Sanders Dewees Bruce

The Thoroughbred Horse

His Origin, how to Breed and how Select him - With the Horse Breeders' Guide

Sanders Dewees Bruce

The Thoroughbred Horse
His Origin, how to Breed and how Select him - With the Horse Breeders' Guide

ISBN/EAN: 9783337144913

Printed in Europe, USA, Canada, Australia, Japan

Cover: Foto ©berggeist007 / pixelio.de

More available books at **www.hansebooks.com**

THOROUGHBRED HORSE.

HIS ORIGIN, HOW TO BREED AND HOW SELECT HIM.

WITH THE

HORSE BREEDERS' GUIDE.

EMBRACING ONE HUNDRED TABULATED PEDIGREES OF THE PRINCIPAL SIRES, WITH FULL PERFORMANCES OF EACH AND BEST OF THEIR GET, COVERING THE SEASON OF 1892.

BY S. D. BRUCE,

Author of the American Stud Book.

PUBLISHED AT THE

Office of the TURF, FIELD AND FARM,

TIMES BUILDING, NEW YORK.

1892.

INDEX

TO THE SIRES OF STALLIONS TABULATED.

INDEX

TO THE DAMS OF STALLIONS TABULATED.

INTRODUCTION.

The author in writing this book and compiling the tabulated pedigrees of the sires to be found in it, has been induced to do so in the hope of their proving useful to those engaged in breeding for the turf, or who are seeking to foster this most popular national amusement. Very many of our largest breeders ignore the lessons taught in the "Stud Book and Racing Calendar." I earnestly commend to all the necessity of observing what sources and what course of breeding have produced the best results in England, which may be most properly called the home of the thoroughbred horse. I have given my own views as to the best mode to successfully breed the race-horse, the best mode to select a stallion and brood mare, and the treatment of the same both in the stud and on the farm. I do not expect that all will agree with me, but the ideas expressed and plans suggested will do away with many of the chances incident to breeding. There is much uncertainty, and always will be, attending the best and most careful mode of breeding, and this opinion is strongly exemplified in the frequent occurrence of one horse being of very high form and an excellent race-horse, and a full brother or sister being only ordinary; yet I differ from a great many in the opinion that breeding depends entirely upon chance. Accidents and other unforeseen causes, some of them so unimportant and abstruse as to escape our attention or come within our knowledge, may prevent the best bred and most promising animal from arriving at its natural size and true shape, and a little difference in conformation, symmetry and constitution may make a decided difference in goodness and speed. The foal may be weak and have a delicate constitution, owing to the dam being starved and exposed to hardships while carrying it, or it may have been improperly reared. This proves that great care and knowledge are necessary in rearing horses for the turf, as well as judgment and attention in selecting mares and stallions from which to breed. The chief points are pure blood, conformation, constitution, racing lineage and hereditary soundness. The nearer we get to true shape with the other points combined, the more certainly we will arrive at excellence. We often find a large horse of good shape and racing symmetry; but where there is one good large one there are a dozen small or medium sized ones. The greater the size, when combined with the good qualities, the greater the excellence and the powers, for a good little horse cannot cope with a great good one. Hence size with constitution, soundness and symmetry constitute the height of perfection. While I advocate and commend pure blood, I am convinced that very often pedigree is the only point at which some breeders look, ignoring altogether shape and action; hence failures. The establishment of reputation by a stallion depends on his having good mares at first, for if he has only bad and indifferent ones the produce will be in low form and a disappointment, and the horse condemned as a failure. To prove this, it is only necessary to cite the Godolphin Arabian and Marske; the former but for the accident of covering Roxana and getting Lath would have died unknown, while Marske, who had been standing for half a guinea and was sold for twenty guineas, after siring Eclipse was sold for one thousand guineas and covered at one hundred guineas. Squirt, the sire of Marske, had been ordered shot and was saved by the intercession of his groom; he afterwards got Marske, Syphon and the celebrated Squirt mare, the dam of Maiden Purity, Pumpkin and other famous horses. That many good race-horses have proved failures as stallions and many poor race-horses successful ones no one can doubt, as instance Lath, who was a famous race-horse but an indifferent stallion, whilst his full brother Cade was an indifferent race-horse but a most excellent stallion. Flying Childers dazzled the world by the splendor of his career on the turf, yet his brother, Bartlett's Childers, proved far the better stallion. According to my notions, no horse can be a good race-horse or a successful stallion that does not possess great symmetry, by which I do not mean beauty, but a show of all those points, such as conformation, length, power and muscle, justly united. Notwithstanding it is often asserted that horses run in all shapes, it cannot be denied that they run better and more frequently where they are well and truly shaped and of tried and approved conformation. A horse may appear to the eye of ordinary judges very plain and ill shaped, and still be in high form, and to the eye of a connoisseur right in all the material points for racing purposes. The machinery properly put together is the point. Chances and accidents are closely allied to all pursuits, and more particularly to breeding for the turf, and luck very frequently predominates over skill and judgment; and whilst breeding cannot be made a certainty or reduced to a science, it should not be left to chance. It cannot be too minutely studied, investigated or attended to in all its branches, and the breeder who selects his mares and stallions with care, attention and judgment as to purity of pedigree, symmetry of form, temper, soundness and constitution, in fact, those possessing all those essential qualities of speed and stoutness, must be more successful than one who breeds at haphazard and pays no regard to these established rules, at the same time crossing and preserving the blood of his mares judiciously, and aiming to remedy the defects, deficient properties and inferior qualities of his mares by the superior conformation, symmetry, admirable properties and brilliant qualities of the stallion, or vice versa. Those who do this will be more likely to produce a more symmetrical, high-typed and successful race-horse than those who pay no attention to these points. Speculative experiments may suit those of large fortunes, but the thinking and judicious breeder, aware of the great expense, constant and unremitting attention necessary for success, will confine himself to the established practice of men whose experience, judgment, attention and success in breeding are worthy of imitation. It is a matter of opinion whether the offspring partake most of the sire or dam. There is no doubt but that they partake of both, though very often more of the one than the other. Some of the mares breed more after themselves, others more after the stallion, then again one foal will partake more after the mare and the next partake most after the horse. It also occasionally happens that the foal will color and mark after the grandsire or grandam or some other more distant cross, and partake of their qualities, hence the necessity of pure blood, conformation and soundness through many generations. To appreciate what care, attention and sound judgment have done for the thoroughbred horse, it is only necessary for us to look at his origin; unquestionably the thoroughbred horse as he now exists in his great perfection originated from a cross of the Arabian, Barb and Turkish horse, the representative scions being the Godolphin and Darley Arabians; the former, generally conceded to be a "Barb," was the sire of Lath, Cade, Regulus, Blank, Babraham, Bajazet, Old England, etc. It is said Mr. Coke brought him from France, and that he had actually drawn a cart in the streets of Paris. Mr. Coke gave him to Mr. Williams, keeper of the St. James Coffee House, by whom he was presented to Lord Godolphin, hence his name. He was teaser to Hobgoblin in the years 1730 and 1731, and, on the latter refusing to cover Roxana by the Bald Galloway, she was bred to the Arabian, and from that cover produced Lath, the first colt ever credited to him, which brought him into prominent notice. He was represented 15 hands high. The Darley Arabian was the property of Mr. Darley, of Buttercramb, near York. He was the sire of the great Flying Childers, supposed to be the fastest horse in the world, sire of Second, Snip, Blacklegs, etc., Bartlett's Childers and Almanzor. Bartlett's Childers never raced, but he was famous as a stallion. He was sire of Squirt (the sire of Marske and Syphon) and the Little Hartley mare, the dam of Janus, Blank, Old England, Slouch, etc. The Curwen Bay Barb was a present to Louis XIV. of France from the Emperor of Morocco, and was brought into England by Mr. Curwen. St. Victor Barb was the sire of the Bald Galloway, and he was the sire of Roxana that brought the Goldophin Arabian into notice. The Compton Barb was sire of Coquette and others. Hutton's Bay Barb was sire of Blacklegs. The Byerly Turk was Capt. Byerly's charger in Ireland during King William's wars in 1689. It is from this horse that the Herod blood originated. An examination of the Stud Book and Racing Calendar will show that nearly the entire family of great race-horses, both in England and America, dates back to the sources mentioned, and that it is extreme folly in our day to expect to improve the present magnificent race-horse by an infusion of the blood of the modern Arabian. We commenced upon the Arab, Barb and Turk with a height not exceeding 15 hands, and have, by judicious crossing, generous diet, discreet and careful handling, built up the most magnificent specimens of the equine race, ranging up to 17 hands in height, with bone, muscle, length, action and all the other grand qualities in proportion. In the early days of breeding, from necessity, very many of the best horses were very much inbred, and even incestuously so, and the question of inbreeding is one which commends itself strongly to mature consideration and examination. According to the maxim that "like produces like," we ought to follow form, blood, speed and other good qualities, and if these good qualities can be maintained and improved by inbreeding, then it must be desirable to a certain extent. Some of the best English and American horses were very much inbred, and the consanguinity of blood did not work deterioration. I am not an advocate of incestuous breeding, but there is no doubt that manly beauty, graceful form and intellectual vigor have resulted from preservation of these high qualities in the human race, where not carried too far; and I can see no reason why the inbreeding of the truest, best bred and best shaped racers can work an injury to the equine race. Still, I think, an out-cross of pure blood, with the essential qualities of a good race-horse, and then back to the superior blood and conformation, is likely to produce the best results. I will give a few examples of inbred modern horses in England, and the same in America, and in doing so will select those of high character which distinguished themselves on the turf and in the stud for instance: The Baron, not only a good race-horse, but sire of Stockwell and Rataplan, was by Birdcatcher by Sir Hercules by Whalebone by Waxy by Pot-8-o-'s by Eclipse. The Baron's dam, Echidna by Economist by Whisker, own brother to Whalebone by Waxy by Pot-8-o-'s by Eclipse; Echidna's dam, Miss Pratt by Blacklock by Whitelock by Hambletonian by King Fergus by Eclipse. Touchstone, a fine race-horse and one of the best stallions that ever lived, was by Camel, son of Whalebone by Waxy by Pot-8-o-'s by Eclipse; his dam, Banter by Master Henry, son of Orville by Beningbrough by King Fergus by Eclipse; second dam, Boadicea by Alexander, son of Eclipse. The great Stockwell was much inbred on his dam's side, Pocahontas by Glencoe, dam Marpessa by Muley, son of Orville by Beningbrough by King Fergus by Eclipse. Muley's dam, Eleanor by Whiskey, son of Saltram by Eclipse. Pocahontas' grandam was Clare by Marmion, son of Whiskey by Saltram by Eclipse; Clare's dam, Harpalice by Gohanna, son of Mercury by Eclipse. Queen Mary, the dam of Blink Bonny, Bonnie Scotland, etc., was much inbred. Gladiator, her sire, was by Partisan, son of Walton by Sir Peter by Highflyer by Herod; her dam by Plenipotentiary, son of Emilius by Orville by Beningbrough by King Fergus by Eclipse; her grandam, Myrrha by Whalebone, son of Waxy by Pot-8-o-'s by Eclipse; her great grandam, Gift by Y. Gohanna, son of Gohanna by Mercury by Eclipse, out of a daughter of Sir Peter, son of Highflyer by Herod. Partisan was out of Parasol by Pot-8-o-'s, son of Eclipse. Pocahontas' best son, Stockwell, was by an inbred horse, and Blink Bonny, Queen Mary's best daughter, was by Melbourne, a horse inbred to Herod and Eclipse. So, if the preservation of good blood through inbreeding in these striking cases has been a success, is it not reasonable to suppose that the same results must follow inbreeding to good blood and true shape with other desirable qualities in this country? Boston was inbred to Diomed, as also his best son, Lexington. Wanderer and Uncas are both much inbred on the dam's side, being out of a grandaughter of Glencoe and tracing to an own sister of the old hero. Glenmore, one of the best race-horses recently on the turf, and whose performances are of the best at all distances, is very much inbred. His dam, Lotta, is by Hunter's Glencoe, son of imp. Glencoe and the Blue filly (Fiatt) by imp. Hedgeford, she out of Lady Thompkins by American Eclipse. Glenmore's grandam, Sally Lewis, is by imp. Glencoe; her dam, Motto by imp. Barefoot, out of Lady Thompkins by Am. Eclipse. Barefoot was by Tramp and Glencoe's dam by Tramp, so that he is, strictly speaking, incestuously bred; yet he was a first class race-horse. Norfolk, a superior race-horse, is inbred to Sir Archy and Diomed. Falsetto is inbred to Lexington, the dam of his sire is by Lexington, and his grandam by Lexington, and he has nine crosses of Diomed. Imp. Eclipse was much inbred; his dam Gaze was by Bay Middleton, son of Sultan and Cobweb by Phantom, son of Walton; Flycatcher, his grandam, was by Godolphin, son of Partisan by Walton, and his great grandam was an own sister to Cobweb by Phantom. Then if the Lexington, Leamington and Glencoe blood is to be preserved to the country, it can only be done by a judicious course of inbreeding, and so uniting the choice of both as to combine and render permanent the qualities possessed by each. Some regard inbreeding as an active cause of degeneration, as unnatural and calculated to develop hereditary diseases. These evils can only spring from abuse of the system. If proper care is taken to exclude the weak and those having hereditary tendencies to disease from participation in reproduction, it may be the means of preserving those estimable qualities so much desired in the equine race, and transmitting them unimpaired to succeeding generations. The greatest success has been achieved by breeding from the nearest affinities of blood, and one should not hesitate to breed a half-brother or sister together where they possess many points of superiority. In the selection of a stallion we want first pure blood, size, substance and conformation with symmetry, not a tall, narrow-chested horse, but one inclined to be thick-set; all coarseness should be avoided, especially in the head, neck, shoulders, ankles and hocks; the eye should be large, clear and bright, with no coarse hair about it, the absence of which is indicative of high breeding in all animals; the jawbones not too massive or heavy, tapering gradually to the nose, free from beefiness; good space between the jawbones for the windpipe; a clear, full, steady eye denotes good temper and enduring qualities, whilst the one showing too much white is generally found in flighty, speedy, non-staying horses. The neck of moderate length, deep and not too thick at the crest or curve; nostrils large, full and roomy. The chest moderately wide, the shoulder blade oblique and inclined backward, with sufficient muscle to cover it, with arms long and muscular, cannon bones short and flat, and the pasterns moderately long, not upright; knee broad and flat, and rather inclined to arch or bend over than backward or calf-kneed, which causes an extra tension on the back tendons; chest deep, extending down between the elbows, which should be straight, inclining neither in nor out, and toes pointed straight forward; body and barrel round and not too long; muscular arched loins, with good length from point of the hip to haunch bone, a slight drooping towards the root of the tail, good length from hip to point of stifle and thence to the hock, which should be broad, finely cut and free from beefiness; the cannon bones flat, tendons detached and well defined, feet strong and hoof not too large or too small: back ribs long, round, and slightly detached from the hip joint. There should be sufficient general length, not to be determined by the length of the back, but the ground the horse covers when standing in a natural position. Good blood is essential and necessary, but good form is superiority. In the selection of brood mares, form is as much a desideratum as in the stallion. First select from the most fashionable blood from running families on both sides with conformation, constitution, good temper and speed. Some prefer large mares, others of medium height. Large mares are not preferable because they are large, but if well and truly shaped, from running blood, there is no objection to size. As a rule, the deep-chested, large-bodied, short-legged mares, with large pelvis, with wide hips and good length, from 15 to 15½ hands high, have proved the best and most successful brood mares. The mare, above all things, should be good-tempered and free from all hereditary defects and disease. It does not follow that a mare which may be blemished from some unforeseen cause may not be as good a brood mare as one entirely sound. Mares in good health will breed until twenty-five or thirty years old, but they require attention, air and exercise, with proper shelter from storms and bad weather, with sufficient food to keep them always in good, strong condition—not beefy fat, as nothing is more fatal to fruitfulness than obesity. Unless kept in good, strong condition, the foals are apt to be weak and weedy at birth. The time to make a race-horse is when the foal is in embyro; in the

Irish vernacular, to make a race-horse you must do so before he is born. Stallions, to do themselves justice, must have plenty of exercise in the open air. If they cannot be trusted in an open paddock, they should be ridden three or four hours each day. Idleness results in indigestion, loss of vigor and flatulence, which often prove fatal. The colts from the day they are foaled should be fed, if the dam does not afford sufficient milk to insure speedy and healthy growth, and broken at weaning time, which should be the last of September or first of October. It is a capital plan to feed the colts in pens for a month or six weeks before weaning them. I am decidedly of the opinion that foals which come the last of March or first of April have full as much advantage as those foaled earlier before the grasses, necessary to afford an ample supply of milk from the dam, arrive at perfection. Those who believe in having early foals should always sow down in the early fall a patch of good rye for the use of the mares and foals. Stallions, mares and colts all require plenty of fresh air and exercise; air and light is life—darkness, death. Horses, and particularly colts, from their natural activity, require more exercise than any other animals, and when properly given it is productive of the most salutary effects. It is the more necessary to colts highly and grossly feed than to those stinted or fed moderately. It enables you to preserve them in a perfect state of health. The food is converted into wholesome nourishment, the circulation of the blood promoted, and all the secretions and discharges facilitated. It invigorates the whole system, gives additional flow to the spirits, adds firmness and strength to the muscles, increases the firmness, texture and growth of the bone, promotes insensible perspiration, assists digestion, prevents flatulence and prepares the system for fresh supplies of aliment. It enables the animal to endure fatigue. In fact, without constant and habitual exercise no animal can enjoy perfect health. High feeding without proper exercise produces many evils, such as indigestion, flatulency, costiveness; the circulation becomes languid, incurable diseases follow, and frequently death terminates the scene. The stallions and mares, treated as I have advised, whose blood is pure and uncontaminated, and whose conformation, strength, activity and vigor are conspicuous in every movement, must impart to their offspring not only sound constitutions, but speed, native fire and energy, which are necessary to support them under the severest exercise of their powers. With their variety of soil, perennial grasses and favorable climate, the Americans should excel all nations in producing the most perfect and the grandest of the equine race. The greatest danger which threatens our ultimate success as the grandest breeding and racing nation of the earth is the constant influx of the rejected and trashy animals imported. What the breeders and racing men of England do not want cannot improve the blood horse of America, but, upon the contrary, work irreparable injury.

THE THOROUGHBRED HORSE;

HIS ORIGIN, HOW TO BREED HIM AND HOW TO SELECT HIM.

THE native country of the horse is not certainly known. According to ancient classic mythology, the beautiful and useful animal originated in the contention of two deities, emulous to confer on the human race the most valuable gift. Minerva, the Goddess of Wisdom, War and the Liberal Arts, created the olive tree, when Neptune, the God of the Sea, Rivers and Fountains, to outrival her, struck the earth with his trident and gave existence to the horse. His noble prowess, bold majestic front, speed of thought and graceful symmetry have been the admiration of men in every age and the theme of poets of all lands. Job, Homer, Virgil, Shakespeare and many others have paid willing tribute of their genius to the elegance of his form, the animation of his spirit and courage. A friend to man, an ornament to the earth, no other animal has contributed so much to the social advancement of the human race. Speed, endurance, strength, elasticity, elegance and power, qualities so varied and valuable, are so thoroughly combined in this admirable animal as to proclaim him without a peer in the sphere of animated nature. In the Bible we are told that as early as 1650 years before the birth of Christ the horse had been domesticated by the Egyptians. When Joseph carried his father's remains from Egypt to Canaan, there went with him both chariots and horsemen. One hundred and fifty years afterwards the horse constituted the principal strength of the Egyptian army. Pharaoh pursued the Israelites with six hundred chariots, and all the chosen horsemen of Egypt. Fifty years after the expulsion of the Israelites from Egypt, and 1450 years before the birth of Christ, the horse was so far naturalized and utilized in Greece that the Olympic Games were instituted, including chariot and horse races. This short digression from our subject is justified in order to show that horses were used in races long before the Christian era. According to Whyte in his history of the British turf, the earliest mention of race-horses, called running horses in those days, was in the ninth century, when Hugh, founder of the royal house of Capet in France, sent horses as a present to King Athelstane, whose sister, Ethelswitha, he was soliciting in marriage. In the reign of William the Conqueror, the Earl of Shrewsbury imported some stallions from Spain into his estate at Powisland; we find their produce celebrated by Drayton, the poet. This is the first well-authenticated step we can find towards the improvement of the breed of horses. These horses were more than probably of Eastern blood. In the reign of Richard I. matches were run for large sums, and swift-running horses were greatly esteemed and were sold for very high prices. There is no data in subsequent reigns, but we find Edward III., 1326, purchasing running horses, and in the ninth year of his reign he received as a present two running horses from the king of Navarre, and they were supposed to be valuable as the king made large pecuniary compensation to the person who brought them. In the reign of Henry VII. the custom of gelding horses came into general practice. Henry VIII., 1509, took great pains to improve the breed of horses, especially in the royal stud; Sir Thomas Chaloner, a writer in the commencement of the reign of Queen

Elizabeth, mentions his having imported horses from Turkey, Naples and Spain. Those horses unquestionably were largely bred from Moorish and Barbary blood. Her successor, James I., in 1602, placed the turf upon a solid basis and made it a permanent institution. From the royal patronage accorded it, and rules framed for its regulation, horse racing had a gradual and uninterrupted progress until it reached its zenith. James I. by his public patronage of the race course, and royal support, which was needed to bring it into good form and repute, should be canonized as the patron saint of the turf. But the improvement of the race-horse originated by him was of much and superior importance. The crossing of foreign blood with the English horse was not new at this period, but the eastern blood had never been tried, and to James I. is due the credit of first attempting that, which, although considered a failure, has resulted in the excellence of the race-horse of the present day. King James purchased of Mr. Markham an Arabian known as the Markham Arabian. This Arabian was a failure, and no horses of note or value descended from him. The king subsequently purchased the Turk known as Place's White Turk. Mr. Place was subsequently Stud-master to the Protector, Oliver Cromwell. Place's White Turk was the sire of Commoner and Wormwood, and the grandam of Cartouch, Grey Ramsden and Windham. The Brimmer mare, dam by Place's White Turk, grandam by Dodsworth out of the Layton Barb mare, produced Bay Layton by D'Arcy's Counsellor, Brown Farewell and her sister by Makeless. The Place White Turk mare, dam by Dodsworth out of the Layton Barb mare, was the dam of the noted Makeless. Mr. Croft's Brimmer mare, the mare by Hautboy, dam of the Pet mare by Wastell's Turk, Trumpet and Gray Layton, by Counsellor. Bay Layton above was the dam of Old Scar by Makeless, Chesnut Layton by Makeless and Scar by Croft's Bay Barb. Brown Farewell was the dam of Guy, Young Greyhound and sister to them by Greyhound, and Miss Partner by Croft's Partner. This sister to Guy was the dam of Bay Bloody Buttocks and Grey Bloody Buttocks (Dairy Maid) by Bloody Buttocks, Little Partner (Vane's), Young Partner, Red Rose and sister to them by Croft's Partner. This sister to Vane's Little Partner was the dam of the Starling mare that produced Leedes by Second and his sister that was the dam of Amaranthus, Gift and the mare by Cub which was foaled in 1762 and imported into America by Gen. Delancy, of New York, prior to the American revolution. Miss Partner above was the dam of Whimsey and Madame by Bloody Buttocks, Toy by Bartlet's Childers, Drowsy by Fox Cub, Torismond by Old Starling, a filly by Roundhead (dam of Juliet by Bustard in Ireland), Young Cade, Miss Cade, Villager and Milksop, all by Cade. A sister to Miss Partner by Croft's Partner was the dam of Sweepstakes and Clio by the Gower stallion, son of the Godolphin Arabian. Clio was the dam of Mr. Wilkinson's Conqueror in Ireland, foaled 1754 by Bajazet. Another sister to Miss Partner was the dam of Dormouse, by the Godolphin Arabian, Mercury, Merryman, and Cadormus by Cade and Bolton, foaled 1752 by Shock, imported into America. Still another sister to Miss Partner was the dam of Miss Roundhead by Roundhead, Changeling and the noted Matchem by Cade, son of the Godolphin Arabian, Young Starling by Starling, etc. Matchem was the best race-horse of his day and famous as a stallion. He got in the twenty-three years he was in the stud 354 winners of $755,485.00. Bay Bloody Buttocks, mentioned above, was the dam of the famous Spinster, better known as Mr. Panton's Widdrington mare by Partner. Spinster was the dam of Spinster and Shepherd's Crab and Fancy by Crab, Mr. Panton's Deputy, afterwards Mr. Vane's Lofty, by Godolphin Arabian, Golden Grove by Blank and others. Spinster by Crab was the dam of a filly by Janus, whose daughter by Skim, son of Traveler, was the dam of Mr. Vernon's Telemachus and Duke of Hamilton's Expectation, both by Herod: Grey Starling by Bolton Starling and others. Fancy by Crab was the dam of Cloudy and Countess by Blank. Cloudy was the dam of Amaranda by Omnium, son of Snap, foaled in 1771. Amaranda was the dam of Joe Andrews (first called Dennis O) by Eclipse, who was the sire of the two good horses Jack Andrews and Dick Andrews, dam by Highflyer. Golden Grove was the dam of Petworth (Gandy) and an own sister by Herod. Petworth was the dam of Rosina by Woodpecker, she, Rosina, was the dam of Hermes by Mercury, son of Eclipse. Expectation, sister to Telemachus, by Herod, was the dam of Orange Girl, Betsey and Anticipation, all by Beningbrough, son of King Fergus by Eclipse. Anticipation was the dam of Maniac by Shuttle (the dam of Harriet by Stripling, Lunatic by Prime Minister, Bedlamite by Welbeck, Lunacy and Bee-in-a-Bonnet by Blacklock, and Young Maniac by Tramp), Offa Dyke by Paynator, Handmaiden (the dam of Inheritor by Lottery and Executrix, by Liverpool) by Walton. A filly by Cervantes, son of Don Quixote, by Eclipse, was the dam of Rebecca by Lottery. Rebecca was the dam of The Provost by The Saddler, Alice Hawthorn by Muley Moloch, Annandale by Touchstone, Fair Helen by Pantaloon, son of Castrel, that was the dam of Lord of the Isles by Touchstone.

Alice Hawthorn was the most famous race-mare of her day, winner of 16 cups, including Doncaster and Goodwood cups, each twice; 18 Queen's plates, and 17 other races, and was the dam of Oulston by Melbourne, a superior racer, Thormanby, winner of the Derby, 1860, by Melbourne or Windhound, Lady Hawthorn by Windhound; Sweet Hawthorn by Sweetmeat, etc. Lady Hawthorn was the dam of Lady Alice Hawthorn by Newminster, she the dam of Hawthorn Bloom by Kettledrum, she the dam of imported Bread Fruit by Breadalbane, Imported Albert by Albert Victor. Lady Alice Hawthorn was also dam of Hawthorndale by Kettledrum, she the dam of Hauteur, winner of the 1,000 Guineas, by Rosicrucian. Lady Hawthorn was the dam of Maybloom, by Newminster, she the dam of Corisande, winner of the Cesarewitch, and her sister, imported Lady Mentmore by King Tom. Lady Mentmore is the dam of Mentmore Lass, winner in England, Rica, a superior race-mare, and Mehallah by Kingfisher, son of Lexington. Another sister to Lady Mentmore and Corisande, Verdure is the dam of imported Judith and Clover by Macaroni, and a sister Vista, she the dam of Bonavista, winner of the 2,000 Guineas in 1892, by Bend'or. Sweet Hawthorn was the dam of Lady Nateby by Van Galen, she the dam of Lady Portland by The Primate, and she the dam of Miss Jummy, winner of the 1,000 Guineas and Oaks, by Petrarch. To Anticipation in the female line trace Inheritor, Bedlamite, Common, winner of the 2,000 Guineas, Derby and St. Leger, Macgregor, winner of the 2,000 Guineas. Our Nell, winner of the Oaks, Manganese, winner of the 1,000 Guineas, Apology, winner of the 1,000 Guineas, Oaks and St. Leger, Mandrake, The Miner, Wenlock, winner of the St. Leger, Kisber, winner of the Grand Prize of Paris and Derby, imported Agenoria and her brother imported Pizarro, sire of Pessara, whose early death was a serious loss to America.

England is entitled to the credit of originating the thoroughbred horse. The breed has spread over the entire intelligent habitable globe wherever the value of beauty in form, utility, speed combined with strength, so indispensably necessary to labor, pastime, war and the turf is valued. To understand his origin we must begin with the sources from which he has descended and trace him from a combination of the Arab, Barb and Turkish horses. It is impossible for any man, a student of the stud book and conversant with the turf, to look at the present high-bred horse, and those of earlier days, without being surprised that the blood which was so highly and justly esteemed should now be unfashionable and little used, but the judicious crossing and intermingling of the blood of Herod, Matchem and Eclipse has built up such a grand and noble specimen of the equine race that the original blood cannot improve it. The three great sources of blood will readily be seen in the following tables:

The Byerly Turk was Captain Byerly's charger in Ireland, in King William's wars in 1689, and founded what is styled and known as the Herod Blood. The Byerly Turk got Jigg out of a Spanker mare. Spanker was by D'Arcy's Yellow Turk, who was the sire, also, of Brimmer. Jigg got Partner out of a sister to Mixbury, by the Curwen Bay Barb, who was a present to Louis XIV. from Muley Ishmael, King of Morocco. The Curwen Bay Barb was brought into England by Mr. Curwen, who procured him, together with the Thoulouse Barb in France, through Count Byram and Count Thoulouse, two natural sons of Louis XIV, the former Master of the Horse, and the latter an Admiral. They proved excellent stallions

Partner got Tartar out of Meliora by Fox, she out of Milkmaid by Sir W. Blacket's Snail. Fox was by Clumsy, son of Wilke's Old Hautboy by the White D'Arcy or Sedbury Turk, out of a Royal mare, brought into England by the Master of the Horse to King Charles II. Tartar got King Herod out of Cypron by Blaze, she out of Selima, a daughter of Bethell's Arabian. Blaze was by Flying Childers, son of the Darley Arabian, dam Betty Leedes by Careless, out of sister to Leedes by Leede's Arabian. Herod got Highflyer, dam Rachel, by Blank, son of the Godolphin Arabian (Barb), out of a daughter of Regulus, son, also, of the Godolphin Arabian. Herod, also, got Fortitude, dam by Snap, son of Snip by Flying Childers, out of Milksop by Cade, son of the Godolphin Arabian. Herod also sired Woodpecker, dam Miss Ramsden by Old Cade, son of the Godolphin Arabian, out of a daughter of the Lonsdale Bay Arabian, etc. Woodpecker sired imp. Buzzard, out of Misfortune by Dux. Highflyer got Sir Peter Teazle, commonly called Sir Peter, dam Papillon by Snap, out of Miss Cleveland by Regulus. Among the famous sons of Sir Peter were Sir Solomon, dam Matron, by Florizel, by Herod, out of Maiden, by Matchem; Hap-hazard, dam Miss Hervey by Eclipse, out of Clio by Young Cade; Stamford, dam Horatia by Eclipse, out of Countess, Delpini's dam by Blank; Walton, dam Arethusa by Dungannon (son of Eclipse and Aspasia by Herod), her dam by Prophet, son of Regulus, out of Virago, Saltram's dam by Snap; Williamson's Ditto, a full brother to Walton; and Sir Oliver, dam Fanny by our imported Diomed, out of Ambrosia by Woodpecker, etc. This is the first class or Herod blood.

The Godolphin Arabian, generally believed to have been a Barb, was first the property of Mr. Coke, who presented him to Mr. Williams, the keeper of the St. James Coffee House, by whom he was presented to Lord Godolphin. The Godolphin Arabian was teazer to Hobgoblin in the years 1730 and 1731, and on the latter refusing to cover Roxana, she was put to the Godolphin Arabian, and from that cover she produced Lath, a superior race-horse, and in 1734 Cade, by the Godolphin Arabian, who took his name from the fact that Roxana died when he was ten days old, and he was brought up on cow's milk. "To Cade, to breed up in softness."

Roxana was by the Bald Galloway, son of the St. Victor Barb, his dam Grey Whynot by Whynot, son of the Fenwick Barb, his grandam a Royal mare; Roxana's dam, sister to Chaunter by the Akaster Turk. Cade got Matchem, foaled 1748, dam sister to Miss Partner by Partner, son of Jigg by the Byerley Turk, out of Brown Farewell of Makeless, son of the Oglethorpe Arabian; Matchem got Conductor, dam by Snap, grandam by the Cullen Arabian, out of Grisewood's Lady Thigh by Croft's Partner; Conductor got Imperator, dam by Herod, grandam Carina by Marske (sire of Eclipse), son of Squirt by Bartlet's Childers; Conductor, also, got Trumpator, dam Brunette by Squirrel, son of Traveler, out of Dove by Matchless, son of the Godolphin Arabian.

Imperator got Pipator, out of Brunette, the dam of Trumpator, and Pipator got Remembrancer out of Queen Mab, sister to Mercury by Eclipse, out of a daughter of Tartar; Trumpator got Sorcerer, dam Young Giantess, by imported Diomed, out of Giantess, by Matchem. This is the second class, or Matchem blood.

The Darley Arabian was the property of Mr. Darley, of Buttercramb, near York. He was thought by some to be a Turk or Syrian horse, but was unquestionably an Arabian of pure blood. He got Bartlet's Childers own brother to Flying Childers, dam Betty Leedes by Careless, son of Spanker, son of the D'Arcy Yellow Turk, out of Sister to Leedes by Leedes Arabian. Bartlet's Childers got Squirt, dam sister to Old Country Wench, by Snake, son of the Lister Turk, out of Grey Wilks by Old Hautboy, son of the White D'Arcy or Sedbury Turk. Squirt got Marske out of the Ruby mare by Blacklegs, son of Hutton's Bay Barb, his grandam by Bay Bolton, son of Grey Hautboy. Marske got Eclipse (O'Kelly's) dam Spiletta by Regulus, son of the Godolphin Arabian, grandam, Mother Western by Smith's son of Snake by Mr. Lister's Snake, out of daughter of the Akaster Turk. Eclipse took his name from the fact that he was foaled during the great eclipse, in 1764. Eclipse got King Fergus, dam Tuting's Polly by Black-and-all-Black (Othello), grandam Fanny by Tartar. Eclipse got Mercury, Volunteer and Queen Mab, dam by Tartar, out of a daughter of Mogul by the Godolphin Arabian. Eclipse got Dungannon, dam Aspasia by Herod, grandam Doris by Blank, out of Helen by Spectator, son of Crab, by Alcock Arabian. Eclipse got Meteor, dam daughter of Merlin, by Second, son of Flying Childers, grandam Mother Pratt, by Marksman, son of the Godolphin Arabian. Eclipse got imported Saltram, dam Virago by Snap, out of a daughter of Regulus. Eclipse got Pot-8-o's, dam Sportsmistress by Warren's Sportsman, son of Cade, grandam Golden Locks by Oroonoko, son of Crab, out of Valiant's dam, by Crab. Volunteer got Eagle and Spread Eagle, out of a daughter of Highflyer, grandam by Engineer, son of Blaze, out of a daughter of Regulus. Mercury got Gohanna and Precipitate, out of a daughter of Herod, grandam Maiden, sister to Pumpkin, by Matchem. King Fergus got Beningbrough, out of a daughter of Herod, grandam Pyrrha, by Matchem, out of Duchess, by Whitenose, son of the Godolphin Arabian. King Fergus got Hambletonian, dam by Highflyer, out of Monimia by Matchem. Saltram got Whiskey, out of Calash by Herod, grandam Teresa by Matchem. Pot-8-o's got Waxy and Worthy, out of Maria by Herod, grandam Lisette by Snap, out of Miss Windsor by the Godolphin Arabian. This constitutes the third class or Eclipse blood. To these three strains of blood trace all the horses of this day and all the best for the last century, and the horses now so successful on the turf in America and England combine these three strains of blood to a greater or less extent on both the paternal and maternal sides.

Taking a retrospect of the winners of the two great fixed events of the English turf (there being really no National fixed Event in America if we except the Futurity and the Realization stakes), the Derby and St. Leger, we find that in the tail male line The Derby, in the 113 years it has been run, has been won by the First line, Herod blood, thirty-five times, commencing with Diomed in the inauguration Derby, 1780. Diomed was imported in 1798, being then 22 years old. He was sold for 50 guineas and subsequently sold to Col. Hoomes, of Virginia, by whom he was imported. He died in Virginia in 1808, being then 31 years old. There is scarcely a really good race-horse of the present day which does not partake of his blood. He was the sire of Sir Archy, out of imp. Castianira by Rockingham, justly styled the Godolphin Arabian of America. Diomed also sired Ball's Florizel, Duroc, the sire of American Eclipse, Potomac, Diomed Eagle, Stump the Dealer, Vingt'un, Hamlintonian Virginius, Peacemaker and a number of superior race-horses and brood mares, dams of winners. Other winners of the Derby imported were Saltram by Eclipse, sire of Whip, also imported, and the Saltram Mare, dam of Timoleon, the sire of Boston; John Bull by Fortitude, son of Herod; Spread Eagle by Volunteer, son of Eclipse; Sir Harry, son of Sir Peter, by Herod; Archduke by Sir Peter; Priam by Emilius; St. Giles by Tramp, by Dick Andrews; Blue Gown by Beadsman died on the passage; Kingcraft by King Tom died on the passage, and St. Blaise by Hermit. All these from Priam down trace on the tail male line to Eclipse, whose descendants in a direct male line won the Derby 72 times, including Iroquois. The descendant of Matchem only six times, and one, the horse Aimwell, who won it in 1785, from the Alcock Arabian through Mark Anthony by Spectator by Crab, his dam a sister to Postmaster by Herod. The other great fixed event, the St. Leger, was inaugurated in 1776, but did not receive its present name until 1778. It was won in 1776 by Allabaculia by Sampson, son of Blaze by Flying Childers, by the Darley Arabian, Sampson's dam by Hip, son of the Curwen Bay Barb, this race has been run one hundred and sixteen years, from 1776 to 1891 inclusive, and has been won 72 times by the descendants of the Eclipse blood in the tail male line, 28 times by the Herod line and 16 times by the Matchems, but the winners were largely inbred to the three great strains of blood. The 2,000 Guineas, another of the important fixed events of the English Turf, was inaugurated in 1809, and has been run annually since that date. It is the important opening three year old event of the year. It is run over the Rowley Mile, one mile 11 yards. It has been won by the descendants of Eclipse, in the tail male line, 45 times, by the Herod line, 27 times and by the Matchem line, 12 times. These three events are open to both colts and fillies, but few of the latter are entered or start in them, hence the winners have generally been colts. The Two Thousand Guineas has been won but four times by fillies, in 1822 by Pastille by Rubens, son of Buzzard; in 1840 by Crucifix by Priam, son of Emilius, in 1878 by Pilgrimage by The Earl or The Palmer, and in 1882 by Shotover by Hermit, son of Newminster. In 1868 Formosa by Buccaneer ran a dead heat and divided the stakes with Moslem. The Derby has been won only three times by fillies, in 1801 by Eleanor by Whiskey, in 1857 by Blink Bonny by Melbourne, and in 1882 by Shotover by Hermit. The St. Leger, run in the fall, has been won more frequently by fillies, for the reason that good ones have oftener started for this event. Allabacnlia by Sampson won it in 1776. Hollandaise by Matchem in 1778. Serina by Goldfinder in 1781. Imperatrix by Alfred in 1782. Omphale by Highflyer in 1784. Cowslip by Highflyer in 1785. Young Flora by Highflyer in 1788. Pewett by Tandem in consequence of a foul on the part of Duke of Hamilton's Laurel colt, in 1789. Paulina by Sir Peter in 1807. Altisidora by Dick Andrews in 1813. Duchess by Cardinal York in 1816. Matilda by Comus in 1827. Queen of Trumps by Velocipede in 1835, She also won the Oaks. Blue Bonnet by Touchstone in 1842. Imperieuse by Orlando in 1857. Sunbeam by Chanticleer in 1858. Caller On by Stockwell in 1861. Achievement by Stockwell in 1867. Formosa by Buccaneer in 1868, Formosa also ran a dead heat with Moslem for the 2,000 Guineas and won the Oaks the same year. Hannah by King Tom in 1871. Hannah also won the Oaks. Marie Stuart by Scottish Chief in 1873. Apology by Adventurer in 1874. Marie Stuart and Apology both won the Oaks. Jannette by Lord Clifden in 1878. Jannette also won the Oaks. Dutch Oven by Dutch Skater in 1882. Seabreeze by Isonomy in 1888. She also won the Oaks. Memoir by St. Simon in 1890. Memoir won the Oaks. The Oaks, the most important fixed event, for fillies only, was established in 1779, the year prior to The Derby. It has been run one hundred and fourteen times, including the year 1892, and has been won 64 times by the descendants of Eclipse in the tail male line, 37 times by the descendants of Herod and 13 times by the descendants of Matchem. The length of the Derby course in 1780 and up to 1784 was one mile. From 1784 up to and including 1871, the distance was one and a half miles. In 1872 and since the start has been from the New High Level Starting Post, the distance being one and a half miles and 29 yards. The Oaks is run over the Derby course. The St. Leger course was originally two miles. In 1813 the distance was changed to one mile 6 furlongs and 193 yards. In 1826 the distance was reduced 61 yards and has remained since that date one mile 6 furlongs and 132 yards. It appears from the statistics above that the tail male line of the Eclipse blood has won a large majority of the fixed events above mentioned, but it must be borne in mind that Matchem was foaled in 1748, King Herod in 1758 and Eclipse in 1764, so that Eclipse had the benefit of the daughters of the two other strains of blood and nearly all of the best and most famous of his sons and grandsons were from Herod's daughters and granddaughters, fortified by the Matchem blood. There were but few males from the Matchem blood compared with the number from the Herod line, and the same may be said of the number of the Herod's compared with the Eclipse's, and nearly all the descendants of Eclipse which have won the great events in the last quarter of a century, are descended from Whalebone through his sons Camel and Sir Hercules. To the former traces Touchstone, and to the latter Stockwell, both being inbred to Herod and Matchem on both the paternal and maternal sides, and the same is true of every grand race-horse that has appeared in any country on the globe.

It was in the reign of the unfortunate Charles I. that the introduction of Eastern blood of much importance took place, notably the Helmsley Turk, brought to England by the Duke of Buckingham, and sold to Mr. Helmsley, hence his name. The Helmsley Turk was the sire of Bustler, whose dam was not known; Hutton's Royal colt, dam a Sedbury Royal mare, and Vixen, out of Dodsworth's dam, a Royal mare. Vixen was the dam of the Old Child mare, by the Gresley Bay Arabian (Bay Roan). The Old Child mare was the dam of a filly, by the Darley Arabian, which produced a filly by Rattle, who was by a son of Sir J. Harpur's Barb, out of a Royal mare. This Rattle mare was the dam of Silvertail, by Heneage's Whitenose, son of the Hall Arabian; Silvertail was the dam of Whimsey, by Cullen Arabian: Sportsman and Cade, imported into South Carolina in 1762 (both by Cade, son of the Godolphin Arabian): Careless and Fearnought by Regulus, son of the Godolphin Arabian.

Fearnought was imported into Virginia in 1761, by Col. John Baylor, and was noted as a sire in this country. Silvertail, also, produced Joan and a sister to her, both by Regulus; Joan was the dam of Rantipole, by Blank, son of the Godolphin Arabian; Columbine, by Eclipse, and others, but from her daughter, Rantipole, she became famous. Rantipole was the dam of Nymph, Trusty and Nimble by Florizel. Nimble was the dam of Niko (winner of the Oaks in 1797), by Alexander, son of Eclipse. Nutmeg by Sir Peter, son of Highflyer, and Donna Clara, by Cesario, by John Bull, son of Fortitude, by Herod; Donna Clara was the dam of a filly by Dick Andrews, son of Joe Andrews, by Eclipse; Donna Maria, by Partisan, son of Walton, and others. Donna Maria's daughter, by Buzzard, son of Blacklock, produced imp. Barbarity, by Simoon, son of Camel, by Whalebone, that gave to America the superior mares Ruthless, Relentless and Remorseless, with Young Eclipse, Barbarian and Devastation, all by imp. Eclipse, son of Orlando, by Touchstone. Niko, through her daughter by Trafalgar, son of Sir Peter by Highflyer, gave us imp. Alarm by Thunderbolt, son of Sorcerer, by Trumpator. Alarm was the dam of Zadora and Di Vernon, by American Eclipse; Lady Alert (Alert), by Eclipse Lightfoot; Clara Howard, Young Alarm and Fanny Elssler, by imp. Barefoot. Clara Howard was a superior race-mare, and the dam of Eliza Butler and Mary Elliott, by imp. Leviathan. Mary Taylor, by imp. Sovereign; Angeline, Star-Spangled Banner, and Banner, by imp. Albion. Angeline was the dam of Lantana, by Capt. Elgee, and Peytona Barry, by Rogers, son of Lexington. Lantana was the dam of Poca Wiley, by Muggins, Larkspar, by Jack Malone, Beersheba, Balance-All, Bran Dance (in England), all by imp. Bonnie Scotland; the good race-mares Liahtunah and Panorama, by John Morgan, and Tantalus, by imp. Great Tom. The reputation of the Helmsley Turk was mainly established and immortalized as the sire of Bustler, whose dam is not stated in the stud book, or by other authorities, but was undoubtedly out of one of the Royal mares. He was the sire of Lord D'Arcy's Blunderbuss, Old Merlin, the sire of Mr. Bethel's Woodcock and Castaway, and Mr. Richardson's Merlin. The Bolton Sweepstakes' great grandam was out of daughter of Bustler. The pedigree of the Ruby mare, by Blacklegs, dam of Marske (sire of Eclipse and imp. Shark), ends in a daughter of Bustler. The Bolton Sweepstakes was the sire of the dam of Whistle jacket, and the grandam of Mr. O'Kelly's famous Old Tartar mare. The Old Tartar mare was the dam of Mercury, Volunteer, Queen Mab, Venus, Jupiter, Adonis and Lily of the Valley, all by Mr. O'Kelly's noted Eclipse. Bolton Sweepstakes was the sire of Turner's Sweepstakes, out of a Bay Bolton mare. The names of Bustler and Hutton's Royal colt are to be found at the end of the pedigrees of many of the most noted horses.

Place's White Turk was imported and owned by Mr. Place, stud-master to Oliver Cromwell when protector, and was the sire of Mr. Croft's Commoner, Mr. Tregonwell's White Turk Mare, out of a natural Barb mare. This White Turk mare was the dam of a mare by the Taffolet Morocco Barb that produced the Byerly Turk mare, which was the dam of Grey Ramsden, by Grey Hautboy, son of Hautboy, by the D'Arcy White Turk, a filly by Spanker, son of the D'Arcy Yellow Turk, one filly each by the Darley Arabian and Newcastle Turk. From the Spanker filly mentioned, descended in direct female line, Wooodpecker, Goldfinder, Waxy Pope, Whalebone, Whisker, Partisan, Rockingham, Glencoe, Bay Middleton, Lord Lyon, Dollar, Trumpeter, Melbourne, Bend'or, Robert the Devil, Uncas, Speculum, Charibert, The Nob, Silvio, Blue Gown, Pellegrino, Trappist, Craig Millar, Mortemer, King of Trumps (by Lord Clifden), The Lambkin, The Confessor (by The Palmer), Paradox, The Bard (by Petrarch), Minting, and many other noted race-horses and stallions.

The Byerly Turk was one of the valuable importations to England during the reign of William and Mary. He was the charger of Capt. Byerly throughout the war of King William in Ireland in 1689. This Turk became one of the principal founders of the thoroughbred horse. To him traces what is generally styled the First Class, or Herod Blood. The Byerly Turk was the sire of Jigg, dam by Spanker, son of the D'Arcy Yellow Turk. The grandam of Jigg is unknown. Jig was the sire of Partner, 1718, out of Sister to Mixbury, by the Curwin Bay Barb, second dam by Curwin's Old Spot, son of the Selaby Turk, third dam by the white-legged Lowther Barb, out of the Old Vintner mare. He was also the sire of Miss Jigg and her sister, out of Partner's dam. Partner got Tartar, 1743, dam Meliora, 1729, by Fox, son of Clumsy by Hautboy, second dam Witty's Milkmaid, 1720, by Snail, son of Whynot, by St. Victor Barb, third dam by Mr. Curwin's Shields Galloway. Tartar was an excellent racer and superior stallion. He was the sire of King Herod, commonly called Herod, 1758, and his sister Thais, dam Cypron, by Blaze, son of Flying Childers, second dam Selima, by Bethell's Arabian, etc. Tartar was also the sire of Beaufremont, and Mr. O'Kelly's mare, dam of Mercury, Volunteer, etc. Herod got Highflyer, 1774, dam Rachel, by Blank, son of the Godolphin Arabian, second dam by Regulus, son of the Godolphin Arabian, out of a daughter of Soreheels, son of Basto, by Byerly Turk, etc. Herod was also the sire of Anvil, Bordeaux, Drone, Florizel (sire of imp. Diomed, winner of the first Derby, 1780), Fortitude, Phenomenon, Postmaster, Telemachus, Woodpecker (sire of Buzzard) and other stallions. He was the sire of many excellent race and successful brood mares, including Calash (dam of Whiskey, by Saltram, and Kite, by Buzzard), Faith (dam of Bobtail), Maid of the Oaks (dam of imp. Druid), Maria (dam of Waxy and Worthy, by Pot-8-o's, Quiz and Wowski by Mentor), Peggy, sister to Postmaster (dam of Peggy, imported to America, and her sisters, Spinnetta and Trumpetta), by Trumpator, son of Conductor, by Matchem, Perdita, Pomona, Sting, sister to Florizel, Tuberose, etc. Herod was a superior race-horse, and as a stallion was inferior to no horse that ever lived. In nineteen years he got 497 winners that won $1,007,525. Highflyer got Sir Peter Teazle, 1784 (commonly called Sir Peter, winner of the Derby, 1787), dam Papillon, by Snap, son of Snip, by Flying Childers, second dam Miss Cleveland, by Regulus, out of Midge, by a son of Bay Bolton, tracing to the Byerly Turk mare, dam of the two True Blues, Lady Teazle and Brown Bess, sisters to Sir Peter. Highflyer was the sire of Noble (winner of the Derby in 1786). Skyscraper (winner of the Derby in 1789). Highflyer was also sire of Omphale and Cowslip, winners of the St. Leger in 1784 and 1785, Spadille and Young Flora, winners of the St. Leger in 1787 and 1788. Highflyer never paid a forfeit and was never beaten. Sir Peter was the best son of Highflyer, and was the best race-horse of his day. He was the sire of imp. Sir Harry, 1795, dam Matron by Alfred, son of Matchem out of Pilot's dam by Marske. Sir Harry won the Derby in 1798, and was imported into Virginia in 1804. He was the sire of Sir Alfred, and Sir Hal, his name is often found in the pedigrees of our best racers. Imp. Robin Redbreast, sire of Sumpter's dam, was by Sir Peter, dam Wren, by Woodpecker. Imp. Archduke, and his brother, Stamford, out of Horatia, sister to Achilles by Eclipse, second dam Countess, Delpini's dam by Blank. Archduke won the Derby in 1799, and was imported by Col. Hoomes into Virginia in 1803. He is best known in America as the sire of the dam of Countess Plater by Virginian who was the dam of the good race-horse Altorf by imp. Fylde and Janey, dam by imp. Stirling that was the dam of the famous Polly Hopkins and Ivanhoe by Virginian (son of Sir Archy) and Hyazim by Sir Archy. Sir Peter was the sire of Sir Solomon (first called Tankersley), dam Matron by Florizel, out of Maiden by Matchem. Sir Solomon, 1796, was the sire of Warrior and Mary Ann, the dam of Windle by Beningbrough. Sir Peter also got Haphazard, 1797, and his sister, dam Miss Hervey by Eclipse, out of Clio by Young Cade. Haphazard was a grand race-horse and the sire of Filho-da-Puta, 1812, dam Mrs. Barnet by Waxy, son of Pot-8-o's, out of daughter of Woodpecker, etc. Filho-da-Puta won the St. Leger in 1815. Filho-da-Puta was the sire of Birmingham that won the St. Leger in 1830, and Colwick, dam Stella by Sir Oliver, son of Sir Peter. Colwick was the sire of Attila, dam Progress by Langar, son of Selim, that won the Derby in 1842. Haphazard was the sire of Antar and Reginald, that won the 2,000 Guineas in 1819 and 1821, and Rowena, winner of the 1,000 Guineas in 1820. Imp. Hedgeford, foaled 1825, was by Filho-da-Puta, dam, Miss Craigie by Orville, son of Beningbrough, second dam Marchioness by Lurcher, son of Dungannon by Eclipse, out of Miss Cogden by Phenomenon, son of Herod, imported into America. Hedgeford was a full brother to Birmingham, and was a good race-horse. He got Duane, a superior race-horse, dam Goodloe Washington by Washington, son of Timoleon by Sir

Archy, Ruffin, dam Duchess of Marlborough by Sir Archy. Ruffin, the property of Jos. G. Boswell, known as the lucky Kentuckian, won the great Gold stakes at Lexington, Kentucky, in 1843, and many other good races. Hedgeford was the sire of Fiatt (Blue, filly), dam, Lady Thompkins by American, Eclipse, tracing back to imp. Duchess, that produced Blue Bell by Chorister, son of imp. Contract by Catton. Blue Bell was the dam of Nellie Gray and Betsey Hunter by Oliver, son of Wagner, and Flight by imp. Leviathan, Julia Mattingly by John Morgan, son of imp. Sovereign, Bessie Lee by Hunter's Lexington, and others. Betsey Hunter is the dam of Kate Mattingly (Creosote dam) by John Morgan, Chassense, Joe Johnson, Aunt Winnie, Judge Wickliffe, etc., nearly all of which raced well and have produced race-horses. Blue Filly was also the dam of Kate Hunter, Lizzy Morgan, Hunters' Glencoe, and others.

Lizzy Morgan left a distinguished progeny through her son, Morgan Scout by John Morgan, Ratan, the dam of Wildfire, Girotlé, Girotla, etc. This family has been much inbred, but always preserved its individuality. Blue Bonnet by imp. Hedgeford, dam Grey Fanny by Bertrand, was one of the best race-mares of her day and a brood-mare that left behind her a history. She was the dam of such good ones, as Little Arthur, Nebula (the dam of Asteroid, the unbeaten son of Lexington, Sue Lewis, Luna, Aster and Aneroid, his sisters and brothers, and Asterisk by Ringgold, son of Boston), Alice Jones (Jonesboro's dam), and The Gloamin, all by imp. Glencoe: Lightning, Loadstone, Thunder and Lancaster, a quartet of superior race-horses, and Bonnet, all by Lexington. Hedgeford was also the sire of the dam of Prunella by imp. Glencoe, she the dam of Sympathy and Lizzie W. by imp. Scythian, son of Orlando by Touchstone, Nellie Grey (the dam of Alroy, Amadis, Aramis, Marie Michon, Bazil, etc.), by Lexington. Other daughters of Hedgeford left good stock, and his early death was a loss to the country. He combined through his sire and dam a large infusion of Herod and Eclipse blood. Maria Black, the daughter of Filho-da-Puta, grandson of Sir Peter, dam by Smolensko, son of Sorcerer by Trumpator, out of a daughter of Sir Peter, figured conspicuously on the turf in America in 1838 and 1839, and won at two, three and four mile heats, in the latter race winning after four heats were run. She was a superior brood-mare, dam of John Black by imp. Trustee, Gertrude by Chorister, son of imp. Contract, Sellie Waters by Glencoe, Madame Bruce by Boston, Frank Waters by Wagner, Bay Leaf by imp. Yorkshire and Tripod by imp. Sovereign. But her name and fame is perpetuated through her daughter Bay Leaf, one of the best brood-mares ever in America. Bay Leaf was the dam of Rubicon, (sent to England), Bay Flower, Beacon, Bayswater, Baywood, Bayonet, Niagara, Bay Rose and Bay Final (a winner in England), all by Lexington, Bingaman and M. A. B. by Asteroid and Bay Bush by Australian. Bay Flower was a good race-mare, and dam of several good performers, including Ivy Leaf (the dam of Bazar by Jack Malone), Bramble, Bonnie Leaf, Bye and Bye (Now or Never's dam), Brambaletta (Merry Monarch's dam), all by imp. Bonnie Scotland, Eclat by Enquirer, Thistle by Great Tom, Niagara and Jackson by Luke Blackburn. Eliza by Filho-da-Puta, dam Zephyrina by Middlethorpe, son of Shuttle by Young Marske, was the dam of the good mare Albine by Jeff Davis, son of Hero by Bertrand jr., and she the dam of the good race-horse Gen. Yorke by Planet. Nameless by Filho-da-Puta was a dam of winners, and The Nun's Daughter by Filho-da-Puta, dam The Nun by Blacklock was the dam of the good race-horse Consol jr. by imp. Consol and of the Stumps mare, by Stumps, son of Whalebone, that produce Pot-8-o's by Glencoe, and Blacklock by O'Meara, son of Glencoe. Stamford, the brother to imp. Archduke was a superior race-horse, and his name appears in many excellent pedigrees. Sir Peter got Walton and his brother Williamson's Ditto, out of Arethusa by Dungannon, son of Eclipse, grandam by Prophet, son of Regulus, out of Virago, Saltram's dam by Snap, grandson of Flying Childers. Walton was a superior race-horse and left a grand reputation as a stallion. He was sire of Phantom, out of Julia (sister to Eleanor) by Whiskey, son of Saltram, grandam Young Giantess, the dam of Sorcerer, by Diomed. Phantom won the Derby in 1811, and Walton's son St. Patrick, dam by Dick Andrews, son of Joe Andrews by Eclipse, won the St. Leger in 1820 and the Gold Cup at York. Phantom got Cedric, out of sister to Repeator by Trumpator, grandam Demirep by Highflyer, who won the

Derby in 1824, and Middleton, dam Web by Waxy, son of Pot-8-o's, out of Penelope by Trumpator, that won the Derby in 1825. But Phantom's best reputation comes through his daughters. He was the sire of Cobweb, out of Filagree by Soothsayer, son of Sorcerer, grandam Web, sister to Whalebone by Waxy. Cobweb won the Oaks and 1,000 Guineas in 1824, and was the dam of Bay Middleton (son of Sultan, by Selim), that won the 2,000 Guineas and the Derby in 1836. Bay Middleton was the sire of the Flying Dutchman (dam Barbelle by Sandbeck, son of Catton), that won both the Derby and St. Leger in 1849, the Ascot Cup in 1850, and defeated the noted Voltigeur in the Match at York, two miles. Bay Middleton was the sire of Hermit, that won the 2,000 Guineas in 1854, Aphrodite the 1,000 Guineas in 1851, Andover the Derby in 1854. He was the sire of Ellen Middleton, dam Myrrha by Malek, son of Blacklock, out of Bessy by Young Gouty, grandson of Sir Peter. Ellen Middleton is famous as the dam of Wild Dayrell, that won the Derby in 1855, by Ion, son of Cain, by Paulowitz, by Sir Paul, by Sir Peter, much inbred to Sir Peter. Wild Dayrell was the sire of Buccaneer, out of Cruiser's dam by Little Red Rover, son of Tramp. Buccaneer was the sire of Kisber, winner of the Derby in 1876, Formosa, winner of the Oaks and St. Leger in 1868, Cadet, See Saw, Brigantine, winner of the Oaks in 1869, and The Rake, out of England's Beauty by Birdcatcher, who was imported after being a failure in England, but died within a year. Walton was also the sire of Partisan, dam Parasol by Pot-8-o's, son of Eclipse, grandam Prunella by Highflyer, son of Herod, out of Promise by Snap. Partisan was the sire of Mameluke, dam Miss Sophia by Stamford, son of Sir Peter, grandam Sophia by Buzzard, son of Woodpecker, by Herod. Mameluke won the Derby in 1827, and is known to America as the sire of imp. Myrtle, the dam of the famous brood-mare Magnolia by imp. Glencoe, that produced such grand race-horses as Kentucky, second to no son of Lexington over any distance of ground; Madeline, the grandam of Iroquois, winner of the Derby, St. Leger and Prince of Wales stakes in 1881. Charley Ball, Princeton, Charles Morgan, Hanover by imp. Yorkshire, Simon Kenton, Madonna and Skedaddle by the same sire, and Gilroy, brother to Kentucky and Victory by Uncle Vic, son of Lexington. Partisan was sire of Glaucus, dam Nanine by Selim, son of Buzzard, and Glaucus was the sire of Refraction, that won the Oaks in 1845, and also won the Ascot Gold Cup from Rockingham. Glaucus was also the sire of The Nob, dam Octave, sister to Oxygen, by Emilius, out of Whizgig by Rubens. The Nob was the sire of The Nabob, dam Hester by Camel, son of Whalebone, out of Monimia by Muley, etc. The Nabob was the sire of Vermouth, dam Vermeille by The Baron, out of Fair Helen by Priam, and Bois Roussel, dam Agar by Sting, son of Slane; also Nutbourne, out of Princess by Merry Monarch. Vermouth got Boiard (winner of the French Derby and Grand Prix of Paris in 1873), dam La Bossue by De Clare, son of Touchstone. The Nabob was sire also of Nutbourne, the sire of imp. Woodlands, dam Whiteface by Turnus. Partisan further handed down the Sir Peter blood through his two sons, Gladiator, dam Pauline by Moses (Derby 1822), out of Quadrille by Selim, and Venison, dam Fawn by Smolensko, son of Sorcerer, out of Jerboa by Gohanna, son of Mercury by Eclipse. Gladiator and Venison finished second and third to Bay Middleton in the Derby in 1836. Partisan was the sire of Patron, winner of the 2,000 Guineas in 1829, Zeal, winner of the 1,000 Guineas in 1821, and Cyprian, winner of the Oaks in 1836. Gladiator went to France, where he became famous as a sire. He got Miss Gladiator, dam Taffrail by Sheet Anchor, son of Lottery by Tramp. Miss Gladiator was the dam of the famous Gladiateur that won the 2,000 Guineas, Derby, St. Leger and Grand Prix of Paris, and was the only horse which ever accomplished the feat. Gladiator was sire also of Sweetmeat, dam Lollypop by Starch or Voltaire, out of Belinda by Blacklock. Gladiator also got Fulvie, the dam of Dutch Skater by Flying Dutchman. Sweetmeat was the sire of Mincemeat and Mincepie, winners of the Oaks in 1854 and 1856, and Macaroni, dam Jocose by Pantaloon, son of Castrel, by imp. Buzzard. Macaroni won the 2,000 Guineas and Derby in 1863, and was the sire of Spinaway, winner of the 1,000 Guineas and Oaks in 1875, Bonny Jean, winner of the Oaks in 1883, Macgregor, Macheath Couronne de Fer, and imp. Macaroon, dam Songstress by Chanticleer, son of Birdcatcher.

Macaroon was imported to America, and is in the Brookdale stud of the late Mr. D. D. Withers. He is sire of the winners Macbeth and Macduff (the latter now in the McGrathiana stud, and highly prized). Cadence, dam Castagnette, by Marsyas, son of Orlando (she the dam of Trill and Cascade by Uncas, son of Lexington and other winners), Favoress, dam imp. Faverdale by The Palmer, in Dixiana stud; Gem, dam Bijou by Australian, in Chesterbrook stud, Pennsylvania, Knicknack, sister to Gem and dam of Bibelot, in Brookdale stud, Paradox, dam Maxim, by Planet, in McGrathiana stud, and Gregory, out of imp. Abundance by Kaiser, a first-class horse. Macaroni also got Bonbon (imported by the late M. H. Sanford), dam Blondella by The Flying Dutchman. She is the dam of the good race-horse Little Ruffin, Petersburgh, and their sisters, Caramel and Tulare, all by Monarchist. Imp. Clover by Macaroni, dam Verdure, by King Tom, son of Harkaway, was the dam of Flageoletta and Aftermath by Rayon D'or. Dauntless by Macaroni, dam Artless by Archy, son of Camel by Whalebone, is the dam of Florimel, Delilah, and King Cole by Ill Used, and Mr. Pelham and Fearless by imp. St. Blaise. Imp. Lady Nar by Macaroni, dam Queen O' Scotts by Blair Athol, is owned by Dr. A. W. McAlester of Missouri. Polenta (in the late Nursery stud) by Macaroni, dam by Lord of the Isles, son of Touchstone, out of Shot by Birdcatcher, is the dam of Farina by Kentucky, Tapioca by Kingfisher, Polonius (in England), Polydora by St. Blaise, and Louise by Ill Used. Imp. Inverness by Macaroni, dam Elfrida by Faugh-a-Ballagh, son of Sir Hercules by Whalebone, is the dam of Short Line, and Garrick by Lexington, Lady Caroline by Australian, Patroness by Pat Malloy, Haidee, Mary K, and Joe Cotton by King Alfonso, the most of which were prominent on the turf, and Joe Cotton was one of the speediest horses that ever faced a starter. His one mile and three-sixteenths in 2.00¼ with 109½ lbs. up was one of the best races ever run, and has not been excelled, the best being 200¼.

Imp. Judith by Macaroni, dam Verdure, by King Tom, out of May Bloom, by Newminster, is in the Algeria stud. Imp. Queen Maud, dam Queen of the Vale by King Tom, out of Agnes by Pantaloon, dam of Queen O' the May, is the dam of Stoney Montgomery and Hawkstone.

Moccasin by Macaroni, dam Madame Strauss by King Tom, out of Jetty Treffz by Melbourne, is a large, fine horse, which never raced. Before going to Canada he got some good horses, such as Vampire, Faith, Amazon, Disdain, etc. He is now in the Edgewater Stud, Kentucky, and should be heard from as a sire. No horse can be better bred.

Macaroni's daughters have been successes in the stud in England. Lily Agnes, dam Polly Agnes by The Cure, son of Physician, son of Brutandorf by Blacklock, is the dam of imp. Rossington by Doncaster, son of Stockwell, and of the grand race-horse Ormonde by Bend'or, son of Doncaster, by Stockwell, and Rouge Rose by Thormanby, the latter and Bend'or Derby winners, and Ormonde, winner of the 2,000 Guineas, Derby, and St. Leger. Gladiator was also the sire of Fitz Gladiator, dam Zarah by Reveller, son of Comus, by Sorcerer. Fitz Gladiator was the sire of Compiègne, dam Maid of Hart by The Provost, son of The Saddler, by Waverly, by Whalebone. Compiègne was the sire of imp. Mortemer, dam Comtesse by The Baron or Nuncio, credited to the latter, who was a son of Plenipotentiary by Emilius. Mortemer was not only a superior race-horse, but the sire of such good ones as Chamant, winner of the 2,000 Guineas; Verneuil, winner of the Grand Prix of Paris, and St. Christophe, winner of the Gold Vase, Gold Cup and Alexandra Plate at Ascot. His success in America was equal to that in Europe, and whenever bred to mares possessing racing blood the produce was good. He got Exile, dam imp. Second Hand by Stockwell, out of Gaiety by Touchstone, also a superior horse, Adonis, dam Alice Ward by Lexington, Wanda and Winfred, out of Minnie Minor by Lexington, Chimera, Cyclops, and Cambyses out of Lizzie Lucas by Australian. Saluda out of Perfection by Leamington, Savanac out of Sly Boots by Rivoli, son of Revenue by imp. Trustee; Heimdel and Housatonic out of Vandalite by Vandal, son of Glencoe.

Parmesan by Sweetmeat, dam Gruyère by Verulam, son of Lottery by Tramp, grandam Jennala by Touchstone. Parmesan is the sire of Cremorne (winner of the Derby, 1872) dam Rigolboche by Rataplan, son of The Baron, by Birdcatcher. Parmesan is sire of imp. Strachino, out of May Bell, by Hetman Platoff. Favonius by Parmesan dam Zephyr by King Tom, son of Harkaway by Economist, won the Derby in 1871. Favonius was not only a race-horse of merit, but is the sire of Sir Bevys (winner of the Derby in 1879), dam Lady Langden by Kettledrum (winner of the Derby in 1861), son of Rataplan. Gladiator is immortalized as the sire of Queen Mary, dam by Plenipotentiary, son of Emilius by Orville, grandam Myrrha by Whalebone, son of Waxy by Pot-8-o's, etc. Queen Mary has given us a grand lot of Stallions and Matrons which have made their mark on both sides of the Atlantic. She produced twenty foals, including Haricot by Mango or Lanercost, Braxey by Moss Trooper, son of Liverpool by Tramp, Bab at the Bowster and imp. Balrownie by Annandale, son of Touchstone, Blooming Heather and Blink Bonny by Melbourne, imp. Bonnie Scotland by Iago, son of Don John by Tramp or Waverly, Bonnyfield, Bonny Breastknot and Bonny Bell by Voltigeur, Broomielaw by Stockwell, Bertie by Newminster. Blinkhoolie by Rataplan, Bertha by Young Melbourne and Bonnie Doon by Rapid Rhone, son of Young Melbourne. None of her sons or daughters, except Blink Bonny winner of the Oaks and Derby, 1857, won any of the great fixed events of the English turf, Bonny Doon is the dam of imp. Bella Donna by Hermit, she the dam of the winner Bel Demonico her first foal. Bonny Bell is the dam of imp. Blythwood and Beauclerc, winner of the Middle Park Plate. Haricot produced Caller Ou by Stockwell, that won the St. Leger in 1861, the Brighton Cup and the Northumberland Plate in 1863 and '64, and was the dam of The Pearl by Newminster. Haricot was the dam of Lady Langden by Kettledrum, that was the dam of Hampton by Lord Clifden. Hampton was the sire of Merry Hampton (Derby winner in 1887), dam Doll Tearsheet by Broomielaw, Queen Mary's son, and Ayrshire (Derby winner in 1888) out of Atalanta by Galopin, son of Vedette. Lady Langden was also the dam of Sir Bevys, Derby winner in 1879 Braxey was the dam of Thrift by Stockwell, and she the dam of Tristan by Hermit. Tristan ran second in many of the big events, won the Gold Cup, two Gold Vases and Hardwicke Stakes. Thrift was also dam of Trapeze by Hermit, and Pursebearer by Scottish Chief. Bernice by Stockwell, out of Braxey, was in the Nursery Stud and was the dam of Bertram and Beatrice, the dam of Bellona, she the dam of Belladonna (Arnica's dam), Belinda, etc. Blink Bonny won the Oaks and Derby, 1857, and was the dam of Blair Athol by Stockwell. Blair Athol won the Derby and St. Leger, and Breadalbane his brother, won the Prince of Wales and Gratwicke Stakes. Blair Athol got Silvio (Derby winner, 1877), out of Silverhair by Kingston, son of Venison. Beauclerc by Rosicrucian, dam Bonny Bell, won the Middle Park Plate. Bella Donna by Hermit dam Bonnie Doon by Rapid Rhone, is in the Nursery Stud. Lady Stockwell by Knowsley, son of Stockwell, out of Bab-at-the-Bowster by Annandale, is in the Walnut Hill Stud in Kentucky, her daughter Salara (imported *in utero*) by Salvator, is proving a good broodmare, as are also her daughters Sophronia by Ten Broeck and Sungleam by King Alfonso. Venison was the sire of The Ugly Buck, winner of the 2,000 Guineas in 1844, Miami winner of the Oaks in 1847 and Clementina who won the 1,000 Guineas in the same year. Venison was also the sire of Alarm, dam Southdown, by Defence son of Whalebone, and of Kingston, dam Queen Ann, by Slane son of Royal Oak by Catton. Kingston was a fine and game horse, he was the sire of Caractacus (winner of the Derby in 1862), Queen Bertha (winner of the Oaks in 1863), Ely, and Silverhair, dam of Silvio (winner of both the Derby and St. Leger), Coimbra, dam of Claremont, imp. Stylites and Imp. Stonehenge in the Brookdale Stud, and a successful sire. Non Pareille, the sister to Silverhair, was in the late Nursery Stud, and is the dam of Matador by Gladiateur, Perfection by Kingfisher and Nina, the dam of Punka, by Kentucky. Imp. Babta by Kingston, dam Alice Lowe, by Defence son of Whalebone, is the dam of Glenelg, (imported *in utero*) by Citadel son of Stockwell, who has established his reputation as one of the most successful stallions in America. Eltham Lass (imported), by Kingston, dam by Pyrrhus the First, son of Epirus by Langar, out of Palmyra by Sultan, was the dam of Kingfisher, winner of the Belmont, Champion and Travers Stakes, and an excellent and successful stallion, Sabina (dam of

Rachiel) Syria, The Widow, (dam of Ollie Glenn, by Hindoo, and Ocean by imp. Billet), King Pin, King Bolt and Majestic (the dam of the superior race-horse Eurus, winner of the Suburban and other important races), all by Lexington. Imp. Charaxus, by Distin, son of Trumpeter by Orlando, is out of Sappho by Kingston and was not only a good race-horse, but is making himself known as a successful stallion.

Sir Paul, own brother to Paulina, winner of the St. Leger in 1807, was by Sir Peter, dam Pewet by Tandem (son of Syphon by Squirt) out of Termagant by Tantrum, son of Cripple by the Godolphin Arabian. Pewet was the dam also of Clinkerina (the dam of Humprey Clinker) by Clinker. Termagant was the dam of Evelina by Highflyer, the dam of Cervantes, Orville and Paulowitz by Sir Paul. Paulowitz was the sire of Archibald (winner of the 2,000 Guineas), and Cain, dam by Paynator, son of Trumpator by Conductor. Cain is famous as the sire of Ion, who ran second in the Derby and St. Leger in 1838 and was the sire of Wild Dayrell and Tadmor, he the sire of Seclusion, Hermit's dam. Wild Dayrell was the sire of Buccaneer, Wild Oats, imp. The Rake and Hurricane winner of the 1,000 Guineas, 1862. Ion was also the sire of Adeline, the grandam of St. Simon by Galopin. St. Simon was disqualified for the great events, the Derby and St. Leger, by the death of Prince Batthyany, but was a superior race-horse, won a large number of events including the Epsom, Ascot and Goodwood Cups, and is regarded as one of the best sires in England. The Byerly Turk blood is not less famous through Herod's other son Woodpecker, foaled in 1773, dam Miss Ramsden by Cade, son of the Godolphin Arabian, grandam by Lonsdale Bay Arabian, third dam sister to Whitefoot, and Bonny Lass, by Bay Bolton, out of a daughter of the Byerly Turk, tracing to a Natural Barb mare. Woodpecker was a superior race-horse and a grand and successful stallion. He was the sire of Bustard, dam Matron by Alfred, imp. Buzzard, dam Misfortune by Dux, imp. Dragon and his daughter are famous in the stud. Ambrosia by Woodpecker, dam Ruth by Blank, was the dam of Fanny (the dam of Sir Oliver, Miss Teazle, Poulton, etc.) by Diomed. Abigail by Woodpecker, dam Firetail by Eclipse was the dam of Zodiac by St. George. Chesnut and Grey Skim were sisters by Woodpecker, dam Silver's dam by Herod. The former, Chesnut Skim, was the dam of Election (winner of the Derby in 1807) by Gohanna; the latter, Grey Skim, was the dam of Skim and Castanea, by Gohanna, that figures in many of our pedigrees. Active by Woodpecker, dam Laura by Whistle-Jacket, son of Mogul, was the dam of imp. Expedition (by Pegasus son of Eclipse), the sire of Honesty (Monmouth Eclipse's dam), Katydid, Expedition mare (dam of American Boy by Sea Gull), Maid of the Oaks (the dam of Medoc, Midas, Gipsy and Cora by American Eclipse), Katy Ann by Ogle's Oscar, to which the Lewis family trace that has produced so many good ones such as Potomac, Sensation, Stratford, Onondaga, Glenmore, Acrobat, John Morgan, Hunter's Lexington, Morgan Scout, Ratan, Girotlé, etc. Matchless, another daughter of Expedition, founded the Morgan family, through her daughter, Lady Morgan by John Richard, son of Sir Archy, another distinguished horse in male line to the Byerly Turk. Woodpecker's daughter Rosina, out of Petworth by Herod, inbred to the Byerly Turk, was the dam of Hermes, by Mercury, son of Eclipse. Wren by Woodpecker out of Papillon (Sir Peter's dam), by Snap, was the dam of Agonistes and imp. Robin Redbreast by Sir Peter, the latter sire of the Robin mare that was the dam of Thornton's Rattler, Sumpter, Childers, Elizabeth and Flirtilla, all by Sir Archy, a horse in the male line to the Byerly Turk. A daughter of Woodpecker, dam Everlasting by Eclipse, was the dam of Fraxinella (grandam of Tramp) by Trentham, and Fractious (the dam of Carthage, Amazon, and Hannibal (this last winner of the Derby in 1803), all by Driver, son of Trentham by Sweepstakes. Woodpecker was the sire of Catharine and Colibri, dam Camilla by Trentham. Catharine was the dam of Golumpus, Hedley, Kate and Wanderer by Gohanna, Sprite by Bobtail, and Worthy by Whalebone. Woodpecker got many other good brood-mares. Woodpecker's son Buzzard, who was imported into Virginia in March, 1805, and died in Kentucky in 1808, left his impression deeply on the stock of this country. He got in England, prior to his importation, Hornby Lass, dam Puzzle by Matchem (she the dam of Morel, Mango, Truffle, and Emperor by Sorcerer), Sophia, dam Huncamunca by Highflyer (she the dam of Miss Sophia by Stamford, Miss Sophia was dam of Master Henry and of Mameluke, winner of Derby in 1827), Vanity, dam Dabchick by Pot-8-o's (she the dam of Conceit and Caprice by Walton), and Coquette by Dick Andrews. Buzzard was the sire of the three brothers Castrel, Selim and Rubens, and their sister, Bronze, dam by Alexander, son of Eclipse, and Rosamond, dam Roseberry by Phoenomenon, she the dam of imp. Barefoot by Tramp. Barefoot was the sire of the good race-mares Motto and Clara Howard, whose descendants have proven the value of this blood on the turf.

Castrel was the sire of Bustard, dam Miss Hap by Shuttle, son of Young Marske. Bustard got Heron, dam by Orville, son of Beningbrough by King Fergus. Heron got the noted race-horse Fisherman, out of Mainbrace by Sheet Anchor, son of Lottery. Fisherman ran one hundred and nineteen races and won sixty-nine. He was exported to Australia where he made a great reputation as a stallion. Heron was also the sire of Snowdrop, dam Fairy by Filho-da-Puta, and she the dam of Gemma di Vergy by Sir Hercules, son of Whalebone, who has made a good reputation as a stallion. Castrel was also the sire of Merlin, dam Miss Newton by Delpini, and of Pantaloon, dam Idalia by Peruvian, son of Sir Peter. Pantaloon was the sire of Sleight of Hand, dam Decoy by Filho-da-Puta, and The Libel, dam Pasquinade, sister to Touchstone by Camel. The Libel is sire of Bribery, dam Splitvote by St. Luke (she the dam of St. Albans, winner of the St. Leger in 1860); also sire of Traducer, dam Arethusa by Elis, that is the sire of imp. Sir Modred and Cheviot (see tables). Pantaloon also sired the brothers, Hobbie Noble, Elthiron and Windhound, out of Phryne by Touchstone. Windhound is admitted to be the sire of Thormanby (winner of the Derby in 1860) dam Alice Hawthorn by Muly Moloch. Thormanby is the sire of Charibert, Violet (the grandam of Melton, winner of the Derby in 1885) and Rouge Rose, the dam of Bend'or winner of the Derby in 1880, he the sire of Ormonde. Thormanby was the sire of Thuringian Prince, dam Eastern Princess by Surplice, and he the sire of the superior race-horse Bird of Freedom, winner of the City and Suburban Handicap, Ascot Cup and Gold Vase; also of Plaudit, he the sire of Balfe (a good race-horse and stallion), dam Bohemia by Weatherbit, Thormanby was the sire of imp. Glengarry, out of Carbine by Rifleman. Legerdemain, sister of Drone, Sleight of Hand and Van Amburgh by Pantaloon, dam Decoy by Filho-da-Puta, is the dam of Toxophilite by Longbow, son of Ithuriel by Touchstone. Toxophilite is the sire of Stray Shot, out of Vaga by Stockwell, she the dam of Torpedo, Martini and the noted Shotover by Hermit. Toxophilite is sire of Musket, out of daughter of West Australian, and he the sire of Petronel, winner of the 2.000 Guineas, 1880. Musket was exported to Australia and has proved highly successful there, as a sire. Pantaloon is the sire of Ghuznee (winner of the Oaks in 1841), her dam Languish by Cain. Ghuznee was the dam of Storm, Escalade, etc., by Touchstone, the latter the dam of Citadel by Stockwell, sire of imp. Glenelg. Pantaloon was the sire of Satirist (winner of the St. Leger in 1841, and the Ascot Gold Vase), out of Sarcasm by Teniers, son of Rubens. Pantaloon was the sire of the dam of imp. Leamington, and Jocose the dam of Macaroni. Selim was the sire of Langar, dam by Walton, out of Young Giantess by Diomed. Langar was the sire of Stockport, Elis (winner of the St. Leger in 1836) and Epirus, out of Olympia by Sir Oliver, son of Sir Peter. Epirus was the sire of Pyrrhus I. (winner of the Derby in 1846), dam Fortress by Defence, son of Whalebone. Langar was sire of Vulture, dam Kite by Bustard, son of Castrel, that was the dam of Orlando (winner of the Derby in 1844), by Touchstone, and of Progress, the dam of Attila (winner of the Derby in 1842), her dam by Blacklock. Selim was the sire of Sultan, foaled in 1816, dam Bacchante by Williamson's Ditto, out of sister to Calomel by Mercury, son of Eclipse. Sultan was the sire of imp. Glencoe, dam Trampoline by Tramp (see table). Glencoe won the 2,000 Guineas in 1834, and was imported to America after the season of 1836. The Glencoe mare, dam Alice by Whalebone, was the dam of Adine by Slane, Torment by Alarm and others, but Glencoe's reputation is lasting as the sire of Pocahontas, dam Marpessa by Muley, son of Orville, out of Clare by Marmion, son of Whiskey by Saltram. Pocahontas was the dam of Stockwell (winner of the 2,000 Guineas and the St. Leger in 1852), Rataplan (third in the St. Leger in 1853, Tradesman's Cup, etc.), both by The Baron, son of Birdcatcher, King Tom, by Harkaway, son of Economist, by Whisker, Ayacanora by Birdcatcher, Knight of Kars by Nutwith, son of Tomboy, Knight of St. Patrick by imp. Knight of St. George, Auricula and Araucaria by Ambrose, son of Touchstone, and Annette by imp. Priam. Stockwell proved the most potent and valuable of all the sires in England. He got The Marquis, Bothwell and Gang Forward (winners of the 2,000 Guineas and The Marquis also won the St. Leger), Lord Lyon (winner of the 2,000 Guineas, Derby and St. Leger), Blair Athol, winner of the Derby and St. Leger, Doncaster, winner of the Derby, Lady Augusta, Repulse and Achievement, winners of the 1,000 Guineas (Achievement also won the St. Leger), Regalia, winner of the Oaks, Savernake and St. Albans, the latter winner of the St. Leger, Caller Ou, the grand mare, winner of St. Leger and a number of Queen's Plates. Blair Athol got Prince Charlie, winner of the 2,000 Guineas, Silvio, winner of both Derby and St. Leger, and Craig Millar, winner of the St. Leger. St. Albans got Julius and Springfield, the latter the sire of Sainfoin, winner of the Derby in 1890. Doncaster got Bend'or, winner of the Derby, and Bend'or sired the great Ormonde, winner of the 2,000 Guineas, Derby and St. Leger, Bertram, the grandson of Stockwell, is the sire of Robert the Devil, winner of the St. Leger. Rataplan was the sire of Kettledrum, winner of the Derby, The Miner, Ben Battle, Blinkhoolie and Cymbal, good stallions. King Tom got Tomato, winner of the 1,000 Guineas, Hannah, winner of the 1,000 Guineas, Oaks and St. Leger, Tormentor and Hippia, Oaks winners, Kingcraft, winner of the Derby, Great Tom, King Ernest, King Ban and Phaeton, imported into America and successful and valuable stallions. Ayacanora is the dam of Chattanooga by Orlando, Cachuca by Voltigeur, the latter dam of imp. Castagnette by Marsyas and grandam of Quicklime, Chattanooga is the sire of Printanniere, the dam of Poulet in France. Araucaria is the dam of Stephanotis by Macaroni, Wellingtonia by Chattanooga, and the grand horse Rayon d'Or, now at the head of the Algeria stud (see table). Of all the stallions subsequent to Diomed, Glencoe made the greatest impression on the blood stock of America. Glencoe got but few males, and at his death, Vandal and Panic were about the only sons alive. The latter had no chance in the stud, and Vandal little, until purchased by the late Gen. Harding and put in the Belle Meade stud, when he got Ventilator and Vanguard, out of Carolin by imp. Scythian, son of Orlando, Mollie Jackson, the grand race-mare and great grandam of Foxhall. Ventilator is sire of Mary Anderson, Drizzle, Airtight, Airplant, Airshaft and others. Vandal also got Vandalite, the best mare of her day, Vanderbilt (sire of Madstone, etc.), out of Melrose by Childe Harold, Voltigeur (sire of Walter H, Frank Ward, etc.), dam Duet by Highlander, son of Glencoe, and Coral, dam of Uncas and Wanderer. But in Vandal's son Virgil, we find his greatest excellence. Virgil's dam was Hymenia by imp. Yorkshire. Virgil was the sire of Vagrant, Vigil, Vera Cruz, Virginius, Fairplay, Virgilian, Santa Anita, Momento, Maurice O. and the grand race-horse and popular stallion Hindoo (sire of Hanover, Sallie McClelland, Buddhist, Mabel Glenn, etc.), and the unbeaten Tremont, sire of Dagonet, a promising two-year old, 1891. Glencoe was the sire of a large number of grand broodmares, including the sisters Mildred and Lightsome, Charmer the grand race-mare, Sally Lewis, Topaz (dam of Wagram, Waterloo, Lodi, Arcola, and Colton), Reel (dam of Lecomte, Prioress, Starke, War Dance, etc.), Magnolia (dam of Kentucky, Gilroy, Victory, Princeton, Skedaddle, Madeline, the grandam of Iroquois, etc.), Novice (dam of Norfolk), Nebula (dam of Asteroid), Kitty Clark (dam of Maiden, La Henderson, etc.), Bonnie Lassie, Blonde, Florine (dam of Idlewild, Aerolite, etc.), Pryor, and his two sisters, one of them the dam of Glycera and Colossus, by imp. Sovereign, the other the dam of Alumina, Sister of Charity, Lord Jersey, Wheatly, Midsummer, etc. A Glencoe mare was the dam of Optimist. Sultan was the sire of Bay Middleton, dam Cobweb by Phantom, son of Walton by Sir Peter, a splendid race-horse, winner of the 2,000 Guineas and Derby, in 1836. He is the sire of Flying Dutchman (winner of the Derby and St. Leger in 1849), and his full brother Vanderdecken. Flying Dutchman was sire of Dutch Skater and Dollar. Bay Middleton was sire also of Andover, dam sister to Aegis, by Defence, winner of the Derby in 1854, sire also of Ennui, dam Blue Devils by Velocipede. Ennui was a superior racer and dam of

Saunterer by Birdcatcher, Loiterer by Stockwell and Bravery by Gameboy. Bravery was the dam of Salamanca by Student, and she the dam of Pero Gomez (winner of the St. Leger in 1869) by Beadsman. Bay Middleton was also sire of Cowl, dam Crucifix by Priam. Cowl was the sire of Miss Sellon, the grandam of Hermit sire of St. Blaise, also Madame Eglantine, the dam of Rosicrucian, and The Palmer, by Beadsman, also of Morgan-la-Faye, dam Miami (Oaks winner, 1847), by Venison, she the dam of Marie Stuart (winner of the Oaks and St. Leger in 1873), by Scottish Chief. But the great reputation of Bay Middleton in the female line is from Ellen Middleton, the dam of the noted Wild Dayrell. Sultan was the sire of Ibrahim (winner of the 2,000 Guineas, 1835), dam sister to Cobweb by Phantom, Augustus (winner of the 2,000 Guineas, 1830), dam Augusta by Woful, Achmet (brother to Bay Middleton), winner of the 2,000 Guineas, 1837, Greenmantle (winner of the Oaks, 1829), dam Dulcinea by Cervantes, son of Don Quixote, and Galata (winner of the 1,000 Guineas and Oaks, 1832), dam Advance by Pioneer.

Sultan was sire also of Hampton and Adana (the latter imported to America), dam Rachel by Whalebone; Beiram, dam Miss Cantley by Stamford; Jereed, dam My Lady, by Comus; Ishmael, dam by Phantom, sire of Star of Erin. Selim was the sire of Turcoman, dam Pope Joan by Waxy.

Rubens, the brother to Castrel and Selim, was a good race-horse, and got a large number of racers. He is particularly distinguished for his daughters. Landscape, dam Iris, by Brush, son of Eclipse, won the Oaks in 1816; Pastille, dam Parasol, by Pot-8-o-'s, son of Eclipse, won the 2,000 Guineas and Oaks in 1822, and Whizgig, dam Penelope, by Trumpator, won the 1,000 Guineas the same year. Whizgig was the dam of Oxygen, winner of the Oaks, and grandam of Trumpeter. Rubens was the sire of the dam of Phosphorus, winner of the Derby in 1837, and his sister May Day, winner of the 1,000 Guineas in 1834. Rubens was the sire of Defiance, the dam of Dangerous, by Tramp (winner of the Derby in 1833), Defence by Whalebone, Design by Tramp, and Delight by Reveller, the last two imported into America, and dams of winners. Rubens got Ruby, dam, by Williamson's Ditto, that was the dam of Coronation (Derby winner in 1841) by Sir Hercules. Rubens was also the sire of Bobadil, dam by Skyscraper, son of Highflyer. Bobadil was the sire of the grandam of the great brood-mare Magnolia. Rubens was also the sire of Teniers, dam Snowdrop by Highland Fling, son of Spadille, and Peter Lely, dam Stella by Sir Oliver, winner of the Gold Cups at Worcester, Warwick, and Leicester; Miniature, the dam of Enamel (winner of the 2,000 Guineas, 1825) by Phantom. Imp. Eliza, dam Little Folly by Highland Fling, was the dam of Emerald (the dam of Topaz), by imp. Leviathan. Rubens sired many other brood-mares that produced winners. The Byerly Turk blood in the male line permeates the pedigrees of all the best race-horses in America, through Diomed, who was sold and imported to Virginia after he was 22 years old, but such was his natural vigor and stamina that he lived to be 31 years old, having been foaled in 1777, and died in 1808 in Virginia. Diomed was by Florizel, son of Herod, by Tartar, by Partner, by Jigg, by Byerly Turk. Diomed's dam was sister to Juno by Spectator, son of Crab by Alcock Arabian, out of Horatia by Blank, son of the Godolphin Arabian. Diomed won the initial Derby in 1780. He also won other races up to four-mile heats, and got some excellent horses and brood-mares before his importation, such as Glaucus and Lais, dam Grace by Snap, Grey Diomed, Robin Grey, Greyhound, all out of Grey Dorimant by Dorimant; Wrangler, imported into Virginia; Champion, the Diomed Mare of Lord Fitzwilliam (dam of Wonder and Miracle by Phoenomenon, son of Herod, Caleb Quotem by Sir Peter, and Cossack by Sir Paul), Diomed Mare (dam of Regulator and Momentilla by Whiskey): Young Giantess, dam Giantess by Matchem, the dam of Sorcerer by Trumpator, Eleanor by Whiskey (winner of the Oaks and the first mare to win the Derby, and the dam of Muley, the sire of imp. Leviathan and Margrave), Julia, Lydia and Cressida, all by Whiskey. Julia was dam of Phantom (winner of the Derby in 1811) by Walton, son of Sir Peter. Lydia was the dam of The Corporal by Orville, and The Drummer by Waxy. Cressida was the dam of imp. Priam by Emilius, Priam got the famous Crucifix (winner of the 2,000 and 1,000 Guineas and Oaks in 1840), the dam of Surplice (winner of the Derby, 1848) by Touchstone; Miss Letty, dam by Orville, son of Beningbrough, winner of the Oaks in 1837, and the dam of Weatherbit by Sheet Anchor, he the sire of Beadsman, winner of the Derby in 1858, and he sire of Rosicrucian (winner of the Prince of Wales and Craven stakes), The Palmer (sire of Jenny Howlett (winner of the Oaks and dam of Chitabob), Gray Palmer, Pellegrino, Pilgrimage), etc. Beadsman also got Blue Gown (winner of the Derby, 1868), out of Bas-Bleu by Stockwell. Priam was also sire of Industry, dam Arachne by Fillo-da-Puta. Industry won the Oaks, 1838, and was the dam of Stitch by Hornsea, Lady Evelyn (winner of the Oaks, 1849), by Don John, and Chevalier d'Industrie by Orlando. Diomed also got Young Noisette, dam Noisette by Squirrel, she the dam of Clermont, Hymen and Navigator by Trumpator, son of Conductor, and Marmion by Whiskey, the latter sire of the grandam of Pocahontas. But it is in America where Diomed left an imperishable name. He was the sire of Sir Archy, the Godolphin Arabian of America. Sir Archy was inbred to Byerly Turk through sire and dam. Diomed was a grandson of Herod, and Castianira was by Rockingham, also grandson of Herod. Castianira raced successfully in this country, as also did her son Sir Archy, who was one of the most successful sires in the world. He was the sire of Bertrand and Pacific, out of Eliza by imp. Bedford, son of Dungannon by Eclipse, both good and successful stallions. Sir Archy Montorio (Big Archy) and Sir William of Transport, out of Transport by Virginius, son of Diomed, Young Lottery, Kosciusko, Saxe-Weimar, Crusader and Phenomenon, out of Lottery by imp. Bedford, she out of imp. Anvilina by Anvil, son of Herod, Director and Virginian out of Meretrix by Magog, Coquette, Arab, Tariff, Brilliant, Eliza and Gen. Brook, out of Bet Bounce by imp. Sir Harry, son of Sir Peter by Highflyer. Coquette, bred to her own sire, Sir Archy, produced Janet (Virginia Lafayette) and Virginia Taylor. Sir Archy got Sir Charles and Janette, dam by imp. Citizen, son of Pacolet by Blank; Phillis, Gohanna, Brunette, Sally Trent and Charlotte Temple, out of Merino Ewe by imp. Jack Andrews, son of Joe Andrews by Eclipse; Thornton's Rattler, Childers, Sumpter, Elizabeth and Flirtilla out of the Robin mare by imp. Robin Redbreast, son of Sir Peter, she out of a mare by imp. Obscurity, son of Eclipse. Flirtilla was bred to her own sire and produced Flirtilla, Jr., and Antoinette. Flirtilla, Jr., bred to Boston (inbred to Diomed through Timoleon and Florizel), produced Ringgold (a good race-horse and for his chances a very successful stallion), sire of Onward, Delaware, Ringmaster, Tipperary, good race-horses, and Goldring, the dam of Bullion, etc. Ringgold covered only a few thoroughbred mares. Flirtilla, Jr., also produced Cassandra by imp. Priam, a good race-mare and the dam of Trojan, Basil and Philo by Mariner, the latter the dam of Lady Blessington by imp. Eclipse, and she the dam of Count D'Orsay, The Countess and The Baroness by Kentucky, Lady Rosebery and Duchess by Kingfisher, and Fairwater by Ill Used. The Baroness is the dam of Badge, by Ill Used. Lady Rosebery is the dam of Lady Primrose, Lord Dalmeny, etc., and Duchess is the dam of the good race-horse Watterson by Great Tom. Sir Archy was the sire of Vanity and Reality, out of a daughter of imp. Medley by Gimcrack, both good race-mares, and Reality was the dam of Johnson's Medley by Sir Hal, son of imp. Sir Harry, Slender and the noted Bonnets o' Blue by Sir Charles. Bonnets o' Blue was the dam of Mariner by Shark, son of American Eclipse and the noted mare Fashion by imp. Trustee. Fashion beat Boston in the famous match four-mile heats over the Union Course, L. I., May 10th, 1842, in 7.32½–7.45, the fastest race ever run in the world to that date. Fashion started in 36 races, 24 of which were four-mile heats, of which she won 21, won eight at three-mile heats, won three at two-mile heats, and lost one, winning 32 out of 36 races.

Sir Archy got Carolinian, Boxer and Lady Randolph, out of Poll by imp. Druid, son of Pot-8-o-'s by Eclipse; Wild Bill (Pilot) and Charles Kemble, dam Maria by Gallatin, son of imp. Bedford and Gabriella, dam Calypso by Bellair, son of imp. Medley. Gabriella, bred to Garrison's Zinganee (son of Sir Archy and Miss Chance by imp. Chance, son of Lurcher by Dungannon), produced George Martin and Miss Martin. George Martin was a superior race-horse at all distances and defeated Hannah Harris and the famous Reel, four-mile heats, in 7.33, 7.43, in 1844. He had previously won the first heat in 7.36½, the race being won in 7.39, 7.51½, by the noted little prodigy Miss Foot by imp. Consol, out of imp. Gabriella by Partisan. Miss Martin was the dam of Tokay by imp. Yorkshire, she the dam of Nannie Butler, King Lear, King Tom, King John and King Henry, good racers by Lexington. Nannie Butler was the dam of Girl of the Period (dam of Gerald, Geraldine and Geranium by imp. Saxon) by Virgil. Nannie Black, sister to Girl of the Period, was the dam of Glene (Jennie June's dam), by Glenelg, Princess Blandina and Brown Princess by Prince Charlie: and Nannie Bay by Glenelg, dam Nannie Butler, is the dam of Temple by Tremont. Gabriella also produced Lize by American Eclipse, son of Duroc by Diomed, that was the grandam of Enquirer, one of the best sons of Leamington, and a very successful stallion. Sir Archy was sire of Betsey Richards (dam of the good race-horse Gano by American Eclipse), John Richards and Rockingham, dam by Rattle, son of imp. Shark by Marske. He also got Marion and Pandora, dam by imp. Citizen. Marion was the sire of the good race-horse John Blount, that beat Boston and Fashion a first heat of four miles in 7.42, and broke down in the second. Marion was the sire of the good race-mare Maria West, out of Ella Crump by imp. Citizen, she the dam of the great race-horse Wagner, by Sir Charles, son of Sir Archy, that beat the noted Grey Eagle the two four-mile heat races run at Louisville, Ky., in 1839, the first one 7.48-7.44, the second one in 7.51-7.43. Grey Eagle won the first heat and broke down in the second. Maria West was dam of Fanny (dam of Millwood and Lithgow by imp. Monarch) by American Eclipse, Trinket by Andrew, son of Sir Charles, Margaret Wood by imp. Priam, Childe Harold by imp. Sovereign and Nuncio by imp. Herald, son of Plenipotentiary by Emilius. Millwood was the dam of Mill Creek and Charleston (a superior race-horse taken to England), by imp. Sovereign, Macon, Congaree and Rosa Bonheur, by imp. Glencoe, the latter, Rosa Bonheur, the dam of Plantagenet, a fine race-horse by Planet. Margaret Wood built up a family of her own. She was the dam of Heraldry by imp. Herald, Wade Hampton by Boston, Star Davis by imp. Glencoe, Maria Wood, Allendale (in Canada), Capt. Travis, Yorktown, Mary Wood and Anna Travis, all by imp. Yorkshire, son of St. Nicholas by Emilius, and Georgia Wood by imp. Knight of St. George, son of Birdcatcher and Maltese by Hetman Platoff. Heraldry was the dam of Margaret West, Balloon, Parachute, Ascension, Sailor and Godard, all by imp. Yorkshire, Petercumalinctum and Sally of the Valley by Austerlitz, son of imp. Yorkshire and Topaz by Glencoe. Balloon was the dam of Revolver (a good race-horse and sire) by Revenue, son of imp. Trustee, The Banshee and True Blue by Lexington, Ballet by Planet, and Buff and Blue by War Dance. The Banshee raced successfully and is the dam of Krupp Gun and others. Ballet is the dam of Balancer, Blue Grass Belle and Modesty (good racers) by War Dance, Thundercloud by imp. Thunderstorm, son of Thunderbolt by Stockwell, Lady Longfellow, and Peg Woffington by Longfellow, Maj. Richards and Katie H. by Ten Broeck. Buff and Blue is the dam of Badge (the dam of Logic, Minnie Elkins and O'Kelly), by Gilroy, Banner by Longfellow, Bonnie Blue by imp. Great Tom, Blue Stocking by Fellowcraft, Ban Cloche, Monita Hardy and Sister Geneva by imp. King Ban. Parachute had no chance in the stud, but was the dam of Parapluie (the dam of Windsail) by Kentucky. Georgia Wood was the dam of Ginger, Ella Wotten, Una and Ginger Pop by War Dance, La Gitana by Uncle Vic, son of Lexington, and Maggie Woods by Endorser, son of Wagner. La Gitana is the dam of Modjeska and Zingara by Tom Bowling, Guitar by St. Martin, son of imp. Phaeton, and Geneva by War Dance. Geneva is the dam of the good race-horse Riley by Longfellow. Sir Archy was the sire of Industry, dam by Ball's Florizel. Industry was the sire of Atalanta, the dam of Boston, Jr. (Cost Johnson), by Boston, and 2d Priam by imp. Priam, whose names appear in many good pedigrees. Sir Archy got the celebrated Sally Hope, dam an imported mare, by Chance, out of Jemima by Phoenomenon. She was the dam of Alice Riggs by imp. Leviathan. Cicero by Sir Archy, dam by imp. Diomed, was the sire of the dam of the famous Trifle by Sir Charles, and she the dam of Miss Peyton (Allendorf's dam), and Gloriana (the dam of Jack Malone and Pat Malloy by Lexington, good race-horses and valuable stallions), both by American Eclipse. Stockholder by Sir Archy, dam by imp. Citizen, was the sire of Betsey

Malone, dam by Potomac, son of imp. Diomed, grandam by imp. Diomed. Betsey was the dam of the three sisters Sleeping Maggie, Jenny Lind and the famous Charmer by imp. Glencoe. Charmer was one of the best race-mares of her day, particularly at long distances. She was the winner of the five heat race at four miles, run at New Orleans in 1851. Rigadoon won the first heat in 7.41, Louis d'Or the second in 7.39, the third was a dead heat between Charmer and Louis d'Or, in 7.47½. Charmer won the fourth heat in 8.05½, and galloped over for the fifth heat. Charmer produced, by imp. Yorkshire, Maggie Mitchell, whose daughter Marian (by Malcolm, son of imp. Bonnie Scotland and Lady Lancaster by imp. Monarch) is the dam of the grand race-horse Emperor of Norfolk and his brothers, all by Norfolk, the invincible son of Lexington, and the crack Yo Tambien (winner 1 1-16 miles 1.45½, fastest on record) by Joe Hooker. Stockholder was also sire of Miss Shepherd (dam Miranda by Top Gallant, son of Diomed), she the dam of Miss Russell by Thornton's Rattler, that was the dam of Sally Russell by Boston. From Sally Russell, through her daughter Miss Russell, came the famous trotting-mare Maud S., and many others. Sally Kirby by Stockholder, dam by Barksdale's Gray Diomed, son of imp. Diomed, was the dam of Princess Ann, Warlock, Black Satin (Ebony) and Sonora Love, all by imp. Leviathan. Princess Ann was the dam of Elizabeth McNairy and Chancellor by imp. Ambassador, Cottage Girl and Augusta by imp. Ainderby, son of Velocipede by Blacklock. Cottage Girl, through her daughters Ninette by imp. Albion, and Ada Cheatham by Lexington, has given us a lot of superior race-horses, such as Planetarium and her sons Font and Bob Fisher, by Fonso, Lida Stanhope by Waverly, etc. The name of Stockholder is found in many of the pedigrees of our best race-horses. Lady Burton by Sir Archy, dam Sultana by the imp. Barb Horse, out of the imp. Barb Mare, presented to President Jefferson by the Bey of Tunis, was the dam of Conter Snapper by imp. Chance, Sidi Hamet by Virginian (son of Sir Archy), and Shylock by Shylock, son of imp. Bedford, that was the grandam of the good race-mare Lilac by imp. Leviathan. Sir Archy was sire of Roanoke, Angelica and Last Chance (out of imp. Lady Bunbury by Trumpator, son of Conductor by Matchem), also of the noted race-horse Timoleon, dam by imp. Saltram, son of Eclipse, out of a daughter of Symme's Wildair, son of imp. Fearnought by Regulus. Timoleon was the sire of the great race-horse Boston, dam sister to Tuckahoe by Ball's Florizel, son of imp. Diomed, out of a daughter of imp. Alderman, son of Pot-8-o-'s by Eclipse. Boston has had no superior as a race-horse, and as a stallion was second to none that ever lived for his chances. It must be remembered that when he retired to the stud, racing and breeding had gone down to nothing. Boston ran forty-five races, won forty, thirty of which were four-mile heats, nine three-mile heats and one two-mile heats. But for an infirmity of temper he would have won some races that he lost. Boston was inbred to Diomed, his grandsire Sir Archy, and Florizel, the sire of his dam being both sons of imp. Diomed. Boston had met all the best horses of his day and beaten them all, until conquered by Fashion, first in a race at Camden, N. J., when she won in 7.42, 7.48. John Blount won the first heat and broke down in the second. This led to the match for $20,000 a side, run over the Union Course, L. I., May 10th, 1842, Fashion winning in 7.32½, 7.45, the fastest race run to that day. Boston was then nine years old and had been in the stud. He was wonderfully successful in the stud, a fact which tends much to weaken the opinion of Count Lehndorff, who says that no stallion, only two removes from a common ancestry, has ever been a success in the stud. Boston got Red Eye, a first-class race-horse, dam Lucy Long by imp. Priam, and Lexington, the best race-horse at all distances ever bred in America, and the Emperor of stallions. Lexington won all his races but one, and was beaten in that one by Lecompte, son of Boston, by the error of Lexington's jockey in pulling up at the end of the third mile in the first heat. The time was 7.26, 7.38¾. Lexington subsequently won his match against Lecompte's time, 7.26, in 7.19¾, and beat Lecompte in 7.23¾, Lecompte being drawn after the first heat, two races unequaled by one and the same horse in the annals of the turf. Lexington's dam, Alice Carneal, was by imp. Sarpedon, son of Emilius, out of Rowena by Sumpter, son of Sir Archy. Lexington is the sire of the unbeaten Norfolk, dam Novice by Glencoe, Asteroid, dam Nebula by Glencoe, Kentucky (one of the best race-horses ever bred in America), Gilroy and Daniel Boone, dam Magnolia by Glencoe, Lightning, Loadstone, Thunder and Lancaster, dam Blue Bonnet by imp. Hedgeford, Beacon, Rubicon, Bay Flower, Bayswater, Baywood, Bayonet, Niagara, Preakness, Bay Rose and Bay Final, dam Bay Leaf by imp. Yorkshire, son of St. Nicholas by Emilius, Sprightly (the dam of Volturno, Elias Lawrence, Avalon and Aretino), Crucifix (the dam of Fairplay and Semper Felix (the latter the dam of Leonatus), Semper Felix is also dam of Semper Vive by Waverly (she the dam of Fordham, Semper Idem, Longstreet's dam, Portchester, etc.), St. Augustine and Latonia (dam of Sena and Lew Weir), etc.). Lexington also got Salina, dam of Salvator (who has the fastest mile on record ever run, 1.35½), Nevada (the dam of Luke Blackburn, Ecuador, etc.), dam Lightsome by imp. Glencoe, Minx, the dam of Minnock, Monitor, Mirth (in England and dam of Bouthillier), Lotta C., etc., dam Mildred, sister to Lightsome by Glencoe; Stamps, sister to Minx (dam of Katie Pearce), Bertha, Louisette, Preciosa, etc.; Maiden and her sister La Henderson, dam Kitty Clark by imp. Glencoe. Maiden was the dam of James A., Parole, Powhattan (the promising sire), Perfection, Pappoose, and Parthenia. La Henderson is the dam of Ferida, who has the best four-mile heat, 7.23½, ever run by a mare, Aella, Planetia, Virgilian, etc. Lexington also got Susan Beane, dam Sally Lewis by imp. Glencoe, she the dam of Stratford, Sensation, Onondaga, Susquehanna, (Potomac's dam), Soubrette, etc.; Ulrica (dam imp. Emilia (Australian's dam) by Young Emilius), the dam of Megara (Spinaway's dam), Memento and Ben Ali by Virgil; Regan (dam Lorette by imp. Sovereign, son of Emilius), the dam of Vigil, Vera Cruz, Jennie B., Jennie D., Relay, etc.; Florence (dam imp. Weatherwitch by Weatherbit, son of Sheet Anchor by Lottery), the dam of the grand race-horse Hindoo, and his sister Florida, the dam of that most excellent race-mare, Firenzi, whose two races at 1½ miles in 2.33 and 2.33½ have never been equaled in the world; Waltz (dam of Hop, Glidelia, 1¾ miles in 3.01, Farandole, Raquet, Kermesse, etc.), dam Schottische by imp. Albion, son of Cain or Actaeon; Susan Ann (dam of Thora (three miles in 5.25½, and the dam of the speedy Yorkville Belle), Henlopen, etc.), dam Roxana by imp. Chesterfield, son of Priam; Finework (dam of Elkhorn Lass, Portland, etc.), dam imp. Filagree by Stockwell; The Nun (sister to Norfolk and dam of Clara D.), Sister Anne (dam of Volante), dam Novice by Glencoe; Mattie Gross (dam of Mate, Whisper, Grenada, etc.), out of Dick Doty's dam by American Eclipse; Arizona (dam of Aranza, Amerique, Louise, etc.), out of imp. Zone by The Cure, son of Physician; Hira (dam of the good race-horse and successful sire, Himyar, Hegiaz, Hi Ban, Highflight, etc.), out of Hegira by imp. Ambassador, son of Plenipotentiary; Lady Motley (dam of Blazes, Lucifer, etc.), her dam, Anna C., by Glencoe; Echo (dam of Report, Duplex, Echoless and Naphtha (the grandam of Dewdrop), etc.), out of Maria Innis by imp. Yorkshire; Item (dam of Dan. Sparling, Bob Woolley, Joe Rhodes, etc.), out of Katona by Voucher, son of Wagner; Woodbine (dam of Bonnie Wood, Belle of the Meade, Boardman, etc.), out of sister to Compromise by Glencoe; Veritas (dam of Maxim, Lady Prewitt, Vera, etc.), out of Verona by imp. Yorkshire; Mary Martin (dam of Santa Anita, Gano's dam), out of Alice Jones by imp. Glencoe; Jury (sister to Susan Ann), dam of Verdict, Acquittal, etc; Mazurka, dam of Tallulah (she the dam of Bribery, Miss Ford's dam), Minnie Brown, Zoo Zoo, Gyptis, Maze, etc., out of Miss Morgan by imp. Yorkshire; Nannie Butler, dam of Girl of the Period, Nannie Black (dam of Brown Princess and Princess Blandina), Nannie Bay (Temple's dam), Vermont, etc., out of Tokay by imp. Yorkshire; Aerolite, dam of Spendthrift, Fellowcraft (four miles, 7.19¾), Miser, Rutherford, Jersey Belle, Platina (dam of Drake Carter, 3 miles, 5.24, Fellowplay, Salile M., Bersan's dam, etc.), out of Florine by Glencoe, and Aerolite's own sister, Idlewild (4 miles in 7.26¼), dam of Wildidle (4 miles, 7.25), Fanchon, etc.; Annie Bush (dam of Bushwhacker, Diana, Virginia Bush, etc), dam Banner by imp. Albion; The Banshee (dam of Krupp Gun, Brigand Belle, Emma E., etc.), dam Balloon by imp. Yorkshire; Beulah (dam of Hilarity, Hoder, Hannibal, etc.), dam Eureka by Glencoe; Blandina, dam of Calash (Eolian's dam), Christina, Sea Foam, etc., dam My Lady by Glencoe; Carrie Atherton (dam of Janet Norton, Duke, Nettie Norton, etc.), out of Glycera by imp. Sovereign; Coquette (dam of Viola, Vivien, Vampire, etc.), out of Susan Harris by Revenue; Cordelia, dam of

Cammie F. (History's dam), Invalid, North Elkhorn, Glendalia (Woodcutter's dam), out of Kitturah by Brawner's Eclipse; Edeny, dam of Lottie Moon (Ida Moon's dam), Eva Sherley, Saratoga, La Favorita (Ballarat's dam), out of Sally Polk by imp. Ambassador; Emily Peyton (dam of Lady Peyton, Bessie Peyton, Fannie Peyton, etc.), out of Sally Roper by imp. Albion; Evadne, dam of Bandana (dam of Banner Bearer, Napa, etc.), Bulwark, Evasive, Tennyson, etc., out of Volga by Glencoe: Geneva, dam of Silvio, Greystone, Bijou (dam of Mag, Sea Fog and Drizzle's dam, etc.), out of Grisette by Glencoe: Glenrose (dam of Glendora, Glendew (Gnenn's dam), Restless, etc.), out of Sally Lewis by Glencoe; Grecian Bend (dam of Modiste, Grecian Maid, La Belle Helene, etc.), out of Miriam by Glencoe: Hester, dam of Springbok (fastest 2¼ miles, 3.56¼), Galanthis, Aspinwall, etc., out of Heads I Say by Glencoe; Impudence, dam of Hindoostan, Wamba, Insolence, Sweetbriar (the dam of Dr. Helmuth, Dr. Hasbrouck, etc.), out of Magnetta by Mahomet: India (dam of Cathcart, Calcutta, Gallifet, etc.), dam Miss Morgan by Yorkshire: Invermore (dam of Ganley, Inverwick, etc.), out of imp. Invercauld by St. Albans; Lady Wallenstein, dam of Wallenstein (a successful racer in England), Witch, etc., out of Louisa by Yorkshire; La Polka (dam of Post Guard, Strathspey, La Esmeralda, Heel and Toe, Los Angeles, etc.), out of Dance by Glencoe; Letola (dam of Unrest, Utility, Cayuga, etc.), out of Capitola by Vandal: Lida, dam of Enquirer (one of the best sons of Leamington), Analysis, Oceanica, The Squaw, etc., out of Laze by American Eclipse; Lilly Duke (dam of Water Lily, Lizzie Dwyer, Red Dress, etc.), out of Lilla by Yorkshire: Madame Dudley, dam of Judith, Belinda, Virgie D. (Hueneme's dam), Vintage, Santa Rita, etc., out of imp. Britannia IV by The Flying Dutchman; Majestic, sister to Kingfisher, dam of the superior race-horse Eurus, out of imp. Eltham Lass by Kingston, and Sabina, full sister to Majestic, and dam of The Widow, Ocean, Ruchiel, Chief Justice, etc.: Marguerite, sister to Blandina, dam of Prude, Katie Creel (Kenwood's dam), Azalia, Macola, etc.; Mary Clark, dam of Spark, Pure Rye, Grey Dawn, etc., out of Eagless by Glencoe; Minnie Minor, dam of Warwick, Minority, John Ledford, Wanda, Winfred, etc., out of Julia by Glencoe; Miss Bassett, sister to Harry Bassett, dam of Ernest, Masher, Plasir, etc., out of Canary Bird by Albion: Mollie Cad, dam of Maj. Pickett, Dolly Cad, Vallisca, etc., out of imp. Weatherwitch by Weatherbit; Mollie Wood, dam of Golden Gate, Altaire, Woodlark, Gascon, etc., out of Hymena by Yorkshire; Mandane, dam of Sunbeam, Lounger, Tonawanda, Blue Wing, The Teal, Blue Garter, etc., out of Sally Bowen by imp. Hooton: Queen Victoria, sister to Duke of Magenta, dam of Queen's Own, Infanta, Jennie Treacy, Powhattan Queen, etc.; Quickstep, sister to Waltz, dam of False-Step, Fernwood, Alfaretta, Quickmarch, etc.; Revolt, dam of Regicide, Renegade, Misrule, etc., out of Sue Washington by Revenue, son of imp. Trustee: Sallie (dam of Solicitor, Nathan Oaks, Experience Oaks, Red Coat, Glenella, etc.), out of Sallie by Buford, son of imp. Nonplus by Catton; Sarong (dam of Aristides, Eliza Adams, Wissahickon, Daisey Hoey, Petersborough etc.), out of Greek Slave by Glencoe; Springbrook, dam of Silent Friend, Full Cry (Checkmate's dam), Verbena (dam of Phil Lee and Michael), etc., out of Emmckfaw, by Mambrino, son of American Eclipse; Squeeze'em, dam of Day Star, Sylph (dam of La Sylphide, Kaloolah, Kinesem, etc.), Palmetto, Helena, etc., out of Skedaddle by Yorkshire; Sultana (dam of Soudan, She, Padisha, etc.), out of Mildred by Glencoe; Ultima, dam of Aurora Raby (dam of Bill Bruce and Charley Gorham), Nuisance, Sunrise, Sonora, Ultimatum, etc., out of Utilla by imp. Margrave, son of Muley; Zephyr (sister to Arizona), dam of Berlin, Westwind, Typhoon, Zetetic, etc., out of imp. Zone by The Cure. Amongst the horses of Lexington's get who have done well in the stud and on the turf, we can mention Norfolk, never beaten, 3 miles in 5.27½-5.29½, and a successful stallion: Asteroid, never beaten, and sire of a number of brood-mares, successful producers, including Mattie W. (dam of Jim Malone, four miles in 7.29½), Eleini, Emeti, Flutter, Lass of Ayr, Lerna, Luna, Modiste, Susan Queen, Waterwitch, etc.: Kentucky, sire of Bertram, Beatrice (Bellona's dam), Woodbine (dam of Forester), The Baroness (Badge's dam), The Countess, Nina, Freebooter; Vauxhall and Foster, both good race-horses, the former sire of Viator, Cloverbrook, Oden, etc.; Concord, sire of Galway, Picolo, etc.: Lever, sire of Leveler, Essillah, La Gloria, Petrel, Mahlstock, Sweetheart etc.; Uncas and

Wanderer, brothers, both good race-horses and stallions; Jack Malone and Pat Malloy, good race-horses and sires: Lightning (sire of Janet, four miles 7.25), a superior race-horse: Kingfisher, sire of a number of good ones in the Nursery stud, including Fillette (dam of Fides), Magnetism (dam of Magnetizer), etc.; Harry Bassett, Merrill, Marion. Ansel, Optimist (sire in France of Mars, a popular stallion, and sire of Jongleur); Preakness, a good sire in England; Chesapeake, Colton, sire of Monday, etc., and many others.

The Herod line is stronger in numbers than the Matchem, but not so numerous as those of the Eclipse line. The best now alive in England are Balfe, a good race-horse by Plaudit, son of Thormanby, dam Bohemia by Weatherbit, son of Sheet Anchor by Lottery. Bird of Freedom (winner of the Ascot Gold Vase, City and Suburban Handicap and other good races), by Thuringian Prince, son of Thormanby, dam Vitula by Arthur Wellesley, son of Melbourne. Bruce (winner of the Grand Prize of Paris), by See Saw, son of Buccaneer by Wild Dayrell, dam Carine by Stockwell. Atlantic (winner of the 2.000 Guineas), by Thormanby, dam of Hurricane by Wild Dayrell. Charibert (winner of the 2,000 Guineas), by Thormanby, dam Gertrude by Saunterer, son of Birdcatcher. Macaroon by Macaroni, dam Margery Daw by Brocket, son of Melbourne. Macheath by Macaroni, dam Heather Bell by Stockwell. Macgregor (winner of the 2,000 Guineas) by Macaroni, dam Necklace by The Fallow Buck. Mask, a great sire, by Carnival, son of Sweetmeat, dam Meteor by Thunderbolt, son of Stockwell. Kisber (winner of the Derby and Grand Prize of Paris) by Buccaneer, dam Mineral by Rataplan. Ossian (winner of the St. Leger) by Salvator, dam Music by Stockwell. Sir Bevys (winner of the Derby) by Favonius, dam Lady Langden by Kettledrum. Salvator (winner of the French Derby and Grand Prize of Paris) by Dollar, son of Flying Dutchman, dam Sauvagine by Ion. Saxifrage by Vertugadin, son of Fitz Gladiator, dam Slapdash by Annandale. Perplexe (winner of the French St. Leger) by Vermouth, son of The Nabob, dam Peripetie by Sting, the last three bred in France. In America we are very strong in the line to Herod, through the sons of Lexington and male descendents of imp. Glencoe. The brothers Lucas and Wanderer, both good race-horses by Lexington, dam Coral by Vandal, son of Glencoe. Tom Ochiltree, sire of Major Domo and a superior race-horse by Lexington, dam Katona by Voucher, son of Wagner. Vauxhall, a good race-horse by Lexington, dam Verona by imp. Yorkshire. Duke of Magenta, a superior racer, by Lexington, dam Magenta by imp. Yorkshire. Monday by Colton, son of Lexington, dam Mollie Jackson by Vandal. Bishop by Shannon, son of Monday, dam Bettie Bishop by imp. Buckden. Argyle by Monday. dam Cuba by imp. Australian. Joe Hooker by Monday, dam Mayflower by imp. Eclipse. Day Star by Star Davis, son of imp. Glencoe, dam Squeez'em by Lexington. Hindoo, a grand race-horse by Virgil, son of Vandal, dam Florence by Lexington. Hanover, a superior race-horse, by Hindoo, dam Bourbon Belle by imp. Bonnie Scotland. Vocalic by Virgil, dam Acoustic by imp. Australian. Ventilator by Vandal, dam Carolin by imp. Scythian. Tremont by Virgil, dam Ann Fief by Alarm, was never beaten and is promising well as a sire. Vanderbilt by Vandal, dam Melrose by Childe Harold, son of imp. Sovereign. Vagabond by Vandal, dam Gem by Childe Harold. Vigil by Virgil, dam Regan by Lexington. Portland by Virgil, dam Finework by Lexington. Isaac Murphy by Virgil, dam Mary Howard by imp. Hartington. Imported Mortemer by Compiegne, son of Fitz Gladiator, dam Comtesse by The Baron or Nuncio. Mortemer won the Gold Cup at Ascot, and his sons are Exile, a superior race-horse, dam imp. Second Hand by Stockwell; Pontico, a good race-horse, dam imp. Agenoria by Adventurer; Cholula, a good racer, dam Fanny Ludlow by imp. Eclipse, son of Orlando; Cyclops, dam Lizzie Lucas by imp. Australian: Wilfred, dam Minnie Minor by Lexington; Heimdel, dam Vandalite by Vandal. Imp. Macaroon by Macaroni, dam Songstress by Chanticleer, son of Birdcatcher. Imp. Moccasin by Macaroni, dam Madame Strauss by King Tom. Imp. Woodlands by Nutbourne, son of The Nabob, dam Whiteface by Turnus. The imported brothers Sir Modred and Cheviot by Traducer, son of The Libel, dam Idalia by Cambuscan. Grinstead by Gilroy, son of Lexington, dam Sister to Ruric by imp. Sovereign. Gano by Grinstead, dam Santa Anita by Virgil. Flood by Norfolk, dam Hennie Farrow by imp. Shamrock.

Emperor of Norfolk and his brother by Norfolk, dam Marian by Malcolm, son of imp. Bonnie Scotland.

King William, through Mr. Marshall, imported the White Barb Chillaby, together with the Barb mare Slugey or Salugy (which means greyhound), in foal to the White Barb. She produced a colt which was named Greyhound, and it is through this horse Greyhound that the blood was valuable. He was the sire of Othello (Duke of Wharton's), Goliah, Sampson and his sisters out of a Curwen Barb mare, Rake and his sister out of Mr. Croft's Pet mare by Wastell's Turk, Desdemona out of Chestnut Thornton by Makeless. The Greyhound mare (sister to Sampson and Goliah) was the dam of Lady Thigh (Grisewood's), Meynell, Craven, Pompey and Jenny Spinner by Croft's Partner, Cloudy and Sally by Forester. Greyhound also got Y. Greyhound, Guy and his sister, out of Brown Farewell by Makeless. The sister to Rake was the dam of Barforth and his sister by Croft's Partner. Grisewood's Lady Thigh was the dam of imported Duchess by Cullen Arabian, and Miss Thigh by Rib, she the dam of Selim by Bajazet. and the Cullen Arabian mare (sister to imp. Duchess) that was the dam of Principessa by Blank, she the dam of Heinel by Squirrel. Heinel's daughter by Woodpecker was the dam of Cobbea by Skyscraper [she the grandam of Cadland (Derby winner, 1828), by Andrew] and of Mrs. Barnet by Waxy, the dam of Filho-da-Puta, St. Leger winner, 1815. Another full sister to Duchess had a filly by Snap, son of Snip, that was the dam of Conductor, Alfred, Georgiana, Dictator, Scorpion, Eyebright and Laurestinus, all by Matchem. Georgiana was the dam of Lady Harriet by Mark Anthony, Grouse by Highflyer, Speculator and Moorcock, brother to Grouse. Eyebright was the dam of Jemima by Phænomenon, that was the dam of the Chance mare, imported into Virginia, that was the dam of Sally Hope by Sir Archy. The sister to Guy by Greyhound was the dam of Bay Bloody Buttocks and Grey Bloody Buttocks by Bloody Buttocks Arabian (which was a grey Arabian with a red mark on his hip), Little Partner (sister to Vane's Little Partner, which was the grandam of the imp. Cub mare), Young Partner and Red Rose by Old Partner. Bay Bloody Buttocks was the dam of Spinster (the Widdrington mare) by Old Partner, son of Jigg. Spinster (the Widdrington mare) was the dam of Spinster and Shepherd's Crab by Crab, son of the Alcock Arabian, Lofty by Godolphin Arabian, Golden Grove by Blank, etc. Golden Grove was the dam of Petworth (Gaudy) by Herod, and she the dam of Quiz by Tandem, Bounty by Mercury, and Rosina by Woodpecker, the latter the dam of Hermes by Mercury, son of Eclipse. Hermes was the sire of Gibside Fairy, the dam of Emma and Maria by Whisker. Emma was the dam of imp. Trustee and his brother Mundig (winner of the Derby, 1835), by Catton, The Ladye of Silverkeld Well by Velocipede, Cotherstone (winner of the Derby, 1813) and Mowerina (West Australian's dam) by Touchstone. Maria was the dam of Euclid, Laura, Equation, Example and Eclipsis by Emilius. Grey Bloody Buttocks ran under the name of Dairymaid, and was the dam of Parker's Lady Thigh by Partner, son of Jigg, Cornforth's Forester by Forester, Smallbones and Squirrel by Traveler, son of Partner. Young Greyhound got the dam of Engineer and Miner, out of a daughter of the Curwen Bay Barb. Engineer was the sire of Mambrino, out of a daughter of Cade by the Godolphin Arabian. Mambrino was the sire of imp. Messenger, out of a daughter of Turf by Matchem, and imp. Mambrina, out of Naylor's Sally by Blank, son of the Godolphin Arabian. Mambrina was the dam of Gallatin, Fairy and Eliza, all by imp. Bedford, and Eliza was the dam of the great race-horse and stallion Bertrand, and Pacific by Sir Archy by imp. Diomed. A sister to Engineer by Young Greyhound produced a filly by Blaze, son of Flying Childers, that was the dam of Sappho, Semele and Seraphina by Blank. Sappho was the dam of Sophonisba by Squirrel and Madcap by Eclipse. Madcap's daughter, Caprice, by Anvil, son of Herod, produced Floranthe by Octavian son of Stripling by Phænomenon by Herod. Floranthe was the dam of Economist by Whisker. Economist was the sire of the wonderful race-horse Harkaway, out of Fanny Dawson by Nabocklish. Harkaway got King Tom out of Pocahontas by imp. Glencoe, and King Tom got imp. King Ernest out of Ernestine by Touchstone, imp. King Ban out of Atlantis by Thormanby, imp. Phaeton out of Merry Sunshine by Storm, son of Touchstone, imp. Great Tom out of Woodcraft by Voltigeur. These horses proved big successes in America, and their daughters are doing equally well in the stud.

The Akaster Turk was brought to England about the year 1700, by whom the authorities do not state. He was the sire of Chaunter, and a full sister, which was the dam of the famous Roxana (the dam of Lath and Cade), and Silverlocks by The Bald Galloway, Terror and Mr. Thwaite's Dun Mare (the dam of Mr. Beaver's Driver). Akaster Turk was also the sire of the dam of William's Squirrel and his full brother, Smith's son of Mr Lister's Snake. Silverlocks was owned by Mr. Coke and was the dam of Lord Portmore's Silvertail and Lord Godolphin's Buffcoat by the Godolphin Arabian, who was imported into Virginia, and died there in 1757. There was also an own sister to Roxana and Silverlocks, which was the dam of Mr. Panton's Cato by Partner, and a sister to Cato was the dam of the Foxhunter mare that produced Molly-long-legs by Babraham. This sister to Cato by Partner was also the dam of imp. Silver (Sylva) by the Belsize Arabian, which was imported into America and was the dam of Sally Painter by Evan's imp. Stirling, also son of the Belsize Arabian, Babraham and a full sister by imp. Fearnought, and Grenadier by True Briton. Sally Painter was the dam of Diana by Clodius, son of imp. Janus. Diana was the dam of Minerva by imp. Obscurity, son of Eclipse. Minerva was the dam of Camilla by Symme's Wildair, she the dam of Priestly by Chanticleer, son of imp. Wildair. Diana was also the dam of Rosetta by imp. Centinel, son of Blank by the Godolphin Arabian. Rosetta was the dam of Narcissa by imp. Shark, which produced Meretrix by Magog, son of Chanticleer, which gave us Director and Virginian, brothers by Sir Archy, son of imp. Diomed, good race-horses and stallions. Julietta by imp. Dare Devil, son of Magnet by Herod, out of Rosetta, was the dam of Nell Saunders and sister by Wilke's Wonder, son of imp. Diomed. From Nell Saunders, and her daughter by Pacolet, son of imp. Citizen by Pacolet, son of Blank, bred to Stockholder, son of Sir Archy, dam by imp. Citizen, came the dam of Compromise the great gelding, and Miss Pattie by imp. Glencoe, Martha Dunn and Invincible by imp. Sovereign, and Laura (the dam of Harry of the West, Lilly Ward and Fanny Cheatham, all superior racers and by Lexington. Miss Pattie was the dam of Woodbine by Lexington and Jessamine by Brown Dick (she the dam of Barnton and others), Madeira by John Morgan, Belle of the Meade, Bonnie Wood and Boardman by Bonnie Scotland. Patty Puff by Pacolet, tracing through Rosy Clack (dam of Tennessee Oscar by Wilke's Wonder), by imp. Saltram, son of Eclipse, to Diana, was the dam of Tennessee Citizen, Byron, Maria Shelby by Old Stockholder, son of Sir Archy, and the three sisters Angora, Celerity and Velocity, by imp. Leviathan, son of Muley by Orville, good racers. To Celerity trace many good race-horses, such as Ann Percy, the dam of Goldring (Bullion's dam) by Ringgold, son of Boston. Martha Washington, Bill Deering (Gamble). Martha Dunn by imp. Sovereign son of Emilius by Orville, produced Lulu Horton by imp. Albion, son of Cain or Actæon, that was the dam of Spartan, Kildare, Pequot, Lute and Killarney by Lexington. Lute was the dam of Virginius by Virgil, a good race-horse, and Glenluine by imp. Glenelg, the latter the dam of Windrush, a superior race-horse, by King Alfonso, Patrimony (the dam of the good but unfortunate Lisimony) by Pat Malloy, Troubadour, the grand race-horse by Lisbon, son of imp. Phaeton by King Tom, that beat The Bard, in the Ocean Stakes, 1½ miles, 1.56¼, and Monmouth Cup, 1¾ miles in 3.04, and also won the Suburban Handicap in 1886. Lulu Horton was also the dam of Amy Farley by Planet, she the dam of Titania (dam of Charity by Lever, and Lina by King Alfonso) by Lexington, Mattie Amelia by King Alfonso, Pathfinder, Terese and Blue Grass, a prominent winner in England, by Pat Malloy.

The Selaby Turk (called also the Marshall Turk) was brought to England by Mr. Marshall, brother to the King's Master of the Stud, in King William's reign. He was the sire of Mr. Curwen's Old Spot, sire of the dam of Mixbury and brother to Mixbury (Chedworth's), the latter the sire of Little Bowes (the dam of Mab by Hobgoblin, Regulus Mixbury and Medusa, dam of Weazle by Regulus, Sophia by Blank, and Young Bowes, the dam of Chedworth's Snap, by Dormouse). The Selaby Turk was also the sire of the grandam of Mr. Croft's Partner, sire of Tartar and grandsire of King Herod. He also got the dam of Old Wyndham by Hautboy, the grandam of Soreheels by Basto, Sloven by Bay Bolton, and Fearnought by Bay Bolton. The Selaby Turk was the sire of the Coppin mare, frequently confounded with the Coffin mare (a mare so called from being concealed in a cellar in Frenchman street, at the time of the Restoration). The Coppin mare's dam was by Place's White Turk. To a daughter of the Coppin mare by Commoner, son of Place's White Turk, traces in a direct female line Imperator by Conductor, Royalist by Saltram (Derby winner, 1783), both imported to America, Actæon by Send, Emilius (Derby winner, 1823) by Orville, Dalesman by King Tom, Barbillon (French) by Pretty Boy, son of Idle Boy by Harkaway, Warminster (imported) by Newminster, imp. Saxon by Beadsman, imp. Camilla by King Tom, the dam of Carolin, Victoria and Medora by Kentucky, Carina (dam of St. Carlo) by Kingfisher, Carita (the dam of King Crab, St. Charles, etc.) by Ill Used, and her sister Clara, the dam of Clarendon, Chatham, etc., imp. Queen Maud by Macaroni, the dam of Queen of the May, Hawkstone, Stoney Montgomery, etc. The Selaby Turk was also sire of the mare, dam by Bustler, grandam by Place's White Turk, out of a daughter of Dodsworth, a natural Barb. This mare was the dam of Windham (Wyndham) by Hautboy. Wyndham was the sire of Miss Wyndham, out of a daughter of Belgrade Turk, who was taken at the siege of Belgrade in Turkey from the Bashaw of that place, and sold by the Minister of Prince Lorraine in London to Sir M. Wyvill's. He was the sire of Young Belgrade and of the dam of Miss Wyndham by Wyndham, who produced a mare by Brother to Bolton Fearnought, that was the dam of Beaufremont by Tartar, and the Cade mare that was the dam of Lardella by Young Marske, to which traces imp. Frolicksome Fanny (the dam of Tom Corwin by imp. Emancipation, Miss Margrave by imp. Margrave, and Nina by Boston) by Lottery. Nina was a superior race-mare and a grand brood-mare, dam of Planet (a superior race-horse), Exchequer, Ninette (dam of Diavolo and Grace Darling by Jonesboro'), Orion by Revenue, Conductor by Engineer, Oriana by Deucalion, son of Revenue, Ripley by Jeff Davis, Eugenia (dam of Euterpe, Eunice and Eusebia by Vauxhall, son of Lexington, and Eula B. by Dickens,) by Eugene, son of Revenue, Ecliptic, the dam of Sunbeam (Constantina and Winston's dam) by imp. Leamington, Baby and North Anna by imp. Strachino, son of Parmesan, Catalpa and Virgo by Abd-el-Kader (son of imp. Australian,) by imp. Eclipse, Algerine, winner of the Belmont stakes, and his sister Algeria by Abd-el-Kader. Algerine is now in the Algeria stud, Erie, Pennsylvania.

The St. Victor Barb was sent to England by Monsieur St. Victor, of France. He was the sire of the celebrated Bald Galloway, his dam Grey Whynot by Old Whynot (son of the Fenwick Barb), grandam a Royal mare. The Bald Galloway was sire of Buckhunter (afterwards the Carlisle gelding) and Old Lady, a sister to Buckhunter, Lord Portmore's Snake and Daffodil, Grey Robinson, out of sister to Old Country Wench by Snake. Grey Robinson was the dam of the famous Regulus by the Godolphin Arabian. Roxana by The Bald Galloway, dam sister to Chaunter by the Akaster Turk, was the dam of Lath and Cade, by the Godolphin Arabian, and to her the Godolphin owes his success in the stud and his immortality. Roxana was also the dam of Roundhead by Flying Childers. Silverlocks, sister to Roxana, was the dam of Buffcoat by the Godolphin Arabian imported into Virginia. The sister to Regulus, out of Grey Robinson, was the dam of Polly by Blank, Coquette (the dam of Camilla, Glider, Driver, Darter and Cocker by Trentham) by the Compton Barb, imp. Emilia by Young Emilius (dam of imp. Australian and Ulrica), out of Persian by Whisker, imp. Rosalind by Barnton, dam sister to Bay Rosalind by Orlando, imp. Adriana by Blinkhoolie, dam Essence by St. Albans, imp. Santa Lucia (the dam of Aurelia, Torso, etc.) by Lord Lyon, dam Lady Margaret by Honiton, imp. Star Actress by Siderolite, dam The Tinted Venus by Macaroni, all trace through Coquette by The Compton Barb and sister to Regulus to Grey Robinson. The filly by Snap, son of Snip, out of the sister to Regulus, in 1775, produced in Ireland, Heroine by Hero, son of Cade. Heroine, in 1781, produced Tom Turf by Lennox, son of Bustard by Crab, and Young Heroine, the great grandam of Birdcatcher by Bagot, son of Herod. Another sister to Roxana produced Cato by Partner, and his sister that was the dam of imp. Silver (Sylva) by

Belsize Arabian. St. Victor Barb was also sire of Cupid, own brother to Bald Galloway, and his sister Points, out of Grey Whynot by Old Whynot, out of a Royal mare. In 1715 Points foaled the filly Flying Whigg by Williams' Woodstock Arabian. Flying Whigg was the dam of the Large Hartley mare by Mr. Hartley's Blind-Horse, son of the Holderness Turk, Little Hartley mare by Bartlet's Childers, and three fillies by Hip, son of the Curwen Bay Barb. The Large Hartley mare was the dam of imp. Selima by The Godolphin Arabian, and her brothers Babraham, Mogul and Marlborough, sent to Ireland, also a roan filly by Hip. It is to this latter Hip mare that many of our best importations trace their origin, including imp. Maud by Stockwell, the dam of Maudina (the dam of Galway, Piccolo, Elizabeth, Cloverbrook, Oden, Minerva, Empress, Maumee, Marauder, etc.) by imp. Australian, Attraction (the dam of Magnetism, Magnetizer's dam) by imp. Balrownie, Alarm by imp. Eclipse, one of the best and speediest horses of his day and a successful stallion. To this Hip mare also traces imp. Jenny Mills by Whisker. She was the dam of Ambassador, imported *in utero*, by English Plenipotentiary, a good stallion, Liatunah and Sewanee, good racers, by imp. Ainderby, son of Velocipede. To the same identical source traces also imp. Buttercup by The Primate, son of St. Albans by Stockwell, her dam Buttercrambe by Ben Webster, son of Barnton, brother to Voltigeur. She is the dam of a number of winners. The Little Hartley mare was the dam of Janus, Blank and Old England by the Godolphin Arabian, and Miss Meredith by Cade, son of the Godolphin Arabian. To Miss Meredith trace imp. Boaster, foaled in 1795, by Dungannon, son of Eclipse, imp. Driver, foaled in 1806, by Driver, son of Trentham, and Gabriel, foaled in 1790, by Dorimant, son of Otho by Moses. These horses are found in many of our best racing pedigrees.

The Oglethorpe Arabian was the property of Sir Thomas Oglethorpe of Scotland. Little is known of his history. He was the sire of Makeless, whose dam was Trumpet's dam by Place's White Turk, second dam by Dodsworth, out of the Layton Barb Mare. Makeless was not only a distinguished race-horse in his day, but was greatly esteemed as a stallion, being the sire of Sir Ralph Milbank's famous Black mare, the dam of Hartley's Blind Horse by the Holderness Turk, and Gipsy by King William's Black No-tongued Barb. From the latter in direct female line descend Sir Archy (the Godolphin Arabian of America), Highland Mary, Hephestion, Castania, Virgo, Noli-me-Tangere, etc. Makeless was also the sire of Sir M. Pierson's Makeless mare, dam by Brimmer, grandam by D'Arcy's Diamond (son of the Helmsley Turk), out of an own sister to Old Merlin by Bustler. This Makeless mare was the dam of Bay Bolton (Brown Lusty), and a sister to Bay Bolton by Grey Hautboy, Bay Lusty and Lamprie by the same stallion, and the Luggs mare by the Darley Arabian. It is from this sister to Bay Bolton that Dr. Syntax and Little Red Rover descend in direct female line. The Oglethorpe Arabian was also sire of Bold Frampton. The Blind Horse was bred by Leonard Hartley, Esq., near Richmond, Yorkshire. He was sire of Miss Neasham and the Large Hartley mare, she the dam of Babraham, Mogul, Marlborough and Selima, all by the Godolphin Arabian. Selima was imported into the United States and founded a distinguished family of race-horses. Makeless was the sire of Old Scarborough mare, dam by Brimmer, and she was the dam of Scarborough Colt by Tifter, a filly by Belgrade Turk which was the dam of Miss Wyndham by Wyndham, and a filly by Bay Bolton that was the dam of Young Belgrade by Belgrade Turk. Miss Wyndham was the dam of Beaufremont by Tartar, and Mr. Hutton's mare by Cade, she the dam of Lardella by Young Marske, from whence in direct female line descend Comus by Sorcerer, Slane by Royal Oak, Young Melbourne by Melbourne, North Lincoln by Pylades, son of Surplice, and Sefton by Speculum. Makeless was also the sire of Bay Farewell and Brown Farewell, the latter the dam of Guy, Young Greyhound and full sister to these two by Greyhound. This latter mare was the dam of Bay Bloody Buttocks, Grey Bloody Buttocks (Dairymaid) by Bloody Buttocks, Little Partner, Young Partner and Red Rose by Croft's Partner. Makeless was also sire of Bay and Chestnut Thornton, dam Old Thornton by Brimmer, grandam by Dicky Pierson (son of Dodsworth), out of the Burton Barb mare, a Natural Barb mare of Mr. Burton's. It is from this Burton Barb mare that descend in a direct female line the following horses: The Ancaster Starling, Mungo by Sampson, Waverley by Whalebone, Safeguard, Whiskey, Selim, Camillus, Ardrossan, Castrel, Rubens, Catton, Whitelock, Blacklock, Bobadil, Sir Hercules, Irish Drone, Don John, Hetman Platoff, Harkaway, Wonersley, Cremorne, Voltigeur, Surplice, Cowl, Ithuriel, King of Trumps (by Velocipede), Teddington, The Marquis (by Stockwell), St. Albans, Lord Clifden, Skirmisher, Lecturer, Camballo, Martyrdom, Mask, Pell Mell, Carnelion, Thurio, Umpire (by King Tom), Althotas, Brag, Sheen, etc.

The following imported horses are also descendants in the female line from the same mare: Abjer, Ainderby, Billet, Brilliant, Chance, Claret, Clifden, Contract, Dion, Eagle, Spread Eagle (winner of the Derby), Gouty, Granby, Harkforward, Hurrah, Jolly Roger, Langford, Margrave (winner of St. Leger), Don John (winner of St. Leger), Merry Tom, Nonplus, North Star, Pantaloon (by Herod), Phenomenon (winner of St. Leger), Restless, Skylark, St. George, The Tester, Thunderstorm, Yorkshire, and Waverly (by imp. Australian), bred in Kentucky.

The Darley Arabian was the property of Mr. Darley, of Buttercramb, near York, England, to whom he was presented by his brother, residing abroad. He was a Bedouin horse of the family called Keheilan-Ras-el-Fedawi, imported in the latter end of Queen Anne's reign, about 1702. He was the sire of Almanzor (his dam the dam of Terror by Akaster Turk, and Graham's Champion by Harpur's Arabian, Aleppo, Whitelegs and Smockface, full brothers to Almanzor), Brisk, out of Coneyskins' dam by Jigg, and Cupid, both good horses, Daedalus, a fleet horse, Skipjack, Manica, imp. Bulle Rock, Dart, Whistlejacket, good plate horses, and Whimsey and Kitty Burdett, out of the Young Child mare, good plate mares and producers. The great reputation of the Darley Arabian comes from his son Flying Childers (commonly called Childers), and his full brother, Bartlett's Childers, who was at first called Bleeding Childers, from his frequently bleeding from the nose. Flying Childers, foaled in 1715, was by the Darley Arabian, dam Betty Leedes by Old Careless, son of Spanker, and a Barb mare, second dam sister to Leedes by Leede's Arabian, third dam by Spanker, fourth dam Old Morocco mare (Old Peg, Spanker's own dam) by Morocco Barb, fifth dam, old Bald Peg by an Arabian out of a Barb mare. Childers started against the best horses of his time and was never beaten. In the year 1721 it is said that Childers ran a trial against Almanzor and Brown Betty, carrying 128 lbs., over the Round Course at Newmarket, 3 miles, 6 furlongs and 93 yards, in 6 minutes and 40 seconds. It was also said that he moved 82½ feet in a second of time, which is nearly at the rate of a mile in one minute. He ran over the Beacon Course, four miles, one furlong and 138 yards, in 7 minutes and 30 seconds. His stride covered 25 feet. These performances are not reliable, but he was unquestionably one of the fleetest horses that ever lived. He was not only a superior racer, but a wonderfully valuable stallion. He was the sire of Plaistow, Blacklegs, Second, Snip, Commoner, Blaze, Winall, Spanking Roger, Lord Manner's Poppet, Fleec'em, Steady, Young Miss Belvoir (the dam of Miss, Redcap and Duchess by Lath, and Foxhunter by Blank). He also got two sisters to Young Miss Belvoir, one of which was the dam of Allworthy, Cypher and Pytho by Crab, Feather by the Godolphin Arabian, Horatius and Horatia by Blank, and a filly by Panton's Arabian. The other sister was the dam of the Ancaster Blossom and Godolphin Blossom, both by Crab, the former the dam of Blossom, Jilt and Bragg by the Godolphin Arabian, Gaudy, Gamesome, Grizzle and Chrysolite by Blank. Childers was also the sire of Bay Basto and Brown Basto, sisters to Snip. Bay Basto was the dam of Chance, Crazy and Gipsy by Lath, and Chaunter by Cade. Brown Basto was the dam of Hector, Cassandra and Lady by Blank. Sister to Blaze, she the dam of Whitenose by the Godolphin Arabian. Another daughter was the dam of Little Driver by Beaver's Driver. Bartlett's Childers was not raced, but was an excellent stallion, sire of Smales' Childers, Œdipus, Grey Childers, Squirt (the sire of Marske and Syphon), Coughing Polly, Little Hartley mare (the dam of Janus, Blank and Old England by the Godolphin Arabian, Trimmer and Shakespeare by Hobgoblin, Miss Meredith and Slouch by Cade), the dam of Wyvill's Volunteer, from which mare in direct female line descend Waxy, Paynator, Thunderbolt (by Sorcerer),

Smolensko, Colsterdale, Ellington, Ashton, Souvenir, Poussin, Bertram, etc. The mare out of sister to the Two True Blues, dam of Middleton's Squirrel, Midge, Thwackum and Camilla by a son of Bay Bolton, Miss Belsea and a sister by Regulus, son of the Godolphin Arabian. Miss Belsea was the dam of Wildman's Snap, Hyæna, Timidity, Rosebud, Harpy, Elfrida and others, all by Snap, son of Snip, by Flying Childers. Hyæna was the dam of Eagle, Everlasting and Lavinia by Eclipse, and Evander by Highflyer. The sister to Miss Belsea by Regulus was the dam of Fribble by Snip, in Ireland, the mares Snapdragon, Curiosity, Angelica and a sister by Snap, and Cypher by Squirrel. Curiosity was the dam of Misfortune by Dux (son of Matchem by Cade), that produced the great horse Buzzard, imported into America by Col. Hoomes, of Virginia, in 1805. Few horses in his day were superior to Buzzard, and he left behind him a distinguished progeny in England, being the sire of the three brothers, Selim, Castrel and Rubens, and their sister, Bronze. Buzzard was also the sire of Rosamond, the dam of imp. Barefoot by Tramp. From Barefoot, through his daughter Motto (the dam of Nannie Lewis, Sally Lewis and My Lady, all by imp. Glencoe) a distinguished family of horses descend. Sally Lewis was the dam of John Morgan by imp. Sovereign, Hunter's Lexington, Glenrose, Susan Beane and Acrobat by Lexington, Linda Lewis by Oliver, son of Wagner, and Lotta by Hunter's Glencoe, son of imp. Glencoe, and Fiatt (The Blue Filly) by imp. Hedgeford. John Morgan and Hunter's Lexington were good race-horses and were successful in the stud. Susan Beane was the dam of Stratford, Sensation, Onondaga, Sioux and Susquehanna, the last one the dam of the grand race-horse Potomac (winner of the Futurity and Realization Stakes) and Chesapeake, a good horse. Susan Beane was also the dam of Soubrette and Sir Hercules by Alarm.

Glenrose is the dam of Glendew, she the dam of the fleet Guenne by imp. Glengarry, and Restless by imp. Mortemer, a real good one. Linda Lewis is the dam of Kisba, Quiver, Withrow (by Longbow, son of Longfellow) and the superior race-horse and promising stallion Linden by Longfellow. Lotta is the dam of the fine race-horse Glenmore by imp. Glen Athol, son of Blair Athol by Stockwell. Glenmore was a good race-horse at all distances in fast time, his three heats in four miles in 7.29½, 7.30½, 7.31, being the fastest and best ever run in the world. Buzzard was the sire of Hephestion, out of imp. Castianira (the dam of Sir Archy) by Rockingham, son of Highflyer by Herod. Hephestion was a good race-horse and quite a successful stallion. Buzzard's daughters were a great success in America. One out of The Fawn by Craig's Alfred was the dam of Old Court, Lady Adams by Whipster, Huntress by Cherokee, Lady Fortune, the greatgrandam of Longfellow, by Brimmer or Blue Beard, and the noted Woodpecker by Bertrand. Woodpecker was the sire of Grey Eagle, the competitor of Wagner in the two famous races at Louisville, Ky., run in the fall of 1839. Another daughter of Buzzard, out of Arminda by imp. Medley, son of Gimcrack by Cripple, son of the Godolphin Arabian, was the dam of Grey Fanny by Bertrand, son of Sir Archy. Grey Fanny was the dam of Grey Medoc by Medoc, son of American Eclipse, that was the best race-horse of his day. He won the great race, four-mile heats, at New Orleans, in 1841, the first heat, a dead heat with Altorf, in 7.35, second heat won by Altorf in 8.19, and the third and fourth heats won by Grey Medoc in 7.42, 8.17. Grey Fanny also produced Kate Aubrey, a good mare by American Eclipse, and the noted Blue Bonnet by imp. Hedgeford. Blue Bonnet won a number of good races, and retired in the stud in her ninth year. She was a success in the stud, being the dam of Little Arthur, Nebula (the dam of the unbeaten Asteroid, Sue Lewis, Luna, and Aneroid by Lexington, and Asterisk by Ringgold, son of Boston), Alice Jones (the dam of Jonesboro, Mary Martin, Annette and others by Lexington), all by imp. Glencoe), and the four noted brothers Lightning, Thunder, Loadstone and Lancaster, with their sister Bonnet by Lexington. Nearly all those mentioned raced successfully, and those tried in the stud produced well. Mary Martin was the dam of Santa Anita (Gano's dam), Venezuela and Vulpine by Virgil. Lightning, Loadstone and Lancaster were all superior race-horses, and Lightning for his chances a successful stallion. Bonnet is the dam of Steinbok, Chimneysweep, etc. Another daughter of Buzzard was the dam of Nell by Orphan, son of Ball's Florizel by imp. Diomed, that was the dam of Eliza Jenkins by Sir William of Transport, son of Sir Archy. Eliza was the dam of Pressure, Hawkeye, Thos. R. Roots (all good horses in their day), Gold Eagle and Tulip by Grey Eagle, Louisa M. Berry and Clipper by American Eclipse. Nell was also the dam of the mare by Sir Archy Montorio (Big Archy) that produced the excellent race-horse Creath by imp. Tranby. The Eclipse mare, out of Nell was the dam of the superior race-horse Dick Doty by Boston, whose race of four miles at Lexington, Ky., in 1853, 7.37½, 7.16½, was the fastest race at the distance run in Kentucky up to that date. This Eclipse mare was also the dam of Louisa by imp. Yorkshire, son of St. Nicholas by Emilius, and Mattie Gross by Lexington. Louisa was the dam of Jessamine Porter (the dam of Donnybrook, Ida Wells by Leamington, and Maggie Hunt by King Ernest). Mattie Gross was the dam of Tasmania, Jennie June (the dam of Dan K.), Baronet and Mate (a superior race-horse), all by imp. Australian, Whisper, a good race-horse and stallion, by Planet, son of Revenue by imp. Trustee, Mattie Glenn (the dam of Bonanza) by imp. Glen Athol, and others.

Darley Arabian got Flying Childers and his brother, Bartlett's Childers, out of Betty Leedes by Old Careless (son of Spanker and a Barb mare), grandam by Leedes Arabian, great grandam by Spanker (son of the D'Arcy Yellow Turk), out of Old Morocco mare (Old Peg), Spanker's own dam, by Lord Fairfax's Morocco Barb. Flying Childers got Snip out of sister to Soreheels by Basto (son of Byerly Turk), she out of Partner's dam by the Curwen Bay Barb.

The Darley Arabian founded the third division, or Eclipse line. Snip was the sire of Snap, dam sister to Slipby by Fox, out of Gipsy by Bay Bolton. Snap was the sire of Latham's Snap, 1759, dam Dormouse's dam by Cade, Chedworth's Snap, 1760, dam Young Bowes by Dormouse, Snapdragon, Curiosity, and Angelica out of Fribble's dam by Regulus. Snapdragon was the dam of Paymaster by Blank, and Rarity by Matchem, she the dam of Maid of the Oaks by Herod, and Radish by Pot-8-o's, and filly by Gimcrack, she the dam of Constable's imported mare by Pot-8-o's, which was the dam of Miller's Damsel by imp. Messenger, she the dam of American Eclipse. Curiosity was the dam of Pantaloon and Young Pantaloon by Matchem, Misfortune (dam of imp. Buzzard by Dux), son of Matchem, Justice and Faith (dam of Bobtail by Eclipse) by Herod. Angelica was the dam of Flora (Spadille's dam) by Squirrel, and a filly (dam of Sir George) by Chrysolite. Amanda, the dam of imp. Medley, was by Snap. Caroline by Snap was the dam of Pomona by Herod. Laïs by Diomed, dam Grace by Snap, was the dam of Spoliator and Thaïs by Trumpator, and Pamela by Whiskey. Hyæna by Snap was the dam of Eagle, Everlasting and Lavinia by Eclipse, and Evander by Highflyer. Lisette by Snap was the dam of Maria (the dam of Waxy and Worthy by Pot-8-o's, Quiz and Wowski by Mentor) by Herod. Middlesex by Snap was the dam of imp. Seagull by Woodpecker. Papillon by Snap, dam Miss Cleveland by Regulus, was the dam of the famous Highflyer, and his sister Lady Teazle by Sir Peter, Wren and Wagtail by Woodpecker. Penultima by Snap was the dam of imp. Baronet by Vertumnus, son of Eclipse. Promise by Snap was the dam of Peppermint and the famous Prunella by Highflyer. From Prunella are descended some of the most famous horses known to the turf, including Penelope (the dam of Wire, Web, Whalebone, Woful, Whisker, etc.) by Trumpator. Virago by Snap was the dam of imp. Saltram and Speranza by Eclipse. A daughter of Snap, dam sister to Regulus by the Godolphin Arabian, was the dam in Ireland of Heroine by Hero, son of Cade, and Heroine was the dam of Tom Turf, foaled 1781, by Lennox, son of Bustard by Crab, and of Young Heroine by Bagot, the grandam of Irish Birdcatcher. A daughter of Snap, dam sister to imp. Duchess, from which descended the Lexington family, and the Maid of Oaks family, out of Lady Thigh by Croft's Partner, was the dam of Conductor, sire of Trumpator, Alfred, Georgiana, etc., all by Matchem. The Snap mare out of Warwickshire Wag's dam by Marlborough was the dam of imp. Shark by Marske. The Snap mare dam by the Gower Stallion was the dam of Postmaster, imp. Peggy and their sister, from which have descended in the female line such good horses as Bustard (by Buzzard), Hampton (by Lord Clifden), Blair Athol and all the Queen Mary family, Petrarch, Pretender, Sir Bevys, Tristan, Beauclerc, Rotherhill, etc. From Peggy in

America, through Peggy her daughter by imp. Bedford, have come a superior lot of race-horses, and there are many now on the turf in direct female line from her. A daughter of Snap out of Chalkstone's dam by Shepherd's Crab was the dam of imp. Gabriel by Dorimant. The blood can only be found in the female line, as there is no stallion in the male line from Snap. To Bartlett's Childers, own brother to Flying Childers, we look for the founder of the Darley Arabian line. He was first distinguished by the name of "Bleeding Childers," from his frequent bleeding from his nose. He was never trained for racing but used as stallion. He was the sire of Squirt, foaled in 1732, dam sister to Old Country Wench by Snake, son of the Lister Turk, grandam Grey Wilkes by Hautboy, out of Miss Betty D'Arcy Pet mare, daughter of a Sedbury Royal mare. Squirt was a superior race-horse, and when a stallion in Sir Harry Harpur's stud was ordered to be shot, but when Miles Thistlewaite was leading him to the Dog Kennel, he was begged off by Sir Harry's groom, after which he became famous as the sire of Marske, foaled in 1750, dam the Ruby mare by Hutton's Blacklegs, grandam by Bay Bolton out of a daughter of Fox Cub, etc., Syphon, dam by Patriot, son of Bay Bolton, grandam by Crab, out of sister to Sloven by Bay Bolton, etc., and Mr. Pratt's famous Squirt mare, her dam Lot's dam by Mogul, out of Camilla by Bay Bolton. This Squirt mare was the dam of Conundrum, Rauthos, Ænigma, Riddle, Miss Timms, Pumpkin, Maiden, Rasselas and Purity, all by Matchem, son of Cade. Marske became famous as the sire of Mr. O'Kelly's Eclipse, foaled during the Eclipse in 1764, from which he took his name. Eclipse was never beaten and founded a family of his own. His dam was Spiletta by Regulus, son of the Godolphin Arabian, grandam Mother Western by Smith's son of Snake, own brother to William's Squirrel by Lister Snake out of a daughter of Lord D'Arcy's Old Montagu.

Eclipse was the sire of Pot-8-o'-s, foaled 1773, dam Sportsmistress by Warren's Sportsman, son of Cade by the Godolphin Arabian, grandam Golden Locks by Oroonoko, son of Crab, great grandam by Crab, fourth dam by Croft's Partner, son of Jigg, out of Thwaite's Dun mare by Akaster Turk. Another son of Eclipse was King Fergus, foaled 1775, dam Tuting's Polly by Black-and-All-Black, son of Crab, out of Fanny by Tartar, tracing through Old Peg by Lord Fairfax's Morocco Barb to a Barb mare. Eclipse also got Joe Andrews, dam Amaranda by Omnium, son of Snap, grandam Cloudy by Blank, etc., Mercury, 1778, Volunteer, 1780, Jupiter, Venus and Queen Mab, dam by Tartar (sire of Herod), son of Partner, grandam by Mogul, brother to Babraham, great-grandam by Sweepstakes and tracing to the old Vintner mare, Alexander, 1782, Don Quixote, 1784, and Xantippe, dam Grecian Princess by William's Forester, son of Croft's Forester by Hartley's Blind Horse, grandam by Coalition colt, son of the Godolphin Arabian, tracing through Charming Molly by Second to a Royal mare. Eclipse sired Dungannon, 1780, dam Aspasia by Herod, grandam Doris by Blank, Soldier, 1779, dam Miss Spindleshanks by Omar, son of the Godolphin Arabian, grandam by Old Starling, etc., and imported Saltram, 1780, dam Virago by Snap, out of daughter of Regulus. Eclipse sired three Derby winners, Young Eclipse, 1781, Saltram, 1783, Serjeant, 1784, and one winner of the Oaks, Annette, in 1787. No St. Leger winner is credited to him. Pot-8-o'-s was a superior race-horse and a valuable and successful stallion. He was the sire of imp. Alderman, dam Lady Bolingbroke by Squirrel, Waxy, foaled 1790, winner of the Derby, and Worthy, 1795, dam Maria by Herod, out of Lisette by Snap. Waxy was the sire of Whalebone, Whisker, Woful, Web, Wire and Wilful, dam Penelope by Trumpator, son of Conductor by Matchem, out of Prunella by Highflyer. Whalebone won the Derby in 1810, and Whisker won it in 1815. Whalebone was the sire of Sir Hercules, dam Peri by Wanderer, son of Gohanna by Mercury by Eclipse, grandam Thalestris by Alexander, son of Eclipse out of Rival by Sir Peter. Whalebone also got Camel, 1822, dam by Selim, son of Buzzard, out of Maiden by Sir Peter, etc., Waverley, 1817, dam Margaretta by Sir Peter, out of sister to Cracker by Highflyer, and Defence, 1824, dam Defiance by Rubens, son of Buzzard out of Little Folly by Highland Fling, son of Spadille. Sir Hercules was the sire of the brothers, Irish Birdcatcher, 1833, and Faugh-a-Ballagh, 1841, dam Guiccioli by Bob Booty, son of Chanticleer by Woodpecker, grandam Flight, by Irish Escape, son of Commodore by Tom Tug by Herod. Irish Birdcatcher was the sire of The Baron, dam Echidna by Economist, son of Whisker, grandam Miss Pratt by Blacklock, out of Gadabout by Orville, son of Beningbrough, etc. The Baron won the St. Leger in 1845 and was the sire of Stockwell, winner of the 2,000 Guineas and St. Leger in 1852, and his brother, Rataplan, dam Pocahontas by Glencoe, out of Marpessa by Muley, she out of Clare by Marmion, son of Whiskey by Saltram. Stockwell was a superior race-horse and as a stallion was the grandest success ever known, and richly deserved the sobriquet of "Emperor of Stallions." Stockwell was the sire of The Marquis (winner of the 2,000 Guineas and St. Leger in 1862), dam Cinizelli by Touchstone, Lord Lyon winner of the 2,000 Guineas, Derby and St. Leger in 1866), dam Paradigm by Paragone, son of Touchstone, Blair Athol (winner of the Derby and St. Leger in 1864), dam Blink Bonny by Melbourne, St. Albans (winner of the St. Leger in 1860), dam Bribery by The Libel, son of Pantaloon, Bothwell (winner of the 2,000 Guineas, 1871), dam Katharine Logie by The Flying Dutchman, Achievement (winner of the 1,000 Guineas and St. Leger in 1867), dam Paradigm, Caller Ou (winner of the St. Leger in 1861, the Northumberland Plate twice, Brighton Cup and 31 Queen's Plates), dam Haricot, by Mango or Lanercost, Gang Forward (winner of the 2,000 Guineas in 1873), dam Lady May by Orlando, Doncaster (winner of the Derby in 1873), dam Marigold by Teddington. Doncaster got Bend'or (winner of the Derby in 1880), dam Rouge Rose by Thormanby, and Bend'or got Ormonde (winner of the 2,000 Guineas, Derby and St. Leger in 1886), dam Lily Agnes by Macaroni, and Bonavista (winner of 2,000 Guineas, 1892), dam Vista by Macaroni.

Stockwell also sired Lady Augusta (winner of the 1,000 Guineas in 1863), dam Meeanee by Touchstone, and Repulse (winner of the 1,000 Guineas in 1866), dam Sortie by Melbourne. Rataplan, whilst he did not win any of the great events, was a grand race-horse and a stout one. He was the sire of Kettledrum (winner of the Derby in 1861), dam Hybla by The Provost, and Blinkhoolie, dam Queen Mary by Gladiator. Blinkhoolie was the sire of Wisdom, dam Aline by Stockwell, and Wisdom is the sire of Surefoot (winner of the 2,000 Guineas in 1890), out of daughter of Galopin, and Sir Hugo (winner of the Derby in 1892), out of Manœuvre by Lord Clifden. Kettledrum was the sire of Lady Langden, out of Haricot by Mango or Lanercost. Lady Langden is the dam of Hampton by Lord Clifden, son of Newminster. Hampton is the sire of Merry Hampton (winner of the Derby in 1887), dam Doll Tearsheet by Broomielaw, and Ayrshire (winner of the 2,000 Guineas and Derby in 1888), dam Atalanta by Galopin. Hampton is also sire of Reve d'Or (winner of the 1,000 Guineas and the Oaks in 1887). Rataplan is the sire of Mandragora, The Miner and Mineral, dam Manganese (winner of the 1,000 Guineas) by Irish Birdcatcher. Mineral is the dam of Wenlock (winner of the St. Leger in 1872) by Lord Clifden, and the sister to Wenlock was dam of Geologist by Sterling. Mineral also produced Kisber (winner of the Derby in 1876, and the Grand Prize of Paris same year,) by Buccaneer. Mandragora by Rataplan is the dam of Mandrake by Weatherbit, Agility and Apology (the latter winner of the 1,000 Guineas, Oaks and St. Leger in 1874), both by Adventurer, son of Newminster. Honeymoon by Rataplan was the dam of Cradle and Cylinder by See Saw. The Baron was also sire of Costa out of Catharine Hayes by Lanercost. Costa got some good brood mares. Birdcatcher also got Augur, dam Nickname by Ishmael, son of Sultan; Daniel O'Rourke (winner of the Derby in 1852), dam Forget-me-not by Hetman Platoff, son of Brutandorf; Gamekeeper, dam Swallow by Lanercost; Warlock (winner of the St. Leger in 1856), dam Elphine by Emilius. Warlock was the sire of Tynedale. Saunterer by Irish Birdcatcher, dam Ennui by Bay Middleton, was sire of Gamos (Oaks winner in 1870), Gertrude, she the dam of Charibert (winner of the 2,000 Guineas in 1879) by Thormanby, and Childeric by Scottish Chief. Knight of St. George by Irish Birdcatcher, dam Maltese by Hetman Platoff, won the St. Leger in 1854, and was sire of the Knight of St. Patrick out of Pocahontas by Glencoe, Ada B., etc., in England. Knight of St. George had little or no chance in the stud in America, owing to the Civil War, but he was sire of Breana (grandam of The Bard, Tibbie Dunbar, etc). Edina (dam of Minnie Mc, Elemi, etc.), Eliza Davis (dam of War Song, she the dam of Eole, St. Saviour, Eon, etc.), Fairy (dam of Grey Nun, Artist, etc.), Georgia Wood (dam of Ginger, La Gitana, Una, etc.), Heliotrope (dam of Marchioness, Mariposa, etc.), Jenny H. (dam of Jentling, Ada S., Toledo, etc.), Lass of Sidney (dam of Tubman, Mary Constant, Phyllis, Mary P., Queechy, etc.), Mishap (dam of Crockford, Perhaps (the dam of Chance, Reckon, Doubt, etc.), and Misfortune (dam of Abbotsford, Syntax, etc). Knight of St. George's name appears in many of our best pedigrees. Wormersley by Irish Birdcatcher, dam Cinizelli by Touchstone, was the sire of Codrington, Gen. Williams, and Wanona, the grandam of Robert the Devil, out of a daughter of Hampton by Sultan. Oxford by Irish Birdcatcher, dam Honey Dear by Plenipotentiary, is the sire of Sterling, he the sire of Isonomy, the sire of Seabreeze (winner of the Oaks and St. Leger in 1888), and the great Common (winner of the 2,000 Guineas, Derby and St. Leger in 1891). Sterling is the sire also of Paradox, Enterprise and Enthusiast, winners of the 2,000 Guineas, and Harvester, who ran a dead heat in the Derby in 1884, and divided the stakes with St. Gatien. Sterling is also sire of Geologist, Fernandez, Poste Restante, the imp. horses Loyalist, Top Gallant and Stalwart. Birdcatcher was sire also of imp. Mickey Free. Faugh-a-Ballagh (winner of the St. Leger in 1844) was the sire of Ethelbert, dam Espoir by Liverpool. [Ethelbert was the sire of Isolina, grandam of Isonomy.] Constance, dam Milkmaid by Glaucus, and she, Constance, was the dam of Bertram, sire of Robert the Devil. It is to Faugh-a-Ballagh's son Leamington that America is indebted for this blood. Leamington's dam was by Pantaloon, son of Castrel by Buzzard, out of Daphne by Laurel, son of Blacklock. Leamington made his first season in America in 1866, at the Bosque Bonita stud, and from that season came Enquirer, dam Lida by Lexington, a first-class race-horse and a successful stallion, Longfellow, the grand race-horse and popular sire, dam Nantura by Brawner's Eclipse, son of American Eclipse, Lyttleton, a good race-horse, dam Fanny Holton by Lexington, and Lynchburg, dam Neutrality by Revenue. These were followed by Eolus, dam Fanny Washington by Revenue, a capital race-horse and one of the most successful stallions in America, and subsequently came Reform, Hyder Ali, James A., Rhadamanthus, the unbeaten Sensation and his brothers Stratford and Onondaga, Parole, Bob Woolley, Aristides, Harold, Fernclifte, Blazes, the great Iroquois, and a number of most excellent mares, including Susquehanna (dam of Potomac), Francesca, Katie Pearce (dam of Lizzie S., Ballard, Katrina, etc.), Spinaway, Spark, Jaconet (dam of Sir Dixon, Belvidere, etc.), Mary Buckley, dam of Stonebuck, Perfection (dam of Saluda), Preciosa, Syria, dam of Dalsyrian, etc., Theodocia, The Squaw, Wissahickon, Wyandotte, etc. It will be observed that the best sires of all these both on the turf and in the stud are from those in both male and female line to the Eclipse and Herod lines of blood. Sir Hercules was also the sire of Coronation (winner of the Derby in 1841), dam Ruby by Rubens, Gemma di Vergy, sire of Roccoco, dam Snowdrop by Heron, son of Bustard by Castrel, Gunboat and Lifeboat, out of Yard Arm by Sheet Anchor, son of Lottery by Tramp, and Robert de Gorham, dam Duvernay by Emilius. Whalebone was the sire of Camel, dam by Selim, son of Buzzard, out of Maiden by Sir Peter. Camel was the sire of Touchstone, dam Banter by Master Henry, son of Orville by Beningbrough. Camel was one of the most noted of Whalebone's sons, as from his sons Touchstone, Launcelot and Simoon, a host of distinguished race-horses and brood mares have descended. Touchstone won the St. Leger in 1834. He got Blue Bonnet (winner of the St. Leger in 1845), Orlando (winner of the Derby in 1844.) In this Derby Running Rein, who came in first, was subsequently proven to be a four year old Maccabeus, and he was disqualified.

Orlando's dam was Vulture by Langar, son of Selim by Buzzard. Orlando was the sire of Teddington (winner of the Derby, 1851), Imperieuse (winner of 1,000 Guineas and the St. Leger, 1857), Fazzoletto, Fitz Roland and Diophantus, winners of the 2,000 Guineas, also of Trumpeter, sire of Distin. Sire also of Plutus, the sire of Flageolet, that got Rayon d'Or, Zut and Beaumnet; Marsyas, sire of George Frederick (Derby winner, 1874), dam Malibran by Whisker. Orlando was also the sire of Little Lady, dam of Camballo, winner of the 2,000 Guineas, 1875, Bay Celia, dam of The Duke, Laura, dam of Petrarch (winner of 2,000 Guineas and St. Leger, 1876), and Doralice the dam of Speculum (winner of the City and Suburban, and sire of Sefton, winner of the Derby, 1878). Teddington was the sire of Marigold, dam of Doncaster, winner of the Derby, 1873, and he the sire of Bend'or, winner of Derby, 1880. Fitz Roland was the sire of Miss Roland, dam of Craig Millar, winner of the St. Leger, 1875. Touchstone was the most famous and valuable of all Camel's sons. He was the sire of Cotherstone, winner of the 2,000 Guineas and Derby, 1843, Flatcatcher and Nunnykirk, winners of the 2,000 Guineas in 1848 and 1849, Lord of the Isles, winner of the same event, 1855, Surplice, winner of the Derby and St. Leger, 1848, Mendicant, winner of 1,000 Guineas and the Oaks, 1846. Newminster, winner of the St. Leger, 1851, who was out of the famous Bee's-wing by Dr. Syntax, son of Paynator by Trumpator by Conductor by Matchem, grandam by Ardrossan, son of John Bull by Fortitude by Herod, great-grandam Lady Eliza, by Whitworth, son of Agonistes by Sir Peter, thus tracing to the Matchem and Herod blood. Newminster got Adventurer, dam Palma by Emilius. Adventurer won the City and Suburban and was the sire of Pretender, winner of the 2,000 Guineas and Derby, 1869, Apology, winner of the 1,000 Guineas, Oaks and St. Leger, 1874, Wheel of Fortune, winner of the 1,000 Guineas and Oaks, 1879, imp. Blythewood, a sire of winners in this country, and Pizzaro, now dead, a horse that promised success. Newminster is immortalized as the sire of Hermit, dam Seclusion by Tadmor, son of Ion by Cain by Paulowitz by Sir Paul by Sir Peter, grandam Miss Sellon by Cowl, son of Bay Middleton by Sultan by Selim by Buzzard, largely inbred to the Herod blood. Hermit won the Derby in 1867 and was one of the most successful and popular stallions ever in England. Hermit was sire of Trappist, Shotover (winner of the Derby and 2,000 Guineas, 1882), Trapeze and Tristan, out of Thrift by Stockwell. Tristan was beaten in the Derby and St. Leger by Iroquois, and in the Grand Prize of Paris by Foxhall, but he was a superior race-horse. Hermit also got Torpedo, brother to Shotover, Peter, the grand race-horse, and his brother Timothy, out of Lady Masham, Clairvaux, Thebais, and St. Marguerite, out of Devotion by Stockwell. Thebais won the 1,000 Guineas and Oaks, 1881, and St. Margnerite the 1,000 Guineas in 1882. He was also sire of The Abbot, Charon, Marden and Nautilus out of Barchettina by Pelion, son of Ion. Hermit's son, St. Blaise, winner of the Derby, 1883, dam Fusee by Marsyas, son of Orlando, grandam Vesuvienne by Gladiator, son of Partisan, out of Venus by Sir Hercules, has created a sensation. He has been one of the most phenomenal successes in the stud which ever came to this country. He is inbred to Touchstone. Hermit was grandson of Touchstone and Marsyas, the sire of St. Blaise's dam was also grandson of Touchstone, and he has a large infusion of Eclipse blood in both sire and dam's side through Whalebone and Whisker. St. Blaise is the sire of La Tosca, St. Florian, Clarendon and many others. Any and all strains of blood suited him. Newminster was sire of Lord Clifden out of The Slave by Melbourne. Lord Clifden won the St. Leger in 1863, and was the sire of Petrarch, winner of the 2,000 Guineas and St. Leger in 1876 (and he the sire of of Busybody and Miss Jummy (winners of 1,000 Guineas and Oaks in 1884 and 1886), Hawthornden and Wenlock, winners of the St. Leger, 1870 and 1872, Janette, winner of the Oaks and St. Leger in 1878. Newminster is also sire of Cambuscan, out of The Arrow by Slane, son of Royal Oak by Catton. Cambuscan is sire of Camballo, winner of the 2,000 Guineas, 1875. Touchstone was sire of Lord of the Isles (dam Fair Helen by Pantaloon), he the sire of Scottish Chief, he the sire of Childeric, Violet Melrose (the dam of Melton), Marie Stuart, winner of the Oaks and St. Leger, 1873, King of the Forest, imp. Kantaka (out of Seclusion, Hermit's dam), and Dundee, out of Marmalade by Sweetmeat, winner of six out of seven races at two years old. Touchstone also got Ithuriel, sire of Longbow, he the sire of Toxopholite, the sire of Musket, Maxim's sire, and Bay Archer. Whalebone was also sire of Waverley, out of Margaretta by Sir Peter. Waverley was the sire of Don John (winner of St. Leger, 1838), he the sire of Maid of Masham, the grandam of Peter and Timothy, also of Iago, sire of Bonnie Scotland. The Saddler, by Waverley, out of Castrellina by Castrel, was the sire of Inheritress (dam of the Heiress), and Hybla (the dam of Kettledrum), by The Saddler's son. The Provost, produced also Maid of Hart, dam of Compiegne, sire of Mortemer.

Whalebone was the sire of Defence, dam Defiance by Rubens, son of Buzzard. Defence was the sire of Diversion, the grandam of Rosicrucian and The Palmer, and of

Protection, the grandam of See Saw. His best son was the Emperor, dam daughter of Reveller by Comus. He was the sire of Monarque, dam Poetess by Royal Oak, son of Catton. Monarque was the sire of Gladiateur, dam Miss Gladiator. Gladiateur was the winner of the 2,000 Guineas, Derby, St. Leger and Grand Prize of Paris in 1865. Gladiateur was the sire of Grand Coup, Highborn, Lord Gough and Lady Gladys. Monarque was the sire of La Favorita, the dam of Flageolet, sire of imp. Rayon d'Or. Monarque was sire also of Trocadero, sire of Fra Diavolo, Hospodar and Consul, sire of Archiduc, Farfadet and Nougat in France, and of Reine, winner of the 1,000 Guineas and the Oaks in 1872. Whisker, the brother to Whalebone, was the sire of Economist, dam Floranthe by Octavian, son of Phænomenon by Herod, grandam Caprice by Anvil, son of Herod, out of Madcap by Eclipse. He was sire of Harkaway, a grand race-horse up to any weight. Harkaway was the sire of Idle Boy, dam Iole by Sir Hercules; he the sire of Tomboy. Harkaway's great reputation comes through his son King Tom, whose dam was old Pocahontas, dam of Stockwell, Rataplan, etc. Upon him depends almost entirely the transmission of the blood of this wonderful horse in the male line. King Tom was the sire of Kingcraft, winner of the Derby in 1870, and his brother, imp. Great Tom, out of Woodcraft by Voltigeur, King Lud, winner of the Cesarewitch stakes, out of Qui Vive, sister to Vedette; Skylark, winner of the Queen's Vase at Ascot; imp. King Ernest, out of Ernestine by Touchstone; imp. King Ban, dam Atlantis by Thormanby; imp. Phaeton, dam Merry Sunshine by Storm, son of Touchstone. King Tom got a number of choice brood mares. Of these St. Angela, dam Adeline by Ion, was the dam of St. Simon. Empress was the dam of Chislehurst. Premature was dam of Poste Restante. Princess was dam of Royal Hampton, Prince Royal and His Highness. A sister to King Alfred by King Tom was the dam of Enterprise (winner of the 2,000 Guineas in 1887). Zephyr by King Tom was the dam of Favonius (winner of the Derby in 1871). A daughter of King Tom is the dam of the promising young stallion and sire imp. King Galop. Euxine was dam of Town Moor. Whisker was also the sire of Peron, dam Darioletta by Amadis, out of Selima by Selim. Whisker was also the sire of The Colonel (winner of the St. Leger in 1828), out of My Lady's dam by Delpini, and of Memnon (winner of the St. Leger in 1825), dam Manuella by Dick Andrews. Whisker's best daughter was Emma, dam of imp. Trustee and of Cotherstone (winner of the 2,000 Guineas and Derby in 1843), of Mundig (winner of the Derby in 1835), and of Mowerina, the dam of West Australian. The Colonel was the sire of Chatham, who sired The Governess, winner of the 1,000 Guineas and Oaks in 1858. Economist was also the sire of Echidna, out of Miss Pratt by Blacklock, she the dam of The Baron, etc. King Fergus was the sire of Young Traveller, winner of the St. Leger in 1791, Beningbrough, winner of the St. Leger in 1794, and Hambletonian, winner of the St. Leger in 1795, dam by Highflyer, son of Herod, out of Monimia by Matchem. Hambletonian was the sire of Whitelock, dam Rosalind by Phenomenon, son of Herod, grandam Atalanta by Matchem, out of Lass of the Mill by Oroonoko, son of Crab.

Whitelock, 1803, was the sire of Blacklock, 1814, dam by Coriander, son of Pot-8-o's, grandam Wildgoose by Highflyer, out of Coheiress by Pot-8-o's. Blacklock was the sire of Brutandorf, 1821, dam Mandane by Pot-8-o's, out of Young Camilla by Woodpecker. Brutandorf was the sire of Physician, dam Primette by Prime Minister, son of Sancho, out of Miss Paul by Sir Paul. Physician was the sire of The Cure, dam Morsel by Mulatto, son of Catton. The Cure was the sire of Polly Agnes and Little Agnes, out of Miss Agnes by Birdcatcher. Polly Agnes is the dam of Lily Agnes (the dam of Ormonde and imp. Rossington) by Macaroni. Little Agnes is the dam of Tibthorpe, winner of the Steward's Cup, by Voltigeur. Brutandorf was also sire of Hetman Platoff, who got The Cossack and Neasham. Hetman Platoff is the sire of Maltese (dam of imp. Knight of St. George), Cossack Maid (grandam of Balfe), the grandam of Solon, which mare is great grandam of Barcaldine. Alexina, the dam of Tomboy, Muscovite, sire of Vauban (winner of the 2,000 Guineas in 1867), and of Siberia, (1,000 Guineas in 1865), Bunch, Trappist's dam, etc. Blacklock was the sire of imp. Belshazzar, dam Manuella by Dick Andrews (and he, Belshazzar, was the sire of some good ones in America), of imp. Tranby, the sire of Vandal's dam, Young Blacklock

(Watt's), sire of Magpie, out of Kitten by Sligo, Waxy, and Velocipede, dam by Juniper, son of Whiskey, out of a daughter of Sorcerer. Velocipede was the sire of Amato (winner of the Derby in 1838), Queen of Trumps (winner of the Oaks and St. Leger in 1835), King of Trumps, Hornsea, etc., whose names appear in many good pedigrees, and of Voltaire, dam by Phantom, son of Walton by Sir Peter.

Voltaire was the sire of Barnton (sire of Fandango), and his brother Voltigeur, (winner of the Derby, St. Leger and Doncaster Cup in 1850), out of Martha Lynn by Mulatto, Charles XII (winner of the St. Leger in 1839), and Tearaway. Voltigeur has perpetuated the Blacklock blood. He is the sire of Bonny Bell (dam of Blantyre, Beauclerc, etc.), out of Queen Mary, Dulcibella (dam of Onslow, Ma Belle, etc.), Beatrice, dam of Prestonpans, Makeshift (dam of Pell Mell and Helter Skelter by Young Melbourne), Our Mary Ann (dam of imp. Doncaster Lass by Doncaster), Qui Vive (dam of imp. Felucca), and Vedette, out of Mrs. Ridgway by Birdcatcher. Voltigeur was also the sire of Skirmisher, The Ranger, imp. Hartington, Tibthorpe, Buckstone, Cavendish and John Davis. Vedette was the best one of all his get. He won the 2,000 Guineas, was not in the Derby or St. Leger, won the Doncaster Cup two consecutive years, and many other races. Vedette was the sire of Gardevisure (winner of the Cambridgeshire), out of Paradigm by Paragone, Speculum (winner of the City and Suburban Handicap), out of Doralice by Orlando, son of Touchstone, and Galopin (winner of the Derby in 1875), out of Flying Duchess by The Flying Dutchman, and it is through these two horses that the Blacklock blood has become so famous in this day. Speculum was the sire of Rosebery (winner of the Cesarewitch and Cambridgeshire Handicaps), out of Ladylike by Newminster, Sefton (winner of the Derby in 1878), out of Lady Sefton by West Australian, son of Melbourne, Flotsam and Jetsam, out of Flotilla by Kingston, Hagioscope, dam Sophia by Macaroni, he the sire of Queen's Birthday, Amphion by Speculum or Rosebery, out of Suicide by Hermit, a superior race-horse and a popular stallion. Penton, dam Progress by Thormanby, he the sire of Penelope, a good mare. Galopin is the sire of Donovan (winner of the Derby and St. Leger in 1889), and Modwena, out of Mowerina by Scottish Chief, grandam Stockings by Stockwell, Galliard, dam Mavis by Macaroni, out of Merlette by The Baron, etc. Galliard won the 2,000 Guineas, Prince of Wales, Ascot, Chesterfield and other stakes. St. Simon is by Galopin, dam St. Angela by King Tom, out of Adeline by Ion, etc. Donovan and Galliard are young in the stud but must prove successful. It is worthy of note that Galopin is much inbred, being by Vedette, son of Voltigeur by Voltaire, his dam Flying Duchess by The Flying Dutchman, whilst his grandam Merope is by Voltaire, son of Blacklock, and his great grandam is Velocipede's dam by Juniper. Galopin's son Donovan won the Derby and St. Leger in 1889, and Semolina, Donovan's half sister, won the 1,000 Guineas in 1890, she being by St. Simon, son of Galopin. The Blacklock blood is represented in this country through descendants of Belshazzar, Tranby, the sire of Tranbyana, the dam of Big Boston. Belle Lewis by Glencoe, and the Tranby mare, dam of Vandal, Volga and Levity, and their descendants, King Galop by Galopin, dam by King Tom, out of Sunshine by Thormanby, a very fine horse and superior breeder. Beningbrough by King Fergus, dam by Herod, grandam Pyrrha by Matchem, out of Duchess by Whitenose, son of the Godolphin Arabian, is another strong link in the Darley Arabian line. Beningbrough got Orville, 1799, dam Evelina by Highflyer, out of Termagant by Tantrum, son of Cripple by the Godolphin Arabian. Orville won the St. Leger in 1802 and was the sire of Emilius, dam Emily by Stamford, son of Sir Peter, out of daughter of Whiskey, son of Saltram by Eclipse. Emilius was the sire of Euclid, dam Maria, sister to Emma by Whisker, he the sire of Bantam, out of Lady Fanny by Dr. Syntax. Emilius was the sire of Plenipotentiary, winner of the Derby in 1834, out of Harriet by Pericles, son of Evander by Delpini. Plenipotentiary was the sire of Herald, dam imp. Delphine by Whisker, that won the second heat in the great Peyton stake won by Peytona at Nashville, Tenn., October 10, 1843. Herald was the sire of Heraldry, dam of Balloon, Ascension, Parachute, etc. From Balloon a large number of good race-horses have descended. Plenipotentiary was sire of Potentia (winner of the 1,000 Guineas in 1841) and of Poison (winner of the Oaks in 1843). Plenipotentiary was a valuable sire in the female line in England. He is the sire of the dam of Queen Mary, of Delhi, the great grandam of Bend'or, of Honey Dear, the dam of Oxford, sire of Sterling, etc., of Monstrosity, the dam of The Ugly Buck (winner of the 2,000 Guineas in 1844). Priam by Emilius, dam Cressida, sister to Eleanor by Whiskey, son of Saltram by Eclipse, grandam Young Giantess by Diomed, out of Giantess by Matchem, won the Derby in 1830 and was imported to America in 1837. Priam was the sire of Crucifix (winner of the 2,000 Guineas, 1,000 Guineas and Oaks in 1840), out of Octaviana by Octavian, son of Stripling by Phenomenon, Miss Letty (winner of the Oaks in 1837), out of a daughter of Orville, Industry (winner of the Oaks in 1838) and Scamander, her brother, out of Arachne by Filho-da-Puta. Crucifix's great reputation rests upon being the dam of Cowl by Bay Middleton. Cowl was the sire of Miss Sellon, out of Belle Dame by Belshazzar, and Miss Sellon is the grandam of the great and popular sire Hermit by Newminster. Cowl was also sire of Madame Eglantine, the dam of Rosicrucian and The Palmer, out of Diversion by Defence. Crucifix was also the dam of Surplice (winner of the Derby in 1848) by Touchstone. Surplice was sire of Eastern Princess, the dam of imp. Prince Charlie (winner of the 2,000 Guineas in 1872), and of Flax, the dam of Queen Bertha (winner of the Oaks in 1863). Miss Letty was the dam of Weatherbit by Sheet Anchor, son of Lottery by Tramp, and My Dear by Bay Middleton. Weatherbit was the sire of Beadsman (winner of the Derby in 1858), out of Mendicant by Touchstone. Beadsman was sire of Blue Gown (winner of the Derby in 1868), and Pero Gomez (winner of the St. Leger in 1869). Beadsman was also sire of Rosicrucian and The Palmer, the latter sire of Jenny Howlet (winner of the Oaks in 1880), dam of Chitabob, both valuable and successful stallions. Weatherbit is also sire of Brown Bread, out of Brown Agnes by West Australian, he the sire of Toastmaster, a very fine race-horse. Priam got in this country Cassandra, out of Flirtilla jr. by Sir Archy. Cassandra was the dam of Trojan, Basil and Philo by Mariner, son of Shark by American Eclipse. Philo was the dam of Lady Blessington by imp. Eclipse, son of Orlando, she the dam of The Baroness (Badge's dam) by Kentucky, Lady Rosebery and Duchess by Kingfisher, and Fairwater by Ill Used in the Nursery Stud. Priam got a number of other good brood mares in America, such as Cornelia, dam Bay Maria by Eclipse, Creusa, dam of the good race-mare Bostona, dam Miss Valentine by imp. Valentine; Delta, dam of Gazelle by Albion, Talma by Glencoe, Duette by Highlander, son of Glencoe, out of Gamma by Pacific; Emily Thomas, dam of Pasta out of Nancy Norman by Tom Tough; Margaret Wood, dam of Star Davis, Capt. Travis, Georgia Wood, etc., out of Maria West by Marion; Eudora, dam of Kate Hays, Martha Worsham, Sally Woodward, etc , out of Bet Bosley, jr., by imp. Bluster; Fidelity, dam of Faith, Felicity, etc., out of Maria Louisa by Monsieur Tonson; Little Rose, dam of Red Rose by Glencoe, out of Tuberose by Arab; Little Trick, dam of Odd Trick by Lexington, etc., out of Bet Bosley, Jr.; Lucy Long, dam of the great race-horse Red Eye by Boston, out of Polly Franklin by Shakespeare, and The Queen, dam of Deucalion, out of imp. Delphine by Whisker. Monarch, own brother to The Queen, was an excellent race horse and left some good brood mares, including Bellamira (dam of Dic Clapperton, Embry's Lexington, Concord, etc.), Breeze, dam of the good race-mare Albine, Castinet, dam of Highlander by Glencoe, Young Fashion (dam of Surprise, Scotland, Lady Fashion, Columbia, Wananita, Bonnie Kate, etc.), out of the great race-mare Fashion by imp. Trustee, Lady Lancaster (dam of Malcolm, Ontario, Lady Amanda, etc.), out of Lady Canton by imp. Trauby, Millwood (dam of Charleston, Rosa Bonheur, Mill Creek, etc.), out of Fanny by American Eclipse. Emilius was also the sire of imp. Sarpedon, dam Icaria by The Flyer, son of Vandyke, Jr. by Walton. Sarpedon was the sire of Alice Carneal (Lexington's dam), out of Rowena by Sumpter, and Ariel, dam Lancess by Lance. Sarpedon got some good race-horses but left no descendant in the male line. Sovereign by Emilius, dam Fleur-de-Lis by Bourbon, son of Sorcerer by Trumpator, was imported when a yearling and never raced. He was the sire of a large number of good racers, such as Rurie, Charleston (sent to England), Berry, Mogul and Invincible. He got a large number of good brood mares, and his blood is transmitted in the female line to very many of our best horses. Daughters of Sovereign are sister to Rurie (dam of Grinstead), Ann Dunn and Prioress (winner in this country and England), out of Reel, Alexandra (dam of Clemmie G.), Carrie Cage (dam of Betty Belle, Eppie L., etc.), Dixie (dam of Herzog, Florac, Aureola, War Reel, Hearsay, etc.), dam St. Mary by Hamlet, Edith (dam of Edina, Stonewall Jackson, etc.), Fanny Wells (dam of Jils Johnson, Rena B., Early Light, etc.), dam Reel, Fleur-de-Lis (dam of Bourbon, Censor and Columbia), out of Maria West by Marion, Glycera (dam of Knighthood, Carrie Atherton, Sue Ryder, The Gleaner, etc.), out of sister to Prior by Glencoe, Iodine (dam of Iodine and Iona by Lexington, Arnica, Collodion, etc.), out of daughter of Stockholder, Kate Anderson, out of Chloe Anderson by Rodolph, Lady Barry, Lady Sherrod, Lorette (dam of Mishap, Regan, Sophy Badderly, Lorena, etc.), out of Mary Ogden by Thornhill, son of Glencoe, Mamona, dam of good ones, Martha Dunn, dam of Lulu Horton, Miss Carter (dam of Hildegarde, Evelyn Carter, etc.), Miss Odom (dam of Lucy Brooks, Mary Wylie, etc.), Monica, dam of Juanita, Nebraska (dam of Jack the Barber, Syren, etc.), Nora (dam of Bombay, Persia, etc.), Polly (dam of Ooltowah, Mary Wylie, etc.), Princess Royal, Sallie, Rosemary and many others. Emilius was the sire of St. Nicholas, he the sire of imp. Yorkshire, that made his impression on the stock of America through his sons and daughters, many of the latter being great successes in the stud. Orville was also sire of Muley, 1810, dam Eleanor, the first mare that ever won the Derby and the Oaks, by Whiskey, out of Young Giantess by Diomed. Muley was the sire of Muley Moloch, and his sister imp. Britannia, out of Nancy by Dick Andrews. Muley Moloch was the sire of Alice Hawthorn, the famous racer and dam of Oulston, Lady Hawthorn, Sweet Hawthorn, Thormanby, winner of the Derby in 1860, etc., out of Rebecca by Lottery. Britannia left a distinguished progeny through her daughters La Victime, Variation and Verona.

Muley got Morisco, out of Aquilina by Eagle. Morisco is credited with Taurus, out of Catharine by Soothsayer. Taurus got Turnus, out of Clarissa by Defence, and Turnus was the sire of Whiteface (the dam of imp. Woodlands, quite successful in the stud); also of Butterfly, winner of the Oaks, 1860. Muley was also sire of imp. Leviathan, dam by Windle, son of Beningbrough, grandam by Anvil, son of Herod, and imp. Margrave, dam by Election, son of Gohanna by Mercury, out of Fair Helen by Hambletonian, son of King Fergus. Leviathan was a good race-horse and a valuable stallion, sire of The Poney, Othello, Capt. McHeath, John R. Grymes, etc. His blood is preserved through his daughters, of whom Cotillion was the dam of Dance, Schottische, Gallopade, Jr., and O'Meara, all by Glencoe. Dance was the dam of Schottische (she the dam of Waltz, Tarantella, Gorlitza, etc.), by imp. Albion, La Polka by Lexington, etc. Crucifix (dam of Gov. Poindexter and Giantess) was by Leviathan, out of Virginia by Sir Archy. Giantess was the dam of Aduella, the famous Peytona and Union, by imp. Glencoe. Leviathan also got Dolphin, dam by Sir Archy, and Emerald (the dam of Topaz), out of imp. Eliza by Rubens, son of Buzzard. Topaz was the dam of Lodi, Waterloo, Austerlitz and Wagram by imp. Yorkshire, Colton and Arcola by Lexington, and Rivoli by Revenue. Fandango by Leviathan was the dam of Judith (Edith's dam), Rigadoon, Jeanette, Grisette and Hornpipe by imp. Glencoe. Grisette is dam of Geneva, Greyling [Gen. Ewell] by Lexington, and Quickstep by imp. Australian. Flight by Leviathan, out of Charlotte Hamilton by Sir Charles, was the dam of Oliver by Wagner, Hegira and Mecca by imp. Ambassador, Mahomet and Medina by imp. Sovereign. Betsey Coody and Sally Jones were sisters to Flight, the latter being the dam of Noty Price by Cost Johnson, Tancred, etc. Gaslight by Leviathan was the dam of Budelight, Beaconlight, etc., out of Pigeon by Pacolet. Jane Mitchell by Leviathan, dam by Stockholder, was the dam of Territier and Ella Temple. Other mares by Leviathan were Jeannetteau, the dam of Arrow by Boston and grandam of Optimist, who raced successfully in England and was valuable as a stallion in France, Laura (dam of Kansas by Sovereign, Lilly Ward, Harry of the West, Fanny Cheatham and Judge Durrell by Lexington, and Lerna by Asteroid), Martha Malone (dam of La Bacchante and Julia H. by Glencoe), out of Tachecana by Bertrand, Leviathan mare (dam of Satellite by imp. Albion, that was sent to England, and Fanny McAlister by O'Meara), out of imp. Anna Maria by Truffle, Polly Elliott (dam of Ada Tevis, Peytona Barry, etc.), out of

Caledonia by Jerry, son of Pacolet, Parasina (dam of Mary Hadley, Florence Nightingale and Blazella by O'Meara, Mary Wynne by Jack Malone, and Mattie Seviere by Blacklock), Princess Ann (dam of Elizabeth McNairy by Ambassador, Cottage Girl and Augusta by imp. Ainderby, son of Velocipede), Waxlight (dam of Torchlight and Scintilla by imp. Glencoe), out of Pigeon by Pacolet, Beeswing, Sarah Bladen, and many others. Margrave won the St. Leger and was a very successful stallion. Nearly all his get were males. He was the sire of Blue Dick, Brown Dick, that ran three miles at New Orleans in 5.30¾, 5.28, Tom Payne, Alamode, Doubloon, Florin, etc. The few mares that he left have proved the value of the blood in the stud.

Countess (the dam of Katona by Voucher), was by Margrave, out of a daughter of American Eclipse. Katona was the dam of Bettie West by Colton, Metarie and Joke by Star Davis, Item (Dan Sparling's dam), Kadi, the superior race-horse Tom Ochiltree by Lexington, Governess by Planet, etc. Crisis, dam of Donerail, was by Margrave, out of Susette by Aratus. Eleanor Margrave and Emma Wright were by Margrave, out of Fanny Wright by Silverheels, son of Oscar by imp. Gabriel. Eleanor Margrave was the dam of Rurica (dam of Harkaway) by Ruric, and Margretta by Lexington. Emma Wright was the dam of Laura Farris and Miss Doyle by Lexington, Mollie Jackson by Vandal, the winner of the great three-mile race at Louisville, Ky., in 5.35½, 5.34¾, 5.28¾, Sherrod winning the second heat. Mollie Jackson was the dam of Monday by Colton, Long Branch and Fanny Ludlow by imp. Eclipse, and Doubt by imp. Eclipse or Cavalier, son of imp. Eclipse. Fanny Ludlow was the dam of Jamaica by Lexington, and she the dam of Foxhall, that won the Grand Prize of Paris, the Cesarewitch and Cambridgeshire Handicaps, etc. Fanny G. by Margrave, out of Lancess by Lance, was the dam of Endorser by Wagner, Flora G. and Lady Dan Bryant by Lexington, Emma Maratta by Vandal, Estella by imp. Australian, and Fanny B. by Asteroid. Estella is the dam of Alma Mater, that produced the two great trotting stallions Alcantara and Alcyone. Lady Margrave (the dam of Undine by imp. Yorkshire) was by Margrave, out of Lady Adams by Whipster. Undine was the dam of Uncle Vic by Lexington, Australia by imp. Australian, Margaret Hunter by Margrave was the dam of Kate Bateman by imp. Yorkshire, Sallie by imp. Sovereign, Greek Slave by Glencoe, Stumps by Lexington, etc. Sallie was the dam of Eagle by Vandal, Long Nine by Lightning, O'Malley by imp. Mickey Free, La Reve (dam of Strychnine), and Lampi by Lightning. Greek Slave was the dam of Black Slave (the dam of Maratana and Anne Augusta by Leamington, Topsy by Ill Used, Black Gal and Bessie B. by Tom Ochiltree) by imp. Eclipse, son of Orlando, Sarong (the dam of Aristides, Wissahickon, Daisy Hoey, etc.) by Lexington, Petty and Tibitha by Tipperary, son of Ringgold by Boston. Margrave mare, dam Mistletoe by Cherokee, son of Sir Archy, was the dam of Mary Churchill by Alexander Churchill, son of imp. Zinganee. Capitola (dam of King Alfonso and Belle Barclay by imp. Phaeton, Letola by Lexington, Hospodar by Longfellow) by Vandal, La Grande Duchess (dam of Aniella, Talisman, etc.) and Versailles by Vandal. Mary Churchill was the dam of Eva Douglass and Florence Wallace by Vandal, Hamburg by Lexington, Astell (Miss Nailer's dam) by Lexington or Asteroid. A large number of good race-horses trace to this family. Margravine, the dam of the superior race-horse Jas. A. Connolly, was by Margrave, out of Gabriella by Sir Archy. The Margrave mare, out of Rosalie Somers by Sir Charles, was the dam of Josephine R. Rowan, and Rebecca T. Price by The Colonel. Rebecca T. Price is the dam of China by Lexington, the superior race-horse Mahlstick and Sweetheart by Lever, and Netty Brady by Jack Malone. Utilla by Margrave, dam Too Soon by Sir Leslie, was the dam of Young Utilla by Hurricane (Warfield's Florizel), (she the dam of Tasmania and Teetotal by Melbourne, Jr., son of Knight of St. George, Gaberlunzie by Bonnie Scotland, etc.), Ulverston and Ultima (the dam of Naisance, Sunrise, Sonora, Ultimatum, etc.), by Lexington. Victoire by Margrave, dam Argentile by Bertrand, was the dam of Zaidee by Belshazzar, Vandalia and Bijou by Vandal, Lilla (the dam of Lemonade, Judge Curtis, Chillicothe, and Lilly Duke, the last the dam of Water Lily and Lizzie Dwyer) by Lexington, Vesta by Asteroid, Leonore, etc. There is no horse in the male line from either Leviathan or Margrave. Muley was sire of Little Wonder (winner of the Derby in 1840) and of Vespa (winner of the Oaks in 1833). His daughter Marpessa was dam of the unparalleled Pocahontas, and of her half brother Idas (winner of the 2,000 Guineas in 1845.)

Joe Andrews (Dennis O.), 1778, by Eclipse, was a good race-horse, dam Amaranda by Omnium, son of Snap, out of Cloudy by Blank, son of the Godolphin Arabian. Joe Andrews was the sire of Dick Andrews, 1797, and imp. Jack Andrews, 1794, dam by Highflyer, son of Herod, grandam by Cardinal Puff, son of Babraham by the Godolphin Arabian. Dick Andrews was the sire of Tramp, 1810, dam by Gohanna, son of Mercury by Eclipse, grandam Fraxinella by Trentham, son of Sweepstakes (Turner's), great grandam by Woodpecker, son of Herod, out of Everlasting by Eclipse. Tramp got Liverpool, 1828, dam by Whisker, out of Mandane by Pot-8-o-'s, etc., imp. Zingance, 1825, out of Folly by Young Drone, and Lottery, 1820, dam Mandane by Pot-8-o-'s, out of Young Camilla by Woodpecker, etc. Liverpool got Lanercost, dam Otis by Bustard, son of Buzzard, out of Gayhurst's dam by Election.

Liverpool got Van Tromp and De Ruyter, dam Barbelle by Sandbeck, son of Catton, Loup-garou, dam Moonbeam by Tomboy, Colsterdale and Ellerdale (Ellington's dam), dam by Tomboy, out of Tesane by Whisker. Tramp got also Little Red Rover, dam Miss Syntax, sister to Dr. Syntax by Paynator, he the sire of the dam of Buccaneer. Lottery was the sire of Inheritor, dam Hand Maiden by Walton, son of Sir Peter, Verulam, dam Wire by Whisker, and Sheet Anchor, dam Morgiana by Muley. Sheet Anchor got Weatherbit, out of Miss Letty by Priam, and Collingwood, out of Kalmia by Magistrate. Weatherbit's blood comes to us through imp. Cicily Jopson (dam of Waverly, Phœbe Mayflower, Katrine, Flash of Lightning, etc.), her dam Cestrea by Faugh-a-Ballagh; imp. Weatherwitch (dam of Weatherby, Little Mack, Florence (the last the dam of Hindoo and grandam of the grand mare Firenzi), Waterwitch, Fonso, the promising sire, Fonwitch, etc. Tramp was the sire of Trampoline, dam of Glencoe, winner of the 2,000 Guineas in 1834, Miss Rose, imp. Yorkshire's dam, out of daughter of Sancho by Don Quixote, son of Eclipse. Tramp was also sire of Charlotte West and Tarantula (winners of the 1,000 Guineas in 1830 and 1833), Lady Moore Carew, the dam of Mendicant (winner of the 1.000 Guineas and Oaks in 1846), and of Fury, dam of Our Nell (winner of the Oaks in 1842). Tramp was also sire of St. Giles (winner of the Derby in 1832), Dangerous (winner of the Derby in 1833), and Barefoot (winner of the St. Leger in 1823), who was imported to America and made his impression on the stock of this country. Mercury, 1778, by Eclipse, dam by Tartar, son of Partner, grandam by Mogul, son of the Godolphin Arabian. Mercury got Gohanna, 1790, dam by Herod, out of Maiden by Matchem. Gohanna got Wanderer, out of Catherine by Woodpecker. Wanderer was sire of Peri, dam of Sir Hercules. Gohanna also got the dam of Moses (sire of Pauline, dam of Gladiator), Golumpus, out of Catherine by Woodpecker, sire of the grandam of Melbourne, imp. Trustee, and Mundig (winner of the Derby in 1835), out of Emma by Whisker. Trustee was a good and successful stallion in America, sire of Revenue, dam Rosalie Somers by Sir Charles. Revenue got Planet, Exchequer, Orion and Ninette, out of Nina by Boston. Trustee also got the noted gelding Reube, out of Minstrel by Medoc, John Black, out of imp. Maria Black by Filho-da-Puta, Hiddlestone, dam by Bullock's Muckle John. Trustee was the sire of the noted race-mare Fashion, that defeated Boston in the great match. Through Fashion's daughter Young Fashion there are some good ones. Trustee also got Ann Stevens (dam of Kate and Frank Boston), out of imp. Caprice by Muley, and the noted Levity, out of Vandal's dam by imp. Tranby. Levity was the dam of Ruric by imp. Sovereign, the noted producing sisters Lightsome and Mildred by imp. Glencoe, Brenna, the dam of Brademante (The Bard's dam), Ella Hankins and Athalaric (Tom Martin's dam) by Gilroy, Lever and Legatee by Lexington. Catton was also the sire of imp. Gallopade, out of Camillina by Camillus, son of Hambletonian by King Fergus. Gallopade was the dam of Fandango and Cotillion by imp. Leviathan, Cracovienne, Quadrille, Hornpipe and Reel by imp. Glencoe. Reel is the dam of Prioress, Starke, Lecompte, that defeated Lexington, War Dance, etc. The blood is found in the pedigrees of many of our very best horses.

Catton was the sire of Royal Oak, out of a daughter of Smolensko. Royal Oak got Slane out of a daughter of Orville. Slane got Queen Anne, dam of Kingston, out of Garcia by Octavian, The Arrow, dam of Cambuscan, out of Southdown by Defence, The Merry Monarch, winner of the Derby, 1845, out of The Margravine by Little John, son of Octavius, Conyngham (winner of the 2,000 Guineas, 1847). Catton also got Mulatto, out of Desdemona by Orville. Mulatto got Martha Lynn, dam of Voltigeur, out of Leda by Filho-da-Puta. Mulatto also got Bloomsbury, winner of the Derby, 1839, out of Arcot Lass by Ardrossan. Catton got Sandbeck out of Orvillina, sister to Orville, by Beningbrough. Sandbeck got Redshank, sire of Ellen Horne, the grandam of Bend'or, and Barbelle, the dam of The Flying Dutchman, out of Darioletta by Amadis, son of Don Quixote by Eclipse. Catton also got imp. Contract out of Helen by Hambletonian, and imp. Nonplus, out of Miss Garforth by Walton. Saltram by Eclipse, dam Virago by Snap, won the Derby, 1783, and was imported to America in 1800. He got Whiskey, sire of the dams of imp. Priam, Phantom and Muley, in England. Saltram got the dam of Timoleon (sire of Boston), Jenny Cockracy, and Betsey Haxall, out of a daughter of Symme's Wildair. A daughter of Saltram dam by imp. Medley, was the dam of Sir Hal by imp. Sir Harry. Saltram got Rosy Clack, out of Camilla by Melzar, son of imp. Medley, she the dam of Tennessee Oscar and Rosetta by Wilke's Wonder, son of imp. Diomed, Patty Puff and her sister, dam of Betsey Saunders by Pacolet. The Darley Arabian male line is represented in England by such good ones as Ayrshire (Derby, 1888) by Hampton, dam Atalanta by Galopin, Bend'or (Derby, 1880) by Doncaster (Derby, 1873), dam Rouge Rose by Thormanby, Bendigo by Ben Battle, dam Hasty Girl by Lord Gough, Brag by Struan, dam Bounce by Flatterer, Donovan (Derby and St. Leger, 1889) by Galopin (Derby, 1875), dam Mowerina by Scottish Chief, Common (2,000 Guineas, Derby and St. Leger, 1891) by Isonomy, dam Thistle by Scottish Chief, Galliard (2,000 Guineas, 1883) by Galopin, dam Mavis by Macaroni, Galopin (Derby, 1875) by Vedette, dam Flying Duchess by The Flying Dutchman, Hagioscope by Speculum, dam Sophia by Macaroni, Minting by Lord Lyon, dam Mint Sauce by Young Melbourne, Muncaster by Doncaster, dam Windermere by Macaroni, Lowland Chief by Lowlander, dam Bathilde by Stockwell, Peter by Hermit, dam Lady Masham by Brother to Strafford, Petrarch by Lord Clifden dam Laura by Orlando. Pero Gomez (St. Leger, 1869) by Beadsman, dam Salamanca by Student, Petronel (2,000 Guineas, 1880) by Musket, dam Crytheia by Hesperus, Ormonde (2,000 Guineas, Derby and St. Leger, 1886) by Bend'or, dam Lily Agnes by Macaroni, Saraband by Muncaster, dam Highland Fling by Scottish Chief, Silvio (Derby and St. Leger, 1877) by Blair Athol, dam Silverhair by Kingston, Sefton (Derby, 1878) by Speculum, dam Lady Sefton by West Australian, St. Gatien (dead heat with Harvester and walked over for the Derby, 1884) by Rotherhill or The Rover, dam Saint Editha by Kingley Vale, Springfield by St. Albans, dam Viridis by Marsyas, Surefoot (2,000 Guineas, 1890) by Wisdom, dam by Galopin, out of Miss Foote by Orlando, The Bard by Petrarch dam Magdalene by Syrian, St. Simon by Galopin, dam St. Angela by King Tom, Timothy by Hermit, dam Lady Masham by Brother to Stafford, Tristan by Hermit, dam Thrift by Stockwell, Wenlock (St. Leger, 1872) by Lord Clifden, dam Mineral by Rataplan, Wisdom by Blinkhoolie, dam Aline by Stockwell, Merry Hampton (Derby, 1887) by Hampton, dam Doll Tearsheet by Broomielaw, Sainfoin (Derby, 1890) by Springfield, dam Sanda by Wenlock, the two last untried in the stud. In America we are full as strong in this blood, not only in imported horses, many of them successes in the stud, but in native bred horses who were first-class race-horses and are popular and successful stallions. Imported St. Blaise (Derby, 1883) by Hermit, dam Fusee by Marsyas, Rayon d'Or (St. Leger, 1879) by Flageolet, dam Araucaria by Ambrose, Dalnacardoch (dead) by Rataplan, dam Mayonaise by Teddington, Blythewood by Adventurer, dam Bonny Bell by Voltigeur, The Jacobite by Prince Charlie, dam sister to Little Lady by Orlando, Uhlan by The Ranger, dam La Mechante by Turnus, Deceiver by Wenlock, dam Boot and Saddle by Trumpeter, Stonehenge by Blair Athol, dam Coimbra by Kingston. Charaxus by Distin, dam Sappho by Kingston, Maxim by Musket, dam Realization by Vespasian. Kantaka by Scottish Chief, dam Seclusion (Hermit's dam) by Tadmor, Saxon by Beadsman, dam Girasol by Asteroid, Galore by Galopin, dam Lady Maura by Macaroni. King Galop by Galopin, dam by King Tom, out of Sunshine by

Thormanby. Great Tom by King Tom, dam Woodcraft by Voltigeur. Top Gallant and Stalwart brothers by Sterling, dam Sea Mark by Adventurer. Loyalist by Sterling, dam Casuistry by The Miner. The Sailor Prince by Albert Victor, dam Hermita by Hermit. Laureate by Petrarch, dam Macaria by Macaroni. Dandie Dinmont by Silvio, dam Meg Merrilies by Macgregor. Glenelg by Citadel, dam Babta by Kingston. Eothen by Hampton, dam Sultana by Oxford. Rossington by Doncaster, dam Lily Agnes (Ormonde's dam) by Macaroni. Wagner by Prince Charlie, dam Duchess of Malfi by Elland. Kyrle Daly by Artillery, dam Colleen Rhue by Gemma di Vergy. Siddartha by Pero Gomez, dam The Pearl by Newminster. Many of these are not Derby or St. Leger winners but very highly bred and successful stallions. Amongst those bred in America, and who possess racing qualities sufficient to class them with the best performers in England, are the following: Salvator by Prince Charlie, dam Salina by Lexington (winner of the fastest mile ever run in the world, 1.35¼, and 1¼ miles with 122 lbs. in 2.05), Longfellow, the equal of any sire in the world, by imp. Leamington, dam Nantura by Brawner's Eclipse, Leonatus by Longfellow, dam Semper Felix by imp. Phaeton, winner of ten stakes at 3 years old, and a successful stallion. Luke Blackburn by imp. Bonnie Scotland, dam Nevada by Lexington, winner of 22 out of 24 races at 3 years old, and a successful stallion. Eolus, a superior race-horse and successful stallion by imp. Leamington, dam Fanny Washington by Revenue. Enquirer, a superior race-horse and successful stallion by imp. Leamington, dam Lida by Lexington, the brothers Onondaga, Stratford and Sensation (unbeaten) by imp. Leamington, dam Susan Beane by Lexington. Inspector B. and Getaway, good racers, by Enquirer, dam Colossa by Colossus, Iroquois (winner of the Derby and St. Leger in 1881) by imp. Leamington, dam Maggie B. B. by imp. Australian, Lisbon by imp. Phaeton, dam imp. Lady Love by Stockwell or Caterer, Troubadour by Lisbon, dam Glenluine by imp. Glenelg, Powhattan (brother to Parole) by imp. Leamington, dam Maiden by Lexington, Alarm by imp. Eclipse, dam imp. Maud by Stockwell. Panique and Wawekus (brothers) by Alarm, dam Maggie B. B. by imp. Australian, Falsetto by Enquirer, dam Farfaletta by imp. Australian, The Bard, a superior horse, by Longfellow, dam Brademante by War Dance, Blue Eyes by Enquirer, dam Buchu by Planet, Linden by Longfellow, dam Linda Lewis by Oliver, Longstreet by Longfellow, dam Semper Idem by imp. Glen Athol, Belvidere and Sir Dixon (brothers) by imp. Billet, dam Jaconet by imp. Leamington, Himyar by Alarm, dam Hira by Lexington, Aramis by imp. Phaeton, dam Nellie Grey by Lexington, Aretino by Aramis, dam Sprightly by Lexington, Elias Lawrence by imp. Billet, dam Sprightly by Lexington, Whisper by Planet, dam Mattie Gross by Lexington, Bramble by imp. Bonnie Scotland, dam Ivy Leaf by imp. Australian, Foxhall by King Alfonso, dam Jamaica by Lexington (winner of the Grand Prize of Paris, etc.), Forester by imp. Ill Used, dam Woodbine by Censor or Kentucky, King Alfonso by imp. Phaeton, dam Capitola by Vandal, Pontiac by Pero Gomez, dam imp. Agenoria by Adventurer, Jils Johnson by Longfellow, dam Fanny Wells by imp. Sovereign, His Highness by imp. Ill Used, dam imp. Princess by King Tom, a grand good one, with many others which possess on the female side a large infusion of the coveted Herod blood. In fact, our American horses possess more of the Herod blood through Glencoe, who sired England's most noted matron, Pocahontas, dam of Stockwell, Rataplan, King Tom, etc., than can be found in the world.

The Alcock Arabian was imported into England about 1720, by whom it is not stated, but from his name by Mr. Alcock. He was noted as the sire of Crab, dam sister to Sorcheels by Basto, 2d dam (Partner's dam), sister to Mixbury Galloway by the Curwen Bay Barb, 3d dam by Curwen's Old Spot, 4th dam by White-legged Lowther Barb, 5th dam the old Vintner mare. This old Vintner mare was the property of Mr. Curwen, of Workington, and was said by Mr. Croft to have been the best bred mare and best racer of her day. She was the dam (in addition to the filly above mentioned) of a filly by Pulleine's Chestnut Arabian, which was the dam of the filly by Blossom, son of Curwen's Bay Barb. To the old Vintner mare trace in direct female line Croft's Partner, Sorcheels, Crab, Snip, Pacolet, Syphon, Soldier, Mercury, Master Bagot, St. Giles, Boudrow, Dick Andrews, Remembrancer, John o' Gaunt, Pompey, Muley Moloch, Nutwith, Lambton, The Confessor, Prism, Warlock, Ratan, Springy Jack, Daniel O'Rourke, Wingrave, Outfit, Bendigo, Peter, Peregrine Tertius, etc.

This Arabian also got Mr. Alcock's Spot, the Duke of Ancaster's Gentleman, and the mare, dam Flash by the Curwen Bay Barb, out of a natural Barb mare, that was the dam of Dismal by the Godolphin Arabian, Whitefoot by Bolton Whitefoot, and Trifle by Fox. Crab was a superior race-horse, and as a stallion in Mr. Panton's stud was noted as the sire of many good race-horses, stallions and brood mares.

The Godolphin Arabian. This, the most noted of all the Eastern sires, was a dark bay or brown horse, foaled in 1724, and was by many regarded as a Barb. He was brought from France by Mr. Coke, and it is said that he actually drew a cart in the streets of Paris. Be this true or not, he unquestionably contributed more to the breed of thoroughbreds than any stallion either before or since his time. Mr. Coke presented him to Mr. Williams, keeper of the St. James Coffee House, by whom he was presented to Lord Godolphin, whose property he remained until his death at Gogmagog, in 1753. It is true that every superior race-horse of the present day, either in America, France or England, partakes largely of his blood. The Godolphin was teazer to Hobgoblin, son of Aleppo by the Darley Arabian and Mr. Brewster's old Hautboy mare, in 1730 and 1731. On Hobgoblin's refusal to cover Roxana, she was bred to the Godolphin Arabian, and from that cover she produced Lath, the first horse the Arabian ever got. Lath, a bay horse foaled in 1732 (named from his flat sides), was a superior race-horse, the best one of his day. Cade, his own brother, was a bay horse, foaled in 1734, and was named Cade from the fact that he was brought up on cow's milk, his dam dying when he was only ten days old, "To cade, to bring up in softness." Cade was an indifferent race-horse, but a very successful stallion, which success is attributed not only to his superior blood, but to his grand symmetry in form. Roxana, the dam of Lath, Cade, and Roundhead by Devonshire Childers, was owned by Lord Godolphin. She was by the Bald Galloway, dam sister to Chaunter by the Akaster Turk, 2d dam by the Leedes Arabian, 3d dam by Spanker, 4th dam old Morocco mare (Old Peg, Spanker's dam) by Morocco Barb, 5th dam Old Bald Peg by an Arabian, out of a Barb mare. Lath got Hector, Miss (the dam of Fairy) by Shepherd's Crab, Redcap, and Duchess, dam Young Miss Belvoir by Flying Childers. Duchess, was the dam of Brimstone by Devonshire Chestnut Arabian, and she the dam of Squib by Spectator. A daughter of Lath, out of Brown Basto (sister to Snip), was the dam of Omar by the Godolphin Arabian, Crazy by Lath, dam Bay Basto (sister to Snip), was the dam of Virago and two sisters by Panton Arabian, and the bay horse Protector, imported into America and here called Lath, by Shepherd's Crab. Virago was the dam of Hollandaise by Matchem (winner of St. Leger, 1778), and the grey filly Gunilda, imported to America by Star, son of Regulus by the Godolphin Arabian. Gunilda (called also Virago) was the dam of Marske, Mask and Virago by imp. Shark, son of Marske by Squirt. From Virago by Shark, many excellent race-horses have descended, including Phœnix Belle, Cora Lynn, Miss Carter, Lilac, Joe Daniels, etc. Miss Carter is the dam of Hildegarde, Isse, Poteen, etc. Crazy was also dam of Quill and sister by Feather, son of the Godolphin Arabian. This Feather mare, sister to Quill, was the dam of Anvil by Herod, who was sire of the grandam of imp. Leviathan.

Lath was also the grandsire of Bonny Lass by Snip. Cade was much more distinguished as a stallion. He got Matchem (1748), dam sister to Miss Partner by Croft's Partner, 2d dam Brown Farewell by Makeless, 3d dam by Brimmer, 4th dam by Place's White Turk, 5th dam by Dodsworth, out of the Layton Violet Barb mare, Changeling, a brother and a full sister out of the same mare. Matchem was a superior race-horse, and was a very successful stallion, being the sire of 354 winners in 23 years that won in money $759,850.00, in addition to cups.

Matchem got Conductor (1767), Alfred, Georgiana, Eyebright, etc., out of a daughter of Snap by Snip by Flying Childers, 2d dam by the Cullen Arabian, 3d dam Grisewood's Lady Thigh by Croft's Partner, 4th dam by Greyhound, tracing through the old Montagu mare to a Royal mare. Conductor got Imperator (1773), dam by Herod, 2d dam Carina by Marske (sire of Eclipse), out of a daughter of Blank, son of the Godolphin Arabian, tracing through Miss Hip by Oysterfoot (Arabian) to the noted Coppin mare. Conductor also got Trumpator (1782), Cat, Cantator, etc., dam Brunette by Squirrel, son of Traveler by Partner, 2d dam Dove by Matchless, son of the Godolphin Arabian, tracing through Look-at-me-Lads by Grasshopper, son of the Byerly Turk, to the Oldfield mare, said to have been very highly bred. Imperator got Pipator (1786), out of Brunette, the dam of Trumpator by Squirrel. Pipator got Remembrancer (winner of St. Leger, 1803) out of Queen Mab, sister to Mercury by Eclipse. Trumpator got Paynator (1791), dam by Mark Anthony, son of Spectator by Crab, 2d dam Signora by Snap, grandson of Flying Childers, out of Miss Windsor by the Godolphin Arabian. Trumpator also got Sorcerer (1796), dam Young Giantess by Diomed, 2d dam Giantess by Matchem, tracing through a sister to Roxana (dam of Lath and Cade), to the old Morocco mare (Spanker's dam) by Morocco Barb. Paynator got Dr. Syntax (1811), dam by Beningbrough, son of King Fergus by Eclipse, 2d dam Jenny Mole by Carbuncle, grandson of the Godolphin Arabian, out of a daughter of Prince T'Quassaw, grandson of the Darley Arabian. Dr. Syntax got The Doctor (sire of Malcolm), dam by Lottery, out of Elizabeth by Walton, and the noted Bee's-wing, dam of Newminster, out of a mare by Ardrossan, son of John Bull, out of Lady Eliza by Whitworth, son of Agonistes.

Paynator got the dam of Cain by Paulowitz, out of a Delpini mare. Sorcerer got Comus, dam Houghton Lass by Sir Peter, son of Highflyer by Herod, out of Alexina by King Fergus, son of Eclipse. Comus got Humphrey Clinker, out of Clinkerina by Clinker, son of Sir Peter. Humphrey Clinker got Melbourne, out of daughter of Cervantes, son of Don Quixote by Eclipse, grandam by Golumpus, son of Gohanna by Mercury. Melbourne got West Australian (winner of the 2,000 Guineas, Derby and St. Leger, 1853), dam Mowerina by Touchstone, out of Emma by Whisker, son of Waxy, Blink Bonny (winner of the Oaks and Derby, 1857), dam of Blair Athol (winner of Derby and St. Leger, 1864), and Breadalbane, out of Queen Mary by Gladiator, Canezon (winner of 1,000 Guineas, 1848), out of Madame Pelerine by Velocipede, son of Blacklock. Canezon was the dam of Fazzoletto by Orlando, Cape Flyaway by The Flying Dutchman, etc. Melbourne got Cymba (winner of the Oaks, 1848), out of Skiff by Sheet Anchor, and she was the grandam of Marden, Charon and The Abbot. Marchioness (winner of the Oaks, 1855) by Melbourne, out of Cinizelli by Touchstone, was sent to Australia and was the dam of Angler by Fisherman, winner of the Victoria Derby, St. Leger and Ballarat Derby, and sire of Onyx and Robinson Crusoe. Melbourne was also sire of Mentmore Lass (winner of the 1,000 Guineas, 1853), out of Emerald by Defence. Mentmore Lass was the dam of Hannah (winner of the 1,000 Guineas, Oaks and St. Leger, 1871). West Australian was the sire of Brown Agnes (dam of Brown Bread), out of Miss Agnes by Birdcatcher, Lady Sefton (dam of Sefton, winner of the Derby, 1878), out of Clarissa by Pantaloon, Solon, sire of Barcaldine, Philammon, etc., out of Darling's dam by Birdcatcher, and the dam of Musket by Toxophilite. Musket is sire of imp. Maxim, Martini-Henry and Petronel (winner of the 2,000 Guineas, six Queen's Plates and the Doncaster Cup). West Australian also got Summerside (winner of the Oaks, 1859), dam of Printanniere (Poulet's dam), out of Ellerdale by Lanercost. West Australian also got imp. Australian (dam Emilia (imp.) by Young Emilius), who was not only a superior race-horse but one of the best and most popular stallions ever in America, sire of Wildidle (four miles, 7.25¼), out of Idlewild (4 miles, 7.26½) by Lexington, Fellowcraft (4 miles, 7.19½), Spendthrift, Miser, Rutherford, etc., out of Aerolite by Lexington, Baden Baden, Helmbold, etc., out of Lavender by Wagner, Abd-el-Kader, Abd-el-Koree, Wilful, Silent Friend, Mate, Atilla, Leinster, Farfaletta (the dam of Falsetto), Maggie B. B., dam of Iroquois (winner of both the Derby and St Leger), Ivy Leaf (dam of Bramble, Brambaletta, etc.), and Lizzie Lucas, dam of Lytton, Cambyses, etc. Melbourne was also sire of Sir Tatton Sykes (winner of St. Leger, 1846), dam by Margrave (imp.), Prime Minister (dam by Sheet Anchor), who got Knight of the Garter (winner of the Tradesman Plate, Chester, 1869), out of Rosa Bonheur by Touchstone. Knight of the Garter was the sire of Przedswit, out of The Jewel by

Stockwell. Melbourne also got Young Melbourne, dam Clarissa by Pantaloon. son of Castrel. out of a daughter of Glencoe. Young Melbourne got Gen Peel and Statesman, dam a daughter of Orlando, out of Brown Bess by Camel. Gen. Peel won the 2,000 Guineas, 1864, and was second in both the Derby and St. Leger to Blair Athol. Young Melbourne also got New Holland (winner of the Goodwood Cup), out of Bonnceaway by Zuyder Zee, son of Orlando, Brother to Strafford, out of daughter of Gameboy, Rapid Rhone, dam by Lanercost or Retriever, Pell Mell, dam Makeshift by Voltigeur, he the sire of Carlton, out of Bonny Spec by Speculum. Young Melbourne showed best through his daughters, Mint Sauce, dam of Minting by Lord Lyon (winner of the Grand Prize of Paris), The Lambkin, (winner of the St. Leger, 1884) by Camballo, and his sister Minthe (winner of the 1,000 Guineas, 1889), out of Sycee by Marsyas, Adelaide, dam of Peregrine (winner of the 2,000 Guineas, 1881) by Pero Gomez, dam by Teddington, Faraway, (dam of Rafaello), out of Maid of Masham by Don John, Patchwork (dam of Quilt, Crosspatch, etc.), out of Makeshift by Voltigeur; Wheatear (dam of Skylark, winner of the Gold Vase at Ascot, by King Tom, and Harvester, who ran a dead heat and divided the Derby with St. Gatien, 1884), out of Swallow by Cotherstone. Humphrey Clinker was also sire of Bran, dam Velvet by Oiseau, out of Wire, sister to Whalebone by Waxy. Bran was the sire of Our Nell, winner of the Oaks, 1842. Humphrey Clinker was sire of Rockingham (winner of St. Leger, 1833), The Biddy (dam of Priestess by The Doctor) out of Idalia by Peruvian. The Biddy was dam of The Gem by Touchstone, and she the dam of Regalia (winner of the Oaks, 1865) by Stockwell. The Priestess was the dam of Dulcibella by Voltigeur, and she the dam of Idalia by Cambuscan, that produced imp. Sir Modred and imp. Cheviot. Comus was sire of Reveller (winner of St. Leger, 1818), out of Rosette by Beningbrough. Reveller was the sire of imp. Delight, out of Defiance by Rubens. Delight was the dam of Hagar and Matilda Bynum by imp. Glencoe, and Shamrock (imported *in utero*) by St. Patrick, son of Walton by Sir Peter. Shamrock was the sire of Hennie Farrow, out of Ida by imp. Belshazzar, son of Blacklock, she the dam of Mayflower by imp. Eclipse, that produced Joe Hooker, the popular sire, and California by Monday, Warwick and Annie Laurie by Hubbard. Mimi, the sister to Mayflower, is dam of Mask, Mikado and Kinglike by King Ernest, the latter a very promising stallion. Ballerina by imp. Balrownie, out of Hennie Farrow, is the dam of Ballinette (dam of Clara W., Alta and Del Norte) by Monday or Young Eclipse, and Countess Zeika (dam of Ed Corrigan, Judge Post, etc. by Norfolk. Reveller was the sire of Zillah, out of Morisca by Morisco, son of Muley. Zillah was the dam of Prairie Bird by Touchstone, and she the dam of England's Beauty by Birdcatcher, she the dam of imp. The Rake (sire of Dundee) by Wild Dayrell, Silverhair (the dam of Silvio, Silvester and Fetterlock) by Kingston, Rose of Kent (the dam of Sycee by Marsyas) by Kingston. Sycee was the dam of Mint Sauce, dam of Minting, winner of the Grand Prize of Paris, and The Lambkin, winner of the St. Leger, 1864. Prairie Bird was also dam of Vitula by Arthur Wellesley, the dam of Bird of Freedom (winner of City and Suburban Handicap). Comus was the sire of imp. My Lady out of The Colonel's dam by Delpini, she the dam of imp. Delphine by Whisker, that was the dam of imp. Monarch by Priam, The Queen, full sister to Monarch and imp. Herald by Plenipotentiary. Sorcerer was the sire of Smolensko (winner of the 2,000 Guineas and Derby, 1813), dam Wowski by Mentor, son of Justice by Herod, out of Maria, dam of Waxy by Herod. Clari by Smolensko, dam by Precipitate, son of Mercury by Eclipse, was the dam of Chat, Chatterer, Chatterbox and Chitchat by Magpie in Ireland. Smolensko was the sire of Jerry (winner of the St. Leger, 1824), out of Louisa by Orville. Jerry was the sire of Tomboy, out of Bee's-wing's dam by Ardrossan. Tomboy was the sire of Nutwith, out of daughter of Comus. Nutwith (winner of St. Leger, 1843) was the sire of Knight of Kars, out of Pocahontas by Glencoe, and of Gameboy, out of Lady Moore Carew by Tramp. Smolensko was also the sire of The Fawn, out of Jerboa by Gohanna, and The Fawn was the dam of Venison by Partisan, son of Walton. Venison was the sire of Kingston, out of Queen Anne by Slane, son of Royal Oak by Catton, Royal Oak's dam by Smolensko. Kingston was the sire of imp. Blue Mantle, Silverhair (dam of Silvio, winner of Derby and St. Leger, 1877), imp. Non Pareille, Man-at-Arms, Alarm, etc. Smolensko's blood is represented in America through imp. Maria Black by Filho-da-Puta, whose dam was by Smolensko out of a daughter of Sir Peter. Maria Black was the dam of Bay Leaf by imp. Yorkshire, the dam of Beacon, Bayswater, Bayonet, Baywood Preakness, Bay Final, Bay Flower and Niagara by Lexington, Bingaman and M. A. B. by Asteroid, and Bay Bush by imp. Australian. Bay Flower is the dam of Ivy Leaf, she the dam of Bramble, Brambaletta (Merry Monarch's dam), Bye and Bye (Now or Never's dam), Bonnie Leaf, all by imp. Bonnie Scotland, Bazaar by Jack Malone, and Niagara, a speedy, good race-horse, by Luke Blackburn. The grandam of Gallopade, that founded the great Reel and Dance families, is a daughter of Smolensko. Sorcerer is also sire of Soothsayer (winner of the St. Leger, 1811), dam Goldenlocks by Delpini, son of Highflyer, out of Violet by imp. Shark, son of Marske. Soothsayer was the sire of Filagree, out of Web, sister to Wire and Whalebone by Waxy. Filagree was the dam of Cobweb (winner of the 1,000 Guineas and Oaks) by Phantom, sire of Walton and imp. Riddlesworth (winner of the 2,000 Guineas, 1831) by Emilius, and of Charlotte West (winner of the 1,000 Guineas, 1830) by Tramp. Cobweb was the dam of Bay Middleton by Sultan, son of Selim by imp. Buzzard. Bay Middleton was a grand race-horse (winner of the 2,000 Guineas and Derby, 1836) and successful stallion, being the sire of Cowl, out of Crucifix by Priam, Hesperus, out of Plenary by Emilius, Ellen Middleton (dam of Wild Dayrell, winner of Derby, 1855), out of Myrrha by Malek, son of Blacklock, Britannia IV., imported (dam of Madame Dudley and Brown Prince, sent to England, by Lexington, Brigand by Planet, Booty by Asteroid, etc). Madame Dudley is the dam of Judith and Belinda by Glenelg, Vintage (the dam of Ruby Royal, Flavilla, etc.) by Virgil. Santa Rita and Virgie D. (sisters to Vintage), the latter dam of Hueneme by Billet. Booty is the dam of Violator by King Ban, Estelle by Himyar, etc. Bay Middleton was also sire of Eunice, dam of Saunterer, out of Blue Devils by Velocipede; the dam of King Alfred, and grandam of Enterprise, out of West Country Lass by Venison; Bay Missy, grandam of Scottish Chief, out of Camilla by Young Phantom; The Flying Dutchman (winner of Derby and St. Leger, 1849), sire of Flying Duchess, Galopin's dam, out of Barbelle by Sandbeck, son of Catton, and of Dutch Skater, out of Fulvie by Gladiator, with other good ones; My Dear (grandam of Oxford by Birdcatcher), out of Miss Letty by Priam; Hesperus the grandam of Petronel, out of Plenary by Emilius; the sister to Cobweb by Phantom produced Flycatcher by Godolphin, son of Partisan, who was the grandam of imp. Eclipse, who got Alarm, Ruthless, Relentless, Remorseless and many other good ones. Soothsayer was sire also of Tiresias (winner of the Derby, 1819), out of Pledge by Waxy.

The Godolphin Arabian also got Blossom, Jilt and Bragg, dam Ancaster Blossom by Crab, son of the Alcock Arabian, grandam by Flying Childers, out of Miss Belvoir by Grey Grantham, etc. Jilt was the dam of Arab by Panton Arabian, Jilt, Hornet, Bashful, Caligula and Bauble, all by Blank. Cripple was own brother to Cygnet and Creeper by the Godolphin Arabian, dam Blossom, sister to Ancaster Blossom by Crab. Cripple was the sire of the great Gimcrack, a superior race-horse and the sire of imp. Medley, foaled in 1776. Medley was a fine race-horse and sire of Grey Diomed (foaled in 1786), dam by Sloe, son of Partner, grandam by imp. Vampire, out of imp. Calista by Forester; Bellair (Taylor's), foaled 1786, and Calypso, foaled 1793, dam Selima (Young) by Yorick, grandam Black Selima by imp. Fearnought, out of imp. Selima by the Godolphin Arabian; Boxer, foaled 1790, dam by imp. Fearnought; Quicksilver (foaled 1783), dam by Symme's Wildair; Melzar, foaled 1791, dam by Symme's Wildair, grandam by imp. Vampire, out of imp. Kitty Fisher by Cade, son of the Godolphin Arabian; Lamplighter, dam by imp. Lonsdale, by Jolly Roger, son of Gower stallion, out of imp. Kitty Fisher by Cade; Grey Medley (Barry's of Tennessee), dam by True Whig, grandam Circe by Ariel, out of imp. Lady Northumberland; the Medley mare (owned by Marmaduke Johnson of Virginia), dam of Miss Jefferson by imp. Diomed, Carolina by imp. Saltram, Carolinian by imp. True Blue, and the famous sisters, Vanity and Reality by Sir Archy, the latter dam of Johnson's

Medley by Sir Hal, Slender and Bonnets o' Blue by Sir Charles, the latter the dam of the noted Fashion (that defeated Boston in the great match, four-mile heats, run over the Union Course, L. I., May 10th, 1842, in 7.32½, 7.45, the best race run in America to that date), Mariner by Shark, etc. Gimcrack was also sire of imp. Clockfast, and of the dam of Constable's imp. Pot-8-o-'s mare, grandam of American Eclipse. Godolphin Arabian was the sire of Blank, bay horse, foaled 1740, dam Little Hartley mare by Bartlett's Childers, 2d dam Flying Whigg by William's Woodstock Arabian, 3d dam Points, own sister to the Bald Galloway by the St. Victor Barb, 4th dam Grey Whynot by Whynot, son of the Fenwick Barb, out of a Royal mare. The Little Hartley mare was also the dam of Janus and Old England, and a filly by the Godolphin Arabian, which filly was the dam of Merlin by Second, Babraham Blank by Babraham, etc. Blank was not successful as a race-horse but was an exceedingly valuable stallion, and sire of Chatsworth out of Fairy by Shepherd's Crab; Chrysolite, Gaudy and Gamesome, out of Blossom by Panton's Crab; Pacolet out of Whiteneck by Panton's Crab, son of Alcock Arabian; Paymaster out of Snap Dragon by Snap, son of Snip; Tatler out of Spectator's dam by Partner. Blank was the sire of Countess, out of a daughter of Rib, son of Crab. Countess was the dam of Horatia and Achilles by Eclipse, Delpini by Highflyer, etc. Horatia was the dam of Stamford by Sir Peter, and his brothers, the Derby winners, Paris, 1806, and Archduke, 1799, imp. into Virginia. Blank also got Dizzy (out of Dizzy by Driver, son of the Wynn Arabian), she the dam of Grey Dorimant by Dorimant, son of Otho: Golden Grove, the dam of Petworth by Herod out of Spinster, the Widdrington mare, by Partner: Helen (the dam of Riot and Brim by Squirrel) out of Mab by Crab; Horatia, dam of Juno by Spectator, and her sister (the dam of imp. Diomed), out of a daughter of Flying Childers; Julia, dam of Promise by Snap, and Princess by Herod, out of Spectator's dam by Partner; Lilly, dam of Hippona by Snap, Jocasta and Drone by Herod, out of Peggy by Cade; Principessa, dam of Heinel by Squirrel, out of a daughter of the Cullen Arabian.

Little Hartley mare was the dam of Tortoise and his sister by Whitefoot, son of Bay Bolton. This Whitefoot mare was the dam of the Gower stallion by the Godolphin Arabian. The Gower stallion was the sire of Little David, Lord Gower's Sweepstakes (sire of Young Sweepstakes, Trentham, Whitelegs, Corsair, Doctor, etc.), Clio (dam of Conqueror by Bajazet), etc. The Godolphin Arabian got Regulus, and his sister (the dam of Folly by Blank, and Coquette by the Compton Barb, she the dam of Camilla, Glider, etc., by Trentham), dam Grey Robinson by the Bald Galloway, 2d dam sister to Old Country Wench by Snake, son of the Lister Turk, 3d dam Grey Wilkes, sister to Clumsy by Hautboy, 4th dam Miss Betty D'Arcy's Pet mare, daughter of a Sedbury Royal mare. Regulus was a bay horse foaled in 1739. Under the name of Sweetlips he won a number of races, and at six years old he won eight Royal Plates and one £50 Plate. He was never beaten, and was superior to any horse of his day. Regulus was noted as a stallion, and was the sire of Careless, Cato, Morwick Ball, Adolphus, Elephant, South, Turk, Prophet, Vampire (imported), etc. He was also the sire of A la Greeque, dam of Pontac and sister by Marske, Balloon, Emma and Maria by Telemachus. Regulus also got Spiletta, out of Mother Western by Smith's son of Snake, own brother to William's Squirrel by Lister's Snake, out of an Akaster Turk mare. Spiletta was the dam of O'Kelly's Eclipse and his sister Proserpine and brother Garrick (Hyperion), etc. The Godolphin Arabian was also sire of Babraham, Mogul, Marlborough, and imported Selima, out of the Large Hartley mare by Mr. Hartley's Blind Horse, son of the Holderness Turk. Godolphin Arabian also sired Bajazet (foaled in 1740), dam by Whitefoot, grandam by Leedes, out of the Moonah Barb mare.

Rachel, dam of the noted Highflyer by Herod, and Mark Anthony by Spectator, was out of a daughter of Regulus by the Godolphin Arabian; Rantipole, dam of Nymph and Nimble by Florizel, son of Herod, out of sister to Careless by Regulus: Rutilia, sister to Rachel, dam of Jocasta and Harmony by Herod; Sally, dam of imp. St. George, out of Poppet by Black Chance, son of Hutton's Bay Barb; Sappho, the dam of Sophonisba by Squirrel, and Madcap by Eclipse, out of a daughter of

Blaze, son of Flying Childers; Sophia, dam of Jewel by Squirrel, and Cricket by Herod, out of Lord Leigh's Diana by Second; Toy, sister to Pacolet, dam of Marigold by Herod, out of Whiteneck by Crab. The Matchem blood in the male line is almost extinct in England, and has but few reputable representatives. The best now living is Barcaldine (a fine race-horse) by Solon, son of West Australian by Melbourne, dam Ballyroe by Belladrum, son of Stockwell, out of Bon Accord by Adventurer; Morion (winner of Ascot Gold Cup) by Barcaldine, dam Chaplet by Beadsman, out of Madame Eglantine by Cowl; Carlton (winner of the Doncaster Cup) by Pell Mell, son of Young Melbourne by Melbourne, dam Bonny Spec by Speculum, out of Bonny May by Newminster; Kilwarlin (won the St. Leger in 1887) by Arbitrator, son of Solon by West Australian, dam Hasty Girl by Lord Gough, son of Gladiateur, out of Irritation by King of Trumps, and Plebeian by Joskin, son of West Australian, dam Queen Elizabeth by Autocrat, son of Bay Middleton, out of Bay Rosalind by Orlando. In America we are remarkably strong in this blood, especially from the Melbourne line. Imp. Darebin, a superior race-horse, by The Peer, son of Melbourne, dam Larline by Traducer, son of The Libel by Pantaloon, out of Mermaid by King Tom; Wilful by imp. Australian, son of Melbourne, dam imp. Pussy by Diophantus, son of Orlando, out of Agapemone by Bay Middleton, a good race-horse and successful stallion for his chances; the American-bred Springbok, winner of the fastest 2¼ miles in the world in 3.56½, by imp. Australian dam Hester by Lexington, out of Heads I Say by Glencoe; the brothers Spendthrift, sire of Kingston, Fellowcraft, 4 miles in 7.19¾, and Miser, sire of Yorkville Belle by imp. Australian, dam Ærolite by Lexington, out of Florine by Glencoe; Wildidle, 4 miles in 7.25½, by imp. Australian, dam Idlewild, 4 miles in 7.26¼, by Lexington, out of Florine by Glencoe; Duke of Montrose, a good race-horse and stallion, by Waverly, son of imp. Australian, dam Kelpie by imp. Bonnie Scotland, out of Sister to Rurie by imp. Sovereign; Harry O'Fallon, sire of Terra Cotta, etc., by imp. Australian, dam imp. Sunny South by Birdcatcher, out of Equal by The Cure, purely English bred; Strathmore, a fine race-horse, by Waverly, son of imp. Australian, dam Brenna by imp. Knight of St. George, son of Birdcatcher, out of Levity by imp. Trustee; Atilla (winner of Travers' stake) by imp. Australian, dam Ultima by Lexington, out of Utilla by imp. Margrave.

The Compton Barb (the Sedley Grey Arabian) was brought to England by Mr. Compton, and became the property of Sir Charles Sedley, after which he was known as the Sedley Grey Arabian. He was principally noted as the sire of Coquette, out of sister to Regulus by the Godolphin Arabian. Coquette was the dam of Camilla, Glider, Driver, Darter, Cocker, and two unnamed fillies by Trentham, son of Sweepstakes by the Gower stallion. Camilla was the dam of Colibri, Catherine, Sophia, Crazy, Poetess and Humming Bird by Woodpecker, son of Herod, Ragged Jack by Highflyer, and Jerboa by Gohanna. Catherine was a famous brood mare, dam of Golumpus, Hedley, Kate, Wanderer and Vagabond by Gohanna, son of Mercury by Eclipse, Slipper by Precipitate, Sprite by Bobtail, son of Precipitate, by Mercury, Rectory by Octavius, and Worthy by Whalebone. Sprite was the dam of Variety by Selim or Soothsayer, Amiable by Orville, and Spite by Tiresias. Humming Bird was the dam of Mouse (Vittoria's dam), Hanna and Fan, all by Gohanna. Jerboa was a noted successful brood mare. She was the dam of Lacerta by Zodiac, that produced The Little Known by Muley, Little Wonder by the same (winner of the Derby in 1840), Fawn by Smolensko, she the dam of Venison by Partisan. Venison is the sire of Kingston, out of Queen Anne by Slane. Kingston was the sire of Queen Bertha (the grandam of Childeric and Charibert), out of Flax by Surplice, Silverhair and imp. Non Pareille, out of England's Beauty by Birdcatcher. Silverhair was the dam of Silvester and of Silvio, that won both the Derby and St. Leger, 1877, the latter by Blair Athol. Kingston also got Man-at-Arms, out of Paradigm by Paragone, Alarm out of Southdown by Defence, sire of the grandam of Petrarch, winner of the St. Leger, 1876. The Kingston blood is highly esteemed for stoutness.

After the Restoration, and when King Charles II. had been firmly settled on the throne, he sent Sir John Fenwick (Master of the Horse and prominent on the turf) abroad to procure a number of high bred mares and stallions, and the mares so purchased were called Royal mares, one of which was in foal and produced the noted Dodsworth, who was the sire of Dicky Pierson, sometimes called Son of Dodsworth. Both of these names are found in many of the old pedigrees. Dodsworth got the grandam of Dyer's Dimple, and the latter was the sire of Sophonisba, out of sister to Little George by Curwen's Bay Barb. Dodsworth was sire of the grandam of Grasshopper (Ancaster's), and the great grandam of Spider by the Lonsdale Arabian. Dodsworth also sired the great grandam of Bay Layton by D'Arcy's Counsellor, and Brown Farewell by Makeless, the latter the dam of Miss Partner by Partner, the dam of Young Cade by Cade, son of the Godolphin Arabian, and the sister to Miss Partner (who was the dam of the noted Matchem, that founded the Matchem line of blood, and Changeling his brother) by Cade, son of the Godolphin Arabian. Another sister to Miss Partner was the dam of Dormouse by the Godolphin Arabian, and imp. Bolton by Shock. Sir John Fenwick also brought from abroad the Barb horse known as the Fenwick Barb, which was the sire of Whynot, out of a Royal mare, and she was the dam of the Bald Galloway, Cupid, and Points by the St. Victor Barb. Points was the dam of Flying Whigg by William's Woodstock Arabian, she the dam of the Large Hartley and Little Hartley mares. The former was the dam of imp. Selima, that founded the Mollie Jackson family, Babraham, Mogul and Marlborough, all by the Godolphin Arabian. The latter was the dam of Janus, Blank and Old England by the Godolphin Arabian, Tortoise by Whitefoot, and Slouch by Cade. Another daughter of Whynot was the dam of Bay Wilkinson and his sister by Snake, and a filly by Hutton's Grey Barb, that was the dam of Grisewood's Partner by Old Partner. Through the descendants of the Bald Galloway, Grisewood's Partner and Selima, the blood of the Fenwick Barb has been transmitted to many of the best horses of this day.

After the death of Charles II., the Royal mares were dispersed throughout the Kingdom, and their female descendants were called Royal mares. The Sedbury Royal mare was the dam of Hutton's Royal colt by the Helmsley Turk, and great grandam of Old Country Wench, and her sister by Lister's Snake, D'Arcy Royal mare, dam of Hautboy by D'Arcy White Turk, the Blunderbuss Royal mare, dam of Old Royal by Holderness Turk, and grandam of Hartley's Blind Horse by the same Turk, D'Arcy's Royal mare, dam of Brimmer by the D'Arcy Yellow Turk, Christopher D'Arcy's Royal mare, to which traces the dam of Rachel by Blank, she the dam of Mark Anthony by Spectator, and Highflyer by Herod, the D'Arcy Black-legged Royal mare, whose daughter by Lord Oxford's Dun Arabian was the dam of Miss Slamerkin by Young True Blue, she the dam of Bustard, Othello (Black-and-all-Black) and Oroonoko, all by Crab, and Duchess by Whitenose. There are other Royal mares, but the larger number of them were in the Sedbury stud of the D'Arcy family. Lord D'Arcy imported the White Turk (called also Sedbury Turk), who was the sire of Old Hautboy (Wilke's). He got Grey Hautboy, the sire of Bay Bolton, out of daughter of Makeless, Wyndham, the dam of Lister's Snake, the dams of Almanzor and Aleppo by Darley Arabian, Terror by Akaster Turk, and Champion by Sir J. Harpur's Arabian. D'Arcy Yellow Turk got the noted Spanker, out of Old Peg (old Morocco mare) by the Morocco Barb, out of Old Bald Peg by an Arabian, out of a Barb mare.

The Curwen Bay Barb and Thoulouse Barb were imported in the reign of King Charles II. They were a present from Muley Ishmael, King of Morocco, to Louis XIV. of France. They were brought from Barbary when Count of Thoulouse was Lord High Admiral of the Fleet and Count Byram was Master of the Horse to the Monarch. Mr. Curwen, of Workington, got possession of these Barbs and brought them to England. The Curwen Bay Barb, from a daughter of Old Spot, son of the Selaby Turk, out of a mare by the White-legged Lowther Barb, out of the old Vintner mare, got Mixbury, his two sisters and Whiteneck. One of these sisters was the dam of Miss Jigg and Croft's Partner by Jigg, Soreheels and his sister by Basto, and the other was the dam of Sloven by Bay Bolton. Miss Jigg was the dam of Monkey by

Lonsdale Bay Arabian. The sister to Soreheels was the dam of the noted Crab by Alcock Arabian, Blacklegs, Bay Basto, Second, Hip, Snip and Brown Basto, all by Flying Childers. The Curwen Bay Barb was famous also as the sire of Brocklesby and Brocklesby Betty, out of Mr. Lister's Hobby mare by his Turk, grandam Piping Peg. Brocklesby Betty was famous on the turf, and was the dam of the grey mare Brocklesby by Greyhound, and she the dam of Grey Brocklesby by Bloody Buttocks, and Bay Brocklesby by Partner. It is to this line of blood to which trace in direct female line the following good horses: Octavius, Bob Booty, Cannon Ball, Bretby, St. Martin, imported Mickey Free, Chanticleer, Solon, Tom King, Xenophon, Lowland Chief, Zealot, Hagioscope, Barcaldine, Ossian, Ambergris, Signorina, imp. Filagree and her daughters, imp. Ga-elle and her descendants.

In the reign of James II. the Stradling, or Lister Turk, was brought to England. This Turk was taken in 1686, at the siege of Buda, by the Duke of Berwick, who sold him to Mr. Lister, of Lincolnshire, in whose stud he became famous under the name of the Lister Turk. He got Brisk and Coneyskins, out of a daughter of Jigg, son of the Byerly Turk, grandam of Bolton Sweepstakes Mr. Lister's Snake (the sire of William's Squirrel and his own brother, Smith's son of Snake) out of a daughter of the Akaster Turk, the Hobby mare, dam of Brocklesby Betty, and her sister, the dam of Mr. Pelham's Hip by Curwen Bay Barb. Through these, and particularly Coneyskins and Mr. Lister's Snake, the Lister Turk must be regarded as one of the Eastern progenitors of the thoroughbred horse.

Lonsdale Bay Arabian was the property of Lord Lonsdale. He got Lord Lonsdale's Monkey, Spider, Sultan, etc., and Mr. Panton's Veteran. He also got the dam of Miss Ramsden by Cade, out of sister to Bonny Lass by Bay Bolton. Miss Ramsden was noted as the dam of Woodpecker, Quicksand, Wormwood and Whipcord, all by Herod. Woodpecker was famous as the sire of imp. Buzzard, out of Misfortune by Dux, son of Matchem. Buzzard sired the three brothers, Selim, Castrel and Rubens. Woodpecker also sired a large number of valuable brood mares, including Ambrosia (the dam of Fanny by Diomed, Cripple by Saltram, etc.), Colibri (dam of Princess by Precipitate, Cardinal Beaufort and Canopus by Gohanna), Rosina (dam of Hermes by Mercury), Wren (the dam of imp. Robin Redbreast, Agonistes, Tiney and Milo, all by Sir Peter), and the dams of Fractious by Mercury, Fraxinella by Trentham, and Petworth by Precipitate.

The Honeywood Arabian (known by the name of William's Turk) was imported to England by Sir John Williams in the reign of King William. Sir John sold him to Mr. Honeywood, hence his name. In 1709 Mr. Honeywood bred his Byerly Turk mare to this Arabian, and the grey colt, foaled in 1710, was named True Blue, and won the King's Plate at York in 1718, in which year the Byerly Turk mare produced Young True Blue, his full brother, who was the best plate horse of his day. There were two sisters to the two True Blues. It is to these two sisters of the True Blues to which trace, in direct female line, nearly all the best race-horses and stallions for more than one hundred years, including Sir Peter, imp. Buzzard, Skyscraper, Spadille, Champion, Tramp, Master Henry, Bizarre, imp. Chateau Margaux, Rector, Velocipede, The Saddler, Glaucus, Lanercost, Galaor, Sleight of Hand, Flatcatcher, Jeremy Diddler. Pyrrhus the First, Van Tromp, The Flying Dutchman, Stockwell, Rataplan, King Tom, Hobbie Noble, Galopin, Chattanooga, Windhound, Kettledrum, Knight of the Garter, Mr. Winkle, Statesman, Ivan, Artillery, Plum Pudding, Toxophilite, Knowsley, Tim Whiffler, Lacydes, Musket, Victor by Vindex, Cardinal York, Paul Jones, Victorious, Tomahawk, Gen. Peel, Vermouth, Favonius, Bruce, Vertugadin, Master Kildare, Wellingtonia (in France), Pride of Prussia, Border Minstrel, Quicklime, Highland Chief, Maribyrnong (in Australia), imp. Rayon d'Or, imp. Robin Redbreast, imp. King Ernest, imp. Escape (Horns), imp. Double Duke, imp. Oscar, imp. Eothen, imp. Consternation, imp. Lochiel, imp. Lapdog, imp. Mr. Pickwick, imp. Roman, imp. Glengarry, imp. Dalnacardoch, imp. King Ban, American Eclipse, Viator, Reform, the imp. mares Castagnette, Cyclone, Angela, Periwinkle, Athenais, Teardrop and many others of note.

A number of Barb mares were imported to England and were designated by the names of their owners. The Burton Barb mare, property of Mr. Burton, said to have been a natural Barb mare, had a colt, St. Martin by Spanker, and a filly by Dodsworth. This filly by Dodsworth was the dam of Morgan's Dun by a son of the Helmsley Turk, and an own sister to him, which produced a filly by Hautboy. The Burton Barb mare had also a filly by Dicky Pierson, son of Dodsworth. This Dicky Pierson mare was the dam of Old Thornton and her sister by Brimmer, son of the D'Arcy Yellow Turk. Old Thornton was the dam of Chestnut Thornton by Makeless, she the dam of Lusty Thornton by Croft's Bay Barb, Desdemona by Greyhound, and Woodcock Thornton by Woodcock, son of Bustler by the Helmsly Turk. Lusty Thornton was the dam of Brown Woodcock by Woodcock, and Ringbone by Partner. The latter was the dam of Ancaster Starling, Madame Flirt, Miss Starling and Miss Starling Junior, all by Bolton (Old) Starling, a filly by Bloody Buttocks (Arabian), a filly by Regulus, Barforth and a sister by Forester, son of Hartley's Blind Horse, Legacy by Cade, and the chestnut filly Mary Grey, foaled in 1742, and imported to America, by Roundhead, son of Flying Childers. The sister to Old Thornton produced Bay Thornton, who had a filly by Croft's Bay Barb, and four fillies by Greyhound. Brown Woodcock had Emma, Miss Doe and two sisters by Greyhound, Miss Barforth and her sister (dam of Miss Makeless) by Croft's Partner, etc. This sister to Miss Barforth was also dam of Roger of the Vale (Jolly Roger), imported to America, by Roundhead, son of Flying Childers, Brown Slipby by Slipby, and Brown Starling by Bolton Starling. It is from this Burton Barb mare that has descended in direct female line Phenomenon, Mungo, Dick, Waverley, Safeguard, Whiskey, Selim, Castrel, Rubens, Camillus, Ardrossan, Catton, Whitelock, Blacklock, Bobadil, St. Giles, Sir Hercules, Irish Drone, Margrave (imported), Don John, Hetman Platoff, Harkaway, Womersley, Cremorne, Voltigeur, Surplice, Cowl, Ithuriel, King of Trumps (by Velocipede), Teddington, The Marquis (by Stockwell), St. Albans, Lord Clifden, Mask, Skirmisher, Lecturer, Camballo, Martyrdom, Pell Mell, Carnelion, Thurio, Umpire (Tom King), Althotas, Brag, Sheen, Gold, imp. Ainderby, imp. Billet, imp. Chance, imp. Claret, imp. Contract, imp. Eagle, imp. Hurrah, imp. Nonplus, imp. Skylark, imp. Spread Eagle, imp. St. George, imp. Thunderstorm and imp. Yorkshire.

A Royal mare of Capt. Appleyard's, bred to Brimmer (or, as some reports allege, to Son of Brimmer), produced a filly, which, bred to Bethell's Castaway, son of Merlin, produced a filly, which filly was the dam of a filly by Bethell's Arabian and one by Commoner (Wharton's), the latter was the dam of Grey Childers by Flying Childers. This Castaway mare, above mentioned, was the dam of Bald Charlotte, first called Lady Legs, by Old Royal, son of the Holderness Turk. Bald Charlotte's daughter by the Somerset Arabian was the dam of imp. Kitty Fisher by Cade, she the dam of Wormsly's King Herod, Cripple Fearnought, Young Kitty Fisher, etc., by imp. Fearnought, son of Regulus by the Godolphin Arabian. It is through Young Kitty Fisher and her daughter by Symme's Wildair that the noted race-horse and stallion Boston traces to this family.

The Dick Burton Barb mare produced Ruffler and his sister by a son of Brimmer (and mare by the Chesterfield Arabian). This sister to Ruffler was the dam of a filly by Darley Arabian, she the dam of a filly by Fox, son of Clumsey by Hautboy, which produced a filly by Blaze, son of Flying Childers. The Blaze mare was the dam of Riot and Red Rose by Regulus, and Riot was the dam of Flora (the dam of Ruler by Young Marske) by Lofty. It is to this Dick Burton mare that Ruler, Wanton, Fazzoletto, Cape Flyaway, The Speaker (by Filbert), Wallace and others trace in direct female line.

The natural Barb mare of Mr. Tregonwell, bred to Place's White Turk, produced a filly, which, bred to the Taffolet, or Morocco Barb, produced also a filly, and this filly, bred to the Byerly Turk, produced a filly which was the dam of Grey Ramsden by Grey Hautboy, a filly by Spanker, one by the Darley Arabian, and one by the Newcastle Turk. The Darley Arabian mare was the dam of Merry Andrew by Fox, Paragon by Snip, a filly by Lonsdale Bay Arabian, and a filly by Partner (Croft's). This Partner mare was the dam of Naylor by Cade, son of the Godolphin Arabian,

Spectator by Crab, son of the Alcock Arabian, Tatler, Julia and two sisters of Julia by Blank, son of the Godolphin Arabian. Naylor was the dam of Contest, imp. Centinel (foaled 1758, and imported into South Carolina), Nancy and others, all by Blank. Julia was the dam of Prize and Promise by Snap, and Princess by Herod. Promise was the dam of Patience, Peeress and Pallas by Herod, Prude, Peppermint and Prunella by Highflyer, and a filly by Pot-8-o's. Prunella was the dam of Penelope (Wire's dam), and Pawn, both by Trumpator, Parasol (Partisan's dam), by Pot-8-o's, Pioneer by Whiskey, Waxy Pope (Pope), winner of the Derby in 1809, Pledge (Tiresias' dam), Pope Joan and Prudence, all by Waxy, son of Pot 8-o's by Eclipse, and Piquet by Sorcerer. It is to this Tregonwell Barb mare that the following horses trace in direct female line: Woodpecker, Goldfinder, Waxy Pope, Whalebone, Whisker, Partisan, Turcoman, Rockingham, Bay Middleton, John Davis, Fitz Roland, Lord Lyon, Dollar, Trumpeter, Nutbourne, The Nob, Verulam, The Cossack, Melbourne, Bran, Gunboat, Bend'or, Robert the Devil, Brocket, Uncas, Fetterlock, Silvio, Speculum, Charibert, Childeric, Blue Gown, Ceruleus, Favo, Pellegrino, Trappist, Craig Millar, Silvester, Orphelin, King of Trumps (by Lord Clifden), The Lambkin, Glen Arthur, Controversy, Town Moor, Privateer, Victor Chief, The Confessor (by The Palmer), Paradox, The Bard (by Petrarch), Minting, imp. Glencoe, imp. Eclipse (Morris'), imp. Riddlesworth, imp. Athlete, imp. Scythian, imp. Blue Mantle, imp. Glen Lyon, imp. Tubal Cain, Belmont (of California), Uncas, and Wanderer by Lexington, and a number of good producing mares.

The mares which have founded families in America are a study to lovers and admirers of racing lineage. One of the most prominent of these, whose descendants have always been famous on the turf and in the stud, was the Cub mare. She was a bay mare, foaled in 1762, by Cub, son of Fox by Clumsey by old Hautboy, her dam Amaranthus' dam by Second, son of Flying Childers, grandam sister to Vane's Little Partner by Partner, son of Jigg by the Byerly Turk, great grandam sister to Guy by Greyhound (son of King William's White Barb, Chillaby, out of Slugey, a natural Barb mare), she out of Brown Farewell by Makeless, son of the Oglethorpe Arabian, tracing to the noted Layton Barb mare. This Cub mare was imported along with Wildair by Gen. Jas. DeLancey, of New York, in 1764. Wildair was repurchased and sent back to England in 1773. Gen. DeLancey bred several foals by Wildair out of this Cub mare. Hero, Moulton, Ruth Black Eyes, Bashaw (foaled 1767), and Maria Slamerkin, better known as Old Slamerkin or Miss Slamerkin, and it is from this last mare that so many of the best race-horses in our day derive their origin. Old Slamerkin had (in 1788) a filly by imp. Obscurity, son of Eclipse, Moll by imp. Figure, Fair American by Lloyd's Traveler (this line is nearly extinct), and Dido by imp. Bay Richmond, son of Feather by the Godolphin Arabian.

The Obscurity mare of 1788 was the dam of a filly called the Robin mare by imp. Robin Redbreast, who produced Thornton's Rattler, 1816; Childers, 1817; Sumpter, 1818; Elizabeth, 1819, and Flirtilla, 1820—all by Sir Archy. The horses were good stallions, and in the female line from Flirtilla have come a large number of superior race-horses. Flirtilla produced Tom Kimball, 1832, and Mary Wynn, 1833, by American Eclipse, the former sent to Canada; Antoinette, 1829, and Flirtilla, Jr., 1828, by her own sire, Sir Archy; Gaslight by imp. Chateau Margaux. Flirtilla, Jr., produced, 1836, Olympius by American Eclipse; Ringgold, a superior race-horse and good stallion, 1842, by Boston, and Cassandra, 1838, by imp. Priam. It is to this last that we look for the descendants of this line. Cassandra produced Trojan, 1847; Basil, 1848; Philo, 1851, all by Mariner, son of Shark by American Eclipse; and Prophet, 1856, by imp. Monarch. Philo produced Constantine, 1858, by Revenue, son of imp. Trustee, and Lady Blessington, 1861, by imp. Eclipse, son of Orlando by Touchstone. Lady Blessington produced Count D'Orsay, 1870, a good race-horse, by Kentucky; The Countess, 1871, by Kentucky; The Baroness, 1874, by Kentucky; Lord Lytton, 1876, by Kentucky or Kingfisher, 1878, Lady Rosebery, 1878, Fairwater, 1880, Duchess, 1881, and Lord Beaconsfield, 1882, all by Kingfisher. The Countess was the dam of Falco, Le Petit Duc, Contessa, Grand Duke and Peeress by imp. Ill Used. These all raced well, and Contessa is the dam of the winner Midget by King Alfonso. The Baroness is the dam of Barouella (Music's dam) by imp. Matador, son of Gladiateur, Ballerina and Badge (a very superior race-horse and winner in fast time), both by imp. Ill Used; Barcarola and Berceuse, both by Falsetto Lady Rosebery was the dam of Lady Primrose, Lady Margaret, and Lord Dalmeny by imp. Ill Used. These were all good performers on the turf and valuable in the stud. Fairwater is the dam of Gracie by imp. Siddartha, Katie by imp. Macaroon, and Laughing Water (a good two-year old) by imp. Dalnacardoch. Duchess is the dam of Watterson, a good race-horse, by imp. Great Tom.

Old Slamerkin produced Moll (Young Cub) by imp. Figure. Moll was the dam of Cockfighter and Esopus, good horses, by imp. Grey Highlander; Major Moll and Lady Jane by imp. Obscurity; Jane Hunt, 1796, and her sister by Wade Hampton's Paragon, son of imp. Flimnap, son of South by Regulus. It is from Jane Hunt that this line descends to us. She was the dam, 1802, of Indiana by Butler's Columbus, son of imp. Pantaloon by Herod; Shepherdess, 1803, by imp. Speculator, son of Dragon by Woodpecker by Herod; Sally Sneed, 1807, by imp. Buzzard; Tiger, 1812, and Grecian Princess, 1816, by Cook's or Blackburn's Whip, son of imp. Whip. Indiana was the dam of Hannah Harris (Paragon), 1807, by imp. Buzzard. Hannah Harris produced Aurora, 1847, by Aratus, son of Director by Sir Archy. Aurora produced Lancess (Miss Lancess), 1835, by Lance, son of American Eclipse. Lancess raced successfully, and was the dam in 1844 of Ariel by imp. Sarpedon, 1845, Fanny G. by imp. Margrave, son of Muley, and, 1849, Emma Taylor by imp. Glencoe. Ariel produced Corinne, 1852, by imp. Glencoe, and she in 1861 produced Virginia by Revenue, son of Imp. Trustee. Virginia in 1868 produced Jennie C. by Norfolk, 1875, Lottery, and, 1879, sister to Lottery by Monday, son of Colton by Lexington, 1883. Virginius by Longfield, and in 1884 Bagatelle by Jim Brown, son of Foster by Lexington. Jennie C. is the dam of Mark L., 1875, sister to Mark L., 1879, and Prudence, 18-5, all by Monday, and Jessie C., 1887, by Bishop, son of Shannon by Monday. Sister to Lottery is the dam of Loto, Corona and other fillies by Norfolk, son of Lexington. Fanny G. was a speedy, good race-mare, and the dam of Liz Mardis, 1853, by imp. Glencoe; Endorser, 1856, a fine race horse, by Wagner; Flora G., 1859, Falcon, 1860, Lady Dan Bryant, 1862, all by Lexington; Emma Maratta, 1865, and Violette, 1867, both by Vandal; Estella, 1866, by imp. Australian. Estella is the dam of Alma Mater, the dam of the trotters Alcantara, record 2.23; Alcyone, 2.27; Alicia, 2.30; Almater, 2.29½, and Arbiter, 2.22¾, and Fanny B., 1870, by Asteroid. Liz Mardis was the dam of Leon, 1864, Sir Rufus, 1868, and in 1873 of a bay filly, all by Lexington; Coquina, 1870, by imp. Australian; Dame Winnie, 1871, by Planet (she the dam of Bridget Ella, 1878, by Wanderer, son of Lexington, and the trotters Big Jim, record 2 23¼, Palo Alto 2.08¾, and Gertrude, Russell, 2.23½), and a bay filly, 1873, by Asteroid. Leon produced chestnut filly, 1869, by Colossus, and Galt Reporter, 1870, by same; Lottie, 1874, by King Tom, son of Lexington; Mecca, 1882, by imp. Billet, and Hindoostane, 1884, by Hindoo. Lottie by King Tom is the dam of Winnamucca, Ballet and Bassanio by imp. Billet, and Montpelier by Hindoo. Estella is also the dam of bay filly, Timbrel, 1879, by imp. Saxon, Miss Lady, 1882, by Falsetto, Welcome, 1883, and Stella, 1885, both by imp. Mortemer, and Ethelda, 1884, by imp. Moccasin, son of Macaroni; the last named dam of Ethelair by imp. Pizarro. Emma Taylor produced Eva Bulwer (Miss Bulwer), 1857, by Bulwer, son of Grey Eagle, she the dam of Ruth Ryan and Jewel by Lodi, son of imp. Yorkshire. This stock is owned in California. Emma Taylor is also the dam of Montezuma, 1859, Strideaway, 1858, and Trampolette, 1860, all by Billy Cheatham, son of Cracker by Boston. Trampolette produced Julia Morgan, 1877, Mercedes (Vivandière), 1876, and Gipsy, 1879, all by Lodi. Mercedes is the dam of Mamie T. and her sister by Wheatly, son of War Dance. These are owned in Rancho del Paso stud, California.

Hannah Harris also produced Jenny Slamerkin, 1823, by Tiger, he by Whip, out of Jane Hunt, the grandam of Hannah Harris. Jenny Slamerkin produced Miss Obstinate, 1829, by Sumpter, and The Mule and her sister by Medoc, son of American Eclipse. Miss Obstinate greatly increased the reputation of this family. She was the dam of Morgiana, 1834, by Thornton's Rattler (brother to Sumpter). Mary

Morris, 1836, by Medoc, Ann Innis, 1838, by American Eclipse, Solo, 1850, Marchioness, 1851, and Kitty Clark, 1853, all by imp. Glencoe. Mary Morris was famous on the turf, and was the dam of chestnut mare by imp. Envoy, son of Memnon by Whisker, Wild Irishman, 1850, Frankfort, 1851 (both excellent race-horses), by imp. Glencoe, and Grey Mare by Grey Eagle, son of Woodpecker. The Envoy mare produced Gleneona, 1855, by imp. Glencoe, who in 1861 produced Elkhorna by Lexington, she the dam of Butler, 1870, and Arcturus, 1872, by Planet, Farfaletta, 1867, and Elkhorn, 1871, by imp. Australian. Farfaletta produced Pirouette, 1873, by Harvey Villian, son of Asteroid, Felicia, 1874, by imp. Phaeton, Falsetto, 1876 (a grand race-horse and one of the popular stallions of the day), and his sisters, Fortuna, 1875, and Fac Simile, 1881, by Enquirer, Farfadette, 1886, by imp. Siddartha. Pirouette produced Hattie Harris, 1880, Wood River, 1881, Avondale, 1882, Gem of the Mountains, 1883, and Susie S., 1884, all by Marmaduke, son of Enquirer. Hattie Harris is the dam of the winners Spectator by Springbok, and Cornie Buckingham by Powhattan. The remainder of this branch of the family are in the Fleetwood stud, Kentucky, and have bred winners.

The Grey Eagle mare produced Eagless, 1855, by imp. Glencoe, American (Volscian), 1857, by Vandal, Priscilla, 1859, by imp. Scythian, Annette, 1860, by Lexington, Jenny H., 1861, by imp. Knight of St. George, Della, 1866, by Rifleman, son of Glencoe, and, in 1867, a grey filly by Lodi. Eagless produced Luna, 1863, and Mat Eagless, 1864, by Swigert's Lexington, son of Lexington, Mary Clark, 1868, and Lexington Belle, 1875, by Lexington, Grey Planet, 1869, Steel Eyes, 1871, and Eaglet, 1876, all by Planet, and, in 1870, Lizzie Lucas by imp. Australian. They nearly all raced successfully. Luna was the dam of Glenmary, 1874 (a twin), by imp. Glenelg, Mamie R., 1877, Minnie D., 1878, both by Longfellow, and Nannie R., 1879, by Virgil. Glenmary is the dam of Glen Ban and Catharine Ban by imp. King Ban, and bay mare Loraine by Lisbon. Nannie R. left the filly Cheekie, by Longfellow. Mat Eagless produced Juanita by Wanderer, and she produced Nita, Fanita, Carroll, Blue Vail, Hazel Hurst and others. Mary Clark produced Preciosa, 1876, Spark, 1878 (both winners), by imp. Leamington, Pure Rye, 1883, Grey Dawn, 1885, and Old Rye, 1886, winners, by imp. Billet. Lizzie Lucas produced Cerise by imp. Moccasin (she the dam of Cherry Stone by King Bolt, and Wild Cherry by Wilful, and the great two-year old Morello by Eolus, winner of 11 out of 14 races this year), Chimera (dam of Chimere by Iroquois), Cyclops, Cambyses, Cleola, Cleofas and Veva by imp. Mortemer, and Leonora by imp. Sir Modred. Eaglet produced Olanthe, Laura Garrison and Spring Eagle (Sadie Mc) by Springbok. Annette went to California. She was the dam of Oliver Cromwell, Charley Chase and Camilla Urso by Lodi, Carrie C. (Covey) by Monday, and Jim Farley by Hubbard, and the trotter Ansel, 2.20, in Mr. Robert Bonner's stud. Camilla Urso was the dam of Raven by Monday, and Armada by Crichton, son of Glencoe (she dam of Meda Howard by Sacramento, son of Rifleman, and Hazel by Balboa, son of Norfolk), Miss Camilla (Mozelle) by Monday, and brown mare, 1880, by Shannon. Carrie Covey is the dam of Marin by Thad. Stevens, son of Langford. Marin is the dam of Reata by Milner, and Marinette by imp. Partisan, son of Tomahawk, Catalina by Wheatly, Sonoma by Longfield, and Bessie Barnes by imp. Darebin. Jenny H. produced Jenny Black, 1867, and Jenny Patterson, 1869, by imp. Australian, Jentling and Saturn by Planet; Daraley, Ada S. Manette and Toledo by Asteroid, Rosa R. and Useful by Pat Malloy. Jentling is the dam of Vassar by Virgil (she the dam of Vellum, Mary Porter, Perkins, alias Ben Hur, and Paris by imp. Billet, and a bay filly, 1888, by the same sire), Gentle by Baywood, and bay filly, 1888, by imp. Rotherhill. Ada S. produced two fillies, Leona and Surprise, by Big Fellow, son of War Dance. Manette produced Maryetta by Red Bluff, son of imp. Australian, Emma Gene and sister by Storey, son of Monarchist. Useful is the dam of Patmos, Laura Ford, etc., by Longfellow. The Mule produced Our Gal, 1850, Jean D'Arc, 1851, and Margaret Morris, 1852, by imp. Glencoe, the last of which was the dam of Harper, Marquis, Marsellaise, Mary Morris and Morris, all by Lexington. Mary Morris was the dam of Lizzie Gant (Bob Forsyth's dam), Annie Lexington (dam of Brevet by Brigadier, and Bonnie Annie by Scotlander), both by Exchequer, son of Revenue and Bonnie Park by Bonnie Scotland, she the dam of East Lake and Bonnie May by Enquirer, and Bracken by Bramble.

Kitty Clark, 1853, by Glencoe produced Charlotte Buford, 1857, Ann Clark, 1859, Dr. Lindsey, 1861, Maiden, 1863, La Henderson, 1865, Uncle Nick, 1866, and Utica, 1870, all by Lexington, Rosa Clark, 1864, and Kith, 1873, both by imp. Australian, and Glendolen by imp. Glen Athol. Charlotte Buford was a good mare and the dam of Charade, Culloden and Florence I. by imp. Australian, and Ida May by Planet. Florence I. is the dam of Florence Payne by Blarneystone, Florence Lamertine by Tom Bowling, Florence Cook by War Dance, Powhattan II. by imp. Glenelg, Florio by Virgil, Florence Fonso by King Alfonso, Florimer by imp. Rayon d'Or, etc. A majority of these were good race-horses, and Florio is the dam of the winner Franco. Ida May is the dam of Flotsam (Pumice's dam) by Waverly, Jetsam (the dam of Venture) by imp. Strachino, son of Parmesan, and Chissie May, etc., by Clipses. Ann Clark produced Aspire, a bay filly by Escape, son of imp. King of Cymry, Mrs Winslow by imp. Australian, and bay filly by Crichton, son of Glencoe. Aspire is dam of Alice Murphy (dam of Donatello and Alice Racine) by Jim Murphy, son of West Roxbury, and Josie (dam of Oleander by Aliunde, and Secret by Tennyson) by Ramadan. Mrs. Winslow is the dam of Methilde and Mateal, by Restless, son of Voucher. Maiden was the best filly of her year, and winner of a number of good stakes, including the Travers at Saratoga. She was the dam of Lady Washington, 1870, by Beacon, son of Lexington, James A., Parole (gelded, and one of the best race-horses of his day, winner in both America and in England), Mineola, Perfection, Pawnee (gelded), Pappoose, Paw Paw and Powhattan, the successful stallion, all by imp. Leamington, and Parthenia by Alarm. Lady Washington is the dam of Combination by Crockford, son of Lightning, Aileen by Amadis, son of Rivoli, Martha Washington by Volturno, son of imp. Billet, etc. Perfection is the dam of Smilax by imp. Saxon, Perception by imp. Mortemer, Saluda by imp. Mortemer, Perfecto by Duke of Magenta, etc. Pappoose is dam of Parolina by Duke of Montrose. Parthenia is dam of mare Rizpah by imp. Mortemer, and mare Persistence by imp. Sir Modred. La Henderson was the dam of Kitty Claborn by Jerome Edger, Planetia by Planet, Virgilian (a good race-horse), and Vanguard by Virgil, the noted Ferida (four miles 7.23½, fastest ever run by a mare), Aella, a superior race-mare, and Ferona, a winner, all by imp. Glenelg. Planetia is the dam of Placid (Playfellow's dam) by Enquirer, Virginia (Lilly M.) by Virgil, Fostress (Question's dam), by Foster, Plantress and Ned Cook by Flood, and others. Ferida is the dam of Fay, Medje, Flitaway, Carrie C. and a filly, etc. by Sensation. Aella is dam of Prince George, a winner, Thrill, and three colts by Sensation, the unbeaten.

Rosa Clark was the dam of Lilly Clark, Lulu Baxter and Judge Spofford by Watson, son of Lexington, Magnolia Clark, Clark and Ophelia Spofford by imp. Canwell, son of Stockwell, Barclay and Fronie Louise by imp. Glengarry, son of Thormanby. Lilly Clark is the dam of Goliad, Beulah Clark, Mattie Jackson and Olive Varcoe by Heretog, son of imp. Australian. Magnolia Clark is dam of Palmetto (Bamboo's dam) by imp. Glenelg, Emma Mack by Ten Broeck, Ae-le-ta and Kossuth by Longfellow. Glendolen is the dam of Alsace by Aramis, Atticus by Aristides, and filly by Punster. Kith is the dam of Kinsman, a winner, by Monarchist, Vic by Virgil, and Gold Fish by Prince Charlie or Virgil.

Ann Innis by American Eclipse was the dam of Hebron, a good race-horse, Vivandiere and Jim Barton, a good winner, by Grey Eagle, son of Woodpecker, Maria Innis by imp. Yorkshire, and Swigert's Lexington and La Grande Duchesse by Lexington. Maria Innis was the dam of Bulletin, Echo and Kathleen by Lexington, Sovereign Eclipse by imp. Sovereign, Australian Chief, Lagonda and Emma by imp. Australian, Select by Swigert's Lexington, and Meteor by Asteroid. Echo was the dam of the geldings Report and Duplex, good race-horses, Nemesis and Naptha by imp. Eclipse, Echoless and Annie Hall by imp. Leamington, Court Hampton by imp. Hampton Court, Eccola (Rose's dam) and her sister by imp. King Ernest. Echoless was the dam of Resound and Peanuts by imp. Hampton Court, Glen Echo and Re-Echo by imp. Glenelg. Nemesis was the dam of Vengeance and sister by

Censor, son of Lexington, Rhadamanthus, a speedy, good race-horse and a large winner, and his sister Wyandotte (the dam of El Trinidad, Lady, Lassie, Ottawa, etc.) both by imp. Leamington. Retribution (dam of Arab and Sir William), Eachus and Mink by Reform, Acoustic by imp. Australian and Brushwood by imp. Woodlands. Acoustic is the dam of Vocalic (a good racer and successful sire), Alida and Fancy by Virgil, Music, Acasia, etc., by imp. Glenelg. Naptha was the dam of Golden Gate by imp. Leamington, Explosion and Gleam by imp. Hampton Court, son of Young Melbourne, Ten Drop by Ten Broeck, The Jewess by Saracen, son of Leamington, Miss Hattie by Falsetto, etc. Golden Gate was the dam of Laura Winston, Faralone and Oro by Norfolk. Laura Winston is the dam of Miss Harker, Snowdrop, Broncho and Snowball by Joe Hooker. Explosion is the dam of Dangerous by Saxon, Disdain (dam of Omega, Grand Duke, Arrogance, etc.) by imp. Moccasin, Dione, Daruna and Dolores by imp. Mortemer. Dew Drop, the grand race-mare that sold for $29,000 (dead), and her sister, Firework by Falsetto, and Rainbow by Iroquois. Gleam is the dam of Glitter by imp. Stylites, Fidelia and Flemington by Falsetto, etc. Kathleen is the dam of Kennesaw, a good horse, Kathairon (dam of Kathleen by Shannon, Keepsake by Flood, and Kenneth by Joe Hooker) by Harry of the West, son of Lexington, John Happy and the superior race-horse George Kinney by imp. Bonnie Scotland, Hetty, Kennebeck, Khartoum and Pullman by imp. Glengarry, Kern by Ten Broeck, Katuna by Bramble, etc. Emma by imp. Australian was dam of Judge Murray by Vauxhall, Maggie Duffy by Charley Howard, J. O. Nay by Fellowcraft, Charity by Sensation, etc. La Grande Duchesse was the dam of Motella, Aniella and Christine by imp. Australian, Elsie by imp. Bonnie Scotland, Marmion by Planet, Talisman and Uberto by Virgil. Motella was the dam of Misadie and Miss Ella by Enquirer, the latter dam of Lita, alias Lyda (the dam of Jum Jum, Michael and O. K. by Regent) by imp. True Blue, son of Vedette, Maretzeck and Mariet by King Alfonso, Mendelssohn by imp. Buckden, Whisperine (dam of Whisban) by Whisper, Miramar (dam of Iola) by Longfellow, Metal by Blue Eyes, and Maga by imp. Siddartha. Aniella is the dam of Allegro (the dam of Aria, Lizzie L. and Conrad O. by Whisper), Anniban by King Ban, Annona and Anadem by Blue Eyes. Elsie is the dam of Etna (dam of Outset, Outstep, Outbound and Outspeed) by King Alfonso, Egeria by Lever, Rumor and Eola by Whisper, Eros II. by Lisbon, etc. Christine is the dam of Gypsy by War Dance, Bonnie Australian, Vice Regent, Montana Regent, Julia L. and High Tariff by Longfellow, and Beth Broeck by Ten Broeck. Gypsy is the dam of Pat Regent, Don Regent and King Regent by Regent, and Poet Scout, a superior race-horse, by Longfellow, Gypsy Girl and Lorna Doone by imp. Sir Modred.

Sally Sneed by imp. Buzzard was the dam of Grey Goose by Pacolet. Grey Goose was the dam of John Bascomb, a famous racer, and his two sisters, Lady Huntsville and Hannah Harris by Bertrand, son of Sir Archy, Pilot and Glance by Wild Bill, son of Sir Archy. Lady Huntsville had two fillies by imp. Consol by Lottery. One of these fillies was the dam of Betty Wharton by Othello, son of imp. Leviathan, and she the dam of the good race-horse and stallion, Rebel, and his sister, Miss Socks by Socks, son of imp. Albion, Anna Bishop by Lytle, son of Jim Allen (she the dam of Betty Bedford by Rupee, and Ellen Lane by Tar River), West Maria (Anna Hurst's dam) by Gen. Burleson, and Last Chance by Regent. Glance was the dam of Little Mistress by imp. Shamrock, she the dam of Little Miss by imp. Sovereign, she the dam of Letty by imp. Australian, the dam of Grace and Lettina by Harry Bassett, Letretia and her sister by Tom Ochiltree, and Refund by Sensation. Little Miss was also dam of Uga by King Alfonso, and she the dam of Ultimo by Red Bluff, Ianthe by Storey, etc.

Grecian Princess, 1816, sister to Tiger by Blackburn's Whip, was the dam, 1829, of Ann Merry by Sumpter, Titus (Cripple) by Medoc, etc. Ann Merry was the dam of Mirabeau and Cub (a superior race-mare) by Medoc, and Barilla by Lieut. Bassinger. Cub was the dam of Sportsmistress by American Eclipse, Monte (a good racer) and Martha Buford by Wagner, a filly by imp. Trustee, Fanny Fern and Fanny Campbell by imp. Glencoe. Sportsmistress was dam of Arnold Harris by Whalebone,

Charles Harris by Wagner, Susan Harris by Revenue, and Alfred Harris by imp. Sovereign. Susan Harris produced Zadoc by Marion, and Coquette by Lexington, the latter the dam of Vivien by imp. Leamington, Viola by imp. Eclipse, Variety, Vixen and Maid of Athens by imp. Saxon, Dolinka by imp. Mortemer, and Coquette by Iroquois. Vivien is the dam of Comanche by Leinster, Young Bazar and Aunt Jane by Bazar, owned in California. Viola is the dam of Mamie M. by imp. Saxon (she the dam of Lille and Lancer by King Alfonso), Blaze Duke by Duke of Magenta, Prosperity by Onondaga, etc. The Trustee mare, out of Cub, produced a filly by 2d Altorf, which was the dam of Lizzie Lewis and Stella Price by Epsilon, son of Pacific. Lizzie Lewis was the dam of Varina Bradford by Abu Becker, son of Mahomet by imp. Sovereign, she the dam of May Curl and Peerless by Bill Bass. Stella Price was dam of Plover Wing by imp. Stone Plover, son of Cotherstone, she the dam of Lady Longstreet by Gen. Longstreet (she the dam of Kate by Frogtown, Bella Golden and Atossa by Bowie, son of Leamington), Nancy by St. Valentine, Golden Plover and Minnichee by Chariton, son of Pat Malloy, and Tooan-Tuh by Frogtown.

Martha Buford by Wagner was the dam of Mantua by Lexington, Martha Buford by Ringgold, son of Boston, and Maud by imp. Australian. The last named produced Bonnie Scotia and Paper Maker by imp. Bonnie Scotland, and Eurydice by imp. Leamington. Eurydice had a filly, Astoria, by imp. Athlete. Fanny Fern was the dam of Eugene by Revenue, Oleata by Lexington, etc. Oleata was the dam of Emily Fuller, Pollywog, Osseo and Oleaster by imp. Eclipse, son of Orlando, Olitipa (one of the speediest mares ever foaled) by imp. Leamington, and Odessa by imp. Glenlyon, son of Stockwell. Emily Fuller is the dam of Prosper, Pantaloon, Olive and Persuasion by imp. Leamington, Florence E. by Reform, Xalapa and Brussels by imp. Billet, Hannah Moberly by Hindoo, etc. Nearly all these race well. Olitipa is the dam of Olivia and Orator by imp. Ill Used, Mary C. (Ophelia) by Kingfisher, St. Omer and St. Olga by imp. St. Blaise. Olive is dam of Olivie by Tom Ochiltree, Oliver Twist, All Fours and Oliver by imp. The Jacobite, son of Prince Charlie. Odessa is dam of Udessa by imp. Uhlan. Fanny Campbell produced two fillies, a chestnut filly, 1831, and Gentle Annie by Monte, son of Wagner, that were owned by George Treat, California.

Major Moll by imp. Obscurity was the dam of Patrick Highlander by imp. Brown Highlander, and a filly by Wade Hampton's Paragon. This Paragon mare was the dam of a filly by imp. Buzzard, and a filly by Copperbottom, son of Brilliant. This Copperbottom mare was the dam of John Henry by Aratus, and Hunter's Duke of Bedford. The Buzzard mare was the dam, 1826, of Mary Hunter and, 1828, of Kitty Hunter by Harris' Paragon, son of Blackburn's Whip. Kitty Hunter was the dam of a filly by Hephestion, Ann Stewart, a good racer, by American Eclipse, and a filly by Goode's Arab, son of Sir Archy. This blood is only preserved through these two mares. The Goode's Arab mare was the dam of Geneva, a good race mare, by Medoc, she the dam of Madame House by imp. Leviathan. The latter was the dam of Maria and Carrie Cage by imp. Sovereign, Villir Allen (the dam of Pearl by Buckmaster) by Hiawatha, son of imp. Albion, and Epizootic by Rogers, son of Lexington. Maria was the dam of Pennyroyal by imp. Albion, she the dam of Wild Addie and Lulu L. (Parepa Rosa) by Caledonia, son of Bonnie Scotland. Carrie Cage was the dam of a filly by Jack Malone, Miss Cheatham by Bill Cheatham, son of imp. Albion, a filly by Bulletin, son of Lexington, Betty Bell and Eppie L. by Surprise, son of imp. Bonnie Scotland. Miss Cheatham is the dam of a filly by Prolific, Cousin Cheatham and Third Cousin by Uncle Vic, and Amelia Johnson by Uncle Tom. Betty Bell is the dam of Bettie L. (Lillie McElwain), Vodena's dam, and Brigade by imp. Brigadier, and Bagatelle by imp. Ill Used. Eppie L. is the dam of Peekskill and a filly, 1888, by imp. Billet, Lizzie B. and Purity by Hindoo. Epizootic is the dam of Sweetheart by Jack of Diamonds, and she the dam of Va Tout by Faustus, and Omen by Vandal, Jr., son of Vandal.

Mary Hunter was the dam of Mary Streshley by John Henry, Miss Singleton by Kosciusko, Streshley, Webster, Emerine Price and sister to them by Medoc, and Lady Lovell by Sir Lovell. Mary Streshley produced Neoma Ferguson, John Horace,

Udica and a sister to them by Medoc. This Medoc mare produced a filly by Brawner's Eclipse, son of American Eclipse, and she produced Waxy by Lexington, that is the grandam of the famous trotting mare Sunol, record 2.08¼, the fastest ever made. Fair American by Lloyd's Traveler produced a filly by imp. Grey Highlander, whose daughter by imp. Expedition was the dam of Rosalinda by Ogle's Oscar. Rosalinda was the dam of Sally Van Dyke by Henry, Fleta by Johnson's Medley, son of Sir Hal, she the dam of St. Louis by Altorf, and St. Charles by American Eclipse. Sally Van Dyke produced a filly by Hugh L. White, son of imp. Leviathan. Oscaloosa is the dam of Olio by Memnon, son of Epsilon, and she the dam of the Rebel mare, dam of Envenom and Eight to Seven by Enquirer.

Dido by imp. Bay Richmond founded what is now known as the Morgan family. Dido produced Prizefighter by imp. Expedition, and Zelipha by imp. Messenger. Zelipha produced two fillies, through which the blood is perpetuated, viz., Aurora by imp. Honest John, son of Sir Peter, and Honesty by imp. Expedition, son of Pegasus by Eclipse. Aurora produced a filly by Sir Solomon, son of imp. Tickle Toby by Alfred by Matchem. This filly was the dam of Matchless by imp. Expedition, she the dam of Lady Morgan by John Richards, son of Sir Archy. Lady Morgan was the dam of Sally Morgan by imp. Emancipation, son of Whisker, and Polly Rowe by Eutaw Shark. Sally Morgan was the dam of Mary Blueskin by Grey Eagle, a good race-mare, Miss Morgan and Bondy by imp. Yorkshire, Panic by imp. Glencoe, and Ada C. and Dolly Morgan by Revenue. Miss Morgan was the dam of Mazurka, Mignonette (Minnie Milton), and India by Lexington, Mattie Morgan and Ave Maria by imp. Australian, Addie Hart and Peggy Morgan by Asteroid. Mazurka was the dam of Minnie Brown, Gyptis, Zoo Zoo, Typhoon and Rebecca, all by imp. Australian, Tallulah, Carlina and Maze by Planet, Zalite (dam of Lela May) by Lisbon, Bull's Head, etc. by King Alfonso. Minnie Brown was the dam of Medina by Watson, Elissa (dam of Lizzie K. and Lela K.) by Enquirer, Eleanor C. by Powhattan, The Don by Lisbon, etc. Tallulah is the dam of Barbary, Blithesome, Bonnie Meade and Bribery, all by imp. Bonnie Scotland, Tarantula (Centipede and Tara Blackburn's dam) and Termagant by imp. Great Tom, Tally-Ho and Bettina by Bramble, King Iro by Iroquois, and Talluda by Enquirer. Barbary, a winner, is the dam of Robert Bruce, Tip Top, Rhadamantha (the dam of Little Girl), Miss Maud and Sir Abner. Blithesome is the dam of Leipsic, Kee-Vee-Na and Wautauga, the two last by Enquirer. Bonnie Meade was a winner, and is the dam of Trophy (in Rancho-del Paso stud) by imp. Great Tom, Exile, Eukonia, Nellie B., Kempton, Reporter and Basil Duke by Enquirer, Dictum and Belle Meade by Iroquois. Bribery is the dam of Lida C. (Enguerande), Miss Ford, Zuleika, Boodler, Blue Dress, Baldwin, Tom Paxton, etc., by Enquirer. Gyptis, a winner, is the dam of Annie C. (Godiva) by imp. Saxon (she the dam of Violet, Vanity, Mamie R. and Ittie Reno by Vanguard), Golconda by imp. Glenlyon, Moula and Tyrone by imp. Mortemer, Magenta and Fabian by Duke of Magenta. Maze is the dam of Quandary (the dam of Uncertainty by Virgil or Monarchist) and Gleam by imp. Glenelg. Zoo Zoo, a winner, is the dam of Zamora by imp. Saxon, son of Beadsman, Zamor and Zobel by imp. Mortemer, and Zulu by imp. Pizarro. Zicka is the dam of Leather Stocking by imp. Moccasin, Astoria by imp. Kantaka, Zara (Lulu S.) by imp. Mortemer, and Pagan by imp. Pizarro.

Mattie Morgan was the dam of Alberta by Watson (she the dam of Mamie K. by Ballinkeel, Glen Ellen by imp. Glengarry, and Briganzee by imp. Brigadier, son of Monarque), Matagorda, a winner, by imp. Glengarry, she the dam of Enright by Enquirer, Right, Outright, etc. by Onondaga. Ave Maria was the dam of Martica by Daniel Boone, and a filly by imp. Buckden. Martica is the dam of Bric-a-Brac (dam of Miss Ryan and Algonquin by imp. Bonnie Scotland), Evelyn and sister by Enquirer, Mountain Deer by Iroquois, etc. India is the dam of Cathcart by Lisbon, Calcutta by King Alfonso, and Gallifet by Falsetto. Peggy Morgan is the dam of Austraoid (the dam of Ubet, Dornix, Mary Tindall, Assiniboine, etc.) by Red Bluff, Petite by Storey, etc. Addie Hart is the dam of Orange Girl, Beck and Bias by Bertram, son of Kentucky, and Brentano by imp. Great Tom. Dolly Morgan was the dam of Jennie Morgan,

Ambush and Hayti by imp. Australian, Stampede, a fine race horse by War Dance, Aurelius by Asteroid, Pearl Thorn, Bob Miles, Frankfort and a sister by Pat Malloy, and Despenia by King Alfonso. Jennie Morgan was the dam of Minnie T. Morgan (the dam of Gen. Monroe and Fosteral) by Enquirer, Glenival by imp. Glenelg, and Scully by Tom Bowling. Hayti is the dam of Hope, Exeter and Hazel Kirk by Lever, Roger Eastman and Virgilina by Virgil, Haytienne, Genie, Jack Batchelor, My Chick, etc. by imp. Glenelg. Pearl Thorn is the dam of Pearl Set, Pearl Rivers and Pearl Top by Falsetto. Ada C. is the dam of Pill Box, Ballot Box, Abbey, Conner, Treasure Box, Spring Box, Narcola and Lucky Baldwin by Norfolk, and Surinam by Joe Hooker. Abbey is the dam of La Scala, Porter Ashe, etc., by Joe Hooker, and Osric, a winner, by imp. Cheviot. Polly Rowe has only one line left, through her daughter Mary Good by Bill Cheatham. She produced Belle of Rockhill by Lynchburgh, son of imp. Leamington, she the dam of Mark Master and Mamie B. by Abd-el-Koree, and Daisy by Westminster.

Honesty by imp. Expedition was the dam of Tormentor, Flagellator, Gen. Jackson and Monmouth Eclipse, all good horses and winners. Monmouth Eclipse was by American Eclipse. He stood in Kentucky and died there. Honesty also produced Helen Mar by Thornton's Rattler, and Ethelinda by Marshal Bertrand, son of Marshal Duroc by Duroc. Ethelinda produced a filly by imp. Barefoot, which was the dam of Esta by Bolivar, son of Thornton's Rattler. Esta was the dam of Nellie Graves by imp. Eclipse, she the dam of Invergarry and Kate Florence by imp. Leamington, and Kate Florence, the dam of Florrie (Ione) by Pat Malloy, Gralucca by Kingfisher, and Milesia (the dam of Silvy) by Milesian. It therefore appears that a mare foaled 130 years ago, through one daughter by imp. Wildair, has left scores of descendants who are now racing and winning. All have speed and stamina, and when any of the mares in the female line to Old Slamerkin are bred to horses of pure pedigrees and racing lineage they produce race-horses. An examination of the Barb mares will show the distinguished race-horses and stallions tracing to the Layton Barb mare, to which the DeLancey Cub mare traces.

Duchess (Bland's), called also Diamond, was foaled about the year 1755 or '56 and imported to America by Gen. Alexander Spottswood. She was by the Cullen Arabian (as also was Jack of Diamonds, imported at the same time), her dam Lady Thigh (Grisewood's) by Croft's Partner, son of Jigg by Byerly Turk, grandam by Greyhound, a natural Barb, great-grandam Sophonisba's dam by the Curwen Bay Barb, great-great-grandam by D'Arcy's Chestnut Arabian, tracing through daughters of Whiteshirt, Lord D'Arcy's old Montagu, Hautboy and Brimmer to a Royal mare. Duchess was the dam of Apollo and his sister by imp. Traveler, son of Partner, a filly by Junius, filly by imp. Jack of Diamonds, True Whig by Fitzhugh's Regulus, son of imp. Fearnought, and a filly by imp. Regulus, son of Regulus by the Godolphin Arabian. This blood is transmitted to us through her two daughters, by Jack of Diamonds and imp. Regulus.

The Jack of Diamonds mare, bred to Ariel (son of imp. Traveler by Partner and imp. Selima by the Godolphin Arabian), produced a filly which was bred to imp. Fearnought and produced a filly which was bred to imp. Highflyer, son of Highflyer by Herod. This Highflyer filly, bred to Melzar, son of imp. Medley, produced Maria, and she produced in 1817 the bay mare Lady Grey by Robin Grey, son of imp. Royalist by Saltram by Eclipse. Lady Grey in 1821 produced Lucy by Orphan, son of Ball's Florizel by imp. Diomed, in 1823 bay mare Maria by Davis' Hamiltonian, son of Tayloe's Hamiltonian by imp. Diomed, and in 1826 chestnut mare Rowena by Sumpter, son of Sir Archy. Lucy was the dam of Blacknose by Medoc, a fine race-horse, Dick Chinn by Sumpter, and Lucilla by Trumpator, son of Sir Solomon. Lucilla was the dam of Carlotta by imp. Trustee, two bay fillies by imp. Tranby, son of Blacklock, and a bay filly by Old Boston. The Tranby mare, foaled in 1838, was the dam of Alaric, a fine race-horse by Mirabeau, Levity by imp. Trustee, Atala, a superior horse, by Boston, Vandal and Volga by imp. Glencoe, and Vistula by imp. Scythian. Vandal was one of the best race-horses of his day (winner of 3-mile heats in 5.36½, 5.33).

Levity was the dam of Ruric, a superior horse, and his sister by imp. Sovereign, Lightsome and Mildred by imp. Glencoe, Brenna by imp. Knight of St. George, son of Birdcatcher, Lever and Legatee, a good racer, by Lexington. Lightsome was the dam of Sprightly, Lexicon, Crucifix, Salina and Nevada by Lexington, Linda by imp. Scythian, son of Orlando, Lively, by Ringgold, son of Boston, Emmet by Asteroid, and Gulnare by imp. Glen Athol. Sprightly was bred to trotters for years, and then to imp. Billet, by whom she produced Volturno, a grand race-horse, and Elias Lawrence, a good one and winner at 3 miles at three years old in 5.28¼, Lady Way by imp. Eclipse, Avalon and Aretino (the last a grand race-horse) by Aramis, son of imp. Phaeton. Lady Way is the dam of Fox Glove by Lever, and Lady Wayward by Virgil. Linda had no chance in the stud, but was the dam of Linda Green (Yeteve's dam) by imp. Australian, and Neptunus, a good race-horse, by Planet. Crucifix was the dam of Fairplay, a good racer, by Virgil, Semper Felix by imp. Phaeton, Cruciform, Crusader and Blanche J. by imp. Australian, Quito (Cripple), The Judge (Cordova), Cortez, St. Augustine, a good horse, and La Cruz by King Alfonso, Maria D. by Lisbon, Cross Fire and St. Albans by Falsetto. Semper Felix is the dam of Semper Vive by Waverly, son of imp. Australian, Semitone (Kate J.) by imp. Buckden, Leonatus, a grand race-horse and successful stallion, by Longfellow, Latonia (Button's dam) by imp. Billet, Felicia and Semper Paratis by Stratford, etc. Semper Vive is the dam of Semper Idem and Semper Aurum by imp. Glen Athol, Fordham, Folsom and Westchester, all good race-horses, by Falsetto. Semper Idem is the dam of Long Glen, Longstreet (one of the best race-horses in America and winner of 16 out of 18 races in 1891, including match 1¼ miles in 2.07½, defeating Tenny), Longford, a good race-horse, Long Beach, etc., by Longfellow. Semitone is dam of Ella Pinkerton by Longfellow, and Semaphore, a winner, by Onondaga. Blanche J. was a good mare and the dam of Blandona (the dam of Verona, Donna and Katona) by Longfellow, Sea Shell (Long Shore's dam) and Australian Ban by imp. King Ban, Blantyre by Lelaps, etc. Maria D. is the dam of Lucy B. and Victorine by Onondaga, Mark S. (Moxie) and Ed Bell by Duke of Montrose. Salina was the dam of Duchess, Saltpetre and Saltillo by imp. Glenelg, Virlein by Virgil, Prince Charlie (Priam) and Salvator by imp. Prince Charlie. Salvator ran a mile in 1.35½, beat Tenny a match 1¼ miles in 2.05, and won the Champion Stakes, 1½ miles, in 2.35½. Nevada was the dam of Nova Zembla, Sierra Nevada, Greenland and Greengage by imp. Glengarry, the phenomenal three-year old in his year, Luke Blackburn by imp. Bonnie Scotland, Ecuador, a good horse, by Enquirer, Bradford and Bravo, both winners, by Bramble, and Neva by Longfellow. Nova Zembla is dam of Nisa and her sister by Shannon, and Superba by Flood. Sierra Nevada was the dam of Mountain Range, Monte Rosa's dam, and Wamsutta by Longfellow, Sutro by Enquirer, etc. Gulnare is the dam of Costello and the good filly Satisfaction by Sensation, etc.

Mildred was the dam of Zoe and Mittie by imp. Scythian (the latter dam of Young Australian by imp. Australian, and Gen. Longstreet by Asteroid II.), Ringlet by Ringgold, she the dam of Reckless and Restless by Voucher, Debby, Jim Lupe, Debbie (Flora) and Epsom by Derby, son of imp. Eclipse, Maximilian, Wandering, J. J. Healy and Waterbury (Rambler) by Wanderer. Mildred produced Nora Worth by imp. Eclipse, she the dam of Nero, Rocket, Roxella, North West, Startle, Trumps, Darkness, Renown, Force and Finality by West Roxbury, and Adventure 2d by Amadis, the majority of which were good racers. Rocket is the dam of Rocket II. by imp. Intruder (she the dam of Minnie W. and Lady Sumner), Penelope, Nannie S., Leo by Pantaloon, Ericson and Cora A. by imp. Eric. Roxella was the dam of Molly Malloy (Queen Malloy) by Pat Malloy, Fleet and Roxella 2d by imp. Intruder, and Amadias by Amadis. Molly Malloy is the dam of Pequot, Anna S., Birdette and Al by Pantaloon, and Voltalloy by Voltigeur, Dick Malloy and Sam Murphy by imp. Eric, son of Pretender. Fleet is dam of Athlete by Amadis, and Lilly and others by Pantaloon. Darkness is the dam of Fleetness and Interloper by imp. Intruder, etc. Fleetness is dam of Alert by Amadis, Dave S., Millie C. and Sam by Pantaloon. Minx is the dam of Minnock, the great gelding Monitor, Lotta C., Mirth and Romeo, the two last in England. Mirth was the dam of Bouthillier, a winner in England last year. Minnock is the dam of Salviator, Lady B., Minnarette, Selma D. by Springbok, and Onondagas by Onondaga, etc. Lotta C. is the dam of Roulette (Sefton's dam), Illisee and Lizzie M. by Baden Baden. Stamps was the dam of Katie Pearce by imp. Leamington, she the dam of Lizzie S. (Extra's dam) by Wanderer, Ballard by imp. Billet, Katrina by Virgil, and Dal by imp. Mortemer. Bertha, Loch Tanna (in England), Louisette, Preciosa, Billy Pinkerton by imp. Glenelg, and Valleria by imp. Glenelg or Virgil, unquestionably by the latter. Bertha is the dam of Reina Victoria by imp. Saxon, Delphine and Saucy Lass by imp. Mortemer, and Mohican by Iroquois. Valleria is the dam of Baylight, Laura Stone and Versatile by imp. Rayon d'Or. Sultana is the dam of Sovrana (Inah's dam) and Sultan by imp. Ill Used, Soudan by imp. Matador, Saladin by Nonesuch or imp. Ill Used, she by imp. Glenelg, Padishah, Semiramis and Caliph by imp. St. Blaise.

Brenna was the dam of Tibbie Dunbar by imp. Bonnie Scotland, she the dam of Voltigeur, Lena Dunbar (dam of Lena's First), Clifton Bell, Capt. Kidd, Ed Smith, Canny Scott, Broadchurch (Bradshaw) and Lady Evangeline by Leinster, and Lizzie Dunbar, the dam of May Dunbar, by Bazar. Lady Evangeline is the dam of Valiant by Monday, Oceanica, Floodtide, Evelina and a sister by Flood. Brenna was also the dam of Harie by imp. Australian, a good race-horse, Eclair (Factory Girl) by Lightning, Ella Hankins by Gilroy, Britzka by imp Phaeton, Brademante by War Dance, Athalaric by Gilroy, Strathmore, Kinkead and Sir Walter by Waverly, Brenda by imp. Thunderstorm, Long Stop by Longfellow, etc. Ella Hankins is the dam of Amadine by Amadis, she the dam of Hilarity by Punster, and Polly S. by imp. Pizarro, and Oriana by Onondaga, bay filly by Pizarro, etc. Athalaric is the dam of Tom Martin, a superior horse, by Longfellow, Belle of Brooklyn and Aethel by Stratford. Sister to Ruric by imp. Sovereign was the dam of Kelpie by imp. Bonnie Scotland, Ermengarde, Hilderic (Jim Connor) and Genseric by Lightning, Grinstead, a fine race-horse and successful stallion, Tilford and Viceroy, a good horse, by Gilroy, Pat Farrell and Hypatia by Waverly. Kelpie was the dam of Miss Austine and Janet (winner at four miles in 7.25) by Lightning, Maxwell by Gilroy, a filly and Duke of Montrose, a good racer and successful stallion, by Waverly, Annette and Kelp by imp. Strachino, Ovation, Overture and Dalgetty by Onondaga. Miss Austine is the dam of Austina by Waverly. Bridesmaid by imp. Bonnie Scotland, Aconite (Cryolete's dam) by Ten Broeck, Teuton, a good one, by Ten Broeck, and Faraday, a superior colt, by Himyar. Janet is the dam of Sinaloa, Volunteer (Duffer), Sonora's dam by Ten Broeck, Spaldie by imp. Blue Mantle, La Mascotte, Golden Reel and others by Spendthrift. Annette is the dam of Shoemaker, Ortowin, etc., by Onondaga. Kelp is the dam of Kempie, a winner, by St. Martin, Keepsake, Onaway, Upward, Ondawa, etc., by Onondaga. Ermengarde is the dam of Brocade, Bingen, Baltic and Bedotte by imp. Bonnie Scotland, Belle Walsh (Tuscumbia) by imp. Great Tom, Bellevue (Blackburn and Belvidere) and sister by Bramble. Brocade is the dam of Bertha B. by imp. Buckden, the gelding Burch, War Dress by War Dance, Long Cloth and Jennie B. by Longfellow, Annie C. by Enquirer, and Dodo by Falsetto. Bertha B. is the dam of Lizzie Baker by Longfellow, Judge Murray by Ten Broeck, Tom Vaughn (George H.) by King Alfonso, Pat Kelly by Pat Malloy, etc. Bedotte is the dam of Still Alarm, Goosequill by Pardee, etc. Hypatia is the dam of Slickaway by St. Martin, Hypocrite by Longfellow, Deceit by Ten Broeck, Escapade by Onondaga, etc. Volga was the dam of Rurica by Ruric, Eminence, Barney Williams, a good horse, Valdine, Invoice and Evadne by Lexington, Jenny McKinney and Alice Mitchell by Planet, Tecalco by imp. Glen Athol, etc.

Evadne was the dam of Evasive by imp. Canwell, son of Stockwell, Blink Bonnie, Buckingham, Bulwark, and Bandana by imp. Bonnie Scotland, Evasive (Bruce's dam) by John Morgan, Tennyson by imp. Great Tom, Eve Effingham by Enquirer, Montague by imp. Mortemer, Eblis, etc., by Iroquois. Blink Bonnie was the dam of Saxony by imp. Saxon, and Tullahoma by Great Tom. Saxony is the dam of Erebus and Once a Week, geldings, Tatian by Great Tom, etc. Tullahoma is the dam of the good mare Tulla Blackburn by Luke Blackburn, Tammany by Iroquois, etc. Bandana is the dam of Banner Bearer (gelded), Napa and Equipoise (Equality's dam) by Enquirer, Talisman by imp. Great Tom, Saranac, Emblem and Red Banner, that is one of the best two-year old fillies in the West, by Iroquois, etc. Jenny McKinney was the dam of Ellen Alice by Barney Williams, Claudia by War Dance, Buccaneer by imp. Buckden, Poet by Longfellow, Omaha and Vassar by Tom Ochiltree. Ellen Alice is the dam of Vigiline and Purdy by Virgil, and Vigiline is the dam of Vigilant, Vigilette by Eolus, while Claudia is the dam of May D. by Voltigeur, son of Vandal, Umpire and his sister by Versailles, son of Vandal, Catalong by imp. Rayon d'Or, etc. Invoice is the dam of Fatima by Abd-el-Kader, and is dead. Tecalco is the dam of Tolu by Ten Broeck, Rock-and-Rye by Virgilian, Tonique by Sensation, Hot Scotch by imp. Hopeful, The Wandering Jew by Tom Ochiltree, etc.

Vistula had no chance in the stud, but was the dam of Kentucky Belle by Lexington, she the dam of Pride, Vis-a-Vis and Himalaya, a good race-horse, by Virgil. Pride is the dam of Glitter and Pretence by imp. Glenelg. Maria by Davis Hamiltonian was the dam of Catharine Ogle by Trumpator, Tom Benton and Catherine by Bertrand, the latter the dam of the superior race-horse Alexander Churchill by imp. Zingonee. Rowena by Sumpter was the dam of Celestian, a good racer, and Merry Gold by Sir Leslie, Alice Carneal by imp. Sarpedon, a good race-mare, Sarah Moreton by Sidi Hamet, a superior mare, and Boswig by imp. Hedgeford. Merry Gold was the dam of Hebe and Vulcan (Tom Nichols) by Berthune, son of Sidi Hamet, Cyclone and Magelone by Vandal, and Almeria by Jack Lane, son of imp. Yorkshire. There is no report of Hebe since 1875. Cyclone was the dam of Red Wing by imp. Balrownie, Cycloid by imp. Lapidist, and a bay filly by Kentucky. Magelone was the dam of Lily of the Valley by Lightning, and Van Billet by imp. Billet. Lily of the Valley was the dam of Kenilworth by Waverly, Frank Martin by St. Martin, Blue Times by imp. Blue Mantle, and Greynell by imp. Glenelg, the latter the dam of a filly by imp. Deceiver, a colt by Hindoo, etc.

Alice Carneal founded a family of her own. She was a good mare at all distances, was the dam of Miss Trustee by imp. Trustee, Fance and Grey Alice by Chorister, son of imp. Contract by Catton, Lexington (the renowned racer, winner of the time match, four miles, in 7.19¾, and afterwards defeating Lecompte in 7.23¾, the fastest two heats of four miles ever run by one horse) by Boston, Didie (Maid of Orleans), Release and Rescue by Berthune, Waxy, a fine race-horse, by Buford, Lavender by Wagner, Umpire by Lecompte, and Annette by imp. Scythian, the last two sent to England. Miss Trustee was the dam of Florizel (Hurricane) by imp. Belshazzar, Mystery by Euric, a bay filly by Ringgold, and Miss Trustee by J. C. Breckenrige, son of Lexington. Mystery was the dam of a bay filly by Revolver, Dil Wiggins, Winchester, and a chestnut filly by imp. Hurrah. Fance was the dam of Bazil by imp. Sovereign, Anne Tarlton and Lady Vandal by Vandal. Anne Tarlton was the dam of Charity by imp. Hurrah, Minor Jackson by Dick Jackson, Victress and Vestvalli by Victory, son of Uncle Vic. Victress is the dam of Vanguard, a good racer, Victrix by Longfellow, and a colt by Stratford. Vestvalli is the dam of Wanda West, Veracity (Festina's dam), Vestella, Vesper Bells and Valedictory by Wanderer. Wanda West is the dam of Speedwest by Spendthrift, and Imogene by Powhattan. Vestella is dam of Mahomedan by Hindoo, etc. Grey Alice was the dam of Pickaway by Wagner, and Nellie Gwynn by imp. Bonnie Scotland. Nellie Gwynn was the dam of Louise Gwynn and Faith by imp. Hurrah. Louise Gwynn is the dam of Grace Lee by imp. Kyrle Daly, and Hector by Grenada. Release was the dam of Carrie P. by Norbourne, son of Lecompte, Hetty C. by Uncle Jeff, and Night Rose by imp. Knight of St. George. Carrie P. is the dam of Cape Sterling, Cape Henry, Cedar Grove and Carrie Race by Cape Race, son of Lexington. Night Rose was the dam of Nightmare by imp. Australian, Voltagne by imp. Glen Athol, Fred Fogle by Planet, Haddisco by Pat Malloy, Johnny E. (Pocassett) and filly by Powhattan. Rescue was the dam of Abu Beeker by Mahomet, Relief (Maid of the Cave) by Ringgold, Relief by Star Davis, Remorse by imp. Eclipse, Abd-el-Kader, Abd-el-Koree, a good race-horse, and Rigmarole by imp. Australian. Rigmarole was the dam of Regale by Bulletin, Rigmal by Harry of the West, Riglin and Bessie Franklin by imp. Glengarry, Carmen (Blue

Jeans) by Gorlitza, son of Lexington, Merry Dance by War Dance, etc. Riglin is the dam of Rowland, Regina and Royal by Shannon, Chancellor by Monday, Gladys by Flood, etc. Bessie Franklin is the dam of Gen. Early and I. O. by Erlanger, son of Enquirer. Carmen is the dam of Blue Streak and Blue Wind by Bingaman, son of Asteroid. Lavender was the dam of Rock by Ruric, Nannie Letcher, Helmbold, a superior race-horse, Bob Shelton, a winner, Lava, Barricade, a winner, and Baden Baden, a superior horse, by imp. Australian, Lax by imp. Scythian, George Wallace by Asteroid, Buchu by Planet, and Glen Wood by imported Glen Athol. Nannie Letcher was the dam of Corinne by Planet (she the dam of Rapidity, Sweetheart, America and Lady Rapture by imp. Rapture), Mignonne (dam of Rapture's Daughter), and Emma G. (dam of Hermit by Hurrah) by imp. Phaeton. Lax was the dam of Sea Breeze by King Lear, she the dam of Slow Breeze by Restless, she the dam of Golden Slipper by McCreery, and she the dam of Le Premier by Lucifer, Verge d'Or by Virginius, and Crystallite by Joquita, Bragelone and Neppie Moore by Baywood, the latter the dam of Trosach by Terror, son of Alarm, Virgie by Virgil, she the dam of Ona by Onondaga, Virgin and Virgin 2d by Milner, son of imp. Leamington, and Release by imp. Strachino.

The Regulus mare out of Duchess produced a filly by True Whig, which produced a filly by Gallant, son of imp. Fearnought. This filly produced a filly by Rockingham, son of Partner by imp. Traveler, which Rockingham mare was the dam of Annette by imp. Shark, son of Marske, sire of English Eclipse. Annette was the dam of Nancy Air by imp. Bedford by Dungannon by Eclipse, and Maid of the Oaks by imp. Spread Eagle, son of Volunteer by Eclipse. Nancy Air was the dam of Transport, Poor Girl and Young Nancy Air by Virginius son of imp. Diomed, and Seagull by Sir Archy. Transport was the dam of Sir Archy Monterio (Big Archy), and Sir William of Transport, by Sir Archy, Bertrand Jr. and Julia by Bertrand, son of Sir Archy, and Clara Fisher by Virginius. The female line from Nancy Air has became extinct.

Maid of the Oaks was the dam of Maid of the Oaks and Lady of the Lake by Hickory, son of imp. Whip, Marshal Duroc and Cinderella by Duroc, son of imp. Diomed, Forest Maid by Ratray, son of imp. Clifden, Young Maid of the Oaks by imp. Expedition, son of Pegasus by Eclipse, Orphan Boy by American Eclipse, and Lionella by imp. Cœur de Lion. Lady of the Lake was the dam of Lady Jackson, Maryland Eclipse, and Misfortune by American Eclipse. The line from here is extinct. Cinderella was the dam of Modesty by imp. Expedition (she the dam of Jane Maria and Alpine by Henry (this line is extinct), Betsey Richard (Young) by John Richards (line extinct), and Lalla Rookh by Ogle's Oscar, son of imp. Gabriel. Lalla Rookh was the dam of Hamilton Moore, Tom Moore and Mark Moore by American Eclipse, Henry Moore, Jim Moore and No Moore by Henry, son of Sir Archy, Cerito by imp. Priam and Hinda by Monmouth Eclipse, the latter the dam of Fanny by Shark, Jr., son of Shark by American Eclipse, she the dam of White Oak and Warbeshaw by Young Langford. This female line is extinct. Cinderella also produced Cornelia Conover and Celeste by Henry. Celeste left nothing in the female line. Cornelia Conover was the dam of Shark, Jr. by Shark, Heiress and Grace Darling by imp. Trustee, and the latter was the dam of Countess by Truxton, son of imp. Barefoot. Countess was the dam of Nettie by Kenneth, son of imp. Yorkshire, and Nellie by Touchstone, son of imp. Lapidist. Nettie was the dam of Exotic by Extra, son of Endorser, Fanny Wizer by Terror, son of Ruric, she the dam of Stilton by imp. Strachino, etc. Nellie is the dam of Una by Gen. Custer, son of Donerail, and Bothwell by Montrose, son of Duke of Montrose. Young Maid of the Oaks was the dam of Miss Walton by Mendoza, brother to Sir Walter, Katy Ann by Ogle's Oscar, son of imp. Gabriel, Highland Mary, the noted and popular race-horse and stallion Medoc, Cora and Gipsy by American Eclipse, Coronet by Henry, and a mare by Gohanna that was dam of Fanny Bouton by imp. Trustee. Miss Walton was the dam of Goliah, Floranthe and Zela by American Eclipse, Dosoris by Henry, and Moustache by imp. Trustee. This family line is extinct.

Katy Ann was the dam of Lady Thompkins by American Eclipse, Maid of Miami by Orphan Boy, Fleur de Lis by imp. Valentine, son of Magistrate, and Butterfly by imp. Skylark. Lady Thompkins was the dam of Tranbyana by imp. Tranby, Motto by imp. Barefoot, Sally Ann by imp. Trustee, Fiatt (the Blue filly) by imp. Hedgeford, and a filly by Cadmus. Tranbyana was the dam of Belle Lewis, Bay Dick and Maggie Cragg by imp. Glencoe, a filly by Wagner, Big Boston by Boston, and Altoona by Vincent Nolte, son of American Eclipse. Belle Lewis was the dam of Neil Robinson, a good race-horse, by Wagner, Bonnie (dam of Susie Allen), Luther by Lexington, Rurica by Rurie, and several foals by Trotter. This line has run out. Maggie Cragg was the dam of Carrie Cosby by Oliver, son of Wagner, and she the dam of May D. (May Beverly), Lady Cripple, Docility and Stella by Hunter's Lexington, Creola by Creedmore, son of Asteroid, and Wickland by Judge Wickliffe. May Beverly was the dam of Maid of Woodland by John Morgan, Proctor Knott by King Pynn and Actress by Longbow. Maid of Woodlands was dam of Birdie Ray by King Pynn, Maid of Richland by Voltigeur, etc. Stella is the dam of Frank Ward, a good one, Stelletta, Larry Farrell, Sam Ross. Crissie, etc., by Voltigeur, son of Vandal. Creola is the dam of Cretonia, Trown, etc. by Voltigeur.

Motto was the dam of Red Oak, Nannie Lewis, a real good one (dam of Aldabaran), Sally Lewis, Gov. Wickliffe and My Lady by imp. Glencoe, Melita Hunter by Lexington, and Raglan by imp. Scythian. Sally Lewis was the dam of John Morgan, a superior race-horse, by imp. Sovereign, Minnie Lewis by Revenue, Lotta by Hunter's Glencoe, Linda Lewis by Oliver, Hunter's Lexington, Glenrose, Susan Beane and Acrobat, a good racer, by Lexington, and Planivoo by Planet. Nannie Lewis left nothing in the female line. Minnie Lewis was the dam of Lorentia by imp. Australian, Melrose and Alf Riley by Asteroid, Mollie McGinley by imp. Glen Athol, Pinback by King Pynn, and Zuma by Longbow. Lorentia was the dam of Willie Hinkle by Rogers, son of Lexington, King Cole by King Lear, Anemone by King Pynn, son of imp. Australian, Leesburg by Lever, McBowling by Tom Bowling. Four Tens by Ten Broeck, Laureate and Unlucky by Leonatus. Willie Hinkle is the dam of Bessie Hinckley by imp. Buckden, Weeks by Ferncliff, Hanley S. and her sister by Erdenheim. Anemone is the dam of Delaware (Dell) by Little Jim, son of Planet, and she the dam of Bullrush by Stampede. Zuma is the dam of Marion C. and her sister by Harry O'Fallon. Lotta was the dam of Vixen by Planet, Glenmore (the superior race-horse, winner of the great four mile race, three heats, 7.29½, 7.30¼, 7.31, the fastest three heats ever made in a race), Greenfield, Lottery and Gladiator by imp. Glen Athol, and a filly by Powhattan, Linda Lewis was the dam of Nazareth by Asteroid, Kioba and Quiver by Longbow, Linden, a superior race-horse and very promising stallion, Withrow by Longfellow, and Linnie by Macduff. Glenrose was the dam of Glendora and Glendew by imp. Glengarry, Biloxi and Buckthorn by imp. Bonnie Scotland, and Restless, a good mare (dam of Repose), by imp. Mortemer. Glendora is the dam of Emblem (dam of Lottie S.), Endora by Enquirer and Beaver by Bramble. Glendew is the dam of Glendair by Norfolk. Guardsman by Monday, Guenne, Geoffrey and Glenlivet by Flood. Susan Beane was the dam of Stratford (Leamington II.), a superior race-horse, the unbeaten Sensation, Onondaga, a good racer and sire, Susquehanna, a good mare, and Sioux. all by imp. Leamington, Soubrette and Sir Hercules by Alarm. Susquehanna is the dam of Sussex by imp. Ill Used, Susanna by Kingfisher, Chesapeake, a winner, the great Potomac, and Schuylkill by imp. St. Blaise. Soubrette is the dam of Darlington by Iroquois, John Lakeland by Runnymede, etc.

My Lady was the dam of Onward, a superior race-horse, by Ringgold, Blandina and Marguerite by Lexington, Maggie Welch by Asteroid, Clifton by Planet, a filly by King Alfonso, and Fingal by Lisbon. Blandina was the dam of Calash by imp. Phaeton, Red Bluff by imp. Australian, Christina, Bresica (Polina's dam) and Ulsie by King Alfonso, Sea Foam by Lisbon, False Note, Keynote and Falsita by Falsetto. Calash is the dam of Albemarle and Tar Heel by Abd-el-Koree, Elite, Etha, Eolian, a very speedy good racer, Ethie, Clark, Ethus, Cantey, Cante and Caleche, all by Eolus.

Elite is the dam of the winners Capstone, Hetty Carey (Katie C.) and Annie G. by imp. Charaxus. Elsie is the dam of the good fillies Bandusia, Emma G., etc., by Bend'Or, son of imp. Buckden. Marguerite was the dam of Foddie Pope by Planet, she the dam of Katura II. by Silent Friend, and she the dam of Melodine by Melodist, Prude, Katie Creel (Kenwood's dam), Azalea, Macola and Nettie by King Alfonso, and Lady Athol by imp. Glen Athol. Azalea is the dam of Gazalea and Pinkie T. by imp. Glengarry. Melita Hunter left nothing in the female line. Sally Ann by imp. Trustee was the dam of Harper by imp. Glencoe, Sandhurst by Lexington, and Kate Quinn by Wagner, she the dam of Herman by Vincente Nolte, Quincy by Ruric, filly by Lexington, one by Ulverston, Wallaby Jack and his sister by imp. Australian, Asteroid II. by Asteroid, colt Australian by Young Australian.

Fiatt (the Blue filly) was the dam of Blue Bell by Chorister, son of imp. Contract by Catton, Kate Hunter, Liz Morgan, Hunter's Glencoe and a sister by imp. Glencoe. Blue Bell was the dam of Lunette and Cygnet by Big Boston, Grey Don by Hornpipe, son of imp. Glencoe, Nellie Grey and Betsey Hunter by Oliver, son of Wagner, Julia Mattingly, Songstress and Songbird by John Morgan, Cruiskeen, Bessie Lee, Specie Payment and Blue Bird (Breastpin's dam) by Hunter's Lexington. Lunette was dam of Rebel Morgan by John Morgan. There is no female from either her or Cygnet. Nellie Grey was the dam of Sallie Newton and Nannie May (dam of Sue Willie) by John Morgan, Kate Adams, Lizzie Hayden and Betty Magruder by Hunter's Lexington, Lunette by Longfellow, Malise by Waverly, Lilly Grey by Judge Wickliffe, Ellen Terry (Lillie Smith) and a sister by Longbow. Sallie Newton was the dam of Modesty (Ruth) by Revolver (she the dam of Blush, Laura Keene and Ruth by Ramadan), Eldorado and Little Criss (the dam of Bonnie S. by Planter, Jim Clare by Granger, Ilma B. and her sister by John Harper), both by Revolver, Capt. Flagherty, Lilly B., Autumn Leaf by imp. Hurrah, Sallie Harper and a sister by John Harper, and Frank E. by Probability. Betty Magruder is the dam of Becky B. (Becky Bye) and Maggie P. by Longbow, Lucius by Harry O'Fallon, and St. Bridget by Leonatus. Betsey Hunter was the dam of Kate Mattingly by John Morgan, Chasseuse, Joe Johnson, Fanny Johnson, Aunt Winnie and Judge Wickliffe by Hunter's Lexington, Inquire by Enquirer, Katharine M. by imp. Glen Athol, and Miss Hunter (Annex) by King Alfonso. Kate Mattingly was the dam of Hybla by Hunter's Lexington (she the dam of King Pippin by King Pynn, Huntress by Bob Woolley, Susie Forbes and a sister by Rebel), Long Girl by Longfellow, Herbert (Tom Plunkett) by imp. Glenelg, Matinee and Miss Mattie by imp. Buckden, and The Maid by Erdenheim. Chasseuse was the dam of Dilsey by Hunter's Lexington (she the dam of McMurtry by Miser, Mamie Cole by Harry O'Fallon, and Miss Gibson by imp. King Ban), Little Planet by Planet, Florence D. (Longchase) by Longfellow (she the dam of Debut by Pat Malloy, and G. W. C. by Little Ruffin). Fanny Johnson is the dam of Sallie Jennings by Ortolan, son of Donerail, Frank B. by Longfellow, Via (Lolion's dam), Full Moon, Virago and Nina Darnell by Vagabond, son of Vandal. Sallie Jennings is the dam of Carolina by Harry Lamar, Vagantes by Vagabond, Bonnie Blaze by George Kinney, Hattie O'Fallon and Jim Guest by Harry O'Fallon. Julia Mattingly was the dam of Fanny Mattingly, Belle of Nelson and Belle of Maywood by Hunter's Lexington, Wanderess by Wanderer, and Alarmist (Fright's dam) by Alarm. Fanny Mattingly is the dam of Glendale by imp. Glenelg, Equity, a sister, and Mirabeau by Enquirer, Turnbull by imp. Great Tom, Minnie L. and other foals by Iroquois. Equity is the dam of Justice by imp. Woodlands, and a filly and colt by imp. Dalnacardoch. Belle of Nelson is dam of Marion by Malise, and Belle Noyes by Grenada. Belle of Maywood is dam of Branch (Stiletto) and Mayrose by Duke of Montrose, the great race-horse Tenny and Maywood by imp. Rayon d'Or, and Rafter by imp. Kantaka. Songbird is the dam of Bob May and May Bird by Hunter's Lexington. Songstress was the dam of Julia Jackson (Pat Clayton's dam) and Little Jack by Hunter's Lexington. Bessie Lee was the dam of Orinoca by Ortolan, Procrastination by Bulwark, Vicinity and Dot by Vagabond. Lizzy Morgan was the dam of Mollie Fisher by imp. Knight of St. George, Sally Morgan by Revenue, Morgan Scout and Ellen C.

by John Morgan, Ratan by Lexington, Rosy Morgan by Planet, and Minnie B. by imp. Glen Athol. Mollie Fisher was the dam of Morgiana by John Morgan, Nellie Knight, Maud Hampton and Lida Knight by Hunter's Lexington, Pearl by War Dance, and Lamar by King Pynn. Morgiana was the dam of Jewel by War Dance, Pauline Lee by Lee Paul, son of Lexington, Conn by Daniel Boone, Moriana by Longbow, Harry Fields, Donnell and a sister by Harry O'Fallon. Jewel is the dam of Duette by King Alfonso, Leah Sobèl by imp. Buckden, Ban Dance and Jewel Ban by imp. King Ban, Jewel II. and Mr. Appleby by Himyar, and a chestnut filly by Alarm. Duette is the dam of Opera and Oratorio by Harry O'Fallon, Seraphina by imp. Siddartha, and Blue Maid by Blue Eyes. Pearl was the dam of Kossuth by Londonderry, and Banka by imp. King Ban. Maud Hampton is the dam of Creton by Judge Wickliffe, the grand race horses Ban Fox, King Fox, King Thomas, Ethel and Mabel by imp. King Ban, a colt by imp. St. Blaise, and a filly by imp. Sir Modred. Creton is the dam of Jennie S. by Fellowcraft, Clean Heels by Duke of Montrose, and Ragnarok by imp. Deceiver.

Sally Morgan was the dam of Maggie Morgan and Charity Morgan (King Cyrus' dam) by Hunter's Lexington, Jack Trigg by Lightning, Longbow, Sozodont, Longway and Miss Nelson by Longfellow, Vrill by imp. Glen Athol, and Enterprise by Enquirer. Maggie Morgan was the dam of Janette by John Morgan, Tolima by imp. Glen Athol, Orelia by Waverly, Anxiety by Alarm, La Cigale by King Alfonso, and Vender by Virgil. Tolima is the dam of Curtolima (Kitestring's dam) by Judge Curtis, Maggie Bruce by Bill Bruce, Glen Fox by Caligula, Strathlene by imp. Strachino, Countess and Pentolima by imp. Newcourt, and Ogestoh by Lisbon. Orelia is the dam of Ram Lal (Eugene Brodie), Mr. Isaacs and Yolande by imp. Glen Athol, and Laura Gie by Falsetto. Anxiety is the dam of Remsen by King Alfonso, and Lizzie Sauter by Falsetto. La Cigale is the dam of Maj. Mulford by Longfellow. Vrill was dam of Merter by Revolver, she the dam of Dew Drop by Chillicothe, son of Lexington, and Cyclon by Chillicothe. Sozodont is the dam of Doolittle by King Alfonso, Ursula by Duke of Montrose, So So by Longfield, and a filly by imp. Darebin. Ratan was dam of Cuba, Lancewood, Girollé and Girolla by imp. Leamington, and Wildfire by Wildidle (she the dam of Edgewood by Enquirer, Ithaca, and a sister by imp. Saxon). Rosy Morgan was the dam of Maggie D. (the dam of Joe Harris and Twang), and Guadalupe by Pilgrim, Sam Bennett (Rebel Scout) and Nettie (Miss Hawkins) by Rebel, and Morgan Scout, Jr., by Silent Friend, son of imp. Australian. Nettie is the dam of Rebel Planet by Wade Hampton, son of Rodolph by Gen. Twiggs, Charleie by Blue Dick, Waneta and Ella Trent by Plusetto.

Fleur de Lis by imp. Valentine was the dam of Buckeye Lass by Admiral, son of imp. Barefoot. Buckeye Lass was the dam of Woodnymph by Wagner, Monarch, Jr., by imp. Monarch, Romance by imp. Scythian, Ella Boston by Lexington, and Waxy (Bertha) by imp. Cruiser. Romance was the dam of Prelude by imp. Cruiser, Nellie Boston by Frank Boston, son of Lexington, Roscola by Daniel the Prophet, and Grafton by Virginian, son of Ringgold. Prelude was the dam of Tom Boston by Frank Boston, Lithe and Preface by imp. Hurrah. Lithe is the dam of Lithsome and her sister by Barnton. Nellie Boston is the dam of Fannie Shields, Maggie Shields (Peruna's dam), Woodbine, Valentine and Woodstock, Jr., by Woodstock, Cynthia Rose by Red Eagle, son of Grey Eagle, and Glen Daly by imp. Kyrle Daly. Fannie Shields was the dam of Sattie Seymour by imp. Hurrah, Good Friday by imp. Kyrle Daly, Belle of Licking and Siddie Mitchell by Grenada. Waxy was the dam of Wexford (Turenne), Dan O'Hara, Bertha May, Mattie G., Harry Wexford and Lady Langtry by Woodstock, son of imp. Australian (the latter the dam of Katie Mac, Benedict and Nivalis by Bertram, son of Kentucky), and Lolah (dam of Susy T.) by Northumberland, brother to Norfolk. Cora by American Eclipse was the dam of 2d Priam and Lucy Toland by imp. Priam, 3d Boston by Cost Johnson, son of Boston, and a chestnut mare by Wagner, she the dam of Ellen Fall and Adele by imp. Albion, Balmoral by Lexington, and Jesse Joiner by Bill Cheatham. Ellen Fall was the dam of Josaphine Saunders by Hiawatha (she the dam of Chili McIntosh, Lawproof and a sister by Victory), and Watsona by Watson, she the dam of Allman and Vina by Cœlebs, son of Pat Malloy. Adele was the dam of Fanny Moore by imp. Bonnie Scotland, Gladys by Ringmaster, son of Ringgold, Ranch-Wai-Me, Rotha, Blasco and Wah-ta-Wah by J. C. Breckenridge, son of Lexington, Wah-O-Naisa by Plowman, Fair Count, Wedding Day and Finale by Dudley, son of Planet, and Wahnooka by Bob Woolley. Fannie Moore was the dam of Lady Woodford by Woodford, son of Ringgold, Spiteful by Edgecomb, son of Lexington, Harmony and Nettie Martin by Concord, son of Lexington. Lady Woodford was the dam of Paris Green, Bonnie Bruce and Bonnie Lee by Scottish Eclipse, son of imp. Bonnie Scotland, Mollie Bawn, Morna Woodford and Sallie Hagen by Faustus, Yolande II. by Aliunde, and Hagen by Pat Farrell. Paris Green was the dam of Faustina (dam of Spanish Fly and Dick Whittington) by Faustus, and Gretna Green by Moscow, son of War Dance. Bonnie Bruce was the dam of Froila and Bobby Beach by Faustus. Bonnie Lee is the dam of Fausterman, Jack Harding and Bessie Briggs by Faustus. Spiteful is the dam of Leila Belle, Mary Bet. Chaney and Ella S. by Concord. Old Mike and George Morrow by Duke of Kent. Leila Bell is dam of Ella Rogers by Duke of Kent, and Mrs. Evans by Ten Stone. Mary Bet. Chaney is dam of Effaline and West Anna by Duke of Kent. Nettie Martin is dam of Nettie Kent, Nannie P. and Capt. Ruby by Duke of Kent, and Gracie M. by Ten Stone. Gladys is the dam of Cultivator by Plowman, Miss-Go-Lively (Hatty Cheeney's dam), Oakleaf, Raton and Yelta by Dudley, and Joe. O. Set by Vocalic. Ranch-Wai-Me is the dam of Zotum, Hatto, Alveda, Zi-Zi, Dover and a sister by Dudley. Wah-ta-Wah is the dam of Eleanore, Toolooah and Count Dudley by Dudley. Wah-O-Naisa is the dam of Wan-o-sae-na, Hard to Tell (Duster's dam) and Nekama by Dudley. Balmoral was the dam of Miss Alma by Bill Cheatham, she the dam of Annie Malloy by Nightshade, Lizzie McDonald by Escape, and Bitter Sweet by Crichton, son of imp. Glencoe, she the dam of Icewitch by Iceberg. Lochiel and Lew Joslyn by imp. Lochiel, son of Sweetmeat.

Gipsy, sister to Cora, was the dam of Niagara and his sister by imp. Trustee, Sunbeam by imp. Langford, Young Gipsy by imp. Mercer. Pryor and his sisters, Nos. 1 and 2, by imp. Glencoe. Pryor was a superior race-horse and died in England. The blood of this branch of the Duchess family is transmitted through the two sisters of Pryor and Young Gipsy. The latter was the dam of Bosio by imp. Eclipse, Feathers and Gipsy Lass by Revenue. Bosio was the dam of Rivulet by Rivoli, she the dam of Clematis (Rill) by Creedmore, son of Asteroid, Cornucopia by Virgil, a filly by Flood, and Modesto by Monday. Bosio also produced Heron and Emma Hunter by Lever. Emma Hunter is the dam of Tom Stacey by Lisbon, Walter T. (Elk Hunter) by Falsetto, Stranger and Avon Lass by Stratford. Sister to Pryor No. 1 was the dam of Glycera and Colossus, both good racers, by imp. Sovereign, Susie Spears by imp. Arab Fysaul, Lute, sent to England, and Saratoga by imp. Knight of St. George. Glycera was the dam of Knighthood and Sue Ryder by imp. Knight of St. George, Merry Bird by imp. Mickey Free, Carrie Atherton by Lexington, Mary Edith by imp. Fazzoletto, Warpath, The Gleaner and Cordelia by War Dance. Sue Ryder was the dam of Ranee, Bonne Bouche and Remus by Reform, Alistair by Alarm, and Cracksman by imp. Woodlands. Ranee is the dam of Maggie Ahrens, Woodranie, Register and Summer Coon by imp. Woodlands, and Waiting (Miss Doloris) by imp. Dalnacardoch. Merry Bird was the dam of Annie Richards by imp. Bonnie Scotland, she the dam of Boomerang, Festus, Nodaway Queen, etc., by Faustus. Carrie Atherton was the dam of Janet Norton and Duke by imp. Leamington, Nortonian by Panic, Glenroy and Nettie Norton by imp. Glenelg, Buttress by Barricade, Marvine (Canteen's dam) and Mont d'Or by imp. Mortemer. Janet Norton is the dam of Free Love and Keyser by Luke Blackburn. Mary Edith had a filly by War Dance, and Cordelia was the dam of Mark Daly by imp. Kyrle Daly, and Polaris by imp. Thunderstorm. Susie Spears was dam of Mamie M. by War Dance, she the dam of Eastover (Elkridge's dam) by John Payne, and Sue Casey by Hermit, son of Warwick. Saratoga was the dam of Sweet Sixteen, Monmouth, Simoon, Congressman and Claudina by War Dance. Sweet Sixteen was the dam of Limo by Limestone, son of War Dance, and she the dam of Princess Limo by imp. Prince Charlie, and Limo Brook by Ten Broeck.

etc. Simoon is the dam of Moonlight and Moonshine by Princeton, son of Oakland, Half Moon and Rosa Maybud by imp. Strachino. Sister to Pryor No. 2 was the dam of Reporter and Summerside by Lexington, both sent to England, Alumina by imp. Eclipse, John Kilgour by imp. Mickey Free, Sister of Charity and Buckshot by imp. Knight of St. George, Wheatly, Lord Jersey and Midsummer by War Dance. Alumina was the dam of Leona and sister by War Dance, Mettle and Milan by Melbourne, Jr., Alchemy, Fairfield, Amalgam, Oxygen and Virginia Keene by Bullion, son of War Dance. Leona is the dam of Golden Era by Bullion, Emma Hancock, Dick Hancock and Stella Walker by Outcast, and Swifter by Hyder Ali. Sister of Charity was the dam of Pimlico, Warsaw, Sister of Mercy, John R. Swiney and Sisterly by War Dance. Sister of Mercy was the dam of Heck by Leamington or Reform, Bonnie Chief, Pardee, Compassion and Devotee by Alarm, Iuka, Tessa K. and Sister Marie by Reform, Merci and Bashford by imp. Woodlands. Iuka is dam of Caress by imp. Woodlands, and Alarming by Alarm. Midsummer was the dam of Billion by Bullion, Summer Storm (Amos A.'s dam) by King Faro or imp. Thunderstorm, Trinkitat (the dam of Daisy F. and Balgowan by imp. Thunderstorm), Outlaw and Eugenia by Outcast, and a filly by Geo. Kinney. This Duchess family is one of the most famous racing and producing ones to be found in the stud-book.

Peggy, bay mare foaled in 1788, imp. by Col. John Tayloe, Virginia, was by Trumpator, son of Conductor by Matchem, dam Peggy, sister to Postmaster by Herod, grandam by Snap, son of Snip by Flying Childers, great grandam by Gower Stallion, son of the Godolphin Arabian, out of a daughter of Childers. Peggy produced Britannia by English Pegasus, Peggy by imp. Bedford, son of Dungannon by Eclipse, Fairy by Highflyer, and others. A sister to Peggy and Aimator was bred to Sir Peter Teazle, son of Highflyer by Herod, then crossed on Young Gohanna, Whalebone and Plenipotentiary, three of the tail-male line to Eclipse, and then to Gladiator, a Herod, and we have thus the great brood mare Queen Mary, the dam of Balrownie, Bonnie Scotland, Broomielaw, Haricot (the dam of Cramond and Caller-Ou, winner of 34 Queen's Plates and the St. Leger, Brighton and York Cups), Braxey (dam of imp. Bernice), and Thrift (the dam of Tristan, La Dauphine and Kate Hampton). Blink Bonny, a winner of the Derby and Oaks and dam of Borealis, Blair Athol, winner of the Derby and St. Leger, and Breadalbane, winner of the Prince of Wales and Gratwicke stakes. Other famous winners in the family are Sir Bevys (Derby), Hampton (Epsom Gold Cup, Doncaster and Goodwood Cups and Goodwood stakes), Broomielaw (Dee stakes and Chesterfield stakes), Blinkhoolie (Ascot Gold Vase and Alexandra Plate), Good Hope (Vienna Derby), Beauclerc (Middle Park Plate), and other famous sires and matrons. Peggy by imp. Bedford was the dam of Trumpetta by Hephestion, son of imp. Buzzard and imp. Castianira, the dam of Sir Archy, by Rockingham. This is the only female credited to her. Trumpetta produced Little Peggy by Gallatin, son of imp. Bedford, imp. Mambrina by Mambrino, and two fillies by Kosciusko, son of Sir Archy. The two last have disappeared, leaving nothing behind them. Little Peggy became the property of Dr. Elisha Warfield, of The Meadows, near Lexington, the breeder of the famous Lexington. She had four fillies, one by Sir Archy Montorio (Big Archy), one, Electra, by Sidi Hamet, son of Virginian, one, Too Soon, by Sir Leslie, son of Bertrand, and Cherry Bird by Celestion, son of Sir Leslie and Rowena, the dam of Alice Carneal, Lexington's dam. The first two left no female produce. Too Soon produced Buford by imp. Nonplus, and Garret Davis by Glencoe, both good race-horses, Primrose by Sidi Hamet, Utilla by imp. Margrave, Peggy by Boston, and Kaloolah by imp. Yorkshire. Primrose left no thoroughbred foals, and the same is true of Kaloolah. Utilla produced Ashland by Mambrino Chief, Wind and Pit-a-Pat by Berthune, a filly by imp. Consternation, Young Utilla by Hurricane (Warfield's Florizel), Ulverston and Ultima by Lexington. Young Utilla produced several good race-horses, including Monomania (Tasmania), Dr. Butler and Teetotal by Melbourne, Jr., Gaberlunzie by Bonnie Scotland, Hattie by Australian, and several foals by Pimlico. Ultima produced Aurora Raby and Attila by imp. Australian, the latter of which won at a mile, over Jerome Park course, in 1.44½, one mile and a furlong in 1.58, and the Travers stake in 1874, one-and-three-quarter miles in 2.08¾, after a dead heat with Acrobat in 2.09½. Ultima also produced Nuisance, Sunrise and Totilla by Planet, Sonora, Ultimatum and Alma by King Alfonso, Pleasantry and Cleopatra by Falsetto, etc. Aurora Raby was a good race-mare and the dam of Bill Bruce, one of the best sons of Enquirer, Respond by the same sire, and the noted little gelding Charley Gorham by Blarneystone. Nuisance produced several good ones. Sunrise has had no chance in the stud, and the others are virtually untried, with the exception of Monomania, who has produced several good ones, such as Monitress, Scissors and Monopoly by War Dance, Mammonist, Monopole, Monogram, Matrimony and Mona, all by imp. Buckden and good performers, Gilt Edge and Lida Laroy by Gilroy. Peggy by Boston was the dam of Annie Laurie by Vincente Nolte, son of American Eclipse, Bridget by Paddy Burns, son of Grey Eagle, Princess Royal by imp. Sovereign, and Mathilde by imp. Scythian. Annie Laurie produced a number of good ones, such as Kate McDonald by imp. Mickey Free (the dam of Trouble, the best horse over timber ever in America), and Pest by imp. Phaeton (the latter a good one), Locust Post, Flying Locust and others by Melbourne, Jr., Annie Butler by Ulverston, she the dam of Mary Howard by imp. Hartington, that produced the speedy mare Pearl Jennings, Mary Payne and Ada Rees by Lelaps, son of imp. Leamington, Sally Howard, dam of Mary Ellis, by King Lear, son of Lexington, Isaac Murphy by Virgil, etc. Annie Laurie is also dam of Experiment by imp. Phaeton, Venturia, and Mamie O. by Virgil, and Locust Bloom by Tom Bowling, the latter the dam of Chilhowee by King Ban, Elknoe by Longfellow, etc. Venturia is the dam of some good ones. The family is a racing one, and has produced race-horses uniformly when properly bred.

Imp. Selima was a bay mare, foaled 1746, bred by Lord Craven, by the Godolphin Arabian, dam Large Hartley mare by Mr. Hartley's Blind Horse, grandam Flying Whigg by William's Woodstock Arabian, great grandam Grey Whynot by Whynot, son of the Fenwick Barb, out of a Royal mare. She was own sister to Babraham, Mogul and Marlborough. Many of the best horses and successful stallions in America and England trace in direct female line to this family, such as Blank, Babraham, Gower Stallion, Merlin by Second, Sweetbriar, Hornsea, Amsterdam, Soothsayer, Jerry, Harvester, Skylark, etc. In America imp. Ambassador by Plenipotentiary, imp. Buckden, imp. Driver, imp. Gabriel, imp. Mendosa, and the following bred in this country, Alarm, Picolo,* Oden, Cloverbrook, the famous Grey Eagle, Robin Grey (sire of Lady Grey), Hanover, Foxhall, etc., descend from the same source. Selima was the dam of Partner, Ariel and a sister by imp. Traveler, son of Croft's Partner, Stella, Selin and Ebony by imp. Othello, Black Selima by imp. Fearnought, etc.

The Traveler mare, sister to Partner and Ariel, was the dam of Peg by imp. Juniper, she the dam of a filly by imp. Othello, she the dam of the filly by Leonidas, son of Lloyd's Traveler, and she the dam of filly by Hall's Union, son of imp. Slim, who was the dam of Pandora by Grey Diomed. Pandora was the dam of Daisy Cropper and Silverheels by Ogle's Oscar, Aurora and Pocahontas by Lloyd's Vingt'un. Aurora was the dam of Fanny Wright by Silverheels, and she the dam of Tiberius by imp. Priam, Emma Wright and Eleanor Margrave by imp. Margrave. Emma Wright was the dam of Laura Farris and Miss Doyle by Lexington, Jerseyman and the noted Mollie Jackson (who won at all distances and won the famous three-mile race in 5.35¼, 5.34½, 5. 28¾, which has never been equalled) by Vandal, Earring by Ringgold, Woodstock by imp. Australian, Geo. Wilkes and Emet by Astoria. The blood is best preserved in this line through Mollie Jackson, who was the dam of Monday by Colton, son of Lexington, Fanny Ludlow, Temptation and Long Branch by imp. Eclipse, and Doubt by imp. Eclipse or Cavalier, son of imp. Eclipse. Fanny Ludlow was the dam of Jamaica by Lexington, Cedric, Cherokee, Chipola and Choctaw by imp. Saxon, Clare by imp. Glenlyon, Cholula, Catalime and Courisande by imp. Mortemer, and Alexina by Duke of Magenta. Jamaica is the dam of the grand race-horse Foxhall, Potosi, American Girl, Alexander and Abaca by King Alfonso, Octopus and Joanna by Falsetto. Doubt is the dam of Mollie Jackson, Jr., by Battle Axe, and Stockton by Spendthrift.

Stella by imp. Othello was the dam of Harmony (dam of Hall's Union) and a sister by imp. Figure, Primrose and Thistle by imp. Dove. The sister to Harmony founded the family through which Exchange by Richmond, son of imp. Sir Harry, traces. Exchange was the dam of Mary Selden by Sussex, and Glorvina by Industry, son of Sir Archy. Glorvina was the dam of Laura Spillman by Wagner, and she the dam of Nora, Williamsburg, Blue Bells and Daylight by Lightning, Mary Clay by Revenue, and Busy Bee by War Dance, the latter dam of Violet by Viley, and Busy Body by Day Star. Mary Clay was the dam of Sanford by imp. Glenelg, Lancaster and Sue Finnie by imp. Buckden, the latter the dam of Bettie C., Ofalece, Trimble and Alf Allen by Harry O'Fallon. Primrose by imp. Dove was the dam of Cades by Wormsley's King Herod, she the dam of Hornet and Sting by imp. Diomed, and Lady Bolingbroke by imp. Pantaloon. Lady Bolingbroke was the dam of Cælia by Symme's Wildair, Desdemona and Virginia by imp. Dare Devil, Lavinia, Superior, Primrose, Wrangler and Lady Chesterfield by imp. Diomed. Lady Chesterfield was the dam of the mare by Sir Archy that was the dam of Ophelia by Wild Medley, and she the dam of the superior race-horse Grey Eagle, and Falcon by Woodpecker, son of Bertrand by Sir Archy. Falcon was the dam of Lilla D. and Ella D. by Vandal, Alert by Lexington, and Scythia by imp. Scythian. Ella D. was the dam of Paris and Paris Belle by Lexington, and Bourbon Belle (the dam of the grand race-horse Hanover and Houston by Hindoo) by imp. Bonnie Scotland. Alert was the dam of Vandalla (Glencaire's dam) by Vandal, Tolona by imp. Phaeton, Harper by Longfellow, Kenton by imp. Glenelg, Richmond, Lookout and Active by Virgil. Scythia was the dam of Ida and Parisa by Paris, son of Lexington, etc. Ida is dam of Virgilia and Idaho by Virgil, Idler by imp. Glenelg, and Irene by Leveller. Parisa is dam of Swamp Fox by Fonso. Cælia by Symme's Wildair was the dam of Fair Forester by imp. Chance, and Rosemary by imp. Diomed. . Fair Forester was the dam of Polly Peacham by John Richards, son of Sir Archy. She produced a filly by imp. Monarch which was the dam of Mary O'Neil and Fawn by St. Louis, son of Altorf. Mary O'Neil was the dam of Maggie O'Neil, and she the dam of Tillie W. by Joe Daniels, and Maggie O. by Shannon. Fawn was the dam of Bettie by Gen. Longstreet, she the dam of Mary Shaw by Gov. Bowie, and Bernice by imp. Intruder. Polly by imp. Sovereign, out of Polly Peacham, was the dam of Susan Gillespie by Engineer, son of Revenue, and Kitty Guild by Hiawatha. Susan Gillespie was the dam of Alice Thompkins by imp. Glengarry, Miss Malloy and Brunette (Omega) by Pat Malloy. There are a few mares from Rosemary: only Eula by Muggins, Leona King and Lady Corinne. Virginia by imp. Dare Devil was the dam of Virginia by Sir Archy, she the dam of Caledonia by Jerry, son of Pacelot, Dolphin and the good race-mare Giantess by imp. Leviathan. Caledonia was the dam of Polly Elliott by imp. Leviathan, she the dam of Peytona Barry by Rogers, Ada Tevis and Tenth Legion by imp. Albion. Ada Tevis was the dam of Blanche Rousseau by Commodore, son of Boston, Charlotte Thompson by imp. Mickey Free, and Johnetta by Austerlitz. Blanche Rousseau is the dam of Lucy Wallace and Reel Dance by War Dance, Margaret Haight by Gilroy, La Blanche and Susie L. by Vocalic. Lucy Wallace is the dam of Alma by Himyar, Martha and Luminous by imp. Rayon d'Or. Reel Dance is the dam of Cotillion and Minuet by imp. Rayon d'Or, and Fandango by imp. Kantaka. Charlotte Thompson was the dam of the great gelding Barnum by imp. Bonnie Scotland, Sallie Gardner by Vandal. and Miss Muggins by Muggins. Sallie Gardner is the dam of Gardey and Sallie G. by Flood, and Miss Muggins is the dam of Guitar by Grit, Philopena and Miss Blonde (Peppermint) by Plenipo, son of imp. Bonnie Scotland. Johnetta was the dam of Melita and two sisters by Muggins, son of Jack Malone. Melita is the dam of Von Moltke by Vermont, son of Vandal, Miss Harding by imp. Great Tom, Egmont by Enquirer, Little Emily (Brillianteen) and Bendigo by Luke Blackburn, and Moi Miloi by Bramble. Dolphin was the dam of Minnow and Whale by Voucher, son of Wagner. Minnow was the dam of a mare by Revenue, she the dam of a filly by Lightning, she the dam of Adele by imp. Australian. she the dam of Banda by King Ban, Adelina by Macduff, and fillies by Linden.

Giantess was the dam of Aduella, Union, Glencoe, Jr., and the grand mare Peytona (winner of the great Peyton stake at Nashville. 1843), all by imp. Glencoe.

Peytona was the dam of two fillies by imp. Ruby, son of Emilius, and Transylvania by imp. Arab Massoud. One of the Ruby mares was the dam of Matilda by imp. Sovereign, she the dam of Miss Tilton by Daniel Boone, and Satilla by imp. Buckden. Miss Tilton is the dam of the gelding G. W. Cook, Minnie by Congressman, Misfit by imp. Siddartha, etc. Transylvania was the dam of Palm Tree by Lexington, Date Tree by imp. Micky Free, Blanchette and Dewberry by imp. Knight of St. George, Limestone, Lime Tree and Limit by War Dance. Date Tree was the dam of Melodist by Melbourne, Jr,, Wah-ta-Wah by War Dance (she the dam of Olemus, Snook and Lass of Tamales), Arabia and Clara L. by imp. Thunderstorm, and Lou Monette by imp. Strachino. Dewberry was the dam of J. H. Haverly by War Dance, Thundercloud by imp. Thunderstorm, and Mary Berry by Bullion or Outcast. Lime Tree is the dam of Limey by imp. Thunderstorm, Limestreet by Streatham, Belle of Mediapolis and Limey Long by Lead On, son of imp. Leamington. Limit is the dam of Easter III. by imp. Thunderstorm, and Limitone by Lead On. Some of the best race-horses, brood mares and stallions in England and America trace to this line of blood.

The Popinjay mare was foaled 1808, and imported by Hon. John Randolph, of Virginia. She was by Popinjay, son of imp. Buzzard, dam Bourbon's dam by Precipitate, son of Mercury by Eclipse, grandam Young Tiffany by Highflyer, son of Herod, out of Tiffany by Eclipse, tracing through Young Hag by Skim, Hag by Crab, Ebony by Flying Childers, and Old Ebony by Basto, to the Massey mare by Mr. Massey's Black Barb, to which trace in direct female line Trentham, Florizel (sire of imp. Diomed), Election, Quiz, Royal Oak, Mulatto, Defence, The Emperor, The Palmer, Rosicrucian, Dutch Skater, Doncaster, Gladiateur, Hermit (sire of imp. St. Blaise, Tristan, Peter, etc.), imp. Flimnap, imp. Barefoot, imp. Shamrock, imp. Speculator, imp. Eliza (the grandam of the great producing mare Topaz by Glencoe), imp. Kantaka, imp. Ocean Queen (dam of Tourmaline, Seadrift, Fairy Queen, etc.). This Popinjay mare was the dam of Parrot, Woodlark and Macaw by Randolph's Roanoke, son of Sir Archy. Parrot was the dam of Mary Thomas by imp. Consol, and Sally Hunter by American Eclipse. Mary Thomas was the dam of Odd Stocking, Magnolia and Mary Ogden by Thornhill, son of imp. Glencoe, and Thetis by imp. Sovereign. Odd Stocking's line is extinct. Magnolia was the dam of Kate Jewell by Wagner, several fillies by imp. Sovereign, and Wreath by Brown Dick. Kate Jewell was the dam of Tennessean by Lightning, and a filly by War Dance. Mary Ogden was the dam of Lorette by imp. Sovereign, and she the dam of Mishap by imp. Knight of St. George, Regan by Lexington, Charette by Lightning, Ada A., Lerona and Nannie B. by Asteroid, Sophy Badderly by imp. Australian, and Loriot and Peaceful by Planet. Mishap was the dam of King Benazet and Crockford by Lightning, Misfortune, Mischief and Misdeal by Gilroy, Perhaps by imp. Australian, Marimon by Waverly, Sophy and her sister by imp. Strachino, and Mischief by imp. Thunderstorm. Misfortune is the dam of Merrimack and Abbotsford (Mistake) by Waverly, who raced successfully in England, Miss Dance (Netroma's dam) by War Dance, Syntax by St. Martin, Hopeful and Glen Fortune by imp. Glenelg, Adversity by Spendthrift, King Fortune by imp. King Ban, and Long Fortune by Longfellow. Perhaps is the dam of Perplex by Gilroy, Warrior and Chance (a good horse) by War Dance, Doubt by St Martin, Probability and Dilemma by Onondaga, Reckon by imp. Pizarro, and Perblaze by imp. St. Blaise. Sophy is the dam of Repose (Pelham's dam) by Gilroy, and Morning Glory by Fonso. Regan was the dam of Regal, Virgil, Vera Cruz, Virgilette, Relay, Hawley, Vengeance and Printer (Last Chance) by Vigil, Jennie D. and Jennie B. by imp. Glenelg. Jennie D. is the dam of Grismer, Verano, Janova and La Cienega by Grinstead. Jennie B. is the dam of Ganymede, Honduras and a sister by Grinstead. Relay is the dam of Donohue, etc. by Stratford. Ada A. was the dam of Patsey Duffy, Frank Rhoads, Lizzie P., Nubia, Ab. Stemler, Rosa G. (Mary B.), (Tom O'Hara's dam), Ysabel (dam of Warsaw and Florence by Warwick), Lady Leinster and Leon, all by Leinster. Lizzie P. is the dam of Idaline Cotton by Jim Brown, Louis P. by Joe Hooker, Varona, etc., by Young Bazar, Sophy Badderly was the dam of Toulie by Gilroy, Olivette by Alarm, Peytona Barry by imp. Strachino,

Bob Pate by St. Martin, Aunt Em and Olive by Onondaga. Peytona Barry was the dam of Endie Mills, Salvanos (Bob Thomas), and Sweetheart by Enquirer. Thetis was the dam of Themis by Austerlitz, son of imp. Yorkshire. Themis was the dam of Dacota by Lightning, Thesia by Calvin, son of Tipperary. Charlotte by imp. Prince Charlie, The Sheriff by imp. King Ban, Ben's Pet by Himyar, and filly by imp. Deceiver. Thesia was the dam of Blunette by imp. Glenelg, and Ramona by Ten Broeck. This is one of the valuable families to perpetuate and improve racing blood. The Popinjay mare was not only fashionably bred, but from one of the best and most distinguished sources in the stud book, and one which has demonstrated its value even to this day.

One of the latest importatons to which a large number of race-horses owe their origin was Myrtle, imported by Messrs. Kirkman, Nashville, Tenn. Myrtle was by Mameluke (winner of the Derby in 1827, and second in the St. Leger the same year), son of Partisan by Walton and Miss Sophia by Stamford, dam Bobadilla (winner of the Gold Cup at Ascot and Drawing-Room Stakes at Goodwood in 1828) by Bobadil, son of Rubens by imp. Buzzard, grandam Pythoness by Sorcerer, son of Trumpator by Conductor, out of Princess by Sir Peter, tracing through the best racing blood to the Layton Barb mare. Myrtle is credited with but two foals, Dr. John and Magnolia by imp. Glencoe.

Magnolia became the property of Dr. Mercer, of New Orleans, who presented her to the Hon. Henry Clay, of Kentucky, who gave her to his son, the late John M. Clay, of the Ashland stud. She produced a number of good ones, including Magic, Madonna, Princeton, Charles Morgan, Hanover, Simon Kenton and Skedaddle, all by imp. Yorkshire (son of St. Nicholas by Emilus, and imp. Miss Rose by Tramp), Charley Ball by Wagner, Daniel Boone, a good race-horse, the noted Cup horse Kentucky and Gilroy by Lexington, Victory by Uncle Vic, son of Lexington, and Madeline by Old Boston. Madeline was the dam of Magdalena by Bob Letcher, son of Medoc, Madostone (ran as Kate Duncan) by imp. Stone Plover, Income by Revenue, son of imp. Trustee, and Maggie B. B. by imp. Australian, and Madelina by Lightning. Magdalena was the dam of Myra by Tite Eye (son of Herr's Boston). Madostone was the dam of Manton by Orion (son of Revenue), Bide-a-Blink and Bee-in-the-Bonnet by imp. Bonnie Scotland. Bee-in-the-Bonnet was the dam of Mike's Pride (Pike's Pride), Jaubert (Star Ban), Bonnie Ban, French Park, a good race-horse, all by imp. King Ban, Reverta and Craftie by Fellowcraft, Pike's Peak (Pixy) by Lelaps, and Senator Hearst by Longfellow. Income was the dam of Goldmine by imp. Australian, Coupon by Lightning, Incommode, a fine race-horse, by Melbourne, Jr., son of imp. Knight of St. George, Wavelet (dam of Penelope, Tidal Wave and Briar Root) and Clan Alpine by Waverly, Interest (Pickup's dam) by St. Martin, Warda by Long Taw, Runaway by imp. Highlander, and Outcome by Forester. Goldmine was the dam of Pearl (dam of Grayson, and Bijou (the dam of Kilkenny by Geo. Kinney), both by War Path, son of War Dance. Maggie B. B. (winner of the Sequel stakes) was the dam of Lord Clive, Pera (the dam of Petition, Padre's dam, Pericolor, Darya, Gretina), Magnum Bonum, Jaconet, Harold, Iroquois, winner of the Derby, St. Leger, Prince of Wales, and St. James' Palace stakes, and Francesca, all by imp. Leamington, Red and Blue (dam of the fine race-mare Sallie McClelland), Panique and Wawekus by Alarm, Okemas, Onas, Hutoka and Homoopathy by Reform, Hypatia and Flatlands by imp. Woodlands. Jaconet is the dam of The Niece (dam of Lydia Belle, Alarm Bell and La Colonia), Mattie T. (dam of Amy Davenport and Ethel S.) by imp. Billet, Belvidere and Sir Dixon, good race-horses, and Blacklock by imp. Billet, Cambric (Lisric's dam) and Intrepid by Hindoo. Francesca is the dam of Franconia by imp. Billet, Frisco and Halloo by Hindoo. Madelina was the dam. of Ollie C. by Alarm, she the dam of the winner Sir Arthur by Leonatus.

Skedaddle was the dam of Saucebox and Scramble by Star Davis, son of imp. Glencoe, Slyboots by Rivoli, son of Revenue, Squeez'em by Lexington. Florence B. by Tom Bowling, Joe S. by King Lear, The Slashes by imp. Glenelg. Saucebox was the dam of Sylvia Springer and Music Box by imp. Buckden, and Safety by Buchanan. Sylvia Springer is the dam of Judith by Ten Broeck. Hot Spring by Rebel, Camilla by

Erdenheim, and Silvana by imp. Silvermine. Scramble is the dam of Alice J., Ruth Howard, Remember and Maggie J. by Ballinkeel, son of Asteroid, Outscramble by Outcast, and Sam Morse by Leonatus. Alice J. is the dam of Alice D. (the dam of Pittipat by Pat Molloy), by Add, son of Revolver. Glenmound, Glendale and Glencliffe by Glenmore, etc. Remember is the dam of Counterfeit by Glenmore, and Release by Tom Ochiltree or imp. Hopeful. Slyboots is the dam of the superior race-horse Leveller by Lever, son of Lexington, Sly Dance, Dakota and Sachem by War Dance, son of Lexington, Savanac and Volunteer by imp. Mortemer. Sly Dance is the dam of Pavanne by Iroquois, Sunnyside by imp. Mortemer, Portia by Joe Daniels, son of imp. Australian. Squeez'em is the dam of Day Star (winner of the Blue Ribbon stakes and Kentucky Derby) by Star Davis, Sylph and Tom Naven by imp. Glenelg, Palmetto by Virgil, Recluse, Fardello, and Helena by imp. Mortemer, Eliza Rooney by Baden Baden, etc. Sylph is the dam of John K. by War Dance, La Sylphide by Fellowcraft, Kaloolah, Philip D. and Kincsem by Longfellow, Sylvia and Sea Mew (Logica's dam) by imp. King Ban. Sylvia is dam of Fauna by Longfellow. La Sylphide is the dam of Semper Fidele, a grand race-mare, by Longfellow, Semper Rex by Lelaps, Semper Lex by Falsetto, and Semper Cara by King Alfonso. The family is a great racing and producing one. Iroquois won the Derby and St. Leger in England, a feat accomplished by few horses, and is now regarded as one of the most valuable and potent sires in the country. Kentucky was a first-class race-horse, and was, for his chances, a grand success in the stud. The family trace their origin to the Layton Barb mare, to which trace in direct female line Young Greyhound, Cartouch, Shepherd's Crab, Matchem, Young Cade, Drone (by Herod), Blucher, Sir Walter Raleigh, The Ugly Buck, Clairvaux, Simoom, Audubon, Philammon, St. Honorat, and many other good ones.

BREEDING.

In breeding for the Turf purity of blood, racing lineage and true conformation are the desirable and requisite qualities, and a combination of these qualities must improve the mechanical structure which nature conferred upon the horse, designed to give him strength and fleetness, a combination seldom found in the animal world, and which greatly enhances his utility to the human race. These powers, great in an uncultivated state, are susceptible of much increase and improvement by judicious attention to breeding and crossing. It is an axiom in breeding (and this is particularly applicable to the horse) that the character, quality and form of the offspring is governed in a large degree by the qualities we wish to propagate. These should be found in the parents, so that by coupling together a male and female endowed with the qualities we wish to be transmitted to the foals, we acquire the power of moulding the produce to our ideas. The principle that parents of all breeds and families of whatever species produce young ones analogous to themselves, has been acted on for ages. The parents should be selected with the greatest discrimination, so that by uniting in their offspring the superiorities possessed by both stallion and mare, the improvement is almost sure to follow. Those who breed for the turf must remember and act upon the value of the precept, and use the utmost caution and judgment in selecting the sire and dam, so as to combine, improve and render permanent the good qualities possessed by each. In selecting the parents, independent of conformation, pedigree and racing lineage, it is all important that both be sound, that is, free from any hereditary disease or taint. There are certain diseases which are transmissable from parent to offspring, but they are few. Nearly all the diseases to which the horse is liable, or blemishes, either in certain diseases of the eyes, spavin, ringbones or roaring, are traceable to some cause not hereditary. Many of our best and most successful stallions have been blind, but not from constitutional causes, and the same is true as to roarers, this defect not showing itself in their offspring, whilst others, sound in all respects, have bred those with these defects, evidently from some cause unknown. It is a mooted question as to what age brood mares should be allowed to commence breeding. If they have been well raised, and are not deficient in energy and muscular development, they can be safely bred to the stallion at three years old. It is a notable fact that more valuable horses for general uses and more good race-horses have been bred from mares well advanced in life than from those of immature age, and brood mares properly cared for retain the functions of production to an advanced age. Alice Carneal foaled Lexington at the age of 11. He was by far her best one, and she had her last foal at the age of 23. Reel foaled Lecompte, her best, when 12, Starke at 17, and War Dance when 21 years old. Boston, by odds the best foal of his dam, was foaled when his mother was 19. Pocahontas foaled Stockwell at 12, and Rataplan when 13 years old. Blue Bonnet threw Lightning at 18, Thunder at 19, Loadstone at 22, and Lancaster when 24 years old, all good race-horses. Salina threw Salvator at 18 years old, and imp. Princess produced His Highness at 16 years old. It is necessary in breeding race horses that the muscular system of the stallion should be toned up by abundant exercise, and if the stallion will not exercise when turned into his paddock, he should be ridden. Exercise contributes to his health and muscular energy, and if he is deficient in these qualities at the time of copulation, it is but reasonable to suppose that the foal must suffer constitutionally in a like respect. For a similar reason it is obviously injurious to the stamina of the foal if either parent shall have suffered from insufficient nourishing food or other debilitating causes. No animal can be well formed or strong that has been stinted in the ample supply of good nourishing food necessary to its full development and growth, from the earliest period of its existence to the complete formation of all its parts. The old saying is that the time to commence to breed the race-horse and insure success is to begin "before he is born." The dam is the fountain, and if she is neglected and starved the produce which she is carrying, with the one at her side, will not only show it, but prove it subsequently. It is but reasonable that such should be the case when the system is taxed to contribute to the formation of the embryo, as well as to sustain life and supply the foal at her side with nourishment. Mares should be kept in good, strong condition, not swine fat, but fat enough for health and strength, and to contribute the same to her foals. The foal will be larger, healthier, stronger, and the hair have a glossy and healthy appearance. Many breeders are of the opinion that a mare that has never raced is better to breed from than one that has raced. Statistics do not support this position. Some of the most eminent English breeders, notably Lord Falmouth, bred from no mares that were not tried, and few breeders were more successful or bred a larger number of winners of the important racing events. It is no disadvantage to race mares a reasonable time, and now that the system of heat racing has been abolished, and horses are not subjected to the long and severe preparation, exhausting the vital powers to a great extent, no reasonable racing or training can injure mares for breeding purposes. Alice Carneal, the dam of the immortal Lexington, ran two, three and four mile heats. Reel, the dam of Lecompte, Starke, Prioress, etc., ran four mile heats. Brademante, dam of The Bard, ran heat races. Idlewild, dam of Wildidle, won at four miles. Nantura, the dam of Longfellow, raced and won. Toucques, dam of La Tosca, raced until four years old. Salina, dam of Salvator, raced. Other successful brood mares that raced are Nevada, the dam of Luke Blackburn, Florence, dam of Hindoo and grandam of Firenzi, Aerolite, dam of Spendthrift, Fellowcraft, Miser, etc., Ann Fief, dam of the unbeaten Tremont, Maria Black, dam of Bay Leaf, the dam of Bayflower, Preakness, Bayonet, etc., Topaz, dam of Wagram, Lodi, Colton, etc., Fanny Washington, dam of Eolus, etc., War Song, dam of Eole, St. Saviour, Eon, etc., Nina, dam of Planet, Exchequer, etc., Arizona, dam of Aranza, Maggie B. B., dam of Iroquois, Panique, etc., Jamaica, dam of Foxhall, Balloon, dam of True Blue, The Banshee, etc., imp. Maud, dam of Alarm, Maudina, dam of Piccolo, Oden, Cloverbrook, Marauder, etc., Minx, dam of Monitor, Susan Beane, dam of Sensation, Onondaga, Stratford, etc., Rescue, dam of Abd-el-Kader, Abd-el-Koree, etc., Sarong, dam of Aristides, Volga, dam of Barney Williams, Levity, dam of Ruric, Legatee, Lever, etc., Ivy Leaf, dam of Bazar, Bramble, Niagara, Brambaletta, etc., Magnolia, dam of Princeton, Charley Ball, Magic, etc., Maiden, dam of Parole, James A., Perfection, Powhattan, etc., Magenta, dam of Duke of Magenta, La Henderson, dam of Ferida, Aella, etc., Farfaletta, dam of Falsetto, Felucia, etc., Fanny Holton, dam of Ten Broeck, Ulrica, dam of Ben Ali, Megara, Memento, etc., Waltz, dam of Glidelia,

Farandole, etc., Sultana, dam of She, Padisha, etc., Susquehanna, dam of Potomac, Chesapeake, etc., Belle of Maywood, dam of Tenny. There are many others that could be added to this list, none of which were severely trained and raced at heats beyond their fifth year. It should be an established axiom with breeders that no thoroughbred animal with hereditary defects should be used in the stud. But the question is, what are hereditary defects? Spindling bone, chronic weakness and infirmities of constitution are the most pronounced hereditary defects. There are defects arising from accidents, such as affections of the wind or respiratory organs, in consequence of distemper or severe colds, ringbone or spavins from injury or over strains, which should not be termed or considered hereditary defects, and which are not transmissable from parent to offspring. But to reach the acme in breeding the thoroughbred horse, no constitutionally unsound mares should be used for the stud or breeding purposes. Then how is this fact to be ascertained? The most practical test of soundness of limbs, digestive organs, constitution. temper and nervous organization is in training and racing on public tracks. Mares which are trained and raced in their two and three-year old form, when fit, even with moderate success, unless some constitutional defect develops, must be regarded as sound; and such mares, if properly mated, should breed sound progeny, and such as should race successfully. Accidents or unforeseen causes, of course, may intervene and prevent success. We have sometimes seen mares, from racing families, which have raced successfully, and retired to all appearances sound and with constitutions unimpaired, that proved failures in the stud. This is doubtless owing to the fact that they were raced too long, and the vital powers exhausted from the effects, and from the long and severe preparation for the class of races which they ran. One or two notable instances can be cited. Fashion a very highly shaped mare, well bred and from a racing family, was a failure in the stud. It must be remembered that she raced until eleven years old, running 36 races, of which she won 32. Of these 24 were at four mile heats, of which she won 21, 8 were three mile heats, all won, and 4 two mile heats, of which she lost one. She was bred the first three seasons to her half-brother Mariner by Shark, producing two fillies and one colt. Her best produce was Dangerous, her last (foaled when she was 22 years old), by Bonnie Scotland. Her daughter Young Fashion by imp. Monarch has done well in the stud, as also her daughters A la mode and Etiquette by Mariner. There are others which were famous on the turf and trained until aged that were failures in the stud.

"Breeding In and In." This expression is used to specify the practice of breeding animals successively from the same race and blood, with similar form and symmetry, and closely allied in blood lines. It has always been regarded as an active cause of degeneracy, as unnatural and calculated to engender and perpetuate hereditary diseases. These evils, in my opinion, can only arise from the abuse of the system. If proper attention and care be taken to exclude the weak and diseased from participating in reproduction, it will be the means of improving and continuing for an unlimited time those perfections and beauties which are to be highly prized. "Breeding In and In" bestows the power of improving and perpetuating the excellences of any breed of animals and transmitting those excellences unimpaired to succeeding generations, and for this reason should not be too readily disregarded and condemned. The application of these principles to generation of our useful domestic animals has produced the swift and beautiful race-horse, the disposition to fatten in cattle, and the fine-woolled breeds of sheep; and that which has been originated and established by art in breeding, can be continued and improved by the same means. By judiciously selecting the parents, both male and female, we can establish new varieties, and by continuing to breed from such improved and perfected stock we will be able to perpetuate the excellent qualities. The evils which often arise in the human race from the union of those too closely allied in blood may be altogether avoided and obviated in the brute creation, by wholly debarring those tainted with hereditary diseases, or such as are deficient in constitutional stamina, from participation in reproduction. The higher the intelligence the more marked the deterioration, where there is a taint of hereditary disease or lack of constitutional stamina. In opposition to this method practised in propagating the useful animal, the horse, may be cited the practice of uniting sire and dam, differing more or less in physical proportion, with the object of obtaining a mean of their qualities, which process is denominated crossing. If followed with good judgment and just discrimination, it is very often a source of improvement, but the contrary course, where the junction of extreme varieties and violation of the propriety of nature has been permitted, has led to the propagation of an animal of inferior character. The improvement of the race-horse arose from a happy combination of Eastern blood, represented by the Arabian, Barb and Turkish horse, and it is to the union of these bloods that we trace his present excellence. The union of the blood conferred upon the possessors wonderful stamina, great speed and the power of resisting and enduring extraordinary feats of muscular exertion without fatigue, exceeding all other breeds in these estimable qualities. In the infancy of racing improvement seemed to mark every fresh infusion of Eastern blood. The Godolphin Arabian, Darley Arabian, Curwen Barb, Byerly Turk, Lister Turk, Place's White Turk, Compton Barb and the Royal mares originated a progeny celebrated in the annals of the turf. But the infusion of new Eastern blood is now more likely to degenerate than to improve the thoroughbred horse. Our modern race-horse has become such a grand and distinct improvement from the source that originated him that he excels it in speed, stamina and staying qualities, and that any further admixture of Eastern blood would fail to benefit the horse. It has often been the custom to cross from one blood into another, and to breed from a stallion or mare of inferior shape, symmetry and goodness in preference to breed, as it is styled, in and in. Whilst I do not advocate incestuous breeding, I believe that crossing out to good blood and form, and then back to the same good blood and symmetry, if the animals used in reproduction are as indicated above, must result in improvement. Our very best horses from the earliest dates have been bred in and in, often incestuously so, without deterioration, and I am of the opinion that we should keep in good blood, where there is symmetry, constitution and no hereditary disease, even incestuously so, rather than exchange it for inferior blood and form, under the idea that it is necessary to cross at all events into distinct blood. But as good and bad qualities descend, if we go from superior blood to inferior, and from high form to bad, the stock must degenerate. I do not assert that breeding in and in is superior or preferable to any other mode of breeding, but I am confident there are not those physical objections to it, which are commonly entertained. An improvement of any race can only be expected from crossing with superior symmetry of form and better blood, and it is only in such cases that we decidedly recommend it. I desire to support my views by facts, and leave the breeder to draw his own conclusions.

I will only cite examples of famous horses. Flying Childers was bred in and in. He was the fleetest and most famous horse of his day, by the Darley Arabian, dam Betty Leedes by Old Careless, son of Spanker, by the D'Arcy Yellow Turk, grandam sister to Leedes by Leedes Arabian, out of daughter of Spanker, son of D'Arcy Yellow Turk, she out of old Morocco mare, Spanker's own dam.

Highflyer, the most famous race-horse of his day, and a grand and successful stallion, was by Herod, dam Rachel by Blank, grandam by Regulus. Blank and Regulus were both sons of the Godolphin Arabian.

Old Fox, an excellent racer and sire of Meliora, the dam of Tartar, the sire of Herod, was by Clumsey, son of Hautboy, dam Bay Peg by Leedes Arabian, out of Young Bald Peg by the same Arabian.

Priestess, the most famous mare of her day, was by Matchless, son of Cade, her dam by the Gower stallion, grandam by Regulus. Cade, Gower stallion and Regulus were all by the Godolphin Arabian.

The celebrated Gimcrack, sire of imp. Medley, was by Cripple, dam by Grisewood's Partner, grandam by Croft's Partner, the sire of Grisewood's Partner.

Imp. Shark, the most noted race-horse of his day, and a very successful stallion, was by Marske, son of Squirt by Bartlett's Childers, his dam by Snap, son of Snip by Flying Childers, own brother to Bartlett's Childers.

The celebrated O'Kelly's Eclipse, who was never beaten or paid a forfeit, was much inbred to Lister Turk and Hautboy, and much of his excellence was attributed to this fact.

Diomed (winner of the first Derby ever run), imported to America when twenty-two years old, and the most valuable stallion ever imported, was much inbred. Florizel, his sire, was by Herod, dam daughter of Cygnet, son of the Godolphin Arabian. His dam Sister to Juno was by Spectator, whose dam was by Partner, grandsire of Herod. Diomed's grandam Horatia was by Blank, son of the Godolphin Arabian. Herod, Florizel and Horatia were all inbred to the Darley Arabian through the two Childers.

Imp. Glencoe, whose daughter Pocahontas was the grandest and most successful brood mare in England, and whose son Vandal made such an impression on the stock of this country through his sons, and whose daughters are the dams and grandams of many of the most famous race-horses and stallions in America, was much inbred to Herod, Eclipse and Childers, having no less than eleven direct crosses of Herod blood, seven of Eclipse, and the same to Childers, through Snap.

Imp. Leamington, the grand and successful stallion, was much inbred to Herod and Eclipse. He had no less than twenty-five crosses of the former and fourteen of the latter blood.

Imp. Phaeton, a phenomenal success in the stud in America, is very much inbred to the same blood, Herod and Eclipse, through Buzzard's sons, Castrel and Selim; and through the best sources, Whalebone, Whisker and Beningbrough.

Imported Priam by Emilius (winner of the Derby, 1823), and sire of Crucifix (Winner of the 2,000 and 1,000 Guineas and Oaks, 1840), Miss Letty and Industry (winners of the Oaks, 1837-1838), was inbred to Whiskey, son of imp. Saltram. He was the highest priced horse imported up to his day, costing, $17,500 in England.

Imported Billet, an excellent race-horse and popular stallion, was inbred to Filho-da-Puta. Voltigeur, his sire, was by Voltaire, dam Martha Lynn by Mulatto, out of Leda by Filho-da-Puta, Billet's dam Calcutta by Flatcatcher, whose dam, Decoy, was by Filho-da-Puta, and he has any number of crosses to Eclipse, through the best sources, Pot-8-o-'s and King Fergus, backed by a large infusion of Herod blood from the best and most potent sources.

Galopin, the most popular stallion in England (winner of the Derby in 1875), sire of Galliard (winner of the 2,000 Guineas, 1883), Donovan (winner of the Derby and St. Leger, 1889), St. Simon (a superior race-horse, sire of Semolina, winner of the 1,000 Guineas, 1890, Memoir, winner of the Oaks and St Leger in 1890, etc.), is inbred to Voltaire, being by Vedette, son of Voltigeur by Voltaire, dam Flying Duchess by The Flying Dutchman, out of Merope by Voltaire, who is also inbred to Blacklock.

Barcaldine, the best living representative of the Melbourne blood in England, and a very popular stallion, sire of Mimi (winner of the Oaks), is inbred to Birdcatcher. He is by Solon, son of West Australian by Melbourne, Solon's dam, Darling's dam by Birdcatcher. Barcaldine's dam Ballyroe by Belladrum, grandam Bon Accord by Adventurer, out of Darling's dam by Birdcatcher.

Isonomy, the grand race-horse, and who was one of the most popular stallions, sire of Seabreeze, winner of the Oaks, Common, winner of the 2,000 Guineas, Derby and St. Leger, is inbred to Birdcatcher, being by Sterling, son of Oxford by Birdcatcher, dam Isola Bella by Stockwell, son of The Baron by Birdcatcher, and traces four times direct to Sir Hercules.

Wellingtonia, a superior race-horse and sire of Clover, winner of the Prix du Jockey Club (French Derby), is inbred to old Pocahontas. He is by Chattanooga, son of Orlando by Touchstone, Chattanooga's dam Ayacanora by Birdcatcher, out of Pocahontas by Glencoe. Wellingtonia's dam Araucaria was by Ambrose, son of Touchstone, out of Pocahontas by Glencoe.

Merry Hampton, winner of the Derby in 1887, was inbred to Pocahontas and Queen Mary on both sides of sire and dam. Hampton, his sire, is by Lord Clifden, son of Newminster by Touchstone, his dam Lady Langden by Kettledrum, son of Rataplan, who was by The Baron out of Pocahontas, grandam Haricot by Mango or Lanercost, out of Queen Mary. Doll Tearsheet, Merry Hampton's dam, is by Broomielaw, son of Stockwell, own brother to Rataplan, by The Baron out of Pocahontas, his dam Queen Mary by Gladiator.

St. Blaise (imported) is by Hermit, dam Fusee by Marsyas, son of Orlando by Touchstone. Hermit was by Newminster, son of Touchstone, and he has a number of crosses to Waxy, son of Pot-8-o's by Eclipse, through the brothers Whalebone and Whisker. St. Blaise is the most popular stallion in America. He got Potomac, winner of the Futurity and Realization stakes, out of Susquehanna (a mare inbred to Sir Archy and Diomed), La Tosca, Chesapeake, etc. He sold at the dissolution of the Nursery Stud for $100,000.

Imported Ill Used, a superior race-horse (winner of the Kenner and Sequel stakes at Saratoga, both two miles), sire of Forester, a superior race-horse, Bellona, Woodflower, Jack of Hearts, Florimel, Delilah, Badge and the grand two-year-old, His Highness, was inbred to Birdcatcher. He was by Breadalbane, brother to Blair Athol, son of Stockwell by The Baron by Birdcatcher, dam Ellermire by Chanticleer, son of Birdcatcher, and he traces to Waxy many times, through the brothers Whalebone and Whisker.

Imp. Galore is by Galopin, inbred to Voltaire, dam Lady Maura by Macaroni, grandam by Stockwell, great grandam by The Flying Dutchman. Galopin's dam was by The Flying Dutchman. Galore was a superior race-horse, but has yet to be tried as stallion.

Imp. Rayon d'Or, the grand race-horse and popular stallion, is inbred to Touchstone, and has almost numberless crosses to Eclipse through the brothers Whalebone and Whisker and through Whiskey, also a large number direct to Herod.

Imp. Deceiver (Padlock), now one of the popular stallions of America, is inbred to Touchstone, and has two crosses of Glencoe.

Imp. Mortemer, a superb race-horse and sire of St. Christophe, winner of the Grand Prize of Paris, Verneuil, winner of the Ascot Gold Vase, Gold Cup and Alexandra Plate, Chamant, winner of the 2,000 Guineas, 1877, Exile, Wanda, Pontico and other good ones in America, is inbred to Partisan, son of Walton by Sir Peter, and has a large infusion of Eclipse blood on both sides from the best sources.

Many of the most famous producing mares in England were closely inbred. Mandragora (dam of Mandrake, Apology and Agility) and her sister Mineral (dam of Wenlock, St. Leger winner, 1872, Schwindler and Kisber, Derby winner, 1876) were inbred to Birdcatcher.

Marpessa, the dam of Boarding-School Miss, Pocahontas (the dam of Stockwell, Rataplan, King Tom, Ayacanora and Araucaria, dam of Rayon d'Or), is inbred to Whiskey, son of imp. Saltram by Eclipse. Her blood is to be found in nearly all the best race-horses of our day.

Seclusion, the dam of the great Hermit, sire of St. Blaise, Marden, Peter, Tristan, etc., is inbred to Sultan, son of Selim by imp. Buzzard.

Isoline (dam of Isola Bella, Isonomy's dam, and of St. Christophe) was inbred to Sir Hercules by Whalebone.

Bay Celia, the dam of The Earl and The Duke, by Orlando, son of Touchstone by Camel, dam Hersey by Glaucus, out of Hester by Camel, was inbred to Camel.

Mowerina, own sister to Cotherstone, was inbred to Waxy and Penelope. Her sire Touchstone, was a great grandson of Whalebone by Waxy, out of Penelope. Emma, her dam, and also the dam of imp. Trustee and Mundig, was by Whisker, own brother to Whalebone by Waxy, out of Penelope by Trumpator.

Many of our imported matrons were also inbred. Imp. Barbarity by Simoom, son of Camel by Whalebone and Sea-Breeze by Paulowitz, son of Sir Paul by Sir Peter, dam by Buzzard, son of Blacklock, out of Donna Maria by Partisan, son of Walton by Sir Peter, was inbred to Sir Peter. Bred to imp. Eclipse, son of Orlando, by Touchstone by Camel, out of Gaze by Bay Middleton, inbred to Phantom (son of Walton by Sir Peter). Barbarity produced those good race-mares, Relentless, Remorseless and Ruthless, with Barbarian and Devastation. Imp. Britannia by Muley, son of Orville by

Beningbrough, dam Nancy by Dick Andrews, son of Joe Andrews by Eclipse, out of Spitfire by Beningbrough, son of King Fergus by Eclipse, was inbred to Beningbrough and Eclipse, and produced to imp. Belshazzar (a horse inbred to Eclipse) Verifier, La Verité, Vandyke and La Victime, and by imp. Yorkshire (a horse inbred to Eclipse) Verona, the dam of Vauxhall, Foster, Veritas, etc. A number of superior race-horses are from Verona's daughters. Imp. Princess, dam of Prince Royal and His Highness, is inbred to Sultan. Her sire, King Tom, is a great grandson of Sultan through Glencoe. Her dam is great granddaughter of Sultan through Bay Middleton.

Sir Archy, justly styled the Godolphin Arabian of America, was inbred to Herod. He was by Diomed, son of Florizel by Herod, dam imp. Castianira by Rockingham, son of Highflyer by Herod, and had no less than seven direct crosses of the Godolphin Arabian through his dam.

Boston was the best race-horse of his day. He ran 45 races and won 40, 30 of which were four mile heats, 9 three mile heats, and one two mile heats. He was the best stallion of his day, sire of the immortal Lexington, Lecompte, Red Eye, Tally-Ho, Ringgold, Arrow, Bostona, Nina and many others, and was closely inbred to imp. Diomed. He was by Timoleon, son of Sir Archy by Diomed, dam sister to Tuckahoe, by Ball's Florizel, son of imp. Diomed. Timoleon's dam was by imp. Saltram, son of Eclipse. Boston's grandam was by imp. Alderman, son of Pot-8-o-'s by Eclipse.

The great Lexington, the only horse that ever ran four miles better than 7.25 on two occasions (7.19¾, 7.23¾), was inbred. He was by Boston, grandson of Sir Archy by imp. Diomed. Lexington's dam was Alice Carneal by imp. Sarpedon (a horse inbred to Eclipse), grandam Rowena by Sumpter, son of Sir Archy, and he had a large infusion of Eclipse blood with the Herod on both sides.

Jack Malone and Pat Malloy, own brothers, both superior race-horses, full of substance and quality, were much inbred. They were got by Lexington (himself inbred to Sir Archy by imp. Diomed), dam Gloriana by American Eclipse, son of Duroc by imp. Diomed, grandam Trifle by Sir Charles, son of Sir Archy by imp. Diomed, she out of a daughter of Cicero, son of Sir Archy. Both were successful stallions, and many of their daughters are grand good brood mares.

Longfellow, the grand race-horse and stallion, sire of The Bard, Leonatus, Longstreet, etc., is much inbred on his dam's side. Nantura, his dam, was by Brawner's Eclipse (grandson of Duroc by Diomed, his dam by Henry, son of Sir Archy by Diomed, grandam, Young Romp by Duroc, son of Diomed), Longfellow's grandam, Quiz by Bertrand, son of Sir Archy by Diomed.

The Bard by Longfellow, a very superior race-horse and promising stallion, is inbred to Sir Hercules, having three crosses of that blood, one through Faugh-a-Ballagh and two through Birdcatcher, the own brother to Faugh-a-Ballagh, and is also inbred to Diomed through Lexington, sire of War Dance and Nantura, dam of Longfellow.

Salvator, the great and popular race-horse, whose one and a quarter miles in 2.05, and one mile in 1.35½, has never been equalled by one horse, is inbred to Glencoe. Stockwell, the grandsire of Prince Charlie (Salvator's sire), was out of Pocahontas by Glencoe, and Salvator's dam, Salina, is out of Lightsome by Glencoe.

Enquirer, one of the best sons of Leamington as a race-horse and stallion, is much inbred on his dam's side. Lida, his dam, was by Lexington, son of Boston, by Timoleon by Sir Archy, his grandam, Lize, by American Eclipse, grandson of Diomed, his great grandam, Gabriella, by Sir Archy by Diomed.

Eolus, one of the best and most popular and successful stallions in America, is inbred on the dam's side. Revenue, the sire of his dam, was out of Rosalie Somers by Sir Charles, son of Sir Archy, his grandam Mischief by Virginian, son of Sir Archy, Eolus' grandam Sarah Washington by Garrison's Zingance, son of Sir Archy, his great grandam Stella by Contention, he by Sir Archy.

Falsetto, a superior race-horse, winner of the Phœnix Hotel Stakes, Lexington, Ky., Clark Stakes, Louisville, Ky., Travers and Kenner Stakes at Saratoga, N. Y., is inbred to Lexington and Boston on both sire and dam's sides. His sire, Enquirer, was out of Lida by Lexington, and his dam, Farfaletta, is by imp. Australian, out of Elkhorna by Lexington, son of Boston, both inbred to imp. Diomed, as was also the dam of Enquirer. All the best of his get, Dew-Drop, Jennie Treacy, Fordham, Kenwood and Gascon, were from mares by Lexington or granddaughters of Lexington.

Foxhall, that went to England, and whom William Day said was the best race-horse he ever trained, winner of the Grand Prize of Paris, Cesarewitch and Cambridgeshire Handicaps and Ascot Gold Cup, is inbred to Vandal. King Alfonso, his sire, is out of Capitola by Vandal. Jamaica, his dam, out of Fanny Ludlow by imp. Eclipse, she out of Mollie Jackson by Vandal.

Hanover, the superior race-horse, winner of all his races at two years old, and 20 out of 27 stakes at three years old, and now regarded as a promising stallion, is inbred to Vandal, son of Glencoe, being by Hindoo, son of Virgil by Vandal, dam Bourbon Belle by imp. Bonnie Scotland, out of Ella D. by Vandal.

Lisbon, a good race-horse, sire of Troubadour, Ripple, Grimaldi, etc., is much inbred to Pocahontas. Imp. Phaeton, his sire, was by King Tom, son of Harkaway and Pocahontas by Glencoe, his dam imp. Lady Love by Stockwell or Caterer, son of Stockwell; Stockwell by The Baron, dam Pocahontas.

Norfolk, one of the best sons of Lexington, both as race-horse and stallion, was closely inbred to Sir Archy and imp. Diomed. Boston and Lexington were inbred to Sir Archy and Diomed. Novice, the dam of Norfolk, was by imp. Glencoe, dam Chloe Anderson by Rodolph, son of Sir Archy Montorio, son of Sir Archy, grandam Belle Anderson by Sir William of Transport, own brother to Sir Archy Montorio. Norfolk won two of the most remarkable races on record, two-mile heats in 3.37½, 3.38½, three-mile heats in 5.27½, 5.29½.

Many of our best producing native mares are much inbred. Coral, the dam of Uncas and Wanderer, good race-horses and successful stallions, was inbred to Sultan. She was by Vandal, son of Glencoe by Sultan, dam by Cotherstone, grandam by Slane out of Gleneairne, own sister to Glencoe by Sultan.

La Henderson, the dam of Virgilian, Ferida, Aella, Ferona, etc., is inbred to Sir Archy and imp. Diomed. She is by Lexington, dam Kitty Clark by imp. Glencoe, out of Miss Obstinate by Sumpter, son of Sir Archy. Maiden, her sister, is the dam of Parole, James A., Powhattan, Pappoose, Parthenia, etc.

Mattie Gross, the dam of the fine race-horses Mate, Whisper and Grenada, was inbred to Sir Archy and Diomed. She was by Lexington, dam by American Eclipse, son of Duroc by Diomed, grandam by Orphan, son of Ball's Florizel by Diomed, Orphan's dam by Diomed.

Fanny Holton, dam of Ten Broeck, whose five races are unequalled in the world (mile 1.39¾, two miles 3.27½, 2⅝ miles, 4.58½, three miles 5.26½, four miles 7.15¾), was much inbred to Sir Archy and Diomed. She was by Lexington, dam Nantura by Brawner's Eclipse, son of American Eclipse by Duroc by Diomed, Brawner's Eclipse out of daughter of Henry by Sir Archy, grandam sister to American Eclipse's dam by Duroc, son of Diomed, Nantura's dam Quiz by Bertrand, son of Sir Archy.

Aerolite, my beau ideal as a brood mare, and her sister Idlewild, were inbred to Sir Archy and Diomed. They were by Lexington, dam Florine by imp. Glencoe, out of Melody by Medoc, son of American Eclipse by Duroc by Diomed. Aerolite, bred to imp. Australian (himself inbred to Whisker), produced Spendthrift, Fellowcraft, Miser, Rutherford, etc., all good horses and stallions. Idlewild ran four miles herself in 7.26¼, and her son Wildidle by Australian won at four miles in 7.25½. Her daughter Fanchon is the dam of winners.

Katona, the dam of the superior race-horse and good stallion Tom Ochiltree, Kadi, Partnership, etc., is inbred to Muley. She was by Voucher, son of Wagner, out of imp. Britannia by Muley, her dam by imp. Margrave by Muley, son of Orville by Beningbrough. Her sons by Lexington are inbred to Sir Archy and Diomed.

I have quoted examples enough of both English and American horses to establish my opinion that inbreeding is not injurious, and that, if properly carried out, it does not lead to degeneracy. Being in possession of good blood, symmetry (by which I mean all the elements of good conformation) and good temper, there can be no question that crossing out and then back to the same good blood, symmetry and temper, the shape and good qualities are preserved and perpetuated, and the produce never degenerates.

CONFORMATION.

The race-horse, or rather the thoroughbred horse, from his symmetry, power, graceful action, light, elastic form, speed and stamina, is beyond question the highest type of the equine family. He possesses more of the beauties of formation than any animal which approximates to him in size and shape; he excels all others in the intelligent expression of his countenance and shape of his head, and in the simplicity, compactness and completness of his digestive organs, the great leverage of his hocks and hind quarters, and the wonderful mechanism displayed in the construction of his legs and feet.

BONE.

The bones are the passive agents of locomotion and are the levers upon which the muscles act; they support and sustain the body and act as protectors to the softer and more vital parts. The projections, roughened surfaces of the bones are the points for attachment of the muscles. As these projections act with lever power, they consequently cannot be too long; length of leverage, mechanically speaking, diminishes the intensity of the muscular action. This is notably applicable to the spinous and transverse processes of the vertebral column, the point of the elbow (ulna) and projecting point of the hocks (os calcis). Largeness of bone is not an invariable evidence of strength, as the density of material and closeness of structure gives greater specific gravity to the bone of the thoroughbred than is found in common stock. Joints are composed of the articulating surfaces of the bones, and are the means of extensive motion. In order that flexure may be as free as possible, and to avoid occurrence of disease from friction, which sometimes ensues, the bones at the joints should have considerable expansion, so as to divide and distribute the weight as much as possible over an extensive surface. It is an established law in mechanics that a simple joint does not possess great strength or admit of much motion. In animals, nature, strength and accelerated motion is aimed at, has made the joint a compound one, or a series of joints. This is the case in the human foot, which is arched, and where strength is insured by the peculiar arrangement of the bones, enabling them to bear the entire weight of the body, while the elasticity is maintained by the play between each separate joint. It is the same, practically speaking, in the knee, hock and foot of the horse, each being composed of a series of joints. In the hock the motion is restricted to two bones, the others being employed in sustaining the weight of the body and receiving the lever pressure of the point of the hock, and imparting a spring to counteract the concussive effects. The motion of each individual joint is in itself insignificant, but the effect of the series combined is sufficient to give strength and protect the parts.

MUSCLE.

The muscles are the active motive powers of the animal machine, and in most cases give indication of a horse's real energy. They are the true organs of motion and are distinct from the other tissues of the body by their texture and vital contractility. Muscles are made up of fibres, arranged in bundles, proportioned in quantity to the required action. The effects of their contraction are conveyed to the distant parts by rope-like tendons in which they terminate. The larger muscles are provided with correspondingly larger tendons, hence the relative size of the tendons is indicative of muscular strength. The muscles perform various functions in the animal economy, the most important of which is the result of their contraction on the bones, thus producing locomotion. The effect of the contraction of muscle is the drawing together of the points of attachment, increasing its diameter and decreasing its length. Hence, in proportion to the length of the muscular fibres, will be their contractile power. In all cases where speed is the desired quality, long muscles are indispensable, providing for the complete flexion and extension of the limbs, thus giving the greatest possible length to the stride. In order that this extensive contraction should be continued with ease and without fatigue, a due proportion of vital tonicity of the muscular fibre is indispensable. No animal can be strong unless the muscles are bulky and well developed. Power depends upon the amount of the contractile fibre. Exercise develops the muscles, enlarges the outline of the limbs and increases and intensifies the ease and force of the contraction. Tonicity of the muscular system constitutes the leading element of condition, which is judged by the firmness of the muscles, absence of unnecessary connective fatty tissue between the bundles and between the muscles and the skin, showing a prominent, distinct, clean and well-defined appearance of each muscle beneath the integuments.

THE HEAD.

No part of the horse more definitely illustrates the superiority of blood than the head. A cold, cross or coarse one is apparent in its contour. It is a ready evidence of equine aristocracy and is an unmistakable indication of the spirit and temper. The head should be clean and bloodlike, and of a length in proportion to the size of the frame, divested of all unnecessary covering of flesh, and should articulate with the neck in a free and graceful manner. The bones of the lower jaw not too heavy, with sufficient width to admit of flexure towards the throat without interruption of respiration. The line drawn from the foretop to the muzzle cannot be too straight. A curvature outward will indicate sluggishness, and if inward is likely to impede respiration. The cranium or skull should be wide, as containing the sensorium, and the face from the eyes down in due proportion to constitute a well-formed head. If narrow between the eyes, the horse is apt to be obstinate and sluggish. The eyes should be full, clear and nearly circular, color rather dark, as those showing too much white are inclined to be flighty and unreliable. If the eyeball is sunk into the orbit or has a corrugated appearance of the lids it indicates some latent disease. The ears are the index to the temper of the horse, telegraph his will and give warning of his intentions. I like fair-sized ears, rather long than short, not too wide apart, lively in motion and erect. They give animation and expression to the head and indicate vigor and endurance. The lips are prehensile. They collect the food and direct its course to the teeth. As an evidence of blood they must be thin, and as an evidence of vigor they must be firm against the teeth and gums. If relaxed and pendulous it is a sure criterion of the lack of muscular energy.

THE NECK AND CREST.

A well developed neck, properly shaped, is a sure indication of strength; a thin, poorly developed one shows weakness and want of constitution. The lower surface, that is, the portion from the point of the shoulder to the larynx or gullet, cannot be too short, while the length from the top of the withers to the poll of the head should be more than double that of the former, because there will be less resistance in the air passages and rapid respiration much easier. This formation is affected more or less by the proper conformation of the other portions of the frame, more particularly of the shoulders. The point of the shoulder being pushed forward and the apex of the shoulder blade being well inclined back into the backbone, lengthens the upper surface of the neck, and the scapula at this point pushed forward shortens the lower surface to the neck and lessens the space between the breast and the larynx. It is not often that you find a good neck on a bad animal, or a bad shaped neck on a good one. A good neck is strong, deep and broad, running into the shoulders smoothly, and formed as above described. A thin, ewe neck or a light, tapering, peacocky one is a sure indication of weakness, not only in this part, but of the whole system.

THE SHOULDER.

The shoulder is composed of two bones, the scapula and the humerus, which are enveloped in strong, powerful muscles. A certain proportion of muscularity is indispensable to the possession of power and continuance of great exertion. The scapula or shoulder blade should be obliquely directed backward, so that the lower end meets the humerus or upper bone of the arm in a slanting direction. This formation not only lessens or obviates concussion, but increases the performing power of the muscles. The shoulder should be well covered with muscle, but not bear the appearance of being overloaded. The scapula and humerus being in too straight a line to each other act to great disadvantage, and the muscles, in order to overcome the malformation, expend a useless amount of contractile force. When these bones meet each other in an oblique direction this force is unnecessary, as the parts are already partially flexed and can act with concentrated effect. The limb is more easily elevated and the shoulder action increased. Youatt says: "The point of the shoulder is projected forward, and therefore the pillars which support the fore part of the horse are likewise placed proportionately forward, and they have less weight to carry. They are exposed to less concussion, and especially concussion in rapid action. The horse is safer, for having less weight lying before the pillars of support, he is not likely to have the centre of gravity thrown before and beyond them by an accidental trip, hence he is not likely to fall, and rides more pleasantly as there is less weight in the hand of the rider." A clean, thin, weak and knify shoulder is not desirable, nor is an upright and overloaded one.

THE CHEST.

The chest contains the chief organs of animal life, those of respiration and circulation. The chest should be of sufficient width to allow free action of the respiratory organs. It should be not too broad, but round rather than narrow and deep. It is by the shape of the chest and its capacity that we judge of the constitutional vigor of the animal. The girth of the chest does not give an infallible idea of its internal dimensions. On the principle that a circle contains more than an ellipse, so the broad chest contains a greater internal area than a deep one, but I do not fancy too wide a chest as you stand before a horse, for in such a formation the frame is not so compact and regular as is required for continued exertion and freedom of action.

THE FORE LEG.

The fore leg is composed of the cannon bone in the front and two small splint bones behind. The cannon bone sustains the bulk of the weight and communicates it to the pastern. A minor part of the weight is transmitted from the knee to the head of the splint bones, which are attached by centres of ossification to the cannon bone, and when in action descend and break or lessen the concussive force, and ascend to facilitate the extension of the legs. Thus by playing up and down as the foot is alternately raised and placed upon the ground, they bestow elasticity and prevent concussion. By over exertion or injury this part becomes inflamed and the extremities are unnaturally joined to the cannon bone, which is called splint and produces lameness. The cannon and splint bones cannot be too short, relatively speaking, nor the radius or bone of the arm too long. They should be broad and flat as you look at them sideways, the muscles running full into the knees; the legs should stand perpendicular and the elbows as near in a line with the stifle joints as possible, pointing neither in nor out. If inward, the action is jarring, stiff and weak, owing to the concussion being more direct. If outward, the action will incline to be sprawling and ineffective. But if compelled to choose between either, I should prefer the in-elbowed horse to the out-elbowed. It is of vital importance that the arm be muscular, and, as an indication of speed, of good length. The radius or upright bone and the ulna, which forms the elbow, become consolidated into one when the animal matures. The knee is formed by a treble series of joints, and corresponds to the wrist in a human being. In speedy animals it should be long, broad and flat in front, but not thick when viewed laterally. The six different bones composing it are each covered above and below by a thick coating of cartilage connected together by strong ligaments, separated by fluids and surrounded by membranes. Thus concussion is shared by all of them and by their peculiar connection it is rendered harmless. The seventh bone, called the trapezium from its quadrangular shape, is placed behind the knee, but bears no portion of weight. Two of the flexor muscles proceed from the bone of the arm and are inserted into it, and being thus thrown off of the limb, act more directly and with considerable power. As the terminous of the muscles descend on the limbs, they are tied down by strong ligamentous bands, particularly in the neighborhood of the joint. The extensor tendons, which lie principally in front of the cannon bone, are prevented from starting, strengthened and assisted in their action, but the flexor tendons which are at the back of the leg would be liable to friction and their motion impeded were they bound down too tightly. The trapezium bone prevents the ring-like ligament from pressing too closely on the main flexor tendons, whilst it leaves them room to play and has a capsule filled with sinovia, which oozes out slowly in the course of the tendons and supplies them with a fluid which prevents much injurious friction. The arm, therefore, being composed of bone and muscle, and from the knee downward made up of tendons and ligaments, cannot be proportionately too long or the space between the knee and the fetlock joint too short or broad. The tendons should enter the posterior portion of the knee boldly. The leg, as you look at it laterally, as a whole should not appear to be made up of a series of joints, but should present the appearance of gradually tapering from the elbow to the fetlock joint. Looking at the conformation of the whole leg and not at any parts of it, I would advise the discarding of a calf-kneed horse, or what is called back at the knees. They are usually speedy, but no matter how well shaped the leg in other respects, the strain on the legs in such conformation must be severe and cannot long be endured without injury. Lord George Bentinck, who nominally reduced racing to a science, tested horses with arched knees and tried them over different courses. His experience was that on level and inclined courses there was little or no difference, but over uphill courses the superiority of the arch-kneed horse was very marked. Of the two formations the arched knee is preferable and will stand longer, that is, if naturally so, not from contraction caused by disease.

THE FETLOCK AND PASTERN.

Four bones compose the fetlock joint: the upper and large pastern, the lower and small pastern, and two sessamoid bones. From the back part of the cannon or shank bone arises a strong ligament, called by veterinarians the suspensory or main ligament, which is of great elasticity, and which passes down the leg, bifurcates and connects with the two sessamoid bones at the back of the pastern joint. This tendon elongates with pressure and contracts when the weight is removed, and thus by its elasticity assists the flexor muscles in flexing the pastern joint. When overtaxed this tendon is broken through, and the horse then, in turf parlance, is said to be broken down. When thoroughly so, the point of the fetlock comes to the ground, and the horse is unfit for great exertion or racing afterwards. The pasterns should not be too upright, nor is an excess of obliquity desirable, and they should be more oblique behind than in front. Nature, having made such extensive provision for elasticity in the formation of the front limbs, economises as much as possible to increase the strength of the parts below. There is, therefore, less necessity for great obliquity of the pasterns in the front legs than in the hind ones. Elasticity and strength may be regarded as antagonistic forces. As one predominates the other diminishes; hence, where strength is demanded, a more upright pastern is indicated than where elasticity is required. It may be put down as an axiom, that a horse is unworthy of the name of a race-horse that does not possess good and properly formed fore-legs, and a material point in such formation is an oblique, large pastern. If the pasterns are too short, or upright and stiff, they indicate a want of speed and tax both foot and leg unduly.

THE FEET.

The true shape of the foot is the segment of a cylinder. The front part of the wall is called the toe, the sides the quarters, the posterior part the heels. The under surface of the foot is composed of the frog, bars and sole. It should be high at the heels, very concave underneath, wide at the bar, with well-developed frog, which is both elastic and firm, enabling it to resist the greatest shock without injury to the foot. There should be no enlargement about the coronet, and the horn should be of healthy color and of even and regular form. If rings appear of a dirty reddish hue, there is some hidden disease or inflammation.

THE TRUNK OR BODY.

The trunk or body should be of good length, with proportionate depth of girth or true ribs, but more circular than flat and deep. See that the girth is properly formed, with sufficient width through the heart, and that the large muscles, which extend forward and outward the ribs that expand the chest, move the shoulder blade in action, contract the humerus, brace the muscles of the back and aid the muscles of the haunch in propelling the body. The requisite development of these muscles or thickness between the knees is a great sign of strength of limb and wind. As a rule, horses of such formation will stay longer and carry more weight than others. Circularity of carcass with a great development of the lateral muscles interferes the least with movement of the animal, and provides room for healthy digestion. Herring-bellied or light-bodied horses are, as a rule, bad feeders, and a horse which is a bad feeder is seldom or never a reliable race-horse.

BACK AND RIBS.

The false, standing or back ribs will form a gradual decline from the girth, and should not appear to swell out too far, as viewed from the rear or front. The inter-costal muscles between the ribs are connected with the diaphragm internally and externally, and with the muscles of the back. The internal and external muscles of the ribs are beautifully and advantageously formed for action, the fibres running in adverse direction and crossing each other give additional strength to united action; that is, in contracting and drawing the ribs closer together in connection with the diaphragm, the muscle extending from the thorax or breast to the termination of the loins by distinct fibres by which the viscera is forced backwards and inspiration is perfected. The muscles of the abdomen contract and, converging the ribs, push back the diaphragm and expel the air from the lungs. I do not like a spare-ribbed horse or one deficient in the muscles which act on the respiratory organs, as such horses seldom exhibit ability to perform satisfactorily. The large muscle along the back-bone is of vital importance, as it is connected with the muscles of the loins, bracing all the muscles of the back and aiding in working the shoulder. No horse can ever be a good one, able to carry weight or perform any extraordinary exertion, which is deficient in muscular development of the back. It has been observed that the best backed horses are the best performers. The back should be straight and rather hollow than roached, as in the latter formation the horse is inclined to over-reach, carry his head low, and possess little or no elasticity of the spine. A little depression just behind the withers and then an almost straight line to the loins is the form most desirable for beauty and strength.

THE LOINS.

The proper formation of the loins is material in all horses, and more especially in the race-horse. They must be broad, well rounded and muscular. The strength of the loin, and more especially of the hind quarter, depends much on this. In formation the bones of the loin are larger than those of the back, and a more dovetailed union exists between them than those of the back. Every provision is made for strength. The union of the back and loins should be carefully scrutinized, for there is frequently a depression between them at the junction, which shows an imperfection in just formation of the spine, and is an evidence of weakness.

THE CROUP AND TAIL.

The setting on of the tail is an indication of breeding in the horse, and all those possessing aristocracy of blood, display it in the graceful manner in which the tail is attached to the body. The croup should curve very little and should not be too long. A small, contracted and protruding sphincter of the anus denotes superior muscular energy in the animal, and is an evidence of constitutional vigor.

THE HIND QUARTERS.

The hind quarters differ from the fore ones in the fact that there is less elasticity in their connections. Nature, in order to give greater efficiency to the muscles of the hind quarters, has attached the hind extremities to the frame by direct bony articulations, which is not the case in the thoracic extremities. The functions of the anterior extremities is to support and direct, whilst that of the posterior limbs is to propel the frame.

The haunch is composed of three bones. The ilium is principally concerned in the formation of the haunch. The ischium or hip-bone is behind and below the ilium.

The pubis unites with the two former below and behind. The loins, to the root of the tail, should be slightly rounded, the haunch bones then will be more oblique and will produce a corresponding obliquity in the thigh bone, in which direction the muscles act most advantageously. This shape of the haunch is characteristic of the thoroughbred horse. Width of the haunch is a matter of great consequence, as it evidently affords more room for the attachment of muscles, but not so wide as to cause the horse to be ragged hipped, though many of our ragged hipped horses have been both fleet and strong, whilst few of the narrow hipped ones possess strength. If the loins are broad and the horse well ribbed up, the points of the hip can scarcely be too far apart. In all quadrupeds celebrated for speed we find the hind legs remarkable for their length, and the muscles of the loins and quarters for their great size and power. Speed depends almost entirely on the length of the hind extremities and possession of sufficient power to propel the carcass with vigor and rapidity. If they are deficient in length of leverage or muscular energy, the most perfect formation of the other portion of the body cannot correct the fault. The three bones composing the haunch, ilium, ischium and pubis on each side, form an arch for the support and protection of the posterior parts of the body and their contents. These three bones become partially consolidated, and form at their point of union a cup for the reception of the head of the true thigh bone (femur), forming the cavity of the hip joint, which is a ball-and-socket joint, and known by the name of whirl or round bone, which is deeply covered with muscles. From the hip joint the thigh bone slants forward to meet the lower or second thigh bone (tibia), which, with the patella, a bone corresponding to the knee pan in the human race, constitutes the stifle. The lower or second thigh bone is direcred backward, forming a considerable angle with the true thigh bone, giving the muscles acting on the parts a mechanical advantage which they would not otherwise possess, and throwing the weight on them in such an oblique direction as to prevent concussion. These two bones, the femur and tibia, should be long, for the reason that it gives them double mechanical power by lengthening the muscular fibre and increasing the leverage at which the muscles act. As the length of the extremities above the knees and hocks is increased, that below these points is diminished. If the horse has short quarters and shoulders, the columns of support must be elongated somewhere, which is usually below the knee and hock. If he has a deep forehand and long quarters, the cannon bones are diminished in length, a formation much desired and esteemed.

THE HIND LEG, FETLOCKS AND PASTERNS.

There is little or no difference in the mechanical construction of these parts from those of the fore extremities, the anatomy of the parts below the knee and hock. The cannon or shank bone is longer and the splint bone larger in the hind extremities. The pasterns are more oblique and deservedly so. The elastic play of the upper parts of the fore limbs being much greater, a less obliquity of pastern is requisite. The hind limbs being attached to the frame by bone, it is necessary that the pasterns be much more oblique.

THE HOCK.

The proper conformation of the hock is of exceeding importance. It is composed of ten bones, between some of which there is extensive motion, while between others the motion is inconsiderable, but there is much strain on all of them during progression. It is scarcely less important than the thigh. If there is malformation or weakness here, the muscles of the thigh will be weakened in their force of action or perform their action with feeble results. The os calcis or point of the hock should stand out boldly from the other bones, as the action and leverage of the tendons will be in exact proportion to the distance; hence the hock should be broad as you look at it from the sides, and should be free from all beefiness between the os calcis and other bones as well as behind and below the calcis, and taper gradually into the shank without a tied or wasted appearance at the lower extremity, and should be straight in preference to sickled or cow hocked. The os calcis or point of the hock is the most important feature of this joint. It projects outward and upward from the other bones and gives attachment, according to its length and direction, to the tendons of the most powerful muscles flexing the limb. The strongest and best wearing hocks are straight. The importance of a proper shaped hock is the more marked from its complicated structure and great strain in its work, which renders it the seat of lameness. The os calcis too long means an excess of leverage, hence curbs, an important anatomical feature in the race-horse, who from the nature of his work throws unusual strain on these parts.

BROOD MARES.

In purchasing mares to breed from it is wise to pay particular attention to the following points:

1. In the management of young mares be careful that they are not overloaded with flesh and gross, nor kept too thin, and hence weak. Either extreme is inimical to the production of sound, healthy and strong foals.

2. In selecting a mare for your stud, examine the stud-book and see if her dam and grandam have uniformly bred winners, or are descended from a family which is noted as producing race-horses and winners. The more superior and sound the race-horses from the immediate and near relations, the better and more desirable she must be. If, however, she is the only one of her family from a number of her brothers and sisters tried that has proved a good performer, her acquisition is not so desirable unless she has extraordinary qualities to commend her.

3. Examine carefully in regard to flightiness of temper, nervousness, organs of respiration and digestion, shape and soundness of feet, objectionable habits—such as wind sucking and cribbing. All these must be taken into consideration; also the strong or weak points which are peculiar to her family that may be transmitted.

4. Before buying a mare or mares that have bred foals, in addition to the qualities of her produce, satisfy yourself as to the state of her uterus. Look for outward signs of rupture, and, when trotted or cantered, for a blubbering noise, which would indicate rupture extending to its internal parts. Such an animal is worthless, and if she should chance to have a foal it would be an accident.

5. Examine carefully the bag or udder. See if it is naturally developed and both sides can be equally used. The condition of the last foal will show whether the mare is a good nurse, and the quality of the nourishment which it received from the dam. The quality and quantity of the mother's milk will solve the question whether the foal will grow into a race-horse or develop into a miserable weakling. Small and puny foals develop with wonderful rapidity at the side of a mare which supplies good, nourishing, rich milk. Whilst others, whose foals have size, shape and quality, go backward and melt away like butter in the sun. It is advisable, with good mares from racing families that produce winners, to take away the foals soon after they drop, and give them to some mare known to be a good nurse. Or, if the mother is an indifferent milker, which is readily ascertained by the colt not filling himself, give the best fresh milk, slightly sweetened. He will take it first from the finger being inserted in the milk, but will soon drink it as a calf. No colt need be despaired of, either from a bad mother or if the mother dies.

Sensitive mares do not easily submit to exchange of foals, but with care and patience you will have no difficulty. All mares know their own foals by smell, as one can readily see by observation, when collected together. Rub into the coats of the foals to be exchanged a little aniseed oil for a few days, until the mares become accustomed to it. This will prepare for the deception. Separate the mares from the foals and keep them away until the pressure of the milk in the bag become distressing. The foals being exchanged, the mares are returned and the change will be successful.

Avoid purchasing and reject all mares which the stud book reports show have been repeatedly barren, slipped their foals or produced twins.

One of the most important requisites in a brood mare is symmetry of conformation. She should be long, deep in the chest, with large, roomy pelvis, with breadth of hip, in order that the foal have ample room and space for its development. Leggy, light and short-bodied mares cannot be expected to produce large foals or good shaped ones. Other not glaring imperfections can be better overlooked than these faults in conformation.

Many good judges do not like mares with too coarse and luxuriant growth of hair in the mane and especially with full and bushy tails at the root. Foals by such stallions or out of such mares generally lack quality and are deficient in energy. A rat-tailed horse is not a beauty to the eye, but it is seldom that you find a bad horse with a rat tail.

With mares possessing the qualifications mentioned, with marked development of muscles and bone, added to symmetry in conformation, which every animal should possess which is intended to propagate its species, we may hope that the foals will possess similar good qualities. I particularly like horses and mares which possess generic character, unmistakeably expressed, in either sex. I condemn all mares of studdish appearance, either in shape or manners, especially when examination is necessary to determine whether they are mares or geldings. The more quality a mare possesses the more marked will be the evidence of her sex.

Avoid the system, ruinously adopted and followed by many, of breeding from mares or stallions merely because they were good. Consider the essential qualifications of the race-horse. See that the animal is eminently gifted with certain inestimable qualities, such as temper, handiness, immense nervous action and indomitable courage. These may and often do outweigh great defects in conformation, but unless sure of such grand qualities and peculiarities being handed down to their progeny in equal ratio, do not regard them beyond their true worth.

It is worse than useless to purchase the stoutest, best bred mares from racing families, and expect to breed race-horses by fattening them in rich pastures until November, and then starve and half feed until April, under the false idea that it makes the colts hardier and tough. It must be remembered that, for every one living through it, two die, and the survivors must have iron constitutions, and with proper food and care would have proved better, sounder and possess more size and substance. Abundant nourishment tends to the formation of bone and muscle; want of it leads to emaciation, weakness and death.

SUMMING UP (NECESSARILY A REPETITION).

It is generally claimed and admitted by all writers upon the thoroughbred horse that he owes his origin to three great ancestors or sources: The Darley Arabian, Byerly Turk and the Godolphin Arabian. Although this theory is accepted as true, it is misleading and not sustained by facts. The first class, as arranged in the stud book, is that of the Darley Arabian. If the Arabian alone was entitled to the credit of founding this line, what becomes of the female line and what credit is due to the mare Betty Leedes that first brought the Darley Arabian into notice? Betty Leedes, the dam of Flying Childers and Bartlett's Childers, was by old Careless, son of Spanker and a Barb mare, her dam sister to Leedes (called Cream Cheeks) by the Leedes Arabian, grandam by Spanker, great grandam old Morocco mare (Old Peg), Spanker's own dam by Lord Fairfax's Morocco Barb, out of Old Bald Peg, who was by an Arabian, out of a Barb mare. There were other race-horses in the family prior to the advent of Flying Childers, but the great excellence of the family in the male line is through the two Childers, especially through Bartlett's Childers. Flying Childers was said to have been the fleetest horse in the world. It is said that in the year 1721 he ran a trial against Almanzor and Brown Betty over the Round Course at Newmarket, 3 miles 4 furlongs 139 yards, carrying 128 lbs., in 6 minutes and 40 seconds. He ran over the Beacon Course, 4 miles 1 furlong 173 yards, in 7.30. It is not believed that these statements of time are true, as there were no watches in those days by which time could be taken. He got Snip, sire of Snap. Bartlett's Childers founded the Eclipse line. He got Squirt, dam Shock's dam by Lister's Snake, grandam Grey Wilkes, sister to Clumsey by Old Hautboy, tracing to a Sedbury Royal mare. So little was Squirt esteemed as a stallion that he was ordered to be shot, but was begged off by his groom while being led out to the dog kennel. He afterward got Marske, Syphon and the noted old Squirt mare, the dam of Pumpkin, Maiden and Purity. Marske got imp. Shark, dam by Snap, son of Snip, a grand race-horse and successful stallion. From the cover of Marske, Spiletta produced

Eclipse and his sister Proserpine. Spiletta was by Regulus, son of the Godolphin Arabian, out of Mother Western by Smith's son of Mr. Lister's Snake, own brother to William's Squirrel, out of a daughter of the Akaster Turk. Eclipse got Pot-8-o-'s, King Fergus, Mercury, Alexander, Don Quixote, Saltram, etc. It is to these sons of Eclipse nearly all of the best race-horses of our day in the male line from Eclipse trace their origin. Pot-8-o-'s, out of a mare by Warren's Sportsman, son of Cade by the Godolphin Arabian, got Waxy, out of Penelope by Trumpator, son of Conductor by Matchem.

Waxy got Whalebone, Whisker, Woful and their sisters, Web and Wire, out of Prunella by Highflyer, son of Herod. The commingling of this blood with that of the Byerly Turk and Godolphin Arabian is what gave us the grand race-horses in this line. Whalebone got Sir Hercules, out of Peri by Wanderer, son of Gohanna by Mercury, and Sir Hercules got the brothers Irish, Birdcatcher and Faugh-a-Ballagh, out of Guiccioli by Bob Booty, son of Chanticleer by Woodpecker by Herod.

Stockwell by The Baron, son of Birdcatcher by Sir Hercules by Whalebone by Waxy by Pot-8-o's by Eclipse, dam Pocahontas by Glencoe by Sultan, son of Selim by Buzzard by Woodpecker by Herod, grandam Marpessa by Muley son of Orville by Beningbrough by King Fergus by Eclipse, out of Clare by Marmion, son of Whiskey by Saltram by Eclipse. Stockwell got Blair Athol (Derby and St. Leger winner, 1864), dam Blink Bonny (Derby and Oaks winner, 1857) by Melbourne, son of Humphrey Clinker by Comus, out of Queen Mary (dam of Bonnie Scotland, etc.) by Gladiator, son of Partisan.

Blair Athol got Andred, out of Wonderful by Voltigeur; Clanronald, out of Isilia by Newminster; Ethus, out of Theresa by Touchstone; Tangible, out of Touch-me-Not by Touchstone; Craig Millar (winner of the St. Leger, 1875), out of Miss Roland by Fitz Roland, son of Orlando by Touchstone; Prince Charlie (winner of the 2,000 Guineas, 1872, and imported to America), dam Eastern Princess by Surplice, son of Touchstone, out of Tomyris by Sesostris, son of Slane by Royal Oak. Prince Charlie got Prestonpans, dam Beatrice by Voltigeur, out of Bribery by The Libel; Salvator (winner of the Suburban Handicap, 1¼ miles, 2.06⅘, and the Match 1¼ in 2.05, and the fastest mile ever run in the world, 1.35½), dam Salina by Lexington, out of Lightsome by imp. Glencoe; Prince Fonso, out of May Wilson by King Alfonso; imp. Wagner, dam Duchess of Malfi by Elland, out of The Duchess by St. Albans, he the sire of Minnehaha III.; imp. The Jacobite, dam sister to Little Lady by Orlando, out of Volley (own sister to Voltigeur) by Voltaire. Blair Athol also got Silvio (winner of the Derby and St. Leger, 1879), out of Silverhair by Kingston. Blair Athol got Struan, out of Terrific by Touchstone, and Struan is sire of Brag, out of Bounce by Flatterer. Silvio is the sire of imp. Silvermine, out of Nuneaton by Orlando, and imp. Dandy Dinmont, out of Meg Merrilies by Macgregor.

Blair Athol got imp. Glen Athol, dam Greta by Voltigeur, out of Mountain Flower by Ithuriel, son of Touchstone.

Glen Athol got Glenmore, dam Lotta by Hunter's Glencoe, out of Sally Lewis by imp. Glencoe. Glenmore was a superior race-horse, winner of the best four-mile heat race of three heats ever run, 7.29½, 7.30½, 7.31, and was, for his chances, quite a success in the stud.

Blair Athol got imp. Stonehenge (dam Coimbra by Kingston, out of Calcavella by Birdcatcher, Son of Sir Hercules) and his full brother Claremont, the sire of imp. Esher, out of Una by Dusk.

Blair Athol also got imp. Child of the Mist, owned by Mr. M. Daly, out of Ma Belle by Lord Clifden.

Stockwell got Breadalbane, own brother to Blair Athol. Breadalbane got imp. Ill Used, dam Ellermire by Chanticleer, son of Birdcatcher, out of Ellerdale by Lanercost, son of Liverpool by Tramp. Ill Used was a good race-horse and is sire of Badge, dam The Baroness by Kentucky, His Highness, dam imp. Princess (Royal Hampton's dam) by King Tom, out of Mrs Lincoln by North Lincoln, son of Pylades by Surplice, Forester, dam Woodbine by Censor or Kentucky, out of imp. Fleur-des-Champs by Newminster, Magnetizer, dam Magnetism by Kingfisher, out of Attraction by imp. Balrownie, son of Annandale by Touchstone, and Nonesuch, dam imp. Non Pareille by Kingston, out of England's Beauty by Birdcatcher. Stockwell is also sire of Citadel, dam Sortie by Melbourne, out of Escalade by Touchstone. Citadel is best known as the sire of imp. Glenelg, dam Babta by Kingston, son of Venison, out of Alice Lowe by Defence, son of Whalebone. Glenelg got Post Guard (Gen. Phillips) and Strathspey, dam La Polka by Lexington, Dan Sparling, dam Item by Lexington, Dry Monopole, dam Peru by imp. Glengarry, son of Thormanby. Grenadier, out of Alala by War Dance, Little Minch, out of Goldstone by Loadstone, son of Lexington, Tristan, out of Traviata by Tom Bowling, son of Lexington, Sir Joseph, out of Susie Linwood by Judge Leonard, son of Zero by Boston. Glenelg is the sire of a vast number of valuable brood mares producing winners and the superior mare Firenzi. Stockwell got Broomielaw, out of Queen Mary by Gladiator, and Asteroid, out of Teetotum by Touchstone. Asteroid got Siderolite, out of Aphrodite by Bay Middleton. Stockwell got The Marquis (winner of the 2,000 Guineas and St. Leger, 1862), out of Cinizelli by Touchstone, Rustic, out of Village Lass by Pyrrhus the First, Caterer, dam Selina by Orlando, The Duke, out of Bay Celia by Orlando. The Duke got Merry Duchess, Blandford, out of Auricula by Ambrose, and Bertram, out of Constance by Faugh-a-Ballagh, son of Sir Hercules. Bertram got Robert the Devil (winner of the St. Leger, 1880), out of Cast Off by Promised Land, son of Jericho by Jerry by Smolensko, Lord Lyon (winner of the 2,000 Guineas, Derby and St. Leger, 1866), out of Paradigm by Paragone, son of Touchstone. Lord Lyon got Minting, out of Mint Sauce by Young Melbourne, and Touchet, out of Lady Audley, sister to Buccaneer by Wild Dayrell. Stockwell also got St. Albans and Savernake, dam Bribery by The Libel, son of Pantaloon by Castrel, out of Splitvote by St. Luke, son of Bedlamite by Welbeck by Soothsayer.

St. Albans (winner of the St. Leger, 1860) got Julius, out of Julie by Orlando; Springfield, out of Viridis by Marsyas, son of Orlando. Springfield got Sainfoin (winner of the Derby, 1890), out of Sanda by Wenlock.

St. Albans also got Martyrdom, out of Eulogy by Euclid, son of Emilius; Mr. Winkle, dam Peri by Birdcatcher; Silvester, out of Silverhair by Kingston, and St. Ronan, out of Elspeth by Birdcatcher.

Stockwell got Lord Ronald, out of Edith by Newminster, grandam Deiadamia by Pyrrhus I., son of Epirus by Langar.

Lord Ronald got Master Kildare, out of Silk by Plum Pudding, son of Sweetmeat.

Master Kildare got Melton (winner of the Derby and St. Leger, 1885), out of Violet Melrose by Scottish Chief.

Stockwell got Thunderbolt, dam Cordelia by Red Deer, son of Venison, out of Emilia by Young Emilius. Thunderbolt got Vulcan, out of Alarum by Alarm; Lemnos, out of Laura by Orlando; Reverberation, out of The Golden Horn by Harkaway, and imp. Thunderstorm, out of Killarney by Newminster.

Stockwell got Doncaster (winner of the Derby, 1873), dam Marigold, by Teddington, son of Orlando, out of sister to Singapore by Ratan, son of Buzzard by Blacklock. Doncaster got Bend'or (winner of the Derby, 1880, and Epsom Gold Cup), dam Rouge Rose by Thormanby, grandam Ellen Horne by Redshank, son of Sandbeck by Catton, out of Delhi by Plenipotentiary, son of Emilius, Muncaster, dam Windermere by Macaroni, out of Miss Agnes by Birdcatcher. Muncaster got Saraband, out of Highland Fling by Scottish Chief, and imp. Rossington out of Lily Agnes by Macaroni. Bend'or got Ormonde (winner of the 2,000 Guineas, Derby and St. Leger, 1886), dam Lily Agnes by Macaroni, grandam Polly Agnes by The Cure, son of Physician by Brutandorf by Blacklock, out of Miss Agnes by Birdcatcher, and Orvieto, out of Napoli by Macaroni. Bend'or got Bonavista, winner of the 2,000 Guineas, 1892, out of Vista by Macaroni, and Kendal, out of Windermere by Macaroni. Stockwell got Uneas, dam Nightingale by Mountain Deer, son of Touchstone, out of Clarinda by Sir Hercules. Uneas got Prism, dam Rainbow by Yorkminster, son of Newminster, out of Blue Bonnet by Young Melbourne.

Rataplan was own brother to Stockwell by The Baron, dam Pocahontas by

Glencoe. Rataplan got Blinkhoolie, dam Queen Mary (Blink Bonny and Bonnie Scotland's dam) by Gladiator. Blinkhoolie got Wisdom, dam Aline by Stockwell, grandam Jeu d'Esprit by Flatcatcher, son of Touchstone, out of Extempore by Emilius.

Wisdom sired Surefoot (winner of the 2,000 Guineas and Eclipse stakes, 1891, defeating the great Common), dam by Galopin, grandam Miss Foote by Orlando, out of Gossamer by Birdcatcher, and Veracity, out of Vanish by Honiton. Rataplan got Ben Battle, dam Young Alice by Young Melbourne, out of Sweet Hawthorn by Sweetmeat. Ben Battle got Bendigo, out of Hasty Girl by Lord Gough, son of Gladiateur.

Rataplan got The Miner and Mineral, dam Manganese by Birdcatcher, out of Moonbeam by Tomboy, son of Jerry by Smolensko, Kettledrum, dam Hybla by The Provost, son of The Saddler, out of Otisina by Liverpool, son of Tramp. Kettledrum got Lady Langden, the dam of Hampton and Sir Bevys, out of Hariot by Mango or Lanercost. Rataplan also got imp. Dalnacardoch out of Mayonaise by Teddington.

Birdcatcher got Oxford, out of Honey Dear by Plenipotentiary, in the male line to Eclipse. Oxford got Sterling, out of Whisper by Flatcatcher, in the male line to Eclipse. Sterling got Energy, Harvester, Paradox and Isonomy, the latter out of a daughter of Stockwell. Isonomy got Common, the winner of the triple event in England in 1891, out of Thistle by Scottish Chief, a horse in the male line to Eclipse. Sterling is also sire of imp. Loyalist, brother to Paradox, Stalwart and imp. Top Gallant, the two last brothers, tracing through Sea Mark, a daughter of Adventurer, to Pewet, the dam of Sir Paul, Paulina and Clinkerina, the dam of Humphrey Clinker, sire of Melbourne, and Pewet was half sister to Evelina, the dam of Orville, Orvillina, dam of Sandbeck, Cervantes, sire of Melbourne's dam, Paulowitz, sire of Cain, that got Ion, the sire of Wild Dayrell. Birdcatcher also got imported Knight of St. George, Warlock (sire of Tyndale), Saunterer, imp. Mickey Free and Womersley. Faugh-a-Ballagh got imp. Leamington, out of a daughter of Pantaloon, a horse in the male line to Herod. Leamington got Longfellow, out of Nantura by Brawner's Eclipse, Herod line, sire of The Bard, out of Brademante by War Dance, Herod line. Longford and Longstreet, out of Semper Idem by Glen Athol, son of Blair Athol; Linden, a grand race-horse and promising sire. Enquirer, out of Lida by Lexington, Herod line, a good race-horse and popular sire. Powhattan, out of Maiden by Lexington. Eolus, out of Fanny Washington by Revenue, son of Trustee, Eclipse line. Eolus is one of the most popular and successful stallions in America. Hyder Ali, Sensation, Onondaga, Stratford, Aristides, out of daughters of Lexington, Iroquois, out of Maggie B. B. by Australian, Matchem line. Lelaps, out of imp. Pussy by Diophantus. Then there are Falsetto, Blue Eyes, McCreery, Faustus and Inspector B., good stallions nearly all, except Falsetto in the Eclipse line. Whalebone got Camel, out of daughter of Selim, Herod line.

Touchstone by Camel, son of Whalebone by Waxy by Pot-8-o's by Eclipse, dam Banter by Master Henry, son of Orville, by Beningbrough by King Fergus by Eclipse, grandam Boadicea by Alexander, son of Eclipse, out of Brunette by Amaranthus, son of Old England by the Godolphin Arabian.

Touchstone got Artillery, dam Jeanette by Birdcatcher, son of Sir Hercules, grandam Perdita by Langar, son of Selim, out of Delenda by Gohanna, son of Mercury by Eclipse.

Artillery was the sire of imp. Kyrle Daly, dam Colleen Rhue by Gemma di Vergy, son of Sir Hercules.

Touchstone sired Claret and Mountain Deer, dam Mountain Sylph by Belshazzar, son of Blacklock. Claret got Blarney, and Mountain Deer got The Druid and many good mares.

Touchstone got Assaults and Storm, out of Ghuznee by Pantaloon, and Annandale, dam Rebecca by Lottery, out of a daughter of Cervantes son of Don Quixote by Eclipse.

Annandale got imp. Balrownie, out of Queen Mary by Gladiator. Balrownie got West Roxbury, out of Columbia by imp. Glencoe, she out of Fleur de Lis by imp.

Sovereign. West Roxbury was a good race-horse and quite a successful stallion. Balrownie also got Fleetwing, out of Rhoda by imp. Glencoe.

Touchstone got Cotherstone, dam Emma by Whisker, son of Waxy. Cotherstone (winner of the Derby and 2,000 Guineas, 1843) was the sire of imp. Cairngorm, out of Glenluce by Slane, she the dam of Pilgrim by Lexington and Coral by Vandal, the dam of Uncas and Wanderer by Lexington.

Touchstone got Ithuriel, dam Verbena by Velocipede, son of Blacklock, out of Rosalba by Milo, son of Sir Peter. Ithuriel got Longbow, out of Miss Bowe by Catton. Longbow got Toxophilite, out of Legerdemain by Pantaloon.

Toxophilite got Musket, out of a daughter of West Australian by Melbourne, and Stray Shot (the dam of Shotover and Torpedo by Hermit), out of Vaga by Stockwell.

Musket got Petronel (winner of the 2,000 Guineas, 1880), out of Crytheia by Hesperus, son of Bay Middleton, and imp. Maxim (in Rancho del Paso stud), dam Realization by Vespasian, son of Newminster.

Touchstone got Surplice (winner of the Derby, 1848), out of Crucifix by imp. Priam, who sired several mares of note, Eastern Princess, Flax, etc., also Vindex, dam Garland by Langar, out of Cast Steel by Whisker. Vindex got Victor, out of a daughter of Scroggins, son of Tramp.

Victor got Valour, out of daughter of Mount Zion, son of Mountain Deer.

Touchstone got Lord of the Isles, dam Fair Helen by Pantaloon, out of Rebecca by Lottery.

Lord of the Isles got Scottish Chief and Macdonald, out of Miss Ann by The Little Known, son of Muley, she out of Bay Missy by Bay Middleton.

Scottish Chief got Childeric, out of Gertrude by Saunterer, Fitz James, out of Hawthorn Bloom by Kettledrum, King of the Forest, out of Lioness by Fandango, Pursebearer, out of Thrift by Stockwell, imp. Donald A., out of imp. Algebra by Diophantus, imp. Kantaka, out of Seclusion (Hermit's dam) by Tadmor, and Dundee, out of Marmalade by Sweetmeat.

Touchstone got Newminster, dam Beeswing by Dr. Syntax, son of Paynator by Trumpator by Conductor by Matchem, grandam by Ardrossan, son of John Bull by Fortitude by Herod, out of Lady Eliza by Whitworth, son of Agonistes by Sir Peter by Highflyer by Herod.

Newminster got Cambuscan, out of The Arrow by Slane. Cambuscan got Camballo (winner of the 2,000 Guineas, 1875), out of Little Lady by Orlando, son of Touchstone, and Onslow, out of Dulcibella by Voltigeur.

Newminster got Adventurer (winner of City and Suburban), dam Palma by Emilius, out of Francesca by Partisan.

Adventurer got Ishmael, out of Lena by Stockwell, Privateer, out of La Favorita by Thormanby, Glen Arthur, out of Maid of the Glen by Kingston, Pretender, out of Ferina by Venison, and imp. Pizarro, out of Milliner by Rataplan, she out of Manganese by Birdcatcher, and imp. Blythewood, out of Bonny Bell by Voltigeur. Pizarro's early death was a serious loss to the country.

Newminster got Cathedral, out of Stolen Moments by Melbourne. Cathedral got Landmark, out of Miss Agnes by Birdcatcher.

Newminster got Hermit (winner of the Derby, 1867), dam Seclusion by Tadmor, son of Ion by Cain, grandam Miss Sellon by Cowl, son of Bay Middleton by Sultan, out of Belle Dame by Belshazzar, son of Blacklock.

Hermit got The Abbot, Charon and Marden, out of Barchettina by Pelion, son of Ion, Clairvaux, Thebais and St. Marguerite, out of Devotion by Stockwell, Peter and Timothy, out of Lady Masham by Brother to Strafford, son of Young Melbourne, Tristan and Trapeze, out of Thrift by Stockwell.

Trappist, out of Bunch by Muscovite, son of Hetman Platoff, Whitefriar, out of Lady Blanche by Thunderbolt, Edward the Confessor, out of Princess of Wales by Stockwell, Retreat, out of Quick March by Rataplan, Zealot, out of Zelle by Stockwell, Southampton, out of Preface by Stockwell, imp. Mr. Pickwick (dead), out of Tomato by King Tom, sire of Picknicker, Facial B., Ida Pickwick, etc., Friar Tuck (now in California), out of Romping Girl by Wild Dayrell, and imp. St. Blaise, out of Fusee by Marsyas, son of Orlando, sire of St. Florian, Admiral, St. Pancras, Clarendon, Padishah, Potomac, Chesapeake, La Tosca, etc.

Newminster got Lord Clifden, dam The Slave by Melbourne, grandam Volley (sister to Voltigeur) by Voltaire, out of Martha Lynn by Mulatto, son of Catton.

Lord Clifden got Petrarch (winner of the 2,000 Guineas and St. Leger, 1876), and imp. Rotherhill, out of Laura (the dam of imp. Laureate) by Orlando. Wenlock (winner of St. Leger, 1872) and his sister Siluria (dam of Geologist), out of Mineral by Rataplan, Hampton, and imp. Abingdon and imp. Deceiver, out of Boot-and-Saddle by Trumpeter, son of Orlando, out of Lady Langden by Kettledrum, Hawthornden (sent to New South Wales), out of Bonny Blink by The Flying Dutchman, and Winslow, out of Creslow by King Tom. Hampton is the sire of Merry Hampton (winner of the Derby, 1889), out of Doll Tearsheet by Broomielaw, Ayrshire (winner of the 2,000 Guineas and Derby in 1888), out of Atalanta by Galopin, Royal Hampton, out of imp. Princess (dam of Prince Royal and His Highness) by King Tom, Imp. Kingston, out of Last Love by Annandale, and imp. Eothen, out of Sultana by Oxford.

Wenlock got Quicklime, out of Duvernay by Beadsman. Newminster got Strathconan, out of Souvenir by Chanticleer, son of Birdcatcher. He was the sire of Scot Guard, Strathavon and Strathern, winner of the Royal Hunt Cup, and of Buchanan, out of Flurry by Young Melbourne.

Touchstone got Orlando, dam Vulture by Langar, son of Selim, out of Kite by Bustard, son of Castrel.

Orlando got Chattanooga, out of Ayacanora by Birdcatcher. Chattanooga got Wellingtonia, out of Araucaria by Ambrose, son of Touchstone. Wellingtonia got Clover (winner of the French Derby), out of Princess Catherine by Prince Charlie. Orlando got Marsyas (out of Malibran by Whisker), who sired George Frederick (winner of the Derby in 1874) and Albert Victor, out of Princess of Wales by Stockwell. George Frederick was the sire of Beau Brummell. Albert Victor was the sire of Victor Emanuel and imp. Albert, the latter out of Hawthorn Bloom by Kettledrum, and imp. The Sailor Prince, out of Hermita by Hermit.

Orlando got Orestes (dam by Bay Middleton), he the sire of Crest (out of Lady Louisa by Touchstone), who was the sire of Hollywood, out of Furze Chat by King Tom.

Orlando got Teddington, out of Miss Twickenham by Rockingham, son of Humphrey Clinker.

Teddington got Moulsey, out of Sabra by Pantaloon, and Marigold (the dam of Doncaster and imp. Lady Molesworth), out of Delilah by Thormanby.

Orlando got Trumpeter, out of Cavatina by Redshank, son of Sandbeck. Trumpeter got Queen's Messenger, out of Queen Bertha by Kingston, and Distin, out of Miss Bowser by Hesperus. Distin got Avontes, out of Avondale by Ratan, and imp. Charaxus, out of Sappho by Kingston. Charaxus is a very promising sire.

Trumpeter got Plutus, dam by Planet, son of Bay Middleton, out of Alice Bray by Venison.

Plutus got Flageolet, out of La Favorita by Monarque. Flageolet got imp. Rayon d'Or, out of Araucaria by Ambrose, son of Touchstone, Zut, out of Regalia by Stockwell, and Beauminet, out of Beauty by Knowsley.

Orlando got imp. Scythian, out of Scythia by Hetman Platoff, and imp. Eclipse, out of Gaze by Bay Middleton, he sire of Alarm, he sire of Himyar.

Whalebone got Waverley, out of Margaretta by Sir Peter. Waverley got Don John, out of a Comus mare, and Don John got Iago, out of a Selim mare, and he got imp. Bonnie Scotland, out of Queen Mary by Gladiator, Herod line. Whalebone also got Defence, out of Rubens mare, and he got The Emperor, out of a daughter of Reveller by Comus. Whisker got Economist, out of Floranthe by Octavian, Herod line. Economist was the sire of Echidna and Harkaway, the latter out of Fanny Dawson by Nabocklish, in the Herod line.

King Tom by Harkaway, son of Economist by Whisker by Waxy by Pot-8-o-'s by Eclipse, dam Pocahontas (dam of Stockwell, Rataplan, etc.) by imp. Glencoe, son of Sultan by Selim by Buzzard by Woodpecker by Herod, grandam Marpessa by Muley. King Tom got imp Kingcraft (dead), winner of the Derby, 1870, and imp. Great Tom, in the Belle Meade Stud, Tenn., out of Woodcraft by Voltigeur, King Lud, out of Qui Vive (sister to Vedette) by Voltigeur, Skylark, out of Wheatear by Young Melbourne, Dalesman (dam Agnes by Pantaloon), he the sire of Lowlander, out of Lufra by Windhound, son of Pantaloon, and he the sire of Lowland Chief, out of Bathilde by Stockwell. King Tom also got King Alfred, dam by Bay Middleton, Restitution, dam by Slane, King o' Scots, out of Katharine Logie by The Flying Dutchman. He got Bruce, out of Recluse by Newcastle.

King Tom got imp. King Ban, out of Atlantis by Thormanby (died in Dixiana Stud, Ky., March, 1888), imp King Ernest (dead), dam Ernestine by Touchstone, in the Brookdale Stud, Monmouth Co., N. J. Imp. Phaeton (died 1874), dam Merry Sunshine by Storm.

Phaeton got Ten Broeck (1 mile, 1.39¾; 2 miles, 3.27½; 3 miles, 5.26½, and 4 miles, 7.15¾, performances never equalled by one and the same horse), out of Fanny Holton by Lexington. King Alfonso, out of Capitola by Vandal, he the sire of Foxhall (that won the Grand Prize of Paris, Cesarewitch and Cambridgeshire Handicaps), out of Jamaica by Lexington, and Lisbon, out of imp. Lady Love by Stockwell or Caterer.

King Fergus was by Eclipse, dam Creeping Polly by Black-and-All-Black, son of Crab by Alcock Arabian. King Fergus got Hambletonian, out of a daughter of Highflyer by Herod, grandam by Matchem. Hambletonian got Whitelock, out of Rosalind by Phænomenon, son of Herod. Whitelock got Blacklock, out of daughter of Coriander by Pot-8-o-'s. Blacklock got Brutandorf, sire of Physician, sire of The Cure and Hetman Platoff, imp. Belshazzar, Velocipede, sire of King of Trumps, Hornsea, etc.; Voltaire, out of daughter of Phantom by Walton, Herod line.

Voltigeur by Voltaire, son of Blacklock by Whitelock by Hambletonian by King Fergus by Eclipse, dam Martha Lynn by Mulatto, son of Catton by Golumpus by Gohanna by Mercury by Eclipse, grandam Leda by Filho-da-Puta, son of Haphazard by Sir Peter by Highflyer by Herod, out of Treasure by Camillus, son of Hambletonian by King Fergus,

Voltigeur got Skirmisher and The Ranger, dam by Gardham, son of Falcon by Bustard by Castrel, out of daughter of Langar. Skirmisher got Kaiser, dam Regina by King Tom.

The Ranger got imp. Uhlan, dam La Mechante by Turnus, son of Taurus. Uhlan was a superior race-horse, a successful stallion in France, and is now in Chesterbrook Stud, Pennsylvania. Voltigeur got John Davis, dam Jamaica by Liverpool, son of Tramp. John Davis got John Day, out of Breakwater by Buccaneer, and Fortissimo, out of Vocalist by Trumpeter. Voltigeur got Tibthorpe, dam Little Agnes by The Cure, son of Physician. Tibthorpe got Thurio (winner of the Grand Prize of Paris), out of Verona by Orlando.

Voltigeur got Vedette (winner of the 2,000 Guineas and two Doncaster Cups), dam by Birdcatcher, out of Nan Darrell by Inheritor, son of Lottery.

Vedette got Speculum (winner of the City and Suburban), dam Doralice by Alarm or Orlando, out of Preserve by Emilius.

Speculum got Castlereagh, out of Lady Trespass by Birdcatcher, Sefton (winner of the Derby, 1878), out of Lady Sefton by West Australian, Rosebery (winner of the Cesarewitch and Cambridgeshire Handicaps), out of Ladylike by Newminster, and Hagioscope, out of Sophia by Macaroni, he the sire of Queen's Birthday, out of Matilda by Beauclerc.

Vedette got Galopin (winner of the Derby, 1875), dam Flying Duchess by The Flying Dutchman, out of Merope by Voltaire, son of Blacklock.

Galopin got Galliard (winner of the 2,000 Guineas, 1883), out of Mavis by Macaroni, St. Simon, out of St. Angela by King Tom, and Donovan (winner of the Derby and St. Leger, 1889), out of Mowerina by Scottish Chief.

Galopin got imp. Galore, a superior race-horse, out of Lady Maura by Macaroni, in the Bowling Brook Stud, Md , and imp. King Galop, dam by King Tom, out of Sunshine by Thormanby, now in Iroquois Stud, Lexington, Kentucky.

St. Simon got Memoir (winner of the Oaks and St. Leger, 1890) and La Fleche (winner of the 1,000 Guineas and Oaks, 1892), out of Quiver by Toxophilite, Semolina (winner of the 1,000 Guineas, 1890), out of Mowerina, Masetto, out of Lady Abbess by Cathedral, in Dixiana Stud, Kentucky, and imp. Simon Magus, out of Wheel of Fortune by Adventurer, now in the Rancocas Stud, New Jersey.

Voltigeur got imp. Hartington (winner of the Cesarewitch), out of Countess of Burlington by Touchstone, and imp. Billet, dam Calcutta by Flatcatcher, son of Touchstone, out of Miss Martin by St. Martin, son of Actæon. Billet was the sire of Volturno, Elias Lawrence, Miss Woodford, Barnes, Runnymede, Belvidere and other good ones.

Beningbrough by King Fergus, dam by King Herod, was the sire of Orville, out of Evelina by Highflyer by Herod. Orville was the sire of Emilius, out of Emily by Stamford, son of Sir Peter. Emilius was sire of Plenipotentiary, sire of imp. Herald, imp. Priam, out of Cressida by Whiskey, son of imp. Saltram, imp. Sovereign, out of Fleur de Lis by Bourbon (Matchem line), sire of Ruric, etc. Imp. Sarpedon, out of Icaria by The Flyer, Herod line. Sarpedon was sire of Alice Carneal, dam of Lexington. St. Nicholas, out of Seamew by Send, son of Beninbrough. St. Nicholas was sire of imp. Yorkshire, out of imp. Miss Rose by Tramp, a valuable stallion. Send by Beningbrough was the sire of Actæon, that got imp. Albion, out of a daughter of Comus or Blacklock. Orville got Muley, out of a daughter of Whiskey. Muley sired imp. Margrave and Leviathan, but there are no stallions in the male line from either of these, though the blood is found in many good and successful brood-mares. Muley also got Muley Moloch, the sire of the grand brood-mare imp. Britannia.

Joe Andrews by Eclipse, dam by Omnium, son of Snap by Snip, got Dick Andrews, out of Highflyer mare, who got Tramp, out of a mare by Gohanna, son of Mercury. Tramp got imp. Zinganee, Lanercost, out of Otis by Bustard, son of imp. Buzzard, Lottery, out of Mandane by Pot-8-o-'s. Tramp got Sheet Anchor, out of a Muley mare.

Weatherbit by Sheet Anchor, son of Lottery by Tramp by Dick Andrews by Joe Andrews by Eclipse, dam Miss Letty by imp. Priam, son of Emilius by Orville by Beningbrough by King Fergus by Eclipse, grandam by Orville, out of daughter of imp. Buzzard, son of Woodpecker by Herod.

Weatherbit got Brown Bread, out of Brown Agnes by West Australian. Brown Bread got Toastmaster, out of Mayoress by The Marquis, Hilarious, out of Hygeia by Knight of Kars, and Sweetbread, out of Peffar by Adventurer, also the imp. mares Bordelaise, Bread and Honey, Dry Toast, Miss Gwilt and Undercrust.

Weatherbit got Mandrake, out of Mandragora by Rataplan, and Beadsman (Derby winner, 1858), dam Mendicant by Touchstone, out of Lady Moore Carew by Tramp.

Beadsman got Blue Gown (Derby winner, 1868), and Cœruleus, out of Bas Bleu by Stockwell. Beadsman also got imp. Saxon, out of Girasol by Asteroid, a good sire and winner of the fastest Belmont Stake ever run, The Palmer and Rosicrucian, out of Madame Eglantine by Cowl. The Palmer got Grey Palmer, out of Eller by Chanticleer, and Pellegrino, out of Lady Audley by Macaroni. Rosicrucian got Geheimniss (winner of the Oaks, 1882), Hauteur, (winner of the 1,000 Guineas, 1883).

Mercury by Eclipse, dam by Tartar, sire of King Herod. Mercury got Gohanna, dam by Herod, grandam Maiden by Matchem. Gohanna got Election, out of Chestnut Skim by Woodpecker, son of Herod and Golumpus, out of Catherine by Woodpecker. Golumpus got Catton, out of Lucy Gray by Timothy, son of Delpini by Highflyer. Catton got Royal Oak, sire of Slane, out of Orville mare, imp. Nonplus, out of a Walton mare, imp. Trustee, out of Emma by Whisker. Trustee sired Revenue, out of Rosalie Somers by Sir Charles. Revenue got Planet and Exchequer, out of Nina by Boston and Rivoli, out of Topaz by Glencoe. Planet got Hubbard, out of Minnie Mansfield by Glencoe, Dudley, out of Regan by Lexington, and Whisper, out of Mattie Gross by Lexington. Catton also got Mulatto, out of Desdemona by Orville. He was the sire of Bloomsbury and Martha Lynn, the dam of Voltigeur. Catton got Gallopade, dam of Reel. Saltram by Eclipse, dam by Snap, son of Snip by

Flying Childers, got Whiskey out of Calash by Herod. Whiskey was the sire of dams of imp. Priam, Phantom and Muley, all good horses. Saltram also got the dams of Timoleon, the sire of Boston, Sir Hal and imp. Whip. The descendants of the Darley Arabian or Eclipse line in America are as good as any to be found in the world, as we have two sons of King Tom now alive, King Ernest and Great Tom, and in the male line Lisbon, King Alfonso, St. Martin, Aramis, Fonso, Grenada and Foxhall. Ill Used and his sons Forester, Badge and the grand race-horse His Highness, as good as any in the world. Leamington and his sons, Sensation, Onondaga, Stratford, Enquirer, and his sons Falsetto, Faustus, Blue Eyes, Inspector B., etc. Longfellow and his sons, The Bard, Jils Johnson, Long Taw, Leonatus, Longstreet and Linden. Eolus and his sons, St. Saviour, Eon, Russell. Aristides, Bob Woolley, Hyder Ali, Pantaloon, Lelaps, Outcast, Powhattan, Blazes, Reform, Warwick, etc. There is not any descendant from Sir Hercules through Faugh-a-Ballagh equal to them. Billet's sons, from the Blacklock line, Barnes, Runnymede, Belvidere, Elias Lawrence, King Galop, Galore, imp. Simon Magus, Masetto. Touchstone line, imp. St. Blaise, imp. Uhlan, Plevna, Bead'or, Buchanan, imp. Deceiver, imp. Kingston, Rayon d'Or, Blythewood, Alarm, Himyar, Panique, Wawekus, Gabriel, imp. Charaxus, etc. In the Whalebone line through Waverley and Bonnie Scotland's sons, Lake Blackburn, Bramble, Plenipo, Barnton, Lochiel, Spindrift, etc. Birdcatcher line, Glenelg, Ill-Used, Dalnacardoch, Glen Athol, imp. Wagner, The Jacobite, and the great Salvator. In the Tramp line, imp. Saxon and imp. Siddartha, so that the Eclipse line is very strongly represented by a number of first-class horses descended from racing and producing families.

The Byerly Turk line. The Byerly Turk was brought into England in the reign of William and Mary by Captain Byerly, whose charger he was during the whole of King William's wars in 1689 in Ireland. This noted horse and his owner, Captain Byerly, narrowly escaped capture on the banks of the Boyne in the last brilliant charge made by the Irish Dragoons, which broke the English lines and retrieved the fortunes of the day for the unfortunate James. The Byerly Turk escaped unscathed, and became one of the most noted of the founders of the English race-horse. The Byerly Turk got Basto, 1702, dam Bay Peg by Leedes Arabian, grandam Young Bald Peg, also by Leedes Arabian, great grandam old Morocco mare (Old Peg), Spanker's dam by Lord Fairfax's Morocco Barb, out of Old Bald Peg by an Arabian, out of a Barb mare. Byerly Turk got Jigg, out of a daughter of Spanker by the D'Arcy Yellow Turk. It will be seen that there was good blood to found the female line in this class, as well as in the first class, to the Darley Arabian. Basto got Sorcheels and his sister, out of sister to Mixbury by Curwen's Bay Barb, grandam by Curwen's Old Spot, son of the Selaby Turk, great grandam by the White-legged Lowther Barb, out of the old Vintner mare. This sister to Soreheels was the dam of Crab by the Alcock Arabian, Blacklegs, Second, Hip, Snip, Bay Basto and Brown Basto, all by Flying Childers. Jigg got Partner, 1718, out of the dam of Soreheels by Basto as above, and with Partner the cross becomes famous in turf history. Partner got Tartar, dam Meliora by Fox, son of Clumsy by Old Hautboy, grandam Milkmaid by Blackett's Snail, son of Whynot by the Fenwick Barb, out of a mare by Shields Galloway, the most famous Galloway of his day. Partner also got Sedbury, out of Old Montague by D'Arcy's Woodcock and Old Traveller, out of sister to Spinner by Almanzor, son of the Darley Arabian. Partner got imp. Traveler, out of Bay Bloody Buttocks. The great excellence of this blood comes down to us through Tartar's son King Herod, commonly called Herod, dam Cypron by Blaze, son of Flying Childers, grandam Selima by Bethell's Arabian, great grandam by Graham's Champion, son of Harpur's Arabian, fourth dam by Darley Arabian, out of a daughter of Old Merlin, son of Bustler. Herod got Florizel, dam by Cygnet, son of the Godolphin Arabian, grandam by Young Cartouch, son of Cartouch, son of the Bald Galloway, tracing through Ebony by Flying Childers to the Massey mare by Mr. Massey's black Barb. Florizel got imp. Diomed, dam sister to Juno by Spectator, son of Crab, grandam Horatia by Blank, great grandam by Flying Childers, tracing through Miss Belvoir by Grey Grantham and Betty Percival to an imported Barb mare.

Diomed (winner of the Derby, 1780) by Florizel by Herod by Tartar by Partner by Jigg by Byerly Turk, dam sister to Juno by Spectator by Crab by Alcock Arabian, out of Horatia by Blank, son of the Godolphin Arabian.

His sons, Sir Archy, dam imp. Castianira by Rockingham by Highflyer by Herod, out of Tabitha by Trentham.

Ball's Florizel, dam by imp. Shark by Marske, sire of O'Kelly's Eclipse, out of daughter of Harris' imp. Eclipse by Partner.

Duroc, dam Amanda by Grey Diomed by imp. Medley by Gimcrack, out of a daughter of Virginia Cade. Potomac, Hamiltonian, Virginius, Vingt'un, etc.

His daughters in England, Young Giantess, dam Giantess by Matchem, the dam of Sorcerer by Trumpator, Eleanor by Whiskey, the first mare that ever won the Oaks and Derby and the dam of Muley by Orville, Julia by Whiskey, dam of Phantom (winner of the Derby, 1811) by Walton, Cressida by Whiskey (dam of Antar by Haphazard, winner of the 2,000 Guineas, 1819), and imp. Priam (winner of the Derby, 1830, and two Goodwood Cups) by Emilius, and a Walton mare, dam of Langar and his brother Nicolo (winner of the 2.000 Guineas, 1823) by Selim.

In America the dams of Henry, Cicero, Shylock, Duchess of Marlborough, Lady Chesterfield, Haine's Maria, etc.

Sir Archy got Rattler, Sumpter, Childers and Flirtilla, dam by imp. Robin Redbreast, son of Sir Peter by Highflyer by Herod, out of daughter of imp. Obscurity, son of Eclipse.

Sir Archy Montorio and Sir Archy of Transport, out of Transport by Virginius, son of imp. Diomed.

Marion, Stockholder and Sir Charles, out of daughters of imp. Citizen, son of Pacolet by Blank by Godolphin Arabian.

The brothers Pacific and Bertrand, dam Eliza by imp. Bedford, son of Dungannon by Eclipse, out. of imp. Mambrina by Mambrino.

Virginian and Director, dam Meretrix by Magog, son of Chanticleer by imp. Wildair.

Saxe Weimar, Kosciusko and Crusader, dam Lottery by imp. Bedford, out of imp. Anvilina by Anvil, son of Herod.

Contention, dam by imp. Dare Devil, son of Magnet by Herod.

Zinganee (Garrison's), dam Miss Chance by imp. Chance, son of Lurcher by Dungannon by Eclipse.

Cicero, dam by imp. Diomed.

Arab, Brilliant, Tariff, Gen. Brooke, the good mares Coquette and Eliza Riley, dam Bet Bounce by imp. Sir Harry, son of Sir Peter.

John Richards, dam by Rattle, son of imp. Shark, Sea Gull, dam Nancy Air by imp. Bedford, Industry, dam by Ball's Florizel.

Timoleon, dam by imp. Saltram, son of Eclipse, out of a daughter of Symmes' Wildair, son of imp. Fearnought by Regulus by the Godolphin Arabian.

Timoleon got Boston (winner of 34 four-mile heat and 9 three-mile heat races), dam sister to Tuckahoe by Ball's Florizel, son of Diomed, grandam by imp. Alderman, son of Pot-8-o-'s by Eclipse, out of daughter of imp. Clockfast.

Boston got Lexington (4 miles, 7.19¾ and 7.23¾), dam Alice Carneal by imp. Sarpedon, out of Rowena by Sumpter, Lecompte (4 miles, 7.26, 7.38¾), dam Reel by imp. Glencoe, out of imp. Gallopade by Catton.

Red Eye, dam Lucy Long by imp. Priam, Commodore, dam Rosalie Somers by Sir Charles, Cracker, dam by Lance, son of American Eclipse, Ringgold, dam Flirtilla, Jr., by Sir Archy, Tally Ho, dam Nancy Norman by Tom Tough The noted mares Nina (dam of Planet, Exchequer, etc.), dam imp. Frolicksome Fanny by Lottery, Madeline, grandam of Iroquois, dam Magnolia by imp. Glencoe, Bostona, dam Creusa by imp. Priam, Goldpin, dam imp. Goldwire by Whalebone.

Lexington got Kentucky (4 miles, 7.31½), Daniel Boone and Gilroy, dam Magnolia by imp. Glencoe, Monarchist (3 miles, 5.34½; four miles, 7.33¾, both over Jerome Park old track), dam Mildred by imp. Glencoe, Norfolk (3 mile heats in 5.27½, 5.29½), dam Novice by imp. Glencoe, Asteroid (never beaten), dam Nebula by imp. Glencoe, Tom Ochiltree, dam Katona by Voucher, son of Wagner, Loadstone, Lightning, Thunder and Lancaster, dam Blue Bonnet by imp. Hedgeford, Harry Bassett and Charley Howard, dam Canary Bird by imp. Albion.

Rubicon, Beacon, Bayswater, Baywood, Bayonet and Preakness brothers, dam Bay Leaf by imp. Yorkshire, son of St. Nicholas by Emilius.

Colton, dam Topaz by imp. Glencoe, Jack Malone and Pat Malloy, dam Gloriana by American Eclipse.

Vauxhall and Foster, dam Verona by imp. Yorkshire, out of imp. Britannia by Muley.

Hunter's Lexington, dam Sally Lewis by imp. Glencoe.

Wanderer and Uncas ($1\frac{3}{4}$ miles in 2.21$\frac{3}{4}$), dam Coral by Vandal, out of imp. Cairngorm by Cotherstone, War Dance, dam Reel by imp. Glencoe, Kingfisher, dam imp. Eltham Lass by Kingston.

Merrill and Marion, dam Miriam by imp. Glencoe.

Duke of Magenta, dam Magenta by imp. Yorkshire.

Lever and Legatee, out of Levity by imp. Trustee by Catton, and the following noted mares, Sprightly, dam of Volturno, Susan Beane, dam of Sensation, Onondaga, Stratford, Susquehanna (Potomac's dam), Jamaica (dam of Foxhall), Nevada (dam of Luke Blackburn), Ærolite, dam of Spendthrift, Miser, Fellowcraft (4 miles, 7.19$\frac{1}{2}$), Rutherford, etc., Idlewild (4 miles, 7.26$\frac{1}{4}$), dam of Wildidle (4 miles, 7.25$\frac{1}{2}$), Maiden, dam of Parole, Powhattan, James A., etc., Hester, dam of Springbok ($2\frac{1}{2}$ miles, 3.56), Florence, dam of Hindoo and grandam of Firenzi.

Hira, dam of Himyar, Arizona, dam of Aranza, Regan, dam of Vigil, Vera Cruz, etc. Salina, dam of Salvator (1 mile, 1.35$\frac{1}{2}$), Minx, dam of Monitor, Waltz, dam of Glidelia, La Polka, dam of Post Guard (Gen. Philips), Heel and Toe, Strathspey, etc. Kathleen, dam of Geo. Kinney, Sarong, dam of Aristides, Crucifix, dam of Fair Play, Semper Felix (dam of Leonatus and great grandam of Longstreet), Mattie Gross, dam of Grenada, Mate, etc. Nannie Holton, dam of Ten Broeck, Squeez'em, dam of Day Star, Sylph, etc. Stamps, dam of Katie Pearce, Louisette, Preciosa, etc. La Henderson, dam of Ferida (4 miles, 7.23$\frac{1}{2}$), Aella, Virgilian, Ferona, etc. Susan Ann, dam of Thora. Lute, dam of Virginius and grandam of Troubadour, Minnie Minor, dam of Warwick, Wanda, John Ledford, etc.

Herod got Woodpecker out of Miss Ramsden by Cade, son of Godolphin Arabian. Woodpecker got imp. Buzzard, dam Misfortune by Dux, son of Matchem. Buzzard got the three brothers Castrel, Selim and Rubens, dam by Alexander, son of Eclipse, grandam by Highflyer, son of Herod. Castrel got Bustard, sire of Heron, out of Miss Hap by Shuttle, Pantaloon, out of Idalia by Peruvian, son of Sir Peter by Highflyer. Pantaloon got The Libel, out of Pasquinade, sister to Touchstone by Camel, Windhound, Hobbie Noble, The Reiver and Elthiron, out of Phryne by Touchstone, Sleight of Hand, out of Decoy by Filho-da-Puta, in Herod line. The Libel got Traducer out of Arethusa by Elis. Traducer got imp. Sir Modred and Cheviot, out of Idalia by Cambuscan, son of Newminster. Windhound got Thormanby out of Alice Hawthorn by Muley Moloch, Eclipse line. Thormanby got Plaudit out of Plausible by Springy Jack, Charibert, out of Gertrude by Saunterer, imp. Glengarry, out of Carbine by Rifleman, Rouge Rose, the dam of Bend'or, and Violet, the grandam of Melton. Plaudit, his son, got Balfe out of Bohemia by Weatherbit, Eclipse line. Selim was the sire of Langar, out of a daughter of Walton, son of Sir Peter. Langar got Elis and Epirus out of Olympia by Sir Oliver, the latter the sire of Pyrrhus the First, out of Fortress by Defence. Langar got Vulture, the dam of Orlando, out of Kite by Bustard, son of Castrel. Selim got Sultan, out of Bacchante by Williamson's Ditto, son of Sir Peter by Highflyer.

Glencoe (winner of the 2,000 Guineas, 1834, the Ascot Gold and Goodwood Cups) by Sultan, son of Selim by (imp.) Buzzard by Woodpecker by Herod, dam Trampoline by Tramp, son of Dick Andrews by Joe Andrews by Eclipse, grandam Web by Waxy, son of Pot-8-o-s by Eclipse, out of Penelope by Trumpator, son of Conductor by Matchem.

Glencoe got Panic, dam Sally Morgan by imp. Emancipation, son of Whisker by Waxy.

Star Davis, dam Margaret Wood by imp. Priam. Star Davis got Day Star, dam Squeez'em by Lexington.

Glencoe also got Vandal, dam by imp. Tranby, son of Blacklock by Whitelock by Hambletonian by King Fergus by Eclipse, grandam Lucilla by Trumpator, out of Lucy by Orphan, son of Ball's Florizel.

Wild Irishman and Frankfort out of Mary Morris by Medoc. Highflyer and Basvecchi, dam Castinet by imp. Monarch, son of Priam

Glencoe got a number of superior brood mares, such as Magnolia, dam of Kentucky, Daniel Boone, Gilroy and Madeline, the grandam of Iroquois.

Reel, dam of Lecompte, Starke, Prioress, War Dance, Fanny Wells, etc.

Topaz, dam of Austerlitz, Wagram, Waterloo, Lodi, Colton, etc. Novice, dam of Norfolk. Kitty Clark, dam of Maiden (Parole and Powhattan's dam), and La Henderson, Ferida's dam.

Florine, dam of Idlewild (Wildidle's dam) and Ærolite, dam of Spendthrift, Miser, Fellowcraft, etc.

Nebula, dam of Asteroid, Asterisk, etc.

Lightsome, dam of Sprightly, Nevada, Salina and Crucifix.

Mildred, dam of Monarchist, Minx (Monitor's dam), Stamps, Nora Worth, etc.

Sally Lewis, dam of John Morgan, Hunter's Lexington, Susan Beane (dam of Sensation, Onondaga, Stratford, etc).

Vandal got Virgil, dam Hymenia by imp. Yorkshire. Ventilator, dam Carolin by imp. Scythian. Voltigeur, dam Duet by Highlander. Vanderbilt, dam Melrose by Childe Harold. Vagabond, dam Gem by Childe Harold, son of imp. Sovereign.

Virgil got Hindoo, dam Florence by Lexington, out of imp. Weatherwitch by Weatherbit.

Hindoo got Hanover, dam Bourbon Belle by imp. Bonnie Scotland.

Virgil got Tremont, dam Ann Fief by Alarm. Vocalic, dam Acoustic by imp. Australian. Ben Ali, dam Ulrica by Lexington. Vigil, dam Regan by Lexington. Virgilian, dam La Henderson by Lexington. Portland, dam Finework by Lexington. Isaac Murphey, dam Mary Howard by imp. Hartington.

Vanderbilt got Madstone, dam Nina Turner by Hiawatha, son of imp. Albion.

Vagabond got Judge Morrow, out of Moonlight by Joe Johnson, son of Hunter's Lexington.

Rubens was the sire of Bobadil, out of a daughter of Skyscraper; Peter Lely, out of Stella by Sir Oliver; Teniers, out of Snowdrop by Highland Fling, and Defiance, the dam of Defence, out of Little Folly by Highland Fling.

Woodpecker got imp. Dragon, out of Juno by Spectator, and Dragon got imp. Speculator, dam by Herod. Woodpecker was also the sire of Chanticleer, out of a daughter of Eclipse, she out of Rosebud by Snap. Chanticleer got Bob Booty, out of Ierne by Bagot, son of Herod. Bob Booty was the sire of Guicciol:, the dam of Birdcatcher and Faugh-a-Ballagh. Woodpecker also got Colibri and Catharine, out of Camilla by Trentham. Catharine was the dam of Wanderer and Golumpus by Gohanna, son of Mercury by Eclipse. Herod got Highflyer, dam Rachel by Blank, son of the Godolphin Arabian, out of a daughter of Regulus by the Godolphin Arabian, tracing to a Royal mare. Highflyer never paid a forfeit or was beaten. He was the sire of Delpini, out of Countess by Blank. Delpini was the sire of Timothy, out of Cora by Matchem. Delpini was also the sire of Goldenlocks, the dam of Soothsayer, out of Violet by Shark, and of Seymour, out of Bay Javelin by Javelin, son of Eclipse. Highflyer's most celebrated son was Sir Peter, foaled 1784, dam Papillon by Snap, son of Snip by Flying Childers, grandam Miss Cleveland by Regulus, son of Godolphin Arabian, tracing through the Byerly Turk mare, dam of the two True Blues to the Burton Barb mare, as conjectured. Sir Peter got Walton and Williamson's Ditto, winner of the Derby in 1803, out of Arethusa by Dungannon, son of Eclipse. Sir Harry (imported), winner of the Derby in 1798, out of Matron by Alfred, son of Matchem. Stamford, imp. Archduke, winner of the Derby in 1799, and Paris, winner of the Derby in 1806, out of Horatia, sister to Achilles by Eclipse. Imp. Robin Redbreast, out of Wren by Woodpecker. Walton got Phantom, winner of the Derby in

1811, out of Julia by Whiskey, and Partisan, out of Parasol by Pot-8-o-'s. Phantom was the sire of Cobweb, dam of Bay Middleton, and also of the dam of Voltaire, sire of Voltigeur, and his blood is found in many of our best pedigrees.

Bay Middleton by Sultan, son of Selim by imp. Buzzard, dam Cobweb by Phantom, son of Walton by Sir Peter, grandam Filagree by Soothsayer, son of Sorcerer by Trumpator by Conductor by Matchem, out of Web by Waxy, son of Pot-8-o-'s by Eclipse.

His sons:

Autocrat, dam Empress by Emilius, son of Orville by Beningbrough by King Fergus, out of Mangelwurzel by Merlin, son of Castrel by imp. Buzzard.

Cowl, dam Crucifix by Priam (imported), son of Emilius, grandam Octaviana by Octavian, son of Stripling by Phenomenon by Herod.

Cowl is sire of The Confessor, dam Forest Fly by Mosquito, son of Emilius, out of Walfruna by Velocipede, son of Blacklock. Sire also of Miss Sellon, grandam of Hermit by Newminster, and Kautaka by Scottish Chief (in Algeria Stud), out of Belle Dame by Belshazzar, son of Blacklock.

Cowl is also sire of Madame Eglantine (dam of Rosicrucian and The Palmer by Beadsman), out of Diversion by Defence, son of Whalebone.

The Flying Dutchman (dam Barbelle by Sandbeck, son of Catton by Golumpus, grandam Darioletta by Amadis, son of Don Quixote by Eclipse), was the sire of Amsterdam, dam Urania by Idle Boy, son of Satan by Lottery by Tramp, grandam Venus by Langar, out of Vesta by Governor, son of Trumpator. He was also sire of Dollar, dam Payment by Slane, son of Royal Oak by Catton by Golumpus, grandam Receipt by imp. Rowton, son of Oiseau by Camillus by Hambletonian.

Dollar is sire of the French horse Salvator, dam Sauvagine by Ion, son of Cain by Paulowitz, out of Cuckoo by Elis, son of Langar, and Fontainebleau, dam Finlande by Ion, out of Fraudulent by Venison, son of Partisan by Walton.

The Flying Dutchman got also Dutch Skater, dam Fulvie by Gladiator, son of Partisan, out of Boutique by Giges, son of imp. Priam; Ellington (winner of the Derby in 1856), dam Ellerdale by Lanercost, son of Liverpool by Tramp, out of a mare by Tom Boy, son of Jerry by Smolensko; Jarnac, dam Joliette by Surplice, son of Touchstone by Camel, out of Jessamine by Paragone by Touchstone; Ignoramus, 1854, dam Ignorance by The Little Known, son of Muley by Orville, out of Bohemienne by Confederate, son of Comus by Sorcerer; Tom Bowline, 1857, dam Miss Bowe by Catton, son of Golumpus by Gohanna, out of Tranby's dam by Orville.

Dutch Skater got Insulaire, dam Green Sleeves by Beadsman, son of Weatherbit by Sheet Anchor, out of Mrs. Quickley by Longbow.

Gladiator by Partisan, son of Walton, dam Pauline by Moses, son of Whalebone or Seymour by Delpini, out of Quadrille by Selim, son of imp. Buzzard. Gladiator got Fitz Gladiator, dam Zarah by Reveller, son of Comus by Sorcerer, out of sister to Wouvermans by Rubens, son of imp. Buzzard.

Fitz Gladiator got Compiègne, out of Maid of Hart by The Provost, son of The Saddler by Waverley, out of Martha Lynn (Voltigeur's dam) by Mulatto. Compiègne was sire of imp. Mortemer, dam Comtesse by The Baron or Nuncio, the latter son of Plenipotentiary by Emilius, out of Eusebia by Emilius.

Mortemer was sire of St. Christophe (winner of the Grand Prize of Paris), out of Isoline by Ethelbert; Verneuil (winner of the Gold Vase, Gold Cup and Alexandra Plate at Ascot), out of Regalia by Stockwell; Chamant (winner of 2,000 Guineas), out of Araucaria by Ambrose, in France; Exile (a grand race-horse in America), out of imp. Second Hand by Stockwell; Pontiac, out of imp. Agenoria by Adventurer; Winfred, out of Minnie Minor by Lexington; Cyclops, Cambyses and Cleofas, out of Lizzie Lucas by imp. Australian; Cholula (Stiegen) and Cataline, out of Fanny Ludlow by imp. Eclipse; Heimdal, Housatonic and Iconoclast, out of Vandalite by Vandal; Savanac, out of Sly Boots by Rivoli, son of Revenue, with Wanda, Chimera, Restless, Heva, Saluda and Unrest. Fitz Gladiator was also sire of Orphelin, out of Echelle by Sting. Gladiator sired Ventre-St.-Gris, out of Belle de Nuit by Young Emilius.

Gladiator also got Sweetmeat, dam Lollypop by Starch or Voltaire, the latter son of Blacklock, out Belinda by Blacklock.

Sweetmeat got Macaroni, dam Jocose by Pantaloon, son of Castrel by imp. Buzzard, out of Banter (Touchstone's dam) by Master Henry, son of Orville.

Macaroni got Couronne de Fer, out of Miss Agnes by Birdcatcher; Maximilian, dam Duchess by St. Albans; Macgregor, out of Necklace by The Fallow Buck; imp. Maccaroon, dam Songstress by Chanticleer, son of Birdcatcher, out of Mrs. Carter by Humphrey, son of Sandbeck; imp. Moccasin, dam Madame Strauss by King Tom, he the sire of Vampire, Amazon, Disdain, etc. Maccaroon is the sire of Macduff, out of Jersey Lass by imp. King Ernest, son of King Tom; Macbeth, out of Jersey Belle (sister to Spendthrift and Fellowcraft) by imp. Australian; Gregory, dam imp. Abundance by Kaiser, son of Skirmisher by Voltigeur; Masher, dam Miss Bassett (sister to Harry Bassett) by Lexington.

Sweetmeat got Parmesan, dam Gruyere by Verulam, son of Lottery by Tramp, out of Jennala by Touchstone.

Parmesan got Favonius (winner of the Derby, 1871), out of Zephyr by King Tom; Cremorne (winner of the Derby, 1872), out of Rigolboche by Rataplan; Hopbloom, out of Cognisaunce by Stockwell; D'Estournel, out of a daughter of Chanticleer, son of Birdcatcher; Strachino, out of Lady Blanche by Voltigeur; imp. Strachino, out of May Belle by Hetman Platoff.

Favonius got Sir Bevys (Derby winner, 1879), dam Lady Langden by Kettledrum, son of Rataplan, out of Haricot by Mango or Lanercost.

Sweetmeat got Carnival, dam Volatile by Buckthorn, son of Venison. Carnival got Mask, out of Meteor by Thunderbolt, son of Stockwell, and Scobell, out of Lady Sophie by King Tom.

Sweetmeat got Saccharometer, out of Defamation by Iago, he the sire of Cucumber, out of Aminette by Birdcatcher, and Vanderdecken, out of Stolen Moments by Melbourne, Lozenge, out of Down-with-the-Dust by Star of Erin, son of Ishmael by Sultan.

Partisan was also the sire of Glaucus, out of Nanine by Selim. Glaucus got The Nob and he got The Nabob, and The Nabob got Vermouth and Bois Roussel. Vermouth got Boiard. The Nabob also got Nutbourne and Nutbourne got imp. Woodlands, out of Whiteface by Turnus. Glaucus was also sire of Milkmaid, the grandam of Bertram, sire of Robert the Devil.

Sir Peter was sire of Sir Paul and Paulina, out of Pewet by Tandem, son of Syphon. Sir Paul was the sire of Paulowitz, out of Evelina by Highflyer. Paulina won the St. Leger in 1807 and was the dam of Galatea by Amadis, the dam of Soldier's Joy, both distinguished brood mares. Paulowitz got Cain, out of a daughter of Paynator by Trumpator. Cain got Ion, out of Margaret by Edmund, and Ion got Wild Dayrell, out of Ellen Middleton by Bay Middleton.

Wild Dayrell by Ion, son of Cain by Paulowitz by Sir Paul by Sir Peter by Highflyer by Herod, dam Ellen Middleton by Bay Middleton by Sultan by Selim by Buzzard by Woodpecker by Herod, grandam Myrrha by Malek, son of Blacklock by Whitelock by Hambletonian by King Fergus by Eclipse, out of Bessy by Young Gouty, son of Gouty by Sir Peter.

Wild Dayrell got Buccaneer, out of daughter of Little Red Rover, son of Tramp. Buccaneer got Kisber (winner of the Derby in 1876), out of Mineral by Rataplan. Kisber is the sire of imp. Viola (Victory's dam), out of Parma by Parmesan. Buccaneer got See Saw, out of Margery Daw by Brocket, son of Melbourne. See Saw got Bruce (winner of the Grand Prize of Paris), out of Carine by Stockwell. Buccaneer also got imp. Felucca, out of Revival by Newminster. Felucca is the dam of Felicia, Feu Follet (dam of St. Florian), etc.

Wild Dayrell got Bold Dayrell, out of Rosina by The Emperor, son of Defence. Bold Dayrell is sire of imp. Lady Darrell and imp. Richmond, out of Miss Harriot by Kettledrum.

Wild Dayrell also got Wild Moor and Wild Oats, out of The Golden Horn by Harkaway, Guy Darrell, out of Reginella by King Tom, and imp. The Rake, out of

England's Beauty by Birdcatcher, sire of Dundee. Wild Moor is sire of imp. Wild Blossom, a good brood mare.

Ion was also the sire of Tadmor, out of Palmyra by Sultan, son of Selim. Tadmor was the sire of Seclusion, the dam of the great Hermit by Newminster. Tadmor also sired Palm, the grandam of Petronel by Musket. Sir Peter got Sir Oliver, out of Fanny by Diomed, and Haphazard, out of Miss Hervey by Eclipse, and he got Filho-da-Puta, out of Mrs. Barnet by Waxy. Filho-da-Puta got Colwick, out of Stella by Sir Oliver, imp. Maria Black, the dam of Bay Leaf, one of the grandest brood mares ever in America, and Leda, the grandam of Voltigeur, and imp. Hedgford, out of Miss Craigie by Orville. Hedgford was a success in this country, sire of Rattin, Duane, Blue Bonnet, dam of Loadstone, Lightning, Thunder, Lancaster, etc., and Fiatt (the Blue filly), the dam of Hunter's Glencoe, Kate Hunter, Liz Morgan and Blue Bells, to which Tenny traces. Herod also got Phaenomenon, out of Frenzy by Eclipse, who got Anvil, that sired the grandam of Economist and Stripling, the sire of Octavian.

Phaenomenon was also sire of Rosalind, dam of Whitelock. Herod got Drone, out of Lily by Blank, Bagot, out of Marotte by Matchem, and Maria, the dam of the noted Waxy by Pot-8-o-'s, out of Lisette by Snap. The Byerly Turk class is becoming extinct in England; the best representatives are found in America. There is nothing in the male line from Glencoe in Europe, while we have Hindoo, Hanover, Tremont, Vocalic, Ventilator, Vanderbilt, and his son Madstone, Day Star, Vagabond. From Diomed, through Boston and his son Lexington, Vauxhall, Uncas, Wanderer, Duke of Magenta, Monday, Joe Hooker, Grinstead and a few others. It is the interest of America to preserve this blood. The representatives of it now alive were the equals on the turf of any ever bred in Europe.

The Godolphin Arabian. The Godolphin Arabian was said to have drawn a cart in the streets of Paris, France. He was purchased and brought to England by Mr. Coke in 1730, who gave him to Mr. Williams, keeper of the St. James' Coffee House, who presented him to Lord Godolphin. Whether he was an Arabian or a Barb is a disputed point. He was a brown bay, about fifteen hands high, with some white on the off hind heel. It is generally admitted that he contributed more to improve the thoroughbred horse than any stallion either before or since his time. He accidentally became famous. He was teaser to Hobgoblin in 1730 and 1731, and upon the latter refusing to cover Roxana, she was bred to the Godolphin Arabian; from that cover Roxana produced in 1732 Lath, the first horse the Arabian ever got. Roxana was by the Bald Galloway, son of St. Victor Barb, dam Grey Whynot by old Whynot, son of Fenwick Barb, out of a Royal mare, Roxana's dam, sister to Chaunter by Akaster Turk, grandam by the Leedes Arabian, great grandam by Spanker, out of the old Morocco mare, Spanker's own dam by Lord Fairfax's Morocco Barb, she out of Old Bald Peg by an Arabian, out of a Barb mare. It will thus be seen that Lath was not a chance horse, his dam being as well bred as any mare of her day, and a race-mare of merit, and Lath was a superior race-horse himself, regarded as second only to Flying Childers. In 1734 Roxana produced Cade, the own brother to Lath, and, although he was not celebrated on the turf, it is to him that we are indebted for the Matchem line of blood. Cade lost his dam at foaling, and was brought up by hand. He was a horse of high conformation, and proved one of the best of stallions. Cade got Matchem, 1748, and Changeling out of an own sister to Miss Partner by Croft's Partner, son of Jigg by Byerly Turk, grandam Brown Farewell by Makeless (son of the Oglethorpe Arabian), tracing to the Layton Barb mare. Matchem was distinguished on the turf and became famous in the stud. He sired in the twenty-three years he was in the stud 354 winners of $755,485, in addition to cups, etc. Matchem got Conductor, 1767, Alfred, 1770, Georgiana, 1771, Eyebright, 1777, and Laurestinus, 1778, all out of a daughter of Snap by Snip, grandam by Cullen Arabian, out of Grisewood's Lady Thigh by Croft's Partner, etc. Conductor got Imperator, 1776, out of the dam of imp. Royalist by Herod, grandam Carina by Marske, sire of Eclipse. Imperator got Trumpator, 1782, and Pipator, 1786, out of Brunette by Squirrel, son of Traveler by Partner, grandam Dove by Matchless, son of the Godolphin Arabian. Alfred got imp. Tickle Toby, out of Cælia by Herod, who got Sir Solomon, out of Vesta by

Dreadnought, son of Meade's Celer by imp. Janus, and Sir Solomon got Trumpator, out of a daughter of Hickory, son of imp. Whip by Saltram. Cade also got imp. Wildair, 1753, out of a daughter of Steady by Flying Childers. So highly was Wildair valued in England, that he was repurchased and sent back to England, but he left behind him in America Bashaw, Sim's Wildair and Old Slamerkin. From the latter very many of the best race-horses of our day and many excellent stallions, including Thornton's Rattler, Sumpter and Childers, have descended.

Trumpator got Sorcerer, out of Young Giantess by Diomed, grandam Giantess by Matchem, and Paynator, dam by Mark Anthony, son of Spectator. Paynator got Dr. Syntax, out of a daughter of Beningbrough, son of King Fergus by Eclipse. Dr. Syntax was the sire of the noted mare Beeswing, the dam of Newminster by Touchstone. Newminster is sire of Herod (sire of St. Blaise), Lord Clifden and Adventurer, the sire of Ishmael and imp. Pizarro. Paynator was also the sire of the dam of Cain, sire of Ion, the sire of Tadmor and Wild Dayrell. Tadmor is sire of Seclusion, Hermit's dam. Sorcerer was the sire of Comus, out of Houghton Lass by Sir Peter. Comus got Humphrey Clinker, out of Clinkerina by Clinker, son of Sir Peter.

Melbourne, 1834, by Humphrey Clinker, son of Comus by Sorcerer by Trumpator by Conductor by Matchem, dam by Cervantes, son of Don Quixote by Eclipse, grandam by Golumpus, son of Gohanna by Mercury by Eclipse, out of sister to Zodiac by St. George, son of Highflyer by Herod.

Melbourne got Young Melbourne, dam Clarissa by Pantaloon, son of Castrel, out of daughter of (imp.) Glencoe. Young Melbourne got Pell Mell (out of Makeshift by Voltigeur), he the sire of Carlton, out of Bonny Spec by Speculum; Strafford and Brother to Strafford, out of daughter of Gameboy, Statesman and Gen. Peel, dam by Orlando, Rapid Rhone and Brother, out of mare by Lanercost or Retriever.

Melbourne got The Prime Minister (out of Pantalonade by Pantaloon), he the sire of Knight of the Garter, out of Rosa Bonheur by Touchstone; Arthur Wellesley, out of Lady Barbara by Launcelot: West Australian, dam Mowerina by Touchstone, out of Emma (dam of imp. Trustee) by Whisker.

West Australian won the 2,000 Guineas, Derby and St. Leger in 1853. He got The Wizard, dam by The Cure, Ruy Blas (out of Rosati by Gladiator), he the sire of Mourle, Reluisant and Financier; Joskin (out of Peasant Girl by The Major, son of Sheet Anchor), he the sire of Plebeian, dam Queen Elizabeth by Autocrat; Solon, dam by Birdcatcher, grandam by Hetman Platoff, out of Whim by Drone. Solon got Philammon, out of Satanella by Wild Dayrell, and Barcaldine, out of Ballyroe by Belladrum. Barcaldine is sire of Morion, out of Chaplet by Beadsman.

West Australian got imp. Australian, dam imp. Emilia by Young Emilius, out of Persian by Whisker. Australian got Spendthrift, Fellowcraft, Miser and Rutherford, out of Aerolite by Lexington, Helmbold, Baden Baden and Barricade, out of Lavender by Wagner, Springbok and Aspinwall, out of Hester by Lexington, Wildidle, out of Idlewild by Lexington, Wilful, out of imp. Pussy by Diophantus, and Harry O'Fallon, out of imp. Sunny South by Birdcatcher.

Melbourne is also the sire of The Slave, dam of Lord Clifden, Canezou, Cymba, Marchioness, and of The Peer, out of Cinizelli by Touchstone. The Peer is the sire of imp. Darebin, out of Lurline by Traducer, in the Rancho del Paso Stud of Mr. J. B. Haggin. Humphrey Clinker was the sire of imp. Clink, the dam of Ann Watson, she the dam of Rhynodyne and Watson by Glencoe. Sorcerer got Smolensko, out of Wowski by Mentor, son of Justice by Herod. Smolensko got Jerry, out of Louisa by Orville, and Jerry got Tomboy, out of Beeswing's dam by Ardrossan. Tomboy got Gameboy, out of Lady Moore Carew by Tramp, and Nutwith, out of a daughter of Comus, and Nutwith got Knight of Kars, out of Pocahontas by Glencoe. Sorcerer also got Bourbon, out of a daughter of Precipitate by Mercury, son of Eclipse, and Bourbon was the sire of Fleur de Lis, the dam of imp. Sovereign. Smolensko is the sire of the dam of the great mare imp. Maria Black, dam of Bayleaf by imp. Yorkshire, and the grandam of imp. Gallopade, the dam of the noted Reel by Glencoe. From these mares some of our most famous race-horses, stallions and brood mares owe their origin. Sorcerer also got Soothsayer, out of Golden Locks by Delpini. Soothsayer got Tiresias, winner of the Derby, out of Pledge by Waxy, and Filagree, the dam of imp. Riddlesworth, and Cobweb, the dam of Bay Middleton.

The Godolphin Arabian also got the Gower Stallion, out of sister to Tortoise by Whitefoot, Blank, Old England and Janus, out of the Little Hartley mare, by Bartlett's Childers. Babraham, Mogul, Marlborough and imp. Selima, out of the Large Hartley mare by Hartley's Blind Horse, son of the Holderness Turk. Imp. Selima was the dam of Ariel, Partner and their sister, by Moreton's imp. Traveler, son of Croft's Partner; Stella and Ebony by imp. Othello; Black Selima by imp. Fearnought, son of Regulus. It is from one of the above fillies, sister to Ariel and Partner, that we are indebted for the famous race-mare Mollie Jackson, whose three-mile heat race in 5.35½, 5.34¾, 5.28¾ is the best in the world, and she is the dam of Monday by Colton, and he the sire of Joe Hooker. Her daughter, Fanny Ludlow by imp. Eclipse, is the dam of Jamaica by Lexington, and she the dam of the superior race-horse Foxhall, that won the Grand Prize of Paris, the Cesarewitch and Cambridgeshire Handicaps. Blank got Rachel, the dam of the noted Highflyer by Herod, and imp. Centinel, out of Naylor by Cade, son of the Godolphin Arabian. Blank got Pacolet, out of Whiteneck by Crab. Pacolet got imp. Citizen, out of Princess by Turk, son of Regulus, he the sire of the dam of Marion. Imp. Citizen got Pacolet, out of Mary Grey by Tippoo Saib. Pacolet was sire of Monsieur Tonson, Sir Henry Tonson and Sir Richard Tonson, out of Madame Tonson by Top Gallant. Janus got imp. Janus out of a daughter of Fox, and Janus got Meade's Celer, out of Brandon by imp. Aristotle, son of the Cullen Arabian. The name of Celer figures in many of our best racing pedigrees. The Godolphin Arabian got Cripple out of Blossom by Crab, who got Gimcrack, out of Miss Elliot by Grisewood's Partner, son of Croft's Partner by Jigg. Gimcrack got imp. Medley, out of Arminda by Snap, son of Snip, and imp. Clockfast, out of Miss Ingram by Regulus. Clockfast was the sire of the great grandam of Boston, while Medley was the sire of Gray Diomed, who got Amanda, the dam of Duroc, sire of American Eclipse, and Melzar, out of Kitty Fisher by Symmes' Wildair. Melzar got Maria, out of a daughter of imp. Highflyer, son of Highflyer by Herod, and Maria was the dam of Lady Grey by Robin Grey, son of imp. Royalist by imp. Saltram by Eclipse. It is to Lady Gray that Lexington, Vandal, and all the Levity family, including Luke Blackburn, Salvator, Leonatus, Longstreet and other noted horses, trace their origin. The Godolphin Arabian got Regulus, out of Grey Robinson by the Bald Galloway. Regulus was famous on the turf, and was never beaten. Regulus got Careless and imp. Fearnought, out of Silvertail by Heneage's Whitenose, son of the Hall Arabian, and South, out of a daughter of Soreheels, son of Jigg. South got imp. Flimnap, out of a daughter of Cygnet, son of the Godolphin Arabian. Regulus got Dragon, out of Whimsey by Cullen Arabian, and Dragon got imp. St. George, out of Sally by Blank. His name appears in many of our best pedigrees. Regulus got Spiletta, the dam of Eclipse, out of Mother Western by Smith's son of Snake, and the mare grandam of Herod's great son Highflyer. Imp. Fearnought got Symmes' Wildair, out of a daughter of imp. Jolly Roger. Wildair was the sire of the grandam of Timoleon, Fitzhugh's Regulus, Godolphin and Specimen, out of imp. Jenny Dismal by Dismal, also son of the Godolphin Arabian. Apollo, Harris' Eclipse, Dandridge's Fearnought, etc. The Godolphin line is only represented on the other side of the Atlantic by the male descendants of Melbourne, of which the best representative is Barcaldine, and his son Morion, out of Chaplet by Beadsman.

In America we have a number of very highly bred race-horses and successful stallions in the male line to the Godolphin Arabian, through the same Melbourne line. Wilful, a good race-horse and sire of winners, out of imp. Pussy by Diophantus, son of Orlando. The brothers Spendthrift (sire of Kingston, Lamplighter, etc.), Fellowcraft, Rutherford and Miser, the latter sire of Yorkville Belle, Harry O'Fallon, sire of Terra Cotta, and other good ones, out of imp. Sunny South by Birdcatcher; Duke of Montrose, out of Kelpie by Bonnie Scotland; Strathmore, out of Brenna by Knight of St. George; Leinster, out of Luileme by Lexington; Wildidle (4 miles, 7.25½), out of Idlewild (4 miles, 7.26¼) by Lexington; Algerine, out of Nina by Boston; Springbok, out of Hester by Lexington; imp. Darebin, out of Lurline by

Traducer. There is then no necessity to send abroad for trash to improve this line of blood. In fact, any student of the stud book, or others who have read and digested the book on the thoroughbred horse, must be convinced that, whether we go to the Eclipse, Herod or Matchem lines of blood, we have as good in each and every class as can be found in the world.

THE

BREEDERS' HAND BOOK AND GUIDE.

ABBREVIATIONS.

Am.................................American.
Anc.................................Ancaster, that means Duke of Ancaster's.
Ar.................................Arabian.
Babr.................................Babraham.
Bolt.................................Bolton.
Br. or Broth.........................Brother.
Dau.................................Daughter.
Fl. Childers.........................Flying Childers.
God.................................Godolphin.
h.................................Horse.
H. C..................................Hampton Court.
Ir.................................Irish.
m.................................Mare.
Sir Peter T... ....Sir Peter Teazle.
T.................................Turk.
T. Sweepstakes... ....Turner's Sweepstakes.
W's Ditto.............................Williamson's Ditto.
Y.................................Young.
b............. ....bay.
bl. or blk.............................black.
br.................................brown.
ch.................................chestnut.
ro.................................roan.

‡ Means double parentage, which is given on foot line of page.

§ See for foot line of page.

by [Selim or] Soothsayer..............means has double parentage, and Soothsayer is accepted as sire.

**Diomed* ....by imported Diomed

*Diomed**.................. ....got by Diomed in England before his exportation.

† *Fisherman*..........................means imported from England to Australia.

* [The names in the last four instances are only given as example, and the marks hold good for every other animal.]

The top line on the pages of pedigrees extends the tail male line, which ceases on the 7th cross, to the Byerly Turk, Darley Ar. or God. Ar.

The tables give the pedigree full in 5 crosses; after the 5th cross is given the pedigree of the respective animal in the female line to the 9th cross.

IMPORTED ABINGDON.

The property of J. D. Lucas, Goodwood Stud, St. Louis, Mo.

ABINGDON by Wenlock (winner of the St. Leger in 1872), best son of Lord Clifden, was foaled in 1887, bred by Lord Norreys, dam Lady Langden (dam of Hampton, winner of the Doncaster Cup, Goodwood Cup and Epsom Gold Cups, and Sir Bevys, winner of the Derby) by Kettledrum, 2d dam Haricot (dam of Caller Ou, winner of the St. Leger, York and Brighton Cups and 34 Queen's Plates) by Mango or Lanercost, she out of the famous Queen Mary by Gladiator, the dam of Bonnie Scotland, Blink Bonny, that won the Derby and Oaks and was the dam of Borealis, Blair Athol, winner of the Derby and St. Leger, and Breadalbane, winner of the Prince of Wales and Gratwicke Stakes and sire of imp. Ill Used. Other famous winners in the family are Sir Bevys, winner of the Derby: Hampton, Epsom Gold Cup, Doncaster and Goodwood Cups and Goodwood Stakes; Broomielaw, Dee Stakes and Chesterfield Stakes; Blinkhoolie, Ascot Gold Vase and Alexandra Plate; Freeman, Great Northern Handicap twice, Goodwood Stakes, Chesterfield Cup and Alexandra Plate; Good Hope, Vienna Derby; Beauclerc, Middle Park Plate. Thrift, the dam of Tristan, winner of the Ascot Gold Cup and Hardwicke Stakes three times, for which $40,000 was refused, is from the same family. The family is potent in great race-horses, prolific brood mares and successful sires. Abingdon started only once, ran unplaced in the Hastings Plate at Newmarket, won by Heaume. He is a grandly bred horse and from one of the most distinguished racing and producing families in the world.

ABINGDON.

BAY HORSE, FOALED 1887; BRED BY LORD NORREYS, ENGLAND.

WENLOCK, b. h., 1869.	Lord Clifden, b. h., 1860.	Newminster, b. h., 1848.	Touchstone, br. h., 1831.	Camel, br. h., 1822.	Whalebone 1807 (Waxy 1790)—Dau. 1812 of Selim 1802 (Buzzard* 1787)—Maiden 1801 by Sir Peter T. 1784 (Highflyer 1774)—Dau. 1788 of Phenomenon 1780 (Herod 1758)—Matron 1782 by Florizel 1768 (Herod 1758).
				Banter, br. m., 1826.	Master Henry 1815 (Orville 1799)—Boadicea 1807 by Alexander 1782 (Eclipse 1764)—Brunette by Amaranthus 1766 (Old England 1741)—Mayfly 1771 by Matchem 1748 (Cade 1734)—Dau. 1755 of Starling 1738 (Starling 1727).
			Bee's-wing, b. m., 1833.	Dr. Syntax, br. h., 1811.	Paynator 1791 (Trumpator 1782)—Dau. of Beningbrough 1791 (King Fergus 1775)—Jenny Mole 1787 by Carbuncle 1772 (Babr. Blank 1758)—Dau. 1773 of Prince T'Quassaw 1751 (Snip 1736)—Sultana 1759 by Regulus 1739 (God. Ar. 1724).
				Tomboy's dam, ch. m., 1817.	Ardrossan 1809 (John Bull 1789)—Lady Eliza 1813 by Whitworth 1805 (Agonistes 1797)—Dau. 1793 of Spadille 1784 (Highflyer 1774)—Sylvia 1783 by Y. Marske 1771 (Marske 1750)—Ferret 1765 by Broth. to Silvio (Cade 1734).
		The Slave, b. m., 1852.	Melbourne, br. h., 1834.	Humphrey Clinker, b. h., 1822.	Comus 1809 (Sorcerer 1796)—Clinkerina 1812 by Clinker 1805 (Sir Peter T. 1784)—Pewet 1786 by Tandem 1773 (Syphon 1750)—Termagant by Tantrum 1760 (Cripple 1750)—Cantatrice by Sampson 1745 (Blaze 1733).
				Daughter of, b. m., 1825.	Cervantes 1806 (Don Quixote 1784)—Dau. 1818 of Golumpus 1802 (Gohanna 1790)—Dau. 1810 of Paynator 1791 (Trumpator 1782)—Dau. 1802 of St. George 1789 (Highflyer 1774)—Abigail 1788 by Woodpecker 1773 (Herod 1758).
			Volley, b. m., 1845.	Voltaire, br. h. 1826.	Blacklock 1814 (Whitelock 1803)—Dau. 1816 of Phantom 1808 (Walton 1799)—Dau. 1802 of Overton 1788 (King Fergus 1775)—Dau. 1796 of Walnut 1786 (Highflyer 1774)—Dau. 1786 of Ruler 1777 (Y. Marske 1771).
				Martha Lynn, br. m., 1837.	Mulatto 1823 (Catton 1809)—Leda 1824 by Filho da Puta 1812 (Haphazard 1797)—Treasure 1809 by Camillus 1803 (Hambletonian 1792)—Dau. 1804 of Hyacinthus 1797 (Coriander 1786)—Flora 1789 by King Fergus 1775 (Eclipse 1764).
	Mineral, ch. m., 1863.	Rataplan, ch. h., 1850.	The Baron, ch. h., 1842.	Ir. Birdcatcher, ch. h., 1833.	Sir Hercules 1826 (Whalebone 1807)—Guiccioli 1823 by Bob Booty 1804 (Chanticleer 1787)—Flight 1809 by Ir. Escape 1802 (Commodore 1793)—Y. Heroine by Bagot 1780 (Herod 1758)—Heroine 1775 by Hero 1753 (Cade 1734).
				Echidna, b. m., 1838.	Economist 1825 (Whisker 1812)—Miss Pratt 1825 by Blacklock 1814 (Whitelock 1803)—Gadabout 1812 by Orville 1799 (Beningbrough 1791)—Minstrel 1803 by Sir Peter T. 1784 (Highflyer 1774)—Matron 1782 by Florizel 1768 (Herod 1758).
			Pocahontas, b. m., 1837.	Glencoe,* ch. h., 1831.	Sultan 1816 (Selim 1802)—Trampoline 1825 by Tramp 1810 (Dick Andrews 1797)—Web 1808 by Waxy 1790 (Pot-8-o's 1773)—Penelope 1798 by Trumpator 1782 (Conductor 1767)—Prunella 1788 by Highflyer 1774 (Herod 1758).
				Marpessa, b. m., 1830.	Muley 1810 (Orville 1799)—Clare 1824 by Marmion 1806 (Whiskey 1789)—Harpalice 1814 by Gohanna 1790 (Mercury 1778)—Amazon 1799 by Driver 1783 (Trentham 1766)—Fractious 1792 by Mercury 1778 (Eclipse 1764).
		Manganese, ch. m., 1853.	Ir. Birdcatcher, ch. h., 1833.	Sir Hercules, blk. h., 1826.	Whalebone 1807 (Waxy 1790)—Peri 1822 by Wanderer 1811 (Gohanna 1790)—Thalestris 1809 by Alexander 1782 (Eclipse 1764)—Rival 1800 by Sir Peter T. 1784 (Highflyer 1774)—Hornet 1790 by Drone 1777 (Herod 1758).
				Guiccioli, ch. m., 1823.	Bob Booty 1804 (Chanticleer 1787)—Flight 1809 by Ir. Escape 1802 (Commodore 1793)—Y. Heroine by Bagot 1780 (Herod 1758)—Heroine 1775 by Hero 1753 (Cade 1734)—Dau. 1766 of Snap 1750 (Snip 1736).
			Moonbeam, ch. m., 1838.	Tomboy, b. h., 1829.	Jerry 1821 (Smolensko 1810)—Dau. 1817 of Ardrossan 1809 (John Bull 1789)—Lady Eliza 1813 by Whitworth 1805 (Agonistes 1797)—Dau. 1793 of Spadille 1784 (Highflyer 1774)—Sylvia 1783 by Y. Marske 1771 (Marske 1750).
				Lunatic, b. m., 1818.	Prime Minister 1810 (Sancho 1801)—Maniac 1806 by Shuttle 1793 (Y. Marske 1771)—Anticipation 1802 by Beningbrough 1791 (King Fergus 1775)—Expectation 1779 by Herod 1758 (Tartar 1743)—Dau. 1758 of Skim 1746 (Starling 1727).
LADY LANGDEN (Hampton's dam), br. m., 1868.	Kettledrum, ch. h., 1858.	Rataplan, ch. h., 1850.	The Baron, ch. h., 1842.	Ir. Birdcatcher, ch. h., 1833.	Sir Hercules 1826 (Whalebone 1807)—Guiccioli 1823 by Bob Booty 1804 (Chanticleer 1787)—Flight 1809 by Ir. Escape 1802 (Commodore 1793)—Y. Heroine by Bagot 1780 (Herod 1758)—Heroine 1775 by Hero 1753 (Cade 1734).
				Echidna, b. m., 1838.	Economist 1825 (Whisker 1812)—Miss Pratt 1825 by Blacklock 1814 (Whitelock 1802)—Gadabout 1812 by Orville 1799 (Beningbrough 1791)—Minstrel 1808 by Sir Peter T. 1784 Highflyer 1774)—Matron 1782 by Florizel 1768 (Herod 1758).
			Pocahontas, b. m., 1837.	Glencoe,* ch. h., 1831.	Sultan 1816 (Selim 1802)—Trampoline 1825 by Tramp 1810 (Dick Andrews 1797)—Web 1808 by Waxy 1790 (Pot-8-o's 1773)—Penelope 1798 by Trumpator 1782 (Conductor 1767)—Prunella 1788 by Highflyer 1774 (Herod 1758).
				Marpessa, b. m., 1830.	Muley 1810 (Orville 1799)—Clare 1824 by Marmion 1806 (Whiskey 1789)—Harpalice 1814 by Gohanna 1790 (Mercury 1778)—Amazon 1799 by Driver 1783 (Trentham 1766)—Fractious 1792 by Mercury 1778 (Eclipse 1764).
		Hybla, b. m., 1846.	The Provost, br. h., 1836.	The Saddler, br. h., 1828.	Waverley 1817 (Whalebone 1807)—Castrellina 1823 by Castrel 1801 (Buzzard* 1787)—Dau. 1813 of Waxy 1790 (Pot-8-o's 1773)—Bizarre 1811 by Peruvian 1806 (Sir Peter T. 1784)—Violante 1802 by John Bull 1789 (Fortitude 1777).
				Rebecca, b. m., 1831.	Lottery 1820 (Tramp 1810)—Dau. 1818 of Cervantes 1806 (Don Quixote 1784)—Anticipation 1802 by Beningbrough 1791 (King Fergus 1775)—Expectation 1779 by Herod 1758 (Tartar 1743)—Dau. 1758 of Skim 1746 (Starling 1727).
			Otisina, br. m., 1837.	Liverpool, b. h., 1828.	Tramp 1810 (Dick Andrews 1797)—Dau. 1822 of Whisker 1812 (Waxy 1790)—Mandane 1800 by Pot-8-o's 1773 (Eclipse 1764)—Y. Camilla 1787 by Woodpecker 1773 (Herod 1758)—Camilla 1778 by Trentham 1766 (T. Sweepstakes 1743).
				Otis, b. m., 1820.	Bustard 1801 (Buzzard* 1787)—Dau. 1813 of Election 1804 (Gohanna 1790)—Dau. 1791 of Highflyer 1774 (Herod 1758)—Dau. 1773 of Eclipse 1764 (Marske 1750)—Rosebud 1765 by Snap 1750 (Snip 1736).
	Haricot, br. m., 1847.	[Mango or] Lanercost, br. h., 1835.	Liverpool, b. h., 1828.	Tramp, b. h., 1810.	Dick Andrews 1797 (Joe Andrews 1778)—Dau. 1803 of Gohanna 1790 (Mercury 1778)—Fraxinella 1793 by Trentham 1766 (T. Sweepstakes 1743)—Dau. 1785 of Woodpecker 1773 (Herod 1758)—Everlasting 1775 by Eclipse 1764 (Marske 1750).
				Daughter of, b. m., 1822.	Whisker 1812 (Waxy 1790)—Mandane 1800 by Pot-8-o's 1773 (Eclipse 1764)—Y. Camilla 1787 by Woodpecker 1773 (Herod 1758)—Camilla 1778 by Trentham 1766 (T. Sweepstakes 1743)—Coquette 1765 by Compton Barb.
			Otis, b. m., 1830.	Bustard, b. h., 1801.	Buzzard* 1787 (Woodpecker 1773)—Gipsy 1789 by Trumpator 1782 (Conductor 1767)—Sist. 1774 to Postmaster by Herod 1758 (Tartar 1743)—Dau. of Snap 1750 (Snip 1736)—Dau. of Gower Stallion 1740 (God. Ar. 1724).
				Gayhurst's dam, —, 1813.	Election 1804 (Gohanna 1790)—Dau. 1791 of Highflyer 1774 (Herod 1758)—Dau. 1773 of Eclipse 1764 (Marske 1750)—Rosebud 1765 by Snap 1750 (Snip 1736)—Miss Belsea 1753 by Regulus 1739 (God. Ar. 1724).
		Queen Mary, b. m., 1843.	Gladiator, ch. h., 1833.	Partisan, b. h., 1811.	Walton 1799 (Sir Peter T. 1784)—Parasol 1800 by Pot-8-o's 1773 (Eclipse 1764)—Prunella 1788 by Highflyer 1774 (Herod 1758)—Promise 1768 by Snap 1750 (Snip 1736)—Julia 1756 by Blank 1740 (God. Ar. 1724).
				Pauline, b. m., 1826.	Moses 1819 ([Whalebone or] Seymour 1807)—Quadrille 1815 by Selim 1802 (Buzzard* 1787)—Canary Bird 1806 by [Whiskey or] Sorcerer 1796 (Trumpator 1782)—Canary 1797 by Coriander 1786 (Pot-8-o's 1773)—Miss Green 1787 by Highflyer 1774 (Herod 1758)
			Dau. of, b. m., 1840.	Plenipotentiary, ch. h., 1831.	Emilius 1820 (Orville 1799)—Harriet 1819 by Pericles 1809 (Evander 1801)—Dau. 1812 of Selim 1802 (Buzzard* 1787)—Pipylina 1803 by Sir Peter T. 1784 (Highflyer 1774)—Rally 1790 by Trumpator 1782 (Conductor 1767).
				Myrrha, b. m., 1830.	Whalebone 1807 (Waxy 1790)—Gift gr. m. 1818 by Y. Gohanna 1810 (Gohanna 1790)—Sist. b. m. 1802 to Grazier by Sir Peter T. 1784 (Highflyer 1774)—Sist. b. m. 1788 to Aimator by Trumpator 1782 (Conductor 1767)—Sist. 1774 to Postmaster by Herod 1758 (Tartar 1743)—Dau. of Snap 1750 (Snip 1736)—Dau. of Gower Stallion 1740 (God. Ar. 1724)—Dau. of Childers.

IMPORTED ALBERT.

WINNER OF THE SEATON DELAVAL STAKES AND SECOND TO SATCHEL IN THE RACING STAKES AT GOODWOOD.

The property of J. N. Camden, Jr., Hartland Stud, near Versailles, Woodford County, Kentucky.

ALBERT, bay horse, foaled 1882, bred by Mr. R. Jardine, by Albert Victor, own brother to the Derby winner George Frederick, son of Marsyas by Orlando, dam Hawthorn Bloom, the dam of May Bloom, winner of the May Stakes at Chester and Tyro Stakes at Liverpool, Fitz James, winner of the Prince of Wales Stakes and an excellent sire in England, Heath Bird, winner of the York Cup and Ayr Gold Cup, Lady Gowrie, winner of the Duke of Westminster's Cup, Whin Blossom, winner of the Findon Stakes, Ebor St. Leger and Brighton Stakes, beating the crack Thebais, winner of the Oaks; and Hawthorn, full sister to Albert, that won the Prince of Wales Stakes and Osmaston Nursery Handicap, in which she beat Reve d'Or, winner of the Oaks and 1,000 Guineas, by Kettledrum, winner of the Derby and Doncaster Cup, son of Rataplan, own brother to Stockwell. Hawthorn Bloom is full sister to Hawthorndale, dam of Hauteur, winner of the 1,000 Guineas, and daughter of Lady Alice Hawthorn, she being full sister to Maybloom, the great grandam of Bonavista, winner of the 2,000 Guineas in 1892. Lady Hawthorn, his third dam, is sister to Thormanby (winner of the Derby and Ascot Gold Cup) and Oulston by Windhound, she out of the noted race-mare Alice Hawthorn, winner of 16 Cups, including two Doncaster and two Goodwood Cups, 18 Queen's Plates and 17 other races, tracing to the Layton Barb mare. Albert Victor was himself a superior race-horse, he won the Middle Park Plate, beating Hannah (Oaks and St. Leger winner), Bothwell (2,000 Guineas winner) and others. At three years old won the Newmarket Biennial Stakes, beating Favonius (winner of the Derby), also ran second to Favonius in the Derby, won Ebor St. Leger, and ran second in the St. Leger. At four years old won the Gold Vase at Ascot, beating Sterling and Dutch Skater, won Brighton Cup, 2 miles, and Great Ebor Handicap. He is sire of Maskelyne, that beat Iroquois in Hopeful Stakes, Victor Emanuel, winner of the Chesterfield Cup and Great Ebor Handicap, imp. The Sailor Prince, winner of the Cambridgeshire Handicap, beating Carlton, Melton, St. Honorat, etc., Whitelegs, winner of the Royal Hunt Cup at Ascot, and other good race-horses.

Albert is descended in the male line from Eclipse through his son Pot-8-o-'s and grandson Waxy, and is full of the blood of that extraordinary progenitor of great race-horses He is descended from Eclipse in the direct male line through Pot-8-o-'s, Waxy and Touchstone, a strain of blood which is in by far the greatest repute in England to-day on account of the fact that the most successful race-horses have been so bred, such as Newminster, Scottish Chief, Adventurer, Hermit, Marsyas, Lord Clifden, Strathconan, Trumpeter, Cathedral, Petrarch, Gladiateur, Melton, Paradox, The Bard, etc. Lady Moore Carew mentioned in the fifth generation was a very celebrated mare, and winner of 18 races.

Albert is 16¼ hands high. He has the best of feet and legs, powerful loin, short, strong back, but grand length of body, finely sloping shoulders, and clean, intelligent, blood-like head.

ALBERT.

BAY HORSE, FOALED 1882; BRED BY MR. R. JARDINE, ENGLAND.

ALBERT VICTOR, ch. h., 1868.	Marsyas, ch. h., 1851.	Orlando, b. h., 1841.	Touch-stone, br. h., 1831.	Camel, br. h., 1822. Whalebone 1807 (Waxy 1790)—Dau. 1812 of Selim 1802 (Buzzard* 1787)—Maiden 1801 by Sir Peter T. 1784 (Highflyer 1774)—Dau. 1788 of Phenomenon 1780 (Herod 1758)—Matron 1782 by Florizel 1768 (Herod 1758).
				Banter, br. m., 1826. Master Henry 1815 (Orville 1799)—Boadicea 1807 by Alexander 1782 (Eclipse 1764)—Brunette by Amaranthus 1766 (Old England 1741)—Mayfly 1771 by Matchem 1748 (Cade 1734)—Dau. 1755 of Starling 1738 (Starling 1727).
			Vulture, ch. m., 1833.	Langar, ch. h., 1817. Selim 1802 (Buzzard* 1787)—Dan. 1808 of Walton 1799 (Sir Peter T. 1784)—Y. Giantess 1790 by Diomed* 1777 (Florizel 1768)—Giantess 1769 by Matchem 1748 (Cade 1734)—Molly Long Legs 1753 by Babraham 1740 (God. Ar. 1724).
				Kite, b. m., 1821. Bustard 1813 (Castrel 1801)—Olympia 1815 by Sir Oliver 1800 (Sir Peter T. 1784)—Scotilla 1795 by Anvil 1777 (Herod 1758)—Scota 1783 by Eclipse 1764 (Marske 1750)—Harmony 1775 by Herod 1758 (Tartar 1743).
		Malibran, b. m., 1830.	Whisker, b. h., 1812.	Waxy, b. h., 1790. Pot-8-o-'s 1773 (Eclipse 1764)—Maria 1777 by Herod 1758 (Tartar 1743)—Lisette 1772 by Snap 1750 (Snip 1736)—Miss Windsor 1751 by God. Ar. 1724—Dau. of Y. Belgrade (Belgrade Turk).
				Penelope, b. m., 1798. Trumpator 1782 (Conductor 1767)—Prunella 1788 by Highflyer 1774 (Herod 1758)—Promise 1768 by Snap 1750 (Snip 1736)—Julia 1756 by Blank 1740 (God. Ar. 1724)—Spectator's Dam 1735 by Partner 1718 (Jigg).
			Garcia, br. m., 1823.	Octavian, ch. h., 1807. Stripling 1795 (Phenomenon 1780)—Dau. 1797 of Oberon 1790 (Highflyer 1774)—Sist 1779 to Sharper by Ranthos 1763 (Matchem 1748)—Dau. of T. Sweepstakes 1743 (B. Sweepstakes 1722)—Dau. 1748 of Hutton's Spot 1728 (Hartley's Blind Horse).
				Daughter of, b. m., 1806. Shuttle 1793 (Y. Marske 1771)—Katharine 1798 by Delpini 1781 (Highflyer 1774)—Dau. 1788 of Paymaster 1766 (Blank 1740)—Dau. 1774 of Le Sang 1759 (Changeling 1747)—Dau. 1751 of Rib 1736 (Crab 1722).
	Princess of Wales, ch. m., 1862.	Stockwell, ch. h., 1849.	The Baron, ch. h., 1842.	Ir. Birdcatcher ch. h., 1833. Sir Hercules 1826 (Whalebone 1807)—Guiccioli 1823 by Bob Booty 1804 (Chanticleer 1787)—Flight 1809 by Ir. Escape 1802 (Commodore 1793)—Y. Heroine by Bagot 1780 (Herod 1758)—Heroine 1775 by Hero 1753 (Cade 1734).
				Echidna, b. m., 1838. Economist 1825 (Whisker 1812)—Miss Pratt 1825 by Blacklock 1814 (Whitelock 1803)—Gadabout 1812 by Orville 1799 (Beningbrough 1791)—Minstrel 1803 by Sir Peter T. 1784 (Highflyer 1774)—Matron 1782 by Florizel 1768 (Herod 1758).
			Pocahontas, b. m., 1837.	Glencoe,* ch. h., 1831. Sultan 1816 (Selim 1802)—Trampoline 1825 by Tramp 1810 (Dick Andrews 1797)—Web 1808 by Waxy 1790 (Pot-8-o-'s 1773)—Penelope 1798 by Trumpator 1782 (Conductor 1767)—Prunella 1788 by Highflyer 1774 (Herod 1758).
				Marpessa, b. m., 1830. Muley 1810 (Orville 1799)—Clare 1824 by Marmion 1806 (Whiskey 1789)—Harpalice 1814 by Gohanna 1790 (Mercury 1778)—Amazon 1799 by Driver 1783 Trentham 1766)—Fractious 1792 by Mercury 1778 (Eclipse 1764).
		The Bloomer, b. m., 1870.	Melbourne, br. h., 1834.	Humprey Clinker, b. h., 1822. Comus 1809 (Sorcerer 1796)—Clinkerina 1812 by Clinker 1805 (Sir Peter T. 1784)—Pewet 1786 by Tandem 1773 (Syphon 1750)—Termagant by Tantrum 1760 (Cripple 1750)—Cantatrice by Sampson 1745 (Blaze 1733).
				Daughter of, b. m., 1825. Cervantes 1806 (Don Quixote 1784)—Dau. 1818 of Golumpus 1802 (Gohanna 1790)—Dau. 1810 of Paynator 1791 (Trumpator 1782)—Dau. 1802 of St. George 1789 (Highflyer 1774)—Abigail 1788 by Woodpecker 1773 (Herod 1758).
			Lady Sarah, ch. m., 1841	Velocipede, ch. h., 1825. Blacklock 1814 (Whitelock 1803)—Dau. 1817 of Juniper 1805 (Whiskey 1789)—Dau. 1810 of Sorcerer 1796 (Trumpator 1782)—Virgin 1801 by Sir Peter T. 1784 (Highflyer 1774)—Dau. 1794 of Pot-8-o-'s 1773 (Eclipse 1764).
				Lady Moore Carew, b. m., 1830. Tramp 1810 (Dick Andrews 1797)—Kite 1821 by Bustard 1813 (Castrel 1801)—Olympia 1815 by Sir Oliver 1800 (Sir Peter T. 1784)—Scotilla 1795 by Anvil 1777 (Herod 1758)—Scota 1783 by Eclipse 1764 (Marske 1750).
HAWTHORN BLOOM, br. m., 1861.	Kettledrum, ch. h., 1858.	Rataplan, ch. h., 1850.	The Baron, ch. h., 1842.	Ir. Birdcatcher ch. h., 1833. Sir Hercules 1826 (Whalebone 1807)—Guiccioli 1823 by Bob Booty 1804 (Chanticleer 1787)—Flight 1809 by Ir. Escape 1802 (Commodore 1793)—Y. Heroine by Bagot 1780 (Herod 1758)—Heroine 1775 by Hero 1753 (Cade 1734).
				Echidna, b. m., 1838. Economist 1825 (Whisker 1812)—Miss Pratt 1825 by Blacklock 1814 (Whitelock 1803)—Gadabout 1812 by Orville 1799 (Beningbrough 1791)—Minstrel 1803 by Sir Peter T. 1784 (Highflyer 1774)—Matron 1782 by Florizel 1768 (Herod 1758).
			Pocahontas, b. m., 1837.	Glencoe,* ch. h., 1831. Sultan 1816 (Selim 1802)—Trampoline 1825 by Tramp 1810 (Dick Andrews 1797)—Web 1808 by Waxy 1790 (Pot-8-o-'s 1773)—Penelope 1798 by Trumpator 1782 (Conductor 1767)—Prunella 1788 by Highflyer 1774 (Herod 1758).
				Marpessa, b. m., 1830. Muley 1810 (Orville 1799)—Clare 1824 by Marmion 1806 (Whiskey 1789)—Harpalice 1814 by Gohanna 1790 (Mercury 1778)—Amazon 1799 by Driver 1783 (Trentham 1766)—Fractious 1792 by Mercury 1778 (Eclipse 1764).
		Hybla, b. m., 1846.	The Provost, br. h., 1836.	The Saddler, br. h., 1828. Waverley 1817 (Whalebone 1807)—Castrellina 1822 by Castrel 1801 (Buzzard* 1787)—Dau. 1815 of Waxy 1790 (Pot-8-o-'s 1773)—Bizarre 1811 by Peruvian 1806 (Sir Peter T. 1784)—Violante 1802 by John Bull 1789 (Fortitude 1777).
				Rebecca, b. m., 1831. Lottery 1820 (Tramp 1810)—Dau. 1818 of Cervantes 1806 (Don Quixote 1784)—Anticipation 1802 by Beningbrough 1791 (King Fergus 1775)—Expectation 1779 by Herod 1758 (Tartar 1743)—Dau. 1778 of Skim 1746 (Starling 1727).
			Otisina, br. m., 1837.	Liverpool, b. h., 1828. Tramp 1810 (Dick Andrews 1797)—Dau. 1822 of Whisker 1812 (Waxy 1790)—Mandane 1800 by Pot-8-o-'s 1773 (Eclipse 1764)—Y. Camilla 1787 by Woodpecker 1773 (Herod 1758)—Camilla 1778 by Trentham 1766 (T. Sweepstakes 1743).
				Otis, b. m., 1820. Bustard 1801 (Buzzard* 1787)—Dau. 1813 of Election 1804 (Gohanna 1790)—Dau. 1791 of Highflyer 1774 (Herod 1778)—Dau. 1778 of Eclipse 1764 (Marske 1750)—Rosebud 1765 by Snap 1750 (Snip 1736).
	Lady Alice Hawthorn, b. m., 1859.	Newminster, b. h., 1848.	Touchstone, br. h., 1831.	Camel, br. h., 1822. Whalebone 1807 (Waxy 1790)—Dau. 1812 of Selim 1802 (Buzzard* 1787)—Maiden 1801 by Sir Peter T. 1784 (Highflyer 1774)—Dau. 1788 of Phenomenon 1780 (Herod 1758)—Matron 1782 by Florizel 1768 (Herod 1758).
				Banter, br. m., 1826. Master Henry 1815 (Orville 1799)—Boadicea 1807 by Alexander 1782 (Eclipse 1764)—Brunette by Amaranthus 1766 (Old England 1741)—Mayfly 1771 by Matchem 1748 (Cade 1734)—Dau. 1755 of Starling 1738 (Starling 1727).
			Bee's-wing, b. m., 1833.	Dr. Syntax, br. h., 1811. Paynator 1791 (Trumpator 1782)—Dau. of Beningbrough 1791 (King Fergus 1775)—Jenny Mole 1787 by Carbuncle 1772 (Babr. Blank 1758)—Dau. 1773 of Prince T'Quassaw 1751 (Snip 1736)—Sultana 1759 by Regulus 1739 (God. Ar. 1724).
				Tomboy's dam, ch. m., 1817. Ardrossan 1809 (John Bull 1789)—Lady Eliza 1813 by Whitworth 1805 (Agonistes 1797)—Dau. 1793 of Spadille 1784 (Highflyer 1774)—Sylvia 1783 by Y. Marske 1771 (Marske 1750)—Ferret 1765 by Broth. to Silvio (Cade 1734).
		Lady Hawthorn, b. m., 1854.	Windhound, br. h., 1847.	Pantaloon, ch. h., 1824. Castrel 1801 (Buzzard* 1787)—Idalia 1815 by Peruvian 1806 (Sir Peter T. 1784)—Musidora 1804 by Meteor 1783 (Eclipse 1764)—Maid-of-all-Work 1786 by Highflyer 1774 (Herod 1758)—Dau. 1771 of Syphon 1750 (Squirt 1732).
				Phryne, br. m., 1840. Touchstone 1831 (Camel 1822)—Decoy 1830 by Filho-da-Puta 1812 (Haphazard 1797)—Finesse 1815 by Peruvian 1806 (Sir Peter T. 1784)—Violante 1802 by John Bull 1789 (Fortitude 1777)—Dau. 1788 of Highflyer 1774 (Herod 1758).
			Alice Hawthorn, b. m., 1838.	Muley Moloch, br. h., 1830. Muley 1810 (Orville 1799)—Nancy 1813 by Dick Andrews 1797 (Joe Andrews 1778)—Spitfire 1800 by Beningbrough 1791 (King Fergus 1775)—Dau. of Y. Sir Peter 1775 (Doge 1762)—Dau. of Engineer 1756 (Sampson 1745).
				Rebecca, b. m., 1831. Lottery 1820 (Tramp 1810)—Dau. b. m. 1818 of Cervantes 1806 (Don Quixote 1784—Anticipation ch. m. 1802 by Beningbrough 1791 (King Fergus 1775)—Expectation gr. m. 1779 by Herod 1758 (Tartar 1743)—Dau. 1778 of Skim 1746 (Starling 1727)—Dau. 1751 of Janus 1738 (God. Ar. 1724)—Spinster gr. m. 1743 by Crab 1722 (Alcock Ar.)—Spinster

(The Widdrington mare) ch. m. 1735 by Partner 1718 (Jigg)—Bay Bloody Buttocks b. m. 1729 by Bloody Buttocks—Sist. 1722 to Guy by Greyhound (Chillaby)—Brown Farewell 1710 by Makeless (Oglethorpe Ar.)—Dau. of Brimmer (D'Arcy Yellow Turk)—Dau. of Place's White Turk—Dau. of Dodsworth—Layton Barb mare.

BELVIDERE.

WINNER OF THE RED BANK, REAPER'S, VAN COURTLAND STAKES, WESTCHESTER, LONG BRANCH AND DELAWARE HANDICAPS.

Private stallion in the Preakness Stud of Mr. James Galway, located at the Stonewall Stud of Mr. J. Breckenridge Viley, near Midway, Woodford Co., Ky.

BELVIDERE by imp. Billet, son of Voltigeur, foaled 1884, bred in the Runnymede Stud, Bourbon Co., Ky., dam Jaconet, own sister to Iroquois (winner of the Derby and St. Leger) by imp. Leamington, son of Faugh-a-Ballagh, out of Maggie B. B. by imp. Australian, etc. Billet, his sire, was a good race-horse, winner of the Zetland Stakes at York, Egham Stakes at Egham, and other races, by Voltigeur, dam Calcutta by Flatcatcher, son of Touchstone. Billet was one of the best stallions of our modern importations, sire of Volturno, Miss Woodford, Runnymede, The Pepper, Raceland, Blue Wing and many other winners. Belvidere's dam, Jaconet (Sir Dixon's dam), is full sister to Iroquois, one of the best race-horses, judged by his performances, ever bred in America, and the only horse, except Donovan in 1889, that ever won the Derby, Prince of Wales Stakes and the Doncaster St. Leger. Iroquois was second to Peregrine in the 2,000 Guineas, and beat him in the Derby. Iroquois also won the Chesterfield Stakes at Newmarket, Levant Stakes at Goodwood, the Burwell Stakes, Newmarket Derby and other important events. In these races he beat such good ones as Peregrine, Ishmael, Scobel, Geologist, Tristan, Cameliard, Bal Gal, Lucy Glitters and others. Belvidere, therefore, is descended from one of the best bred and most potent racing families in America. His grandam, Maggie B. B., by imp. Australian, was the dam of Pera, Harold, Panique, Red and Blue (Sally McClelland's dam), etc. Harold had only one colt on the turf, Harefoot, an excellent race-horse, and Harold was own brother to Iroquois, a good index to the potency of the blood. Madeline, his great grandam, was by the great Boston, sire of Lexington, Lecompte, Bostona, Nina and a host of good ones. Magnolia was the dam of Kentucky, the best son of Lexington. Magnolia was also the dam of Daniel Boone, Magic, Charley Ball, Madonna, Skedaddle, Gilroy, Victory, etc. Skedaddle is the dam of Saucebox, Scramble, Florence B., Sly Boots, Joe S., The Slashes and Squeez'em (the dam of Day Star, winner of the Kentucky Derby, Blue Ribbon Stakes and other fast races) and Sylph (the dam of La Sylphide, 1¼ miles in 2.07½, and dam of Semper Fidele), Kaloolah, Kincsem, etc. Sly Boots is the dam of Leveller, Sly Dance, Sachem and Savanac. Myrtle was by Mameluke (winner of the Derby). Bobadilla won the Ascot Gold Cup and Drawing Room Stakes. Belvidere won the Red Bank Stakes for two-year olds, a Free Handicap Sweepstakes, 1⅛ miles, in 1.57½; won Reaper's Stakes, 1$\frac{3}{16}$ miles, 112 lbs., in 2.03; Handicap Sweepstakes, 1$\frac{1}{16}$ miles, 105 lbs., in 1.50½; Van Courtland Stakes, 1½ miles, 108 lbs., in 2.46; Fordham Handicap, 1¼ miles, 115 lbs., in 2.12; Handicap Sweepstakes, 1⅛ miles, 120 lbs., in 1.58½; Westchester Handicap, 1⅜ miles, 120 lbs., in 2.26¼; Long Branch Handicap, 1¼ miles, 118 lbs., in 2.10½; Free Handicap Sweepstakes, 1½ miles, 120 lbs., in 2.43½; Delaware Handicap, 1¼ miles, 120 lbs., in 2.09, and was placed second and third in many excellent races in the best of company. He cannot fail to transmit his great qualities to his progeny. The pedigree is pure and unadulterated to the famous Layton Barb mare.

BELVIDERE.

BAY HORSE, FOALED 1882; BRED IN THE RUNNYMEDE STUD.

*BILLET, b. h., 1865.
Voltigeur, br. h., 1847.
Voltaire, br. h., 1826.
Blacklock, b. h., 1814.
Daughter of, b. m., 1816.
Martha Lynn, br. m., 1837.
Mulatto, b. h., 1823.
Leda, b. m., 1824.
Calcutta, b. m., 1853.
Flatcatcher, br. h., 1845.
Touchstone, br. h., 1831.
Decoy, b. m., 1830.
Miss Martin, br. m., 1844.
St. Martin, br. h., 1833.
Wagtail, br. m., 1829.

Whitelock, b. h., 1803. Hambletonian 1792 (King Fergus 1775)—Rosalind 1788 by Phenomenon 1780 (Herod 1758)—Atalanta 1769 by Matchem 1748 (Cade 1734)—Y. Lass of the Mill 1756 by Oroonoko 1745 (Crab 1722)—Dau. 1744 of Traveler 1735 (Partner 1718).

Daughter of, —, 1799. Coriander 1786 (Pot-8-o-'s 1773)—Wildgoose 1792 by Highflyer 1774 (Herod 1758)—Coheiress 1786 by Pot-8-o-'s 1773 (Eclipse 1764)—Manilla 1777 by Goldfinder 1764 (Snap 1750)—Dau. 1765 of Old England 1741 (God. Ar. 1724).

Phantom, b. h., 1808. Walton 1799 (Sir Peter T. 1784)—Julia 1799 by Whiskey 1789 (Saltram* 1780)—Y. Giantess 1790 by Diomed* 1777 (Florizel 1768)—Giantess 1769 by Matchem 1748 (Cade 1734)—Molly Long Legs 1753 by Babraham 1740 (God. Ar. 1724).

Daughter of, b. m., 1802. Overton 1788 (King Fergus 1775)—Dau. 1793 of Walnut 1786 (Highflyer 1774)—Dau. 1786 of Ruler 1777 (Y. Marske 1771)—Piracantha 1772 by Matchem 1748 (Cade 1734)—Prophetess 1758 by Regulus 1739 (God. Ar. 1724).

Catton, b. h., 1809. Golumpus 1802 (Gohanna 1790)—Lucy Gray 1804 by Timothy 1794 (Delpini 1781)—Lucy 1789 by Florizel 1768 (Herod 1758)—Frenzy 1774 by Eclipse 1764 (Marske 1750)—Dau. of Engineer 1756 (Sampson 1745).

Desdemona, br. m., 1811. Orville 1799 (Beningbrough 1791)—Fanny 1796 by Sir Peter T. 1784 (Highflyer 1774)—Dan. 1788 of Diomed* 1777 (Florizel 1768)—Desdemona 1779 by Marske 1750 (Squirt 1732)—Y. Hag 1761 by Skim 1746 (Starling 1727).

Filho-da-Puta br. h., 1812. Haphazard 1797 (Sir Peter T. 1784)—Mrs. Barnet 1806 by Waxy 1790 (Pot-8-o-'s 1773)—Dau. 1788 of Woodpecker 1773 (Herod 1758)—Heinel 1771 by Squirrel 1754 (Traveler 1735)—Principessa 1762 by Blank 1740 (God. Ar. 1724).

Treasure, ch. m., 1800. Camillus 1803 (Hambletonian 1792)—Dau. 1804 of Hyacinthus 1797 (Coriander 1786)—Flora 1780 by King Fergus 1775 (Eclipse 1764)—Atalanta 1769 by Matchem 1748 (Cade 1734)—Y. Lass of the Mill 1756 by Oroonoko 1745 (Crab 1722).

Camel, br. h., 1822. Whalebone 1807 (Waxy 1790)—Dau. 1812 of Selim 1802 (Buzzard* 1787)—Maiden 1801 by Sir Peter T. 1784 (Highflyer 1774)—Dau. 1788 of Phenomenon 1780 (Herod 1758)—Matron 1782 by Florizel 1768 (Herod 1758).

Banter, br. m., 1826. Master Henry 1815 (Orville 1799)—Boadicea 1807 by Alexander 1782 (Eclipse 1764)—Brunette by Amaranthus 1766 (Old England 1741)—Mayfly 1771 by Matchem 1748 (Cade 1734)—Dau. 1755 of Anc. Starling 1738 (Starling 1727).

Filho-da-Puta br. h., 1812. Haphazard 1797 (Sir Peter T. 1784)—Mrs. Barnet 1806 by Waxy 1790 (Pot-8-o-'s 1773)—Dau. 1788 of Woodpecker 1773 (Herod 1758)—Heinel 1771 by Squirrel 1754 (Traveler 1735)—Principessa 1762 by Blank 1740 (God. Ar. 1724).

Finesse, b. m., 1815. Peruvian 1806 (Sir Peter T. 1784)—Violante 1802 by John Bull 1789 (Fortitude 1777)—Dau. 1788 of Highflyer 1774 (Herod 1758)—Everlasting 1775 by Eclipse 1764 (Marske 1750)—Hyæna 17.2 by Snap 1750 (Snip 1736).

Acteon, ch. h., 1822. Scud 1804 (Beningbrough 1791)—Diana 1808 by Stamford 1794 (Sir Peter T. 1784)—Dau. 1799 of Whiskey 1789 (Saltram* 1780)—Grey Dorimant 1781 by Dorimant 1772 (Otho 1760)—Dizzy 1757 by Blank 1740 (God. Ar. 1724).

Galena, br. m., 1823. Walton 1799 (Sir Peter T. 1784)—Comedy 1817 by Comus 1809 (Sorcerer 1796)—Dau. 1803 of Star* 1785 (Highflyer 1774)—Dau. 1794 of Y. Marske 1771 (Marske 1750)—Emma 1781 by Telemachus 1770 (Herod 1758).

Whisker, b. h., 1812. Waxy 1790 (Pot-8-o-'s 1773)—Penelope 1798 by Trumpator 1782 (Conductor 1767)—Prunella 1788 by Highflyer 1774 (Herod 1758)—Promise 1768 by Snap 1750 (Snip 1736)—Julia 1756 by Blank 1740 (God. Ar. 1724).

Daughter of, br. m., 1814. Sorcerer 1796 (Trumpator 1782)—Dau. 1804 of Sir Solomon 1793 (Sir Peter T. 1784)—Dau. 1794 of Y. Marske 1771 (Marske 1750)—Dau. 1788 of Phenomenon 1780 (Herod 1758)—Calliope 1763 by Slouch 1747 (Cade 1734).

JACONET (full sist. to Iroquois), ch. m., 1867.
*Leamington, br. h., 1853.
Faugh-a-Ballagh, br. h., 1841.
Sir Hercules, blk. h., 1826.
Guiccioli, ch. m., 1823.
Daughter of, b. m., 1841.
Pantaloon, ch. h., 1824.
Daphne, br. m., 1837.
Maggie B. B., ch. m., 1867.
*Australian, ch. h., 1858.
West Australian, b. h., 1850.
Emilia,* b. m., 1840.
Madeline, ch. m., 1862.
Boston, ch. h., 1833.
Magnolia, ch. m., 1841.

Whalebone, br. h., 1807. Waxy 1790 (Pot-8-o-'s 1773)—Penelope 1798 by Trumpator 1782 (Conductor 1767)—Prunella 1788 by Highflyer 1774 (Herod 1758)—Promise 1768 by Snap 1750 (Snip 1736)—Julia 1756 by Blank 1740 (God. Ar. 1724).

Peri, b. m., 1822. Wanderer 1811 (Gohanna 1790)—Thalestris 1809 by Alexander 1782 (Eclipse 1764)—Rival 1800 by Sir Peter T. 1784 (Highflyer 1774)—Hornet 1790 by Drone 1777 (Herod 1758)—Manilla 1777 by Goldfinder 1764 (Snap 1750).

Bob Booty, ch. h., 1804. Chanticleer 1787 (Woodpecker 1773)—Ierne 1790 by Bagot 1780 (Herod 1758)—Dau. of Gamahoe (Bustard 1741)—Patty by Tim 1752 (Squirt 1732)—Miss Patch 1737 by Justice 1725 (H. C. Litton Ar.).

Flight, ch. m., 1809. Ir. Escape 1802 (Commodore 1793)—Y. Heroine by Bagot 1780 (Herod 1758)—Heroine 1775 by Hero 1753 (Cade 1734)—Dau. 1766 of Snap 1750 (Snip 1736)—Sist. 1743 to Regulus by God. Ar. 1724.

Castrel, ch. h., 1801. Buzzard* 1787 (Woodpecker 1773)—Dau. of Alexander 1782 (Eclipse 1764)—Dau. of Highflyer 1774 (Herod 1758)—Dau. of Alfred 1770 (Matchem 1748)—Dau. 1770 of Engineer 1756 (Sampson 1745).

Idalia, ch. m., 1815. Peruvian 1806 (Sir Peter T. 1784)—Musidora 1804 by Meteor 1783 (Eclipse 1764)—Maid of all Work 1786 by Highflyer 1774 (Herod 1758)—Sist. 1771 to Tandem by Syphon 1750 (Squirt 1732)—Dau. of Regulus 1739 (God. Ar. 1724).

Laurel, br. h., 1824. Blacklock 1814 (Whitelock 1803)—Wagtail 1818 by Prime Minister 1810 (Sancho 1801)—Tranby*'s Dam 1812 by Orville 1799 (Beningbrough 1791)—Miss Grimstone 1796 by Weazle 1776 (Herod 1758)—Dau. of Ancaster 1768 (Blank 1740).

Maid of Honor, br. m., 1829. Champion 1812 (Selim 1802)—Etiquette 1820 by Orville 1799 (Beningbrough 1791)—Boadicea 1807 by Alexander 1782 (Eclipse 1764)—Brunette by Amaranthus 1766 (Old England 1741)—Mayfly 1771 by Matchem 1748 (Cade 1734).

Melbourne, br. h., 1834. Humprey Clinker 1822 (Comus 1809)—Dau. 1825 of Cervantes 1806 (Don Quixote 1784)—Dau. 1818 of Golumpus 1802 (Gohanna 1790)—Dau. 1810 of Paynator 1791 (Trumpator 1782)—Circle 1802 by St. George 1789 (Highflyer 1774).

Mowerina, b. m., 1843. Touchstone 1831 (Camel 1822)—Emma 1824 by Whisker 1812 (Waxy 1790)—Gibside Fairy 1811 by Hermes 1790 (Mercury 1778)—Vicissitude 1800 by Pipator 1786 (Imperator 1776)—Beatrice 1791 by Sir Peter T. 1784 (Highflyer 1774).

Y. Emilius, b. h., 1833. Emilius 1820 (Orville 1799)—Shoveler 1816 by Scud 1804 (Beningbrough 1791)—Goosander 1805 by Hambletonian 1792 (King Fergus 1775)—Rally 1790 by Trumpator 1782 (Conductor 1767)—Fancy 1780 (Sist. to *Diomed) by Florizel 1768 (Herod 1758).

Persian, b. m., 1829. Whisker 1812 (Waxy 1790)—Variety 1816 by [Selim or] Soothsayer 1808 (Sorcerer 1796)—Sprite 1807 by Bobtail 1795 (Precipitate* 1787)—Catherine 1795 by Woodpecker 1773 (Herod 1758)—Camilla 1778 by Trentham 1766 (T. Sweepstakes 1743).

Timoleon, ch. h., 1813. Sir Archy 1805 (*Diomed 1777)—Dau. of *Saltram 1780 (Eclipse 1764)—Dau. of Symme's Wildair 1767 (*Fearnought 1755)—Dau. of Tyler's Driver 1760 (*Othello)—Dau. of *Fallower 1761 (Blank 1740).

Sister to Tuckahoe, ch. m., 1814. Ball's Florizel 1801 (*Diomed 1777)—Dau. 1793 of *Alderman 1787 (Pot-8-o-'s 1773)—Dau. of *Clockfast 1774 (Gimcrack 1760)—Dau. of Symme's Wildair 1767 (*Fearnought 1755)—Y. Kitty Fisher 1767 by *Fearnought 1755 (Regulus 1739).

***Glencoe, ch. h., 1831.** Sultan 1816 (Selim 1802)—Trampoline 1825 by Tramp 1810 (Dick Andrews 1797)—Web 1808 by Waxy 1790 (Pot-8-o-'s 1773)—Penelope 1798 by Trumpator 1782 (Conductor 1767)—Prunella 1788 by Highflyer 1774 (Herod 1758).

***Myrtle, ch. m., 1834.** Mameluke 1824 (Partisan 1811)—Bobadilla br. m. 1823 by Bobadil 1813 (Rubens 1805)—Pythoness blk. m. 1813 by Sorcerer 1796 (Trumpator 1782)—Princess blk. m. 1796 by Sir Peter T. 1784 (Highflyer 1774)—Dau. br. m. 1789 of Dungannon 1780 (Eclipse 1764)—Dau. ch. m. 1783 of Turf 1760 (Matchem 1748)—Dau. b. m. 1779 of Herod 1758 (Tartar 1743)—Golden Grove ch. m. 1760 by Blank 1740 (God. Ar. 1724)—Spinster (the Widdrington mare) ch. m. 1735 by Partner 1718 (Jigg)—Bay Bloody Buttocks b. m. 1729 by Bloody Buttocks—Sist. 1722 to Guy by Greyhound (Chillaby)—Brown Farewell br. m. 1710 by Makeless (Oglethorpe Ar.)—Dau. of Brimmer (D'Arcy Yellow Turk)—Dau. of Place's White Turk—Dau. of Dodsworth—Layton Barb mare.

BEND'OR.

WINNER OF THE COKE STAKES, BOARD OF TRADE HANDICAP AT CHICAGO, SUMMER HANDICAP WELTER STAKES AT SARATOGA AND OTHER FAST RACES.

Property of Mr. John B. Ewing, Willamette Stud, near Nashville, Tenn.

BEND'OR by imp. Buckden, son of Lord Clifden by Newminster, bay horse, foaled 1878, bred by Sidney Taylor, Madison Co., Ky., dam Kate Walker by Embry's Lexington, son of Lexington and Bellamira by imp. Monarch, out of Carrie D. by Don Juan, son of Berthune by Sidi Hamet by Virginian by Sir Archy. Bend'or is half brother to Ann Fief, the dam of the unbeaten Tremont, by Alarm. Bend'or was a first-class race-horse; at three years old won the Galt House Handicap, 1½ miles with 95 lbs., in 2.38.

Bend'or as a four year old was one of the best race-horses this country has ever produced on the score of speed and weight carrying capacity. He started in twelve races, of which he won nine. At Louisville he won a purse, 1 1/16 miles, 117 lbs., in 1.49, heats of 1 1/16 miles, 117 lbs., in 1.49¾, 1.49, 1.51½, Bootjack winning the first heat. At Pittsburgh, on a heavy track, he won the Coke Stakes, 1½ miles, 117 lbs., in 2.47½. At Chicago he carried 117 lbs., a 10 lb. penalty, and won the Board of Trade Handicap, 1½ miles in 2.46½, track heavy. At Saratoga he won the Welter Stakes, 1 mile in 1.45, carrying 149 lbs. At same place won 1 mile and 500 yards, 115 lbs., in 2.10½, best time on record to date, beating Ada Glenn, Gen. Monroe, and others, and carrying the top weight; the Summer Handicap, 1½ miles, 118 lbs., in 2.35½, beating the great Thora and other good ones. In the Grand Prize of Saratoga, 1¾ miles, he carried 122 lbs., and was second to Glidelia (5), 116, in 3.01, the fastest race at the distance ever run to that date, At Saratoga he carried 123 lbs., and won a Handicap Sweepstakes, 1½ miles in 2.36¾, beating Gen. Monroe (4), 101 lbs., Ada Glen (6), 102 lbs., and others; a Handicap Sweepstakes, 1 mile and five furlongs, in 2.49, carrying 125 lbs., and beating Creosote (4), 106 lbs., Lida Stanhope (4), 102 lbs., and other first-class horses. This race, considering age and weight carried, is the best on record at the distance, and has only been beaten by ¼ of a second by Exile (4), 115 lbs. Bend'or is a very large, strong horse, and his get all show fine quality. Ann Fief is the dam of the unbeaten Tremont, and two such horses as Bend'or and Tremont from one and the same family is sufficient to establish its producing excellence. All of Bend'or's get that have started are winners. Bandusia won the Belles Stakes, ¾ of a mile, at Sheepshead Bay, in 1.16½, beating Los Angeles and other good ones. Inferno (Iago), Baltimore, A. O. H., Emma J., Goldstep, Golucky, Rheingold, Maid of Altonwood, Even Weight, Tactician, Billet Doux colt, Strategy colt, Equality, Equity, the latter winner of the Titan and Mermaid Stakes, Mamie B. B., Tea Set, Waltz, etc. There is no stallion now in the stud of greater promise.

BEND'OR.

BAY HORSE, FOALED 1878; BRED BY SID. TAYLOR, MADISON COUNTY, KENTUCKY.

- *BUCKDEN, b. h., 1869.
  - Lord Clifden, b. h., 1860.
    - Newminster, b. h., 1848.
      - Touchstone, br. h., 1831.
      - Bee's-wing, b. m., 1833.
    - The Slave, b. m., 1852.
      - Melbourne, br. h., 1834.
      - Volley, b. m., 1845.
  - Consequence, b. m., 1857.
    - Bay Middleton, b. h., 1833.
      - Sultan, b. h., 1816.
      - Cobweb, b. m., 1821.
    - Result, b. m., 1838.
      - Mulatto, b. h., 1823.
      - Problem, b. m., 1823.
- KATE WALKER, br. m., 18 8.
  - Embry's Lexington, b. h., 1858.
    - Lexington, b. h., 1850.
      - Boston, ch. h., 1833.
      - Alice Carneal, b. m., 1836.
    - Bellamira, b. m., 1842.
      - *Monarch, b. h., 1834.
      - Kitty Heath, ch. m., 1833.
  - Carrie D., b. m., 1861.
    - Don Juan, bl. h., 1855.
      - Berthune, b. h., 1837.
      - Ariel, br. m., 1841.
    - Romance, b. m., 1839.
      - Sir Leslie, b. h., 1828.
      - Arian, b. m., 1831.

Camel, br. h., 1822.	Whalebone 1807 (Waxy 1790)—Dau. 1812 of Selim 1802 (Buzzard* 1787)—Maiden 1801 by Sir Peter T. 1784 (Highflyer 1774)—Dau. 1788 of Phenomenon 1780 (Herod 1758)—Matron 1782 by Florizel 1768 (Herod 1758).
Banter, br. m., 1826.	Master Henry 1815 (Orville 1799)—Boadicea 1807 by Alexander 1782 (Eclipse 1764)—Brunette by Amaranthus 1766 (Old England 1741)—Mayfly 1771 by Matchem 1748 (Cade 1734)—Dau. 1755 of Anc. Starling 1738 (Starling 1727).
Dr. Syntax, br. h., 1811.	Paynator 1791 (Trumpator 1782)—Dau. of Beningbrough 1791 (King Fergus 1775)—Jenny Mole 1787 by Carbuncle 1772 (Babe. Blank 1756)—Dau. 1772 of Prince T'Quassaw 1751 (Snip 1736)—Sultana 1759 by Regulus 1739 (God. Ar. 1724).
Daughter of, ch. m., 1817.	Ardrossan 1809 (John Bull 1789)—Lady Eliza 1813 by Withworth 1805 (Agonistes 1797)—Dau. 1793 of Spadille 1784 (Highflyer 1774)—Sylvia 1783 by Y. Marske 1771 (Marske 1750)—Ferret 1765 by Br. to Silvio (Cade 1734).
Humprey Clinker, b. h., 1822.	Comus 1809 (Sorcerer 1796)—Clinkerina 1812 by Clinker 1805 (Sir Peter T. 1784)—Pewet 1786 by Tandem 1773 (Syphon 1770)—Termagant by Tantrum 1760 (Cripple 1750)—Cantatrice by Sampson 1745 (Blaze 1733).
Daughter of, b. m., 1825.	Cervantes 1806 (Don Quixote 1784)—Dau. 1818 of Golumpus 1802 (Gohanna 1790)—Dau. 1810 of Paynator 1791 (Trumpator 1782)—Circle 1802 by St. George 1789 (Highflyer 1774)—Abigail 1788 by Woodpecker 1773 (Herod 1758).
Voltaire, br. h., 1826.	Blacklock 1814 (Whitelock 1803)—Dau. 1816 of Phantom 1808 (Walton 1799)—Dau. 1802 of Overton 1788 (King Fergus 1775)—Dau. 1796 of Walnut 1786 (Highflyer 1774)—Dau. 1786 of Ruler 1777 (Y. Marske 1771).
Martha Lynn, br. m., 1837.	Mulatto 1823 (Catton 1809)—Leda 1824 by Filho-da-Puta 1812 (Haphazard 1797)—Treasure 1809 by Camillus 1803 (Hambletonian 1792)—Dau. 1801 of Hyacinthus 1797 (Coriander 1786).
Selim, ch. h., 1802.	Buzzard* 1787 (Woodpecker 1773)—Dau. of Alexander 1782 (Eclipse 1764)—Dau. of Highflyer 1774 (Herod 1758)—Dau. of Alfred 1770 (Matchem 1748)—Dau. 1770 of Engineer 1756 (Sampson 1745).
Bacchante, br. m., 1809.	W's Ditto 1800 (Sir Peter T. 1784)—Dau. 1791 of Mercury 1778 (Eclipse 1764)—Dau. 1776 of Herod 1758 (Tartar 1743)—Folly 1771 by Marske 1750 (Squirt 1732)—Vixen 1753 by Regulus 1739 (God. Ar. 1724).
Phantom, b. h., 1808.	Walton 1799 (Sir Peter T. 1784)—Julia 1799 by Whiskey 1789 (Saltram* 1780)—Y. Giantess 1790 by Diomed* 1777 (Florizel 1768)—Giantess 1769 by Matchem 1748 (Cade 1734)—Molly Long Legs 1753 by Babraham 1740 (God. Ar. 1724).
Filagree, ch. m., 1815.	Soothsayer 1808 (Sorcerer 1796)—Web 1808 by Waxy 1790 (Pot-8-o-'s 1773)—Penelope 1798 by Trumpator 1782 (Conductor 1767)—Prunella 1788 by Highflyer 1774 (Herod 1758)—Promise 1768 by Snap 1750 (Snip 1736).
Catton, b. h., 1809.	Golumpus 1802 (Gohanna 1790)—Lucy Gray 1804 by Timothy 1794 (Delpini 1781)—Lucy 1789 by Florizel 1768 (Herod 1758)—Frenzy 1774 by Eclipse 1764 (Marske 1750)—Dau. of Engineer 1756 (Sampson 1745).
Desdemona, br. m., 1811.	Orville 1799 (Beningbrough 1791)—Fanny 1796 by Sir Peter T. 1784 (Highflyer 1774)—Dau. 1788 of Diomed* 1777 (Florizel 1768)—Desdemona 1770 by Marske 1750 (Squirt 1732)—Y. Hag 1761 by Skim 1746 (Starling 1727).
Euclid, ch. h., 1836.	Emilius 1820 (Orville 1799)—Maria 1827 by Whisker 1812 (Waxy 1790)—Gibside Fairy 1811 by Hermes 1790 (Mercury 1778)—Vicissitude 1800 by Pipator 1786 (Imperator 1776)—Beatrice 1791 by Sir Peter T. 1784 (Highflyer 1774).
Parthenessa, br. m., 1821.	Cervantes 1806 (Don Quixote 1784)—Marianne 1816 by Sorcerer 1796 (Trumpator 1782)—Thomasina 1804 by Timothy 1794 (Delpini 1781)—Violet 1787 by Shark* 1771 (Marske 1750)—Dau. 1772 of Syphon 1770 (Squirt 1732).
Timoleon, ch. h., 1813.	Sir Archy 1805 (*Diomed 1777)—Dau. of *Saltram 1780 (Eclipse 1764)—Dau. of Symmes' Wildair 1767 (*Fearnought 1755)—Dau. of Driver 1760 (*Othello)—Dau. of *Fallower 1761 (Blank 1740).
Daughter of, ch. m., 1814.	Ball's Florizel 1801 (*Diomed 1777)—Dau. 1790 of *Alderman 1787 (Pot-8-o-'s 1773)—Dau. of *Clockfast 1774 (Gimcrack 1760)—Dau. of Symmes' Wildair 1767 (*Fearnought 1755)—Y. Kitty Fisher 1767 by *Fearnought 1755 (Regulus 1739).
*Sarpedon, br. h., 1828.	Emilius 1820 (Orville 1799)—Icaria 1824 by The Flyer 1814 (Vandyke Jr. 1808)—Parma 1813 by Dick Andrews 1797 (Joe Andrews 1778)—May 1804 by Beningbrough 1791 (King Fergus 1775)—Primrose 1787 by Mambrino 1768 (Engineer 1756).
Rowena, ch. m., 1826.	Sumpter 1818 (Sir Archy 1805)—Lady Grey 1817 by Robin Grey 1805 (*Royalist 1790)—Maria 1802 by Melzar 1791 (*Medley 1776)—Dau. of *Highflyer 1784 (Highflyer 1774)—Dau. of *Fearnought 1755 (Regulus 1739).
Priam,* b. h., 1827.	Emilius 1820 (Orville 1799)—Cressida 1807 by Whiskey 1789 (Saltram* 1780)—Y. Giantess 1790 by Diomed* 1777 (Florizel 1768)—Giantess 1769 by Matchem 1748 (Cade 1734)—Molly Long Legs 1753 by Babraham 1740 (God. Ar. 1724).
Delphine,* br. m., 1825.	Whisker 1812 (Waxy 1790)—My Lady* 1818 by Comus 1809 (Sorcerer 1796)—The Colonel's dam 1802 by Delpini 1781 (Highflyer 1774)—Tipple Cyder 1788 by King Fergus 1775 (Eclipse 1764)—Sylvia 1783 by Y. Marske 1771 (Marske 1750).
Am. Eclipse, ch. h., 1814.	Duroc 1806 (*Diomed 1777)—Miller's Damsel 1802 by *Messenger 1780 (Mambrino 1768)—*Dau. 1792 of Pot-8-o-'s 1773 (Eclipse 1764)—Dau. 1778 of Gimcrack 1760 (Cripple 1750)—Snapdragon 1759 by Snap 1750 (Snip 1736).
Am. Pomona, b. m., 1820.	Sir Alfred 1805 (*Sir Harry 1795)—Dau. of *Sir Harry 1795 (Sir Peter T. 1784)—*Pomona 1803 by Worthy 1795 (Pot-8-o-'s 1773)—Comedy 1799 by Buzzard* 1787 (Woodpecker 1773)—Huncamunca 1787 by Highflyer 1774 (Herod 1758).
Sidi Hamet, b. h., 1825.	Virginian 1815 (Sir Archy 1805)—Lady Burton 1813 by Sir Archy 1805 (*Diomed 1777)—Sultana by *Barb horse Black Sultan)—Barb M. presented 1806 to Pres. Jefferson by Dey of Tunis.
Susette, br. m., 1828.	Aratus 1820 (Director 1811)—Jenny Cockracy 1814 by Potomac 1805 (*Diomed 1777)—Timoleon's dam by *Saltram 1780 (Eclipse 1764)—Dau. of Symmes' Wildair 1767 (*Fearnought 1755)—Dau. of Driver 1760 (*Othello).
Sarpedon,* br. h., 1828.	Emilius 1820 (Orville 1799)—Icaria 1824 by The Flyer 1814 (Vandyke Jr. 1808)—Parma 1813 by Dick Andrews 1797 (Joe Andrews 1778)—May 1804 by Beningbrough 1791 (King Fergus 1775)—Primrose 1787 by Mambrino 1768 (Engineer 1756).
Lancess, b. m., 1835.	Lance 1821 (Am. Eclipse 1814)—Aurora 1827 by Aratus 1820 (Director 1811)—Paragon 1807 by *Buzzard 1787 (Woodpecker 1773)—Indiana 1802 by Columbus (*Pantaloon 1778)—Jane Hunt 1796 by Paragon 1788 (*Flimnap 1765).
Sir William of Transport, b. h., 1820.	Sir Archy 1805 (*Diomed 1777)—Transport 1812 by Virginius 1805 (*Diomed 1777)—(Florizel 1768)—Nancy Air 1799 by *Bedford 1792 (Dungannon 1780)—Annette by *Shark 1771 (Marske 1750)—dau. of Rockingham (Partner 1755).
Miss Dowden, ch. m., 1809.	*Buzzard 1787 (Woodpecker 1773)—Dau. of *Speculator 1795 (Dragon* 1787)—Dau. of Damon (*Fearnought 1755)—Dau. of Fearnought.
Trumpator, br. h., 1823.	Sir Solomon 1805 (*Tickle Toby 1786)—Dau. of Hickory 1804 (*Whip 1794)—Trumpetta 1797 by Trumpator 1782 (Conductor 1767)—Sist. 1786 to Lambinos by Highflyer 1774 (Herod 1758)—Dau. 1775 of Eclipse 1764 (Marske 1750).
Mary Jones, ch. m., 1827.	Kosciusko 1815 (Sir Archy 1805)—Multiflora's grandam by Y. Bedford (*Bedford 1792)—Dau. of Arion (Polyphemus)—Dau. of Brilliant (Burwell's Traveller, son of *Traveller 1747).

IMPORTED BLACK DEAN.

Brother to Geheimniss, owned by Eugene Leigh, La Belle Stud, Yarnallton, Kentucky. Stands for public mares.

BLACK DEAN, own brother to Geheimniss, winner of the Oaks, brown horse, foaled 1886, bred by Mr. John Watson. By Rosicrucian, son of Beadsman, by Weatherbit, dam Nameless, dam of Geheimniss, winner of the Oaks, by Blinkhoolie, son of Rataplan, out of No Name by Teddington, winner of the Derby, Ascot and Doncaster Cups, son of Orlando by Touchstone. Black Dean from an injury never raced but his sister Geheimniss, one of the best mares of her day. As a two year old she started seven times. On the first occasion she ran a dead heat for the Sandown Two Years Old Stakes, which were divided, and the remaining six races she won, beating some of the best animals in training. As a three year old she won the Oaks; ran second in the Fernhill Stakes at Ascot, beaten a head; ran second in the St. Leger, beating Shotover (winner of the 2,000 Guineas, Derby, etc.) and many other high-class animals; and wound up the season by running second at Newmarket, beating Kermesse, who as a two year old was only beaten once, and who also won all her three year old engagements with the exception of the race mentioned. As a four year old, Geheimniss won the Trial Stakes at Ascot, giving 14 lbs. weight to all starters, won the Bunbury Handicap, one mile; ran second in the Steward's Cup at Goodwood, beaten a head, giving the winner 14 lbs.; ran second for the Portland Plate, carrying top weight, giving the winner 10 lbs.; won the Kempton Park Cambridgeshire Trial Handicap, 1¼ miles, carrying 129 lbs.; won the Trial Stakes at Newmarket, carrying 140 lbs., and finished the year by giving Brag 21 lbs. and running second to him for the Brighton Cup. The five year old form of Geheimniss was brilliant. She earned eight brackets, ran second once and was placed once. The races she won were the Westminster Cup, 1¼ miles, at Kempton Park, carrying 143 lbs.; the All-aged Stakes at Ascot, carrying 145 lbs.; the Stockbridge Cup, carrying 132 lbs.; a walk over for the July Cup at Newmarket; walk over for the Bunbury Stakes at same meeting; won the Lennox Stakes at Goodwood, and walked over for the Singleton Stakes at the same place.

Rosicrucian, the sire of Black Dean, is own brother to The Palmer, the sire of Pellegrino and Jenny Howlet, winner of the Oaks, and was a good race-horse; won the Criterion and Troy Stakes, and is the sire of Hauteur, winner of the 1,000 Guineas, and the good stallion Beauclerc. Rosicrucian was regarded by Sir Joseph Hawley as the equal of the Derby winner Blue Gown, and declared to win the Derby with him. His sire, Beadsman (out of Mendicant, winner of the Oaks), won the Derby (sire of Pero Gomez, winner of St. Leger, and imp. Saxon, winner of the fastest Belmont stake ever run), was a son of Weatherbit (by Sheet Anchor), whose dam, Miss Letty by Priam, won the Oaks. Nameless, Black Dean's dam, produced other winners in Friendless and Offenheit. Her sire, Blinkhoolie, was by Rataplan (a full brother to Stockwell), and sire of Kettledrum (winner of the Derby) His dam, Queen Mary, produced imp. Bonnie Scotland and also Blink Bonny, the dam of Blair Athol (winner of the Derby, St. Leger, etc.) and Breadalbane, a good race-horse. No Name, Black Dean's second dam, was a winner, and bred a winner in Bradlaugh. Queen of Beauty, the third dam, produced Cordelia, a winner, and dam of the winners Thunder and Lady Blanche. Vlie, another of Queen of Beauty's daughters, produced the winners First Lord, Black Adder and Amor. Birthday, the fourth dam, produced the winners Filius, Kohinoor, Stockham, Lupellus, Lupus and Vision, while from Honoria, the fifth dam, came the winner Broadsword, and Miss Nutting (dam of the winner Hawk). Any number of other winners have come from the family, including Mr. Haggin's imported stallion Darebin, in addition to Leamington, Touchstone, Launcelot and Macaroni, and the American winners Arizona, Aranza, Glenarm, Amerique, Louise, Palatka, Millie Williams, Zephyr, Cape Race, Armiel, Ella Blackburn, Cyclone, Westwind, Falsehood, etc. Black Dean is a sure and safe foal getter and should succeed in the stud.

Dick Andrews 1797—Joe Andrews 1778 Eclipse 1764—Marske 1750—Squirt 1732—Bartlett's Childers—Darley Arabian 1699.

BLACK DEAN.

BROWN HORSE, FOALED 1886; BRED BY MR. JNO. WATSON, ENGLAND.

ROSICRUCIAN, b. h., 1865.	Beadsman, br. h. 1855.	Weatherbit, br. h., 1842.	Sheet Anchor, br. h., 1832.	Lottery, br. h., 18.0.	Tramp 1810 (Dick Andrews 1797)—Mandane 1800 by Pot-8-o-'s 1773 (Eclipse 1764) Y. Camilla 1787 by Woodpecker 1773 (Herod 1758) Camilla 1778 by Trentham 1766 (T. Sweepstakes 1743)—Coquette 1765 by Compton Barb.
				Morgiana, blk. m., 1820.	Muley 1810 (Orville 1799) Miss Stephenson 1811 [Send or] Sorcerer 1796 (Trumpator 1782) Dau. 1796 of Precipitate* 1787 (Mercury 1778) Dau. 1787 of Woodpecker 1773 (Herod 1758)—Sist. 1773 to Juniper by Snap 1750 (Snip 1736).
			Miss Letty, b. m., 1834.	Priam,* b. h., 1827.	Emilius 1820 (Orville 1799) Cressida 1807 by Whiskey 1789 (Saltram* 1780) Y. Giantess 1790 by Diomed* 1777 (Florizel 1768) Giantess 1769 by Matchem 1748 (Cade 1734)—Molly Long Legs 1753 by Babraham 1740 (God. Ar. 1724).
				Daughter of, br. m., 1815.	Orville 1799 (Beningbrough 1791) Dau. 1803 of Buzzard* 1787 (Woodpecker 1773)—Hornpipe 1773 by Trumpator 1782 (Conductor 1767) Luna 1779 by Herod 1758 (Tartar 1743) Proserpine 1763 by Marske 1750 (Squirt 1732).
		Mendicant, br. m., 1843.	Touchstone, br. h., 1831.	Camel, br. h., 1822.	Whalebone 1807 (Waxy 1790) Dau. 1812 of Selim 1802 (Buzzard* 1787) Maiden 1801 by Sir Peter T. 1784 (Highflyer 1774) Dau. 1788 of Phenomenon 1780 (Herod 1758) Matron 1782 by Florizel 1768 (Herod 1758).
				Banter, br. m., 1826.	Master Henry 1815 (Orville 1799) Boadicea 1807 by Alexander 1782 (Eclipse 1764)—Brunette by Amaranthus 1766 (Old England 1741) Mayfly 1771 by Matchem 1748 (Cade 1734)—Dau. 1755 of Starling 1738 (Starling 1727).
			Lady Moore Carew, b. m., 1830.	Tramp, b. h., 1810.	Dick Andrews 1797 (Joe Andrews 1778)—Dau. 1803 of Gohanna 1790 (Mercury 1778) Fraxinella 1793 by Trentham 1766 (T. Sweepstakes 1743) Dau. 1785 of Woodpecker 1773 (Herod 1758)—Everlasting 1775 by Eclipse 1764 (Marske 1750).
				Kite, b. m., 1821.	Bustard 1813 (Castrel 1801)—Olympia 1815 by Sir Oliver 1800 (Sir Peter T. 1784) Scotilla 1795 by Anvil 1777 (Herod 1758)—Scota 1783 by Eclipse 1764 (Marske 1750) Harmony 1775 by Herod 1758 (Tartar 1743).
	Madame Eglentine, b. m., 1857.	Cowl, b. h., 1842.	Bay Middleton, b. h., 1833.	Sultan, b. h., 1816.	Selim 1802 (Buzzard* 1787) Bacchante 1809 by W's Ditto 1800 (Sir Peter T. 1784) Dau. 1791 of Mercury 1778 (Eclipse 1764)—Dau. 1776 of Herod 1758 (Tartar 1743) Folly 1771 by Marske 1750 (Squirt 1732).
				Cobweb, b. m., 1821.	Phantom 1808 (Walton 1799)—Filagree 1815 by Soothsayer 1808 (Sorcerer 1796)—Web 1808 by Waxy 1790 (Pot-8-o-'s 1773)—Penelope 1798 by Trumpator 1782 (Conductor 1767)—Prunella 1788 by Highflyer 1774 (Herod 1758).
			Crucifix, b. m., 1837.	Priam,* b. h., 1827.	Emilius 1820 (Orville 1799)—Cressida 1807 by Whiskey 1789 (Saltram* 1780) Y. Giantess 1790 by Diomed* 1777 (Florizel 1768)—Giantess 1769 by Matchem 1748 (Cade 1734)—Molly Long Legs 1753 by Babraham 1740 (God. Ar. 1724).
				Octaviana, b. m., 1815.	Octavian 1807 (Stripling 1795)—Dau. 1807 of Shuttle 1793 (Y. Marske 1771)—Zara 1801 by Delpini 1781 (Highflyer 1774)—Flora 1789 by King Fergus 1775 (Eclipse 1764)—Atalanta 1769 by Matchem 1748 (Cade 1734).
		Diversion, ch. m., 1838.	Defence, b. h., 1824.	Whalebone, br. h., 1807.	Waxy 1790 (Pot-8-o-'s 1773)—Penelope 1798 by Trumpator 1782 (Conductor 1767)—Prunella 1788 by Highflyer 1774 (Herod 1758)—Promise 1768 by Snap 1750 (Snip 1736)—Julia 1756 by Blank 1740 (God. Ar. 1724).
				Defiance, ch. m., 1816.	Rubens 1805 (Buzzard* 1787)—Little Folly 1806 by Highland Fling 1798 (Spadille 1784)—Harriet 1799 by Volunteer 1780 (Eclipse 1764)—Dau. 1785 of Alfred 1770 (Matchem 1748)—Magnolia 1771 by Marske 1750 (Squirt 1732).
			Folly, ch. m., 1830.	Middleton, ch. h., 1822.	Phantom 1808 (Walton 1799)—Web 1808 by Waxy 1790 (Pot-8-o-'s 1773)—Penelope 1798 by Trumpator 1782 (Conductor 1767)—Prunella 1788 by Highflyer 1774 (Herod 1758)—Promise 1768 by Snap 1750 (Snip 1736).
				Little Folly, b. m., 1806.	Highland Fling 1798 (Spadille 1784)—Harriet 1799 by Volunteer 1780 (Eclipse 1764)—Dau. 1785 of Alfred 1770 (Matchem 1748)—Magnolia 1771 by Marske 1750 (Squirt 1732)—Dau. 1755 by Babraham 1740 (God. Ar. 1724).
NAMELESS, ch. m., 1874.	Blinkhoolie, b. h., 1864.	Rataplan, ch. h., 1850.	The Baron, ch. h., 1842.	Ir Birdcatcher ch. h., 1833.	Sir Hercules 1826 (Whalebone 1807)—Guiccioli 1823 by Bob Booty 1804 (Chanticleer 1787)—Flight 1809 by Ir. Escape 1802 (Commodore 1793)—Y. Heroine by Bagot 1780 (Herod 1758)—Heroine 1775 by Hero 1753 (Cade 1734).
				Echidna, b. m., 1838.	Economist 1825 (Whisker 1812)—Miss Pratt 1825 by Blacklock 1814 (Whitelock 1803)—Gadabout 1812 by Orville 1799 (Beningbrough 1791)—Minstrel 1803 by Sir Peter T. 1784 (Highflyer 1774)—Matron 1782 by Florizel 1768 (Herod 1758).
			Pocahontas, b. m., 1837.	Glencoe,* ch. h., 1831.	Sultan 1816 (Selim 1802)—Trampoline 1825 by Tramp 1810 (Dick Andrews 1797)—Web 1808 by Waxy 1790 (Pot-8-o-'s 1773)—Penelope 1798 by Trumpator 1782 (Conductor 1767)—Prunella 1788 by Highflyer 1774 (Herod 1758).
				Marpessa, b. m., 1830.	Muley 1810 (Orville 1799)—Clare 1824 by Marmion 1806 (Whiskey 1789)—Harpalice 1814 by Gohanna 1790 (Mercury 1778)—Amazon 1799 by Driver 1783 (Trentham 1766)—Fractious 1792 by Mercury 1778 (Eclipse 1764).
		Queen Mary, b. m., 1843.	Gladiator, ch. h., 1833.	Partisan, b. h., 1811.	Walton 1799 (Sir Peter T. 1784)—Parasol 1800 by Pot-8-o-'s 1773 (Eclipse 1764) Prunella 1788 by Highflyer 1774 (Herod 1758)—Promise 1768 by Snap 1750 (Snip 1736)—Julia 1756 by Blank 1740 (God. Ar. 1724).
				Pauline, b. m., 1826.	Moses 1819 ([Whalebone or] Seymour 1807)—Quadrille 1815 by Selim 1802 (Buzzard* 1787)—Canary Bird 1806 by [Whiskey or] Sorcerer 1796 (Trumpator 1782) Canary 1797 by Coriander 1786 (Pot-8-o-'s 1773)—Miss Green 1787 by Highflyer 1774 (Herod 1758).
			Daughter of, b. m., 1840.	Plenipotentiary, ch. h., 1831.	Emilius 1820 (Orville 1799)—Harriet 1819 by Pericles 1809 (Evander 1801)—Dau. 1812 of Selim 1802 (Buzzard* 1787)—Pipylina 1803 by Sir Peter T. 1784 (Highflyer 1774)—Rally 1790 by Trumpator 1782 (Conductor 1767).
				Myrrha, b. m., 1830.	Whalebone 1807 (Waxy 1790)—Gift 1818 by Y. Gohanna 1810 (Gohanna 1790) Dau. 1802 of Sir Peter T. 1784 (Highflyer 1774)—Dau. 1788 of Trumpator 1782 (Conductor 1767)—Sist. 1774 to Postmaster by Herod 1758 (Tartar 1743).
	No Name, br. m., 1861.	Teddington, ch. h., 1848.	Orlando, b. h., 1841.	Touchstone, br. h., 1831.	Camel 1822 (Whalebone 1807)—Banter 1826 by Master Henry 1815 (Orville 1799)—Boadicea 1807 by Alexander 1782 (Eclipse 1764)—Brunette by Amaranthus 1766 (Old England 1741)—Mayfly 1771 by Matchem 1748 (Cade 1734).
				Vulture, ch. m., 1833.	Langar 1817 (Selim 1802)—Kite 1821 by Bustard 1813 (Castrel 1801)—Olympia 1815 by Sir Oliver 1800 (Sir Peter T. 1784)—Scotilla 1795 by Anvil 1777 (Herod 1758)—Scota 1783 by Eclipse 1764 (Marske 1750).
			Miss Twickenham, ch. m., 1838.	Rockingham, b. h., 1833.	Humprey Clinker 1822 (Comus 1809)—Medora 1813 by Swordsman 1796 (Prizefighter 1784)—Ir. Penelope 1797 by Trumpator 1782 (Conductor 1767)—Peppermint 1787 by Highflyer 1774 (Herod 1758)—Promise 1768 by Snap 1750 (Snip 1736).
				Electress, ch. m., 1819.	Election 1804 (Gohanna 1790)—Dau. 1805 of Stamford 1794 (Sir Peter T. 1784)—Miss Judy 1784 by Alfred 1770 (Matchem 1748)—Manilla 1777 by Goldfinder 1764 (Snap 1750)—Dau. 1766 of Old England 1741 (God. Ar. 1724).
		Queen of Beauty, br. m., 1854.	Melbourne, br. h., 1834.	Humprey Clinker, b. h., 1822.	Comus 1809 (Sorcerer 1796)—Clinkerina 1812 by Clinker 1805 (Sir Peter T. 1784)—Pewet 1786 by Tandem 1773 (Syphon 1750)—Termagant by Tantrum 1760 (Cripple 1750)—Cantatrice by Sampson 1745 (Blaze 1733).
				Daughter of, b. m., 1825.	Cervantes 1806 (Don Quixote 1784)—Dau. 1818 of Golumpus 1802 (Gohanna 1790)—Dau. 1810 of Paynator 1791 (Trumpator 1782) Circle 1802 by St. George 1789 (Highflyer 1774)—Abigail 1788 by Woodpecker 1773 (Herod 1758).
			Birthday, br. m., 1843.	Pantaloon, ch. h., 1824.	Castrel 1801 (Buzzard* 1787)—Idalia 1815 by Peruvian 1806 (Sir Peter T. 1784)—Musidora 1804 by Meteor 1783 (Eclipse 1764)—Maid-of-all-Work 1786 by Highflyer 1774 (Herod 1758)—Dau. 1771 of Syphon 1750 (Squirt 1732).
				Honoria, br. m., 1830.	Camel 1822 (Whalebone 1807)—Maid of Honor br. m. 1829 by Champion 1812 (Selim 1802)—Etiquette br. m. 1820 by Orville 1799 (Beningbrough 1791)—Boadicea b. m. 1807 by Alexander 1782 (Eclipse 1764)—Brunette by Amaranthus 1766 (Old England 1741)—Mayfly b. m. 1771 by Matchem 1748 (Cade 1734)—Dau. blk. m. 1755 of Anc. Starling

1738 (Bolton Starling 1727)—Look-at-me-Lads ch. m. 1731 by Bristol Grasshopper (Byerly Turk)—Dau. of Sir M. Newton's Bay Ar.—Lady Mare by Pert (Ely Turk)—Dau. of St. Martin (Spanker)—Dau. of Sir E. Hale's Ar.—Oldfield Mare.

BRAMBLE.

WINNER OF THE YOUNG AMERICA STAKE NO. 2 AND MAXWELL HOUSE STAKES AT NASHVILLE, SARATOGA AND CONGRESS STAKES AT SARATOGA, BALTIMORE CUP, MONMOUTH CUP, WESTCHESTER CUP, SARATOGA CUP, AND OTHER IMPORTANT RACES.

Property of Eugene Leigh, La Belle Stud, Yarnalton, Ky., and is a public stallion. Terms published annually.

BRAMBLE by imp. Bonnie Scotland, bred by Gen. W. G. Harding, Belle Meade Stud, foaled 1875, dam Ivy Leaf, dam of Bazar, Brambaletta, etc., by imp. Australian, out of Bay Flower, sister to Bayonet, Bayswater, Preakness, etc., by Lexington. Bramble ran seven times as a two-year old; won the Young America Stakes No. 2 at Nashville, three-quarters of a mile, in 1.20, track heavy, won the Saratoga Stakes, at Saratoga, three-quarters of a mile, in 1.17½, was second in two and third in two races, beating Duke of Magenta, Pride of the Village and others. He made his *debut* in his three-year old form in the Withers Stake at Jerome Park, one mile, was second to Duke of Magenta in 1.48, each with 118 lbs., one of the best races ever run, the track being muddy and heavy—Danicheff and Pride of the Village were behind him; beat Simoom, Oriole, etc., one mile, in 1.48½; ran second to Duke of Magenta for the Belmont Stakes, 1½ miles, in 2.43½, Spartan, Danicheff, Bayard and others behind him. Long Branch, won dash of three-quarters of a mile, beating Idalia, Egypt, etc., in 1.17; same place, was beaten a match, one and a quarter miles, by Spartan, in 2.16, evidently badly managed or out of condition. Saratoga, ran second to the Duke of Magenta in Travers Stakes, one and three-quarters miles, in 3.08, a capital race; won dash of one mile, beating Lou Lanier, Garrick, etc., in 1.45; won dash of one mile, beating Idalia, Pique, etc., in 1.47½, track heavy; beat Oriole, Idalia, etc., one and a quarter miles, in 2.17; walked over for purse, one and a half miles, received half purse and entrance money; beat Pique, Pride of the Village, etc., one mile, in 1.47½, carrying 113 lbs; ran second to Duke of Magenta for Kenner Stakes, two miles, in 3.41½, with good ones behind him; won race 1⅛ miles, in 1.58, beating Bonnie Wood, Kate Claxton, etc.; beat Lady d'Arcy, Bonnie Wood, etc.; three-quarters of a mile, in 1.19; beat Patriot, Kate Claxton and others, one mile, in 1.45¾. He was sent to Louisville, where he won the great American Stallion Stakes, 1¾ miles, in 3.14½, beating Cammie F., Warfield, Momentum, Day Star and five others; same place, was second to Warfield, giving him 5 lbs., 1⅛ miles, in 1.56, the fastest race at the distance during the year. Nashville, won the Maxwell House Stakes, mile heats, in 1.43, 1.44, beating Warfield and Bergamot. At four years old started twenty times, won fifteen races, was second in two, third in one and unplaced in two. Bramble at Baltimore won dash of 1¼ miles, defeating Gov. Hampton, Albert and Judith; time, 2.15½; same meeting, won the Baltimore Cup, 2¼ miles, defeating Bushwhacker, Bonnie Wood, Lou Lanier and Bayard, time, 4.01¾. Jerome Park, won dash of one mile, defeating Una, Franklin, Bonnie Wood and three others, time 1.45½. Same meeting won the Westchester Cup, 2¼ miles, defeating his only opponent, Invermore, time, 4.20½. Spring Meeting, Coney Island Jockey Club, at Prospect Park, Seaside Handicap, 2 miles, Willful finished first, Gen. Philips second and Bramble last; time, 3.34¾. Monmouth Park, won the Ocean Stakes, 1⅛ miles, defeating Monitor, Lancewood, Susquehanna and Belinda; time, 1.58¾; won dash 1⅛ miles, defeating Una and Tom Scarlet; time, 2.27. Won the Monmouth Cup, 2¼ miles, defeating Lou Lanier, Gen. Philips and Willful; Bramble carried 5 lbs. penalty for winning the Westchester Cup; time, 4.18½. Saratoga, ran second to Monitor in the All-Aged Stakes, 1¼ miles; time, 2.12¼; won the Saratoga Cup, 2¼ miles, defeating Willful, Lou Lanier, etc.; time, 4.11¼; won dash of 1⅛ miles, defeating Mintzer, Dan Sparling, etc.; time, 2.00; won dash 1 mile, defeating Bennett; time, 1.49; won three-quarters of a mile, defeating Annie Augusta, Bonnie Wood, etc., time, 1.17½; won dash of 1 mile, beating Oriole and Gabriel, time, 1.44½; won dash 1⅛ miles, defeating Gabriel, Dan Sparling, etc., time, 1.58; won the Congress Hall Stakes, heats of three-quarters of a mile, time, 1.19, 1.18; won the Brighton Cup, 2¼ miles, defeating Franklin by a head, time, 4.16½. At 5 years won the Centennial Stakes at Nashville, Tenn., 1⅛ miles, in 1.58, beating Beatitude and others; at 6 years, starting only once; ran second to Ferida; at Baltimore, mile heats in 1.44½, 1.44½, 1.48¾, Bramble won the first heat. This was his last race. Ivy Leaf, the dam of Bramble, was never trained. Bay Flower, his grandam, was a good race-mare at all distances, she beat the famous Rhynodyne three and four mile heats at St. Louis; four miles, in 7.45½, 7.45. Bramble is a rich red bay with large star in

(*Continued on page 120.*)

Pot-8-o-'s 1773—Eclipse 1764—Marske 1750—Squirt 1732—Bartlett's Childers—Darley Arabian 1699.

BRAMBLE.

BAY HORSE, FOALED 1875; BRED IN THE BELLE MEADE STUD, TENNESSEE.

- ***BONNIE SCOTLAND, b. h., 1853.**
  - Iago, b. h., 1843.
    - Don John, b. h., 1835.
      - Waverley, br. h., 1817.
      - Daughter of [illegible], gr. m., 1821.
    - Scandal, b. m., 1832.
      - Selim, ch. h., 1802.
      - Daughter of, br. m., 1816.
  - Queen Mary, b. m., 1843.
    - Gladiator, ch. h., 1833.
      - Partisan, b. h., 1811.
      - Pauline, b. m., 1826.
    - Daughter of, b. m., 1840.
      - Plenipotentiary, ch. h., 1831.
      - Myrrha, b. m., 1830.
- **IVY LEAF, ch. m., 1867.**
  - *Australian, ch. h., 1858.
    - West Australian, b. h., 1850.
      - Melbourne, br. h., 1834.
      - Mowerina, b. m., 1843.
    - Emilia,* b. m., 1840.
      - Y. Emilius, b. h., 1828.
      - Persian, b. m., 1829.
  - Bay Flower, b. m., 1859.
    - Lexington, b. h., 1850.
      - Boston, ch. h., 1833.
      - Alice Carneal, b. m., 1836.
    - Bay Leaf, b. m., 1853.
      - *Yorkshire, b. h., 1834.
      - *Maria Black, br. m., 1834.

Whalebone, br. h., 1807.	Waxy 1790 (Pot-8-o's 1773)—Penelope 1798 by Trumpator 1782 (Conductor 1767)—Prunella 1788 by Highflyer 1774 (Herod 1758)—Promise 1768 by Snap 1750 (Snip 1736)—Julia 1756 by Blank 1740 (God. Ar. 1724).
Margaretta, br. m., 1802.	Sir Peter T. 1784 (Highflyer 1774)—Dau. 1792 of Highflyer 1774 (Herod 1758)—Nutcracker 1767 by Matchem 1748 (Cade 1734)—Miss Starling 1750 by Starling 1727 (Bay Bolton 1705)—Ringbone 1732 by Partner 1718 (Jigg).
Comus, ch. h., 1809.	Sorcerer 1796 (Trumpator 1782)—Epsom Lass 1801 by Sir Peter T. 1784 (Highflyer 1774)—Alexina 1788 by King Fergus 1775 (Eclipse 1764)—Lardella 1780 by Y. Marske 1771 (Marske 1750)—Dau. of Y. Cade 1747 (Cade 1734).
Marciana, gr. m., 1809.	Stamford 1794 (Sir Peter T. 1784)—Marcia 1797 by Coriander 1786 (Pot-8-o's 1773)—Faith 1779 by Pacolet 1763 (Blank 1740)—Atalanta 1769 by Matchem 1748 (Cade 1734)—Y. Lass-of-the-Mill 1756 by Oroonoko 1745 (Crab 1722).
Buzzard,* ch. h., 1787.	Woodpecker 1773 (Herod 1758)—Misfortune 1775 by Dux 1761 (Matchem 1748)—Curiosity 1760 by Snap 1750 (Snip 1736)—Dau. 1749 of Regulus 1739 (God. Ar. 1724)—Dau. of Bartlett's Childers (Darley Ar. 1699).
Daughter of, b. m., —.	Alexander 1782 (Eclipse 1764)—Dau. of Highflyer 1774 (Herod 1758)—Dau. of Alfred 1770 (Matchem 1748)—Dau. 1770 of Engineer 1756 (Sampson 1745)—Bay Malton's dam 1756 by Cade 1734 (God. Ar. 1724).
Haphazard, br. h., 1797.	Sir Peter T. 1784 (Highflyer 1774)—Miss Hervey 1775 by Eclipse 1764 (Marske 1750)—Clio 1760 by Y. Cade 1747 (Cade 1734)—Dau. of Starling 1727 (Bay Bolton 1705)—Dau. of Bartlett's Childers (Darley Ar. 1699).
Princess, b. m., 1799.	Precipitate* 1787 (Mercury 1778)—Colibri 1793 by Woodpecker 1773 (Herod 1758)—Camilla 1778 by Trentham 1766 (T. Sweepstakes 1743)—Coquette 1765 by Compton Barb—Sist. 1743 to Regulus by God. Ar. 1724.
Walton, b. h., 1799.	Sir Peter T. 1784 (Highflyer 1774)—Arethusa 1792 by Dungannon 1780 (Eclipse 1764)—Dau. 1777 of Prophet 1760 (Regulus 1739)—Virago 1764 by Snap 1750 (Snip 1736)—Dau. of Regulus 1739 (God. Ar. 1724).
Parasol, b. m., 1800.	Pot-8-o's 1773 (Eclipse 1764)—Prunella 1788 by Highflyer 1774 (Herod 1758)—Promise 1768 by Snap 1750 (Snip 1736)—Julia 1756 by Blank 1740 (God. Ar. 1724)—Spectator's dam 1735 by Partner 1718 (Jigg).
Moses, b. h., 1819.	[Whalebone or] Seymour 1807 (Delpini 1781)—Dau. 1807 of Gohanna 1790 (Mercury 1778)—Grey Skim 1793 by Woodpecker 1773 (Herod 1758)—Dau. 1780 of Herod 1758 (Tartar 1743)—Y. Hag 1761 by Skim 1746 (Starling 1727).
Quadrille, b. m., 1815.	Selim 1802 (Buzzard* 1787)—Canary Bird 1806 by [Whiskey or] Sorcerer 1796 (Trumpator 1782)—Canary 1797 by Coriander 1786 (Pot-8-o's 1773)—Miss Green 1787 by Highflyer 1774 (Herod 1758)—Harriet 1769 by Matchem 1748 (Cade 1734).
Emilius, b. h., 1820.	Orville 1799 (Beningbrough 1791)—Emily 1810 by Stamford 1794 (Sir Peter T. 1784)—Dau. 1799 of Whiskey 1789 (Saltram* 1780)—Grey Dorimant 1781 by Dorimant 1772 (Otho 1760)—Dizzy 1757 by Blank 1740 (God. Ar. 1724).
Harriet, blk. m., 1819.	Pericles 1809 (Evander 1801)—Dau. 1812 of Selim 1802 (Buzzard* 1787)—Pipylina 1803 by Sir Peter T. 1784 (Highflyer 1774)—Rally 1790 by Trumpator 1782 (Conductor 1767)—Fancy 1780 (Sist. to Diomed*) by Florizel 1768 (Herod 1758).
Whalebone, br. h., 1807.	Waxy 1790 (Pot-8-o's 1773)—Penelope 1798 by Trumpator 1782 (Conductor 1767)—Prunella 1788 by Highflyer 1774 (Herod 1758)—Promise 1768 by Snap 1750 (Snip 1736)—Julia 1756 by Blank 1740 (God. Ar. 1724).
Gift, gr. h., 1818.	Y. Gohanna 1810 (Gohanna 1790)—Dau. 1802 of Sir Peter T. 1784 (Highflyer 1774)—Dau. 1788 of Trumpator 1782 (Conductor 1767)—Sist. 1774 to Postmaster by Herod 1758 (Tartar 1743)—Dau. of Snap 1750 (Snip 1736).
Humphrey Clinker, b. h., 1822.	Comus 1809 (Sorcerer 1796)—Clinkerina 1812 by Clinker 1805 (Sir Peter T. 1784)—Pewet 1786 by Tandem 1773 (Syphon 1750)—Termagant by Tantrum 1760 (Cripple 1750)—Cantatrice by Sampson 1745 (Blaze 1733).
Daughter of, b. m., 1825.	Cervantes 1806 (Don Quixote 1784)—Dau. 1818 of Golumpus 1802 (Gohanna 1790)—Dau. 1810 of Paynator 1791 (Trumpator 1782)—Circe 1802 by St. George 1789 (Highflyer 1774)—Abigail 1788 by Woodpecker 1773 (Herod 1758).
Touchstone, br. h., 1831.	Camel 1822 (Whalebone 1807)—Banter 1826 by Master Henry 1815 (Orville 1799)—Boadicea 1807 by Alexander 1782 (Eclipse 1764)—Brunette by Amaranthus 1766 (Old England 1741)—Mayfly 1771 by Matchem 1748 (Cade 1734).
Emma, ch. m., 1824.	Whisker 1812 (Waxy 1790)—Gibside Fairy 1811 by Hermes 1790 (Mercury 1778)—Vicissitude 1800 by Pipator 1786 (Imperator 1776)—Beatrice 1792 by Sir Peter T. 1784 (Highflyer 1774)—Pyrrha 1771 by Matchem 1748 (Cade 1734).
Emilius, b. h., 1820.	Orville 1799 (Beningbrough 1791)—Emily 1810 by Stamford 1794 (Sir Peter T. 1784)—Dau. 1799 of Whiskey 1789 (Saltram* 1780)—Grey Dorimant 1781 by Dorimant 1772 (Otho 1760)—Dizzy 1757 by Blank 1740 (God. Ar. 1724).
Shoveler, b. m., 1816.	Scud 1804 (Beningbrough 1791)—Goosander 1805 by Hambletonian 1792 (King Fergus 1775)—Rally 1790 by Trumpator 1782 (Conductor 1767)—Fancy 1780 (Sist. to Diomed*) by Florizel 1768 (Herod 1758)—Dau. 1763 of Spectator 1749 (Crab 1722).
Whisker, b. h., 1812.	Waxy 1790 (Pot-8-o's 1773)—Penelope 1798 by Trumpator 1782 (Conductor 1767)—Prunella 1788 by Highflyer 1774 (Herod 1758)—Promise 1768 by Snap 1750 (Snip 1736)—Julia 1756 by Blank 1740 (God. Ar. 1724).
Variety, ch. m., 1816.	[Selim or] Soothsayer 1808 (Sorcerer 1796)—Sprite 1807 by Bobtail 1795 (Precipitate* 1787)—Catherine 1795 by Woodpecker 1773 (Herod 1758)—Camilla 1778 by Trentham 1766 (T. Sweepstakes 1743)—Coquette 1765 by Compton Barb.
Timoleon, ch. h., 1813.	Sir Archy 1805 (*Diomed 1777)—Dau. of *Saltram 1780 (Eclipse 1764)—Dau. of Symmes' Wildair 1767 (*Fearnought 1755)—Dau. of Tyler's Driver 1760 (*Othello)—Dau. of *Fallower 1761 (Blank 1740).
Sister to Tuckahoe, ch. m., 1814.	Ball's Florizel 1801 (*Diomed 1777)—Dau. 1799 of *Alderman 1787 (Pot-8-o's 1773)—Dau. of *Clockfast 1774 (Gimcrack 1760)—Dau. of Symmes' Wildair 1767 (*Fearnought 1755)—Y. Kitty Fisher 1767 by *Fearnought 1755 (Regulus 1739).
*Sarpedon, br. h., 1828.	Emilius 1820 (Orville 1799)—Icaria 1824 by The Flyer 1811 (Vandyke Jr. 1808)—Parma 1813 by Dick Andrews 1797 (Joe Andrews 1778)—May 1804 by Beningbrough 1791 (King Fergus 1775)—Primrose 1787 by Mambrino 1768 (Engineer 1756).
Rowena, ch. m., 1826.	Sumpter 1818 (Sir Archy 1805)—Lady Grey 1817 by Robin Grey 1805 (*Royalist 1790)—Maria by Melzar 1791 (*Medley 1776)—Dau. of *Highflyer 1784 (Highflyer 1774)—Dau. of *Fearnought 1755 (Regulus 1739).
St. Nicholas, br. h., 1827.	Emilius 1820 (Orville 1799)—Sea Mew 1815 by Scud 1804 (Beningbrough 1791)—Goosander 1805 by Hambletonian 1792 (King Fergus 1775)—Rally 1790 by Trumpator 1782 (Conductor 1767)—Fancy 1780 by Florizel 1768 (Herod 1758).
Miss Rose,* b. m., 1826.	Tramp 1810 (Dick Andrews 1797)—Dau. 1810 of Sancho 1801 (Don Quixote 1784)—Blacklock's dam 1799 by Coriander 1786 (Pot-8-o's 1773)—Wildgoose 1792 by Highflyer 1774 (Herod 1758)—Coheiress 1786 by Pot-8-o's 1773 (Eclipse 1764).
Filho-da-Puta, br. h., 1812.	Haphazard 1797 (Sir Peter T. 1784)—Mrs. Barnet 1806 by Waxy 1790 (Pot-8-o's 1773)—Dau. 1788 of Woodpecker 1773 (Herod 1758)—Heinel 1771 by Squirrel 1754 (Traveller 1735)—Principessa 1762 by Blank 1740 (God. Ar. 1724).
Daughter of, —, 1817.	Smolensko 1810 (Sorcerer 1796)—Dau. br. m. 1803 by Sir Peter T. 1784 (Highflyer 1774)—Dau. gr. m. 1785 by Mambrino 1768 (Engineer 1756)—Marigold b. m. 1777 by Herod 1758 (Tartar 1743)—Toy gr. m. 1761 by Blank 1740 (God. Ar. 1724)—Whiteneck gr. m. 1751 by Crab 1722 (Alcock Ar.)—Dau. of God. Ar. 1724—Dau. of Conyer's Ar.—Sist. to Mixbury by Curwen Bay Barb—Dau. of Curwen Spot (Selaby Turk)—Dau. of White-legged Lowther Barb—Vintner mare.

BRAMBLE—Continued.

forehead, 15 hands 2½ inches high. girths 70 inches, length of shoulder 28 inches, around the arm 20 inches, below the knee 8½ inches, from hip to point of hock 36¾ inches, around gaskins 16½ inches, from point of shoulder to point of buttocks 64 inches. He is full of substance and power, and greatly resembles his sire. He is strongly inbred to Herod and Eclipse through the best sources with the Archy and Diomed blood through Lexington, he was speedy, could stay and carry weight.

Bramble, with a limited chance in the stud, is the sire of the following winners, Biggonet, winner of the Withers, etc., Bryson, Brambleton, Bordelaise, Orderly, Barb, Bradford, Rambler, Bertha, Armiel, Bravo, Daisy Woodruff, Katunah, Mora, Romp, Wrestler, Barrister, Ben Harrison, Mayor B., Bettina, Blackthorn, Gracie M., Jakie Toms, Manhattan, Pedestrian, Gambler, Wightman, Ivanhoe II., Nominee, Bellevue, Jack Murray, Briar, Pixy, Lizzie English, Lady Jane, Thorns, Thornless, Bertha, etc.

BUCKMASTER.

(BROTHER TO KIMBALL.)

The property of J. D. Lucas, Goodwood Stud, St. Louis, Mo. Service by private contract.

Buckmaster, bay horse, bred by Capt. Wm. Cottrill, Mobile, Ala., own brother to Kimball by imp. Buckden, son of Lord Clifden, dam Meta H. by Harry of the West, son of Lexington, out of Louisiana B. by Daniel Boone, son of Lexington, thus giving him a double cross of this inestimable blood upon a double cross of imp. Glencoe through the noted mare Magnolia, grandam of Iroquois, and Fanny King, the dam of the great race-horse Brown Dick, son of imp. Margrave by Muley.

His own brother Kimball, as a two-year old, won the Young America Stakes, ⅝ mile, at Nashville, in 1.04¾; won the Alexander Stakes, Louisville, ½ mile, by 1½ lengths, in 49¾ seconds, beating Luke Blackburn, Big Medicine, etc.; ran third to Wallenstein and Fonso for the Tennessee Stakes, Louisville, beating Luke Blackburn Big Medicine, etc., and conceded 5 lbs. to all his opponents; won the Hotel Stakes, ¾ mile, St. Louis, time, 1.16½, by 10 lengths, beating Luke Blackburn, etc.; won the Grand Pacific Hotel Stakes, ¾ mile, Chicago, in 1.18½, beating Luke Blackburn, Mistake, Victory, etc., and ran third for the Criterion Stakes, Chicago, beating Luke Blackburn, Wallenstein, Bye-and-Bye, etc. As a three-year old, he won the Cumberland Stakes, mile heats, at Nashville, in two straight heats, 1.49, 1.47¾; ran second to Fonso for the Kentucky Derby; second for the Merchant Stakes, St. Louis, and second in the Tidal Stakes, Gravesend, N. Y., to Luke Blackburn. Meta H. produced the winners Ruth and Mattie K. Louisiana B., the second dam, had only one foal, viz., Meta H., but from the third dam, Maggie G., came Magnolia, Sally Watson, Bonaventure, Adventure and Buckden Lass. Magnolia produced the winners The Cripple (Odd Fellow), Mollie Merrill (dam of Ira E. Bride), Aleck Ament, Orange Blossom and Myrtle H. Sally Watson threw Buckleo. Peradventure, out of Bonaventure, is the dam of Adventurer, a winner. From Bonaventure came another Adventurer (by Buckden), in addition to Kitty Wells and Bonny Lass (dam of English Lady, winner of the Zoo-Zoo Stakes, etc.), and Adventure is the dam of the winners Lamont and Teddy Venture (a two-year old winner).

Buckmaster did not run until seven years old. He won ⅞ of a mile in 1.21½, with 118 lbs. Won mile heats in 1.47, 1.46. Won at 1½ miles in 2.44, and again in 2.46. Won at a mile in 1.48¼, and half a mile in 0.53. He was second and third in several other races. He is a well-bred horse, tracing to the Barb mare presented to President Jefferson by the Dey of Tunis. The sons of Buckden are doing well in the stud.

BUCKMASTER.

BAY HORSE, FOALED 1876; BRED BY CAPT. WM. COTTRILL, MOBILE, ALA.

*BUCKDEN, b. h., 1869.
- Lord Clifden, b. h., 1860.
  - Newminster, b. h., 1848.
    - Touchstone, br. h., 1831.
    - Beeswing, b. m., 1833.
  - The Slave, b. m., 1852.
    - Melbourne, br. h., 1834.
    - Volley, b. m., 1845.
- Consequence, b. m., 1855.
  - Bay Middleton, b. h., 1833.
    - Sultan, b. h., 1816.
    - Cobweb, b. m., 1821.
  - Result, b. m., 1838.
    - Mulatto, b. h., 1823.
    - Problem, b. m., 1842.

META, H., ch. m., 1870.
- Harry of the West, b. h., 1862.
  - Lexington, b. h., 1850.
    - Boston, ch. h., 1833.
    - Alice Carneal, b. m., 1836.
  - Laura, ch. m., 1846.
    - *Leviathan, ch. h., 1823.
    - Daughter of, b. m., 1833.
- Louisiana B., ch. m., —.
  - Daniel Boone, b. h., 1857.
    - Lexington, b. h., 1850.
    - Magnolia, ch. m., 1841.
  - Maggie G., ch. m., 1856.
    - Brown Dick, br. h., 1850.
    - Sally Riddlesworth, ch. m., 1842.

Camel, br. h., 1822.	Whalebone 1807 (Waxy 1790)—Dau. 1812 of Selim 1802 (Buzzard* 1787)—Maiden 1801 by Sir Peter T. 1784 (Highflyer 1774)—Dau. 1788 of Phenomenon 1780 (Herod 1758)—Matron 1782 by Florizel 1768 (Herod 1758).
Banter, br. m., 1826.	Master Henry 1815 (Orville 1799)—Boadicea 1807 by Alexander 1782 (Eclipse 1764)—Brunette by Amaranthus 1766 (Old England 1741)—Mayfly 1771 by Matchem 1748 (Cade 1734)—Dau. 1755 of Ancaster Starling 1738 (Starling 1727).
Dr. Syntax, br. h., 1811.	Paynator 1791 (Trumpator 1782)—Dau. of Beningbrough 1791 (King Fergus 1775)—Jenny Mole 1787 by Carbuncle 1772 (Babr. Blank 1758)—Dau. 1773 of Prince T'Quassaw 1751 (Snip 1736)—Sultana 1759 by Regulus 1739 (God. Ar. 1724).
Daughter of, ch. m., 1817.	Ardrossan 1809 (John Bull 1789)—Lady Eliza 1813 by Whitworth 1805 (Agonistes 1795)—Dau. 1793 of Spadille 1784 (Highflyer 1774)—Sylvia 1783 by Y. Marske 1771 (Marske 1750)—Ferret 1765 by Br. to Silvio (Cade 1734).
Humprey Clinker, b. h., 1822.	Comus 1809 (Sorcerer 1796)—Clinkerina 1812 by Clinker 1805 (Sir Peter T. 1784)—Pewet 1786 by Tandem 1773 (Syphon 1750)—Termagant by Tantrum 1760—Cripple 1750—Cantatrice by Sampson 1745 (Blaze 1733).
Daughter of, b. m., 1825.	Cervantes 1806 (Don Quixote 1784)—Dau. 1818 of Golumpus 1802 (Gohanna 1790)—Dau. 1810 of Paynator 1791 (Trumpator 1782)—Circle 1802 by St. George 1789 (Highflyer 1774)—Abigail 1788 by Woodpecker 1773 (Herod 1758).
Voltaire, br. h., 1826.	Blacklock 1814 (Whitelock 1803)—Dau. 1816 of Phantom 1808 (Walton 1799)—Dau. 1802 of Overton 1788 (King Fergus 1775)—Dau. 1796 of Walnut 1786 (Highflyer 1774)—Dau. 1786 of Ruler 1777 (Y. Marske 1771).
Martha Lynn, br. m., 1837.	Mulatto 1823 (Catton 1809)—Leda 1824 by Filho-da-Puta 1812 (Haphazard 1797)—Treasure 1809 by Camillus 1803 (Hambletonian 1792)—Dau. 1801 of Hyacinthus 1797 (Coriander 1786).
Selim, ch. h., 1802.	Buzzard* 1787 (Woodpecker 1773)—Dau. of Alexander 1782 (Eclipse 1764)—Dau. of Highflyer 1774 (Herod 1758)—Dau. of Alfred 1770 (Matchem 1748)—Dau. 1770 of Engineer 1756 (Sampson 1745).
Bacchante, br. m., 1809.	W's Ditto 1800 (Sir Peter T. 1784)—Dau. 1791 of Mercury 1778 (Eclipse 1764)—Dau. 1776 of Herod 1758 (Tartar 1743)—Folly 1771 by Marske 1750 (Squirt 1732)—Vixen 1753 by Regulus 1739 (God. Ar. 1724).
Phantom, b. h., 1808.	Walton 1799 (Sir Peter T. 1784)—Julia 1799 by Whiskey 1789 (Saltram* 1780)—Y. Giantess 1790 by Diomed* 1777 (Florizel 1768)—Giantess 1769 by Matchem 1748 (Cade 1734)—Molly Long Legs 1753 by Babraham 1740 (God. Ar. 1724).
Filagree, ch. m., 1815.	Soothsayer 1808 (Sorcerer 1796)—Web 1808 by Waxy 1790 (Pot-8-o's 1773)—Penelope 1798 by Trumpator 1782 (Conductor 1767)—Prunella 1788 by Highflyer 1774 (Herod 1758)—Promise 1768 by Snap 1750 (Snip 1736).
Catton, b. h., 1809.	Golumpus 1802 (Gohanna 1790)—Lucy Gray 1804 by Timothy 1794 (Delpini 1781)—Lucy 1789 by Florizel 1768 (Herod 1758)—Frenzy 1774 by Eclipse 1764 (Marske 1750)—Dau. of Engineer 1756 (Sampson 1745).
Desdemona, br. m., 1811.	Orville 1799 (Beningbrough 1791)—Fanny 1796 by Sir Peter T. 1784 (Highflyer 1774)—Dau. 1788 of Diomed* 1777 (Florizel 1768)—Desdemona 1770 by Marske 1750 (Squirt 1732)—Y. Hag 1761 by Skim 1746 (Starling 1727).
Euclid, ch. h., 1836.	Emilius 1820 (Orville 1799)—Maria 1827 by Whisker 1812 (Waxy 1790)—Gibside Fairy 1811 by Hermes 1790 (Mercury 1778)—Vicissitude 1800 by Pipator 1786 (Imperator 1776)—Beatrice 1791 by Sir Peter T. 1784 (Highflyer 1774).
Parthenessa, br. m., 1821.	Cervantes 1806 (Don Quixote 1784)—Marianne 1816 by Sorcerer 1796 (Trumpator 1782)—Thomasina 1804 by Timothy 1794 (Delpini 1781)—Violet 1787 by Shark* 1771 (Marske 1750)—Dau. 1772 of Syphon 1750 (Squirt 1732).
Timoleon, ch. h., 1813.	Sir Archy 1805 (*Diomed 1777)—Dau. of *Saltram 1780 (Eclipse 1764)—Dau. of Symmes' Wildair 1767 (*Fearnought 1755)—Dau. of Tyler's Driver 1760 (*Othello)—Dau. of *Fallower 1761 (Blank 1740).
Sist. to Tuckahoe, ch. m., 1814.	Ball's Florizel 1801 (*Diomed 1777)—Dau. 1799 of *Alderman 1787 (Pot-8-o's 1773)—Dau. of *Clockfast 1774 (Gimcrack 1760)—Dau. of Symmes' Wildair 1767 (*Fearnought 1755)—Y. Kitty Fisher 1767 by *Fearnought 1755 (Regulus 1739).
*Sarpedon, br. h., 1828.	Emilius 1820 (Orville 1799)—Icaria 1824 by The Flyer 1814 (Vandyke Jr. 1808)—Parma 1813 by Dick Andrews 1797 (Joe Andrews 1778)—May 1804 by Beningbrough 1791 (King Fergus 1775)—Primrose 1787 by Mambrino 1768 (Engineer 1756).
Rowena, ch. m., 1826.	Sumpter 1818 (Sir Archy 1805)—Lady Grey 1817 by Robin Grey 1805 (*Royalist 1790)—Maria 1802 by Melzar 1791 (*Medley 1776)—Dau. of *Highflyer 1784 (Highflyer 1774)—Dau. of *Fearnought 1755 (Regulus 1739).
Muley, b. h., 1810.	Orville 1799 (Beningbrough 1791)—Eleanor 1798 by Whiskey 1789 (Saltram* 1780)—Y. Giantess 1790 by Diomed* 1777 (Florizel 1768)—Giantess 1769 by Matchem 1748 (Cade 1734)—Molly Long Legs 1753 by Babraham 1740 (God. Ar. 1724).
Daughter of, blk. m., 1809.	Windle 1804 (Beningbrough 1791)—Dau. 1798 of Anvil 1777 (Herod 1758)—Virago 1764 by Snap 1750 (Snip 1736)—Dau. of Regulus 1739 (God. Ar. 1724)—Sist. 1750 to Othello by Crab 1722 (Alcock Ar.).
Stockholder, b. h., —.	Sir Archy 1805 (*Diomed 1777)—Dau. of *Citizen 1785 (Pacolet 1763)—Dau. of *Stirling 1791 (Volunteer 1780)—Dau. of *Mousetrap 1787 (Y. Marske 1771)—Dau. of Harris' Eclipse 1770 (*Fearnought 1755).
Daughter of, ch. m., 1822.	Pacolet 1806 (*Citizen 1785)—Nell Saunders 1815 by Wilkes' Wonder 1800 (*Diomed 1777)—Julietta by *Dare Devil 1787 (Magnet 1770)—Rosetta 1785 by *Centinel 1758 (Blank 1740)—Diana by Clodius 1778 (*Janus 1746).
Boston, ch. h., 1833.	Timoleon 1813 (Sir Archy 1805)—Sist. 1814 to Tuckahoe by Ball's Florizel 1801 (*Diomed 1777)—Dau. 1799 of *Alderman 1787 (Pot-8-o's 1773)—Dau. of *Clockfast 1774 (Gimcrack 1760)—Dau. of Symmes' Wildair 1767 (*Fearnought 1755).
Alice Carneal, b. m., 1836.	*Sarpedon 1828 (Emilius 1820)—Rowena 1826 by Sumpter 1818 (Sir Archy 1805)—Lady Grey 1817 by Robin Grey 1805 (*Royalist 1790)—Maria 1802 by Melzar 1791 (*Medley 1776)—Dau. of *Highflyer 1784 (Highflyer 1774).
*Glencoe, ch. h., 1831.	Sultan 1816 (Selim 1802)—Trampoline 1825 by Tramp 1810 (Dick Andrews 1797)—Web 1808 by Waxy 1790 (Pot-8-o's 1773)—Penelope 1798 by Trumpator 1782 (Conductor 1767)—Prunella 1788 by Highflyer 1774 (Herod 1758).
*Myrtle, ch. m., 1834.	Mameluke 1824 (Partisan 1811)—Bobadilla 1825 by Bobadil 1813 (Rubens 1805)—Pythoness 1813 by Sorcerer 1796 (Trumpator 1782)—Princess 1796 by Sir Peter T. 1784 (Highflyer 1774)—Dau. 1789 of Dungannon 1780 (Eclipse 1764).
*Margrave, ch. h., 1829.	Muley 1810 (Orville 1799)—Dau. 1815 of Election 1804 (Gohanna 1790)—Fair Helen 1808 by Hambletonian 1792 (King Fergus 1775)—Helen 1801 by Delpini 1781 (Highflyer 1774)—Rosalind 1788 by Phenomenon 1780 (Herod 1758).
Fanny King, b. m., 1841.	*Glencoe 1831 (Sultan 1816)—Mary Smith 1830 by Sir Richard 1823 (Pacolet 1806)—Dau. of Tenn. Oscar 1815 (Wilkes' Wonder 1800)—Ledbetter M. by Sir Archy 1805 (*Diomed 1777)—Dau. of *Diomed 1777 (Florizel 1768).
*Riddlesworth, ch. h., 1828.	Emilius 1820 (Orville 1799)—Filagree 1815 by Soothsayer 1808 (Sorcerer 1796)—Web 1808 by Waxy 1790 (Pot-8-o's 1773)—Penelope 1798 by Trumpator 1782 (Conductor 1767)—Prunella 1788 by Highflyer 1774 (Herod 1758).
Quacker Girl, ch. m., 1834.	*Leviathan 1823 (Muley 1810)—Sally Burton b. m. 1822 by Sir Archy 1805 (*Diomed 1777)—Dau. of Junius (*Buzzard 1787)—Dau. of *Citizen 1785 (Pacolet 1763)—*Barb Mare presented to President Jefferson by the Dey of Tunis in 1806.

CENTAUR.

(*Late* THE FAVERDALE COLT.)

The property of the Estate of the late D. D. Withers, Brookdale Stud, N. J. Used as private stallion or by contract.

CENTAUR, bay horse, bred in the Brookdale Stud by Mr. D. D. Withers, by the unbeaten Sensation, own brother to Stratford and Onondaga, son of imp. Leamington, dam imp. Faverdale. dam of Frolic, Oneko, Favoress, etc., by The Palmer, own brother to Rosicrucian by Beadsman. The Palmer won the Ascot, Derby, Royal Stakes and Liverpool Autumn Cup, and is the sire of Jenny Howlet, winner of the Oaks, Pellegrino, Pilgrimage (winner of the Dewhurst Plate, the 1,000 and 2,000 Guineas), Grey Palmer, etc. His grandam Georgiana (full sister to Countess of Burlington, the dam of Cavendish and imp. Hartington, winner of the Cesarewitch Handicap) by Touchstone, winner of the St. Leger and one of the most valuable stallions ever bred in England. Touchstone sired Mowerina, the dam of West Australian, winner of the 2,000 Guineas, the Derby and St. Leger. Mowerina, was out of Emma, sister to Caroline, the fourth dam of Centaur by Whisker, third dam Lady Emily by Muley Moloch, son of Muley, and sire of the famous Alice Hawthorn, dam of Thormanby, winner of the Derby, tracing through Gibside Fairy by Hermes, son of Mercury by Eclipse, to the D'Arcy black-legged Royal mare. Centaur was a first-class race-horse, winner at two years old of the Hopeful Stakes, ¾ of a mile, 115 lbs., in 1.16, beating Windfield, Diablo, etc.; was second to French Park in Sequence Stakes, 5 furlongs in 1.04; ran second to Tipstaff in the Sapling Stakes, ¾ of a mile in 1.15½, conceding him 10 lbs.; ran second to Proctor Knott (conceding him 3 lbs.) in Junior Champion Stakes, ¾ of a mile in 1.14, with Fresno, Salvator and Diablo behind him; won the Home Bred Produce Stakes, ¾ of a mile, 115 lbs., in 1.16¼, and the Carteret Handicap, ¾ of a mile, with 121 lbs., in 1.16½, beating Fresno, Fides, etc.; won the Prospect Stakes, ¾ of a mile, 115 lbs., in 1.17½, beating Diablo, Fresno, etc.; won the Algeria Stakes, ¾ of a mile, 120 lbs., in 1.16½, beating Senorita, etc.; was unplaced in Tyro Stakes, won by Reporter. At three years old he was unplaced in the Tidal Stakes, won by Salvator; won Handicap Sweepstake, 1⅜ miles, 112 lbs, in 2.23, beating Le Logos, Inverwick and Reporter; ran third to Salvator and Longstreet in Lorillard Stakes, 1½ miles in 2.37½, and was third to Madstone and Rhono in Newark Stakes, 1 mile in 1.43. He was injured in this race and never ran again. From his grand qualities as a race-horse and high breeding, he should be a grand success as a stallion. He is descended in direct female line from the Black-legged Royal mare, to which also trace the following excellent race-horses and successful stallions, Oroonoko, Saltram (imported), Beningbrough, Williamson's Ditto, Cotherstone, West Australian, Walton, Wild Dayrell, Parmesan, Diophantus, Caterer, Lord Ronald, Ishmael, Wisdom, The Miser, Saxifrage and Harry O'Fallon, in America.

CENTAUR (Ran as FAVERDALE COLT).

BROWN HORSE, FOALED 1886; BRED BY THE LATE MR. D. D. WITHERS, BROOKDALE STUD, NEW JERSEY.

- SENSATION, br. h., 1877.
  - *Leamington, br. h., 1853.
    - Faugh-a-Ballagh, br. h., 1841.
      - Sir Hercules, blk. h., 1826.
      - Guiccioli, ch. m., 1823.
    - Daughter of, b. m., 1841.
      - Pantaloon, ch. h., 1824.
      - Daphne, br. m., 1837.
  - Susan Beane, ch. m., 1863.
    - Lexington, b. h., 1850.
      - Boston, ch. h., 1833.
      - Alice Carneal, b. m., 1836.
    - Sally Lewis, ch. m., 1852.
      - *Glencoe, ch. h., 1831.
      - Motto, ch. m., 1833.
- *FAVERDALE, br. m., 1873.
  - The Palmer, b. h., 1864.
    - Beadsman, br. h., 1855.
      - Weatherbit, br. h., 1842.
      - Mendicant, br. m., 1843.
    - Madame Eglentine, b. m., 1857.
      - Cowl, b. h., 1842.
      - Diversion, ch. m., 1831.
  - Georgiana, b. m., 1851.
    - Touchstone, br. h., 1831.
      - Camel, br. h., 1822.
      - Banter, br. m., 1826.
    - Lady Emily, b. m., 1842.
      - Muley Moloch, br. h., 1830.
      - Caroline, b. m., 1828.

Whalebone, br. h., 1807.	Waxy 1790 (Pot-8-o-'s 1773)—Penelope 1798 by Trumpator 1782 (Conductor 1767)—Prunella 1788 by Highflyer 1774 (Herod 1758)—Promise 1768 by Snap 1750 (Snip 1736)—Julia 1756 by Blank 1740 (God. Ar. 1724).
Peri, b. m., 1822.	Wanderer 1811 (Gohanna 1790)—Thalestris 1809 by Alexander 1782 (Eclipse 1764)—Rival 1800 by Sir Peter T. 1784 (Highflyer 1774)—Hornet 1790 by Drone 1777 (Herod 1758)—Manilla 1777 by Goldfinder 1764 (Snap 1750).
Bob Booty, ch. h., 1804.	Chanticleer 1787 (Woodpecker 1773)—Ierne 1790 by Bagot 1780 (Herod 1758)—Dau. of Gamahoe (Bustard 1741)—Patty by Tim 1752 (Squirt 1732)—Miss Patch 1737 by Justice 1725 (H. C. Litton Ar.).
Flight, ch. m., 1809.	Ir. Escape 1802 (Commodore 1793)—Y. Heroine by Bagot 1780 (Herod 1758)—Heroine 1775 by Hero 1753 (Cade 1734)—Dau. 1766 of Snap 1750 (Snip 1736)—Sist. 1743 to Regulus by God. Ar. 1724.
Castrel, ch. h., 1801.	Buzzard 1787 (Woodpecker 1773)—Dau. of Alexander 1782 (Eclipse 1764)—Dau. of Highflyer 1774 (Herod 1758)—Dau. of Alfred 1770 (Matchem 1748)—Dau. 1770 of Engineer 1755 (Sampson 1745).
Idalia, ch. m., 1815.	Peruvian 1806 (Sir Peter T. 1784)—Musidora 1804 by Meteor 1783 (Eclipse 1764)—Maid-of-all-Work 1786 by Highflyer 1774 (Herod 1758)—Sist. 1771 to Tandem by Syphon 1750 (Squirt 1732)—Dau. of Regulus 1739 (God. Ar. 1724).
Laurel, br. h., 1824.	Blacklock 1814 (Whitelock 1803)—Wagtail 1818 by Prime Minister 1810 (Sancho 1801)—Tranby*'s dam 1812 by Orville 1799 (Beningbrough 1791)—Miss Grimstone 1793 by Wenzle 1776 (Herod 1758)—Dau. of Ancaster 1768 (Blank 1740).
Maid of Honor, br. m., 1829.	Champion 1812 (Selim 1802)—Etiquette 1820 by Orville 1799 (Beningbrough 1791)—Boadicea 1807 by Alexander 1782 (Eclipse 1764)—Brunette by Amaranthus 1766 (Old England 1741)—Mayfly 1771 by Matchem 1748 (Cade 1734).
Timoleon, ch. h., 1813.	Sir Archy 1805 (*Diomed 1777)—Dau. of *Saltram 1780 (Eclipse 1764)—Dau. of Symmes' Wildair 1767 (*Fearnought 1755)—Dau. of Tyler's Driver 1760 (*Othello)—Dau. of *Fallower 1761 (Blank 1740).
Sist. to Tuckahoe, ch. m., 1814.	Ball's Florizel 1801 (*Diomed 1777)—Dau. 1799 of *Alderman 1787 (Pot-8-o-'s 1773)—Dau. of *Clockfast 1774 (Gimcrack 1760)—Dau. of Symmes' Wildair 1767 (*Fearnought 1755)—Y. Kitty Fisher 1767 by *Fearnought 1755 (Regulus 1739).
*Sarpedon, br. h., 1828.	Emilius 1820 (Orville 1799)—Icaria 1824 by The Flyer 1814 (Vandyke Jr. 1808)—Parma 1813 by Dick Andrews 1797 (Joe Andrews 1778)—May 1804 by Beningbrough 1791 (King Fergus 1775)—Primrose 1787 by Mambrino 1768 (Engineer 1756).
Rowena, ch. m., 1826.	Sumpter 1818 (Sir Archy 1805)—Lady Grey 1817 by Robin Grey 1805 (*Royalist 1790)—Maria 1802 by Melzar 1791 (*Medley 1776)—Dau. of *Highflyer 1784 (Highflyer 1774)—Dau. of *Fearnought 1755 (Regulus 1739).
Sultan, b. h., 1816.	Selim 1802 (Buzzard* 1787)—Bacchante 1809 by W's Ditto 1800 (Sir Peter T. 1784)—Dau. 1791 of Mercury 1778 (Eclipse 1764)—Dau. 1776 of Herod 1758 (Tartar 1743)—Folly 1771 by Marske 1750 (Squirt 1732).
Trampoline, ch. m., 1825.	Tramp 1810 (Dick Andrews 1797)—Web 1808 by Waxy 1790 (Pot-8-o-'s 1773)—Penelope 1798 by Trumpator 1782 (Conductor 1767)—Prunella 1788 by Highflyer 1774 (Herod 1758)—Promise 1768 by Snap 1750 (Snip 1736).
*Barefoot, ch. h., 1820.	Tramp 1810 (Dick Andrews 1797)—Rosamond 1798 by Buzzard* 1787 (Woodpecker 1773)—Roseberry 1792 by Phenomenon 1780 (Herod 1758)—Miss West 1777 by Matchem 1748 (Cade 1734)—Sist. to Favourite by Regulus 1739 (God. Ar. 1724).
Lady Thompkins, ch. m., 1831.	Am. Eclipse 1814 (Duroc 1806)—Katy Ann 1825 by Ogles' Oscar 1800 (*Gabriel 1790)—Medoc's dam 1817 by *Expedition 1795 (Pegasus 1784)—Maid of the Oaks 1801 by *Spread Eagle 1792 (Volunteer 1780)—Annette by *Shark 1771 (Marske 1750).
Sheet Anchor, br. h., 1832.	Lottery 1820 (Tramp 1810)—Morgiana 1820 by Muley 1810 (Orville 1799)—Miss Stephenson 1814 by [Scud or] Sorcerer 1796 (Trumpator 1782)—Dau. 1796 of Precipitate* 1787 (Mercury 1778)—Dau. 1787 of Woodpecker 1773 (Herod 1758).
Miss Letty, b. m., 1834.	Priam* 1827 (Emilius 1820)—Dau. 1815 of Orville 1799 (Beningbrough 1791)—Dau. 1800 of Buzzard* 1787 (Woodpecker 1773)—Hornpipe 1793 by Trumpator 1782 (Conductor 1767)—Luna 1779 by Herod 1758 (Tartar 1743).
Touchstone, br. h., 1831.	Camel 1822 (Whalebone 1807)—Banter 1826 by Master Henry 1815 (Orville 1799)—Boadicea 1807 by Alexander 1782 (Eclipse 1764)—Brunette by Amaranthus 1766 (Old England 1741)—Mayfly 1771 by Matchem 1748 (Cade 1734).
Lady Moore Carew, b. m., 1830.	Tramp 1810 (Dick Andrews 1797)—Kite 1821 by Bustard 1813 (Castrel 1801)—Olympia 1815 by Sir Oliver 1800 (Sir Peter T. 1784)—Scotilla 1795 by Anvil 1777 (Herod 1758)—Scota 1783 by Eclipse 1764 (Marske 1750).
Bay Middleton, b. h., 1833.	Sultan 1816 (Selim 1802)—Cobweb 1821 by Phantom 1808 (Walton 1799)—Filagree 1815 by Soothsayer 1808 (Sorcerer 1796)—Web 1808 by Waxy 1790 (Pot-8-o-'s 1773)—Penelope 1798 by Trumpator 1782 (Conductor 1767)
Crucifix, b. m., 1837.	Priam* 1827 (Emilius 1820)—Octaviana 1815 by Octavian 1807 (Stripling 1795)—Dau. 1807 of Shuttle 1793 (Y. Marske 1771)—Zara 1801 by Delpini 1781 (Highflyer 1774)—Flora 1789 by King Fergus 1775 (Eclipse 1764).
Defence, b. h., 1824.	Whalebone 1807 (Waxy 1790)—Defiance 1816 by Rubens 1805 (Buzzard* 1787)—Little Folly 1806 by Highland Fling 1798 (Spadille 1784)—Harriet 1799 by Volunteer 1780 (Eclipse 1764)—Lady Sarah's dam 1785 by Alfred 1770 (Matchem 1748).
Folly, ch. m., 1820.	Middleton 1822 (Phantom 1808)—Little Folly 1806 by Highland Fling 1798 (Spadille 1784)—Harriet 1799 by Volunteer 1780 (Eclipse 1764)—Dau. 1785 of Alfred 1770 (Matchem 1748)—Magnolia 1771 by Marske 1750 (Squirt 1732).
Whalebone, br. h., 1807.	Waxy 1790 (Pot-8-o-'s 1773)—Penelope 1798 by Trumpator 1782 (Conductor 1767)—Prunella 1788 by Highflyer 1774 (Herod 1758)—Promise 1768 by Snap 1750 (Snip 1736)—Julia 1756 by Blank 1740 (God. Ar. 1724).
Daughter of, b. m., 1812.	Selim 1802 (Buzzard* 1787)—Maiden 1801 by Sir Peter T. 1784 (Highflyer 1774)—Dau. 1788 of Phenomenon 1780 (Herod 1758)—Matron 1782 by Florizel 1768 (Herod 1758)—Maiden 1770 by Matchem 1748 (Cade 1734).
Master Henry, b. h., 1815.	Orville 1799 (Beningbrough 1791)—Miss Sophia 1805 by Stamford 1794 (Sir Peter T. 1784)—Sophia 1798 by Buzzard* 1787 (Woodpecker 1773)—Huncamunca 1787 by Highflyer 1774 (Herod 1758)—Cypher 1772 by Squirrel 1754 (Traveller 1735).
Boadicea, b. m., 1807.	Alexander 1782 (Eclipse 1764)—Brunette by Amaranthus 1766 (Old England 1741)—Mayfly 1771 by Matchem 1748 (Cade 1734)—Dau. 1755 of Anc. Starling 1738 (Starling 1727)—Look-at-me-Lads 1731 by Grasshopper (Byerly Turk).
Muley, b. h., 1810.	Orville 1799 (Beningbrough 1791)—Eleanor 1798 by Whiskey 1789 (Saltram* 1780)—Y. Giantess 1790 by Diomed* 1777 (Florizel 1768)—Giantess 1769 by Matchem 1748 (Cade 1734)—Molly Long Legs 1753 by Babraham 1740 (God. Ar. 1724).
Nancy, b. m., 1813.	Dick Andrews 1797 (Joe Andrews 1778)—Spitfire 1800 by Beningbrough 1791 (King Fergus 1775)—Dau. of Y. Sir Peter 1775 (Doge 1762)—Dau. of Engineer 1756 (Sampson 1745)—Dau. of Wilson Arabian.
Whisker, b. h., 1812.	Waxy 1790 (Pot-8-o-'s 1773)—Penelope 1798 by Trumpator 1782 (Conductor 1767)—Prunella 1788 by Highflyer 1774 (Herod 1758)—Promise 1768 by Snap 1750 (Snip 1736)—Julia 1756 by Blank 1740 (God. Ar. 1724).
Gibside Fairy, b. m., 1811.	Hermes 1790 (Mercury 1778)—Vicissitude b. m. 1800 by Pipator 1786 (Imperator 1776)—Beatrice b. m. 1791 by Sir Peter T. 1784 (Highflyer 1774)—Pyrrha b. m. 1771 by Matchem 1748 (Cade 1734)—Duchess b. m. 1748 by Whitenose 1742 (God. Ar. 1724)—Miss Slamerkin b. m. 1729 by Y. True Blue 1718 (Honeywood Ar.)—Dau. of Lord Oxford's Dun Ar.—D'Arcy Blacklegged Royal mare.

CHANCE.

The property of Mr. James B. Clay, Iroquois Stud, Lexington. Stallion advertised for public service each year.

CHANCE, chestnut horse, foaled 1881, bred by the late James A. Grinstead, Walnut Hill Stud, Ky., by War Dance, son of Lexington and the famous Reel, dam of Lecompte (the only horse that ever beat Lexington), Starke, winner of the Goodwood Cup, and Prioress, winner of the Cesarewitch Handicap, etc., by imp. Glencoe, out of imp. Gallopade by Catton. Chance was a good race-horse; won two races out of four starts, and was second in one. Won Colt Stake, ½ mile, 105 lbs., in 0.50½; won purse, 5 furlongs, in 1.05¼. As three-years old won purse, ¾ of a mile, 109 lbs., in 1.17¾; won purse, ¾ of a mile heats, 104 lbs., in 1.18½, 1.18½, track slow; won Falls City Stakes, 1$\frac{1}{16}$ miles, 108 lbs., in 1.55, beating Highflight, Aretino and others, and was placed in other races. Was placed in several good races at four years old. At five years old won purse, 1½ miles, 112 lbs., in 1.57; second to Matinee, 1$\frac{1}{16}$ miles, 108 lbs., in 1.50½; second to Modesty, 1¼ miles, in 2.11; was third in the Tobacco Stakes and two other races. At six years old won 1 mile 70 yards in 1.51¼; won ¾ of a mile in 1.16¼; 7 furlongs in 1.30½, 107 lbs., track heavy; was second in one race and third in three. His dam, Perhaps, is the dam of Perplex, Doubt, Dilemma and the speedy, good race-mare Reckon by imp. Australian, she out of Mishap by imp. Knight of St. George by Irish Birdcatcher. Mishap, the second dam, produced Crockford, Misfortune (dam of Mistake, a good winner in England, Hopeful, Syntax and Miss Dance, Walcott's dam), Sophy (dam of Repose, Pelham's dam) and Mischief (dam of Judge Hughes). Lorette, the third dam, produced Regan, Ada A. (late Ada Allen), Sophy Badderly and Peaceful. Of these Regan was the dam of Vigil, Vera Cruz, Jennie D. (dam of Grismer, Verano and Lijero), Jennie B. (dam of Primero, Ganymede and Honduras), Jolly Sir John, Vengeance and Printer (late Last Chance). Ada A. produced Patsey Duffy, Frank Rhoads, Leon (the two last named each winners of the California Stakes, etc.), Lizzie P. (dam of Idalene Cotton and Louis P.) and Rosa G. (dam of Tom O'Hara and Fanny J.). Sophy Badderley produced the winners Peytona Barry (dam of Bob Thomas and Worth), Bob Pate and Martindale, while from Peaceful, the last of Lorette's daughters, came the winners Montezuma and Steve Jerome. It is a noted racing and producing family in both America and England; from it came Henry Perritt, Mogul, Cazique, Our Ban, in addition to those above, and in England Florizel, Election, Royal Oak, Mulatto, The Palmer, Rosicrucian, Dutch Skater, Doncaster, Gladiateur, Hermit (by Newminster), sire of imp. St. Blaise, etc. Chance should be a success in the stud; he combines the best racing blood and is from a racing and producing family. He is sire of Eagle Bird, a good two year old, and the Narka colt, both winners. He is descended in direct female line from the Massey mare, to which trace Trentham, Florizel, Election, Quiz, Royal Oak, Mulatto, Defence, The Palmer, Rosicrucian, Dutch Skater, Doncaster, Gladiateur, Hermit (by Newminster), The Merry Monarch, Hackthorp and other good ones.

CHANCE.

CHESTNUT HORSE, FOALED 1831; BRED BY JAS. A. GRINSTEAD, LEXINGTON, KY.

PERHAPS, ch. m., 1874.
- Mishap, b. m., 1862.
  - Lorette, b. m., 1854.
    - Mary Ogden, ch. m., .
    - *Sovereign, b. h., 1836.
  - *Knight of St. George, b. h., 1851.
    - Maltese, b. m., 1845.
    - Ir. Birdcatcher, ch. h., 1833.
- *Australian, ch. h., 1858.
  - Emilia,* b. m., 1840.
    - Persian, b. m., 1829.
    - Y. Emilius, b. h., 1835.
  - West Australian, b. h., 1850.
    - Mowerina, b. m., 1843.
    - Melbourne, br. h., 1834.

WAR DANCE, ch. h., 1859.
- Reel, gr. m., 1838.
  - *Galopade, gr. m., 1828.
    - Camillina, gr. m., 1822.
    - Catton, b. h., 1809.
  - *Glencoe, ch. h., 1831.
    - Trampoline, ch. m., 1825.
    - Sultan, b. h., 1816.
- Lexington, b. h., 1850.
  - Alice Carneal, b. m., 1836.
    - Rowena, ch. m., 1826.
    - Sarpedon, br. h., 1828.
  - Boston, ch. h., 1833.
    - Sister to Tuckahoe, ch. m., 1814.
    - Timoleon, ch. h., 1813.

Florizel 1768—Herod 1758 Tartar 1743 Partner 1718 Jigg- Byerly Turk, about 1685.

Sir Archy, b. h., 1805.	*Diomed 1777 (Florizel 1768) *Castianira 1796 by Rockingham 1781 (Highflyer 1774)—Tabitha 1782 by Trentham 1766 (T. Sweepstakes 1743) Dau. of Bosphorus 1754 (Babraham 1740) Dau. of W's Forester 1750 (Forester 1736).
Daughter of, ch. m., —.	*Saltram 1780 (Eclipse 1764)—Dau. of Symmes' Wildair 1767 *Fearnought 1755 Dau. of Tyler's Driver 1761 (*Othello)—Dau. of *Fallower 1761 (Blank 1740) Dau. of *Vampire 1757 (Regulus 1739).
Ball's Florizel, ch. h., 1801.	*Diomed 1777 (Florizel 1768)—Dau. of *Shark 1771 (Marske 1750) Dau. of Harris' Eclipse 1770 (*Fearnought 1755) Dau. of *Fearnought 1755 (Regulus 1739) Dau. of *Jolly Roger 1741 (Roundhead 1733).
Daughter of, b. m., 1799.	*Alderman 1787 (Pot-8-o's 1773)—Dau. of *Clockfast 1771 (Gimcrack 1760) Dau. of Symmes' Wildair 1767 (*Fearnought 1755) Y. Kitty Fisher 1767 by *Fearnought 1755 (Regulus 1739) *Kitty Fisher 1755 by Cade 1734 (God. Ar. 1724).
Emilius, b. h., 1820.	Orville 1799 (Beningbrough 1791) Emily 1810 by Stamford 1794 (Sir Peter T. 1784) Dau. 1799 of Whiskey 1789 (Saltram* 1780) Grey Dorimant 1781 by Dorimant 1772 (Otho 1760) Dizzy 1757 by Blank 1740 (God. Ar. 1724).
Icaria, b. m., 1824.	The Flyer 1814 (Vandyke Jr. 1808) Parma 1813 by Dick Andrews 1797 (Joe Andrews 1778) May 1804 by Beningbrough 1791 (King Fergus 1775) Primrose 1787 by Mambrino 1768 (Engineer 1756) Cricket 1776 by Herod 1758 (Tartar 1743).
Sumpter, ch. h., 1818.	Sir Archy 1805 (*Diomed 1777) Dau. of *Robin Redbreast 1796 (Sir Peter T. 1784) Dau. 1788 of *Obscurity 1777 (Eclipse 1764) Slamerkin 1768 by *Wildair 1753 (Cade 1734) *Cub mare 1762 by Cub 1739 (Fox 1714).
Lady Grey, b. m., 1817.	Robin Grey 1805 (*Royalist 1790) Maria 1802 by Melzar 1791 (*Medley 1776) Dau. of *Highflyer 1784 (Highflyer 1774)—Dau. of *Fearnought 1755 (Regulus 1739) Dau. of Ariel 1756 (*Traveller 1747).
Selim, ch. h., 1802.	Buzzard* 1787 (Woodpecker 1773) Dau. of Alexander 1782 (Eclipse 1764)—Dau. of Highflyer 1774 (Herod 1758) Dau. of Alfred 1770 (Matchem 1748) Dau. 1770 of Engineer 1756 (Sampson 1745).
Bacchante, br. m., 1809.	W's Ditto 1800 (Sir Peter T. 1784)—Dau. 1791 of Mercury 1778 (Eclipse 1764) Dau. 1776 of Herod 1758 (Tartar 1743)- Folly 1771 by Marske 1750 (Squirt 1732)- Vixen 1753 by Regulus 1739 (God. Ar. 1724).
Tramp, b. h., 1810.	Dick Andrews 1797 (Joe Andrews 1778)—Dau. 1803 of Gohanna 1790 (Mercury 1778)—Fraxinella 1793 by Trentham 1766 (T. Sweepstakes 1743)—Dau. 1785 of Woodpecker 1773 (Herod 1758)—Everlasting 1775 by Eclipse 1764 (Marske 1750).
Web, b. m., 1808.	Waxy 1790 (Pot-8-o's 1773)—Penelope 1798 by Trumpator 1782 (Conductor 1767) Prunella 1788 by Highflyer 1774 (Herod 1758)—Promise 1768 by Snap 1750 (Snip 1736)—Julia 1756 by Blank 1740 (God. Ar. 1724).
Gohanna, b. h., 1802.	Gohanna 1790 (Mercury 1778) Catherine 1795 by Woodpecker 1773 (Herod 1758) Camilla 1778 by Trentham 1766 (T. Sweepstakes 1743) Coquette 1765 by Compton Barb—Sist. 1743 to Regulus by God. Ar. 1724.
Lucy Gray, ch. m., 1804.	Timothy 1794 (Delpini 1781)—Lucy 1789 by Florizel 1768 (Herod 1758)- Frenzy 1774 by Eclipse 1764 (Marske 1750)—Dau. of Engineer 1756 (Sampson 1745)- Dau. 1763 of Blank 1740 (God. Ar. 1724).
Camillus, gr. h., 1803.	Hambletonian 1792 (King Fergus 1775)—Faith 1779 by Pacolet 1763 (Blank 1740) Atalanta 1769 by Matchem 1748 (Cade 1734)- Y. Lass-of-the-Mill 1756 by Oroonoko 1745 (Crab 1722) Dau. 1744 of Traveller 1735 (Partner 1718).
Daughter of, b. m., 1818.	Smolensko 1810 (Sorcerer 1796)- Miss Cannon 1811 by Orville 1799 (Beningbrough 1791)—Dau. 1796 of Weathercock 1785 (Highflyer 1774)- Cora 1777 by Matchem 1748 (Cade 1734)—Dau. of Turk 1763 (Regulus 1739).
Humprey Clinker, b. h., 1822.	Comus 1809 (Sorcerer 1796)—Clinkerina 1812 by Clinker 1805 (Sir Peter T. 1784)—Pewet 1786 by Tandem 1773 (Syphon 1750)—Termagant by Tantrum 1760 (Cripple 1750)—Cantatrice by Sampson 1745 (Blaze 1733).
Daughter of, b. m., 1825.	Cervantes 1806 (Don Quixote 1784) Dau. 1818 of Golumpus 1802 (Gohanna 1790)—Dau. 1810 of Paynator 1791 (Trumpator 1782) Circle 1802 by St. George 1789 (Highflyer 1774)—Abigail 1788 by Highflyer 1774 (Herod 1758).
Touchstone, br. h., 1831.	Camel 1822 (Whalebone 1807) Banter 1826 by Master Henry 1815 (Orville 1799) Boadicea 1807 by Alexander 1782 (Eclipse 1764) Brunette by Amaranthus 1766 (Old England 1741)—Mayfly 1771 by Matchem 1748 (Cade 1734).
Emma, ch. m., 1824.	Whisker 1812 (Waxy 1790)—Gibside Fairy 1811 by Hermes 1790 (Mercury 1778)—Vicissitude 1800 by Pipator 1786 (Imperator 1776) Beatrice 1791 by Sir Peter T. 1784 (Highflyer 1774)—Pyrrha 1771 by Matchem 1748 (Cade 1734).
Emilius, b. h., 1820.	Orville 1799 (Beningbrough 1791)- Emily 1810 by Stamford 1794 (Sir Peter T. 1784)—Dau. 1799 of Whiskey 1789 (Saltram* 1780) Grey Dorimant 1781 by Dorimant 1772 (Otho 1760)—Dizzy 1757 by Blank 1740 (God. Ar. 1724).
Shoveler, b. m., 1816.	Scud 1804 (Beningbrough 1791)—Goosander 1805 by Hambletonian 1792 (King Fergus 1775) Rally 1790 by Trumpator 1782 (Conductor 1767) Fancy 1780 by Florizel 1768 (Herod 1758) Sist. 1763 to Juno by Spectator 1749 (Crab 1722).
Whisker, b. h., 1812.	Waxy 1790 (Pot-8-o's 1773)—Penelope 1798 by Trumpator 1782 (Conductor 1767)—Prunella 1788 by Highflyer 1774 (Herod 1758)—Promise 1768 by Snap 1750 (Snip 1736)—Julia 1756 by Blank 1740 (God. Ar. 1724).
Variety, ch. m., 1816.	[Selim or] Soothsayer 1808 (Sorcerer 1796) Sprite 1807 by Bobtail 1795 (Precipitate* 1787)—Catherine 1795 by Woodpecker 1773 (Herod 1758)—Camilla 1778 by Trentham 1766 (T. Sweepstakes 1743)—Coquette 1765 by Compton Barb.
Sir Hercules, blk. h., 1826.	Whalebone 1807 (Waxy 1790)—Peri 1822 by Wanderer 1811 (Gohanna 1790)—Thalestris 1809 by Alexander 1782 (Eclipse 1764) Rival 1800 by Sir Peter T. 1784 (Highflyer 1774)—Hornet 1790 by Drone 1777 (Herod 1758).
Guiccioli, ch. m., 1823.	Bob Booty 1804 (Chanticleer 1787)—Flight 1809 by Ir. Escape 1802 (Commodore 1793)—Y. Heroine by Bagot 1780 (Herod 1758)—Heroine 1775 by Hero 1753 (Cade 1734)—Dau. 1766 of Snap 1750 (Snip 1736).
Hetman Platoff, b. h., 1836.	Brutandorf 1821 (Blacklock 1814) Dau. 1820 of Comus 1809 (Sorcerer 1796)—Marciana 1809 by Stamford 1794 (Sir Peter T. 1784)—Marcia 1797 by Coriander 1786 (Pot-8-o's 1773)—Faith 1779 by Pacolet 1763 (Blank 1740).
Waterwitch, blk. m., 1833.	Sir Hercules 1826 (Whalebone 1807)—Mary Anne 1820 by Waxy Pope 1806 (Waxy 1790)—Witch 1811 by Sorcerer 1796 (Trumpator 1782)—Miss Buckle 1801 by Precipitate* 1787 (Mercury 1778)—Dau. 1788 of Highflyer 1774 (Herod 1758).
Emilius, b. h., 1820.	Orville 1799 (Beningbrough 1791)—Emily 1810 by Stamford 1794 (Sir Peter T. 1784)—Dau. 1799 of Whiskey 1789 (Saltram* 1780)- Grey Dorimant 1781 by Dorimant 1772 (Otho 1760)—Dizzy 1757 by Blank 1740 (God. Ar. 1724).
Fleur-de-Lis, b. m., 1822.	Bourbon 1811 (Sorcerer 1796)—Lady Rachel 1805 by Stamford 1794 (Sir Peter T. 1784)—Y. Rachel 1799 by Volunteer 1780 (Eclipse 1764)—Rachel 1790 by Highflyer 1774 (Herod 1758) Dau. 1771 of Syphon 1750 (Squirt 1732).
Thornhill, ch. h., 1838.	*Glencoe 1831 (Sultan 1816)—*Pickle 1831 by Emilius 1820 (Orville 1799)—Mustard 1824 by Merlin 1815 (Castrel 1801) Morel 1805 by Sorcerer 1796 (Trumpator 1782)—Hornby Lass 1796 by Buzzard* 1787 (Woodpecker 1773).
MaryThomas, b. m., 1837.	*Consol 1828 (Lottery 1820)—Parrot b. m 1825 by Roanoke 1817 (Sir Archy 1805)—Paroquet b. m. 1819 by *Merryfield 1808 (Cockfighter 1796)—*Popinjay mare b. m. 1808 by Popinjay 1797 (Buzzard* 1787)—Bourbon's dam gr. m. 1797 by Precipitate* 1787 (Mercury 1778)—Y. Tiffany gr. m. 1787 by Highflyer 1774 (Herod 1758)—Tiffany gr. m. 1775 by Eclipse 1764 (Marske 1750)—Y. Hag gr. m. 1761 by Skim 1746 (Starling 1727)—Hag gr. m. 1744 by Crab 1722 (Alcock Ar.)—Ebony b. m. 1728 by Flying Childers 1715 (Darley Ar. 1699)—Old Ebony blk. m. 1714 by Basto 1702 (Byerly Turk)—Duke of Rutland's Massey mare by Massey's Black Barb.

IMPORTED CHARAXUS.

WINNER OF THE LIVERPOOL NURSERY STAKES, WORCESTER CUP, TRADESMAN WELTER CUP, CRAVEN STAKES AND OTHER GOOD RACES.

The property of Capt. R. J. Hancock, Ellerslie Stud, Overton, Va. Yearlings sold annually. Service by contract.

CHARAXUS, bay horse, foaled 1876, bred by Mr. J. G. Watkins, by Distin, son of Trumpeter by Orlando by Touchstone, dam Sappho by Kingston, son of Venison, out of Sacrifice by Voltaire, son of Blacklock, etc. Distin, the sire of Charaxus, comes from a racing family. Miss Bowzer, his dam, is the grandam of Craig Millar, winner of the St. Leger and Doncaster Cup, she out of Mangosteen, sister to imp. Mango, winner of the St. Leger, and to Preserve, winner of the 1,000 Guineas, and grandam of Speculum, winner of the Goodwood Cup. Trumpeter, sire of Distin, is the sire of Plutus, grandsire of Rayon d'Or. Distin is the sire of Avontes, a good race-horse and winner of a number of important races, and is the sire of Netheravon, a good winner. Kingston, sire of Charaxus' dam, won the Goodwood Cup and was a popular sire in England, and is the sire of Babta, imp. Glenelg's dam, imp. Eltham Lass, dam of Kingfisher, Caractacus, winner of the Derby, imp. Bluemantle, Queen Bertha, winner of the Oaks and dam of Spinaway, Wheel of Fortune, both winners of the 1,000 Guineas and Oaks. Sacrifice, his grandam, is also grandam of Devotion, the dam of Thebais, winner of the 1,000 Guineas and Oaks, the unbeaten Clairvaux, St. Marguerite, winner of the 1,000 Guineas and dam of Seabreeze, winner of the Oaks and St. Leger by Voltaire, the sire of English Voltigeur, who won the Derby and St. Leger and sired imp. Billet, sire of Barnes, Runnymede, Miss Woodford, etc., and Vedette, winner of the 2,000 Guineas, he the sire of Galopin, winner of the Derby. Galopin is the sire of St. Simon, the most popular stallion in England, Galliard, winner of the 2,000 Guineas, and Donovan, who won both the Derby and St. Leger in 1889, she out of Virginia, the dam of Virago, winner of the 1,000 Guineas, Goodwood and Doncaster Cups, and full sister to Puce, dam of Flea, winner of the 1,000 Guineas and grandam of The Ugly Buck, winner of the 2,000 Guineas, by the St. Leger winner imp. Rowton. Virginia is out of Pucelle by Muley, she out of Medora, winner of the Oaks, and dam of Gulnare, an Oaks winner, by Selim. The pedigree runs back through fashionable channels to the Layton Barb mare, to which trace Falsetto and Iroquois, and from Golden Grove back the pedigree is identical with that of Iroquois. To the same fountain head trace also Powhattan, Parole, Rebel, Virgilian, La Henderson, Ferida, Aella and many others of note on this side and in England. Matchem, Inheritor, Annandale, Ion, Simoom, St. Honorat all descent in direct female line, and are a few of the equine wonders that trace directly to this family. As a race-horse Charaxus was distinguished. At two years old he won the Liverpool Nursery Stakes, in which he beat Strathconan. At three he won the Worcester Welter Cup, 1 mile, 112 lbs, the Shobdon Welter Cup, 1 mile, carrying in the latter race 127 lbs., the Alexandra Handicap, 1½ miles, 114 lbs. At five he won the Herefordshire Cup, the Birmingham Autumn Handicap and the Worcester Autumn Handicap. At six he won Easter Handicap Plate, 1¼ miles, carrying 136 lbs., at Bath, won the County Members' Plate, 1 mile, carrying 145 lbs., a mile Selling Welter Race, carrying 156 lbs., Worcestershire Stakes (Handicap), 1¾ miles, 124 lbs., Worcester Tradesmen's Welter Cup, 1 mile, with 168 lbs., the Shepperton Handicap Stakes, 1¼ miles, 132 lbs., and Her Majesty's Plate, at Lichfield, 2 miles, 137 lbs. At seven he won the Craven Stakes, 1¼ miles, 115 lbs., at Goodwood. From this it will be seen that the career of Charaxus on the turf was very highly successful, and his form and royal pedigree should insure his success as a sire of race-horses. The only ones of his get tried, Capstone, Katie C. (Hetty Cary), Charade, Lester, Annie G., Lallah, Character, Hammie and Ma Belle, are good ones and winners in first-class company.

CHARAXUS.

BAY HORSE, FOALED 1876; BRED BY MR. I. G. WATKINS, ENGLAND.

DISTIN, b. h., 1864.	Trumpeter, ch. h., 1856.	Orlando, b. h., 1841.	Touchstone, br. h., 1831.	Camel, br. h., 1822.	Whalebone 1807 (Waxy 1790)—Dau. 1812 of Selim 1802 (Buzzard* 1787)—Maiden 1801 by Sir Peter T. 1784 (Highflyer 1774)—Dau. 1788 of Phenomenon 1780 (Herod 1758)—Matron 1782 by Florizel 1768 (Herod 1758).
				Banter, br. m., 1826.	Master Henry 1815 (Orville 1799)—Boadicea 1807 by Alexander 1782 (Eclipse 1764)—Brunette by Amaranthus 1766 (Old England 1741)—Mayfly 1771 by Matchem 1748 (Cade 1734)—Dau. 1755 of Anc. Starling 1748 (Old Starling 1727).
			Vulture, ch. m., 1833.	Langar, ch. h., 1817.	Selim 1802 (Buzzard* 1787)—Dau. 1808 of Walton 1799 (Sir Peter T. 1784)—Y. Giantess 1790 by Diomed* 1777 (Florizel 1768)—Giantess 1769 by Matchem 1748 (Cade 1734)—Molly Long Legs 1753 by Babraham 1740 (God. Ar. 1724).
				Kite, b. m., 1821.	Bustard 1813 (Castrel 1801)—Olympia 1815 by Sir Oliver 1800 (Sir Peter T. 1784)—Scotilla 1795 by Anvil 1777 (Herod 1758)—Scota 1783 by Eclipse 1764 (Marske 1750)—Harmony 1775 by Herod 1758 (Tartar 1743).
		Cavatina, ch. m., 1845.	Redshank, b. h., 1833.	Sandbeck, b. h., 1818.	Catton 1809 (Golumpus 1802)—Orvillina 1804 by Beningbrough 1791 (King Fergus 1775)—Evelina 1791 by Highflyer 1774 (Herod 1758)—Termagant by Tantrum 1760 (Cripple 1750)—Cantatrice by Sampson 1745 (Blaze 1733).
				Johanna, ch. m., 1813.	Selim 1802 (Buzzard* 1787)—Dau. of Skyscraper 1786 (Highflyer 1774)—Dau. 1796 of Dragon* 1787 (Woodpecker 1773)—Fidget's dam 1777 by Matchem 1748 (Cade 1734)—Dau. 1770 of Syphon 1750 (Squirt 1732).
			Oxygen, b. m., 1828.	Emilius, b. h., 1820.	Orville 1799 (Beningbrough 1791)—Emily 1810 by Stamford 1794 (Sir Peter T. 1784)—Dau. 1799 of Whiskey 1789 (Saltram* 1780)—Grey Dorimant 1781 by Dorimant 1772 (Otho 1760)—Dizzy 1757 by Blank 1740 (God. Ar. 1724).
				Whizgig, ch. m., 1819.	Rubens 1805 (Buzzard* 1787)—Penelope 1798 by Trumpator 1782 (Conductor 1767)—Prunella 1788 by Highflyer 1774 (Herod 1758)—Promise 1768 by Snap 1750 (Snip 1736)—Julia 1756 by Blank 1740 (God. Ar. 1724).
	Miss Bowzer, b. m., 1856.	Hesperus, b. h., 1839.	Bay Middleton, b. h., 1833.	Sultan, b. h., 1816.	Selim 1802 (Buzzard* 1787)—Bacchante 1809 by W's Ditto 1800 (Sir Peter T. 1784)—Dau. 1791 of Mercury 1778 (Eclipse 1764)—Dau. 1776 of Herod 1758 (Tartar 1743)—Folly 1771 by Marske 1750 (Squirt 1732).
				Cobweb, b. m., 1821.	Phantom 1808 (Walton 1799)—Filagree 1815 by Soothsayer 1808 (Sorcerer 1796)—Web 1808 by Waxy 1790 (Pot-8-o-'s 1773)—Penelope 1798 by Trumpator 1782 (Conductor 1767)—Prunella 1788 by Highflyer 1774 (Herod 1758).
			Plenary, br. m., 1837.	Emilius, b. h., 1820.	Orville 1799 (Beningbrough 1791)—Emily 1810 by Stamford 1794 (Sir Peter T. 1784)—Dau. 1799 of Whiskey 1789 (Saltram* 1780)—Grey Dorimant 1781 by Dorimant 1772 (Otho 1760)—Dizzy 1757 by Blank 1740 (God. Ar. 1724).
				Harriet, blk. m., 1819.	Pericles 1809 (Evander 1801)—Dau. 1812 of Selim 1802 (Buzzard* 1787)—Pipylina 1803 by Sir Peter T. 1784 (Highflyer 1774)—Rally 1790 by Trumpator 1782 (Conductor 1767)—Fancy 1780 (Sist. to *Diomed) by Florizel 1768 (Herod 1758).
		Mangosteen, b. m., 1834.	Emilius, b. h., 1820.	Orville, b. h., 1799.	Beningbrough 1791 (King Fergus 1775)—Evelina 1791 by Highflyer 1774 (Herod 1758)—Termagant by Tantrum 1760 (Cripple 1750)—Cantatrice by Sampson 1745 (Blaze 1733)—Dau. 1757 of Regulus 1739 (God. Ar. 1724).
				Emily, ch. b., 1810.	Stamford 1794 (Sir Peter T. 1784)—Dau. 1799 of Whiskey 1789 (Saltram* 1780)—Grey Dorimant 1781 by Dorimant 1772 (Otho 1760)—Dizzy 1757 by Blank 1740 (God. Ar. 1724)—Anc. Dizzy 1741 by Driver 1727 (Wynn Ar.).
			Mustard, ch. m., 1824.	Merlin, ch. h., 1815.	Castrel 1801 (Buzzard* 1787)—Miss Newton 1804 by Delpini 1781 (Highflyer 1774)—Tipple Cyder 1788 by King Fergus 1775 (Eclipse 1764)—Sylvia 1783 by Y. Marske 1771 (Marske 1750)—Ferret 1765 by Bro. to Silvio (Cade 1734).
				Morel, ch. m., 1805.	Sorcerer 1796 (Trumpator 1782)—Hornby Lass 1796 by Buzzard* 1787 (Woodpecker 1773)—Puzzle 1778 by Matchem 1748 (Cade 1734)—Princess 1769 by Herod 1758 (Tartar 1743)—Julia 1756 by Blank 1740 (God. Ar. 1724).
SAPPHO, br. m., 1859.	Kingston, b. h., 1849.	Venison, br. h., 1833.	Partisan, b. h., 1811.	Walton, b. h., 1799.	Sir Peter T. 1784 (Highflyer 1774)—Arethusa 1792 by Dungannon 1780 (Eclipse 1764)—Dau. 1777 of Prophet 1760 (Regulus 1739)—Virago 1764 by Snap 1750 (Snip 1736)—Dau. of Regulus 1739 (God. Ar. 1724).
				Parasol, b. m., 1800.	Pot-8-o-'s 1773 (Eclipse 1764)—Prunella 1788 by Highflyer 1774 (Herod 1758)—Promise 1768 by Snap 1750 (Snip 1736)—Julia 1756 by Blank 1740 (God Ar. 1724)—Spectator's Dam 1735 by Partner 1718 (Jigg).
			Fawn, br. m., 1823.	Smolensko, blk. h., 1810.	Sorcerer 1796 (Trumpator 1782)—Wowski 1797 by Mentor 1784 (Justice 1774)—Maria 1777 by Herod 1758 (Tartar 1743)—Lisette 1772 by Snap 1750 (Snip 1736)—Miss Windsor 1754 by God. Ar. 1724.
				Jerboa, b. m., 1803.	Gohanna 1790 (Mercury 1778)—Camilla 1778 by Trentham 1766 (T. Sweepstakes 1743)—Coquette 1764 by Compton Barb—Sist. 1743 to Regulus by God. Ar. 1724—Grey Robinson 1723 by Bald Galloway (St. Victor Barb.
		Queen Anne, b. m., 1843.	Slane, b. h., 1833.	Royal Oak, br. h., 1823.	Catton 1809 (Golumpus 1802)—Dau. 1818 of Smolensko 1810 (Sorcerer 1796)—Lady Mary 1809 by Beningbrough 1791 (King Fergus 1775)—Dau. of Highflyer 1774 (Herod 1758)—Sist. to Miss Bell by Marske 1750 (Squirt 1732).
				Daughter of, b. m., 1819.	Orville 1799 (Beningbrough 1791)—Epsom Lass 1803 by Sir Peter T. 1784 (Highflyer 1774)—Alexina 1788 by King Fergus 1775 (Eclipse 1764)—Lardella 1780 by Y. Marske 1771 (Marske 1750) Dau. of Y. Cade 1747 (Cade 1734).
			Garcia, br. m., 1829.	Octavian, ch. h., 1807.	Stripling 1795 (Phenomenon 1780)—Dau. 1797 of Oberon 1790 (Highflyer 1774)—Dau. 1779 of Rauthos 1763 (Matchem 1748)—Dau. of T. Sweepstakes 1743 (B. Sweepstakes 1722)—Sist. 1748 to Careless by Hutton's Spot 1728 (Hartley's Blind Horse).
				Daughter of, b. m., 1803.	Shuttle 1793 (Y. Marske 1771)—Katharine 1798 by Delpini 1781 (Highflyer 1774)—Dau. 1783 of Paymaster 1766 (Blank 1740)—Dau. 1774 of Le Sang 1759 (Changeling 1747)—Dau. 1751 of Rib 1736 (Crab 1722)
	Sacrifice, br. m., 1847.	Voltaire, br. h., 1826.	Blacklock, b. h., 1814.	Whitelock, b. h., 1803.	Hambletonian 1792 (King Fergus 1775)—Rosalind 1788 by Phenomenon 1780 (Herod 1758)—Atalanta 1769 by Matchem 1748 (Cade 1734)—Y. Lass of the Mill 1756 by Oroonoko 1745 (Crab 1722)—Dau. 1744 of Traveller 1735 (Partner 1718).
				Daughter of, — m., 1799.	Coriander 1786 (Pot-8-o-'s 1773)—Wildgoose 1792 by Highflyer 1774 (Herod 1758)—Coheiress 1786 by Pot-8-o'-s 1773 (Eclipse 1764)—Manilla 1777 by Goldfinder 1764 (Snap 1750)—Dau. 1763 of Old England 1741 (God. Ar. 1724).
			Daughter of, b. m., 1816.	Phantom, b. h., 1808.	Walton 1799 (Sir Peter T. 1784)—Julia 1799 by Whiskey 1789 (Saltram* 1780)—Y. Giantess 1790 by Diomed* 1777 (Florizel 1768) Giantess 1769 by Matchem 1748 (Cade 1734)—Molly Long Legs 1753 by Babraham 1740 (God. Ar. 1724).
				Daughter of, b. m., 1802.	Overton 1788 (King Fergus 1775)—Dau. 1793 of Walnut 1786 (Highflyer 1774)—Dau. 1786 of Ruler 1777 (Y. Marske 1771)—Piracantha 1772 by Matchem 1748 (Cade 1724)—Prophetess 1758 by Regulus 1739 (God. Ar. 1724).
		Virginia, ch. m., 1835.	Rowton*, ch. h., 1826.	Oiseau, ch. h., 1809.	Camillus 1803 (Hambletonian 1792)—Dau. 1796 of Ruler 1777 (Marske 1771)—Treecreeper 1789 by Woodpecker 1773 (Herod 1758)—Dau. 1779 of Trentham 1766 (T. Sweepstakes 1743)—Cunegonde 1769 by Blank 1740 (God. Ar. 1724).
				Katherina, b. m., 1817.	Woful 1809 (Waxy 1790)—Landscape 1813 by Rubens 1805 (Buzzard* 1787)—Iris 1795 by Brush 1785 (Eclipse 1764) Dau. of Herod 1758 (Tartar 1743) Doctor's sist. 1774 by Goldfinder 1764 (Snap 1750).
			Pucelle, ch. m., 1821.	Muley, b. h., 1810.	Orville 1799 (Beningbrough 1791)—Eleanor 1798 by Whiskey 1789 (Saltram* 1780)—Y. Giantess 1790 by Diomed* 1777 (Florizel 1768)—Giantess 1769 by Matchem 1748 (Cade 1734)—Molly Long Legs 1753 by Babraham 1740 (God. Ar. 1724).
				Medora, ch. m., 1811.	Selim 1802 (Buzzard* 1787)—Dau. ch. m. 1803 of Sir Harry* 1795 (Sir Peter T. 1784)—Dau. of ch. m. 1793 by Volunteer 1780 (Eclipse 1764)—Dau. of b. m. 1779 Herod 1758 (Tartar 1743)—Golden Grove ch. m. 1760 by Blank 1740 (God. Ar. 1724)—Spinster (the Widdrington mare) ch. m. 1735 by Partner 1718 (Jigg)—Bay Bloody Buttocks b. m. 1729 by Bloody Buttocks—Sist. 1722 to Guy by Greyhound (Chillaby)—Brown Farewell br. m. 1710 by Makeless (Oglethorpe Ar.)—Dau. of Brimmer (D'Arcy Yellow Turk)—Dau. of Place's White Turk—Dau. of Dodsworth—Layton Barb Mare.

IMPORTED CHEVIOT.

WINNER OF THE CANTERBURY DERBY, MIDSUMMER HANDICAP AND CRAVEN STAKES, AUSTRALIA.

Is in the Fairview Stud of Messrs. Charles Reed & Sons, Gallatin, Tenn. Yearlings sold annually.

CHEVIOT, bay horse, brother to Sir Modred, foaled 1879, bred in the Middle Park Stud, New Zealand, by Traducer, dam Idalia by Cambuscan, grandam Dulcibella by Voltigeur, out of Priestess by The Doctor. Imported Sir Modred and Cheviot were both great race-horses in Australia, and both have become decided successes in the stud in America. Cheviot won the Canterbury Derby, Midsummer Handicap, 2 miles, 120 lbs., and Craven Stakes, 1½ miles, 133 lbs., and only Sir Modred, his brother, could beat him for the Canterbury Cup. Cheviot is still at the beginning of his stud career. With the same opportunities, he will take equal rank with his older brother as a sire. Among his get, very limited in number, are the following: Osric (winner of the Expectation Stakes last year), Gonzales (a winner last season and of four races this year), Dr. Ross, Vernon, Bernardo (who has captured five races this year, 6 furlongs in 1.15, 1.15¼ and 1.16¾, each time with 119 lbs., 7 furlongs in 1.26, with 124 lbs., and 1 mile 70 yards in 1.48), Princess, half a mile in 0.48¼, with 115 lbs.), Castanet (4½ furlongs in 0.58¼), Mezzotint (a two-year old winner this year at 5 furlongs), Adolph (another two-year old, that defeated a good field this spring at San Francisco), Prize (a two-year old winner at Washington Park, Chicago, this year from a field of fifteen), and Bridal Veil (winner of Ladies' Stakes, St. Paul, Minn., this year). Cheviot's sire, Traducer, was bred in England, by The Libel (son of Pantaloon) and Arethusa. Traducer was the best stallion that ever stood in New Zealand. From 1867 to 1881 he got nine winners of the Canterbury Derby, the greatest event in that country, viz.: Scandal, Envy, Defamation, Calumny, Trump Card, Natator, Sir Modred, The Dauphin and Cheviot. In three of these years his get ran first, second and third, and in two others first and second. He also got Vanguard (winner of the New Zealand Cup), Welcome Jack (winner of the Canterbury Cup), and Lurline, a great mare, out of imp. Mermaid. Lurline produced Darebin, who won the Victoria Derby of 1881, and the best horse of his time throughout Australia. Traducer was to New Zealand and Australia what Lexington or Leamington was to us, or Hermit was to England. Idalia, dam of Cheviot and Sir Modred, was the only Cambuscan mare imported to the colonies. She produced Betrayer, Sir Modred, Idalium, Cheviot, July, all by Traducer, Liverpool by King of Clubs, chestnut filly by Apremont, and a bay filly by Apremont. Betrayer won the Canterbury Champagne Stakes, Canterbury Cup, Wauganal Cup and Timaru Cup. Sir Modred won the Dunedin Champagne Stakes, Canterbury Champagne Stakes, Dunedin Cup, Canterbury Derby, Timaru Cup, Dunedin Birthday Handicap, Canterbury Cup, Christ Church Plate and the Sydney Great Metropolitan. July ran second for the Dunedin Cup. Liverpool won the Welcome Stakes. Cheviot won the Canterbury Derby, Midsummer Handicap and Craven Stakes, and only Sir Modred, his brother, was able to beat him for the Canterbury Cup. Idalia was the Pocahontas of New Zealand, the queen of its stud. In blood she is rare, being a sister to Onslow (who beat Cremorne, in England, in 1871) by Cambuscan, out of Dulcibella, who won the Cesarewitch in 1860. Thus her grandams are old Bee's-wing (Newminster's dam), Southdown (Alarm's dam), and Martha Lynn (Voltigeur's dam.) Cheviot should be a grand stallion. (See Sir Modred, page 254.)

Herod 1758 Tartar 1743—Partner 1718 Jigg Byerly Turk, about 1685.

CHEVIOT.

BAY HORSE, FOALED 1879; BRED BY THE MIDDLE PARK STUD COMPANY, NEW ZEALAND.

†TRADUCER, br. h., 1857.

The Libel, br. h., 1842.

Pantaloon, ch. h., 1824.

Castrel, ch. h., 1801.

Buzzard,* ch. h., 1787. Woodpecker 1773 (Herod 1758) Misfortune 1775 by Dux 1761 (Matchem 1748) Curiosity 1760 by Snap 1750 (Snip 1736) Dau. 1743 of Regulus 1739 (God. Ar. 1724)—Dau. of Bartlett's Childers (Darley Ar. 1689).

Daughter of, b. m., . Alexander 1782 (Eclipse 1764)—Dau. of Highflyer 1774 (Herod 1758) Dau. of Alfred 1770 (Matchem 1748)—Dau. 1770 of Engineer 1756 (Sampson 1745) Bay Malton's dam 1753 by Cade 1734 (God. Ar. 1724).

Idalia, ch. m., 1815.

Peruvian, b. h., 1806. Sir Peter T. 1784 (Highflyer 1774) Dau. 1788 of Boudrow 1777 (Eclipse 1764) Escape's dam 1768 by Squirrel 1754 (Traveller 1735) Sist. to Lowther's Babraham by Babraham 1740 (God. Ar. 1724) Dau. 1750 of Golden Ball 1735 (Partner 1718).

Musidora, b. m., 1804. Meteor 1783 (Eclipse 1764) Maid-of-all-Work 1786 by Highflyer 1774 (Herod 1758) Sist. 1771 to Tandem by Syphon 1750 (Squirt 1732) Dau. of Regulus 1739 (God. Ar. 1724)—Sist. 1758 to Judgement by Snip 1736 (Flying Childers 1715).

Pasquinade, br. m., 1829.

Camel, br. h., 1822.

Whalebone, br. h., 1807. Waxy 1790 (Pot-8-o-'s 1773)—Penelope 1798 by Trumpator 1782 (Conductor 1767) Prunella 1788 by Highflyer 1774 (Herod 1758) Promise 1768 by Snap 1750 (Snip 1736)—Julia 1756 by Blank 1740 (God. Ar. 1724).

Daughter of, b. m., 1812. Selim 1802 (Buzzard* 1787)—Maiden 1801 by Sir Peter T. 1784 (Highflyer 1774) Dau. 1788 of Phenomenon 1780 (Herod 1758) Matron 1782 by Florizel 1768 (Herod 1758) Maiden 1770 by Matchem 1748 (Cade 1734).

Banter, br. m., 1826.

Master Henry, b. h., 1815. Orville 1799 (Beningbrough 1791)—Miss Sophia 1805 by Stamford 1794 (Sir Peter T. 1784) Sophia 1798 by Buzzard* 1787 (Woodpecker 1773) Huncamunca 1787 by Highflyer 1774 (Herod 1758) Cypher 1772 by Squirrel 1754 (Traveller 1735).

Boadicea, b. m., 1807. Alexander 1782 (Eclipse 1764) Brunette by Amaranthus 1766 (Old England 1741) Mayfly 1771 by Matchem 1748 (Cade 1734)—Dau. 1755 of Anc. Starling 1738 (Starling 1727) Look-at-me-Lads 1731 by Grasshopper (Byerly Turk).

Arethusa, b. m., 1839.

Elis, ch. h., 1833.

Langar, ch. h., 1817.

Selim, ch. h., 1802. Buzzard* 1787 (Woodpecker 1773)—Dau. of Alexander 1782 (Eclipse 1764) Dau. of Highflyer 1774 (Herod 1758)—Dau. of Alfred 1770 (Matchem 1748) Dau. 1770 of Engineer 1756 (Sampson 1745).

Daughter of ch. m., 1808. Walton 1799 (Sir Peter T. 1784)—Y. Giantess 1790 by Diomed* 1777 (Florizel 1768)—Giantess 1769 by Matchem 1748 (Cade 1734) Molly Long Legs 1753 by Babraham 1740 (God. Ar. 1724)—Dau. of Cole's Foxhunter 1727 (Brisk 1711).

Olympia, b. m., 1815.

Sir Oliver, b. h., 1800. Sir Peter T. 1784 (Highflyer 1774) Fanny 1790 by Diomed* 1777 (Florizel 1768)—Ambrosia 1783 by Woodpecker 1773 (Herod 1758) Ruth 1761 by Blank 1740 (God. Ar. 1724)—Sist. 1751 to South by Regulus 1739 (God. Ar. 1724).

Scotilla, b. m., 1795. Anvil 1777 (Herod 1758)—Scota 1783 by Eclipse 1764 (Marske 1750) Harmony 1775 by Herod 1758 (Tartar 1743) Rutilia 1769 by Blank 1740 (God. Ar. 1724) Sist. 1751 to South by Regulus 1739 (God. Ar. 1724).

Languid, b. m., 1831.

Cain, b. h., 1822.

Paulowitz, br. h., 1813. Sir Paul 1803 (Sir Peter T. 1784)—Evelina 1791 by Highflyer 1774 (Herod 1758)—Termagant by Tantrum 1760 (Cripple 1750)—Cantatrice by Sampson 1745 (Blaze 1733)—Dau. 1757 of Regulus 1739 (God. Ar. 1724).

Daughter of, —. —., 1810. Paynator 1791 (Trumpator 1782)—Dau. 1805 of Delpini 1781 (Highflyer 1774)—Dau. 1795 of Y. Marske 1771 (Marske 1750)—Dau. of Dorimond 1758 (Dormouse 1738)—Portia 1758 by Regulus 1739 (God. Ar. 1724).

Lydia, b. m., 1822.

Poulton, b. h., 1805. Sir Peter T. 1784 (Highflyer 1774)—Fanny 1790 by Diomed* 1777 (Florizel 1768)—Ambrosia 1783 by Woodpecker 1773 (Herod 1758)—Ruth 1761 by Blank 1740 (God. Ar. 1724)—Sist. 1751 to South by Regulus 1739 (God. Ar. 1724).

Variety, b. m., 1808. Hyacinthus 1797 (Coriander 1786)—Sist. 1790 to Swordsman by Weazle 1776 (Herod 1758)—Dau. of Turk 1763 (Regulus 1739)—Dau. of Locust 1744 (Crab 1722)—Dau. of 1758 Changeling 1747 (Cade 1734).

---

†IDALIA, b. m., 1870.

Cambuscan, ch. h., 1861.

Newminster, b. h., 1848.

Touchstone, br. h., 1831.

Camel, br. h., 1822. Whalebone 1807 (Waxy 1790)—Dau. 1812 of Selim 1802 (Buzzard* 1787)—Maiden 1801 by Sir Peter T. 1784 (Highflyer 1774)—Dau. 1788 of Phenomenon 1780 (Herod 1758)—Matron 1782 by Florizel 1768 (Herod 1758).

Banter, br. m., 1826. Master Henry 1815 (Orville 1799)—Boadicea 1807 by Alexander 1782 (Eclipse 1764)—Brunette by Amaranthus 1766 (Old England 1741) Mayfly 1771 by Matchem 1748 (Cade 1734)—Dau. 1755 of Anc. Starling 1738 (Old Starling 1727).

Bee's-wing, b. m., 1833.

Dr. Syntax, br. h., 1811. Paynator 1791 (Trumpator 1782)—Dau. of Beningbrough 1791 (King Fergus 1775) Jenny Mole 1787 by Carbuncle 1772 (Babr. Blank 1758) Dau. 1773 of Prince T'Quassaw 1751 (Snip 1736)—Sultana 1759 by Regulus 1739 (God. Ar. 1724).

Daughter of, ch. m., 1817. Ardrossan 1809 (John Bull 1789) Lady Eliza 1813 by Whitworth 1805 (Agonistes 1797)—X. Y. Z.'s dam 1793 by Spadille 1781 (Highflyer 1774) Sylvia 1783 by Y. Marske 1771 (Marske 1750) Ferret 1755 by Broth. to Silvio (Cade 1734).

The Arrow, b. m., 1850.

Slane, b. h., 1833.

Royal Oak, br. h., 1823. Catton 1809 (Golumpus 1802)—Dau. 1818 of Smolensko 1810 (Sorcerer 1796)—Lady Mary 1800 by Beningbrough 1791 (King Fergus 1775)—Dau. of Highflyer 1774 (Herod 1758)—Sist. to Miss Bell by Marske 1750 (Squirt 1732).

Daughter of, b. m., 1819. Orville 1799 (Beningbrough 1791)—Epsom Lass 1803 by Sir Peter T. 1784 (Highflyer 1774) Alexina 1788 by King Fergus 1775 (Eclipse 1764)—Lardella 1780 by Y. Marske 1771 (Marske 1750)—Dau. of Y. Cade 1747 (Cade 1734).

Southdown, b. m., 1835.

Defence, b. h., 1824. Whalebone 1807 (Waxy 1790)—Defiance 1816 by Rubens 1805 (Buzzard* 1787) Little Folly 1806 by Highland Fling 1798 (Spadille 1784) Harriet 1799 by Volunteer 1780 (Eclipse 1764) Dau. 1785 of Alfred 1770 (Matchem 1748).

Feltona, b. m., 1819. X. Y. Z. 1808 (Haphazard 1797)—Janetta 1803 by Beningbrough 1791 (King Fergus 1775)—Dau. 1797 of Drone 1777 (Herod 1758) Contessina 1787 by Y. Marske 1771 (Marske 1750) Tuberose 1772 by Herod 1758 (Tartar 1743).

Dulcibella, br. m., 1857.

Voltigeur, br. h., 1847.

Voltaire, br. h., 1826.

Blacklock, b. h., 1814. Whitelock 1803 (Hambletonian 1792)—Dau. 1799 of Coriander 1786 (Pot-8-o-'s 1773)—Wildgoose 1792 by Highflyer 1774 (Herod 1758) Coheiress 1786 by Pot-8-o-'s 1773 (Eclipse 1764) Manilla 1777 by Goldfinder 1764 (Snap 1750).

Daughter of, b. m., 1816. Phantom 1808 (Walton 1799)—Dau. 1802 of Overton 1788 (King Fergus 1775)—Gratitude's dam 1793 by Walnut 1786 (Highflyer 1774) Dau. 1786 of Ruler 1777 (Y. Marske 1771)—Piracantha 1772 by Matchem 1748 (Cade 1734).

Martha Lynn, br. m., 1837.

Mulatto, b. h., 1823. Catton 1809 (Golumpus 1802)—Desdemona 1811 by Orville 1799 (Beningbrough 1791)—Fanny 1796 by Sir Peter T. 1784 (Highflyer 1774) Dau. 1788 of Diomed* 1777 (Florizel 1768)—Desdemona 1770 by Marske 1750 (Squirt 1732).

Leda, b. m., 1824. Filho-da-Puta 1812 (Haphazard 1797)—Treasure 1809 by Camillus 1803 (Hambletonian 1792)—Dau. 1801 of Hyacinthus 1797 (Coriander 1786) Flora 1789 by King Fergus 1775 (Eclipse 1764) Atalanta 1769 by Matchem 1748 (Cade 1734).

Priestess, bl. m., 1846.

The Doctor, bl. h., 1834.

Dr. Syntax, br. h., 1811. Paynator 1791 (Trumpator 1782)—Dau. of Beningbrough 1791 (King Fergus 1775)—Jenny Mole 1787 by Carbuncle 1772 (Babr. Blank 1758)—Dau. 1773 of Prince T'Quassaw 1751 (Snip 1736) Sultana 1759 by Regulus 1739 (God. Ar. 1724).

Daughter of, br. m., 1828. Lottery 1820 (Tramp 1810)—Elizabeth 1821 by Walton 1799 (Sir Peter T. 1784) Trulla 1815 by Sorcerer 1796 (Trumpator 1782)—Dau. 1796 of Weathercock 1785 (Highflyer 1774)—Cora 1777 by Matchem 1748 (Cade 1734).

The Biddy, ch. m., 1839.

Bran, ch. h., 1831. Humprey Clinker 1822 (Comus 1809)—Velvet 1823 by Oiseau 1809 (Camillus 1803)—Wire 1811 by Waxy 1790 (Pot-8-o-'s 1773) Penelope 1798 by Trumpator 1782 (Conductor 1767) Prunella 1788 by Highflyer 1774 (Herod 1758).

Idalia, ch. m., 1815. Peruvian 1806 (Sir Peter T. 1784)—Musidora b. m. 1804 by Meteor 1783 (Eclipse 1764)—Maid-of-all-Work b. m. 1786 by Highflyer 1774 (Herod 1758)—Sist. b. m. 1771 to Tandem by Syphon 1750 (Squirt 1732)—Dau. b. m. of Regulus 1739 (God. Ar. 1724)—Sist. 1758 to Judgement by Snip 1736 (Flying Childers 1715)—Dau. b. m. 1745 of Cottingham

1735 (Hartley's Blind Horse)—The Warlock Galloway ch. m. 1728 by Snake (Lister Turk)—Old Lady ch. m. by the Bald Galloway (St. Victor Barb)—The Wharton mare by the Carlisle Turk—Dau. of Byerly Turk.

COLONEL CLARK.

Property of Wilson P. Hunt, Glen Echo Stud, Normandy, Mo.

COLONEL CLARK, chestnut horse, foaled 1882, bred by H. C. Shafer, Tennessee, by imp. Glengarry, son of Thormanby, winner of the Derby, by Windhound, dam Kate Fisher by Bill Cheatham, son of imp. Albion, and Lalla Rookh by imp. Glencoe, out of Lizzie Hoover's dam by imp. Sovereign, etc. Glengarry was imported in December, 1866, and so badly injured that he raced only moderately well. Thormanby, his sire, was by Melbourne or Windhound, unquestionably by the latter, son of Pantaloon by Castrel. Thormanby defeated, in the Derby, The Wizard, winner of the 2,000 Guineas, won the Ascot Gold Cup, and was the sire of Charibert and Atlantis, winners of the 2,000 Guineas, and of Rouge Rose, the dam of the grand race-horse Bend'or, winner of the Derby and sire of the unbeaten Ormonde. Glengarry's dam Carbine by Rifleman, winner of the Great Yorkshire Stakes, Convivial and Claret Stakes, and by Touchstone. Colonel Clark's dam, Kate Fisher, is the dam of Kingsland, Joe Mitchell, Lucinda C., Glenfisher, Longfish (winner of the St. Louis Derby) Kitty Cheatham, etc., and by Bill Cheatham. Henrietta, half sister to Colonel Clark, is the dam of the winners Germania, Sample, Fraud and Harry Glenn. Colonel Clark's grandam the dam of Lizzie Hoover by imp. Sovereign, son of Emilius, and the celebrated Fleur de Lis by Bourbon, she out of Katy Fagan by imp. Leviathan, she the dam of Miss Odom (Bay Odom), the dam of Mary Wylie, Godiva, Shanghai, Savoy, etc., and she out of Lucy Brooks by Old Bertrand, son of Sir Archy and Eliza by imp. Bedford. The family race and produce race-horses. Colonel Clark was a good race-horse, winner at one mile in 1.45, and again in 1.51 with 149 lbs., 7 furlongs in 1.28½, and was placed second in a number of good races with good horses behind him. Colonel Clark has been in the stud only a few seasons, but is the sire of the winners, Lottie Mills, Laura B., Lucy Clark, etc.

COLONEL CLARK.

CHESTNUT HORSE, FOALED 1882; BRED BY H. C. SHAFER, TENNESSEE.

*GLENGARRY, br. h., 1866.

Thormanby, ch. h., 1857. — [Melbourne or] Windhound, br. h., 1847. — Alice Hawthorn, b. m., 1838.

Carbine, b. m., 1858. — Rifleman, b. h., 1852. — Troica, br. m., 1845.

Pantaloon, ch. h., 1824. Phryne, br. m., 1840. Muley Moloch, br. h., 1830. Rebecca, b. m., 1831. Touchstone, br. h., 1831. [illegible], ch. m., 1838. Lanercost, br. h., 1835. Siberia, b. m., 1835.

KATE FISHER, ch. m., 1856.

Bill Cheatham, b. h., 1855. — *Albion, blk. h., 1837. — Lalla Rookh, —, 1840.

Lizzie Hoover's dam, b. m., 1853. — *Sovereign, b. h., 1836. — Katy Fagan, b. m., 1833.

[Cain or] Actæon, ch. h., 1822. Panthea, br. m., 1821. *Glencoe, ch. h., 1831. Queen of Diamonds, b. m., 1833. Emilius, b. h., 1820. Fleur-de-Lis, b. m., 1822. *Leviathan, ch. h., 1823. Lucy Brooks, b. m., —.

Woodpecker 1773 Herod 1758 Tartar 1743 Partner 1718 Jigg—Byerly Turk, about 1685.

Castrel, ch. h., 1801.	*Buzzard* 1787 (Woodpecker 1773) Dau. of Alexander 1782 (Eclipse 1764) Dau. of Highflyer 1774 (Herod 1758) Dau. of Alfred 1770 (Matchem 1748) Dau. 1770 of Engineer 1755 (Sampson 1745).
Idalia, ch. m., 1815.	Peruvian 1806 (Sir Peter T. 1784) Musidora 1804 by Meteor 1783 (Eclipse 1764)—Maid-of-all-Work 1786 by Highflyer 1774 (Herod 1758) Dau. 1774 of Syphon 1750 (Squirt 1732) Dau. of Regulus 1739 (God. Ar. 1724).
Touchstone, br. h., 1831.	Camel 1822 (Whalebone 1807) Banter 1826 by Master Henry 1815 (Orville 1799) Boadicea 1807 by Alexander 1782 (Eclipse 1764) Brunette by Amaranthus 1766 (Old England 1741) Mayfly 1771 by Matchem 1748 (Cade 1734).
Decoy, b. m., 1830.	Filho da-Puta 1812 (Haphazard 1797) Finesse 1815 by Peruvian 1806 (Sir Peter T. 1784) Violante 1802 by John Bull 1789 (Fortitude 1777) Dau. 1788 of Highflyer 1774 (Herod 1758) Everlasting 1775 by Eclipse 1764 (Marske 1750).
Muley, b. h., 1810.	Orville 1799 (Beningbrough 1791) Eleanor 1798 by Whiskey 1789 (Saltram 1780) Y. Giantess 1790 by Diomed* 1777 (Florizel 1768)—Giantess 1769 by Matchem 1748 (Cade 1734) Molly Long Legs 1753 by Babraham 1740 (God. Ar. 1724).
Nancy, b. m., 1813.	Dick Andrews 1797 (Joe Andrews 1778) Spitfire 1800 by Beningbrough 1791 (King Fergus 1775) Dau. of Y. Sir Peter 1775 (Dorge 1762) Dau. of Engineer 1756 (Sampson 1745)—Dau. of Wilson Ar.
Lottery, br. h., 1820.	Tramp 1810 (Dick Andrews 1797) Mandane 1800 by Pot-8-o-'s 1773 (Eclipse 1764)—Y. Camilla 1787 by Woodpecker 1773 (Herod 1758) Camilla 1778 by Trentham 1766 (T. Sweepstakes 1743) Coquette 1765 by Compton Barb.
Daughter of, b. m., 1818.	Cervantes 1806 (Don Quixote 1784) Anticipation 1802 by Beningbrough 1791 (King Fergus 1775)—Expectation 1779 by Herod 1758 (Tartar 1743)—Dau. 1758 of Skim 1746 (Starling 1727)—Dau. 1751 of Janus 1738 (God. Ar. 1724).
Camel, br. h., 1822.	Whalebone 1807 (Waxy 1790) Dau. 1812 of Selim 1802 (Buzzard* 1787) Maiden 1801 by Sir Peter T. 1784 (Highflyer 1774) Dau. 1788 of Phenomenon 1780 (Herod 1758) Matron 1782 by Florizel 1768 (Herod 1758).
Banter, br. m., 1826.	Master Henry 1815 (Orville 1799)—Boadicea 1807 by Alexander 1782 (Eclipse 1764) Brunette by Amaranthus 1766 (Old England 1741) Mayfly 1771 by Matchem 1748 (Cade 1734)—Dau. 1755 of Starling 1738 (Starling 1727).
The Colonel, ch. h., 1825.	Whisker 1812 (Waxy 1790)—Dau. 1802 of Delpini 1781 (Highflyer 1774) Tipple Cyder 1788 by King Fergus 1775 (Eclipse 1764) Sylvia 1783 by Y. Marske 1771 (Marske 1750)—Ferret 1765 by Broth. to Silvio (Cade 1734).
Galatea, br. m., 1813.	Amadis 1807 (Don Quixote 1784) Paulina 1804 by Sir Peter T. 1784 (Highflyer 1774)—Pewet 1786 by Tandem 1773 (Syphon 1750) Termagant by Tantrum 1760 (Cripple 1750)—Cantatrice by Sampson 1745 (Blaze 1733).
Liverpool, b. h., 1828.	Tramp 1810 (Dick Andrews 1797) Dau. 1822 of Whisker 1812 (Waxy 1790)—Mandane 1800 by Pot-8-o-'s 1773 (Eclipse 1764)—Y. Camilla 1787 by Woodpecker 1773 (Herod 1758)—Camilla 1778 by Trentham 1766 (T. Sweepstakes 1743).
Otis, b. m., 1820.	Bustard 1801 (Buzzard* 1787)—Dau. 1813 of Election 1804 (Gohanna 1790) Dau. 1791 of Highflyer 1774 (Herod 1758)—Dau. 1778 of Eclipse 1764 (Marske 1750)—Rosebud 1765 by Snap 1750 (Snip 1736).
Brutandorf, b. h., 1821.	Blacklock 1814 (Whitelock 1803) Mandane 1800 by Pot-8-o's 1773 (Eclipse 1764)—Y. Camilla 1787 by Woodpecker 1773 (Herod 1758)—Camilla 1778 by Trentham 1766 (T. Sweepstakes 1743)—Coquette 1765 by Compton Barb.
Daughter of, b. m., 1824.	Blucher 1811 (Waxy 1790)—Opal 1806 by Sir Peter T. 1784 (Highflyer 1774)—Olivia 1786 by Justice 1774 (Herod 1758)—Cypher 1772 by Squirrel 1754 (Traveller 1735)—Fribble's dam 1749 by Regulus 1739 (God. Ar. 1724).
Scud, b. h., 1804.	Beningbrough 1791 (King Fergus 1775)—Eliza 1791 by Highflyer 1774 (Herod 1758)—Augusta 1784 by Eclipse 1764 (Marske 1750)—Dau. of Herod 1758 (Tartar 1743)—Nettle 1763 by Bajazet 1740 (God. Ar. 1724).
Diana, ch. m., 1808.	Orville 1799 (Beningbrough 1791)—Emily 1810 by Stamford 1794 (Sir Peter T. 1784)—Dau. 1790 of Whiskey 1789 (Saltram* 1780) Grey Dorimant 1781 by Dorimant 1772 (Otho 1760)—Dizzy 1757 by Blank 1740 (God. Ar. 1724).
[Comus or] Blacklock, b. h., 1814.	Whitelock 1803 (Hambletonian 1792) Dau. 1790 of Coriander 1786 (Pot-8-o-'s 1773)—Wildgoose 1792 by Highflyer 1774 (Herod 1758)—Coheiress 1786 by Pot-8-o-'s 1773 (Eclipse 1764)—Manilla 1777 by Goldfinder 1764 (Snap 1750).
Manuella, b. m., 1809.	Dick Andrews 1797 (Joe Andrews 1778)—Mandane 1800 by Pot-8-o-'s 1773 (Eclipse 1764)—Y. Camilla 1787 by Woodpecker 1773 (Herod 1758)—Camilla 1778 by Trentham 1766 (T. Sweepstakes 1743)—Coquette 1764 by Compton Barb.
Sultan, b. h., 1816.	Selim 1802 (Buzzard* 1787)—Bacchante 1809 by W's Ditto 1800 (Sir Peter T. 1784) Dau. 1791 of Mercury 1778 (Eclipse 1764)—Dau. 1776 of Herod 1758 (Tartar 1743)—Folly 1771 by Marske 1750 (Squirt 1732).
Trampoline, ch. m., 1825.	Tramp 1810 (Dick Andrews 1797)—Web 1808 by Waxy 1790 (Pot-8-o-'s 1773)—Penelope 1798 by Trumpator 1782 (Conductor 1767) Prunella 1788 by Highflyer 1774 (Herod 1758)—Promise 1768 by Snap 1750 (Snip 1736).
*Leviathan, ch. h., 1830.	Muley 1810 (Orville 1799)—Dau. 1809 of Windle 1804 (Beningbrough 1791)—Dau. 1788 of Anvil 1777 (Herod 1758)—Virago 1764 by Snap 1750 (Snip 1736) Dau. of Regulus 1739 (God. Ar. 1724).
Sally Burton, b. m., 1822.	Sir Archy 1805 (*Diomed 1777)—Dau. of Junius (*Buzzard 1787)—Dau. of *Citizen 1785 (Pacolet 1765)—*Barb mare presented by the Dey of Tunis to President Jefferson in 1806.
Orville, b. h., 1799.	Beningbrough 1791 (King Fergus 1775)—Evelina 1791 by Highflyer 1774 (Herod 1758)—Termagant by Tantrum 1760 (Cripple 1770)—Cantatrice by Sampson 1745 (Blaze 1733)—Dau. 1757 of Regulus 1739 (God. Ar. 1724).
Emily, ch. m., 1810.	Stamford 1794 (Sir Peter T. 1784)—Dau. 1790 of Whiskey 1789 (Saltram* 1780) Grey Dorimant 1781 by Dorimant 1772 (Otho 1760)—Dizzy 1757 by Blank 1740 (God. Ar. 1724)—Dizzy 1741 by Driver 1727 (Wynn Ar.).
Bourbon, b. h., 1811.	Sorcerer 1796 (Trumpator 1782)—Dau. 1797 of Precipitate* 1787 (Mercury 1778)—Y. Tiffany 1787 by Highflyer 1774 (Herod 1758) Tiffany 1775 by Eclipse 1764 (Marske 1750)—Y. Hag 1761 by Skim 1746 (Starling 1727).
Lady Rachel, b. m., 1805.	Stamford 1794 (Sir Peter T. 1784)—Y. Rachel 1789 by Volunteer 1780 (Eclipse 1764)—Rachel 1770 by Highflyer 1774 (Herod 1758)—Dau. 1771 of Syphon 1750 (Squirt 1732)—Dau. of Regulus 1739 (God. Ar. 1724).
Muley, b. h., 1810.	Orville 1799 (Beningbrough 1791)—Eleanor 1798 by Whiskey 1789 (Saltram* 1780)—Y. Giantess 1790 by Diomed* 1777 (Florizel 1768)—Giantess 1769 by Matchem 1748 (Cade 1734)—Molly Long Legs 1753 by Babraham 1740 (God. Ar. 1724).
Daughter of, blk. m., 1809.	Windle 1804 (Beningbrough 1791)—Dau. 1788 of Anvil 1777 (Herod 1758)—Virago 1764 by Snap 1750 (Snip 1736)—Dau. of Regulus 1739 (God. Ar. 1724)—Sist. to Othello 1750 by Crab 1722 (Alcock Ar.).
Bertrand, b. h., 1821.	Sir Archy 1805 (*Diomed 1777)—Eliza 1804 by *Bedford 1792 (Dungannon 1780)—*Mambrina 1785 by Mambrino 1768 (Engineer 1756)—Sist. to Sally by Blank 1740 (God. Ar. 1724) Dau. of Ward 1731 (Crab 1722).
The Paramour filly, b. m., —.	*Whip 1794 (Saltram* 1780)—Dau. of *Stirling 1791 (Volunteer 1780)—Dan. of Telemachus (Meade's Celer 1774)—Meade's thoroughbred mare.

IMPORTED DAREBIN.

The property of Mr. J. B. Haggin, Rancho-del-Paso Stud, Sacramento, Cal. Used only as private stallion. Yearlings sold annually.

DAREBIN, brown horse, foaled 1878, bred by Mr. S. Gardiner, of Bundoora Park, Australia, by The Peer, an English-bred horse, by Melbourne, dam Cinizelli, and full brother to Marchioness, the Oaks winner, by Touchstone. Larline, the dam of Darebin, was one of the best mares that was ever stripped for a race in Australia. She won the Australian Cup of 1875, the Christchurch Plate in 1872 and 1873, the Canterbury Cup of 1872 and 1873, and is the dam of Prometheus, Motea Maroondah, etc., the best race-horses of New Zealand. Darebin was one of the greatest, possibly the greatest, race-horses of the Southern Hemisphere. Traducer, the sire of Lurline, was by The Libel, son of Pantaloon and Pasquinade, sister to Touchstone. Darebin, at two years old, was so big and raw that he did not do much. He ran third for the Ascot Vale Stakes at the V. R. C. Autumn Meeting of 1881, the race being won by Royal Maid. At the same meeting, with 124 lbs., he was second to Coreena, 88 lbs., for the Second Nursery, Tourmaline, 106 lbs., third, and sixteen others behind them. At Adelaide he was unplaced, with 122 lbs., for the Nursery, won by Henrietta, 122 lbs., and the following day won the Two-Year-Old Handicap, 122 lbs., Result, the second horse, having 117 lbs., and Spectre 108 lbs., seven others unplaced. At three years old he began by winning the great Victoria Derby, Santa Claus second, Commotion and several others unplaced. Two days later Darebin essayed a great task. Mr. Dakem, his owner, started him for the Great Melbourne Cup, two miles (handicap), with 96 lbs., won by Zulu, a four-year old with only 80 lbs. on his back. At the same meeting he won the Mares' Produce Stakes, 5 lbs. penalty, beating Monmouth, Commotion, Santa Claus, etc. At the V. R. C. Summer Meeting he ran second to Respite for the Touraville Stakes, four others starting. He was third in the St. Leger to Commotion and Pell Mell; was second to Bathurst in the Grand Stand Stakes; was unplaced with 122 lbs., to Commotion's 119 lbs., for the Town Plate. At Adelaide he won the St. Leger, beating Dutchman. He also won the Birthday Cup of £1,000 with 120 lbs., beating Crystal, 4 years, 107 lbs., Topaz, 3 years, 100 lbs., and eleven others. At four years old Darebin ran unplaced with 135 lbs., to Little Jack, 3 years, 91 lbs., for the Caulfield Cup, twenty-eight starting. At Melbourne he won the Melbourne Stakes, beating Belmont, Larpent and seven others. In the Melbourne Cup he took up 135 lbs. and ran unplaced to Assyrian, a five-year old, with 111 lbs. At the same meeting won the Royal Park Stakes, beating Lord Burghley, Liberator and three others, and was second to Commotion for the Canterbury Plate, beating Smart William and two others. Darebin was purchased by Mr. E. C. Cox for stud purposes, and the great horse crowned a splendid career by landing the Sydney Cup, although he had up 134 lbs. In this race he gave 30 lbs. to Mistaken, who finished second, and 20 lbs. to Willeroo, the third horse, and conceded 47 lbs. to the minimum weight, Musjid, an aged gelding.

At the death of Mr. Cox his stud was sold at auction, and Darebin was held in reserve at 1,000 guineas. He was soon after sold privately to Mr. Alexander Gordon, of the Grange Stud, Ipswich, Queensland, through his agents, Messrs. Ryan, Hammond & Donker. He made the season of 1885 at the Grange Stud. It was at this time that, failing to secure Commotion, Mr. Haggin made an offer of $15,000 for Darebin, which was refused, and Mr. Haggin purchased Sir Modred, the crack racer of New Zealand. His Antipodean purchase pleased him, as the additional purchase of Darebin would indicate.

Darebin is probably nearer to old English Melbourne than any other stallion living, and this alone should make his blood eagerly sought after, as the Melbourne blood, as shown through West Australian, Sir Tatton Sykes, Blink Bonny, Mentmore Lass, Barcaldine, imp. Australian, etc., etc., has done so exceedingly well.

Darebin has had no opportunity in the stud in comparison with Sir Modred, but the number and success of his get which have run demonstrate that he needs only time to make as brilliant a reputation as a sire as he earned as a race-horse. Before leaving Australia he got many good animals, and here he is the sire of Ludwig, Kildeer (mile in 1.37½), Mephisto, Two Lips, Mileties, Bessie W., Blackmailer, Madge L., Garwood, Daring, Julien, Tom Watson, Iris colt, etc.

It is to the Oldfield mare to which trace in direct female line Trumpator (by Conductor), Touchstone, Buccaneer, imp. Leamington, Macaroni, Saccharometer, Wild Oats, Saraband, etc., etc.

DAREBIN.

BROWN HORSE, FOALED 1878; BRED BY MR. S. GARDINER, VICTORIA, NEW SOUTH WALES, AUSTRALIA.

- †THE PEER, b. h., 1855.
  - Melbourne, br. h., 1834.
    - Humphrey Clinker, b. h., 1822.
      - Comus, ch. h., 1809.
      - Clinkerina, br. m., 1812.
    - Daughter of, br. m., 1825.
      - Cervantes, b. h., 1806.
      - Daughter of, br. m., 1818.
  - Cinizelli, b. m., 1842.
    - Touchstone, br. h., 1831.
      - Camel, br. h., 1822.
      - Banter, br. m., 1826.
    - Brocade, br. m., 1831.
      - Pantaloon, ch. h., 1824.
      - Bombasine, br. m., 1817.
- LURLINE, br. m., 1869.
  - †Traducer, b. h., 1857.
    - The Libel, br. h., 1842.
      - Pantaloon, ch. h., 1824.
      - Pasquinade, br. m., 1829.
    - Arethusa, b. m., 1839.
      - Elis, ch. h., 1833.
      - Languid, b. m., 1831.
  - †Mermaid, ch. m., 1860.
    - King Tom, b. h., 1851.
      - Harkaway, ch. h., 1834.
      - Pocahontas, b. m., 1837.
    - Waterwitch, b. m., 1853.
      - The Flying Dutchman, br. h., 1846.
      - Evening Star, br. m., 1839.

Sorcerer, blk. h., 1796.	Trumpator 1782 (Conductor 1767) Y. Giantess 1790 by Diomed* 1777 (Florizel 1768)—Giantess 1769 by Matchem 1748 (Cade 1734) Molly Long Legs 1753 by Babraham 1740 (God. Ar. 1724) Dau. of Foxhunter 1727 (Brisk 1711).
Houghton Lass, b. m., 1801.	Sir Peter T. 1784 (Highflyer 1774) Alexina 1788 by King Fergus 1775 (Eclipse 1764)—Lardella 1780 by Y. Marske 1771 (Marske 1750) Dau. of Y. Cade 1747 (Cade 1734) Beaufremont's dam by Bro. to Fearnought (Bay Bolton 1705).
Clinker, br. h., 1805.	Sir Peter T. 1784 (Highflyer 1774)—Hyale 1795 by Phenomenon 1780 (Herod 1758)—Rally 1790 by Trumpator 1782 (Conductor 1767)—Fancy 1780 by Florizel 1768 (Herod 1758)—Sist. 1763 to Juno by Spectator 1749 (Crab 1722).
Pewet, b. m., 1786.	Tandem 1773 (Syphon 1750)—Termagant by Tantrum 1760 (Cripple 1750) Cantatrice by Sampson 1745 (Blaze 1733) Dau. 1757 of Regulus 1739 (God. Ar. 1724) The Ruby mare (Marske's dam) by Hutton's Blacklegs 1725 (Mulso Bay Turk).
Don Quixote, ch. h., 1781.	Eclipse 1764 (Marske 1750) Grecian Princess 1770 by W. Forester 1770 (Cr. Forester 1736) Dau. of The Coalition Colt 1739 (God. Ar. 1724)—Dau. of Bustard 1711 (Crab 1722)—Charming Molly 1742 by Second 1732 (Flying Childers 1715).
Evelina, br. m., 1791.	Highflyer 1774 (Herod 1758) Termagant by Tantrum 1760 (Cripple 1750) Cantatrice by Sampson 1745 (Blaze 1733) Dau. 1757 of Regulus 1739 (God. Ar. 1724) The Ruby mare (Marske's dam) by Hutton's Blacklegs 1725 (Mulso Bay Turk).
Gohunpus, b. h., 1802.	Gohanna 1790 (Mercury 1778) Catherine 1795 by Woodpecker 1773 (Herod 1758)—Camilla 1778 by Trentham 1766 (T. Sweepstakes 1743)—Coquette 1761 by the Compton Barb—Sist. 1743 to Regulus by God. Ar. 1724.
Daughter of, b. m., 1810.	Paynator 1791 (Trumpator 1782)—Circle 1802 by St. George 1789 (Highflyer 1774)—Abigail 1788 by Woodpecker 1773 (Herod 1758) Firetail 1772 by Eclipse 1764 (Marske 1750) Sist. 1761 to Contest by Blank 1740 (God. Ar. 1724).
Whalebone, br. h., 1807.	Waxy 1790 (Pot-8-o's 1773) Penelope 1798 by Trumpator 1782 (Conductor 1767)—Prunella 1788 by Highflyer 1774 (Herod 1758)—Promise 1768 by Snap 1750 (Snip 1736)—Julia 1756 by Blank 1740 (God. Ar. 1724).
Daughter of, b. m., 1812.	Selim 1802 (Buzzard* 1787) Maiden 1801 by Sir Peter T. 1784 (Highflyer 1774)—Dau. 1788 of Phenomenon 1780 (Herod 1758)—Matron 1782 by Florizel 1768 (Herod 1758)—Maiden 1770 by Matchem 1748 (Cade 1734).
Master Henry, b. h., 1815.	Orville 1799 (Beningbrough 1791)—Miss Sophia 1805 by Stamford 1794 (Sir Peter T. 1784) Sophia 1798 by Buzzard* 1787 (Woodpecker 1773)—Huncamunca 1787 by Highflyer 1774 (Herod 1758)—Cypher 1772 by Squirrel 1754 (Traveller 1735).
Boadicea, b. m., 1807.	Alexander 1782 (Eclipse 1764)—Brunette by Amaranthus 1766 (Old England 1741)—Mayfly 1771 by Matchem 1748 (Cade 1734)—Dau. 1755 of Anc. Starling 1738 (Starling 1727)—Look-at-me-Lads 1731 by Bristol Grasshopper (Byerly T.).
Castrel, ch. h., 1801.	Buzzard* 1787 (Woodpecker 1773)—Dau. of Alexander 1782 (Eclipse 1764) Dau. of Highflyer 1774 (Herod 1758)—Dau. of Alfred 1770 (Matchem 1748)—Dau. 1770 of Engineer 1756 (Sampson 1745).
Idalia, ch. m., 1815.	Peruvian 1806 (Sir Peter T. 1784)—Musidora 1804 by Meteor 1783 (Eclipse 1764)—Maid of all Work 1786 by Highflyer 1774 (Herod 1758)—Sist. 1771 to Tandem by Syphon 1750 (Squirt 1732)—Dau. of Regulus 1739 (God. Ar. 1724).
Thunderbolt, blk. h., 1806.	Sorcerer 1796 (Trumpator 1782)—Wowski 1797 by Mentor 1784 (Justice 1774)—Maria 1777 (Waxy's dam) by Herod 1758 (Tartar 1743)—Lisette 1772 by Snap 1750 (Snip 1736)—Miss Windsor 1754 by God. Ar. 1724.
Delta, br. m., 1810.	Alexander 1782 (Eclipse 1764)—Isis 1805 by Sir Peter T. 1784 (Highflyer 1774)—Ibis 1791 by Woodpecker 1773 (Herod 1758)—Isabella 1783 by Eclipse 1764 (Marske 1750)—Isabel's dam 1775 by Squirrel 1754 (Traveller 1735).
Castrel, ch. h., 1801.	Buzzard* 1787 (Woodpecker 1773)—Dau. of Alexander 1782 (Eclipse 1764)—Dau. of Highflyer 1774 (Herod 1758)—Dau. of Alfred 1770 (Matchem 1748)—Dau. 1770 of Engineer 1756 (Sampson 1745).
Idalia, ch. m., 1815.	Peruvian 1806 (Sir Peter T. 1784)—Musidora 1804 by Meteor 1783 (Eclipse 1764)—Maid-of-all-Work 1786 by Highflyer 1774 (Herod 1758)—Sist. 1771 to Tandem by Syphon 1750 (Squirt 1732)—Dau. of Regulus 1739 (God. Ar. 1724).
Camel, br. h., 1822.	Whalebone 1807 (Waxy 1790)—Dau. 1812 of Selim 1802 (Buzzard* 1787)—Maiden 1801 by Sir Peter T. 1784 (Highflyer 1774)—Dau. 1788 of Phenomenon 1780 (Herod 1758)—Matron 1782 by Florizel 1768 (Herod 1758).
Banter, br. m., 1826.	Master Henry 1815 (Orville 1799)—Boadicea 1807 by Alexander 1782 (Eclipse 1764)—Brunette by Amaranthus 1766 (Old England 1741) Mayfly 1771 by Matchem 1748 (Cade 1734)—Dau. 1755 of Anc. Starling 1738 (Starling 1727).
Langar, ch. h., 1817.	Selim 1802 (Buzzard* 1787)—Dau. 1808 of Walton 1799 (Sir Peter T. 1784)—Y. Giantess 1790 by Diomed* 1777 (Florizel 1768)—Giantess 1769 by Matchem 1748 (Cade 1734)—Molly Long Legs 1753 by Babraham 1740 (God. Ar. 1724).
Olympia, b. m., 1815.	Sir Oliver 1800 (Sir Peter T. 1784)—Scotilla 1795 by Anvil 1777 (Herod 1758)—Scota 1783 by Eclipse 1764 (Marske 1750)—Harmony 1775 by Herod 1758 (Tartar 1743)—Rutilia 1769 by Blank 1740 (God. Ar. 1724).
Cain, b. h., 1822.	Paulowitz 1813 (Sir Paul 1802)—Dau. 1810 of Paynator 1791 (Trumpator 1782)—Dau. 1805 of Delpini 1781 (Highflyer 1774)—Dau. 1795 of Y. Marske 1771 (Marske 1750)—Gentle Kitty 1774 by Silvio 1754 (Cade 1734).
Lydia, b. m., 1822.	Poulton 1805 (Sir Peter T. 1784)—Variety 1808 by Hyacinthus 1797 (Coriander 1786)—Sist. 1790 to Swordsman by Weazle 1776 (Herod 1758)—Dau. of Turk 1763 (Regulus 1739)—Dau. of Locust 1744 (Crab 1722).
Economist, b. h., 1825.	Whisker 1812 (Waxy 1790)—Floranthe 1818 by Octavian 1807 (Stripling 1795)—Caprice 1797 by Anvil 1777 (Herod 1758)—Madcap 1774 by Eclipse 1764 (Marske 1750)—Sappho 1763 by Blank 1740 (God. Ar. 1724).
Fanny Dawson, ch. m., 1823.	Nabocklish 1811 (Rugantino 1803)—Miss Tooley 1808 by Teddy The Grinder 1798 (Asparagus 1787)—Lady Jane 1796 by Sir Peter T. 1784 (Highflyer 1774)—Paulina 1778 by Florizel 1768 (Herod 1758)—Captive 1771 by Matchem 1748 (Cade 1734).
Glencoe,* ch. h., 1831.	Sultan 1816 (Selim 1802)—Trampoline 1825 by Tramp 1810 (Dick Andrews 1797)—Web 1808 by Waxy 1790 (Pot-8-o's 1773)—Penelope 1798 by Trumpator 1782 (Conductor 1767)—Prunella 1788 by Highflyer 1774 (Herod 1758).
Marpessa, b. m., 1830.	Muley 1810 (Orville 1799)—Clare 1824 by Marmion 1806 (Whiskey 1789)—Harpalice 1814 by Gohanna 1790 (Mercury 1778)—Amazon 1799 by Driver 1783 (Trentham 1766)—Fractious 1792 by Mercury 1778 (Eclipse 1764).
Bay Middleton, b. h., 1833.	Sultan 1816 (Selim 1802)—Cobweb 1821 by Phantom 1808 (Walton 1799)—Filagree 1815 by Soothsayer 1808 (Sorcerer 1796)—Web 1808 by Waxy 1790 (Pot-8-o's 1773)—Penelope 1798 by Trumpator 1782 (Conductor 1767).
Barbelle, b. m., 1836.	Sandbeck 1818 (Catton 1809)—Darioletta 1822 by Amadis 1807 (Don Quixote 1784)—Selima 1810 by Selim 1802 (Buzzard* 1787)—Dau. 1794 of Pot-8-o's 1773 (Eclipse 1764)—Editha 1781 by Herod 1758 (Tartar 1743).
Touchstone, br. h., 1831.	Camel 1822 (Whalebone 1807)—Banter 1826 by Master Henry 1815 (Orville 1799)—Boadicea 1807 by Alexander 1782 (Eclipse 1764)—Brunette by Amaranthus 1766 (Old England 1741)—Mayfly 1771 by Matchem 1748 (Cade 1734).
Bertha, ch. m., 1821.	Rubens 1805 (Buzzard* 1787)—Boadicea b. m. 1807 by Alexander 1782 (Eclipse 1764)—Brunette by Amaranthus 1766 (Old England 1741)—Mayfly b. m. 1771 by Matchem 1748 (Cade 1734)—Dau. blk. m. 1755 of Anc. Starling 1738 (Old Starling 1727)—Look-at-me-Lads ch. m. 1731 by Bristol Grasshopper (Byerly Turk)—Dau. of Sir Newton's Bay Ar.—Lady Mare by Old Pert (Ely Turk)—Dau. of St. Martin (Spanker)—Dau. of Sir E. Hale's Turk—Oldfield Mare.

DAY STAR.

WINNER OF THE BLUE RIBBON STAKES AT LEXINGTON, AND KENTUCKY DERBY AT LOUISVILLE, KY., IN 1878.

Will stand for mares at the farm of W. S. Payne, near Lexington, Ky. Terms at opening of season.

DAY STAR, by Star Davis, son of imp. Glencoe, bred by John M. Clay, Ashland Stud, Kentucky, foaled 1875, dam Squeez'em by Lexington, out of the noted Skedaddle by imp. Yorkshire, son of St. Nicholas by Emilius. Day Star started twice as a two-year old, in 1877, was unplaced in the Colt Stakes, three-quarters of a mile, at Lexington, won by Himyar in 1.16¾; ran second to Blue Eyes in the Sanford Stake, 1 mile at Louisville, Ky., in 1.45¼, beating Himyar. As a three-year old, won the Blue Ribbon Stake at Lexington, 1½ miles, in 2.40¼, defeating Leveler, Solicitor, Blue Eyes, Fortuna and six others. Louisville, Ky., won the Kentucky Derby, 1½ miles, in 2.37¼, beating Himyar, Leveler, Solicitor, McHenry and four others; ran dead heat with Solicitor for second place in Clark Stake, 2 miles, won by Leveler in 3.37, beating Blue Eyes. Nashville, Tenn., ran second to Milan, in purse race, 1 mile, 1.45¼, beating Parides. At four years old started eight times, ran a close, good second to the great mare Janet, 4 miles, in 7.25, he carrying 110 lbs. to her 115, he four and she six years old. In this race he beat Charley Bush, Minnoch and two others. He was then purchased by Charles Reed and put over timber. Baltimore, Md., won steeplechase, full course, carrying 154 lbs., beating Faustina and Judith. Saratoga, won steeplechase, about 2¼ miles, with 152 lbs., beating Capt. Franklin, Rose and Faustina; ran third in race on flat, 1½ miles, won by Turenne, in 2.45¾; was unplaced in flat race, 1¾ miles, won by Oden in 2.54½; won hurdle race, 1½ miles, over six hurdles, in 2.53½, beating Frank Short, The Stranger and Rose; ran second to Chimney-sweep, 2 miles, over eight hurdles, in 3.55, three others behind him, and ran third to Pomeroy and Lizzie D. in steeplechase at Baltimore, in the fall, about 1½ miles, in 4.03; he was unplaced in two other races over timber. At six years old started ten times, won five races, was second in three and third in one. Baltimore, ran dead heat and divided purse in steeplechase with Derby, beating Joe Hunt and Surprise. Sheepshead Bay, won steeplechase, short course, in 5.15, beating Highland Fling, Capt. Franklin and two others; won steeplechase over full course, in 5.42, beating Frank Short, Joe Hunt and two others. Jerome Park, won steeplechase over short course in 3.23, beating Trouble, Bertha and three others; won hurdle race, 1¾ miles, over seven hurdles, in 3.29, carrying 149 lbs., beating Ingomar, Frank Short, etc. In all the other races he was beaten by the best horses then on the turf, he carrying his full impost of weight. At seven years old won one race and was third in one. Sheepshead Bay, won steeplechase, short course, in 5.09¼, beating Ohio Boy, Ike Bonham and four others. Day Star is a bright chestnut, with roan flanks, both front ankles white, and one hind foot white half way to hock, and a narrow white stripe down the face; stands full 15½ hands, with great substance and power, is a highly shaped horse and very evenly balanced all over, exhibiting speed, strength and endurance in a remarkable degree. He is descended from one of the greatest and best racing families in America. On the side of his sire we have Wagner, Maria Wood, Heraldry, Wade Hampton, Yorktown, Capt. Travis, Balloon, True Blue and other famous horses. On the dam's side we have Magic, Madonna, Princeton, Daniel Boone, Skedaddle, Kentucky, Madeline and Maggie B. B., the dam of Harold and the great horse Iroquois, winner of both the St. Leger and Derby in England. Thus it is fair to presume that a horse possessing all the qualities of a good race-horse, and descended on both sire and dam's side from one of the best racing families in America, and tracing through the collateral branches to the most famous horses in England, must get race-horses, and we commend Day Star to the breeders of America.

Day Star has covered but few thoroughbred mares, but is sire of the winners Day Dream, Altair, Jack Star, King Star, Celia, winner at 7 furlongs, in 1.28, 1 mile in 1.41, $1\frac{1}{16}$ miles, in 1.49 and 1.49½, 1¼ miles, on muddy track, in 2.10¼.

DAY STAR.

CHESTNUT HORSE, FOALED 1875; BRED BY MR. JOHN M. CLAY, ASHLAND STUD, KENTUCKY.

Herod 1758—Tartar 1743—Partner 1718—Jigg—Byerly Turk, about 1685.

STAR DAVIS, b. h., 1849.	*Glencoe, ch. h., 1831.	Sultan, b. h., 1816.	Selim, ch. h., 1802.	Buzzard,* ch. h., 1787.	Woodpecker 1773 (Herod 1758)—Misfortune 1775 by Dux 1761 (Matchem 1748)—Curiosity 1760 by Snap 1750 (Snip 1736)—Dau. 1749 of Regulus 1739 (God. Ar. 1724)—Dau. of Bartlett's Childers (Darley Ar. 1689).
				Daughter of, b. m., —.	Alexander 1782 (Eclipse 1764)—Dau. of Highflyer 1774 (Herod 1758)—Dau. of Alfred 1770 (Matchem 1748)—Dau. 1770 of Engineer 1756 (Sampson 1745)—Bay Malton's dam 1753 by Cade 1734 (God. Ar. 1724).
			Bacchante, br. m., 1809.	Williamson's Ditto, b. h., 1800.	Sir Peter T. 1784 (Highflyer 1774)—Arethusa 1792 by Dungannon 1780 (Eclipse 1764)—Dau. 1777 of Prophet 1760 (Regulus 1739)—Virago 1764 by Snap 1750 (Snip 1736)—Dau. of Regulus 1739 (God. Ar. 1724).
				Sist. to Calomel, b. m., 1791.	Mercury 1778 (Eclipse 1764)—Dau. 1776 of Herod 1758 (Tartar 1743)—Folly 1771 by Marske 1750 (Squirt 1732)—Vixen 1753 by Regulus 1739 (God. Ar. 1724)—Dau. of Hutton's Spot 1728 (Hartley's Blind Horse).
		Trampoline, ch. m., 1825.	Tramp, b. h., 1810.	Dick Andrews, b. h., 1797.	Joe Andrews 1778 (Eclipse 1764)—Dau. 1790 of Highflyer 1774 (Herod 1758)—Dau. 1777 of Cardinal Puff 1760 (Babraham 1740)—Dau. 1768 of Tatler 1754 (Blank 1740)—Bay Snip 1746 by Snip 1736 (Flying Childers 1715).
				Daughter of, b. m., 1803.	Gohanna 1790 (Mercury 1778)—Fraxinella 1793 by Trentham 1766 (T. Sweepstakes 1743)—Dau. 1785 of Woodpecker 1773 (Herod 1758)—Everlasting 1775 by Eclipse 1764 (Marske 1750)—Hyaena 1762 by Snap 1750 (Snip 1736).
			Web, b. m., 1808.	Waxy, b. h., 1790.	Pot-8-o-'s 1773 (Eclipse 1764)—Maria 1777 by Herod 1758 (Tartar 1743)—Lisette 1772 by Snap 1750 (Snip 1736)—Miss Windsor 1751 by God. Ar. 1724—Dau. of Y. Belgrade (Belgrade Turk).
				Penelope, b. m., 1798.	Trumpator 1782 (Conductor 1767)—Prunella 1788 by Highflyer 1774 (Herod 1758)—Promise 1768 by Snap 1750 (Snip 1736)—Julia 1756 by Blank 1740 (God. Ar. 1724)—Spectator's dam 1735 by Partner 1718 (Jigg).
	Margaret Wood, b. m., 1840.	*Priam, b. h., 1827.	Emilius, b. h., 1820.	Orville, b. h., 1799.	Beningbrough 1791 (King Fergus 1775)—Evelina 1791 by Highflyer 1774 (Herod 1758)—Termagant by Tantrum 1760 (Cripple 1750)—Cantatrice by Sampson 1745 (Blaze 1733)—Dau. 1757 by Regulus 1739 (God. Ar. 1724).
				Emily, ch. m., 1810.	Stamford 1794 (Sir Peter T. 1784)—Dau. 1799 of Whiskey 1789 (Saltram* 1780)—Grey Dorimant 1782 by Dorimant 1772 (Otho 1760)—Dizzy 1757 by Blank 1740 (God. Ar. 1724)—Anc. Dizzy 1741 by Driver 1727 (Wynn Ar.).
			Cressida, b. m., 1807.	Whiskey, b. h., 1789.	Saltram* 1780 (Eclipse 1764)—Calash 1775 by Herod 1758 (Tartar 1743)—Teresa 1767 by Matchem 1748 (Cade 1734)—Brown Regulus 1759 by Regulus 1739 (God. Ar. 1724)—Miss Starling Jr. 1752 by Starling 1727 (Bay Bolton 1705).
				Y. Giantess, b. m., 1790.	Diomed* 1777 (Florizel 1768)—Giantess 1769 by Matchem 1748 (Cade 1734)—Molly Long Legs 1753 by Babraham 1740 (God. Ar. 1724)—Dau. of Cole's Foxhunter 1727 (Brisk 1711)—Sist. to Cato by Partner 1718 (Jigg).
		Maria West, b. m., 1827.	Marion, b. h., 1820.	Sir Archy, b. h., 1805.	*Diomed 1777 (Florizel 1768)—*Castianira 1796 by Rockingham 1781 (Highflyer 1774)—Tabitha 1782 by Trentham 1766 (T. Sweepstakes 1743)—Dau. of Bosphorus 1754 (Babraham 1740)—Dau. of Forester 1750 (Forester 1736).
				Daughter of, b. m., —.	*Citizen 1785 (Pacolet 1763)—Dau. of *Alderman 1787 (Pot-8-o-'s 1773)—Dau. of Ashe's Roebuck (*Sweeper)—Dau. of Haine's King Herod 1770 (*Fearnought 1755)—Dau. of Partner 1755 (*Traveller 1747).
			Ella Crump, br. m., —.	*Citizen, b. h., 1785.	Pacolet 1763 (Blank 1740)—Princess 1774 by Turk 1768 (Regulus 1739)—Fairy Queen 1762 by Y. Cade 1747 (Cade 1734)—Black Eyes 1740 by Crab 1722 (Alcock Ar.)—Warlock Galloway 1728 by Snake (Lister Turk).
				Daughter of, br. m., 1795.	Huntsman 1780 (*Mousetrap 1771)—Dau. of Symmes' Wildair 1767 (*Fearnought 1755)—Dau. of *Fearnought 1755 (Regulus 1739)—Dau. of *Janus 1746 (Janus 1739).
SQUEEZ'EM, ch. m., 1863.	Lexington, b. h., 1850.	Boston, ch. h., 1833.	Timoleon, ch. h., 1813.	Sir Archy, b. h., 1805.	*Diomed 1777 (Florizel 1768)—*Castianira 1796 by Rockingham 1781 (Highflyer 1774)—Tabitha 1782 by Trentham 1766 (T. Sweepstakes 1743)—Dau. of Bosphorus 1754 (Babraham 1740)—Dau. of Forester 1750 (Forester 1736).
				Daughter of ch. m., —.	*Saltram 1780 (Eclipse 1764)—Dau. of Symmes' Wildair 1767 (*Fearnought 1755)—Dau. of Tyler's Driver 1760 (*Othello)—Dau. of *Fallower 1761 (Blank 1740)—Dau. of *Vampire 1757 (Regulus 1739).
			Sist. to Tuckahoe, ch. m., 1814.	Ball's Florizel, ch. h., 1801.	*Diomed 1777 (Florizel 1768)—Dau. of *Shark 1771 (Marske 1750)—Dau. of Harris' Eclipse 1770 (*Fearnought 1755)—Dau. of *Fearnought 1755 (Regulus 1739)—Dau. of *Jolly Roger 1741 (Roundhead 1733)—Dau. of *Sober John 1748 (Rib 1736).
				Daughter of, b. m., 1799.	*Alderman 1787 (Pot-8-o-'s 1773—Dau. of *Clockfast 1774 (Gimcrack 1760)—Dau. of Symmes' Wildair 1767 (*Fearnought 1755)—Y. Kitty Fisher 1767 by *Fearnought 1755 (Regulus 1739)—*Kitty Fisher 1755 by Cade 1734 (God. Ar. 1724).
		Alice Carneal, b. m., 1836.	*Sarpedon, br. h., 1828.	Emilius, b. h., 1820.	Orville 1799 (Beningbrough 1791)—Emily 1810 by Stamford 1794 (Sir Peter T. 1784)—Dau. 1799 of Whiskey 1789 (Saltram* 1780)—Grey Dorimant 1781 by Dorimant 1772 (Otho 1760)—Dizzy 1757 by Blank 1740 (God. Ar. 1724).
				Icaria, b. m., 1824.	The Flyer 1814 (Vandyke Jr. 1808)—Parma 1813 by Dick Andrews 1797 (Joe Andrews 1778)—May 1804 by Beningbrough 1791 (King Fergus 1775)—Primrose 1787 by Mambrino 1768 (Engineer 1756)—Cricket 1776 by Herod 1758 (Tartar 1743).
			Rowena, ch. m., 1826.	Sumpter, ch. h., 1818.	Sir Archy 1805 (*Diomed 1777)—Robin mare by *Robin Redbreast 1796 (Sir Peter T. 1784)—Dau. 1788 of *Obscurity 1777 (Eclipse 1764)—Slamerkin 1768 by *Wildair 1753 (Cade 1734)—*Cub mare 1762 by Cub 1739 (Fox 1714).
				Lady Grey, b. m., 1817.	Robin Grey 1805 (*Royalist 1790)—Maria 1802 by Melzar 1791 (*Medley 1776)—Dau. of *Highflyer 1784 (Highflyer 1774)—Dau of *Fearnought 1755 (Regulus 1739)—Dau. of Ariel 1756 (*Traveler 1747).
	Skedaddle, ch. m., 1860.	*Yorkshire, b. h., 1834.	St. Nicholas, br. h., 1827.	Emilins, b. h., 1820.	Orville 1799 (Beningbrough 1791)—Emily 1810 by Stamford 1794 (Sir Peter T. 1784)—Dau. 1799 of Whiskey 1789 (Saltram* 1780)—Grey Dorimant 1781 by Dorimant 1772 (Otho 1760)—Dizzy 1757 by Blank 1740 (God. Ar. 1424).
				Sea-mew, b. m., 1815.	Scud 1804 (Beningbrough 1791)—Goosander 1805 by Hambletonian 1792 (King Fergus 1775)—Rally 1790 by Trumpator 1782 (Conductor 1767)—Fancy 1780 by Florizel 1768 (Herod 1758)—Sist. 1763 to Juno by Spectator 1749 (Crab 1722).
			Miss Rose*, b. m., 1826.	Tramp, b. h., 1810.	Dick Andrews 1797 (Joe Andrews 1778)—Dau. 1803 of Gohanna 1790 (Mercury 1778)—Fraxinella 1793 by Trentham 1766 (T. Sweepstakes 1743)—Dau. 1785 of Woodpecker 1773 (Herod 1758)—Everlasting 1775 by Eclipse 1764 (Marske 1750).
				Daughter of, ch. m., 1810.	Sancho 1801 (Don Quixote 1784)—Blacklock's Dam 1799 by Coriander 1786 (Pot-8-o-'s 1773)—Wildgoose 1792 by Highflyer 1774 (Herod 1758)—Coheiress 1786 by Pot-8-o-s 1773 (Eclipse 1764)—Manilla 1777 by Goldfinder 1764 (Snap 1750).
		Magnolia, ch. m., 1841.	*Glencoe, ch. h., 1831.	Sultan, b. h., 1816.	Selim 1802 (Buzzard* 1787)—Bacchante 1809 by W's Ditto 1800 (Sir Peter T. 1784)—Dau. 1791 of Mercury 1778 (Eclipse 1764)—Dau. 1776 of Herod 1758 (Tartar 1743)—Folly 1771 by Marske 1750 (Squirt 1732)—Vixen 1753 by Regulus 1739 (God. Ar. 1724).
				Trampoline, ch. m., 1825.	Tramp 1810 (Dick Andrews 1797)—Web 1808 by Waxy 1790 (Pot-8-o-'s 1773)—Penelope 1798 by Trumpator 1782 (Conductor 1767)—Prunella 1788 by Highflyer 1774 (Herod 1758)—Promise 1768 by Snap 1750 (Snip 1736).
			*Myrtle, ch. m., 1834.	Mameluke, b. h., 1824.	Partisan 1811 (Walton 1799)—Miss Sophia 1805 by Stamford 1794 (Sir Peter T. 1784)—Sophia 1798 by Buzzard* 1787 (Woodpecker 1773)—Huncamunca 1787 by Highflyer 1774 (Herod 1758)—Cypher 1772 by Squirrel 1754 (Traveller 1735).
				Bobadilla, br. m., 1825.	Bobadil 1813 (Rubens 1805)—Pythoness blk. m. 1813 by Sorcerer 1796 (Trumpator 1782)—Princess blk. m. 1796 by Sir Peter T. 1784 (Highflyer 1774)—Dau. br. m. 1789 by Dungannon 1780 (Eclipse 1764)—Dau. ch. m. 1783 of Turf 1760 (Matchem 1748)—Dau. b. m. 1779 by Herod 1758 (Tartar 1743)—Golden Grove ch. m. 1760 by Blank 1740 (God Ar. 1724)—Spinster ch. m. 1735 by Partner 1718 (Jigg)—Bay Bloody Buttocks b. m. 1729 by Bloody Buttocks—Sist. 1722 to Guy by Greyhound (Chillaby)—Brown Farewell br. m. 1710 by Makeless (Oglethorpe Ar.)—Dau. of Primmer (D'Arcy Yellow Turk)—Dau. of Place's White Turk—Dau. of Dodsworth—Layton Barb Mare.

IMPORTED DECEIVER.

WINNER OF THE EPSOM GRAND PRIZE.

Property of Messrs. S. C. Lyne & Co. Stands at Larchmont Stock Farm, Windom, Ky. Service by contract.

IMPORTED DECEIVER, bay horse, foaled 1880, bred by Her Majesty Queen Victoria, by Wenlock (winner of the St. Leger), son of Lord Clifden (winner of the St. Leger) by Newminster by Touchstone. His dam Boot and Saddle by Trumpeter, son of Orlando (winner of the Derby) by Touchstone, out of Rinderpest by Alarm, winner of the Cambridgeshire Handicap, son of Venison by Partisan. Wenlock is the sire of Sanda, the dam of Sainfoin, winner of the Derby in 1890. Adine, his (Deceiver's) grandam is by Slane, sire of Conyngham, winner of the 2,000 Guineas, and Merry Monarch, winner of the Derby, and The Princess, winner of the Oaks, son of Royal Oak by Catton. Alarm is sire of Torment, the dam of the Oaks winner Tormentor by King Tom, and dam of Laura, dam of Petrarch, winner of the 2,000 Guineas, St. Leger and Ascot Gold Cup, imp. Laureate and Rotherhill. Deceiver's great grandam is by imp. Glencoe. Deceiver was a good race-horse at two years old, ran third in the Chesterfield Stakes to Galliard and Export, Malibran, Touch-Me-Not behind him; ran second in Hulnaker Stakes to Reine Blanche, eight others behind him, including Blue Rock, and was unplaced in the Prendergast Stakes. At three years old ran third to Grandmaster and Ossian in the Newmarket Craven Stakes with eight others behind him; was unplaced in 2,000 Guineas, won by Galliard; won the Epsom Grand Prize, 1¼ miles, 115 lbs., beating Goldfield, Ossian, Energy and four others; was unplaced in the Ascot Derby; ran second to Galliard in the St. James Palace Stakes, 1 mile, beating Potosi; ran third in the Midsummer Stakes at Newmarket, and was unplaced in the Stewards' Cup at Goodwood and Chesterfield Cup at Ascot; ran second to Montroyd in Wharncliffe Stakes, 1 mile, at Doncaster, with three others behind him; was third in the Free Welter Handicap Plate at Brighton and third in the Ashcombe Handicap at Lewes. At four years old was unplaced in City and Suburban Handicap; ran second to Southampton in Princess of Wales Free Handicap at Sandown Park, with nine behind him; was third in the Epsom Town Plate and unplaced in Royal Stakes at Epsom. Deceiver combines all the best racing blood in his pedigree, and traces to the Childers mare, to which trace in direct female line Bustard (by Buzzard), Hampton (by Lord Clifden), Blair Athol and all the Queen Mary family, Petrarch, Pretender, Sir Bevys, Tristan, Carlton, Beauclerc, Rotherhill and all the Postmaster and Peggy family in America.

Deceiver is a success in the stud. Sire of Brownwood,Ceverton, Cruickshank, Deceit, Doncaster, Ducat, Falstaff, Fay S., Frank Kinny, Florence T., John Berkley, Judge Arkell, Marietta, Mirage, Peddler, Pretender, Purycar D., Ollie, Ragnarock, etc.

DECEIVER (PADLOCK).

BAY HORSE, FOALED 1880; BRED BY HER MAJESTY.

WENLOCK, b. h., 1869.	Lord Clifden, b. h., 1860.	Newminster, b. h., 1848.	Touchstone, br. h., 1831.	Camel, br. h., 1822.	Whalebone 1807 (Waxy 1790)—Dau. 1812 of Selim 1802 (Buzzard 1787)—Maiden 1801 by Sir Peter T. 1784 (Highflyer 1774)—Dau. 1788 of Phenomenon 1780 (Herod 1758)—Matron 1782 by Florizel 1768 (Herod 1758).
				Banter, br. m., 1826.	Master Henry 1815 (Orville 1799)—Boadicea 1807 by Alexander 1782 (Eclipse 1764)—Brunette by Amaranthus 1766 (Old England 1741)—Mayfly 1771 by Matchem 1748 (Cade 1734)—Dau. 1755 of Starling 1738 (Starling 1727).
			Bee's-wing, b. m., 1833.	Dr. Syntax, br. h., 1811.	Paynator 1791 (Trumpator 1782)—Dau. of Beningbrough 1791 (King Fergus 1775)—Jenny Mole 1787 by Carbuncle 1772 (Babr. Blank 1758)—Dau. 1773 of Prince T'Quassaw 1751 (Snip 1736)—Sultana 1759 by Regulus 1739 (God. Ar. 1724).
				Tomboy's dam, ch. m., 1817.	Ardrossan 1809 (John Bull 1789)—Lady Eliza 1813 by Whitworth 1805 (Agonistes 1797)—Dau. 1793 of Spadille 1784 (Highflyer 1774)—Sylvia 1783 by Y. Marske 1771 (Marske 1750)—Ferret 1765 by Broth. to Silvio (Cade 1734).
		The Slave, b. m., 1852.	Melbourne, br. h., 1834.	Humprey Clinker, b. h., 1822.	Comus 1809 (Sorcerer 1796)—Clinkerina 1812 by Clinker 1805 (Sir Peter T. 1784)—Pewet 1786 by Tandem 1773 (Syphon 1750)—Termagant by Tantrum 1760 (Cripple 1750)—Cantatrice by Sampson 1745 (Blaze 1733).
				Daughter of, b. m., 1825.	Cervantes 1806 (Don Quixote 1784)—Dau. 1818 of Golumpus 1802 (Gohanna 1790)—Dau. 1810 of Paynator 1791 (Trumpator 1782)—Circle 1802 by St. George 1789 (Highflyer 1774)—Abigail 1788 by Woodpecker 1773 (Herod 1758).
			Volley, b. m., 1845.	Voltaire, br. h. 1826.	Blacklock 1814 (Whitelock 1803)—Dau. 1816 of Phantom 1808 (Walton 1799)—Dau. 1802 of Overton 1788 (King Fergus 1775)—Dau. 1796 of Walnut 1786 (Highflyer 1774)—Dau. 1786 of Ruler 1777 (Y. Marske 1771).
				Martha Lynn, br. m., 1837.	Mulatto 1823 (Catton 1809)—Leda 1824 by Filho-da-Puta 1812 (Haphazard 1797)—Treasure 1809 by Camillus 1803 (Hambletonian 1792) Dau. 1804 of Hyacinthus 1797 (Coriander 1786)—Flora 1789 by King Fergus 1775 (Eclipse 1764).
	Mineral, ch. m., 1863.	Rataplan, ch. h., 1850.	The Baron, ch. h., 1842.	Ir. Birdcatcher ch. h., 1833.	Sir Hercules 1826 (Whalebone 1807)—Guiccioli 1823 by Bob Booty 1804 (Chanticleer 1787)—Flight 1809 by Ir. Escape 1802 (Commodore 1793)—Y. Heroine by Bagot 1780 (Herod 1758)—Heroine 1775 by Hero 1758 (Cade 1734).
				Echidna, b. m., 1838.	Economist 1825 (Whisker 1812)—Miss Pratt 1825 by Blacklock 1814 (Whitelock 1803)—Gadabout 1812 by Orville 1799 (Beningbrough 1791)—Minstrel 1803 by Sir Peter T. 1784 (Highflyer 1774)—Matron 1782 by Florizel 1768 (Herod 1758),
			Pocahontas, b. m., 1837.	Glencoe,* ch. h., 1831.	Sultan 1816 (Selim 1802)—Trampoline 1825 by Tramp 1810 (Dick Andrews 1797)—Web 1808 by Waxy 1790 (Pot-8-o-'s 1773)—Penelope 1798 by Trumpator 1782 (Conductor 1767)—Prunella 1788 by Highflyer 1774 (Herod 1758).
				Marpessa, b. m., 1830.	Muley 1810 (Orville 1799)—Clare 1824 by Marmion 1806 (Whiskey 1789)—Harpalice 1814 by Gohanna 1790 (Mercury 1778)—Amazon 1799 by Driver 1783 (Trentham 1766)—Fractious 1792 by Mercury 1778 (Eclipse 1764).
		Manganese, ch. m., 1853.	Ir. Birdcatcher, ch. h., 1833.	Sir Hercules, blk. h., 1826.	Whalebone 1807 (Waxy 1790)—Peri 1822 by Wanderer 1811 (Gohanna 1790)—Thalestris 1809 by Alexander 1782 (Eclipse 1764)—Rival 1800 by Sir Peter T. 1784 (Highflyer 1774)—Hornet 1790 by Drone 1777 (Herod 1758).
				Guiccioli, ch. m., 1823.	Bob Booty 1804 (Chanticleer 1787)—Flight 1809 by Ir. Escape 1802 (Commodore 1793)—Y. Heroine by Bagot 1780 (Herod 1758)—Heroine 1775 by Hero 1758 (Cade 1734)—Dau. 1766 of Snap 1750 (Snip 1736).
			Moonbeam, ch. m., 1838.	Tomboy, b. h., 1829.	Jerry 1821 (Smolensko 1810)—Dau. 1817 of Ardrossan 1809 (John Bull 1789)—Lady Eliza 1813 by Whitworth 1805 (Agonistes 1797)—Dau. 1793 of Spadille 1784 (Highflyer 1774)—Sylvia 1783 by Y. Marske 1771 (Marske 1750).
				Lunatic, b. m., 1818.	Prime Minister 1810 (Sancho 1801)—Maniac 1806 by Shuttle 1793 (Y. Marske 1771)—Anticipation 1802 by Beningbrough 1791 (King Fergus 1775)—Expectation 1779 by Herod 1758 (Tartar 1743)—Dau. 1758 of Skim 1746 (Starling 1727).
BOOT AND SADDLE, b. m., 1873.	Trumpeter, ch. h., 1856.	Orlando, b. h., 1841.	Touchstone, br. h., 1831.	Camel, br. h., 1822.	Whalebone 1807 (Waxy 1790)—Dau. 1812 of Selim 1802 (Buzzard* 1787)—Maiden 1801 by Sir Peter T. 1784 (Highflyer 1774)—Dau. 1788 of Phenomenon 1780 (Herod 1758)—Matron 1782 by Florizel 1768 (Herod 1758).
				Banter, br. m., 1826.	Master Henry 1815 (Orville 1799)—Boadicea 1807 by Alexander 1782 (Eclipse 1764)—Brunette by Amaranthus 1766 (Old England 1741)—Mayfly 1771 by Matchem 1748 (Cade 1734)—Dan. 1755 of Anc. Starling 1738 (Starling 1727).
			Vulture, ch. m., 1833.	Langar, ch. h., 1817.	Selim 1802 (Buzzard* 1787)—Dau. 1808 of Walton 1799 (Sir Peter T. 1784)—Y. Giantess 1790 by Diomed* 1777 (Florizel 1768)—Giantess 1769 by Matchem 1748 (Cade 1734)—Molly Long Legs 1753 by Babraham 1740 (God. Ar. 1724).
				Kite, b. m., 1821.	Bustard 1813 (Castrel 1801)—Olympia 1815 by Sir Oliver 1800 (Sir Peter T. 1784)—Scotilla 1795 by Anvil 1777 (Herod 1758)—Scota 1783 by Eclipse 1764 (Marske 1750)—Harmony 1775 by Herod 1758 (Tartar 1748).
		Cavatina, ch. m., 1845.	Redshank, b. h., 1833.	Sandbeck, b. h., 1818.	Catton 1809 (Golumpus 1802)—Orvillina 1804 by Beningbrough 1791 (King Fergus 1775)—Evelina 1791 by Highflyer 1774 (Herod 1758)—Termagant by Tantrum 1760 (Cripple 1750)—Cantatrice by Sampson 1745 (Blaze 1733).
				Johanna, ch. m., 1813.	Selim 1802 (Buzzard* 1787)—Dau. of Skyscraper 1786 (Highflyer 1774)—Dau. 1796 of Dragon* 1787 (Woodpecker 1773)—Fidget's dam 1777 by Matchem 1748 (Cade 1734)—Sist. 1770 to Sweetbriar by Syphon 1750 (Squirt 1732).
			Oxygen, b. m., 1828.	Emilius, b. h., 1820.	Orville 1799 (Beningbrough 1791)—Emily 1810 by Stamford 1794 (Sir Peter T. 1784)—Dau. 1793 of Whiskey 1789 (Saltram* 1780)—Grey Dorimant 1781 by Dorimant 1772 (Otho 1760)—Dizzy 1757 by Blank 1740 (God. Ar. 1724).
				Whizgig, ch. m., 1819.	Rubens 1805 (Buzzard* 1787)—Penelope 1798 by Trumpator 1782 (Conductor 1767)—Prunella 1788 by Highflyer 1774 (Herod 1758)—Promise 1768 by Snap 1750 (Snip 1736)—Julia 1756 by Blank 1740 (God. Ar. 1724).
	Rinderpest, — m., 1860.	Alarm, b. h., 1842.	Venison, br. h., 1833.	Partisan, b. h., 1811.	Walton 1799 (Sir Peter T. 1784)—Parasol 1800 by Pot-8-o-'s 1773 (Eclipse 1764)—Prunella 1788 by Highflyer 1774 (Herod 1758)—Promise 1768 by Snap 1750 (Snip 1736)—Julia 1756 by Blank 1740 (God. Ar. 1724).
				Fawn, br. m., 1823.	Smolensko 1810 (Sorcerer 1796)—Jerboa 1803 by Gohanna 1790 (Mercury 1778)—Camilla 1778 by Trentham 1766 (T. Sweepstakes 1743)—Coquette 1764 by Compton Barb—Sist. 1743 to Regulus by God. Ar. 1724.
			Southdown, b. m., 1836.	Defence, b. h., 1824.	Whalebone 1807 (Waxy 1790)—Defiance 1816 by Rubens 1805 (Buzzard* 1787)—Little Folly 1806 by Highland Fling 1798 (Spadille 1784)—Harriet 1799 by Volunteer 1780 Eclipse 1764)—Dau. 1785 of Alfred 1770 (Matchem 1748).
				Feltona, b. m., 1819.	X. Y. Z. 1808 (Haphazard 1797)—Janetta 1803 by Beningbrough 1791 (King Fergus 1775)—Dau. 1797 of Drone 1777 (Herod 1778)—Contessina 1787 by Y. Marske 1771 (Marske 1750)—Tuberose 1772 by Herod 1758 (Tartar 1748).
		Adine, b. m., 1849.	Slane, b. h., 1833.	Royal Oak, br. h., 1823.	Catton 1809 (Golumpus 1802)—Dau. 1818 of Smolensko 1810 (Sorcerer 1796)—Lady Mary 1800 by Beningbrough 1791 (King Fergus 1775)—Dau. of Highflyer 1774 (Herod 1758)—Dau. of Marske 1750 (Squirt 1732).
				Daughter of, b. m., 1819.	Orville 1799 (Beningbrough 1791)—Epsom Lass 1803 by Sir Peter T. 1784 (Highflyer 1774)—Alexina 1788 by King Fergus 1775 (Eclipse 1764)—Lardella 1780 by Y. Marske 1771 (Marske 1750)—Dau. of Y. Cade 1747 (Cade 1734).
			Daughter of, br. m., 1836.	Glencoe,* ch. h., 1831.	Sultan 1816 (Selim 1802)—Trampoline 1825 by Tramp 1810 (Dick Andrews 1797)—Web 1808 by Waxy 1790 (Pot-8-o-'s 1773)—Penelope 1798 by Trumpator 1782 (Conductor 1767)—Prunella 1788 by Highflyer 1774 (Herod 1758).
				Alea, br. m., 1828.	Whalebone 1807 (Waxy 1790)—Hazardess b. m. 1818 by Haphazard 1797 (Sir Peter T. 1784)—Dau. br. m. 1809 of Orville 1799 (Beningbrough 1791)—Spinetta b. m. 1792 by Trumpator 1782 (Conductor 1767)—Peggy b.m. 1778 by Herod 1758 (Tartar 1743)—Dau. of Snap 1750 (Snip 1736)—Dau. of Gower Stallion 1740 (God. Ar. 1724)—Dau. of Childers.

IMPORTED DONALD A.

Imported in utero. The Property of J. D. Lucas, Goodwood Stud, St. Louis, Mo. Service by private contract.

DONALD A., chestnut horse, foaled 1881, bred by Com. Kittson in the Erdenheim Stud, Pa., by Scottish Chief, son of Lord of the Isles by Touchstone and Miss Ann by The Little Known, son of Muley, dam imp. Algebra by Diophantus, son of Orlando by Touchstone, out of Beatrice by Voltigeur, etc. Scottish Chief, his sire, won the Biennial Stakes at Ascot, and Chesterfield Stakes at two years old. At three years old ran third to Blair Athol and Gen. Peel in the Derby, won the Biennial at Ascot and Ascot Gold Cup, beating Lord Clifden, The Ranger and others. Scottish Chief is the sire of Marie Stuart, winner of the Oaks and St. Leger, Childeric, winner of the Chesterfield, Prince of Wales and Prendergast Stakes, third in the Derby to Sefton and Insulaire, and second to Janette in the St. Leger, and won the Doncaster Stakes and Royal Stakes at Newmarket. Scottish Chief is the sire of the dams of Common, winner of the 2,000 Guineas, Derby and St. Leger, of Donovan, winner of the Derby and St. Leger, Melton, winner of the Derby and St. Leger, and Semolina, winner of the 1,000 Guineas. Algebra, his dam, is by Diophantus, winner of the 2,000 Guineas, grandam Beatrice, dam of Prestonpans, winner, at two years old, of the Alexandra, Prince of Wales, Nursery, Hopeful and Criterion Stakes. At three years old he beat Philammon and Petronel (winner of the 2,000 Guineas) in the Liverpool Autumn Cup, by Voltigeur, winner of the Derby and St. Leger, and sire of Vedette (winner of the 2,000 Guineas). Vedette is the sire of Galopin (Derby winner) and Voltigeur is sire of the dam of Kingcraft (Derby winner) and imp. Great Tom. Bribery, his third dam, is the dam of St. Albans, winner of the St. Leger and half sister to Lady Mary, the dam of Gang Forward, winner of the 2,000 Guineas, by The Libel, out of Splitvote, half sister to Miss Twickenham, the dam of Teddington, winner of the Derby, Ascot and Doncaster Cups, by St. Luke. Donald A. had a chequered career on the turf. He started in thirty-four races and won nine, at ⅝, ¾, 7 furlongs, 1 mile and over hurdles with high weights. Donald A. is a very highly bred horse and from a distinguished racing family in England, and traces through the best racing blood to the Burton Barb mare, from which descended in direct female line Whiskey, Selim, Rubens, Castrel, Catton, Whitelock, Blacklock, Sir Hercules, imp. Margrave, Harkaway, Voltigeur, Surplice, St. Albans, Lord Clifden, Mask, Pell Mell, Brag, Sheen, Gold, imp. Hurrah, imp. Ainderby, imp. Billet, imp. Eagle and Spread Eagle, Waverly (by imp. Australian), etc.

DONALD A.

CHESTNUT HORSE, FOALED 1881 (IMPORTED IN UTERO); BRED BY COM. N. W. KITTSON, ERDENHEIM STUD, PA.

SCOTTISH CHIEF, b. h., 1861.
- Lord of the Isles, b. h., 1852.
  - Touchstone, br. h., 1831.
    - Camel, br. h., 1822.
    - Banter, br. m., 1826.
  - Fair Helen, ch. m., 1843.
    - Pantaloon, ch. h., 1824.
    - Rebecca, b. m., 1831.
- Miss Ann, b. m., 1846.
  - The Little Known, b. h., 1836.
    - Muley, b. h., 1810.
    - Lacerta, b. m., 1816.
  - Bay Missy, b. m., 1842.
    - Bay Middleton, b. h., 1833.
    - Camilla, br. m., 1842.

*ALGEBRA, b. m., 1870.
- Diophantus, ch. h., 1858.
  - Orlando, b. h., 1841.
    - Touchstone, br. h., 1831.
    - Vulture, ch. m., 1833.
  - Equation, b. m., 1839.
    - Emilius, b. h., 1820.
    - Maria, b. m., 1827.
- Beatrice, b. m., 1860.
  - Voltigeur, br. h., 1847.
    - Voltaire, br. h., 1826.
    - Martha Lynn, br. m., 1837.
  - Bribery, ch. m., 1851.
    - The Libel, br. h., 1842.
    - Splitvote, ch. m., 1841.

Whalebone, br. h., 1807.	Waxy 1790 (Pot-8-o-'s 1773) Penelope 1798 by Trumpator 1782 (Conductor 1767) Prunella 1788 by Highflyer 1774 (Herod 1758)—Promise 1768 by Snap 1750 (Snip 1736)—Julia 1756 by Blank 1740 (God. Ar. 1724).
Daughter of, b. m., 1812.	Selim 1802 (Buzzard* 1787) Maiden 1801 by Sir Peter T. 1784 (Highflyer 1774) Dau. 1788 of Phenomenon 1780 (Herod 1758) Matron 1782 by Florizel 1768 (Herod 1758) Maiden 1770 by Matchem 1748 (Cade 1734).
Master Henry, b. h., 1815.	Orville 1799 (Beningbrough 1791) Miss Sophia 1805 by Stamford 1794 (Sir Peter T. 1784) Sophia 1798 by Buzzard* 1787 (Woodpecker 1773) Huncamunca 1787 by Highflyer 1774 (Herod 1758) Cypher 1772 by Squirrel 1754 (Traveller 1735).
Boadicea, b. m., 1807.	Alexander 1782 (Eclipse 1764) Brunette by Amaranthus 1766 (Old England 1741) Mayfly 1771 by Matchem 1748 (Cade 1734) Dau. 1755 of Anc. Starling 1738 (Starling 1727) Look-at-me-Lads 1731 by Brist. Grasshopper (Byerly T.)
Castrel, ch. h., 1801.	Buzzard* 1787 (Woodpecker 1773) Dau. of Alexander 1782 (Eclipse 1764) Dau. of Highflyer 1774 (Herod 1758) Dau. of Alfred 1770 (Matchem 1748) Dau. 1770 of Engineer 1756 (Sampson 1745).
Idalia, ch. m., 1815.	Peruvian 1806 (Sir Peter T. 1784)—Musidora 1804 by Meteor 1783 (Eclipse 1764) Maid-of-all-Work 1786 by Highflyer 1774 (Herod 1758)—Sist. 1771 to Tandem by Syphon 1750 (Squirt 1732) Dau. of Regulus 1739 (God. Ar. 1724).
Lottery, br. h., 1820.	Tramp 1810 (Dick Andrews 1797)—Mandane 1800 by Pot-8-o-'s 1773 (Eclipse 1764)—Y. Camilla 1787 by Woodpecker 1773 (Herod 1758) Camilla 1778 by Trentham 1766 (T. Sweepstakes 1743) Coquette 1765 by Compton Barb.
Daughter of, b. m., 1818.	Cervantes 1806 (Don Quixote 1784)—Anticipation 1802 by Beningbrough 1791 (King Fergus 1775) Expectation 1779 by Herod 1758 (Tartar 1743)—Dau. 1758 of Skim 1746 (Starling 1727)—Dau. 1751 of Janus 1738 (God. Ar. 1724).
Orville, b. h., 1799.	Beningbrough 1791 (King Fergus 1775) Evelina 1791 by Highflyer 1774 (Herod 1758) Termagant by Tantrum 1760 (Cripple 1750) Cantatrice by Sampson 1745 (Blaze 1733) Dau. 1757 of Regulus 1739 (God. Ar. 1724).
Eleanor, b. m., 1798.	Whiskey 1789 (Saltram* 1780) Y. Giantess 1790 by Diomed* 1777 (Florizel 1768) Giantess 1769 by Matchem 1748 (Cade 1734) Molly Long Legs 1753 by Babraham 1740 (God. Ar. 1724) Dau. of Foxhunter 1727 (Brisk 1711).
Zodiac, ch. h., 1801.	St. George 1789 (Highflyer 1774) Abigail 1788 by Woodpecker 1773 (Herod 1758) Firetail 1772 by Eclipse 1764 (Marske 1750) Dau. 1764 of Blank 1740 (God. Ar. 1724)—Naylor 1748 by Cade 1734 (God. Ar. 1724).
Jerboa, b. m., 1803.	Gohanna 1790 (Mercury 1778) Camilla 1778 by Trentham 1766 (T. Sweepstakes 1743)—Coquette 1764 by Compton Barb—Sist. 1719 to Regulus by God. Ar. 1724—Grey Robinson 1723 by Bald Galloway (St. Victor Barb).
Sultan, b. h., 1816.	Selim 1802 (Buzzard* 1787) Bacchante 1809 by W's Ditto 1800 (Sir Peter T. 1784)—Dau. 1791 of Mercury 1778 (Eclipse 1764) Dau. 1776 of Herod 1758 (Tartar 1743) Folly 1771 by Marske 1750 (Squirt 1732).
Cobweb, b. m., 1821.	Phantom 1808 (Walton 1799)—Filagree 1815 by Soothsayer 1808 (Sorcerer 1796)—Web 1808 by Waxy 1790 (Pot 8-o-'s 1773) Penelope 1798 by Trumpator 1782 (Conductor 1767)—Prunella 1788 by Highflyer 1774 (Herod 1758).
Y. Phantom, b. h., 1822.	Phantom 1808 (Walton 1799)—Emmeline 1817 by Waxy 1790 (Pot-8-o-'s 1773)—Sorcery 1808 by Sorcerer 1796 (Trumpator 1782) Cobbea 1802 by Skyscraper 1786 (Highflyer 1774)—Dau. 1788 of Woodpecker 1773 (Herod 1758).
Sister to Speaker, b. m., 1817.	Camillus 1803 (Hambletonian 1792) Dau. 1809 by Sancho 1801 (Don Quixote 1784)—Miss Hornpipe Teazle 1802 by Sir Peter T. 1784 (Highflyer 1774)—Hornpipe 1793 by Trumpator 1782 (Conductor 1767)—Luna 1779 by Herod 1758 (Tartar 1743).
Camel, br. h., 1822.	Whalebone 1807 (Waxy 1790)—Dau. 1812 of Selim 1802 (Buzzard* 1787)—Maiden 1801 by Sir Peter T. 1784 (Highflyer 1774)—Dau. 1788 of Phenomenon 1780 (Herod 1758)—Matron 1782 by Florizel 1768 (Herod 1758).
Banter, br. m., 1826.	Master Henry 1815 (Orville 1799)—Boadicea 1807 by Alexander 1782 (Eclipse 1764)—Brunette by Amaranthus 1766 (Old England 1741)—Mayfly 1771 by Matchem 1748 (Cade 1734)—Dau. 1755 of Starling 1738 (Starling 1727).
Langar, ch. h., 1817.	Selim 1802 (Buzzard* 1787)—Dau. 1808 of Walton 1799 (Sir Peter T. 1784)—Y. Giantess 1790 by Diomed* 1777 (Florizel 1768) Giantess 1769 by Matchem 1748 (Cade 1734)—Molly Long Legs 1753 by Babraham 1740 (God. Ar. 1724).
Kite, b. m., 1821.	Bustard 1813 (Castrel 1801)—Olympia 1815 by Sir Oliver 1800 (Sir Peter T. 1784)—Scotilla 1795 by Anvil 1777 (Herod 1758)—Scota 1783 by Eclipse 1764 (Marske 1750)—Harmony 1775 by Herod 1758 (Tartar 1743).
Orville, b. h., 1799.	Beningbrough 1791 (King Fergus 1775)—Evelina 1791 by Highflyer 1774 (Herod 1758)—Termagant by Tantrum 1760 (Cripple 1750)—Cantatrice by Sampson 1745 (Blaze 1733)—Dau. 1757 of Regulus 1739 (God. Ar. 1724).
Emily, ch. m., 1810.	Stamford 1794 (Sir Peter T. 1784) Dau. 1799 of Whiskey 1789 (Saltram* 1780)—Grey Dorimant 1781 by Dorimant 1772 (Otho 1760)—Dizzy 1757 by Blank 1740 (God. Ar. 1724)—Dizzy 1741 by Driver 1727 (Wynn Ar.).
Whisker, b. h., 1812.	Waxy 1790 (Pot-8-o-'s 1773)—Penelope 1798 by Trumpator 1782 (Conductor 1767)—Prunella 1788 by Highflyer 1774 (Herod 1758)—Promise 1768 by Snap 1750 (Snip 1736)—Julia 1756 by Blank 1740 (God. Ar. 1724).
Gibside Fairy, b. m., 1811.	Hermes 1790 (Mercury 1778) Vicissitude 1800 by Pipator 1786 (Imperator 1776)—Beatrice 1791 by Sir Peter T. 1784 (Highflyer 1774) Pyrrha 1771 by Matchem 1748 (Cade 1734)—Duchess 1748 by Whitenose 1742 (God. Ar. 1724).
Blacklock, b. h., 1814.	Whitelock 1803 (Hambletonian 1792) Dau. 1799 of Coriander 1786 (Pot-8-o-'s 1773)—Wildgoose 1792 by Highflyer 1774 (Herod 1758)—Coheiress 1786 by Pot-8-o-'s 1773 (Eclipse 1764)—Manilla 1777 by Goldfinder 1764 (Snap 1750).
Daughter of b. m., 1816.	Phantom 1808 (Walton 1799)—Dau. 1804 of Overton 1788 (King Fergus 1775) Dau. 1793 of Walnut 1786 (Highflyer 1774)—Dau. 1786 of Ruler 1777 (Y. Marske 1771)—Piracantha 1772 by Matchem 1748 (Cade 1734).
Mulatto, b. h., 1823.	Catton 1809 (Golumpus 1802)—Desdemona 1811 by Orville 1799 (Beningbrough 1791)—Fanny 1796 by Sir Peter T. 1784 (Highflyer 1774)—Dau. 1783 of Diomed* 1777 (Florizel 1768)—Desdemona 1770 by Marske 1750 (Squirt 1732).
Leda, b. m., 1824.	Filho-da-Puta 1812 (Haphazard 1797)—Treasure 1809 by Camillus 1803 (Hambletonian 1792)—Dau. 1804 of Hyacinthus 1797 (Coriander 1786) Flora 1789 by King Fergus 1775 (Eclipse 1764)—Atalanta 1769 by Matchem 1748 (Cade 1734).
Pantaloon, ch. h., 1824.	Castrel 1801 (Buzzard* 1787)—Idalia 1815 by Peruvian 1806 (Sir Peter T. 1784)—Musidora 1804 by Meteor 1783 (Eclipse 1764)—Maid-of-all-Work 1786 by Highflyer 1774 (Herod 1758)—Dau. 1771 of Syphon 1750 (Squirt 1732).
Pasquinade, br. m., 1829.	Camel 1822 (Whalebone 1807)—Banter 1826 by Master Henry 1815 (Orville 1799)—Boadicea 1807 by Alexander 1782 (Eclipse 1764)—Brunette by Amaranthus 1766 (Old England 1741)—Mayfly 1771 by Matchem 1748 (Cade 1734).
St. Luke, ch. m., 1833.	Bedlamite 1823 (Welbeck 1815) Eliza Leeds 1817 by Comus 1809 (Sorcerer 1796)—Helen 1811 by Hambletonian 1792 (King Fergus 1775)—Susan 1800 by Overton 1788 (King Fergus 1775)—Drowsy 1790 by Drone 1777 (Herod 1758).
Electress, ch. m., 1819.	Election 1804 (Gohanna 1790)—Dau. b. m. 1805 of Stamford 1794 (Sir Peter T. 1784)—Miss Judy b. m. 1784 by Alfred 1770 (Matchem 1748)—Manilla br. m. 1777 by Goldfinder 1764 (Snap 1750)—Dau. b. m. 1766 of Old England 1741 (God. Ar. 1724)—Dau. b. m. 1756 by Cullen Ar. 1745—Miss Cade ch. m. 1750 by Cade 1734 (God. Ar. 1724)—Miss Makeless b. m. 1737 by Y. Greyhound 1723 (Greyhound)—Sist. 1731 to Miss Barforth by Partner 1718 (Jigg)—Brown Woodcock br. m. 1717 by Woodcock (Bustler)—Lusty Thornton ch. m. by Croft's Bay Barb (Chillaby)—Chestnut Thornton ch. m. by Makeless (Oglethorpe Ar.)—Old Thornton by Brimmer (D'Arcy Yellow Turk)—Dau. of Dickey Pierson (Dodsworth)—Burton Natural Barb mare.

IMPORTED DUTCH ORGAN.

Property of Wm. Arnette, Sharon Stock Farm, Sharon, Ill. Service fee published annually.

Dutch Organ, bay horse, foaled 1878, bred by Mr. T. Whiting, by Lowlander, son of Dalesman by King Tom and Lufra by Windhound, son of Pantaloon, dam Tymbestere by The Rake, son of Wild Dayrell by Ion, second dam Timbrel by Rataplan, own brother to Stockwell, out of Belsay by Lanercost, son of Liverpool. Lowlander was a good race-horse, winner of the Great Warwickshire Handicap, Royal Hunt Cup at Ascot, Ascot Plate, Newport Cup twice, Stockbridge Cup, Stewards' Cup at Chester, Queen's Stand Plate at Ascot, Lenox Stakes at Goodwood, in which he defeated Trappist, etc., and other good races, and is the sire of Lowland Chief, winner of the Stewards' Cup at Goodwood and Royal Stakes at Epsom. Imp. The Rake, sire of his dam, is sire of the good colt The Reaper. Rataplan, sire of his grandam, is own brother to Stockwell, a superior race-horse, sire of Kettledrum (winner of the Derby) and of the dams of Apology (winner of the 1,000 Guineas, Oaks and St. Leger), Rigolboche (dam of Cremorne, winner of the Derby), and Mineral, dam of Wenlock (winner of the St. Leger), and Kisber (winner of the Derby). Belsay, his third dam, is by Lanercost, sire of Catharine Hayes (winner of the Oaks) Van Tromp (winner of the St. Leger) and Ellerdale, the dam of Ellington (winner of the Derby) and Summerside (winner of the Oaks). Belsay is half sister to the good horse Vindex and Vesta, dam of Sabinus (winner of the Cambridgeshire Handicap and Ascot Gold Cup). Garland, his fourth dam, is half sister to Gaiety, dam of Gamester (winner of the St. Leger). Dutch Organ ran second to Syringe in the first year of the 24th Zetland Biennial Stakes, 5 furlongs, with three behind him. At three years old ran second to Macadam in Silver Bells Handicap, 1 mile, at Paisley; second to Master Rat in Maiden Plate, 1 mile, at Ayr; second to Melodious in Plate, ¾ of a mile, at Ayr, and was third in Athole Handicap, 5 furlongs. At four years old won the Silver Bells Handicap, 1 mile, 122 lbs., at Paisley; won Glasgow Handicap Plate, 1 mile, 94 lbs., at same place, and same day won the Burgh Member's Plate, 1¼ miles, 131 lbs., giving 21 lbs. to Astronomy, who was second. At five years old ran third in National Hurdle Plate, 2 miles, to Ridotto and Attache, 145 lbs., with seven behind him. At Sandown Park won Claremont Steeplechase Handicap, 2 miles, 145 lbs., in field of nine; won Hampton Hurdle Handicap, 2 miles, at Packington, 151 lbs. At Nottingham won Oxton Hurdle Handicap, 2 miles, 155 lbs., defeating field of eight. Dutch Organ is 16 hands high, weighs 1,265 lbs., good bone and excellent disposition.

DUTCH ORGAN.

BAY HORSE, FOALED 1878; BRED BY MR. T. WHITING, ENGLAND.

Parent	Grandparent	Great-grandparent	G.-g.-grandparent	Ancestor	Pedigree
LOWLANDER, ch. h., 1870.	Dalesman, ch. h., 1863.	King Tom, b. h., 1851.	Harkaway, ch. h., 1834.	Economist, b. h., 1825.	Whisker 1812 (Waxy 1790)—Floranthe 1818 by Octavian 1807 (Stripling 1795) Caprice 1797 by Anvil 1777 (Herod 1758)—Madcap 1774 by Eclipse 1764 (Marske 1750) Sappho 1763 by Blank 1740 (God. Ar. 1724).
				Fanny Dawson, ch. m., 1823.	Nabockliish 1811 (Rugantino 1803)—Miss Tooley 1808 by Teddy the Grinder 1798 (Asparagus 1787) Lady Jane 1796 by Sir Peter T. 1784 (Highflyer 1774) Paulina 1778 by Florizel 1768 (Herod 1758) Captive 1771 by Matchem 1748 (Cade 1734).
			Pocahontas, b. m., 1837.	Glencoe,* ch. h., 1831.	Sultan 1816 (Selim 1802)—Trampoline 1825 by Tramp 1810 (Dick Andrews 1797) Web 1808 by Waxy 1790 (Pot-8-o's 1773) Penelope 1798 by Trumpator 1782 (Conductor 1767)—Prunella 1788 by Highflyer 1774 (Herod 1758).
				Marpessa, b. m., 1830.	Muley 1810 (Orville 1799)—Clare 1824 by Marmion 1806 (Whiskey 1789)—Harpalice 1814 by Gohanna 1790 (Mercury 1778) Amazon 1799 by Driver 1783 (Trentham 1766) Fractions 1792 by Mercury 1778 (Eclipse 1764).
		Agnes, br. m., 1848.	Pantaloon, ch. h., 1824.	Castrel, ch. h., 1801.	Buzzard* 1787 (Woodpecker 1773)—Dau. of Alexander 1782 (Eclipse 1764) Dau. of Highflyer 1774 (Herod 1758)—Dau. of Alfred 1770 (Matchem 1748)—Dau. 1770 of Engineer 1756 (Sampson 1745).
				Idalia, ch. m., 1815.	Peruvian 1806 (Sir Peter T. 1784) Musidora 1804 by Meteor 1783 (Eclipse 1764)—Maid-of-all-Work 1786 by Highflyer 1774 (Herod 1758)—Dau. 1771 of Syphon 1750 (Squirt 1732)—Dau. of Regulus 1739 (God. Ar. 1724).
			Black Agnes, blk. m., 18[illegible].	Velocipede, ch. h., 1825.	Blacklock 1814 (Whitelock 1803) Dau. 1817 of Juniper 1805 (Whiskey 1789) Dau. 1810 of Sorcerer 1796 (Trumpator 1782)—Virgin 1801 by Sir Peter T. 1784 (Highflyer 1774) Dau. 1791 of Pot-8-o's 1773 (Eclipse 1764).
				Daughter of, br. m., 1808.	Walton 1799 (Sir Peter T. 1784) Y. Noisette 1789 by Diomed* 1777 (Florizel 1768) Noisette 1774 by Squirrel 1754 (Traveller 1735)—Carina 1766 by Marske 1750 (Squirt 1732) Dau. 1755 of Blank 1740 (God. Ar. 1724).
	Lufra, br. m., 1860.	Windhound, br. h., 1847.	Pantaloon, ch. h., 1824.	Castrel, ch. h., 1801.	Buzzard* 1787 (Woodpecker 1773)—Dau. of Alexander 1782 (Eclipse 1764) Dau. of Highflyer 1774 (Herod 1758)—Dau. of Alfred 1770 (Matchem 1748)—Dau. 1770 of Engineer 1756 (Sampson 1745).
				Idalia, ch. m., 1815.	Peruvian 1806 (Sir Peter T. 1784)—Musidora 1804 by Meteor 1783 (Eclipse 1764) Maid-of-all-Work 1786 by Highflyer 1774 (Herod 1758) Dau. 1771 of Syphon 1750 (Squirt 1732) Dau. of Regulus 1739 (God. Ar. 1724).
			Phryne, br. m., 1840.	Touchstone, br. h., 1831.	Camel 1822 (Whalebone 1807) Banter 1826 by Master Henry 1815 (Orville 1799) Boadicea 1807 by Alexander 1782 (Eclipse 1764) Brunette by Amaranthus 1766 (Old England 1741) Mayfly 1771 by Matchem 1748 (Cade 1734).
				Decoy, b. m., 1830.	Filho-da-Puta 1812 (Haphazard 1797)—Finesse 1815 by Peruvian 1806 (Sir Peter T. 1784)—Violante 1802 by John Bull 1789 (Fortitude 1777)—Dau. 1788 of Highflyer 1774 (Herod 1758)—Everlasting 1775 by Eclipse 1764 (Marske 1750).
		Maud, ch. m., 1854.	Loup-garou, br. h., 1846.	Lanercost, br. h., 1835.	Liverpool 1828 (Tramp 1810) Otis 1820 by Bustard 1801 (Buzzard* 1787) Dau. 1813 of Election 1804 (Gohanna 1790)—Dau. 1791 of Highflyer 1774 (Herod 1758) Chanticleer's dam 1778 by Eclipse 1764 (Marske 1750).
				Moonbeam, ch. m., 1838.	Tomboy 1829 (Jerry 1821) Lunatic 1818 by Prime Minister 1810 (Sancho 1801) Maniac 1806 by Shuttle 1793 (Y. Marske 1771) Anticipation 1802 by Beningbrough 1791 (King Fergus 1775)—Expectation 1779 by Herod 1758 (Tartar 1743).
			Venilia, b. m., 1834.	Velocipede, ch. h., 1825.	Blacklock 1814 (Whitelock 1803) Dau. 1817 of Juniper 1805 (Whiskey 1789) Dau. 1810 of Sorcerer 1796 (Trumpator 1782)—Virgin 1801 by Sir Peter T. 1784 (Highflyer 1774)—Dau. 1791 of Pot-8-o's 1773 (Eclipse 1764).
				Charity, b. m., 1824.	Tramp 1810 (Dick Andrews 1797)—Dau. 1809 of Shuttle 1793 (Y. Marske 1771) Dau. 1797 of Drone 1777 (Herod 1758)—Contessina 1787 by Y. Marske 1771 (Marske 1750)—Tuberose 1772 by Herod 1758 (Tartar 1743).
TYMBESTERE, b. m., 1874.	The Rake*, b. h., 1864.	Wild Dayrell, br. h., 1852.	Ion, br. h., 1835.	Cain, b. h., 1822.	Paulowitz 1813 (Sir Paul 1802)—Dau. 1810 of Paynator 1791 (Trumpator 1782) Dau. 1805 of Delpini 1781 (Highflyer 1774)—Dau. 1795 of Y. Marske 1771 (Marske 1750)—Gentle Kitty 1774 by Silvio 1754 (Cade 1734).
				Margaret, br. m., 1831.	Edmund 1824 (Orville 1799)—Medora 1811 by Selim 1802 (Buzzard* 1787)—Dau. 1803 by Sir Harry* 1795 (Sir Peter T. 1784)—Dau. 1793 of Volunteer 1780 (Eclipse 1764) Dau. 1779 of Herod 1758 (Tartar 1743).
			Ellen Middleton, br. m., 1846.	Bay Middleton, b. h., 1833.	Sultan 1816 (Selim 1802)—Cobweb 1821 by Phantom 1808 (Walton 1799)—Filagree 1815 by Soothsayer 1808 (Sorcerer 1796) Web 1808 by Waxy 1790 (Pot-8-o's 1773)—Penelope 1798 by Trumpator 1782 (Conductor 1767).
				Myrrha, ch. m., 1831.	Malek 1824 (Blacklock 1814)—Bessy 1815 by Y. Gouty 1805 (Gouty 1796)—Grandiflora 1810 by Sir Harry Dimsdale 1800 (Sir Peter T. 1784) Dau. 1799 of Pipator 1786 (Imperator 1776) Dau. 1796 of Phenomenon 1780 (Herod 1758).
		England's Beauty, ch. m., 1850.	Ir. Birdcatcher, ch. h., 1833.	Sir Hercules, blk. h., 1826.	Whalebone 1807 (Waxy 1790)—Peri 1822 by Wanderer 1811 (Gohanna 1790)—Thalestris 1809 by Alexander 1782 (Eclipse 1764)—Rival 1800 by Sir Peter T. 1784 (Highflyer 1774)—Hornet 1790 by Drone 1777 (Herod 1758).
				Guiccioli, ch. m., 1823.	Bob Booty 1804 (Chanticleer 1787)—Flight 1809 by Ir. Escape 1802 (Commodore 1793)—Y. Heroine by Bagot 1780 (Herod 1758) Heroine 1775 by Hero 1753 (Cade 1734)—Dau. 1766 of Snap 1750 (Snip 1736).
			The Prairie Bird, b. m., 1844.	Touchstone, br. h., 1831.	Camel 1822 (Whalebone 1807) Banter 1826 by Master Henry 1815 (Orville 1799) Boadicea 1807 by Alexander 1782 (Eclipse 1764)—Brunette by Amaranthus 1766 (Old England 1741)—Mayfly 1771 by Matchem 1748 (Cade 1734).
				Zillah, ch. m., 1835.	Reveller 1815 (Comus 1809)—Morisca 1826 by Morisco 1819 (Muley 1810)—Waltz 1822 by Election 1804 (Gohanna 1790) Penelope 1798 by Trumpator 1782 (Conductor 1767)—Prunella 1788 by Highflyer 1774 (Herod 1758).
	Timbrel, ch. m., 1864.	Rataplan, ch. h., 1850.	The Baron, ch. h., 1842.	Ir. Birdcatcher, ch. h., 1833.	Sir Hercules 1826 (Whalebone 1807)—Guiccioli 1823 by Bob Booty 1804 (Chanticleer 1787)—Flight 1809 by Ir. Escape 1802 (Commodore 1793)—Y. Heroine by Bagot 1780 (Herod 1758)—Heroine 1775 by Hero 1753 (Cade 1734).
				Echidna, b. m., 1838.	Economist 1825 (Whisker 1812) Miss Pratt 1825 by Blacklock 1814 (Whitelock 1803)—Gadabout 1812 by Orville 1799 (Beningbrough 1791) Minstrel 1803 by Sir Peter T. 1784 (Highflyer 1774)—Matron 1782 by Florizel 1768 (Herod 1758).
			Pocahontas, b. m., 1837.	Glencoe,* ch. h., 1831.	Sultan 1816 (Selim 1802)—Trampoline 1825 by Tramp 1810 (Dick Andrews 1797)—Web 1808 by Waxy 1790 (Pot-8-o's 1773)—Penelope 1798 by Trumpator 1782 (Conductor 1767)—Prunella 1788 by Highflyer 1774 (Herod 1758).
				Marpessa, b. m., 1830.	Muley 1810 (Orville 1799)—Clare 1824 by Marmion 1806 (Whiskey 1789) Harpalice 1814 by Gohanna 1790 (Mercury 1778) Amazon 1799 by Driver 1783 (Trentham 1766)—Fractions 1792 by Mercury 1778 (Eclipse 1764).
		Belsay, b. m., 1848.	Lanercost, br. h., 1835.	Liverpool, b. h., 1828.	Tramp 1810 (Dick Andrews 1797)—Dau. 1822 of Whisker 1812 (Waxy 1790) Mandane 1800 by Pot-8-o's 1773 (Eclipse 1764) Y. Camilla 1787 by Woodpecker 1773 (Herod 1758) Camilla 1778 by Trentham 1766 (T. Sweepstakes 1743).
				Otis, b. m., 1820.	Bustard 1801 (Buzzard* 1787)—Dau. 1813 of Election 1804 (Gohanna 1790)—Dau. 1791 of Highflyer 1774 (Herod 1758)—Dau. 1778 of Eclipse 1764 (Marske 1750)—Rosebud 1765 by Snap 1750 (Snip 1736).
			Garland, br. m., 1835.	Langar, ch. h., 1817.	Selim 1802 (Buzzard* 1787)—Dau. 1808 of Walton 1799 (Sir Peter T. 1784) Y. Giantess 1790 by Diomed* 1777 (Florizel 1768)—Giantess 1769 by Matchem 1748 (Cade 1734—Molly-Long-Legs 1753 by Babraham 1740 (God. Ar. 1724).
				Cast-steel, br. m., 1828.	Whisker 1812 (Waxy 1790)—The Twinkle b. m. 1821 by Walton 1799 (Sir Peter T. 1784)—Dau. gr. m. 1814 of Orville 1799 (Beningbrough 1791)—Lisette gr. m. 1806 by Hambletonian 1792 (King Fergus 1775)—Constantia gr. m. 1796 by Walnut 1786 (Highflyer 1774)—Contessina gr. m. 1787 by Y. Marske 1771 (Marske 1750)—Tuberose gr. m. 1772 by Herod 1758 (Tartar 1743)—Grey Starling gr. m. 1745 by Bolton Starling 1727 (Bay Bolton 1705)—

Coughing Polly b. m. 1736 by Bartlett's Childers (Darley Ar. 1699)—Sist. to Thunderbolt by Wood's Counsellor 1694 (D'Arcy Counsellor)—Dau. of Snake (Lister Turk)—Dau. of Luggs (D'Arcy White Turk)—Dau. of Davill's Old Woodcock.

ECUADOR.

WINNER OF THE LORILLARD STAKE, MONMOUTH PARK, 1884.

Property of Benj. F. Cockrill, Richland Stud, Nashville, Tenn.

ECUADOR, chestnut horse, foaled 1881, bred by Capt. Jas. Franklin, Kennesaw Stud, Tennessee, by Enquirer, son of imp. Leamington and Lida by Lexington, dam Nevada, the dam of the grand race-horse Luke Blackburn, winner of 22 out of 24 races at three years old, Greenland, Nova Zembla, Sierra Nevada, Bravo, etc., by Lexington, grandam Lightsome, dam of Salina (Salvator's dam), Sprightly (dam of Volturno, Elias Lawrence, Aretino, etc.), Crucifix (dam of Semper Felix, Leonatus' dam), and Semper Vive (dam of Fordham and Semper Idem, Longstreet's dam) by imp. Glencoe, out of Levity (dam of Lever, Legatee, Ruric, Mildred, dam of Monarchist), Minx (Monitor's dam), Stamps (dam of Katie Pearce), Sultana (dam of She, Padisha, etc.), Nora Worth (dam of Force, Startle, Renown, etc.), Brenna (dam of Strathmore), Brademante (dam of The Bard), Athalaric (dam of Tom Martin), Sister to Ruric (dam of Grinstead), Kelpie (dam of Janet, 4 miles in 7.25), Duke of Montrose, Miss Austine (dam of Teuton, Faraday, etc.), Ermengarde, dam of Brocade (dam of Burch, Jennie B., Bertha B. (dam of Lizzie Baker, Judge Murray, Pat Kelly, etc.). Levity was out of Vandal's dam by imp. Tranby, that produced Volga (the dam of Barney Williams, Evade, dam of Blink Bonny, the dam of Tullahoma, she the dam of Tulla Blackburn and the great race-horse Tammany, winner of the Withers, Realization and Lorillard Stakes in 1892). Ecuador ran during his turf career 109 races, won at 5 furlongs three-quarters of a mile at two years old, won at mile in 1.45, 141 lbs., won the Lorillard Stake, 1½ miles, 118 lbs., in 2.40¼, won over timber, 1½ miles, in 2.50¾, with 138 lbs., won hurdle race, 1 mile, 134 lbs., in 2.00, won hurdle races, 1½ miles, 160 lbs., and 2 miles, 150 lbs., and 1¼ miles, 163 lbs. He ran in and out of condition and was both a speedy and game horse. Ecuador is one of the nearest inbred Diomed horses living, his sire being out of a Lexington mare and his dam by Lexington, tracing back through Eclipse, a grandson of Diomed, and then to a daughter of the great Sir Archy, the Highflyer of America. On his dam's side he belongs to the great Levity family, the *ne plus ultra* of American pedigrees. This family has produced the greatest stallions of the last half century. Ecuador's dam will ever rank as one of the great matrons of America through her sons, Luke Blackburn, the "nonpareil," Greenland, Nova Zembla, Sierra Nevada, Bravo and others.

Ecuador is young in the stud, is the sire of the winners Eleanor, Forest Rose, Gov. Brown, Majestic, Salvation, Pete, etc.

Pot-8-o-'s 1773 Eclipse 1764—Marske 1750 Squirt 1732 Bartlett's Childers —Darley Arabian 1699.

ECUADOR.

CHESTNUT HORSE, FOALED 1881; BRED BY CAPT. JAS. FRANKLIN, KENNESAW STUD, TENNESSEE.

ENQUIRER, b. h., 1867.
*Leamington, br. h., 1853.
Faugh-a-Ballagh, br. h., 1841.
Sir Hercules, blk. h., 1826.
Guiccioli, ch. m., 1823.
Daughter of, b. m., 1841.
Pantaloon, ch. h., 1824.
Daphne, br. m., 1837.
Lida, b. m., 1858.
Lexington, b. h., 1850.
Boston, ch. h., 1833.
Alice Carneal, b. m., 1836.
Lize, ch. m., —.
American Eclipse, ch. h., 1814.
Gabriella, ch. m., 1833.

NEVADA, b. m., 1869.
Lexington, b. h., 1850.
Boston, ch. h., 1833.
Timoleon, ch. h., 1813.
Sist. to Tuckahoe, ch. m., 1811.
Alice Carneal, b. m., 1836.
*Sarpedon, br. h., 1828.
Rowena, ch. m., 1826.
Lightsome, ch. m., 1854.
*Glencoe, ch. h., 1831.
Sultan, b. h., 1816.
Trampoline, ch. m., 1825.
Levity, b. m., 1845.
*Trustee, ch. h., 1829.
Vandal's dam, b. m., 1837.

Whalebone, br. h., 1807.	Waxy 1790 (Pot-8-o-'s 1773)—Penelope 1798 by Trumpator 1782 (Conductor 1767)—Prunella 1788 by Highflyer 1774 (Herod 1758)—Promise 1768 by Snap 1750 (Snip 1736)—Julia 1756 by Blank 1740 (God. Ar. 1724).
Peri, b. m., 1822.	Wanderer 1811 (Gohanna 1790)—Thalestris 1809 by Alexander 1782 (Eclipse 1764)—Rival 1800 by Sir Peter T. 1784 (Highflyer 1774)—Hornet 1790 by Drone 1777 (Herod 1758)—Manilla 1777 by Goldfinder 1764 (Snap 1750).
Bob Booty, ch. h., 1804.	Chanticleer 1787 (Woodpecker 1773)—Ierne 1790 by Bagot 1780 (Herod 1758)—Dau. of Gamahoe (Bustard 1741)—Patty by Tim 1752 (Squirt 1732)—Miss Patch 1757 by Justice 1725 (H. C. Litton Ar.)
Flight, ch. m., 1809.	Ir. Escape 1802 (Commodore 1793)—Y. Heroine by Bagot 1780 (Herod 1758)—Heroine 1775 by Hero 1753 (Cade 1734)—Dau. 1766 of Snap 1750 (Snip 1736)—Sist. 1753 to Regulus by God. Ar. 1724.
Castrel, ch. h., 1801.	Buzzard* 1787 (Woodpecker 1773)—Dau. of Alexander 1782 (Eclipse 1764)—Dau. of Highflyer 1774 (Herod 1758)—Dau. of Alfred 1770 (Matchem 1748)—Dau. 1770 of Engineer 1756 (Sampson 1745).
Idalia, ch. m., 1815.	Peruvian 1806 (Sir Peter T. 1784)—Musidora 1804 by Meteor 1783 (Eclipse 1764)—Maid-of-all-Work 1786 by Highflyer 1774 (Herod 1758)—Sist. 1771 to Tandem by Syphon 1750 (Squirt 1732)—Dau. of Regulus 1739 (God. Ar. 1724).
Laurel, br. h., 1824.	Blacklock 1814 (Whitelock 1803)—Wagtail 1818 by Prime Minister 1810 (Sancho 1801)—Tranby*'s dam 1812 by Orville 1799 (Beningbrough 1791)—Miss Grimstone 1796 by Weazle 1776 (Herod 1758)—Dau. of Ancaster 1768 (Blank 1740).
Maid of Honor, br. m., 1829.	Champion 1812 (Selim 1802)—Etiquette 1820 by Orville 1799 (Beningbrough 1791)—Boadicea 1807 by Alexander 1782 (Eclipse 1764)—Brunette by Amaranthus 1766 (Old England 1741)—Mayfly 1771 by Matchem 1748 (Cade 1734).
Timoleon, ch. h., 1813.	Sir Archy 1805 (*Diomed 1777)—Dau. of *Saltram 1780 (Eclipse 1764)—Dau. of Symmes' Wildair 1767 (*Fearnought 1755)—Dau. of Tyler's Driver 1760 (*Othello)—Dau. of *Fallower 1761 (Blank 1740).
Sister to Tuckahoe, ch. m., 1814.	Ball's Florizel 1801 (*Diomed 1777)—Dau. 1799 of *Alderman 1787 (Pot-8-o-'s 1773)—Dau. of *Clockfast 1774 (Gimcrack 1760)—Dau. of Symmes' Wildair 1767 (*Fearnought 1755)—Y. Kitty Fisher 1767 by *Fearnought 1755 (Regulus 1739).
*Sarpedon, br. h., 1828.	Emilius 1820 (Orville 1799)—Icaria 1824 by The Flyer 1814 (Vandyke Jr. 1808)—Parma 1813 by Dick Andrews 1797 (Joe Andrews 1778)—May 1804 by Beningbrough 1791 (King Fergus 1775)—Primrose 1787 by Mambrino 1768 (Engineer 1756).
Rowena, ch. m., 1826.	Sumpter 1818 (Sir Archy 1805)—Lady Grey 1817 by Robin Grey 1805 (*Royalist 1790)—Maria 1802 by Melzar 1791 (*Medley 1776)—Dau. of *Highflyer 1784 (Highflyer 1774)—Dau. of *Fearnought 1755 (Regulus 1739).
Duroc, ch. h., 1806.	*Diomed 1777 (Florizel 1768)—Amanda 1800 by Grey Diomed 1786 (*Medley 1776)—Dau. of Va. Cade (Partner 1755)—Dau. of Hickman's Independence 1776 (*Fearnought 1755)—Dau. of *Badger (Bosphorus 1754).
Miller's Damsel, gr. m., 1802.	*Messenger 1780 (Mambrino 1768)—*Pot-8-o-'s mare 1792 by Pot-8-o-'s 1773 (Eclipse 1764)—Dau. 1778 of Gimcrack 1760 (Cripple 1750)—Snapdragon 1759 by Snap 1750 (Snip 1736)—Dau. 1749 of Regulus 1739 (God. Ar. 1724).
Sir Archy, b. h., 1805.	*Diomed 1777 (Florizel 1768)—*Castianira 1796 by Rockingham 1781 (Highflyer 1774)—Tabitha 1782 by Trentham 1766 (T. Sweepstakes 1743)—Dau. of Bosphorus 1754 (Babraham 1740)—Dau. of Forester 1750 (Croft's Forester 1736).
Calypso, b. m., 1805.	Bellair 1786 (*Medley 1776)—Dau. 1798 of *Dare Devil 1787 (Magnet 1770)—Dau. 1792 of Symmes' Wildair 1767 (*Fearnought 1755)—Picadilla by Y. Fearnought 1777 (*Fearnought 1755)—Dau. of Godolphin 1770 (*Fearnought 1755).
Sir Archy, b. h., 1805.	*Diomed 1777 (Florizel 1768)—*Castianira 1796 by Rockingham 1781 (Highflyer 1774)—Tabitha 1782 by Trentham 1766 (T. Sweepstakes 1743)—Dau. of Bosphorus 1754 (Babraham 1740)—Dau. of W's Forester 1750 (Forester 1736).
Daughter of, ch. m., —.	*Saltram 1780 (Eclipse 1764)—Dau. of Symmes' Wildair 1767 (*Fearnought 1755)—Dau. of Tyler's Driver 1760 (*Othello)—Dau. of *Fallower 1761 (Blank 1740)—Dau. of *Vampire 1757 (Regulus 1739).
Ball's Florizel, ch. h., 1801.	*Diomed 1777 (Florizel 1768)—Dau. of *Shark 1771 (Marske 1750)—Dau. of Harris' Eclipse 1770 (*Fearnought 1755)—Dau. of *Fearnought 1755 (Regulus 1739)—Dau. of *Jolly Roger 1741 (Roundhead 1733).
Daughter of, b. m., 1799.	*Alderman 1787 (Pot-8-o-'s 1773)—Dau. of *Clockfast 1774 (Gimcrack 1760)—Dau. of Symmes' Wildair 1767 (*Fearnought 1755)—Y. Kitty Fisher 1767 by *Fearnought 1755 (Regulus 1739)—*Kitty Fisher 1755 by Cade 1734 (God. Ar. 1724).
Emilius, b. h., 1820.	Orville 1799 (Beningbrough 1791)—Emily 1810 by Stamford 1794 (Sir Peter T. 1784)—Dau. 1799 of Whiskey 1789 (Saltram* 1780)—Grey Dorimant 1781 by Dorimant 1772 (Otho 1760)—Dizzy 1757 by Blank 1740 (God. Ar. 1724).
Icaria, b. m., 1824.	The Flyer 1814 (Vandyke Jr. 1808)—Parma 1813 by Dick Andrews 1797 (Joe Andrews 1778)—May 1804 by Beningbrough 1791 (King Fergus 1775)—Primrose 1787 by Mambrino 1768 (Engineer 1756)—Cricket 1776 by Herod 1758 (Tartar 1743).
Sumpter, ch. h., 1818.	Sir Archy 1805 (*Diomed 1777)—Dau. of *Robin Redbreast 1796 (Sir Peter T. 1784)—Dau. 1788 of *Obscurity 1777 (Eclipse 1764)—Slamerkin 1768 by *Wildair 1753 (Cade 1734)—*Cub mare 1762 by Cub 1739 (Fox 1714).
Lady Grey, b. m., 1817.	Robin Grey 1805 (*Royalist 1790)—Maria 1802 by Melzar 1791 (*Medley 1776)—Dau. of *Highflyer 1781 (Highflyer 1774)—Dau. of *Fearnought 1755 (Regulus 1739)—Dau. of Ariel 1756 (*Traveller 1747).
Selim, ch. h., 1802.	Buzzard* 1787 (Woodpecker 1773)—Dau. of Alexander 1782 (Eclipse 1764)—Dau. of Highflyer 1774 (Herod 1758)—Dau. of Alfred 1770 (Matchem 1748)—Dau. 1770 of Engineer 1756 (Sampson 1745).
Bacchante, br. m., 1809.	W's Ditto 1800 (Sir Peter T. 1784)—Dau. 1791 of Mercury 1778 (Eclipse 1764)—Dau. 1776 of Herod 1758 (Tartar 1743)—Folly 1771 by Marske 1750 (Squirt 1732)—Vixen 1753 by Regulus 1739 (God. Ar. 1724).
Tramp, b. h., 1810.	Dick Andrews 1797 (Joe Andrews 1778)—Dau. 1803 of Gohanna 1790 (Mercury 1778)—Fraxinella 1798 by Trentham 1766 (T. Sweepstakes 1743)—Dau. 1785 of Woodpecker 1773 (Herod 1758)—Everlasting 1775 by Eclipse 1764 (Marske 1750).
Web, b. m., 1808.	Waxy 1790 (Pot-8-o-'s 1773)—Penelope 1798 by Trumpator 1782 (Conductor 1767)—Prunella 1788 by Highflyer 1774 (Herod 1758)—Promise 1768 by Snap 1750 (Snip 1736)—Julia 1756 by Blank 1740 (God Ar. 1724).
Catton, b. h., 1809.	Golumpus 1802 (Gohanna 1790)—Lucy Gray 1804 by Timothy 1794 (Delpini 1781)—Lucy 1789 by Florizel 1768 (Herod 1758)—Frenzy 1774 by Eclipse 1764 (Marske 1750)—Dau. of Engineer 1756 (Sampson 1745).
Emma, ch. m., 1824.	Whisker 1812 (Waxy 1790)—Gibside Fairy 1811 by Hermes 1790 (Mercury 1778)—Vicissitude 1800 by Pipator 1786 (Imperator 1776)—Beatrice 1791 by Sir Peter T. 1784 (Highflyer 1774)—Pyrrha 1771 by Matchem 1748 (Cade 1734).
*Tranby, br. h., 1826.	Blacklock 1814 (Whitelock 1803)—Dau. 1812 of Orville 1799 (Beningbrough 1791)—Miss Grimstone 1796 by Weazle 1776 (Herod 1758)—Dau. of Ancaster 1768 (Blank 1740)—Dau. of Damascus Ar. 1754.
Lucilla, b. m., —.	Trumpator 1823 (Sir Solomon 1805)—Lucy b. m. 1821 by Orphan 1810 (Ball's Florizel 1801)—Lady Grey b. m. 1817 by Robin Grey 1805 (*Royalist 1790)—Maria 1802 by Melzar 1791 (*Medley 1776)—Dau. of *Highflyer 1781 (Highflyer 1774)—Dau. of *Fearnought 1755 (Regulus 1739)—Dau. of Ariel 1756 (*Traveller 1747)—Dau. of *Jack of Diamonds (Cullen Ar. 1745)—*Duchess by Cullen Ar. 1745—Lady Thigh ch. m. 1731 by Partner 1718 (Jigg)—Dau. ch. m. 1723 of Greyhound (Chillaby)—Sophonisba's dam by Curwen Bay Barb—Dau. of D'Arcy Chestnut Ar.—Dau. of Whiteshirt—Lord Darcy's Montagu mare by Old Montagu—Dau. of Hautboy (D'Arcy White Turk)—Dau. of Brimmer (D'Arcy Yellow Turk)—Royal mare.

EMPEROR.

The property of Pierre Lorillard, Rancocas Stud, Jobstown, N. J., and is a private stallion.

EMPEROR, bay horse, foaled 1881, bred in the Belle Meade Stud, Tennessee, by Enquirer, son of imp. Leamington and Lida by Lexington; regarded, upon performances and success in the stud, as the best son of Leamington. Emperor's dam Vesperlight, the dam of the superior race-mare Vandalite, that won the Sequel Stakes at Saratoga, 2 miles; the Dixie and Breckenridge Stakes at Baltimore, 2 miles (Dixie in 3.35½, and Breckenridge, with a penalty of 7 lbs., in 3.35, the fastest ever run to that date). Emperor won a Sweepstakes, 1½ miles, at Coney Island, in 1.58, his only start. He is exceptionally well bred, tracing to the imported mare Miss Shipton by the famous Waxy. Vesperlight (his dam) is also the dam of Nellie Ransom, and she the dam of Ferncliff, winner of the Withers Stakes. Vesperlight is also dam of Skylight by Jack Malone, and she the dam of Moonlight, the dam of Judge Morrow. Emperor is virtually untried in the stud, but his son Vestibule won the Galliard Stakes and showed not only speed but ability to stay. Sirocco is a fast and reliable colt.

Emperor's dam, Vesperlight, is by Childe Harold, son of imp. Sovereign by Emilius and Maria West, dam of Wagner by Marion, son of Sir Archy, grandam Budelight by imp. Glencoe, out of Gaslight by imp. Leviathan, son of Muley by Orville, tracing through Miss Shipton by Waxy, son of Pot-8-o-'s by Eclipse, to Maria, the dam of Waxy by Herod, to which trace in direct female line not only Waxy (Derby winner), but Paynator, Thunderbolt (by Sorcerer), Smolensko (2,000 Guineas and Derby), Colsterdale, Ellington (Derby winner), Ashton, Souvenir, Poussin, Bertram (sire of Robert the Devil, winner of the St. Leger), etc. The pedigree is strong in the best racing blood.

The following by him are all winners: Belshazzar, Chartreuse, Cyrus, Fagot, Happy Dan, Kirsch, 1 mile in 1.38, Padre, Pardon, Uncertainty, Wizard, Stalactite, Wang, Darius, Vestibule, Lorimer, Prince Imperial, Illian, Breeze filly, etc.

EMPEROR.

BROWN HORSE, FOALED 1881; BRED IN BELLE MEADE STUD, TENNESSEE.

Pot-8-o-'s 1773 Eclipse 1764—Marske 1750—Squirt 1732 Bartlett's Childers—Darley Arabian 1699.

ENQUIRER, b. h., 1867.

*Leamington, br. h., 1853.

Faugh-a-Ballagh, br. h., 1841.

Sir Hercules, blk. h., 1826.

Whalebone, br. h., 1807.	Waxy 1790 (Pot-8-o-'s 1773) Penelope 1798 by Trumpator 1782 (Conductor 1767) Prunella 1788 by Highflyer 1774 (Herod 1758) Promise 1768 by Snap 1750 (Snip 1736) Julia 1756 by Blank 1740 (God. Ar. 1724).
Peri, b. m., 1822.	Wanderer 1811 (Gohanna 1790) Thalestris 1809 by Alexander 1782 (Eclipse 1764) Rival 1800 by Sir Peter T. 1784 (Highflyer 1774)—Hornet 1790 by Drone 1777 (Herod 1758) Manilla 1777 by Goldfinder 1764 (Snap 1750).

Guiccioli, ch. m., 1823.

Bob Booty, ch. h., 1804.	Chanticleer 1787 (Woodpecker 1773) Ierne 1790 by Bagot 1780 (Herod 1758) Dau. of Gamahoe (Bustard 1741) Patty by Tim 1752 (Squirt 1732) Miss Patch 1737 by Justice 1725 (H. C. Litton Ar.)
Flight, ch. m., 1809.	Ir. Escape 1802 (Commodore 1793) Y. Heroine by Bagot 1780 (Herod 1758)—Heroine 1775 by Hero 1753 (Cade 1734) Dau. 1766 of Snap 1750 (Snip 1736) Sist. 1743 to Regulus by God. Ar. 1724.

Daughter of, b. m., 1841.

Pantaloon, ch. h., 1824.

Castrel, ch. h., 1801.	Buzzard* 1787 (Woodpecker 1773) Dau. of Alexander 1782 (Eclipse 1764) Dau. of Highflyer 1774 (Herod 1758) Dau. of Alfred 1770 (Matchem 1748) Dau. 1770 of Engineer 1756 (Sampson 1745).
Idalia, ch. m., 1815.	Peruvian 1806 (Sir Peter T. 1784)—Musidora 1804 by Meteor 1783 (Eclipse 1764)—Maid-of-all-Work 1786 by Highflyer 1774 (Herod 1758) Sist. 1771 to Tandem by Syphon 1750 (Squirt 1732) Dau. of Regulus 1739 (God. Ar. 1724).

Daphne, br. m., 1837.

Laurel, br. h., 1824.	Blacklock 1814 (Whitelock 1803) Wagtail 1818 by Prime Minister 1810 (Sancho 1801) Tranby*'s dam 1812 by Orville 1799 (Beningbrough 1791) Miss Grimstone 1796 by Weazle 1776 (Herod 1758) Dau. of Ancaster 1768 (Blank 1740).
Maid of Honor, br. m., 1829.	Champion 1812 (Selim 1802) Etiquette 1820 by Orville 1799 (Beningbrough 1791) Boadicea 1807 by Alexander 1782 (Eclipse 1764)—Brunette by Amaranthus 1766 (Old England 1741)—Mayfly 1771 by Matchem 1748 (Cade 1734).

Lida, b. m., 1858.

Lexington, b. h., 1850.

Boston, ch. h., 1833.

Timoleon, ch. h., 1813.	Sir Archy 1805 (*Diomed 1777) Dau. of *Saltram 1780 (Eclipse 1764)—Dau. of Symmes' Wildair 1767 (*Fearnought 1755)—Dau. of Tyler's Driver 1760 (*Othello)—Dau. of *Fallower 1761 (Blank 1740).
Sister to Tuckahoe, ch. m., 1814.	Ball's Florizel 1801 (*Diomed 1777)—Dau. 1790 of *Alderman 1787 (Pot-8-o-'s 1773)—Dau. of *Clockfast 1774 (Gimcrack 1760) Dau. of Symmes' Wildair 1767 (*Fearnought 1755) Y. Kitty Fisher 1767 by *Fearnought 1755 (Regulus 1739).

Alice Carneal, b. m., 1836.

*Sarpedon, br. h., 1828.	Emilius 1820 (Orville 1799)—Icaria 1824 by The Flyer 1814 (Vandyke Jr. 1808) Parma 1813 by Dick Andrews 1797 (Joe Andrews 1778)—May 1804 by Beningbrough 1791 (King Fergus 1775)—Primrose 1787 by Mambrino 1768 (Engineer 1756).
Rowena, ch. m., 1826.	Sumpter 1818 (Sir Archy 1805)—Lady Grey 1817 by Robin Grey 1805 (*Royalist 1790)—Maria 1802 by Melzar 1791 (*Medley 1776)—Dau. of *Highflyer 1784 (Highflyer 1774)—Dau. of *Fearnought 1755 (Regulus 1739).

Lize, ch. m., —.

American Eclipse, ch. h., 1814.

Duroc, ch. h., 1806.	*Diomed 1777 (Florizel 1768)—Amanda 1800 by Grey Diomed 1786 (*Medley 1776)—Dau. of Va. Cade (Partner 1755)—Dau. of Hickman's Independence 1776 (*Fearnought 1755)—Dau. of *Badger (Bosphorus 1754).
Miller's Damsel, gr. m., 1802.	*Messenger 1780 (Mambrino 1768)—*Pot-8-o-'s mare 1792 by Pot-8-o-'s 1773 (Eclipse 1764) Dau. 1778 of Gimcrack 1760 (Cripple 1750) Snapdragon 1759 by Snap 1750 (Snip 1736) Dau. 1749 of Regulus 1739 (God. Ar. 1724).

Gabriella, ch. m., 1826.

Sir Archy, b. h., 1805.	*Diomed 1777 (Florizel 1768)—*Castianira 1796 by Rockingham 1781 (Highflyer 1774)—Tabitha 1782 by Trentham 1766 (T. Sweepstakes 1743)—Dau. of Bosphorus 1754 (Babraham 1740)—Dau. of Forester 1750 (Croft's Forester 1736).
Calypso, b. m., 1805.	Bellair 1786 (*Medley 1776)—Dau. 1798 of *Dare Devil 1787 (Magnet 1770)—Dau. 1792 of Symmes' Wildair 1767 (*Fearnought 1755)—Picadilla by Y. Fearnought 1777 (*Fearnought 1755)—Dau. of Godolphin 1770 (*Fearnought 1755).

VESPERLIGHT, br. m., 1869.

Childe Harold, b. h., 1842.

*Sovereign, b. h., 1836.

Emilius, b. h., 1820.

Orville, b. h., 1799.	Beningbrough 1791 (King Fergus 1775)—Evelina 1791 by Highflyer 1774 (Herod 1758)—Termagant by Tantrum 1760 (Cripple 1750) Cantatrice by Sampson 1745 (Blaze 1733) Dau. 1757 of Regulus 1739 (God. Ar. 1724).
Emily, ch. m., 1810.	Stamford 1794 (Sir Peter T. 1784)—Dau. 1799 of Whiskey 1789 (Saltram* 1780)—Grey Dorimant 1781 by Dorimant 1772 (Otho 1760)—Dizzy 1757 by Blank 1740 (God. Ar. 1724)—Dizzy 1741 by Driver 1727 (Wynn Ar.).

Fleur-de-Lis, b. m., 1822.

Bourbon, b. h., 1811.	Sorcerer 1796 (Trumpator 1782)—Dau. 1797 of Precipitate* 1787 (Mercury 1778)—Y. Tiffany 1787 by Highflyer 1774 (Herod 1758)—Tiffany 1775 by Eclipse 1764 (Marske 1750)—Y. Hag 1761 by Skim 1746 (Starling 1727).
Lady Rachel, b. m., 1805.	Stamford 1794 (Sir Peter T. 1784)—Y. Rachel 1799 by Volunteer 1780 (Eclipse 1764)—Rachel 1790 by Highflyer 1774 (Herod 1758)—Dau. 1771 of Syphon 1750 (Squirt 1732)—Dau. of Regulus 1739 (God. Ar. 1724).

Maria West, b. m., 1827.

Marion, b. h., 1820.

Sir Archy, b. h., 1805.	*Diomed 1777 (Florizel 1768)—*Castianira 1796 by Rockingham 1781 (Highflyer 1774)—Tabitha 1782 by Trentham 1766 (T. Sweepstakes 1743)—Dau. of Bosphorus 1754 (Babraham 1740) Dau. of Forester 1750 (Forester 1736).
Daughter of, b. m., —.	*Citizen 1785 (Pacolet 1763) Dau. of *Alderman 1787 (Pot-8-o-'s 1773)—Dau. of Roebuck (*Sweeper)—Dau. of Haines' King Herod (*Fearnought 1755) Dau. of Partner 1755 (*Traveller 1747).

Ella Crump, br. m., —.

*Citizen, b. h., 1785.	Pacolet 1763 (Blank 1740)—Princess 1774 by Turk 1763 (Regulus 1739) Fairy Queen 1762 by Y. Cade 1747 (Cade 1734)—Black Eyes 1740 by Crab 1722 (Alcock Ar.)—Warlock Galloway 1728 by Snake (Lister Turk).
Daughter of, br. m., 1795.	Huntsman 1789 (*Monsetrap 1771)—Dau. of Symmes' Wildair 1767 (*Fearnought 1755)—Dau. of *Fearnought 1755 (Regulus 1739)—Dau. of *Janus 1746 (Janus 1739).

Budelight, ch. m., 1841.

*Glencoe, ch. h., 1831.

Sultan, b. h., 1816.

Selim, ch. h., 1802.	Buzzard* 1787 (Woodpecker 1773)—Dau. of Alexander 1782 (Eclipse 1764)—Dau. of Highflyer 1774 (Herod 1758)—Dau. of Alfred 1770 (Matchem 1748)—Dau. 1770 of Engineer 1756 (Sampson 1745).
Bacchante, br. m., 1809.	W's Ditto 1800 (Sir Peter T. 1784) Dau. 1791 of Mercury 1778 (Eclipse 1764)—Dau. 1776 of Herod 1758 (Tartar 1743)—Folly 1771 by Marske 1750 (Squirt 1732)—Vixen 1753 by Regulus 1739 (God. Ar. 1724).

Trampoline, ch. m., 1825.

Tramp, b. h., 1810.	Dick Andrews 1797 (Joe Andrews 1778)—Dau. 1803 of Gohanna 1790 (Mercury 1778)—Fraxinella 1793 by Trentham 1766 (T. Sweepstakes 1743)—Dau. 1785 of Woodpecker 1773 (Herod 1758)—Everlasting 1775 by Eclipse 1764 (Marske 1750).
Web, b. m., 1808.	Waxy 1790 (Pot-8-o-'s 1773)—Penelope 1798 by Trumpator 1782 (Conductor 1767)—Prunella 1788 by Highflyer 1774 (Herod 1758)—Promise 1768 by Snap 1750 (Snip 1736)—Julia 1756 by Blank 1740 (God. Ar. 1724).

Gaslight, ch. m., 1835.

***Leviathan, ch. h., 1823.**

Muley, b. h., 1810.	Orville 1799 (Beningbrough 1791)—Eleanor 1798 by Whiskey 1789 (Saltram* 1780) Y. Giantess 1790 by Diomed* 1777 (Florizel 1768)—Giantess 1769 by Matchem 1748 (Cade 1734) Molly Long Legs 1753 by Babraham 1740 (God. Ar. 1724).
Daughter of, blk. m., 1809.	Windle 1804 (Beningbrough 1791)—Dau. 1788 of Anvil 1777 (Herod 1758)—Virago 1764 by Snap 1750 (Snip 1736)—Dau. of Regulus 1739 (God. Ar. 1724)—Sist. 1750 to Othello by Crab 1722 (Alcock Ar.).

Pigeon, —, —.

Pacolet, gr. h., 1806.	*Citizen 1785 (Pacolet 1763)—Mary Grey by Tippoo Saib (Lindsay's Ar.)—Dau. of Goode's Brimmer 1779 (Harris' Eclipse 1770)—Dau. of *Silvereye (Cullen Ar.) Dau. of *Valiant 1756 (Dormouse 1738).
*Miss Shipton, about 1801.	Waxy 1790 (Pot-8-o-'s 1773)—Mother Shipton b. m. 1794 by Anvil 1777 (Herod 1758)—Jemima b. m. 1786 by Satellite 1774 (Eclipse 1764)—Maria (Waxy's dam) b. m. 1777 by Herod 1758 (Tartar 1743) Lisette b. m. 1772 by Snap 1750 (Snip 1736)—Miss Windsor b. m. 1754 by God. Ar. 1724—Sist. to Volunteer by Y. Belgrade (Belgrade Turk)—Dau. of Bartlett's Childers (Darley Ar. 1699)—Dau. of Devonshire Grey Turk—Sist. to Westbury by Curwen Bay Barb—Dau. of Curwen Old Spot (Selaby Turk)—Dau. of Old Woodcock (Bustler).

ENQUIRER.

WINNER OF THE PHŒNIX HOTEL AND CITIZEN STAKES AT LEXINGTON, CONTINENTAL HOTEL AND ROBINS STAKES AT LONG BRANCH, AND THE KENNER STAKES AT SARATOGA, ALL IN 1870.

Property of Gen. W. H. Jackson, Belle Meade Stud, near Nashville, Tenn. Application to Gen. W. H. Jackson, lock box 383, Nashville, Tenn. Yearling sales annually about May 1st.

ENQUIRER by imp. Leamington, son of Faugh-a-Ballagh, bred by H. F. Vissman, Louisville, Ky., dam Lida by Lexington, son of Boston, second dam Lize by American Eclipse, out of Gabriella, dam of George Martin by Garrison's Zinganee, son of Sir Archy. As a race-horse Enquirer ranks as the best son of Leamington, both as race-horse and sire. He started three times as a two-year old in 1869: was unplaced in Sweepstake at Cincinnati (1 mile), won by Hamburg in 1.45; ran unplaced in Willard Hotel Stakes at Louisville (1 mile) in 1.48, won by his stable companion, Lynchburg. Won George Elliot Stake for two-year olds at Memphis (1 mile), beating Hamburg, Lynchburg, Mundane and Kildare, in 1.48. In 1870, when three years old, ran six races and won them all. Lexington, won the Phœnix Hotel Stakes (mile heats) in 1.44¾, 1.44¼, beating Catina and distancing Longfellow in the second heat; same place won Citizens' Stakes (2-mile heats), in 3.35¾, 3.44½, beating Lyttleton: the first mile in the first heat was run in 1.43½. Cincinnati, won Sweepstakes for three-year olds (2-mile heats), in 3.50½, 3.39½, beating Conductor and Restless. Long Branch, won the Continental Hotel Stakes (mile heats), in 1.47, 1.49, 1.51¼, beating Lynchburg (who won the first heat and broke his shoulder at the start of the second), Maggie B. B. (dam of Iroquois), Haric, Susan Ann, dam of Thora, etc. Same place, won the Robin's Stakes (2-mile heats), beating Kingfisher, who won the first heat, and the Major, in 3.56½, 3.54, 4.00, track muddy. Saratoga, won the Kenner Stakes (2 miles), in 3.48¾, beating Hamburg, Telegram and Remorseless, value, $10,250, the most valuable Kenner ever run. Value of winnings at three years old, $20,800. His leg gave way in his four-year old form, and he was put into the stud, but trained again at seven years old, when he started in one race (2 miles) but was unplaced. The family from which he descends is a good racing one. His dam, Lida, and grandam, Lize, were not trained. Eclipse, the sire of his grandam, was the race-horse of his day; he beat all the best horses of his time, and won the great sectional match, the North against the South, run over the Union Course, L. I., May 27, 1823, for $20,000 a side, Eclipse beating Henry, who won the first heat, in 7.37½, 7.49, 8.24. Gabrielle, his great grandam, by the renowned Sir Archy, was a fine race-mare, and dam of the noted Geo. Martin, which beat Reel and Hannah Harris (4-mile heats), in 7.33, 7 43. Yet he has always held a high position as a winning sire. Amongst the best of his get, all winners, are Leander (Searcher), mile in 1.41¾: McWhirter, 2 miles, 3.30½; the fastest race ever run by a three-year old, Lizzie Whipps (2-mile heats), in 3.36½, 3.39; Harkaway (mile heats), in 1.43, 1.42½ (2-mile heats), in 3.39, 3.35½; Solicitor, 2 miles in 3.34½; Bill Bruce, ¾ mile in 1.15¼, mile heats, 1.43¾, 1.41, 2 miles in 3.36½; McHenry, Blue Eyes, 1⅛ miles in 1.55¼, and 2¼ miles in 3.58¾; Goodnight, ⅞ of mile in 1.28¾, mile, 1.42½, and 1⅛ miles, 120 lbs., in 1.55¼: Getaway, 1¼ miles in 2.07¾, fastest on record to that date, 2-mile heats in 3.36¼, 3.35½; Falsetto, 1¼ miles in 2.08¾, 2 miles in 3.35½, with 118 lbs.; Pinafore, 1 mile 500 yards, in 2.11: Inspector B., Egmont, Lizzie McWhirter, Fortuna, Faustus, Reporter, Banner Bearer, Bella B., Miss Ford, Burch, Basil Duke, Ernest, Telie Doe, Kitty T., Marie K., Nellie Van, Boodler, Nelly B., Kee-Vee-Na, Centipede, Ada A., Enterprise, Queer Toy, Envoy, Emperor, Eloise, Embargo, Mirabeau, Executor, Willow, Emigrant, Edisto, Jessica, Endymion, Equipoise, Napa, Everett, Economy, Editor, Erebus, Clay Pete, Endurer, Eight-to-Seven, Baldwin, Argenta, Bob Thomas, Tom Paxton, Late Fogle, Little Phil, Caligula, Revenge, Clarissima, Miss Ella, Mannie Grey, Adeline, Amy Hyde, Lizzie D., Enquiress, Zuleika, Sadie McNairy, Enigma, Drift, Engarita, Florence Anderson, Secret, Respond, etc. Enquirer traces through several crosses of Sir Archy to the great imp. Diomed, winner of the first English Derby, and a horse in which the best stock of America can be traced through uncontaminated sources. There is no better line through which to perpetuate and retain this blood, with the Whalebone and Pantaloon blood inherited through his sire. He is 16¼ hands high: girth, 73 in.: length of shoulder, 29¾ in.; circumference of arm, 22 in.: around the leg below the knee, 9 in.; from point of shoulder to turn of the buttocks, 69 in.: from point of hip to point of hock, 38½ in.; around the gaskins, 18 in., and weighs 1,200 lbs.

ENQUIRER.

BAY HORSE, FOALED 1867; BRED BY H. F. VISSMAN, LOUISVILLE, KY.

					Eclipse 1764—Marske 1750 Squirt 1732 Bartlett's Childers Darley Arabian 1699.
*LEAMINGTON, br. h., 1853.	Faugh-a-Ballagh, br. h., 1841.	Sir Hercules, blk. h., 1826.	Whalebone, br. h., 1807.	Waxy, b. h., 1790.	Pot-8-o-'s 1773 (Eclipse 1764)—Maria 1777 by Herod 1758 (Tartar 1743)—Lisette 1772 by Snap 1750 (Snip 1736)—Miss Windsor 1751 by God. Ar. 1724—Sist. to Volunteer by Y. Belgrade (Belgrade Turk).
				Penelope, b. m., 1798.	Trumpator 1782 (Conductor 1767)—Prunella 1788 by Highflyer 1774 (Herod 1758)—Promise 1768 by Snap 1750 (Snip 1736)—Julia 1756 by Blank 1740 (God. Ar. 1724)—Spectator's dam 1735 by Partner 1718 (Jigg).
			Peri, b. m., 1822.	Wanderer, b. h., 1811.	Gohanna 1790 (Mercury 1778)—Catherine 1795 by Woodpecker 1773 (Herod 1758)—Camilla 1778 by Trentham 1766 (T. Sweepstakes 1743)—Coquette 1764 by Compton Barb—Sist. 1743 to Regulus by God. Ar. 1724.
				Thalestris, blk. m., 1809.	Alexander 1782 (Eclipse 1764)—Rival 1800 by Sir Peter T. 1784 (Highflyer 1774)—Hornet 1790 by Drone 1777 (Herod 1758)—Manilla 1777 by Goldfinder 1764 (Snap 1750)—Dau. 1765 of Old England 1741 (God. Ar. 1724).
		Guiccioli, ch. m., 1823.	Bob Booty, ch. h., 1804.	Chanticleer, ch. h., 1787.	Woodpecker 1773 (Herod 1758)—Dau. 1778 of Eclipse 1764 (Marske 1750)—Rosebud 1765 by Snap 1750 (Snip 1736)—Miss Belsea 1753 by Regulus 1739 (God. Ar. 1724)—Dau. of Bartlett's Childers (Darley Ar. 1699).
				Ierne, ch. m., 1790.	Bagot 1780 (Herod 1758)—Dau. of Gamahoe (Bustard 1741)—Patty by Tim 1752 (Squirt 1732)—Miss Patch 1737 by Justice 1725 (H. C. Litton Ar.)—Ringtail Galloway by Curwen Bay Barb.
			Flight, ch. m., 1809.	Ir. Escape, ch. h., 1802.	Commodore 1793 (Tom Tug 1777)—Moll-in-the-Wad 1791 by Highflyer 1774 (Herod 1758)—Shift 1779 by Sweetbriar 1769 (Syphon 1750)—Black Susan 1762 by Snap 1750 (Snip 1736)—Dau. of Cade 1734 (God. Ar. 1724).
				Y. Heroine, —, —.	Bagot 1780 (Herod 1758)—Heroine 1775 by Hero 1753 (Cade 1734)—Dau. 1756 of Snap 1750 (Snip 1736)—Sist. 1743 to Regulus by God. Ar. 1724—Grey Robinson 1723 by Bald Galloway (St. Victor Barb).
	Daughter of, b. m., 1841.	Pantaloon, ch. h., 1824.	Castrel, ch. h., 1801.	Buzzard,* ch. h., 1787.	Woodpecker 1773 (Herod 1758)—Misfortune 1775 by Dux 1761 (Matchem 1748)—Curiosity 1760 by Snap 1750 (Snip 1736)—Dau. 1749 of Regulus 1739 (God. Ar. 1724)—Dau. of Bartlett's Childers (Darley Ar. 1699).
				Daughter of, b. m., —.	Alexander 1782 (Eclipse 1764)—Dau. of Highflyer 1774 (Herod 1758)—Dau. of Alfred 1770 (Matchem 1748)—Dau. 1770 of Engineer 1756 (Sampson 1745)—Bay Malton's dam 1756 by Cade 1734 (God. Ar. 1724).
			Idalia, ch. m., 1815.	Peruvian, b. h., 1806.	Sir Peter T. 1784 (Highflyer 1774)—Dau. 1788 of Boudrow 1777 (Eclipse 1764)—Dau. 1764 of Squirrel 1754 (Traveller 1735)—Sist. to Lowther's Babraham by Babraham 1740 (God. Ar. 1724)—Dau. 1750 of Golden Ball 1735 (Partner 1718).
				Musidora, b. m., 1804.	Meteor 1783 (Eclipse 1764)—Maid-of-all-Work 1786 by Highflyer 1774 (Herod 1758)—Dau. 1771 of Syphon 1750 (Squirt 1732)—Dau. of Regulus 1739 (God. Ar. 1724)—Sist. 1758 to Judgement by Snip 1736 (Flying Childers 1715).
		Daphne, br. m., 1837.	Laurel, br. h., 1824.	Blacklock, b. h., 1814.	Whitelock 1803 (Hambletonian 1792)—Dau. 1799 of Coriander 1786 (Pot-8-o-'s 1773)—Wildgoose 1792 by Highflyer 1774 (Herod 1758)—Coheiress 1786 by Pot-8-o-'s 1773 (Eclipse 1764)—Manilla 1777 by Goldfinder 1764 (Snap 1750).
				Wagtail, b. m., 1818.	Prime Minister 1810 (Sancho 1801)—Tranby*'s dam 1812 by Orville 1799 (Beningbrough 1791)—Miss Grimstone 1796 by Weazle 1776 (Herod 1758)—Dau. of Ancaster 1768 (Blank 1740)—Dau. of Damascus Ar. 1754.
			Maid of Honor, br. m., 1829.	Champion, ch. h., 1812.	Selim 1802 (Buzzard* 1787)—Podagra 1805 by Gouty 1796 (Sir Peter T. 1784)—Jet 1785 by Magnet 1770 (Herod 1758)—Jewel 1775 by Squirrel 1754 (Traveller 1735)—Sophia 1764 by Blank 1740 (God. Ar. 1724).
				Etiquette, br. m., 1820.	Orville 1799 (Beningbrough 1791)—Boadicea 1807 by Alexander 1782 (Eclipse 1764)—Brunette by Amaranthus 1766 (Old England 1741)—Mayfly 1771 by Matchem 1748 (Cade 1734)—Dau. 1755 of Starling 1738 (Starling 1727).
LIDA, b. m. 1858.	Lexington, b. h., 1850.	Boston, ch. h., 1833.	Timoleon, ch. h., 1813.	Sir Archy, b. h., 1805.	*Diomed 1777 (Florizel 1768)—*Castianira 1796 by Rockingham 1781 (Highflyer 1774)—Tabitha 1782 by Trentham 1766 (T. Sweepstakes 1743)—Dau. of Bosphorus 1754 (Babraham 1740)—Dau. of Forester 1750 (Forester 1736).
				Daughter of, ch. m., —.	*Saltram 1780 (Eclipse 1764)—Dau. of Symmes' Wildair 1767 (*Fearnought 1755)—Dau. of Tyler's Driver 1760 (*Othello)—Dau. of *Fallower 1761 (Blank 1740)—Dau. of *Vampire 1757 (Regulus 1739).
			Sist. to Tuckahoe, ch. m., 1814.	Ball's Florizel, ch. h., 1801.	*Diomed 1777 (Florizel 1768)—Dau. of *Shark 1771 (Marske 1750)—Dau. of Harris' Eclipse 1770 (*Fearnought 1755)—Dau. of *Fearnought 1755 (Regulus 1739)—Dau. of *Jolly Roger 1741 (Roundhead 1733).
				Daughter of, b. m., 1799.	*Alderman 1787 (Pot-8-o-'s 1773)—Dau. of *Clockfast 1774 (Gimcrack 1760)—Dau. of Symmes' Wildair 1767 (*Fearnought 1755)—Y. Kitty Fisher 1767 by *Fearnought 1755 (Regulus 1739)—*Kitty Fisher 1755 by Cade 1734 (God. Ar. 1724).
		Alice Carneal, b. m., 1836.	*Sarpedon, br. h., 1828.	Emilius, b. h., 1820.	Orville 1799 (Beningbrough 1791)—Emily 1810 by Stamford 1794 (Sir Peter T. 1784)—Dau. 1799 of Whiskey 1789 (Saltram* 1780)—Grey Dorimant 1781 by Dorimant 1772 (Otho 1760)—Dizzy 1757 by Blank 1740 (God. Ar. 1724).
				Icaria, b. m., 1824.	The Flyer 1814 (Vandyke Jr. 1808)—Parma 1813 by Dick Andrews 1797 (Joe Andrews 1778)—May 1804 by Beningbrough 1791 (King Fergus 1775)—Primrose 1787 by Mambrino 1768 (Engineer 1756)—Cricket 1776 by Herod 1758 (Tartar 1743).
			Rowena, ch. m., 1826.	Sumpter, ch. h., 1818.	Sir Archy 1805 (*Diomed 1777)—Dau. of *Robin Redbreast 1796 (Sir Peter T. 1784)—Dau. 1788 of *Obscurity 1777 (Eclipse 1764)—Slamerkin 1768 by *Wildair 1753 (Cade 1734)—*Cub mare 1762 by Cub 1739 (Fox 1714).
				Lady Grey, b. m., 1817.	Robin Grey 1805 (*Royalist 1790)—Maria 1802 by Melzar 1791 (*Medley 1776)—Dau. of *Highflyer 1784 (Highflyer 1774)—Dau. of *Fearnought 1755 (Regulus 1739)—Dau. of Ariel 1756 (*Traveller 1747).
	Lize, ch. m., —.	Am. Eclipse, ch. h., 1814.	Duroc, ch. h., 1806.	*Diomed, ch. h., 1777.	Florizel 1768 (Herod 1758)—Dau. 1763 of Spectator 1749 (Crab 1722)—Horatia 1758 by Blank 1740 (God. Ar. 1724)—Dau. of Fl. Childers 1715 (Darley Ar. 1699)—Miss Belvoir by Grey Grantham (Brownlow Turk).
				Amanda, ch. m., 1800.	Grey Diomed 1786 (*Medley 1776)—Dau. of Va. Cade (Partner 1755)—Dau. of Hickman's Independence 1776 (*Fearnought 1755)—Dau. of *Badger (Bosphorus 1754).
			Miller's Damsel, gr. m., 1802.	*Messenger, gr. h., 1780.	Mambrino 1768 (Engineer 1756)—Dau. 1774 of Turf 1760 (Matchem 1748)—Dau. 1761 of Regulus 1739 (God. Ar. 1724)—Dau. 1753 of Starling 1727 (Bay Bolton 1705)—Snap's dam 1739 by Fox 1714 (Clumsey).
				*Daughter of, ch. m., 1792.	Pot-8-o-'s 1773 (Eclipse 1764)—Dau. 1778 of Gimcrack 1760 (Cripple 1750)—Snap Dragon 1759 by Snap 1750 (Snip 1736)—Dau. 1749 of Regulus 1739 (Godolph. Ar. 1724)—Dau. of Bartlett's Childers (Darley Ar. 1699).
		Gabriella, ch. m., 1826.	Sir Archy, b. h., 1805.	*Diomed, ch. h., 1777.	Florizel 1768 (Herod 1758)—Dau. 1763 of Spectator 1749 (Crab 1722)—Horatia 1758 by Blank 1740 (God. Ar. 1724)—Dau. of Fl. Childers 1715 (Darley Ar. 1699)—Miss Belvoir by Grey Grantham (Brownlow Turk).
				*Castianira, br. m., 1796.	Rockingham 1781 (Highflyer 1774)—Tabitha 1782 by Trentham 1766 (T. Sweepstakes 1743)—Dau. of Bosphorus 1754 (Babraham 1740)—Dau. of W's Forester 1750 (Forester 1736)—Dau. of Coalition Colt 1739 (God. Ar. 1724).
			Calypso, b. m., 1805.	Bellair, gr. h., 1786.	*Medley 1776 (Gimcrack 1760)—Y. Selima by Yorick 1760 (*Traveller 1747)—Black Selima 1763 by *Fearnought 1755 (Regulus 1739)—*Selima 1746 by God. Ar. 1724—Large Hartley mare 1729 by Hartley's Blind Horse (Holderness Turk).
				Contention's dam, ch. m., 1798.	*Dare Devil 1787 (Magnet 1770)—Sallard mare ch. m. 1792 by Symmes' Wildair 1767 (*Fearnought 1755)—Picadillia ch. m. by Y. Fearnought 1777 (*Fearnought 1755)—Dau. of Godolphin 1770 (*Fearnought 1755)—Archibald Cary's mare by *Hob or Nob 1747 (Goliah 1730)—Dau. of *Jolly Roger 1741 (Roundhead 1733)—Dau. of *Valiant 1756 (Dormouse 1738)—Dau. of Tryall.

EOLUS.

WINNER OF THE CITY HOTEL STAKES AT BALTIMORE AND OTHER GOOD RACES.

R. J. Hancock, proprietor, Overton P. O., Va. Yearlings will be sold annually.

EOLUS by imp. Leamington, son of Faugh-a-Ballagh, bred by Major Thomas W. Doswell, Bullfield Stud, Hanover Junction, Va., foaled 1868, dam Fanny Washington, dam of Rappahannock, etc , by Revenue, son of imp. Trustee, cnt of Sarah Washington, the dam of Oratrix, Inspector, etc., by Garrison's Zinganee, son of Sir Archy. Fanny Washington, his dam, was one of the best mares of her day; was winner at all distances, 2 mile heats in 3.47¾, 3.40½, 3 mile heats in 5.43, 5.44½, defeating Tar River, and in 5.43¼, 5.33¾; she won 4 mile heats in 7.37, 8.02, defeating La Variete, Endorser and others, and beat the best ones on all the tracks; her full sister, Sue Washington, was also noted as winner at all distances. Revenue, her sire, won at all distances, and was the sire of the great Planet, Exchequer and other good race-horses. Sarah Washington, his grandam, was one of the best mares of her day, winner at all distances; 3 mile heats in 5.42, 5.39½, and beat the famous Lady Clifden, 4 mile heats. She produced Inspector, one of the best race-horses of his day, and Oratrix, that beat the noted Bostona. Eolus did not start as a two-year old, but won three out of the six races in which he started as a three-year old, a club purse, 2 miles, at Saratoga, in 3.38¾; a purse, 1¾ miles, in 3.14½, beating such horses as Tubman, Frogtown and others, and the City Hotel Stakes at Baltimore, 1½ miles, in the mud, in 2.55. He did not start as a four-year old, and only won three races at five years old; 1¼ miles at Baltimore, in 2.43¾; 1¾ miles at Saratoga, in 3.15, and one of 1⅛ miles, in 2.02¼. He was close up in several races in first-class company. When six years old he ran three races, won two, one of 2½ miles, at Baltimore, in 4.39, and one of 2-mile heats, in 3.40, 3.39¼, 3.36¾, Mart Jordan winning the second heat. This third heat is the fastest and best third heat ever run, and stamped him as a race-horse of the first water, possessing both speed and stamina. He had but one representative on the turf in 1881, Eole, and, although much out of condition during the year, he ran creditably in all his races and won some of great merit, notably his 1¾ miles, in 3.07½, 1 mile and 5 furlongs, in 2.49¾, 1 mile and a furlong, with 116 lbs., in 1.58¾, 1 mile and 3 furlongs in 2.24, 1¼ miles in 2.11½, 1½ miles in 2.39½. In 1882 Eole, although saddled with the top weights, won the great Metropolitan Stakes at Jerome Park, 1¼ miles, in 2.41, defeating General Monroe, Monitor, Parole, etc., and the Jockey Club Handicap, 2 miles, in 3.38½, carrying 123 lbs. to Monitor's 122, he being second. In the Coney Island Cup he forced Hindoo to run the first 2 miles in 3.29¾, and the 2¼ miles in 3.58, the fastest race at the distance during the year. Eole won the Monmouth Cup, 2¼ miles, in 4.07¾, beating Girofle and Monitor. He won the Champion Stakes, the Morrissey Stakes at Saratoga, and the Autumn Cup at Coney Island, 3 miles, in 5.26½, in which he defeated Lida Stanhope, Thora, Glenmore and four others. In this race he carried 120 lbs. and conceded all lumps of weight.

Eolus is the sire of the winners Eleve, St. Saviour, Eolist, Eolo, Elkwood, Knight of Ellerslie, Eolian, Calculus, Kathleen, Notus, Eolie, Starling, Spartan, Thorndale, Count-Me-In, Souvenir and the grand two-year old Morello, Diablo, Ninone, Young Grace, Charley Russell, Harry Russell, Eoline, Seedtick, Faustina, Bay Archer, Elkton, John Cavenagh, Albemarle, Amos, Russell, winner at ½ a mile in 46½ and 49¾, ¾ in 1.10, ⅝ in 1.02, Eurus, one mile in 1.41, 1½ miles in 1.55¾, 1¼ miles in 2.08, Royal Garter, Sterling, Raymond G., Little Jim and other winners.

Eolus is a dark bay horse, with star in his forehead and two white hind heels, full 15¾ hands, and weighs 1,100 lbs. He is a large, fine horse, of great substance and power, and in addition to his Waxy blood through Whalebone and Whisker, he has four distinct crosses of Sir Archy on his dam's side, and traces to Eclipse and Herod through his collateral branches. The stock all possess symmetry, constitution and staying powers.

It is to Old Bald Peg in direct female line that the following famous horses trace, Lath, Cade, Flying Childers, imp. Diomed, Sorcerer, imp. Priam, Phantom, Langar, Muley, Plenipotentiary, Grey Medoc, Thunder, Loadstone, Lightning, Lancaster, and all the Blue Bonnet family.

EOLUS.

BAY HORSE, FOALED 1868; BRED BY MAJ. TH. W. DOSWELL, BULLFIELD STUD, VIRGINIA.

*LEAMINGTON, br. h., 1853.
- Faugh-a-Ballagh, br. h., 1841.
  - Sir Hercules, blk. h., 1826.
    - Whalebone, br. h., 1807.
    - Peri, b. m., 1822.
  - Guiccioli, ch. m., 1823.
    - Bob Booty, ch. h., 1804.
    - Flight, ch. m., 1822.
- Daughter of, b. m., 1841.
  - Pantaloon, ch. h., 1821.
    - Castrel, ch. h., 1801.
    - Idalia, ch. m., 1815.
  - Daphne, br. m., 1837.
    - Laurel, br. h., 1824.
    - Maid of Honor, br. m., 1829.

FANNY WASHINGTON, ch. m., 1855.
- Revenue, b. h., 1843.
  - *Trustee, ch. h., 1829.
    - Catton, b. h., 1809.
    - Emma, ch. m., 1824.
  - Rosalie Somers, b. m., 1831.
    - Sir Charles, ch. h., 1816.
    - Mischief, b. m., 1825.
- Sarah Washington, b. m., 1837.
  - Garrison's Zinganee, b. h., 1825.
    - Sir Archy, b. h., 1805.
    - Miss Chance, —, —.
  - Stella, b. m., —.
    - Contention, ch. h., 1815.
    - Daughter of, —, —.

Waxy, b. h., 1790.	Pot-8-o-'s 1773 (Eclipse 1764)—Maria 1777 by Herod 1758 (Tartar 1743)—Lisette 1772 by Snap 1750 (Snip 1736)—Miss Windsor 1751 by God. Ar. 1724—Sist. to Volunteer by Y. Belgrade (Belgrade Turk).
Penelope, b. m., 1798.	Trumpator 1782 (Conductor 1767)—Prunella 1788 by Highflyer 1774 (Herod 1758)—Promise 1768 by Snap 1750 (Snip 1736)—Julia 1756 by Blank 1740 (God. Ar. 1724) Spectator's dam 1735 by Partner 1718 (Jigg).
Wanderer, b. h., 1811.	Gohanna 1790 (Mercury 1778) Catherine 1795 by Woodpecker 1773 (Herod 1758) Camilla 1778 by Trentham 1766 (T. Sweepstakes 1743)—Coquette 1764 by Compton Barb Sist. 1743 to Regulus by God. Ar. 1724.
Thalestris, blk. m., 1809.	Alexander 1782 (Eclipse 1764)—Rival 1800 by Sir Peter T. 1784 (Highflyer 1774) Hornet 1790 by Drone 1777 (Herod 1758)—Manilla 1777 by Goldfinder 1764 (Snap 1750) Dau. 1766 of Old England 1741 (God. Ar. 1724).
Chanticleer, ch. h., 1787.	Woodpecker 1773 (Herod 1758)—Dau. 1778 of Eclipse 1764 (Marske 1750)—Rosebud 1765 by Snap 1750 (Snip 1736)—Miss Belsea 1753 by Regulus 1739 (God. Ar. 1724)—Dau. of Bartlett's Childers (Darley Ar. 1699).
Ierne, ch. m., 1790.	Bagot 1780 (Herod 1758)—Dau. of Gamahoe (Bustard 1741)—Patty by Tim 1752 (Squirt 1732)—Miss Patch 1737 by Justice 1725 (H. C. Litton Ar.)—Ringtail Galloway by Curwen Bay Barb.
Ir. Escape, ch. h., 1802.	Commodore 1793 (Tom Tug 1777)—Moll-in-the-Wad 1791 by Highflyer 1774 (Herod 1751)—Shift 1779 by Sweetbriar 1769 (Syphon 1750)—Black Susan 1762 by Snap 1750 (Snip 1736) Dau. of Cade 1734 (God. Ar. 1724).
Y. Heroine, —, —.	Bagot 1780 (Herod 1758)—Heroine 1775 by Hero 1753 (Cade 1734)—Dau. 1766 of Snap 1750 (Snip 1736)—Sist. 1743 to Regulus by God. Ar. 1724—Grey Robinson 1723 by Bald Galloway (St. Victor Barb).
Buzzard,* ch. h., 1787.	Woodpecker 1773 (Herod 1758)—Misfortune 1775 by Dux 1761 (Matchem 1748)—Curiosity 1760 by Snap 1750 (Snip 1736)—Dau. 1749 of Regulus 1739 (God. Ar. 1724)—Dau. of Bartlett's Childers (Darley Ar. 1699).
Daughter of, b. m., —.	Alexander 1782 (Eclipse 1764)—Dau. of Highflyer 1774 (Herod 1758)—Dau. of Alfred 1770 (Matchem 1748)—Dau. 1770 of Engineer 1756 (Sampson 1745)—Bay Malton's dam 1756 by Cade 1734 (God. Ar. 1724).
Peruvian, b. h., 1806.	Sir Peter T. 1784 (Highflyer 1774)—Dau. 1788 of Bondrow 1777 (Eclipse 1764)—Dau. 1764 of Squirrel 1754 (Traveller 1735)—Sist. to Lowther's Babraham by Babraham 1740 (God. Ar. 1724)—Dau. 1750 of Golden Ball 1735 (Partner 1718).
Musidora, b. m., 1804.	Meteor 1783 (Eclipse 1764)—Maid-of-all-Work 1786 by Highflyer 1774 (Herod 1758)—Dau. 1771 of Syphon 1750 (Squirt 1732)—Dau. of Regulus 1739 (God. Ar. 1724)—Sist. 1758 to Judgement by Snip 1736 (Flying Childers 1715).
Blacklock, b. h., 1814.	Whitelock 1803 (Hambletonian 1792) Dau. 1799 of Coriander 1786 (Pot-8-o-'s 1773) Wildgoose 1792 by Highflyer 1774 (Herod 1758)—Coheiress 1786 by Pot-8-o-'s 1773 (Eclipse 1764)—Manilla 1777 by Goldfinder 1764 (Snap 1750).
Wagtail, b. m., 1818.	Prime Minister 1810 (Sancho 1801)—Tranby*'s dam 1812 by Orville 1799 (Beningbrough 1791) Miss Grimstone 1796 by Weazle 1776 (Herod 1758)—Dau. of Ancaster 1768 (Blank 1740)—Dau. of Damascus Ar. 1754.
Champion, ch. h., 1812.	Selim 1802 (Buzzard* 1787) Podagra 1805 by Gouty 1793 (Sir Peter T. 1784)—Jet 1785 by Magnet 1770 (Herod 1758)—Jewel 1775 by Squirrel 1754 (Traveller 1735)—Sophia 1764 by Blank 1740 (God. Ar. 1724).
Etiquette, br. m., 1820.	Orville 1799 (Beningbrough 1791) Boadicea 1807 by Alexander 1782 (Eclipse 1764)—Brunette by Amaranthus 1766 (Old England 1741)—Mayfly 1771 by Matchem 1748 (Cade 1734)—Dau. 1755 of Starling 1738 (Starling 1727).
Golumpus, b. h., 1802.	Gohanna 1790 (Mercury 1778) Catherine 1795 by Woodpecker 1773 (Herod 1758) Camilla 1778 by Trentham 1766 (T. Sweepstakes 1743)—Coquette 1764 by Compton Barb—Sist. 1743 to Regulus 1739 (God. Ar. 1724).
Lucy Gray, ch. m., 1804.	Timothy 1794 (Delpini 1781)—Lucy 1789 by Florizel 1768 (Herod 1758)—Frenzy 1774 by Eclipse 1764 (Marske 1750)—Dau. of Engineer 1756 (Sampson 1745)—Dau. 1763 of Blank 1740 (God. Ar. 1724).
Whisker, b. h., 1812.	Waxy 1790 (Pot-8-o-'s 1773)—Penelope 1798 by Trumpator 1782 (Conductor 1767)—Prunella 1788 by Highflyer 1774 (Herod 1758) Promise 1768 by Snap 1750 (Snip 1736)—Julia 1756 by Blank 1740 (God. Ar. 1724).
Gibside Fairy, b. m., 1811.	Hermes 1790 (Mercury 1778)—Vicissitude 1800 by Pipator 1786 (Imperator 1776)—Beatrice 1791 by Sir Peter T. 1784 (Highflyer 1774)—Pyrrha 1771 by Matchem 1748 (Cade 1734)—Duchess 1748 by Whitenose 1742 (God. Ar. 1724).
Sir Archy, b. h., 1805.	*Diomed 1777 (Florizel 1768)—*Castianira 1796 by Rockingham 1781 (Highflyer 1774)—Tabitha 1782 by Trentham 1766 (T. Sweepstakes 1743)—Dau. of Bosphorus 1754 (Babraham 1740)—Dau. of Forester 1750 (Forester 1736).
Daughter of, ch. m., —.	*Citizen 1785 (Pacolet 1763)—Dau. of Commutation (Symmes' Wildair 1767)—Dau. of *Dare Devil 1787 (Magnet 1770)—Sally Shark by *Shark 1771 (Marske 1750)—Betsy Pringle by *Fearnought 1755 (Regulus 1739).
Virginian, b. h., 1815.	Sir Archy 1805 (*Diomed 1777)—Meretrix by Magog 1799 (Chanticleer)—Narcissa 1796 by *Shark 1771 (Marske 1750)—Rosetta 1785 by *Centinel 1758 (Blank 1740)—Diana by Clodius 1778 (*Janus 1746).
Grandam of Trifle, gr. m., —.	*Bedford 1792 (Dungannon 1780)—Dau. of Bellair 1786 (*Medley 1776)—Dau. of *Shark 1771 (Marske 1750)—Dau. of Symme's Wildair 1767 (*Fearnought 1755)—Dau. of Lexington 1774 (Washington).
*Diomed, ch. h., 1777.	Florizel 1768 (Herod 1758)—Dau. 1763 of Spectator 1749 (Crab 1722)—Horatia 1758 by Blank 1740 (God. Ar. 1724)—Dau. of Flying Childers 1715 (Darley Ar. 1699)—Miss Belvoir by Grey Grantham (Brownlow Turk).
*Castianira, br. m., 1796.	Rockingham 1781 (Highflyer 1774)—Tabitha 1782 by Trentham 1766 (T. Sweepstakes 1743)—Dau. of Bosphorus 1754 (Babraham 1740)—Dau. of Forester 1750 (Forester 1736)—Dau. of Coalition Colt 1739 (God. Ar. 1724).
*Chance, b. h., 1797.	Lurcher 1789 (Dungannon 1780)—Recovery 1778 by Hyder Ally 1765 (Blank 1740)—Perdita 1769 by Herod 1758 (Tartar 1743) Fair Forester 1753 by Sloe 1740 (Crab 1722)—Dau. 1746 of Forester 1736 (Hartley's Blind Horse).
Daughter of, —, —.	*Eagle 1796 (Volunteer 1780)—Maria 1805 by Bay Yankee (President 1788) Green's Celer mare by Meade's Celer 1774 (*Janus 1746)—Dau. of Partner 1755 (*Traveller 1747)—Dau. of Apollo 1767 (*Traveller 1747).
Sir Archy, b. h., 1805.	*Diomed 1777 (Florizel 1768)—*Castianira 1796 by Rockingham 1781 (Highflyer 1774)—Tabitha 1782 by Trentham 1766 (T. Sweepstakes 1743)—Dau. of Bosphorus 1754 (Babraham 1740)—Dau. of Forester 1750 (Forester 1736).
Daughter of, ch. m., 1798.	*Dare Devil 1787 (Magnet 1770)—Dau. 1792 of Symmes' Wildair 1767 (*Fearnought 1755)—Picadilla by Y. Fearnought 1777 (*Fearnought 1755)—Dau. of Godolphin 1770 (*Fearnought 1755)—Dau. of *Hob or Nob 1747 (Goliah 1730).
*Speculator, b. h., 1795.	Dragon* 1787 (Woodpecker 1773)—Dau. 1780 of Herod 1758 (Tartar 1743)—Florizel's dam 1761 by Cygnet 1753 (God. Ar. 1724)—Dau. of Y. Cartouch 1731 (Cartouch 1717)—Ebony 1728 by Flying Childers 1715 (Darley Ar. 1699)
Pompadour, ch. m., —.	*Valiant 1756 (Dormouse 1738)—*Jenny Cameron by Cuddy 1727 (Fox 1714) Miss Belvoir by Grey Grantham (Brownlow Turk)—Dau. of Paget Turk—Betty Percival by Leedes Ar. Dau. 1690 of Spanker (D'Arcy Yellow Turk)—Old Morocco mare by Morocco Barb—Old Bald Peg by an Arabian—Barb mare.

IMPORTED EOTHEN.

Owned by Nick Finzer, Esq., Louisville, Ky. Service by contract.

IMPORTED EOTHEN, bay horse, foaled 1883. bred in the Yardley Stud, imported by the late D. D. Withers as a yearling. by Hampton, son of Lord Clifden, winner of the Derby, by Newminster, Eothen's dam Sultana by Oxford, son of Birdcatcher by Sir Hercules. out of Besika by Beiram, son of Sultan. Hampton, the sire of Eothen, was a superior horse, winner of the Goodwood and Doncaster Cups, and is half brother to Sir Bevys, winner of the Derby, both being out of Lady Langden by Kettledrum, son of Rataplan, own brother to Stockwell, and out of Haricot by Mango or Lanercost, she out of Queen Mary, dam of Blink Bonny (Bair Athol's dam), Bonnie Scotland, etc. Hampton is a popular and successful stallion in England, sire of Ayrshire, winner of the 2,000 Guineas and Derby, Reve d'Or, winner of the 1,000 Guineas and Oaks, imp. Kingston and the great horse Sheen, that won the Cesarewitch Handicap, carrying 128 lbs., and the only horse that ever won it with such an impost. Eothen's dam Sultana by Oxford, who was the sire of Sterling, who sired the great horse Isonomy. Besika, his grandam, is the dam of Moslem. winner of the 2.000 Guineas, and is half sister to Flying Duchess, the dam of the popular stallion and Derby winner Galopin, she out of Merope, almost full sister to Velocipede and to Estelle, the dam of Picnic, winner of the 1,000 Guineas and the dam of Mayonaise, winner of the 1,000 Guineas. Sultana, the dam of Eothen, is the dam of the winners Turkish Delight (winner of five races at three years old and two races last year), Gulbeyaz (winner of four races at two years old, including Exning Plate), Graceful and Ayrshire Lass. Eothen was imported when a yearling by the late D. D. Withers, but was so badly injured in the voyage to America that he was never trained. The oldest of Eothen's get are now yearlings and are good looking colts, of high finish and much promise. From the Byerly Turk mare, dam of the two True Blues, descend in direct female line the following famous horses, Sir Peter, Buzzard, Tramp, Master Henry, Champion, Galopin, Lanercost, Glaucus, Stockwell, Rataplan, The Flying Dutchman, King Tom, Quicklime, Kettledrum, Toxophilite, Favonius, Windhound, imp. Chateau Margaux, imp. Robin Redbreast, imp. Hampton Court, American Eclipse, imp. Rayon d'Or, Reform, imp. King Ernest, imp. King Ban, imp. Ill-Used, etc.

EOTHEN.

BAY HORSE, FOALED 1883; BRED AT THE YARDLEY STUD, ENGLAND.

HAMPTON, b. h., 1872. — Lord Clifden, b. h., 1860 (Newminster, b. h., 1848 [Touchstone, br. h., 1831; Beeswing, b. m., 1833]; The Slave, b. m., 1852 [Melbourne, br. h., 1834; Volley, b. m., 1845]). Lady Langden, br. m., 1868 (Kettledrum, ch. h., 1858 [Rataplan, ch. h., 1850; Hybla, b. m., 1846]; Haricot, br. m., 1847 [Lanercost, br. h., 1835; [Mangoer] Queen Mary, b. m., 1843]).

SULTANA, ch. m., 1869. — Oxford, ch. h., 1857 (Ir. Birdcatcher, ch. h., 1833 [Sir Hercules, blk. h., 1826; Guiccioli, ch. m., 1823]; Honey Dear, b. m., 1844 [Plenipotentiary, ch. h., 1831; My Dear, b. m., 1841]). Besika, ch. m., 1852 (Beiram, ch. h., 1829 [Sultan, b. h., 1816; Miss Cantley, ch. m., 1812]; Merope, b. m., 1841 [Voltaire, br. h., 1826; Velocipede's dam, 1817]).

Camel, br. h., 1822.	Whalebone 1807 (Waxy 1790) Dau. 1812 of Selim 1802 (Buzzard 1787) Maiden 1801 by Sir Peter T. 1784 (Highflyer 1774) Dau. 1788 of Phenomenon 1780 (Herod 1758) Matron 1782 by Florizel 1768 (Herod 1758).
Banter, br. m., 1826.	Master Henry 1815 (Orville 1799) Boadicea 1807 by Alexander 1782 (Eclipse 1764) Brunette by Amaranthus 1766 (Old England 1741) Mayfly 1771 by Matchem 1748 (Cade 1734) Dau. 1755 of Anc. Starling 1738 (Starling 1727).
Dr. Syntax, br. h., 1811.	Paynator 1791 (Trumpator 1782)- Dau. of Beningbrough 1791 (King Fergus 1775) Jenny Mole 1787 by Carbuncle 1772 (Babraham Blank 1758) Dau. 1773 of Prince T'Quassaw 1751 (Snip 1736) Sultana 1759 by Regulus 1739 (God. Ar. 1724).
Tomboy's dam, ch. m., 1817.	Ardrossan 1809 (John Bull 1789) Lady Eliza 1813 by Whitworth 1805 (Agonistes 1797) Dau. 1793 of Spadille 1784 (Highflyer 1774) Sylvia 1783 by Y. Marske 1771 (Marske 1750) Ferret 1765 by Broth. to Silvio (Cade 1734).
Humprey Clinker, b. h., 1822.	Comus 1809 (Sorcerer 1796) Clinkerina 1812 by Clinker 1805 (Sir Peter T. 1784) Pewet 1786 by Tandem 1773 (Syphon 1750) Termagant by Tantrum 1760 (Cripple 1750) Cantatrice by Sampson 1745 (Blaze 1733).
Daughter of, b. m., 1825.	Cervantes 1806 (Don Quixote 1784) Dau. 1818 of Golumpus 1802 (Gohanna 1790) Dau. 1810 of Paynator 1791 (Trumpator 1782) Circle 1802 by St. George 1789 (Highflyer 1774)—Abigail 1788 by Woodpecker 1773 (Herod 1758).
Voltaire, br. h., 1826.	Blacklock 1814 (Whitelock 1803) Dau. 1816 of Phantom 1808 (Walton 1799) Dau. 1802 of Overton 1788 (King Fergus 1775)—Dau. 1796 of Walnut 1786 (Highflyer 1774) Dau. 1786 of Ruler 1777 (Y. Marske 1771).
Martha Lynn, br. m., 1837.	Mulatto 1823 (Catton 1809) Leda 1824 by Filho-da-Puta 1812 (Haphazard 1797) Treasure 1805 by Camillus 1803 (Hambletonian 1792) Dau. 1801 of Hyacinthus 1797 (Coriander 1786)—Flora 1789 by King Fergus 1775 (Eclipse 1764).
The Baron, ch. h., 1842.	Ir. Birdcatcher 1833 (Sir Hercules 1826)—Echidna 1838 by Economist 1825 (Whisker 1812) Miss Pratt 1825 by Blacklock 1814 (Whitelock 1803) Gadabout 1812 by Orville 1799 (Beningbrough 1791) Minstrel 1803 by Sir Peter T. 1784 (Highflyer 1774).
Pocahontas, b. m., 1837.	Glencoe* 1831 (Sultan 1816)—Marpessa 1830 by Muley 1810 (Orville 1799) Clare 1824 by Marmion 1806 (Whiskey 1789) Harpalice 1814 by Gohanna 1790 (Mercury 1778) Amazon 1799 by Driver 1783 (Trentham 1766).
The Provost, br. h., 1836.	The Saddler 1828 (Waverley 1817) Rebecca 1831 by Lottery 1820 (Tramp 1810) Dau. 1818 of Cervantes 1806 (Don Quixote 1784) Anticipation 1802 by Beningbrough 1791 (King Fergus 1775) Expecation 1779 by Herod 1758 (Tartar 1743).
Otisina, br. m., 1837.	Liverpool 1828 (Tramp 1810)- Otis 1820 by Bustard 1801 (Buzzard* 1787)—Dau. 1813 of Election 1804 (Gohanna 1790) Sist. 1791 to Skysweeper by Highflyer 1774 (Herod 1758) Dau. 1778 of Eclipse 1764 (Marske 1750).
Liverpool, b. h., 1828.	Tramp 1810 (Dick Andrews 1797) Dau. 1822 of Whisker 1812 (Waxy 1790) Mandane 1800 by Pot-8-o-'s 1773 (Eclipse 1764) Y. Camilla 1787 by Woodpecker 1773 (Herod 1758) Camilla 1778 by Trentham 1766 (T. Sweepstakes 1743).
Otis, b. m., 1820.	Bustard 1801 (Buzzard* 1787) Dau. 1813 of Election 1804 (Gohanna 1790) Sist. 1791 to Skysweeper by Highflyer 1774 (Herod 1758)—Dau. 1778 of Eclipse 1764 (Marske 1750)—Rosebud 1765 by Snap 1750 (Snip 1736).
Gladiator, ch. h., 1833.	Partisan 1811 (Walton 1799) Pauline 1823 by Moses 1819 ([Whalebone or] Seymour 1807)- Quadrille 1815 by Selim 1802 (Buzzard* 1787) Canary Bird 1806 by [Whiskey or] Sorcerer 1796 (Trumpator 1782)- Canary 1797 by Coriander 1786 (Pot-8-o's 1773).
Daughter of, b. m., 1840.	Plenipotentiary 1831 (Emilius 1820)—Myrrha 1830 by Whalebone 1807 (Waxy 1790)—Gift 1818 by Y. Gohanna 1810 (Gohanna 1790)—Dau. 1802 of Sir Peter T. 1784 (Highflyer 1774)—Sist. 1788 to Aimator by Trumpator 1782 (Conductor 1767).
Whalebone, br. h., 1807.	Waxy 1790 (Pot-8-o-'s 1773) Penelope 1798 by Trumpator 1782 (Conductor 1767) Prunella 1788 by Highflyer 1774 (Herod 1758)—Promise 1768 by Snap 1750 (Snip 1736)—Julia 1756 by Blank 1740 (God. Ar. 1724).
Peri, b. m., 1822.	Wanderer 1811 (Gohanna 1790)—Thalestris 1809 by Alexander 1782 (Eclipse 1764)—Rival 1800 by Sir Peter T. 1784 (Highflyer 1774)—Hornet 1790 by Drone 1777 (Herod 1758)—Manilla 1777 by Goldfinder 1764 (Snap 1750).
Bob Booty, ch. h., 1804.	Chanticleer 1787 (Woodpecker 1773)—Ierne 1790 by Bagot 1780 (Herod 1758) Dau. of Gamahoe (Bustard 1741)—Patty by Tim 1752 (Squirt 1732)—Miss Patch 1737 by Justice 1725 (H. C. Litton Ar.)
Flight, ch. m., 1809.	Ir. Escape 1802 (Commodore 1793)—Y. Heroine by Bagot 1780 (Herod 1758)—Heroine 1775 by Hero 1753 (Cade 1734)—Dau. 1766 of Snap 1750 (Snip 1736)—Sist. 1743 to Regulus by God. Ar. 1724.
Emilius, b. h., 1820.	Orville 1799 (Beningbrough 1791)—Emily 1810 by Stamford 1794 (Sir Peter T. 1784)—Dau. 1799 of Whiskey 1789 (Saltram* 1780)—Grey Dorimant 1781 by Dorimant 1772 (Otho 1760)—Dizzy 1757 by Blank 1740 (God. Ar. 1724).
Harriet, blk. m., 1819.	Pericles 1809 (Evander 1801)—Dau. 1812 of Selim 1802 (Buzzard* 1787)—Pipylina 1803 by Sir Peter T. 1784 (Highflyer 1774)—Rally 1790 by Trumpator 1782 (Conductor 1767)—Fancy 1780 by Florizel 1768 (Herod 1758).
Bay Middleton, b. h., 1833.	Sultan 1816 (Selim 1802)—Cobweb 1821 by Phantom 1808 (Walton 1799)—Filagree 1815 by Soothsayer 1808 (Sorcerer 1796)- Web 1808 by Waxy 1790 (Pot-8-o-'s 1773)—Penelope 1798 by Trumpator 1782 (Conductor 1767).
Miss Letty, b. m., 1834.	Priam* 1827 (Emilius 1820)—Dau. 1815 of Orville 1799 (Beningbrough 1791)—Dau. 1800 of Buzzard* 1787 (Woodpecker 1773)—Hornpipe 1793 by Trumpator 1782 (Conductor 1767)—Luna 1779 by Herod 1758 (Tartar 1743).
Selim, ch. h., 1802.	Buzzard* 1787 (Woodpecker 1773)—Dau. of Alexander 1782 (Eclipse 1764)—Dau. of Highflyer 1774 (Herod 1758)—Dau. of Alfred 1770 (Matchem 1748)—Dau. 1770 of Engineer 1756 (Sampson 1745).
Bacchante, br. m., 1809.	W.'s Ditto 1800 (Sir Peter T. 1784)—Dau. 1791 of Mercury 1778 (Eclipse 1764)—Dau. 1776 of Herod 1758 (Tartar 1743)—Folly 1771 by Marske 1750 (Squirt 1732)- Vixen 1753 by Regulus 1739 (God. Ar. 1724).
Stamford, br. h., 1794.	Sir Peter T. 1784 (Highflyer 1774)—Horatia 1778 by Eclipse 1764 (Marske 1750)—Countess 1760 by Blank 1740 (God. Ar. 1724)—Dau. of Rib 1736 (Crab 1722)—Dau. of Wynn Arabian.
Sist. to Silver, gr. m., 1790.	Mercury 1778 (Eclipse 1764)—Dau. 1780 of Herod 1758 (Tartar 1743)—Y. Hag. 1761 by Skim 1746 (Starling 1727)—Hag 1744 by Crab 1722 (Alcock Ar.)—Ebony 1728 by Fl. Childers 1715 (Darley Ar. 1699).
Blacklock, b. h., 1814.	Whitelock 1803 (Hambletonian 1792) Dau. 1799 of Coriander 1786 (Pot-8-o-'s 1773)—Wildgoose 1792 by Highflyer 1774 (Herod 1758) Coheiress 1786 by Pot-8-o-'s 1773 (Eclipse 1764)—Manilla 1777 by Goldfinder 1764 (Snap 1750).
Daughter of, b. m., 1816.	Phantom 1808 (Walton 1799)—Dau. 1802 of Overton 1788 (King Fergus 1775)—Dau. 1796 of Walnut 1786 (Highflyer 1774)—Dau. 1786 of Ruler 1777 (Y. Marske 1771)—Piracantha 1772 by Matchem 1748 (Cade 1734).
Juniper, ch. h., 1805.	Whiskey 1789 (Saltram* 1780)—Jenny Spinner 1797 by Dragon* 1787 (Woodpecker 1773)—Dau. 1775 of Eclipse 1764 (Marske 1750) Miss Spindleshanks 1770 by Omar 1752 (God. Ar. 1727)—Dau. 1753 of Starling 1727 (Bay Bolton 1705).
Daughter of, br. m., 1810.	Sorcerer 1796 (Trumpator 1782)—Virgin br. m. 1801 by Sir Peter T. 1784 (Highflyer 1774)—Dau. b. m. 1794 by Pot-8-o-'s 1773 (Eclipse 1764)—Editha b. m. 1781 by Herod 1758 (Tartar 1743)—Elfrida br. m. 1768 by Snap 1750 (Snip 1736)—Miss Belsea b. m. 1758 by Regulus 1739 (God. Ar. 1724)—Dau. of Bartlett's Childers (Darley Ar. 1699)—Dau. gr. m. of Honeywood's Arab.—Dam of the two True Blues by the Byerly Turk.

IMPORTED ESHER.

*Located in the Hartland Stud, near Versailles, Ky. Services at **$50** the season, with privilege of return.*

ESHER, brown horse, foaled 1883, bred by W. B. Van Haansbergen, owned by J. N. Camden, Jr., Hartland Stud, Versailles, Ky., by Claremont, full brother to Mr. D. D. Withers' imp. Stonehenge, a very richly bred horse, and the sire of many winners in England, namely: Candahar, Miss Betty, Netherton, Mrs. Malaprop, Benedict, Verona, Candahar II., Stonefall, Wilkie, Auchencruive, Humility, Sir David, Hersham, Charlotte, Reveille, etc.

Claremont was a very highly tried horse in his work and was a good race-horse. At two years old Claremont started twice, running second to Camballo for the Hurstbourne Stakes, and third to Balfe for the Chesterfield Stakes. At three years old he started only twice, ran unplaced to Camballo for the 2,000 Guineas, and ran Galopin to a length for the Derby, beating Balfe, Camballo and others. At four years old he started six times, won twice. The Severn Cup, 1 mile, 151 lbs. up, beating six others, and the Great Autumn Welter Cup, 1 mile, 151 lbs. up, beating five others.

Blair Athol, Claremont's sire, was by Stockwell (winner of the St. Leger), dam Blink Bonny (winner of the Oaks and Derby) by Melbourne. Blair Athol was a first-class race-horse, winner of the Derby and St. Leger, and a first-class sire in England.

Ellington, who divides the paternity of Una, Esher's dam, won the Derby, and was by The Flying Dutchman, a famous race-horse, and winner of the Derby, St. Leger and other races, and out of Ellerdale by Lanercost, who was the grandam of imp. The Ill'Used.

Dusk was a son of Wild Dayrell (winner of the Derby) and Circassian Maid by Lanercost.

Esher won three races at two years old. At Nottingham, the Little John Plate of 200 sovereigns, about 5 furlongs, Esher, 124 lbs., won by 2 lengths, Wild Notes, 111 lbs., second, Joyous, 117 lbs., third, and seven others. At Northampton, for the Auction Stakes of £245, 5 furlongs, Esher, 114 lbs., won by a length, the Celosia Colt, 114 lbs., second, and Glade, 106 lbs., third, and five others. At Stockbridge, for the Stockbridge Foal Stakes of £430, Bush In, Esher, 126 lbs., won by a head from Modwena, 127 lbs., second, Pampas Grass, 121 lbs., third, and three others.

At four years old, Esher won twice. At Newmarket, for the Seham Plate of 200 sovereigns, 5 furlongs, Esher, 112 lbs., won by three parts of a length from Invention, 129 lbs., second, Pierrepont, 116 lbs., third, and six others. At Lewes, for the Castle Plate of £195, 5 furlongs, Esher, 126 lbs., won by two lengths from Laceman, 123 lbs., second, Indian Star, 123 lbs., third, and four others.

Imp. Esher is a grand looking horse, with plenty of bone, substance and every requisite for a first-class horse and from his breeding is almost sure to get first-class horses.

Esher combines in his pedigree all the popular winning strains of the day. Through his sire he has the Blair Athol, Stockwell, Irish Birdcatcher, Melbourne, Gladiator, Partisan, Venison and Sir Hercules blood, and on his dam's side the Flying Dutchman, Bay Middleton, Sultan, Selim, Lanercost, Irish Birdcatcher, Humphrey Clinker, Comus, Partisan and Whiskey strains. Esher traces directly through the famous Eleanor, the first mare to win the Derby, to Roxana, that first brought the Godolphin Arabian into notice.

ESHER.

BROWN HORSE, FOALED 1883; BRED BY MR. W. B. VAN HAANSBERGEN, ENGLAND.

CLAREMONT (brother *Stonehenge), b. h., 1872.
- Blair Athol, ch. h., 1861.
  - Stockwell, ch. h., 1849.
    - The Baron, ch. h., 1842.
    - Pocahontas, b. m., 1837.
  - Blink Bonny, b. m., 1854.
    - Melbourne, br. h., 1834.
    - Queen Mary, b. m., 1843.
- Coimbra, b. m., 1861.
  - Kingston, b. h., 1849.
    - Venison, br. h., 1833.
    - Queen Anne, b. m., 1843.
  - Calcavella, b. m., 1844.
    - Ir. Birdcatcher, ch. h., 1833.
    - Caroline, b. m., 1836.

UNA, .. 1866.
- [Ellington or] Dusk, br. h., 1859.
  - Wild Dayrell, br. h., 1852.
    - Ion, br. h., 1835.
    - Ellen Middleton, br. m., 1846.
  - Circassian Maid, b. m., 1844.
    - Lanercost, br. h., 1835.
    - Miss Clifton, b. m., 1822.
- Conjecture, b. m., 1856.
  - Augur, ch. h., 1849.
    - Ir. Birdcatcher, ch. h., 1833.
    - Nickiane, ch. m., 1837.
  - Daughter of, b. m., 1839.
    - Bran, ch. h., 1831.
    - Active, b. m., 1820.

Ir Birdcatcher, ch. h., 1833.	Sir Hercules 1826 (Whalebone 1807)—Guiccioli 1823 by Bob Booty 1804 (Chanticleer 1787)—Flight 1809 by Ir. Escape 1802 (Commodore 1793)—Y. Heroine by Bagot 1780 (Herod 1758)—Heroine 1775 by Hero 1753 (Cade 1734).
Echidna, b. m., 1838.	Economist 1825 (Whisker 1812)—Miss Pratt 1825 by Blacklock 1814 (Whitelock 1803)—Gadabout 1812 by Orville 1799 (Beningbrough 1791)—Minstrel 1803 by Sir Peter T. 1784 (Highflyer 1774)—Matron 1782 by Florizel 1768 (Herod 1758).
Glencoe,* ch. h., 1831.	Sultan 1816 (Selim 1802)—Trampoline 1825 by Tramp 1810 (Dick Andrews 1797)—Web 1808 by Waxy 1790 (Pot-8-o's 1773)—Penelope 1798 by Trumpator 1782 (Conductor 1767)—Prunella 1788 by Highflyer 1774 (Herod 1758).
Marpessa, b. m., 1830.	Muley 1810 (Orville 1799)—Clare 1824 by Marmion 1806 (Whiskey 1789)—Harpalice 1814 by Gohanna 1790 (Mercury 1778)—Amazon 1799 by Driver 1783 (Trentham 1766)—Fractious 1792 by Mercury 1778 (Eclipse 1764).
Humprey Clinker, b. h., 1822.	Comus 1809 (Sorcerer 1796)—Clinkerina 1812 by Clinker 1805 (Sir Peter T. 1784)—Pewet 1786 by Tandem 1773 (Syphon 1750)—Termagant by Tantrum 1760 (Cripple 1750)—Cantatrice by Sampson 1745 (Blaze 1733).
Daughter of, b. m., 1825.	Cervantes 1806 (Don Quixote 1784)—Dau. 1818 of Golumpus 1802 (Gohanna 1790)—Dau. 1810 of Paynator 1791 (Trumpator 1782)—Circle 1802 by St. George 1789 (Highflyer 1774)—Abigail 1788 by Woodpecker 1773 (Herod 1758).
Gladiator, ch. h., 1833.	Partisan 1811 (Walton 1799)—Pauline 1826 by Moses 1819 [Whalebone or] Seymour 1807)—Quadrille 1815 by Selim 1802 (Buzzard* 1787)—Canary Bird 1806 by [Whiskey or] Sorcerer 1796 (Trumpator 1782)—Canary 1797 by Coriander 1786 (Pot-8-o's 1773).
Daughter of, b. m., 1840.	Plenipotentiary 1831 (Emilius 1820)—Myrrha 1830 by Whalebone 1807 (Waxy 1790)—Gift 1818 by Y. Gohanna 1810 (Gohanna 1790)—Dau. 1802 of Sir Peter T. 1784 (Highflyer 1774)—Dau. 1788 of Trumpator 1782 (Conductor 1767).
Partisan, b. h., 1811.	Walton 1799 (Sir Peter T. 1784)—Parasol 1800 by Pot-8-o's 1773 (Eclipse 1764)—Prunella 1788 by Highflyer 1774 (Herod 1758)—Promise 1768 by Snap 1750 (Snip 1736)—Julia 1756 by Blank 1740 (God. Ar. 1724).
Fawn, br. m., 1823.	Smolensko 1810 (Sorcerer 1796)—Jerboa 1803 by Gohanna 1790 (Mercury 1778)—Camilla 1778 by Trentham 1766 (T. Sweepstakes 1743)—Coquette 1764 by Compton Barb—Sist. 1743 to Regulus by God. Ar. 1724.
Slane, b. h., 1833.	Royal Oak 1823 (Catton 1809)—Dau. 1819 of Orville 1799 (Beningbrough 1791)—Epsom Lass 1803 by Sir Peter T. 1784 (Highflyer 1774)—Alexina 1788 by King Fergus 1775 (Eclipse 1764)—Lardella 1780 by Y. Marske 1771 (Marske 1750).
Garcia, br. m., 1823.	Octavian 1807 (Stripling 1795)—Dau. 1806 of Shuttle 1793 (Y. Marske 1771)—Katharine 1798 by Delpini 1781 (Highflyer 1774)—Dau. 1786 of Paymaster 1766 (Blank 1740)—Dau. 1774 of Le Sang 1759 (Changeling 1747).
Sir Hercules, blk. h., 1826.	Whalebone 1807 (Waxy 1790)—Peri 1822 by Wanderer 1811 (Gohanna 1790)—Thalestris 1809 by Alexander 1782 (Eclipse 1764)—Rival 1800 by Sir Peter T. 1784 (Highflyer 1774)—Hornet 1790 by Drone 1777 (Herod 1758).
Guiccioli, ch. m., 1823.	Bob Booty 1804 (Chanticleer 1787)—Flight 1809 by Ir. Escape 1802 (Commodore 1793)—Y. Heroine by Bagot 1780 (Herod 1758)—Heroine 1775 by Hero 1753 (Cade 1734)—Dau. 1766 of Snap 1750 (Snip 1736).
Ir. Drone, gr. h., 1823.	Master Robert 1811 (Buffer 1798)—Dau. 1817 of Sir Walter Raleigh 1801 (Waxy 1790)—Miss Tooly 1808 by Teddy the Grinder 1798 (Asparagus 1787)—Lady Jane 1796 by Sir Peter T. 1784 (Highflyer 1774)—Paulina 1788 by Florizel 1768 (Herod 1758).
The Potentate's dam, b. m., 1823.	Don Juan 1814 (Orville 1799)—Moll-in-the-Wad 1810 by Hambletonian 1792 (King Fergus 1775)—Spitfire 1794 by Pipator 1786 (Imperator 1776)—Farewell 1772 by Slope 1782 (Highflyer 1774)—Dau. of Y. Marske 1771 (Marske 1750).
Cain, b. h., 1822.	Paulowitz 1813 (Sir Paul 1802)—Dau. 1810 of Paynator 1791 (Trumpator 1782)—Dau. 1805 of Delpini 1781 (Highflyer 1774)—Dau. 1795 of Y. Marske 1771 (Marske 1750)—Gentle Kitty 1774 by Silvio 1754 (Cade 1734).
Margaret, br. m., 1831.	Edmund 1824 (Orville 1799)—Medora 1811 by Selim 1802 (Buzzard* 1787)—Dau. 1803 by Sir Harry* 1795 (Sir Peter T. 1784)—Dau. 1793 of Volunteer 1780 (Eclipse 1764)—Dau. 1779 of Herod 1758 (Tartar 1743).
Bay Middleton, b. h., 1833.	Sultan 1816 (Selim 1802)—Cobweb 1821 by Phantom 1808 (Walton 1799)—Filagree 1815 by Soothsayer 1808 (Sorcerer 1796)—Web 1808 by Waxy 1790 (Pot-8-o's 1773)—Penelope 1798 by Trumpator 1782 (Conductor 1767).
Myrrha, ch. m., 1831.	Malek 1824 (Blacklock 1814)—Bessy 1815 by Y. Gouty 1805 (Gouty 1796)—Grandiflora 1810 by Sir Harry Dimsdale 1800 (Sir Peter T. 1784)—Dau. 1799 of Pipator 1786 (Imperator 1776)—Dau. 1796 of Phenomenon 1780 (Herod 1758).
Liverpool, b. h., 1828.	Tramp 1810 (Dick Andrews 1797)—Dau. 1822 of Whisker 1812 (Waxy 1790)—Mandane 1800 by Pot-8-o's 1773 (Eclipse 1764)—Y. Camilla 1787 by Woodpecker 1773 (Herod 1758)—Camilla 1778 by Trentham 1766 (T. Sweepstakes 1743).
Otis, b. m., 1820.	Bustard 1801 (Buzzard* 1787)—Dau. 1813 of Election 1804 (Gohanna 1790)—Dau. 1791 of Highflyer 1774 (Herod 1758)—Dau. 1778 of Eclipse 1764 (Marske 1750)—Rosebud 1765 by Snap 1750 (Snip 1736).
Partisan, b. h., 1811.	Walton 1799 (Sir Peter T. 1784)—Parasol 1800 by Pot-8-o's 1773 (Eclipse 1764)—Prunella 1788 by Highflyer 1774 (Herod 1758)—Promise 1768 by Snap 1750 (Snip 1736)—Julia 1756 by Blank 1740 (God. Ar. 1724).
Isis, b. m., 1805.	Sir Peter T. 1784 (Highflyer 1774)—Ibis 1791 by Woodpecker 1773 (Herod 1758)—Isabella 1783 by Eclipse 1764 (Marske 1750)—Dau. 1775 of Squirrel 1754 (Traveller 1735)—Ancaster Nancy 1768 by Blank 1740 (God. Ar. 1724).
Sir Hercules, blk. h., 1826.	Whalebone 1807 (Waxy 1790)—Peri 1822 by Wanderer 1811 (Gohanna 1790)—Thalestris 1809 by Alexander 1782 (Eclipse 1764)—Rival 1800 by Sir Peter T. 1784 (Highflyer 1774)—Hornet 1790 by Drone 1777 (Herod 1758).
Guiccioli, ch. m., 1823.	Bob Booty 1804 (Chanticleer 1787)—Flight 1809 by Ir. Escape 1802 (Commodore 1793)—Y. Heroine by Bagot 1780 (Herod 1758)—Heroine 1775 by Hero 1753 (Cade 1734)—Dau. 1766 of Snap 1750 (Snip 1736).
Ishmael, ch. h., 1830.	Sultan 1816 (Selim 1802)—Sist. 1820 to Cobweb by Phantom 1808 (Walton 1799)—Filagree 1815 by Soothsayer 1808 (Sorcerer 1796)—Web 1808 by Waxy 1790 (Pot-8-o's 1773)—Penelope 1798 by Trumpator 1782 (Conductor 1767).
Misnomer, b. m., 1831.	Merlin 1815 (Castrel 1801)—Dau. of Phantom 1808 (Walton 1799)—Dau. 1817 of Pericles 1809 (Evander 1801)—Mary 1803 by Sir Peter T. 1784 (Highflyer 1774)—Dau. 1788 of Diomed* 1777 (Florizel 1768).
Humprey Clinker, b. h., 1822.	Comus 1809 (Sorcerer 1796)—Clinkerina 1812 by Clinker 1805 (Sir Peter T. 1784)—Pewet 1786 by Tandem 1773 (Syphon 1750)—Termagant by Tantrum 1760 (Cripple 1750)—Cantatrice by Sampson 1745 (Blaze 1733).
Velvet, b. m., 1823.	Oiscan 1803 (Camillus 1803)—Wire 1811 by Waxy 1790 (Pot-8-o's 1773)—Penelope 1798 by Trumpator 1782 (Conductor 1767)—Prunella 1788 by Highflyer 1774 (Herod 1758)—Promise 1768 by Snap 1750 (Snip 1736).
Partisan, b. h., 1811.	Walton 1799 (Sir Peter T. 1784)—Parasol 1800 by Pot-8-o's 1773 (Eclipse 1764)—Prunella 1788 by Highflyer 1774 (Herod 1758)—Promise 1768 by Snap 1750 (Snip 1736)—Julia 1756 by Blank 1740 (God. Ar. 1724).
Eleanor, b. m., 1798.	Whiskey 1789 (Saltram* 1780)—Y. Giantess b. m. 1790 by Diomed* 1777 (Florizel 1768)—Giantess b. m. 1769 by Matchem 1748 (Cade 1734)—Molly Long Legs b. m. 1753 by Babraham 1740 (God. Ar. 1724)—Dau. 1744 of Cole's Foxhunter 1727 (Brisk 1711)—Sist. to Cato (dam of *Silver) by Partner 1718 (Jigg)—Sist. ch. m. 1719 to Roxana

by Bald Galloway (St. Victor Barb)—Sist. to Chaunter by Akaster Turk—Cream Cheeks by Leedes Ar.—Dau. 1690 by Spanker (D'Arcy Yellow Turk)—Old Morocco mare by Morocco Barb—Old Bald Peg by an Arabian—A Barb mare.

EXILE.

WINNER OF THE BROOKLYN HANDICAP, BROOKLYN CUP, RARITAN STAKES, AND OTHER IMPORTANT RACES.

Property of Messrs. Charles Reed & Sons, Fairview Stud, near Gallatin, Tenn. Yearlings sold annually.

EXILE, bay horse, foaled 1882, bred by Mr. P. Lorillard, Rancocas Stud, N. J., by imp. Mortemer, a superior race-horse, won two races at two years old and was second in three, winning the Omnium, nearly a mile, carrying 123 lbs., a remarkable performance for a two-year old. At three years old he won four races out of eight starts, and was second in three and third in one. At four years old he won nine races, running from 1½ miles to 2 miles, carrying from 132 to 153 lbs.; was second in seven, and third in one, being unplaced but once in eighteen races. At five years old he won five out of six starts, carrying from 133 to 151 lbs., in one of these races beating Dutch Skater; same year won the Stockbridge Cup, and the following year won two out of four starts, including the Gold Cup at Ascot, and was pronounced by Admiral Rous to be by seven pounds the best horse in England. He was also a successful sire. His son, St. Christophe, won the Grand Prize of Paris in 1877 and other good races. Verneuil was a winner on the Continent, and in England won the Gold Vase, Gold Cup and Alexandra Plate at Ascot in 1878. Chamant won the 2,000 Guineas in 1877. In this country he sired first-class racers.

Exile's dam imp. Second Hand, winner of the Craven Handicap, by Stockwell, winner of the 2,000 Guineas and St. Leger, grandam Gaiety, the dam of Gamester (St. Leger winner), Garnish and Gadabout (the dam of Scamander). Second Hand is full sister to Doncaster, winner of the Bristol Plate, Chatsworth, Zetland Stakes and Dee stand Cup, etc., great grandam Cast Steel, dam of Galanthus, winner of the Great Yorkshire Handicap, and Garland (the dam of Vindex), and Vesta (the dam of Vespasian and Sabinus, winner of the Cambridgeshire Handicap and Ascot Gold Cup). To Gossamer, another daughter of Cast Steel, traces Surefoot, winner of the 2,000 Guineas. Lisette, the sixth dam, was the dam of Clearwell, winner of the 2,000 Guineas, and Madame Vestris, to which traces Vedette, Retreat, Sir Hugo and King Lud, etc. Exile was a first-class race-horse and is descended from a distinguished racing and producing family. During his racing career he started 73 times, won 30 races, was second in 17 and third in six. At two years old won a sweepstake, ¾ of a mile, 100 lbs., in 1.18, and again ¾ of a mile, 115 lbs., in 1.17¼; was second to Richmond in the Hopeful Stakes, and second to Brookwood in July Stakes. At three years old won four out of nine races; purse, 1⅛ miles, 103 lbs., in 1.57¾; the Raritan Stakes, 1¼ miles, 113 lbs., in 2.12½; a purse, 1⅛ miles, 105 lbs., in 1.58½, and purse, 1¼ miles, 105 lbs., in 2.11¼. At four years old won five out of nine races, was second in three and third in one; won Brooklyn Eagle Stakes, 1¼ miles, 113 lbs., in 2.11½; purse, 7 furlongs, 118 lbs., in 1.29½; purse, 1$\frac{1}{16}$ miles, in 1.52, 133 lbs.; won Welter Handicap Stakes, 1$\frac{3}{16}$ miles, 126 lbs., in 2.03, the fastest race at the distance to that date, 1886; won Handicap Sweepstake, 1⅝ miles, 116 lbs., in 2.48¾, the fastest race ever run at the distance with the weight. At five years old started 14 times, won four races and was second in three; won Bay Ridge Handicap, 1½ miles, 114 lbs., in 2.36, defeating field of 16; won Sweepstake, 1½ miles, 121 lbs., the top weight, in 2.34½; won purse, 1¼ miles, 122 lbs., in 2.11½; won Mayflower Handicap, 1⅜ miles, 118 lbs., in 2.22½, all good races. At six years old ran 21 races, won 10, and was second in 7 and third in one; won Handicap Sweepstake, 1⅛ miles, 117 lbs., in 1.57¼; won purse, 1¼ miles, 122 lbs., in 2.12½; won High Weight Handicap, ¾ of a mile, 128 lbs., in 1.18; won High Weight Handicap Sweepstake, 2 miles, 140 lbs., in 3.36½; won purse, 1⅛ miles, 115 lbs., in 2.00; won Purse, 1⅛ miles, 133 lbs., in 2.03¾; won purse, 1 mile, 122 lbs., in 1.46¾; won Twin City Handicap, 1¼ miles, 117 lbs., in 2.08; won Woodlawn Handicap, 1$\frac{3}{16}$ miles, 116 lbs., in 2.08¾. In these races he beat the best horses of the year. At seven years old won three races out of six, the Brooklyn Handicap, 1¼ miles, 116 lbs., in 2.07½; won purse, 1 mile, 124 lbs., in 1.47; won Twin City Handicap, 1¼ miles, 125 lbs., in 2.09½. At eight years old won the only two races in which he started, a Sweepstakes, 1 mile, 116 lbs., in 1.46¾, and Brooklyn Cup, 1½ miles, 128 lbs., in 2.35½. Exile was able to go a distance, possessed speed and was up to weight, and should be a success in the stud, as he is handsomely bred. From the old Woodcock mare descend in direct female line the following good horses, Skim, Reveller (by Comus), Heron, Alarm, Monarque, Vedette, Cambuscan, Vespasian, Isonomy, Fernandez, King Lud, Lisbon, imp. Woodlands, Pat Malloy, Jack Malone, etc., etc.

Sir Peter T. 1784 Highflyer 1774 Herod 1758—Tartar 1743 Partner 1718 Jigg Byerly Turk, about 1685.

EXILE.

BAY HORSE, FOALED 1882; BRED BY P. LORILLARD, RANCOCAS STUD, NEW JERSEY.

*MORTEMER, ch. h., 1865.

Compiègne, ch. h., 1858.

Fitz Gladiator, ch. h., 1850.

Gladiator, ch. h., 1833.	Partisan, b. h., 1811.	Walton 1799 (Sir Peter T. 1784) Parasol 1800 by Pot-8-o's 1773 (Eclipse 1764) Prunella 1788 by Highflyer 1774 (Herod 1758) Promise 1768 by Snap 1750 (Snip 1736) Julia 1756 by Blank 1740 (God. Ar. 1724).
	Pauline, b. m., 1826.	Moses 1819 ([Whalebone or] Seymour 1807) Quadrille 1815 by Selim 1802 (Buzzard* 1787) Canary Bird 1806 by (Whiskey or) Sorcerer 1796 (Trumpator 1782) Canary 1797 by Coriander 1786 (Pot-8-o's 1773) Miss Green 1787 by Highflyer 1774 (Herod 1758).
Zarah, b. m., 1835.	Reveller, b. h., 1815.	Comus 1809 (Sorcerer 1796) Rosette 1803 by Beningbrough 1791 (King Fergus 1775) Rosamond 1788 by Tandem 1773 (Syphon 1750) Tuberose 1772 by Herod 1758 (Tartar 1743) Grey Starling 1745 by Starling 1727 (Bay Bolton 1705).
	Daughter of, ch. m., 1814.	Rubens 1805 (Buzzard* 1787) Brightonia 1808 by Gohanna 1790 (Mercury 1778) Nutmeg 1800 by Sir Peter T. 1784 (Highflyer 1774) Nimble 1784 by Florizel 1768 (Herod 1758) Rantipole 1769 by Blank 1740 (God. Ar. 1724).

Maid of Hart, br. m., 1846.

The Provost, br. h., 1836.	The Saddler, br. h., 1828.	Waverley 1817 (Whalebone 1807) Castrellina 1823 by Castrel 1801 (Buzzard* 1787) Dau. 1815 of Waxy 1790 (Pot-8-o's 1773) Bizarre 1811 by Peruvian 1806 (Sir Peter T. 1784) Violante 1802 by John Bull 1789 (Fortitude 1777).
	Rebecca, b. m., 1831.	Lottery 1820 (Tramp 1810) Dau. 1818 of Cervantes 1806 (Don Quixote 1784) Anticipation 1802 by Beningbrough 1791 (King Fergus 1775) Expectation 1779 by Herod 1758 (Tartar 1743) Dau. 1758 of Skim 1746 (Starling 1727).
Martha Lynn, br. m., 1837.	Mulatto, b. h., 1823.	Catton 1809 (Golumpus 1802) Desdemona 1811 by Orville 1799 (Beningbrough 1791) Fanny 1796 by Sir Peter T. 1784 (Highflyer 1774) Dau. 1788 of Diomed* 1777 (Florizel 1768) Desdemona 1770 by Marske 1750 (Squirt 1732).
	Leda, b. m., 1824.	Filho-da-Puta 1812 (Haphazard 1797) Treasure 1809 by Camillus 1803 (Hambletonian 1792) Dau. 1804 of Hyacinthus 1797 (Coriander 1786) Flora 1789 by King Fergus 1775 (Eclipse 1764) Atalanta 1769 by Matchem 1748 (Cade 1734).

Contesse, b. m., 1855.

[The Baron or] Nuncio, br. h., 1839.

Plenipotentiary, ch. h., 1831.	Emilius, b. h., 1820.	Orville 1799 (Beningbrough 1791) Emily 1810 by Stamford 1794 (Sir Peter T. 1784) Dau. 1799 of Whiskey 1789 (Saltram* 1780) Grey Dorimant 1781 by Dorimant 1772 (Otho 1760) Dizzy 1757 by Blank 1740 (God. Ar. 1724).
	Harriet, blk. m., 1819.	Pericles 1809 (Evander 1801) Dau. 1812 of Selim 1802 (Buzzard* 1787) Pipylina 1803 by Sir Peter T. 1784 (Highflyer 1774) Rally 1790 by Trumpator 1782 (Conductor 1767) Fancy 1780 (Sist. to Diomed) by Florizel 1768 (Herod 1758).
Ally, b. m., 1818.	Partisan, b. h., 1811.	Walton 1799 (Sir Peter T. 1784) Parasol 1800 by Pot-8-o's 1773 (Eclipse 1764) Prunella 1788 by Highflyer 1774 (Herod 1758) Promise 1768 by Snap 1750 (Snip 1736) Julia 1756 by Blank 1740 (God. Ar. 1724).
	Jest, gr. m., 1810.	Waxy 1790 (Pot-8-o's 1773)—Scotia 1799 by Delpini 1781 (Highflyer 1774)—Dau. 1790 of King Fergus 1775 (Eclipse 1764)—Cælia 1775 by Herod 1758 (Tartar 1743)—Proserpine 1766 by Marske 1750 (Squirt 1732).

Eusebia, ch. m., 1839.

Emilius, b. h., 1820.	Orville, b. h., 1799.	Beningbrough 1791 (King Fergus 1775) Evelina 1791 by Highflyer 1774 (Herod 1758)—Termagant by Tantrum 1760 (Cripple 1750)—Cantatrice by Sampson 1745 (Blaze 1733)—Dau. 1757 of Regulus 1739 (God. Ar. 1724).
	Emily, ch. m., 1810.	Stamford 1794 (Sir Peter T. 1784)—Dau. 1799 of Whiskey 1789 (Saltram* 1780) Grey Dorimant 1781 by Dorimant 1772 (Otho 1760)—Dizzy 1757 by Blank 1740 (God. Ar. 1724)—Anc. Dizzy 1741 by Driver 1727 (Wynn Ar.).
Mangelwurzel, ch. m., 1823.	Merlin, ch. h., 1815.	Castrel 1801 (Buzzard* 1787) Miss Newton 1804 by Delpini 1781 (Highflyer 1774) Tipple Cyder 1788 by King Fergus 1775 (Eclipse 1764) Sylvia 1783 by Y. Marske 1771 (Marske 1750)—Ferret 1765 by Broth. to Silvio (Cade 1734).
	Morel, ch. m., 1805.	Sorcerer 1796 (Trumpator 1782)—Hornby Lass 1796 by Buzzard* 1787 (Woodpecker 1773)—Puzzle 1778 by Matchem 1748 (Cade 1734)—Princess 1769 by Herod 1758 (Tartar 1743) Julia 1756 by Blank 1740 (God. Ar. 1724).

*SECOND HAND, b. m., 1864.

Stockwell, ch. h., 1849.

The Baron, ch. h., 1842.

Ir. Birdcatcher, ch. h., 1833.	Sir Hercules, blk. h., 1826.	Whalebone 1807 (Waxy 1790)—Peri 1822 by Wanderer 1811 (Gohanna 1790)—Thalestris 1809 by Alexander 1782 (Eclipse 1764) Rival 1800 by Sir Peter T. 1784 (Highflyer 1774)—Hornet 1790 by Drone 1777 (Herod 1758).
	Guiccioli, ch. m., 1823.	Bob Booty 1804 (Chanticleer 1787) Flight 1809 by Ir. Escape 1802 (Commodore 1793)—Y. Heroine by Bagot 1780 (Herod 1758) Heroine 1775 by Hero 1758 (Cade 1734)—Dau. 1766 by Snap 1750 (Snip 1736).
Echidna, b. m., 1838.	Economist, b. h., 1825.	Whisker 1812 (Waxy 1790)—Floranthe 1818 by Octavian 1807 (Stripling 1795) Caprice 1797 by Anvil 1777 (Herod 1758)—Madcap 1774 by Eclipse 1764 (Marske 1750)—Sappho 1768 by Blank 1740 (God. Ar. 1724).
	Miss Pratt, b. m., 1825.	Blacklock 1814 (Whitelock 1803)—Gadabout 1812 by Orville 1799 (Beningbrough 1791)—Minstrel 1803 by Sir Peter T. 1784 (Highflyer 1774)—Matron 1782 by Florizel 1768 (Herod 1758)—Maiden 1770 by Matchem 1748 (Cade 1734).

Pocahontas, b. m., 1837.

Glencoe*, ch. h., 1831.	Sultan, b. h., 1816.	Selim 1802 (Buzzard* 1787)—Bacchante 1809 by W's Ditto 1800 (Sir Peter T. 1784)—Dau. 1791 of Mercury 1778 (Eclipse 1764)—Dau. 1776 of Herod 1758 (Tartar 1743)—Folly 1771 by Marske 1750 (Squirt 1732).
	Trampoline, ch. m., 1825.	Tramp 1810 (Dick Andrews 1797)—Web 1808 by Waxy 1790 (Pot-8-o's 1773)—Penelope 1798 by Trumpator 1782 (Conductor 1767)—Prunella 1788 by Highflyer 1774 (Herod 1758)—Promise 1768 by Snap 1750 (Snip 1736).
Marpessa, b. m., 1830.	Muley, b. h., 1810.	Orville 1799 (Beningbrough 1791)—Eleanor 1798 by Whiskey 1789 (Saltram* 1780)—Y. Giantess 1790 by Diomed* 1777 (Florizel 1768)—Giantess 1769 by Matchem 1748 (Cade 1734)—Molly Long Legs 1753 by Babraham 1740 (God. Ar. 1724).
	Clare, b. m., 1824.	Marmion 1806 (Whiskey 1789)—Harpalice 1814 by Gohanna 1790 (Mercury 1778)—Amazon 1799 by Driver 1783 (Trentham 1766)—Fractious 1792 by Mercury 1778 (Eclipse 1764)—Dau. 1785 of Woodpecker 1773 (Herod 1758).

Gaiety, br. m., 1844.

Touchstone, br. h., 1831.

Camel, br. h., 1822.	Whalebone, br. h., 1807.	Waxy 1790 (Pot-8-o's 1773)—Penelope 1798 by Trumpator 1782 (Conductor 1767)—Prunella 1788 by Highflyer 1774 (Herod 1758)—Promise 1768 by Snap 1750 (Snip 1736)—Julia 1756 by Blank 1740 (God. Ar. 1724).
	Daughter of, b. m., 1812.	Selim 1802 (Buzzard* 1787)—Maiden 1801 by Sir Peter T. 1784 (Highflyer 1774)—Dau. 1788 of Phenomenon 1780 (Herod 1758)—Matron 1782 by Florizel 1768 (Herod 1758)—Maiden 1770 by Matchem 1748 (Cade 1734).
Banter, br. m., 1826.	Master Henry, b. h., 1815.	Orville 1799 (Beningbrough 1791) Miss Sophia 1805 by Stamford 1794 (Sir Peter T. 1784)—Sophia 1798 by Buzzard* 1787 (Woodpecker 1773)—Huncamunca 1787 by Highflyer 1774 (Herod 1758)—Cypher 1772 by Squirrel 1754 (Traveller 1735).
	Boadicea, b. m., 1807.	Alexander 1782 (Eclipse 1764)—Brunette by Amaranthus 1766 (Old England 1741)—Mayfly 1771 by Matchem 1748 (Cade 1734)—Dau. 1755 of Anc. Starling 1738 (Starling 1727)—Look-at-me-Lads 1731 by Brist. Grasshopper (Byerly T.)

Cast Steel, br. m., 1828.

Whisker, b. h., 1812.	Waxy, b. h., 1790.	Pot-8-o's 1773 (Eclipse 1764)—Maria 1777 by Herod 1758 (Tartar 1743)—Lisette 1772 by Snap 1750 (Snip 1736) Miss Windsor 1754 by God. Ar. 1724—Sist. to Volunteer by Y. Belgrade (Belgrade Turk).
	Penelope, b. m., 1798.	Trumpator 1782 (Conductor 1767)—Prunella 1788 by Highflyer 1774 (Herod 1758)—Promise 1768 by Snap 1750 (Snip 1736)—Julia 1756 by Blank 1740 (God. Ar. 1724)—Spectator's dam 1735 by Partner 1718 (Jigg).
The Twinkle, b. m., 1821.	Walton, b. h., 1799.	Sir Peter 1784 (Highflyer 1774)—Arethusa 1792 by Dungannon 1780 (Eclipse 1764)—Dau. 1777 of Prophet 1760 (Regulus 1739)—Virago 1764 by Snap 1750 (Snip 1736)—Dau. of Regulus 1739 (God. Ar. 1724).
	Daughter of, gr. m., 1814.	Orville 1799 (Beningbrough 1791)—Lisette gr. m. 1806 by Hambletonian 1792 (King Fergus 1775)—Constantia gr. m. 1796 by Walnut 1786 (Highflyer 1774)—Contessina gr. m. 1787 by Y. Marske 1771 (Marske 1750)—Tuberose gr. m. 1772 by Herod 1748 (Tartar 1743)—Grey Starling gr. m. 1745 by Starling 1727 (Bay Bolton 1705)—Coughing Polly

b. m. 1736 by Bartlett's Childers (Darley Ar. 1699)—Dau. of Counsellor 1694 (D'Arcy Counsellor)—Dau. of Snake (Lister Turk)—Dau. of Luggs (D'Arcy White Turk)—Dau. of David's Woodcock.

FALSETTO.

WINNER OF THE PHŒNIX HOTEL STAKES AT LEXINGTON, CLARK STAKES AT LOUISVILLE, TRAVERS AND KENNER STAKES AT SARATOGA, N. Y., ALL IN 1879.

Located in the Woodburn Stud, A. J. Alexander, proprietor, near Spring Station, Ky. Application to L. Brodhead, Spring Station. Annual sales of yearlings in May.

FALSETTO by Enquirer, son of imp. Leamington, bred by the late Col. J. W. Hunt Reynolds, Fleetwood Stud, Kentucky, foaled 1876. dam Farfaletta, dam of Felicia, Fortuna, etc., by imp. Australian, out of Elkhorna, dam of Elkhorn, Arcturus, etc., by Lexington. Leamington, Enquirer, Australian and Lexington, all noted horses, covering the first four crosses of the blood of Falsetto, are to be found in this book. His dam, Farfaletta, has never produced anything but good ones. Felicia and Fortuna were both good race-mares and creditable performers. Falsetto made his maiden effort at Lexington, Ky., May 10, first day of the Spring meeting, in Phœnix Hotel Stakes, 1¼ miles. Ten horses started, Falsetto won, defeating Ada Glenn by a neck, Scully a good third, balance unplaced; time, 2.08¾. Falsetto's second appearance was in the Kentucky Derby, 1½ miles; he ran second to Lord Murphy, suffering his first and only defeat. Strathmore was third, and time 2.37. Same place, May 27, won the Clark Stakes, 2 miles, defeating Bucktie, Trinidad and six others, time, 3.40½. Saratoga, July 19, he met the hitherto invincible Spendthrift, also Harold, Jericho and Dan Sparling, in the Travers Stakes, 1¾ miles. Falsetto won; Spendthrift was second, Harold third; time, 3.09¼, a very fast race for the track. Saratoga, August 12, won the Kenner Stakes, 2 miles, defeating Spendthrift, Jericho, Monitor and Harold; time 3.35½. Total winnings in his three-year old form, $18,275. Falsetto was purchased and sent to England by Mr. P. Lorillard, he paying $18,000 for him. Giving away in one of his legs in his preparation, he was returned to America and sold to the Woodburn Stud. Falsetto is from one of the great racing families of America, his dam, grandam and great grandam were handled but little, but all have produced winners; Mary Morris, his fifth dam, was a fine race-mare at all distances, and was the dam of the two fine race-horses, Wild Irishman and Frankfort, the great grandam of Parole, and from the same blood in a direct line sprang Rattler, Childers and Sumpter, all good race-horses and stallions. Flirtilla by Sir Archy was out of the same family, and from her come Ringgold, the sire of Onward and Delaware. The imp. Cub mare, from which all this stock emanated, was out of Amaranthus' dam by Second, and they trace through a popular racing family of England to the Layton Barb mare. Falsetto has four of the best racing strains in this country to begin with, and his blood is pure through a long line of distinguished horses. Falsetto is a successful sire. He is inbred to Lexington, and has the Waxy blood on both sides, from Whalebone, winner of the Derby in 1810, to Whisker, winner of the Derby in 1815, and his son Memnon, sire of Envoy, won the St. Leger in 1825. He is also very strongly inbred to Diomed through Lexington, Am. Eclipse and Gabriella. Falsetto is a brown, stands 16 hands ¾ of an inch high, with a blaze in his face and snip on the nose, and four white ankles.

Falsetto is a grand looker and a horse of fine temper. He has been a grand success in the stud. He is the sire of Dewdrop, that sold for $29,000, Jennie T. Rupert, Ceawood, Emotion, Falsehood, Frontenac, Fernwood, Falcon, Falerna, Folsom, Filmore, Fordham, Foxmede, Ferriere, Forty, Flemington, George C., Gallifet, Gascon, Granite, Glee Boy, Henry Owsley, Kenwood, Keynote, June Day, Lizzie Scott, Miracle, Mon Droit, Mulligan, Patron, Portchester, Pearl Set, Poor Jonathan, Ravenhill, School Girl, Shoshone, Sanford, Tenor, Tom Rogers, Umilta, Vulpina, Vandergrift, Willie L., Miss Dixie, Bob L., False B. and many others. To the Layton Barb mare trace in direct female line Young Greyhound, Matchem, Young Cade, Drone (by Herod), Blucher, Audubon, Kentucky, Daniel Boone, Gilroy, the great Iroquois, Powhattan, and all the descendants of the imp. Cub mare.

FALSETTO.

BAY HORSE, FOALED 1876; BRED BY COL. J. W. HUNT REYNOLDS, FLEETWOOD STUD, KENTUCKY.

ENQUIRER, b. h., 1867. — Leamington, br. h., 1853. — Faugh-a-Ballagh, br. h., 1841. — Sir Hercules, blk. h., 1826. — Guiccioli, ch. m., 1823. — Daughter of, b. m., 1841. — Pantaloon, ch. h., 1824. — Daphne, br. m., 1837. — Lida, b. m., 1858. — Lexington, b. h., 1850. — Boston, ch. h., 1833. — Alice Carneal, b. m., 1836. — Lize, ch. m., 1851. — American Eclipse, ch. h., 1814. — Gabriella, ch. m., 1835.

FARFALETTA, b. m., 1867. — *Australian, ch. h., 1858. — West Australian, b. h., 1850. — Melbourne, br. h., 1834. — Mowerina, b. m., 1843. — Emilia,* b. m., 1840. — Y. Emilius, b. h., 1828. — Persian, b. m., 1829. — Elkhorna, b. m., 1861. — Lexington, b. h., 1850. — Boston, ch. h., 1833. — Alice Carneal, b. m., 1836. — Glencona, ch. m., 1855. — *Glencoe, ch. h., 1831. — Daughter of, ch. m., ...

Whalebone, br. h., 1807. Waxy 1790 (Pot-8-o-'s 1773) Penelope 1798 by Trumpator 1782 (Conductor 1767) Prunella 1788 by Highflyer 1774 (Herod 1758) Promise 1768 by Snap 1750 (Snip 1736) Julia 1756 by Blank 1740 (God. Ar. 1724).

Peri, b. m., 1822. Wanderer 1811 (Gohanna 1790) Thalestris 1809 by Alexander 1782 (Eclipse 1764) Rival 1800 by Sir Peter T. 1784 (Highflyer 1774)—Hornet 1790 by Drone 1777 (Herod 1758) Manilla 1777 by Goldfinder 1764 (Snap 1750).

Bob Booty, ch. h., 1804. Chanticleer 1787 (Woodpecker 1773) Ierne 1790 by Bagot 1780 (Herod 1758) Dau. of Gamahoe (Bustard 1771) Patty by Tim 1752 (Squirt 1732) Miss Patch 1787 by Justice 1725 (H. C. Litton Ar.)

Flight, ch. m., 1809. Ir. Escape 1802 (Commodore 1793) Y. Heroine by Bagot 1780 (Herod 1758)—Heroine 1775 by Hero 1753 (Cade 1734) Dau. 1766 of Snap 1750 (Snip 1736) Sist. 1743 to Regulus by God. Ar. 1724.

Castrel, ch. h., 1801. Buzzard* 1787 (Woodpecker 1773) Dau. of Alexander 1782 (Eclipse 1764) Dau. of Highflyer 1774 (Herod 1758) Dau. of Alfred 1770 (Matchem 1748) Dau. 1770 of Engineer 1756 (Sampson 1745).

Idalia, ch. m., 1815. Peruvian 1806 (Sir Peter T. 1784)—Musidora 1804 by Meteor 1783 (Eclipse 1764)—Maid-of-all-Work 1786 by Highflyer 1774 (Herod 1758) Sist. 1774 to Tandem by Syphon 1750 (Squirt 1732) Dau. of Regulus 1739 (God. Ar. 1724).

Laurel, br. h., 1824. Blacklock 1814 (Whitelock 1803) Wagtail 1818 by Prime Minister 1810 (Sancho 1801) Tranby*'s dam 1812 by Orville 1799 (Beningbrough 1791) Miss Grimstone 1796 by Weazle 1776 (Herod 1758) Dau. of Ancaster 1768 (Blank 1740).

Maid of Honor, br. m., 1829. Champion 1812 (Selim 1802) Etiquette 1820 by Orville 1799 (Beningbrough 1791) Boadicea 1807 by Alexander 1782 (Eclipse 1764)—Brunette by Amaranthus 1766 (Old England 1741) Mayfly 1771 by Matchem 1748 (Cade 1734).

Timoleon, ch. h., 1813. Sir Archy 1805 (*Diomed 1777)—Dau. of *Saltram 1780 (Eclipse 1764)—Dau. of Symmes' Wildair 1767 (*Fearnought 1755)—Dau. of Tyler's Driver 1760 (*Othello) Dau. of *Fallower 1761 (Blank 1740).

Sister to Tuckahoe, ch. m., 1814. Ball's Florizel 1801 (*Diomed 1777)—Dau. 1799 of *Alderman 1787 (Pot-8-o-'s 1773)—Dau. of *Clockfast 1774 (Gimcrack 1760)—Dau. of Symmes' Wildair 1767 (*Fearnought 1755) Y. Kitty Fisher 1767 by *Fearnought 1755 (Regulus 1739).

***Sarpedon, br. h., 1828.** Emilius 1820 (Orville 1799)—Icaria 1824 by The Flyer 1814 (Vandyke Jr. 1808) Parma 1813 by Dick Andrews 1797 (Joe Andrews 1778)—May 1804 by Beningbrough 1791 (King Fergus 1775)—Primrose 1787 by Mambrino 1768 (Engineer 1756).

Rowena, ch. m., 1826. Sumpter 1818 (Sir Archy 1805)—Lady Grey 1817 by Robin Grey 1805 (*Royalist 1790)—Maria 1802 by Melzar 1791 (*Medley 1776)—Dau. of *Highflyer 1784 (Highflyer 1774)—Dau. of *Fearnought 1755 (Regulus 1739).

Duroc, ch. h., 1806. *Diomed 1777 (Florizel 1768)—Amanda 1800 by Grey Diomed 1786 (*Medley 1776)—Dau. of Va. Cade (Partner 1755)—Dau. of Hickman's Independence 1776 (*Fearnought 1755)—Dau. of *Badger (Bosphorus 1754).

Miller's Damsel, gr. m., 1802. *Messenger 1780 (Mambrino 1768)—*Pot-8-o-'s mare 1792 by Pot-8-o-'s 1773 (Eclipse 1764)—Dau. 1778 of Gimcrack 1760 (Cripple 1750)—Snapdragon 1759 by Snap 1750 (Snip 1736)—Dau. 1749 of Regulus 1739 (God. Ar. 1724).

Sir Archy, b. h., 1805. *Diomed 1777 (Florizel 1768)—*Castianira 1796 by Rockingham 1781 (Highflyer 1774)—Tabitha 1782 by Trentham 1766 (T. Sweepstakes 1743)—Dau. of Bosphorus 1754 (Babraham 1740)—Dau. of Forester 1750 (Croft's Forester 1736).

Calypso, b. m., 1805. Bellair 1786 (*Medley 1776)—Dau. 1798 of *Dare Devil 1787 (Magnet 1770)—Dau. 1792 of Symmes' Wildair 1767 (*Fearnought 1755)—Picadilla by Y. Fearnought 1777 (*Fearnought 1755) Dau. of Godolphin 1770 (*Fearnought 1755).

Humprey Clinker, b. h., 1822. Comus 1809 (Sorcerer 1796)—Clinkerina 1812 by Clinker 1805 (Sir Peter T. 1784)—Pewet 1786 by Tandem 1773 (Syphon 1750)—Termagant by Tantrum 1760 (Cripple 1750)—Cantatrice by Sampson 1745 (Blaze 1733).

Daughter of, b. m., 1825. Cervantes 1806 (Don Quixote 1784)—Dau. 1818 of Golumpus 1802 (Gohanna 1790)—Dau. 1810 of Paynator 1791 (Trumpator 1782)—Circle 1802 by St. George 1789 (Highflyer 1774)—Abigail 1788 by Woodpecker 1773 (Herod 1758).

Touchstone, br. h., 1831. Camel 1822 (Whalebone 1807)—Banter 1826 by Master Henry 1815 (Orville 1799)—Boadicea 1807 by Alexander 1782 (Eclipse 1764)—Brunette by Amaranthus 1766 (Old England 1741)—Mayfly 1771 by Matchem 1748 (Cade 1734).

Emma, ch. m., 1824. Whisker 1812 (Waxy 1790)—Gibside Fairy 1811 by Hermes 1790 (Mercury 1778)—Vicissitude by Pipator 1786 (Imperator 1776)—Beatrice 1791 by Sir Peter T. 1784 (Highflyer 1774)—Pyrrha 1771 by Matchem 1748 (Cade 1734).

Emilius, b. h., 1820. Orville 1799 (Beningbrough 1791)—Emily 1810 by Stamford 1794 (Sir Peter T. 1784)—Dau. 1799 of Whiskey 1789 (Saltram* 1780)—Grey Dorimant 1781 by Dorimant 1772 (Otho 1760)—Dizzy 1757 by Blank 1740 (God. Ar. 1724).

Shoveler, b. m., 1816. Scud 1804 (Beningbrough 1791)—Goosander 1805 by Hambletonian 1792 (King Fergus 1775)—Rally 1790 by Trumpator 1782 (Conductor 1767)—Fancy 1780 by Florizel 1768 (Herod 1758)—Dau. 1763 of Spectator 1749 (Crab 1722).

Whisker, b. h., 1812. Waxy 1790 (Pot-8-o-'s 1773)—Penelope 1798 by Trumpator 1782 (Conductor 1767) Prunella 1788 by Highflyer 1774 (Herod 1758)—Promise 1768 by Snap 1750 (Snip 1736)—Julia 1756 by Blank 1740 (God. Ar. 1724).

Variety, ch. m., 1816. [Selim or] Soothsayer 1808 (Sorcerer 1796)—Sprite 1807 by Bobtail 1795 (Precipitate* 1787)—Catherine 1795 by Woodpecker 1773 (Herod 1758)—Camilla 1778 by Trentham 1766 (T. Sweepstakes 1743)—Coquette 1765 by Compton Barb.

Timoleon, ch. h., 1813. Sir Archy 1805 (*Diomed 1777)—Dau. of *Saltram 1780 (Eclipse 1764)—Dau. of Symmes' Wildair 1767 (*Fearnought 1755)—Dau. of Tyler's Driver 1760 (*Othello)—Dau. of *Fallower 1761 (Blank 1740).

Sister to Tuckahoe, ch. m., 1814. Ball's Florizel 1801 (*Diomed 1777)—Dau. 1799 of *Alderman 1787 (Pot-8-o-'s 1773)—Dau. of *Clockfast 1774 (Gimcrack 1760)—Dau. of Symmes' Wildair 1767 (*Fearnought 1755) Y. Kitty Fisher 1767 by *Fearnought 1755 (Regulus 1739).

***Sarpedon, br. h., 1828.** Emilius 1820 (Orville 1799)—Icaria 1824 by The Flyer 1814 (Vandyke Jr. 1808)—Parma 1813 by Dick Andrews 1797 (Joe Andrews 1778)—May 1801 by Beningbrough 1791 (King Fergus 1775)—Primrose 1787 by Mambrino 1768 (Engineer 1756).

Rowena, ch. m., 1826. Sumpter 1818 (Sir Archy 1805)—Lady Grey 1817 by Robin Grey 1805 (*Royalist 1790)—Maria 1802 by Melzar 1791 (*Medley 1776)—Dau. of *Highflyer 1784 (Highflyer 1774)—Dau. of *Fearnought 1755 (Regulus 1739).

Sultan, b. h., 1816. Selim 1802 (Buzzard* 1787)—Bacchante 1809 by W's Ditto 1800 (Sir Peter T. 1784) Dau. 1791 of Mercury 1778 (Eclipse 1764)—Dau. 1776 of Herod 1758 (Tartar 1743)—Folly 1771 by Marske 1750 (Squirt 1732).

Trampoline, ch. m., 1825. Tramp 1810 (Dick Andrews 1797)—Web 1808 by Waxy 1790 (Pot-8-o-'s 1773)—Penelope 1798 by Trumpator 1782 (Conductor 1767)—Prunella 1788 by Highflyer 1774 (Herod 1758)—Promise 1768 by Snap 1750 (Snip 1736).

***Envoy, b. h., 1833.** Memnon 1822 (Whisker 1812)—Zarina 1827 by Morisco 1819 (Muley 1810)—Ina 1821 by Smolensko 1810 (Sorcerer 1796)—Morgiana 1804 by Coriander 1786 (Pot-8-o-'s 1773)—Fairy 1782 by Highflyer 1774 (Herod 1758).

Mary Morris, b. m., 1836. Medoc 1829 (Am. Eclipse 1814)—Miss Obstinate br. m. 1820 by Sumpter 1818 (Sir Archy 1805)—Jenny Slamerkin b. m. 1823 by Tiger 1812 (Cook's Whip 1804)—Paragon b. m. 1807 by *Buzzard 1787 (Woodpecker 1773)—Indiana b. m. 1802 by Columbus (*Pantaloon 1778)—Jane Hunt b. m. 1796 by Hampton's Paragon 1788 (*Flimnap 1765)—Moll b. m. by *Figure 1757 (Grey Figure 1747)—Slamerkin b. m. 1768 by *Wildair 1753 (Cade 1734)—*Cub mare b. m. 1762 by Cub 1739 (Fox 1714)—Amaranthus' dam b. m. 1755 by Second 1732 (Flying Childers 1715)—Dau. 1746 by Starling 1727 (Bay Bolton 1705)—Sist. b. m. 1740 to Vane's Little Partner by Partner 1718 (Jigg)—Sist. 1722 to Guy by Greyhound (Chillaby)—Brown Farewell br. m. 1710 by Makeless (Oglethorpe Ar.)—Dau. of Brimmer (D'Arcy Yellow Turk)—Dau. of Place's White Turk—Dau. of Dodsworth—Layton Barb mare.

FARANDOLE.

The Property of Ed. S. Gardner, Avondale Stud, Saundersville, Tenn. Application for service to E. S. Gardner.

FARANDOLE, chestnut horse, foaled 1881, bred in the Kennesaw Stud, Tenn., owned by E. S. Gardner, Avondale Stud, Tenn., by Enquirer, son of imp. Leamington, dam Waltz by Lexington, out of Schottische by imp. Albion. Farandole only ran three races; was second to Joquita in Maxwell House Stake, at Nashville, 1 mile, in 1.48; second to Billy Gilmore in the Kennesaw Stake, 1½ miles, in 2.09¾, and second to the same in the Woodlawn Stakes, 1½ miles, in 2.01½, the track was very heavy and slow in both these races. His sire, Enquirer, is second to no son of Leamington as a race-horse and stallion. Waltz, his dam was a good race-mare and is from one of the best racing and producing families in the world. She is the dam of Hop (the dam of Kirkman, Carson, Fond-du-Lac, etc.), Glidelia, she having the fastest race at 1¾ miles ever run up to 1891, Racket, Bolero (a winner in England), Gladiola, Raquet, Kermesse, etc. Schottische is the dam of Ballinkeel, Cachuca, Reveller, Quickstep (the dam of Falsestep, Fernwood, etc.). Dance, his great grandam, was the dam of La Polka, the dam of Post Guard (Gen. Phillips), Strathspey, Heel-and-Toe, Los Angeles, etc. Imp. Gallopade was the dam of the renowned Reel, and she the dam of Lecompte, the only horse that ever beat Lexington, Starke, winner of the Goodwood Cup, Prioress, winner of the Cesarewitch, War Dance and Fanny Wells, Nellie Grey by Commodore, Jils Johnson and a host of fine race-horses are from this family, which is a famous English one, tracing to Piping Peg, to which trace in direct female line Octavius, Bob Booty, Cannon Ball, St. Martin, Mickey Free, Chanticleer, Solon, Xenophon, Zealot, Barcaldine, Lowland Chief, Ossian, Ambergris, Lecompte, Starke, War Dance, Jils Johnson, etc. Farandole combines the best racing blood on both sides, and should be a success in the stud. He is the sire of the winners Annie Brown, Billy Smith, Douglas, Dixie V., Free Trade, Fringe, Indigo, La Mothe, Melia P. These are about all of his get yet trained.

FARANDOLE.

CHESTNUT HORSE, FOALED 1881; BRED BY L. & T. E. KITTSON, ERDENHEIM STUD, PENNSYLVANIA.

Pot-8-o-'s 1773 Eclipse 1764—Marske 1750 Squirt 1732 Bartlett's Childers Darley Arabian 1689.

ENQUIRER, b. h., 1867.

- Leamington, br. h., 1853.
  - Faugh-a-Ballagh, br. h., 1841.
    - Sir Hercules, blk. h. 1826.
    - Guiccioli, ch. m., 1823.
  - Daughter of, b. m., 1841.
    - Pantaloon, ch. h., 1824.
    - Daphne, br. m., 1837.
- Lida, b. m., 1858.
  - Lexington, b. h., 1850.
    - Boston, ch. h., 1833.
    - Alice Carneal, b. m., 1836.
  - Lize, ch. m.,
    - American Eclipse, ch. h., 1814.
    - Gabriella, ch. m., 1831.

Whalebone, br. h., 1807.	Waxy 1790 (Pot-8-o-'s 1773) Penelope 1798 by Trumpator 1782 (Conductor 1767) Prunella 1788 by Highflyer 1774 (Herod 1758) Promise 1768 by Snap 1750 (Snip 1736) Julia 1756 by Blank 1740 (God. Ar. 1724).
Peri, b. m., 1822.	Wanderer 1811 (Gohanna 1790) Thalestris 1809 by Alexander 1782 (Eclipse 1764) Rival 1800 by Sir Peter T. 1784 (Highflyer 1774)—Hornet 1790 by Drone 1777 (Herod 1758) Manilla 1777 by Goldfinder 1764 (Snap 1750).
Rob Roy, ch. h., 1814.	Chanticleer 1787 (Woodpecker 1773) Ierne 1790 by Bagot 1780 (Herod 1758) Dau. of Gamahoe (Bustard 1741) Patty by Tim 1752 (Squirt 1732) Miss Patch 1757 by Justice 1725 (H. C. Litton Ar.)
Flight, ch. m., 1809.	Ir. Escape 1802 (Commodore 1793) Y. Heroine by Bagot 1780 (Herod 1758)—Heroine 1775 by Hero 1753 (Cade 1734) Dau. 1766 of Snap 1750 (Snip 1736) Sist. 1743 to Regulus by God. Ar. 1724.
Castrel, ch. h., 1801.	Buzzard* 1787 (Woodpecker 1773) Dau. of Alexander 1782 (Eclipse 1764) Dau. of Highflyer 1774 (Herod 1758) Dau. of Alfred 1770 (Matchem 1748) Dau. 1770 of Engineer 1756 (Sampson 1745).
Idalia, ch. m., 1815.	Peruvian 1806 (Sir Peter T. 1784)—Musidora 1804 by Meteor 1783 (Eclipse 1764)—Maid-of-all-Work 1786 by Highflyer 1774 (Herod 1758) Sist. 1771 to Tandem by Syphon 1750 (Squirt 1732) Dau. of Regulus 1739 (God. Ar. 1724).
Laurel, br. h., 1824.	Blacklock 1814 (Whitelock 1803) Wagtail 1818 by Prime Minister 1810 Sancho 1801—Tranby's dam 1812 by Orville 1799 (Beningbrough 1791) Miss Grimstone 1796 by Weazle 1776 (Herod 1758) Dau. of Ancaster 1768 (Blank 1740).
Maid of Honor, br. m., 1829.	Champion 1812 (Selim 1802) Etiquette 1820 by Orville 1799 (Beningbrough 1791) Boadicea 1807 by Alexander 1782 (Eclipse 1764)—Brunette by Amaranthus 1766 (Old England 1741) Mayfly 1771 by Matchem 1748 (Cade 1734).
Timoleon, ch. h., 1813.	Sir Archy 1805 (*Diomed 1777) Dau. of *Saltram 1780 (Eclipse 1764)—Dau. of Symmes' Wildair 1767 (*Fearnought 1755)—Dau. of Tyler's Driver 1760 (*Othello) Dau. of *Fallower 1761 (Blank 1740).
Sister to Tuckahoe, ch. m., 1814.	Ball's Florizel 1801 (*Diomed 1777) Dau. 1799 of *Alderman 1787 (Pot-8-o-'s 1773)—Dau. of *Clockfast 1771 (Gimcrack 1760) Dau. of Symmes' Wildair 1767 (*Fearnought 1755) Y. Kitty Fisher 1767 by *Fearnought 1755 (Regulus 1739).
*Sarpedon, br. h., 1828.	Emilius 1820 (Orville 1799)—Icaria 1824 by The Flyer 1814 (Vandyke Jr. 1808) Parma 1813 by Dick Andrews 1797 (Joe Andrews 1778)—May 1804 by Beningbrough 1791 (King Fergus 1775)—Primrose 1787 by Mambrino 1768 (Engineer 1756).
Rowena, ch. m., 1826.	Sumpter 1818 (Sir Archy 1805) Lady Grey 1817 by Robin Grey 1805 (*Royalist 1790)—Maria 1802 by Melzar 1791 (*Medley 1776)—Dau. of *Highflyer 1784 (Highflyer 1774)—Dau. of *Fearnought 1755 (Regulus 1739).
Duroc, ch. h., 1806.	*Diomed 1777 (Florizel 1768) Amanda 1800 by Grey Diomed 1786 (*Medley 1776) Dau. of Va. Cade (Partner 1755)—Dau. of Hickman's Independence 1776 (*Fearnought 1755)—Dau. of *Badger (Bosphorus 1754).
Miller's Damsel, gr. m., 1802.	*Messenger 1780 (Mambrino 1768) *Pot-8-o-'s mare 1782 by Pot-8-o-'s 1773 (Eclipse 1764) Dau. 1778 of Gimcrack 1760 (Cripple 1750)—Snapdragon 1759 by Snap 1750 (Snip 1736) Dau. 1749 of Regulus 1739 (God. Ar. 1724).
Sir Archy, b. h., 1805.	*Diomed 1777 (Florizel 1768)—*Castianira 1796 by Rockingham 1781 (Highflyer 1774)—Tabitha 1782 by Trentham 1766 (T. Sweepstakes 1743)—Dau. of Bosphorus 1754 (Babraham 1740)—Dau. of Forester 1750 (Croft's Forester 1736).
Calypso, b. m., 1805.	Bellair 1786 (*Medley 1776)—Dau. 1798 of *Dare Devil 1787 (Magnet 1770) Dau. 1792 of Symmes' Wildair 1767 (*Fearnought 1755)—Picadilla by Y. Fearnought 1777 (*Fearnought 1755)—Dau. of Godolphin 1770 (*Fearnought 1755).

WALTZ, ch. m., 1867.

- Schottische, ch. m., 1859.
  - *Albion, blk. h., 1837.
    - [Cain or] Actæon, ch. h., 1822.
    - Panthea, br. m., 1831.
  - Dance, ch. m., 1853.
    - *Glencoe, ch. h., 1831.
    - Cotillion, ch. m., 1837.
- Lexington, b. h., 1850.
  - Boston, ch. h., 1833.
    - Timoleon, ch. h., 1813.
    - Sist. to Tuckahoe, ch. m., 1814.
  - Alice Carneal, b. m., 1836.
    - *Sarpedon, br. h., 1828.
    - Rowena, ch. m., 1826.

Sir Archy, b. h., 1805.	*Diomed 1777 (Florizel 1768)—*Castianira 1796 by Rockingham 1781 (Highflyer 1774)—Tabitha 1782 by Trentham 1766 (T. Sweepstakes 1743)—Dau. of Bosphorus 1754 (Babraham 1740)—Dau. of W's Forester 1750 (Forester 1736).
Daughter of, ch. m., —.	*Saltram 1780 (Eclipse 1764) Dau. of Symmes' Wildair 1767 (*Fearnought 1755)—Dau. of Tyler's Driver 1760 (*Othello) Dau. of *Fallower 1761 (Blank 1740) Dau. of *Vampire 1757 (Regulus 1739).
Ball's Florizel, ch. h., 1801.	*Diomed 1777 (Florizel 1768)—Dau. of *Shark 1771 (Marske 1750)—Dau. of Harris' Eclipse 1770 (*Fearnought 1755)—Dau. of *Fearnought 1755 (Regulus 1739)—Dau. of *Jolly Roger 1741 (Roundhead 1733).
Daughter of, b. m., 1799.	*Alderman 1787 (Pot-8-o-'s 1773) Dau. of *Clockfast 1774 (Gimcrack 1760)—Dau. of Symmes' Wildair 1767 (*Fearnought 1755)—Y. Kitty Fisher 1767 by *Fearnought 1755 (Regulus 1739)—*Kitty Fisher 1755 by Cade 1734 (God. Ar. 1724).
Emilius, b. h., 1820.	Orville 1799 (Beningbrough 1791)—Emily 1810 by Stamford 1794 (Sir Peter T. 1784)—Dau. 1799 of Whiskey 1789 (Saltram* 1780)—Grey Dorimant 1781 by Dorimant 1772 (Otho 1760) Dizzy 1757 by Blank 1740 (God. Ar. 1724).
Icaria, b. m., 1824.	The Flyer 1814 (Vandyke Jr. 1808)—Parma 1813 by Dick Andrews 1797 (Joe Andrews 1778)—May 1804 by Beningbrough 1791 (King Fergus 1775) Primrose 1787 by Mambrino 1768 (Engineer 1756)—Cricket 1776 by Herod 1758 (Tartar 1743).
Sumpter, ch. h., 1818.	Sir Archy 1805 (*Diomed 1777) Dau. of *Robin Redbreast 1793 (Sir Peter T. 1784) Dau. 1788 of *Obscurity 1777 (Eclipse 1764)—Slamerkin 1768 by *Wildair 1753 (Cade 1734)—*Cub mare 1762 by Cub 1739 (Fox 1714).
Lady Grey, b. m., 1817.	Robin Grey 1805 (*Royalist 1790) Maria 1802 by Melzar 1791 (*Medley 1776)—Dau. of *Highflyer 1784 (Highflyer 1774) Dau. of *Fearnought 1755 (Regulus 1739)—Dau. of Ariel 1756 (*Traveller 1747).
Sead, b. h., 1804.	Beningbrough 1791 (King Fergus 1775) Eliza 1791 by Highflyer 1774 (Herod 1758)—Augusta 1781 by Eclipse 1764 (Marske 1750)—Hardwicke's dam by Herod 1758 (Tartar 1743)—Nettle 1763 by Bajazet 1740 (God. Ar. 1724).
Diana, ch. m., 1808.	Stamford 1794 (Sir Peter T. 1784) Dau. 1799 of Whiskey 1789 (Saltram* 1780)—Grey Dorimant 1781 by Dorimant 1772 (Otho 1760)—Dizzy 1757 by Blank 1740 (God. Ar. 1724)—Anc. Dizzy 1741 by Driver 1727 (Wynn Ar.).
[Comus or] Blacklock, b. h., 1814.	Whitelock 1803 (Hambletonian 1792) Dau. 1799 of Coriander 1786 (Pot-8-o-'s 1773)—Wildgoose 1792 by Highflyer 1774 (Herod 1758) Coheiress 1786 by Pot-8-o-'s 1773 (Eclipse 1764)—Manilla 1777 by Goldfinder 1764 (Snap 1750).
Mandella, b. m., 1809.	Dick Andrews 1797 (Joe Andrews 1778)—Mandane 1800 by Pot-8-o-'s 1773 (Eclipse 1764)—Y. Camilla 1787 by Woodpecker 1773 (Herod 1758)—Camilla 1778 by Trentham 1766 (T. Sweepstakes 1743)—Coquette 1765 by Compton Barb.
Sultan, b. h., 1816.	Selim 1802 (Buzzard* 1787)—Bacchante 1809 by W's Ditto 1800 (Sir Peter T. 1784)—Dau. 1791 of Mercury 1778 (Eclipse 1764)—Dau. 1776 of Herod 1758 (Tartar 1743)—Folly 1771 by Marske 1750 (Squirt 1732).
Trampoline, ch. m., 1825.	Tramp 1810 (Dick Andrews 1797)—Web 1808 by Waxy 1790 (Pot-8-o-'s 1773) Penelope 1798 by Trumpator 1782 (Conductor 1767)—Prunella 1788 by Highflyer 1774 (Herod 1758) Promise 1768 by Snap 1750 (Snip 1736).
*Leviathan, ch. h., 1823.	Muley 1810 (Orville 1799)—Dau. 1809 of Windle 1804 (Beningbrough 1791) Dau. 1788 of Anvil 1777 (Herod 1758)—Virago 1764 by Snap 1750 (Snip 1736) Dau. of Regulus 1739 (God. Ar. 1724)—Sist. 1750 to Othello by Crab 1722 (Alcock Ar.).
*Galopade, gr. m., 1828.	Catton 1809 (Golumpus 1802)—Camillina gr. m. 1822 by Camillus 1803 (Hambletonian 1792)—Dau. b. m. 1818 of Smolensko 1810 (Sorcerer 1796)—Miss Cannon b. m. 1811 by Orville 1799 (Beningbrough 1791)—Dau. b. m. 1796 of Weathercock 1785 (Highflyer 1774)—Cora ro. m. 1777 by Matchem 1748 (Cade 1734)—Dau. of Turk 1763 (Regulus 1739)—Dau. 1764 of Cub 1739 (Fox 1714)—Dau. 1750 of Allworthy 1744 (Crab 1722)—Dau. of Bolton Starling 1727 (Bay Bolton 1705)—Dairymaid b. m. 1737 by Bloody Buttocks—Bay Brocklesby b. m. 1731 by Partner 1718 (Jigg)—Brocklesby gr. m. 1721 by Greyhound (Chillaby)—Brocklesby Betty ch. m. 1711 by Curwen Bay Barb—Leedes Hobby mare by Lister Turk, out of Piping Peg.

FELLOWCRAFT.

THE FIRST HORSE TO BEAT LEXINGTON'S FOUR MILE TIME, 7.19¾, WHICH HE DID AT SARATOGA, AUGUST 20, 1874, RUNNING IN 7.19½.

Fellowcraft is in the Dixiana Stud. Application to Maj. B. G. Thomas, proprietor, Lexington, Ky. Sales annually.

FELLOWCRAFT by imp. Australian, son of West Australian, bred by A. J. Alexander, Woodburn Stud, Ky., foaled 1870, dam Aerolite, the dam of Mozart, Rutherford, Miser, Spendthrift, etc., by Lexington, 2d dam Florine, the dam of Idlewild by imp. Glencoe, out of Melody by Medoc, son of Am. Eclipse. As a two-year old Fellowcraft started five times and won one race; was unplaced in Hopeful Stakes, one-half mile; unplaced in a dash of three-quarters of a mile; won a dash of three-quarters of a mile in 1.20, beating Marsyas and three others. At Saratoga was unplaced in the Flash Stakes and Saratoga Stakes. As a three-year old he started nine times and won one race. Jerome Park, unplaced in the Belmont Stakes; second to Survivor in a dash of 1¼ miles, run in 2.15¾; unplaced in the Weatherby Stakes, 1½ miles; was third to Tom Bowling and Springbok in the Jerome Stakes, 2 miles; was third in the Grand National Handicap 2¼ miles, won by Preakness in 4.08¾, with Harry Basset second. At Saratoga, unplaced in the Travers Stakes, 1¾ miles, won by Tom Bowling in 3.09¾; second in the Sequel Stakes, won by imp. Ill Used in 3.46½, and walked over for a Sweepstakes for three-year olds, 2 miles. When four years old he started twelve times and won three races. Jerome Park, unplaced in the Fordham Handicap, the Jockey Club Handicap, and in a race of mile heats, won by Kitty Pease in 1.43¾, 1.44; third in purse race, 1¾ miles; third in Handicap Sweepstakes, 2½ miles, won by Kitty Pease in 4.35½. At Long Branch, was unplaced in a race of two-mile heats; won purse, $1,000, 4 miles. in 7.43; was second in a race of two-mile heats, won by Vandalite in 3.49, 3.37. At Saratoga, was third in a dash of three miles. won by Springbok in 5.42½. with Preakness second; won dash of 1½ miles in 2.42½, beating Katie Pease, The Governess and Wanderer; was second in a dash of 2¼ miles, won by Wanderer in 4.00½; and won dash of 4 miles in 7.19½, beating Wanderer and Katie Pease. This was the best race, at the distance, ever run up to that date. The first mile was run in 1.47½, two miles in 3.38, three miles in 5.29½, and the four miles in 7.19½. This closed the turf career of Fellowcraft. Since his retirement, with a limited advantage, he has sired some fast horses. Knight Templar won three-quarters of a mile in 1.14, three-quarter mile heats in 1.16¾, 1.16½, and 1 mile in 1.42, all of which are creditable. Blue Lodge won heats of three-quarters of a mile in 1.20, 1.19¾, beating Brambaletta, Pilgrimage, Maggie C. and others; ran dead heat and divided Citizens' Plate with Long Taw, at St. Louis, 1⅛ miles, in 1.55¾. Blue Lodge won some good races in 1882. Fellowcraft is a rich chestnut, with star and two white hind feet, is well formed and a fine blood-like specimen of a horse, 16 hands high. He is handsomely bred, was a good race-horse, with great speed and the capacity to stay over a distance of ground. He is strongly inbred to Diomed and imp. Medley, son of Gimcrack by Cripple by the Godolphin Arabian; he possesses the Waxy blood through Whisker and Whalebone, and numerous crosses to Herod and Eclipse, the pedigree terminating in the blood of Medley by Gimcrack.

Fellowcraft is the sire of the winners Boaz, Dr. Jekyl, Jim Murphy, My Fellow, Lady Reel, Harrodsburgh, Woodcraft, Abby Craft. Fellowship, Simrock, Shibboleth. Hansa, Fannie S., Capt. Brown, Hydra, Free Mason, My Craft, Timberland, Tom Tough and others.

Trumpator 1782 - Conductor 1767—Matchem 1748 Cade 1731 Godolphin Arabian 1724.

FELLOWCRAFT.

CHESTNUT HORSE, FOALED 1870; BRED IN THE WOODBURN STUD, SPRING STATION, KY.

*AUSTRALIAN, ch. h., 1858.	West Australian, b. h., 1850.	Melbourne, br. h., 1834.	Humphrey Clinker, b. h., 1822.	Comus, ch. h., 1809. Sorcerer 1796 (Trumpator 1782)—Epsom Lass 1803 by Sir Peter T. 1784 (Highflyer 1774)—Alexina 1788 by King Fergus 1775 (Eclipse 1764) Lardella 1780 by Y. Marske 1771 (Marske 1750) Dau. of Y. Cade 1747 (Cade 1731).
				Clinkerina, br. m., 1812. Clinker 1805 (Sir Peter T. 1784) Pewet 1786 by Tandem 1773 (Syphon 1750) Termagant by Tantrum 1760 (Cripple 1750) Cantatrice by Sampson 1745 (Blaze 1733) Dau. 1757 of Regulus 1739 (God. Ar. 1724).
			Daughter of, b. m., 1825.	Cervantes, b. h., 1806. Don Quixote 1784 (Eclipse 1764) Evelina 1791 by Highflyer 1774 (Herod 1758) Termagant by Tantrum 1760 (Cripple 1750) Cantatrice by Sampson 1745 (Blaze 1733) Dau. 1757 of Regulus 1739 (God. Ar. 1724).
				Daughter of, br. m., 1818. Golumpus 1802 (Gohanna 1790) Dau. 1810 of Paynator 1791 (Trumpator 1782) Circle 1802 by St. George 1789 (Highflyer 1774) Abigail 1788 by Woodpecker 1773 (Herod 1758) Firetail 1773 by Eclipse 1764 (Marske 1750).
		Mowerina, b. m., 1843.	Touchstone, br. h., 1831.	Camel, br. h., 1822. Whalebone 1807 (Waxy 1790) Dau. 1812 of Selim 1802 (Buzzard* 1787) Maiden 1801 by Sir Peter T. 1784 (Highflyer 1774) Dau. 1788 of Phenomenon 1780 (Herod 1758)—Matron 1782 by Florizel 1768 (Herod 1758).
				Banter, br. m., 1826. Master Henry 1815 (Orville 1799) Boadicea 1807 by Alexander 1782 (Eclipse 1764) - Brunette by Amaranthus 1766 (Old England 1741) Mayfly 1771 by Matchem 1748 (Cade 1731)—Dau. 1755 of Anc. Starling 1738 (Starling 1727).
			Emma, ch. m., 1824.	Whisker, b. h., 1812. Waxy 1790 (Pot-8-o-'s 1773) Penelope 1798 by Trumpator 1782 (Conductor 1767) Prunella 1788 by Highflyer 1774 (Herod 1758) Promise 1768 by Snap 1750 (Snip 1736) - Julia 1756 by Blank 1740 (God. Ar. 1724).
				Gibside Fairy, b. m., 1811. Hermes 1790 (Mercury 1778) Vicissitude 1800 by Pipator 1786 (Imperator 1776)—Beatrice 1791 by Sir Peter T. 1784 (Highflyer 1774) Pyrrha 1771 by Matchem 1748 (Cade 1734) - Duchess 1748 by Whitenose 1742 (God. Ar. 1724).
	Emilia,* b. m., 1840.	Y. Emilius, b. h., 1833.	Emilius, b. h., 1820.	Orville, b. h., 1799. Beningbrough 1791 (King Fergus 1775) Evelina 1791 by Highflyer 1774 (Herod 1758)—Termagant by Tantrum 1760 (Cripple 1750)—Cantatrice by Sampson 1745 (Blaze 1733)—Dau. 1757 of Regulus 1739 (God. Ar. 1724).
				Emily, ch. m., 1810. Stamford 1794 (Sir Peter T. 1784) Dau. 1799 of Whiskey 1789 (Saltram* 1780) Grey Dorimant 1781 by Dorimant 1772 (Otho 1760) Dizzy 1757 by Blank 1740 (God. Ar. 1724) Dizzy 1741 by Driver 1727 (Wynn Ar.).
			Shoveler, b. m., 1816.	Scud, b. h., 1804. Beningbrough 1791 (King Fergus 1775) Eliza 1791 by Highflyer 1774 (Herod 1758)—Augusta 1784 by Eclipse 1764 (Marske 1750)—Hardwicke's dam by Herod 1758 (Tartar 1743) - Nettle 1763 by Bajazet 1740 (God. Ar. 1724).
				Goosander, b. m., 1805. Hambletonian 1792 (King Fergus 1775)—Rally 1790 by Trumpator 1782 (Conductor 1767)—Fancy 1780 by Florizel 1768 (Herod 1758)—Dau. 1763 of Spectator 1749 (Crab 1722)—Horatia 1758 by Blank 1740 (God. Ar. 1724).
		Persian, b. m., 1829.	Whisker, b. h., 1812.	Waxy, b. h., 1790. Pot-8-o-'s 1773 (Eclipse 1764)—Maria 1777 by Herod 1758 (Tartar 1748)—Lisette 1772 by Snap 1750 (Snip 1736) Miss Windsor 1754 by God. Ar. 1724—Sist. to Volunteer by Y. Belgrade (Belgrade Turk).
				Penelope, b. m., 1798. Trumpator 1782 (Conductor 1767)—Prunella 1788 by Highflyer 1774 (Herod 1758)—Promise 1768 by Snap 1750 (Snip 1736)—Julia 1756 by Blank 1740 (God. Ar. 1724) - Spectator's dam 1735 by Partner 1718 (Jigg).
			Variety, ch. m., 1816.	[Selim or] Soothsayer, ch. h., 1808. Sorcerer 1796 (Trumpator 1782) —Golden Locks 1793 by Delpini 1781 (Highflyer 1774)—Violet 1787 by Shark* 1771 (Marske 1750)—Dau. 1772 of Syphon 1750 (Squirt 1732) - Charlotte 1756 by Blank 1740 (God. Ar. 1724).
				Sprite, ch. m., 1807. Bobtail 1795 (Precipitate* 1787)—Catherine 1795 by Woodpecker 1773 (Herod 1758)—Camilla 1778 by Trentham 1766 (T. Sweepstakes 1743) - Coquette 1764 by Compton Barb—Sist. 1743 to Regulus by God. Ar. 1724.
AEROLITE (Sist. to Idlewild), ch. m., 1862.	Lexington, b. h., 1850.	Boston, ch. h., 1833.	Timoleon, ch. h., 1813.	Sir Archy, b. h., 1805. *Diomed 1777 (Florizel 1768)—*Castianira 1796 by Rockingham 1781 (Highflyer 1774)—Tabitha 1782 by Trentham 1766 (T. Sweepstakes 1743)—Dau. of Bosphorus 1754 (Babraham 1740)—Dau. of W's Forester 1750 (Forester 1736).
				Daughter of, ch. m., —. *Saltram 1780 (Eclipse 1764)—Dau. of Symmes' Wildair 1767 (*Fearnought 1755)- Dau. of Tyler's Driver 1760 (*Othello)—Dau. of *Fallower 1761 (Blank 1740) Dau. of *Vampire 1757 (Regulus 1739).
			Sist. to Tuckahoe, ch. m., 1814.	Ball's Florizel, ch. h., 1801. *Diomed 1777 (Florizel 1768)—Dau. of *Shark 1771 (Marske 1750)—Dau. of Harris's Eclipse 1770 (*Fearnought 1755) - Dau. of *Fearnought 1755 (Regulus 1739)—Dau. of *Jolly Roger 1741 (Roundhead 1733).
				Daughter of, b. m., 1799. *Alderman 1787 (Pot-8-o-'s 1773) Dau. of *Clockfast 1771 (Gimcrack 1760)—Dau. of Symmes' Wildair 1767 (*Fearnought 1755)—Y. Kitty Fisher by *Fearnought 1755 (Regulus 1739)—*Kitty Fisher 1755 by Cade 1734 (God. Ar. 1724).
		Alice Carneal, b. m., 1836.	*Sarpedon, br. h., 1828.	Emilius, b. h., 1820. Orville 1799 (Beningbrough 1791)—Emily 1810 by Stamford 1794 (Sir Peter T. 1784)—Dau. 1799 of Whiskey 1789 (Saltram* 1780) Grey Dorimant 1781 by Dorimant 1772 (Otho 1760) Dizzy 1757 by Blank 1740 (God. Ar. 1724).
				Icaria, b. m., 1824. The Flyer 1814 (Vandyke Jr. 1808)- Parma 1813 by Dick Andrews 1797 (Joe Andrews 1778)—May 1804 by Beningbrough 1791 (King Fergus 1775)—Primrose 1787 by Mambrino 1768 (Engineer 1756) - Cricket 1776 by Herod 1758 (Tartar 1743).
			Rowena, ch. m., 1826.	Sumpter, ch. h., 1818. Sir Archy 1805 (*Diomed 1777)—Dau. of *Robin Redbreast 1796 (Sir Peter T. 1784)—Dau. 1788 of *Obscurity 1777 (Eclipse 1764)—Slamerkin 1768 by *Wildair 1753 (Cade 1734)—*Cub mare 1762 by Cub 1739 (Fox 1714).
				Lady Grey, b. m., 1817. Robin Grey 1805 (*Royalist 1790)—Maria 1802 by Melzar 1791 (*Medley 1773)—Dau. of *Highflyer 1784 (Highflyer 1774)- Dau. of *Fearnought 1755 (Regulus 1739)- Dau. of Ariel 1753 (*Traveller 1747).
	Florine, ch. m., 1854.	*Glencoe, ch. h., 1831.	Sultan, b. h., 1816.	Selim, ch. h., 1802. Buzzard* 1787 (Woodpecker 1773)—Dau. of Alexander 1782 (Eclipse 1764)—Dau. of Highflyer 1774 (Herod 1758)—Dau. of Alfred 1770 (Matchem 1748)—Dau. 1770 of Engineer 1756 (Sampson 1745).
				Bacchante, br. m., 1809. W's Ditto 1800 (Sir Peter T. 1784)—Dau. 1791 of Mercury 1778 (Eclipse 1764)—Dau. 1776 of Herod 1758 (Tartar 1743)—Folly 1771 by Marske 1750 (Squirt 1732)—Vixen 1753 by Regulus 1739 (God. Ar. 1724).
			Trampoline, ch. m., 1825.	Tramp, b. h., 1810. Dick Andrews 1797 (Joe Andrews 1778)—Dau. 1803 of Gohanna 1790 (Mercury 1778)—Fraxinella 1793 by Trentham 1766 (T. Sweep-takes 1743)—Dau. 1785 of Woodpecker 1773 (Herod 1758)—Everlasting 1775 by Eclipse 1764 (Marske 1750).
				Web, b. m., 1808. Waxy 1790 (Pot-8-o-'s 1773) —Penelope 1798 by Trumpator 1782 (Conductor 1767) —Prunella 1788 by Highflyer 1774 (Herod 1758)—Promise 1768 by Snap 1750 (Snip 1736)—Julia 1756 by Blank 1740 (God Ar. 1724).
		Melody, ch. m., 1837.	Medoc, ch. h., 1829.	Am. Eclipse, ch. h., 1814. Duroc 1806 (*Diomed 1777) —Miller's Damsel 1802 by *Messenger 1780 (Mambrino 1768)—*Dau. 1792 of Pot-8-o-'s 1773 (Eclipse 1764) - Dau. 1778 of Gimcrack 1760 (Cripple 1750) Snapdragon 1759 by Snap 1750 (Snip 1736).
				Y. Maid of the Oaks, ch. m., 1817. *Expedition 1795 (Pegasus 1784)—Maid of the Oaks 1801 by *Spread Eagle 1792 (Volunteer 1780)—Annette by *Shark 1771 (Marske 1750)—Dau. of Rockingham (Partner 1755) - Dau. of True Whig (Fitzhugh's Regulus 1765).
			Rodolph's dam, br. m., 1824.	Haxall's Moses, b. h., 1816. *Sir Harry 1795 (Sir Peter T. 1784) *Mermaid by Waxy 1790 (Pot-8-o-'s 1773) *Promise 1803 by Buzzard* 1787 (Woodpecker 1773)—Dau. 1795 of Precipitate* 1787 (Mercury 1778)—Lady Harriet 1773 by Mark Anthony 1767 (Spectator 1749).
				Daughter of, —, —. Cook's Whip 1804 (*Whip 1794) Hart's Maria by Craig's Alfred (*Medley 1776)—The Pryor mare [a good race-mare] by Tayloe's Bellair 1786 (*Medley 1776)—Thoroughbred mare of Mr. T. D. Owings.

IMPORTED GALORE.

WINNER OF THE HARRINGTON, MICHAELMAS, PAYNE STAKES, AND MADE A DEAD HEAT FOR THE ST. JAMES' PALACE STAKES.

The property of R. W. Walden, Bolingbroke Stud, Middlebury, Md.

GALORE, brown horse, foaled 1885, bred by the Duke of Portland, by Galopin, son of Vedette, dam Lady Maura by Macaroni, son of Sweetmeat, out of Noblesse by Stockwell. Galopin, the sire of Galore, won the Derby in 1875, Hyde Park Plate, New Stakes, Fernhill Stakes and Newmarket Derby. Galopin is the sire of the unbeaten St. Simon, Galliard (winner of the 2,000 Guineas), Donovan (winner of the Derby and St. Leger), Atalanta (dam of Ayrshire, winner of the 2,000 Guineas and Derby), and of the dam of Surefoot (winner of the 2,000 Guineas), his dam, Lady Maura, like the dam of Galliard, is by Macaroni, sire also of Macgregor (winner of the 2,000 Guineas), Spinaway and Camelia (winners of the 1,000 Guineas and Oaks), Bonny Jean (winner of the Oaks), Vista, dam of Bonavista (winner of the 2,000 Guineas), and Lily Agnes, dam of Ormonde (winner of the 2,000 Guineas, Derby and St Leger). Noblesse, his grandam, is by the emperor of stallions, Stockwell, and out of Brown Duchess (winner of the Oaks) by The Flying Dutchman, winner of the Derby and St. Leger.

Galore is an extraordinary fine horse and combines in his pedigree all of the most fashionable strains of English blood. He was a superior race-horse, as his performances show.

As a two-year old Galore ranked with the best of his year, never did he contend for a prize except with youngsters of the highest class. He won Harrington Stakes at the Derby Spring meeting, Arrandale (a winner) second, and Seabreeze (the winner of the Oaks and St. Leger the following year) third, with seven others behind. The Michaelmas Stakes at Sandown Park was also placed to his credit, such subsequent winners as St. Symphorien, Blanton, Nunwick, Juggler, Bartizan, etc., being defeated. In the Severals Plate at Newmarket he ran second to Sandal (winner of the Halnaker Stakes, Devonshire Nursery), Galore carrying 134 lbs. and the latter 131 lbs., Hark, Father Confessor and King Galop being in the rear. A year later Galore proved his two-year old form was no fluke by winning the Hastings Plate at Newmarket (distance 1 mile), with such flyers as Hazlehatch, Sheen, Anarch, Sky Pilot, Caerlaverock and three others, those named being all in the winning list, and prominent this year (1889). He won the Payne Stakes at Newmarket (one mile and two furlongs), with Patchouli, Scene Shifter and three others behind him. The Derby was the next effort, but he could get no nearer than fourth to Ayrshire, Crowberry and Van Dieman's Land; Orbit, Chillington and three others were behind him. In the Prince of Wales' Stakes at Ascot he ran second to Ossory, who accompanied Galore on his voyage across the Atlantic, but died during the journey; both these colts carried 125 lbs., giving weight to nearly all their seven opponents. They met again in the St. James' Palace Stakes, each carrying 126 lbs., but so close did they finish that the judges could not separate them, a dead heat being the result and stake divided. In the Zetland Plate, Sheen just managed to avenge the defeat which he had previously suffered, but in the Hamilton Park Stakes, over $1\frac{3}{4}$ miles, eight opponents were easily disposed of.

None of Galore's get have yet raced, but from his high breeding and excellence as a race-horse he should be a success in the stud.

Galore descends in direct female line from the Royal mare, to which Conductor, O'Kelly's Eclipse, Voltaire, Oxford, Kingston, Scottish Chief, Adventurer, Lexington, Vandal, Leonatus, Longstreet, Salvator and other good ones trace.

GALORE.

BROWN HORSE, FOALED 1885; BRED BY THE DUKE OF PORTLAND, ENGLAND.

GALOPIN, b. h., 1872.	Vedette, br. h., 1854.	Voltigeur, br. h., 1847.	Voltaire, br. h., 1826.	Blacklock, b. h., 1814.	Whitelock 1803 (Hambletonian 1792) Dau. 1790 of Coriander 1786 (Pot-8-o's 1773) Wildgoose 1792 by Highflyer 1774 (Herod 1758)—Coheiress 1786 by Pot-8-o's 1773 (Eclipse 1764) Manilla 1777 by Goldfinder 1764 (Snap 1750).
				Daughter of, b. m., 1816.	Phantom 1808 (Walton 1799)—Dau. 1802 of Overton 1788 (King Fergus 1775) Dau. 1796 of Walnut 1786 (Highflyer 1774) Dau. 1786 of Ruler 1777 (Y. Marske 1771) Piracantha 1772 by Matchem 1748 (Cade 1734).
			Martha Lynn, br. m., 1837.	Mulatto, b. h., 1823.	Catton 1809 (Golumpus 1802)—Desdemona 1811 by Orville 1799 (Beningbrough 1791)—Fanny 1756 by Sir Peter T. 1784 (Highflyer 1774) Dau. 1788 of Diomed 1777 (Florizel 1768) Desdemona 1770 by Marske 1750 (Squirt 1732).
				Leda, b. m., 1824.	Filho-da-Puta 1812 (Haphazard 1797) Treasure 1803 by Camillus 1803 (Hambletonian 1792) Dau. 1804 of Hyacinthus 1797 (Coriander 1786) Flora 1789 by King Fergus 1775 (Eclipse 1764) Atalanta 1769 by Matchem 1748 (Cade 1734).
		Mrs. Ridgway, b. m., 1849.	Birdcatcher, ch. h., 1833.	Sir Hercules, blk. h., 1826.	Whalebone 1807 (Waxy 1790) Peri 1822 by Wanderer 1811 (Gohanna 1790) Thalestris 1809 by Alexander 1782 (Eclipse 1764) Rival 1800 by Sir Peter T. 1784 (Highflyer 1774)—Hornet 1790 by Drone 1777 (Herod 1758).
				Guiccioli, ch. m., 1823.	Bob Booty 1804 (Chanticleer 1787)—Flight 1809 by Ir. Escape 1802 (Commodore 1793) Y. Heroine by Bagot 1780 (Herod 1758)—Heroine 1775 by Hero 1753 (Cade 1734) Dau. 1766 of Snap 1750 (Snip 1736).
			Nan Darell, gr. m., 1844.	Inheritor, blk. h., 1831.	Lottery 1820 (Tramp 1810)—Hand Maiden 1817 by Walton 1799 (Sir Peter T. 1784) Anticipation 1802 by Beningbrough 1791 (King Fergus 1775) Expectation 1799 by Herod 1758 (Tartar 1743)—Dau. 1758 of Skim 1746 (Starling 1727).
				Nell, gr. m., 1831.	Blacklock 1814 (Whitelock 1803) Madame Vestris 1818 by Comus 1809 (Sorcerer 1796)—Lisette 1806 by Hambletonian 1792 (King Fergus 1775) Constantia 1796 by Walnut 1786 (Highflyer 1774)—Contessina 1787 by Y. Marske 1771 (Marske 1750).
	Flying Duchess, b. m., 1853.	The Flying Dutchman, br. h., 1846.	Bay Middleton, b. h., 1833.	Sultan, b. h., 1816.	Selim 1802 (Buzzard* 1787) Bacchante 1809 by W's Ditto 1800 (Sir Peter T. 1784) Dau. 1794 of Mercury 1778 (Eclipse 1764) Dau. 1776 of Herod 1758 (Tartar 1743) Folly 1771 by Marske 1750 (Squirt 1732).
				Cobweb, b. m., 1821.	Phantom 1808 (Walton 1799) Filagree 1815 by Soothsayer 1808 (Sorcerer 1796) Web 1808 by Waxy 1790 (Pot-8-o's 1773) Penelope 1798 by Trumpator 1782 (Conductor 1767)—Prunella 1788 by Highflyer 1774 (Herod 1758).
			Barbelle, b. m., 1836.	Sandbeck, b. h., 1818.	Catton 1803 (Golumpus 1802)—Orvillina 1804 by Beningbrough 1791 (King Fergus 1775)—Evelina 1791 by Highflyer 1774 (Herod 1758)—Termagant by Tantrum 1760 (Cripple 1750)—Cantatrice by Sampson 1745 (Blaze 1733).
				Darioletta, br. m., 1822.	Amadis 1807 (Don Quixote 1784)—Selima 1810 by Selim 1802 (Buzzard* 1787) Dau. 1791 of Pot-8-o's 1773 (Eclipse 1764)—Editha 1781 by Herod 1758 (Tartar 1743) Elfrida 1768 by Snap 1750 (Snip 1736).
		Merope, b. m., 1841.	Voltaire, br. h., 1826.	Blacklock, b. h., 1814.	Whitelock 1803 (Hambletonian 1792)—Dau. 1790 of Coriander 1786 (Pot-8-o's 1773)—Wildgoose 1792 by Highflyer 1774 (Herod 1758)—Coheiress 1786 by Pot-8-o's 1773) (Eclipse 1764) Manilla 1777 by Goldfinder 1764 (Snap 1750).
				Daughter of, b. m., 1816.	Phantom 1808 (Walton 1799)—Dau. 1802 of Overton 1788 (King Fergus 1775) Dau. 1796 of Walnut 1786 (Highflyer 1774)—Dau. 1786 of Ruler 1777 (Y. Marske 1771)—Piracantha 1772 by Matchem 1748 (Cade 1734).
			Velocipede's dam, 1817.	Juniper, ch. h., 1805.	Whiskey 1789 (Saltram* 1780)—Jenny Spinner 1797 by Dragon* 1787 (Woodpecker 1773)—Dau. 1775 of Eclipse 1764 (Marske 1750) Miss Spindleshanks 1770 by Omar 1752 (God. Ar. 1724)—Dau. 1753 of Starling 1727 (Bay Bolton 1705).
				Daughter of, br. m., 1810.	Sorcerer 1796 (Trumpator 1782)—Virgin 1801 by Sir Peter T. 1784 (Highflyer 1774)—Dau. 1794 of Pot-8-o's 1773 (Eclipse 1764)—Editha 1781 by Herod 1758 (Tartar 1743)—Elfrida 1768 by Snap 1750 (Snip 1736).
LADY MAURA, b. m., 1879.	Macaroni, b. h., 1860.	Sweetmeat, br. h., 1842.	Gladiator, ch. h., 1833.	Partisan, b. h., 1811.	Walton 1799 (Sir Peter 1784)—Parasol 1800 by Pot-8-o's 1773 (Eclipse 1764)—Prunella 1788 by Highflyer 1774 (Herod 1758)—Promise 1768 by Snap 1750 (Snip 1736)—Julia 1756 by Blank 1740 (God. Ar. 1724).
				Pauline, b. m., 1826.	Moses 1819 ([Whalebone or] Seymour 1807)—Quadrille 1815 by Selim 1802 (Buzzard* 1787)—Canary Bird 1806 by [Whiskey or] Sorcerer 1796 (Trumpator 1782)—Canary 1797 by Coriander 1786 (Pot-8-o's 1773) Miss Green 1787 by Highflyer 1774 (Herod 1758).
			Lollypop, blk. m., 1836.	[Starch or] Voltaire, br. h., 1826.	Blacklock 1814 (Whitelock 1803)—Dau. 1816 of Phantom 1808 (Walton 1799)—Dau. 1802 of Overton 1788 (King Fergus 1775) Dau. 1796 of Walnut 1786 (Highflyer 1774)—Dau. 1786 of Ruler 1777 (Y. Marske 1771).
				Belinda, br. m., 1825.	Blacklock 1814 (Whitelock 1803)—Wagtail 1818 by Prime Minister 1810 (Sancho 1801)—Tranby*'s dam 1812 by Orville 1799 (Beningbrough 1791)—Miss Grimstone 1796 by Weazle 1776 (Herod 1758)—Dau. of Ancaster 1768 (Blank 1740).
		Jocose, b. m., 1843.	Pantaloon, ch. h., 1824.	Castrel, ch. h., 1801.	Buzzard* 1787 (Woodpecker 1773)—Dau. of Alexander 1782 (Eclipse 1764)—Dau. of Highflyer 1774 (Herod 1758)—Dau. of Alfred 1770 (Matchem 1748)—Dau. 1770 of Engineer 1756 (Sampson 1745).
				Idalia, ch. m., 1815.	Peruvian 1806 (Sir Peter T. 1784)—Musidora 1804 by Meteor 1783 (Eclipse 1764)—Maid-of-all-Work 1786 by Highflyer 1774 (Herod 1758)—Dau. 1771 of Syphon 1750 (Squirt 1732)—Dau. of Regulus 1739 (God. Ar. 1724).
			Banter, br. m., 1826.	Master Henry, b. h., 1815.	Orville 1799 (Beningbrough 1791)—Miss Sophia 1805 by Stamford 1794 (Sir Peter T. 1784)—Sophia 1798 by Buzzard* 1787 (Woodpecker 1773)—Huncamunca 1787 by Highflyer 1774 (Herod 1758)—Cypher 1772 by Squirrel 1754 (Traveller 1735).
				Boadicea, b. m., 1807.	Alexander 1782 (Eclipse 1764)—Brunette by Amaranthus 1766 (Old England 1741)—Mayfly 1771 by Matchem 1748 (Cade 1734) Dau. 1755 of Anc. Starling 1738 (Starling 1727) Look-at-me-Lads 1731 by Brist. Grasshopper (Byerly T.).
	Noblesse, b. m., 1868.	Stockwell, ch. h., 1849.	The Baron, ch. h., 1842.	Ir. Birdcatcher ch. h., 1833.	Sir Hercules 1826 (Whalebone 1807)—Guiccioli 1823 by Bob Booty 1804 (Chanticleer 1787)—Flight 1809 by Ir. Escape 1802 (Commodore 1793)—Y. Heroine by Bagot 1780 (Herod 1758) Heroine 1775 by Hero 1753 (Cade 1734).
				Echidna, b. m., 1838.	Economist 1825 (Whisker 1812)—Miss Pratt 1825 by Blacklock 1814 (Whitelock 1803)—Gadabout 1812 by Orville 1799 (Beningbrough 1791)—Minstrel 1803 by Sir Peter T. 1784 (Highflyer 1774)—Matron 1782 by Florizel 1768 (Herod 1758).
			Pocahontas, b. m., 1837.	Glencoe,* ch. h., 1831.	Sultan 1816 (Selim 1802)—Trampoline 1825 by Tramp 1810 (Dick Andrews 1797)—Web 1808 by Waxy 1790 (Pot-8-o's 1773)—Penelope 1798 by Trumpator 1782 (Conductor 1767)—Prunella 1788 by Highflyer 1774 (Herod 1758).
				Marpessa, b. m., 1830.	Muley 1810 (Orville 1799)—Clare 1824 by Marmion 1806 (Whiskey 1789)—Harpalice 1814 by Gohanna 1790 (Mercury 1778)—Amazon 1799 by Driver 1783 (Trentham 1766)—Fractious 1792 by Mercury 1778 (Eclipse 1764).
		Brown Duchess, br. m., 1858.	The Flying Dutchman, br. h., 1846.	Bay Middleton, b. h., 1833.	Sultan 1816 (Selim 1802)—Cobweb 1821 by Phantom 1808 (Walton 1799) Filagree 1815 by Soothsayer 1808 (Sorcerer 1796)—Web 1808 by Waxy 1790 (Pot-8-o's 1773)—Penelope 1798 by Trumpator 1782 (Conductor 1767).
				Barbelle, b. m., 1836.	Sandbeck 1818 (Catton 1803)—Darioletta 1822 by Amadis 1807 (Don Quixote 1784)—Selima 1810 by Selim 1802 (Buzzard* 1787)—Dau. 1791 of Pot-8-o's 1773 (Eclipse 1764)—Editha 1781 by Herod 1758 (Tartar 1743).
			Espoir, b. m., 1841.	Liverpool, b. h., 1828.	Tramp 1810 (Dick Andrews 1797)—Dau. 1822 of Whisker 1812 (Waxy 1790)—Mandane 1800 by Pot-8-o's 1773 (Eclipse 1764)—Y. Camilla 1787 by Woodpecker 1773 (Herod 1758)—Camilla 1778 by Trentham 1766 (T. Sweepstakes 1743).
				Esperance, b. m., 1836.	Lapdog* 1823 (Whalebone 1807)—Grisette ch. m. 1825 by Merlin 1815 (Castrel 1801)—Coquette ch. m. 1814 by Dick Andrews 1797 (Joe Andrews 1778)—Vanity b. m. 1803 by Buzzard* 1787 (Woodpecker 1773)—Dabchick br. m. 1798 by Pot-8-o's 1773 (Eclipse 1764)—Drab b. m. 1791 by Highflyer 1774 (Herod 1758)—Hebe b. m. 1774 by Chrysolite 1763 (Blank 1740)—Proserpine (sist. to Eclipse) b. m. 1766 by Marske 1750 (Squirt 1732)—Spiletta b. m. 1749 by Regulus 1739 (God. Ar. 1724)—Mother Western by Easby Snake (Snake)—Montagu mare by D'Arcy's old Montagu—Dau. of Hautboy (D'Arcy White Turk)—Dau. of Brimmer (D'Arcy Yellow Turk)—Royal mare.

IMPORTED GREAT TOM.

WINNER OF ST. JAMES' PALACE STAKES AT ASCOT, AND DONCASTER STAKES AT DONCASTER, IN 1876.

Property of Gen. W. H. Jackson, Belle Meade Stud, Nashville, Tenn. Annual sale of yearlings about May 1st.

GREAT TOM by King Tom, son of Harkaway, bred by Lord Falmouth, foaled 1873, dam Woodcraft, the dam of Kingcraft, Andred and Anderida by Voltigeur, winner of the Derby, St. Leger and Doncaster Cup, out of Falcon and Ostreger's dam by Venison. Great Tom started once at two years old: at Newmarket First October Meeting, for the Boscawen Post Stakes of 100 sovereigns each, for two-year olds, three-quarters of a mile (six subs.); won by Twine the Plaiden, 119 lbs., Great Tom, 122 lbs., unplaced. At three years old started nine times; won two, was second three times, third twice, and unplaced in two. Newmarket Craven Meeting for Post Sweepstakes of 100 sovereigns each, half forfeit, for three year olds, Ditch mile, 7 furlongs and 220 yards, Wild Tommy, 122 lbs., was first, Great Tom, 122 lbs., second, Coltness, 118 lbs., third, won by a head, a length and a half between second and third. At Newmarket, for Two Thousand Guineas Stakes, Rowley mile, 1 mile and 17 yards, won by Petrach, Great Tom was unplaced. At Epsom, for the Derby 1½ miles, won by Kisber, Forerunner second, Julius Cæsar third, Great Tom was unplaced, fifteen started. At Ascot, for the Prince of Wales Stakes, 1 mile and 5 furlongs, Petrarch, 127 lbs., first, Great Tom, 115 lbs., second, Julius Cæsar, 122 lbs., third, won by a length. St. James' Palace Stakes, old mile (12 subs.), Great Tom, 115 lbs., and Glacis, 115 lbs., ran a dead heat. Great Tom afterwards walked over, and the stakes were divided. At Doncaster, for the Doncaster Stakes, 1½ miles (36 subs.), Great Tom, 122 lbs., was first, Coltness, 127 lbs., second, Morning Star, 122 lbs., third, won by two lengths, three lengths between second and third. At Newmarket, for the Triennial Produce Stakes, A. F., 1 mile 2 furlongs 73 yards, won by Camembert, Twine the Plaiden second, Great Tom third. Beaufort Post Stakes, Rowley mile (5 subs.), won by Twine the Plaiden, Great Tom second, Wild Tommy third, Camembert fourth. Newmarket Derby, 1½ miles, Lord Falmouth's Skylight, 123 lbs., first, Moulin, 123 lbs., second, Great Tom, 123 lbs., third, won by a length, two lengths between second and third.

At four years old he started six times and won once, second once, third twice, and unplaced twice. At Newmarket First October Meeting for the Triennial Produce Stakes, Ditch In, 2 miles and 105 yards, Augusta was first, Great Tom second, Footstep third. The Cesarewitch Handicap Stakes, 2¼ miles and 28 yards, won by Hilarious, 3 years, 88 lbs., Great Tom was unplaced. The Champion Stakes, Across the Flat, 1¼ miles and 73 yards, Springfield, 4 years, 130 lbs., first, Silvio, 3 years, 118 lbs., second, Great Tom, 4 years, 130 lbs., third. Winding-up Handicap, Rowley mile, 1 mile and 17 yards, Great Tom, 3 years, 117 lbs., was first, Augusta, 4 years, 105 lbs., second, Sheldrake, 3 years, 91 lbs., third, won by three lengths, a head between second and third. At Shrewsbury, for the Queen's Plate, about 2½ miles, Sheldrake, 3 years old, 117 lbs., was first, Redoubt, 3 years old, 117 lbs., second, Great Tom, 4 years old, 133 lbs., third. Was unplaced for the Hawkstone Welter Cup, New mile, won by Speculation, 3 years old, 134 lbs.

King Tom, sire of Great Tom, was by Harkaway, the best race-horse of his day. He was not in either the Derby or St. Leger, but won 21 out of the 28 races in which he started up to the end of his four-year old year, including eight King's Plates, the Goodwood Cup and the Royal Whip, beating all the best horses on the Irish and English turf, carrying the top weights. He also won the Goodwood Cup when five years old with 130 lbs.

Great Tom is a rich, golden chestnut, with a white stripe down his face, running down between his nostrils, and two white feet behind, half way to the hock. He measures 16 hands 2 inches high; around the girth, 77 inches; length of shoulder, 32 inches; around the arm, 22 inches; below the knee, 9 inches; from point of hip to point of hock, 40 inches; circumference of gaskin, 19 inches; from point of shoulder to point of buttocks, 69 inches, and weighs 1,305 lbs. Among the best of his get are Tyrant (winner of the Withers and Belmont), General Harding (the best two-year old of 1883, winner of nine out of ten races), Tennyson, Telie Doe (with the fastest seven furlongs on record, 1.27¼, up to that time), Thackeray, Trombone, Treasurer, Tatton, Standiford Keller, Tabitha, Whizgig, Tartar, Thistle, Belle Pate, Cremorne, Swift, Watterson, Captain Wagener, Calcium, Dutchman, Hilda, Mayor Nolan, Tarquin, John C., Tally Ho, Hildegarde, Brentano, Honeymoon, Great Dixon, Jack Richelieu, Little Billie, Ella, Tullahoma, Tocsin, Tallapoosa (Proctor Knott's dam), Tom Boy, Tea Rose, Hattie S. Bliss, Tomahawk, Toribera, Tantrum, Graduate, Great Guns, Simon Pure, Tangent, Jase Phillips, Soprano, Gipsy and many others.

GREAT TOM.

CHESTNUT HORSE, FOALED 1873; BRED BY LORD FALMOUTH, ENGLAND.

KING TOM, b. h., 1851.	Harkaway, ch. h., 1834.	Economist, b. h., 1825.	Whisker, b. h., 1812.	Waxy, b. h., 1790.	Pot-8-o-'s 1773 (Eclipse 1764)—Maria 1777 by Herod 1758 (Tartar 1743) Lisette 1772 by Snap 1750 (Snip 1736) Miss Windsor 1754 by God. Ar. 1724 Sist. to Volunteer by Y. Belgrade (Belgrade Turk).
				Penelope, b. m., 1798.	Trumpator 1782 (Conductor 1767)—Prunella 1788 by Highflyer 1774 (Herod 1758) Promise 1768 by Snap 1750 (Snip 1736) Julia 1756 by Blank 1740 (God. Ar. 1724) Spectator's dam 1735 by Partner 1718 (Jigg).
			Floranthe, b. m., 1818.	Octavian, ch. h., 1807.	Stripling 1795 (Phenomenon 1780) Dau. 1787 of Oberon 1790 (Highflyer 1774) Sist. 1779 to Sharper by Ranthos 1763 (Matchem 1748)—Dau. of T. Sweepstakes 1743 (B. Sweepstakes 1722) Sist. 1718 to Careless by Spot 1728 (Hartley's Blind Horse).
				Caprice, b. m., 1797.	Anvil 1777 (Herod 1758) Madcap 1774 by Eclipse 1764 (Marske 1750) Sappho 1763 by Blank 1740 (God. Ar. 1724) Dau. 1750 of Blaze 1733 (Fl. Childers 1715 Dau. of Y. Greyhound 1723 (Greyhound).
		Fanny Dawson, ch. m., 1823.	Nabocklish, ch. h., 1811.	Rugantino, ch. h., 1804.	Commodore 1793 (Tom Tug 1777) Moll-in-the-Wad 1791 by Highflyer 1774 (Herod 1758) Shift 1779 by Sweetbriar 1769 (Syphon 1750) Black Susan 1762 by Snap 1750 (Snip 1736)—Dau. of Cade 1731 (God. Ar. 1724).
				Butterfly, ch. m., 1804.	Master Bagot 1787 (Bagot 1780) Dau. of Bagot 1780 (Herod 1758)—Mother Brown 1771 by Trunnion 1747 (Cade 1734) Dau. of Old England 1741 (God. Ar. 1724)—Jenny Jessamy 1748 by Starling 1727 (Bay Bolton 1705).
			Miss Tooly, b. m., 1808.	Teddy the Grinder, b. h., 1798.	Asparagus 1787 (Pot-8-o-'s 1773) Stargazer 1782 by Highflyer 1774 (Herod 1758) Miss West 1777 by Matchem 1748 (Cade 1731) Sist. to Favourite by Regulus 1739 (God. Ar. 1724)—Y. Ebony 1752 by Crab 1722 (Alcock Ar.).
				Lady Jane, b. m., 1796.	Sir Peter T. 1784 (Highflyer 1774)—Paulina 1778 by Florizel 1768 (Herod 1758) Captive 1771 by Matchem 1748 (Cade 1734) Calliope 1763 by Slouch 1747 (Cade 1734)—Y. Lass-of-the-Mill 1776 by Oroonoko 1745 (Crab 1722).
	Pocahontas, b. m., 1837.	Glencoe, ch. h., 1831.	Sultan, b. h., 1816.	Selim, ch. h., 1802.	Buzzard* 1787 (Woodpecker 1773)—Dau. of Alexander 1782 (Eclipse 1764) Dau. of Highflyer 1774 (Herod 1758) Dau. of Alfred 1770 (Matchem 1748)—Dau. 1770 of Engineer 1756 (Sampson 1745).
				Bacchante, br. m., 1809.	W's Ditto 1800 (Sir Peter T. 1784) Dau. 1791 of Mercury 1778 (Eclipse 1764 Dau. 1776 of Herod 1758 (Tartar 1743)—Folly 1771 by Marske 1750 (Squirt 1732)—Vixen 1773 by Regulus 1739 (God. Ar. 1724).
			Trampoline, ch. m., 1825.	Tramp, b. h., 1810.	Dick Andrews 1797 (Joe Andrews 1778)—Dau. 1803 of Gohanna 1790 (Mercury 1778)—Fraxinella 1793 by Trentham 1766 (T. Sweepstakes 1743) Dau. 1785 of Woodpecker 1773 (Herod 1758)—Everlasting 1775 by Eclipse 1764 (Marske 1750).
				Web, b. m., 1808.	Waxy 1790 (Pot-8-o-'s 1773) Penelope 1798 by Trumpator 1782 (Conductor 1767) Prunella 1788 by Highflyer 1774 (Herod 1758) Promise 1768 by Snap 1750 (Snip 1736)—Julia 1756 by Blank 1740 (God. Ar. 1724).
		Marpessa, b. m., 1830.	Muley, b. h., 1810.	Orville, b. h., 1799.	Beningbrough 1791 (King Fergus 1775) Evelina 1791 by Highflyer 1774 (Herod 1758) Termagant by Tantrum 1760 (Cripple 1750)—Cantatrice by Sampson 1745 (Blaze 1733) Dau. 1757 of Regulus 1739 (God. Ar. 1724).
				Eleanor, b. m., 1798.	Whiskey 1789 (Saltram* 1780)—Y. Giantess 1790 by Diomed* 1777 (Florizel 1768) Giantess 1769 by Matchem 1748 (Cade 1734) Molly Long Legs 1753 by Babraham 1740 (God. Ar. 1724)—Dau. of Cole's Foxhunter 1727 (Brisk 1711).
			Clare, b. m., 1824.	Marmion, b. h., 1806.	Whiskey 1789 (Saltram* 1780) Y. Noisette 1798 by Diomed* 1777 (Florizel 1768)—Noisette 1774 by Squirrel 1754 (Traveller 1735)—Carina 1766 by Marske 1750 (Squirt 1732) Dau. 1755 of Blank 1740 (God. Ar. 1724).
				Harpalice, b. m., 1814.	Gohanna 1790 (Mercury 1778)—Amazon 1790 by Driver 1783 (Trentham 1766)—Fractious 1792 by Mercury 1778 (Eclipse 1764)—Dau. 1785 of Woodpecker 1773 (Herod 1758)—Everlasting 1775 by Eclipse 1764 (Marske 1750).
WOODCRAFT, b. m., 1861.	Voltigeur, br. h., 1847.	Voltaire, br. h., 1826.	Blacklock, b. h., 1814.	Whitelock, b. h., 1803.	Hambletonian 1792 (King Fergus 1775)—Rosalind 1788 by Phenomenon 1780 (Herod 1758)—Atalanta 1769 by Matchem 1748 (Cade 1734)—Y. Lass-of-the-Mill 1756 by Oroonoko 1745 (Crab 1722)—Dau. 1744 of Traveller 1735 (Partner 1718).
				Daughter of, —, 1799.	Coriander 1786 (Pot-8-o-'s 1773)—Wildgoose 1792 by Highflyer 1774 (Herod 1758)—Coheiress 1786 by Pot-8-o-'s 1773 (Eclipse 1764)—Camilla 1777 by Goldfinder 1764 (Snap 1750)—Dau. 1766 of Old England 1741 (God. Ar. 1724).
			Daughter of, b. m., 1816.	Phantom, b. h., 1808.	Walton 1799 (Sir Peter T. 1784)—Julia 1799 by Whiskey 1789 (Saltram* 1780)—Y. Giantess 1790 by Diomed* 1777 (Florizel 1768)—Giantess 1769 by Matchem 1748 (Cade 1734)—Molly Long Legs 1753 by Babraham 1740 (God. Ar. 1724).
				Daughter of, b. m., 1802.	Overton 1788 (King Fergus 1775)—Dau. 1796 of Walnut 1786 (Highflyer 1774)—Dau. 1786 of Ruler 1777 (Y. Marske 1771)—Piracantha 1772 by Matchem 1748 (Cade 1731)—Prophetess 1758 by Regulus 1739 (God. Ar. 1724).
		Martha Lynn, br. m., 1837.	Mulatto, b. h., 1823.	Catton, b. h., 1809.	Golumpus 1802 (Gohanna 1790) Lucy Gray 1804 by Timothy 1794 (Delpini 1781)—Lucy 1789 by Florizel 1768 (Herod 1758)—Frenzy 1774 by Eclipse 1764 (Marske 1750)—Dau. of Engineer 1756 (Sampson 1745).
				Desdemona, br. m., 1811.	Orville 1799 (Beningbrough 1791) Fanny 1796 by Sir Peter T. 1784 (Highflyer 1774)—Dau. 1788 of Diomed* 1777 (Florizel 1768)—Desdemona 1770 by Marske 1750 (Squirt 1732)—Y. Hag 1761 by Skim 1746 (Starling 1727).
			Leda, b. m., 1824.	Filho-da-Puta, br. h., 1812.	Haphazard 1797 (Sir Peter T. 1784) Mrs. Barnet 1806 by Waxy 1790 (Pot-8-o-'s 1773)—Dau. 1788 of Woodpecker 1773 (Herod 1758)—Heinel 1771 by Squirrel 1754 (Traveller 1735)—Principessa 1762 by Blank 1740 (God. Ar. 1724).
				Treasure, ch. m., 1809.	Camillus 1803 (Hambletonian 1792)—Dau. 1804 of Hyacinthus 1797 (Coriander 1786)—Flora 1789 by King Fergus 1775 (Eclipse 1764)—Atalanta 1769 by Matchem 1748 (Cade 1734)—Y. Lass-of-the-Mill 1756 by Oroonoko 1745 (Crab 1722).
	Daughter of b. m., 1849.	Venison, br. h., 1833.	Partisan, b. h., 1811.	Walton, b. h., 1799.	Sir Peter T. 1784 (Highflyer 1774)—Arethusa 1792 by Dungannon 1780 (Eclipse 1764)—Dau. 1777 of Prophet 1760 (Regulus 1739)—Virago 1764 by Snap 1750 (Snip 1736)—Dau. of Regulus 1739 (God. Ar. 1724).
				Parasol, b. m., 1800.	Pot-8-o-'s 1773 (Eclipse 1764)—Prunella 1788 by Highflyer 1774 (Herod 1758)—Promise 1768 by Snap 1750 (Snip 1736)—Julia 1756 by Blank 1740 (God. Ar. 1724)—Spectator's dam 1735 by Partner 1718 (Jigg).
			Fawn, br. m., 1823.	Smolensko, blk. h., 1810.	Sorcerer 1796 (Trumpator 1782)—Wowski 1797 by Mentor 1784 (Justice 1774)—Maria 1777 (Waxy's dam) by Herod 1758 (Tartar 1743)—Lisette 1772 by Snap 1750 (Snip 1736)—Miss Windsor 1754 by God. Ar. 1724.
				Jerboa, b. m., 1803.	Gohanna 1790 (Mercury 1778)—Camilla 1778 by Trentham 1766 (T. Sweepstakes 1743)—Coquette 1764 by Compton Barb—Sist. 1743 to Regulus by God. Ar. 1724—Grey Robinson 1723 by Bald Galloway (St. Victor Barb).
		Wedding Day, b. m., 1842.	Camel, br. h., 1822.	Whalebone, br. h., 1807.	Waxy 1790 (Pot-8-o-'s 1773)—Penelope 1798 by Trumpator 1782 (Conductor 1767)—Prunella 1788 by Highflyer 1774 (Herod 1758)—Promise 1768 by Snap 1750 (Snip 1736)—Julia 1756 by Blank 1740 (God. Ar. 1724).
				Daughter of, b. m., 1812.	Selim 1802 (Buzzard* 1787)—Maiden 1801 by Sir Peter T. 1784 (Highflyer 1774)—Dau. 1788 of Phenomenon 1780 (Herod 1758)—Matron 1782 by Florizel 1768 (Herod 1758)—Maiden 1770 by Matchem 1748 (Cade 1731).
			Margellina, b. m., 1826.	Whisker, b. h., 1812.	Waxy 1790 (Pot-8-o-'s 1773)—Penelope 1798 by Trumpator 1782 (Conductor 1767)—Prunella 1788 by Highflyer 1774 (Herod 1758)—Promise 1768 by Snap 1750 (Snip 1736)—Julia 1756 by Blank 1740 (God. Ar. 1724).
				Manuella, b. m., 1809.	Dick Andrews 1797 (Joe Andrews 1778)—Mandane ch. m. 1800 by Pot-8-o-'s 1773 (Eclipse 1764)—Y. Camilla b. m. 1787 by Woodpecker 1773 (Herod 1758)—Camilla b. m. 1778 by Trentham 1766 (T. Sweepstakes 1743)—Coquette b. m. 1764 by Compton Barb—Sist. b. m. 1743 to Regulus by God. Ar. 1724—Grey Robinson gr. m. 1723 by Bald Galloway (St. Victor Barb)—Sist. to Old Country Wench by Snake (Lister Turk)—Grey Wilkes gr. m. by Hautboy (D'Arcy White Turk)—Miss D'Arcy's Pet mare out of a Sedbury Royal mare.

GRINSTEAD.

WINNER OF THE CHAMPAGNE STAKES AT JEROME PARK, MANSION HOUSE STAKES AT LONG BRANCH, ALL-AGED SWEEPSTAKES AND SUMMER HANDICAP AT SARATOGA.

The Property of E. J. Baldwin, Savannah, Los Angelos Co., California.

GRINSTEAD by Gilroy, son of Lexington, bred by Jas. A. Grinstead, Walnut Hill Stud, Ky., foaled 1871, dam sister to Ruric by imp. Sovereign, son of Emilius, out of Levity, dam of Lever, Mildred, Legatee, etc., by imp. Trustee. Grinstead, as a two-year old, started in two races, of which he won one. At Jerome Park, the Champagne Stakes for two-year olds, ¾ of a mile, in 1.17¾, beating Dublin, Weathercock and seven others. At Saratoga, was second in the Kentucky Stakes for two-year olds, 1 mile, won by Battle-Axe in 1.45½. Three years old, started twelve times and won three races. Jerome Park, won a dash of 1¼ miles in 2.40¾, beating Kadi and seven others, among which were Harry Bassett, Whisper and Spindrift. Won a dash of 1¾ miles in 3.10, beating Shylock, Lyttleton, Wildidle, Kadi and others. Baltimore, Md., won a compensation purse, mile heats, in 1.45½, 1.45½, beating six others, among which were Aaron Pennington and Survivor. Jerome Park, was second in the Belmont Stakes, 1½ miles, won by imp. Saxon, in 2.39½. Among the starters were Aaron Pennington, Elkhorn, Rutherford, Brigand and Reform. Saratoga, was second in the Sequel Stakes for three-year olds, 2 miles, won by Vandalite, in 3.40¾. Was third in a dash of 1 mile and a furlong, won by Madge, in 1.57¼. Was third in a dash of 2 miles, won by Culpepper, in 3.40¼. Jerome Park, was second in a handicap, 1½ miles, won by Skylock, in 2.38. Four years old, started in eleven races, of which he won seven. Jerome Park, won a race of heats of 1 mile and a furlong, in 1.59¾, 2.01¾, beating Rhadamanthus and Fadladeen. Monmouth Park, won a race of mile heats, in 1.45¼, 1.45, 1.47. D'Artagnan won first heat and Grinstead was fifth. Won Mansion House Stakes for four-year olds, 2½ miles, in 4.40½, beating Rutherford, Stampede and two others. Saratoga, won Sweepstakes for all ages, 1¼ miles, 2.08¾, beating Springbok, Olitipa and Mate. Won the Summer Handicap, 2 miles, in 3.37½, beating Wildidle and Mattie W. Won a dash of 1½ miles in 2.40, beating Aaron Pennington, Scratch and two others. Jerome Park, was third in the Jockey Club Handicap, 2 miles, in 3.38½, won by Wildidle. Was third in the Woodburn Stakes for four-year olds, 2½ miles, won by Aaron Pennington, in 4.36½. Saratoga, was third in the Saratoga Cup, 2¼ miles, Springbok and Preakness running a dead heat for first place, in 3.56¼, which still stands the fastest time on record at the distance. Reno, Nevada, won race, mile heats, three in five, in 1.47, 1.46½, 1.46½. Bay District Course, Cal, was second in the Wise Plate, a post stake, 4 miles, won by Wildidle, in 7.25½. As a five-year old started once and was unplaced. Grinstead is a blood bay horse, full 16 hands, of great substance and power, with excellent shoulders, good body and strong, well placed back, hips and loin. He was a good, honest race-horse and comes from a grand racing family, and a host of good ones have descended from it. Grinstead has covered only a few mares. He sired Gano, winner of the Conner Stakes, Grisner, Bonita, Goliah, Ed. McGinnis, Geraldine, Glen Almond, Ganymede, Gladiator, Honduras, Cleopatra, Clio, Costa Rica, Hemet, Puente, Santiago, Sinaloa II., Esperanza, Winona, Salonica, El Rayo, Santa Anna and many other winners. It is to the Royal mare through old Montagu mare that trace in direct female line Conductor, Eclipse, Sterling, Weatherbit, Oxford, Kingston, Scottish Chief, Adventurer, Lexington, Vandal, Leonatus, Salvator, Onondaga, Stratford, Sensation and all the Duchess family.

GRINSTEAD.

BAY HORSE, FOALED 1873; BRED BY JAS. A. GRINSTEAD, WALNUT HILL STUD, KENTUCKY.

Parents	Grandparents	Great-grandparents	Great-great-grandparents	Great-great-great-grandparents	Further ancestry
GILROY, b. h., 1862.	Lexington, b. h., 1850.	Boston, ch. h., 1833.	Timoleon, ch. h., 1813.	Sir Archy, b. h., 1805.	*Diomed 1777 (Florizel 1768)—Castianira 1796 by Rockingham 1781 (Highflyer 1774)—Tabitha 1782 by Trentham 1766 (T. Sweepstakes 1743)—Dau. of Bosphorus 1754 (Babraham 1740)—Dau. of W's Forester 1750 (Forester 1736).
				Daughter of, ch. m.	Saltram 1780 (Eclipse 1764)—Dau. of Symmes' Wildair 1767 (*Fearnought 1755)—Dau. of Tyler's Driver 1761 (Othello)—Dau. of *Fallower 1761 (Blank 1740)—Dau. of Vampire 1757 (Regulus 1739).
			Sister to Tuckahoe, ch. m., 1814.	Ball's Florizel, ch. h., 1801.	*Diomed 1777 (Florizel 1768)—Dau. of Shark 1771 (Marske 1750)—Dau. of Harris' Eclipse 1770 (*Fearnought 1755)—Dau. of *Fearnought 1755 (Regulus 1739)—Dau. of *Jolly Roger 1741 (Roundhead 1733).
				Daughter of, b. m., 1799.	Alderman 1787 (Pot-8-o's 1773)—Dau. of *Clockfast 1774 (Gimcrack 1760)—Dau. of Symmes' Wildair 1767 (*Fearnought 1755)—Y. Kitty Fisher 1767 by *Fearnought 1755 (Regulus 1739)—*Kitty Fisher 1755 by Cade 1734 (God. Ar. 1724).
		Alice Carneal, b. m., 1836.	*Sarpedon, br. h., 1828.	Emilius, b. h., 1820.	Orville 1799 (Beningbrough 1791)—Emily 1810 by Stamford 1794 (Sir Peter T. 1784)—Dau. 1799 of Whiskey 1789 (Saltram 1780)—Grey Dorimant 1781 by Dorimant 1772 (Otho 1760)—Dizzy 1757 by Blank 1740 (God. Ar. 1724).
				Icaria, b. m., 1824.	The Flyer 1814 (Vandyke Jr. 1808)—Parma 1813 by Dick Andrews 1797 (Joe Andrews 1778)—May 1801 by Beningbrough 1791 (King Fergus 1775)—Primrose 1787 by Mambrino 1768 (Engineer 1756)—Cricket 1776 by Herod 1758 (Tartar 1743).
			Rowena, ch. m., 1826.	Sumpter, ch. h., 1818.	Sir Archy 1805 (*Diomed 1777)—Dau. of *Robin Redbreast 1796 (Sir Peter T. 1784)—Dau. 1788 of *Obscurity 1777 (Eclipse 1764)—Slamerkin 1768 by *Wildair 1753 (Cade 1734)—*Cub mare 1762 by Cub 1739 (Fox 1714).
				Lady Grey, b. m., 1817.	Robin Grey 1805 (*Royalist 1790)—Maria 1802 by Melzar 1791 (*Medley 1776)—Dau. of *Highflyer 1784 (Highflyer 1774)—Dau. of *Fearnought 1755 (Regulus 1739)—Dau. of Ariel 1756 (*Traveller 1747).
	Magnolia, ch. m., 1841.	*Glencoe, ch. h., 1831.	Sultan, b. h., 1816.	Selim, ch. h., 1802.	Buzzard* 1787 (Woodpecker 1773)—Dau. of Alexander 1782 (Eclipse 1764)—Dau. of Highflyer 1774 (Herod 1758)—Dau. of Alfred 1770 (Matchem 1748)—Dau. 1770 of Engineer 1756 (Sampson 1745).
				Bacchante, br. m., 1809.	W's Ditto 1800 (Sir Peter T. 1784)—Dau. 1791 of Mercury 1778 (Eclipse 1764)—Dau. 1776 of Herod 1758 (Tartar 1743)—Folly 1771 by Marske 1750 (Squirt 1732)—Vixen 1753 by Regulus 1739 (God. Ar. 1724).
			Trampoline, ch. m., 1825.	Tramp, b. h., 1810.	Dick Andrews 1797 (Joe Andrews 1778)—Dau. 1803 of Gohanna 1790 (Mercury 1778)—Fraxinella 1798 by Trentham 1766 (T. Sweepstakes 1743)—Dau. 1785 of Woodpecker 1773 (Herod 1758)—Everlasting 1775 by Eclipse 1764 (Marske 1750).
				Web, b. m., 1808.	Waxy 1790 (Pot-8-o's 1773)—Penelope 1798 by Trumpator 1782 (Conductor 1767)—Prunella 1788 by Highflyer 1774 (Herod 1758)—Promise 1768 by Snap 1750 (Snip 1736)—Julia 1756 by Blank 1740 (God. Ar. 1724).
		*Myrtle, ch. m., 1831.	Mameluke, b. h., 1824.	Partisan, b. h., 1811.	Walton 1799 (Sir Peter T. 1784)—Parasol 1800 by Pot-8-o's 1773 (Eclipse 1764)—Prunella 1788 by Highflyer 1774 (Herod 1758)—Promise 1768 by Snap 1750 (Snip 1736)—Julia 1756 by Blank 1740 (God. Ar. 1724).
				Miss Sophia, b. m., 1805.	Stamford 1794 (Sir Peter T. 1784)—Sophia 1798 by Buzzard* 1787 (Woodpecker 1773)—Huncamunca 1787 by Highflyer 1774 (Herod 1758)—Cypher 1772 by Squirrel 1754 (Traveller 1735)—Dau. 1749 of Regulus 1739 (God. Ar. 1724).
			Bobadilla, br. m., 1824.	Bobadil, ch. h., 1813.	Rubens 1805 (Buzzard* 1787)—Dau. 1795 of Skyscraper 1786 (Highflyer 1774)—Isabel 1787 by Woodpecker 1773 (Herod 1758)—Dau. 1775 of Squirrel 1754 (Traveller 1735)—Nancy 1767 by Blank 1740 (God. Ar. 1724).
				Pythoness, blk. m., 1813.	Sorcerer 1796 (Trumpator 1782)—Princess 1796 by Sir Peter T. 1784 (Highflyer 1774)—Dau. 1780 of Dungannon 1780 (Eclipse 1764)—Dau. 1783 of Turf 1760 (Matchem 1748)—Dau. 1779 of Herod 1758 (Tartar 1743).
SISTER TO RURIC, b. m., 1862.	*Sovereign, b. h., 1836.	Emilius, b. h., 1820.	Orville, b. h., 1799.	Beningbrough, b. h., 1791.	King Fergus 1775 (Eclipse 1764)—Dau. 1780 of Herod 1758 (Tartar 1743)—Pyrrha 1771 by Matchem 1748 (Cade 1734)—Dutchess 1748 by Whitenose 1742 (God. Ar. 1724)—Miss Slamerkin 1729 by Y. True Blue 1718 (Honeywood's Ar.).
				Evelina, br. m., 1791.	Highflyer 1774 (Herod 1758)—Termagant by Tantrum 1760 (Cripple 1750)—Cantatrice by Sampson 1745 (Blaze 1733)—Dau. 1757 of Regulus 1739 (God. Ar. 1724)—Ruby mare (Marske's dam) by Blacklegs 1725 (Hutton's Bay Turk).
			Emily, ch. m., 1810.	Stamford, br. h., 1794.	Sir Peter T. 1784 (Highflyer 1774)—Horatia 1778 by Eclipse 1764 (Marske 1750)—Countess (Delpini's dam) 1760 by Blank 1740 (God. Ar. 1724)—Dau. of Rib 1736 (Crab 1722)—Dau. of Wynn Ar.
				Daughter of, ch. m., 1799.	Whiskey 1789 (Saltram* 1780)—Grey Dorimant 1781 by Dorimant 1772 (Otho 1760)—Dizzy 1757 by Blank 1740 (God. Ar. 1724)—Dizzy 1741 by Driver 1727 (Wynn Ar.)—Dau. of Smiling Tom 1721 (Conyers' Ar.).
		Fleur-de-Lis, b. m., 1822.	Bourbon, b. h., 1811.	Sorcerer, blk. h., 1793.	Trumpator 1782 (Conductor 1767)—Y. Giantess 1790 by Diomed* 1777 (Florizel 1768)—Giantess 1769 by Matchem 1748 (Cade 1734)—Molly Long Legs 1753 by Babraham 1740 (God. Ar. 1724)—Dau. of Foxhunter 1727 (Brisk 1711).
				Daughter of, gr. m., 1797.	Precipitate* 1787 (Mercury 1778)—Y. Tiffany 1787 by Highflyer 1774 (Herod 1758)—Tiffany 1773 by Eclipse 1764 (Marske 1750)—Y. Hag 1761 by Skim 1746 (Starling 1727)—Hag 1744 by Crab 1722 (Alcock Ar.)
			Lady Rachel, b. m., 1803.	Stamford, br. h., 1794.	Sir Peter T. 1784 (Highflyer 1774)—Horatia 1778 by Eclipse 1764 (Marske 1750)—Countess (Delpini's dam) 1760 by Blank 1740 (God. Ar. 1724)—Dau. of Rib 1736 (Crab 1722)—Dau. of Wynn Ar.
				Y. Rachel, b. m., 1790.	Volunteer 1780 (Eclipse 1764)—Rachel 1790 by Highflyer 1774 (Herod 1758)—Dau. 1771 of Syphon 1750 (Squirt 1732)—Dau. of Regulus 1739 (God. Ar. 1724)—Sist. 1758 to Judgement by Snip 1736 (Flying Childers 1715).
	Levity, b. m., 1845.	*Trustee, ch. h., 1829.	Catton, b. h., 1809.	Golumpus, b. h., 1802.	Gohanna 1790 (Mercury 1778)—Catherine 1795 by Woodpecker 1773 (Herod 1758)—Camilla 1778 by Trentham 1766 (T. Sweepstakes 1743)—Coquette 1765 by Compton Barb—Sist. 1748 to Regulus by God. Ar. 1724.
				Lucy Gray, ch. m., 1804.	Timothy 1794 (Delpini 1781)—Lucy 1789 by Florizel 1768 (Herod 1758)—Frenzy 1774 by Eclipse 1764 (Marske 1750)—Dau. of Engineer 1756 (Sampson 1745)—Dau. 1763 of Blank 1740 (God. Ar. 1724).
			Emma, ch. m., 1824.	Whisker, b. h., 1812.	Waxy 1790 (Pot-8-o's 1773)—Penelope 1798 by Trumpator 1782 (Conductor 1767)—Prunella 1788 by Highflyer 1774 (Herod 1758)—Promise 1768 by Snap 1750 (Snip 1736)—Julia 1756 by Blank 1740 (God. Ar. 1724).
				Gibside Fairy, b. m., 1811.	Hermes 1790 (Mercury 1778)—Vicissitude 1800 by Pipator 1786 (Imperator 1776)—Beatrice 1791 by Sir Peter T. 1784 (Highflyer 1774)—Pyrrha 1771 by Matchem 1748 (Cade 1734)—Dutchess 1748 by Whitenose 1742 (God. Ar. 1724).
		Vandal's dam, b. m., 1837.	*Tranby, br. h., 1826.	Blacklock, b. h., 1814.	Whitelock 1803 (Hambletonian 1792)—Dau. 1799 of Coriander 1786 (Pot-8-o's 1773)—Wildgoose 1792 by Highflyer 1774 (Herod 1758)—Coheiress 1786 by Pot-8-o's 1773 (Eclipse 1764)—Manilla 1777 by Goldfinder 1764 (Snap 1750).
				Miss Bowe's dam, br. m., 1812.	Orville 1799 (Beningbrough 1791)—Miss Grimstone 1796 by Weazle 1776 (Herod 1758)—Dau. of Ancaster 1768 (Blank 1740)—Dau. of Damascus Ar. 1754—Dau. of Sampson 1745 (Blaze 1733).
			Lucilla, b. m., [illegible].	Trumpator, br. h., 1823.	Sir Solomon 1805 (*Tickle Toby 1786)—Dau. of Hickory 1804 (*Whip 1794)—*Trumpetta 1797 by Trumpator 1782 (Conductor 1767)—Sist. 1786 to Lambinos by Highflyer 1774 (Herod 1758)—Dau. 1775 of Eclipse 1764 (Marske 1750).
				Lucy, b. m., 1821.	Orphan 1810 (Ball's Florizel 1801)—Lady Grey b. m. 1817 by Robin Grey 1805 (*Royalist 1790)—Maria 1802 by Melzar 1791 (*Medley 1776)—Dau. of *Highflyer 1784 (Highflyer 1774)—Dau. of *Fearnought 1755 (Regulus 1739)—Dau. of Ariel 1756 (*Traveller 1747)—Dau. of *Jack of Diamonds (Cullen Ar. 1745)—*Duchess by Cullen Ar. 1745—Lady Thigh ch. m. 1731 by Partner 1718 (Jigg)—Dau. ch. m. 1723 of Greyhound (Chillaby)—Sophonisba's dam by Curwen Bay Barb—Dau. of D'Arcey Chestnut Ar.—Dau. of Whiteshirt—Montagu mare by D'Arcey Old Montagu—Dau. of Hautboy (D'Arcey White Turk)—Dau. of Brimmer (D'Arcey Yellow Turk)—Royal mare.

HANOVER.

WINNER OF THE HOPEFUL, JULY AND SAPLING STAKES AT TWO YEARS OLD, TWENTY OUT OF TWENTY-SEVEN RACES AT THREE YEARS OLD, INCLUDING WITHERS, BELMONT, SWIFT, TIDAL, LORILLARD, STOCKTON, STEVENS, BROOKLYN DERBY, CONEY ISLAND DERBY, EMPORIUM, CHAMPION AND OTHER STAKES.

The property of Milton Young, McGrathiana Stud, Lexington, Ky.

HANOVER, chestnut horse, foaled 1884, bred in Runnymede Stud, by Hindoo, son of Virgil, dam Bourbon Belle by imp. Bonnie Scotland, out of Ella D. by Vandal, son of imp. Glencoe. Hanover was a grand race-horse. At two years old won the Hopeful Stakes, ¾ of a mile, 115 lbs., in 1.15¾, beating Fitzroy, Roi d'Or and three others; won July Stakes, ¾ of a mile, 120 lbs., in 1.15½, defeating Oneko, Roi d'Or, etc.; won Sapling Stakes, ⅝ of a mile, 1.15, defeating Kingston, Austriana and four others. All these races were run at Monmouth Park. As a three-year old, started 27 times, won 20 races. At Brooklyn won the Carlton Stakes, 1 mile, 118 lbs., in 1.43¼; the Brookdale Handicap, 1⅛ miles, 111 lbs., in 1.54½, Dry Monopole, Oriflame and others behind him; won Brooklyn Derby, 1½ miles, 118 lbs., in 2.43½, beating Dunbine and Bronzomarte. Jerome Park, won Withers Stakes, 1 mile, 118 lbs., in 1.46½, Stockton, Belvidere etc., behind him; won Belmont Stakes, 1½ miles, 118 lbs., beating Oneko, in 2.43½. Coney Island, won Swift Stakes, 7 furlongs, 118 lbs., beating Kingston and Firenzi, in 1.32, track muddy; won the Tidal Stakes, 1 mile, 118 lbs., beating Kingston and Oneko, in 1.41¾; won Coney Island Derby, 1½ miles, 118 lbs., beating Oneko and Hypasia in 2.44½, track heavy; won the Emporium Stakes, 1½ miles, 128 lbs., in 2.35¾, beating Dunbine, Oneko and five others; won the Spindrift Stakes, 1¼ miles, 125 lbs., beating Stockton, 111½ lbs., and Raveller, 112 lbs., in 2.11½. Monmouth Park, won the Lorillard Stakes, 1½ miles, 118 lbs., beating Almy, in 2.40½; won Stockton Stakes, 1¼ miles, 125 lbs., beating Kingdom, 118 lbs, in 2.13; won Barnegat Stakes, 1½ miles, 125 lbs., beating Miss Ford, 113 lbs., and Kingdom, 110 lbs., in 2.41¼; won Stevens Stakes, 1⅝ miles, 123 lbs., beating Almy, 114 lbs., in 2.56; ran second to Laggard, 111 lbs., in the Raritan Stakes, Hanover, 128 lbs., in 2.14, Bradford, 109 lbs., third; won Champion Stakes, 1½ miles, 109 lbs., beating Firenzi, 100 lbs., and Volante, 121 lbs., in 2.38; ran third to Laggard, 118 lbs., and Firenzi, 120 lbs., Hanover, 125 lbs., in the Omnibus Stakes, 1½ miles, in 2.44, track muddy, Kingston and Esquimau behind him; was beaten by Banburg, 110 lbs., in Choice Stakes, 1½ miles, in 2.45 Hanover, 118 lbs. Saratoga, won United States Hotel Stakes, 1½ miles, 118 lbs., beating Waxy and others, in 2.38½. Brooklyn Fall Meeting, was unplaced in first special, 1¼ miles, won by Volante in 2.12; second won special, 1⅛ miles, 114 lbs., in 1.57½, Favor (5), 123 lbs., Banburg and Grisette behind him. Jerome Park Fall Meeting, beaten by Firenzi, 122 lbs., Hanover 125 lbs., in Jerome Stakes, 1¾ miles, in 3.09¾. Baltimore, won Breckenridge Stakes, 1¾ miles, 123 lbs., in 2.52½, beating Glenmound, 118 lbs., and Stockton, 118 lbs.; won Dixie Stakes, 2 miles, 123 lbs., beating Glenmound, 118 lbs., in 3.51¾, track heavy; ran second to Eolian, 118 lbs., 7 furlongs, in 1.28¾, two behind him; ran second to Eolian (4), 123 lbs., Hanover (3), 114 lbs., 1 mile, in 1.43, four others behind him; won purse, 1 mile, 112 lbs., in 1.41¾, Orvid, 96 lbs., and four others behind him. At four years old ran three times, was second to The Bard in Brooklyn Handicap, 1¼ miles, each 125 lbs., in 2.13, Exile, Volante and seven others behind him; ran second to Richmond (6), 112 lbs., Hanover (4), 125 lbs., in Brookdale Handicap, 1⅛ miles, in 1.58, Dry Monopole, Eurus and Fitzroy behind him; ran third to The Bard (5), 123 lbs., Volante (6), 124 lbs., Hanover (4), 118 lbs., in Brooklyn Cup, 1½ miles, in 2.42½, Fenelon (4), 118 lbs., fourth. At five years old started 17 times, won 9 races, was second in 6, third in 1 and unplaced in 1. Brooklyn, was unplaced in Brooklyn Handicap, 1¼ miles, won by Exile in 2.07½; won purse, 7 furlongs, 124 lbs., in 1.28½, beating Guarantee, Eolian and Eolo; ran second to Diablo in St. James Hotel Stakes, 1¼ miles, in 2.09½, beating Elkwood; was second to Marauder (4), 122 lbs., Hanover (5), 127 lbs., in Brooklyn Cup, 1½ miles, in 2.42¼, Elkwood, Eurus and Glen Echo behind him; won Sweepstake, ½ a mile, 125 lbs., in 0.48¾, seven behind him. Coney Island Jockey Club, won Sweepstake, 1⅛ miles, 114 lbs., in 1.55, three behind him; won Sweepstake, 5½ furlongs, 124 lbs., in 1.06⅗, eight behind him; won Coney Island Cup, 1½ miles, 116 lbs., in 2.35⅗, beating Firenzi and three others; won Coney Island Stakes, 1⅛ miles, 119 lbs., in 1.54⅘, Kingston, 121 lbs., and three others behind him. Saratoga, won the California Stakes, 1 mile, 124 lbs., in 1.43½, three behind him; ran second to Los Angeles (4), 112 lbs., in Excelsior Stakes, 1¼ miles, Hanover (5), 124 lbs., in 2.13, Montrose third; won Merchants' Stakes, 1⅝ miles, 124 lbs., beating Montrose and Elkwood.

(Continued on page 170.)

HANOVER.

CHESTNUT HORSE, FOALED 1884; BRED BY MESSRS. CLAY & WOODFORD, RUNNYMEDE STUD, KENTUCKY.

HINDOO, b. h., 1878.
- Virgil, br. h., 1864.
  - Vandal, b. h., 1850.
    - *Glencoe, ch. h., 1831.
    - Daughter of, b. m., 1837.
  - Hymenia, b. m., 1851.
    - *Yorkshire, b. h., 1834.
    - Little Peggy, ch. m., 1837.
- Florence, ch. m., 1869.
  - Lexington, b. h., 1850.
    - Boston, ch. h., 1833.
    - Alice Carneal, b. m., 1836.
  - *Weatherwitch, ch. m., 1854.
    - Weatherbit, br. h., 1842.
    - Daughter of, br. m., 1833.

BOURBON BELLE, b. m., 1869.
- *Bonnie Scotland, b. h., 1853.
  - Iago, b. h., 1843.
    - Don John, b. h., 1835.
    - Scandal, b. m., 1832.
  - Queen Mary, b. m., 1843.
    - Gladiator, ch. h., 1833.
    - Daughter of, b. m., 1840.
- Ella D., b. m., 1857.
  - Vandal, b. h., 1850.
    - *Glencoe, ch. h., 1831.
    - Daughter of, b. m., 1837.
  - Falcon, b. m., 1839.
    - Woodpecker, b. h., 1828.
    - Ophelia, gr. m., 1825.

Sultan, b. h., 1816.	Selim 1802 (Buzzard* 1787)- Bacchante 1809 by W's Ditto 1800 (Sir Peter T. 1784) Dau. 1791 of Mercury 1778 (Eclipse 1764)—Dau. 1776 of Herod 1758 (Tartar 1743)--Folly 1771 by Marske 1750 (Squirt 1732).
Trampoline, ch. m., 1825.	Tramp 1810 (Dick Andrews 1797)—Web 1808 by Waxy 1790 (Pot-8-o-'s 1773) Penelope 1798 by Trumpator 1782 (Conductor 1767) Prunella 1788 by Highflyer 1774 (Herod 1758) Promise 1768 by Snap 1750 (Snip 1736).
*Tranby, br. h., 1826.	Blacklock 1814 Whitelock 1803) Dau. 1812 of Orville 1799 (Beningbrough 1791) Miss Grimstone 1796 by Weazle 1776 (Herod 1758) Dau. of Ancaster 1768 (Blank 1740)—Dau. of Damasc. Ar. 1754.
Lucilla, b. m., —.	Trumpator 1823 (Sir Solomon 1805) Lucy 1821 by Orphan 1810 (Ball's Florizel 1801)—Lady Grey 1817 by Robin Grey 1805 (*Royalist 1790)—Maria 1802 by Melzar 1791 (*Medley 1776)—Dau. of *Highflyer 1784 (Highflyer 1774).
St. Nicholas, br. h., 1827.	Emilius 1820 (Orville 1799) Sea-Mew 1815 by Scud 1804 (Beningbrough 1791) Goosander 1805 by Hambletonian 1792 (King Fergus 1775)—Rally 1790 by Trumpator 1782 (Conductor 1767) Fancy 1780 (Sist. to Diomed) by Florizel 1768 (Herod 1758).
Miss Rose,* b. m., 1820.	Tramp 1810 (Dick Andrews 1797) Dau. 1810 of Sancho 1801 (Don Quixote 1784) Blacklock's Dam 1799 by Coriander 1786 (Pot-8-o-'s 1773) Wildgoose 1792 by Highflyer 1774 (Herod 1758)—Coheiress 1786 by Pot-8-o-'s 1773 (Eclipse 1764)
Cripple, b. h., 1836.	Medoc 1829 (Am. Eclipse 1814) Grecian Princess 1816 by Cook's Whip 1804 (*Whip 1794) Jane Hunt 1796 by Paragon 1788 (*Flimnap 1765) Moll by *Figure 1757 (Grey Figure 1747)—Slamerkin 1768 by *Wildair 1753 (Cade 1734).
Peggy Stewart, b. m., 1824.	Cook's Whip 1804 (*Whip 1794)—Mary Bedford 1816 by Duke of Bedford (*Bedford 1792)—Dau. 1813 of *Speculator 1795 (Dragon* 1787) Dau. 1805 of *Dare Devil 1787 (Magnet 1770)—*Trumpetta 1797 by Trumpator 1782 (Conductor 1767).
Timoleon, ch. h., 1813.	Sir Archy 1805 (*Diomed 1777)- Dau. of *Saltram 1780 (Eclipse 1764) Dau. of Symmes' Wildair 1767 (*Fearnought 1755)—Dau. of Driver 1760 (*Othello) Dau. of *Fallower 1761 (Blank 1740).
Daughter of, ch. m., 1814.	Ball's Florizel 1801 (*Diomed 1777) Dau. 1799 of *Alderman 1787 (Pot-8-o-'s 1773) Dau. of *Clockfast 1774 (Gimcrack 1760)—Dau. of Symmes' Wildair 1767 (*Fearnought 1755)—Y. Kitty Fisher 1767 by *Fearnought 1755 (Regulus 1739).
*Sarpedon, br. h., 1828.	Emilius 1820 (Orville 1799)—Icaria 1824 by The Flyer 1814 (Vandyke Jr. 1808) Parma 1813 by Dick Andrews 1797 (Joe Andrews 1778)—May 1804 by Beningbrough 1791 (King Fergus 1775)—Primrose 1787 by Mambrino 1768 (Engineer 1756).
Rowena, ch. m., 1826.	Sumpter 1818 (Sir Archy 1805)- Lady Grey 1817 by Robin Grey 1805 (*Royalist 1790)—Maria 1802 by Melzar 1791 (*Medley 1776)—Dau. of *Highflyer 1784 (Highflyer 1774)—Dau. of *Fearnought 1755 (Regulus 1739).
Sheet Anchor, br. h., 1832.	Lottery 1820 (Tramp 1810)—Morgiana 1820 by Muley 1810 (Orville 1799)—Miss Stephenson 1814 by [Seud or] Sorcerer 1796 (Trumpator 1782)—Sist. 1794 to Petworth by Precipitate* 1787 (Mercury 1778)—Dau. 1787 of Woodpecker 1773 (Herod 1758).
Miss Letty, b. m., 1834.	Priam* 1827 (Emilius 1820)—Dau. 1815 of Orville 1799 (Beningbrough 1791)- Dau. 1800 of Buzzard* 1787 (Woodpecker 1773)—Hornpipe 1793 by Trumpator 1782 (Conductor 1767) —Luna 1779 by Herod 1758 (Tartar 1743).
Ir. Birdcatcher, ch. h., 1833.	Sir Hercules 1826 (Whalebone 1807)—Guiccioli 1823 by Bob Booty 1804 (Chanticleer 1787)—Flight 1809 by Ir. Escape 1802 (Commodore 1793)—Y. Heroine by Bagot 1780 (Herod 1758)—Heroine 1775 by Hero 1758 (Cade 1734).
Colocynth, b. m., 1840.	Physician 1829 (Brutandorf 1821) Camelina 1823 by Whalebone 1807 (Waxy 1790)—Dau. 1812 of Selim 1802 (Buzzard* 1787)—Maiden 1801 by Sir Peter T. 1784 (Highflyer 1774)—Dau. 1788 of Phenomenon 1780 (Herod 1758).
[Tramp or] Waverley, br. h., 1817.	Whalebone 1809 (Waxy 1790)—Margaretta 1802 by Sir Peter T. 1784 (Highflyer 1774)—Dau. 1792 of Highflyer 1774 (Herod 1758)—Nutcracker 1767 by Matchem 1748 (Cade 1734)—Miss Starling 1759 by Starling 1727 (Bay Bolton 1705).
Daughter of, gr. m., 1821.	Comus 1809 (Sorcerer 1796)—Marc'ana 1809 by Stamford 1794 (Sir Peter T. 1784)—Marcia 1797 by Coriander 1786 (Pot-8-o-'s 1773)—Faith 1779 by Pacolet 1763 (Blank 1740)—Atalanta 1769 by Matchem 1748 (Cade 1734).
Selim, ch. h., 1802.	Buzzard* 1787 (Woodpecker 1773)—Dau. of Alexander 1782 (Eclipse 1764)—Dau. of Highflyer 1774 (Herod 1758)—Dau. of Alfred 1770 (Matchem 1748)—Dau. 1770 of Engineer 1756 (Sampson 1745).
Daughter of, br. m., 1816.	Haphazard 1797 (Sir Peter T. 1784)—Princess 1799 by Precipitate* 1787 (Mercury 1778)—Colibri 1793 by Woodpecker 1773 (Herod 1758)—Camilla 1778 by Trentham 1766 (T. Sweepstakes 1743)—Coquette 1765 by Compton Barb.
Partisan, b. h., 1811.	Walton 1799 (Sir Peter T. 1784)—Parasol 1800 by Pot-8-o-'s 1773 (Eclipse 1764)—Prunella 1788 by Highflyer 1774 (Herod 1758)—Promise 1768 by Snap 1750 (Snip 1736)—Julia 1756 by Blank 1740 (God. Ar. 1724).
Pauline, b. m., 1826.	Moses 1819 ([Whalebone or] Seymour 1807)—Quadrille 1815 by Selim 1802 (Buzzard* 1787)—Canary Bird 1806 by [Whiskey or] Sorcerer 1796 (Trumpator 1782)—Canary 1797 by Coriander 1786 (Pot-8-o-'s 1773)—Miss Green 1787 by Highflyer 1774 (Herod 1758).
Plenipotentiary, ch. h., 1831.	Emilius 1820 (Orville 1799) Harriet 1819 by Pericles 1809 (Evander 1801)—Dau. 1812 of Selim 1802 (Buzzard* 1787)—Pipylina 1803 by Sir Peter T. 1784 (Highflyer 1774)—Rally 1790 by Trumpator 1782 (Conductor 1767).
Myrrha, b. m., 1830.	Whalebone 1807 (Waxy 1790)—Gift 1818 by Y. Gohanna 1810 (Gohanna 1790)—Dau. 1802 of Sir Peter T. 1784 (Highflyer 1774)—Dau. 1788 of Trumpator 1782 (Conductor 1767)—Sist. 1774 to Postmaster by Herod 1758 (Tartar 1743).
Sultan, b. h., 1816.	Selim 1802 (Buzzard* 1787)—Bacchante 1809 by W's Ditto 1800 (Sir Peter T. 1784)—Dau. 1791 of Mercury 1778 (Eclipse 1764)—Dau. 1776 of Herod 1758 (Tartar 1743)—Folly 1771 by Marske 1750 (Squirt 1732).
Trampoline, ch. m., 1825.	Tramp 1810 (Dick Andrews 1797)—Web 1808 by Waxy 1790 (Pot-8-o-'s 1773)—Penelope 1798 by Trumpator 1782 (Conductor 1767)—Prunella 1788 by Highflyer 1774 (Herod 1758)—Promise 1768 by Snap 1750 (Snip 1736).
*Tranby, br. h., 1826.	Blacklock 1814 (Whitelock 1803)—Dau. 1812 of Orville 1799 (Beningbrough 1791)—Miss Grimstone 1796 by Weazle 1776 (Herod 1758)—Dau. of Ancaster 1768 (Blank 1740)—Dau. of Damasc. Ar. 1754.
Lucilla, b. m., —.	Trumpator 1823 (Sir Solomon 1805)—Lucy 1821 by Orphan 1810 (Ball's Florizel 1801)—Lady Grey 1817 by Robin Grey 1805 (*Royalist 1790)—Maria 1802 by Melzar 1791 (*Medley 1776)—Dau. of *Highflyer 1784 (Highflyer 1774).
Bertrand, b. h., 1821.	Sir Archy 1805 (*Diomed 1777)—Eliza 1801 by *Bedford 1792 (Dungannon 1780)—*Mambrina 1783 by Mambrino 1768 (Engineer 1756)—Sist. to Sally by Blank 1740 (God. Ar. 1724)—Dau. of Ward 1731 (Crab 1722).
Daughter of, ch. m., 1809.	*Buzzard 1787 (Woodpecker 1773)—The Faun by Craig's Alfred (*Medley 1776)—Shepherdess by Wormsley's King Herod (*Fearnought 1755)—Dau. of *Traveler 1747 (Partner 1718)—Dau. of *Whittington (Cullen Ar.)
Wild Medley, gr. h., —.	Mendoza 1796 (Boxer 1790)—Dau. of *Pensacola [a Spanish horse]—Dau. of *Whittington (Cullen Ar.)—Dau. of Lindsay's Ar.—Dau. of *Fearnought 1755 (Regulus 1739)—Dau. of *Ranger.
Daughter of, b. m., 1821.	Sir Archy 1805 (*Diomed 1777)—Lady Chesterfield b. m. 1801 by *Diomed 1777 (Florizel 1768)—Lady Bolingbroke b. m. by *Pantaloon 1778 (Herod 1758)—Cades by Wormsley's King Herod (*Fearnought 1755)—Primrose b. m. 1764 by *Dove 1756 (Y. Cade 1747)—Tasker's Stella b. m. 1758 by *Othello (Crab 1722)—*Selima b. m. 1746 by God. Ar.

1724—Large Hartley mare ch. m. 1729 by Hartley's Blind Horse (Holderness Turk)—Flying Whigg 1715 by William's Woodstock Ar.—Points by St. Victor Barb—Grey Whynot by Old Whynot (Fenwick Barb)—Royal mare.

HANOVER.—Continued.

Morris Park, ran second, 119 lbs., to Senorita (3), 106 lbs., in New Rochellé Stakes, 1 3/8 miles, in 2.03, Taragon and Raceland behind him; ran third to Raceland (4), 120 lbs., Firenzi (5), 125 lbs., Hanover (5), 126 lbs., in N. Y. Jockey Club Handicap, 1¼ miles, in 2.09¾, ten others behind him; ran second to Tenny (3), 115 lbs., Hanover (5), 122 lbs., in Electric Stakes, ¾ of a mile, in 1.12, seven others behind him; won the Express Stakes, 7 furlongs, 129 lbs., in 1.29, Defaulter (4), 115 lbs., and four others behind him; ran second to Volunteer II., each (5) 122 lbs., in Sweepstakes, 5½ furlongs, in 1.06¾, with five others behind him. This closed his turf career, having won in purses and stakes $118,000. Hanover traces back to imp. Selima by the Godolphin Arabian, thence to a Royal mare, to which trace in direct female line Blank, Babraham, Gower Stallion, Merlin (by Second), Hornsea, Rattle, Amsterdam Jerry, Jacques, Tearaway, Foxhall, Monday, Harvester, etc. He has a double cross of Glencoe, both from Vandal, and should succeed in the stud.

HARRY O'FALLON.

Application to L. B. Fields & Bro., Fields Place, Lexington, Ky.

Harry O'Fallon by imp. Australian, bred by James O'Fallon, St. Louis, Mo., foaled 1869, dam imp. Sunny South, dam of Ozark by Irish Birdcatcher, son of Sir Hercules, second dam Equal by The Cure, out of Equation, the dam of Diophantus, Archimedes, Artesian, etc. by Emilius. Australian is in this volume. Sunny South, Harry's dam, was the dam of the good race-horse Ozark and others, she by the great Birdcatcher, sire of The Baron, who got Stockwell and Rataplan. Birdcatcher sired Habena and Manganese, winners of the 1,000 Guineas, Songstress, winner of the Oaks, Daniel O'Rourke, winner of the Derby, The Baron, Knight of St. George and Warlock, winners of the St. Leger, Chanticleer, Early Bird, Saunterer, Exact and other good race-horses. The Cure, sire of his grandam, was a fine race-horse; he won the Champagne Stakes at Doncaster, the Dee Stakes at Chester, the Suffolk Stakes at Newmarket, the North Derby Stakes, and was second to Faugh-a-Ballagh in the St. Leger, 1844. He sired Little Agnes, Dictator and others. Harry's great grandam was the dam of Exact, Diophantus, winner of the 2,000 Guineas, and others. Maria, his fourth dam, was the dam of Euclid, who was third in the Derby and ran a dead heat with Charles XII. for the St. Leger, Extempore, winner of the 1,000 Guineas, and other good ones. Harry O'Fallon ran only two races, when out of condition, and was unplaced. There is no more stoutly-bred horse in America than Harry O'Fallon. He has double crosses of Whalebone and Whisker, the famous sons of Waxy, and double crosses of the speedy Emilius, and a cross of Touchstone, with any number to Eclipse and Herod through his collateral branches. Harry has sired a fair number of good horses which have been trained, and they have all run creditably; amongst them, Krupp Gun, Sam Ecker, Expectation, Harry Doswell, Jack of Spades and John Davis, the latter a winner of the Citizens' Plate, 1¾ miles, at St. Louis, 104 lbs., in 3.06¾, 1¼ miles, in 2.11, the Garden City Cup, at Chicago, 2¼ miles, in 4.30½, track very muddy, 1 mile and 500 yards, 112 lbs., in 2.13¾, Northwestern Stakes, at Chicago, 117 lbs., in 3.06½, and the October Handicap, 111 lbs., in 2.09, and the winners Carrie, Harvard, Hiram Wood, Jack Cocks, O'Fellus, Revoke, Chandler, Keystone, Marion C., Miss Flood, Ofalece, Owen Golden, Walker, The Deacon, McDonald, Revival, Heron, Harry Weldon, Carrie A., Callie, Kosciusko K., Santa Lena, Leadville, Tristram, Split Silk, Alonzo, Bob Toombs, Cup Bearer, Little Annie, Shiloh, Torrent, Powers, Dud Hughes, Tom Jones, Terra Cotta, one of the cracks of his year, with many others. Harry is inbred to Waxy through Whalebone and Whisker, both Derby winners, and has many crosses to Eclipse, Herod and Matchem. Few stallions that ever lived have done as well with his chances. He traces in direct female line to the same Royal mare to which Oroonoko, imp. Saltram, Beningbrough, Williamson's Ditto, Cotherstone, West Australian, Walton, Wild Dayrell, Parmesan, Diophantus, Orest, Caterer, Lord Ronald, Ishmael, Wisdom, The Miser, Saxifrage, etc., trace.

HARRY O'FALLON.

CHESTNUT HORSE, FOALED 1869; BRED BY MR. J. O. FALLON, ST. LOUIS, MO.

- *SUNNY SOUTH, ch. m., 1858.
  - Equal, b. m., 1849.
    - Equation, b. m., 1839.
      - Maria, b. m., 1827.
      - Emiline, b. h., 1820.
    - The Cure, b. h., 1841.
      - Morsel, b. m., 1836.
      - Physician, b. h., 1829.
  - Ir. Birdcatcher, ch. h., 1833.
    - Guiccioli, ch. m., 1823.
      - Flight, ch. m., 1809.
      - Bob Booty, ch. h., 1804.
    - Sir Hercules, blk. h., 1826.
      - Peri, b. m., 1822.
      - Whalebone, br. h., 1807.
- *AUSTRALIAN, ch. h., 1858.
  - Emilia,* b. m., 1840.
    - Persian, b. m., 1829.
      - Variety, ch. m., 1816.
      - Whisker, b. h., 1812.
    - Y. Emilius, b. h., 1833.
      - Shoveler, b. m., 1816.
      - Emilius, b. h., 1820.
  - West Australian, b. h., 1850.
    - Mowerina, b. m., 1843.
      - Emma, ch. m., 1824.
      - Touchstone, br. h., 1831.
    - Melbourne, br. h., 1834.
      - Daughter of, b. m., 1825.
      - Humphrey Clinker, b. h., 1822.

Comus, ch. h., 1809.	Sorcerer 1796 (Trumpator 1782) Epsom Lass 1803 by Sir Peter T. 1784 (Highflyer 1774) Alexina 1788 by King Fergus 1775 (Eclipse 1764) Lardella 1780 by Y. Marske 1771 (Marske 1750) Dau. of Y. Cade 1747 (Cade 1734).
Clinkerina, br. m., 1812.	Clinker 1805 (Sir Peter T. 1784) Pewet 1786 by Tandem 1773 (Syphon 1750) Termagant by Tantrum 1760 (Cripple 1750) Cantatrice by Sampson 1745 (Blaze 1733) Dau. 1757 of Regulus 1739 (God. Ar. 1724).
Cervantes, b. h., 1806.	Don Quixote 1784 (Eclipse 1764) Evelina 1791 by Highflyer 1774 (Herod 1758) Termagant by Tantrum 1760 (Cripple 1750) Cantatrice by Sampson 1745 (Blaze 1733) Dau. 1757 of Regulus 1739 (God. Ar. 1724).
Daughter of, br. m., 1818.	Golumpus 1802 (Gohanna 1790) Dau. 1810 of Paymator 1791 (Trumpator 1782) Circle 1802 by St. George 1789 (Highflyer 1774) Abigail 1788 by Woodpecker 1773 (Herod 1758) Firetail 1772 by Eclipse 1764 (Marske 1750).
Camel, br. h., 1822.	Whalebone 1807 (Waxy 1790) Dau. 1812 of Selim 1802 (Buzzard 1787) Maiden 1801 by Sir Peter T. 1784 (Highflyer 1774) Dau. 1788 of Phenomenon 1780 (Herod 1758)—Matron 1782 by Florizel 1768 (Herod 1758).
Banter, br. m., 1826.	Master Henry 1815 (Orville 1799) Boadicea 1807 by Alexander 1782 (Eclipse 1764) Brunette by Amaranthus 1766 (Old England 1741) Mayfly 1771 by Matchem 1748 (Cade 1734)—Dau. 1755 of Anc. Starling 1738 (Starling 1727).
Whisker, b. h., 1812.	Waxy 1790 (Pot-8-o-'s 1773) Penelope 1798 by Trumpator 1782 (Conductor 1767) Prunella 1788 by Highflyer 1774 (Herod 1758) Promise 1768 by Snap 1750 (Snip 1736) Julia 1756 by Blank 1740 (God. Ar. 1724).
Gibside Fairy, b. m., 1811.	Hermes 1790 (Mercury 1778) Vicissitude 1800 by Pipator 1786 (Imperator 1776)—Beatrice 1791 by Sir Peter T. 1784 (Highflyer 1774) Pyrrha 1771 by Matchem 1748 (Cade 1734) Duchess 1748 by Whitenose 1742 (God. Ar. 1724).
Orville, b. h., 1799.	Beningbrough 1791 (King Fergus 1775) Evelina 1791 by Highflyer 1774 (Herod 1758) Termagant by Tantrum 1760 (Cripple 1750)—Cantatrice by Sampson 1745 (Blaze 1733) Dau. 1757 of Regulus 1739 (God. Ar. 1724).
Emily, ch. m., 1810.	Stamford 1794 (Sir Peter T. 1784) Dau. 1799 of Whiskey 1789 (Saltram* 1780) Grey Dorimant 1781 by Dorimant 1772 (Otho 1760) Dizzy 1757 by Blank 1740 (God. Ar. 1724) Dizzy 1741 by Driver 1727 (Wynn Ar.).
Scud, b. h., 1804.	Beningbrough 1791 (King Fergus 1775) Eliza 1791 by Highflyer 1774 (Herod 1758) Augusta 1784 by Eclipse 1764 (Marske 1750)—Hardwicke's dam by Herod 1758 (Tartar 1743) Nettle 1763 by Bajazet 1740 (God. Ar. 1724).
Goosander, b. m., 1805.	Hambletonian 1792 (King Fergus 1775)—Rally 1790 by Trumpator 1782 (Conductor 1767)—Fancy 1780 by Florizel 1768 (Herod 1758)—Dau. 1763 of Spectator 1749 (Crab 1722)—Horatia 1758 by Blank 1740 (God. Ar. 1724).
Waxy, b. h., 1790.	Pot-8-o-'s 1773 (Eclipse 1764)—Maria 1777 by Herod 1758 (Tartar 1743)—Lisette 1772 by Snap 1750 (Snip 1736) Miss Windsor 1754 by God. Ar. 1724—Sist. to Volunteer by Y. Belgrade (Belgrade Turk).
Penelope, b. m., 1798.	Trumpator 1782 (Conductor 1767) Prunella 1788 by Highflyer 1774 (Herod 1758)—Promise 1768 by Snap 1750 (Snip 1736) Julia 1756 by Blank 1740 (God. Ar. 1724)—Spectator's dam 1735 by Partner 1718 (Jigg).
[Selim or] Soothsayer, ch. h., 1808.	Sorcerer 1796 (Trumpator 1782) Golden Locks 1798 by Delpini 1781 (Highflyer 1774)—Violet 1787 by Shark* 1771 (Marske 1750) Dau. 1772 of Syphon 1750 (Squirt 1732)—Charlotte 1756 by Blank 1740 (God. Ar. 1724).
Sprite, ch. m., 1807.	Bobtail 1795 (Precipitate* 1787)—Catherine 1795 by Woodpecker 1773 (Herod 1758)—Camilla 1778 by Trentham 1766 (T. Sweepstakes 1743) Coquette 1765 by Compton Barb—Sist. 1743 to Regulus by God. Ar. 1724.
Waxy, b. h., 1790.	Pot-8-o-'s 1773 (Eclipse 1764)—Maria 1777 by Herod 1758 (Tartar 1743)—Lisette 1772 by Snap 1750 (Snip 1736)—Miss Windsor 1754 by God. Ar. 1724—Dau. of Y. Belgrade (Belgrade Turk).
Penelope, b. m., 1798.	Trumpator 1782 (Conductor 1767)—Prunella 1788 by Highflyer 1774 (Herod 1758)—Promise 1768 by Snap 1750 (Snip 1736)—Julia 1756 by Blank 1740 (God. Ar. 1724) Spectator's dam 1735 by Partner 1718 (Jigg).
Wanderer, b. h., 1811.	Gohanna 1790 (Mercury 1778)—Catherine 1795 by Woodpecker 1773 (Herod 1758) Camilla 1778 by Trentham 1766 (T. Sweepstakes 1743) Coquette 1765 by Compton Barb.—Sist. 1743 to Regulus by God. Ar. 1724.
Thalestris, blk. m., 1809.	Alexander 1782 (Eclipse 1764)—Rival 1800 by Sir Peter T. 1784 (Highflyer 1774)—Hornet 1790 by Drone 1777 (Herod 1758)—Manilla 1777 by Goldfinder 1764 (Snap 1750)—Dau. 1766 of Old England 1741 (God. Ar. 1724).
Chanticleer, ch. h., 1787.	Woodpecker 1773 (Herod 1758)—Dau. 1778 of Eclipse 1764 (Marske 1750)—Rosebud 1765 by Snap 1750 (Snip 1736)—Miss Belsea 1753 by Regulus 1739 (God. Ar. 1724)—Dau. of Bartlett's Childers (Darley Ar. 1703).
Ierne, ch. m., 1790.	Bagot 1780 (Herod 1758)—Dau. of Gamahoe (Bustard 1741)—Patty by Tim 1752 (Squirt 1732)—Miss Patch 1737 by Justice 1725 (H. C. Litton Ar.)—Ringtail Galloway by Curwen Bay Barb.
Ir. Escape, ch. h., 1802.	Commodore 1793 (Tom Tug 1777)—Moll in the Wad 1791 by Highflyer 1774 (Herod 1758) Shift 1779 by Sweetbriar 1769 (Syphon 1750)—Black Susan 1762 by Snap 1750 (Snip 1736) Dau. of Cade 1734 (God. Ar. 1724).
Y. Heroine, —, —.	Bagot 1780 (Herod 1758)—Heroine 1775 by Hero 1758 (Cade 1734)—Dau. 1766 of Snap 1750 (Snip 1736)—Sist. 1743 to Regulus by God. Ar. 1724—Grey Robinson 1723 by Bald Galloway (St. Victor Barb.)
Brutandorf, b. h., 1821.	Blacklock 1814 (Whitelock 1803)—Mandane 1800 by Pot-8-o-'s 1773 (Eclipse 1764)—Y. Camilla 1787 by Woodpecker 1773 (Herod 1758)—Camilla 1778 by Trentham 1766 (T. Sweepstakes 1743)—Coquette 1765 by Compton Barb.
Primette, b. m., 1820.	Prime Minister 1810 (Sancho 1801)—Miss Paul 1811 by Sir Paul 1802 (Sir Peter T. 1784)—Miss Dunnington 1807 by Shuttle 1793 (Y. Marske 1771)—Miss Grimstone 1795 by Weazle 1776 (Herod 1758)—Dau. of Ancaster 1768 (Blank 1740).
Mulatto, b. h., 1823.	Catton 1809 (Golumpus 1802)—Desdemona 1811 by Orville 1799 (Beningbrough 1791) Fanny 1796 by Sir Peter T. 1784 (Highflyer 1774) Dau. 1788 of Diomed* 1777 (Florizel 1768)—Desdemona 1770 by Marske 1750 (Squirt 1732).
Linda, b. m., 1825.	Waterloo 1814 (Walton 1799)—Cressida 1807 by Whiskey 1789 (Saltram* 1780)—Y. Giantess 1790 by Diomed* 1777 (Florizel 1768)—Giantess 1769 by Matchem 1748 (Cade 1734)—Molly Long Legs 1753 by Babraham 1740 (God. Ar. 1724).
Orville, b. h., 1799.	Beningbrough 1791 (King Fergus 1775) Evelina 1791 by Highflyer 1774 (Herod 1758)—Termagant by Tantrum 1760 (Cripple 1750) Cantatrice by Sampson 1745 (Blaze 1733)—Dau. 1757 of Regulus 1739 (God. Ar. 1724).
Emily, ch. m., 1810.	Stamford 1794 (Sir Peter T. 1784)—Dau. 1799 of Whiskey 1789 (Saltram* 1780)—Grey Dorimant 1781 by Dorimant 1772 (Otho 1760)—Dizzy 1757 by Blank 1740 (God. Ar. 1724)—Dizzy 1741 by Driver 1727 (Wynn Ar.).
Whisker, b. h., 1812.	Waxy 1790 (Pot-8-o-'s 1773)—Penelope 1798 by Trumpator 1782 (Conductor 1767) Prunella 1788 by Highflyer 1774 (Herod 1758)—Promise 1768 by Snap 1750 (Snip 1736)—Julia 1756 by Blank 1740 (God. Ar. 1724).
Gibside Fairy, b. m., 1811.	Hermes 1790 (Mercury 1778)—Vicissitude b. m. 1800 by Pipator 1786 (Imperator 1776)—Beatrice b. m. 1791 by Sir Peter T. 1784 (Highflyer 1774)—Pyrrha b. m. 1771 by Matchem 1748 (Cade 1734)—Duchess b. m. 1748 by Whitenose 1742 (God. Ar. 1724)—Miss Slamerkin b. m. 1720 by Y. True Blue 1714 (Honeywood Ar.)—Dau. of Oxford Dun Arabian—D'Arcy Blacklegged Royal mare.

HIMYAR.

WINNER OF THE BELLE MEADE STAKES FOR TWO-YEAR OLDS AT LOUISVILLE, BELLE MEADE STAKES, No. 1, AT NASHVILLE, PHŒNIX HOTEL STAKES AT LEXINGTON, THE MERCHANTS' AND TURF STAKES AT LOUISVILLE, KY., AND JANUARY STAKES AT ST. LOUIS.

In the Dixiana Stud, Maj. B. G. Thomas, proprietor, near Lexington, Ky. By private contract. Sales annually in May.

HIMYAR by Alarm, son of imp. Eclipse, bred and owned by Maj. B. G. Thomas, Dixiana Stud, Lexington, Ky., foaled 1875, dam Hira by Lexington, out of Hegira by imp. Ambassador. Himyar made his *debut* as a two-year old at Lexington, Ky., in the Colt and Filly Stakes, 5 furlongs, running third to Pomeroy, in 1.04½. Won the Colt Stakes, three-quarters of a mile, in 1.16¾, beating Blue Eyes, Leveler, etc. Fall, won the Colt and Filly Stakes, 1 mile, in 1.44½, beating Leveler, Blue Eyes and five others. Louisville, won the Belle Meade Stakest, hree-quarters of a mile, in 1.16½, beating Leveler and four others. Was third in the Sanford Stakes to Blue Eyes and Day Star, 1 mile, in 1.45. Himyar, three years old, started in five races, won three and was second in two. Nashville, May 4, won Belle Meade Stake No. 1, 1½ miles, in 2.43, track heavy, beating Bergamot, Artful and Glenmore. Lexington, Ky., May 16, won Phœnix Hotel Stakes, 1¾ miles, in 3.22½, track deep in mud, beating McHenry, J. R. Swinney and Solicitor. Louisville, Ky., May 21, ran second to Day Star in Kentucky Derby, 1½ miles, in 2.37¼, the fastest Derby ever run to that date, beating Leveler, Solicitor, McHenry and four others. St Louis, June 4, won January Stakes, mile heats, in 1.42½, 1.43¼, beating Leveler, Kate Claxton and McHenry. Lexington, September 13, was second in Elkhorn Stakes, 2 miles, in 4.04¾, won by Cammie F., track very muddy. Four years old, started four times, won all of his races. Lexington, won a dash of 1 mile in 1.51¾, track muddy, defeating a field of four; walked over for Club Purse, 2 miles. Louisville, won a Sweepstake for all ages, 1½ miles, beating Goodnight, Long Taw and four others, in 1.56; won Club Purse, 2 miles, in 3.35, beating Long Taw and two others. At five years old started twelve times. Lexington, Ky., was third to Mendelssohn and Verdict in Sweepstake for all ages, 1¼ miles, in 2.08. Won Club Purse, mile heats, in 1.42½, 1.44½, beating Warfield, Victim and four others. Louisville, was unplaced in Louisville Cup, 2¼ miles, in 4.20½, won by Volturno, track very muddy; won the Merchants' Stake, 1½ miles, in 1.55¼, beating Blue Eyes, One Dime and four others. St. Louis, was second to L'Argentine, 1¾ miles, in 2.10¼; was unplaced in Citizens' Plate, 1½ miles, dead heat and stakes divided between Long Taw and Blue Lodge, in 1.55¼. Lexington Fall Meeting, was third to Renown and Montreal, 1 mile, in 1.49¾, track heavy; was second to Renown in Club Purse, 1½ miles, no time, five others behind him. Louisville, won Turf Stakes, 1½ miles, in 1.57½, beating Renown and three others; was second to Bowling Green in Club Purse, 1¼ miles, in 2.42½, three others behind him; won Club Purse, 1½ miles, in 1.54¾, defeating Montreal and Mattie Walker. Nashville, was beaten by Jim Malone in Citizens' Stake, mile heats, in 1.46½, 1.47½, 1.53¾, Himyar won first heat. Six years old, Lexington, Ky., was unplaced in Club Purse, 1 mile, won by Patti in 1.44½. The family is a distinguished racing one. Hira, his dam, was a fine race-mare, Hegira, his grandam, had no equal for speed in her day. She ran 2 miles at New Orleans, in 1850, in 3.34½, the fastest time made to that day. Mecca, her full sister, was also a fine race-mare. Flight, his great grandam, was a superior race-mare and dam of Oliver, a very distinguished race-horse, by Wagner, winner of three-mile heats in 5.38½, 5.38½. Shylock, Cicero, Pedlar, etc., all distinguished in their day, descended from the family. Himyar is a light bay with a long star and both hind feet white. He has a neat, bony head, broad between the eyes, good neck, splendid, well-set, oblique shoulders, with abundance of muscle over the shoulder blade, great depth of girth, good body, but rather short back ribs, which makes his flank look light, with broad, strong hips, immense stifle and second thigh, and the muscle running tapering to excellent hocks, with sound feet and legs. Taking him altogether, he is a very fine, well-shaped horse, with most wonderful muscular power behind.

Himyar, as his high racing lineage warranted, is a success in the stud. He is the sire of the speedy Correction, Georgia, Faraday, a prominent stake winner. Other winners (some of them extra good ones) by Himyar are Estelle, Lou Dudley, Quesal (a superior mare), Johnnie Winkle, Greenwich (an excellent stake winner), Hamilton, Jewel H., Helen, Fannie S., Miss Williams, Ella T., Lady Mary, Henry Tyler, Wah Jim, Mackintosh, Reve d'Or, Mary D., Rosa H., Rorka, Bush Bolt, Himlex, Ben's Pet, Dago, The Hero, Sir Edward, Lord Tom Himyar, The Ironmaster, etc.

From the Barton Barb mare descend in direct female line the good horses Whiskey, Selim, Castrel, Rubens, Catton, Blacklock, Sir Hercules, Margrave (imported), Harkaway, Voltigeur, Surplice, Cowl, Teddington, St. Albans, Lord Clifden, Brag, Sheen, Gold, imp. Billet, Major Domo, Defaulter, etc., etc.

Waxy 1790—Pot-8-o's 1773 Eclipse 1764—Marske 1750 Squirt 1732- Bartlett's Childers—Darley Ar. 1689.

HIMYAR.

BAY HORSE, FOALED 1875; BRED BY MAJ. B. G. THOMAS, DIXIANA STUD, KENTUCKY.

Sire / Dam	Grandparents	Great-grandparents	Great-great-grandparents		
ALARM, b. h., 1869.	*Eclipse, b. h., 1855.	Orlando b. h., 1841.	Touchstone, br. h., 1831.	Camel, br. h., 1822.	Whalebone 1807 (Waxy 1790) Dau. 1812 of Selim 1802 (Buzzard* 1787)—Maiden 1801 by Sir Peter T. 1784 (Highflyer 1774)- Dau. 1788 of Phenomenon 1780 (Herod 1758 Matron 1782 by Florizel 1768 (Herod 1758).
				Banter, br. m., 1826.	Master Henry 1815 (Orville 1799)— Boadicea 1807 by Alexander 1782 (Eclipse 1764) Brunette by Amaranthus 1766 (Old England 1741) Mayfly 1771 by Matchem 1748 (Cade 1734)- Dau. 1755 of Starling 1738 (Starling 1727).
			Vulture, ch. m., 1833.	Langar, ch. h., 1817.	Selim 1802 (Buzzard* 1787) Dau 1808 of Walton 1799 (Sir Peter T. 1784) Y. Giantess 1790 by Diomed* 1777 (Florizel 1768)—Giantess 1769 by Matchem 1748 (Cade 1734) Molly Long Legs 1753 by Babraham 1740 (God. Ar. 1724).
				Kite, b. m., 1821.	Bustard 1813 (Castrel 1801) Olympia 1815 by Sir Oliver 1800 (Sir Peter T. 1784) Scotilla 1795 by Anvil 1777 (Herod 1758)—Scota 1783 by Eclipse 1764 (Marske 1750)— Harmony 1775 by Herod 1758 (Tartar 1743).
		Gaze, b. m., 1842.	Bay Middleton, b. h., 1833.	Sultan, b. h., 1816.	Selim 1802 (Buzzard* 1787) Bacchante 1809 by W's Ditto 1800 (Sir Peter T. 1784) Dau. 1791 of Mercury 1778 (Eclipse 1764) Dau. 1776 of Herod 1758 (Tartar 1743) Folly 1771 by Marske 1750 (Squirt 1732).
				Cobweb, b. m., 1821.	Phantom 1808 (Walton 1799)—Filagree 1815 by Soothsayer 1808 (Sorcerer 1796 —Web 1808 by Waxy 1790 (Pot 8-o's 1773) Penelope 1798 by Trumpator 1782 (Conductor 1767) Prunella 1788 by Highflyer 1774 (Herod 1758).
			Flycatcher, ch. m., 1831.	Godolphin, b. h., 1818.	Partisan 1811 (Walton 1799)—Ridicule 1810 by Shuttle 1793 (Y. Marske 1771) Dau. 1792 of Dungannon 1780 (Eclipse 1764)—Letitia 1783 by Highflyer 1774 (Herod 1758) Dau. of Matchem 1748 (Cade 1734).
				Sister to Cobweb, ch. m., 1820.	Phantom 1808 (Walton 1799)— Filagree 1815 by Soothsayer 1808 (Sorcerer 1796)— Web 1808 by Waxy 1790 (Pot-8-o's 1773) Penelope 1798 by Trumpator 1782 (Conductor 1767) Prunella 1788 by Highflyer 1774 (Herod 1758).
	*Maud, b. m., 1859.	Stockwell, ch. h., 1849.	The Baron, ch. h., 1842.	Ir. Birdcatcher ch. h., 1833.	Sir Hercules 1826 (Whalebone 1807)- Guiccioli 1823 by Bob Booty 1804 (Chanticleer 1787)—Flight 1809 by Ir. Escape 1802 (Commodore 1793)—Y. Heroine by Bagot 1780 (Herod 1758) Heroine 1775 by Hero 1753 (Cade 1734).
				Echidna, b. m., 1838.	Economist 1825 (Whisker 1812)—Miss Pratt 1825 by Blacklock 1814 (Whitelock 1803)— Gadabout 1812 by Orville 1799 (Beningbrough 1791) -Minstrel 1803 by Sir Peter T. 1784 (Highflyer 1774)—Matron 1782 by Florizel 1768 (Herod 1758).
			Pocahontas, b. m., 1837.	Glencoe,* ch. h., 1831.	Sultan 1816 (Selim 1802) Trampoline 1825 by Tramp 1810 (Dick Andrews 1797) Web 1808 by Waxy 1790 (Pot-8-o's 1773)—Penelope 1798 by Trumpator 1782 (Conductor 1767)—Prunella 1788 by Highflyer 1774 (Herod 1758).
				Marpessa, b. m., 1830.	Muley 1810 (Orville 1799)—Clare 1824 by Marmion 1806 (Whiskey 1789)—Harpalice 1814 by Gohanna 1790 (Mercury 1778)- Amazon 1799 by Driver 1783 (Trentham 1766)—Fractions 1792 by Mercury 1778 (Eclipse 1764).
		Countess of Albemarle, b. m., 1847.	Lanercost, br. h., 1835.	Liverpool, b. h., 1828.	Tramp 1810 (Dick Andrews 1797) Dau. 1822 of Whisker 1812 (Waxy 1790)- Mandane 1800 by Pot-8-o's 1773 (Eclipse 1764) Y. Camilla 1787 by Woodpecker 1773 (Herod 1758)- Camilla 1778 by Trentham 1766 (T. Sweepstakes 1743).
				Otis, b. m., 1820.	Bustard 1801 (Buzzard* 1787) Dau. 1813 of Election 1804 (Gohanna 1790)- Dau. 1791 of Highflyer 1774 (Herod 1758)—Dau. 1778 of Eclipse 1764 (Marske 1750)—Rosebud 1765 by Snap 1750 (Snip 1736).
			Sist. to Hornsea, b. m., 1832.	Velocipede, ch. h., 1825.	Blacklock 1814 (Whitelock 1803) Dau. 1817 of Juniper 1805 (Whiskey 1789)- Dau. 1810 of Sorcerer 1796 (Trumpator 1782) Virgin 1801 by Sir Peter T. 1784 (Highflyer 1774)—Dau. 1794 of Pot-8-o's 1773 (Eclipse 1764).
				Daughter of, b. m., 1820.	Cerberus 1802 (Gohanna 1790)—Miss Cranfield 1803 by Sir Peter T. 1784 (Highflyer 1774)- Dau. 1796 of Pegasus 1784 (Eclipse 1764)—Dau. 1783 of Paymaster 1766 (Blank 1740)—Pomona 1775 by Herod 1758 (Tartar 1743).
HIRA, ch. m., 1864.	Lexington, b. h., 1850.	Boston, ch. h., 1833.	Timoleon, ch. h., 1813.	Sir Archy, b. h., 1805.	*Diomed 1777 (Florizel 1768)—*Castianira 1796 by Rockingham 1781 (Highflyer 1774)—Tabitha 1784 by Trentham 1766 (T. Sweepstakes 1743)—Dau. of Bosphorus 1754 (Babraham 1740)—Dau. of Forester 1730 (Forester 1730).
				Daughter of, ch. m., —.	*Saltram 1780 (Eclipse 1764)—Dau. of Symmes' Wildair 1767 (*Fearnought 1755)—Dau. of Tyler's Driver 1760 (*Othello)—Dau. of *Fallower 1761 (Blank 1740)—Dau. of *Vampire 1757 (Regulus 1739).
			Sist. to Tuckahoe, ch. m., 1814.	Ball's Florizel, ch. h., 1801.	*Diomed 1777 (Florizel 1768)—Dau. of *Shark 1771 (Marske 1750)—Dau. of Harris' Eclipse 1770 (*Fearnought 1755)—Dau. of *Fearnought 1755 (Regulus 1739)—Dau. of *Jolly Roger 1741 (Roundhead 1733).
				Daughter of, b. m., 1799.	*Alderman 1787 (Pot-8-o's 1773)—Dau. of *Clockfast 1774 (Gimcrack 1760)—Dau. of Symmes' Wildair 1767 (*Fearnought 1755)—Y. Kitty Fisher 1767 by *Fearnought 1755 (Regulus 1739)—*Kitty Fisher 1755 by Cade 1734 (God. Ar. 1724).
		Alice Carneal, b. m., 1836.	*Sarpedon, br. h., 1828.	Emilius, b. h., 1820.	Orville 1799 (Beningbrough 1791)—Emily 1810 by Stamford 1794 (Sir Peter T. 1784)—Dau. 1799 of Whiskey 1789 (Saltram* 1780)—Grey Dorimant 1781 by Dorimant 1772 (Otho 1760)—Dizzy 1757 by Blank 1740 (God Ar. 1724).
				Icaria, b. m., 1824.	The Flyer 1814 (Vandyke Jr. 1808)—Parma 1813 by Dick Andrews 1797 (Joe Andrews 1778)—May 1804 by Beningbrough 1791 (King Fergus 1775)—Primrose 1787 by Mambrino 1768 (Engineer 1756)—Cricket 1776 by Herod 1758 (Tartar 1743).
			Rowena, ch. m., 1826.	Sumpter, ch. h., 1818.	Sir Archy 1805 (*Diomed 1777)—Dau. of *Robin Redbreast 1790 (Sir Peter T. 1784)—Dau. 1789 of *Obscurity 1777 (Eclipse 1764)—Slamerkin 1768 by *Wildair 1753 (Cade 1734)—*Cub mare 1762 by Cub 1739 (Fox 1714).
				Lady Grey, b. m., 1817.	Robin Grey 1805 (*Royalist 1790)—Maria 1802 by Melzar 1791 (*Medley 1776)—Dau. of *Highflyer 1784 (Highflyer 1774)—Dau of *Fearnought 1755 (Regulus 1739)—Dau. of Ariel 1756 (*Traveller 1747).
	Hegira, ch. m., 1846.	*Ambassador, ch. h., 1833.	Plenipotentiary, ch. h., 1831.	Emilius, b. h., 1820.	Orville 1799 (Beningbrough 1791)—Emily 1810 by Stamford 1794 (Sir Peter T. 1784)—Dau. 1799 of Whiskey 1789 (Saltram* 1780)—Grey Dorimant 1781 by Dorimant 1772 (Otho 1760)—Dizzy 1757 by Blank 1740 (God. Ar. 1724).
				Harriet, blk. m., 1819.	Pericles 1809 (Evander 1801) Dau. 1812 of Selim 1802 (Buzzard* 1787) Pipylina 1803 by Sir Peter T. 1784 (Highflyer 1774)—Rally 1790 by Trumpator 1782 (Conductor 1767)—Fancy 1780 by Florizel 1768 (Herod 1758).
			*Jenny Mills, ch. m., 1825.	Whisker, b. h., 1812.	Waxy 1790 (Pot-8-o's 1773)- Penelope 1798 by Trumpator 1782 (Conductor 1767) Prunella 1788 by Highflyer 1774 (Herod 1758) - Promise 1768 by Snap 1750 (Snip 1736)—Julia 1756 by Blank 1740 (God. Ar. 1724).
				Hornsea's dam, b. m., 1820.	Cerberus 1802 (Gohanna 1790) Miss Cranfield 1803 by Sir Peter T. 1784 (Highflyer 1774)- Dau. 1796 of Pegasus 1784 (Eclipse 1764)- Dau. 1783 of Paymaster 1766 (Blank 1740) -Pomona 1775 by Herod 1758 (Tartar 1743).
		Flight, ch. m., 1837.	*Leviathan, ch. h., 1823.	Muley, b. h., 1810.	Orville 1799 (Beningbrough 1791)—Eleanor 1798 by Whiskey 1789 (Saltram* 1780)—Y. Giantess 1790 by Diomed* 1777 (Florizel 1768)—Giantess 1769 by Matchem 1748 (Cade 1734)—Molly Long Legs 1753 by Babraham 1740 (God. Ar. 1724).
				Daughter of, blk. m., 1809.	Windle 1804 (Beningbrough 1791) Dau. 1788 of Anvil 1777 (Herod 1758)—Virago 1764 by Snap 1750 (Snip 1736) Dau. of Regulus 1739 (God. Ar. 1724)—Sist. 1750 to Othello by Crab 1722 (Alcock Ar.).
			Charlotte Hamilton, ch. m., 1826.	Sir Charles, ch. h., 1816.	Sir Archy 1805 (*Diomed 1777)—Dau. of *Citizen 1785 (Pacolet 1763)— Dau. of Commutation (Symmes' Wildair 1767)—Dau. of *Dare Devil 1787 (Magnet 1770)—Sally Shark by *Shark 1771 (Marske 1750).
				Lady of the Lake, b. m., 1812.	*Sir Harry 1795 (Sir Peter T. 1784)—Shylock's dam by *Diomed 1777 (Florizel 1768)—Dau. of *St. George 1771 (Dragon 1757)—Dau. of *Fearnought 1755 (Regulus 1739)—Dau. of *Jolly Roger 1741 (Roundhead 1733)—*Mary Grey gr. m. 1742 by Roundhead 1733 (Fl. Childers 1715)—Ringbone 1732 by Partner 1718 (Jigg)—Lusty Thornton ch. m. by Croft's Bay Barb (Chillaby)—Chestnut Thornton ch. m. by Makeless (Oglethorpe Ar.)—Old Thornton by Brimmer (D'Arcy Yellow Turk)—Dau. of Dicky Pierson (Dodsworth)—Burton Barb mare.

HINDOO.

WINNER OF SEVEN OUT OF NINE STAKES AT TWO YEARS OLD, WINNER OF THE KENTUCKY DERBY AND CLARK STAKES AT LOUISVILLE, BLUE RIBBON STAKES AT LEXINGTON, TIDAL STAKES AND CONEY ISLAND DERBY AT SHEEPSHEAD BAY, OCEAN STAKES, JERSEY DERBY AND LORILLARD STAKES AT LONG BRANCH, THE TRAVERS AND KENNER STAKES AT SARATOGA, ALL IN 1881; LOUISVILLE CUP AND TURF STAKES AND CONEY ISLAND CUP, 1882.

Property of the Runnymede Stud, near Paris, Ky. Application to Messrs. Clay & Woodford, Paris, Ky. Annual sales of yearlings in May.

HINDOO by Virgil, son of Vandal, bred by D. Swigert, Elmendorf Stud, Ky., foaled 1878, dam Florence by Lexington, son of Boston, out of imp. Weatherwitch, dam of Little Mack and Mollie Cad by Weatherbit, son of Sheet Anchor. Hindoo was the sensation two-year old of his year, started in nine races, won seven, was second in one, third in one. Hindoo, at Lexington, Ky., won the Colt and Filly Stakes for two-year olds, three-quarters of a mile, in 1.17¾, beating Alfambra, Brambaletta and seven others. Louisville, Ky., won Alexander Stakes, half a mile, in 50 seconds, beating Banter, Maretzek and eight others; won the Tennessee Stakes for two-year olds, three-quarters of a mile, in 1.16, beating Brambaletta, Ripple, Bootjack and five others. St. Louis, won the Juvenile Stakes for two-year olds, three-quarters of a mile, in 1 17¼, beating Voltague and six others; won the Jockey Club Stakes for two-year olds, 1 mile, in 1.44, beating Lelex, Voltague and Enniskillen, a fast and good race. Chicago, won the Criterion Stakes for two-year olds, three-quarters of a mile, in 1.15, beating Ripple, Greenland and three others. This was the fastest three-quarters ever run by a two-year old to that date. Won the Tremont Hotel Stakes for two-year olds, 1 mile, in 1.48, beating Lizzie S., Ripple and Moses. In all these races, except the Tennessee Stakes, he carried 100 lbs., in the Tennessee, 105 lbs. Saratoga, ran third to Crickmore and Bonnie Lizzie in the Windsor Hotel Stakes for two-year olds 5 furlongs, in 1.05, track heavy, beating Thora and four others; ran second to Thora in the Day Boat Line Stakes, three-quarters of a mile, in 1.17½, beating Bonnie Lizzie and three others. In these races he carried 110 lbs. He had just changed hands and did not run up to his previous form.

Three years old, won eighteen out of twenty races. Lexington, Ky., won Blue Ribbon Stakes, 1½ miles, in 2 38, beating Getaway, Bend'or, Creosote and four others. Louisville, Ky., won Kentucky Derby, 1½ miles, in 2.40, beating Lelex, Alfambra and three others; won Clark Stakes, 1¼ miles, in 2.10¾, beating Alfambra, Bootjack, Bend'or and Sligo. Jerome Park, won a dash of 1 mile and a furlong, in 2.02½, beating Sir Hugh, Jack of Hearts and Rob Roy; won a dash of 1 mile and 3 furlongs, in 2.34, track heavy. Sheepshead Bay, won Tidal Stakes, 1 mile, in 1.43½, beating Crickmore and Saunterer; won the Coney Island Derby, 1½ miles, in 2.47½, beating Baltic. Monmouth Park, won Ocean Stakes for all ages, 1 mile and a furlong, in 1.57, beating Monitor, Glidelia, etc.; won Lorillard Stakes, 1½ miles, in 2.39¾., beating Crickmore and Saunterer; walked over for a Sweepstakes of $1,000 each, with $2,000 added. Saratoga, won the Travers Stakes, 1¾ miles, in 3.07½, beating Eole, Getaway and Compensation; won Sequel Stakes, 1¾ miles, in 3.21, track heavy. In this race Hindoo carried 123 lbs., including a penalty of five pounds, and beat Greenland and Valentino; won United States Hotel Stakes, 1½ miles, in 2 36, beating Crickmore, Bonfire and Gladiola; won Kenner Stakes, 2 miles, in 3.32, a fast race. Monmouth Park, won Champion Stakes for all ages, 1½ miles, in 2.39, beating Monitor and Parole; won Jersey St. Leger, 1¾ miles, in 3.18, carrying 123 lbs., beating Bona Fide. Sheepshead Bay, won race of mile heats, in 1.42¼, 1.45½, beating Sir Hugh; won a dash of a mile, in 1.42, beating Sir Hugh and Edendary; was second in the Brighton Beach Purse, 1½ miles, won by Crickmore, in 2.36½; was third in the September Handicap, 1¾ miles, won by Crickmore, in 3.03, Hindoo carrying 123 lbs.

As a four-year old won five races out of six. Louisville, Ky., won Louisville Cup, 2¼ miles, in 3.57¾, beating Checkmate, Glidelia, Lida Stanhope and Blazes. Won Merchants' Stakes for all ages, 1 mile and a furlong, in 1.59½, beating Checkmate, Runnymede and Creosote; won Turf Stakes, 1¼ miles, in 2.08½, carrying 122 lbs., beating Checkmate (aged), 123 lbs., and Creosote (4), 114 lbs. Sheepshead, won Coney Island Stakes for three-year olds and upwards, 1 mile and a furlong, in 1.57¾, Barrett the only starter; won Coney Island Cup, 2¼ miles, in 3.58, beating Eole and Parole. This was the best race run at the distance during the year, and compares favorably with similar races. There is no question but that Hindoo was the best

(*Continued on page 176.*)

HINDOO.

BAY HORSE, FOALED 1878; BRED BY D. SWIGERT, ELMENDORF STUD, KENTUCKY.

VIRGIL, br. h., 1864.
Vandal, b. h., 1850. — Hymenia, b. m., 1851.
Glencoe, ch. h., 1831. — Alarie's dam, b. m., 182?. — *Yorkshire, b. h., 1834. — Little Peggy, ch. m., 1837.
Sultan, b. h., 1816. — Trampoline, ch. m., 1825. — Tranby, br. h., 1826. — Lucilla, b. m., —. — St. Nicholas, br h., 1827. — Miss Rose, b. m., 1823. — Cripple, b. h., 1826. — Peggy Stewart, b. m., 1821.

FLORENCE, ch. m., 1869.
Lexington, b. h., 1850. — *Weather-Witch, ch. m., 1858.
Boston, ch. h., 1833. — Alice Carneal, b. m., 1836. — Weatherbit, br. h., 1842. — Daughter of, br. m., 1832.
Timoleon, ch. h., 1813. — Sist. to Tuckahoe, ch. m., 1814. — *Sarpedon, br. h., 1828. — Rowena, ch. m., 1826. — Sheet Anchor, br. h., 1832. — Miss Letty, b. m., 1834. — Ir. Birdcatcher, ch. h., 1833. — Colocynth, b. m., 1840.

Woodpecker 1773—Herod 1758—Tartar 1743—Partner 1718—Jigg—Byerly Turk, about 1685.

Selim, ch. h., 1802.	Buzzard 1787 (Woodpecker 1773)—Dau. of Alexander 1782 (Eclipse 1764)—Dau. of Highflyer 1774 (Herod 1758)—Dau. of Alfred 1770 (Matchem 1748)—Dau. 1770 of Engineer 1756 (Sampson 1745).
Bacchante, br. m., 1809.	W's Ditto 1800 (Sir Peter T. 1784)—Dau. 1791 of Mercury 1778 (Eclipse 1764)—Dau. 1776 of Herod 1758 (Tartar 1743)—Folly 1771 by Marske 1750 (Squirt 1732)—Vixen 1753 by Regulus 1739 (God. Ar. 1724).
Tramp, b. h., 1810.	Dick Andrews 1797 (Joe Andrews 1778)—Dau. 1803 of Gohanna 1790 (Mercury 1778)—Fraxinella 1793 by Trentham 1766 (T. Sweepstakes 1743)—Dau. 1785 of Woodpecker 1773 (Herod 1758)—Everlasting 1775 by Eclipse 1764 (Marske 1750).
Web, b. m., 1808.	Waxy 1790 (Pot-8-o's 1773)—Penelope 1798 by Trumpator 1782 (Conductor 1767)—Prunella 1788 by Highflyer 1774 (Herod 1758)—Promise 1768 by Snap 1750 (Snip 1736)—Julia 1756 by Blank 1740 (God. Ar. 1724).
Blacklock, b. h., 1814.	Whitelock 1803 (Hambletonian 1792)—Dau. 1799 of Coriander 1786 (Pot-8-o's 1773)—Wildgoose 1792 by Highflyer 1774 (Herod 1758)—Coheiress 1786 by Pot-8-o's 1773 (Eclipse 1764)—Manilla 1777 by Goldfinder 1764 (Snap 1750).
Miss Bowe's dam, br. m., 1812.	Orville 1799 (Beningbrough 1791)—Miss Grimstone 1796 by Weazle 1776 (Herod 1758)—Dau. of Ancaster 1768 (Blank 1740)—Dau. of Damascus Ar. 1754—Dau. of Sampson 1745 (Blaze 1733).
Davenport's Trumpator, br. h., 1824.	Sir Solomon 1805 (*Tickle Toby 1786)—Dau. of Hickory 1801 (*Whip 1794)—Trumpetta 1797 by Trumpator 1782 (Conductor 1767)—Sist. 1786 to Lambinos by Highflyer 1774 (Herod 1758)—Dau. 1775 of Eclipse 1764 (Marske 1750).
Lucy, b. m., 1821.	Orphan 1810 (Ball's Florizel 1801)—Lady Grey 1817 by Robin Grey 1805 (*Royalist 1790)—Maria 1802 by Melzar 1791 (*Medley 1776)—Dau. of *Highflyer 1784 (Highflyer 1774)—Dau. of *Fearnought 1755 (Regulus 1739).
Emilius, b. h., 1820.	Orville 1799 (Beningbrough 1791)—Emily 1810 by Stamford 1794 (Sir Peter T. 1784)—Dau. 1799 of Whiskey 1789 (*Saltram 1780)—Grey Dorimant 1781 by Dorimant 1772 (Otho 1760)—Dizzy 1757 by Blank 1740 (God. Ar. 1724).
Sea Mew, b. m., 1815.	Scud 1804 (Beningbrough 1791)—Goosander 1805 by Hambletonian 1792 (King Fergus 1775)—Rally 1790 by Trumpator 1782 (Conductor 1767)—Fancy 1780 by Florizel 1768 (Herod 1758)—Dau. 1763 of Spectator 1749 (Crab 1722).
Tramp, b. h., 1810.	Dick Andrews 1797 (Joe Andrews 1778)—Dau. 1803 of Gohanna 1790 (Mercury 1778)—Fraxinella 1793 by Trentham 1766 (T. Sweepstakes 1743)—Dau. 1785 of Woodpecker 1773 (Herod 1758)—Everlasting 1775 by Eclipse 1764 (Marske 1750).
Daughter of, ch. m., 1810.	Sancho 1801 (Don Quixote 1784)—Blacklock's dam 1799 by Coriander 1786 (Pot-8-o's 1773)—Wildgoose 1792 by Highflyer 1774 (Herod 1758)—Coheiress 1786 by Pot-8-o's 1773 (Eclipse 1764)—Manilla 1777 by Goldfinder 1764 (Snap 1750).
Medoc, ch. h., 1829.	Am. Eclipse 1814 (Duroc 1806)—Y. Maid-of-the-Oaks 1817 by *Expedition 1795 (Pegasus 1784)—Maid-of-the-Oaks 1801 by *Spread Eagle 1792 (Volunteer 1780)—Annette by *Shark 1771 (Marske 1750)—Dau. of Rockingham (Partner 1755).
Grecian Princess, b. m., 1816.	Cook's Whip 1804 (*Whip 1794)—Jane Hunt 1796 by Hampton's Paragon 1788 (*Flimnap 1765)—Moll by *Figure 1757 (Figure 1747)—Slamerkin 1768 by *Wildair 1753 (Cade 1734)—*Cub mare 1762 by Cub 1759 (Fox 1714).
Cook's Whip, b. h., 1804.	*Whip 1794 (Saltram* 1780)—Speckleback 1790 by Randolph's Celer (Meade's Celer 1774)—Old Speckleback by Meade's Celer 1774 (*Janus 1745)—Dau. of *Sober John 1748 (Rib 1736)—Thoroughbred mare of Mr. Wade.
Mary Bedford, b. m., 1816.	Duke of Bedford (*Bedford 1792)—Dau. 1818 of *Speculator 1795 (Dragon* 1787)—Dau. 1806 of *Dare Devil 1787 (Magnet 1770)—*Trumpetta 1797 by Trumpator 1782 (Conductor 1767)—Sist. 1786 to Lambinos by Highflyer 1774 (Herod 1758).
Sir Archy, b. h., 1805.	*Diomed 1777 (Florizel 1768)—*Castianira 1796 by Rockingham 1781 (Highflyer 1774)—Tabitha 1782 by Trentham 1766 (T. Sweepstakes 1743)—Dau. of Bosphorus 1754 (Babraham 1740)—Dau. of Forester 1770 (Croft's Forester 1736).
Daughter of, ch. m., —.	*Saltram 1780 (Eclipse 1764)—Dau. of Symmes' Wildair 1767 (*Fearnought 1755)—Dau. of Tyler's Driver 1760 (*Othello)—Dau. of *Fallower 1761 (Blank 1740)—Dau. of *Vampire 1757 (Regulus 1739).
Ball's Florizel, ch. h., 1801.	*Diomed 1777 (Florizel 1768)—Dau. of *Shark 1771 (Marske 1750)—Dau. of Harris' Eclipse 1770 (*Fearnought 1755)—Dau. of *Fearnought 1755 (Regulus 1739)—Dau. of *Jolly Roger 1741 (Roundhead 1733).
Daughter of, b. m., 1799.	*Alderman 1787 (Pot-8-o's 1773)—Dau. of *Clockfast 1774 (Gimcrack 1760)—Dau. of Symmes' Wildair 1767 (*Fearnought 1755)—Y. Kitty Fisher 1767 by *Fearnought 1755 (Regulus 1739)—*Kitty Fisher 1755 by Cade 1734 (God. Ar. 1724).
Emilius, b. h., 1820.	Orville 1799 (Beningbrough 1791)—Emily 1810 by Stamford 1794 (Sir Peter T. 1784)—Dau. 1799 of Whiskey 1789 (Saltram* 1780)—Grey Dorimant 1781 by Dorimant 1772 (Otho 1760)—Dizzy 1757 by Blank 1740 (God Ar. 1724).
Icaria, b. m., 1824.	The Flyer 1814 (Vandyke Jr. 1808)—Parma 1813 by Dick Andrews 1797 (Joe Andrews 1778)—May 1804 by Beningbrough 1791 (King Fergus 1775)—Primrose 1787 by Mambrino 1768 (Engineer 1756)—Cricket 1776 by Herod 1758 (Tartar 1743).
Sumpter, ch. h., 1818.	Sir Archy 1805 (*Diomed 1777)—Dau. of *Robin Redbreast 1796 (Sir Peter T. 1784)—Dau. 1788 of *Obscurity 1777 (Eclipse 1764)—Slamerkin 1768 by *Wildair 1753 (Cade 1734)—*Cub mare 1762 by Cub 1759 (Fox 1714).
Lady Grey, b. m., 1817.	Robin Grey 1805 (*Royalist 1790)—Maria 1802 by Melzar 1791 (*Medley 1776)—Dau. of *Highflyer 1784 (Highflyer 1774)—Dau of *Fearnought 1755 (Regulus 1739)—Dau. of Ariel 1751 (*Traveller 1747).
Lottery, br. h., 1820.	Tramp 1810 (Dick Andrews 1797)—Mandane 1800 by Pot-8-o's 1773 (Eclipse 1764)—Y. Camilla 1787 by Woodpecker 1773 (Herod 1758)—Camilla 1778 by Trentham 1766 (T. Sweepstakes 1743)—Coquette 1765 by Compton Barb.
Morgiana, blk. m., 1820.	Muley 1810 (Orville 1799)—Miss Stephenson 1814 by [Scud or] Sorcerer 1796 (Trumpator 1782)—Sist. 1796 to Petworth by Precipitate* 1787 (Mercury 1778)—Dau. 1787 of Woodpecker 1773 (Herod 1758)—Dau. 1773 of Snap 1750 (Snip 1736).
Priam,* b. h., 1827.	Emilius 1820 (Orville 1799)—Cressida 1807 by Whiskey 1789 (Saltram* 1780)—Y. Giantess 1790 by Diomed* 1777 (Florizel 1768)—Giantess 1769 by Matchem 1748 (Cade 1734)—Molly Long Legs 1753 by Babraham 1740 (God. Ar 1724).
Daughter of, br. m., 1815.	Orville 1799 (Beningbrough 1791)—Dau. 1800 of Buzzard* 1787 (Woodpecker 1773)—Hornpipe 1793 by Trumpator 1782 (Conductor 1767)—Luna 1779 by Herod 1758 (Tartar 1743)—Proserpine 1766 by Marske 1750 (Squirt 1732).
Sir Hercules, blk. h., 1826.	Whalebone 1807 (Waxy 1790)—Peri 1822 by Wanderer 1811 (Gohanna 1790)—Thalestris 1809 by Alexander 1782 (Eclipse 1764)—Rival 1800 by Sir Peter T. 1784 (Highflyer 1774)—Hornet 1790 by Drone 1777 (Herod 1758).
Guiccioli, ch. m., 1823.	Bob Booby 1804 (Chanticleer 1787)—Flight 1809 by Ir. Escape 1802 (Commodore 1793)—Y. Heroine by Bagot 1780 (Herod 1758)—Heroine 1775 by Hero 1753 (Cade 1734)—Dau. 1766 of Snap 1750 (Snip 1736).
Physician, b. h., 1829.	Brutandorf 1821 (Blacklock 1814)—Primette 1820 by Prime Minister 1810 (Sancho 1801)—Miss Paul 1811 by Sir Paul 1802 (Sir Peter T. 1784)—Miss Dunnington 1807 by Shuttle 1793 (Y. Marske 1771)—Miss Grimstone 1796 by Weazle 1776 (Herod 1758).
Camelina, b. m., 1823.	Whalebone 1807 (Waxy 1790)—Dau. b. m. 1812 of Selim 1802 (Buzzard 1787)—Maiden b. m. 1801 by Sir Peter T. 1784 (Highflyer 1774)—Dau. ch. m. 1788 of Phenomenon 1780 (Herod 1758)—Matron b. m. 1782 by Florizel 1768 (Herod 1758)—Maiden ch. m. 1770 by Matchem 1748 (Cade 1734)—Pratt's Old Squirt mare ch. m. 1750 by Squirt 1732

(Bartlett's Childers)—Lot's dam by Mogul (God. Ar. 1724)—Camilla ch. m. 1722 by Bay Bolton 1705 (Grey Hautboy)—Starling's dam by Son of Brownlow Turk—Old Lady by Pulleine's Chestnut Ar.—Dau. of Rockwood—Dau. of Helmsley Turk.

HINDOO—Continued.

race-horse which has appeared in this country for more than a decade. It is doubtful if any horse in England could have beaten him in the great classic events. Hindoo is a grandly bred horse, his blood combining all the best strains in England and America. His sire is by the best son of Glencoe, and he has a large infusion of Diomed blood through both sides, together with a double cross of imp. Buzzard, fortified by the stout crosses Whalebone, Herod and Eclipse on the dam's side. Hindoo is a dark bay, 15 hands 3½ inches high, with large star in his forehead, right hind foot white to pastern.

Hindoo is a grand success in the stud, sire of the great Hanover, Jim Gore, Merry Monarch, one of the best two-year olds of 1891, Hindoocraft, Buddhist, Houston, Heathen, Chief Justice, Newburg, Montpelier, Prospect, Dungarven, War Plot, Woodford, Black Knight, Bowen, Hyacinth, Headlight, Cortland, Col. Zeb Ward, Hawkstone, Heatherton, Brahmin, Bright Light, Hempstead, O. P. B., Hannibal, Indus, Speth, Kehama, Tar and Tartar, Montpelier, Halcyon and Sir Curtis. Among his daughters are Sallie McClelland, Catalpa, Mabel Glenn, Hindoo Rose, Hindoo Lass, Hindoogam, Hispania, Madolin, Nina Archer, Sallie Taylor, Hinda, Sunbeam, Lady Hindoo, Lizzie B., Lillian Lindsay, Servia, Franchise, Lizzie C., Miss Thomas, Madumma, Daylight, Centella, Hannah Moberly, Lizette, Matilda, Natalie, etc.

It is to the Helmsley Turk mare to which trace in direct female line Old Starling, Pumpkin, Gohanna, Camel, The Baron, sire of Stockwell and Rataplan, etc.

INSPECTOR B. (ENVOLEUR).

WINNER OF THE BELMONT, TIDAL, LORILLARD, TRAVERS, IROQUOIS AND OTHER GREAT STAKES.

In the Belle Meade Stud, Gen. W. H. Jackson, proprietor, Nashville, Tenn. Terms published annually.

INSPECTOR B., bay horse, foaled 1883, bred in Belle Meade Stud, owned by Gen. W. H. Jackson, by Enquirer, son of imp. Leamington, dam Colossa by Colossus, son of imp. Sovereign, out of Rurica by Rurie, son of imp. Sovereign.

Inspector B. is a very muscular animal, of the most symmetrical formation, having great length, good size, grand substance, and stands on good short legs. His racing lines are as nearly perfect as possible, and at the stud he should equal his first class performances on the turf. He commenced his career as a two-year old at Jerome Park, where he won a ½ mile race by two lengths, in 0.50¼, beating Dry Monopole, etc. Ran third to Dew Drop and Dry Monopole, in a ⅝ mile race at Sheepshead Bay, beating six others; won a ¾ mile race, conceding weight to all his opponents, which included The Bard and five others. At Coney Island, he ran second to Dew Drop, in the Great Eastern Handicap, beating a field of nineteen; ran second to The Bard in the Bouquet Stakes, ⅞ mile, giving him 4 lbs., behind him were Ban Fox, Blue Wing, Pure Rye and eight others, nearly all in receipt of weight. At Coney Island, won a ¾ mile Sweepstakes by a length, in 1.17, beating Biggonet, Winfred, Bessie B. and others. At Jerome Park he ran second to Dew Drop in the Champagne Stakes, and wound up his two-year old season by winning a ¾ mile Sweepstakes at the same place, carrying top weight.

At three he won a Sweepstakes, 1¼ miles, with top weight, in 2.11; won the Belmont Stakes, 1½ miles, in 2.41, beating The Bard, Linden, Saxony, etc.; won the Tidal Stakes, 1 mile, in 1.46¾, beating Biggonet, Quito, Portland, etc; ran second to Ban Fox in the Coney Island Derby, 1½ miles, in 2.38¾, beating Elkwood; won the Stud Stakes, 1⅜ miles, in 2.23¾, beating The Bard, Ban Fox and Buffalo; won the Lorillard Stakes, 1½ miles, in 2.40; won the Harvest Handicap, 1¼ miles, at the same place, carrying top weight; won the Travers Stakes, 1¾ miles, in 3.10¼; won the Iroquois Stakes, 1 mile 500 yards, in 2.16¼, and the United States Hotel Stakes, 1½ miles, in 2.12¾.

At five he won at 1⅛ miles in 1.55¾; a mile dash on the turf in 1.45¾; 7 furlongs in 1.29, with top weight; was second to Bradford, 6 furlongs, in 1.14¾, giving the winner 4 lbs.; was second to Swift, 1⅛ miles, in 1.54½, giving the winner 9 lbs.

At six he won the Brookdale Handicap, 1⅛ miles, in 1.55, beating Tea Tray, etc.; won a 1 1/16 miles race; won the Lawnview Handicap, 1⅛ miles, in a canter, beating Eurus, etc.; won a 1 1/16 mile handicap in 1.48¾; walked over for a 1⅛ miles race; won a Sweepstakes, 1 3/16 miles, beating Elkwood, Eurus, etc., in 2.03, and won the Bay Ridge Handicap, 1½ miles, with 122 lbs. up, time 2.35½, beating Tea Tray, Terra Cotta, etc.

Inspector B. descends in direct female line from the same Royal mare to which Blank, Babraham, Gower Stallion, Hornsea, Soothsayer, Jerry, Harvester, Skylark, Foxhall, Mollie Jackson and all the imp. Selima family trace.

INSPECTOR B.

BAY HORSE, FOALED 1883; BRED IN THE BELLE MEADE STUD, TENNESSEE.

- ENQUIRER, b. h., 1867.
  - *Leamington, br. h., 1853.
    - Faugh-a-Ballagh, br. h., 1841.
      - Sir Hercules, blk. h., 1826.
      - Guiccioli, ch. m., 1823.
    - Daughter of, b. m., 1841.
      - Pantaloon, ch. h., 1824.
      - Daphne, br. m., 1837.
  - Lida, b. m., 1858.
    - Lexington, b. h., 1850.
      - Boston, ch. h., 1833.
      - Alice Carneal, b. m., 1836.
    - Lize, ch. m., .
      - Am. Eclipse, ch. h., 1814.
      - Gabriella, ch. m., 1826.
- COLOSSA, b. m., 1869.
  - Colossus, b. h., 1856.
    - *Sovereign, b. h., 1836.
      - Emilius, b. h., 1820.
      - Fleur-de-lis, b. m., 1822.
    - Sist. to Pryor No. 1, ch. m., 1850.
      - *Glencoe, ch. h., 1831.
      - Gipsey, ro. m., 1832.
  - Rurica, b. m., 1861.
    - Ruric, b. h., 1851.
      - *Sovereign, b. h., 1836.
      - Levity, b. m., 1845.
    - Eleanor Margrave, gr. m., 1846.
      - *Margrave, ch. h., 1829.
      - Fanny Wright, gr. m., 1827.

Whalebone, br. h., 1807.	Waxy 1790 (Pot-8-o-'s 1773) Penelope 1798 by Trumpator 1782 (Conductor 1767) - Prunella 1788 by Highflyer 1774 (Herod 1758) Promise 1768 by Snap 1750 (Snip 1736)—Julia 1756 by Blank 1740 (God. Ar. 1724).
Peri, b. m., 1822.	Wanderer 1811 (Gohanna 1790) Thalestris 1809 by Alexander 1782 (Eclipse 1764)—Rival 1800 by Sir Peter T. 1784 (Highflyer 1774) Hornet 1790 by Drone 1777 (Herod 1758) - Manilla 1777 by Goldfinder 1764 (Snap 1750).
Bob Booty, ch. h., 1804.	Chanticleer 1787 (Woodpecker 1773)—Ierne 1790 by Bagot 1780 (Herod 1758) Dau. of Gamahoe (Bustard 1741) Patty by Tim 1752 (Squirt 1732) Miss Patch 1737 by Justice 1725 (H. C. Litton Ar.).
Flight, ch. m., 1809.	Ir. Escape 1802 (Commodore 1793) Y. Heroine by Bagot 1780 (Herod 1758) Heroine 1775 by Hero 1753 (Cade 1734) Dau. 1766 of Snap 1750 (Snip 1736) Sist. 1743 to Regulus by God. Ar. 1724.
Castrel, ch. h., 1801.	Buzzard* 1787 (Woodpecker 1773) Dau. of Alexander 1782 (Eclipse 1764) Dau. of Highflyer 1774 (Herod 1758) Dau. of Alfred 1770 (Matchem 1748) Dau. 1770 of Engineer 1756 (Sampson 1745).
Idalia, ch. m., 1815.	Peruvian 1806 (Sir Peter T. 1784) Musidora 1804 by Meteor 1783 (Eclipse 1764)—Maid-of-all-Work 1786 by Highflyer 1774 (Herod 1758) Dau. 1771 of Syphon 1750 (Squirt 1732)—Dau. of Regulus 1739 (God. Ar. 1724).
Laurel, br. h., 1824.	Blacklock 1814 (Whitelock 1803) Wagtail 1818 by Prime Minister 1810 (Sancho 1801) - Tranby*'s dam 1812 by Orville 1799 (Beningbrough 1791) Miss Grimstone 1796 by Weazle 1776 (Herod 1758)—Dau. of Ancaster 1768 (Blank 1740).
Maid of Honor, br. m., 1829.	Champion 1812 (Selim 1802)—Etiquette 1820 by Orville 1799 (Beningbrough 1791) - Boadicea 1807 by Alexander 1782 (Eclipse 1764) Brunette by Amaranthus 1766 (Old England 1741)—Mayfly 1771 by Matchem 1748 (Cade 1734).
Timoleon, ch. h., 1813.	Sir Archy 1805 (*Diomed 1777) Dau. of *Saltram 1780 (Eclipse 1764)—Dau. of Symmes' Wildair 1767 (*Fearnought 1755)—Dau. of Tyler's Driver 1760 (*Othello)—Dau. of *Fallower 1761 (Blank 1740).
Sister to Tuckahoe, ch. m., 1814.	Ball's Florizel 1801 (*Diomed 1777)—Dau. 1799 of *Alderman 1787 (Pot-8-o-'s 1773)—Dau. of *Clockfast 1774 (Gimcrack 1760) Dau. of Symmes' Wildair 1767 (*Fearnought 1755) - Y. Kitty Fisher 1767 by *Fearnought 1755 (Regulus 1739).
*Sarpedon, br. h., 1828.	Emilius 1820 (Orville 1799) - Icaria 1824 by The Flyer 1814 (Vandyke Jr. 1808)—Parma 1813 by Dick Andrews 1797 (Joe Andrews 1778) May 1801 by Beningbrough 1791 (King Fergus 1775)—Primrose 1787 by Mambrino 1768 (Engineer 1756).
Rowena, ch. m., 1826.	Sumpter 1818 (Sir Archy 1805) Lady Grey 1817 by Robin Grey 1805 (*Royalist 1790)—Maria 1802 by Melzar 1791 (*Medley 1776)—Dau. of *Highflyer 1784 (Highflyer 1774)—Dau. of *Fearnought 1755 (Regulus 1739).
Duroc, ch. h., 1806.	*Diomed 1777 (Florizel 1768) - Amanda 1800 by Grey Diomed 1786 (*Medley 1776)—Dau. of Va. Cade (Partner 1755) Dau. of Hickman's Independence 1776 (*Fearnought 1755) - Dau. of *Badger (Bosphorus 1754).
Miller's Damsel, gr. m., 1802.	*Messenger 1780 (Mambrino 1768) - *Pot-8-o-'s Mare 1792 by Pot-8-o-'s 1773 (Eclipse 1764) Dau. 1778 of Gimcrack 1760 (Cripple 1750) Snapdragon 1759 by Snap 1750 (Snip 1736)—Dau. 1749 of Regulus 1739 (God. Ar. 1724).
Sir Archy, b. h., 1805.	*Diomed 1777 (Florizel 1768)—*Castianira 1796 by Rockingham 1781 (Highflyer 1774)—Tabitha 1782 by Trentham 1766 (T. Sweepstakes 1743)—Dau. of Bosphorus 1754 (Babraham 1740)—Dau. of Forester 1750 (Forester 1736).
Calypso, b. m., 1803.	Bellair 1786 (*Medley 1776)—Dau. 1798 of *Dare Devil 1787 (Magnet 1770)—Dau. 1792 of Symmes' Wildair 1767 (*Fearnought 1755)—Picadilla by Y. Fearnought 1777 (*Fearnought 1755)—Dan. of Godolphin 1770 (*Fearnought 1755).
Orville, b. h., 1799.	Beningbrough 1791 (King Fergus 1775)—Evelina 1791 by Highflyer 1774 (Herod 1758)—Termagant by Tantrum 1760 (Cripple 1750)—Cantatrice by Sampson 1745 (Blaze 1733)—Dau. 1757 of Regulus 1739 (God. Ar. 1724).
Emily, ch. m., 1810.	Stamford 1794 (Sir Peter T. 1784) Dau. 1799 of Whiskey 1789 (Saltram* 1780)—Grey Dorimant 1781 by Dorimant 1772 (Otho 1760)—Dizzy 1757 by Blank 1740 (God. Ar. 1724).—Dizzy 1741 by Driver 1727 (Wynn Ar.).
Bourbon, b. h., 1811.	Sorcerer 1796 (Trumpator 1782)—Dau. 1797 of Precipitate* 1787 (Mercury 1778)—Y. Tiffany 1787 by Highflyer 1774 (Herod 1758)—Tiffany 1775 by Eclipse 1764 (Marske 1750) Y. Hag 1761 by Skim 1746 (Starling 1727).
Lady Rachel, b. m., 1805.	Stamford 1794 (Sir Peter T. 1784)—Y. Rachel 1799 by Volunteer 1780 (Eclipse 1764)—Rachel 1790 by Highflyer 1774 (Herod 1758)—Sist. 1771 to Tandem by Syphon 1750 (Squirt 1732)—Dau. of Regulus 1739 (God. Ar. 1724).
Sultan, b. h., 1816.	Selim 1802 (Buzzard* 1787)—Bacchante 1803 by W's Ditto 1800 (Sir Peter T. 1784)—Dau. 1791 of Mercury 1778 (Eclipse 1764)—Dau. 1776 of Herod 1758 (Tartar 1743)—Folly 1771 by Marske 1750 (Squirt 1732).
Trampoline, ch. m., 1825.	Tramp 1810 (Dick Andrews 1797)—Web 1808 by Waxy 1790 (Pot-8-o-'s 1773)—Penelope 1798 by Trumpator 1782 (Conductor 1767) Prunella 1788 by Highflyer 1774 (Herod 1758)—Promise 1768 by Snap 1750 (Snip 1736).
American Eclipse, ch. h., 1814.	Duroc 1806 (*Diomed 1777) Miller's Damsel 1802 by *Messenger 1780 (Mambrino 1768)—*Dau. 1792 of Pot-8-o-'s 1773 (Eclipse 1764)—Dau. 1778 of Gimcrack 1760 (Cripple 1750)—Snapdragon 1759 by Snap 1750 (Snip 1736).
Y. Maid of the Oaks, ch. m., 1817.	*Expedition 1795 (Pegasus 1781)—Maid-of-the-Oaks 1801 by *Spread Eagle 1792 (Volunteer 1780)—Annette by *Shark 1771 (Marske 1750) Dau. of Rockingham (Partner 1755)—Dau. of True Whig (Fitzhugh's Regulus 1765).
Emilius, b. h., 1820.	Orville 1799 (Beningbrough 1791)—Emily 1810 by Stamford 1794 (Sir Peter T. 1784)—Dau. 1799 of Whiskey 1789 (Saltram* 1780)—Grey Dorimant 1781 by Dorimant 1772 (Otho 1760)—Dizzy 1757 by Blank 1740 (God. Ar. 1724).
Fleur-de-Lis, b. m., 1822.	Bourbon 1811 (Sorcerer 1796) Lady Rachel 1805 by Stamford 1794 (Sir Peter T. 1784)—Y. Rachel 1799 by Volunteer 1780 (Eclipse 1764)—Rachel 1790 by Highflyer 1774 (Herod 1758)—Dau. 1771 of Syphon 1750 (Squirt 1732).
*Trustee, ch. h., 1829.	Catton 1809 (Golumpus 1802)—Emma 1824 by Whisker 1812 (Waxy 1790)—Gibside Fairy 1811 by Hermes 1790 (Mercury 1778) Vicissitude 1800 by Pipator 1786 (Imperator 1776)—Beatrice 1791 by Sir Peter T. 1784 (Highflyer 1774).
Vandal's dam. b. m., 1837.	*Tranby 1826 (Blacklock 1814) Lucilla by Trumpator 1823 (Sir Solomon 1805)—Lucy 1821 by Orphan 1810 (Ball's Florizel 1801)—Lady Grey 1817 by Robin Grey 1805 (*Royalist 1790)—Maria 1802 by Melzar 1791 (*Medley 1776).
Muley, b. h., 1810.	Orville 1799 (Beningbrough 1791)—Eleanor 1798 by Whiskey 1789 (Saltram* 1780)—Y. Giantess 1790 by Diomed* 1777 (Florizel 1768) Giantess 1769 by Matchem 1748 (Cade 1734) Molly Long Legs 1753 by Babraham 1740 (God. Ar. 1724).
Daughter of, ch. m., 1815.	Election 1804 (Gohanna 1790) Fair Helen 1808 by Hambletonian 1792 (King Fergus 1775) Helen 1801 by Delpini 1781 (Highflyer 1774)—Rosalind 1788 by Phenomenon 1780 (Herod 1758)—Atalanta 1769 by Matchem 1748 (Cade 1734).
Silverheels, gr. h., 1815.	Ogle's Oscar 1800 (*Gabriel 1790)—Pandora by Grey Diomed 1786 (*Medley 1776)—Floretta's dam by Hall's Union 1777 (*Slim 1768)—Dau. of Leonidas (Lloyd's Traveller)—Dau. of *Othello (Crab 1722).
Aurora, gr. m., 1814.	Lloyd's Vingt'un (*Diomed 1777) Pandora gr. m. by Grey Diomed 1786 (*Medley 1776)—Floretta's dam by Hall's Union 1777 (*Slim 1768)—Dau. of Leonidas (Lloyd's Traveller)—Dau. of *Othello (Crab 1722)—Dau. of *Juniper 1752 (Babraham 1740)—Dau. 1757 of *Traveller 1747 (Partner 1718)—*Selima b. m. 1746 by God. Ar. 1724—Large Hartley mare ch. m. 1729 by Hartley's Blind Horse (Holderness Turk)—Flying Whigg 1715 by Woodstock Ar.—Points by St. Victor Barb—Grey Whynot by Whynot (Fenwick Barb)—Royal mare.

IROQUOIS.

WINNER OF THE CHESTERFIELD STAKES AT NEWMARKET AND LEVANT STAKES AT GOODWOOD, 1880. THE BURWELL STAKES, DERBY AT EPSOM, PRINCE OF WALES STAKES, DONCASTER ST. LEGER, NEWMARKET DERBY, AND OTHER IMPORTANT EVENTS IN ENGLAND.

Property of Gen. W. H. Jackson, Belle Meade Stud, near Nashville, Tenn. Application to Gen. W. H. Jackson, lock box 383, Nashville, Tenn. Yearling sales annually about June 20th, in New York.

IROQUOIS by imp. Leamington, son of Faugh-a-Ballagh, bred by Mr. A. Welch, Erdenheim Stud, foaled 1878, dam Maggie B. B., dam of Harold by imp. Australian, second dam Madeline by Old Boston, out of Magnolia, dam of Kentucky by imp. Glencoe. As a two-year-old Iroquois started 12 times, won 4 races, was second in two and unplaced in six. He made his bow to the public by winning the Newmarket two-year old Plate, 5 furlongs, defeating Herman, Kuhleborn and three others; his second success was winning the two-year old stakes at Epsom, 5 furlongs, beating Eliacin. At the Newmarket July Meeting ran Bal Gal to a head for the July Stakes, 5 furlongs 136 yards, having Neophite, Thebais and seven others behind him; same meeting won the Chesterfield Stakes, 5 furlongs, beating Panique, Voluptuary and seven others: at Goodwood won the Levant Stakes, 5 furlongs, defeating Isola Madre, Canace and three others; same meeting ran second to Wandering Nun in the Findon Stakes, ¾ of a mile, with Albion, Ishmael and Worthing behind him; he was unplaced in his other races. As a three-year old made his first appearance by running second to Peregrine in the 2,000 Guineas, having Don Fulano, Camiliard, Scobel and nine others behind him; won the Newmarket Stakes, Ditch mile, beating Lennoxlove; Newmarket Second Spring Meeting, walked over for the Burwell Stakes, Abingdon mile; Epsom Summer Meeting, won the 102d renewal of the Derby Stakes, 1½ miles, defeating Peregrine, Town Moor, Scobel, Geologist, St. Louis, Don Fulano, Tristan and seven others. Ascot won the Prince of Wales Stakes, 1⅝ miles, defeating Geologist, Great Carle and four others; same place won the St. James Palace Stakes, Old mile, beating Leon, his only opponent; Doncaster September Meeting, won the Doncaster St. Leger Stakes, 1 mile 6 furlongs 132 yards, defeating Geologist, Lucy Glitters, St. Louis, Falkirk, Bal Gal and nine others; Newmarket Second October Meeting, was third to Bend'or, and Scobel in Champion Stakes, across the flat, 1 mile 2 furlongs and 73 yards, Buchanan, Falkirk, Muriel and Fiddler behind him; won the Newmarket Derby, 1½ miles, beating Ishmael, Lennoxlove and Lord Chelmsford. Thus starting in 9 races, running second for the 2,000 Guineas, third in the Champion Stakes, and winning the Derby and St. Leger, a feat only accomplished ten times in the one hundred and twelve years they have been run: Champion in 1800, Surplice in 1818, The Flying Dutchman in 1849, Voltigeur in 1850, West Australian in 1853, Blair Athol in 1864, Gladiateur in 1865, Lord Lyon in 1866, Silvio in 1877, Iroquois in 1881, Melton in 1885, Ormonde in 1886, Donovan in 1889 and Common in 1891. Not one of all these won so many good races in the same year. Iroquois is a brown, 16 hands high, with white stripe down the face and white around the coronet of the left fore foot; he has a well placed oblique shoulder, good barrel, fine hip and loin and sound, good legs and feet. He is descended from the best bred and most successful racing families in England and America. Maggie B. B. is the dam of Pera (a producing dam), Harold, Panique, Jaconet (dam of Sir Dixon and Belvidere), Red and Blue (dam of Sallie McClelland), and Homeopathy. Madeline, the dam of Maggie B. B., was by the great Boston, sire of Lexington, Lecompte, Bostona, Nina and a host of good ones. Magnolia produced Kentucky, the best son of Lexington, Daniel Boone, Magic, Charley Ball, Madonna, Skedaddle, Gilroy, Victory, etc. Skedaddle is the dam of Saucebox, Scramble,

(Continued on page 160)

Eclipse 1764—Marske 1750—Squirt 1732—Bartlett's Childers—Darley Arabian 1699.

IROQUOIS.

BROWN HORSE, FOALED 1878; BRED BY MR. A. WELCH, ERDENHEIM STUD, PENNSYLVANIA.

*LEAMINGTON, br. h., 1853.	Faugh-a-Ballagh, br. h., 1841.	Sir Hercules, blk. h., 1826.	Whalebone, br. h., 1807.	Waxy, b. h., 1790.	Pot-8-o-'s 1773 (Eclipse 1764)—Maria 1777 by Herod 1758 (Tartar 1743)—Lisette 1772 by Snap 1750 (Snip 1736)—Miss Windsor 1751 by God. Ar. 1724—Sist. to Volunteer by Y. Belgrade (Belgrade Turk).
				Penelope, b. m., 1798.	Trumpator 1782 (Conductor 1767)—Prunella 1788 by Highflyer 1774 (Herod 1758)—Promise 1768 by Snap 1750 (Snip 1736)—Julia 1756 by Blank 1740 (God. Ar. 1724)—Spectator's dam 1735 by Partner 1718 (Jigg).
			Peri, b. m., 1822.	Wanderer, b. h., 1811.	Gohanna 1790 (Mercury 1778)—Catherine 1795 by Woodpecker 1773 (Herod 1758)—Camilla 1778 by Trentham 1766 (T. Sweepstakes 1743)—Coquette 1764 by Compton Barb—Sist. 1743 to Regulus by God. Ar. 1724.
				Thalestris, blk. m., 1809.	Alexander 1782 (Eclipse 1764)—Rival 1800 by Sir Peter T. 1784 (Highflyer 1774)—Hornet 1790 by Drone 1777 (Herod 1758)—Manilla 1777 by Goldfinder 1764 (Snap 1750)—Dau. 1766 of Old England 1741 (God. Ar. 1724).
		Guiccioli, ch. m., 1823.	Bob Booty, ch. h., 1804.	Chanticleer, ch. h., 1787.	Woodpecker 1773 (Herod 1758)—Dau. 1778 of Eclipse 1764 (Marske 1750)—Rosebud 1765 by Snap 1750 (Snip 1736)—Miss Belsea 1753 by Regulus 1739 (God. Ar. 1724)—Dau. of Bartlett's Childers (Darley Ar. 1699).
				Ierne, ch. m., 1790.	Bagot 1780 (Herod 1758)—Dau. of Gamahoe (Bustard 1741)—Patty by Tim 1752 (Squirt 1732)—Miss Patch 1737 by Justice 1725 (H. C. Litton Ar.)—Ringtail Galloway by Curwen Bay Barb.
			Flight, ch. m., 1809.	Ir. Escape, ch. h., 1802.	Commodore 1793 (Tom Tug 1777)—Moll-in-the-Wad 1791 by Highflyer 1774 (Herod 1758)—Shift 1779 by Sweetbriar 1769 (Syphon 1750)—Black Susan 1762 by Snap 1750 (Snip 1736)—Dau. of Cade 1734 (God. Ar. 1724).
				Y. Heroine, ...	Bagot 1780 (Herod 1758)—Heroine 1775 by Hero 1753 (Cade 1734)—Dau. 1766 of Snap 1750 (Snip 1736)—Sist. 1743 to Regulus by God. Ar. 1724—Grey Robinson 1723 by Bald Galloway (St. Victor Barb).
	Daughter of, b. m., 1841.	Pantaloon, ch. h., 1824.	Castrel, ch. h., 1801.	Buzzard,* ch. h., 1787.	Woodpecker 1773 (Herod 1758)—Misfortune 1775 by Dux 1761 (Matchem 1748)—Curiosity 1760 by Snap 1750 (Snip 1736)—Dau. 1749 of Regulus 1739 (God. Ar. 1724)—Dau. of Bartlett's Childers (Darley Ar. 1699).
				Daughter of, b. m., —.	Alexander 1782 (Eclipse 1764)—Dau. of Highflyer 1774 (Herod 1758)—Dau. of Alfred 1770 (Matchem 1748)—Dau. 1770 of Engineer 1755 (Sampson 1745)—Bay Malton's dam 1756 by Cade 1734 (God. Ar. 1724).
			Idalia, ch. m., 1815.	Peruvian, b. h., 1806.	Sir Peter T. 1784 (Highflyer 1774)—Dau. 1788 of Boudrow 1777 (Eclipse 1764)—Dau. 1764 of Squirrel 1754 (Traveller 1735)—Sist. to Lowther's Babraham by Babraham 1740 (God. Ar. 1724)—Dau. 1750 of Golden Ball 1735 (Partner 1718).
				Musidora, b. m., 1804.	Meteor 1783 (Eclipse 1764)—Maid-of-all-Work 1786 by Highflyer 1774 (Herod 1758)—Dau. 1771 of Syphon 1750 (Squirt 1732)—Dau. of Regulus 1739 (God. Ar. 1724)—Sist. 1758 to Judgement by Snip 1736 (Flying Childers 1715).
		Daphne, br. m., 1837.	Laurel, br. h., 1824.	Blacklock, b. h., 1814.	Whitelock 1803 (Hambletonian 1792)—Dau. 1799 of Coriander 1786 (Pot-8-o-'s 1773)—Wildgoose 1792 by Highflyer 1774 (Herod 1758)—Coheiress 1786 by Pot-8-o-'s 1773 (Eclipse 1764)—Manilla 1777 by Goldfinder 1764 (Snap 1770).
				Wagtail, b. m., 1818.	Prime Minister 1810 (Sancho 1801)—Tranby*'s dam 1812 by Orville 1799 (Beningbrough 1791)—Miss Grimstone 1796 by Weazle 1776 (Herod 1758)—Dau. of Ancaster 1768 (Blank 1740)—Dau. of Damascus Ar. 1754.
			Maid of Honor, br. m., 1829.	Champion, ch. h., 1812.	Selim 1802 (Buzzard* 1787)—Podagra 1805 by Gouty 1796 (Sir Peter T. 1784)—Jet 1785 by Magnet 1770 (Herod 1758)—Jewel 1775 by Squirrel 1754 (Traveller 1735)—Sophia 1764 by Blank 1740 (God. Ar. 1724).
				Etiquette, br. m., 1820.	Orville 1799 (Beningbrough 1791)—Boadicea 1807 by Alexander 1782 (Eclipse 1764)—Brunette by Amaranthus 1766 (Old England 1741)—Mayfly 1771 by Matchem 1748 (Cade 1734)—Dau. 1755 of Starling 1733 (Starling 1727).
MAGGIE B. B., ch. m., 1867.	*Australian, ch. h., 1858.	West Australian, b. h., 1850.	Melbourne, br. h., 1834.	Humphrey Clinker, b. h., 1822.	Comus 1809 (Sorcerer 1796)—Clinkerina 1812 by Clinker 1805 (Sir Peter T. 1784)—Pewet 1786 by Tandem 1773 (Syphon 1750)—Termagant by Tantrum 1760 (Cripple 1750)—Cantatrice by Sampson 1745 (Blaze 1733).
				Daughter of, b. m., 1825.	Cervantes 1806 (Don Quixote 1784)—Dau. 1818 of Golumpus 1802 (Gohanna 1790)—Dau. 1810 of Paynator 1791 (Trumpator 1782)—Circle 1802 by St. George 1789 (Highflyer 1774)—Abigail 1788 by Woodpecker 1773 (Herod 1758).
			Mowerina, b. m., 1843.	Touchstone, br. h., 1831.	Camel 1822 (Whalebone 1807)—Banter 1826 by Master Henry 1815 (Orville 1799)—Boadicea 1807 by Alexander 1782 (Eclipse 1764)—Brunette by Amaranthus 1766 (Old England 1741)—Mayfly 1771 by Matchem 1748 (Cade 1734).
				Emma, ch. m., 1824.	Whisker 1812 (Waxy 1790)—Gibside Fairy 1811 by Hermes 1790 (Mercury 1778)—Vicissitude 1800 by Pipator 1786 (Imperator 1776)—Beatrice 1791 by Sir Peter T. 1784 (Highflyer 1774)—Pyrrha 1771 by Matchem 1748 (Cade 1734).
		Emilia,* b. m., 1840.	Y. Emilius, b. h., 1833.	Emilius, b. h., 1820.	Orville 1799 (Beningbrough 1791)—Emily 1810 by Stamford 1794 (Sir Peter T. 1784)—Dau. 1799 of Whiskey 1789 (Saltram* 1780)—Grey Dorimant 1781 by Dorimant 1772 (Otho 1760)—Dizzy 1757 by Blank 1740 (God. Ar. 1724).
				Shoveler, b. m., 1816.	Scud 1804 (Beningbrough 1791)—Goosander 1805 by Hambletonian 1792 (King Fergus 1775)—Rally 1790 by Trumpator 1782 (Conductor 1767)—Fancy 1780 by Florizel 1768 (Herod 1758)—Dau. 1773 of Spectator 1749 (Crab 1722).
			Persian, b. m., 1829.	Whisker, b. h., 1812.	Waxy 1790 (Pot-8-o-'s 1773)—Penelope 1798 by Trumpator 1782 (Conductor 1767)—Prunella 1788 by Highflyer 1774 (Herod 1758)—Promise 1768 by Snap 1750 (Snip 1736)—Julia 1756 by Blank 1740 (Godolph. Ar. 1724).
				Variety, ch. m., 1816.	[Selim or] Soothsayer 1808 (Sorcerer 1796)—Sprite 1807 by Bobtail 1795 (Precipitate* 1787)—Catherine 1795 by Woodpecker 1773 (Herod 1758)—Camilla 1778 by Trentham 1766 (T. Sweepstakes 1743)—Coquette 1764 by Compton Barb.
	Madeline, ch. m., 1849.	Boston, ch. h., 1833.	Timoleon, ch. h., 1813.	Sir Archy, b. h., 1805.	*Diomed 1777 (Florizel 1768)—*Castianira 1796 by Rockingham 1781 (Highflyer 1774)—Tabitha 1782 by Trentham 1766 (T. Sweepstakes 1743)—Dau. of Bosphorus 1754 (Babraham 1740)—Dau. of W.'s Forester 1750 (Forester 1736).
				Daughter of, ch. m., —.	*Saltram 1780 (Eclipse 1764)—Dau. of Symmes' Wildair 1767 (*Fearnought 1755)—Dau. of Tyler's Driver 1760 (*Othello)—Dau. of *Fallower 1761 (Blank 1740)—Dau. of *Vampire 1757 (Regulus 1739).
			Sist. to Tuckahoe, ch. m., 1814.	Ball's Florizel, ch. h., 1801.	*Diomed 1777 (Florizel 1768)—Dau. of *Shark 1771 (Marske 1750)—Dau. of Harris' Eclipse 1770 (*Fearnought 1755)—Dau. of *Fearnought 1755 (Regulus 1739)—Dau. of *Jolly Roger 1741 (Roundhead 1733).
				Daughter of, b. m., 1790.	*Alderman 1787 (Pot-8-o-'s 1773)—Dau. of *Clockfast 1774 (Gimcrack 1760)—Dau. of Symmes' Wildair 1767 (*Fearnought 1755)—Y. Kitty Fisher 1767 by *Fearnought 1755 (Regulus 1739)—*Kitty Fisher 1755 by Cade 1734 (God. Ar. 1724).
		Magnolia, ch. m., 1841.	*Glencoe, ch. h., 1831.	Sultan, b. h., 1816.	Selim 1802 (Buzzard* 1787)—Bacchante 1809 by W.'s Ditto 1800 (Sir Peter T. 1784)—Dau. 1791 of Mercury 1778 (Eclipse 1764)—Dau. 1776 of Herod 1758 (Tartar 1743)—Folly 1771 by Marske 1750 (Squirt 1732).
				Trampoline, ch. m., 1825.	Tramp 1810 (Dick Andrews 1797)—Web 1808 by Waxy 1790 (Pot-8-o-'s 1773)—Penelope 1798 by Trumpator 1782 (Conductor 1767)—Prunella 1788 by Highflyer 1774 (Herod 1758)—Promise 1768 by Snap 1750 (Snip 1736).
			*Myrtle, ch. m., 1834.	Mameluke, b. h., 1824.	Partisan 1811 (Walton 1799)—Miss Sophia 1805 by Stamford 1794 (Sir Peter T. 1784)—Sophia 1798 by Buzzard* 1787 (Woodpecker 1773)—Huncamunca 1787 by Highflyer 1774 (Herod 1758)—Cypher 1772 by Squirrel 1754 (Traveller 1735).
				Bobadilla, br. m., 1825.	Bobadil 1813 (Rubens 1805)—Pythoness blk. m. 1813 by Sorcerer 1796 (Trumpator 1782)—Princess blk. m. 1796 by Sir Peter T. 1784 (Highflyer 1774)—Dau. br. m. 1789 of Dungannon 1780 (Eclipse 1764)—Dau. ch. m. 1783 of Turf 1760 (Matchem 1748)—Dau. b. m. 1779 of Herod 1758 (Tartar 1743)—Golden Grove ch. m. 1760 by Blank 1740 (Godolph. Ar. 1724)—Spinster, the Widdrington mare, ch. m. 1735 by Partner 1718 (Jigg)—Bay Bloody Buttocks b. m. 1729 by Bloody Buttocks—Sist. 1722 to Guy by Greyhound (Chillaby)—Brown Farewell br. m. 1710 by Makeless (Oglethorpe Ar.)—Dau. of Brimmer (D'Arcy Yellow Turk)—Dau. of Place's White Turk—Dau. of Dodsworth—Layton Barb mare.

IROQUOIS—Continued.

Florence B., Sly Boots, Joe S., The Slashes and Squeez'em (dam of Day Star, winner of the Kentucky Derby, Blue Ribbon Stakes and other races), Sylph (dam of John K., Kaloolah, Philip D. and Kinesem), and La Sylphide, 1¼ miles in 2.07¼, she the dam of Semper Fidele. Sly Boots is the dam of Leveler, Sly Dance, Sachem, Savanac and Volunteer II.

Iroquois is a grand success in the stud, sire of Tammany, winner of the Great Eclipse Stakes, the Criterion Stakes at Monmouth, the 6 furlongs in both races being run in 1.12½, the Lorillard Stakes, 1¾ miles, in 2.20½; Realization Stakes, 1⅝ miles, in 2.51¾; Withers Stakes, 1 mile, in 1.40. Addie was the "crack" filly of the West, winner of eleven races. Other winners by Iroquois are Rancocas, Mohican, White Wings, Mountain Deer, McKeever, Heyday, Laura Doxey, Guildura, Annie Irwin, Beansey, Darlington, Carteret, Bengal, Algonquin, Eblis, Dictum, Emblem, Irregular, Julius Sax, Kiawah, Pan Handle, Ted Gregg, Malachi, Alice D., Claudine, Nellie Pearl, Minnie L., Iowa, Josie M., Mayflower, Piquante, Rainbow, Starlight, Helen Nichols (unbeaten), Huron, Marguerite, St. Croix, Queen Enid, Interior, Red Banner, Bennett Young, Edith Belmont, Cayuga, a rare good one, Haydee, Nettie B., Princess Loraine, G. W. Johnson, unbeaten and $22,000 refused for him. He will stand at the head of the list of winning stallions in 1892.

JULIEN.

Owned by Messrs. Smith and Hughes.

JULIEN, bay horse, foaled 1886, bred by Hon. W. L. Scott, Algeria Stud, Erie, Pa., by imp. Rayon D'or, dam imp. Clover by Macaroni, son of Sweetmeat, grandam Verdure by King Tom, son of Harkaway, etc. Macaroni, the sire of Clover, won the 2,000 Guineas and Derby, and was the sire of the dams of Pilgrimage, winner of the 1,000 and 2,000 Guineas, Ormonde, winner of the 2,000 Guineas, Derby and St. Leger, Galliard, winner of the 2,000 Guineas, Busybody, winner of the 1,000 Guineas and Oaks, and Bonavista, winner of 2,000 Guineas, and her dam, full sister to Clover, the dam of Julien. Verdure, the grandam of Julien, was a winner in England, and own sister to Corisande, that won the Cesarewitch Handicap in a field of twenty-seven at three years old, carrying 110 lbs., the Chesterfield, Granby, New, Coronation and Grande Duke Michael Stakes, by King Tom, sire of imp. Phaeton, imp. King Ban, imp. Great Tom and his brother, Kingcraft, winner of the Derby, and of the dams of the Derby winner Favonius, and Enterprise, winner of the 2,000 Guineas. Maybloom, Julien's third dam, was full sister to Lady Alice Hawthorn, grandam of Hauteur, winner of the 1,000 Guineas, and out of Lady Hawthorn, full sister to Thormanby, winner of the Derby and Ascot Gold Cup. Julien was a good race-horse, winner at 5 furlongs, and placed second and third in several good and fast races. Rayon d'Or, his sire, is from the Pocahontas family, that produced Stockwell, Rataplan, King Tom and other good ones, and his dam from the great Hawthorn family, tracing to the Layton Barb mare.

JULIEN.

BAY HORSE, FOALED 1886; BRED BY W. L. SCOTT, ALGERIA STUD, PENNSYLVANIA.

*RAYON-D'OR, ch. h., 1876.
- Flageolet, ch. h., 1870.
  - Plutus, b. h., 1863.
    - Trumpeter, ch. h., 1856.
    - Daughter of, b. m., 1853.
  - La Favorite, b. m., 1863.
    - Monarque, b. h., 1852.
    - Constance, ch. m., 1848.
- Araucaria, b. m., 1862.
  - Ambrose, blk. h., 1849.
    - Touchstone, br. h., 1831.
    - Annette, b. m., 1835.
  - Pocahontas, b. m., 1837.
    - Glencoe*, ch. h., 1831.
    - Marpessa, b. m., 1830.

*JUDITH, b. m., 1877.
- Macaroni, b. h., 1860.
  - Sweetmeat, br. h., 1842.
    - Gladiator, ch. h., 1833.
    - Lollypop, blk. m., 1836.
  - Jocose, b. m., 1843.
    - Pantaloon, ch. h., 1824.
    - Banter, br. m., 1826.
- Verdure, ch. m., 1867.
  - King Tom, b. h., 1851.
    - Harkaway, ch. h., 1834.
    - Pocahontas, b. m., 1837.
  - May Bloom, b. m., 1861.
    - Newminster, b. h., 1848.
    - Lady Hawthorn, b. m., 1854.

Orlando, b. h., 1844. Touchstone 1831 (Camel 1822) Vulture 1833 by Langar 1817 (Selim 1802) Kite 1821 by Bustard 1813 (Castrel 1801) Olympia 1815 by Sir Oliver 1800 (Sir Peter T. 1784) Scotilla 1795 by Anvil 1777 (Herod 1758).

Cavatina, ch. m., 1845. Red-hank 1833 (Sandbeck 1818) Oxygen 1828 by Emilius 1820 (Orville 1799)—Whizgig 1819 by Rubens 1805 (Buzzard* 1787)—Penelope 1798 by Trumpator 1782 (Conductor 1767) Prunella 1788 by Highflyer 1774 (Herod 1758).

Planet, b. h., 1844. Bay Middleton 1833 (Sultan 1816) Plenary 1817 by Emilius 1820 (Orville 1799) Harriet 1819 by Pericles 1809 (Evander 1801)—Dau. 1812 of Selim 1802 (Buzzard 1787)—Pipylina 1803 by Sir Peter T. 1784 (Highflyer 1774).

Alice Bray, b. m., 1848. Venison 1833 (Partisan 1811) Darkness 1837 by Glencoe* 1831 (Sultan 1816) Fanny 1827 by Whisker 1812 (Waxy 1790) Dau. 1813 of Camillus 1803 (Hambletonian 1792)—Dau. 1804 of Precipitate* 1787 (Mercury 1778).

[The Baron-Stag or] The Emperor, ch. h., 1841. Defence 1824 (Whalebone 1807) Dau. 1821 of Reveller 1815 (Comus 1809)—Design* 1827 by Tramp 1810 (Dick Andrews 1797)—Defiance 1816 by Rubens 1805 (Buzzard* 1787)—Little Folly 1806 by Highland Fling 1798 (Spadille 1784).

Poetess, b. m., 1848. Royal Oak 1823 (Catton 1809)—Ada 1824 by Whisker 1812 (Waxy 1790) Anna Bella 1807 by Shuttle 1793 (Y. Marske 1771) Dau. 1797 of Drone 1777 (Herod 1758)- Contessina 1787 by Y. Marske 1771 (Marske 1750).

Gladiator, ch. h., 1833. Partisan 1811 (Walton 1799)—Pauline 1826 by Moses 1819 ([Whalebone or] Seymour 1807)—Quadrille 1815 by Selim 1802 (Buzzard* 1787) Canary Bird 1806 by [Whiskey or] Sorcerer 1796 (Trumpator 1782) Canary 1797 by Coriander 1786 (Pot-8-o's 1773).

Lanterne, b. m., 1841. Hercule 1830 (Rainbow 1808)—Elvira 1823 by Eryx 1816 (Milo 1802) Coral 1816 by Orville 1799 (Beningbrough 1791) Fairing 1807 by Waxy 1790 (Pot-8-o's 1773)—Rattle 1793 by Trumpator 1782 (Conductor 1767).

Camel, br. h., 1822. Whalebone 1807 (Waxy 1790) Dau. 1812 of Selim 1802 (Buzzard* 1787)—Maiden 1801 by Sir Peter T. 1784 (Highflyer 1774)—Dau. 1788 of Phenomenon 1780 (Herod 1758) Matron 1782 by Florizel 1768 (Herod 1758).

Banter, br. m., 1826. Master Henry 1815 (Orville 1799)—Boadicea 1807 by Alexander 1782 (Eclipse 1764) Brunette by Amaranthus 1766 (Old England 1741) Mayfly 1771 by Matchem 1748 (Cade 1734)—Dau. 1755 of Starling 1738 (Starling 1727).

Priam,* b. h., 1827. Emilius 1820 (Orville 1799)—Cressida 1807 by Whiskey 1789 (Saltram* 1780) Y. Giantess 1790 by Diomed* 1777 (Florizel 1768)—Giantess 1769 by Matchem 1748 (Cade 1734)—Molly Long Legs 1753 by Babraham 1740 (God. Ar. 1724).

The Potentate's dam, b. m., 1823. Don Juan 1814 (Orville 1799)—Moll-in-the-Wad 1810 by Hambletonian 1792 (King Fergus 1775)- Spitfire 1799 by Pipator 1786 (Imperator 1776)—Farewell 1792 by Slope 1782 (Highflyer 1774)—Dau. of Y. Marske 1771 (Marske 1750).

Sultan, b. h., 1816. Selim 1802 (Buzzard* 1787) Bacchante 1809 by W's Ditto 1800 (Sir Peter T. 1784) Dau. 1791 of Mercury 1778 (Eclipse 1764)—Dau. 1776 of Herod 1758 (Tartar 1743)—Folly 1771 by Marske 1750 (Squirt 1732).

Trampoline, ch. m., 1825. Tramp 1810 (Dick Andrews 1797)—Web 1808 by Waxy 1790 (Pot-8-o's 1773)—Penelope 1798 by Trumpator 1782 (Conductor 1767)—Prunella 1788 by Highflyer 1774 (Herod 1758)—Promise 1768 by Snap 1750 (Snip 1736).

Muley, b. h., 1810. Orville 1799 (Beningbrough 1791)—Eleanor 1798 by Whiskey 1789 (Saltram* 1780) Y. Giantess 1790 by Diomed* 1777 (Florizel 1768)—Giantess 1769 by Matchem 1748 (Cade 1734)—Molly Long Legs 1753 by Babraham 1740 (God. Ar. 1724).

Clare, b. m., 1824. Marmion 1806 (Whiskey 1789)—Harpalice 1814 by Gohanna 1790 (Mercury 1778)—Amazon 1799 by Driver 1783 (Trentham 1766)- Fractious 1792 by Mercury 1778 (Eclipse 1764)—Dau. 1785 of Woodpecker 1773 (Herod 1758).

Partisan, b. h., 1811. Walton 1799 (Sir Peter T. 1784)—Parasol 1800 by Pot-8-o's 1773 (Eclipse 1764)—Prunella 1788 by Highflyer 1774 (Herod 1758)—Promise 1768 by Snap 1750 (Snip 1736)—Julia 1756 by Blank 1740 (God. Ar. 1724).

Pauline, b. m., 1826. Moses 1819 ([Whalebone or] Seymour 1807)—Quadrille 1815 by Selim 1802 (Buzzard* 1787)—Canary Bird 1806 by [Whiskey or] Sorcerer 1796 (Trumpator 1782)—Canary 1797 by Coriander 1786 (Pot-8-o's 1773)—Miss Green 1787 by Highflyer 1774 (Herod 1758).

[Starch or] Voltaire, br. h., 1826. Blacklock 1814 (Whitelock 1803)—Dau. 1816 of Phantom 1808 (Walton 1799) Dau. 1802 of Overton 1788 (King Fergus 1775)—Dau. 1796 of Walnut 1786 (Highflyer 1774)—Dau. 1783 of Ruler 1777 (Y. Marske 1771).

Belinda, br. m., 1825. Blacklock 1814 (Whitelock 1803)—Wagtail 1818 by Prime Minister 1810 (Sancho 1801)—Tranby*'s dam 1812 by Orville 1799 (Beningbrough 1791) Miss Grimstone 1796 by Weazle 1776 (Herod 1758) -Dau. of Ancaster 1768 (Blank 1740).

Castrel, ch. h., 1801. Buzzard* 1787 (Woodpecker 1773) Dau. of Alexander 1782 (Eclipse 1764) Dau. of Highflyer 1774 (Herod 1758)—Dau. of Alfred 1770 (Matchem 1748)—Dau. 1770 of Engineer 1756 (Sampson 1745).

Idalia, ch. m., 1815. Peruvian 1806 (Sir Peter T. 1784)—Musidora 1804 by Meteor 1783 (Eclipse 1764) Maid-of-all-Work 1786 by Highflyer 1774 (Herod 1758)—Dau. 1771 of Syphon 1750 (Squirt 1732)—Dau. of Regulus 1739 (God. Ar. 1724).

Master Henry, b. h., 1815. Orville 1799 (Beningbrough 1791) Miss Sophia 1805 by Stamford 1794 (Sir Peter T. 1784) -Sophia 1798 by Buzzard* 1787 (Woodpecker 1773)—Huncamunca 1787 by Highflyer 1774 (Herod 1758)—Cypher 1772 by Squirrel 1754 (Traveller 1735).

Boadicea, b. m., 1807. Alexander 1782 (Eclipse 1764)—Brunette by Amaranthus 1766 (Old England 1741)—Mayfly 1771 by Matchem 1748 (Cade 1734)—Dau. 1755 of Starling 1738 (Starling 1727)—Look-at-me-Lads 1731 by Grasshopper (Byerly Turk).

Economist, b. h., 1825. Whisker 1812 (Waxy 1790)—Floranthe 1818 by Octavian 1807 (Stripling 1795)—Caprice 1797 by Anvil 1777 (Herod 1758)—Madcap 1774 by Eclipse 1764 (Marske 1750)—Sappho 1768 by Blank 1740 (God. Ar. 1724).

Fanny Dawson, ch. m., 1823. Nabocklish 1811 (Rugantino 1803)—Miss Tooly 1808 by Teddy the Grinder 1798 (Asparagus 1787)—Lady Jane 1796 by Sir Peter T. 1784 (Highflyer 1774)—Paulina 1778 by Florizel 1768 (Herod 1758)—Captive 1771 by Matchem 1748 (Cade 1734).

Glencoe,* ch. h., 1831. Sultan 1816 (Selim 1802)—Trampoline 1825 by Tramp 1810 (Dick Andrews 1797)- Web 1808 by Waxy 1790 (Pot-8-o's 1773)—Penelope 1798 by Trumpator 1782 (Conductor 1767)- Prunella 1788 by Highflyer 1774 (Herod 1758).

Marpessa, b. m., 1830. Muley 1810 (Orville 1799)—Clare 1824 by Marmion 1806 (Whiskey 1789)—Harpalice 1814 by Gohanna 1790 (Mercury 1778)—Amazon 1799 by Driver 1783 (Trentham 1766)—Fractious 1792 by Mercury 1778 (Eclipse 1764).

Touchstone, br. h., 1831. Camel 1822 (Whalebone 1807) Banter 1826 by Master Henry 1815 (Orville 1799) Boadicea 1807 by Alexander 1782 (Eclipse 1764)—Brunette by Amaranthus 1766 (Old England 1741)—Mayfly 1771 by Matchem 1748 (Cade 1734).

Bee's-wing, b. m., 1833. Dr. Syntax 1811 (Paynator 1791)—Dau. 1817 by Ardrossan 1809 (John Bull 1789)—Lady Eliza 1813 by Whitworth 1805 (Agonistes 1797)—Dau. 1793 of Spadille 1784 (Highflyer 1774)—Sylvia 1783 by Y. Marske 1771 (Marske 1750).

Windhound, br. h., 1847. Pantaloon 1824 (Castrel 1801)—Phryne 1840 by Touchstone 1831 (Camel 1822)—Decoy 1830 by Filho-da-Puta 1812 (Haphazard 1797)—Finesse 1815 by Peruvian 1806 (Sir Peter T. 1784)—Violante 1802 by John Bull 1789 (Fortitude 1777).

Alice Hawthorn, b. m., 1838. Muley Moloch 1830 (Muley 1810)—Rebecca b. m. 1831 by Lottery 1820 (Tramp 1810)—Dau. b. m. 1818 of Cervantes 1806 (Don Quixote 1784)—Anticipation ch. m. 1802 by Beningbrough 1791 (King Fergus 1775)—Expectation gr. m. 1779 by Herod 1758 (Tartar 1743)—Dau. 1758 of Skim 1746 (Starling 1727)—Dau. 1751 of Janus 1738 (God. Ar. 1724)—Spinster gr. m. 1743 by Crab 1722 (Alcock Ar.)—Spinster (the Widdrington mare) ch. m. 1735 by Partner 1718 Jigg)—Bay Bloody Buttocks b. m. 1729 by Bloody Buttocks—Sist. 1722 to Guy by Greyhound (Chillaby)—Brown Farewell br. m. 1710 by Makeless (Oglethorpe Ar.)—Dau. of Brimmer (D'Arcy Yellow Turk)—Dau. of Place's White Turk—Dau. of Dodsworth—Layton Barb mare.

IMPORTED KANTAKA.

Owned in the Algeria Stud of the late Hon. W. L. Scott, Erie, Pa.

KANTAKA, chestnut horse, 15¾ hands high, by Scottish Chief, son of Lord of the Isles by Touchstone, was bred by Mr. W. Blenkiron, foaled 1880, dam Seclusion (Hermit's dam) by Tadmor, son of Ion by Cain, grandam Miss Sellon by Cowl, son of Bay Middleton by Sultan.

Kantaka is of the blood royal, being a half brother of the great horse Hermit, who won the Derby and was one of the best sires that ever lived, as is evidenced by the great performances of Peter, Shotover, St. Blaise, Tristan, Thebais, Clairvaux, St. Marguerite, Queen Adelaide, etc. Mr. P. Lorillard paid $5,000 for Kantaka when a yearling, and in his trials he showed great speed and promised well, but he met with an accident and never started in a race. His breeding is of the choicest quality and his sire, Scottish Chief, was himself not only a first-class race-horse, but as a sire of race-horses is one of the most popular stallions of England. He won the Ascot Gold Cup and other good races, and as a sire he has proven his prepotency in having to his credit a good list of winners, among them Marie Stuart, winner of the Oaks and St. Leger, and one of the best mares that ever appeared in England, Thistle, dam of Common, winner of the 2,000 Guineas, Derby and St. Leger, Mowerina, dam of Donovan, winner of the Derby and St. Leger, Semolina, winner of the 1,000 Guineas, Violet Melrose, dam of Melton, winner of the Derby and St. Leger, etc. Lord of the Isles, sire of Scottish Chief, was winner of the 2,000 Guineas, son of the great Touchstone, and on the side of his dam he traces back to stout Herod blood, which is well fortified with the Sultan through Bay Middleton, one of the best race-horses and sires of his day. The remarkable success of Hermit as a sire proves the prepotency of the blood of the family to which he belongs. Newminster, the sire of Hermit, was a son of Touchstone, and Scottish Chief, on his sire's side, is also a grandson of Touchstone. Cowl, sire of Miss Sellon, the grandam of Kantaka, is a son of Bay Middleton and the famous Crucifix by imp. Priam. The grandam of Scottish Chief is by Bay Middleton, from which it will be seen that Kantaka and Hermit are much more closely allied than simply half brothers, for the same strains of blood are conspicuous on the side of their sires. The capacity, therefore, to transmit racing qualities is abundant in Kantaka, and a horse of his rich breeding must prove himself a successful sire eventually.

Kantaka is still young in the stud, but has to his credit Canteen, G. B. Boyden, Dock Wick, Rafter, Irish Dan by Kantaka or Algerine, Tipstaff by Rayon d'Or or Kantaka, Clotho (a very speedy two-year old last season), Flagrant, Guilty, Great Scott, Gold Digger, Knapsack, Ko-Ko, Loyalty, Nubian, Chatter, Cascade, Lizzie Gwynne, Piccalilli, Queen d'Or, Crocus, Lebanon, Wyoming, Wallace, Village Maid and other good ones.

It is to the Massey mare in direct female line descend and trace their origin the following good horses and stallions: Trentham, Florizel, Election, Quiz, Royal Oak, Mulatto, Defence, The Palmer, Rosicrucian, Dutch Skater, Doncaster, Gladiateur Hermit by Newminster and many other noted ones.

KANTAKA.

CHESTNUT HORSE, FOALED 1880; BRED BY MR. W. BLENKIRON, ENGLAND.

SCOTTISH CHIEF, b. h., 1861.

Lord of the Isles, b. h., 1852. — Touchstone, br. h., 1831. — Camel, br. h., 1822. / Banter, br. m., 1826.
Lord of the Isles, b. h., 1852. — Fair Helen, ch. m., 1843. — Pantaloon, ch. h., 1824. / Rebecca, b. m., 1831.
Miss Ann, b. m., 1846. — The Little Known, b. h., 1836. — Muley, b. h., 1810. / Lacerta, b. m., 1816.
Miss Ann, b. m., 1846. — Bay Missy, b. m., 1842. — Bay Middleton, b. h., 1833. / Camilla, br. m., 1822.

Pot-8-o's 1773 Eclipse 1764 - Marske 1750 Squirt 1732 Bartlett's Childers Darley Arabian 1693.

Whalebone, br. h., 1807.	Waxy 1790 (Pot-8-o's 1773) Penelope 1798 by Trumpator 1782 (Conductor 1767) Prunella 1788 by Highflyer 1774 (Herod 1758)—Promise 1768 by Snap 1750 (Snip 1736)—Julia 1756 by Blank 1740 (God. Ar. 1724).
Daughter of, b. m., 1812.	Selim 1802 (Buzzard 1787) Maiden 1801 by Sir Peter T. 1784 (Highflyer 1774) Dau. 1788 of Phenomenon 1780 (Herod 1758) Matron 1782 by Florizel 1768 (Herod 1758) - Maiden 1770 by Matchem 1748 (Cade 1734).
Master Henry, b. h., 1815.	Orville 1799 (Beningbrough 1791) Miss Sophia 1805 by Stamford 1794 (Sir Peter T. 1784) Sophia 1798 by Buzzard 1787 (Woodpecker 1773) Huncamunca 1787 by Highflyer 1774 (Herod 1758) Cypher 1772 by Squirrel 1754 (Traveller 1735).
Boadicea, b. m., 1807.	Alexander 1782 (Eclipse 1764) Brunette by Amaranthus 1766 (Old England 1741) Mayfly 1771 by Matchem 1748 (Cade 1734) Dau. 1755 of Anc. Starling 1738 (Starling 1727) Look-at-me-Lads 1731 by Brist. Grasshopper (Byerly T.)
Castrel, ch. h., 1801	Buzzard 1787 (Woodpecker 1773) Dau. of Alexander 1782 (Eclipse 1764) Dau of Highflyer 1774 (Herod 1758) Dau. of Alfred 1770 (Matchem 1748) Dau. 1770 of Engineer 1756 (Sampson 1745).
Idalia, ch. m., 1815.	Peruvian 1806 (Sir Peter T. 1784) Musidora 1804 by Meteor 1783 (Eclipse 1764) Maid-of-all-Work 1786 by Highflyer 1774 (Herod 1758) Sist. 1771 to Tandem by Syphon 1750 (Squirt 1732) Dau. of Regulus 1739 (God. Ar. 1724).
Lottery, br. h., 1820.	Tramp 1810 (Dick Andrews 1797)—Mandane 1800 by Pot-8-o's 1773 Eclipse 1764 Y. Camilla 1787 by Woodpecker 1773 (Herod 1758) Camilla 1778 by Trentham 1766 (T. Sweepstakes 1743) Coquette 1765 by Compton Barb.
Daughter of, b. m., 1818.	Cervantes 1806 (Don Quixote 1784)—Anticipation 1802 by Beningbrough 1791 (King Fergus 1775) Expectation 1779 by Herod 1758 (Tartar 1743)- Dau. 1758 of Skim 1746 (Starling 1727) Dau. 1751 of Janus 1738 (God. Ar. 1724).
Orville, b. h., 1799.	Beningbrough 1791 (King Fergus 1775) Evelina 1791 by Highflyer 1774 (Herod 1758) Termagant by Tantrum 1760 (Cripple 1750) Cantatrice by Sampson 1745 (Blaze 1733) Dau. 1757 of Regulus 1739 (God. Ar. 1724).
Eleanor, b. m., 1798.	Whiskey 1789 (Saltram* 1780) Y. Giantess 1790 by Diomed* 1777 (Florizel 1768) - Giantess 1769 by Matchem 1748 (Cade 1734) - Molly Long Legs 1753 by Babraham 1740 (God. Ar. 1724)—Dau. of Foxhunter 1727 (Brisk 1711).
Zodiac, ch. h., 1801.	St. George 1789 (Highflyer 1774)—Abigail 1788 by Woodpecker 1773 (Herod 1758) Firetail 1772 by Eclipse 1764 (Marske 1750)—Dau. 1764 of Blank 1740 (God. Ar. 1724)—Naylor 1748 by Cade 1734 (God. Ar. 1724).
Jerboa, b. m., 1803.	Gohanna 1790 (Mercury 1778) Camilla 1778 by Trentham 1766 (T. Sweepstakes 1743)—Coquette 1764 by Compton Barb—Sist. 1743 to Regulus by God. Ar. 1724—Grey Robinson 1723 by Bald Galloway (St. Victor Barb).
Sultan, b. h., 1816.	Selim 1802 (Buzzard* 1787) Bacchante 1809 by W's Ditto 1800 (Sir Peter T. 1784) Dau. 1791 of Mercury 1778 (Eclipse 1764) Dau. 1776 of Herod 1758 (Tartar 1743) Folly 1771 by Marske 1750 (Squirt 1732).
Cobweb, b. m., 1821.	Phantom 1808 (Walton 1799)—Filagree 1815 by Soothsayer 1808 (Sorcerer 1796)—Web 1808 by Waxy 1790 (Pot-8-o's 1773)- Penelope 1798 by Trumpator 1782 (Conductor 1767) Prunella 1788 by Highflyer 1774 (Herod 1758).
Y. Phantom, b. h., 1822.	Phantom 1808 (Walton 1799)—Emmeline 1817 by Waxy 1790 (Pot-8-o's 1773)—Sorcery 1808 by Sorcerer 1796 (Trumpator 1782) Cobbéa 1802 by Skyscraper 1786 (Highflyer 1774)—Dau. 1788 of Woodpecker 1773 (Herod 1758).
Sister to Speaker, b. m., 1817.	Camillus 1803 (Hambletonian 1792)- Dau. 1809 by Sancho 1801 (Don Quixote 1784)—Miss Hornpipe Teazle 1802 by Sir Peter T. 1784 (Highflyer 1774)—Hornpipe 1793 by Trumpator 1782 (Conductor 1767)—Luna 1779 by Herod 1758 (Tartar 1743).

SECLUSION (Hermit's dam), b. m., 1857.

Tadmor, br. h., 1846. — Ion, br. h., 1835. — Cain, b. h., 1822. / Margaret, br. m., 1831.
Tadmor, br. h., 1846. — Palmyra, br. m., 1838. — Sultan, b. h., 1816. / Hester, br. m., 1832.
Miss Sellon, b. m., 1851. — Cowl, b. h., 1842. — Bay Middleton, b. h., 1833. / Crucifix, b. m., 1837.
Miss Sellon, b. m., 1851. — Belle Dame, ch. m., 1839. — Belshazzar*, ch. h., 1830. / Ellen, b. m., 1831.

Paulowitz, br. h., 1813.	Sir Paul 1802 (Sir Peter T. 1784)—Evelina 1791 by Highflyer 1774 (Herod 1758)—Termagant by Tantrum 1760 (Cripple 1750)—Cantatrice by Sampson 1745 (Blaze 1733)—Dau. 1757 of Regulus 1739 (God. Ar. 1724).
Daughter of, —, 1810.	Paynator 1791 (Trumpator 1782)—Dau. 1805 of Delpini 1781 (Highflyer 1774)—Dau. 1795 by Y. Marske 1771 (Marske 1750)—Gentle Kitty 1774 by Silvio 1754 (Cade 1734)—Dau. of Dorimond 1758 (Dormouse 1738).
Edmund, b. h., 1824.	Orville 1799 (Beningbrough 1791)—Emmeline 1817 by Waxy 1790 (Pot-8-o's 1773)—Sorcery 1808 by Sorcerer 1796 (Trumpator 1782)—Cobbéa 1802 by Skyscraper 1786 (Highflyer 1774)—Dau. 1788 of Woodpecker 1773 (Herod 1758).
Medora, ch. m., 1811.	Selim 1802 (Buzzard* 1787)—Dau. 1803 of Sir Harry* 1795 (Sir Peter T. 1784)—Dau. 1793 of Volunteer 1780 (Eclipse 1764)—Dau. 1779 of Herod 1758 (Tartar 1743)—Golden Grove 1760 by Blank 1740 (God. Ar. 1724).
Selim, ch. h., 1802.	Buzzard* 1787 (Woodpecker 1773)—Dau. of Alexander 1782 (Eclipse 1764)—Dau. of Highflyer 1774 (Herod 1758)—Dau. of Alfred 1770 (Matchem 1748)—Dau. 1770 by Engineer 1756 (Sampson 1745).
Bacchante, br. m., 1809.	W's Ditto 1800 (Sir Peter T. 1784)—Dau. 1791 of Mercury 1778 (Eclipse 1764)—Dau. 1776 of Herod 1758 (Tartar 1743)—Folly 1771 by Marske 1750 (Squirt 1732)—Vixen 1753 by Regulus 1739 (God. Ar. 1724).
Camel, br. h., 1822.	Whalebone 1807 (Waxy 1790)—Dau. 1812 of Selim 1802 (Buzzard* 1787)—Maiden 1801 by Sir Peter T. 1784 (Highflyer 1774)—Dau. 1788 of Phenomenon 1780 (Herod 1758)—Matron 1782 by Florizel 1768 (Herod 1758).
Monimia, b. m., 1821.	Muley 1810 (Orville 1799)—Dau. 1796 of Precipitate* 1787 (Mercury 1778)—Dau. 1787 of Woodpecker 1773 (Herod 1758)—Dau. 1773 of Snap 1750 (Snip 1736)—Y. Marske's dam 1759 by Blank 1740 (God. Ar. 1724).
Sultan, b. h., 1816.	Selim 1802 (Buzzard* 1787)—Bacchante 1809 by W's Ditto 1800 (Sir Peter T. 1784)—Dau. 1791 of Mercury 1778 (Eclipse 1764)—Dau. 1776 of Herod 1758 (Tartar 1743)—Folly 1771 by Marske 1750 (Squirt 1732).
Cobweb, b. m., 1821.	Phantom 1808 (Walton 1799)—Filagree 1815 by Soothsayer 1808 (Sorcerer 1796)—Web 1808 by Waxy 1790 (Pot-8-o's 1773)—Penelope 1798 by Trumpator 1782 (Conductor 1767)—Prunella 1788 by Highflyer 1774 (Herod 1758).
Priam,* b. h., 1827.	Emilius 1820 (Orville 1799)—Cressida 1807 by Whiskey 1789 (Saltram* 1780)—Y. Giantess 1790 by Diomed* 1777 (Florizel 1768)—Giantess 1769 by Matchem 1748 (Cade 1734)—Molly Long Legs 1753 by Babraham 1740 (God. Ar. 1724).
Octaviana, b. m., 1815.	Octavian 1807 (Stripling 1795)—Dau. 1807 of Shuttle 1793 (Y. Marske 1771)—Zara 1801 by Delpini 1781 (Highflyer 1774)—Flora 1789 by King Fergus 1775 (Eclipse 1764)—Atalanta 1769 by Matchem 1748 (Cade 1734).
Blacklock, b. h., 1814.	Whitelock 1803 (Hambletonian 1792)—Dau. 1799 of Coriander 1786 (Pot-8-o's 1773)—Wildgoose 1792 by Highflyer 1774 (Herod 1758)—Coheiress 1786 by Pot-8-o's 1773 (Eclipse 1764)—Manilla 1777 by Goldfinder 1764 (Snap 1750).
Manuella, b. m., 1809.	Dick Andrews 1797 (Joe Andrews 1778)—Mandane 1800 by Pot-8-o's 1773 (Eclipse 1764)—Y. Camilla 1787 by Woodpecker 1773 (Herod 1758)—Camilla 1778 by Trentham 1766 (T. Sweepstakes 1743) Coquette 1765 by Compton Barb.
Starch, br. h., 1819.	Waxy Pope 1806 (Waxy 1790)—Miss Staveley 1805 by Shuttle 1793 (Y. Marske 1771)—Dau. 1793 of Drone 1777 (Herod 1758) Dau. 1777 of Matchem 1748 (Cade 1734)—Jocasta 1767 by Forester 1749 (Forester 1736).
Cuirass, ch. m., 1823.	Oiseau 1809 (Camillus 1803) Castanea ch. m. 1803 by Gohanna 1790 (Mercury 1778)—Grey Skim gr. m. 1793 by Woodpecker 1773 (Herod 1758)—Silver's dam gr. m. 1780 by Herod 1758 (Tartar 1743)—Y. Hag gr. m 1761 by Skim 1746 (Starling 1727)—Hag gr. m. 1741 by Crab 1722 (Alcock Ar.)—Ebony b. m. 1728 by Flying Childers 1715 (Darley Ar. 1699)—Ebony blk. m. 1714 by Basto 1702 (Byerly Turk)—Duke of Rutland's Massey mare by Massey Bay Barb.

KING ALFONSO.

WINNER OF THE KENTUCKY ST. LEGER, TOBACCO STAKES AND GALT HOUSE STAKES AT LOUISVILLE, KY., AND THE LINCK'S HOTEL STAKES AT NASHVILLE, TENN., ALL IN 1875.

In the Woodburn Stud, Spring Station, Ky., A. J. Alexander, proprietor. Application to L. Brodhead. Annual sales of yearlings in May.

KING ALFONSO by imp. Phaeton, son of King Tom, bred by Warren Viley, Esq., Stonewall Stud, near Midway, Ky., foaled 1872, dam Capitola, dam of Belle Barclay and Hospodar by Vandal, son of Glencoe, out of Margravine, Versaille's dam by imp. Margrave.

King Alfonso's turf career was short but brilliant. He made his *debut* at Lexington, Ky., September 6, 1875, in a Sweepstakes for three-year olds, 1 mile and 1 furlong, 90 lbs. on each, Bob Woolley won in 1.54, the fastest race ever run at the distance to that date, King Alfonso was a close second, with Katie Pearce, Ten Broeck, Elemi and Redman behind him; September 9, was unplaced in Sweepstakes, 1 mile and 5 furlongs, won by Ten Broeck in 2.49¼, the fastest race ever run at that distance to that date, Bob Woolley being second. Louisville, Ky., September 20, won the Kentucky St. Leger, 2 miles, in 3.34½, beating Ten Broeck, second, Verdegris, third, Geo. Graham, Voltigeur and five others; September 24, won the Tobacco Stakes, mile heats, in 1.44½, 1.45¼, beating Gyptis, Misdeal and three others; the next day, September 25, won Galt House Stakes, 2 mile heats, in 3.34, 3.40½, 3.49, beating George Graham, who won the first heat by a head from King Alfonso, the last quarter of the heat being run in 24 seconds, Emma C., Add and two others. He closed the season and his turf career by winning the Linck's Hotel Stakes, at Nashville, mile heats, in 1.45, 1.47¾, beating Misdeal, Asterlite and three others. He was trained the spring of 1876, when four years old, and showed himself a better horse than ever, but in a very fast trial struck his leg and was thrown out of training, and was sold to A. J. Alexander and placed in the Woodburn Stud.

The family has always, as a racing one, ranked one of the best in America. The Albert mare produced the famous Tiger gelding and Black-eyed Susan, a superior race-mare, dam of Dick Singleton, Plato, Mistletoe, Dick Johnson, Catharine, etc. Dick Singleton defeated the great Collier, four mile heats; Mistletoe won the match between Kentucky and Tennessee. Mistletoe was the dam of Evergreen by imp. Glencoe, who produced Goodwood, Maggiore, Verbena, Glendower and others, and Margravine, the dam of Mary Churchill, Tourist, Capitola, The Grand Dutch S., Versailles, etc. Mary Churchill was the dam of Hamburg, Wade Hampton, winner of the Sequel Stakes at Long Branch and Saratoga in 1872, by Asteroid, and Nettie Viley, she the dam of Conductor, Mirah, etc. The Grand Dutch S. was the dam of Yandell and Lizzie Whips, both fine race-nags.

King Alfonso is sire of Fonso, winner of the Phœnix Hotel Stakes, 1¼ miles, in 2.10¾, beating Kinkead, Luke Blackburn and others, and Kentucky Derby, 1½ miles, in 2.37½. Grenada, the winner of the Potomac Stakes, 1½ miles, in 2.39½, and Preakness Stakes, both at Baltimore, Belmont, Jerome and Lorillard Stakes at Jerome Park, Coney Island Derby at Sheepshead Bay, and Dixie Stakes at Baltimore, 2 miles, in 3.38, winning $32,000 as three-year old. At four years old, won four out of six starts; Peyton Handicap at Baltimore, mile heats, in 1.43½, 1.43¼, 1.47½, Oden won second heat. Other winners by him are Quito; Lavacca; Dodette, Ladies' Stake at Chicago. Telemachus, 1¼ miles, in 2.09½, Go Forth, Alfambra, May Wilson, Mollie Brown, 5 furlongs in 1.02, Edison, Duke of Montalban, Mary S. Lost Cause, Windrush, 2 miles in 3.33½, and again in 3.36½, Bayadere, Bob Johnson, Infanta, winner of Elizabeth Stakes at Long Branch, Issie, winner of Ashland Oaks, Lexington, Ky., Katie Creel, winner of the Kentucky Oaks at Louisville, 1½ miles, 2.39, and Illinois Oaks, Chicago. Arno, Lady Prewitt, The Judge (Cordova), Lorena, Mary S., Olivette, Vera, Don Fulano and Golden Gate, raced creditably in England, and Foxhall won two out of three stakes in which he started at two years old, beating the best colts of the year and giving weight. He won the Grand Prize of Paris in 1881, and is the second horse that ever won the Cesarewitch and Cambridge Handicaps in England in one and the same year. The first was Rosebery, a four-year old, carrying in the Cesarewitch 103 lbs., and in Cambridgeshire 117 lbs. Foxhall, a three-year old, carried in the first 110, and in the latter 126 lbs., beating with the top weight Lucy Glitters (3), 91, Tristan (3), 107, Mistake (4), 102, Wallenstein (4), 97, and twenty-seven others. He won the Ascot Gold Cup in 1882. Other winners by him are Bessie June, Joe Cotton, King Lee.

(Continued on page 180.)

Pot-8-o-'s 1773 Eclipse 1764 Marske 1750—Squirt 1732—Bartlett's Childers—Darley Arabian 1689.

KING ALFONSO.

BAY HORSE, FOALED 1872; BRED BY MR. WARREN VILEY, STONEWALL STUD, KENTUCKY.

*PHAETON, b. h., 1865.	King Tom, b. h., 1851.	Harkaway, ch. h., 1834.	Economist, b. h., 1825.	Whisker, b. h., 1812.	Waxy 1790 (Pot-8-o-'s 1773)—Penelope 1798 by Trumpator 1782 (Conductor 1767)—Prunella 1788 by Highflyer 1774 (Herod 1758)—Promise 1768 by Snap 1750 (Snip 1736)—Julia 1756 by Blank 1740 (God. Ar. 1724).
				Floranthe, b. m., 1818.	Octavian 1807 (Stripling 1795)—Caprice 1797 by Anvil 1777 (Herod 1758)—Madcap 1774 by Eclipse 1764 (Marske 1750)—Sappho 1763 by Blank 1740 (God. Ar. 1724)—Dau. 1750 of Blaze 1733 (Flying Childers 1715).
			Fanny Dawson, ch. m., 1823.	Nabocklish, ch. h., 1811.	Rugantino 1804 (Commodore 1793)—Butterfly 1801 by Master Bagot 1787 (Bagot 1780)—Dau. of Bagot 1780 (Herod 1758)—Mother Brown 1771 by Trunnion 1747 (Cade 1734)—Dau. of Old England 1741 (God. Ar. 1724).
				Miss Tooley, b. m., 1808.	Teddy-the-Grinder 1798 (Asparagus 1787)—Lady Jane 1796 by Sir Peter T. 1784 (Highflyer 1774)—Paulina 1778 by Florizel 1768 (Herod 1758)—Captive 1771 by Matchem 1748 (Cade 1734)—Calliope 1763 by Slouch 1747 (Cade 1734).
		Pocahontas, b. m., 1837.	Glencoe*, ch. h., 1831.	Sultan, b. h., 1816.	Selim 1802 (Buzzard* 1787)—Bacchante 1809 by W's Ditto 1800 (Sir Peter T. 1784)—Dau. 1791 of Mercury 1778 (Eclipse 1764)—Dau. 1776 of Herod 1758 (Tartar 1743)—Folly 1771 by Marske 1750 (Squirt 1732).
				Trampoline, ch. m., 1825.	Tramp 1810 (Dick Andrews 1797)—Web 1808 by Waxy 1790 (Pot-8-o-'s 1773)—Penelope 1798 by Trumpator 1782 (Conductor 1767)—Prunella 1788 by Highflyer 1774 (Herod 1758)—Promise 1768 by Snap 1750 (Snip 1736).
			Marpessa, b. m., 1830.	Muley, b. h., 1810.	Orville 1799 (Beningbrough 1791)—Eleanor 1798 by Whiskey 1789 (Saltram* 1780)—Y. Giantess 1790 by Diomed* 1777 (Florizel 1768)—Giantess 1769 by Matchem 1748 (Cade 1734)—Molly Long Legs 1753 by Babraham 1740 (God. Ar. 1724).
				Clare, b. m., 1824.	Marmion 1806 (Whiskey 1789)—Harpalice 1814 by Gohanna 1790 (Mercury 1778)—Amazon 1799 by Driver 1783 (Trentham 1766)—Fractious 1792 by Mercury 1778 (Eclipse 1764)—Dau. 1785 of Woodpecker 1773 (Herod 1758).
	Merry Sunshine, ch. m., 1855.	Storm, b. h., 1848.	Touchstone, br. h., 1831.	Camel, br. h., 1822.	Whalebone 1807 (Waxy 1790)—Dau. 1812 of Selim 1802 (Buzzard* 1787)—Maiden 1801 by Sir Peter T. 1784 (Highflyer 1774)—Dau. 1788 of Phenomenon 1780 (Herod 1758)—Matron 1782 by Florizel 1768 (Herod 1758).
				Banter, br. m., 1826.	Master Henry 1815 (Orville 1799)—Boadicea 1807 by Alexander 1782 (Eclipse 1764)—Brunette by Amaranthus 1766 (Old England 1741)—Mayfly 1771 by Matchem 1748 (Cade 1734)—Dau. 1755 of Anc. Starling 1738 (Old Starling 1727).
			Ghuznee, b. m., 1838.	Pantaloon, ch. h., 1824.	Castrel 1801 (Buzzard* 1787)—Idalia 1815 by Peruvian 1806 (Sir Peter T. 1784)—Musidora 1801 by Meteor 1783 (Eclipse 1764)—Maid-of-all-Work 1786 by Highflyer 1774 (Herod 1758)—Dau. 1771 of Syphon 1750 (Squirt 1732).
				Languish, b. m., 1830.	Cain 1822 (Paulowitz 1813)—Lydia 1822 by Poulton 1805 (Sir Peter T. 1784)—Variety 1808 by Hyacinthus 1797 (Coriander 1786)—Sist. 1790 to Swordsman by Weazle 1776 (Herod 1758)—Dau. of Turk 1763 (Regulus 1739).
		Daughter of, b. m., 1849.	Falstaff, br. h., 1842.	Touchstone, br. h., 1831.	Camel 1822 (Whalebone 1807)—Banter 1826 by Master Henry 1815 (Orville 1799)—Boadicea 1807 by Alexander 1782 (Eclipse 1764)—Brunette by Amaranthus 1766 (Old England 1741)—Mayfly 1771 by Matchem 1748 (Cade 1734).
				Decoy, b. m., 1830.	Filho-da-Puta 1812 (Haphazard 1797)—Finesse 1815 by Peruvian 1806 (Sir Peter T. 1784)—Violante 1802 by John Bull 1789 (Fortitude 1777)—Dau. 1788 of Highflyer 1774 (Herod 1758)—Everlasting 1775 by Eclipse 1764 (Marske 1750).
			Sist to Pompey, b. m., 1843.	Emilius, b. h., 1820.	Orville 1799 (Beningbrough 1791)—Emily 1810 by Stamford 1794 (Sir Peter T. 1784)—Dau. 1799 of Whiskey 1789 (Saltram* 1780)—Grey Dorimant 1781 by Dorimant 1772 (Otho 1760)—Dizzy 1757 by Blank 1740 (God. Ar. 1724).
				Variation, b. m., 1827.	Bustard 1813 (Castrel 1801)—Johanna Southcothe 1811 by Beningbrough 1791 (King Fergus 1775)—Lavinia 1802 by Pipator 1786 (Imperator 1776)—Dau. 1790 of Highflyer 1774 (Herod 1758)—Dau. 1777 of Cardinal Puff 1760 (Babraham 1740).
CAPITOLA, b. m., 1858.	Vandal, b. h., 1850.	*Glencoe, ch. h., 1831.	Sultan, b. h., 1816.	Selim, ch. h., 1802.	Buzzard* 1787 (Woodpecker 1773)—Dau. of Alexander 1782 (Eclipse 1764)—Dau. of Highflyer 1774 (Herod 1758)—Dau. of Alfred 1770 (Matchem 1748)—Dau. 1770 of Engineer 1756 (Sampson 1745).
				Bacchante, br. m., 1809.	W's Ditto 1800 (Sir Peter T. 1784)—Dau. 1791 of Mercury 1778 (Eclipse 1764)—Dau. 1776 of Herod 1758 (Tartar 1743)—Folly 1771 by Marske 1750 (Squirt 1732)—Vixen 1753 by Regulus 1739 (God. Ar. 1724).
			Trampoline, ch. m., 1825.	Tramp, b. h., 1810.	Dick Andrews 1797 (Joe Andrews 1778)—Dau. 1803 of Gohanna 1790 (Mercury 1778)—Fraxinella 1793 by Trentham 1766 (T. Sweepstakes 1743)—Dau. 1785 of Woodpecker 1773 (Herod 1758)—Everlasting 1775 by Eclipse 1764 (Marske 1750).
				Web, b. m., 1808.	Waxy 1790 (Pot-8-o-'s 1773)—Penelope 1798 by Trumpator 1782 (Conductor 1767)—Prunella 1788 by Highflyer 1774 (Herod 1758)—Promise 1768 by Snap 1750 (Snip 1736)—Julia 1756 by Blank 1740 (God. Ar. 1724).
		Alaric's dam, b. m., 1837.	*Tranby, br. h., 1826.	Blacklock, b. h., 1814.	Whitelock 1803 (Hambletonian 1792)—Dau. 1799 of Coriander 1786 (Pot-8-o-'s 1773)—Wildgoose 1792 by Highflyer 1774 (Herod 1758)—Coheiress 1786 by Pot-8-o-'s 1773 (Eclipse 1764)—Manilla 1777 by Goldfinder 1764 (Snap 1750).
				Miss Bowe's dam, br. m., 1812.	Orville 1799 (Beningbrough 1791)—Miss Grimstone 1796 by Weazle 1776 (Herod 1758)—Dau. of Ancaster 1768 (Blank 1740)—Dau. of Damascus Ar. 1754—Dau. of Sampson 1745 (Blaze 1733).
			Lucilla, b. m., —.	Trumpator, br. h., 1823.	Sir Solomon 1805 (*Tickle Toby 1786)—Dau. of Hickory 1801 (*Whip 1794)—*Trumpetta 1797 by Trumpator 1782 (Conductor 1767)—Sist. 1786 to Lambinos by Highflyer 1774 (Herod 1758)—Dau. 1775 of Eclipse 1764 (Marske 1750).
				Lucy, b. m., 1821.	Orphan 1810 (Ball's Florizel 1801)—Lady Grey 1817 by Robin Grey 1805 (*Royalist 1790)—Maria 1802 by Melzar 1791 (*Medley 1776)—Dau. of *Highflyer 1784 (Highflyer 1771)—Dau. of *Fearnought 1755 (Regulus 1739).
	Daughter of, br. m., 1848.	*Margrave, ch. h., 1829.	Muley, b. h., 1810.	Orville, b. h., 1799.	Beningbrough 1791 (King Fergus 1775)—Evelina 1791 by Highflyer 1774 (Herod 1758)—Termagant by Tantrum 1760 (Cripple 1750)—Cantatrice by Sampson 1745 (Blaze 1733)—Dau. 1757 of Regulus 1739 (God. Ar. 1724).
				Eleanor, b. m., 1798.	Whiskey 1789 (Saltram* 1780)—Y. Giantess 1790 by Diomed* 1777 (Florizel 1768)—Giantess 1769 by Matchem 1748 (Cade 1734)—Molly Long Legs 1753 by Babraham 1740 (God. Ar. 1724)—Dau. of Coles' Foxhunter 1727 (Brisk 1711).
			Daughter of, ch. m., 1815.	Election, ch. h., 1804.	Gohanna 1790 (Mercury 1778)—Chestnut Skim 1794 by Woodpecker 1773 (Herod 1758)—Silver*'s dam 1780 by Herod 1758 (Tartar 1743)—Y. Hag 1761 by Skim 1746 (Starling 1727)—Hag 1744 by Crab 1722 (Alcock Ar.).
				Fair Helen, gr. m., 1808.	Hambletonian 1792 (King Fergus 1775)—Helen 1801 by Delpini 1781 (Highflyer 1774)—Rosalind 1788 by Phenomenon 1780 (Herod 1758)—Atalanta 1769 by Matchem 1748 (Cade 1734)—Y. Lass-of-the-Mill 1756 by Oroonoko 1745 (Crab 1722).
		Mistletoe, b. m., 1842.	Cherokee, b. h., 1829.	Sir Archy, b. h., 1805.	*Diomed 1777 (Florizel 1768)—*Castianira 1796 by Rockingham 1781 (Highflyer 1774)—Tabitha 1782 by Trentham 1766 (T. Sweepstakes 1743)—Dau. of Bosphorus 1754 (Babraham 1740)—Dau. of Forester 1750 (Forester 1736).
				Roxana, ch. m., —.	Hephestion 1808 (*Buzzard 1787)—Roxana 1798 by *Marplot 1781 (Highflyer 1774)—Juliet by *Mexican 1775 (Snap 1750)—Dau. of *Friar 1750 (South 1750).
			Black Eyed Susan, br. m., 1821.	Tiger, b. h., 1812.	Cook's Whip 1804 (*Whip 1794)—Jane Hunt 1796 by Hampton's Paragon 1788 (*Flimnap 1765)—Moll by *Figure 1757 (Grey Figure 1747)—Slamerkin 1768 by *Wildair 1753 (Cade 1734)—*Cub mare 1762 by Cub 1739 (Fox 1714).
				Daughter of, blk. m., —.	Albert 1798 (Americus 1793)—Dau. of Algerine (Dey of Algiers)—Dau. of Grey Alfred 1799 (Lindsay's Ar.)—Dau. of Americus 1793 (*Shark 1771)—Dau. of *Medley 1776 (Gimcrack 1760)—Dau. of *Shark 1771 (Marske 1750)—Dau. of *Fearnought 1755 (Regulus 1739).

KING ALFONSO—Continued.

King Alfonso is a red bay, 16 hands, with a star in his forehead, and is one of the truest and best shaped horses in the world. His head is plain but well shaped, and set upon a good, strong muscular neck, with wide, deep throttle, the shoulders oblique, broad and well placed, and covered with suitable muscle; the chest is well shaped and full, with great depth of girth, the body full and round, with the finest back, hip and loin ever put on a horse, being broad, well rounded and slightly curved. He has great length from the point of the hip to the whirlbone, thence to the point of stifle and hocks, being full of strength and muscle, which run into broad, powerful hocks, the legs and feet being sound and good. His temper is of the best. In addition to his double cross of Glencoe, he has the Waxy blood through Web, Whisker and Whalebone, and is inbred to Sir Archy and imp. Shark, with many crosses of Herod and Eclipse in the collateral branches.

IMPORTED KING GALOP.

Located in the Iroquois Stud, Jas. B. Clay, proprietor, Lexington, Ky. Terms annually. Yearling sales in May each year.

KING GALOP, bay horse, foaled 1885, bred by Mr. W. Barrow, imported by S. S. Howland, New York, by Galopin, son of Vedette, dam by King Tom, son of Harkaway, out of Sunshine by Thormanby, etc. King Galop's breeding is gilt-edged. His sire Galopin, winner of the Derby in 1875, is the sire of Galliard, St. Simon, Oberon, Donovan, Corrie Roy and Modwena, a remarkable galaxy of winners for so young a horse. As a three-year old St. Simon won all of his races, which included the Epsom, Ascot and Newcastle Gold Cups and the Goodwood Cup, and all of his engagements, five in number, at two years of age. Galliard won the Chesterfield and Prince of Wales' Stakes at York, 2,000 Guineas, St. James Palace Stakes and Prince of Wales' Stakes at Ascot, and was third for the Derby in 1883. As a two-year old Modwena won nine out of twelve races. Corrie Roy won the Cesarewitch, Goodwood Stakes, Great Ebor Handicap, and other important races. Oberon won the Select Stakes, was second to Ormonde for the Champion Stakes at Newmarket, and won the Lincolnshire Handicap and other races. Donovan won both the Derby and St. Leger in 1889. These and winners like Fulmen, Sunshine, Little Sister, Arbaces, Galvanic, Gallopin Queen, etc., have established Galopin, who is by Vedette, out of Flying Duchess by The Flying Dutchman, as one of the most successful sires in England. King Galop's dam is by King Tom, who was by Harkaway, out of Pocahontas (dam of Stockwell, "Emperor of Stallions," Rataplan, Knight of Kars and Knight of St. Patrick) by imp. Glencoe. King Tom is the sire of imp. Great Tom (sire of General Harding, Tyrant, Telie Doe and of the dam of Proctor Knott), of imp. Phaeton (sire of Ten Broeck and King Alfonso), and of imp. King Ban (sire of King Fox and French Park). King Galop's second dam was Sunshine by Thormanby, winner of the Derby in 1867.

Sunshine was a wonder on the turf. As a two-year old she was only defeated once, and then ran second. She won the Woodcote Stakes, at Epsom; the July Stakes, at Newmarket; the North of England Biennial Stakes, at York; the Champagne Stakes, at Doncaster; a Sweepstakes, at Newmarket; walked over for the Bretby Stakes, at Newmarket; the Troy Stakes, at Stockbridge, and a Sweepstakes, at Newmarket. The last time she ran as a two-year old she met her only defeat in being beaten a head by Frivolity for the Middle Park Plate, at Newmarket, giving the winner 3 lbs., and beating Kingcraft (winner of the Derby), Hawthornden (winner of the St. Leger) and thirteen others. As a three-year old Sunshine only started three times; she won the Coronation Stakes at Ascot, ran second to Gamos in the Oaks, and finished second to Normanby in the Ascot Triennial Stakes. In addition to Sunray and Napoli, Sunshine produced at the stud the winners Palermo and Newfield.

Sunbeam by the noted Chanticleer, King Galop's third dam, was another great race-mare, and as a two-year old was never unplaced. She won the Corporation Stakes; the Half Holiday Nursery Handicap, at Manchester; the Grand Stand Plate, same place; a Plate, at Newmarket, beating a field of nineteen; the Nursery Stakes, at Liverpool; ran second to Conductor for a Nursery Plate, at Liverpool; second to

(Continued on page 188.)

Hambletonian 1792 King Fergus 1775—Eclipse 1761 Marske 1750 Squirt 1732 Bartlett's Childers Darley Arabian 1689.

KING GALOP.

BAY HORSE, FOALED 1885; BRED BY MR. W. BARROW, ENGLAND.

GALOPIN, b. h., 1872.	Vedette, br. h., 1854.	Voltigeur, br. h., 18[illegible].	Voltaire, br. h., 1826.	Blacklock, b. h., 1814.	Whitelock 1803 (Hambletonian 1792) Dau 1799 of Coriander 1786 (Pot-8-o's 1773) Wildgoose 1792 by Highflyer 1774 (Herod 1758) Coheiress 1786 by Pot-8-o's 1773 (Eclipse 1764) Manilla 1777 by Goldfinder 1764 (Snap 1750).
				Daughter of, b. m., 1816.	Phantom 1808 (Walton 1799)—Dau. 1802 of Overton 1788 (King Fergus 1775) Dau. 1796 of Walnut 1786 (Highflyer 1774) Dau. 1786 of Ruler 1777 (Y. Marske 1771) Piracantha 1772 by Matchem 1748 (Cade 1734).
			Martha Lynn, br. m., 1837.	Mulatto, b. h., 1823.	Catton 1809 (Golumpus 1802)—Desdemona 1811 by Orville 1799 (Beningbrough 1791)—Fanny 1796 by Sir Peter T. 1784 (Highflyer 1774) Dau. 1788 of Diomed 1777 (Florizel 1768) Desdemona 1770 by Marske 1750 (Squirt 1732).
				Leda, b. m., 1824.	Filho-da-Puta 1812 (Haphazard 1797) Treasure 1803 by Camillus 1803 (Hambletonian 1792) Dau. 1804 of Hyacinthus 1797 (Coriander 1786) Flora 1789 by King Fergus 1775 (Eclipse 1761) Atalanta 1769 by Matchem 1748 (Cade 1734).
		Mrs. Ridgway, b. m., 1849.	Birdcatcher, ch. h., 1833.	Sir Hercules, blk. h., 1826.	Whalebone 1807 (Waxy 1790) Peri 1822 by Wanderer 1811 (Gohanna 1790) Thalestris 1809 by Alexander 1782 (Eclipse 1764) Rival 1800 by Sir Peter T. 1784 (Highflyer 1774)—Hornet 1790 by Drone 1777 (Herod 1758).
				Guiccioli, ch. m., 1823.	Bob Booty 1804 (Chanticleer 1787)—Flight 1809 by Ir. Escape 1802 (Commodore 1793) Y. Heroine by Bagot 1780 (Herod 1758)—Heroine 1775 by Hero 1753 (Cade 1734)—Dau. 1766 of Snap 1750 (Snip 1736).
			Nan Darrell, gr. m., 1844.	Inheritor, blk. h., 1831.	Lottery 1820 (Tramp 1810)—Hand Maiden 1817 by Walton 1799 (Sir Peter T. 1784) Anticipation 1802 by Beningbrough 1791 (King Fergus 1775) Expectation 1779 by Herod 1758 (Tartar 1743)—Dau. 1758 of Skim 1746 (Starling 1727).
				Nell, gr. m., 1831.	Blacklock 1814 (Whitelock 1803) Madame Vestris 1818 by Comus 1809 (Sorcerer 1796)—Lisette 1806 by Hambletonian 1792 (King Fergus 1775) Constantia 1796 by Walnut 1786 (Highflyer 1774)—Contessina 1787 by Y. Marske 1771 (Marske 1750).
	Flying Duchess, b. m., 1853.	The Flying Dutchman, br. h., 1846.	Bay Middleton, b. h., 1833.	Sultan, b. h., 1816.	Selim 1802 (Buzzard* 1787) Bacchante 1809 by W.'s Ditto 1800 (Sir Peter T. 1784) Dau. 1791 of Mercury 1778 (Eclipse 1764) Dau. 1776 of Herod 1758 (Tartar 1743) Folly 1771 by Marske 1750 (Squirt 1732).
				Cobweb, b. m., 1821.	Phantom 1808 (Walton 1799) Filagree 1815 by Soothsayer 1808 (Sorcerer 1796) Web 1808 by Waxy 1790 (Pot-8-o's 1773) Penelope 1798 by Trumpator 1782 (Conductor 1767)—Prunella 1788 by Highflyer 1774 (Herod 1758).
			Barbelle, b. m., 1836.	Sandbeck, b. h., 1818.	Catton 1809 (Golumpus 1802)—Orvillina 1804 by Beningbrough 1791 (King Fergus 1775)—Evelina 1791 by Highflyer 1774 (Herod 1758)—Termagant by Tantrum 1760 (Cripple 1750)—Cantatrice by Sampson 1745 (Blaze 1733).
				Darioletta, br. m., 1822.	Amadis 1807 (Don Quixote 1784) Selima 1810 by Selim 1802 (Buzzard* 1787)—Dau. 1791 of Pot-8-o's 1773 (Eclipse 1761)—Editha 1781 by Herod 1758 (Tartar 1742) Elfrida 1768 by Snap 1750 (Snip 1736).
		Merope, b. m., 1841.	Voltaire, br. h., 1826.	Blacklock, b. h., 1814.	Whitelock 1803 (Hambletonian 1792)—Dau. 1799 of Coriander 1786 (Pot-8-o's 1773)—Wildgoose 1792 by Highflyer 1774 (Herod 1758)—Coheiress 1786 by Pot-8-o's 1773 (Eclipse 1764)—Manilla 1777 by Goldfinder 1764 (Snap 1750).
				Daughter of, b. m., 1816.	Phantom 1808 (Walton 1799)—Dau. 1802 of Overton 1788 (King Fergus 1775)—Dau. 1796 of Walnut 1786 (Highflyer 1774)—Dau. 1786 of Ruler 1777 (Y. Marske 1771)—Piracantha 1772 by Matchem 1748 (Cade 1734).
			Velocipede's dam, 1817.	Juniper, ch. h., 1805.	Whiskey 1789 (Saltram* 1780)—Jenny Spinner 1797 by Dragon* 1787 (Woodpecker 1773)—Dau. 1775 of Eclipse 1764 (Marske 1750) Miss Spindleshanks 1770 by Omar 1752 (God. Ar. 1724)—Dau. 1773 of Starling 1727 (Bay Bolton 1705).
				Daughter of, br. m., 1810.	Sorcerer 1796 (Trumpator 1782)—Virgin 1801 by Sir Peter T. 1784 (Highflyer 1774)—Dau. 1794 of Pot-8-o's 1773 (Eclipse 1764)—Editha 1781 by Herod 1758 (Tartar 1743)—Elfrida 1768 by Snap 1750 (Snip 1736).
DAUGHTER OF, b. m., 1877.	King Tom, b. h., 1851.	Harkaway, ch. h., 1834.	Economist, b. h., 1825.	Whisker, b. h., 1812.	Waxy 1790 (Pot-8-o's 1773)—Penelope 1798 by Trumpator 1782 (Conductor 1767)—Prunella 1788 by Highflyer 1774 (Herod 1758) Promise 1768 by Snap 1750 (Snip 1736)—Julia 1756 by Blank 1740 (God. Ar. 1724).
				Floranthe, b. m., 1818.	Octavian 1807 (Stripling 1795)—Caprice 1797 by Anvil 1777 (Herod 1758)—Madcap 1774 by Eclipse 1764 (Marske 1750)—Sappho 1763 by Blank 1740 (God. Ar. 1724)—Dau. 1750 of Blaze 1733 (Flying Childers 1715).
			Fanny Dawson, ch. m., 1823.	Nabocklish, ch. h., 1811.	Rugantino 1801 (Commodore 1793) Butterfly 1801 by Master Bagot 1787 (Bagot 1780) Dau. of Bagot 1780 (Herod 1758) Mother Brown 1771 by Trunnion 1747 (Cade 1734)—Dau. of Old England 1741 (God. Ar. 1724).
				Miss Tooly, b. m., 1808.	Teddy the Grinder 1798 (Asparagus 1787) Lady Jane 1796 by Sir Peter T. 1784 (Highflyer 1774)—Paulina 1778 by Florizel 1768 (Herod 1758)—Captive 1771 by Matchem 1748 (Cade 1734) Calliope 1763 by Slouch 1747 (Cade 1734).
		Pocahontas, b. m., 1837.	Glencoe*, ch. h., 1831.	Sultan, b. h., 1816.	Selim 1802 (Buzzard* 1787)—Bacchante 1809 by W.'s Ditto 1800 (Sir Peter T. 1784)—Dau. 1791 of Mercury 1778 (Eclipse 1764)—Dau. 1776 of Herod 1758 (Tartar 1743)—Folly 1771 by Marske 1750 (Squirt 1732)—Vixen 1753 by Regulus 1739 (God. Ar. 1724).
				Trampoline, ch. m., 1825.	Tramp 1810 (Dick Andrews 1797)—Web 1808 by Waxy 1790 (Pot-8-o's 1773)—Penelope 1798 by Trumpator 1782 (Conductor 1767)—Prunella 1788 by Highflyer 1774 (Herod 1758)—Promise 1768 by Snap 1750 (Snip 1736).
			Marpessa, b. m., 1830.	Muley, b. h., 1810.	Orville 1799 (Beningbrough 1791) Eleanor 1798 by Whiskey 1789 (Saltram* 1780) Y. Giantess 1790 by Diomed* 1777 (Florizel 1768)—Giantess 1769 by Matchem 1748 (Cade 1734)—Molly Long Legs 1753 by Babraham 1740 (God. Ar. 1724).
				Clare, b. m., 1824.	Marmion 1806 (Whiskey 1789)—Harpalice 1814 by Gohanna 1790 (Mercury 1778)—Amazon 1799 by Driver 1783 (Trentham 1766)—Fractions 1792 by Mercury 1778 (Eclipse 1764)—Dau. 1785 of Woodpecker 1773 (Herod 1758).
	Sunshine, b. m., 18[illegible]7.	Thormanby, ch. h., 1857.	[Melb'rne or] Windhound, br. h., 1847.	Pantaloon, ch. h., 1824.	Castrel 1801 (Buzzard* 1787)—Idalia 1815 by Peruvian 1806 (Sir Peter T. 1784)—Musidora 1804 by Meteor 1783 (Eclipse 1764)—Maid-of-all-Work 1786 by Highflyer 1774 (Herod 1758)—Dau. 1771 of Syphon 1750 (Squirt 1732).
				Phryne, br. m., 1840.	Touchstone 1831 (Camel 1822)—Decoy 1830 by Filho-da-Puta 1812 (Haphazard 1797)—Finesse 1815 by Peruvian 1806 (Sir Peter T. 1784)—Violante 1802 by John Bull 1789 (Fortitude 1777)—Dau. 1788 of Highflyer 1774 (Herod 1758).
			Alice Hawthorn, b. m., 1838.	Muley Moloch br. h., 1830.	Muley 1810 (Orville 1799)—Nancy 1813 by Dick Andrews 1797 (Joe Andrews 1787)—Spitfire 1800 by Beningbrough 1791 (King Fergus 1775)—Dau. of Y. Sir Peter 1775 (Doge 1762)—Dau. of Engineer 1756 (Sampson 1745).
				Rebecca, b. m., 1831.	Lottery 1820 (Tramp 1810)—Dau. 1818 of Cervantes 1806 (Don Quixote 1784)—Anticipation 1802 by Beningbrough 1791 (King Fergus 1775)—Expectation 1779 by Herod 1758 (Tartar 1743) Dau. 1758 of Skim 1746 (Starling 1727).
		Sunbeam, b. m., 1855.	Chanticleer, gr. h., 1843.	Ir Birdcatcher ch. h., 1833.	Sir Hercules 1826 (Whalebone 1807) Guiccioli 1823 by Bob Booty 1804 (Chanticleer 1787) Flight 1809 by Ir. Escape 1802 (Commodore 1793)—Y. Heroine by Bagot 1780 (Herod 1758) Heroine 1775 by Hero 1753 (Cade 1734).
				Whim, gr. m., 1832.	Ir. Drone 1823 (Master Robert 1811) Kiss 1827 by Waxy Pope 1806 (Waxy 1790)—Dau. 1810 of Champion 1797 (Pot-8-o's 1773) Brown Fanny 1799 by Maximin 1785 (Evergreen 1769)—Dau. 1782 of Highflyer 1774 (Herod 1758).
			Sunflower, b. m., 1847.	Bay Middleton, b. h., 1833.	Sultan 1816 (Selim 1802)—Cobweb 1821 by Phantom 1808 (Walton 1799) Filagree 1815 by Soothsayer 1808 (Sorcerer 1796)—Web 1808 by Waxy 1790 (Pot-8-o's 1773)—Penelope 1798 by Trumpator 1782 (Conductor 1767).
				Io, ch. m., 1833.	Taurus 1826 ([Phantom or] Morisco 1819) Problem ch. m. 1823 by Merlin 1815 (Castrel 1801)—Pawn b. m. 1808 by Trumpator 1782 (Conductor 1767)—Prunella b. m. 1788 by Highflyer 1774 (Herod 1758)—Promise br. m. 1768 by Snap 1750 (Snip 1736)—Julia b. m. 1756 by Blank 1740 (God. Ar. 1724)—Spectator's dam b. m. 1735 by Partner

1718 (Jigg)—Bonny Lass b. m. 1723 by Bay Bolton 1705 (Grey Hautboy)—Dau. of Darley Ar. 1699 Dau. of Byerly Turk—Dau. of Taffolet Barb Dau. of Place's White Turk—Natural Barb mare of Mr. Tregonwell's.

IMPORTED KING GALOP—Continued.

Perfection for the Findon Stakes, at Goodwood; second to Gildermire for the Rawcliffe Produce Stakes, at York, and ran third in another race. As a three-year old she won the Palatine Stakes, at Chester; the Biennial Stakes, at Bath; the Coronation Stakes, at Ascot; a Sweepstakes, at Goodwood; the Chesterfield Cup, the same place; the St. Leger, at Doncaster; walked over for the Don Stakes and the Select Stakes, and ran second to Stapleton for a Handicap Plate. As a four-year old Sunbeam won a Sweepstakes at Goodwood, over the Cup course, and ran second to Zuyder Zee for the Chesterfield Cup, beating twenty-one opponents. In addition to Sunshine (King Galop's dam), Sunbeam produced amongst others Merry Sunshine (winner of the Zetland Stakes), and the grand race-horse Phœbus, who, as a three-year old, also captured the Zetland Stakes, in addition to the Column Produce Stakes, at Newmarket, the Bennington Stakes, the Beaufort Produce Stakes, etc.

King Galop promises well in the stud. He traces in the female line to the Royal mare, from which descend in direct line Woodpecker, Whalebone, Whisker, Partisan, imp. Glencoe, Melbourne, Bend'or, Robert the Devil, Bay Middleton, imp. Mortemer and all the Prunella family.

KINGLIKE.

WINNER OF THE NURSERY STAKES AND HARVEST HANDICAP.

Is a private stallion in the Brookdale Stud of the estate of D. D. Withers. Yearlings sold annually.

KINGLIKE, bay horse, foaled 1880, bred in the Brookdale Stud, by imp. King Ernest, son of King Tom and Ernestine by Touchstone, dam Mimi by imp. Eclipse, son of Orlando, grandam Hennie Farrow by imp. Shamrock, etc. Kinglike was an excellent race-horse, won the Nursery Stakes at Jerome Park at two years old, in a good field of twelve colts. At three years old won purse at Monmouth Park, 1¼ miles, in 2.12, defeating Parole, Fair Count and Heel-and-Toe; won Handicap Sweepstakes, 1⅛ miles, 116 lbs., in 1.58½; ran third to Pizarro and Geo. Kinney in Ocean Stakes, 1½ miles, in 1.56¼; was third to Renegade and Gonfalon, in Barnegat Stakes, 1½ miles, in 2.39¾; second to Little Minch in Newark Stakes, 1 mile, in 1.46¾, conceding him 5 lbs., with five others behind him; ran third to Breeze and Rica, 1 mile, in 1.43, with five others behind him; second to Pizarro, 1 mile, in 1.43¾, with four others behind him. At four years old ran second to Miss Woodford in Coney Island Stakes, 1⅛ miles, in 1-56¼, beating Miss Brewster; won Handicap Sweepstakes, 1$\frac{3}{16}$ miles, beating Jack of Hearts, Aranza, and six others, time 2.03¼; won purse, 1¼ miles, in 2.15½, beating Vibrator; won Harvest Handicap, 1¼ miles, in 2.11¼, defeating Free Gold, Rica, Monitor, Barnes and five others; ran third to Drake, Carter and Heel-and-Toe, in Monmouth Handicap, 1½ miles, in 2.37½, Louisette, Monitor and Barnes behind him; won Handicap Sweepstakes, 1¼ miles, in 2.11¼, Monitor and three others behind him; was third to Tom Martin and Ferg Kyle in Twin City Handicap, 1¼ miles, in 2.10¼, fifteen others behind him. He is descended from a grand racing family. From it came Electra, dam of Rebecca, Marian, Winifred, Elizabeth (dam of Eminence), Hidalgo, El Dorado, Yolande (dam of Beverwyck) and Fitz James, also Hennie Farrow, Ida, Madam Bosley. Hennie Farrow being the dam of Privateer, Ballerina (dam of Ballet Girl, Balinette, Countess Zicka and Lou Spencer), Mayflower (dam of Joe Hooker, Wild Rose, Warwick and Annie Laurie). Mimi (dam of Mark, Mikado, Viceroy and Stately), the successful sire Shannon, the good race-horse and sire Flood, and the great race-mare Mollie McCarty, dam of Fallen Leaf and Mollie McCarty's Last.

Kinglike has only covered a few mares, but is the sire of those good horses Charley Post, Text and the Laura Gould colt, a winner.

KINGLIKE.

BAY HORSE, FOALED 1880; BRED BY MR. D. D. WITHERS, BROOKDALE STUD, NEW JERSEY.

Sire: *KING ERNEST, b. h., 1869.
- King Tom, b. h., 1851.
  - Harkaway, ch. h., 1834.
    - Economist, b. h., 1825.
    - Fanny Dawson, ch. m., 1823.
  - Pocahontas, b. m., 1837.
    - Glencoe*, ch. h., 1831.
    - Marpessa, b. m., 1830.
- Ernestine, b. m., 1850.
  - Touchstone, br. h., 1831.
    - Camel, br. h., 1822.
    - Banter, br. m., 1826.
  - Lady Geraldine, br. m., 1836.
    - The Colonel, ch. h., 1825.
    - Nurse, b. m., 1831.

Dam: MIMI, br. m., 1869.
- *Eclipse, b. h., 1855.
  - Orlando, b. h., 1841.
    - Touchstone, br. h., 1831.
    - Vulture, ch. m., 1833.
  - Gaze, b. m., 1842.
    - Bay Middleton, b. h., 1833.
    - Flycatcher, ch. m., 1831.
- Hennie Farrow, b. m., 1853.
  - *Shamrock, ch. h., 1835?
    - St. Patrick, ch. h., 1817.
    - Delight?, b. m., 1828?
  - Ida, b. m., 1837?
    - *Belshazzar, ch. h., 1830.
    - Madam Bosley, gr. m., —

Whisker, b. h., 1812.	Waxy 1790 (Pot-8-o-'s 1773)—Penelope 1798 by Trumpator 1782 (Conductor 1767)—Prunella 1788 by Highflyer 1774 (Herod 1758)—Promise 1768 by Snap 1750 (Snip 1736)—Julia 1756 by Blank 1740 (God. Ar. 1724).
Floranthe, b. m., 1818.	Octavian 1807 (Stripling 1795)—Caprice 1797 by Anvil 1777 (Herod 1758)—Madcap 1774 by Eclipse 1764 (Marske 1750)—Sappho 1763 by Blank 1740 (God. Ar. 1724)—Dau. 1750 of Blaze 1733 (Flying Childers 1715).
Nabocklish, ch. h., 1811.	Rugantino 1804 (Commodore 1793)—Butterfly 1804 by Master Bagot 1787 (Bagot 1780)—Dau. of Bagot 1780 (Herod 1758)—Mother Brown 1771 by Trunnion 1747 (Cade 1734)—Dau. of Old England 1741 (God. Ar. 1724).
Miss Tooly, b. m., 1808.	Teddy-the-Grinder 1798 (Asparagus 1787)—Lady Jane 1796 by Sir Peter T. 1784 (Highflyer 1774)—Paulina 1778 by Florizel 1768 (Herod 1758)—Captive 1774 by Matchem 1748 (Cade 1734)—Calliope 1763 by Slouch 1747 (Cade 1734).
Sultan, b. h., 1816.	Selim 1802 (Buzzard* 1787)—Bacchante 1809 by W's Ditto 1800 (Sir Peter T. 1784)—Dau. 1791 of Mercury 1778 (Eclipse 1764)—Dau. 1776 of Herod 1758 (Tartar 1743)—Folly 1771 by Marske 1750 (Squirt 1732).
Trampoline, ch. m., 1825.	Tramp 1810 (Dick Andrews 1797)—Web 1808 by Waxy 1790 (Pot-8-o-'s 1773)—Penelope 1798 by Trumpator 1782 (Conductor 1768)—Prunella 1788 by Highflyer 1774 (Herod 1758)—Promise 1768 by Snap 1750 (Snip 1736).
Muley, b. h., 1810.	Orville 1799 (Beningbrough 1791)—Eleanor 1798 by Whiskey 1789 (Saltram* 1780)—Y. Giantess 1790 by Diomed* 1777 (Florizel 1768)—Giantess 1769 by Matchem 1748 (Cade 1734)—Molly Long Legs 1753 by Babraham 1740 (God. Ar. 1724).
Clare, b. m., 1824.	Marmion 1806 (Whiskey 1789)—Harpalice 1814 by Gohanna 1790 (Mercury 1778)—Amazon 1799 by Driver 1783 (Trentham 1766)—Fractious 1792 by Mercury 1778 (Eclipse 1764)—Dau. 1785 of Woodpecker 1773 (Herod 1758).
Whalebone, br. h., 1807.	Waxy 1790 (Pot-8-o-'s 1773)—Penelope 1798 by Trumpator 1782 (Conductor 1767)—Prunella 1788 by Highflyer 1774 (Herod 1758)—Promise 1768 by Snap 1750 (Snip 1736)—Julia 1756 by Blank 1740 (God. Ar. 1724).
Daughter of, b. m., 1812.	Selim 1802 (Buzzard* 1787)—Maiden 1801 by Sir Peter T. 1784 (Highflyer 1774)—Dau. 1788 of Phenomenon 1780 (Herod 1758)—Matron 1782 by Florizel 1768 (Herod 1758)—Maiden 1770 by Matchem 1748 (Cade 1734).
Master Henry, b. h., 1815.	Orville 1799 (Beningbrough 1791)—Miss Sophia 1805 by Stamford 1794 (Sir Peter T. 1784)—Sophia 1798 by Buzzard* 1787 (Woodpecker 1773)—Huncamunca 1787 by Highflyer 1774 (Herod 1758)—Cypher 1772 by Squirrel 1754 (Traveller 1735).
Boadicea, b. m., 1807.	Alexander 1782 (Eclipse 1764)—Brunette by Amaranthus 1766 (Old England 1741)—Mayfly 1771 by Matchem 1748 (Cade 1734)—Dau. 1755 of Anc. Starling 1738 (Starling 1727)—Look-at-me-Lads 1731 by Grasshopper (Byerly Turk).
Whisker, b. h., 1812.	Waxy 1790 (Pot-8-o-'s 1773)—Penelope 1798 by Trumpator 1782 (Conductor 1767)—Prunella 1788 by Highflyer 1774 (Herod 1758)—Promise 1768 by Snap 1750 (Snip 1736)—Julia 1756 by Blank 1740 (God. Ar. 1724).
Daughter of, b. m., 1802.	Delpini 1781 (Highflyer 1774)—Tipple Cyder 1788 by King Fergus 1775 (Eclipse 1764)—Sylvia 1783 by Y. Marske 1771 (Marske 1750)—Ferret 1765 by Broth. to Silvio (Cade 1734)—Dau. of Regulus 1739 (God. Ar. 1724).
Neptune, b. h., 1825.	Tiresias 1816 (Soothsayer 1808)—Rivulet 1813 by Rubens 1805 (Buzzard* 1787)—Dau. 1793 of Pot-8-o-'s 1773 (Eclipse 1764)—Huncamunca 1787 by Highflyer 1774 (Herod 1758)—Cypher 1772 by Squirrel 1754 (Traveller 1735).
Otis, b. m., 1820.	Bustard 1801 (Buzzard* 1787)—Dau. 1813 of Election 1804 (Gohanna 1790)—Dau. 1791 of Highflyer 1774 (Herod 1758)—Dau. 1778 of Eclipse 1764 (Marske 1750)—Rosebud 1765 by Snap 1750 (Snip 1736).
Camel, br. h., 1822.	Whalebone 1807 (Waxy 1790)—Dau. 1812 of Selim 1802 (Buzzard* 1787)—Maiden 1801 by Sir Peter T. 1784 (Highflyer 1774)—Dau. 1788 of Phenomenon 1780 (Herod 1758)—Matron 1782 by Florizel 1768 (Herod 1758).
Banter, br. m., 1826.	Master Henry 1815 (Orville 1799)—Boadicea 1807 by Alexander 1782 (Eclipse 1764)—Brunette by Amaranthus 1766 (Old England 1741)—Mayfly 1771 by Matchem 1748 (Cade 1734)—Dau. 1755 of Anc. Starling 1738 (Starling 1727).
Langar, ch. h., 1817.	Selim 1802 (Buzzard* 1787)—Dau. 1808 of Walton 1799 (Sir Peter T. 1784)—Y. Giantess 1790 by Diomed* 1777 (Florizel 1768)—Giantess 1769 by Matchem 1748 (Cade 1734)—Molly Long Legs 1753 by Babraham 1740 (God. Ar. 1724).
Kite, b. m., 1821.	Bustard 1813 (Castrel 1801)—Olympia 1815 by Sir Oliver 1800 (Sir Peter T. 1784)—Scotilla 1795 by Anvil 1777 (Herod 1758)—Scota 1783 by Eclipse 1764 (Marske 1750)—Harmony 1775 by Herod 1758 (Tartar 1743).
Sultan, b. h., 1816.	Selim 1802 (Buzzard* 1787)—Bacchante 1809 by W's Ditto 1800 (Sir Peter T. 1784)—Dau. 1791 of Mercury 1778 (Eclipse 1764)—Dau. 1776 of Herod 1758 (Tartar 1743)—Folly 1771 by Marske 1750 (Squirt 1732).
Cobweb, b. m., 1821.	Phantom 1808 (Walton 1799)—Filagree 1815 by Soothsayer 1808 (Sorcerer 1796)—Web 1808 by Waxy 1790 (Pot-8-o-'s 1773)—Penelope 1798 by Trumpator 1782 (Conductor 1767)—Prunella 1788 by Highflyer 1774 (Herod 1758).
Godolphin, b. h., 1818.	Partisan 1811 (Walton 1799)—Ridicule 1810 by Shuttle 1793 (Y. Marske 1771)—Dau. 1792 of Dungannon 1780 (Eclipse 1764)—Letitia 1783 by Highflyer 1774 (Herod 1758)—Dau. of Matchem 1748 (Cade 1734).
Sister to Cobweb, ch. m., 1820.	Phantom 1808 (Walton 1799)—Filagree 1815 by Soothsayer 1808 (Sorcerer 1796)—Web 1808 by Waxy 1790 (Pot-8-o-'s 1773)—Penelope 1798 by Trumpator 1782 (Conductor 1767)—Prunella 1788 by Highflyer 1774 (Herod 1758).
Walton, b. h., 1799.	Sir Peter T. 1784 (Highflyer 1774)—Arethusa 1792 by Dungannon 1780 (Eclipse 1764)—Dau. 1777 of Prophet 1760 (Regulus 1739)—Virago 1764 by Snap 1750 (Snip 1736)—Dau. of Regulus 1739 (God. Ar. 1724).
Daughter of, b. m., 1810.	Dick Andrews 1797 (Joe Andrews 1778)—Dau. 1800 of Trumpator 1782 (Conductor 1767)—Dau. 1793 of Highflyer 1774 (Herod 1758)—Othœthea 1772 by Otho 1760 (Moses 1746)—Dau. 1764 of Snap 1750 (Snip 1736).
Reveller, b. h., 1815.	Comus 1809 (Sorcerer 1796)—Rosette 1803 by Beningbrough 1791 (King Fergus 1775)—Rosamond 1788 by Tandem 1773 (Syphon 1750)—Tuberose 1772 by Herod 1758 (Tartar 1743)—Grey Starling 1745 by Starling 1727 (Bay Bolton 1705).
Defiance, ch. m., 1816.	Rubens 1805 (Buzzard* 1787)—Little Folly 1806 by Highland Fling 1798 (Spadille 1784)—Harriet 1799 by Volunteer 1780 (Eclipse 1764)—Dau. 1785 of Alfred 1770 (Matchem 1748)—Magnolia 1771 by Marske 1750 (Squirt 1732).
Blacklock, b. h., 1814.	Whitelock 1803 (Hambletonian 1792)—Dau. 1799 of Coriander 1786 (Pot-8-o-'s 1773)—Wildgoose 1792 by Highflyer 1774 (Herod 1758)—Coheiress 1786 by Pot-8-o-'s 1773 (Eclipse 1764)—Manilla 1777 by Goldfinder 1764 (Snap 1750).
Manuella, b. m., 1809.	Dick Andrews 1797 (Joe Andrews 1778)—Mandane 1800 by Pot-8-o-'s 1773 (Eclipse 1764)—Y. Camilla 1787 by Woodpecker 1773 (Herod 1758)—Camilla 1778 by Trentham 1766 (T. Sweepstakes 1743)—Coquette 1765 by Compton Barb.
Sir Richard, gr. h., 1824.	Pacolet 1806 (*Citizen 1785)—Madam Tonson 1814 by Top Gallant (Gallatin 1799)—Dau. of Barry's Grey Medley (*Medley 1776)—Dau. of *Oscar (Y. Snip 1750)—Dau. of *Fearnought 1755 (Regulus 1739).
Nancy Nichol, —, —.	*Eagle 1796 (Volunteer 1780)—Bet Bosley ch. m. 1815 by Wilkes' Wonder 1800 (*Diomed 1777)—Simmes' br. m. by Chanticleer (Symmes' Wildair 1767)—Dau. of *Stirling 1791 (Volunteer 1780)—Dau. of Clodius 1775 (*Janus 1746)—Dau. of *Silvereye (Cullen Ar.)—Dau. of *Jolly Roger 1741 (Roundhead 1733)—Dau. of *Partner (Partner 1718)—Dau. of *Monkey 1725 (Lonsdale Bay Ar.)—Imported mare from the Stud of Mr. Harrison of Brandon.

IMPORTED KINGSTON.

Owned by Messrs. S. Sanford & Son, Hurricane Stud, Amsterdam, N. Y. Yearlings sold annually.

KINGSTON, chestnut horse, foaled 1882, bred by Mr. S. Mason, imported 1887, by Hampton, son of Lord Clifden (winner of the St. Leger) and Lady Langden, dam of Sir Bevys by Kettledrum (winner of the Derby), son of Rataplan, own brother to Stockwell, out of Haricot (dam of that marvellous mare Caller Ou), which was out of the famous Queen Mary (dam of imp. Bonnie Scotland, Blink Bonny, winner of the Derby and Oaks, and dam of Blair Athol, winner of the Derby, St. Leger, etc.). Hampton was a first-class race-horse, winner of the Goodwood and Doncaster Cups, in addition to the Northumberland Plate, and also the Goodwood and Metropolitan Stakes, etc. At the stud his services have always been in the greatest demand, and he has stood at the head of the winning stallions in England, where his book was full three years in advance. He has sired Merry Hampton (winner of the Derby), Ayrshire (winner of the 2,000 Guineas, Derby, etc.), Reve d'Or (winner of the Oaks, etc.), Royal Hampton (winner of the City and Suburban, etc.), Gay Hampton (a good stake winner) and hosts of other good ones. Last Love, Kingston's dam, produced the winners Courtier by King of Trumps, Hymen and Bonaparte by Voltigeur, and Mandeville by Mandrake. Her sire, Annandale, a son of Touchstone, ran second in the Derby to Merry Monarch, beating Weatherbit, Idas, Mentor, John Davis and twenty-five others. He also ran second to Weatherbit in a Sweepstakes of 300 sovereigns each, 3½ miles, at Goodwood; ran second to Queen Mab in the Ascot Stakes, beating twelve other cracks; won Her Majesty's Plate at Chester, about 2½ miles; won the Queen's Plate, 3 miles, at Shrewsbury; won the Caledonian Handicap, 2 miles, carrying top weight, in addition to other good performances. He was a half brother to one of the greatest of all race mares, viz., Alice Hawthorn (dam of Thormanby, winner of the Derby, and also of those grand performers Oulston, Findon, Lady Hawthorn, Conwold and Sweet Hawthorn, whose daughter, Rebecca, was a star of the first magnitude, and another of her daughters, Fleur d'Oranger, also proved a winner, and produced the winner Grallah). Fair Helen, another of Annandale's half sisters, produced Lord of the Isles, who won the 2,000 Guineas and sired the great Scottish Chief, etc.

Executrix, the grandam of Kingston, was by Liverpool (sire of Lanercost), and a half sister to the great race-horse Inheritor, winner of many gold cups, King's and Queen's Plates. Kingston only started in one race at Kempton Park, when he ran into a fence and was permanently injured in his shoulders. He is a horse of fine size and great substance, handsomely bred from a great racing family.

From the Layton Barb mare have descended in direct female line Young Greyhound, Cartouch, Matchem, Young Cade, Drone by Herod, Audubon, Philammon, St. Honorat, Kentucky, Daniel Boone, Gilroy, Iroquois and all the Magnolia family.

Waxy 1790—Pot-8-o's 1773—Eclipse 1764—Marske 1750—Squirt 1732—Bartlett's Childers—Darley Arabian 1699.

KINGSTON.

CHESTNUT HORSE, FOALED 1882; BRED BY MR. S. MASON, ENGLAND.

HAMPTON, b. h., 1872.

Lord Clifden, b. h., 1860.

Newminster, b. h., 1848.

Touchstone, br. h., 1831.

Camel, br. h., 1822. — Whalebone 1807 (Waxy 1790)—Dau. 1812 of Selim 1802 (Buzzard 1787)—Maiden 1801 by Sir Peter T. 1784 (Highflyer 1774)—Dau. 1786 of Phenomenon 1780 (Herod 1758)—Matron 1782 by Florizel 1768 (Herod 1758).

Banter, br. m., 1826. — Master Henry 1815 (Orville 1799)—Boadicea 1807 by Alexander 1782 (Eclipse 1764)—Brunette by Amaranthus 1766 (Old England 1741)—Mayfly 1771 by Matchem 1748 (Cade 1734)—Dau. 1755 of Anc. Starling 1738 (Starling 1727).

Bee's-wing, b. m., 1833.

Dr. Syntax, br. h., 1811. — Paynator 1791 (Trumpator 1782)—Dau. of Beningbrough 1791 (King Fergus 1775)—Jenny Mole 1787 by Carbuncle 1772 (Babraham Blank 1758)—Dau. 1773 of Prince T'Quassaw 1751 (Snip 1736)—Sultana 1772 by Regulus 1739 (God. Ar. 1724).

Tomboy's dam, ch. m., 1817. — Ardrossan 1809 (John Bull 1789)—Lady Eliza 1813 by Whitworth 1805 (Agonistes 1795)—Dau. 1793 of Spadille 1784 (Highflyer 1774)—Sylvia 1783 by Y. Marske 1771—Marske 1750)—Ferret 1765 by Broth. to Silvio (Cade 1734).

The Slave, b. m., 1852.

Melbourne, br. h., 1834.

Humphrey Clinker, b. h., 1822. — Comus 1809 (Sorcerer 1796)—Clinkerina 1812 by Clinker 1805 (Sir Peter T. 1784)—Pewet 1786 by Tandem 1773 (Syphon 1750)—Termagant by Tantrum 1760 (Cripple 1750)—Cantatrice by Sampson 1745 (Blaze 1733).

Daughter of, b. m., 1825. — Cervantes 1806 (Don Quixote 1784)—Dau. 1818 of Golumpus 1802 (Gohanna 1790)—Dau. 1810 of Paynator 1791 (Trumpator 1782)—Circle 1802 by St. George 1789 (Highflyer 1774)—Abigail 1788 by Woodpecker 1773 (Herod 1758).

Volley, b. m., 1845.

Voltaire, br. h., 1826. — Blacklock 1814 (Whitelock 1803)—Dau. 1816 of Phantom 1808 (Walton 1799)—Dau. 1802 of Overton 1788 (King Fergus 1775)—Dau. 1796 of Walnut 1786 (Highflyer 1774)—Dau. 1786 of Ruler 1777 (Y. Marske 1771).

Martha Lynn, br. m., 1837. — Mulatto 1823 (Catton 1809)—Leda 1824 by Filho-da-Puta 1812 (Haphazard 1797)—Treasure 1810 by Camillus 1803 (Hambletonian 1792)—Dau. 1801 of Hyacinthus 1797 (Coriander 1786)—Flora 1789 by King Fergus 1775 (Eclipse 1764).

Lady Langden, br. m., 1868.

Kettledrum, ch. h., 1858.

Rataplan, ch. h., 1850.

The Baron, ch. h., 1842. — Irish Birdcatcher 1833 (Sir Hercules 1826)—Echidna 1838 by Economist 1825 (Whisker 1812)—Miss Pratt 1825 by Blacklock 1814 (Whitelock 1803)—Gadabout 1812 by Orville 1799 (Beningbrough 1791)—Minstrel 1803 by Sir Peter T. 1784 (Highflyer 1774).

Pocahontas, b. m., 1837. — Glencoe* 1831 (Sultan 1816)—Marpessa 1830 by Muley 1819 (Orville 1799)—Clare 1824 by Marmion 1806 (Whiskey 1789)—Harpalice 1814 by Gohanna 1790 (Mercury 1778)—Amazon 1799 by Driver 1783 (Trentham 1766).

Hybla, b. m., 1845.

The Provost, br. h., 1833. — The Saddler 1828 (Waverley 1817)—Rebecca 1831 by Lottery 1820 (Tramp 1810)—Dau. 1818 of Cervantes 1806 (Don Quixote 1784)—Anticipation 1802 by Beningbrough 1791 (King Fergus 1775)—Expectation 1779 by Herod 1758 (Tartar 1743).

Otisina, br. m., 1837. — Liverpool 1828 (Tramp 1810)—Otis 1820 by Bustard 1801 (Buzzard* 1787)—Dau. 1813 of Election 1804 (Gohanna 1790)—Sist. 1791 to Skysweeper by Highflyer 1774 (Herod 1758)—Dau. 1778 of Eclipse 1764 (Marske 1750).

Haricot, br. m., 1847.

Lanercost, br. h., 1835.

Liverpool, b. h., 1828. — Tramp 1810 (Dick Andrews 1797)—Dau. 1822 of Whisker 1812 (Waxy 1790)—Mandane 1800 by Pot-8-o's 1773 (Eclipse 1764)—Y. Camilla 1787 by Woodpecker 1773 (Herod 1758)—Camilla 1778 by Trentham 1766 (T. Sweepstakes 1743).

Otis, b. m., 1820. — Bustard 1801 (Buzzard* 1787)—Dau. 1813 of Election 1804 (Gohanna 1790)—Sist. 1791 to Skysweeper by Highflyer 1774 (Herod 1758)—Dau. 1778 of Eclipse 1764 (Marske 1750)—Rosebud 1765 by Snap 1750 (Snip 1736).

Queen Mary, [Ma-goor] b. m., 1843.

Gladiator, ch. h., 1833. — Partisan 1811 (Walton 1799)—Pauline 1826 by Moses 1819 ([Whalebone or] Seymour 1807)—Quadrille 1815 by Selim 1802 (Buzzard* 1787)—Canary Bird 1806 by [Whiskey or] Sorcerer 1796 (Trumpator 1782)—Canary 1797 by Coriander 1786 (Pot-8-o's 1773).

Daughter of, b. m., 1840. — Plenipotentiary 1831 (Emilius 1820)—Myrrha 1830 by Whalebone 1807 (Waxy 1790)—Gift 1818 by Y. Gohanna 1810 (Gohanna 1790)—Dau. 1802 of Sir Peter T. 1784 (Highflyer 1774)—Sist. 1788 to Aimator by Trumpator 1782 (Conductor 1767).

LAST LOVE b. m., 1870.

Annandale, br. h., 1842.

Touchstone, br. h., 1831.

Camel, br. h., 1822.

Whalebone, br. h., 1807. — Waxy 1790 (Pot-8-o's 1773)—Penelope 1798 by Trumpator 1782 (Conductor 1767)—Prunella 1788 by Highflyer 1774 (Herod 1758)—Promise 1768 by Snap 1750 (Snip 1736)—Julia 1756 by Blank 1740 (God. Ar. 1724).

Daughter of, b. m., 1812. — Selim 1802 (Buzzard* 1787)—Maiden 1801 by Sir Peter T. 1784 (Highflyer 1774)—Dau. 1786 of Phenomenon 1780 (Herod 1758)—Matron 1782 by Florizel 1768 (Herod 1758)—Maiden 1779 by Matchem 1748 (Cade 1734).

Banter, br. m., 1826.

Master Henry, b. h., 1815. — Orville 1799 (Beningbrough 1791)—Miss Sophia 1805 by Stamford 1794 (Sir Peter T. 1784)—Sophia 1798 by Buzzard* 1787 (Woodpecker 1773)—Huncamunca 1787 by Highflyer 1774 (Herod 1758)—Cypher 1772 by Squirrel 1754 (Traveller 1735).

Boadicea, b. m., 1807. — Alexander 1782 (Eclipse 1764)—Brunette by Amaranthus 1766 (Old England 1741)—Mayfly 1771 by Matchem 1748 (Cade 1734)—Dau. 1755 of Anc. Starling 1738 (Starling 1727)—Look-at-me-Lads 1731 by Brist. Grasshopper (Byerly T.).

Rebecca, b. m., 1831.

Lottery, br. h., 1820.

Tramp, b. h., 1810. — Dick Andrews 1797 (Joe Andrews 1778)—Dau. 1803 of Gohanna 1790 (Mercury 1778)—Fraxinella 1793 by Trentham 1766 (T. Sweepstakes 1743)—Dau. 1785 of Woodpecker 1773 (Herod 1758)—Everlasting 1775 by Eclipse 1764 (Marske 1750).

Mandane, ch. m., 1800. — Pot-8-o's 1773 (Eclipse 1764)—Y. Camilla 1787 by Woodpecker 1773 (Herod 1758)—Camilla 1778 by Trentham 1766 (T. Sweepstakes 1743)—Coquette 1765 by the Compton Barb—Sist. 1743 to Regulus by God. Ar. 1724.

Daughter of, b. m., 1818.

Cervantes, b. h., 1806. — Don Quixote 1784 (Eclipse 1764)—Evelina 1791 by Highflyer 1774 (Herod 1758)—Termagant by Tantrum 1760 (Cripple 1750)—Cantatrice by Sampson 1745 (Blaze 1733)—Dau. 1757 of Regulus 1739 (God. Ar. 1724).

Anticipation, ch. m., 1802. — Beningbrough 1791 (King Fergus 1775)—Dau. 1758 of Skim 1746 (Starling 1727)—Dau. 1751 of Janus 1738 (God. Ar. 1724)—Spinster 1743 by Crab 1722 (Alcock Ar.)—Spinster (the Widdrington mare) 1735 by Partner 1718 (Jigg).

Executrix, b. m., 1836.

Liverpool, b. h., 1828.

Tramp, b. h., 1810.

Dick Andrews, b. h., 1797. — Joe Andrews 1778 (Eclipse 1764)—Dau. 1790 of Highflyer 1774 (Herod 1758)—Dau. 1777 of Cardinal Puff 1760 (Babraham 1740)—Dau. 1768 of Tatler 1754 (Blank 1740)—Bay Snip 1743 by Snip 1736 (Fl. Childers 1715).

Daughter of, b. m., 1803. — Gohanna 1790 (Mercury 1778)—Fraxinella 1793 by Trentham 1766 (T. Sweepstakes 1743)—Dau. 1785 of Woodpecker 1773 (Herod 1758)—Everlasting 1775 by Eclipse 1764 (Marske 1750)—Hyæna 1762 by Snap 1750 (Snip 1736).

Daughter of, b. m., 1821.

Whisker, b. h., 1812. — Waxy 1790 (Pot-8-o's 1773)—Penelope 1798 by Trumpator 1782 (Conductor 1767)—Prunella 1788 by Highflyer 1774 (Herod 1758)—Promise 1768 by Snap 1750 (Snip 1736)—Julia 1756 by Blank 1740 (God. Ar. 1724).

Mandane, ch. m., 1800. — Pot-8-o's 1773 (Eclipse 1764)—Y. Camilla 1787 by Woodpecker 1773 (Herod 1758)—Camilla 1778 by Trentham 1766 (T. Sweepstakes 1743)—Coquette 1765 by the Compton Barb—Sist. 1743 to Regulus by God. Ar. 1724.

Hand Maiden, b. m., 1817.

Walton, b. h., 1799.

Sir Peter T., br. h., 1784. — Highflyer 1774 (Herod 1758)—Papillon 1769 by Snap 1750 (Snip 1736)—Miss Cleveland 1758 by Regulus 1739 (God. Ar. 1724)—Midge by a Son of (Bay Bolton 1705)—Dau. of Bartlett's Childers (Darley Ar. 1699).

Arethusa, ch. m., 1792. — Dungannon 1780 (Eclipse 1764)—Dau. 1777 of Prophet 1760 (Regulus 1739)—Virago 1764 by Snap 1750 (Snip 1736)—Dau. of Regulus 1739 (God. Ar. 1724)—Sist. 1730 to Othello by Crab 1722 (Alcock Ar.).

Anticipation, ch. m., 1802.

Beningbrough, b. h., 1791. — King Fergus 1775 (Eclipse 1764)—Dau. 1780 of Herod 1758 (Tartar 1743)—Pyrrha 1771 by Matchem 1748 (Cade 1734)—Duchess 1748 by Whitenose 1742 (God. Ar. 1724)—Miss Slamerkin 1739 by Y. True Blue 1718 (Honeywood Ar.).

Expectation, gr. m., 1779. — Herod 1758 (Tartar 1743)—Dau. 1758 of Skim 1746 (Starling 1727)—Dau. 1751 of Janus 1738 (God. Ar. 1724)—Spinster gr. m. 1743 by Crab 1722 (Alcock Ar.)—Spinster (the Widdrington mare) ch. m. 1735 by Partner 1718 (Jigg)—Bay Bloody Buttocks b. m. 1729 by Bloody Buttocks—Sist. 1722 to Guy by Greyhound (Chillaby)—Brown Farewell br. m. 1710 by Makeless (Oglethorpe Ar.)—Dau. of Brimmer (D'Arcy Yellow Turk)—Dau. of Place's White Turk—Dau. of Dodsworth—Layton Barb Mare.

KNIGHT OF ELLERSLIE.

The property of L. O. Appleby, Monmouth County, N. J.

KNIGHT OF ELLERSLIE, bay horse, foaled 1881, bred by Capt. R. J. Hancock, Ellerslie Stud, Va., by Eolus, son of imp. Leamington and Fanny Washington by Revenue, son of imp. Trustee, dam Lizzie Hazlewood by Scathelock, son of imp. Eclipse, grandam War Song, the dam of Eole, St. Saviour, Eon, etc., by the great War Dance, tracing through Eliza Davis, dam of War Cry, to the noted imp. Melrose by Melbourne, she the dam of Target (the dam of Creedmoor, Britomarte, Rifle, etc.), Moss Rose (the dam of Hard Wood, Ben Wade, etc.), Melbourne Jr., John Porter, etc. Lizzie Hazlewood's colts all race, she being the dam of Knight of Ellerslie, Thomasia, Charlie Dreux, winner of the Van Courtland Stakes, Westchester Handicap, defeating the noted Firenzi, Charley Arnold, etc.

Knight of Ellerslie was a superior race-horse. When three years old won a purse, 1 mile, with 95 lbs., in 1.49½; won Army and Navy Stakes, 1⅛ miles, 118 lbs., on heavy track, in 2.01½; won Vernal Stakes, 1 mile, 110 lbs., in 1.45; won Preakness Stakes, 1½ miles, in 2.39½, 110 lbs., and was second to Panique in the Belmont Stakes, 1½ miles, in 2.42, with Himalaya and Vocalic behind him. He was unplaced in the Emporium, in which race he was injured and did not start again.

Knight of Ellerslie descends in direct female line from a sister to Old Merlin by Bustler, to which trace Dr. Syntax, the sire of the noted Bee's-wing, Little Red Rover, sire of the dam of Buccaneer.

KNIGHT OF ELLERSLIE.

CHESTNUT HORSE, FOALED 1881; BRED BY MR. R. J. HANCOCK, ELLERSLIE STUD, VIRGINIA.

EOLUS, b. h., 1868.
*Leamington, br. h., 1853. — Faugh-a-Ballagh, br. h., 1841. — Sir Hercules, blk. h., 1826. Guiccioli, ch. m., 1823.
Daughter of, b. m., 1841. — Pantaloon, ch. h., 1824. Daphne, br. m., 1837.
Fanny Washington, ch. m., 1855. — Revenue, b. h., 1843. — *Trustee, ch. h., 1829. Rosalie Somers, b. m., 1831.
Sarah Washington, b. m., 1837. — Garrison's Zingance, b. h., 1828. Stella, b. m., —.

LIZZIE HAZLEWOOD, ch. m., 1874.
Scathelock, ch. h., 1867. — *Eclipse, b. h., 1855. — Orlando, b. h., 1841. Gaze, b. m., 1842.
Fanny Washington, ch. m., 1855. — Revenue, b. h., 1843. Sarah Washington, b. m., 1837.
War Song, ch. m., 1867. — War Dance, ch. h., 1859. — Lexington, b. h., 1850. Reel, gr. m., 1838.
Eliza Davis, b. m., 1862. — *Knight of St. George, b. h., 1851. *Melrose, b. m., 1845.

Whalebone, br. h., 1807.	Waxy 1790 (Pot-8-o-'s 1773)—Penelope 1798 by Trumpator 1782 (Conductor 1767)—Prunella 1788 by Highflyer 1774 (Herod 1758)—Promise 1768 by Snap 1750 (Snip 1736)—Julia 1756 by Blank 1740 (God. Ar. 1724).
Peri, b. m., 1822.	Wanderer 1811 (Gohanna 1790)—Thalestris 1809 by Alexander 1782 (Eclipse 1764)—Rival 1800 by Sir Peter T. 1784 (Highflyer 1774)—Hornet 1790 by Drone 1777 (Herod 1758)—Manilla 1777 by Goldfinder 1764 (Snap 1750).
Bob Booty, ch. h., 1804.	Chanticleer 1787 (Woodpecker 1773)—Ierne 1790 by Bagot 1780 (Herod 1758)—Dau. of Gamahoe (Bustard 1741)—Patty by Tim 1752 (Squirt 1732)—Miss Patch 1737 by Justice 1725 (H. C. Litton Ar.).
Flight, ch. m., 1809.	Ir. Escape 1802 (Commodore 1793)—Y. Heroine by Bagot 1780 (Herod 1758)—Heroine 1775 by Hero 1753 (Cade 1734)—Dau. 1766 of Snap 1750 (Snip 1736)—Sist. 1743 to Regulus by God. Ar. 1724.
Castrel, ch. h., 1801.	Buzzard* 1787 (Woodpecker 1773)—Dau. of Alexander 1782 (Eclipse 1764)—Dau. of Highflyer 1774 (Herod 1758)—Dau. of Alfred 1770 (Matchem 1748)—Dau. 1770 of Engineer 1756 (Sampson 1745).
Idalia, ch. m., 1815.	Peruvian 1806 (Sir Peter T. 1784)—Musidora 1804 by Meteor 1783 (Eclipse 1764)—Maid-of-all-Work 1786 by Highflyer 1774 (Herod 1758)—Dau. 1771 of Syphon 1750 (Squirt 1732)—Dau. of Regulus 1739 (God. Ar. 1724).
Laurel, br. h., 1824.	Blacklock 1814 (Whitelock 1803)—Wagtail 1818 by Prime Minister 1810 (Sancho 1801)—Tranby*'s dam 1812 by Orville 1799 (Beningbrough 1791)—Miss Grimstone 1796 by Weazle 1776 (Herod 1758)—Dau. of Ancaster 1768 (Blank 1740).
Maid of Honor, br. m., 1829.	Champion 1812 (Selim 1802)—Etiquette 1820 by Orville 1799 (Beningbrough 1791)—Boadicea 1807 by Alexander 1782 (Eclipse 1764)—Brunette by Amaranthus 1766 (Old England 1745)—Mayfly 1771 by Matchem 1748 (Cade 1734).
Catton, b. h., 1809.	Golumpus 1802 (Gohanna 1790)—Lucy Gray 1804 by Timothy 1794 (Delpini 1781)—Lucy 1789 by Florizel 1768 (Herod 1758)—Frenzy 1774 by Eclipse 1764 (Marske 1750)—Dau. of Engineer 1756 (Sampson 1745).
Emma, ch. m., 1824.	Whisker 1812 (Waxy 1790)—Gibside Fairy 1811 by Hermes 1790 (Mercury 1778)—Vicissitude 1800 by Pipator 1786 (Imperator 1776)—Beatrice 1791 by Sir Peter T. 1784 (Highflyer 1774)—Pyrrha 1771 by Matchem 1748 (Cade 1734).
Sir Charles, ch. h., 1816.	Sir Archy 1805 (*Diomed 1777)—Dau. of *Citizen 1785 (Pacolet 1763)—Dau. of Commutation (Symmes' Wildair 1767)—Dau. of *Dare Devil 1787 (Magnet 1770)—Sally Shark by *Shark 1771 (Marske 1750).
Mischief, b. m., 1826.	Virginian 1815 (Sir Archy 1805)—Dau. of *Bedford 1792 (Dungannon 1780)—Dau. of Bellair 1786 (*Medley 1776)—Dau. of *Shark 1771 (Marske 1750)—Dau. of Symmes' Wildair 1767 (*Fearnought 1755).
Sir Archy, b. h., 1805.	*Diomed 1777 (Florizel 1768)—*Castianira 1796 by Rockingham 1781 (Highflyer 1774)—Tabitha 1782 by Trentham 1766 (T. Sweepstakes 1743)—Dau. of Bosphorus 1754 (Babraham 1740)—Dau. of Forester 1750 (Forester 1736).
Miss Chance, —. —.	*Chance 1797 (Lurcher 1789)—Dau. of *Eagle 1796 (Volunteer 1780)—Maria 1805 by Bay Yankee (President 1788)—Green's Celer mare by Meade's Celer 1774 (*Janus 1746)—Dau. of Partner 1755 (*Traveller 1747).
Contention, ch. h., 1815.	Sir Archy 1805 (*Diomed 1777)—Dau. 1798 of *Dare Devil 1787 (Magnet 1770)—Dau. 1792 of Symmes' Wildair 1767 (*Fearnought 1755)—Piccadilly by Y. Fearnought 1777 (*Fearnought 1755)—Dau. of Godolphin 1770 (*Fearnought 1755).
Daughter of, —. —.	*Speculator 1795 (Dragon* 1787)—Pompadour by *Valiant 1756 (Dormouse 1738)—*Jenny Cameron by Cuddy 1727 (Fox 1714)—Miss Belvoir by Grey Grantham (Brownlow Turk)—Dau. of Paget Turk.
Touchstone, br. h., 1831.	Camel 1822 (Whalebone 1807)—Banter 1826 by Master Henry 1815 (Orville 1799)—Boadicea 1807 by Alexander 1782 (Eclipse 1764)—Brunette by Amaranthus 1766 (Old England 1741)—Mayfly 1771 by Matchem 1748 (Cade 1734).
Vulture, ch. m., 1833.	Langar 1817 (Selim 1802)—Kite 1821 by Bustard 1813 (Castrel 1801)—Olympia 1815 by Sir Oliver 1800 (Sir Peter T. 1784)—Scotilla 1795 by Anvil 1777 (Herod 1758)—Scota 1783 by Eclipse 1764 (Marske 1750).
Bay Middleton, b. h., 1833.	Sultan 1816 (Selim 1802)—Cobweb 1821 by Phantom 1808 (Walton 1799)—Filagree 1815 by Soothsayer 1808 (Sorcerer 1796)—Web 1808 by Waxy 1790 (Pot-8-o-'s 1773)—Penelope 1798 by Trumpator 1782 (Conductor 1767).
Flycatcher, ch. m., 1831.	Godolphin 1818 (Partisan 1811)—Sist. 1820 to Cobweb by Phantom 1808 (Walton 1799)—Filagree 1815 by Soothsayer 1808 (Sorcerer 1796)—Web 1808 by Waxy 1790 (Pot-8-o-'s 1773)—Penelope 1798 by Trumpator 1782 (Conductor 1767).
*Trustee, ch. h., 1829.	Catton 1809 (Golumpus 1802)—Emma 1824 by Whisker 1812 (Waxy 1790)—Gibside Fairy 1811 by Hermes 1790 (Mercury 1778)—Vicissitude 1800 by Pipator 1786 (Imperator 1776)—Beatrice 1791 by Sir Peter T. 1784 (Highflyer 1774).
Rosalie Somers, b. m., 1831.	Sir Charles 1816 (Sir Archy 1805)—Mischief 1826 by Virginian 1815 (Sir Archy 1805)—Dau. of *Bedford 1792 (Dungannon 1780)—Dau. of Bellair 1786 (*Medley 1776)—Dau. of *Shark 1771 (Marske 1750).
Garrison's Zingance, b. h., 1828.	Sir Archy 1805 (*Diomed 1777)—Miss Chance by *Chance 1797 (Lurcher 1789)—Dau. of *Eagle 1796 (Volunteer 1780)—Maria 1805 by Bay Yankee (President 1788)—Green's Celer mare by Meade's Celer 1774 (*Janus 1746).
Stella, b. m., —.	Contention 1815 (Sir Archy 1805)—Dau. of *Speculator 1795 (Dragon* 1787)—Pompadour by *Valiant 1756 (Dormouse 1738)—*Jenny Cameron by Cuddy 1727 (Fox 1714)—Miss Belvoir by Grey Grantham (Brownlow Turk).
Boston, ch. h., 1833.	Timoleon 1813 (Sir Archy 1805)—Dau. 1814 of Ball's Florizel 1801 (*Diomed 1777)—Dau. 1799 of *Alderman 1787 (Pot-8-o-'s 1773)—Dau. of *Clockfast 1774 (Gimcrack 1760)—Dau. of Symmes' Wildair 1767 (*Fearnought 1755).
Alice Carneal, b. m., 1836.	*Sarpedon 1828 (Emilius 1820)—Rowena 1826 by Sumpter 1818 (Sir Archy 1805)—Lady Grey 1817 by Robin Grey 1805 (*Royalist 1790)—Maria 1802 by Melzar 1791 (*Medley 1776)—Dau. of *Highflyer 1784 (Highflyer 1774).
*Glencoe, ch. h., 1831.	Sultan 1816 (Selim 1802)—Trampoline 1825 by Tramp 1810 (Dick Andrews 1797)—Web 1808 by Waxy 1790 (Pot-8-o-'s 1773)—Penelope 1798 by Trumpator 1782 (Conductor 1767)—Prunella 1788 by Highflyer 1774 (Herod 1758).
*Galopade, gr. m., 1828.	Catton 1809 (Golumpus 1802)—Camillina 1822 by Camillus 1803 (Hambletonian 1792)—Dau. 1818 of Smolensko 1810 (Sorcerer 1796)—Miss Cannon 1811 by Orville 1799 (Beningbrough 1791)—Dau. 1793 of Weathercock 1785 (Highflyer 1774).
Ir. Birdcatcher, ch. h., 1833.	Sir Hercules 1826 (Whalebone 1807)—Guiccioli 1823 by Bob Booty 1804 (Chanticleer 1787)—Flight 1809 by Ir. Escape 1802 (Commodore 1793)—Y. Heroine by Bagot 1780 (Herod 1758)—Heroine 1775 by Hero 1753 (Cade 1734).
Maltese, b. m., 1845.	Hetman Platoff 1836 (Brutandorf 1821)—Waterwitch 1833 by Sir Hercules 1826 (Whalebone 1807)—Mary Anne 1820 by Waxy Pope 1806 (Waxy 1790)—Witch 1811 by Sorcerer 1796 (Trumpator 1782)—Miss Buckle 1804 by Precipitate* 1787 (Mercury 1778).
Melbourne, br. h., 1834.	Humphrey Clinker 1822 (Comus 1809)—Dau. 1825 of Cervantes 1806 (Don Quixote 1784)—Dau. 1818 of Golumpus 1802 (Gohanna 1790)—Dau. 1810 of Paynator 1791 (Trumpator 1782)—Circle 1802 by St. George 1789 (Highflyer 1774).
Clarkia, b. m., 1840.	Muley Moloch 1830 (Muley 1810)—Sist. ch. m. 1832 to Righton by Palmerin 1816 (Amadis 1807)—Oceana ch. m. 1816 by Cerberus 1802 (Gohanna 1790)—Dr. Syntax's dam by Beningbrough 1791 (King Fergus 1775)—Jenny Mole b. m. 1787 by Carbuncle 1772 (Babraham Blank 1758)—Dau. b. m. 1773 by Prince T'Quassaw 1751 (Snip 1736)—Sultana 1759 by Regulus 1739 (God. Ar. 1724)—Dau. of Partner 1718 (Jigg)—Dau. of Gallant's Smiling Tom 1724 (Conyer Ar.)—Traveller's dam b. m. by Almanzor 1713 (Darley Ar. 1699)—Sist. to Bay Bolton by Grey Hautboy (Hautboy)—Dau. of Makeless (Oglethorpe Ar.)—Dau. of Brimmer (D'Arcy Yellow Turk)—Dau. of D'Arcy Diamond (Helmsley Turk)—Sist. to Merlin by Bustler (Helmsley Turk).

IMPORTED LAUREATE.

Imported and owned by Messrs. Stephen Sanford & Sons, Hurricane Stud, Amsterdam, N. Y. Yearlings sold annually.

LAUREATE, brown horse, foaled 1879, bred by Mr. Gosden, by Rosicrucian, son of Beadsman and Madame Eglentine by Cowl, son of Bay Middleton, dam Laura, the dam of Petrarch, winner of the 2,000 Guineas, St. Leger and Ascot Gold Cup, Protomartyr, Fraulein (winner of the Doncaster Cup), Lemnos, imp Rotherhill, etc., by Orlando, son of Touchstone, second dam Torment, the dam of Nana Sahib, Adrasta, Tormentor (Oaks winner), etc., by Alarm, son of Venison; third dam by imp. Glencoe, etc.

Rosicrucian, the sire of Laureate, won the Criterion and Troy Stakes, and is the sire of Hauteur, winner of the 1,000 Guineas, Geheimniss, winner of the Oaks, imp. Black Dean and Beauclerc, and is full brother to The Palmer. Orlando, the sire of Laura, won the Derby and was the sire of Little Lady (dam of Camballo), Lady Mary (dam of Gang Forward), and of the dam of Gen. Peel, all winners of the 2,000 Guineas, also of Doralice (dam of Speculum, winner of the Goodwood Cup).

Laureate was a high class race-horse. He won the Craven Stakes, 1 mile, 117 lbs., beating Marden, the Twenty-third Biennial at Stockbridge, 1½ miles, 124 lbs., and the Singleton Stakes at Goodwood, 1 mile, 117 lbs., and was second in the Chesterfield Cup, 1¼ miles, won by Vibration, second in the Scarborough Stakes, 1 mile, won by Alban, giving him 12 lbs. He is the sire of the winners Poet Laureate, Lullaby, Queen Laura, Stokesley, Frank Patros, Lagunos and Guinevere. Bonnie Scotland was a success in America, and the descendants of imp. Peggy from the same family are noted in this country and in England. Laureate traces in direct female line to the Childers mare, to which traces Bustard (by Buzzard), Hampton (by Lord Clifden), Blair Athol and all the Queen Mary family, Pretender, Sir Bevys, Tristan, Carlton, Crafton, Bonnie Scotland, Bernice and Bella Donna in the Nursery Stud, and all the Peggy family in America.

LAUREATE.

BROWN HORSE, FOALED 1879; BRED BY MR. GOSDEN, ENGLAND.

ROSICRUCIAN, b. h., 1865.
Beadsman, br. h., 1855.
Weatherbit, br. h., 1842.
Sheet Anchor, br. h., 1832.
Miss Letty, b. m., 1834.
Mendicant, br. m., 1843.
Touchstone, br. h., 1831.
Lady Moore Carew, b. m., 1830.
Madame Eglentine, b. m., 1857.
Cowl, b. h., 1842.
Bay Middleton, b. h., 1833.
Crucifix, b. m., 1837.
Diversion, ch. m., 1838.
Defence, b. h., 1824.
Folly, ch. m., 1830.
LAURA, b. m., 1860.
Orlando, b. h., 1841.
Touchstone, br. h., 1831.
Camel, br. h., 1822.
Banter, br. m., 1826.
Vulture, ch. m., 1833.
Langar, ch. h., 1817.
Kite, b. m., 1821.
Torment, b. m., 1850.
Alarm, b. h., 1842.
Venison, br. h., 1833.
Southdown, b. m., 1836.
Daughter of, br. m., 1837.
Glencoe*, ch. h., 1831.
Alea, br. m., 1828.

Lottery, br. h., 1820.	Tramp 1810 (Dick Andrews 1797)—Mandane 1800 by Pot-8-o-'s 1773 (Eclipse 1764) Y. Camilla 1787 by Woodpecker 1773 (Herod 1758) Camilla 1778 by Trentham 1766 (T. Sweepstakes 1743) Coquette 1765 by Compton Barb.
Morgiana, blk. m., 1820.	Muley 1810 (Orville 1799) Miss Stephenson 1814 [Scud or] Sorcerer 1796 (Trumpator 1782) Dau. 1796 of Precipitate* 1787 (Mercury 1778) Dau. 1787 of Woodpecker 1773 (Herod 1758)—Sist. 1773 to Juniper by Snap 1750 (Snip 1736).
Priam,* b. h., 1827.	Emilius 1820 (Orville 1799) Cressida 1807 by Whiskey 1789 (Saltram* 1780) Y. Giantess 1790 by Diomed* 1777 (Florizel 1768) Giantess 1769 by Matchem 1748 (Cade 1734) - Molly Long Legs 1753 by Babraham 1740 (God. Ar. 1724).
Daughter of, br. m., 1815.	Orville 1799 (Beningbrough 1791) Dau. 1800 of Buzzard* 1787 (Woodpecker 1773) - Hornpipe 1793 by Trumpator 1782 (Conductor 1767) Luna 1779 by Herod 1758 (Tartar 1743) Proserpine 1765 by Marske 1750 (Squirt 1732).
Camel, br. h., 1822.	Whalebone 1807 (Waxy 1790) Dau. 1812 of Selim 1802 (Buzzard* 1787) Maiden 1801 by Sir Peter T. 1784 (Highflyer 1774) Dau. 1788 of Phenomenon 1780 (Herod 1758) Matron 1782 by Florizel 1768 (Herod 1758).
Banter, br. m., 1826.	Master Henry 1815 (Orville 1799) Boadicea 1807 by Alexander 1782 (Eclipse 1764)- Brunette by Amaranthus 1766 (Old England 1741) Mayfly 1771 by Matchem 1748 (Cade 1734) Dau. 1755 of Starling 1738 (Starling 1727).
Tramp, b. h., 1810.	Dick Andrews 1797 (Joe Andrews 1778) Dau. 1803 of Gohanna 1790 (Mercury 1778) - Fraxinella 1793 by Trentham 1766 (T. Sweepstakes 1743) Dau. 1785 of Woodpecker 1773 (Herod 1758)- Everlasting 1775 by Eclipse 1764 (Marske 1750).
Kite, b. m., 1821.	Bustard 1813 (Castrel 1801) Olympia 1815 by Sir Oliver 1800 (Sir Peter T. 1784) Scotilla 1795 by Anvil 1777 (Herod 1758) Scota 1783 by Eclipse 1764 (Marske 1750) Harmony 1775 by Herod 1758 (Tartar 1743).
Sultan, b. h., 1816.	Selim 1802 (Buzzard* 1787) Bacchante 1809 by W's Ditto 1800 (Sir Peter T. 1784) Dau. 1791 of Mercury 1778 (Eclipse 1764) - Dau. 1776 of Herod 1758 (Tartar 1743) Folly 1771 by Marske 1750 (Squirt 1732).
Cobweb, b. m., 1821.	Phantom 1808 (Walton 1799) - Filagree 1815 by Soothsayer 1808 (Sorcerer 1796) Web 1808 by Waxy 1790 (Pot-8-o-'s 1773) Penelope 1798 by Trumpator 1782 (Conductor 1767)—Prunella 1788 by Highflyer 1774 (Herod 1758).
Priam,* b. h., 1827.	Emilius 1820 (Orville 1799) Cressida 1807 by Whiskey 1789 (Saltram* 1780)—Y. Giantess 1790 by Diomed* 1777 (Florizel 1768) - Giantess 1769 by Matchem 1748 (Cade 1734)—Molly Long Legs 1753 by Babraham 1740 (God. Ar. 1724).
Octaviana, b. m., 1815.	Octavian 1807 (Stripling 1795)—Dau. 1807 of Shuttle 1793 (Y. Marske 1771)—Zara 1801 by Delpini 1781 (Highflyer 1774)—Flora 1789 by King Fergus 1775 (Eclipse 1764)—Atalanta 1769 by Matchem 1748 (Cade 1734).
Whalebone, br. h., 1807.	Waxy 1790 (Pot-8-o-'s 1773) Penelope 1798 by Trumpator 1782 (Conductor 1767)- Prunella 1788 by Highflyer 1774 (Herod 1758)—Promise 1768 by Snap 1750 (Snip 1736)—Julia 1756 by Blank 1740 (God. Ar. 1724).
Defiance, ch. m., 1816.	Rubens 1805 (Buzzard* 1787)—Little Folly 1806 by Highland Fling 1798 (Spadille 1784)—Harriet 1799 by Volunteer 1780 (Eclipse 1764)—Dau. 1785 of Alfred 1770 (Matchem 1748) Magnolia 1771 by Marske 1750 (Squirt 1732).
Middleton, ch. h., 1822.	Phantom 1808 (Walton 1799) Web 1808 by Waxy 1790 (Pot-8-o-'s 1773) - Penelope 1798 by Trumpator 1782 (Conductor 1767) Prunella 1788 by Highflyer 1774 (Herod 1758)—Promise 1768 by Snap 1750 (Snip 1736).
Little Folly, b. m., 1806.	Highland Fling 1798 (Spadille 1784)—Harriet 1799 by Volunteer 1780 (Eclipse 1764)—Dau. 1785 of Alfred 1770 (Matchem 1748) Magnolia 1771 by Marske 1750 (Squirt 1732)—Dau. 1755 by Babraham 1740 (God. Ar. 1724).
Whalebone, br. h., 1807.	Waxy 1790 (Pot-8-o-'s 1773)—Penelope 1798 by Trumpator 1782 (Conductor 1767)—Prunella 1788 by Highflyer 1774 (Herod 1758)—Promise 1768 by Snap 1750 (Snip 1736)—Julia 1756 by Blank 1740 (God. Ar. 1724).
Daughter of, b. m., 1812.	Selim 1802 (Buzzard* 1787)—Maiden 1801 by Sir Peter T. 1784 (Highflyer 1774)—Dau. 1788 of Phenomenon 1780 (Herod 1758)—Matron 1782 by Florizel 1768 (Herod 1758)—Maiden 1770 by Matchem 1748 (Cade 1734).
Master Henry, b. h., 1815.	Orville 1799 (Beningbrough 1791)—Miss Sophia 1805 by Stamford 1794 (Sir Peter T. 1784)—Sophia 1798 by Buzzard* 1787 (Woodpecker 1773)—Huncamunca 1787 by Highflyer 1774 (Herod 1758)—Cypher 1772 by Squirrel 1754 (Traveller 1735).
Boadicea, b. m., 1807.	Alexander 1782 (Eclipse 1764) Brunette by Amaranthus 1766 (Old England 1741)—Mayfly 1771 by Matchem 1748 (Cade 1734) - Dau. 1755 of Anc. Starling 1728 (Starling 1727)—Look-at-me-Lads 1731 by Brist. Grasshopper (Byerly T.).
Selim, ch. h., 1802.	Buzzard* 1787 (Woodpecker 1773) - Dau. of Alexander 1782 (Eclipse 1764) Dau. of Highflyer 1774 (Herod 1758)—Dau. of Alfred 1770 (Matchem 1748)—Dau. 1770 of Engineer 1756 (Sampson 1745).
Daughter of, ch. m., 1808.	Walton 1799 (Sir Peter T. 1784) Y. Giantess 1790 by Diomed* 1777 (Florizel 1767)—Giantess 1769 by Matchem 1748 (Cade 1734) - Molly Long Legs 1753 by Babraham 1740 (God. Ar. 1724) - Dau. of Cole's Foxhunter 1727 (Brisk 1711).
Bustard, b. h., 1813.	Castrel 1801 (Buzzard* 1787)—Miss Hap 1806 by Shuttle 1793 (Y. Marske 1771)—Dau. 1798 of Sir Peter T. 1784 (Highflyer 1774)—Miss Hervey 1775 by Eclipse 1764 (Marske 1750)—Clio 1760 by Y. Cade 1747 (Cade 1734).
Olympia, b. m., 1815.	Sir Oliver 1800 (Sir Peter T. 1784)—Scotilla 1795 by Anvil 1777 (Herod 1758)—Scota 1783 by Eclipse 1764 (Marske 1750)—Harmony 1775 by Herod 1758 (Tartar 1743) Rutilia 1769 by Blank 1740 (God. Ar. 1724).
Partisan, b. h., 1811.	Walton 1799 (Sir Peter 1784)—Parasol 1800 by Pot-8-o's 1773 (Eclipse 1764)—Prunella 1788 by Highflyer 1774 (Herod 1758)—Promise 1768 by Snap 1750 (Snip 1736)—Julia 1756 by Blank 1740 (God. Ar. 1724).
Fawn, br. m., 1823.	Smolensko 1810 (Sorcerer 1796)—Jerboa 1803 by Gohanna 1790 (Mercury 1778) Camilla 1778 by Trentham 1766 (T. Sweepstakes 1743)—Coquette 1764 by Compton Barb—Sist. 1743 to Regulus by God. Ar. 1724.
Defence, b. h., 1824.	Whalebone 1807 (Waxy 1790)—Defiance 1816 by Rubens 1805 (Buzzard* 1787)—Little Folly 1806 by Highland Fling 1798 (Spadille 1784)—Harriet 1799 by Volunteer 1780 (Eclipse 1764) Dau. 1785 of Alfred 1770 (Matchem 1748).
Feltona, b. m., 1819.	X. Y. Z. 1808 (Haphazard 1797)—Janetta 1803 by Beningbrough 1791 (King Fergus 1775)—Dau. 1797 of Drone 1777 (Herod 1758)—Contessina 1787 by Y. Marske 1771 (Marske 1750) - Tuberose 1772 by Herod 1758 (Tartar 1743).
Sultan, b. h., 1816.	Selim 1802 (Buzzard* 1787)—Bacchante 1809 by W's Ditto 1800 (Sir Peter T. 1784)—Dau. 1791 of Mercury 1778 (Eclipse 1764)—Dau. 1776 of Herod 1758 (Tartar 1743)—Folly 1771 by Marske 1750 (Squirt 1732).
Trampoline, ch. m., 1825.	Tramp 1810 (Dick Andrews 1797)—Web 1808 by Waxy 1790 (Pot-8-o-'s 1773) Penelope 1798 by Trumpator 1782 (Conductor 1767) - Prunella 1788 by Highflyer 1774 (Herod 1758)—Promise 1768 by Snap 1750 (Snip 1736).
Whalebone, br. h., 1807.	Waxy 1790 (Pot-8-o-'s 1773)—Penelope 1798 by Trumpator 1782 (Conductor 1767)—Prunella 1788 by Highflyer 1774 (Herod 1758)—Promise 1768 by Snap 1750 (Snip 1736)—Julia 1756 by Blank 1740 (God. Ar. 1724).
Hazardess, b. m., 1818.	Haphazard 1797 (Sir Peter T. 1784)—Dau. br. m. 1800 of Orville 1799 (Beningbrough 1791)—Spinetta b. m. 1792 by Trumpator 1782 (Conductor 1767) Peggy b. m. 1778 by Herod 1758 (Tartar 1743)—Dau. of Snap 1750 (Snip 1736)—Dau. of Gower Stallion 1740 (God. Ar. 1724)—Dau. of Childers.

LEONATUS.

In the Runnymede Stud, Messrs. Clay & Woodford, proprietors, Paris, Ky. Yearlings sold annually in May.

LEONATUS, bay horse, foaled 1880, bred by J. Henry Miller, Lexington, Ky., by Longfellow, son of imp. Leamington, dam Semper Felix by imp. Phaeton, son of King Tom by Harkaway, grandam Crucifix by Lexington, out of Lightsome by imp. Glencoe. Leonatus is descended from the most celebrated racing and producing families in the Stud Book. Semper Felix, his dam, is the dam of the winners Warder, Leopold and Felicia II. and of Semper Vive and Latonia, the last named the dam of Sena and Lew Weir, while Semper Vive is the dam of Fordham, Folsom, Portchester and Semper Idem, the latter the dam of the great Longstreet and Longford. Crucifix, the second dam of Leonatus, is one of Lexington's best daughters. Besides Semper Felix she produced the winners Fairplay (a very good horse), St. Albans, the stake winners Quito and St. Augustine, and Blanche J., a producer as well as winner, being the dam of Sea Shell and Blantyre, the former the dam of Long Shore and Toss Up. Maria D., another daughter of Crucifix, is the dam of Mark S. and Ed. Bell. Without going back to Lightsome and Levity, it will be seen that it is a family of the grandest excellence.

Leonatus was a superior race-horse, started only once as two-year old; ran second in Maiden Stakes, ¾ of a mile, 105 lbs., in 1.22; at three years old won all his stakes, defeating the best three-year olds of his year, including Drake Carter; won the Blue Ribbon, 1½ miles, 105 lbs., in 2.38½; won the Kentucky Derby, 1½ miles, 105 lbs., in 2.43; won the Tobacco Stakes, mile heats, 110 lbs., in 1.46¼, 1.47¼; won the Woodburn Stakes, 1¼ miles, 110 lbs., in 2.00¼; won the Hindoo Stakes, 1½ miles, 110 lbs., in 2.40¼; won the Ripple Stakes, 1½ miles, muddy track, 110 lbs., in 2.30½; won the Himyar Stakes, 1¾ miles, 115 lbs., in 2.27¼; won the Dearborn Stakes, 1¾ miles, 115 lbs., in 3.12¼; won the Green Stakes, 1¼ miles, 115 lbs., in 1.58; won the Illinois Derby, 1½ miles, 115 lbs., in 2.51.

Leonatus is a success in the stud upon all classes of mares. He is the sire, among others, of Lord Peyton, a winner of seven races as a two-year old, National, Lemoine II., Birthday, Lake Breeze, Dore, Ed. Eshelby, Ed. Leonard, Judge Hughes, Leontine, Boyle Rhodes, Go-as-You-Please, Miracle, Lagadere, LEONAWELL, Florence Slaughter, Leader II., Leo, Leo Eu, Leumbria, Perlid, Ned. Pendleton, Red Light, REX, Sam Morse, Unlucky, Davidson, Labrador, Later On, Little Prince, Serenader, Vexation, Ja Ja, Leta B., Ollie Glenn, Ethel S., Etruria, Jeannerette, Nerina, Sir Arthur, Leonata, Warsaw, etc. He is descended in direct female line through old Montagu mare from the Royal mare to which traces Conductor, O'Kelly's Eclipse, Filho-da-Puta, Voltaire, Sterling, Oxford, Springfield, Marsyas, Adventurer, Prince Charlie, Lexington, Vandal, Monarchist, Volturno, Luke Blackburn, Salvator, The Bard, Strathmore, Tom Martin, Grinstead, Duke of Montrose and other noted horses.

LEONATUS.

BAY HORSE, FOALED 1880; BRED BY J. HENRY MILLER, KENTUCKY.

Parents	Grandparents	Great-grandparents	4th generation	5th generation	Pedigree
LONGFELLOW, br. h., 1867.	*Leamington, br. h., 1853.	Faugh-a-Ballagh, br. h., 1841.	Sir Hercules, blk. h., 1826.	Whalebone, br. h., 1807.	Waxy 1790 (Pot-8-o's 1773)—Penelope 1798 by Trumpator 1782 (Conductor 1767)—Prunella 1788 by Highflyer 1774 (Herod 1758)—Promise 1768 by Snap 1750 (Snip 1736)—Julia 1756 by Blank 1740 (God. Ar. 1724).
				Peri, b. m., 1822.	Wanderer 1811 (Gohanna 1790)—Thalestris 1809 by Alexander 1782 (Eclipse 1764)—Rival 1800 by Sir Peter T. 1784 (Highflyer 1774)—Hornet 1790 by Drone 1777 (Herod 1758)—Manilla 1777 by Goldfinder 1764 (Snap 1750).
			Guiccioli, ch. m., 1823.	Bob Booty, ch. h., 1804.	Chanticleer 1787 (Woodpecker 1773)—Ierne 1790 by Bagot 1780 (Herod 1758)—Dau. of Gamahoe (Bustard 1741)—Patty by Tim 1752 (Squirt 1732)—Miss Patch 1757 by Justice 1735 (H. C. Litton Ar.).
				Flight, ch. m., 1809.	Ir. Escape 1802 (Commodore 1793)—Y. Heroine by Bagot 1780 (Herod 1758)—Heroine 1775 by Hero 1753 (Cade 1734)—Dau. 1766 of Snap 1750 (Snip 1736)—Sist. 1743 to Regulus by God. Ar. 1724.
		Daughter of, b. m., 1841.	Pantaloon, ch. h., 1824.	Castrel, ch. h., 1801.	Buzzard* 1787 (Woodpecker 1773)—Dau. of Alexander 1782 (Eclipse 1764)—Dau. of Highflyer 1774 (Herod 1758)—Dau. of Alfred 1770 (Matchem 1748)—Dau. 1770 of Engineer 1756 (Sampson 1745).
				Idalia, ch. m., 1815.	Peruvian 1806 (Sir Peter T. 1784)—Musidora 1804 by Meteor 1783 (Eclipse 1764)—Maid-of-all-Work 1786 by Highflyer 1774 (Herod 1758)—Dau. 1771 of Syphon 1750 (Squirt 1732)—Dau. of Regulus 1739 (God. Ar. 1724).
			Daphne, br. m., 1837.	Laurel, br. h., 1824.	Blacklock 1814 (Whitelock 1803)—Wagtail 1818 by Prime Minister 1810 (Sancho 1801)—Tranby*'s dam 1812 by Orville 1799 (Beningbrough 1791)—Miss Grimstone 1796 by Weazle 1776 (Herod 1758)—Dau. of Ancaster 1768 (Blank 1740).
				Maid of Honor, br. m., 1829.	Champion 1812 (Selim 1802)—Etiquette 1820 by Orville 1799 (Beningbrough 1791)—Boadicea 1807 by Alexander 1782 (Eclipse 1764)—Brunette by Amaranthus 1766 (Old England 1741)—Mayfly 1771 by Matchem 1748 (Cade 1734).
	Nantura, b. m., 1855.	Brawner's Eclipse, ch. h., 1829.	Am. Eclipse, ch. h., 1814.	Duroc, ch. h., 1806.	*Diomed 1777 (Florizel 1768)—Amanda 1800 by Grey Diomed 1786 (*Medley 1776)—Dau. of Va. Cade (Partner 1755)—Dau. of Hickman's Independence 1776 (*Fearnought 1755)—Dau. of *Badger (Bosphorus 1754).
				Miller's Damsel, gr. m., 1802.	*Messenger 1780 (Mambrino 1768)—*Dau. 1792 of Pot-8-o's 1773 (Eclipse 1764)—Dau. 1778 of Gimcrack 1760 (Cripple 1750)—Snapdragon 1759 by Snap 1750 (Snip 1736)—Dau. 1749 of Regulus 1739 (God. Ar. 1724).
			Daughter of, ch. m., 1834.	Henry, ch. h., 1819.	Sir Archy 1805 (*Diomed 1777)—Dau. of *Diomed 1777 (Florizel 1768)—Bellona by Bellair 1786 (*Medley 1776)—Indian Queen by Pilgrim 1774 (*Fearnought 1755)—Dau. of *Valiant 1756 (Dormouse 1738).
				Y. Romp, ch. m., 1815.	Duroc 1806 (*Diomed 1777)—Old Romp 1805 by *Messenger 1780 (Mambrino 1768)—*Dau. 1792 of Pot-8-o's 1773 (Eclipse 1764)—Dau. 1778 of Gimcrack 1760 (Cripple 1750)—Snapdragon 1759 by Snap 1750 (Snip 1736).
		Quiz, b. m., 1836.	Bertrand, b. h., 1821.	Sir Archy, b. h., 1805.	*Diomed 1777 (Florizel 1768)—*Castianira 1796 by Rockingham 1781 (Highflyer 1774)—Tabitha 1782 by Trentham 1766 (T. Sweepstakes 1743)—Dau. of Bosphorus 1754 (Babraham 1740)—Sist. to Grecian Princess by Forester 1750 (Forester 1736).
				Eliza, ch. m., 1804.	*Bedford 1792 (Dungannon 1780)—*Mambrina 1785 by Mambrino 1768 (Engineer 1756)—Sist. to Sally by Blank 1740 (God. Ar. 1724)—Dau. of Ward 1731 (Crab 1722)—Dau. of Merlin (Bustler).
			Lady Fortune, ch. m.,	[Brimmer or] Blue Beard, ch. h., 1802.	*Stirling 1791 (Volunteer 1780)—Brilliant 1779 by *Matchem (Bosphorus 1754)—*Dau. of Brilliant 1750 (Crab 1722)—Dau. of Tartar 1743 (Partner 1718)—Dau. of Son of Flying Childers 1715.
				Woodpecker's dam, ch. m., 1809.	*Buzzard 1787 (Woodpecker 1773)—The Faun by Craig's Alfred (*Medley 1776)—Dau. of *Traveller 1747 (Partner 1718)—Dau. of *Whittington (Cullen Ar.)—Dau. of *Childers (Blaze 1733).
SEMPER FELIX, b. m., 1872.	*Phaeton, b. h., 1865.	King Tom, b. h., 1851.	Harkaway, ch. h., 1834.	Economist, b. h., 1825.	Whisker 1812 (Waxy 1790)—Floranthe 1818 by Octavian 1807 (Stripling 1795)—Caprice 1797 by Anvil 1777 (Herod 1758)—Madcap 1774 by Eclipse 1764 (Marske 1750)—Sappho 1763 by Blank 1740 (God. Ar. 1724).
				Fanny Dawson, ch. m., 1823.	Nabocklish 1811 (Rugantino 1803)—Miss Tooley 1808 by Teddy-the-Grinder 1798 (Asparagus 1787)—Lady Jane 1796 by Sir Peter T. 1784 (Highflyer 1774)—Paulina 1778 by Florizel 1768 (Herod 1758)—Captive 1771 by Matchem 1748 (Cade 1734).
			Pocahontas, b. m., 1837.	Glencoe,* ch. h., 1831.	Sultan 1816 (Selim 1802)—Trampoline 1825 by Tramp 1810 (Dick Andrews 1797)—Web 1808 by Waxy 1790 (Pot-8-o's 1773)—Penelope 1798 by Trumpator 1782 (Conductor 1767)—Prunella 1788 by Highflyer 1774 (Herod 1758).
				Marpessa, b. m., 1830.	Muley 1810 (Orville 1799)—Clare 1824 by Marmion 1806 (Whiskey 1789)—Harpalice 1814 by Gohanna 1790 (Mercury 1778)—Amazon 1799 by Driver 1783 (Trentham 1766)—Fractious 1792 by Mercury 1778 (Eclipse 1764).
		Merry Sunshine, ch. m., 1855.	Storm, b. h., 1848.	Touchstone, br. h., 1831.	Camel 1822 (Whalebone 1807)—Banter 1826 by Master Henry 1815 (Orville 1799)—Boadicea 1807 by Alexander 1782 (Eclipse 1764)—Brunette by Amaranthus 1766 (Old England 1741)—Mayfly 1771 by Matchem 1748 (Cade 1734).
				Ghuznee, b. m., 1838.	Pantaloon 1824 (Castrel 1801)—Languish 1830 by Cain 1822 (Paulowitz 1813)—Lydia 1822 by Poulton 1805 (Sir Peter T. 1784)—Variety 1808 by Hyacinthus 1797 (Coriander 1786)—Dau. 1790 of Weazle 1776 (Herod 1758).
			Daughter of, b. m., 1849.	Falstaff, br. h., 1842.	Touchstone 1831 (Camel 1822)—Decoy 1830 by Filho-da-Puta 1812 (Haphazard 1797)—Finesse 1815 by Peruvian 1806 (Sir Peter T. 1784)—Violante 1802 by John Bull 1789 (Fortitude 1777)—Dau. 1788 of Highflyer 1774 (Herod 1758).
				Sister to Pompey, b. m., 1843.	Emilius 1820 (Orville 1799)—Variation 1827 by Bustard 1813 (Castrel 1801)—Johanna Southcote 1811 by Beningbrough 1791 (King Fergus 1775)—Lavinia 1802 by Pipator 1786 (Imperator 1776)—Dick Andrews' dam 1790 by Highflyer 1774 (Herod 1758).
	Crucifix, b. m., 1866.	Lexington, b. h., 1850.	Boston, ch. h., 1833.	Timoleon, ch. h., 1813.	Sir Archy 1805 (*Diomed 1777)—Dau. of *Saltram 1780 (Eclipse 1764)—Dau. of Symmes' Wildair 1767 (*Fearnought 1755)—Dau. of Tyler's Driver 1760 (*Othello)—Dau. of *Fallower 1761 (Blank 1740).
				Sister to Tuckahoe, ch. m., 1814.	Ball's Florizel 1801 (*Diomed 1777)—Dau. 1799 of *Alderman 1787 (Pot-8-o's 1773)—Dau. of *Clockfast 1774 (Gimcrack 1760)—Dau. of Symmes' Wildair 1767 (*Fearnought 1755)—Y. Kitty Fisher 1767 by *Fearnought 1755 (Regulus 1739).
			Alice Carneal, b. m., 1836.	*Sarpedon, br. h., 1828.	Emilius 1820 (Orville 1799)—Icaria 1824 by The Flyer 1814 (Vandyke Jr. 1808)—Parma 1813 by Dick Andrews 1797 (Joe Andrews 1778)—May 1804 by Beningbrough 1791 (King Fergus 1775)—Primrose 1787 by Mambrino 1768 (Engineer 1756).
				Rowena, ch. m., 1826.	Sumpter 1818 (Sir Archy 1805)—Lady Grey 1817 by Robin Grey 1805 (*Royalist 1790)—Maria 1802 by Melzar 1791 (*Medley 1776)—Dau. of *Highflyer 1784 (Highflyer 1774)—Dau. of *Fearnought 1755 (Regulus 1739).
		Lightsome, ch. m., 1855.	*Glencoe, ch. h., 1831.	Sultan, b. h., 1816.	Selim 1802 (Buzzard* 1787)—Bacchante 1809 by W's Ditto 1800 (Sir Peter T. 1784)—Dau. 1791 of Mercury 1778 (Eclipse 1764)—Dau. 1776 of Herod 1758 (Tartar 1743)—Folly 1771 by Marske 1750 (Squirt 1732).
				Trampoline, ch. m., 1825.	Tramp 1810 (Dick Andrews 1797)—Web 1808 by Waxy 1790 (Pot-8-o's 1773)—Penelope 1798 by Trumpator 1782 (Conductor 1767)—Prunella 1788 by Highflyer 1774 (Herod 1758)—Promise 1768 by Snap 1750 (Snip 1736).
			Levity, b. m., 1845.	*Trustee, ch. h., 1829.	Catton 1809 (Golumpus 1802)—Emma 1824 by Whisker 1812 (Waxy 1790)—Gibside Fairy 1811 by Hermes 1790 (Mercury 1778)—Vicissitude 1800 by Pipator 1786 (Imperator 1776)—Beatrice 1791 by Sir Peter T. 1784 (Highflyer 1774).
				Vandal's dam, b. m., 1837.	*Tranby 1826 (Blacklock 1814)—Lucilla b. m. by Trumpator 1823 (Sir Solomon 1805)—Lucy b. m. 1821 by Orphan 1810 (Ball's Florizel 1801)—Lady Grey b. m. 1817 by Robin Grey 1805 (*Royalist 1790)—Maria 1802 by Melzar 1791 (*Medley 1776)—Dau. of *Highflyer 1784 (Highflyer 1774)—Dau. of *Fearnought 1755 (Regulus 1739)—Dau. of Ariel 1756 (*Traveller 1747)—Dau. of *Jack of Diamonds (Cullen Ar. 1745)—Duchess by Cullen Ar. 1745—Grosewood's Lady Thigh ch. m. 1731 by Partner 1718 (Jigg)—Sist. ch. m. 1723 to Sampson by Greyhound (Chillaby)—Sophonisba's dam by Curwen Bay Barb—Dau. of D'Arcy Chestnut Ar.—Dau. of Whiteshirt—Old Montagu mare by D'Arcy Old Montagu—Dau. of Hautboy (D'Arcy White Turk)—Dau. of Brimmer (D'Arcy Yellow Turk)—Royal mare.

LINDEN.

WINNER OF THE NEWARK AND CITIZENS' STAKES, SHREWSBURY AND GRAND NATIONAL HANDICAPS.

The property of James Galway, Preakness Stud, located in the Stonewall Stud, Breckenridge Viley, Midway, Ky. Terms annually.

LINDEN, bay horse, foaled 1883, bred by S. P. Lancaster, Ky., by Longfellow, son of imp. Leamington, dam Linda Lewis by Oliver, son of Wagner, grandam Sally Lewis, dam of John Morgan, Hunter's Lexington, Lotta (Glenmore's dam), Susan Beane (dam of Sensation, Stratford, Onondaga and Susquehanna, dam of Potomac) by imp. Glencoe. My Lady, the full sister to Sally Lewis, was the dam of Blandina by Lexington, she the dam of Calash, which produced Elite, Eolian and Ulsie, dam of Bandusia.

Linden did not race at two years old. At three years old ran second to Bandala in a Free Handicap, $1\frac{1}{16}$ miles, in 1.55; ran third to Inspector B. and The Bard, in the Belmont Stakes, $1\frac{1}{2}$ miles, in 2.41; was third to Dew Drop and Quito in Stockton Stakes, $1\frac{1}{4}$ miles, in $2.13\frac{3}{4}$; was unplaced in Stevens Stakes, $1\frac{5}{8}$ miles, won by Dew Drop in $2.54\frac{1}{2}$; was third to Charity and The Bard in the Raritan Stakes $1\frac{1}{4}$ miles, in 2.14, Pontico and Winifred behind him; won the Newark Stakes, 1 mile, 105 lbs., in $1.44\frac{1}{2}$, with nine others behind him; was unplaced in the Omnibus Stakes, $1\frac{1}{2}$ miles won by The Bard in 2.39; ran second to Elkwood in Kenner Stakes, 2 miles, in $3.34\frac{1}{2}$, three others behind him; was third to The Bard and Ferona in September Stakes, $1\frac{3}{4}$ miles, in 3.05; was unplaced in the Bridge Handicap, $1\frac{1}{2}$ miles, won by Jim Gray in $2.39\frac{1}{2}$; and in Sheepshead Bay Stakes, $1\frac{1}{4}$ miles, won by Blue Wing in 2.08. At four years old won Free Handicap Sweepstakes, $1\frac{3}{16}$ miles, 110 lbs., in $2.13\frac{1}{2}$, beating Phil Lee and Enigma; was unplaced in the Suburban, Bay Ridge, Fourth of July and Long Branch Handicaps; won Free Handicap Sweepstakes, $1\frac{3}{8}$ miles, in 2.26, beating Housatonic and nine others; won Shrewsbury Handicap, $1\frac{1}{2}$ miles, in $2.39\frac{1}{2}$, beating Grimaldi and seven others: was unplaced in the Midsummer Handicap, won by Dry Monopole; was second to Dry Monopole in Navesink Handicap, $1\frac{1}{2}$ miles, in $2.42\frac{1}{2}$, beating Eurus; won Free Handicap Sweepstakes, 110 lbs., $1\frac{3}{8}$ miles, in $2.25\frac{1}{2}$, with four others behind him; won Handicap Sweepstakes, $1\frac{1}{2}$ miles, in 2.44, 110 lbs., and another, $1\frac{3}{8}$ miles, 115 lbs., in $2.26\frac{1}{4}$; won Grand National Handicap, $1\frac{1}{2}$ miles, 117 lbs., in $2.42\frac{3}{4}$, Volante second, Eurus third, Laggard, etc.; won Free Handicap, $1\frac{1}{2}$ miles, 120 lbs., in $2.39\frac{1}{2}$, Volante second, Ben Ali and others in the field: won Citizen Stakes in a canter, 110 lbs., $1\frac{1}{2}$ miles, in $2.40\frac{1}{2}$, Firenzi, Dunbine and six others behind him. At five years old won Handicap, $1\frac{1}{8}$ miles, 122 lbs., in $1.56\frac{1}{2}$, six behind him: was unplaced in City Handicap: was second to his stable companion Belvidere in Fordham Handicap, $1\frac{1}{4}$ miles, 128 lbs., in $2.11\frac{1}{2}$, five behind him; was unplaced in Suburban, won by Elkwood: and ran third to Terra Cotta and Firenzi in Sheepshead Bay Handicap, $1\frac{1}{8}$ miles, in 1.53, five behind him.

Linden is from a grand racing and producing family, one that has made its mark upon the turf in England and America. He is young in the stud, but is the sire of the winner Lustre and other promising two-year olds. He is descended in female line from the same Royal mare to which traces Eclipse, Voltaire, Sterling, Oxford, Kingston, Scottish Chief, Lexington, Vandal, The Bard, Leonatus, Sensation, Salvator, Longstreet, Potomac and all the Duchess family.

LINDEN.

BROWN HORSE, FOALED 1883; BRED BY MR. S. I. LANCASTER, KENTUCKY.

LONGFELLOW, br. h., 1867.

Nantura, b. m., 1855. — *Leamington, br. h., 1853.

Quiz, b. m., 1836. — Brawner's Eclipse, ch. h., 1839. — Daughter of, b. m., 1841. — Faugh-a-Ballagh, br. h., 1841.

Lady Fortune, ch. m., —. — Bertrand, b. h., 1821. — Daughter of, ch. m., 1834. — Am Eclipse, ch. h., 1814. — Daphne, br. m., 1837. — Pantaloon, ch. h., 1824. — Guiccioli, ch. m., 1823. — Sir Hercules, blk. h., 1826.

LINDA LEWIS, ch. m., 1864.

Sally Lewis, ch. m., 1852. — Oliver, ch. h., 1844.

Motto, ch. m., 1839. — *Glencoe, ch. h., 1831. — Flight, ch. m., 1837. — Wagner, ch. h., 1831.

Lady Thompkins, ch. m., 1831. — *Barefoot, ch. h., 1820. — Trampoline, ch. m., 1825. — Sultan, b. h., 1816. — Charlotte Hamilton, ch. m., 1826. — *Leviathan, ch. h., 1823. — Maria West, b. m., 1827. — Sir Charles, ch. h., 1816.

Pot-8-o's 1773 Eclipse 1764—Marske 1750 Squirt 1732 Bartlett's Childers- Darley Arabian 1699.

Whalebone, br. h., 1807.	Waxy 1790 (Pot-8-o's 1773) Penelope 1798 by Trumpator 1782 (Conductor 1767) Prunella 1788 by Highflyer 1774 (Herod 1758) Promise 1768 by Snap 1750 Snip 1736) Julia 1756 by Blank 1740 (God. Ar. 1724).
Peri, b. m., 1822	Wanderer 1811 (Gohanna 1790) Thalestris 1809 by Alexander 1782 (Eclipse 1764) Rival (1800 by Sir Peter T. 1784 Highflyer 1774) Hornet 1790 by Drone 1777 (Herod 1758) Manilla 1777 by Goldfinder 1761 (Snap 1750).
Bob Booty, ch. h., 1804.	Chanticleer 1787 (Woodpecker 1773 Ierne 1790 by Bagot 1780 (Herod 1758) Dau. of Gamahoe (Bustard 1741) Patty by Tim 1752 Squirt 1732) Miss Patch 1757 by Justice 1755 (H. C. Litton Ar.).
Flight, ch. m., 1809.	Ir. Escape 1802 (Commodore 1793) Y. Heroine by Bagot 1780 (Herod 1758) Heroine 1775 by Hero 1753 (Cade 1734) Dau. 1766 of Snap 1750 (Snip 1736) Sist. 1743 to Regulus by God. Ar. 1724.
Castrel, ch. h., 1801.	Buzzard* 1787 (Woodpecker 1773)—Dau. of Alexander 1782 (Eclipse 1764) Dau. of Highflyer 1774 (Herod 1758) Dau. of Alfred 1770 (Matchem 1748) Dau. 1770 of Engineer 1756 (Sampson 1745).
Idalia, ch. m., 1815.	Peruvian 1806 (Sir Peter T. 1784)—Musidora 1804 by Meteor 1783 (Eclipse 1764)—Maid-of-all-Work 1786 by Highflyer 1774 (Herod 1758) Dau. 1771 of Syphon 1750 (Squirt 1732)—Dau. of Regulus 1739 (God. Ar. 1724).
Laurel, br. h., 1824.	Blacklock 1814 (Whitelock 1803) Wagtail 1818 by Prime Minister 1810 (Sancho 1801) Tranby's dam 1812 by Orville 1799 (Beningbrough 1791) Miss Grimstone 1796 by Weazle 1776 (Herod 1758) Dau. of Ancaster 1768 (Blank 1740).
Maid of Honor, br. m., 1829.	Champion 1812 Selim 1802) Etiquette 1820 by Orville 1799 (Beningbrough 1791) Boadicea 1807 by Alexander 1782 (Eclipse 1764) Brunette by Amaranthus 1766 (Old England 1741)—Mayfly 1771 by Matchem 1748 (Cade 1734).
Duroc, ch. h., 1806.	*Diomed 1777 (Florizel 1768) Amanda 1800 by Grey Diomed 1786 (*Medley 1776) Dau. of Va. Cade (Partner 1755) Dau. of Hickman's Independence 1776 (*Fearnought 1755) Dau. of *Badger (Bosphorus 1754).
Miller's Damsel, gr. m., 1802.	*Messenger 1780 (Mambrino 1768)—*Dau. 1792 of Pot-8-o's 1773 (Eclipse 1764)- Dau. 1778 of Gimcrack 1760 (Cripple 1750) Snapdragon 1759 by Snap 1750 (Snip 1736) Dau. 1749 of Regulus 1739 (God. Ar. 1724).
Henry, ch. h., 1819.	Sir Archy 1805 (*Diomed 1777) Dau. of *Diomed 1777 (Florizel 1768)- Bellona by Bellair 1786 (*Medley 1776)- Indian Queen by Pilgrim 1774 (*Fearnought 1755) Dau. of *Valiant 1756 (Dormouse 1738).
Y. Romp, ch. m., 1815.	Duroc 1806 (*Diomed 1777)- Old Romp 1805 by *Messenger 1780 (Mambrino 1768) - *Dau. 1792 of Pot-8-o's 1773 (Eclipse 1764) - Dau. 1778 of Gimcrack 1760 (Cripple 1750)—Snapdragon 1759 by Snap 1750 (Snip 1736).
Sir Archy, b. h., 1805.	*Diomed 1777 (Florizel 1768)—*Castianira 1796 by Rockingham 1781 (Highflyer 1774) - Tabitha 1782 by Trentham 1766 (T. Sweepstakes 1743)- Dau. of Bosphorus 1754 (Babraham 1740)—Sist. to Grecian Princess by Forester 1750 (Forester 1736).
Eliza, ch. m., 1804.	*Bedford 1792 (Dungannon 1780)—*Mambrina 1785 by Mambrino 1768 (Engineer 1756)—Sist. to Sally by Blank 1740 (God. Ar. 1724)—Dau. of Ward 1731 (Crab 1722)—Dan. of Merlin (Bustler).
[Brimmer or] Blue Beard, ch. h., 1802.	*Stirling 1791 (Volunteer 1780)—Brilliant 1779 by *Matchem (Bosphorus 1754) *Dau. of Brilliant 1750 (Crab 1722)- Dau. of Tartar 1743 (Partner 1718)- Dau. of Son of Flying Childers 1715.
Woodpecker's dam, ch. m., 1800.	*Buzzard 1787 (Woodpecker 1773)—The Faun by Craig's Alfred (*Medley 1776)—Dau. of *Traveller 1747 (Partner 1718)—Dau. of *Whittington (Cullen Ar.)- Dau. of *Childers (Blaze 1733).
Sir Archy, b. h., 1805.	*Diomed 1777 (Florizel 1768) *Castianira 1796 by Rockingham 1781 (Highflyer 1774) —Tabitha 1782 by Trentham 1766 (T. Sweepstakes 1743)—Dan. of Bosphorus 1754 (Babraham 1740)—Dau. of W's Forester 1750 (Croft's Forester 1736).
Daughter of, ch. m., —.	*Citizen 1785 (Pacolet 1763)—Dau. of Commutation (Symmes' Wildair 1767)—Dau. of *Dare Devil 1787 (Magnet 1770)- Sally Shark by *Shark 1771 (Marske 1750) Betsey Pringle by *Fearnought 1755 (Regulus 1739).
Marion, b. h., 1820.	Sir Archy 1805 (*Diomed 1777) - Dau. of *Citizen 1785 (Pacolet 1763)- Dau. of *Alderman 1787 (Pot-8-o's 1773)—Dau. of Ashe's Roebuck (*Sweeper)—Dau. of Hayne's King Herod (*Fearnought 1755).
Ella Crump, br. m., —.	*Citizen 1785 (Pacolet 1763)—Dau. 1795 of Huntsman 1789 (*Mousetrap 1771)- Dan. of Symmes' Wildair 1767 (*Fearnought 1755)—Dan. of *Fearnought 1755 (Regulus 1739) —Dau. of *Janus 1746 (Janus 1739)
Muley, b. h., 1810.	Orville 1799 (Beningbrough 1791)—Eleanor 1798 by Whiskey 1789 (Saltram* 1780)—Y. Giantess 1790 by Diomed* 1777 (Florizel 1768)—Giantess 1769 by Matchem 1748 (Cade 1734)- Molly Long Legs 1753 by Babraham 1740 (God. Ar. 1724).
Daughter of, blk. m., 1809.	Windle 1804 (Beningbrough 1791)—Dau. 1788 of Anvil 1777 (Herod 1758)—Virago 1764 by Snap 1750 (Snip 1736)—Dau. of Regulus 1739 (God. Ar. 1724)—Sist. 1750 to Othello by Crab 1722 (Alcock Ar.)
Sir Charles, ch. h., 1816.	Sir Archy 1805 (*Diomed 1777)—Dau. of *Citizen 1785 (Pacolet 1763)—Dau. of Commutation (Symmes' Wildair 1767)—Dau. of *Dare Devil 1787 (Magnet 1770)—Sally Shark by *Shark 1771 (Marske 1750).
Lady of the Lake, b. m., 1812.	*Sir Harry 1795 (Sir Peter T. 1784)—Shylock's dam by *Diomed 1777 (Florizel 1768)—Dau. of *St. George 1771 (Dragon 1757)—Dau. of *Fearnought 1755 (Regulus 1739)—Dau. of *Jolly Roger 1741 (Roundhead 1733).
Selim, ch. h., 1802.	Buzzard* 1787 (Woodpecker 1773)—Dau. of Alexander 1782 (Eclipse 1764)—Dau. of Highflyer 1774 (Herod 1758) - Dau. of Alfred 1770 (Matchem 1748)—Dau. 1770 of Engineer 1756 (Sampson 1745).
Bacchante, br. m., 1809.	W's Ditto 1800 (Sir Peter T. 1784)—Dau. 1791 of Mercury 1778 (Eclipse 1764)—Dau. 1776 of Herod 1758 (Tartar 1743)—Folly 1771 by Marske 1750 (Squirt 1732)- Vixen 1753 by Regulus 1739 (God. Ar. 1724).
Tramp, b. h., 1810.	Dick Andrews 1797 (Joe Andrews 1778)—Dau. 1803 of Gohanna 1790 (Mercury 1778)—Fraxinella 1793 by Trentham 1766 (T. Sweepstakes 1743)—Dau. 1785 of Woodpecker 1773 (Herod 1758)—Everlasting 1775 by Eclipse 1764 (Marske 1750).
Web, b. m., 1808.	Waxy 1790 (Pot-8-o's 1773)—Penelope 1798 by Trumpator 1782 (Conductor 1767)—Prunella 1788 by Highflyer 1774 (Herod 1758)—Promise 1768 by Snap 1750 (Snip 1736)—Julia 1756 by Blank 1740 (God. Ar. 1724).
Tramp, b. h., 1810.	Dick Andrews 1797 (Joe Andrews 1778)—Dau. 1803 of Gohanna 1790 (Mercury 1778)—Fraxinella 1793 by Trentham 1766 (T. Sweepstakes 1743)—Dau. 1785 of Woodpecker 1773 (Herod 1758)—Everlasting 1775 by Eclipse 1764 (Marske 1750).
Rosamond, ch. m., 1798.	Buzzard* 1787 (Woodpecker 1773)—Roseberry 1792 by Phenomenon 1780 (Herod 1758)—Miss West 1777 by Matchem 1748 (Cade 1734)—Dau. of Regulus 1739 (God. Ar. 1724)—Y. Ebony 1749 by Crab 1722 (Alcock Ar.)
Am. Eclipse, ch. h., 1814.	Duroc 1806 (*Diomed 1777)—Miller's Damsel 1802 by *Messenger 1780 (Mambrino 1768) —*Pot-8-o's Mare 1792 by Pot-8-o's 1773 (Eclipse 1764)—Dau. 1778 of Gimcrack 1760 (Cripple 1750) Snap Dragon 1759 by Snap 1750 (Snip 1736).
Katy Ann, ch. m., 1825.	Ogle's Oscar 1800 (*Gabriel 1790)—Y. Maid of the Oaks ch. m. 1817 by *Expedition 1795 (Pegasus 1784)—Maid of the Oaks ch. m. 1801 by *Spread Eagle 1793 (Volunteer 1780)—Annette by *Shark 1771 (Marske 1750)—Dau. of Rockingham (Partner 1755)—Dau. of True Whig (Fitzhugh's Regulus 1765)—Dau. of Baylor's Gallant 1770 (*Fearnought 1755)—Dau. of Burwell's *Regulus 1747 (Regulus 1739)—*Duchess by Cullen Arabian 1745—Lady Thigh

ch. m. 1731 by Partner 1718 (Jigg)—Dau. ch. m. 1723 of Greyhound (Chillaby)—Sophonisba's dam by Curwen Bay Barb.—Dau. of D'Arcy Chestnut Ar. Dau. of Whiteshirt—D'Arcy's Montagu Mare by Montagu—Dau. of Hautboy (D'Arcy White Turk)—Dau. of Brimmer (D'Arcy Yellow Turk)—Royal Mare.

LISBON.

WINNER OF THE ALEXANDER STAKES AT LOUISVILLE, KENTUCKY, 1876.

In the Woodburn Stud, Spring Station, Ky. Application to L. Brodhead, Spring Station, Ky. Annual sales of yearlings in May.

LISBON by imp. Phaeton, son of King Tom, bred by Richard Ten Broeck, foaled 1874, dam imp. Lady Love by Stockwell or Caterer, son of Stockwell, second dam Chere Amie by Sweetmeat, out of Blame by Touchstone. His dam, Lady Love, was a winner in England at two years old, and won the Stewards' Cup at Chester at four years old. Chere Amie, his grandam, was also a race-mare and winner. Caterer, the reputed sire of Lady Love, was a good race-horse. He won the Findon Stakes at Goodwood when two years old, defeating Wingrave, Alvediston, Kaiser and others, and was second to the Marquis in the 2,000 Guineas with Alvediston, Wingrave and thirteen others behind him. Sweetmeat, the sire of his grandam, was a superior race-horse, won the Queen's Vase at Ascot, and the Doncaster Cup in 1845. He was the sire of Mincepie and Mincemeat, both Oaks winners, and Macaroni, winner of the Derby and 2,000 Guineas, and second in the list of winning sires in 1882. Touchstone, the sire of his great grandam Blame, won the St. Leger, the Ascot Gold Cup in 1836–37, and the Doncaster Cup in 1835–36, and was second to no sire of his day. Lisbon started three times as a two-year old and won one race. At Louisville, Ky., won the Alexander Stakes, half-mile, in 0.49¾, beating McWhirter and seven others. Was third in the Colt Stakes at Lexington, Ky., three-quarters of a mile, King Faro and McWhirter running dead heat for first place, in 1.18¾. In this race Lisbon beat Allen Pinkerton, Headlight, Planetarian and Edinburg. Was third in the Belle Meade Stakes at Louisville, three-quarters of a mile, in 1.17, won by McWhirter, Baden-Baden second. As a three-year old he started only twice, and was unplaced in both of his races. Lisbon is the sire of Ripple, who won the Long Branch Handicap, 1¼ miles, in 2.10¼, beating Checkmate, Parole and Sir Hugh; also the the Shrewsbury Handicap, 1¾ miles, in 3.09¾. In this race he carried a penalty of 5 lbs. and 3 lbs. overweight. He unfortunately died that winter. Troubadour (that defeated The Bard), Fagan, Pat Sheedy, Cathcart, Grimaldi, an extra good one, Quincey, Clatter, Leland, Sea Foam, Artless, Lisland, Lisimony, Lizzie Glenn, Amos A., Indiana, Jack Rose, Steve Jerome, Sly Lisbon, Portuguese, Farine, Glenoid, Trovatore, Lisbon Maid, Glenoid, Liselg, Leoculus, Mamie B., all winners. Lisbon is a red bay, full 15½ hands high, with a blaze face and four white ankles. He is a horse of immense substance and power, and in conformation second to none, is very muscular, like all of Phaeton's sons, and great constitution. In addition to his double cross of Glencoe through King Tom and Stockwell, he has a double cross of Touchstone on each side, eight crosses of Waxy through Economist and Emma by Whisker and Camel and Sir Hercules by Whalebone, fortified by the blood of Herod through the famous Highflyer, and Eclipse through Pot-8-o's and King Fergus. The blood is worth preserving through the granddaughters of Phaeton and daughters of Great Tom, King Ban and King Ernest.

Lisbon descends in direct female line from the Woodcock mare, to which trace Reveller (by Comus), Heron, Alarm, Monarque, Vedette, Cambuscan, Vespasian, Isonomy, King Lud, Exile and other good ones.

LISBON.

BAY HORSE, FOALED 1874; BRED BY RICHARD TEN BROECK, KENTUCKY.

*PHAETON, b. h., 1865.	King Tom, b. h., 1851.	Harkaway, ch. h., 1834.	Economist, b. h., 1825.	Whisker, b. h., 1812.	Waxy 1790 (Pot-8-o-'s 1773)—Penelope 1798 by Trumpator 1782 (Conductor 1767)—Prunella 1788 by Highflyer 1774 (Herod 1758)—Promise 1768 by Snap 1750 (Snip 1736)—Julia 1756 by Blank 1740 (God. Ar. 1724).
				Floranthe, b. m., 1818.	Octavian 1807 (Stripling 1795)—Caprice 1797 by Anvil 1777 (Herod 1758)—Madcap 1774 by Eclipse 1764 (Marske 1750)—Sappho 1763 by Blank 1740 (God. Ar. 1724)—Dau. 1750 of Blaze 1733 (Flying Childers 1715).
			Fanny Dawson, ch. m., 1823.	Nabocklish, ch. h., 1811.	Rugantino 1804 (Commodore 1793)—Butterfly 1801 by Master Bagot 1787 (Bagot 1780)—Dau. of Bagot 1780 (Herod 1758)—Mother Brown 1771 by Trunnion 1747 (Cade 1734)—Dau. of Old England 1741 (God. Ar. 1724).
				Miss Tooley, b. m., 1808.	Teddy-the-Grinder 1798 (Asparagus 1787)—Lady Jane 1796 by Sir Peter T. 1784 (Highflyer 1774)—Paulina 1778 by Florizel 1768 (Herod 1758)—Captive 1771 by Matchem 1748 (Cade 1734)—Calliope 1763 by Slouch 1747 (Cade 1734).
		Pocahontas, b. m., 1837.	Glencoe*, ch. h., 1831.	Sultan, b. h., 1816.	Selim 1802 (Buzzard* 1787)—Bacchante 1809 by W's Ditto 1800 (Sir Peter T. 1784)—Dau. 1791 of Mercury 1778 (Eclipse 1764)—Dau. 1776 of Herod 1758 (Tartar 1743)—Folly 1771 by Marske 1750 (Squirt 1732).
				Trampoline, ch. m., 1825.	Tramp 1810 (Dick Andrews 1797)—Web 1808 by Waxy 1790 (Pot-8-o-'s 1773)—Penelope 1798 by Trumpator 1782 (Conductor 1767)—Prunella 1788 by Highflyer 1774 (Herod 1758)—Promise 1768 by Snap 1750 (Snip 1736).
			Marpessa, b. m., 1830.	Muley, b. h., 1810.	Orville 1799 (Beningbrough 1791)—Eleanor 1798 by Whiskey 1789 (Saltram* 1780)—Y. Giantess 1790 by Diomed* 1777 (Florizel 1768)—Giantess 1769 by Matchem 1748 (Cade 1734)—Molly Long Legs 1753 by Babraham 1740 (God. Ar. 1724).
				Clare, b. m., 1824.	Marmion 1806 (Whiskey 1789)—Harpalice 1814 by Gohanna 1790 (Mercury 1778)—Amazon 1799 by Driver 1783 (Trentham 1766)—Fractious 1792 by Mercury 1778 (Eclipse 1764)—Dau. 1785 of Woodpecker 1773 (Herod 1758).
	Merry Sunshine, ch. m., 1855.	Storm, b. h., 1848.	Touchstone, br. h., 1831.	Camel, br. h., 1822.	Whalebone 1807 (Waxy 1790)—Dau. 1812 of Selim 1802 (Buzzard* 1787)—Maiden 1801 by Sir Peter T. 1784 (Highflyer 1774)—Dau. 1788 of Phenomenon 1780 (Herod 1758)—Matron 1782 by Florizel 1768 (Herod 1758).
				Banter, br. m., 1826.	Master Henry 1815 (Orville 1799)—Boadicea 1807 by Alexander 1782 (Eclipse 1764)—Brunette by Amaranthus 1766 (Old England 1741)—Mayfly 1771 by Matchem 1748 (Cade 1734)—Dau. 1755 of Anc. Starling 1738 (Old Starling 1727).
			Ghuznee, b. m., 1838.	Pantaloon, ch. h., 1824.	Castrel 1801 (Buzzard* 1787)—Idalia 1815 by Peruvian 1806 (Sir Peter T. 1784)—Musidora 1804 by Meteor 1783 (Eclipse 1764)—Maid-of-all-Work 1786 by Highflyer 1774 (Herod 1758)—Dau. 1771 of Syphon 1750 (Squirt 1732).
				Languish, b. m., 1830.	Cain 1822 (Paulowitz 1813)—Lydia 1822 by Poulton 1805 (Sir Peter T. 1784)—Variety 1808 by Hyacinthus 1797 (Coriander 1786)—Sist. 1790 to Swordsman by Weazle 1776 (Herod 1758)—Dau. of Turk 1763 (Regulus 1739).
		Daughter of, b. m., 1849.	Falstaff, br. h., 1842.	Touchstone, br. h., 1831.	Camel 1822 (Whalebone 1807)—Banter 1826 by Master Henry 1815 (Orville 1799)—Boadicea 1807 by Alexander 1782 (Eclipse 1764)—Brunette by Amaranthus 1766 (Old England 1741)—Mayfly 1771 by Matchem 1748 (Cade 1734).
				Decoy, b. m., 1830.	Filho-da-Puta 1812 (Haphazard 1797)—Finesse 1815 by Peruvian 1806 (Sir Peter T. 1784)—Violante 1802 by John Bull 1789 (Fortitude 1777)—Dau. 1788 of Highflyer 1774 (Herod 1758)—Everlasting 1775 by Eclipse 1764 (Marske 1750).
			Sist. to Pompey, b. m., 1843.	Emilius, b. h., 1820.	Orville 1799 (Beningbrough 1791)—Emily 1810 by Stamford 1794 (Sir Peter T. 1784)—Dau. 1799 of Whiskey 1789 (Saltram* 1780)—Grey Dorimant 1781 by Dorimant 1772 (Otho 1760)—Dizzy 1757 by Blank 1740 (God. Ar. 1724).
				Variation, b. m., 1827.	Bustard 1813 (Castrel 1801)—Johanna Southcotte 1811 by Beningbrough 1791 (King Fergus 1775)—Lavinia 1802 by Pipator 1786 (Imperator 1776)—Dau. 1790 of Highflyer 1774 (Herod 1758)—Dau. 1777 of Cardinal Puff 1760 (Babraham 1740).
*LADY LOVE, br. m., 1865.	[Stockwell or] Caterer, b. h., 1859.	Stockwell, ch. h., 1849.	The Baron, ch. h., 1842.	Ir Birdcatcher ch. h., 1833.	Sir Hercules 1826 (Whalebone 1807)—Guiccioli 1823 by Bob Booty 1804 (Chanticleer 1787)—Flight 1809 by Ir. Escape 1802 (Commodore 1793)—Y. Heroine by Bagot 1780 (Herod 1758)—Heroine 1775 by Hero 1753 (Cade 1734).
				Echidna, b. m., 1838.	Economist 1825 (Whisker 1812)—Miss Pratt 1825 by Blacklock 1814 (Whitelock 1803)—Gadabout 1812 by Orville 1799 (Beningbrough 1791)—Minstrel 1803 by Sir Peter T. 1784 (Highflyer 1774)—Matron 1782 by Florizel 1768 (Herod 1758).
			Pocahontas, b. m., 1837.	Glencoe,* ch. h., 1831.	Sultan 1816 (Selim 1802)—Trampoline 1825 by Tramp 1810 (Dick Andrews 1797)—Web 1808 by Waxy 1790 (Pot-8-o-'s 1773)—Penelope 1798 by Trumpator 1782 (Conductor 1767)—Prunella 1788 by Highflyer 1774 (Herod 1758).
				Marpessa, b. m., 1830.	Muley 1810 (Orville 1799)—Clare 1824 by Marmion 1806 (Whiskey 1789)—Harpalice 1814 by Gohanna 1790 (Mercury 1778)—Amazon 1799 by Driver 1783 (Trentham 1766)—Fractious 1792 by Mercury 1778 (Eclipse 1764).
		Selina, b. m., 1851.	Orlando, b. h., 1841.	Touchstone, br. h., 1831.	Camel 1822 (Whalebone 1807)—Banter 1826 by Master Henry 1815 (Orville 1799)—Boadicea 1807 by Alexander 1782 (Eclipse 1764)—Brunette by Amaranthus 1766 (Old England 1741)—Mayfly 1771 by Matchem 1748 (Cade 1734).
				Vulture, ch. m., 1833.	Langar 1817 (Selim 1802)—Kite 1821 by Bustard 1813 (Castrel 1801)—Olympia 1815 by Sir Oliver 1800 (Sir Peter T. 1784)—Scotilla 1795 by Anvil 1777 (Herod 1758)—Scota 1783 by Eclipse 1764 (Marske 1750).
			Ladye of S. W.,§ ch. m., 1839.	Velocipede, ch. h., 1825.	Blacklock 1814 (Whitelock 1803)—Dau. 1817 of Juniper 1805 (Whiskey 1789)—Dau. 1810 of Sorcerer 1796 (Trumpator 1782)—Virgin 1801 by Sir Peter T. 1784 (Highflyer 1774)—Dau. 1794 of Pot-8-o-'s 1773 (Eclipse 1764).
				Emma, ch. m., 1824.	Whisker 1812 (Waxy 1790)—Gibside Fairy 1811 by Hermes 1790 (Mercury 1778)—Vicissitude 1800 by Pipator 1786 (Imperator 1776)—Beatrice 1791 by Sir Peter T. 1784 (Highflyer 1774)—Pyrrha 1771 by Matchem 1748 (Cade 1734).
	Chère Amie, br. m., 1857.	Sweetmeat, br. h., 1842.	Gladiator, ch. h., 1833.	Partisan, b. h., 1811.	Walton 1799 (Sir Peter T. 1784)—Parasol 1800 by Pot-8-o-'s 1773 (Eclipse 1764)—Prunella 1788 by Highflyer 1774 (Herod 1758)—Promise 1768 by Snap 1750 (Snip 1736)—Julia 1756 by Blank 1740 (God. Ar. 1724).
				Pauline, b. m., 1826.	Moses 1819 ([Whalebone or] Seymour 1807)—Quadrille 1815 by Selim 1802 (Buzzard* 1787)—Canary Bird 1806 by (Whiskey or) Sorcerer 1796 (Trumpator 1782)—Canary 1797 by Coriander 1786 (Pot-8-o-'s 1773)—Miss Green 1787 by Highflyer 1774 (Herod 1758).
			Lollypop, blk. m., 1836.	[Starch or] Voltaire, br. h., 1826.	Blacklock 1814 (Whitelock 1803)—Dau. 1816 of Phantom 1808 (Walton 1799)—Dau. 1802 of Overton 1788 (King Fergus 1775)—Dau. 1796 of Walnut 1786 (Highflyer 1774)—Dau. 1786 of Ruler 1777 (Y. Marske 1771).
				Belinda, br. m., 1825.	Blacklock 1814 (Whitelock 1803)—Wagtail 1818 by Prime Minister 1810 (Sancho 1801)—Tranby*'s dam 1812 by Orville 1799 (Beningbrough 1791)—Miss Grimstone 1796 by Weazle 1776 (Herod 1758)—Dau. of Ancaster 1768 (Blank 1740).
		Blame, b. m., 1851.	Touchstone, br. h., 1831.	Camel, br. h., 1822.	Whalebone 1807 (Waxy 1790)—Dau. 1812 of Selim 1802 (Buzzard* 1787)—Maiden 1801 by Sir Peter T. 1784 (Highflyer 1774)—Dau. 1788 of Phenomenon 1780 (Herod 1758)—Matron 1782 by Florizel 1768 (Herod 1758).
				Banter, br. m., 1826.	Master Henry 1815 (Orville 1799)—Boadicea 1807 by Alexander 1782 (Eclipse 1764)—Brunette by Amaranthus 1766 (Old England 1741)—Mayfly 1771 by Matchem 1748 (Cade 1724)—Dau. 1755 of Anc. Starling 1738 (Starling 1727).
			Maid of Lyme, br. m., 1813.	Tomboy, b. h., 1829.	Jerry 1821 (Smolensko 1810)—Dau. 1817 of Ardrossan 1809 (John Bull 1789)—Lady Eliza 1813 by Whitworth 1805 (Agonistes 1797)—Dau. 1798 of Spadille 1784 (Highflyer 1774)—Sylvia 1783 by Y. Marske 1771 (Marske 1750).
				Lapwing, br. m., 1837.	Bustard 1813 (Castrel 1801)—Dau. b. m. 1830 by Muley 1810 (Orville 1799)—Rosanne b. m. 1811 by Dick Andrews 1797 (Joe Andrews 1778)—Rosette br. m. 1803 by Beningbrough 1791 (King Fergus 1775)—Rosamond b. m. 1788 by Tandem 1773 (Syphon 1750)—Tuberose gr. m. 1772 by Herod 1758 (Tartar 1743)—Grey Starling gr. m. 1745 by Bolton Starling 1727 (Bay Bolton 1705)—Coughing Polly b. m. 1736 by Bartlett's Childers (Darley Ar. 1699)—Counsellor mare by Wood's Counsellor 1694 (D'Arcy Counsellor)—Dau. of Snake (Lister Turk)—Dau. of Luggs (D'Arcy White Turk)—Dau. of Davill's Old Woodcock.

§ The Ladye of the Silverkeld Well.

LOGIC.

The property of Mrs. J. M. Clay, Ashland Stud, Lexington, Ky. Service by contract. Yearlings sold annually.

LOGIC, bay horse, foaled 1886, bred by J. M. Clay, by Longfellow, son of imp. Leamington, dam Badge by Gilroy, son of Lexington, out of Buff and Blue by War Dance, son of Lexington and Reel by imp. Glencoe. Logic is a fine large horse over 17 hands in height and was a horse of great promise, but badly handled and injured early by an accident. He started only a few times, ran second to Robin Hood, 1 mile, in 1.43¼; ran third to Come-to-Taw and Girondes, 1 mile, in 1.42¾; ran third in Twin City Derby, 1¼ miles, in 2.10. His sire is the most popular stallion in America, and no son of Longfellow is from a family more noted as a racing and producing one. Buff and Blue was the dam of Badge (Logic and Minnie Elkins' dam), Ban Cloche, Blue Stocking (Barefoot's dam), Monita Hardy, Sister Geneva, etc. La Gitana, from the family, is the dam of Geneva (Riley's dam) by War Dance, Glitter, Daisy, Guilford, Pommery Sec, Appomatox, etc., Revolver, a good race-horse and stallion. The Banshee, from this family, won the Travers Stakes, Westchester Cup, etc., and was the dam of Krupp Gun, Brigand Belle, etc. Heraldry, the fourth dam, was a good race-mare, dam of Parachute, Ascension, Sally of the Valley (Essillah's dam), etc., by imp. Herald, winner of the second heat in the great Peyton Stakes run at Nashville, 1843, he by English Plenipotentiary, son of Emilius by Orville; fifth dam Margaret Woods, winner of the Trial Stakes at Nashville, 1843, dam of Wade Hampton, Star Davis, Maria Wood, Capt. Travis, La Gitana, etc., she by imp. Priam (winner of the Derby), son of Emilius and Cressida by Whiskey; sixth dam Maria West by Marion, son of Sir Archy. Maria West was a good race-mare and dam of the noted Wagner, who defeated Gray Eagle in the two famous four-mile heat races run at Louisville, Ky., in 1839.

Logic has a double cross each of Lexington and Glencoe blood and three crosses of the famous Emilius blood, one through Yorkshire, son of St. Nicholas by Emilius, one through Priam, winner of the Derby, and one through imp. Herald, son of Plenipotentiary (Lion of Doncaster) by Emilius.

Logic's oldest foals, now sucklings, are rare good ones and will race successfully.

Pot-8-o's 1773 - Eclipse 1764—Marske 1750 Squirt 1732 Bartlett's Childers - Darley Arabian 1689.

LOGIC.

BAY HORSE, FOALED 1886; BRED BY JOHN M. CLAY, ASHLAND STUD, KENTUCKY.

LONGFELLOW, br. h., 1867.	*Leamington, br. h., 1853.	Faugh-a-Ballagh, br. h., 1841.	Sir Hercules, blk. h., 1826.	Whalebone, br. h., 1807.	Waxy 1790 (Pot-8-o's 1773) Penelope 1798 by Trumpator 1782 (Conductor 1767) Prunella 1788 by Highflyer 1774 (Herod 1758) Promise 1768 by Snap 1750 (Snip 1736) Julia 1756 by Blank 1740 (God. Ar. 1724).
				Peri, b. m., 1822.	Wanderer 1811 (Gohanna 1790) Thalestris 1809 by Alexander 1782 (Eclipse 1764) Rival 1800 by Sir Peter T. 1784 (Highflyer 1774) Hornet 1790 by Drone 1777 (Herod 1758) Manilla 1777 by Goldfinder 1764 (Snap 1750).
			Guiccioli, ch. m., 1823.	Bob Booty, ch. h., 1804.	Chanticleer 1787 (Woodpecker 1773) Ierne 1790 by Bagot 1780 (Herod 1758) Dau. of Gamahoe (Bustard 1741) Patty by Tim 1752 (Squirt 1732) Miss Patch 1737 by Justice 1725 (H. C. Litton Ar.).
				Flight, ch. m., 1809.	Ir. Escape 1802 (Commodore 1793) Y. Heroine by Bagot 1780 (Herod 1758) Heroine 1775 by Hero 1753 (Cade 1734) Dau. 1766 of Snap 1750 (Snip 1736) Sist. 1743 to Regulus 1739 (God. Ar. 1724).
		Daughter of, b. m., 1841.	Pantaloon, ch. h., 1824.	Castrel, ch. h., 1801.	Buzzard* 1787 (Woodpecker 1773)—Dau. of Alexander 1782 (Eclipse 1764) Dau. of Highflyer 1774 (Herod 1758) Dau. of Alfred 1770 (Matchem 1748) Dau. 1779 of Engineer 1756 (Sampson 1745).
				Idalia, ch. m., 1815.	Peruvian 1806 (Sir Peter T. 1784)—Musidora 1804 by Meteor 1783 (Eclipse 1764)—Maid-of-all-Work 1786 by Highflyer 1774 (Herod 1758) Dau. 1771 of Syphon 1750 (Squirt 1732)—Dau. of Regulus 1739 (God. Ar. 1724).
			Daphne, br. m., 1837.	Laurel, br. h., 1824.	Blacklock 1814 (Whitelock 1803) Wagtail 1818 by Prime Minister 1810 (Sancho 1801) Tranby's dam 1812 by Orville 1799 (Beningbrough 1791) Miss Grimstone 1796 by Weazle 1776 (Herod 1758) Dau. of Ancaster 1768 (Blank 1740).
				Maid of Honor, br. m., 1829.	Champion 1812 (Selim 1802) Etiquette 1820 by Orville 1799 (Beningbrough 1791) Boadicea 1807 by Alexander 1782 (Eclipse 1764) Brunette by Amaranthus 1766 (Old England 1741)—Mayfly 1771 by Matchem 1748 (Cade 1734).
	Nantura, b. m., 1855.	Brawner's Eclipse, ch. h., 1829.	Am. Eclipse, ch. h., 1814.	Duroc, ch. h., 1806.	*Diomed 1777 (Florizel 1768) Amanda 1800 by Grey Diomed 1786 (*Medley 1776) Dau. of Va. Cade (Partner 1755)—Dau. of Hickman's Independence 1776 (*Fearnought 1755) Dau. of *Badger (Bosphorus 1754).
				Miller's Damsel, gr. m., 1802.	*Messenger 1780 (Mambrino 1768)—*Dau. 1792 of Pot-8-o's 1773 (Eclipse 1764)—Dau. 1778 of Gimcrack 1760 (Cripple 1750)—Snapdragon 1759 by Snap 1750 (Snip 1736) Dau. 1749 of Regulus 1739 (God. Ar. 1724).
			Daughter of, ch. m., 1834.	Henry, ch. h., 1819.	Sir Archy 1805 (*Diomed 1777) Dau. of *Diomed 1777 (Florizel 1768)—Bellona by Bellair 1786 (*Medley 1776)—Indian Queen by Pilgrim 1774 (*Fearnought 1755)—Dau. of *Valiant 1756 (Dormouse 1738).
				Y. Romp, ch. m., 1815.	Duroc 1806 (*Diomed 1777)—Old Romp 1803 by *Messenger 1780 (Mambrino 1768) *Dau. 1792 of Pot-8-o's 1773 (Eclipse 1764)—Dau. 1778 of Gimcrack 1760 (Cripple 1750)—Snapdragon 1759 by Snap 1750 (Snip 1736).
		Quiz, b. m., 1836.	Bertrand, b. h., 1821.	Sir Archy, b. h., 1805.	*Diomed 1777 (Florizel 1768)—*Castianira 1796 by Rockingham 1781 (Highflyer 1774) Tabitha 1782 by Trentham 1766 (T. Sweepstakes 1743)—Dau. of Bosphorus 1754 (Babraham 1740) Sist. to Grecian Princess by Forester 1750 (Forester 1736).
				Eliza, ch. m., 1804.	*Bedford 1792 (Dungannon 1780)—*Mambrina 1785 by Mambrino 1768 (Engineer 1756)—Sist. to Sally by Blank 1740 (God. Ar. 1724)—Dau. of Ward 1731 (Crab 1722)—Dau. of Merlin (Bustler).
			Lady Fortune, ch. m.,	[Brimmer or] Blue Beard, ch. h., 1802.	*Stirling 1791 (Volunteer 1780)—Brilliant 1779 by *Matchem (Bosphorus 1754) *Dau. of Brilliant 1750 (Crab 1722)—Dau. of Tartar 1743 (Partner 1718)—Dau. of Son of Flying Childers 1715.
				Woodpecker's dam, ch. m., 1809.	*Buzzard 1787 (Woodpecker 1773)—The Faun by Craig's Alfred (*Medley 1776)—Dau. of *Traveller 1747 (Partner 1718)—Dau. of *Whittington (Cullen Ar.) Dau. of *Childers (Blaze 1733).
BADGE, b. m., 1879.	Gilroy, b. h., 1862.	Lexington, b. h., 1850.	Boston, ch. h., 1833.	Timoleon, ch. h., 1813.	Sir Archy 1805 (*Diomed 1777)—Dau. of *Saltram 1780 (Eclipse 1764)—Dau. of Symmes' Wildair 1767 (*Fearnought 1755)—Dau. of Tyler's Driver 1760 (*Othello)—Dau. of *Fallower 1761 (Blank 1740).
				Sister to Tuckahoe, ch. m., 1814.	Ball's Florizel 1801 (*Diomed 1777)—Dau. 1799 of *Alderman 1787 (Pot-8-o's 1773)—Dau. of *Clockfast 1774 (Gimcrack 1760)—Dau. of Symmes' Wildair 1767 (*Fearnought 1755)—Y. Kitty Fisher 1767 by *Fearnought 1755 (Regulus 1739).
			Alice Carneal, b. m., 1836.	*Sarpedon, br. h., 1828.	Emilius 1820 (Orville 1799)—Icaria 1824 by The Flyer 1814 (Vandyke Jr. 1808) Parma 1813 by Dick Andrews 1797 (Joe Andrews 1778)—May 1804 by Beningbrough 1791 (King Fergus 1775)—Primrose 1787 by Mambrino 1768 (Engineer 1756).
				Rowena, ch. m., 1826.	Sumpter 1818 (Sir Archy 1805)—Lady Grey 1817 by Robin Grey 1805 (*Royalist 1790)—Maria 1802 by Melzar 1791 (*Medley 1776)—Dau. of *Highflyer 1784 (Highflyer 1774)—Dau. of *Fearnought 1755 (Regulus 1739).
		Magnolia, ch. m., 1841.	*Glencoe, ch. h., 1831.	Sultan, b. h., 1816.	Selim 1802 (Buzzard* 1787)—Bacchante 1809 by W's Ditto 1800 (Sir Peter T. 1784) Dau. 1791 of Mercury 1778 (Eclipse 1764)—Dau. 1776 of Herod 1758 (Tartar 1743)—Folly 1771 by Marske 1750 (Squirt 1732).
				Trampoline, ch. m., 1825.	Tramp 1810 (Dick Andrews 1797)—Web 1808 by Waxy 1790 (Pot-8-o's 1773)—Penelope 1798 by Trumpator 1782 (Conductor 1767) Prunella 1788 by Highflyer 1774 (Herod 1758)—Promise 1768 by Snap 1750 (Snip 1736).
			*Myrtle, ch. m., 1834.	Mameluke, b. h., 1824.	Partisan 1811 (Walton 1799) Miss Sophia 1805 by Stamford 1794 (Sir Peter T. 1784)—Sophia 1798 by Buzzard* 1787 (Woodpecker 1773)—Huncamunca 1787 by Highflyer 1774 (Herod 1758)—Cypher 1772 by Squirrel 1754 (Traveller 1735).
				Bobadilla, br. m., 1825.	Bobadil 1813 (Rubens 1805)—Pythoness 1818 by Sorcerer 1796 (Trumpator 1782)—Princess 1796 by Sir Peter T. 1784 (Highflyer 1774)—Dau. 1789 of Dungannon 1780 (Eclipse 1764) Dau. 1783 of Turf 1760 (Matchem 1748).
	Buff and Blue, b. m., 1873.	War Dance, ch. h., 1859.	Lexington, b. h., 1850.	Boston, ch. h., 1833.	Timoleon 1813 (Sir Archy 1805)—Dau. 1814 of Ball's Florizel 1801 (*Diomed 1777)—Dau. 1799 of *Alderman 1787 (Pot-8-o's 1773)—Dau. of *Clockfast 1774 (Gimcrack 1760)—Dau. of Symmes' Wildair 1767 (*Fearnought 1755).
				Alice Carneal, b. m., 1836.	*Sarpedon 1828 (Emilius 1820)—Rowena 1826 by Sumpter 1818 (Sir Archy 1805)—Lady Grey 1817 by Robin Grey 1805 (*Royalist 1790)—Maria 1802 by Melzar 1791 (*Medley 1776)—Dau. of *Highflyer 1784 (Highflyer 1774).
			Reel, gr. m., 1838.	*Glencoe, ch. h., 1831.	Sultan 1816 (Selim 1802)—Trampoline 1825 by Tramp 1810 (Dick Andrews 1797)—Web 1808 by Waxy 1790 (Pot-8-o's 1773)—Penelope 1798 by Trumpator 1782 (Conductor 1767)—Prunella 1788 by Highflyer 1774 (Herod 1758).
				*Galopade, gr. m., 1828.	Catton 1809 (Golumpus 1802)—Camillina 1822 by Camillus 1803 (Hambletonian 1792)—Dau. 1818 of Smolensko 1810 (Sorcerer 1796)—Miss Cannon 1811 by Orville 1799 (Beningbrough 1791) Dau. 1796 of Weathercock 1785 (Highflyer 1774).
		Balloon, b. m., 1852.	*Yorkshire, b. h., 1834.	St. Nicholas, br. h., 1827.	Emilius 1820 (Orville 1799) Sea Mew 1815 by Scud 1804 (Beningbrough 1791)—Goosander 1805 by Hambletonian 1792 (King Fergus 1775) Rally 1790 by Trumpator 1782 (Conductor 1767)—Fancy 1780 by Florizel 1768 (Herod 1758).
				Miss Rose,* b. m., 1823.	Tramp 1810 (Dick Andrews 1797)—Dau. 1810 of Sancho 1801 (Don Quixote 1784)—Dau. 1799 of Coriander 1786 (Pot-8-o's 1773)—Wildgoose 1792 by Highflyer 1774 (Herod 1758)—Coheiress 1783 by Pot-8-o's 1773 (Eclipse 1764).
			Heraldry, ch. m., 1846.	*Herald, ch. h., 1839.	Plenipotentiary 1831 (Emilius 1820)—*Delphine 1825 by Whisker 1812 (Waxy 1790) My Lady* 1818 by Comus 1809 (Sorcerer 1796) Dau. 1802 of Delpini 1781 (Highflyer 1774)—Tipple Cyder 1788 by King Fergus 1775 (Eclipse 1764).
				Margaret Wood, b. m., 1840.	*Priam 1827 (Emilius 1820)—Maria West b. m. 1827 by Marion 1820 (Sir Archy 1805)—Ella Crump br. m. by *Citizen 1785 (Pacolet 1763)—Dau. br. m. 1795 of Huntsman 1789 (*Mousetrap 1771)—Dau. of Symmes' Wildair 1767 (*Fearnought 1755)—Dau. of *Fearnought 1755 (Regulus 1739)—Dau. of *Janus 1746 (Janus 1739).

LONGFELLOW.

WINNER OF THE PRODUCE STAKES AT LEXINGTON, KY., OHIO STAKES AT CINCINNATI, AND CITIZENS' STAKES AT NASHVILLE. 1870, MONMOUTH CUP AT LONG BRANCH, 1871 AND 1872, SARATOGA CUP IN 1871.

Owned by F. B. Harper, Nantura Stud, near Midway, Woodford Co., Ky.
Seasons full at $1,000.

LONGFELLOW by imp. Leamington, son of Faugh-a-Ballagh, bred by John Harper, foaled 1867, dam Nantura, dam of Extra, Fanny Holton, etc., by Brawner's Eclipse, son of Am. Eclipse, out of Quiz, sister to Queen Mary, by Bertrand, son of Sir Archy. Longfellow was a fast, good race-horse, made his *debut* at Lexington, Ky., May 16, 1870, in Phœnix Stakes, mile heats, for three-year olds, won by Enquirer in 1.44¾, 1.44½, Catina second and Longfellow third and distanced. Same place, Sept. 16, won Produce Stakes for three-year olds, 2 mile heats, in 3.43¼, 3.44, beating Twinkle, etc. Cincinnati, Ohio, Sept. 30, won Ohio Stakes for three-year olds, 2 mile heats, in 3.37½, 3.55½, beating Pilgrim. Memphis, Tenn., Oct. 25, won Post Stakes, 2 mile heats, in 3.40¾, 3.40, beating Defender, Morgan Scout, etc. Nashville, Oct. 12, won Citizens' Stake, 2 mile heats, in 3.41¾, 3.41, beating and distancing Morgan Scout, etc. Lexington, Ky., May 24, 1871, walked over for purse $400, 2 mile heats. Long Branch, July 5, won Monmouth Cup, 2½ miles, in 4.41¼, beating Helmbold, Regards and Preakness. Saratoga, July 14, won Saratoga Cup, 2¼ miles, in 4.02¾, beating Kingfisher, the first mile was run in 1.40; Aug. 19, walked over for purse of $800, 2¼ miles; Aug. 23, was beaten by Helmbold, 4 miles, in 7.49¼. Lexington, Ky., Sept. 12, won club purse, 2 mile heats, in 3.38¾, 3.41¼. Lexington, Ky., May 15, 1872, won purse, 1½ miles, in 2.41½, beating Metella and Anreola; May 18, won Woolley Stake, 3 miles, in 6.00¾, beating Metella, Talaria, etc. Long Branch, July 2, won Monmouth Cup, 2½ miles, in 4.34; July 4, won purse, mile heats, three best in five, in 1.56¼, 1.54, 1.43¾, beating Susan Ann, Metella and Arizona. Saratoga, July 16, was beaten by Harry Bassett for the Cup, 2¼ miles, in 3.59, the fastest race to that date. The first mile was run in 1.44, the first two miles in 3.30. Longfellow pulled up lame, and with this race closed his turf career, starting in sixteen races, won fourteen, lost two, and won $15,000 in money. Nantura, his dam, was a very superior mare and good winner, Brawner's Eclipse, her sire, was a fine race-horse, and much inbred to imp. Diomed, Quiz, his grandam, was a fine race-mare. Bertrand, her sire, son of Sir Archy, was one of the best race-horses and stallions of his day.

Longfellow stands nearly 17 hands high, with white stripe down his face, with left hind foot white above the ankle, and a little white on the coronet of the right hind foot. He has a neat ear, broad forehead, with Roman nose, long, well-shaped neck, running into strong, well-shaped shoulders, great depth of girth, rather light flank, good strong hips, with good length of whirlbone and stifle, and thence to the hock; good legs and feet. The following are the best of his get: Thora, a first-class mare, her 3 miles at Saratoga when three years old, with 99 lbs., in 5.25¼, was a remarkable race; also her Pimlico Stake, 2½ miles, in 3.47¼. Other good ones are Edinburg, Irish King, Jils Johnson, Leonard, Lilly R., Long Taw (Dave Moore), Mollie Merrill, Gen. Pike, John Harper, Georgia A., Odd Fellow, Tom Montague, Bell Boy, Coquena (Miss Nailer), Longitude, Long Girl, Wildmoor, Wildfellow, Freeland, Hospodar, Fellowplay, Minnie D., Lucy Long, Mary Corbett, King Dutchman, Long Girl, Annie G., Hypocrite, Clay Stockton, Kaloolah, Longstreet, Riley, Poet Scout, Barefoot, Rainbow, Lew Weir, Lannes, Semper Fidele, Peg Woffington, Irish Chief, Cassius and many others. In addition to his being a son of Leamington, he is much inbred on the dam's side to Sir Archy and Diomed, well fortified with the blood of Herod and Eclipse, upon that of Partner and Regulus, son of the Godolphin Arabian. Longfellow traces in direct female line to the Merlin mare, to which traces Herod (King Herod), Coronation, Knight of St. George, Promised Land, Carlton, imp. Glenelg, etc.

LONGFELLOW.

BROWN HORSE, FOALED 1867; BRED BY MR. JOHN HARPER, NEAR MIDWAY, KY.

*LEAMINGTON, br. h., 1853.

- **Faugh-a-Ballagh, br. h., 1841.**
  - **Sir Hercules, blk. h., 1826.**
    - **Whalebone, br. h., 1807.**
      - **Waxy, b. h., 1790.** Pot-8-o-'s 1773 (Eclipse 1764)—Maria 1777 by Herod 1758 (Tartar 1743)—Lisette 1772 by Snap 1750 (Snip 1736)—Miss Windsor 1751 by God. Ar. 1724—Sist. to Volunteer by Y. Belgrade (Belgrade Turk).
      - **Penelope, b. m., 1798.** Trumpator 1782 (Conductor 1767)—Prunella 1788 by Highflyer 1774 (Herod 1758)—Promise 1768 by Snap 1750 (Snip 1736)—Julia 1756 by Blank 1740 (God. Ar. 1724)—Spectator's dam 1735 by Partner 1718 (Jigg).
    - **Peri, b. m., 1822.**
      - **Wanderer, b. h., 1811.** Gohanna 1790 (Mercury 1778) Catherine 1795 by Woodpecker 1773 (Herod 1758) Camilla 1778 by Trentham 1766 (T. Sweepstakes 1743)—Coquette 1764 by Compton Barb Sist. 1743 to Regulus by God. Ar. 1724.
      - **Thalestris, blk. m., 1809.** Alexander 1782 (Eclipse 1764)—Rival 1800 by Sir Peter T. 1784 (Highflyer 1774) Hornet 1790 by Drone 1777 (Herod 1758)—Manilla 1777 by Goldfinder 1764 (Snap 1750)—Dau. 1766 of Old England 1741 (God. Ar. 1724).
  - **Guiccioli, ch. m., 1823.**
    - **Bob Booty, ch. h., 1804.**
      - **Chanticleer, ch. h., 1787.** Woodpecker 1773 (Herod 1758)—Dau. 1778 of Eclipse 1764 (Marske 1750)—Rosebud 1765 by Snap 1750 (Snip 1736)—Miss Belsea 1753 by Regulus 1739 (God. Ar. 1724)—Dau. of Bartlett's Childers (Darley Ar. 1699).
      - **Ierne, ch. m., 1790.** Bagot 1780 (Herod 1758)—Dau. of Gamahoe (Bustard 1741)—Patty by Tim 1752 (Squirt 1732)—Miss Patch 1737 by Justice 1725 (H. C. Litton Ar.)—Ringtail Galloway by Curwen Bay Barb.
    - **Flight, ch. m., 1809.**
      - **Ir. Escape, ch. h., 1802.** Commodore 1793 (Tom Tug 1777)—Moll-in-the-Wad 1791 by Highflyer 1774 (Herod 1758)—Shift 1779 by Sweetbriar 1769 (Syphon 1750)—Black Susan 1762 by Snap 1750 (Snip 1736) Dau. of Cade 1734 (God. Ar. 1724).
      - **Y. Heroine, —.** Bagot 1780 (Herod 1758)—Heroine 1775 by Hero 1758 (Cade 1734)—Dau. 1766 of Snap 1750 (Snip 1736)—Sist. 1748 to Regulus by God. Ar. 1724 Grey Robinson 1723 by Bald Galloway (St. Victor Barb).
- **Daughter of, b. m., 1841.**
  - **Pantaloon, ch. h., 1824.**
    - **Castrel, ch. h., 1801.**
      - **Buzzard,* ch. h., 1787.** Woodpecker 1773 (Herod 1758)—Misfortune 1775 by Dux 1761 (Matchem 1748)—Curiosity 1760 by Snap 1750 (Snip 1736)—Dau. 1749 of Regulus 1739 (God. Ar. 1724)—Dan. of Bartlett's Childers (Darley Ar. 1699).
      - **Daughter of, b. m., —.** Alexander 1782 (Eclipse 1764)—Dau. of Highflyer 1774 (Herod 1758)—Dau. of Alfred 1770 (Matchem 1748)—Dau. 1770 of Engineer 1756 (Sampson 1745)—Bay Malton's dam 1756 by Cade 1734 (God. Ar. 1724).
    - **Idalia, ch. m., 1815.**
      - **Peruvian, b. h., 1806.** Sir Peter T. 1784 (Highflyer 1774)—Dau. 1788 of Boudrow 1777 (Eclipse 1764)—Dau. 1764 of Squirrel 1754 (Traveller 1735)—Sist. to Lowther's Babraham by Babraham 1740 (God. Ar. 1724)—Dau. 1759 of Golden Ball 1735 (Partner 1718).
      - **Musidora, b. m., 1804.** Meteor 1783 (Eclipse 1764)—Maid-of-all-Work 1786 by Highflyer 1774 (Herod 1758)—Dau. 1771 of Syphon 1750 (Squirt 1732)—Dau. of Regulus 1739 (God. Ar. 1724)—Sist. 1758 to Judgement by Snip 1736 (Flying Childers 1715).
  - **Daphne, br. m., 1837.**
    - **Laurel, br. h., 1824.**
      - **Blacklock, b. h., 1814.** Whitelock 1803 (Hambletonian 1792) Dau. 1799 of Coriander 1786 (Pot-8-o-'s 1773)—Wildgoose 1792 by Highflyer 1774 (Herod 1758)—Coheiress 1786 by Pot-8-o-'s 1773 (Eclipse 1764)—Manilla 1777 by Goldfinder 1764 (Snap 1750).
      - **Wagtail, b. m., 1818.** Prime Minister 1810 (Sancho 1801)—Tranby's dam 1812 by Orville 1799 (Beningbrough 1791) Miss Grimstone 1796 by Weazle 1776 (Herod 1758)—Dau. of Ancaster 1768 (Blank 1740)—Dau. of Damascus Ar. 1754.
    - **Maid of Honor, br. m., 1829.**
      - **Champion, ch. h., 1812.** Selim 1802 (Buzzard* 1787)—Podagra 1805 by Gouty 1793 (Sir Peter T. 1784)—Jet 1785 by Magnet 1770 (Herod 1758)—Jewel 1775 by Squirrel 1754 (Traveller 1735)—Sophia 1774 by Blank 1740 (God. Ar. 1724).
      - **Etiquette, br. m., 1820.** Orville 1799 (Beningbrough 1791)—Boadicea 1807 by Alexander 1782 (Eclipse 1764)—Brunette by Amaranthus 1766 (Old England 1741)—Mayfly 1771 by Matchem 1748 (Cade 1734)—Dau. 1755 of Starling 1738 (Starling 1727).

NANTURA, b. m., 1855.

- **Brawner's Eclipse, ch. h., 1839.**
  - **American Eclipse, ch. h., 1814.**
    - **Duroc, ch. h., 1806.**
      - ***Diomed, ch. h., 1777.** Florizel 1768 (Herod 1758)—Dau. 1763 of Spectator 1749 (Crab 1722) Horatia 1758 by Blank 1740 (God. Ar. 1724)—Dau. of Fl. Childers 1715 (Darley Ar. 1699)—Miss Belvoir by Grey Grantham (Brownlow Turk).
      - **Amanda, ch. m., 1800.** Grey Diomed 1786 (*Medley 1776)—Dau. of Va. Cade (Partner 1755)—Dau. of Hickman's Independence 1776 (*Fearnought 1755)—Dau. of *Badger (Bosphorus 1754.)
    - **Miller's Damsel, gr. m., 1802.**
      - ***Messenger, gr. h., 1780.** Mambrino 1768 (Engineer 1756)—Dau. 1774 of Turf 1760 (Matchem 1748) Dau. 1761 of Regulus 1739 (God. Ar. 1724)—Dau. 1753 of Starling 1727 (Bay Bolton 1705)—Snap's dam 1739 by Fox 1714 (Clumsy).
      - ***Pot-8-o-'s mare, ch. m., 1792.** Pot-8-o-'s 1773 (Eclipse 1764)—Dau. 1778 of Gimcrack 1760 (Cripple 1750)—Snap Dragon 1759 by Snap 1750 (Snip 1736)—Dau. 1749 of Regulus 1739 (God. Ar. 1724)—Dau. of Bartlett's Childers (Darley Ar. 1699)
  - **Daughter of, ch. m., 1834.**
    - **Henry, ch. h., 1819.**
      - **Sir Archy, b. h., 1805.** *Diomed 1777 (Florizel 1768)—*Castianira 1796 by Rockingham 1781 (Highflyer 1774)—Tabitha 1782 by Trentham 1766 (T. Sweepstakes 1743)—Dau. of Bosphorus 1754 (Babraham 1740) Dau. of Forester 1750 (Forester 1736).
      - **Daughter of, ch. m., —.** *Diomed 1777 (Florizel 1768) Bellona by Bellair 1786 (*Medley 1776) Indian Queen by Pilgrim 1774 (*Fearnought 1755)—Dau. of *Valiant 1756 (Dormouse 1738)—Dau. of *Janus 1746 (Janus 1738).
    - **Y. Romp, ch. m., 1815.**
      - **Duroc, ch. h., 1806.** *Diomed 1777 (Florizel 1768)—Amanda 1800 by Grey Diomed 1786 (*Medley 1776) Dau. of Va. Cade (Partner 1755)—Dau. of Independence 1776 (*Fearnought 1755) Dau. of *Badger (Bosphorus 1754).
      - **Old Romp, ch. m., 1805.** *Messenger 1780 (Mambrino 1768)—*Dau. 1792 of Pot-8-o-'s 1773 (Eclipse 1764)—Dau. 1778 of Gimcrack 1760 (Cripple 1750)—Snap Dragon 1759 by Snap 1750 (Snip 1736)—Dau. 1749 of Regulus 1739 (God. Ar. 1724).
- **Quiz, b. m., 1836.**
  - **Bertrand, b. h., 1821.**
    - **Sir Archy, b. h., 1805.**
      - ***Diomed, ch. h., 1777.** Florizel 1768 (Herod 1758) Dau. 1763 of Spectator 1749 (Crab 1722) Horatia 1758 by Blank 1740 (God. Ar. 1724)—Dau. of Fl. Childers 1715 (Darley Ar. 1699) Miss Belvoir by Grey Grantham (Brownlow Turk).
      - ***Castianira, br. m., 1796.** Rockingham 1781 (Highflyer 1774)—Tabitha 1782 by Trentham 1766 (T. Sweepstakes 1743) Dau. of Bosphorus 1754 (Babraham 1740) Dau. of W's Forester 1750 (Forester 1736)—Dau. of Coalition Colt 1739 (God. Ar. 1724).
    - **Eliza, ch. m., 1804.**
      - ***Bedford, b. h., 1792.** Dungannon 1780 (Eclipse 1764)—Fairy 1782 by Highflyer 1774 (Herod 1758) Fairy Queen 1762 by Y. Cade 1747 (Cade 1734)—Routh's Black Eyes 1741 by Crab 1722 (Alcock Ar.)—Warlock Galloway 1728 by Snake (Lister Turk).
      - ***Mambrina, ch. m., 1785.** Mambrino 1768 (Engineer 1756)—Sist. to Sally by Blank 1740 (God. Ar. 1724)—Dau. of Ward 1731 (Crab 1722)—Dau. of Merlin (Bustler) Lady Mare by Duke of Ancaster's Pert (Ely Turk).
  - **Lady Fortune,‡ ch. m., —.**
    - **Blue Beard, ch. h., 1802.**
      - ***Stirling, b. h., 1791.** Volunteer 1780 (Eclipse 1764)—Harriet 1783 by Highflyer 1774 (Herod 1758)—Dau. 1767 of Y. Cade 1747 (Cade 1734)—Childerkin 1749 by Second 1732 (Flying Childers 1715)—Snap's dam 1739 by Fox 1714 (Clumsey).
      - **Brilliant, ch. m., 1779.** *Matchem (Bosphorus 1754)—*Dau. of Brilliant 1750 (Crab 1722)—Dau. of Tartar 1743 (Partner 1718)—Dau. of a Son of Flying Childers 1715 Bushy Molly 1717 by H. C. Litton Ar.
    - **Woodpecker's dam, ch. m., 1802.**
      - ***Buzzard, ch. h., 1787.** Woodpecker 1773 (Herod 1758)—Misfortune 1775 by Dux 1761 (Matchem 1748)—Curiosity 1760 by Snap 1750 (Snip 1736)—Dau. 1749 of Regulus 1739 (God. Ar. 1724)—Dau. of Bartlett's Childers (Darley Ar. 1699).
      - **The Fawn, b. m., —.** Craig's Alfred (*Medley 1776)—Shepherdess by Wormsley's King Herod (*Fearnought 1755)—Dau. of Moreton's *Traveller 1747 (Partner 1718)—Dau.* of Whittington (Cullen Ar.) Dau. of *Childers (Blaze 1733)—*Dau. 1738 of Babraham 1740 (God. Ar. 1724) Dau. blk. m. 1752 of Starling 1727 (Bay Bolton 1705)—Salome blk. m. 1733 by Bethell's Arabian—Dau. of Graham's Champion 1707 (Harpur's Ar.)—Dau. of Darley Ar. 1699 Dau. of Merlin (Bustler).

‡ Lady Fortune is by Brimmer or Blue Beard. Blue Beard's pedigree is accepted upon Mr. J. W. Fenwick's statement.

IMPORTED LOYALIST.

Owned in Belle Meade Stud, Gen. W. H. Jackson, proprietor, Nashville, Tenn. Yearling sales annually in May.

LOYALIST is a bay horse, foaled 1885, bred in the Yardley Stud. Loyalist is a very highly bred horse, full brother to Paradox, being by Sterling, out of Casuistry by The Miner, Blair Athol's conqueror out of Lady Caroline by Orlando. Loyalist never started, but his brother Paradox was a first-class race-horse. As a two-year old he made his maiden essay in the Middle Park Plate, for which he ran a dead heat with Royal Hampton for third place, behind Melton (the Derby winner) and Xaintrailles, nine others finishing in the rear. The only other race in which he started as a two-year old was the Dewhurst Plate, which he won by three lengths, turning the tables on Xaintrailles and also beating Lonely (winner of the Oaks), Farewell (winner of the 1,000 Guineas) and six others. As a three-year old Paradox began by winning the 2,000 Guineas; he ran second to Melton in the Derby, beaten a head only. After that he never knew defeat. He was then shipped to France, where he captured the Grand Prize of Paris with the greatest ease. Having returned to England, he won the Sussex Stakes at Goodwood, giving Royal Hampton, who was second, 9 lbs., and conceding more weight to the other four runners. At Newmarket he won the Champion Stakes from four others with 100 to 12 laid on him. His racing career closed with his victory in a handicap across the flat at Newmarket, in which he gave from 11 to 34 lbs. to his antagonists. Sterling, his sire, was a great race-horse, and sired many cracks, including Isonomy, Enterprise, winner of the 2,000 Guineas, and Enthusiast (winner of the 2,000 Guineas, etc.), Harvester (who ran a dead heat in the Derby with St. Gatien), and many other good ones. Casuistry, Loyalist's dam, only started as a two-year old, when she won a Maiden Plate at Epsom, beating Isolina, Beddington, Broad Arrow, Fanfare, Mexico and seven others. She is a full sister to that good race-horse, Controversy, who won hosts of races, including the Lincoln Handicap, the Liverpool Cup, carrying 124 lbs., the Edinburgh Gold Cup, carrying 130 lbs. and beating Craig Millar (winner of the St. Leger, etc.) and Thunder, winner of the City and Suburban, etc., and he also beat the speedy Lowlander in a match. Casuistry is a daughter of The Miner, winner of the Great Yorkshire Stakes, where he beat Blair Athol. He is full brother to Mineral (that raced under Mr. Ten Broeck's colors), the dam of Kisber (winner of the Derby and Grand Prize of Paris), and of Wenlock (winner of the St. Leger). Her dam, Lady Caroline, is by Orlando (winner of the Derby, and sire of Teddington, the winner of the Derby, the 2,000 Guineas winners, Fazzoletto, Diophantus and Fitz Roland, and of Imperieuse, the winner of the 1,000 Guineas and St. Leger). Out of daughters of Orlando came Camballo, Gang Forward, General Peel, winners of the 2,000 Guineas, Petrarch, winner of the 2,000 Guineas and St. Leger. Not to forget that imp. Eclipse, a highly valued cross for speed, is son of Orlando. Lady Caroline threw other winners besides Casuistry and Controversy, her list including the stake winners Childe Harold, Lady Stately and Convert, and is out of a Stockwell mare, Lady Blanche, a combination of blood that produced Gang Forward (winner of the 2,000 Guineas), and she is a daughter of Clementine, winner of the 1,000 Guineas, and half sister to Bay Middleton (winner of the 2,000 Guineas and Derby), and Achmet (winner of the 2,000 Guineas) by Venison. Clementine's dam was that famous mare Cobweb (winner of the 1,000 Guineas and Oaks) by Phantom. All the sons of Sterling are getting racers abroad. Isonomy is the sire of Common, winner last year of the 2,000 Guineas, Derby and St. Leger, Seabreeze, etc. There is no better bred son of Sterling than Loyalist, not even Isonomy himself, Enterprise, Enthusiast, Doubloon or Fernandez.

Loyalist descends in direct line from the same Royal mare to which traces Woodpecker, Waxy Pope, Whalebone, Whisker, Partisan, Bay Middleton, imp. Glencoe, Melbourne, Robert the Devil, Bend'or, imp. Eclipse (Morris), Uncas, Wanderer and all the Prunella family.

Pot-8-o's 1773 Eclipse 1764—Marske 1750 Squirt 1732—Bartlett's Childers Darley Arabian 1699.

LOYALIST.

BAY HORSE, FOALED 1885; BRED AT THE YARDLEY STUD, ENGLAND.

STERLING, b. h., 1868.	Oxford, ch. h., 1857.	Ir. Birdcatcher, ch. h., 1833.	Sir Hercules, blk. h., 1826.	Whalebone, br. h., 1807.	Waxy 1790 (Pot-8-o's 1773)—Penelope 1798 by Trumpator 1782 (Conductor 1767 Prunella 1788 by Highflyer 1774 (Herod 1758) Promise 1768 by Snap 1750 (Snip 1736) Julia 1756 by Blank 1740 (God. Ar. 1724).
				Peri, b. m., 1822.	Wanderer 1811 (Gohanna 1790)—Thalestris 1809 by Alexander 1782 (Eclipse 1764) Rival 1800 by Sir Peter T. 1784 (Highflyer 1774) Hornet 1790 by Drone 1777 (Herod 1758)—Manilla 1777 by Goldfinder 1764 (Snap 1750).
			Guiccioli, ch. m., 1823.	Bob Booty, ch. h., 1804.	Chanticleer 1787 (Woodpecker 1773)—Ierne 1790 by Bagot 1780 (Herod 1758) Dau. of Gamahoe (Bustard 1740) Patty by Tim 1752 (Squirt 1732) Miss Patch 1737 by Justice 1725 (H. C. Litton Ar.).
				Flight, ch. m., 1809.	Ir. Escape 1802 (Commodore 1793) Y. Heroine by Bagot 1780 (Herod 1758—Heroine 1775 by Hero 1753 (Cade 1734)—Dau. 1766 of Snap 1750 (Snip 1736) Sist. 1743 to Regulus by God. Ar. 1724.
		Honey Dear, b. m., 1844.	Plenipotentiary, ch. h., 1831.	Emilius, b. h., 1820.	Orville 1799 (Beningbrough 1791) Emily 1810 by Stamford 1794 (Sir Peter T. 1784) Dau. 1799 of Whiskey 1789 (Saltram 1780)—Grey Dorimant 1781 by Dorimant 1772 (Otho 1760)—Dizzy 1757 by Blank 1740 (God. Ar. 1724).
				Harriet, blk. m., 1819.	Pericles 1809 (Evander 1801)—Dau. 1812 of Selim 1802 (Buzzard* 1787)—Pipylina 1803 by Sir Peter T. 1784 (Highflyer 1774)—Rally 1790 by Trumpator 1782 (Conductor 1767)—Fancy 1780 by Florizel 1768 (Herod 1758).
			My Dear, b. m., 1841.	Bay Middleton, b. h., 1833.	Sultan 1816 (Selim 1802) Cobweb 1821 by Phantom 1808 (Walton 1799) Filagree 1815 by Soothsayer 1808 (Sorcerer 1796)—Web 1808 by Waxy 1790 (Pot-8-o's 1773)—Penelope 1798 by Trumpator 1782 (Conductor 1767).
				Miss Letty, b. m., 1834.	Priam* 1827 (Emilius 1820)—Dau. 1815 of Orville 1799 (Beningbrough 1791)—Dau. 1800 of Buzzard* 1787 (Woodpecker 1773) Hornpipe 1793 by Trumpator 1782 (Conductor 1767)—Luna 1779 by Herod 1758 (Tartar 1743).
	Whisper, b. m., 1857.	Flatcatcher, br. h., 1845.	Touchstone, br. h., 1831.	Camel, br. h., 1822.	Whalebone 1807 (Waxy 1790) Dau. 1812 of Selim 1802 (Buzzard* 1787) Maiden 1801 by Sir Peter T. 1784 (Highflyer 1774)—Dau. 1788 of Phenomenon 1780 (Herod 1758—Matron 1782 by Florizel 1768 (Herod 1758).
				Banter, br. m., 1826.	Master Henry 1815 (Orville 1799) Boadicea 1807 by Alexander 1782 (Eclipse 1764)—Brunette by Amaranthus 1766 (Old England 1741) Mayfly 1771 by Matchem 1748 (Cade 1734)—Dau. 1775 of Anc. Starling 1738 (Starling 1727).
			Decoy, b. m., 1830.	Filho-da-Puta br. h., 1812.	Haphazard 1797 (Sir Peter T. 1784)—Mrs. Barnet 1806 by Waxy 1790 (Pot-8-o's 1773)—Dau. 1788 of Woodpecker 1773 (Herod 1758) Heinel 1771 by Squirrel 1754 (Traveller 1735) Principessa 1762 by Blank 1740 (God. Ar. 1724).
				Finesse, b. m., 1815.	Peruvian 1806 (Sir Peter T. 1784) Violante 1802 by John Bull 1789 (Fortitude 1777) Dau. 1788 of Highflyer 1774 (Herod 1758) Everlasting 1775 by Eclipse 1764 (Marske 1750) Hyæna 1762 by Snap 1750 (Snip 1736).
		Silence, b. m., 1848.	Melbourne, br. h., 1834.	Humphrey Clinker, b. h., 1822.	Comus 1809 (Sorcerer 1796)—Clinkerina 1812 by Clinker 1805 (Sir Peter T. 1784) Pewet 1786 by Tandem 1773 (Syphon 1750)—Termagant by Tantrum 1760 (Cripple 1750)—Cantatrice by Sampson 1745 (Blaze 1733).
				Daughter of, b. m., 1825.	Cervantes 1806 (Don Quixote 1784) Dau. 1818 of Golumpus 1802 (Gohanna 1790) Dau. 1819 of Paynator 1791 (Trumpator 1782)—Circle 1802 by St. George 1789 (Highflyer 1774)—Abigail 1788 by Woodpecker 1773 (Herod 1758).
			Secret, ch. m., 1841.	Hornsea, ch. h., 1832.	Velocipede 1825 (Blacklock 1814) Dau. 1820 of Cerberus 1802 (Gohanna 1790) Miss Crandeld 1803 by Sir Peter T. 1784 (Highflyer 1774)—Dau. 1796 of Pegasus 1784 (Eclipse 1764) Dau. 1783 of Paymaster 1766 (Blank 1740).
				Solace, b. m., 1830.	Longwaist 1821 (Whalebone 1807) Dulcamara 1818 by Waxy 1790 (Pot-8-o's 1773)—Witchery 1814 by Sorcerer 1796 (Trumpator 1782)—Cobbea 1802 by Skyscraper 1786 (Highflyer 1774) Dau. 1788 of Woodpecker 1773 (Herod 1758).
CASUISTRY, b. m., 1876.	The Miner, ch. h., 1861.	Rataplan, ch. h., 1850.	The Baron, ch. h., 1842.	Ir. Birdcatcher ch. h., 1833.	Sir Hercules 1826 (Whalebone 1807)—Guiccioli 1823 by Bob Booty 1804 (Chanticleer 1787)—Flight 1809 by Ir. Escape 1802 (Commodore 1793)—Y. Heroine by Bagot 1780 (Herod 1758)—Heroine 1775 by Hero 1753 (Cade 1734).
				Echidna, b. m., 1838.	Economist 1825 (Whisker 1812) Miss Pratt 1825 by Blacklock 1814 (Whitelock 1803)—Gadabout 1812 by Orville 1799 (Beningbrough 1791)—Minstrel 1803 by Sir Peter T. 1784 (Highflyer 1774) Matron 1782 by Florizel 1768 (Herod 1758).
			Pocahontas, b. m., 1837.	Glencoe,* ch. h., 1831.	Sultan 1816 (Selim 1802) Trampoline 1825 by Tramp 1810 (Dick Andrews 1797)—Web 1808 by Waxy 1790 (Pot-8-o's 1773)—Penelope 1798 by Trumpator 1782 (Conductor 1767)—Prunella 1788 by Highflyer 1774 (Herod 1758).
				Marpessa, b. m., 1830.	Muley 1810 (Orville 1799)—Clare 1824 by Marmion 1806 (Whiskey 1789)—Harpalice 1814 by Gohanna 1790 (Mercury 1778)—Amazon 1799 by Driver 1783 (Trentham 1766)—Fractions 1794 by Mercury 1778 (Eclipse 1764).
		Manganese, ch. m., 1853.	Ir. Birdcatcher, ch. h., 1833.	Sir Hercules, blk. h., 1826.	Whalebone 1807 (Waxy 1790)—Peri 1822 by Wanderer 1811 (Gohanna 1790)—Thalestris 1809 Alexander 1782 (Eclipse 1764)—Rival 1800 by Sir Peter T. 1784 (Highflyer 1774)—Hornet 1790 by Drone 1777 (Herod 1758).
				Guiccioli, ch. m., 1823.	Bob Booty 1804 (Chanticleer 1787)—Flight 1809 by Ir. Escape 1802 (Commodore 1793)—Y. Heroine by Bagot 1780 (Herod 1758)—Heroine 1775 by Hero 1753 (Cade 1734) Dau. 1766 of Snap 1750 (Snip 1736).
			Moonbeam, ch. m., 1838.	Tomboy, b. h., 1829.	Jerry 1821 (Smolensko 1810)—Dau. 1817 of Ardrossan 1809 (John Bull 1789)—Lady Eliza 1813 by Whitworth 1805 (Agonistes 1797)—Dau. 1793 of Spadille 1784 (Highflyer 1774)—Sylvia 1783 by Y. Marske 1771 (Marske 1750).
				Lunatic, b. m., 1818.	Prime Minister 1810 (Sancho 1801)—Maniac 1803 by Shuttle 1793 (Y. Marske 1771)—Anticipation 1802 by Beningbrough 1791 (King Fergus 1775)—Expectation 1779 by Herod 1758 (Tartar 1743) Dau. 1758 of Skim 1746 (Starling 1727).
	Lady Caroline, b. m., 1861.	Orlando, b. h., 1841.	Touchstone, br. h., 1831.	Camel, br. h., 1822.	Whalebone 1807 (Waxy 1790) Dau. 1812 of Selim 1802 (Buzzard* 1787)—Maiden 1801 by Sir Peter T. 1784 (Highflyer 1774)—Dau. 1788 of Phenomenon 1780 (Herod 1758)—Matron 1782 by Florizel 1768 (Herod 1758).
				Banter, br. m., 1826.	Master Henry 1815 (Orville 1799)—Boadicea 1807 by Alexander 1782 (Eclipse 1764)—Brunette by Amaranthus 1766 (Old England 1741)—Mayfly 1771 by Matchem 1748 (Cade 1734)—Dau. 1775 of Anc. Starling 1738 (Starling 1727).
			Vulture, ch. m., 1833.	Langar, ch. h., 1817.	Selim 1802 (Buzzard* 1787) Dan. 1808 of Walton 1799 (Sir Peter T. 1784)—Y. Giantess 1790 by Diomed* 1777 (Florizel 1768)—Giantess 1769 by Matchem 1748 (Cade 1734)—Molly Long Legs 1753 by Babraham 1740 (God. Ar. 1724).
				Kite, b. m., 1821.	Bustard 1813 (Castrel 1801)—Olympia 1815 by Sir Oliver 1800 (Sir Peter T. 1784) Scotilla 1795 by Anvil 1777 (Herod 1758)—Scota 1783 by Eclipse 1764 (Marske 1750) Harmony 1775 by Herod 1758 (Tartar 1743).
		Lady Blanche, b. m., 1856.	Stockwell, ch. h., 1849.	The Baron, ch. h., 1842.	Ir. Birdcatcher 1833 (Sir Hercules 1826)—Echidna 1838 by Economist 1825 (Whisker 1812)—Miss Pratt 1825 by Blacklock 1814 (Whitelock 1803)—Gadabout 1812 by Orville 1799 (Beningbrough 1791) Minstrel 1803 by Sir Peter T. 1784 (Highflyer 1774).
				Pocahontas, b. m., 1837.	Glencoe* 1831 (Sultan 1816)—Marpessa 1830 by Muley 1810 (Orville 1799) Clare 1824 by Marmion 1806 (Whiskey 1789) Harpalice 1814 by Gohanna 1790 (Mercury 1778)—Amazon 1799 by Driver 1783 (Trentham 1766).
			Clementina, b. m., 1844.	Venison, br. h., 1833.	Partisan 1811 (Walton 1799)—Fawn 1823 by Smolensko 1810 (Sorcerer 1796)—Jerboa 1803 by Gohanna 1790 (Mercury 1778)—Camilla 1778 by Trentham 1766 (T. Sweepstakes 1743)—Coquette 1764 by Compton Barb.
				Cobweb, b. m., 1821.	Phantom 1808 (Walton 1799)—Filagree ch. m. 1815 by Soothsayer 1808 (Sorcerer 1796)—Web b. m. 1808 by Waxy 1790 (Pot-8-o's 1773)—Penelope b. m. 1798 by Trumpator 1782 (Conductor 1767)—Prunella b. m. 1788 by Highflyer 1774 (Herod 1758—Promise br. m. 1768 by Snap 1750 (Snip 1736)—Julia b. m. 1756 by Blank 1740 (God. Ar. 1724)—Spectator's dam b. m. 1735 by Partner 1718 (Jigg)—Bonny Lass b. m. 1723 by Bay Bolton 1705 (Grey Hautboy)—Dau. of Darley Ar. 1699—Dau. of Byerly Turk—Dau. of Taffolet Barb—Dau. of Place's White Turk—Natural Barb mare of Mr. Tregonwell's.

LUKE BLACKBURN.

WINNER OF THE TIDAL, LONG ISLAND ST. LEGER AND GREAT CHALLENGE STAKES AT SHEEPSHEAD BAY, OCEAN AND CHAMPION STAKES AT LONG BRANCH, UNITED STATES HOTEL STAKES AND GRAND UNION PRIZE AT SARATOGA, KENTUCKY ST. LEGER AND GREAT AMERICAN STALLION STAKES AT LOUISVILLE, KY.

In the Belle Meade Stud. Application to Gen. W. H. Jackson, proprietor, lock box 383, Nashville, Tenn. Annual sale of yearlings about May 1st.

LUKE BLACKBURN by imp. Bonnie Scotland, bred by Messrs. J. & A. C. Franklin, Sumner Co., Tenn., foaled 1877, dam Nevada by Lexington, out of Lightsome, dam of Salina, Sprightly and Crucifix by imp. Glencoe. In 1879, as a two-year old, Blackburn ran with indifferent success. He started thirteen times, won two races, was second in six, third in one and unplaced in four. Lexington, May 12, ran second in the Colt Stakes, ½ mile, time, 0.50. Louisville, Ky., May 21, ran second to Kimball in the Alexander Stakes, ½ mile, time, 0.49¾, eight started. St. Louis, Mo., June 18, again finished second to Kimball in the Hotel Stakes, ¾ of a mile, time, 1.16½. Chicago, Ill., June 23, ran second to Kimball (third time) in Grand Pacific Hotel Stakes, ¾ of a mile, time 1.18½, Wargentine finished third. Saratoga, Aug. 2, ran second to Lucy George, ⅝ of a mile, time, 1.04¾; Giroflé and Cassatt behind him; same place, Aug. 12, ran second to Grenada in the Windsor Hotel Stakes, ⅝ of a mile, time, 1.03½, seven started. Autumn Meeting, Coney Island Jockey Club, Prospect Park, Sept. 6, won his maiden race, the Breeze Purse, ¾ of a mile, defeating Giroflé, Queen's Own and four others, time, 1.17½. Brighton Beach Fair Grounds, N. Y., Sept. 18, won the Ocean Stakes, ¾ of a mile, defeating Elias Lawrence (Late Bilstein), Quito and three others, time, 1.18¼. After this race Luke Blackburn was retired, and became the property of Dwyer Bros., who gave $2,500 for him.

In the year 1880 Luke Blackburn started in twenty-four races, lost two and won twenty-two. His first race was at Lexington, Ky., when he ran third to Fonso and Kinkead in the Phœnix Hotel Stakes, 1¼ miles, won by Fonso in 2.08¾, Blackburn sick and was unfit to run. Won a dash of ⅞ of a mile at Jerome Park in 1.18, defeating Checkmate and three others; won a dash of 1⅛ miles in 1.58, beating Scotilla, Checkmate and two others; won a race of 1 mile and 3 furlongs, in 2.28¼, beating Scotilla and two others; defeated Monitor, 1½ miles, in 2.39½, and won the Handicap Sweepstakes, 1¼ miles, in 2.13.

All these races were run at Jerome Park during the Spring meeting. Sheepshead Bay, June 19, won the Tidal Stake, 1 mile, in 1.45, defeating Kimball, Kitty J. and Grenada; June 22, won the Coney Island Handicap, 1 mile and 3 furlongs, in 2.24½, beating Duke of Montrose and Vagrant; June 26, was beaten in purse race, 1¼ miles, in 2.12, by Duke of Montrose, Blackburn fell in the first quarter and unseated his jockey. Long Branch, July 3, won the Ocean Stakes, 1 mile and a furlong, in 2.03½, beating Duke of Montrose and Harold; July 10, won Sweepstakes, 1¼ miles, in 2.11½, beating, with 110 lbs. up, Duke of Montrose, 105 lbs., and Grenada, 110 lbs. Saratoga, July 16, won the All-aged Sweepstakes, 1⅛ miles, in 2.11⅝, beating Checkmate and Volturno; July 24, won mile and a furlong, in 1.58, beating Gabriel and Giroflé; July 27, won dash of a mile, in 1.43½, beating Turfman; July 31, won Summer Handicap, 1½ miles, in 2.39, carrying 110 lbs., conceding Juanita 20, Gen. Philips 23 and Ada Glenn 17 lbs. Saratoga, Aug. 5, won United States Hotel Stakes for three-year olds, 1½ miles, in 2.41, beating Ferncliff, Oden, etc.; Aug. 7, won the Grand Union Hotel Prize, handicap, 116 lbs. up, 1¾ miles, in 3.07, beating One Dime, 110 lbs., Glenmore, 118 lbs., and three others. In this race he conceded all from 20 to 36 lbs.; Aug. 12, won the Kenner Stakes, 2 miles, in 3.35½, beating Glidelia and Oden. Long Branch, Aug. 17, won the Champion Stakes, 1½ miles, in 2.34, beating Monitor, Uncas, Grenada and Report, in the fastest and best race ever run at the distance to that date. Sheepshead Bay, Sept. 4, won the Great Challenge Stakes for all ages, 1½ miles, in 2.38, beating Monitor, Uncas and One Dime; Sept. 9, won the Long Island St. Leger, 1¾ miles, in 4.07½; Sept. 14, won a match race for $5,000, beating Uncas, each carrying 108 lbs., 1½ miles, in 2.42½. Louisville, Ky., Sept. 27, won the Kentucky St. Leger, 2 miles, in 3.42, beating Kinkead; Sept. 30, won the Great American Stallion Stakes, 1¾ miles, in 3.04, beating Kimball and Big Medicine. In this race he injured the quarter of one of his fore feet badly.

At four years old started in two races. Jerome Park, won club purse, 1 mile, in

(*Continued on page 210.*)

LUKE BLACKBURN.

BAY HORSE, FOALED 1877; BRED BY MESSRS. J. & A. C. FRANKLIN, SUMNER CO., TENNESSEE.

Sire/Dam	Grandparents	Great-grandparents	Great-great-grandparents		Ancestry
*BONNIE SCOTLAND, b. h., 1853.	Iago, b. h., 1843.	Don John, b. h., 1835.	Waverley, br. h., 1817.	Whalebone, br. h., 1807.	Waxy 1790 (Pot-8-o-'s 1773) Penelope 1798 by Trumpator 1782 (Conductor 1767) Prunella 1788 by Highflyer 1774 (Herod 1758) Promise 1768 by Snap 1750 (Snip 1736)—Julia 1756 by Blank 1740 (God. Ar. 1724).
				Margaretta, br. m., 1802.	Sir Peter T. 1784 (Highflyer 1774) Dau. 1792 of Highflyer 1774 (Herod 1758) Nutcracker 1767 by Matchem 1748 (Cade 1734) Miss Starling 1759 by Starling 1727 (Bay Bolton 1705) Ringbone 1732 by Partner 1718 (Jigg).
			Daughter of [Tramp or] Comus, gr. m., 1821.	Comus, ch. h., 1809.	Sorcerer 1796 (Trumpator 1782) Epsom Lass 1801 by Sir Peter T. 1784 (Highflyer 1774)—Alexina 1788 by King Fergus 1775 (Eclipse 1764)—Lardella 1780 by Y. Marske 1771 (Marske 1750)—Dau. of Y. Cade 1747 (Cade 1734).
				Marciana, gr. m., 1809.	Stamford 1794 (Sir Peter T. 1784)—Marcia 1797 by Coriander 1786 (Pot-8-o-'s 1773)—Faith 1779 by Pacolet 1763 (Blank 1740)—Atalanta 1769 by Matchem 1748 (Cade 1734)—Y. Lass-of-the-Mill 1756 by Oroonoko 1745 (Crab 1722).
		Scandal, b. m., 1832.	Selim, ch. h., 1802.	Buzzard,* ch. h., 1787.	Woodpecker 1773 (Herod 1758)—Misfortune 1775 by Dax 1761 (Matchem 1748) Curiosity 1760 by Snap 1750 (Snip 1736) Dau. 1749 of Regulus 1739 (God. Ar. 1724)—Dau. of Bartlett's Childers (Darley Ar. 1699).
				Daughter of, b. m., —.	Alexander 1782 (Eclipse 1764) Dau. of Highflyer 1774 (Herod 1758) Dau. of Alfred 1770 (Matchem 1748) Dau. 1770 of Engineer 1756 (Sampson 1745) Bay Malton's dam 1756 by Cade 1734 (God. Ar. 1724).
			Daughter of, br. m., 1816.	Haphazard, br. h., 1797.	Sir Peter T. 1784 (Highflyer 1774)—Miss Hervey 1775 by Eclipse 1764 (Marske 1750)—Clio 1760 by Y. Cade 1747 (Cade 1734)—Dau. of Starling 1727 (Bay Bolton 1705)—Dau. of Bartlett's Childers (Darley Ar. 1699).
				Princess, b. m., 1799.	Precipitate* 1787 (Mercury 1778) Colibri 1793 by Woodpecker 1773 (Herod 1758)—Camilla 1778 by Trentham 1766 (T. Sweepstakes 1743)—Coquette 1765 by Compton Barb—Sist. 1743 to Regulus by God. Ar. 1724.
	Queen Mary, b. m., 1843.	Gladiator, ch. h., 1833.	Partisan, b. h., 1811.	Walton, b. h., 1799.	Sir Peter T. 1784 (Highflyer 1774) Arethusa 1792 by Dungannon 1780 (Eclipse 1764)—Dau. 1777 of Prophet 1760 (Regulus 1739)—Virago 1764 by Snap 1750 (Snip 1736)—Dau. of Regulus 1739 (God. Ar. 1724).
				Parasol, b. m., 1800.	Pot-8-o-'s 1773 (Eclipse 1764)—Prunella 1788 by Highflyer 1774 (Herod 1758) Promise 1768 by Snap 1750 (Snip 1736) Julia 1756 by Blank 1740 (God. Ar. 1724)—Spectator's dam 1735 by Partner 1718 (Jigg).
			Pauline, b. m., 1826.	Moses, b. h., 1819.	(Whalebone or) Seymour 1807 (Delpini 1781)—Dau. 1807 of Gohanna 1790 (Mercury 1778)—Grey Skim 1793 by Woodpecker 1773 (Herod 1758) Dau. 1780 of Herod 1758 (Tartar 1743)—Y. Hag 1761 by Skim 1746 (Starling 1727).
				Quadrille, b. m., 1815.	Selim 1802 (Buzzard* 1787) Canary Bird 1806 by [Whiskey or] Sorcerer 1796 (Trumpator 1782)—Canary 1797 by Coriander 1786 (Pot-8-o-'s 1773)—Miss Green 1787 by Highflyer 1774 (Herod 1758) Harriet 1769 by Matchem 1748 (Cade 1734).
		Daughter of, b. m., 1840.	Plenipotentiary, ch. h., 1831.	Emilius, b. h., 1820.	Orville 1799 (Beningbrough 1791)—Emily 1810 by Stamford 1794 (Sir Peter T. 1784)—Dau. 1799 of Whiskey 1789 (Saltram* 1780)—Grey Dorimant 1781 by Dorimant 1772 (Otho 1760)—Dizzy 1757 by Blank 1740 (God. Ar. 1724).
				Harriet, blk. m., 1819.	Pericles 1809 (Evander 1801)—Dau. 1812 of Selim 1802 (Buzzard* 1787) Pipylina 1803 by Sir Peter T. 1784 (Highflyer 1774)—Rally 1790 by Trumpator 1782 (Conductor 1767)—Fancy 1780 (Sist. to Diomed*) by Florizel 1768 (Herod 1758).
			Myrrha, b. m., 1830.	Whalebone, br. h., 1807.	Waxy 1790 (Pot-8-o-'s 1773)—Penelope 1798 by Trumpator 1782 (Conductor 1767)—Prunella 1788 by Highflyer 1774 (Herod 1758)—Promise 1768 by Snap 1750 (Snip 1736)—Julia 1756 by Blank 1740 (God. Ar. 1724).
				Gift, gr. m., 1818.	Y. Gohanna 1810 (Gohanna 1790)—Dau. 1802 of Sir Peter T. 1784 (Highflyer 1774) Dau. 1788 of Trumpator 1782 (Conductor 1767)—Sist. 1774 to Postmaster by Herod 1758 (Tartar 1743) Dau. of Snap 1750 (Snip 1736).
NEVADA, b. m., 1869.	Lexington, b. h., 1850.	Boston, ch. h., 1833.	Timoleon, ch. h., 1813.	Sir Archy, b. h., 1805.	*Diomed 1777 (Florizel 1768)—*Castianira 1796 by Rockingham 1781 (Highflyer 1774)—Tabitha 1782 by Trentham 1766 (T. Sweepstakes 1743)—Dau. of Bosphorus 1754 (Babraham 1740) Dau. of W's Forester 1750 (Forester 1736).
				Daughter of, ch. m., —.	*Saltram 1780 (Eclipse 1764)—Dau. of Symmes' Wildair 1767 (*Fearnought 1755) Dau. of Tyler's Driver 1760 (*Othello)—Dau. of *Fallower 1761 (Blank 1740) Dau. of *Vampire 1757 (Regulus 1739).
			Daughter of, ch. m., 1814.	Ball's Florizel, ch. h., 1801.	*Diomed 1777 (Florizel 1768)—Dau. of *Shark 1771 (Marske 1750)—Dau. of Harris' Eclipse 1770 (*Fearnought 1755)—Dau. of *Fearnought 1755 (Regulus 1739)—Dau. of *Jolly Roger 1741 (Roundhead 1733).
				Daughter of, b. m., 1799.	*Alderman 1787 (Pot-8-o-'s 1773)—Dau. of *Clockfast 1774 (Gimcrack 1760)—Dau. of Symmes Wildair 1767 (*Fearnought 1755)—Y. Kitty Fisher 1767 by *Fearnought 1755 (Regulus 1739) *Kitty Fisher 1755 by Cade 1734 (God. Ar. 1724).
		Alice Carneal, b. m., 1836.	*Sarpedon, br. h., 1828.	Emilius, b. h., 1820.	Orville 1799 (Beningbrough 1791)—Emily 1810 by Stamford 1794 (Sir Peter T. 1784)—Dau. 1799 of Whiskey 1789 (Saltram* 1780)—Grey Dorimant 1781 by Dorimant 1772 (Otho 1760)—Dizzy 1757 by Blank 1740 (God. Ar. 1724).
				Icaria, b. m., 1824.	The Flyer 1814 (Vandyke Jr. 1808)—Parma 1813 by Dick Andrews 1797 (Joe Andrews 1778)—May 1804 by Beningbrough 1791 (King Fergus 1775) Primrose 1787 by Mambrino 1764 (Engineer 1756)—Cricket 1776 by Herod 1758 (Tartar 1743).
			Rowena, ch. m., 1826.	Sumpter, ch. h., 1818.	Sir Archy 1805 (*Diomed 1777)—Dau. of *Robin Redbreast 1796 (Sir Peter T. 1784)—Dau. 1788 of *Obscurity 1777 (Eclipse 1764)—Slamerkin 1768 by *Wildair 1753 (Cade 1734)—*Cub mare 1762 by Cub 1739 (Fox 1714).
				Lady Grey, b. m., 1817.	Robin Grey 1805 (*Royalist 1790)—Maria 1802 by Melzar 1791 (*Medley 1776)—Dau. of *Highflyer 1784 (Highflyer 1774)—Dau. of *Fearnought 1755 (Regulus 1739)—Dau. of Ariel 1756 (*Traveller 1747).
	Lightsome, ch. m., 1854.	*Glencoe, ch. h., 1831.	Sultan, b. h., 1816.	Selim, ch. h., 1802.	Buzzard* 1787 (Woodpecker 1773)—Dau. of Alexander 1782 (Eclipse 1764)—Dau. of Highflyer 1774 (Herod 1758)—Dau. of Alfred 1770 (Matchem 1748)—Dau. 1770 of Engineer 1756 (Sampson 1745).
				Bacchante, br. m., 1809.	W's Ditto 1800 (Sir Peter T. 1784)—Dau. 1791 of Mercury 1778 (Eclipse 1764)—Dau. 1776 of Herod 1758 (Tartar 1743)—Folly 1771 by Marske 1750 (Squirt 1732)—Vixen 1753 by Regulus 1739 (God. Ar. 1724).
			Trampoline, ch. m., 1825.	Tramp, b. h., 1810.	Dick Andrews 1797 (Joe Andrews 1778)—Dau. 1803 of Gohanna 1790 (Mercury 1778)—Fraxinella 1793 by Trentham 1766 (T. Sweepstakes 1743)—Dau. 1785 of Woodpecker 1773 (Herod 1758)—Everlasting 1775 by Eclipse 1764 (Marske 1750).
				Web, b. m., 1808.	Waxy 1790 (Pot-8-o-'s 1773)—Penelope 1798 by Trumpator 1782 (Conductor 1767)—Prunella 1788 by Highflyer 1774 (Herod 1758)—Promise 1768 by Snap 1750 (Snip 1736)—Julia 1756 by Blank 1740 (God. Ar. 1724).
		Levity, b. m., 1845.	*Trustee, ch. h., 1829.	Catton, b. h., 1809.	Golumpus 1802 (Gohanna 1790)—Lucy Gray 1804 by Timothy 1794 (Delpini 1781)—Lucy 1789 by Florizel 1768 (Herod 1758)—Frenzy 1774 by Eclipse 1764 (Marske 1750)—Dau. of Engineer 1756 (Sampson 1745).
				Emma, ch. m., 1824.	Whisker 1812 (Waxy 1790) Gibside Fairy 1811 by Hermes 1790 (Mercury 1778) Vicissitude 1800 by Pipator 1786 (Imperator 1776)—Beatrice 1791 by Sir Peter T. 1784 (Highflyer 1774) Pyrrha 1771 by Matchem 1748 (Cade 1734).
			Vandal's dam, b. m., 1837.	*Tranby, br. h., 1826.	Blacklock 1814 (Whitelock 1803)—Dau. 1812 of Orville 1799 (Beningbrough 1791) Miss Grimstone 1796 by Weazle 1776 (Herod 1758)—Dau. of Ancaster 1768 (Blank 1740)—Dau. of Damascus Ar. 1754.
				Lucilla, b. m., —.	Trumpator 1823 (Sir Solomon 1805)—Lucy b. m. 1821 by Orphan 1810 (Ball's Florizel 1801)—Lady Grey b. m. 1817 by Robin Grey 1805 (*Royalist 1790) Maria 1802 by Melzar 1791 (*Medley 1776) Dau. of *Highflyer 1784 (Highflyer 1774)—Dau. of *Fearnought 1755 (Regulus 1739)—Dau. of Ariel 1756 (*Traveller 1747)—Dau. of *Jack of Diamonds (Cullen Ar. 1745)—*Duchess by Cullen Ar. 1745 Lady Thigh ch. m. 1731 by Partner 1718 (Jigg) Dau. ch. m. 1728 of Greyhound (Chillaby)—Sophonisba's dam by Curwen Bay Barb—Dau. of D'Arcy Chestnut Ar.—Dau. of Whiteshirt—Lord Darcy's Montagu mare by Old Montagu—Dau. of Hautboy (D'Arcy White Turk) Dau. of Brimmer (D'Arcy Yellow Turk) Royal mare.

LUKE BLACKBURN.—Continued.

1.45, beating Topsy and Potomac. Was unplaced in Coney Island Cup, won by Glenmore in 3.58¾. This closed his racing career.

Luke Blackburn is a light bay, with large star and two white ankles behind: he stands, fairly measured, 15 hands and 2½ inches high. Blackburn is a horse of rare symmetry; measures 69½ in. around the girth: length of shoulder, 27½ in.: around the arm, 19 in.; below the knee, 8 in.; from point of shoulder to point of buttocks, 64 in.; from hip to point of hock, 38 in.; around gaskins, 17 inches.

His pedigree is a happy combination of speedy and stout crosses, being much inbred to Diomed through his best son Sir Archy, and from the family which produced Lexington and Vandal. If high form and racing ability, with faultless action coupled with high breeding and racing descent are tokens of excellence, Luke Blackburn is the peer of any horse living and must prove an invaluable stallion.

He is the sire of the following winners: American Lady, Proctor Knott (winner of the Futurity Stakes), Niagara, Bendigo, Lithbert, Rebellion, Minnie Palmer, Nina W., Taviston, Lemon, Lemon Blossom, Lakewood, Tom Elliott, Cerberus, La Grippe, Tulla Blackburn, Chicora, Luke Alexander, Uncle Bob, Worth, Thad Rowe, Belle Blackburn, Fort Worth and many other winners.

IMPORTED MACCAROON.

Maccaroon is a private stallion in the Brookdale Stud, Estate of D. D. Withers, proprietor, Monmouth County, near Holmdel, N. J.

MACCAROON by Macaroni, son of Sweetmeat, bred by Mr. T. T. Parker, foaled 1871, dam Songstress by Chanticleer, son of Irish Birdcatcher, out of Mrs. Carter by Humphrey, son of Sandbeck by Catton. Maccaroon made his first appearance as a two-year old in the Nursery Stakes, 1 mile, won by Rutherford in 1.47¾, and was unplaced; same meeting, ran second to Asteroid, Sue Washington Colt, three-quarters of a mile, in 1.18¼, beating Atilla, Lava and Weathercock. Won a Club Purse for maiden two-year olds, three-quarters of a mile, in 1.17½, beating Weathercock, Mattie W., Mary Buckley and Juliana. As a three-year old started four times; was unplaced in Withers Stakes, 1 mile, won by Dublin in 1.50, track heavy: was unplaced in Club Purse, 1 mile, won by Atilla, in 1.44½. Long Branch, was unplaced in Jersey Derby, 1½ miles, won by Brigand in 2.44¾: was third in Robins Stake, 2 miles, won by Acrobat, in 3.42, beating Bannerette. Macaroni, his sire, won the 2,000 Guineas and the Derby, and is the sire of Spinaway, the winner of 1,000 Guineas and Oaks, Camelia, winner of the 1,000 Guineas, sire also of Lamphrey, Ramsbury, Vermicelli, Gloucester, Bonny Jean and Macheath, the best two-year old in England in 1882, winner of eight out of eleven prominent events in which he started. Macaroni was second to Hermit on the list of winning sires in 1882. Chanticleer, the sire of his dam, was son of Birdcatcher, own brother to Faugh-a-Ballagh, and sired Habena and Manganese, winner of the 1,000 Guineas, Songstress, winner of the Oaks, Daniel O'Rourke, winner of the Derby, Knight of St. George, Warlock and The Baron, all St. Leger winners. The Baron was sire of the great Stockwell and Rataplan. Maccaroon has not sired many colts which have been trained. Nancy, at two years old, won Homebred Stakes at Jerome Park, three-quarters of a mile, in 1.19, beating Oden, Juanita and others. She also won Monmouth Oaks, beating Glidelia, Bye-and-Bye and others. Macbeth and Macduff, both winners at two years old. Macduff won Champagne Stakes at Jerome Park, and in 1882 won a Handicap Purse at Jerome, 1½ miles, in 1.59, defeating a field of eight, and the Free Handicap, 1¼ miles, in 2.41, 107 lbs., beating Gen. Monroe (4), 105 lbs., and four others. Barbara, Maroon, Electricity, Frank Mullins, Gregory, a superior colt, Manda, Rival, Masher, Helen Block, Cactus, Roonette and some other good ones.

Maccaroon is a red chestnut, 16 hands and ¼ of an inch high, has a small star and snip, and weighs 1,180 lbs. He is a fine formed horse, of good temper, and only needed a chance to distinguish himself in the stud. His daughters should be invaluable in the stud, and his son, Macduff, is a success as a stallion.

MACAROON.

CHESTNUT HORSE, FOALED 1871; BRED BY MR. T. T. PARKER, ENGLAND.

Highflyer 1774 Herod 1758—Tartar 1743 Partner 1718 Jigg Byerly Turk, about 1685.

MACARONI, b. h., 1860.

Sweetmeat, br. h., 1842.

Gladiator, ch. h., 1833.

Partisan, b. h., 1811.

Walton, b. h., 1799. — Sir Peter T. 1784 (Highflyer 1774) Arethusa 1792 by Dungannon 1780 (Eclipse 1764) Dau. 1777 of Prophet 1750 (Regulus 1739) Virago 1764 by Snap 1750 (Snip 1736) Dau. of Regulus 1739 (God. Ar. 1724).

Parasol, b. m., 1800. — Pot-8-o-'s 1773 (Eclipse 1764)—Prunella 1788 by Highflyer 1774 (Herod 1758) Promise 1768 by Snap 1750 (Snip 1736) Julia 1756 by Blank 1740 (God. Ar. 1724) Spectator's dam 1735 by Partner 1718 (Jigg).

Pauline, b. m., 1826.

Moses, b. h., 1819. — [Whalebone or] Seymour 1807 (Delpini 1781) Dau. 1807 of Gohanna 1790 (Mercury 1778) Grey Skim 1793 by Woodpecker 1773 (Herod 1758) Dau. 1780 of Herod 1758 (Tartar 1743) Y. Hag 1761 by Skim 1746 (Starling 1727).

Quadrille, b. m., 1815. — Selim 1802 (Buzzard* 1787) Canary Bird 1806 by [Whiskey or] Sorcerer 1796 (Trumpator 1782) Canary 1797 by Coriander 1786 (Pot-8-o-'s 1773) Miss Green 1787 by Highflyer 1774 (Herod 1758) Harriet 1769 by Matchem 1748 (Cade 1734).

Lollypop, blk. m., 1836.

[Starch or] Voltaire, br. h., 1826.

Blacklock, b. h., 1814. — Whitelock 1803 (Hambletonian 1792) Dau. 1790 of Coriander 1786 (Pot-8-o-'s 1773) Wildgoose 1792 by Highflyer 1774 (Herod 1758) Coheiress 1786 by Pot-8-o-'s 1773 (Eclipse 1764) Manilla 1777 by Goldfinder 1764 (Snap 1750).

Daughter of, b. m., 1816. — Phantom 1808 (Walton 1799)—Dau. 1802 of Overton 1788 (King Fergus 1775) Dau. 1796 of Walnut 1786 (Highflyer 1774) Dau. 1786 of Ruler 1777 (Y. Marske 1771) Piracantha 1772 by Matchem 1748 (Cade 1734).

Belinda, br. m., 1825.

Blacklock, b. h., 1814. — Whitelock 1803 (Hambletonian 1792) Dau. 1790 of Coriander 1786 (Pot-8-o-'s 1773) Wildgoose 1792 by Highflyer 1774 (Herod 1758)—Coheiress 1786 by Pot-8-o-'s 1773 (Eclipse 1764)—Manilla 1777 by Goldfinder 1764 (Snap 1750).

Wagtail, b. m., 1818. — Prime Minister 1810 (Sancho 1801) —Tranby*'s dam 1812 by Orville 1799 (Beningbrough 1791) Miss Grimstone 1796 by Weazle 1776 (Herod 1758)—Dau. of Ancaster 1768 (Blank 1740)—Dau. of Damascus Ar. 1754.

Jocose, b. m., 1843.

Pantaloon, ch. h., 1824.

Castrel, ch. h., 1801.

Buzzard,* ch. h., 1787. — Woodpecker 1773 (Herod 1758) Misfortune 1775 by Dux 1761 (Matchem 1748) Curiosity 1760 by Snap 1750 (Snip 1736) Dau. 1749 of Regulus 1739 (God. Ar. 1724)—Dau. of Bartlett's Childers (Darley Ar. 1699).

Daughter of, b. m., —. — Alexander 1782 (Eclipse 1764)—Dau. of Highflyer 1774 (Herod 1758) Dau. of Alfred 1770 (Matchem 1748)—Dau. 1770 of Engineer 1756 (Sampson 1745)—Bay Malton's dam 1756 by Cade 1734 (God. Ar. 1724).

Idalia, ch. m., 1815.

Peruvian, b. h., 1806. — Sir Peter T. 1784 (Highflyer 1774)—Dau. 1788 of Boudrow 1777 (Eclipse 1764) Dau. 1768 of Squirrel 1754 (Traveller 1735) Sist. to Lowther's Babraham by Babraham 1740 (God. Ar. 1724)—Dau. 1750 of Golden Ball 1735 (Partner 1718).

Musidora, b. m., 1804. — Meteor 1783 (Eclipse 1764)—Maid-of-all-Work 1786 by Highflyer 1774 (Herod 1758)—Dau. 1771 of Syphon 1750 (Squirt 1732)—Dau. of Regulus 1739 (God. Ar. 1724)—Sist. 1758 to Judgement by Snip 1736 (Flying Childers 1715).

Banter, br. m., 1826.

Master Henry, b. h., 1815.

Orville, b. h., 1799. — Beningbrough 1791 (King Fergus 1775)—Evelina 1791 by Highflyer 1774 (Herod 1758)—Termagant by Tantrum 1760 (Cripple 1750)—Cantatrice by Sampson 1745 (Blaze 1733)—Dau. 1757 of Regulus 1739 (God. Ar. 1724).

Miss Sophia, b. m., 1805. — Stamford 1794 (Sir Peter T. 1784)—Sophia 1798 by Buzzard* 1787 (Woodpecker 1773)—Huncamunca 1787 by Highflyer 1774 (Herod 1758)—Cypher* 1772 by Squirrel 1754 (Traveller 1735)—Fribble's dam 1749 by Regulus 1739 (God. Ar. 1724).

Boadicea, b. m., 1807.

Alexander, ch. h., 1782. — Eclipse 1764 (Marske 1750)—Grecian Princess 1770 by W's Forester 1750 (Forester 1736)—Dau. of Coalition Colt 1759 (God. Ar. 1724)—Dau. of Bustard 1741 (Crab 1722)—Charming Molly 1742 by Second 1732 (Flying Childers 1715).

Brunette, —, —. — Amaranthus 1766 (Old England 1741)—Mayfly 1771 by Matchem 1748 (Cade 1734)—Dau. 1755 of Anc. Starling 1738 (Starling 1727)—Look-at-me-Lads 1731 by Grasshopper (Byerly Turk)—Dau. of Sir Newton's Bay Ar.

SONGSTRESS, b. m., 1860.

Chanticleer, gr. h., 1843.

Ir. Birdcatcher, ch. h., 1833.

Sir Hercules, blk. h., 1826.

Whalebone, br. h., 1807. — Waxy 1790 (Pot-8-o-'s 1773)—Penelope 1798 by Trumpator 1782 (Conductor 1767)—Prunella 1788 by Highflyer 1774 (Herod 1758)—Promise 1768 by Snap 1750 (Snip 1736)—Julia 1756 by Blank 1740 (God. Ar. 1724).

Peri, b. m., 1822. — Wanderer 1811 (Gohanna 1790)—Thalestris 1809 by Alexander 1782 (Eclipse 1764)—Rival 1800 by Sir Peter T. 1784 (Highflyer 1774)—Hornet 1790 by Drone 1777 (Herod 1758) Manilla 1777 by Goldfinder 1764 (Snap 1750).

Guiccioli, ch. m., 1823.

Bob Booty, ch. h., 1804. — Chanticleer 1787 (Woodpecker 1773)—Ierne 1790 by Bagot 1780 (Herod 1758)—Dau. of Gamahoe (Bustard 1741)—Patty by Tim 1752 (Squirt 1732)—Miss Patch 1737 by Justice 1725 (H. C. Litton Ar.).

Flight, ch. m., 1809. — Ir. Escape 1802 (Commodore 1793)—Y. Heroine by Bagot 1780 (Herod 1758)—Heroine 1775 by Hero 1759 (Cade 1734)—Dau. 1766 of Snap 1750 (Snip 1736)—Sist. 1743 to Regulus 1739 (God. Ar. 1724).

Whim, gr. m., 1832.

Ir. Drone, gr. h., 1823.

Master Robert, gr. h., 1811. — Buffer 1798 (Prizefighter 1784)—Spinster 1805 by Shuttle 1793 (Y. Marske 1771)—Dau. 1797 of Sir Peter T. 1784 (Highflyer 1774)—Bab 1787 by Bourdeaux 1774 (Herod 1758)—Speranza 1778 by Eclipse 1764 (Marske 1750).

Daughter of, br. m., 1817. — Sir Walter Raleigh 1801 (Waxy 1790)—Miss Tooly 1808 by Teddy-the-Grinder 1798 (Asparagus 1787)—Lady Jane 1796 by Sir Peter T. 1784 (Highflyer 1774)—Paulina 1778 by Florizel 1768 (Herod 1758)—Captive 1771 by Matchem 1748 (Cade 1734).

Kiss, b. m., 1827.

Waxy Pope, b. h., 1806. — Waxy 1790 (Pot-8-o-'s 1773)—Prunella 1788 by Highflyer 1774 (Herod 1758)—Promise 1768 by Snap 1750 (Snip 1736)—Julia 1756 by Blank 1740 (God. Ar. 1724)—Spectator's dam 1735 by Partner 1718 (Jigg).

Daughter of, br. m., 1810. — Champion 1797 (Pot-8-o-'s 1773)—Brown Fanny 1799 by Maximin 1785 (Evergreen 1769)—Dau. 1782 of Highflyer 1774 (Herod 1758)—Trinket 1767 by Matchem 1748 (Cade 1734)—Miss Elliot by Grisewood's Partner 1730 (Partner 1718).

Mrs. Carter, —, 1850.

Humphrey, b. h., 1835.

Sandbeck, b. h., 1818.

Catton, b. h., 1809. — Golumpus 1802 (Gohanna 1790)—Lucy Gray 1804 by Timothy 1794 (Delpini 1781)—Lucy 1789 by Florizel 1768 (Herod 1758)—Frenzy 1774 by Eclipse 1764 (Marske 1750)—Dau. of Engineer 1756 (Sampson 1745).

Orvillina, b. m., 1804. — Beningbrough 1791 (King Fergus 1775)—Evelina 1791 by Highflyer 1774 (Herod 1758)—Termagant by Tantrum 1760 (Cripple 1750)—Cantatrice by Sampson 1745 (Blaze 1733)—Dau. 1757 of Regulus 1739 (God. Ar. 1724).

Oceana, ch. m., 1816.

Cerberus, ch. h., 1802. — Gohanna 1790 (Mercury 1778)—Dau. 1780 of Herod 1758 (Tartar 1743)—Desdemona 1770 by Marske 1750 (Squirt 1732)—Y. Hag 1761 by Skim 1746 (Starling 1727)—Hag 1744 by Crab 1722 (Alcock Ar.).

Dr. Syntax's dam, —, —. — Beningbrough 1791 (King Fergus 1775)—Jenny Mole 1787 by Carbuncle 1772 (Babr. Blank 1758)—Dau. 1773 of Prince T'Quassaw 1751 (Snip 1736)—Sultana 1759 by Regulus 1739 (God. Ar. 1724)—Dau. of Partner 1718 (Jigg).

Daughter of, ab. 1835.

Margrave*, ch. h., 1829.

Muley, b. h., 1810. — Orville 1799 (Beningbrough 1791)—Eleanor 1798 by Whiskey 1789 (Saltram* 1780) Y. Giantess 1790 by Diomed* 1777 (Florizel 1768)—Giantess 1769 by Matchem 1748 (Cade 1734)—Molly Long Legs 1753 by Babraham 1740 (God. Ar. 1724).

Daughter of, ch. m., 1815. — Election 1804 (Gohanna 1790)—Fair Helen 1808 by Hambletonian 1792 (King Fergus 1775) Helen 1801 by Delpini 1781 (Highflyer 1774)—Rosalind 1788 by Phenomenon 1780 (Herod 1758)—Atalanta 1769 by Matchem 1748 (Cade 1734).

Elastic's dam, b. m., 1816.

Thunderbolt, blk. h., 1806. — Sorcerer 1796 (Trumpator 1782)—Wowski 1797 by Mentor 1784 (Justice 1774)—Maria 1777 by Herod 1758 (Tartar 1743)—Lisette 1772 by Snap 1750 (Snip 1736)—Miss Windsor 1754 by God. Ar. 1724.

Daughter of, b. m., 1810. — Sancho 1801 (Don Quixote 1784)—Miss Teazle b. m. 1799 by Sir Peter T. 1784 (Highflyer 1774)—Fanny ch. m. 1790 by Diomed* 1777 (Florizel 1768)—Ambrosia b. m. 1783 by Woodpecker 1773 (Herod 1758)—Ruth 1761 by Blank 1740 (God. Ar. 1724)—Sist. 1751 to South by Regulus 1739 (God. Ar. 1724)—Dau. of Soreheels (Basto 1702)—Dau. blk. m. of Makeless (Oglethorne Ar.)—Christopher D'Arcy Royal mare by Blunderbuss (Bustler)—Old Grey Royal by D'Arcy White Turk—Dau. of D'Arcy Yellow Turk—Royal mare.

MANCHACA.

The property of Leonidas Cartwright, Fairview Stud, San Augustine, Texas.

MANCHACA, black horse, foaled 1881, bred by R. N. Weisiger, by Rebel, son of Socks by imp. Albion, dam Lemonade by imp. Leamington, grandam Cricket by Daniel Boone, out of the famous race-mare Lilly Ward by Lexington.

Manchaca never raced but is from a racing and producing family, is 16½ hands high, off hind pastern white, beautiful head and neck, with the best of feet and legs, is symmetrically formed and of high finish. Rebel, his sire, was a good race-horse and successful stallion. Lemonade, his dam, is a daughter of imp. Leamington, whose sons and daughters have been very successful in the stud. From his daughters have come Potomac, Sir Dixon, Belvidere, Chesapeake, Strideaway, Carley B., Solid Silver, Britannic, Lizzie S., Redstone, Katrina, Buckstone, Ovid, Tecumseh, Hindu, Spinalong, Dalsyrian and numerous other high class winners. Lemonade also produced Springlight (dam of Charlie Ford) and Incoleame, who was a winner both at two and three years old. Cricket, the second dam, is the dam of Krick and Critic, the former one of the best two-year olds of her year and the dam of the winners Krikina and Koran. Lily Ward, the third dam, was a first-class race mare and produced Ascension, Mary Faris, Hattie Huger, Buckette and Emily Coster. Ascension produced the Kentucky Derby winner Ascender. Buckette produced the winner Jim Beaman, and Emily Coster gave us the two-year old winner Clementine. Laura, the fourth dam, produced the superior race-horses Harry of the West and Judge Durell and the producing mares Fanny Cheatham and Lerna. Fanny Cheatham is the dam of Fanny Brown, Belle Palmer and Memento. Lerna produced Nellie Booker (dam of Ten Booker, Oarsman, Bootmaker and The Sheriff), and the winners Ida K. (dam of Ida Pickwick), Sister, John Y., Newcastle, Frontenac, Falerna and Elsie S. The Stockholder mare, the fifth dam, was a success in the stud, producing Compromise, Martha Dunn (dam of Lulu Horton), Invincible and Iodine. It is a grand family, from which have come Glenluine, Troubadour, Lizzie Krepps, Virginius, Kildare, Amy Farley, Mattie Amelia, Blue Grass (a great horse in England), Ballarat, Windrush, Puzzle, Punster, Sue Wynne, Billy Gilmore, and any number of other good animals.

Manchaca is descended in direct female line through Old Bald Peg and the Barb mare, to which trace Lath, Cade, Flying Childers, Diomed (imp.), Sorcerer, Priam (imp.), Phantom, Langar, Muley, Plenipotentiary, Troubadour, Harry of the West and other good ones.

MANCHACA.

BAY HORSE, FOALED 1881; BRED BY CAPT. W. COTTRILL, ALABAMA.

- **REBEL**, br. h., 1861.
  - Socks, ch. h., 1855.
    - *Albion, blk. h., 1837.
      - [illegible] Actæon, ch. h., 1822.
      - Panthea, br. m., 1821.
    - Daughter of,
      - *Ainderby, ch. h., 1832.
      - Prima, ch. m., 1838.
  - Betty Wharton.
    - Othello, blk. h., 1832.
      - *Leviathan, ch. h., 1823.
      - Sally Barton, b. m., 1822.
    - Daughter of, ch. m., 1839.
      - *Consol, b. h., 1828.
      - Lady Huntsville, b. m., 1828.
- **LEMONADE**, ch. m., 1873.
  - *Leamington, br. h., 1853.
    - Faugh-a-Ballagh, br. h., 1841.
      - Sir Hercules, blk. h., 1826.
      - Guiccioli, ch. m., 1823.
    - Daughter of, b. m., 1841.
      - Pantaloon, ch. h., 1824.
      - Daphne, br. m., 1824.
  - Cricket, b. m., 1867.
    - Daniel Boone, b. h., 1856.
      - Lexington, b. h., 1850.
      - Magnolia, ch. m., 1841.
    - Lilly Ward, ch. m., 1858.
      - Lexington, b. h., 1850.
      - Laura, ch. m., 1846.

Send, b. h., 1801.	Beningbrough 1791 (King Fergus 1775) Eliza 1791 by Highflyer 1774 (Herod 1758)—Augusta 1784 by Eclipse 1764 (Marske 1750)—Hardwicke's dam by Herod 1758 (Tartar 1743) Nettle 1763 by Bajazet 1740 (God. Ar. 1724).
Diana, ch. m., 1808.	Stamford 1794 (Sir Peter T. 1784) Dau. 1799 of Whiskey 1789 (Saltram* 1780) Grey Dorimant 1781 by Dorimant 1772 (Otho 1760) Dizzy 1757 by Blank 1740 (God. Ar. 1724) Anc. Dizzy 1741 by Driver 1727 (Wynn Ar.).
[Connor or] Blacklock, b. h., 1814.	Whitelock 1803 (Hambletonian 1792) Dau. 1799 of Coriander 1786 (Pot-8-o's 1773) Wildgoose 1792 by Highflyer 1774 (Herod 1758) Coheiress 1786 by Pot-8-o's 1773 (Eclipse 1764) Manilla 1777 by Goldfinder 1764 (Snap 1750).
Manuella, b. m., 1809.	Dick Andrews 1797 (Joe Andrews 1778) Mandane 1800 by Pot-8-o's 1773 (Eclipse 1764)—Y. Camilla 1787 by Woodpecker 1773 (Herod 1758)—Camilla 1778 by Trentham 1766 (T. Sweepstakes 1743) Coquette 1765 by Compton Barb.
Velocipede, ch. h., 1825.	Blacklock 1814 (Whitelock 1803)—Dau. 1817 of Juniper 1805 (Whiskey 1789)—Dau. 1810 of Sorcerer 1796 (Trumpator 1782)—Virgin 1801 by Sir Peter T. 1784 (Highflyer 1774)—Dau. 1794 of Pot-8-o's 1773 (Eclipse 1764).
Kate, ch. m., 1823.	Catton 1809 (Golumpus 1802)—Miss Garforth 1819 by Walton 1799 (Sir Peter T. 1784)—Dau. 1806 of Hyacinthus 1797 (Coriander 1786)—Zara 1801 by Delpini 1781 (Highflyer 1774)—Flora 1789 by King Fergus 1775 (Eclipse 1764).
Corsica, b. h., 1832.	John Richards 1819 (Sir Archy 1805)—Selima 1817 by Top Gallant 1800 (*Diomed 1777) Jack Bull 1805 by *Gabriel 1790 (Dorimant 1772)—Active 1785 by Chatham 1776 (Fitzhugh's Regulus 1765) Shepherdess 1777 by *Slim 1768 (Babraham 1740).
Lady Archiana, gr. m., 1810.	Sir Archy 1805 (*Diomed 1777)—Pandora II. 1825 by Silverheels 1815 (Ogle's Oskar 1805)—Equa 1815 by *Chance 1796 (Lurcher 1789)—Dau. of Republican President 1798 (*Highflyer 1781)—Dau. of *Ranger (Regulus 1739).
Muley, b. h., 1810.	Orville 1799 (Beningbrough 1791)—Eleanor 1798 by Whiskey 1789 (Saltram* 1780)—Y. Giantess 1790 by Diomed* 1777 (Florizel 1768)—Giantess 1769 by Matchem 1748 (Cade 1734)—Molly Long Legs 1753 by Babraham 1740 (God. Ar. 1724).
Daughter of, blk. m., 1809.	Windle 1804 (Beningbrough 1791) Dau. 1788 of Anvil 1777 (Herod 1758)—Virago 1764 by Snap 1750 (Snip 1736) Dau. of Regulus 1739 (God. Ar. 1724)—Sist. 1750 to Othello by Crab 1722 (Alcock Ar.).
Sir Archy, b. h., 1805.	*Diomed 1777 (Florizel 1768)—*Castianira 1796 by Rockingham 1781 (Highflyer 1774)—Tabitha 1782 by Trentham 1766 (T. Sweepstakes 1743)—Dau. of Bosphorus 1754 (Babraham 1740)—Dau. of Forester 1750 (Forester 1736).
Daughter of,	Junius (*Buzzard 1787) Dau. of *Citizen 1785 (Pacolet 1763)—*Barb mare presented 1806 to President Jefferson by the Dey of Tunis.
Lottery, br. h., 1820.	Tramp 1810 (Dick Andrews 1797)—Mandane 1800 by Pot-8-o's 1773 (Eclipse 1764)—Y. Camilla 1787 by Woodpecker 1773 (Herod 1758)—Camilla 1778 by Trentham 1766 (T. Sweepstakes 1743)—Coquette 1765 by Compton Barb.
Daughter of, b. m., 1813.	Cerberus 1802 (Gohanna 1790)—Miss Newton 1804 by Delpini 1781 (Highflyer 1774)—Tipple Cyder 1788 by King Fergus 1775 (Eclipse 1764)—Sylvia 1783 by Y. Marske 1771 (Marske 1750) Ferret 1765 by Broth. to Silvio (Cade 1734).
Bertrand, b. h., 1821.	Sir Archy 1805 (*Diomed 1777)—Eliza 1804 by *Bedford 1792 (Dungannon 1780)—*Mambrina 1785 by Mambrino 1768 (Engineer 1756)—Sist. to Sally by Blank 1740 (God. Ar. 1724)—Dau. of Ward 1731 (Crab 1722).
Grey Goose, gr. m., 1818.	Pacolet 1806 (*Citizen 1785)—Sally Sneed 1807 by *Buzzard 1787 (Woodpecker 1773)—Jane Hunt 1796 by Hampton's Paragon 1788 (*Flimnap 1765)—Moll by *Figure 1757 (Figure 1747) Slamerkin 1768 by *Wildair 1753 (Cade 1734).
Whalebone, br. h., 1807.	Waxy 1790 (Pot-8-o's 1773) Penelope 1798 by Trumpator 1782 (Conductor 1767) Prunella 1788 by Highflyer 1774 (Herod 1758) Promise 1768 by Snap 1750 (Snip 1736)—Julia 1756 by Blank 1740 (God. Ar. 1724).
Peri, b. m., 1822.	Wanderer 1811 (Gohanna 1790)—Thalestris 1809 by Alexander 1782 (Eclipse 1764)—Rival 1800 by Sir Peter T. 1784 (Highflyer 1774)—Hornet 1790 by Drone 1777 (Herod 1758)—Manilla 1777 by Goldfinder 1764 (Snap 1750).
Bob Booty, ch. h., 1804.	Chanticleer 1787 (Woodpecker 1773)—Ierne 1790 by Bagot 1780 (Herod 1758) Dau. of Gamahoe (Bustard 1741)—Patty by Tim 1752 (Squirt 1732)—Miss Patch 1737 by Justice 1725 (H. C. Litton Ar.)
Flight, ch. m., 1809.	Ir. Escape 1802 (Commodore 1793)—Y. Heroine by Bagot 1780 (Herod 1758)—Heroine 1775 by Hero 1753 (Cade 1734)—Dau. 1766 of Snap 1750 (Snip 1736)—Sist. 1743 to Regulus by God. Ar. 1724.
Castrel, ch. h., 1801.	Buzzard* 1787 (Woodpecker 1773)—Dau. of Alexander 1782 (Eclipse 1764)—Dau. of Highflyer 1774 (Herod 1758)—Dau. of Alfred 1770 (Matchem 1748)—Dau. 1770 of Engineer 1756 (Sampson 1745).
Idalia, ch. m., 1815.	Peruvian 1806 (Sir Peter T. 1784)—Musidora 1804 by Meteor 1783 (Eclipse 1764)—Maid-of-all-Work 1783 by Highflyer 1774 (Herod 1758)—Dau. 1771 of Syphon 1750 (Squirt 1732)—Dau. of Regulus 1739 (God. Ar. 1724).
Laurel, br. h., 1824.	Blacklock 1814 (Whitelock 1803)—Wagtail 1818 by Prime Minister 1810 (Sancho 1801)—Tranby*'s dam 1812 by Orville 1799 (Beningbrough 1791)—Miss Grimstone 1796 by Weazle 1776 (Herod 1758)—Dau. of Ancaster 1768 (Blank 1740).
Maid of Honor, br. m., 1829.	Champion 1812 (Selim 1802)—Etiquette 1820 by Orville 1799 (Beningbrough 1791)—Boadicea 1807 by Alexander 1782 (Eclipse 1764)—Brunette by Amaranthus 1766 (Old England 1741)—Mayfly 1771 by Matchem 1748 (Cade 1734).
Boston, ch. h., 1833.	Timoleon 1813 (Sir Archy 1805)—Sist. 1814 to Tuckahoe by Ball's Florizel 1801 (*Diomed 1777)—Dau. 1799 of *Alderman 1787 (Pot-8-o's 1773)—Dau. of *Clockfast 1774 (Gimcrack 1760)—Dau. of Symmes' Wildair 1767 (*Fearnought 1755).
Alice Carneal, b. m., 1836.	Sarpedon 1828 (Emilius 1820)—Rowena 1826 by Sumpter 1818 (Sir Archy 1805) Lady Grey 1817 by Robin Grey 1805 (*Royalist 1790)—Maria 1802 by Melzar 1791 (*Medley 1776)—Dau. of *Highflyer 1784 (Highflyer 1774).
*Glencoe, ch. h., 1831.	Sultan 1816 (Selim 1802)—Trampoline 1825 by Tramp 1810 (Dick Andrews 1797)—Web 1808 by Waxy 1790 (Pot-8-o's 1773)—Penelope 1798 by Trumpator 1782 (Conductor 1767)—Prunella 1788 by Highflyer 1774 (Herod 1758).
*Myrtle, ch. m., 1834.	Mameluke 1824 (Partisan 1811)—Bobadilla 1825 by Bobadil 1813 (Rubens 1805)—Pythoness 1813 by Sorcerer 1796 (Trumpator 1782)—Princess 1796 by Sir Peter T. 1784 (Highflyer 1774)—Dau. 1789 of Dungannon 1780 (Eclipse 1764).
Boston, ch. h., 1833.	Timoleon 1813 (Sir Archy 1805)—Sist. 1814 to Tuckahoe by Ball's Florizel 1801 (*Diomed 1777)—Dau. 1799 of *Alderman 1787 (Pot-8-o's 1773)—Dau. of *Clockfast 1774 (Gimcrack 1760)—Dau. of Symmes' Wildair 1767 (*Fearnought 1755).
Alice Carneal, b. m., 1836.	*Sarpedon 1828 (Emilius 1820) Rowena 1826 by Sumpter 1818 (Sir Archy 1805) Lady Grey 1817 by Robin Grey 1805 (*Royalist 1790)—Maria 1802 by Melzar 1791 (*Medley 1776)—Dau. of *Highflyer 1784 (Highflyer 1774).
*Leviathan, ch. h., 1823.	Muley 1810 (Orville 1799)—Dau. 1809 of Windle 1804 (Beningbrough 1791)—Dau. 1788 of Anvil 1777 (Herod 1758)—Virago 1764 by Snap 1750 (Snip 1736)—Dau. of Regulus 1739 (God. Ar. 1724).
Daughter of, b. m., 1829.	Stockholder (Sir Archy 1805)—Dau. ch. m. 1822 of Pacolet 1806 (*Citizen 1785)—Nell Saunders ch. m. 1815 by Wilkes' Wonder 1800 (*Diomed 1777)—Julietta by *Dare Devil 1787 (Magnet 1770)—Rosetta ch. m. 1785 by *Centinel 1758 (Blank 1740)—Diana blk. m. by Clodius 1773 (*Janus 1746)—Sally Painter by *Stirling 1762 (Bellsize Ar.* 1756)—*Silver (ab. 1762) by Bellsize Ar.* 1756—Sist. to Cato by Partner 1718 (Jigg)—Sist. ch. m. 1719 to Roxana by Bald Galloway (St. Victor Barb)—Sist. to Chaunter by Akaster Turk—Cream Cheeks by Leedes Ar.—Dau. 1690 of Spanker (D'Arcy Yellow Turk)—Old Morocco mare by Fairfax's Morocco Barb—Old Bald Peg by an Arabian—A Barb mare.

IMPORTED MASETTO.

Owned by J. C. Smith. In the Dixiana Stud, Major B. G. Thomas, proprietor, Lexington. Terms advertised annually. Sales of yearlings annually.

MASETTO, bay horse, foaled 1888, bred by Mr. C. D. Rose, by St. Simon, the unbeaten, dam Lady Abbess by Cathedral, grandam Lady Sophie by King Tom, out of Bridle by The Saddler, etc. St. Simon did not start in the Derby and St. Leger, in consequence of the death of his breeder and nominator Prince Batthyany, but he was never beaten during his turf career; won the Ascot Gold Cup, Goodwood Cup and Epsom Gold Cup. St. Simon is the sire of Memoir, winner of the Oaks and St. Leger, Semolina, winner of the 1,000 Guineas, La Flèche, winner of the 1,000 Guineas and Oaks and second in the Derby. His dam, Lady Abbess, is the dam of Bel Demonio, winner of the Gerard Plate, The Rous Plate, Great Eastern Railway Handicap, Empress Prize and Wellington Plate, and by Cathedral, winner of the Manchester Tradesman Cup, son of Newminster, winner of the St. Leger. Lady Abbess is full sister to Exeter, winner of the Hardwicke Stakes, out of Lady Sophie by King Tom, sire of imp. King Ban, Imp. Great Tom, imp. Phaeton and Kingcraft, winner of the Derby, also of Hannah, winner of the 1,000 Guineas, Oaks and St. Leger, Hippia and Tormentor, winners of the Oaks, sire of the dam of Enterprise (winner of the 2,000 Guineas) and Zephyr, dam of Favonius (winner of the Derby). Bridle was the dam of Habena (winner of the 1,000 Guineas), her sire, The Saddler, is the sire of Sorello (winner of the 1,000 Guineas). Masetto as a two-year old ran unplaced in the Stud Produce Stakes at Newmarket; received forfeit from Simon Magus in Produce Match at Newmarket; was third to Keroual and Barbatello in Bradford Plate at Leicester, and unplaced in Maiden Plate at Newmarket. Masetto is a finely bred horse, has a double cross of King Tom and is from a good racing and producing family. He is descended in direct female line from the old Vintner mare, to which trace Partner (Croft's), Soreheels, Crab, Snip, Pacolet, Mercury, Dick Andrews, Muley Moloch, Daniel O'Rourke, Bendigo, Peter, Peregrine, etc., in England; Bramble, Preakness, Bayswater, Bayonet, Vauxhall, Muggins, Bishop and all the Allegrante, Bay Leaf and Virago family.

Whitelock 1803-Hambletonian 1792-King Fergus 1775-Eclipse 1764-Marske 1750-Squirt 1732-Bart. Childers-Darley Ar. 1699.

MASETTO.

BAY HORSE, FOALED 1888; BRED BY MR. C. D. ROSE, ENGLAND.

- **ST. SIMON, b. h., 1881.**
  - Galopin, b. h., 1872.
    - Vedette, br. h., 1854.
      - Voltigeur, br. h., 1847.
        - Voltaire, br. h., 1826. — Blacklock 1814 (Whitelock 1803)—Dau. 1816 of Phantom 1808 (Walton 1799)—Dau. 1802 of Overton 1788 (King Fergus 1775)—Dau. 1796 of Walnut 1786 (Highflyer 1774)—Dau. 1786 of Ruler 1777 (Y. Marske 1771).
        - Martha Lynn, br. m., 1837. — Mulatto 1823 (Catton 1809)—Leda 1824 by Filho-da-Puta 1812 (Haphazard 1797)—Treasure 1809 by Camillus 1803 (Hambletonian 1792)—Dau. 1801 of Hyacinthus 1797 (Coriander 1786)—Flora 1789 by King Fergus 1775 (Eclipse 1764).
      - Mrs. Ridgway, br. m., 1849.
        - Ir. Birdcatcher, ch. h., 1833. — Sir Hercules 1826 (Whalebone 1807)—Guiccioli 1823 by Bob Booty 1804 (Chanticleer 1787)—Flight 1809 by Ir. Escape 1802 (Commodore 1793)—Y. Heroine by Bagot 1780 (Herod 1758)—Heroine 1775 by Hero 1753 (Cade 1734).
        - Nan Darell, gr. m., 1844. — Inheritor 1831 (Lottery 1820)—Nell 1831 by Blacklock 1814 (Whitelock 1803)—Madame Vestris 1818 by Comus 1809 (Sorcerer 1796)—Lisette 1806 by Hambletonian 1792 (King Fergus 1775)—Constantia 1796 by Walnut 1786 (Highflyer 1774).
    - Flying Duchess, b. m., 1853.
      - The Flying Dutchman, br. h., 1846.
        - Bay Middleton, b. h., 1833. — Sultan 1816 (Selim 1802)—Cobweb 1821 by Phantom 1808 (Walton 1799)—Filagree 1815 by Soothsayer 1808 (Sorcerer 1796)—Web 1808 by Waxy 1790 (Pot-8-o's 1773)—Penelope 1798 by Trumpator 1782 (Conductor 1767).
        - Barbelle, b. m., 1836. — Sandbeck 1818 (Catton 1809)—Darioletta 1822 by Amadis 1807 (Don Quixote 1784)—Selima 1810 by Selim 1802 (Buzzard* 1787)—Dau. 1791 of Pot-8-o's 1773 (Eclipse 1764)—Editha 1781 by Herod 1758 (Tartar 1743).
      - Merope, b. m., 1841.
        - Voltaire, br. h., 1826. — Blacklock 1814 (Whitelock 1803)—Dau. 1816 of Phantom 1808 (Walton 1799)—Dau. 1802 of Overton 1788 (King Fergus 1775)—Dau. 1796 of Walnut 1786 (Highflyer 1774)—Dau. 1786 of Ruler 1777 (Y. Marske 1771).
        - Velocipede's dam, —, 1817. — Juniper 1805 (Whiskey 1789)—Dau. 1810 of Sorcerer 1796 (Trumpator 1782)—Virgin 1801 by Sir Peter T. 1784 (Highflyer 1774)—Dau. 1794 of Pot-8-o's 1773 (Eclipse 1764)—Editha 1781 by Herod 1758 (Tartar 1743).
  - St. Angela, b. m., 1865.
    - King Tom, b. h., 1851.
      - Harkaway, ch. h., 1834.
        - Economist, b. h., 1825. — Whisker 1812 (Waxy 1790)—Floranthe 1818 by Octavian 1807 (Stripling 1795)—Caprice 1797 by Anvil 1777 (Herod 1758)—Madcap 1774 by Eclipse 1764 (Marske 1750)—Sappho 1763 by Blank 1740 (God. Ar. 1724).
        - Fanny Dawson, ch. m., 1823. — Nabocklish 1811 (Rugantino 1803)—Miss Tooley 1808 by Teddy-the-Grinder 1798 (Asparagus 1787)—Lady Jane 1796 by Highflyer 1774 (Herod 1758)—Paulina 1778 by Florizel 1768 (Herod 1758)—Captive 1771 by Matchem 1748 (Cade 1734).
      - Pocahontas, b. m., 1837.
        - Glencoe,* ch. h., 1831. — Sultan 1816 (Selim 1802)—Trampoline 1825 by Tramp 1810 (Dick Andrews 1797)—Web 1808 by Waxy 1790 (Pot-8-o's 1773)—Penelope 1798 by Trumpator 1782 (Conductor 1767)—Prunella 1788 by Highflyer 1774 (Herod 1758).
        - Marpessa, b. m., 1830. — Muley 1810 (Orville 1799)—Clare 1824 by Marmion 1806 (Whiskey 1789)—Harpalice 1814 by Gohanna 1790 (Mercury 1778)—Amazon 1799 by Driver 1783 (Trentham 1766)—Fractious 1792 by Mercury 1778 (Eclipse 1764).
    - Adeline, b. m., 1851.
      - Ion, br. h., 1835.
        - Cain, b. h., 1822. — Paulowitz 1813 (Sir Paul 1802)—Dau. 1810 of Paynator 1791 (Trumpator 1782)—Dau. 1805 of Delpini 1781 (Highflyer 1774)—Dau. 1795 of Y. Marske 1771 (Marske 1750)—Gentle Kitty 1774 by Silvio 1754 (Cade 1734).
        - Margaret, br. m., 1831. — Edmund 1824 (Orville 1799)—Medora 1811 by Selim 1802 (Buzzard* 1787)—Dau. 1803 of Sir Harry* 1795 (Sir Peter T. 1784)—Dau. 1793 of Volunteer 1780 (Eclipse 1764)—Dau. 1779 of Herod 1758 (Tartar 1743).
      - Little Fairy, ch. m., 1841.
        - Hornsea, ch. h., 1832. — Velocipede 1825 (Blacklock 1814)—Dau. 1820 of Cerberus 1802 (Gohanna 1790)—Miss Cranfield 1803 by Sir Peter T. 1784 (Highflyer 1774)—Dau. 1796 of Pegasus 1784 (Eclipse 1764)—Dau. 1788 of Paymaster 1766 (Blank 1740).
        - Lacerta, b. m., 1816. — Zodiac 1801 (St. George 1789)—Jerboa 1803 by Gohanna 1790 (Mercury 1778)—Camilla 1778 by Trentham 1766 (T. Sweepstakes 1743)—Coquette 1765 by the Compton Barb—Sist. 1743 to Regulus by God. Ar. 1724.
- **LADY ABBESS, br. m., 1875.**
  - Cathedral, ch. h., 1861.
    - Newminster, b. h., 1848.
      - Touchstone, br. h., 1831.
        - Camel, br. h., 1822. — Whalebone 1807 (Waxy 1790)—Dau. 1812 of Selim 1802 (Buzzard* 1787)—Maiden 1801 by Sir Peter T. 1784 (Highflyer 1774)—Dau. 1788 of Phenomenon 1780 (Herod 1758)—Matron 1782 by Florizel 1768 (Herod 1758)
        - Banter, br. m., 1826. — Master Henry 1815 (Orville 1799)—Boadicea 1807 by Alexander 1782 (Eclipse 1764)—Brunette by Amaranthus 1766 (Old England 1741)—Mayfly 1771 by Matchem 1748 (Cade 1734)—Dau. 1755 of Anc. Starling 1738 (Old Starling 1727).
      - Bee's-wing, b. m., 1833.
        - Dr. Syntax, br. h., 1811. — Paynator 1791 (Trumpator 1782)—Dau. of Beningbrough 1791 (King Fergus 1775)—Jenny Mole 1787 by Carbuncle 1772 (Babraham Blank 1753)—Dau. 1773 of Prince T'Quassaw 1751 (Snip 1736)—Sultana 1759 by Regulus 1739 (God. Ar. 1724).
        - Daughter of, ch. m., 1817. — Ardrossan 1809 (John Bull 1789)—Lady Eliza 1813 by Whitworth 1805 (Agonistes 1797)—Dau. 1793 of Spadille 1784 (Highflyer 1774)—Sylvia 1783 by Y. Marske 1771 (Marske 1750)—Ferret 1765 by Broth. to Silvio (Cade 1734).
    - Stolen Moments, b. m., 1852.
      - Melbourne, br. h., 1834.
        - Humprey Clinker, b. h., 1822. — Comus 1809 (Sorcerer 1796)—Clinkerina 1812 by Clinker 1805 (Sir Peter T. 1784)—Pewet 1786 by Tandem 1773 (Syphon 1750)—Termagant by Tantrum 1760 (Cripple 1750)—Cantatrice by Sampson 1745 (Blaze 1733).
        - Daughter of, b. m., 1825. — Cervantes 1806 (Don Quixote 1784)—Dau. 1818 of Golumpus 1802 (Gohanna 1790)—Dau. 1810 of Paynator 1791 (Trumpator 1782)—Circle 1802 by St. George 1789 (Highflyer 1774)—Abigail 1788 by Woodpecker 1773 (Herod 1758).
      - Lady Elizabeth, ch. m., 1845.
        - Sleight-of-Hand, br. h., 1833. — Pantaloon 1824 (Castrel 1801)—Decoy 1830 by Filho-da-Puta 1812 (Haphazard 1797)—Finesse 1815 by Peruvian 1806 (Sir Peter T. 1784)—Violante 1802 by John Bull 1789 (Fortitude 1777)—Dau. 1788 of Highflyer 1774 (Herod 1758).
        - Dam of Sir Tatton Sykes, ch. m., 1836. — Margrave* 1829 (Muley 1810)—Patty Primrose 1830 by Confederate 1821 (Comus 1809)—Sybil 1822 by Interpreter 1815 (Soothsayer 1808)—Galatea 1816 by Amadis 1807 (Don Quixote 1784)—Paulina 1804 by Sir Peter T. 1784 (Highflyer 1774).
  - Lady Sophie, b. m., 1863.
    - King Tom, b. h., 1851.
      - Harkaway, ch. h., 1834.
        - Economist, b. h., 1825. — Whisker 1812 (Waxy 1790)—Floranthe 1818 by Octavian 1807 (Stripling 1795)—Caprice 1797 by Anvil 1777 (Herod 1758)—Madcap 1774 by Eclipse 1764 (Marske 1750)—Sappho 1763 by Blank 1740 (God. Ar. 1724).
        - Fanny Dawson, ch. m., 1823. — Nabocklish 1811 (Rugantino 1803)—Miss Tooley 1808 by Teddy-the-Grinder 1798 (Asparagus 1787)—Lady Jane 1796 by Sir Peter T. 1784 (Highflyer 1774)—Paulina 1778 by Florizel 1768 (Herod 1758)—Captive 1771 by Matchem 1748 (Cade 1734).
      - Pocahontas, b. m., 1837.
        - Glencoe * ch. h., 1831. — Sultan 1816 (Selim 1802)—Trampoline 1825 by Tramp 1810 (Dick Andrews 1797)—Web 1808 by Waxy 1790 (Pot-8-o's 1773)—Penelope 1798 by Trumpator 1782 (Conductor 1767)—Prunella 1788 by Highflyer 1774 (Herod 1758).
        - Marpessa, b. m., 1830. — Muley 1810 (Orville 1799)—Clare 1824 by Marmion 1806 (Whiskey 1789)—Harpalice 1814 by Gohanna 1790 (Mercury 1778)—Amazon 1799 by Driver 1783 (Trentham 1766)—Fractious 1792 by Mercury 1778 (Eclipse 1764).
    - Bridle, br. m., 1841.
      - The Saddler, br. h., 1828.
        - Waverley, br. h., 1817. — Whalebone 1807 (Waxy 1790)—Margaretta 1802 by Sir Peter T. 1784 (Highflyer 1774)—Dau. 1792 of Highflyer 1774 (Herod 1758)—Nutcracker 1767 by Matchem 1748 (Cade 1734)—Miss Starling 1750 by Bolt. Starling 1727 (Bay Bolton 1705).
        - Castrellina, br. m., 1823. — Castrel 1801 (Buzzard* 1787)—Dau. 1815 of Waxy 1790 (Pot-8-o's 1773)—Bizarre 1811 by Peruvian 1806 (Sir Peter T. 1784)—Violante 1802 by John Bull 1789 (Fortitude 1777)—Sist. 1788 to Skyscraper by Highflyer 1774 (Herod 1758).
      - Moneda, ch. m., 1833.
        - Taurus, ch. h., 1826. — [Phantom or] Morisco 1819 (Muley 1810)—Katherine 1821 by Soothsayer 1808 (Sorcerer 1796)—Quadrille 1815 by Selim 1802 (Buzzard* 1787)—Canary Bird 1806 by [Whiskey or] Sorcerer 1796 (Trumpator 1782)—Canary 1797 by Coriander 1786 (Pot-8-o's 1773).
        - Mona, b. m., 1819 — Partisan 1811 (Walton 1799)—Miltonia b. m. 1804 by Patriot 1790 (Rockingham 1781)—Miss Muston b. m. 1790 by King Fergus 1775 (Eclipse 1764)—Columbine 1781 by Espersykes 1775 (Matchem 1748)—Dau. b. m. 1771 of Babraham Blank 1758 (Babraham 1740)—Tipsy b. m. 1750 by Bolt. Starling 1725 (Bay Bolton 1705)—Switch br. m. 1739 by Lonsdale Bay Ar.—Dau. br. m. 1720 of Cyprus Ar.—Crab's dam by Basto 1702 (Byerly Turk)—Partner's dam by Curwen Bay Barb—Dau. of Curwen Old Spot (Selaby Turk)—Dau. of White-legged Lowther Barb—Old Vintner mare.

IMPORTED MAXIM.

The property of J. B. Haggin, Rancho-del-Paso Stud, Sacramento, Cal. Yearlings sold annually.

MAXIM, brown horse, foaled 1884, by Musket, son of Toxophilite, dam Realization by Vespasian, son of Newminster; second dam Hopeful Duchess by Flying Dutchman; third dam Espoir by Liverpool, out of Esperance by Lapdog.

Maxim is regarded as the most valuable horse ever bred in New Zealand, and is the first son of Musket imported to America. Musket (winner of Ascot Stakes and Alexandra Plate) is the sire of Petronel, winner of the 2,000 Guineas and Doncaster Cup, Martini-Henry, winner of the Victoria Derby and Melbourne Cup, Carbine, the winner of the Melbourne Cup in 1890, Nordenfeldt, winner of the Victoria Derby, and other good ones. Vespasian, the sire of Realization, Maxim's dam, was one of the best horses of his day, winner of Weston, Rawcliff Stakes and Cimbolton Cup; was full brother to Sabinus, winner of the Cambridgeshire Handicap and Ascot Gold Cup, and by Newminster. Hopeful Duchess, his grandam, is full sister to Brown Duchess, winner of the Oaks, by The Flying Dutchman, winner of the Derby and St. Leger, sire of Flying Duchess, dam of Galopin, winner of the Derby, Bonny Blink, dam of Hawthornden (winner of the St. Leger), Katherine Logie, dam of Bothwell (winner of the 2,000 Guineas).

Maxim's turf career was a glorious one. The first day he was stripped in public was Dec. 27, 1886, when he started for the Middlepark Stakes, and won by a length from Gipsy King, Fair Nell and three others being behind the pair mentioned. The next day he gave weight and a beating to Gipsy King, Kimberley and two others in the Nursery Handicap, doing the five furlongs in 1.04, winning very easily. He was then shipped to Melbourne, where, though riding under the disadvantages entailed by a lengthy trip and late arrival, he forced the celebrated Abercorn to extend himself in the Produce Stakes, and was beaten by only a short head. Going back to his native land for the Canterbury Autumn Meeting, Maxim started a red hot favorite for the Champagne-Stakes and won by a length, doing the six furlongs in the good time of 1.15, and putting down such notable ones as Sextant, Gipsy King, Sultan and Cruchfield. Of these four, three subsequently became record holders, and the fourth (Sultan) captured the Dunedin Cup and other leading events. On the second day of the C. J. C. referred to, the course was fetlock-deep in mud and Sextant, having 5 lbs. the best of the weights and an advantage at the start, beat him only a length. As a three-year old Maxim won all his races. These were the Hawke's Bay Guineas, in which he beat Lady Florin, Bangle, Waterfall and The Orator; the Flying Handicap, in which he was left at the post and had apparently no chance at all till nearing the straight; the Spring Handicap at the same meeting, beating Rivulet, Pearl and five others; and then coming back to Canterbury he made an exhibition of Gipsy King, Sextant and Cruchfield in the Derby, one and a half miles, run in 2.40, and distanced Nelson in the Canterbury Cup, doing the two miles and a quarter in 4.01¼. It will be seen that Maxim started in ten races, of which he won eight and was second in two. This is really first-class, and when it is remembered that there are undeniable reasons for each of his defeats, and that he never met any but the very best horses, it will be realized that the bare figures do not do the horse justice. All who saw him race speak of his wonderful power and his far-reaching stride, and those who have seen him since agree that he is the most magnificent looking stallion ever bred in New Zealand. The Australians say that it is a thousand pities that such a splendid horse should leave New Zealand.

Maxim is descended in direct female line through the old Montagu mare to the Royal mare, to which traces Conductor, Eclipse (O'Kelly's), Voltaire, Sterling, Weatherbit, Springfield, Scottish Chief, Adventurer, Prince Charlie, in England; Lexington, Vandal, The Bard, Leonatus, Salvator, Longstreet, Sensation, Onondaga, Stratford, Tenny, Luke Blackburn, and all the Duchess family in America.

MAXIM.

BROWN HORSE, FOALED 1884; BRED BY THE NEW ZEALAND STUD CO., NEW ZEALAND.

- †MUSKET, br. h., 1867.
  - Toxophilite, b. h., 1855.
    - Longbow, b. h., 1849.
      - Ithuriel, br. h., 1841.
      - Miss Bowe, b. m., 1834.
    - Legerdemain, b. m., 1846.
      - Pantaloon, ch. h., 1824.
      - Decoy, b. m., 1830.
  - Daughter of, b. m., 1857.
    - West Australian, b. h., 1850.
      - Melbourne, br. h., 1834.
      - Mowerina, b. m., 1843.
    - Brown Bess, br. m., 1844.
      - Camel, br. h., 1822.
      - Daughter of, b. m., 1829.
- †REALIZATION, b. m., 1875.
  - Vespasian, b. h., 1863.
    - Newminster, b. h., 1848.
      - Touchstone, br. h., 1831.
      - Bee's-wing, b. m., 1833.
    - Vesta, ch. m., 1857.
      - Stockwell, ch. h., 1849.
      - Garland, br. m., 1835.
  - Hopeful Duchess, br. m., 1859.
    - The Flying Dutchman, br. h., 1846.
      - Bay Middleton, b. h., 1833.
      - Barbelle, b. m., 1836.
    - Espoir, b. m., 1841.
      - Liverpool, b. h., 1828.
      - Esperance, b. m., 1836.

Touchstone, br. h., 1831.	Camel 1822 (Whalebone 1807) Banter 1826 by Master Henry 1815 (Orville 1799) Boadicea 1807 by Alexander 1782 (Eclipse 1764) Brunette by Amaranthus 1766 (Old England 1741) Mayfly 1771 by Matchem 1748 (Cade 1734).
Verbena, ch. m., 1832.	Velocipede 1825 (Blacklock 1814) Rosalba 1811 by Milo 1802 (Sir Peter T. 1784) Sist. 1804 to Selim by Buzzard* 1787 (Woodpecker 1773) Dau. of Alexander 1782 (Eclipse 1764) Dau. of Highflyer 1774 (Herod 1758).
Catton, b. h., 1809.	Golumpus 1802 (Gohanna 1790) Lucy Gray 1804 by Timothy 1794 (Delpini 1781)—Lucy 1780 by Florizel 1768 (Herod 1758)—Frenzy 1774 by Eclipse 1764 (Marske 1750) Dau. of Engineer 1756 (Sampson 1745).
Tranby's dam, br. m., 1812.	Orville 1799 (Beningbrough 1791) Miss Grimstone 1796 by Weazle 1776 (Herod 1758) Dau. of Ancaster 1768 (Blank 1740) Dau. of Damascus Ar. 1754—Dau. of Sampson 1745 (Blaze 1733).
Castrel, ch. h., 1801.	Buzzard* 1787 (Woodpecker 1773)—Dau. of Alexander 1782 (Eclipse 1764)—Dau. of Highflyer 1774 (Herod 1758)—Dau. of Alfred 1770 (Matchem 1748) Dau. 1770 of Engineer 1756 (Sampson 1745).
Idalia, ch. m., 1815.	Peruvian 1806 (Sir Peter T. 1784)—Musidora 1804 by Meteor 1783 (Eclipse 1764)—Maid-of-all-Work 1786 by Highflyer 1774 (Herod 1758) Dau. 1774 of Syphon 1750 (Squirt 1732)—Dau. of Regulus 1739 (God. Ar. 1724).
Filho-da-Puta, br. h., 1812.	Haphazard 1797 (Sir Peter T. 1784)—Mrs. Barnet 1806 by Waxy 1790 (Pot-8-o-'s 1773)—Dau. 1788 of Woodpecker 1773 (Herod 1758) Heinel 1774 by Squirrel 1754 (Traveller 1735)—Principessa 1762 by Blank 1740 (God. Ar. 1724).
Finesse, b. m., 1815.	Peruvian 1806 (Sir Peter T. 1784) Violante 1802 by John Bull 1789 (Fortitude 1777)—Dau. 1788 of Highflyer 1774 (Herod 1758) Everlasting 1775 by Eclipse 1764 (Marske 1750)—Hyæna 1762 by Snap 1750 (Snip 1736).
Humphrey Clinker, b. h., 1822.	Comus 1809 (Sorcerer 1796)—Clinkerina 1812 by Clinker 1805 (Sir Peter T. 1784) Pewet 1786 by Tandem 1773 (Syphon 1750) Termagant by Tantrum 1760 (Cripple 1750)—Cantatrice by Sampson 1745 (Blaze 1733).
Daughter of, b. m., 1825.	Cervantes 1806 (Don Quixote 1784) Dau. 1818 of Golumpus 1802 (Gohanna 1790) Dau. 1810 of Paynator 1791 (Trumpator 1782)—Circle 1802 by St. George 1789 (Highflyer 1774)—Abigail 1788 by Woodpecker 1773 (Herod 1758).
Touchstone, br. h., 1831.	Camel 1822 (Whalebone 1807) Banter 1826 by Master Henry 1815 (Orville 1799)—Boadicea 1807 by Alexander 1782 (Eclipse 1764)—Brunette by Amaranthus 1766 (Old England 1741)—Mayfly 1771 by Matchem 1748 (Cade 1734).
Emma, ch. m., 1824.	Whisker 1812 (Waxy 1790) Gibside Fairy 1811 by Hermes 1790 (Mercury 1778)—Vicissitude 1800 by Pipator 1786 (Imperator 1776)—Beatrice 1791 by Sir Peter T. 1784 (Highflyer 1774) Pyrrha 1771 by Matchem 1748 (Cade 1734).
Whalebone, br. h., 1807.	Waxy 1790 (Pot-8-o-'s 1773) Penelope 1798 by Trumpator 1782 (Conductor) 1767 Prunella 1788 by Highflyer 1774 (Herod 1758)—Promise 1768 by Snap 1750 (Snip 1736) Julia 1756 by Blank 1740 (God. Ar. 1724).
Daughter of, b. m., 1812.	Selim 1802 (Buzzard* 1787) Maiden 1801 by Sir Peter T. 1784 (Highflyer 1774) Dau. 1788 of Phenomenon 1780 (Herod 1758)—Matron 1782 by Florizel 1768 (Herod 1758)—Maiden 1770 by Matchem 1748 (Cade 1734).
Brutandorf, b. h., 1821.	Blacklock 1814 (Whitelock 1803)—Mandane 1800 by Pot-8-o's 1773 (Eclipse 1764)—Y. Camilla 1787 by Woodpecker 1773 (Herod 1758)—Camilla 1778 by Trentham 1766 (T. Sweepstakes 1743)—Coquette 1765 by Compton Barb.
Mrs. Cruickshanks, ch. m., 1823.	Welbeck 1815 (Soothsayer 1808) Tramp's dam 1803 by Gohanna 1790 (Mercury 1778)—Fraxinella 1791 by Trentham 1766 (T. Sweepstakes 1743)—Dau. 1785 of Woodpecker 1773 (Herod 1758) Everlasting 1775 by Eclipse 1764 (Marske 1750).
Camel, br. h., 1822.	Whalebone 1807 (Waxy 1790)—Dau. 1812 of Selim 1802 (Buzzard* 1787)—Maiden 1801 by Sir Peter T. 1784 (Highflyer 1774)—Dau. 1788 of Phenomenon 1780 (Herod 1758)—Matron 1782 by Florizel 1768 (Herod 1758).
Banter, br. m., 1826.	Master Henry 1815 (Orville 1799)—Boadicea 1807 by Alexander 1782 (Eclipse 1764)—Brunette by Amaranthus 1766 (Old England 1741) Mayfly 1771 by Matchem 1748 (Cade 1734)—Dau. 1755 of Anc. Starling 1738 (Starling 1727).
Dr. Syntax, br. h., 1811.	Paynator 1791 (Trumpator 1782)—Dau. of Beningbrough 1791 (King Fergus 1775)—Jenny Mole 1787 by Carbuncle 1772 (Babraham Blank 1758)—Dau. 1773 of Prince T'Quassaw 1751 (Snip 1736)—Sultana 1759 by Regulus 1739 (God. Ar. 1724).
Tomboy's dam, ch. m., 1817.	Ardrossan 1809 (John Bull 1789)—Lady Eliza 1813 by Whitworth 1805 (Agonistes 1797)—Dau. 1793 of Spadille 1784 (Highflyer 1774) Sylvia 1783 by Y. Marske 1771 (Marske 1750)—Ferret 1765 by Broth. to Silvio (Cade 1734).
The Baron, ch. h., 1842.	Ir. Birdcatcher 1833 (Sir Hercules 1826)—Echidna 1838 by Economist 1825 (Whisker 1812)—Miss Pratt 1825 by Blacklock 1814 (Whitelock 1803) Gadabout 1812 by Orville 1799 (Beningbrough 1791)—Minstrel 1803 by Sir Peter T. 1784 (Highflyer 1774).
Pocahontas, b. m., 1837.	Glencoe* 1831 (Sultan 1816)—Marpessa 1830 by Muley 1810 (Orville 1799)—Clare 1824 by Marmion 1806 (Whiskey 1789)—Harpalice 1814 by Gohanna 1790 (Mercury 1778)—Amazon 1799 by Driver 1783 (Trentham 1766).
Langar, ch. h., 1817.	Selim 1802 (Buzzard* 1787)—Dau. 1808 of Walton 1799 (Sir Peter T. 1784)—Y. Giantess 1790 by Diomed* 1777 (Florizel 1768)—Giantess 1769 by Matchem 1748 (Cade 1734)—Molly Long Legs 1753 by Babraham 1740 (God. Ar. 1724).
Cast Steel, br. m., 1828.	Whisker 1812 (Waxy 1790)—The Twinkle 1821 by Walton 1799 (Sir Peter T. 1784)—Dau. 1814 of Orville 1799 (Beningbrough 1791)—Lisette 1803 by Hambletonian 1792 (King Fergus 1775)—Constantia 1793 by Walnut 1786 (Highflyer 1774).
Sultan, b. h., 1816.	Selim 1802 (Buzzard* 1787)—Bacchante 1809 by W's Ditto 1800 (Sir Peter T. 1784)—Dau. 1791 of Mercury 1778 (Eclipse 1764)—Dau. 1776 of Herod 1758 (Tartar 1743)—Folly 1771 by Marske 1750 (Squirt 1732).
Cobweb, b. m., 1821.	Phantom 1808 (Walton 1799)—Filagree 1815 by Soothsayer 1808 (Sorcerer 1796)—Web 1808 by Waxy 1790 (Pot-8-o's 1773) Penelope 1798 by Trumpator 1782 (Conductor 1767)—Prunella 1788 by Highflyer 1774 (Herod 1758).
Sandbeck, b. h., 1818.	Catton 1809 (Golumpus 1802) Orvillina 1804 by Beningbrough 1791 (King Fergus 1775)—Evelina 1791 by Highflyer 1774 (Herod 1758) Termagant by Tantrum 1760 (Cripple 1750)—Cantatrice by Sampson 1745 (Blaze 1733).
Darioletta, br. m., 1822.	Amadis 1807 (Don Quixote 1784)—Selima 1810 by Selim 1802 (Buzzard* 1787)—Dau. 1794 of Pot-8-o's 1773 (Eclipse 1764)—Editha 1781 by Herod 1758 (Tartar 1743)—Elfrida 1768 by Snap 1750 (Snip 1736).
Tramp, b. h., 1810.	Dick Andrews 1797 (Joe Andrews 1778)—Dau. 1803 of Gohanna 1790 (Mercury 1778)—Fraxinella 1791 by Trentham 1766 (T. Sweepstakes 1743)—Dau. 1785 of Woodpecker 1773 (Herod 1758)—Everlasting 1775 by Eclipse 1764 (Marske 1750).
Daughter of, b. m., 1822.	Whisker 1812 (Waxy 1790)—Mandane 1800 by Pot-8-o-'s 1773 (Eclipse 1764) Y. Camilla 1787 by Woodpecker 1773 (Herod 1758) Camilla 1778 by Trentham 1766 (T. Sweepstakes 1743)—Coquette 1765 by Compton Barb.
Lapdog,* b. h., 1823.	Whalebone 1807 (Waxy 1790)—Dau. 1812 of Canopus 1803 (Gohanna 1790)—Dau. 1804 of Y. Woodpecker 1794 (Woodpecker 1773)—Fractious 1792 by Mercury 1778 (Eclipse 1764)—Dau. 1785 of Woodpecker 1773 (Herod 1758).
Grisette, ch. m., 1825.	Merlin 1815 (Castrel 1801)—Coquette ch. m. 1814 by Dick Andrews 1797 (Joe Andrews 1778)—Vanity b. m. 1803 by Buzzard* 1787 (Woodpecker 1773)—Dabchick br. m. 1798 by Pot-8-o-'s 1773 (Eclipse 1764)—Drab b. m. 1791 by Highflyer 1774 (Herod 1758)—Hebe b. m. 1774 by Chrysolite 1763 (Blank 1740)—Proserpine b. m. 1766 by Marske 1750

(Squirt 1732)—Spiletta b. m. 1749 by Regulus 1739 (God. Ar. 1724)—Mother Western by Easby Snake (Snake)—Montagu mare by D'Arcy's Old Montagu—Dau. of Hautboy (D'Arcy White Turk)—Dau. of Brimmer (D'Arcy Yellow Turk)—Royal mare.

MISER.

BROTHER TO SPENDTHRIFT.

The property of Chas. Reed & Sons, Fairview Stud, Gallatin, Tenn. Yearlings sold annually.

MISER, chestnut horse, foaled 1877, bred in the Woodburn Stud, by imp. Australian, son of West Australian by Melbourne, dam Aerolite, sister to Idlewild by Lexington, out of Florine by imp. Glencoe.

Miser is a full brother to Spendthrift, Fellowcraft (4 miles, in 7.19½) and Rutherford, all of which were first-class race-horses. Since placed in the stud they have become famous as sires of race-horses. Rutherford sired Lucky B., Molly McCarthy's Last, Estrella and Harry Rose, while Spendthrift's great doings at the stud is shown by such horses as Kingston, Stockton and Lamplighter, 1¼ miles, in 2.32¾, best on record. The dam of Miser, Aerolite, is one of the most successful in turf annals, for her racing excellence and for the uniform and almost invariable transmission to her descendants of first-class racing qualities. She produced Mozart, Spendthrift, Fellowcraft, Jersey Belle, Platina, Trafalgar and Rutherford. Florine, the second dam, was the dam of Idlewild, the best race-mare of her day (4 miles, in 7.26¼) and of Legal Tender. Idlewild was the dam of Wild Idle, a great horse (4 miles, in 7.25½), and now a distinguished sire. Miser is one of the most promising young sires in this country.

In 1889 only two of Miser's get started, and each of them were winners, viz.: English Lady and Cadaverous. English Lady was a very high class filly, winner of the Zoo Zoo Stakes, ¾ of a mile, at Latonia, by three lengths, in 1.17 beating Dilemma, Sis O., Lee, Ballyhoo, Kitty Cheatham and Heartsease, all of which were winners. Three other good races also fell to her share during the season. In 1890 she proved herself the champion three-year old filly of the West, by winning the Strauss Handicap, 1 mile, at Lexington, by three lengths, in 1.42; a Handicap, 1⅛ miles, at Louisville, by four lengths, on a heavy track, in 1.58½; the Kentucky Oaks, 1½ miles, by three lengths, in 2.42¼, and the Latonia Oaks, 1¼ miles, by three lengths, in 2.08¾; a mile race at Saratoga, in 1.42; the Kenner Stakes, 1¾ miles, beating Sir John, etc., in 3.14½, and 1⅛ miles Handicap, with top weight, in 2.02½. Miser is the sire of that good performer Lady Washington, winner of the Senorita Stakes, ½ mile, in 0.52, at Memphis, beating Monte Rosa, Annie Brown and seven others; the Nantura Stakes, ⅝ of a mile, Lexington, in a field of nine, in 1.05½, the Breeders' Stakes, ⅝ of a mile, in 1.04¾, and the Hurstbourne Stakes, ⅝ of a mile, Louisville, in 1.05½. Adventurer was another two-year old son of Miser that won races, and was a winner last year. Buckhound, Markstone, Milo, McMurtry, Rough Diamond, Mendicant, Poverty, Mark S., Jennie Cottrill and Penurious are also among his winners, and it is safe to say that everything by him wins, for his representatives are noted for quality rather than quantity, his opportunities having been very limited to within the past year or two.

His best representative, however, is the fleet and game filly Yorkville Belle, winner last year of the Hollywood Handicap, the Autumn Stakes, the Prospect Stakes, the Willow Stakes, the Fashion Stakes and the Nursery Stakes, and second in the Expectation, fourth in the Junior Champion and second in the Futurity, proving herself the best of her age and sex. This year she has won the Gazelle, Ladies', Mermaid, West End Hotel and Omnibus Stakes. It is a grand racing family, having both speed and stamina; Drake Carter, Rutherford, Con Cregan Sallie M., Bersan, Mozart, Trafalgar, Kingcraft, Macbeth, Macduff, Favor, Legal Tender, Melody, Rodolph and a host of others come from it.

Trumpator 1782—Conductor 1767 Matchem 1748 Cade 1734—Godolphin Arabian 1724.

MISER.

CHESTNUT HORSE, FOALED 1877; BRED IN THE WOODBURN STUD, SPRING STATION, KY.

*AUSTRALIAN, ch. h., 1858.	West Australian, b. h., 1850.	Melbourne, br. h., 1834.	Humphrey Clinker, b. h., 1822.	Comus, ch. h., 1809. Sorcerer 1796 (Trumpator 1782)—Eprom Lass 1801 by Sir Peter T. 1784 (Highflyer 1774)—Alexina 1788 by King Fergus 1775 (Eclipse 1764)—Lardella 1780 by Y. Marske 1771 (Marske 1750)—Dau. of Y. Cade 1747 (Cade 1734).
				Clinkerina, br. m., 1812. Clinker 1805 (Sir Peter T. 1784)—Pewet 1786 by Tandem 1773 (Syphon 1750)—Termagant by Tantrum 1760 (Cripple 1750)—Cantatrice by Sampson 1745 (Blaze 1733)—Dau. 1757 of Regulus 1739 (God. Ar. 1724).
			Daughter of, b. m., 1825.	Cervantes, b. h., 1806. Don Quixote 1784 (Eclipse 1764)—Evelina 1791 by Highflyer 1774 (Herod 1758)—Termagant by Tantrum 1760 (Cripple 1750)—Cantatrice by Sampson 1745 (Blaze 1733)—Dau. 1757 of Regulus 1739 (God. Ar. 1724).
				Daughter of, br. m., 1818. Golumpus 1802 (Gohanna 1790)—Dau. 1810 of Paynator 1791 (Trumpator 1782)—Circle 1802 by St. George 1789 (Highflyer 1774)—Abigail 1788 by Woodpecker 1773 (Herod 1758)—Firetail 1772 by Eclipse 1764 (Marske 1750).
		Mowerina, b. m., 1843.	Touchstone, br. h., 1831.	Camel, br. h., 1822. Whalebone 1807 (Waxy 1790)—Dau. 1812 of Selim 1802 (Buzzard* 1787)—Maiden 1801 by Sir Peter T. 1784 (Highflyer 1774)—Dau. 1788 of Phenomenon 1780 (Herod 1758)—Matron 1782 by Florizel 1768 (Herod 1758).
				Banter, br. m., 1826. Master Henry 1815 (Orville 1799)—Boadicea 1807 by Alexander 1782 (Eclipse 1764)—Brunette by Amaranthus 1766 (Old England 1741)—Mayfly 1771 by Matchem 1748 (Cade 1734)—Dau. 1755 of Anc. Starling 1738 (Starling 1727).
			Emma, ch. m., 1824.	Whisker, b. h., 1812. Waxy 1790 (Pot-8-o-'s 1773)—Penelope 1798 by Trumpator 1782 (Conductor 1767)—Prunella 1788 by Highflyer 1774 (Herod 1758)—Promise 1768 by Snap 1750 (Snip 1736)—Julia 1756 by Blank 1740 (God. Ar. 1724).
				Gibside Fairy, b. m., 1811. Hermes 1790 (Mercury 1778)—Vicissitude 1800 by Pipator 1786 (Imperator 1776)—Beatrice 1791 by Sir Peter T. 1784 (Highflyer 1774)—Pyrrha 1771 by Matchem 1748 (Cade 1734)—Duchess 1748 by Whitenose 1742 (God. Ar. 1724).
	Emilia,* b. m., 1840.	Y. Emilius, b. h., 1828.	Emilius, b. h., 1820.	Orville, b. h., 1799. Beningbrough 1791 (King Fergus 1775)—Evelina 1791 by Highflyer 1774 (Herod 1758)—Termagant by Tantrum 1760 (Cripple 1750)—Cantatrice by Sampson 1745 (Blaze 1733)—Dau. 1757 of Regulus 1739 (God. Ar. 1724).
				Emily, ch. m., 1810. Stamford 1794 (Sir Peter T. 1784)—Dau. 1799 of Whiskey 1789 (Saltram* 1780)—Grey Dorimant 1781 by Dorimant 1772 (Otho 1760)—Dizzy 1757 by Blank 1740 (God. Ar. 1724)—Dizzy 1741 by Driver 1727 (Wynn Ar.).
			Shoveler, b. m., 1816.	Scud, b. h., 1804. Beningbrough 1791 (King Fergus 1775)—Eliza 1791 by Highflyer 1774 (Herod 1758)—Augusta 1784 by Eclipse 1764 (Marske 1750)—Hardwicke's dam by Herod 1758 (Tartar 1743)—Nettle 1763 by Bajazet 1740 (God. Ar. 1724).
				Goosander, b. m., 1805. Hambletonian 1792 (King Fergus 1775)—Rally 1790 by Trumpator 1782 (Conductor 1767)—Fancy 1780 by Florizel 1768 (Herod 1758)—Dau. 1763 of Spectator 1749 (Crab 1722)—Horatia 1758 by Blank 1740 (God. Ar. 1724).
		Persian, b. m., 1829.	Whisker, b. h., 1812.	Waxy, b. h., 1790. Pot-8-o-'s 1773 (Eclipse 1764)—Maria 1777 by Herod 1758 (Tartar 1743)—Lisette 1772 by Snap 1750 (Snip 1736)—Miss Windsor 1754 by God. Ar. 1724—Sist. to Volunteer by Y. Belgrade (Belgrade Turk).
				Penelope, b. m., 1798. Trumpator 1782 (Conductor 1767)—Prunella 1788 by Highflyer 1774 (Herod 1758)—Promise 1768 by Snap 1750 (Snip 1736)—Julia 1756 by Blank 1740 (God. Ar. 1724)—Spectator's dam 1735 by Partner 1718 (Jigg).
			Variety, ch. m., 1816.	[Selim or] Soothsayer, ch. h., 1808. Sorcerer 1796 (Trumpator 1782)—Golden Locks 1793 by Delpini 1781 (Highflyer 1774)—Violet 1787 by Shark* 1771 (Marske 1750)—Dau. 1772 of Syphon 1750 (Squirt 1732)—Charlotte 1756 by Blank 1740 (God. Ar. 1724).
				Sprite, ch. m., 1807. Bobtail 1795 (Precipitate* 1787)—Catherine 1795 by Woodpecker 1773 (Herod 1758)—Camilla 1778 by Trentham 1766 (T. Sweepstakes 1743)—Coquette 1764 by Compton Barb—Sist. 1743 to Regulus by God. Ar. 1724.
AEROLITE (Sist. to Idlewild), ch. m., 1862.	Lexington, b. h., 1850.	Boston, ch. h., 1833.	Timoleon, ch. h., 1813.	Sir Archy, b. h., 1805. *Diomed 1777 (Florizel 1768)—*Castianira 1796 by Rockingham 1781 (Highflyer 1774)—Tabitha 1782 by Trentham 1766 (T. Sweepstakes 1743)—Dau. of Bosphorus 1754 (Babraham 1740)—Dau. of W's Forester 1750 (Forester 1736).
				Daughter of, ch. m., —. *Saltram 1780 (Eclipse 1764)—Dau. of Symmes' Wildair 1767 (*Fearnought 1755)—Dau. of Tyler's Driver 1760 (*Othello)—Dau. of *Fallower 1761 (Blank 1740)—Dau. of *Vampire 1757 (Regulus 1739).
			Sist. to Tuckahoe, ch. m., 1814.	Ball's Florizel, ch. h., 1801. *Diomed 1777 (Florizel 1768)—Dau. of *Shark 1771 (Marske 1750)—Dau. of Harris's Eclipse 1770 (*Fearnought 1755)—Dau. of *Fearnought 1755 (Regulus 1739)—Dau. of *Jolly Roger 1741 (Roundhead 1733).
				Daughter of, b. m., 1799. *Alderman 1787 (Pot-8-o-'s 1773)—Dau. of *Clockfast 1774 (Gimcrack 1760)—Dau. of Symmes' Wildair 1767 (*Fearnought 1755)—Y. Kitty Fisher by *Fearnought 1755 (Regulus 1739)—*Kitty Fisher 1755 by Cade 1734 (God. Ar. 1724).
		Alice Carneal, b. m., 1836.	*Sarpedon, br. h., 1828.	Emilius, b. h., 1820. Orville 1799 (Beningbrough 1791)—Emily 1810 by Stamford 1794 (Sir Peter T. 1784)—Dau. 1799 of Whiskey 1789 (Saltram* 1780)—Grey Dorimant 1781 by Dorimant 1772 (Otho 1760)—Dizzy 1757 by Blank 1740 (God. Ar. 1724).
				Icaria, b. m., 1824. The Flyer 1814 (Vandyke Jr. 1808)—Parma 1813 by Dick Andrews 1797 (Joe Andrews 1778)—May 1804 by Beningbrough 1791 (King Fergus 1775)—Primrose 1787 by Mambrino 1768 (Engineer 1756)—Cricket 1776 by Herod 1758 (Tartar 1743).
			Rowena, ch. m., 1826.	Sumpter, ch. h., 1818. Sir Archy 1805 (*Diomed 1777)—Dau. of *Robin Redbreast 1796 (Sir Peter T. 1784)—Dau. 1788 of *Obscurity 1777 (Eclipse 1764)—Slamerkin 1768 by *Wildair 1753 (Cade 1734)—*Cub mare 1762 by Cub 1739 (Fox 1714).
				Lady Grey, b. m., 1817. Robin Grey 1805 (*Royalist 1790)—Maria 1802 by Melzar 1791 (*Medley 1773)—Dau. of *Highflyer 1784 (Highflyer 1774)—Dau. of *Fearnought 1755 (Regulus 1739)—Dau. of Ariel 1756 (*Traveller 1747).
	Florine, ch. m., 1854.	*Glencoe, ch. h., 1831.	Sultan, b. h., 1816.	Selim, ch. h., 1802. Buzzard* 1787 (Woodpecker 1773)—Dau. of Alexander 1782 (Eclipse 1764)—Dau. of Highflyer 1774 (Herod 1758)—Dau. of Alfred 1770 (Matchem 1748)—Dau. 1770 of Engineer 1756 (Sampson 1745).
				Bacchante, br. m., 1809. W's Ditto 1800 (Sir Peter T. 1784)—Dau. 1791 of Mercury 1778 (Eclipse 1764)—Dau. 1776 of Herod 1758 (Tartar 1743)—Folly 1771 by Marske 1750 (Squirt 1732)—Vixen 1753 by Regulus 1739 (God. Ar. 1724).
			Trampoline, ch. m., 1825.	Tramp, b. h., 1810. Dick Andrews 1797 (Joe Andrews 1778)—Dau. 1803 of Gohanna 1790 (Mercury 1778)—Fraxinella 1798 by Trentham 1766 (T. Sweepstakes 1743)—Dau. 1785 of Woodpecker 1773 (Herod 1758)—Everlasting 1775 by Eclipse 1764 (Marske 1750).
				Web, b. m., 1808. Waxy 1790 (Pot-8-o-'s 1773)—Penelope 1798 by Trumpator 1782 (Conductor 1767)—Prunella 1788 by Highflyer 1774 (Herod 1758)—Promise 1768 by Snap 1750 (Snip 1736)—Julia 1756 by Blank 1740 (God Ar. 1724).
		Melody, ch. m., 1837.	Medoc, ch. h., 1829.	Am. Eclipse, ch. h., 1814. Duroc 1806 (*Diomed 1777)—Miller's Damsel 1802 by *Messenger 1780 (Mambrino 1768)—*Dau. 1792 of Pot-8-o-'s 1773 (Eclipse 1764)—Dau. 1778 of Gimcrack 1760 (Cripple 1750)—Snapdragon 1759 by Snap 1750 (Snip 1736).
				Y. Maid of the Oaks, ch. m., 1817. *Expedition 1795 (Pegasus 1784)—Maid of the Oaks 1801 by *Spread Eagle 1792 (Volunteer 1780)—Annette by *Shark 1771 (Marske 1750)—Dau. of Rockingham (Partner 1755)—Dau. of True Whig (Fitzhugh's Regulus 1765).
			Rodolph's dam, br. m., 1824.	Haxall's Moses, b. h., 1816. *Sir Harry 1795 (Sir Peter T. 1784)—*Mermaid by Waxy 1790 (Pot-8-o-'s 1773)—*Promise 1803 by Buzzard* 1787 (Woodpecker 1773)—Dau. 1795 of Precipitate* 1787 (Mercury 1778)—Lady Harriet 1773 by Mark Anthony 1767 (Spectator 1749).
				Daughter of, —, —. Cook's Whip 1804 (*Whip 1794)—Hart's Maria by Craig's Alfred (*Medley 1776)—The Pryor mare [a good race-mare] by Tayloe's Bellair 1786 (*Medley 1776)—Thoroughbred mare of Mr. T. D. Owings.

IMPORTED MOCCASIN.

Owned in the Edgewater Stud, F. S. Ashbrook, secretary, Cynthiana, Ky.

MOCCASIN by Macaroni, son of Sweetmeat by Gladiater, bred by Mr. H. W. Deacon, foaled 1874 and imported by Mr. P. Lorillard when a yearling, dam Madame Strauss, dam of Rawcliff and Ormelie by King Tom, son of Harkaway, out of Jetty Treffz by Melbourne, son of Humphrey Clinker by Comus. Macaroni, the sire of Moccasin, won the 2,000 Guineas and the Derby in 1863, and is the sire of Spinaway, winner of the 1,000 Guineas and Oaks, Camelia, winner of the 1,000 Guineas, and of Macheath, the best two-year old in England in 1882, and favorite for the Derby, Lilly Agnes, dam of Ormonde. Sweetmeat, Macaroni's sire, was by Gladiator, second to Bay Middleton in the Derby. Sweetmeat was a noted race-horse, ran twenty-four times as a two and three-year old and met but one defeat. He won the Doncaster Cup and other important events. Moccasin has a double cross of that popular blood Pantaloon, which is famous for speed and staying qualities, and is a true racing strain. Pantaloon's son, Van Amburgh, was second to Coronation in the Derby, and Satirist, another son, won the St. Leger, beating Coronation: he is sire of Ghuznee, winner of the Oaks, and of Cardinal Puff, Elthiron, Hobbie Noble, The Reiver, second to West Australian in the St. Leger, Hernandez, winner of the 2,000 Guineas, and of Windhound, the sire of Thormanby. Leamington was out of a daughter of Pantaloon. King Tom, the sire of his dam, is too well known to need comment, sufficient to say he was by Harkaway, out of Pochahontas, the dam of Stockwell and Rataplan by Glencoe. His grandam is by Melbourne, sire of West Australian, winner of the double events, Derby and St. Leger, Blink Bonny, winner of the Oaks and Derby, and the dam of Blair Athol, winner of the Derby and St. Leger, Sir Tatton Sykes, winner of the St. Leger, Cymba and Marchioness, winner of the Oaks. The Melbournes are remarkable for size, substance and soundness, large bone and racing shape. Moccasin never raced, but his first colts, all out of mares the refuse of his owner, were very reputable performers. Vampire and Faith won six races out of twenty-four starts in 1881. Vampire three, carried 140 lbs., and ran a mile in 1.46¼. The two-year old P. Lorillard won four races out of seven starts in 1882. Amazon and Disdain are both creditable performers, Disdain won Optional Stakes. Other winners by him are Leather Stocking, Mayona, Mohawk, Mokana, Mamood, etc. He is the sire of Cerise, which is the dam of Morello, one of the best two-year olds ever seen, winner of the Futurity, 1892, in 1.12½, as a sick horse. The Macaroni blood is most popular in England, and he is the only Macaroni horse accessible to the breeding public in America. He is a dark bay or brown, with great substance and power, fine temper and full 16 hands high. He should be invaluable to perpetuate the King Tom strain of blood through the descendents of Phaeton, King Ban and Great Tom. He is descended in direct female line from the Royal mare, to which traces Highflyer, Don Quixote, Alexander, Stumps, Colwick, Orlando, Beadsman, Albert Victor, George Frederick, Galliard, in England; Martini-Henry, Nordenfeldt, in Australia; Macaroon (imp.), etc.

MOCCASIN.

BAY HORSE, FOALED 1874; BRED BY MR. H. W. DEACON, ENGLAND.

Highflyer 1774 Herod 1758—Tartar 1743—Partner 1718 Jigg Byerly Turk, about 1685.

MACARONI, b. h., 1860.	Sweetmeat, br. h., 1842.	Gladiator, ch. h., 1833.	Partisan, b. h., 1811.	Walton, b. h., 1799. Sir Peter T. 1784 (Highflyer 1774) Arethusa 1792 by Dungannon 1780 (Eclipse 1764)—Dau. 1777 of Prophet 1760 (Regulus 1739) Virago 1764 by Snap 1750 (Snip 1736) Dau. of Regulus 1739 (God. Ar. 1724).
				Parasol, b. m., 1800. Pot-8-o-'s 1773 (Eclipse 1764)—Prunella 1788 by Highflyer 1774 (Herod 1758) Promise 1768 by Snap 1750 (Snip 1736) Julia 1756 by Blank 1740 (God. Ar. 1724) Spectator's dam 1735 by Partner 1718 (Jigg).
			Pauline, b. m., 1826.	Moses, b. h., 1819. [Whalebone or] Seymour 1807 (Delpini 1781) Dau 1807 of Gohanna 1790 (Mercury 1778) Grey Skim 1793 by Woodpecker 1773 (Herod 1758) Dau. 1780 of Herod 1758 (Tartar 1743) Y. Hag 1761 by Skim 1746 (Starling 1727).
				Quadrille, b. m., 1815. Selim 1802 (Buzzard* 1787) Canary Bird 1806 by [Whiskey or] Sorcerer 1796 (Trumpator 1782) Canary 1797 by Coriander 1786 (Pot-8-o-'s 1773) Miss Green 1787 by Highflyer 1774 (Herod 1758) Harriet 1769 by Matchem 1748 (Cade 1734).
	Lollypop, blk. m., 1836.	[Sancho or] Waxy mare, br. h., 1826.		Blacklock, b. h., 1814. Whitelock 1803 (Hambletonian 1792) Dau. 1799 of Coriander 1786 (Pot-8-o-'s 1773) Wildgoose 1792 by Highflyer 1774 (Herod 1758) Coheiress 1786 by Pot-8-o-'s 1773 (Eclipse 1764) Manilla 1777 by Goldfinder 1764 (Snap 1750).
				Daughter of, b. m., 1816. Phantom 1808 (Walton 1799)—Dau. 1802 of Overton 1788 (King Fergus 1775) Dau. 1796 of Walnut 1786 (Highflyer 1774) Dau. 1786 of Ruler 1777 (Y. Marske 1771) Piracantha 1772 by Matchem 1748 (Cade 1734).
		Belinda, br. m., 1825.		Blacklock, b. h., 1814. Whitelock 1803 (Hambletonian 1792) Dau. 1799 of Coriander 1786 (Pot-8-o-'s 1773)—Wildgoose 1792 by Highflyer 1774 (Herod 1758)—Coheiress 1786 by Pot-8-o-'s 1773 (Eclipse 1764)—Manilla 1777 by Goldfinder 1764 (Snap 1750).
				Wagtail, b. m., 1818. Prime Minister 1810 (Sancho 1801)—Trauby*'s dam 1812 by Orville 1799 (Beningbrough 1791) Miss Grimstone 1796 by Weazle 1776 (Herod 1758)—Dau. of Ancaster 1768 (Blank 1740)—Dau. of Damascus Ar. 1754.
	Jocose, b. m., 1843.	Pantaloon, ch. h., 1824.	Castrel, ch. h., 1801.	Buzzard,* ch. h., 1787. Woodpecker 1773 (Herod 1758) Misfortune 1775 by Dux 1761 (Matchem 1748) Curiosity 1760 by Snap 1750 (Snip 1736) Dau. 1740 of Regulus 1739 (God. Ar. 1724)—Dau. of Bartlett's Childers (Darley Ar. 1699).
				Daughter of, b. m., —. Alexander 1782 (Eclipse 1764)—Dau. of Highflyer 1774 (Herod 1758)—Dau. of Alfred 1770 (Matchem 1748)—Dau. 1770 of Engineer 1756 (Sampson 1745) Bay Malton's dam 1756 by Cade 1734 (God. Ar. 1724).
			Idalia, ch. m., 1815.	Peruvian, b. h., 1806. Sir Peter T. 1784 (Highflyer 1774)—Dau. 1788 of Boudrow 1777 (Eclipse 1764) Dau. 1768 of Squirrel 1754 (Traveller 1735) Sist. to Lowther's Babraham by Babraham 1740 (God. Ar. 1724)—Dau. 1750 of Golden Ball 1735 (Partner 1718).
				Musidora, b. m., 1804. Meteor 1783 (Eclipse 1764)—Maid-of-all-Work 1786 by Highflyer 1774 (Herod 1758) Dau. 1771 of Syphon 1750 (Squirt 1732)—Dau. of Regulus 1739 (God. Ar. 1724)—Sist. 1758 to Judgement by Snip 1736 (Flying Childers 1715).
		Banter, br. m., 1826.	Master Henry, b. h., 1815.	Orville, b. h., 1799. Beningbrough 1791 (King Fergus 1775)—Evelina 1791 by Highflyer 1774 (Herod 1758)—Termagant by Tantrum 1760 (Cripple 1750)—Cantatrice by Sampson 1745 (Blaze 1733)—Dau. 1757 of Regulus 1739 (God. Ar. 1724).
				Miss Sophia, b. m., 1805. Stamford 1794 (Sir Peter T. 1784)—Sophia 1798 by Buzzard* 1787 (Woodpecker 1773)—Huncamunca 1787 by Highflyer 1774 (Herod 1758) Cypher 1772 by Squirrel 1754 (Traveller 1735) Fribble's dam 1749 by Regulus 1739 (God. Ar. 1724).
			Boadicea, b. m., 1807.	Alexander, ch. h., 1782. Eclipse 1764 (Marske 1750)—Grecian Princess 1770 by W's Forester 1750 (Forester 1736)—Dau. of Coalition Colt 1739 (God. Ar. 1724)—Dau. of Bustard 1741 (Crab 1722)—Charming Molly 1742 by Second 1732 (Flying Childers 1715).
				Brunette, —, —. Amaranthus 1766 (Old England 1741)—Mayfly 1771 by Matchem 1748 (Cade 1734)—Dau. 1755 of Anc. Starling 1738 (Starling 1727)—Look-at-me-Lads 1731 by Grasshopper (Byerly Turk)—Dau. of Sir Newton's Bay Ar.
MADAME STRAUSS, b. m., 1866.	King Tom, b. h., 1851.	Harkaway, ch. h., 1834.	Economist, b. h., 1825.	Whisker, b. h., 1812. Waxy 1790 (Pot-8-o-'s 1773)—Penelope 1798 by Trumpator 1782 (Conductor 1767) Prunella 1788 by Highflyer 1774 (Herod 1758)—Promise 1768 by Snap 1750 (Snip 1736)—Julia 1756 by Blank 1740 (God. Ar. 1724).
				Floranthe, b. m., 1818. Octavian 1807 (Stripling 1795)—Caprice 1797 by Anvil 1777 (Herod 1758)—Madcap 1774 by Eclipse 1764 (Marske 1750)—Sappho 1763 by Blank 1740 (God. Ar. 1724)—Dau. 1750 of Blaze 1733 (Flying Childers 1715).
			Fanny Dawson, ch. m., 1823.	Nabocklish, ch. h., 1811. Rugantino 1804 (Commodore 1793)—Butterfly 1804 by Master Bagot 1787 (Bagot 1780)—Dau. of Bagot 1780 (Herod 1758)—Mother Brown 1771 by Trunnion 1747 (Cade 1734)—Dau. of Old England 1741 (God. Ar. 1724).
				Miss Tooley, b. m., 1808. Teddy-the-Grinder 1798 (Asparagus 1787)—Lady Jane 1796 by Sir Peter T. 1784 (Highflyer 1774)—Paulina 1778 by Florizel 1768 (Herod 1758)—Captive 1771 by Matchem 1748 (Cade 1734)—Calliope 1763 by Slouch 1747 (Cade 1734).
		Pocahontas, b. m., 1837.	Glencoe*, ch. h., 1831.	Sultan, b. h., 1816. Selim 1802 (Buzzard* 1787)—Bacchante 1809 by W's Ditto 1800 (Sir Peter T. 1784) Dau. 1791 of Mercury 1778 (Eclipse 1764)—Dau. 1776 of Herod 1758 (Tartar 1743) Folly 1771 by Marske 1750 (Squirt 1732).
				Trampoline, ch. m., 1825. Tramp 1810 (Dick Andrews 1797) Web 1808 by Waxy 1790 (Pot-8-o-'s 1773) Penelope 1798 by Trumpator 1782 (Conductor 1767)—Prunella 1788 by Highflyer 1774 (Herod 1758)—Promise 1768 by Snap 1750 (Snip 1736).
			Marpessa, b. m., 1830.	Muley, b. h., 1810. Orville 1799 (Beningbrough 1791)—Eleanor 1798 by Whiskey 1789 (Saltram* 1780)—Y. Giantess 1790 by Diomed* 1777 (Florizel 1768) Giantess 1769 by Matchem 1748 (Cade 1734)—Molly Long Legs 1753 by Babraham 1740 (God. Ar. 1724).
				Clare, b. m., 1824. Marmion 1806 (Whiskey 1789)—Harpalice 1814 by Gohanna 1790 (Mercury 1778)—Amazon 1799 by Driver 1783 (Trentham 1766)—Fractious 1792 by Mercury 1778 (Eclipse 1764)—Dau. 1785 of Woodpecker 1773 (Herod 1758).
	Jetty Treffz, br. m., 1851.	Melbourne, br. h., 1834.	Humphrey Clinker, b. h., 1822.	Comus, ch. h., 1809. Sorcerer 1796 (Trumpator 1782)—Houghton Lass 1801 by Sir Peter T. 1784 (Highflyer 1774)—Alexina 1788 by King Fergus 1775 (Eclipse 1764) Lardella 1780 by Y. Marske 1771 (Marske 1750) Dau. of Y. Cade 1747 (Cade 1734).
				Clinkerina, br. m., 1812. Clinker 1805 (Sir Peter T. 1784)—Pewet 1786 by Tandem 1773 (Syphon 1750)—Termagant by Tantrum 1760 (Cripple 1750)—Cantatrice by Sampson 1745 (Blaze 1733) Dau. 1757 of Regulus 1739 (God. Ar. 1724).
			Daughter of, b. m., 1825.	Cervantes, b. h., 1806. Don Quixote 1784 (Eclipse 1764)—Evelina 1791 by Highflyer 1774 (Herod 1758)—Termagant by Tantrum 1760 (Cripple 1750)—Cantatrice by Sampson 1745 (Blaze 1733)—Dau. 1757 of Regulus 1739 (God. Ar. 1724).
				Daughter of, br. m., 1818. Golumpus 1802 (Gohanna 1790) Dau. 1810 of Paynator 1791 (Trumpator 1782)—Circle 1802 by St. George 1789 (Highflyer 1774)—Abigail 1788 by Woodpecker 1773 (Herod 1758)—Firetail 1772 by Eclipse 1764 (Marske 1750.)
		Ellen Loraine, br. m., 1845.	The Lord Major, ch. h., 1836.	Pantaloon, ch. h., 1824. Castrel 1801 (Buzzard* 1787)—Idalia 1815 by Peruvian 1806 (Sir Peter T. 1784) Musidora 1804 by Meteor 1783 (Eclipse 1764)—Maid-of-all-Work 1786 by Highflyer 1774 (Herod 1758)—Dau. 1771 of Syphon 1750 (Squirt 1732)
				Honeymoon, b. m., 1830. Filho-da-Puta 1812 (Haphazard 1797)—Hybla 1821 by Rubens 1805 (Buzzard* 1787)—Larissa 1813 by Trafalgar 1802 (Sir Peter T. 1784)—Meteora 1802 by Meteor 1783 (Eclipse 1764)—Maid-of-all-Work 1786 by Highflyer 1774 (Herod 1758).
			Lady Mary, br. m., 1838.	Voltaire, br. h., 1826. Blacklock 1814 (Whitelock 1803)—Dau. 1816 of Phantom 1808 (Walton 1799)—Dau. 1802 of Overton 1788 (King Fergus 1775)—Dau. 1796 of Walnut 1786 (Highflyer 1774) Dau. 1786 of Ruler 1777 (Y. Marske 1771).
				Lady Moore Carew, b. m., 1830. Tramp 1810 (Dick Andrews 1797)—Kite b. m. 1821 by Bustard 1813 (Castrel 1801)—Olympia b. m. 1815 by Sir Oliver 1800 (Sir Peter T. 1784) Scotilla b. m. 1795 by Anvil 1777 (Herod 1758)—Scota b. m. 1783 by Eclipse 1764 (Marske 1750)—Harmony 1775 by Herod 1758 (Tartar 1743)—Rutilia b. m. 1769 by Blank 1740 (God. Ar. 1724) Sist. 1751 to South by Regulus 1739 (God. Ar. 1724)—Dau. of Soreheels (Basto 1702) Dau. blk. m. by Makeless (Oglethorpe Ar.)—Christopher D'Arcy's Royal mare by Blunderbuss (Bustler)—Old Grey Royal by D'Arcy White Turk—Dau. of D'Arcy Yellow Turk—Royal mare.

ONEKO.

The property of W. S. Barnes, Melbourne Stud, Lexington, Ky. Sales annually.

ONEKO, brown horse, foaled 1884, bred in the Brookdale Stud, by Uncas, son of Lexington and Coral by Vandal, dam imp. Faverdale, dam of Centaur (Faverdale colt) by The Palmer, own brother to Rosicrucian by Beadsman. The Palmer won the Ascot Derby, Royal Stakes and Liverpool Autumn Cup, and is the sire of Jenny Howlet, winner of the Oaks, Pellegrino, Pilgrimage (winner of the Dewhurst Plate, the 1,000 and 2,000 Guineas), Grey Palmer, etc. His grandam Georgiana is (full sister to Countess of Burlington, the dam of Cavendish and imp. Hartington, winner of the Cesarewitch Handicap) by Touchstone, winner of the St. Leger and one of the most valuable stallions ever bred in England. Touchstone sired Mowerina, the dam of West Australian, winner of the 2,000 Guineas, the Derby and St. Leger. Mowerina was out of Emma, sister to Caroline, the fourth dam of Centaur by Whisker, third dam Lady Emily by Muley Moloch, son of Muley and sire of the famous Alice Hawthorn, dam of Thormanby, winner of the Derby, tracing through Gibside Fairy by Hermes, son of Mercury by Eclipse, to the D'Arcy Blacklegged Royal mare.

Oneko commenced his turf career at two years old by running third to Tremont and Queen of Elizabeth for the Foam Stakes, five-eighths of a mile, at Sheepshead Bay, beating Roi d'Or, Nat Goodwin, Alaric and four others. At Monmouth Park, in the July Stakes, three-quarters of a mile, ran second to the mighty Hanover, beaten by the shortest of heads, beating Roi d'Or, Fitzroy and Reveller, time, 1.15½; at Monmouth, in the Criterion Stakes, three-quarters of a mile, he finished fourth to Milton, Firenzi and Fitzroy, beating Strideaway, Flageoletta, Cambyses, Racquet, Puzzle and four others. Oneko began his three-year old career by running third to Tarbouche in a 1¼ mile race at Jerome Park, beating Strideaway and Lizzie Baker. He was beaten by the great Hanover in the Belmont Stakes, 1½ miles, in 2.43½. At Brooklyn ran third in a mile race to Fitzroy and Stuyvesant, in 1.43, beating Dunbine, Young Luke, Alaric and Valentine; was third in the Tidal Stakes, at Sheephead Bay, 1 mile, in 1.41⅘, to Hanover and Kingston. For the Coney Island Derby, 1½ miles, in 2.44½, Oneko again finished second to Hanover, Hypasia third; and ran second to Dunbine, beaten a neck, in the Beacon Stakes, 1$\frac{3}{16}$ miles, 2.03, defeating Connemara, Fitzroy, Bradford, Reveller, Prodigal, etc. Oneko closed his three-year old year by running third in the Emporium to Hanover and Dunbine, 1½ miles, in 2.35¾, the defeated ones being Stockton, Wilfred, Brynwood, Banbridge and Shasta.

With Hanover out of the way Oneko would have been a first-class stake winner, and he must be placed in that class. The breeding of Oneko is excellent. His sire, Uncas, was a grand race-horse and a successful sire. Faverdale, Oneko's dam, produced the winners Frolic and Centaur (Faverdale colt), the latter one of the best horses of his day. Myopia, a daughter of Faverdale, produced the winner Melinda and Briganza, dam of the two-year old winner J. L. Knight. His grandam, Georgiana by Touchstone, was a successful stud matron, throwing a number of winners in Measure for Measure, Farnsfield, Sunset, Ironmaster, Split the Difference and Cœur de Lion. His fifth dam, Gibside Fairy, was the grandam of imp. Trustee (Levity's sire), Cotherstone (winner of the 2,000 Guineas, Derby, etc.), Mundig (winner of the Derby, etc.), and West Australian (winner of the 2,000 Guineas, Derby, St. Leger, etc.), and the great grandam of Parmesan, sire of the Derby winners Cremorne and Favonius.

He is descended in direct female line from the Blacklegged Royal mare, to which traces the following excellent race-horses and successful stallions: Oroonoko, Saltram (imp.), Beningbrough, Williamson's Ditto, Cotherstone, West Australian, Walton, Wild Dayrell, Parmesan, Diophantus, Caterer, Lord Ronald, Ishmael, Wisdom, The Miser, Saxifrage and Harry O'Fallon, in America.

ONEKO.

BAY HORSE, FOALED 1884; BRED BY THE LATE MR. D. D. WITHERS, BROOKDALE STUD, NEW JERSEY.

UNCAS, b. h., 1875	Lexington, b. h., 1850.	Boston, ch. h., 1833.	Timoleon, ch. h., 1813.	Sir Archy, b. h., 1805.	*Diomed 1777 (Florizel 1768) Castianira 1796 by Rockingham 1781 (Highflyer 1774)—Tabitha 1782 by Trentham 1766 (T. Sweepstakes 1743) Dau. of Bosphorus 1754 (Babraham 1740) Dau. of Forester 1750 (Forester 1736).
				Daughter of, ch. m., .	*Saltram 1780 (Eclipse 1764) Dau. of Symmes' Wildair 1767 (Fearnought 1755) Dau. of Tyler's Driver 1760 (*Othello)- Dau. of *Fallower 1761 (Blank 1740) Dau. of Vampire 1757 (Regulus 1739).
			Sist. to Tuckahoe, ch. m., 1814.	Ball's Florizel, ch. h., 1801.	*Diomed 1777 (Florizel 1768) Dau. of *Shark 1771 (Marske 1750) Dau. of Harris' Eclipse 1770 (*Fearnought 1755) Dau. of *Fearnought 1755 (Regulus 1739)—Dau. of *Jolly Roger 1741 (Roundhead 1733).
				Daughter of, b. m., 1799.	*Alderman 1787 (Pot-8-o's 1773) Dau. of *Clockfast 1774 (Gimcrack 1760) Dau. of Symmes' Wildair 1767 (*Fearnought 1755) -Y. Kitty Fisher 1767 by *Fearnought 1755 (Regulus 1739) *Kitty Fisher 1755 by Cade 1734 (God. Ar, 1724).
		Alice Carneal, b. m., 1836.	*Sarpedon, br. h., 1828.	Emilius, b. h., 1820.	Orville 1799 (Beningbrough 1791) Emily 1810 by Stamford 1794 (Sir Peter T. 1784) Dau. 1799 of Whiskey 1789 (Saltram* 1780) Grey Dorimant 1781 by Dorimant 1772 (Otho 1760) Dizzy 1757 by Blank 1740 (God. Ar. 1724).
				Icaria, b. m., 1824.	The Flyer 1814 (Vandyke Jr, 1808) Parma 1813 by Dick Andrews 1797 (Joe Andrews 1778) May 1801 by Beningbrough 1791 (King Fergus 1775) Primrose 1787 by Mambrino 1768 (Engineer 1756)—Cricket 1776 by Herod 1758 (Tartar 1743).
			Rowena, ch. m., 1826.	Sampter, ch. h., 1818.	Sir Archy 1805 (*Diomed 1777) Robin mare by *Robin Redbreast 1796 (Sir Peter T. 1784) Dau. 1788 of *Obscurity 1777 (Eclipse 1764) - Slamerkin 1768 by *Wildair 1753 (Cade 1734) *Cub mare 1762 by Cub 1739 (Fox 1714).
				Lady Grey, b. m., 1817.	Robin Grey 1805 (*Royalist 1790)—Maria 1802 by Melzar 1791 (*Medley 1776)—Dau. of *Highflyer 1784 (Highflyer 1774) - Dau. of *Fearnought 1755 (Regulus 1739)—Dau. of Ariel 1753 (*Traveller 1747).
	Coral, br. m., 1864.	Vandal, b. h., 1850.	*Glencoe, ch. h., 1831.	Sultan, b. h., 1816.	Selim 1802 (Buzzard* 1787) Bacchante 1809 by W's Ditto 1800 (Sir Peter T. 1784) Dau. 1791 of Mercury 1778 (Eclipse 1764)- Dau. 1776 of Herod 1758 (Tartar 1743) Folly 1771 by Marske 1750 (Squirt 1732).
				Trampoline, ch. m., 1825.	Tramp 1810 (Dick Andrews 1797)—Web 1808 by Waxy 1790 (Pot-8-o's 1773) Penelope 1798 by Trumpator 1782 (Conductor 1767) Prunella 1788 by Highflyer 1774 (Herod 1758) Promise 1768 by Snap 1750 (Snip 1736).
			Alaric's dam, b. m., 1837.	*Tranby, br. h., 1826.	Blacklock 1814 (Whitelock 1803) Dau. 1812 of Orville 1799 (Beningbrough 1791) Miss Grimstone 1796 by Weazle 1776 (Herod 1758)- Dau. of Ancaster 1768 (Blank 1740) Dau. of Damascus Ar. 1754.
				Lucilla, b. m., .	Trumpator 1823 (Sir Solomon 1805)—Lucy 1821 by Orphan 1810 (Ball's Florizel 1801)—Lady Grey 1817 by Robin Grey 1805 (*Royalist 1790) Maria 1802 by Melzar 1791 (*Medley 1776)—Dau. of *Highflyer 1784 (Highflyer 1774).
		*Cairngorm, b. m., 1859.	Cotherstone, b. h., 1840.	Touchstone, br. h., 1831.	Camel 1822 (Whalebone 1807) - Banter 1826 by Master Henry 1815 (Orville 1799)—Boadicea 1807 by Alexander 1782 (Eclipse 1764)—Brunette by Amaranthus 1766 (Old England 1741)—Mayfly 1771 by Matchem 1748 (Cade 1734).
				Emma, ch. m., 1824.	Whisker 1812 (Waxy 1790)—Gibside Fairy 1811 by Hermes 1790 (Mercury 1778)—Vicissitude 1800 by Pipator 1786 (Imperator 1776)—Beatrice 1791 by Sir Peter T. 1784 (Highflyer 1774)—Pyrrha 1771 by Matchem 1748 (Cade 1734).
			Glenluce, b. m., 1849.	Slane, b. h., 1833.	Royal Oak 1823 (Catton 1809) Dau. 1819 of Orville 1799 (Beningbrough 1791) Epsom Lass 1803 by Sir Peter T. 1784 (Highflyer 1774)—Alexina 1788 by King Fergus 1775 (Eclipse 1764)—Lardella 1780 by Y. Marske 1771 (Marske 1750).
				Glencairne, b. m., 1838.	Sultan 1816 (Selim 1802)—Trampoline 1825 by Tramp 1810 (Dick Andrews 1797)—Web 1808 by Waxy 1790 (Pot-8-o's 1773)—Penelope 1798 by Trumpator 1782 (Conductor 1767)—Prunella 1788 by Highflyer 1774 (Herod 1758).
*FAVERDALE, br. m., 1873.	The Palmer, b. h., 1864.	Beadsman, br. h., 1855.	Weatherbit, br. h., 1842.	Sheet Anchor, br. h., 1832.	Lottery 1820 (Tramp 1810)—Morgiana 1820 by Muley 1810 (Orville 1799)—Miss Stephenson 1814 by [Scud or] Sorcerer 1796 (Trumpator 1782)—Dau. 1796 of Precipitate* 1787 (Mercury 1778)—Dau. 1787 of Woodpecker 1773 (Herod 1758).
				Miss Letty, b. m., 1834.	Priam* 1827 (Emilius 1820)—Dau. 1815 of Orville 1799 (Beningbrough 1791) Dau. 1800 of Buzzard* 1787 (Woodpecker 1773)—Hornpipe 1793 by Trumpator 1782 (Conductor 1767) Luna 1779 by Herod 1758 (Tartar 1743).
			Mendicant, br. m., 1843.	Touchstone, br. h., 1831.	Camel 1822 (Whalebone 1807) Banter 1826 by Master Henry 1815 (Orville 1799)—Boadicea 1807 by Alexander 1782 (Eclipse 1764)—Brunette by Amaranthus 1766 (Old England 1741)—Mayfly 1771 by Matchem 1748 (Cade 1734).
				Lady Moore Carew, b. m., 1830.	Tramp 1810 (Dick Andrews 1797)—Kite 1821 by Bustard 1813 (Castrel 1801)—Olympia 1815 by Sir Oliver 1800 (Sir Peter T. 1784)—Scotilla 1795 by Anvil 1777 (Herod 1758)—Scota 1783 by Eclipse 1764 (Marske 1750).
		Madame Eglentine, b. m., 1857.	Cowl, b. h., 1842.	Bay Middleton, b. h., 1833.	Sultan 1816 (Selim 1802) Cobweb 1821 by Phantom 1808 (Walton 1799)—Filagree 1815 by Soothsayer 1808 (Sorcerer 1796) - Web 1808 by Waxy 1790 (Pot-8-o's 1773) - Penelope 1798 by Trumpator 1782 (Conductor 1767)
				Crucifix, b. m., 1837.	Priam* 1827 (Emilius 1820)—Octaviana 1815 by Octavian 1807 (Stripling 1795)—Dau. 1807 of Shuttle 1793 (Y. Marske 1771) Zara 1801 by Delpini 1781 (Highflyer 1774)—Flora 1789 by King Fergus 1775 (Eclipse 1764).
			Diversion, ch. m., 1838.	Defence, b. h., 1824.	Whalebone 1807 (Waxy 1790)—Defiance 1816 by Rubens 1805 (Buzzard* 1787)—Little Folly 1806 by Highland Fling 1798 (Spadille 1784)—Harriet 1799 by Volunteer 1780 (Eclipse 1764) Lady Sarah's dam 1785 by Alfred 1770 (Matchem 1748).
				Folly, ch. m., 1830.	Middleton 1822 (Phantom 1808)—Little Folly 1806 by Highland Fling 1798 (Spadille 1784) Harriet 1799 by Volunteer 1780 (Eclipse 1764)—Dau. 1785 of Alfred 1770 (Matchem 1748)—Magnolia 1771 by Marske 1750 (Squirt 1732).
	Georgiana, b. m., 1851.	Touchstone, br. h., 1831.	Camel, br. h., 1822.	Whalebone, br. h., 1807.	Waxy 1790 (Pot-8-o's 1773) Penelope 1798 by Trumpator 1782 (Conductor 1767)—Prunella 1788 by Highflyer 1774 (Herod 1758)—Promise 1768 by Snap 1750 (Snip 1736)—Julia 1756 by Blank 1740 (God. Ar. 1724).
				Daughter of, b. m., 1812.	Selim 1802 (Buzzard* 1787) Maiden 1801 by Sir Peter T. 1784 (Highflyer 1774)—Dau. 1788 of Phenomenon 1780 (Herod 1758)—Matron 1782 by Florizel 1768 (Herod 1758)—Maiden 1770 by Matchem 1748 (Cade 1734).
			Banter, br. m., 1826.	Master Henry, b. h., 1815.	Orville 1799 (Beningbrough 1791)- Miss Sophia 1805 by Stamford 1794 (Sir Peter T. 1784) Sophia 1798 by Buzzard* 1787 (Woodpecker 1773)—Huncamunca 1787 by Highflyer 1774 (Herod 1758) Cypher 1772 by Squirrel 1754 (Traveller 1735).
				Boadicea, b. m., 1807.	Alexander 1782 (Eclipse 1764) Brunette by Amaranthus 1766 (Old England 1741)—Mayfly 1771 by Matchem 1748 (Cade 1734)- Dau. 1755 of Anc. Starling 1738 (Starling 1727) Look-at-me-Lads 1731 by Grasshopper (Byerly Turk).
		Lady Emily, b. m., 1842.	Muley Moloch, br. h., 1830.	Muley, b. h., 1810.	Orville 1799 (Beningbrough 1791)- Eleanor 1798 by Whiskey 1789 (Saltram* 1780) Y. Giantess 1790 by Diomed* 1777 (Florizel 1768)—Giantess 1769 by Matchem 1748 (Cade 1734) Molly Long Legs 1753 by Babraham 1740 (God. Ar. 1724).
				Nancy, b. m., 1813.	Dick Andrews 1797 (Joe Andrews 1778) Spitfire 1800 by Beningbrough 1791 (King Fergus 1775)—Dau. of Y. Sir Peter 1775 (Doge 1762)—Dau. of Engineer 1756 (Sampson 1745)—Dau. of Wilson Arabian.
			Caroline, b. m., 1828.	Whisker, b. h., 1812.	Waxy 1790 (Pot-8-o's 1773)—Penelope 1798 by Trumpator 1782 (Conductor 1767)—Prunella 1788 by Highflyer 1774 (Herod 1758)—Promise 1768 by Snap 1750 (Snip 1736)—Julia 1756 by Blank 1740 (God. Ar. 1724).
				Gibside Fairy, b. m., 1811.	Hermes 1790 (Mercury 1778)—Vicissitude b. m. 1800 by Pipator 1786 (Imperator 1776)—Beatrice b. m. 1791 by Sir Peter T. 1784 (Highflyer 1774) Pyrrha b. m. 1771 by Matchem 1748 (Cade 1734)—Duchess b. m. 1748 by Whitenose 1742 (God. Ar. 1724)—Miss Slamerkin b. m. 1729 by Y. True Blue 1718 (Honeywood Ar.)—Dau. of Lord Oxford's Dun Ar.—D'Arcy Blacklegged Royal mare.

ONONDAGA.

WINNER OF THE JUVENILE STAKES AT JEROME PARK, JULY STAKES AT LONG BRANCH, AND KENTUCKY STAKES AT SARATOGA, 1881.

Will be used as private stallion in the McGrathiana Stud, near Lexington, Fayette County, Ky., Milton Young, proprietor. Annual sales of yearlings in May.

ONONDAGA by imp. Leamington, son of Faugh-a-Ballagh, bred by Mr. A. Welsh, Erdenheim Stud, Pennsylvania, foaled 1879, dam Susan Beane, dam of Stratford, Sensation, etc., by Lexington out of Sally Lewis, dam of John Morgan, Hunter's Lexington, Acrobat, etc., by imp. Glencoe. Onondaga as a two-year old started in nine races, of which he won four. Jerome Park, won the Juvenile Stakes, half a mile, in 0.50½, beating Gerald, Memento and twelve others. Sheepshead Bay, won match, three-quarters of a mile, in 1.15½, beating Sachem. Monmouth Park, won the July Stakes for two-year olds, 5 furlongs, in 1.02½, beating Gerald, Memento and three others. Saratoga, won the Kentucky Stakes for two-year olds, three-quarters of a mile, in 1.16, beating Nightcap and Glenarm. Sheepshead Bay, was second in the Surf Stakes for two-year olds, 5 furlongs, won by Marsh Redon (Stonehenge-Julietta Colt) in 1.03½. At Monmouth Park, was third in the August Stakes for two-year olds, three-quarters of a mile, won by Marsh Redon in 1.16, and was unplaced in his other races. Onondaga, complaining in his legs, was retired. He is from one of the great racing families of America, is own brother to Sensation, the invincible horse of his year, his dam is own sister to Acrobat and Hunter's Lexington, and half-sister to John Morgan; all were good race-horses, and, for their chances, successful sires. The pedigree is well fortified with Archy and Diomed blood, upon the Eclipse and Herod, and traces through the dam of Medoc to the family which gave Lexington and other famous horses to the country. Onondaga is a chestnut, 15¾ hands, large star in his forehead and snip on the nose, with white hairs in his flanks, left hind foot white above the ankle; he is a good-shaped horse, and was speedy; the family is a racing one, and he is a success in the stud, sire of the following winners: Bixby, Blithesome, Long Branch, Ovation, St. Clair, The Crow, Tocor, Black Pansy, Bootmaker, Blessed, Blessing, Button, Goodloe, Ocean Wave, Once Again, Oregon, Hopeful, Onaway, Outright, Milton, Delemma, Ambulance, The Sheriff, Chimes, Orsman, Unadaga, Hub S., Curt Gunn (Locohatchee), Yale '91, Julio, Portlaw, Busteed, Narina, Derfagilla, Yazoo, Content and many others.

Onondaga traces in direct female line through the old Montagu mare to the Royal mare, to which traces Conductor, O'Kelly's Eclipse, Sterling, Weatherbit, Oxford, Kingston, Scottish Chief, Adventurer, Lexington, Vandal, Leonatus, Duke of Montrose, Grinstead, The Bard, Strathmore, Luke Blackburn, Salvator, Volturno, etc.

ONONDAGA.

CHESTNUT HORSE, FOALED 1879; BRED BY MR. A. WELCH, CHESTNUT HILL STUD, PENNSYLVANIA.

*LEAMINGTON, br. h., 1853.
Faugh-a-Ballagh, br. h., 1841. Daughter of, b. m., 1841.
Sir Hercules, blk. h., 1826. Guiccioli, ch. m., 1823. Pantaloon, ch. h., 1824. Daphne, br. m., 1837.
Whalebone, br. h., 1807. Peri, b. m., 1822. Bob Booty, ch. h., 1804. Flight, ch. m., 1809. Castrel, ch. h., 1801. Idalia, ch. m., 1815. Laurel, br. h., 1824. Maid of Honor, br. m., 1829.

Waxy, b. h., 1790. Pot-8-o-'s 1773 (Eclipse 1764)—Maria 1777 by Herod 1758 (Tartar 1743)—Lisette 1772 by Snap 1750 (Snip 1736)—Miss Windsor 1751 by God. Ar. 1724—Sist. to Volunteer by Y. Belgrade (Belgrade Turk).

Penelope, b. m., 1798. Trumpator 1782 (Conductor 1767)—Prunella 1788 by Highflyer 1774 (Herod 1758)—Promise 1768 by Snap 1750 (Snip 1736)—Julia 1756 by Blank 1740 (God. Ar. 1724)—Spectator's dam 1735 by Partner 1718 (Jigg).

Wanderer, b. h., 1811. Gohanna 1790 (Mercury 1778)—Catherine 1795 by Woodpecker 1773 (Herod 1758)—Camilla 1778 by Trentham 1766 (T. Sweepstakes 1743)—Coquette 1764 by Compton Barb—Sist. 1743 to Regulus by God. Ar. 1724.

Thalestris, blk. m., 1809. Alexander 1782 (Eclipse 1764)—Rival 1800 by Sir Peter T. 1784 (Highflyer 1774)—Hornet 1790 by Drone 1777 (Herod 1758)—Manilla 1777 by Goldfinder 1764 (Snap 1750)—Dau. 1763 of Old England 1741 (God. Ar. 1724).

Chanticleer, ch. h., 1787. Woodpecker 1773 (Herod 1758)—Dau. 1778 of Eclipse 1764 (Marske 1750)—Rosebud 1765 by Snap 1750 (Snip 1736)—Miss Belsea 1753 by Regulus 1739 (God. Ar. 1724)—Dau. of Bartlett's Childers (Darley Ar. 1699).

Ierne, ch. m., 1790. Bagot 1780 (Herod 1758)—Dau. of Gamahoe (Bustard 1741)—Patty by Tim 1752 (Squirt 1732)—Miss Patch 1737 by Justice 1725 (H. C. Litton Ar.)—Ringtail Galloway by Curwen Bay Barb.

Dr. Escape, ch. h., 1802. Commodore 1793 (Tom Tug 1777)—Moll-in-the-Wad 1791 by Highflyer 1774 (Herod 1758)—Shift 1779 by Sweetbriar 1769 (Syphon 1750)—Black Susan 1762 by Snap 1750 (Snip 1736)—Dau. of Cade 1734 (God. Ar. 1724).

Y. Heroine, [illegible]. Bagot 1780 (Herod 1758)—Heroine 1775 by Hero 1753 (Cade 1734)—Dau. 1766 of Snap 1750 (Snip 1736)—Sist. 1743 to Regulus by God. Ar. 1724—Grey Robinson 1723 by Bald Galloway (St. Victor Barb).

Buzzard,* ch. h., 1787. Woodpecker 1773 (Herod 1758)—Misfortune 1775 by Dux 1761 (Matchem 1748)—Curiosity 1760 by Snap 1750 (Snip 1736)—Dau. 1749 of Regulus 1739 (God. Ar. 1724)—Dau. of Bartlett's Childers (Darley Ar. 1699).

Daughter of, b. m., —. Alexander 1782 (Eclipse 1764)—Dau. of Highflyer 1774 (Herod 1758)—Dau. of Alfred 1770 (Matchem 1748)—Dau. 1770 of Engineer 1756 (Sampson 1745)—Bay Malton's dam 1755 by Cade 1734 (God. Ar. 1724).

Peruvian, b. h., 1806. Sir Peter T. 1784 (Highflyer 1774)—Dau. 1788 of Boudrow 1777 (Eclipse 1764)—Dau. 1764 of Squirrel 1754 (Traveller 1735)—Sist. to Lowther's Babraham by Babraham 1740 (God. Ar. 1724)—Dau. 1750 of Golden Ball 1735 (Partner 1718).

Musidora, b. m., 1804. Meteor 1783 (Eclipse 1764)—Maid-of-all-Work 1786 by Highflyer 1774 (Herod 1758)—Dau. 1771 of Syphon 1750 (Squirt 1732)—Dau. of Regulus 1739 (God. Ar. 1724)—Sist. 1758 to Judgement by Snip 1736 (Flying Childers 1715).

Blacklock, b. h., 1814. Whitelock 1803 (Hambletonian 1792)—Dau. 1799 of Coriander 1786 (Pot-8-o-'s 1773)—Wildgoose 1792 by Highflyer 1774 (Herod 1758)—Coheiress 1786 by Pot-8-o-'s 1773 (Eclipse 1764)—Manilla 1777 by Goldfinder 1764 (Snap 1750).

Wagtail, b. m., 1818. Prime Minister 1810 (Sancho 1801)—Tranby's dam 1812 by Orville 1799 (Beningbrough 1791)—Miss Grimstone 1796 by Weazle 1776 (Herod 1758)—Dau. of Ancaster 1768 (Blank 1740)—Dau. of Damascus Ar. 1754.

Champion, ch. h., 1812. Selim 1802 (Buzzard* 1787)—Podagra 1805 by Gouty 1793 (Sir Peter T. 1784)—Jet 1785 by Magnet 1770 (Herod 1758)—Jewel 1775 by Squirrel 1754 (Traveller 1735)—Sophia 1764 by Blank 1740 (God. Ar. 1724).

Etiquette, br. m., 1820. Orville 1799 (Beningbrough 1791)—Boadicea 1807 by Alexander 1782 (Eclipse 1764)—Brunette by Amaranthus 1766 (Old England 1741)—Mayfly 1771 by Matchem 1748 (Cade 1734)—Dau. 1755 of Starling 1738 (Starling 1727).

SUSAN BEANE, ch. m., 1868.
Lexington, b. h., 1850. Sally Lewis, ch. m., 1852.
Boston, ch. h., 1833. Alice Carneal, b. m., 1836. *Glencoe, ch. h., 1831. Motto, ch. m., 1839.
Timoleon, ch. h., 1813. Sist. to Tuckahoe, ch. m., 1814. *Sarpedon, br. h., 1828. Rowena, ch. m., 1826. Sultan, b. h., 1816. Trampoline, ch. m., 1825. *Barefoot, ch. h., 1820. Lady Thompkins, ch. m., 1821.

Sir Archy, b. h., 1805. *Diomed 1777 (Florizel 1768)—*Castianira 1796 by Rockingham 1781 (Highflyer 1774)—Tabitha 1784 by Trentham 1766 (T. Sweepstakes 1743)—Dau. of Bosphorus 1754 (Babraham 1740)—Dau. of Forester 1750 (Forester 1736).

Daughter of, ch. m., —. *Saltram 1780 (Eclipse 1764)—Dau. of Symmes' Wildair 1767 (*Fearnought 1755)—Dau. of Tyler's Driver 1760 (*Othello)—Dau. of *Fallower 1761 (Blank 1740)—Dau. of *Vampire 1757 (Regulus 1739).

Ball's Florizel, ch. h., 1801. *Diomed 1777 (Florizel 1768)—Dau. of *Shark 1771 (Marske 1750)—Dau. of Harris' Eclipse 1770 (*Fearnought 1755)—Dau. of *Fearnought 1755 (Regulus 1739)—Dau. of *Jolly Roger 1741 (Roundhead 1733).

Daughter of, b. m., 1799. *Alderman 1787 (Pot-8-o-'s 1773)—Dau. of *Clockfast 1774 (Gimcrack 1760)—Dau. of Symmes' Wildair 1767 (*Fearnought 1755)—Y. Kitty Fisher 1767 by *Fearnought 1755 (Regulus 1739)—*Kitty Fisher 1755 by Cade 1734 (God. Ar. 1724).

Emilius, b. h., 1820. Orville 1799 (Beningbrough 1791)—Emily 1810 by Stamford 1794 (Sir Peter T. 1784)—Dau. 1799 of Whiskey 1789 (Saltram* 1780)—Grey Dorimant 1781 by Dorimant 1772 (Otho 1760)—Dizzy 1757 by Blank 1740 (God. Ar. 1724).

Icaria, b. m., 1824. The Flyer 1814 (Vandyke Jr. 1808)—Parma 1813 by Dick Andrews 1797 (Joe Andrews 1778)—May 1804 by Beningbrough 1791 (King Fergus 1775)—Primrose 1787 by Mambrino 1768 (Engineer 1756)—Cricket 1776 by Herod 1758 (Tartar 1743).

Sumpter, ch. h., 1818. Sir Archy 1805 (*Diomed 1777)—Dau. of *Robin Redbreast 1796 (Sir Peter T. 1784)—Dau. 1788 of *Obscurity 1777 (Eclipse 1764)—Slamerkin 1768 by *Wildair 1753 (Cade 1734)—*Cub mare 1762 by Cub 1739 (Fox 1714).

Lady Grey, b. m., 1817. Robin Grey 1805 (*Royalist 1790)—Maria 1802 by Melzar 1791 (*Medley 1776)—Dau. of *Highflyer 1784 (Highflyer 1774)—Dau. of *Fearnought 1755 (Regulus 1739)—Dau. of Ariel 1756 (*Traveller 1747).

Selim, ch. h., 1802. Buzzard* 1787 (Woodpecker 1773)—Dau. of Alexander 1782 (Eclipse 1764)—Dau. of Highflyer 1774 (Herod 1758)—Dau. of Alfred 1770 (Matchem 1748)—Dau. 1770 of Engineer 1756 (Sampson 1745).

Bacchante, br. m., 1809. W's Ditto 1800 (Sir Peter T. 1784)—Dau. 1791 of Mercury 1778 (Eclipse 1764)—Dau. 1776 of Herod 1758 (Tartar 1743)—Folly 1771 by Marske 1750 (Squirt 1732)—Vixen 1753 by Regulus 1739 (God. Ar. 1724).

Tramp, b. h., 1810. Dick Andrews 1797 (Joe Andrews 1778)—Dau. 1803 of Gohanna 1790 (Mercury 1778)—Fraxinella 1798 by Trentham 1766 (T. Sweepstakes 1743)—Dau. 1785 of Woodpecker 1773 (Herod 1758)—Everlasting 1775 by Eclipse 1764 (Marske 1750).

Web, b. m., 1808. Waxy 1790 (Pot-8-o-'s 1773)—Penelope 1798 by Trumpator 1782 (Conductor 1767)—Prunella 1788 by Highflyer 1774 (Herod 1758)—Promise 1768 by Snap 1750 (Snip 1736)—Julia 1756 by Blank 1740 (God Ar. 1724).

Tramp, b. h., 1810. Dick Andrews 1797 (Joe Andrews 1778)—Dau. 1803 of Gohanna 1790 (Mercury 1778)—Fraxinella 1798 by Trentham 1766 (T. Sweepstakes 1743)—Dau. 1785 of Woodpecker 1773 (Herod 1758)—Everlasting 1775 by Eclipse 1764 (Marske 1750).

Rosamond, ch. m., 1798. Buzzard* 1787 (Woodpecker 1773)—Roseberry 1792 by Phenomenon 1780 (Herod 1758)—Miss West 1777 by Matchem 1748 (Cade 1734)—Sist. to Favourite by Regulus 1739 (God. Ar. 1724)—Y. Ebony 1742 by Crab 1722 (Alcock Ar.).

Am. Eclipse, ch. h., 1814. Duroc 1806 (*Diomed 1777)—Miller's Damsel 1802 by *Messenger 1780 (Mambrino 1768)—*Mare 1792 by Pot-8-o-'s 1773 (Eclipse 1764)—Dau. 1778 of Gimcrack 1760 (Cripple 1750)—Snapdragon 1759 by Snap 1750 (Snip 1736).

Katy Ann, ch. m., 1825. Ogle's Oscar 1800 (*Gabriel 1790)—Y. Maid-of-the-Oaks ch. m. 1817 by *Expedition 1795 (Pegasus 1784)—Maid-of-the-Oaks ch. m. 1801 by *Spread Eagle 1792 (Volunteer 1780)—Annette by *Shark 1771 (Marske 1750)—Dau. of Rockingham (Partner 1755)—Dau. of True Whig (Fitzhugh's Regulus 1765)—Dau. of Gallant 1770 (*Fearnought 1755)—Dau. of *Regulus 1747 (Regulus 1739)—*Duchess by Cullen Ar.—Lady Thigh ch. m. 1731 by Partner 1718 (Jigg)—Dau. ch. m. 1723 by Greyhound (Chillaby)—Sophonisba's dam by Curwen Bay Barb—Dau. of D'Arcy Chestnut Ar.—Dau. of Whiteshirt—Montagu mare by D'Arcy's old Montagu—Dau. of Hautboy (D'Arcy White Turk)—Dau. of Brimmer (D'Arcy Yellow Turk)—Royal mare.

IMPORTED PARAMATTA.

In the Dixiana Stud, Major B. G. Thomas, Proprietor, Lexington, Ky. Yearlings sold annually.

PARAMATTA, bay horse, foaled 1886, bred by Mr. A. Town, Australia, imported March, 1888, by Cheviot, son of Adventurer and Grey Stocking by Stockwell, dam Scraps by Lord of the Hills, son of Touchstone, out of Ragpicker by Fisherman, etc. Paramatta has a treble cross of Touchstone through his sire and dam, and this upon the Stockwell is a choice combination. His dam, Scraps, is the dam of the winners Morceau and Gen. Gordon; her sire, Lord of the Hills, is own brother to Lord of the Isles, winner of the 2,000 Guineas, out of Ragpicker by (Engl.) Fisherman, one of the stoutest and gamest horses that ever ran in England. He started one hundred and twenty times, won seventy races; amongst them were twenty-six Queen's Plates, the Queen's Vase at Ascot and Gold Cup twice, the Cumberland Plate, etc. Fisherman is sire of Angler, Maribyrnong, etc.

Ragpicker, his grandam, is full sister to Sylvia, winner of V. R. C. Oaks, and dam of Goldsborough, winner of the A. J. C. St. Leger; of Robin Hood, winner of the V. R. C. Derby and A. J. C. St. Leger; of Martini-Henry, winner of V. R. C. Derby, 1½ miles, in 2.39, Melbourne Cup, 2 miles, in 3.30, the fastest times on record, carrying 7 st. 6 lbs., the heaviest weight ever carried by a three-year old to victory in the latter race, and of the V. R. C. St. Leger, 1¾ miles, in 3.24½.

Chrysolite, half sister to Sylvia, was the dam of Lapidist, winner of the V. R. C. Derby, 1873, and of Vancluse, winner of the V. R. C. Oaks, 1882, and the dam of Onyx, which produced that great race-horse Nordenfeldt, winner of the Victoria Derby in 1885. Sylvia was out of Juliet, great grandam of Paramatta, and dam of Charon, winner of the Victoria Derby, 1869, is regarded as the Levity of Australia. Juliet is closely bred in blood to Orlando, winner of the Derby. Paramatta was a good race-horse, but had little chance on the turf; did not race until his fourth year; was winner at 1 mile, 112 lbs., in 1.48½; won purse, 6½ furlongs, 115 lbs., in 1.28½, defeating seven good ones, track muddy; ran second to Michael in Handicap, 1 mile, in 1.41, three behind him; won three good races of 1½ miles each, in good company, and was placed in several good races in good company.

Paramatta is a symmetrically shaped horse and excellent temper, combines in his pedigree the best racing blood in the world and is from a distinguished racing family. He descends in direct female line from the same Royal mare to which trace South, Highflyer, Don Quixote, Alexander, Peter Lely, Elis, Stumps, Colwick, Orlando, Beadsman, Albert Victor, Geo. Frederick, Galliard, in England; Martini-Henry Goldsborough, Nordenfeldt, etc., in Australia.

Waxy 1790—Pot-8-o's 1773 Eclipse 1764 Marske 1750 Squirt 1732 Bartlett's Childers—Darley Arabian 1689.

PARAMATTA.

BAY HORSE, FOALED 1886; BRED BY MR. A. TOWN, NEW SOUTH WALES, AUSTRALIA.

Parent	Grandparent	Great-grandparent	Great-great-grandparent	Ancestor	Pedigree
†CHEVIOT, b. h., 1877.	Adventurer, b. h., 1859.	Newminster, b. h., 1848.	Touchstone, br. h., 1831.	Camel, br. h., 1822.	Whalebone 1807 (Waxy 1790) Dau. 1812 of Selim 1802 (Buzzard 1787)—Maiden 1801 by Sir Peter T. 1784 (Highflyer 1774) Dau. 1788 of Phenomenon 1780 (Herod 1758) Matron 1782 by Florizel 1768 (Herod 1758).
				Banter, br. m., 1826.	Master Henry 1815 (Orville 1799) Boadicea 1807 by Alexander 1782 (Eclipse 1764) Brunette by Amaranthus 1766 (Old England 1741) Mayfly 1771 by Matchem 1748 (Cade 1734) Dau. 1755 of Anc. Starling 1738 (Starling 1727).
			Bee's-wing, b. m., 1833.	Dr. Syntax, br. h., 1811.	Paynator 1791 (Trumpator 1782) Dau. of Beningbrough 1791 (King Fergus 1775) Jenny Mole 1787 by Carbuncle 1772 (Babr. Blank 1758) Dau. 1773 of Prince T'Quassaw 1751 (Snip 1736)—Sultana 1759 by Regulus 1739 (God. Ar. 1724).
				Tomboy's dam, ch. m., 1817.	Ardrossan 1809 (John Bull 1789) Lady Eliza 1813 by Whitworth 1805 (Agonistes 1797) Dau. 1793 of Spadille 1784 (Highflyer 1774) Sylvia 1783 by Y. Marske 1771 (Marske 1750) Ferret 1765 by Broth. to Silvio (Cade 1734).
		Palma, br. m., 1840.	Emilius, b. h., 1820.	Orville, b. h., 1799.	Beningbrough 1791 (King Fergus 1775)—Evelina 1791 by Highflyer 1774 (Herod 1758) Termagant by Tantrum 1760 (Cripple 1750) Cantatrice by Sampson 1745 (Blaze 1733) Dau. 1757 of Regulus 1739 (God. Ar. 1724).
				Emily, ch. m., 1810.	Stamford 1794 (Sir Peter T. 1784) Dau. 1799 of Whiskey 1789 (Saltram 1780)—Grey Dorimant 1781 by Dorimant 1772 (Otho 1760)—Dizzy 1757 by Blank 1740 (God. Ar. 1724)—Dizzy 1741 by Driver 1727 (Wynn Ar.).
			Francesca, br. m., 1829.	Partisan, b. h., 1811.	Walton 1799 (Sir Peter T. 1784)—Parasol 1800 by Pot-8-o's 1773 (Eclipse 1764) Prunella 1788 by Highflyer 1774 (Herod 1758)—Promise 1768 by Snap 1750 (Snip 1736) Julia 1756 by Blank 1740 (God. Ar. 1724).
				Daughter of, br. m., 1815.	Orville 1799 (Beningbrough 1791)—Dau. 1800 of Buzzard* 1787 (Woodpecker 1773) Hornpipe 1793 by Trumpator 1782 (Conductor 1767)—Luna 1779 by Herod 1758 (Tartar 1743)—Proserpine 1766 (Sist. to Eclipse) by Marske 1750 (Squirt 1732).
	Grey Stocking, ch. m., 1863.	Stockwell, ch. h., 1849.	The Baron, ch. h., 1842.	Ir. Birdcatcher ch. h., 1833.	Sir Hercules 1826 (Whalebone 1807) Guiccioli 1823 by Bob Booty 1804 (Chanticleer 1787)—Flight 1809 by Ir. Escape 1802 (Commodore 1793)—Y. Heroine by Bagot 1780 (Herod 1758) Heroine 1775 by Hero 1753 (Cade 1734).
				Echidna, b. m., 1838.	Economist 1825 (Whisker 1812)—Miss Pratt 1825 by Blacklock 1814 (Whitelock 1803)—Gadabout 1812 by Orville 1799 (Beningbrough 1791) Minstrel 1803 by Sir Peter T. 1784 (Highflyer 1774)—Matron 1782 by Florizel 1768 (Herod 1758).
			Pocahontas, b. m., 1837.	Glencoe,* ch. h., 1831.	Sultan 1816 (Selim 1802)—Trampoline 1825 by Tramp 1810 (Dick Andrews 1797)—Web 1808 by Waxy 1790 (Pot-8-o's 1773)—Penelope 1798 by Trumpator 1782 (Conductor 1767)—Prunella 1788 by Highflyer 1774 (Herod 1758).
				Marpessa, b. m., 1830.	Muley 1810 (Orville 1799)—Clare 1824 by Marmion 1806 (Whiskey 1789) Harpalice 1814 by Gohanna 1790 (Mercury 1778)—Amazon 1799 by Driver 1783 (Trentham 1766)—Fractious 1792 by Mercury 1778 (Eclipse 1764).
		Heroine, ch. m., 1855.	Neasham, b. h., 1848.	Hetman Platoff, b. h., 1833.	Brutandorf 1821 (Blacklock 1814)—Dau. 1820 of Comus 1809 (Sorcerer 1796)—Marciana 1809 by Stamford 1794 (Sir Peter T. 1784)—Marcia 1797 by Coriander 1786 (Pot-8-o's 1773)—Faith 1779 by Pacolet 1763 (Blank 1740).
				Wasp, ch. m., 1839.	Muley Moloch 1830 (Muley 1810)—Dau. 1836 of Emilius 1820 (Orville 1799) Bee-in-a-Bonnet 1825 by Blacklock 1814 (Whitelock 1803) Maniac 1806 by Shuttle 1793 (Y. Marske 1771)—Anticipation 1802 by Beningbrough 1791 (King Fergus 1775).
			Maid of Saragossa, b. m., 1841.	Jereed, b. h., 1834.	Sultan 1816 (Selim 1802) My Lady* 1818 by Comus 1809 (Sorcerer 1796) Dau. 1802 of Delpini 1781 (Highflyer 1774)—Tipple Cyder 1788 by King Fergus 1775 (Eclipse 1764)—Sylvia 1783 by Y. Marske 1771 (Marske 1750).
				Sister to *Alnderby, ch. m., 1831.	Velocipede 1825 (Blacklock 1814)—Kate 1823 by Catton 1809 (Golumpus 1802)—Miss Garforth 1819 by Walton 1799 (Sir Peter T. 1784)—Dau. 1806 of Hyacinthus 1797 (Coriander 1786)—Zara 1801 by Delpini 1781 (Highflyer 1774).
SCRAPS, br. m., 1870.	†Lord of the Hills, br. h., 1854.	Touchstone, br. h., 1821.	Camel, br. h., 1822.	Whalebone, br. h., 1807.	Waxy 1790 (Pot-8-o's 1773)—Penelope 1798 by Trumpator 1782 (Conductor 1767)—Prunella 1788 by Highflyer 1774 (Herod 1758)—Promise 1768 by Snap 1750 (Snip 1736)—Julia 1756 by Blank 1740 (God. Ar. 1724).
				Daughter of, b. m., 1812.	Selim 1802 (Buzzard* 1787) Maiden 1801 by Sir Peter T. 1784 (Highflyer 1774)—Dau. 1788 of Phenomenon 1780 (Herod 1758)—Matron 1782 by Florizel 1768 (Herod 1758)—Maiden 1770 by Matchem 1748 (Cade 1734).
			Banter, br. m., 1826.	Master Henry, b. h., 1815.	Orville 1799 (Beningbrough 1791)—Miss Sophia 1805 by Stamford 1794 (Sir Peter T. 1784)—Sophia 1798 by Buzzard* 1787 (Woodpecker 1773)—Hunca-munca 1787 by Highflyer 1774 (Herod 1758) Cypher 1772 by Squirrel 1754 (Traveller 1735).
				Boadicea, b. m., 1807.	Alexander 1782 (Eclipse 1764)—Brunette by Amaranthus 1766 (Old England 1741)—Mayfly 1771 by Matchem 1748 (Cade 1734)—Dau. 1755 of Anc. Starling 1738 (Starling 1727)—Look-at-me-Lads 1731 by Bristol Grasshopper (Byerly Turk).
		Fair Helen, ch. m., 1843.	Pantaloon, ch. h., 1824.	Castrel, ch. h., 1801.	Buzzard* 1787 (Woodpecker 1773)—Dau. of Alexander 1782 (Eclipse 1764) Dau. of Highflyer 1774 (Herod 1758)—Dau. of Alfred 1770 (Matchem 1748)—Dau. 1770 of Engineer 1755 (Sampson 1745).
				Idalia, ch. m., 1815.	Peruvian 1806 (Sir Peter T. 1784)—Musidora 1804 by Meteor 1783 (Eclipse 1764)—Maid-of-all-Work 1786 by Highflyer 1774 (Herod 1758)—Sist. 1771 to Tandem by Syphon 1750 (Squirt 1732) Dau. of Regulus 1739 (God. Ar. 1724).
			Rebecca, b. m., 1831.	Lottery, br. h., 1820.	Tramp 1810 (Dick Andrews 1797)—Mandane 1800 by Pot-8-o's 1773 (Eclipse 1764)—Y. Camilla 1787 by Woodpecker 1773 (Herod 1758)—Camilla 1778 by Trentham 1766 (T. Sweepstakes 1743)—Coquette 1765 by Compton Barb.
				Daughter of, b. m., 1818.	Cervantes 1806 (Don Quixote 1784)—Anticipation 1802 by Beningbrough 1791 (King Fergus 1775)—Expectation 1779 by Herod 1758 (Tartar 1743)—Dau. 1758 of Skim 1746 (Starling 1727)—Dau. 1751 of Janus 1738 (God. Ar. 1724).
	Ragpicker, br. m., 1865.	†Fisherman, br. h., 1853.	Heron, br. h., 1833.	Bustard, b. h., 1813.	Castrel 1801 (Buzzard* 1787) Miss Hap 1806 by Shuttle 1793 (Y. Marske 1771) Dau. 1798 of Sir Peter T. 1784 (Highflyer 1774) Miss Hervey 1775 by Eclipse 1764 (Marske 1750)—Clio 1760 by Y. Cade 1747 (Cade 1734).
				Daughter of, b. m., 1827.	Orville 1799 (Beningbrough 1791)—Rosanne 1811 by Dick Andrews 1797 (Joe Andrews 1778)—Rosetta 1803 by Beningbrough 1791 (King Fergus 1775) Rosamond 1788 by Tandem 1773 (Syphon 1750)—Tuberose 1772 by Herod 1758 (Tartar 1743).
			Mainbrace, br. m., 1844.	Sheet Anchor, br. h., 1832.	Lottery 1820 (Tramp 1810) Morgiana 1820 by Muley 1810 (Orville 1799)—Miss Stephenson 1814 by [Scud or] Sorcerer 1796 (Trumpator 1782)—Dau. 1796 of Precipitate* 1787 (Mercury 1778)—Dau. 1787 of Woodpecker 1773 (Herod 1758).
				Daughter of, b. m., 1830.	Bay Middleton 1833 (Sultan 1816)—Nitocris 1829 by Whisker 1812 (Waxy 1790) Manuella 1809 by Dick Andrews 1797 (Joe Andrews 1778) Mandane 1800 by Pot-8-o's 1773 (Eclipse 1764)—Y. Camilla 1787 by Woodpecker 1773 (Herod 1758).
		†Juliet, br. m., 1851.	Touchstone, br. h., 1831.	Camel, br. h., 1822.	Whalebone 1807 (Waxy 1790)—Dau. 1812 of Selim 1802 (Buzzard* 1787) Maiden 1801 by Sir Peter T. 1784 (Highflyer 1774)—Dau. 1788 of Phenomenon 1780 (Herod 1758)—Matron 1782 by Florizel 1768 (Herod 1758).
				Banter, br. m., 1826.	Master Henry 1815 (Orville 1799)—Boadicea 1807 by Alexander 1782 (Eclipse 1764) Brunette by Amaranthus 1766 (Old England 1741)—Mayfly 1771 by Matchem 1748 (Cade 1734)—Dau. 1755 of Anc. Starling 1738 (Starling 1727).
			Lancashire Witch, ch. m., 1842.	Tomboy, b. h., 1829.	Jerry 1821 (Smolensko 1810) Dau. 1817 of Ardrossan 1809 (John Bull 1789)—Lady Eliza 1813 by Whitworth 1805 (Agonistes 1797)—Dau. 1793 of Spadille 1784 (Highflyer 1774)—Sylvia 1783 by Y. Marske 1771 (Marske 1750).
				Kite, b. m., 1821.	Bustard 1813 (Castrel 1801)—Olympia b. m. 1815 by Sir Oliver 1800 (Sir Peter T. 1784)—Scotilla b. m. 1795 by Anvil 1777 (Herod 1758)—Scota b. m. 1783 by Eclipse 1764 (Marske 1750)—Harmony 1775 by Herod 1758 (Tartar 1743)—Rutilia b. m. 1769 by Blank 1740 (God. Ar. 1724)—Sist 1751 to South by Regulus 1739 (God. Ar. 1724) Dau. of

Soreheels (Basto 1702)—Dau. blk. m. of Makeless (Oglethorpe Ar.)—Christ. D'Arcy's Royal mare by Blunderbuss (Bustler)—Old Grey Royal by D'Arcy White Turk—Dau. of D'Arcy Yellow Turk Royal mare.

PHIL WARREN.

The property of Leonidas Cartwright, San Augustine, Texas. Terms of service advertised annually.

PHIL WARREN, chestnut horse, foaled 1886, bred by Philip Warren, Illinois, by Dakota, son of imp. Billet and Experience Oaks by imp. Bonnie Scotland, dam Annie Booth by Marion, son of Lexington and Miriam by imp. Glencoe, grandam Lady Vic by Uncle Vic, son of Lexington, great grandam Bonnie Jean by Bonnie Laddie, etc., thus giving him three crosses of Lexington and two of imp. Glencoe blood from most excellent sources. Dakota, from an injury, never raced. His dam, Experience Oaks, was a good and successful race-mare and dam of Volcano and Bonnie Oaks; his grandam, Sallie, was the dam of the good race-horses Nathan Oaks, Solicitor and John Mullins. Isola was a great race-mare, produced Olio, the dam of Biddy Malone and Duffy. Susette was another famous race-mare and dam of Berthune by Sidi Hamet, of Crisis, that produced Donerail. Jenny Cockracy was the best mare of her day and dam of Maid of Lodi, Creeping Kate and Betsey Harrison, the dam of that good horse Denmark by imp. Hedgeford. War Over, Yum Yum, Mollie Saffle, Connemara, Tarragon, Owen Bowling, War Duke, Warfellow, Impromptu and Palisade are from the same good family, which traces to the imp. Saltram mare, dam of the good race-horse Timoleon, sire of the immortal Boston, that sired Lexington, Lecompte, Salina, Salvator's dam, etc.

On the dam's side Phil Warren traces to the same family that produced Bonnie Lassie, Belle Anderson (dam of Zenith and Ripple), Chloe Anderson, dam of Novice (she the dam of the unbeaten Norfolk, sire of El Rio Rey, Emperor of Norfolk, etc.). It is a racing and producing family, as evidenced by the form of such horses as Asteroid (never beaten), Freeland, Long Knight, Clara D., Sister Anne, Maria Stuart, Volante, Bombay, Sovereign Pat, Grimaldi and a host of others which have descended from it.

Phil Warren is a bright chestnut horse with both hind feet and off front ankles white, a snip on nose and stripe down his face. Is 15¾ hands high, with good, fine head and neck and great muscular development. His pedigree is a rare combination of good racing blood.

PHIL WARREN.

CHESTNUT HORSE, FOALED 1886; BRED BY PHILIP WARREN, ILLINOIS.

Hambletonian 1792—King Fergus 1775—Eclipse 1764—Marske 1750—Squirt 1732—Bartlett's Childers—Darley Arabian 1699.

DAKOTA, b. h., 1878.	*Billet, b. h., 1865.	Voltigeur, br. h., 1847.	Voltaire, br. h., 1826.	Blacklock, b. h., 1814.	Whitelock 1803 (Hambletonian 1792)—Dau. 1799 of Coriander 1786 (Pot-8-o's 1773)—Wildgoose 1792 by Highflyer 1774 (Herod 1758)—Coheiress 1786 by Pot-8-o's 1773 (Eclipse 1764)—Manilla 1777 by Goldfinder 1764 (Snap 1750).
				Daughter of, b. m., 1816.	Phantom 1808 (Walton 1799)—Dau. 1802 of Overton 1788 (King Fergus 1775)—Dau. 1796 of Walnut 1786 (Highflyer 1774)—Dau. 1785 of Ruler 1777 (Y. Marske 1771)—Paracantha 1772 by Matchem 1748 (Cade 1734).
			Martha Lynn, br. m., 1837.	Mulatto, b. h., 1823.	Catton 1809 (Golumpus 1802)—Desdemona 1811 by Orville 1799 (Beningbrough 1791)—Fanny 1796 by Sir Peter T. 1784 (Highflyer 1774)—Dau. 1788 of Diomed* 1777 (Florizel 1768)—Desdemona 1770 by Marske 1750 (Squirt 1732).
				Leda, b. m., 1824.	Filho-da-Puta 1812 (Haphazard 1797)—Treasure 1809 by Camillus 1803 (Hambletonian 1792)—Dau. 1801 of Hyacinthus 1797 (Coriander 1786)—Flora 1789 by King Fergus 1775 (Eclipse 1764)—Atalanta 1769 by Matchem 1748 (Cade 1734).
		Calcutta, b. m., 1853.	Flatcatcher, br. h., 1845.	Touchstone, br. h., 1831.	Camel 1822 (Whalebone 1807)—Banter 1826 by Master Henry 1815 (Orville 1799)—Boadicea 1807 by Alexander 1782 (Eclipse 1764)—Brunette by Amaranthus 1766 (Old England 1741)—Mayfly 1771 by Matchem 1748 (Cade 1734).
				Decoy, b. m., 1830.	Filho-da-Puta 1812 (Haphazard 1797)—Finesse 1815 by Peruvian 1806 (Sir Peter T. 1784)—Violante 1802 by John Bull 1789 (Fortitude 1777)—Dau. 1788 of Highflyer 1774 (Herod 1758)—Everlasting 1775 by Eclipse 1764 (Marske 1750).
			Miss Martin, br. m., 1844.	St. Martin, br. h., 1833.	Actæon 1822 (Scud 1804)—Galena 1824 by Walton 1799 (Sir Peter T. 1784)—Comedy 1817 by Comus 1809 (Sorcerer 1796)—Dau. 1803 of Star* 1785 (Highflyer 1774)—Dau. 1794 of Y. Marske 1771 (Marske 1750).
				Wagtail, br. m., 1829.	Whisker 1812 (Waxy 1790)—Dau. 1814 of Sorcerer 1796 (Trumpator 1782)—Dau. 1808 of Sir Solomon 1796 (Sir Peter T. 1784)—Dau. 1794 of Y. Marske 1771 (Marske 1750)—Dau. 1788 of Phenomenon 1780 (Herod 1758).
	Experience Oaks, b. m., 1869.	*Bonnie Scotland, b. h., 1853.	Iago, b. h., 1843	Don John, b. h., 1835.	[Tramp or] Waverley 1817 (Whalebone 1807)—Dau. 1820 of Comus 1809 (Sorcerer 1796)—Marciana 1809 by Stamford 1794 (Sir Peter T. 1784)—Marcia 1797 by Coriander 1786 (Pot-8-o's 1773)—Faith 1779 by Pacolet 1763 (Blank 1740).
				Scandal, b. m. 1822.	Selim 1802 (Buzzard* 1787)—Dau. 1816 of Haphazard 1797 (Sir Peter T. 1784)—Princess 1799 by Precipitate* 1787 (Mercury 1778)—Colibri 1793 by Woodpecker 1773 (Herod 1758)—Camilla 1778 by Trentham 1766 (T. Sweepstakes 1743).
			Queen Mary, b. m., 1843.	Gladiator, ch. h., 1833.	Partisan 1811 (Walton 1799)—Pauline 1826 by Moses 1819 ([Whalebone or] Seymour 1807)—Quadrille 1815 by Selim 1802 (Buzzard* 1787)—Canary Bird 1806 by [Whiskey or] Sorcerer 1796 (Trumpator 1782)—Canary 1797 by Coriander 1786 (Pot-8-o's 1773).
				Daughter of, b. m., 1840.	Plenipotentiary 1831 (Emilius 1820)—Myrrha 1830 by Whalebone 1807 (Waxy 1790)—Gift 1818 by Y. Gohanna 1810 (Gohanna 1790)—Dau. 1802 of Sir Peter T. 1784 (Highflyer 1774)—Dau. 1788 of Trumpator 1782 (Conductor 1767).
		Sallie, b. m., 1863.	Lexington, b. h., 1850.	Boston, ch. h., 1833.	Timoleon 1813 (Sir Archy 1805)—Sist. 1814 to Tuckahoe by Ball's Florizel 1801 (*Diomed 1777)—Dau. 1799 of *Alderman 1787 (Pot-8-o's 1773)—Dau. of *Clockfast 1774 (Gimcrack 1760)—Dau. of Symmes' Wildair 1767 (*Fearnought 1755).
				Alice Carneal, b. m., 1836.	*Sarpedon 1828 (Emilius 1820)—Rowena 1826 by Sumpter 1818 (Sir Archy 1805)—Lady Grey 1817 by Robin Grey 1805 (*Royalist 1790)—Maria 1802 by Melzar 1791 (*Medley 1776)—Dau. of *Highflyer 1784 (Highflyer 1774).
			Sallie, ch. m., 1851.	Buford, ch. h., 1843.	*Nonplus 1821 (Catton 1809)—Too Soon 1835 by Sir Leslie 1828 (Sir William 1820)—Little Peggy 1821 by Gallatin 1799 (*Bedford 1792)—Trumpetta 1816 by Hephestion 1807 (*Buzzard 1787)—Peggy 1803 by *Bedford 1792 (Dungannon 1780).
				Isola, b. m., 1839.	Bertrand 1821 (Sir Archy 1805)—Susette 1830 by Aratus 1820 (Director 1811)—Jenny Cockracy 1814 by Potomac 1805 (*Diomed 1777)—Timoleon's dam by *Saltram 1780 (Eclipse 1764)—Dau. of Symmes' Wildair 1767 (*Fearnought 1755).
ANNIE BOOTH, ch. m., 1879.	Marion, ch. h., 1864.	Lexington, b. h., 1850.	Boston, ch. h., 1833.	Timoleon, ch. h., 1813.	Sir Archy 1805 (*Diomed 1777)—Dau. of *Saltram 1780 (Eclipse 1764)—Dau. of Symmes' Wildair 1767 (*Fearnought 1755)—Dau. of Tyler's Driver 1760 (*Othello)—Dau. of *Fallower 1761 (Blank 1740).
				Sister to Tuckahoe, ch. m., 1814.	Ball's Florizel 1801 (*Diomed 1777)—Dau. 1799 of *Alderman 1787 (Pot-8-o's 1773)—Dau. of *Clockfast 1774 (Gimcrack 1760)—Dau. of Symmes' Wildair 1767 (*Fearnought 1755)—Y. Kitty Fisher 1767 by *Fearnought 1755 (Regulus 1739).
			Alice Carneal, b. m., 1836.	*Sarpedon, br. h., 1828.	Emilius 1820 (Orville 1799)—Icaria 1824 by The Flyer 1814 (Vandyke Jr. 1808)—Parma 1813 by Dick Andrews 1797 (Joe Andrews 1778)—May 1804 by Beningbrough 1791 (King Fergus 1775)—Primrose 1787 by Mambrino 1768 (Engineer 1756).
				Rowena, ch. m., 1826.	Sumpter 1818 (Sir Archy 1805)—Lady Grey 1817 by Robin Grey 1805 (*Royalist 1790)—Maria 1802 by Melzar 1791 (*Medley 1776)—Dau. of *Highflyer 1784 (Highflyer 1774)—Dau. of *Fearnought 1755 (Regulus 1739).
		Miriam, ch. m., 1852.	*Glencoe, ch. h., 1831.	Sultan, b. h., 1816.	Selim 1802 (Buzzard* 1787)—Bacchante 1809 by W's Ditto 1800 (Sir Peter T. 1784)—Dau. 1791 of Mercury 1778 (Eclipse 1764)—Dau. 1776 of Herod 1758 (Tartar 1743)—Folly 1771 by Marske 1750 (Squirt 1732).
				Trampoline, ch. m., 1825.	Tramp 1810 (Dick Andrews 1797)—Web 1808 by Waxy 1790 (Pot-8-o's 1773)—Penelope 1798 by Trumpator 1782 (Conductor 1767)—Prunella 1788 by Highflyer 1774 (Herod 1758)—Promise 1768 by Snap 1750 (Snip 1736).
			Minerva Anderson, ch. m., 1837.	*Luzborough, br. h., 1820.	W's Ditto 1800 (Sir Peter T. 1784)—Picton's dam 1814 by Dick Andrews 1797 (Joe Andrews 1778)—Eleanor 1798 by Whiskey 1789 (Saltram* 1780)—Y. Giantess 1790 by Diomed* 1777 (Florizel 1768)—Giantess 1769 by Matchem 1748 (Cade 1734).
				Daughter of, —, —.	Sir Charles 1816 (Sir Archy 1805)—Famous quarter mare by Bess Brimmer (Blue Beard 1802)—[Miriam is grandam of Duke of Magenta].
	Lady Vic, ch. m., 1872.	Uncle Vic, ch. h., 1857.	Lexington, b. h., 1850.	Boston, ch. h., 1833.	Timoleon 1813 (Sir Archy 1805)—Sist. 1814 to Tuckahoe by Ball's Florizel 1801 (*Diomed 1777)—Dau. 1799 of *Alderman 1787 (Pot-8-o's 1773)—Dau. of *Clockfast 1774 (Gimcrack 1760)—Dau. of Symmes' Wildair 1767 (*Fearnought 1755).
				Alice Carneal, b. m., 1836.	*Sarpedon 1828 (Emilius 1820)—Rowena 1826 by Sumpter 1818 (Sir Archy 1805)—Lady Grey 1817 by Robin Grey 1805 (*Royalist 1790)—Maria 1802 by Melzar 1791 (*Medley 1776)—Dau. of *Highflyer 1784 (Highflyer 1774).
			Undine, ch. m., 1853.	*Yorkshire, b. h., 1834.	St. Nicholas 1827 (Emilius 1820)—*Miss Rose 1823 by Tramp 1810 (Dick Andrews 1797)—Dau. 1810 of Sancho 1801 (Don Quixote 1784)—Blacklock's dam 1799 by Coriander 1786 (Pot-8-o's 1773)—Wildgoose 1792 by Highflyer 1774 (Herod 1758).
				Lady Margrave, ch. m., 1847.	*Margrave 1829 (Muley 1810)—Lady Adams 1825 by Whipster (Cook's Whip 1801)—Woodpecker's dam 1807 by *Buzzard 1787 (Woodpecker 1773)—The Faun by Craig's Alfred (*Medley 1776)—Shepherdess by W's King Herod (*Fearnought 1755).
		Bonnie Jean, ch. m., 1861.	Bonnie Laddie, ch. h., 1853.	*Glencoe, ch. h., 1831.	Sultan 1816 (Selim 1802)—Trampoline 1825 by Tramp 1810 (Dick Andrews 1797)—Web 1808 by Waxy 1790 (Pot-8-o's 1773)—Penelope 1798 by Trumpator 1782 (Conductor 1767)—Prunella 1788 by Highflyer 1774 (Herod 1758).
				Magdalen, b. m., 1840.	Medoc 1829 (Am. Eclipse 1814)—Keph's dam 1825 by Sumpter 1818 (Sir Archy 1805)—Dau. 1811 of Lewis's Eclipse (*Diomed 1777)—Hart's Maria by Craig's Alfred (*Medley 1776)—The Pryor Mare by Bellair 1786 (*Medley 1776).
			Lux, ch. m., 1842.	Wagner, ch. h., 1834.	Sir Charles 1816 (Sir Archy 1805)—Maria West 1827 by Marion 1820 (Sir Archy 1805)—Ella Crump by *Citizen 1785 (Pacolet 1763)—Dau. 1795 of Huntsman 1789 (*Mousetrap 1771)—Dau. of Symmes' Wildair 1767 (*Fearnought 1755).
				Butterfly, ch. m., 1825.	Sumpter 1818 (Sir Archy 1805)—Dau. of *Buzzard 1787 (Woodpecker 1773)—Dau. of Dandridge's Fearnought (*Fearnought 1755)—Dau. of *Janus 1746 (Janus 1739)—[The Dandridge Fearnought mare was a good race-mare about the year 1790].

IMPORTED PONTIAC.

The property of J. O. Donner, Ramapo Stud, New Jersey.

PONTIAC, black horse, imported *in utero* by Mr. P. Lorillard, foaled 1881, is by Pero Gomez, winner of the St. Leger, son of Beadsman and Salamanca by Student, son of Chatham. Agenoria, his dam, is own sister in blood to Agility and Apology. The latter won the 1,000 Guineas, Oaks, St. Leger and Ascot Gold Cup. Milliner, his grandam, is own sister to Mandragora, the dam of Mandrake (winner of the Doncaster Cup), Agility and Apology, and to Mineral, the dam of Wenlock (winner of the St. Leger), Schwindler and Kisber (winner of the Derby and Grand Prize of Paris). Milliner is also dam of imp. Pizarro, a superior race-horse. Moonbeam by Tomboy was the dam of Loupgaron, Moonshine, etc. The family is one of the best racing ones in England. Pontiac won the Suburban Handicap, 1¼ miles, 102 lbs., in 2.09½; the Consolation Sweepstakes, the same distance, 109 lbs., in 2.12; the Free Handicap, 1 mile, at Monmouth, with 120 lbs., in 1.43¾; 1⅛ miles, 111 lbs., in 1.57; the Passaic Stakes, 6 furlongs, 118 lbs., in 1.17; the Eatontown Stakes, 1 mile, 118 lbs., in 1.43½; the Manhattan Handicap, 1¼ miles, 121 lbs., in 2.14, and the Free Handicap Sweepstakes, $1\frac{5}{16}$ miles, in 2.26½, with 123 lbs. At five years old won the Coney Island Stakes, 1⅛ miles, 122 lbs., in 1.56¼; won purse, ¾ of a mile, 120 lbs., in 1.14½, and won purse, ¾ of a mile, 120 lbs., in 1.15½. He did not run in 1887. In 1888 he won three out of five races; Sweepstakes, ¾ of a mile, 116 lbs., in 1.18¾; won purse, ¾ of a mile, 120 lbs., in 1.17¼; won Sweepstakes, $1\frac{1}{16}$ miles, 123 lbs., on the turf, in 1.51¼, beating Exile and three others. He was placed second or third in a large majority of the races in which he was beaten, and met and defeated the best and speediest horses on the turf during his career. Adventurer, the sire of his dam Agenoria, won the City and Suburban Handicap and was the sire of Pretender, winner of the 2,000 Guineas and Derby, Apology, winner of the 1,000 Guineas, Oaks and St. Leger, Wheel of Fortune, winner of the 1,000 Guineas and Oaks. Rataplan, the sire of his grandam, was own brother to Stockwell and a superior race-horse, winner of 42 out of the 71 races in which he started, including Tradesman's Cup, at Manchester, and was the sire of Kettledrum, winner of the Derby. Manganese, his third dam, won the 1,000 Guineas, and was by the great Birdcatcher, sire of The Baron (St. Leger winner), imp. Knight of St. George (St. Leger winner), Daniel O'Rourke (Derby winner), and Mrs. Ridgeway, his daughter, was the dam of Vedette (winner of the 2,000 Guineas) and sire of the popular Galopin (winner of the Derby).

Pontiac is a fine, symmetrically formed horse and from a very distinguished racing and producing family. He should be a success in the stud. He is descended in direct female line from the Layton Barb mare, to which traces Young Greyhound, Matchem, imp. Drone (by Herod), Young Cade, Blucher, Philammon, Simoon, imp. Comus, Kentucky, Gilroy, Daniel Boone, the great Iroquois, Parole, Powhattan and all the descendants of the imp. Cub mare and Magnolia family.

PONTIAC.

BLACK HORSE, FOALED 1881 (IMPORTED IN UTERO); BRED BY P. LORILLARD, RANCOCAS STUD, NEW JERSEY.

					Dick Andrews 1797 Joe Andrews 1778—Eclipse 1764—Marske 1750 Squirt 1732 Bartlett's Childers Darley Arabian 1689.
PERO GOMEZ, br. h., 1865.	Beadsman, br. h., 1855.	Weatherbit, br. h., 1842.	Sheet Anchor, br. h., 1832.	Lottery, br. h., 1820.	Tramp 1810 (Dick Andrews 1797) Mandane 1800 by Pot-8-o's 1773 (Eclipse 1764) Y. Camilla 1787 by Woodpecker 1773 (Herod 1758) Camilla 1778 by Trentham 1766 (T. Sweepstakes 1743)—Coquette 1765 by Compton Barb.
				Morgiana, blk. m., 1820.	Muley 1810 (Orville 1799) Miss Stephenson 1814 by [Scud or] Sorcerer 1796 (Trumpator 1782) Dau. 1796 of Precipitate* 1787 (Mercury 1778)—Dau. 1787 of Woodpecker 1773 (Herod 1758) Sist. 1773 to Juniper by Snap 1750 (Snip 1736).
			Miss Letty, b. m., 1834.	Priam,* b. h., 1827.	Emilius 1820 (Orville 1799) Cressida 1807 by Whiskey 1789 (Saltram 1780) Y. Giantess 1790 by Diomed* 1777 (Florizel 1768)—Giantess 1769 by Matchem 1748 (Cade 1734)—Molly Long Legs 1753 by Babraham 1740 (God. Ar. 1724).
				Daughter of, br. m., 1815.	Orville 1799 (Beningbrough 1791) Dau. 1800 of Buzzard* 1787 (Woodpecker 1773) Hornpipe 1793 by Trumpator 1782 (Conductor 1767) Luna 1779 by Herod 1758 (Tartar 1743) Proserpine 1766 by Marske 1750 (Squirt 1732).
		Mendicant, br. m., 1843.	Touchstone, br. h., 1831.	Camel, br. h., 1822.	Whalebone 1807 (Waxy 1790) Dau. 1812 of Selim 1802 (Buzzard* 1787) Maiden 1801 by Sir Peter T. 1784 (Highflyer 1774) Dau. 1788 of Phenomenon 1780 (Herod 1758)—Matron 1782 by Florizel 1768 Herod 1758).
				Banter, br. m., 1826.	Master Henry 1815 (Orville 1799) Boadicea 1807 by Alexander 1782 (Eclipse 1764)—Brunette by Amaranthus 1766 (Old England 1741)—Mayfly 1771 by Matchem 1748 (Cade 1734)—Dau. 1775 of Anc. Starling 1738 (Starling 1727).
			Lady Moore Carew, b. m., 1830.	Tramp, b. h., 1810.	Dick Andrews 1797 (Joe Andrews 1778) Dau. 1803 of Gohanna 1790 (Mercury 1778)—Fraxinella 1793 by Trentham 1766 (T. Sweepstakes 1743) Dau. 1785 of Woodpecker 1773 (Herod 1758) Everlasting 1775 by Eclipse 1764 (Marske 1750).
				Kite, b. m., 1821.	Bustard 1813 (Castrel 1801)—Olympia 1815 by Sir Oliver 1800 (Sir Peter T. 1784) Scotilla 1795 by Anvil 1777 (Herod 1758) Scota 1783 by Eclipse 1764 (Marske 1750) Harmony 1775 by Herod 1758 (Tartar 1743).
	Salamanca, b. m., 1859.	Student, ch. h., 1851.	Chatham, ch. h., 1839.	The Colonel, ch. h., 1825.	Whisker 1812 (Waxy 1790) Dau. 1802 of Delpini 1781 (Highflyer 1774) Tipple Cyder 1788 by King Fergus 1775 (Eclipse 1764) Sylvia 1783 by Y. Marske 1771 (Marske 1750)—Ferret 1765 by Broth. to Silvio (Cade 1734).
				Hester, br. m., 1832.	Camel 1822 (Whalebone 1807) Monimia 1821 by Muley 1810 (Orville 1799) Dau. 1796 of Precipitate* 1787 (Mercury 1778)—Dau. 1787 of Woodpecker 1773 (Herod 1758)—Sist. 1773 to Juniper by Snap 1750 (Snip 1736).
			Daughter of, b. m., 1840.	Laurel, br. h., 1824.	Blacklock 1814 (Whitelock 1803)—Wagtail 1818 by Prime Minister 1810 (Sancho 1801)—Tranby*'s dam 1812 by Orville 1799 (Beningbrough 1791) Miss Grimstone 1796 by Weazle 1776 (Herod 1758) Dau. of Ancaster 1768 (Blank 1740).
				Flight, br. m., 1831.	Velocipede 1825 (Blacklock 1814)—Miss Wilkes 1818 by Octavian 1807 (Stripling 1795)—Dau. 1807 of Remembrancer 1800 (Sir Peter T. 1784) Mary 1797 by Y. Marske 1771 (Marske 1750) Gentle Kitty 1774 by Silvio 1754 (Cade 1734).
		Bravery, br. m., 1853.	Gameboy, blk. h., 1842.	Tomboy, b. h., 1829.	Jerry 1821 (Smolensko 1810)—Dau. 1817 of Ardrossan 1809 (John Bull 1789) Lady Eliza 1813 by Whitworth 1805 (Agonistes 1797)—Dau. 1793 of Spadille 1784 (Highflyer 1774)—Sylvia 1783 by Y. Marske 1771 (Marske 1750).
				Lady Moore Carew, b. m., 1830.	Tramp 1810 (Dick Andrews 1797)—Kite 1821 by Bustard 1813 (Castrel 1801)—Olympia 1815 by Sir Oliver 1800 (Sir Peter T. 1784)—Scotilla 1795 by Anvil 1777 (Herod 1758)—Scota 1783 by Eclipse 1764 (Marske 1750).
			Ennui, br. m., 1843.	Bay Middleton, b. h., 1833.	Sultan 1816 (Selim 1802)—Cobweb 1821 by Phantom 1808 (Walton 1799)—Filagree 1815 by Soothsayer 1808 (Sorcerer 1796) Web 1808 by Waxy 1790 (Pot-8-o's 1773)—Penelope 1798 by Trumpator 1782 (Conductor 1767).
				Blue Devils, ch. m., 1837.	Velocipede 1825 (Blacklock 1814)—Care 1825 by Woful 1809 (Waxy 1790)—Dau. 1819 of Rubens 1805 (Buzzard* 1787)—Tippitywitchet 1808 by Waxy 1790 (Pot-8-o's 1773)—Hare 1794 by Sweetbriar 1769 (Syphon 1750).
*AGENORIA, br. m., 1876.	Adventurer, b. h., 1859.	Newminster, b. h., 1848.	Touchstone, br. h., 1831.	Camel, br. h., 1822.	Whalebone 1807 (Waxy 1790)—Dau. 1812 of Selim 1802 (Buzzard* 1787)—Maiden 1801 by Sir Peter T. 1784 (Highflyer 1774)—Dau. 1788 of Phenomenon 1780 (Herod 1758)—Matron 1782 by Florizel 1768 (Herod 1758).
				Banter, br. m., 1826.	Master Henry 1815 (Orville 1799)—Boadicea 1807 by Alexander 1782 (Eclipse 1764)—Brunette by Amaranthus 1766 (Old England 1741) Mayfly 1771 by Matchem 1748 (Cade 1734)—Dau. 1755 of Anc. Starling 1738 (Starling 1727).
			Bee's-wing, b. m., 1833.	Dr. Syntax, br. h., 1811.	Paynator 1791 (Trumpator 1782) Dau. of Beningbrough 1791 (King Fergus 1775) Jenny Mole 1787 by Carbuncle 1772 (Babr. Blank 1778)—Dau. 1773 of Prince T'Quassaw 1751 (Snip 1736)—Sultana 1759 by Regulus 1739 (God. Ar. 1724).
				Tomboy's dam, ch. m., 1817.	Ardrossan 1809 (John Bull 1789)—Lady Eliza 1813 by Whitworth 1805 (Agonistes 1797)—Dau. 1798 of Spadille 1784 (Highflyer 1774)—Sylvia 1783 by Y. Marske 1771 (Marske 1750)—Ferret 1765 by Broth. to Silvio (Cade 1734).
		Palma, br. m., 1840.	Emilius, b. h., 1820.	Orville, b. h, 1799.	Beningbrough 1791 (King Fergus 1775)—Evelina 1791 by Highflyer 1774 (Herod 1758)—Termagant by Tantrum 1760 (Cripple 1750)—Cantatrice by Sampson 1745 (Blaze 1733)—Dau. 1757 of Regulus 1739 (God. Ar. 1724).
				Emily, ch. m., 1810.	Stamford 1794 (Sir Peter T. 1784)—Dau. 1799 of Whiskey 1789 (Saltram* 1780)—Grey Dorimant 1781 by Dorimant 1772 (Otho 1760)—Dizzy 1757 by Blank 1740 (God. Ar. 1724)—Anc. Dizzy 1741 by Driver 1727 (Wynn Ar.).
			Francesca, br. m., 1827.	Partisan, b. h., 1811.	Walton 1799 (Sir Peter T. 1784)—Parasol 1800 by Pot-8-o's 1773 (Eclipse 1764)—Prunella 1788 by Highflyer 1774 (Herod 1758)—Promise 1768 by Snap 1750 (Snip 1736)—Julia 1756 by Blank 1740 (God. Ar. 1724).
				Miss Letty's dam, br. m., 1815.	Orville 1799 (Beningbrough 1791)—Dan. 1800 of Buzzard* 1787 (Woodpecker 1773)—Hornpipe 1793 by Trumpator 1782 (Conductor 1767)—Luna 1779 by Herod 1758 (Tartar 1743)—Proserpine 1766 by Marske 1750 (Squirt 1732).
	Milliner, ch. m., 1869.	Rataplan, ch. h., 1850.	The Baron, ch. h., 1842.	Ir Birdcatcher ch. h., 1833.	Sir Hercules 1826 (Whalebone 1807)—Guiccioli 1823 by Bob Booty 1804 (Chanticleer 1787)—Flight 1809 by Ir. Escape 1802 (Commodore 1793)—Y. Heroine by Bagot 1780 (Herod 1758)—Heroine 1775 by Hero 1753 (Cade 1734).
				Echidna, b. m., 1838.	Economist 1825 (Whisker 1812)—Miss Pratt 1825 by Blacklock 1814 (Whitelock 1803)—Gadabout 1812 by Orville 1799 (Beningbrough 1791)—Minstrel 1803 by Sir Peter T. 1784 (Highflyer 1774)—Matron 1782 by Florizel 1768 (Herod 1758).
			Pocahontas, b. m., 1837.	Glencoe,* ch. h., 1831.	Sultan 1816 (Selim 1802)—Trampoline 1825 by Tramp 1810 (Dick Andrews 1797)—Web 1808 by Waxy 1790 (Pot-8-o's 1773)—Penelope 1798 by Trumpator 1782 (Conductor 1767)—Prunella 1788 by Highflyer 1774 (Herod 1758).
				Marpessa, b. m., 1830.	Muley 1810 (Orville 1799)—Clare 1824 by Marmion 1806 (Whiskey 1789)—Harpalice 1814 by Gohanna 1790 (Mercury 1778)—Amazon 1799 by Driver 1783 (Trentham 1766)—Fractious 1792 by Mercury 1778 (Eclipse 1764).
		Manganese, ch. m., 1853.	Ir. Birdcatcher, ch. h., 1833.	Sir Hercules, blk. h., 1826.	Whalebone 1807 (Waxy 1790) Peri 1822 by Wanderer 1811 (Gohanna 1790)—Thalestris 1809 by Alexander 1782 (Eclipse 1764)—Rival 1800 by Sir Peter T. 1784 (Highflyer 1774)—Hornet 1790 by Drone 1777 (Herod 1758).
				Guiccioli, ch. m., 1823.	Bob Booty 1804 (Chanticleer 1787)—Flight 1809 by Ir. Escape 1802 (Commodore 1793)—Y. Heroine by Bagot 1780 (Herod 1758)—Heroine 1775 by Hero 1753 (Cade 1734)—Dau. 1766 of Snap 1750 (Snip 1736).
			Moonbeam, ch. m., 1838.	Tomboy, b. h., 1829.	Jerry 1821 (Smolensko 1810)—Dau. 1817 of Ardrossan 1809 (John Bull 1789)—Lady Eliza 1813 by Whitworth 1805 (Agonistes 1797)—Dau. 1798 of Spadille 1784 (Highflyer 1774)—Sylvia 1783 by Y. Marske 1771 (Marske 1750).
				Lunatic, b. m., 1818.	Prime Minister 1810 (Sancho 1801)—Maniac ch. m. 1806 by Shuttle 1793 (Y. Marske 1771)—Anticipation ch. m 1802 by Beningbrough 1791 (King Fergus 1775)—Expectation gr. m. 1779 by Herod 1758 (Tartar 1743)—Dau. 1758 of Skim 1746 (Starling 1727)—Dau. 1751 of Janus 1738 (God. Ar. 1724)—Spinster gr. m. 1743 by Crab 1722 (Alcock Ar.)—Spinster (the Widdrington mare) ch. m. 1735 by Partner 1718 (Jigg)—Bay Bloody Buttocks b. m. 1729 by Bloody Buttocks—Sist. 1722 to Guy by Greyhound (Chillaby)—Brown Farewell br. m. 1710 by Makeless (Oglethorpe Ar.)—Dau. of Brimmer (D'Arcy Yellow Turk)—Dau. of Place's White Turk—Dau of Dodsworth—Layton Barb mare.

POST GUARD (GEN. PHILLIPS).

The property of Messrs. Stephen Sanford and Sons, Hurricane Stud, Amsterdam, N. Y. Yearlings sold annually.

POST GUARD (first called and raced under the name of Gen. Phillips), chestnut horse, foaled 1874, bred in the Preakness Stud, by Mr. M. H. Sanford, by imp. Glenelg, son of Citadel by Stockwell, dam La Polka, dam of Strathspey, Heel-and-Toe, Los Angeles, etc., by Lexington, grandam Dance, dam of Schottische, Redowa, etc., by imp. Glencoe, great grandam Cotillion, dam of Gallopade, Jr., O'Meara, etc., by imp. Leviathan, out of imp. Gallopade, dam of Reel (the dam of War Dance, Lecompte, Starke, Prioress, Fanny Wells, etc.), by Catton, son of Golumpus by Gohanna. Under the name of Gen. Phillips, at three years old, started in nine races, won eight; mile heats in 1.48, 1.45½; 1¼ miles, in 2.12; 1½ miles, in 2.42, and 1¾ miles, in 3.15½. At four years old started fifteen times, won six races, was second in one and third in four; won at mile heats, 1¼ miles, 1½ miles, 1¾ miles, and the Westchester Cup, 2¼ miles, 118 lbs., beating Parole and St. James, and was placed in several fast races. At five years old started eighteen times, won three races, was second in three and third in four; won Handicap Sweepstake, 2 miles, 115 lbs., in 2 39½; won purse, 1¾ miles, 112 lbs., in 3.06½; won Grand National Handicap, 2¼ miles, 108 lbs., defeating the noted Ferida, in 4.07½. At six years old started twenty times, won three races, was second in four and third in six; won purse, 1¼ miles, 108 lbs., in 1.56½; won purse, 1¼ miles, 118 lbs., in 2.41½; won purse, 2¼ miles, 105 lbs., in 3.52½. In his seventh, eighth, ninth and tenth years he started in twenty-four races, principally hurdle and steeple-chases, of which he won twelve, carrying from 136 to 168 lbs. He was a horse of fine speed and a game one, capable of running four-mile heats.

Post Guard is fashionably bred and from one of the best racing and producing families in the world. Lou Lanier, from it, was one of the best race-mares of her day at all distances. Lady Hardaway by Commodore, son of Boston, from it, is the dam of Miss Hardaway and Mercedes, the latter the dam of Barnes, Runnymede, Lydia, etc. Gallopade, Jr., own sister to Dance, was the dam of Schottische (she the dam of Waltz, Ballinkeel and Reveller). Waltz is the dam of Glidelia, she having the fastest record until 1891 at 1¾ miles, 3.01, and of Hop, the dam of Carson, Fond du Lac, Kirkman, etc. Imp. Gallopade was the dam of the renowned Reel, and she the dam of Lecompte, the only horse that ever beat Lexington; Starke, winner of the Goodwood Cup; Prioress, winner of the Cesarewitch; War Dance and Fanny Wells. Nellie Grey, by Commodore, Jils Johnson, and a host of fine race-horses are from this family.

Post Guard is almost untried in the stud, but is the sire of the winners, Guard, Rear Guard, Guardiana, Our Maggie, Post Master, Some More, Miss Aggie, Zam Post, etc. He is descended in direct female line from Old Piping Peg, to which traces Bob Booty, Mickey Free, Chanticleer, Solon, Xenophon, Barcaldine, Lowland Chief, War Dance, Lecompte and all the Reel family.

POST GUARD (Gen. Philips).

CHESTNUT HORSE, FOALED 1874; BRED BY MR. M. H. SANFORD, PREAKNESS STUD, KENTUCKY.

- *GLENELG, b. h., 1866.
  - Citadel, ch. h., 1859.
    - Stockwell, ch. h., 1849.
      - The Baron, ch. h., 1842.
      - Pocahontas, b. m., 1837.
    - Sortie, b. m., 1851.
      - Melbourne, br. h., 1834.
      - Escalade, b. m., 1846.
  - *Babta, b. m., 1858.
    - Kingston, b. h., 1849.
      - Venison, br. h., 1833.
      - QueenAnne, b. m., 1843.
    - Alice Lowe, ch. m., 1841.
      - Defence, b. h., 1824.
      - Pet, b. m., 1836.
- LA POLKA, b. m., 1865.
  - Lexington, b. h., 1850.
    - Boston, ch. h., 1833.
      - Timoleon, ch. h., 1813.
      - Sist. to Tuckahoe, ch. m., 1814.
    - Alice Carneal, b. m., 1836.
      - *Sarpedon, br. h., 1828.
      - Rowena, ch. m., 1826.
  - Dance, ch. m., 1853.
    - *Glencoe, ch. h., 1831.
      - Sultan, b. h., 1816.
      - Trampoline, ch. m., 1825.
    - Cotillion, ch. m., 1837.
      - *Leviathan, ch. h., 1823.
      - *Galopade, gr. m., 1823.

The Baron, ch. h., 1842.	Ir. Birdcatcher, ch. h., 1833.	Sir Hercules 1826 (Whalebone 1807) Guiccioli 1823 by Bob Booty 1801 (Chanticleer 1787) Flight 1809 by Ir. Escape 1802 (Commodore 1793) Y. Heroine by Bagot 1780 (Herod 1758) Heroine 1775 by Hero 1753 (Cade 1734).
	Echidna, b. m., 1838.	Economist 1825 (Whisker 1812) Miss Pratt 1825 by Blacklock 1814 (Whitelock 1803) Gadabout 1812 by Orville 1799 (Beningbrough 1791) Minstrel 1803 by Sir Peter T. 1784 (Highflyer 1774) Matron 1782 by Florizel 1768 (Herod 1758).
Pocahontas, b. m., 1837.	Glencoe,* ch. h., 1831.	Sultan 1816 (Selim 1802)—Trampoline 1825 by Tramp 1810 (Dick Andrews 1797)—Web 1808 by Waxy 1790 (Pot 8 o-'s 1773) Penelope 1798 by Trumpator 1782 (Conductor 1767) Prunella 1788 by Highflyer 1774 (Herod 1758).
	Marpessa, b. m., 1830.	Muley 1810 (Orville 1799) Clare 1824 by Marmion 1806 (Whiskey 1789) Harpalice 1814 by Gohanna 1790 (Mercury 1778)—Amazon 1799 by Driver 1783 (Trentham 1766) Fractions 1792 by Mercury 1778 (Eclipse 1764).
Melbourne, br. h., 1834.	Humprey Clinker, b. h., 1822.	Comus 1809 (Sorcerer 1796) Clinkerina 1812 by Clinker 1805 (Sir Peter T. 1784)—Pewet 1786 by Tandem 1773 (Syphon 1750) Termagant by Tantrum 1760 (Cripple 1750)—Cantatrice by Sampson 1745 (Blaze 1733).
	Daughter of, b. m., 1825.	Cervantes 1806 (Don Quixote 1784) Dau. 1818 of Golumpus 1802 (Gohanna 1790)—Dau. 1810 of Paynator 1791 (Trumpator 1782) Circle 1802 by St. George 1789 (Highflyer 1774) Abigail 1788 by Woodpecker 1773 (Herod 1758).
Escalade, b. m., 1846.	Touchstone, br. h., 1831.	Camel 1822 (Whalebone 1807)—Banter 1826 by Master Henry 1815 (Orville 1799) Boadicea 1807 by Alexander 1782 (Eclipse 1764)—Brunette by Amaranthus 1766 (Old England 1741) Mayfly 1771 by Matchem 1748 (Cade 1734).
	Ghuznee, b. m., 1838.	Pantaloon 1824 (Castrel 1801)—Languish 1830 by Cain 1822 (Paulowitz 1813)—Lydia 1822 by Poulton 1805 (Sir Peter T. 1784) Variety 1808 by Hyacinthus 1797 (Coriander 1786)—Dan. 1790 of Weazle 1776 (Herod 1758).
Venison, br. h., 1833.	Partisan, b. h., 1811.	Walton 1799 (Sir Peter T. 1784) Parasol 1800 by Pot 8 o-'s 1773 (Eclipse 1764) Prunella 1788 by Highflyer 1774 (Herod 1758) Promise 1768 by Snap 1750 (Snip 1736)—Julia 1756 by Blank 1740 (God. Ar. 1724).
	Fawn, br. m., 1823.	Smolensko 1810 (Sorcerer 1796) Jerboa 1803 by Gohanna 1790 (Mercury 1778)—Camilla 1778 by Trentham 1766 (T. Sweepstakes 1743) Coquette 1764 by Compton Barb—Sist. 1743 to Regulus by God. Ar. 1724.
QueenAnne, b. m., 1843.	Slane, b. h., 1833.	Royal Oak 1823 (Catton 1809)—Dau. 1819 of Orville 1799 (Beningbrough 1791) Epsom Lass 1803 by Sir Peter T. 1784 (Highflyer 1774)—Alexina 1788 by King Fergus 1775 (Eclipse 1764)—Lardella 1780 by Y. Marske 1771 (Marske 1750).
	Garcia, br. m., 1823.	Octavian 1807 (Stripling 1795)—Dau. 1806 of Shuttle 1793 (Y. Marske 1771) Katharine 1798 by Delpini 1781 (Highflyer 1774)—Dau. 1786 of Paymaster 1766 (Blank 1740)—Dan. 1774 of Le Sang 1759 (Changeling 1747).
Defence, b. h., 1824.	Whalebone, br. h., 1807.	Waxy 1790 (Pot-8-o-'s 1773)—Penelope 1798 by Trumpator 1782 (Conductor 1767)—Prunella 1788 by Highflyer 1774 (Herod 1758)—Promise 1768 by Snap 1750 (Snip 1736)—Julia 1756 by Blank 1740 (God. Ar. 1724).
	Defiance, ch. m., 1816.	Rubens 1805 (Buzzard* 1787) Little Folly 1806 by Highland Fling 1798 (Spadille 1784)—Harriet 1799 by Volunteer 1780 (Eclipse 1764)—Dau. 1785 of Alfred 1770 (Matchem 1748)—Magnolia 1771 by Marske 1750 (Squirt 1732).
Pet, b. m., 1836.	Gainsborough, br. h., 1813.	Rubens 1805 (Buzzard* 1787) Tiney 1801 by Sir Peter T. 1784 (Highflyer 1774)—Wren 1783 by Woodpecker 1773 (Herod 1758)—Papillon 1769 by Snap 1750 (Snip 1736)—Miss Cleveland 1758 by Regulus 1739 (God. Ar. 1724).
	Daughter of, b. m., 1818.	Topsy Turvey 1805 (St. George 1789) Agnes 1805 by Shuttle 1793 (Y. Marske 1771)—Dau. 1788 of Highflyer 1774 (Herod 1758)—Dau. 1779 of Goldfinder 1764 (Snap 1750)—Lady Bolingbroke 1766 by Squirrel 1754 (Traveller 1735).
Timoleon, ch. h., 1813.	Sir Archy, b. h., 1805.	*Diomed 1777 (Florizel 1768)—*Castianira 1796 by Rockingham 1781 (Highflyer 1774)—Tabitha 1782 by Trentham 1766 (T. Sweepstakes 1743)—Dau. of Bosphorus 1754 (Babraham 1740)—Dau. of Forester 1730 (Forester 1736).
	Daughter of, ch. m., —.	*Saltram 1780 (Eclipse 1764)—Dau. of Symmes' Wildair 1767 (*Fearnought 1755)—Dau. of Tyler's Driver 1760 (*Othello)—Dau. of *Fallower 1761 (Blank 1740)—Dau. of *Vampire 1757 (Regulus 1739).
Sist. to Tuckahoe, ch. m., 1814.	Ball's Florizel, ch. h., 1801.	*Diomed 1777 (Florizel 1768)—Dau. of *Shark 1771 (Marske 1750)—Dau. of Harris' Eclipse 1770 (*Fearnought 1755)—Dau. of *Fearnought 1755 (Regulus 1739)—Dau. of *Jolly Roger 1741 (Roundhead 1733).
	Daughter of, b. m., 1799.	*Alderman 1787 (Pot-8-o-'s 1773)—Dau. of *Clockfast 1774 (Gimcrack 1760)—Dau. of Symmes' Wildair 1767 (*Fearnought 1755)—Y. Kitty Fisher 1767 by *Fearnought 1755 (Regulus 1739)—*Kitty Fisher 1755 by Cade 1734 (God. Ar. 1724).
*Sarpedon, br. h., 1828.	Emilius, b. h., 1820.	Orville 1799 (Beningbrough 1791)—Emily 1810 by Stamford 1794 (Sir Peter T. 1784)—Dau. 1799 of Whiskey 1789 (Saltram* 1780) Grey Dorimant 1781 by Dorimant 1772 (Otho 1760) Dizzy 1757 by Blank 1740 (God. Ar. 1724).
	Icaria, b. m., 1824.	The Flyer 1814 (Vandyke Jr. 1808)—Parma 1813 by Dick Andrews 1797 (Joe Andrews 1778)—May 1804 by Beningbrough 1791 (King Fergus 1775)—Primrose 1787 by Mambrino 1768 (Engineer 1756)—Cricket 1776 by Herod 1758 (Tartar 1743).
Rowena, ch. m., 1826.	Sumpter, ch. h., 1818.	Sir Archy 1805 (*Diomed 1777)—Dau. of *Robin Redbreast 1796 (Sir Peter T. 1784)—Dau. 1788 of *Obscurity 1777 (Eclipse 1764)—Slamerkin 1768 by *Wildair 1753 (Cade 1734)—*Cub mare 1762 by Cub 1739 (Fox 1714).
	Lady Grey, b. m., 1817.	Robin Grey 1805 (*Royalist 1790)—Maria 1802 by Melzar 1791 (*Medley 1776)—Dau. of *Highflyer 1781 (Highflyer 1774)—Dau. of *Fearnought 1755 (Regulus 1739)—Dau. of Ariel 1756 (*Traveller 1747).
Sultan, b. h., 1816.	Selim, ch. h., 1802.	Buzzard* 1787 (Woodpecker 1773)—Dau. of Alexander 1782 (Eclipse 1764)—Dau. of Highflyer 1774 (Herod 1758)—Dau. of Alfred 1770 (Matchem 1748)—Dau. 1770 of Engineer 1756 (Sampson 1745).
	Bacchante, br. m., 1809.	W's Ditto 1800 (Sir Peter T. 1784)—Dau. 1791 of Mercury 1778 (Eclipse 1764)—Dau. 1776 of Herod 1758 (Tartar 1743) Folly 1771 by Marske 1750 (Squirt 1732)—Vixen 1753 by Regulus 1739 (God. Ar. 1724).
Trampoline, ch. m., 1825.	Tramp, b. h., 1810.	Dick Andrews 1797 (Joe Andrews 1778) Dau. 1803 of Gohanna 1790 (Mercury 1778)—Fraxinella 1793 by Trentham 1766 (T. Sweepstakes 1743)—Dau. 1785 of Woodpecker 1774 (Herod 1758)—Everlasting 1775 by Eclipse 1764 (Marske 1750).
	Web, b. m., 1808.	Waxy 1790 (Pot-8-o-'s 1773)—Penelope 1798 by Trumpator 1782 (Conductor 1767)—Prunella 1788 by Highflyer 1774 (Herod 1758)—Promise 1768 by Snap 1750 (Snip 1736)—Julia 1756 by Blank 1740 (God. Ar. 1724).
*Leviathan, ch. h., 1823.	Muley, b. h., 1810.	Orville 1799 (Beningbrough 1791)—Eleanor 1798 by Whiskey 1789 (Saltram* 1780)—Y. Giantess 1790 by Diomed* 1777 (Florizel 1768)—Giantess 1769 by Matchem 1748 (Cade 1734) Molly Long Legs 1753 by Babraham 1740 (God. Ar. 1724).
	Daughter of, blk. m., 1809.	Windle 1804 (Beningbrough 1791)—Dau. 1788 of Anvil 1777 (Herod 1758)—Virago 1764 by Snap 1750 (Snip 1736)—Dau. of Regulus 1739 (God. Ar. 1724)—Sist. 1750 to Othello by Crab 1722 (Alcock Ar.).
*Galopade, gr. m., 1823.	Catton, b. h., 1809.	Golumpus 1802 (Gohanna 1790)—Lucy Gray 1804 by Timothy 1794 (Delpini 1781)—Lucy 1789 by Florizel 1768 (Herod 1758)—Frenzy 1774 by Eclipse 1764 (Marske 1750)—Dau. of Engineer 1756 (Sampson 1745).
	Camillina, gr. m., 1822.	Camillus 1803 (Hambletonian 1792)—Dau. b. m. 1819 by Smolensko 1810 (Sorcerer 1796)—Miss Cannon b. m. 1811 by Orville 1799 (Beningbrough 1791)—Dau. b. m. 1796 of Weathercock 1785 (Highflyer 1774)—Cora ro. m. 1777 by Matchem 1748 (Cade 1734)—Dau. of Turk 1763 (Regulus 1739)—Dau. 1764 of Cub 1739 (Fox 1714)—Dau. 1750 of Allworthy 1744 (Crab 1722)—Dau. of Starling 1727 (Bay Bolton 1705)—Dairymaid b. m. 1737 by Bloody Buttocks—Bay Brocklesby b. m. 1731 by Partner 1718 (Jigg)—Brocklesby gr. m. 1721 by Greyhound (Chillaby)—Brocklesby Betty ch. m. 1711 by Curwen Bay Barb—Hobby mare by Lister Turk out of Piping Peg.

POWHATTAN.

OWN BROTHER TO PAROLE.

In the Woodburn Stud, A. J. Alexander, proprietor, near Spring Station, Woodford Co., Ky. Application to L. Brodhead, Spring Station, Ky. Annual sales of yearlings in May.

POWHATTAN by imp. Leamington, son of Faugh-a-Ballagh, bred by Mr. A. Welch, Erdenheim Stud, Pa., foaled 1879, dam Maiden, dam of Parole by Lexington, son of Boston, out of Kitty Clark by imp. Glencoe. Powhattan started in only one public race, that at Long Branch; meeting with an accident, he was unplaced. He was considered quite promising, and ran a very fast trial at Jerome Park in the spring. Parole, his own brother, has been one of the noted popular favorites of the turf. After racing in his three, four and five-year old form in this country, he was sent to England, and won the Newmarket Handicap, last 1½ miles of the Beacon Course, beating the great Isonomy, Lina and three others. He won the City and Suburban Handicap, 1¼ miles, 119 lbs., beating Ridotto, Cradle and fifteen others, and the great Metropolitan Stakes (handicap), 2¼ miles, 124 lbs., beating Castlereagh; the Great Cheshire Handicap Stakes, 1¼ miles, carrying 134 lbs., defeating Reefer, Sir Joseph, Ridotto and four others, and also won the Epsom Gold Cup, 1½ miles, 125 lbs., beating Alchemist and Primrose. The family has always been a successful racing one. Maiden, his dam, won the Travers Stakes at Saratoga, the grandam and great grandam were all winners, and from the stock have come such horses as Mary Morris, Wild Irishman, Frankfort, Falsetto, Felicia, Fortuna, Sumpter, Childers, Rattler, Ringgold, Flirtilla, Mary Wynn, Tom Kimball, Cassandra, The Countess, Turco, John Bascomb, Helen Mar, Ann Merry, Tiger, Ann Innis, Hebron, Meteor, George Kinney, Perfection, Saluda, Perception, Peregal, Oolah and, in the early days of racing, old Slamerkin. Then, on the score of blood and racing lineage, Powhattan should commend himself to the public. Powhattan is a solid bay, with large star, left fore and right hind ankles white, his head is bony and expressive, with large, full eyes, high at the withers like his noted brother, good barrel, great depth of girth, with good loin and fine coupling, and has the Lexington hind leg; he is 16½ hands high. Whilst he has not the racing record of his great brother, he is a Leamington from a racing family on both sides, and as nearly all the Leamingtons are successes in the stud, he should be. Flying Childers electrified the world by his brilliant racing career, but his brother, Bartlett's Childers excelled him in the stud, as also did Cade excel his famous racing brother Lath. Powhattan is much inbred to Sir Archy and Diomed, and traces back without a flaw through Amaranthus, dam to the Layton Barb mare.

Powhattan is the sire of the winners Duhme, Poteen, Pocassett, Cherry Blossom, Burlington, an extra good one, E. D., Eda, Flossie, Indian Princess, Palisade, Fenn P., Pilgrim, Capt. Jack, Imogene, Palestine, Melanie, Mary Conroy, Pennyroyal, The Pooky, Elsie S., Old Pepper, Cornie Buckingham, The Distiller, Pow Wow, Court Lore, Banjo, Manhassett, Llewellyn, Padlock, etc.

Powhattan descends in direct female line from the Layton Barb mare, to which trace Young Greyhound, Matchem, Young Cade, Drone (by Herod), Blucher, Simoom, Audubon, Philammon, St. Honorat; etc.

POWHATTAN.

BROWN HORSE, FOALED 1879; BRED BY MR. A. WELCH, CHESTNUT HILL STUD, PENNSYLVANIA.

*LEAMINGTON, br. h., 1853.	Faugh-a-Ballagh, br. h., 1841.	Sir Hercules, blk. h., 1826.	Whalebone, br. h., 1807.	Waxy, b. h., 1790.	Pot-8-o-'s 1773 (Eclipse 1764)—Maria 1777 by Herod 1758 (Tartar 1743)—Lisette 1772 by Snap 1750 (Snip 1736)—Miss Windsor 1751 by God. Ar. 1724—Sist. to Volunteer by Y. Belgrade (Belgrade Turk).
				Penelope, b. m., 1798.	Trumpator 1782 (Conductor 1767)—Prunella 1788 by Highflyer 1774 (Herod 1758)—Promise 1768 by Snap 1750 (Snip 1736)—Julia 1756 by Blank 1740 (God. Ar. 1724)—Spectator's dam 1735 by Partner 1718 (Jigg).
			Peri, b. m., 1822.	Wanderer, b. h., 1811.	Gohanna 1790 (Mercury 1778)—Catherine 1795 by Woodpecker 1773 (Herod 1758)—Camilla 1778 by Trentham 1766 (T. Sweepstakes 1743)—Coquette 1764 by Compton Barb—Sist. 1743 to Regulus by God. Ar. 1724.
				Thalestris, blk. m., 1809.	Alexander 1782 (Eclipse 1764)—Rival 1800 by Sir Peter T. 1784 (Highflyer 1774)—Hornet 1790 by Drone 1777 (Herod 1758)—Manilla 1777 by Goldfinder 1764 (Snap 1750)—Dau. 1766 of Old England 1741 (God. Ar. 1724).
		Guiccioli, ch. m., 1823.	Bob Booty, ch. h., 1804.	Chanticleer, ch. h., 1787.	Woodpecker 1773 (Herod 1758)—Dau. 1778 of Eclipse 1764 (Marske 1750)—Rosebud 1765 by Snap 1750 (Snip 1736)—Miss Belsea 1753 by Regulus 1739 (God. Ar. 1724)—Dau. of Bartlett's Childers (Darley Ar. 1699).
				Ierne, ch. m., 1790.	Bagot 1780 (Herod 1758)—Dau. of Gamahoe (Bustard 1741)—Patty by Tim 1752 (Squirt 1732)—Miss Patch 1737 by Justice 1725 (H. C. Litton Ar.)—Ringtail Galloway by Curwen Bay Barb.
			Flight, ch. m., 1801.	Ir. Escape, ch. h., 1802.	Commodore 1793 (Tom Tug 1777)—Moll-in-the-Wad 1791 by Highflyer 1774 (Herod 1758)—Shift 1779 by Sweetbriar 1769 (Syphon 1750)—Black Susan 1762 by Snap 1750 (Snip 1736)—Dau. of Cade 1734 (God. Ar. 1724).
				Y. Heroine, [illegible]	Bagot 1780 (Herod 1758)—Heroine 1775 by Hero 1753 (Cade 1734)—Dau. 1766 of Snap 1750 (Snip 1736)—Sist. 1743 to Regulus by God. Ar. 1724—Grey Robinson 1723 by Bald Galloway (St. Victor Barb).
	Daughter of, b. m., 1841.	Pantaloon, ch. h., 1824.	Castrel, ch. h., 1801.	Buzzard,* ch. h., 1787.	Woodpecker 1773 (Herod 1758)—Misfortune 1775 by Dux 1761 (Matchem 1748)—Curiosity 1760 by Snap 1750 (Snip 1736)—Dau. 1740 of Regulus 1739 (God. Ar. 1724)—Dau. of Bartlett's Childers (Darley Ar. 1699).
				Daughter of, b. m., —.	Alexander 1782 (Eclipse 1764)—Dau. of Highflyer 1774 (Herod 1758)—Dau. of Alfred 1770 (Matchem 1748)—Dau. 1770 of Engineer 1756 (Sampson 1745)—Bay Malton's dam 1754 by Cade 1734 (God. Ar. 1724).
			Idalia, ch. m., 1815.	Peruvian, b. h., 1806.	Sir Peter T. 1784 (Highflyer 1774)—Dau. 1788 of Boudrow 1777 (Eclipse 1764)—Dau. 1764 of Squirrel 1754 (Traveller 1735)—Sist. to Lowther's Babraham by Babraham 1740 (God. Ar. 1724)—Dau. 1750 of Golden Ball 1735 (Partner 1718).
				Musidora, b. m., 1804.	Meteor 1783 (Eclipse 1764)—Maid-of-all-Work 1786 by Highflyer 1774 (Herod 1758)—Dau. 1771 of Syphon 1750 (Squirt 1732)—Dau. of Regulus 1739 (God. Ar. 1724)—Sist. 1758 to Judgement by Snip 1736 (Flying Childers 1715).
		Daphne, br. m., 1837.	Laurel, br. h., 1824.	Blacklock, b. h., 1814.	Whitelock 1803 (Hambletonian 1792)—Dau. 1799 of Coriander 1786 (Pot-8-o-'s 1773)—Wildgoose 1792 by Highflyer 1774 (Herod 1758)—Coheiress 1786 by Pot-8-o-'s 1773 (Eclipse 1764)—Manilla 1777 by Goldfinder 1764 (Snap 1750).
				Wagtail, b. m., 1818.	Prime Minister 1810 (Sancho 1801)—Tranby*'s dam 1812 by Orville 1799 (Beningbrough 1791)—Miss Grimstone 1796 by Weazle 1776 (Herod 1758)—Dau. of Ancaster 1768 (Blank 1740)—Dau. of Damascus Ar. 1754.
			Maid of Honor, br. m., 1829.	Champion, ch. h., 1812.	Selim 1802 (Buzzard* 1787)—Podagra 1805 by Gouty 1796 (Sir Peter T. 1784)—Jet 1785 by Magnet 1770 (Herod 1758)—Jewel 1775 by Squirrel 1754 (Traveller 1735)—Sophia 1764 by Blank 1740 (God. Ar. 1724).
				Etiquette, br. m., 1820.	Orville 1799 (Beningbrough 1791)—Boadicea 1807 by Alexander 1782 (Eclipse 1764)—Brunette by Amaranthus 1766 (Old England 1741)—Mayfly 1771 by Matchem 1748 (Cade 1734)—Dau. 1755 of Starling 1738 (Starling 1727).
MAIDEN, b. m., 1862.	Lexington, b. h., 1850.	Boston, ch. h., 1833.	Timoleon, ch. h., 1813.	Sir Archy, b. h., 1805.	*Diomed 1777 (Florizel 1768)—*Castianira 1796 by Rockingham 1781 (Highflyer 1774)—Tabitha 1782 by Trentham 1766 (T. Sweepstakes 1743)—Dau. of Bosphorus 1754 (Babraham 1740)—Dau. of Forester 1750 (Forester 1736).
				Daughter of, ch. m., —.	*Saltram 1780 (Eclipse 1764)—Dau. of Symmes' Wildair 1767 (*Fearnought 1755)—Dau. of Tyler's Driver 1760 (*Othello)—Dau. of *Fallower 1761 (Blank 1740)—Dau. of *Vampire 1757 (Regulus 1739).
			Sist. to Tuckahoe, ch. m., 1814.	Ball's Florizel, ch. h., 1801.	*Diomed 1777 (Florizel 1768)—Dau. of *Shark 1771 (Marske 1750)—Dau. of Harris' Eclipse 1770 (*Fearnought 1755)—Dau. of *Fearnought 1755 (Regulus 1739)—Dau. of *Jolly Roger 1741 (Roundhead 1733).
				Daughter of, b. m., 1799.	*Alderman 1787 (Pot-8-o-'s 1773)—Dau. of *Clockfast 1774 (Gimcrack 1760)—Dau. of Symmes' Wildair 1767 (*Fearnought 1755)—Y. Kitty Fisher 1767 by *Fearnought 1755 (Regulus 1739)—*Kitty Fisher 1755 by Cade 1734 (God. Ar. 1724).
		Alice Carneal, b. m., 1836.	*Sarpedon, br. h., 1828.	Emilius, b. h., 1820.	Orville 1799 (Beningbrough 1791)—Emily 1810 by Stamford 1794 (Sir Peter T. 1784)—Dau. 1799 of Whiskey 1789 (Saltram* 1780)—Grey Dorimant 1781 by Dorimant 1772 (Otho 1760)—Dizzy 1757 by Blank 1740 (God. Ar. 1724).
				Icaria, b. m., 1824.	The Flyer 1814 (Vandyke Jr. 1808)—Parma 1813 by Dick Andrews 1797 (Joe Andrews 1778)—May 1804 by Beningbrough 1791 (King Fergus 1775)—Primrose 1787 by Mambrino 1768 (Engineer 1756)—Cricket 1776 by Herod 1758 (Tartar 1743).
			Rowena, ch. m., 1826.	Sumpter, ch. h., 1818.	Sir Archy 1805 (*Diomed 1777)—Dau. of *Robin Redbreast 1796 (Sir Peter T. 1784)—Dau. 1788 of *Obscurity 1777 (Eclipse 1764)—Slamerkin 1768 by *Wildair 1753 (Cade 1734)—*Cub mare 1762 by Cub 1739 (Fox 1714).
				Lady Grey, b. m., 1817.	Robin Grey 1805 (*Royalist 1790)—Maria 1802 by Melzar 1791 (*Medley 1776)—Dau. of *Highflyer 1784 (Highflyer 1774)—Dau. of *Fearnought 1755 (Regulus 1739)—Dau. of Ariel 1753 (*Traveller 1747).
	Kitty Clark, b. m., 1853.	*Glencoe, ch. h., 1831.	Sultan, b. h., 1816.	Selim, ch. h., 1802.	Buzzard* 1787 (Woodpecker 1773)—Dau. of Alexander 1782 (Eclipse 1764)—Dau. of Highflyer 1774 (Herod 1758)—Dau. of Alfred 1770 (Matchem 1748)—Dau. 1770 of Engineer 1756 (Sampson 1745).
				Bacchante, br. m., 1809.	W's Ditto 1800 (Sir Peter T. 1784)—Dau. 1791 of Mercury 1778 (Eclipse 1764)—Dau. 1776 of Herod 1758 (Tartar 1743)—Folly 1771 by Marske 1750 (Squirt 1732)—Vixen 1753 by Regulus 1739 (God. Ar. 1724).
			Trampoline, ch. m., 1825.	Tramp, b. h., 1810.	Dick Andrews 1797 (Joe Andrews 1778)—Dau. 1803 of Gohanna 1790 (Mercury 1778)—Fraxinella 1793 by Trentham 1766 (T. Sweepstakes 1743)—Dau. 1785 of Woodpecker 1773 (Herod 1758)—Everlasting 1775 by Eclipse 1764 (Marske 1750).
				Web, b. m., 1808.	Waxy 1790 (Pot-8-o-'s 1773)—Penelope 1798 by Trumpator 1782 (Conductor 1767)—Prunella 1788 by Highflyer 1774 (Herod 1758)—Promise 1768 by Snap 1750 (Snip 1736)—Julia 1756 by Blank 1740 (God Ar. 1724).
		Miss Obstinate, br. m., 1829.	Sumpter, ch. h., 1818.	Sir Archy, b. h., 1805.	*Diomed 1777 (Florizel 1768)—*Castianira 1796 by Rockingham 1781 (Highflyer 1774)—Tabitha 1782 by Trentham 1766 (T. Sweepstakes 1743)—Dau. of Bosphorus 1754 (Babraham 1740)—Dau. of Forester 1750 (Forester 1736).
				Robin mare, ch. m., —.	*Robin Redbreast 1796 (Sir Peter T. 1784)—Dau. 1788 of *Obscurity 1777 (Eclipse 1764)—Slamerkin 1768 by *Wildair 1753 (Cade 1734)—*Cub mare 1762 by Cub 1739 (Fox 1714)—Dau. 1755 of Second 1732 (Flying Childers 1715).
			Jenny Slamerkin, b. m., 1821.	Tiger, b. h., 1812.	Cook's Whip 1804 (*Whip 1794)—Jane Hunt 1795 by Paragon 1788 (*Flimnap 1765)—Moll by *Figure 1757 (Grey Figure 1747)—Slamerkin 1768 by *Wildair 1753 (Cade 1734)—*Cub mare 1762 by Cub 1739 (Fox 1714).
				Paragon, b. m., 1807.	*Buzzard 1787 (Woodpecker 1773)—Indiana b. m. 1802 by Columbus (Pantaloon 1778)—Jane Hunt b. m. 1796 by Hampton's Paragon 1788 (*Flimnap 1765)—Moll b. m. by imp. Figure 1757 (Grey Figure 1747)—Slamerkin b. m. 1768 by *Wildair 1753 (Cade 1734)—*Cub mare b. m. 1762 by Cub 1739 (Fox 1714)—Amaranthus' dam b. m. 1755 by Second 1732 (Flying Childers 1715)—Sist. 1746 to Leedes by Starling 1727 (Bay Bolton 1705)—Sister b. m. 1740 to Vane's Little Partner by Partner 1718 (Jigg)—Sist. 1722 to Guy by Greyhound (Chillaby)—Brown Farewell br. m. 1710 by Makeless (Oglethorpe Ar.)—Dau. of Brimmer (D'Arcy Yellow Turk)—Dau. of Place's White Turk—Dau. of Dodsworth—Layton Barb mare

IMPORTED RAYON-D'OR.

WINNER OF THE CLEARWELL AND LEVANT STAKES IN 1878, ST. JAMES' PALACE STAKES, THE GREAT CHALLENGE STAKES, CHAMPION STAKES AND DONCASTER ST. LEGER IN 1879.

In the Algeria Stud, Erie, Pa., the property of the Estate of Mr. W. L. Scott, as private stallion.

RAYON-D'OR (Ray of Gold) by **Flageolet**, son of Plutus, bred in Dangu Stud, France, foaled 1876, dam Araucaria, dam of Chamant, Camelia, etc., by Ambrose, son of Touchstone, out of Pocahontas, dam of Stockwell, Rataplan, King Tom, etc., by imp. Glencoe. Rayon-d'Or as a two-year old won the Levant Stakes, half a mile, at Goodwood, carrying 122 lbs., beating Flavius, Galantha and two others. At Doncaster won a Sweepstakes, at three-quarters of a mile, with 129 lbs., beating Charibert and Reconciliation. At the Newmarket Second October Meeting won the Clearwell Stakes, carrying 131 lbs., Ringleader and Bay Archer, 122 lbs. each, second and third, and three others: won the Glasgow Stakes at the Newmarket Houghton Meeting, three-quarters of a mile, beating Ringleader and Glencairn. Rayon-d'Or was much fancied, both for the 2,000 Guineas and Derby, but his running was a disappointment. He was third to Charibert and Cadogan for the 2,000 Guineas, and was unplaced for the Derby, won by Sir Bevys. At Ascot was third for the Prince of Wales Stakes to Wheel of Fortune and Adventure; on the third day he won the St. James' Palace Stakes, with 122 lbs., over the severe "Old mile," beating Charibert, Ruperra and seven others. At Goodwood Rayon d'Or won the Sussex Stakes, 1 mile, beating Ruperra, Leap Year and Exeter, and the great Doncaster St. Leger, Ruperra second and Exeter third, and fourteen others unplaced, including the winner of the Derby, Sir Bevys. Rayon-d'Or "walked over" for the Zetland Stakes, Mr. Crawfurd's Pell Mell colt saving his stake. At the Newmarket First October Meeting won the great Foal Stakes, with 131 lbs., beating Discord, Palmbearer and three others, but at the same meeting was beaten by Bay Archer for the Newmarket St. Leger, he yielding 7 lbs., after which Rayon-d'Or won the Select Stakes, Rowley mile, beating Discord and three others; the Champion Stakes, across the flat, 1 mile 2 furlongs and 73 yards, beating Placida, the Oaks winner, Exeter and five others, and the Great Challenge Stakes, Bretby Course, 6 furlongs, beating Lollypop, Placida, Parole and others. He closed his three-year-old career by running third for a handicap "Across the Flat" to Out of Bounds, who carried 110 lbs. to Rayon-d'Or's 126 lbs. His earnings for the year amounted to $87,735. As a four-year old Rayon-d'Or won the Prix du Cadran, 2 miles and 5 furlongs, beating Zut, Salteador and others, and the Prix Rainbow, 3 miles and a furlong, beating Zut and Clocher. He was sent to England, where, on the 16th of April, at Newmarket, he "walked over" for the Post Stakes (the two middle miles). At the Newmarket First Spring Meeting he walked over for the Prince of Wales Stakes (the Cesarewitch Course). At Ascot won the Rous Memorial Stake over the New mile, carrying 132 lbs., but in running for the Hardwicke Stakes, at the same meeting, over the Swinley Course (1¼ miles) he was beaten a head by Exeter, to whom he was giving 10 lbs.. This race closed his turf career.

Flageolet, his sire, was a superior race-horse. Won the Hopeful, Rutland, Forlorn, Rowley Mile, Burwell, 5 furlongs, and Criteron Stakes. As three-year old won the Goodwood Cup, in which he beat both Favonius and Cremorne, the Derby winners of 1871 and 1872; won the Grand Duke Michael Stakes, at Newmarket First October Meeting: won a free Handicap Sweepstakes across the flat, and the Jockey Club Cup at the Newmarket Houghton Meeting. At four years old won the Claret Stakes at Newmarket, England.

An analysis of his tabulated pedigree will show that he is richly and fashionably bred. He has a double cross of Glencoe, a triple cross of Diomed, a double cross of Touchstone, fortified by the blood of Whalebone, doubled in upon the Herod and Eclipse blood on both sides to the Byerly mare, dam of the two True Blues. In color he is a rich true chestnut, with a large, rather faint star in his forehead, standing 16 hands 3¼ inches in height, and a horse of true symmetry.

Rayon-d'Or is the sire of the winners Roi d'Or, Daphne, Defense, Bronzomarte, Quotation, Ransom, Laura Stone, Belle d'Or, Tudor, Tea Tray, Somerset, Tenny, Flageoletta, Arundel, Lady Hemphill, Sparling, Pat Oakley, Quibbler, Julien, Louis d'Or, Alchemist, Harbor Lights, Bellair, Chaos, Chateau d'Or, Coldstream, Cotillion, Gendarme, Gipsy Queen, Gipsy King, Jake Miller, Leighton, Marauder, Ossa, Centaur, Void, Crawfish, Amulet, Torchlight, Quito, Banquet, Bolero, Franco, Miss Ransom, John Atwood, Minuet, Aftermath, Sterling, Miss Olive, The Doctor, Florimor, Exclusion, Martha, Cutalong, Paradox, Canteen, Iona, Kingsford, Ayala, Alcalde, Versatile, Coxswain, Monopolist, Entre, Louise d'Or and Maywood. (See imp. Eothen, page 150.)

RAYON-D'OR.

CHESTNUT HORSE, FOALED 1876; BRED BY M. C. J. LEFÈVRE, FRANCE.

FLAGEOLET, ch. h., 1870.

Plutus, b. h., 1862. — La Favorite, b. m., 1863.

Trumpeter, ch. h., 1856. — Daughter of, b. m., 1853. — Monarque,‡ b. h., 1852. — Constance, ch. m., 1848.

Orlando, b. h., 1841. — Cavatina, ch. m., 1845. — Planet, b. h., 1841. — Alice Bray, b. m., 1818. — The Emperor, ch. h., 1841. — Poetess, b. m., 1838. — Gladiator, ch. h., 1833. — Lanterne, b. m., 1841.

Touchstone, br. h., 1831. Camel 1822 (Whalebone 1807) Banter 1826 by Master Henry 1815 (Orville 1799)—Boadicea 1807 by Alexander 1782 (Eclipse 1764) Brunette by Amaranthus 1766 (Old England 1741) Mayfly 1771 by Matchem 1748 (Cade 1734).

Vulture, ch. m., 1833. Langar 1817 (Selim 1802) Kite 1821 by Bustard 1813 (Castrel 1801) Olympia 1815 by Sir Oliver 1800 (Sir Peter T. 1784)—Scotilla 1795 by Anvil 1777 (Herod 1758) Scota 1783 by Eclipse 1764 (Marske 1750).

Redshank, b. h., 1833. Sandbeck 1818 (Catton 1809) Johanna 1813 by Selim 1802 (Buzzard* 1787) Dau. of Skyscraper 1786 (Highflyer 1774) Dau. 1796 of Dragon* 1787 (Woodpecker 1773)—Fidget's dam 1777 by Matchem 1748 (Cade 1734).

Oxygen, b. m., 1828. Emilius 1820 (Orville 1799) Whizgig 1819 by Rubens 1805 (Buzzard* 1787)—Penelope 1798 by Trumpator 1782 (Conductor 1767) Prunella 1788 by Highflyer 1774 (Herod 1758) Promise 1768 by Snap 1750 (Snip 1736)

Bay Middleton, b. h., 1833. Sultan 1816 (Selim 1802)—Cobweb 1821 by Phantom 1808 (Walton 1799) Filagree 1815 by Soothsayer 1808 (Sorcerer 1796)—Web 1808 by Waxy 1790 (Pot-8-o-'s 1773)—Penelope 1798 by Trumpator 1782 (Conductor 1767).

Plenary, br. m., 1837. Emilius 1820 (Orville 1799)—Harriet 1819 by Pericles 1809 (Evander 1801)—Dau. 1812 of Selim 1802 (Buzzard* 1787) Pipylina 1803 by Sir Peter T. 1784 (Highflyer 1774)—Rally 1790 by Trumpator 1782 (Conductor 1767).

Venison, br. h., 1833. Partisan 1811 (Walton 1799)—Fawn 1823 by Smolensko 1810 (Sorcerer 1796) Jerboa 1803 by Gohanna 1790 (Mercury 1778) Camilla 1778 by Trentham 1766 (T. Sweepstakes 1743)—Coquette 1764 by Compton Barb.

Darkness, ch. m., 1837. Glencoe* 1831 (Sultan 1816)—Fanny 1827 by Whisker 1812 (Waxy 1790) Dau. 1813 of Camillus 1803 (Hambletonian 1792)—Dau. 1804 of Precipitate* 1787 (Mercury 1778)—Mendoza*'s dam 1783 by Paymaster 1766 (Blank 1740).

Defence, b. h., 1824. Whalebone 1807 (Waxy 1790)—Defiance 1816 by Rubens 1805 (Buzzard* 1787) Little Folly 1806 by Highland Fling 1798 (Spadille 1784) Harriet 1799 by Volunteer 1780 (Eclipse 1764)—Dau. 1785 of Alfred 1770 (Matchem 1748).

Daughter of, ch. m., 1831. Reveller 1815 (Comus 1809)—Design* 1827 by Tramp 1810 (Dick Andrews 1797)—Defiance 1816 by Rubens 1805 (Buzzard* 1787) Little Folly 1806 by Highland Fling 1798 (Spadille 1784) Harriet 1799 by Volunteer 1780 (Eclipse 1764).

Royal Oak, br. h., 1823. Catton 1809 (Golumpus 1802) Dau. 1818 of Smolensko 1810 (Sorcerer 1796)—Lady Mary 1800 by Beningbrough 1791 (King Fergus 1775)—Dau. of Highflyer 1774 (Herod 1758) Sist. to Miss Bell by Marske 1750 (Squirt 1732).

Ada, b. m., 1824. Whisker 1812 (Waxy 1790) Anna Bella 1807 by Shuttle 1793 (Y. Marske 1771)—Dau. 1797 of Drone 1777 (Herod 1758)—Contessina 1787 by Y. Marske 1771 (Marske 1750)—Tuberose 1772 by Herod 1758 (Tartar 1743).

Partisan, b. h., 1811. Walton 1799 (Sir Peter T. 1784)—Parasol 1800 by Pot-8-o-'s 1773 (Eclipse 1764)—Prunella 1788 by Highflyer 1774 (Herod 1758)—Promise 1768 by Snap 1750 (Snip 1736)—Julia 1756 by Blank 1740 (God. Ar. 1724).

Pauline, b. m., 1826. Moses 1819 [Whalebone or] Seymour 1807 (Delpini 1781) Quadrille 1815 by Selim 1802 (Buzzard* 1787) Canary Bird 1806 by [Whiskey or] Sorcerer 1796 (Trumpator 1782)—Canary 1797 by Coriander 1786 (Pot-8-o-'s 1773)—Miss Green 1787 by Highflyer 1774 (Herod 1758).

Hercule, ch. h., 1830. Rainbow 1808 (Walton 1799)—Aimable 1822 by Election 1804 (Gohanna 1790)—Dau. 1808 of Y. Whiskey 1801 (Whiskey 1789)—Dau. 1800 of Walnut 1786 (Highflyer 1774)—Bay Javelin 1794 by Javelin 1772 (Eclipse 1764).

Elvira, b. m., 1829. Eryx 1816 (Milo 1802)—Coral 1816 by Orville 1799 (Beningbrough 1791)—Fairing 1807 by Waxy 1790 (Pot-8-o-'s 1773)—Rattle 1793 by Trumpator 1782 (Conductor 1767)—Fancy 1780 (Sist. to Diomed*) by Florizel 1768 (Herod 1758).

ARAUCARIA, b. m., 1862.

Ambrose, blk. h., 1849. — Pocahontas, b. m., 1837.

Touchstone, br. h., 1831. — Annette, b. m., 1835. — Glencoe,* ch. h., 1831. — Marpessa, b. m., 1830.

Camel, br. h., 1822. — Banter, br. m., 1826. — Priam*, b. h., 1827. — The Potentate's dam, b. m., 1823. — Sultan, b. h., 1816. — Trampoline, ch. m., 1825. — Muley, b. h., 1810. — Clare, b. m., 1824.

Whalebone, br. h., 1807. Waxy 1790 (Pot-8-o-'s 1773)—Penelope 1798 by Trumpator 1782 (Conductor 1767)—Prunella 1788 by Highflyer 1774 (Herod 1758)—Promise 1768 by Snap 1750 (Snip 1736)—Julia 1756 by Blank 1740 (God. Ar. 1724).

Daughter of, b. m., 1812. Selim 1802 (Buzzard* 1787)—Maiden 1801 by Sir Peter T. 1784 (Highflyer 1774)—Dau. 1788 of Phenomenon 1780 (Herod 1758)—Matron 1782 by Florizel 1768 (Herod 1758)—Maiden 1770 by Matchem 1748 (Cade 1734).

Master Henry, b. h., 1815. Orville 1799 (Beningbrough 1791)—Miss Sophia 1805 by Stamford 1794 (Sir Peter T. 1784)—Sophia 1798 by Buzzard* 1787 (Woodpecker 1773)—Huncamunca 1787 by Highflyer 1774 (Herod 1758)—Cypher 1772 by Squirrel 1754 (Traveller 1735).

Boadicea, b. m., 1807. Alexander 1782 (Eclipse 1764)—Brunette by Amaranthus 1766 (Old England 1741)—Mayfly 1771 by Matchem 1748 (Cade 1724)—Dau. 1755 of Starling 1758 (Starling 1727) Look-at-me-Lads 1731 by Brist. Grasshopper (Byerly Turk).

Emilius, b. h., 1820. Orville 1799 (Beningbrough 1791)—Emily 1810 by Stamford 1794 (Sir Peter T. 1784)—Dau. 1793 of Whiskey 1789 (Saltram* 1780)—Grey Dorimant 1781 by Dorimant 1772 (Otho 1760)—Dizzy 1757 by Blank 1740 (God. Ar. 1724).

Cressida, b. m., 1807. Whiskey 1789 (Saltram* 1780)—Y. Giantess 1790 by *Diomed 1777 (Florizel 1768)—Giantess 1769 by Matchem 1748 (Cade 1724)—Molly Long Legs 1753 by Babraham 1740 (God. Ar. 1724)—Dau. of Foxhunter 1727 (Brisk 1711).

Don Juan, b. h., 1814. Orville 1799 (Beningbrough 1791)—Peterea 1804 by Sir Peter T. 1784 (Highflyer 1774)—Mary Grey 1784 by Friar 1768 (Hero 1753)—Timante by Tim 1752 (Squirt 1732)—Noblessa 1769 by Gamahoe (Bustard 1741).

Moll-in-the-Wad, br. m., 1810. Hambletonian 1792 (King Fergus 1775)—Spitfire 1799 by Pipator 1786 (Imperator 1776)—Farewell 1792 by Slope 1782 (Highflyer 1774)—Dau. of Y. Marske 1771 (Marske 1750)—Dau. of Broth. to Silvio (Cade 1734).

Selim, ch. h., 1802. Buzzard* 1787 (Woodpecker 1773) Dau. of Alexander 1782 (Eclipse 1764) Dau. of Highflyer 1774 (Herod 1758)—Dau. of Alfred 1770 (Matchem 1748)—Dau. 1770 of Engineer 1756 (Sampson 1745).

Bacchante, br. m., 1809. W's Ditto 1800 (Sir Peter T. 1784)—Dau. 1791 of Mercury 1778 (Eclipse 1764)—Dau. 1776 of Herod 1758 (Tartar 1743)—Folly 1771 by Marske 1750 (Squirt 1732) Vixen 1753 by Regulus 1739 (God. Ar. 1724).

Tramp, b. h., 1810. Dick Andrews 1797 (Joe Andrews 1778)—Dau. 1803 of Gohanna 1790 (Mercury 1778)—Fraxinella 1793 by Trentham 1766 (T. Sweepstakes 1743)—Dau. 1785 of Woodpecker 1773 (Herod 1758)—Everlasting 1775 by Eclipse 1764 (Marske 1750).

Web, b. m., 1808. Waxy 1790 (Pot-8-o-'s 1773)—Penelope 1798 by Trumpator 1782 (Conductor 1767)—Prunella 1788 by Highflyer 1774 (Herod 1758) Promise 1768 by Snap 1750 (Snip 1736)—Julia 1756 by Blank 1740 (God. Ar. 1724).

Orville, b. h., 1799. Beningbrough 1791 (King Fergus 1775)—Evelina 1791 by Highflyer 1774 (Herod 1758)—Termagant by Tantrum 1760 (Cripple 1750)—Cantatrice by Sampson 1745 (Blaze 1733)—Dau. 1757 of Regulus 1739 (God. Ar. 1724).

Eleanor, b. m., 1798. Whiskey 1789 (Saltram* 1780)—Y. Giantess 1790 by Diomed* 1777 (Florizel 1768)—Giantess 1769 by Matchem 1748 (Cade 1734)—Molly Long Legs 1753 by Babraham 1740 (God. Ar. 1724)—Dau. of Cole's Foxhunter 1727 (Brisk 1711).

Marmion, b. h., 1806. Whiskey 1789 (Saltram* 1780)—Y. Noisette 1798 by Diomed* 1777 (Florizel 1768)—Noisette 1774 by Squirrel 1754 (Traveller 1735)—Carina 1766 by Marske 1750 (Squirt 1732)—Thunder's dam 1755 by Blank 1740 (God. Ar. 1724).

Harpalice, b. m., 1814. Gohanna 1790 (Mercury 1778)—Amazon b. m. 1799 by Driver 1783 (Trentham 1766)—Fractious b. m. 1792 by Mercury 1778 (Eclipse 1764)—Dau. b. m. 1785 of Woodpecker 1773 (Herod 1758)—Everlasting b. m. 1775 by Eclipse 1764 (Marske 1750) Hyæna br. m. 1762 by Snap 1750 (Snip 1736)—Miss Belsea b. m. 1753 by Regulus 1739 (God. Ar. 1724)—

Dau. of Bartlett's Childers (Darley Ar. 1699)—Dau. gr. m. by Honeywood's Ar.—Dam of the two True Blues by Byerly Turk.

‡ Monarque is by The Baron, Sting or The Emperor. Emperor's pedigree is accepted.

IMPORTED ROSSINGTON.

The property of Mr. F. B. Harper, Nantura Stud, Midway. Terms of service advertised annually.

ROSSINGTON, chestnut horse, foaled 1881, bred by Duke of Westminster, by Doncaster (son of Stockwell), dam of Lily Agnes (dam of Ormonde, Farewell and Ossory) by Macaroni; grandam Polly Agnes (dam Jessie Agnes) by The Cure; great grandam Miss Agnes (dam of Muncaster and Kendal) by Birdcatcher, out of Agnes by Clarion.

Rossington started in three races. At two years old ran third to Katerfelto and Timbrel in Two-Year Old Plate at Lichfield, with seven others behind him: ran third in the Rutland Stakes to Raffaello and Sinbad, with Pantaloon behind him: was unplaced in the Studley Castle Handicap Nursery Plate, 1 mile, won by Scottish Earl. Rossington is nearly full brother to the unbeaten Ormonde, the latter being by Bend'or, son of Doncaster by Stockwell, whilst Rossington is by Doncaster, who won the Derby and ran second for the St. Leger in 1873, and sired Ben d'or, the Derby winner of 1880, his first year in the stud.

The most important fact in connection with Rossington is that he is one of the great "Agnes family," the most successful English racing family of the present generation. His grandam, Polly Agnes by The Cure, was a crack performer, and in the stud produced Rural Dean by Cathedral, Lily Agnes by Macaroni, Fleur-de-Lis by Mandrake, Tiger Lily by Macaroni, Jessie Agnes by Macaroni, Bay Agnes by Speculum. Lily Agnes was a mare of immense speed, winning, among her many races, the Doncaster Cup, Great Ebor Handicap, with 120 lbs., and the Northumberland Plate of 1874 as a three-year old. As a brood-mare Lily Agnes proved a great treasure; she produced Narcissus by Speculum, Eastern Lily by Speculum, Rossington by Doncaster, Farewell (winner of the 1,000 Guineas of 1885) by Doncaster, Ormonde (winner of the Criterion Stakes and Dewhurst Plate of 1885, and the 2,000 Guineas, Derby and St. Leger of 1886, etc.) by Bend'or, and Ossory (winner of the Criterion, 1887) by Bend'or.

On the sire's side of the pedigree Rossington's grandsire, Macaroni, won the 2,000 Guineas and Derby of 1863. His sire, Sweetmeat, has shown his prepotency remarkably in the Derby. He begot Macaroni (Derby winner), Parmesan, sire of two Derby winners, Cremorne and Favonius. The last-named got another Derby winner in Sir Bevys. Macaroni got three Oaks winners in Spinaway, Camelia and Bonnie Jean; a 2,000 Guineas winner in Macgregor, who got one in Scot-Free. Teddington, maternal grandsire of Doncaster, won the Derby in 1851, and was one of the gamest horses which ever ran, as his memorable race with Kingston, Stockwell and others for the Emperor's Plate showed.

Rossington is 16 hands high and a good-shaped horse. In the stud he has not yet had a fair trial, but is the sire of the winners Emma Primrose, Ida Rossington, Rally, Fred Lee (dam Mamie O.), Unadilla, Belfast, Coquette, Lollie, Emin Bey, Sabine, Frank Eiseman, Sarah Remey, etc. He is descended in direct female line from sister to Stripling by Hutton's Spot, to which trace Burgundy, Glenmasson, Ambrose, Dundee, Brown Bread, Van Amburgh, Muncaster, Kendal, Ormonde, imp. Stonehenge and all the Agnes family.

Waxy 1790 Pot-8-o's 1773—Eclipse 1764 Marske 1750 Squirt 1732 Bartlett's Childers Darley Arabian 1689.

ROSSINGTON.

CHESTNUT HORSE, FOALED 1881; BRED BY DUKE OF WESTMINSTER, ENGLAND.

DONCASTER, ch. h., 1870.

- Stockwell, ch. h., 1849.
  - The Baron, ch. h., 1842.
    - Ir. Birdcatcher, ch. h., 1833.
    - Echidna, b. m., 1838.
  - Pocahontas, b. m., 1837.
    - Glencoe, ch. h., 1831.
    - Marpessa, b. m., 1830.
- Marigold, ch. m., 1860.
  - Teddington, ch. h., 1848.
    - Orlando, b. h., 1841.
    - MissTwickenham, ch. m., 1838.
  - Daughter of, b. m., 1852.
    - Ratan, ch. h., 1841.
    - Daughter of, b. m., 1844.

LILY AGNES (Ormonde's dam), b. m., 1871.

- Macaroni, b. h., 1860.
  - Sweetmeat, br. h., 1842.
    - Gladiator, ch. h., 1833.
    - Lollypop, blk. m., 1833.
  - Jocose, b. m., 1843.
    - Pantaloon, ch. h., 1824.
    - Banter, br. m., 1826.
- Polly Agnes, b. m., 1865.
  - The Cure, b. h., 1841.
    - Physician, b. h., 1829.
    - Morsel, b. m., 1836.
  - Miss Agnes, br. m., 1850.
    - Ir Birdcatcher, ch. h., 1833.
    - Agnes, br. m., 1841.

Sir Hercules, blk. h., 1826.	Whalebone 1807 (Waxy 1790)—Peri 1822 by Wanderer 1811 (Gohanna 1790)—Thalestris 1809 by Alexander 1782 (Eclipse 1764)—Rival 1800 by Sir Peter T. 1784 (Highflyer 1774)—Hornet 1790 by Drone 1777 (Herod 1758).
Guiccioli, ch. m., 1823.	Bob Booty 1804 (Chanticleer 1787)—Flight 1809 by Ir. Escape 1802 (Commodore 1793)—Y. Heroine by Bagot 1780 (Herod 1758)—Heroine 1775 by Hero 1753 (Cade 1734)—Dau. 1766 of Snap 1750 (Snip 1736).
Economist, b. h., 1825.	Whisker 1812 (Waxy 1790)—Floranthe 1818 by Octavian 1807 (Stripling 1795)—Caprice 1797 by Anvil 1777 (Herod 1758)—Madcap 1774 by Eclipse 1764 (Marske 1750)—Sappho 1763 by Blank 1740 (God. Ar. 1724).
Miss Pratt, b. m., 1825.	Blacklock 1814 (Whitelock 1803)—Gadabout 1812 by Orville 1799 (Beningbrough 1791)—Minstrel 1803 by Sir Peter T. 1784 (Highflyer 1774)—Matron 1782 by Florizel 1768 (Herod 1758)—Maiden 1770 by Matchem 1748 (Cade 1734).
Sultan, b. h., 1816.	Selim 1802 (Buzzard 1787)—Bacchante 1809 by W's Ditto 1800 (Sir Peter T. 1784)—Dau. 1791 of Mercury 1778 (Eclipse 1764)—Dau. 1776 of Herod 1758 (Tartar 1743)—Folly 1771 by Marske 1750 (Squirt 1732).
Trampoline, ch. m., 1825.	Tramp 1810 (Dick Andrews 1797)—Web 1808 by Waxy 1790 (Pot-8-o's 1773)—Penelope 1798 by Trumpator 1782 (Conductor 1767)—Prunella 1788 by Highflyer 1774 (Herod 1758)—Promise 1768 by Snap 1750 (Snip 1736).
Muley, b. h., 1810.	Orville 1799 (Beningbrough 1791)—Eleanor 1798 by Whiskey 1789 (Saltram* 1780)—Y. Giantess 1790 by Diomed* 1777 (Florizel 1768)—Giantess 1769 by Matchem 1748 (Cade 1734)—Molly Long Legs 1753 by Babraham 1740 (God. Ar. 1724).
Clare, b. m., 1824.	Marmion 1806 (Whiskey 1789)—Harpalice 1814 by Gohanna 1790 (Mercury 1778)—Amazon 1799 by Driver 1783 (Trentham 1766)—Fractions 1792 by Mercury 1778 (Eclipse 1764)—Dau. 1785 of Woodpecker 1773 (Herod 1758).
Touchstone, br. h., 1831.	Camel 1822 (Whalebone 1807)—Banter 1826 by Master Henry 1815 (Orville 1799)—Boadicea 1807 by Alexander 1782 (Eclipse 1764)—Brunette by Amaranthus 1766 (Old England 1741)—Mayfly 1771 by Matchem 1748 (Cade 1734).
Vulture, ch. m., 1833.	Langar 1817 (Selim 1802)—Kite 1821 by Bustard 1813 (Castrel 1801)—Olympia 1815 by Sir Oliver 1800 (Sir Peter T. 1784)—Scotilla 1795 by Anvil 1777 (Herod 1758)—Scota 1783 by Eclipse 1764 (Marske 1750).
Rockingham, b. h., 1830.	Humprey Clinker 1822 (Comus 1809)—Medora 1813 by Swordsman 1796 (Prizefighter 1784)—Ir. Penelope 1797 by Trumpator 1782 (Conductor 1767)—Peppermint 1787 by Highflyer 1774 (Herod 1758)—Promise 1768 by Snap 1750 (Snip 1736).
Electress, ch. m., 1819.	Election 1804 (Gohanna 1790)—Dau. 1805 of Stamford 1794 (Sir Peter T. 1784)—Miss Judy 1784 by Alfred 1770 (Matchem 1748)—Manilla 1777 by Goldfinder 1764 (Snap 1750)—Dau. 1756 of Old England 1741 (God. Ar. 1724).
Buzzard, b. h., 1821.	Blacklock 1814 (Whitelock 1803)—Miss Newton 1804 by Delpini 1781 (Highflyer 1774)—Tipple Cyder 1788 by King Fergus 1775 (Eclipse 1764)—Sylvia 1783 by Y. Marske 1771 (Marske 1750)—Ferret 1765 by Broth. to Silvio (Cade 1734).
Daughter of, b. m., 1831.	Picton 1819 (Smolensko 1810)—Dau. 1824 of Selim 1802 (Buzzard* 1787)—Dau. 1804 of Pipator 1786 (Imperator 1776)—Queen Mab 1785 by Eclipse 1764 (Marske 1750)—Mercury's dam by Tartar 1743 (Partner 1718).
Melbourne, br. h., 1834.	Humprey Clinker 1822 (Comus 1809)—Dau. 1823 of Cervantes 1805 (Don Quixote 1784)—Dau. 1818 of Golumpus 1802 (Gohanna 1790)—Dau. 1810 of Paynator 1791 (Trumpator 1782)—Circle 1802 by St. George 1789 (Highflyer 1774).
Lisbeth, b. m., 1838.	Phantom 1808 (Walton 1799)—Elizabeth 1823 by Rainbow 1808 (Walton 1799)—Belvoirina 1813 by Stamford 1794 (Sir Peter T. 1784)—Sist. 1790 to Silver by Mercury 1778 (Eclipse 1764)—Dau. 1780 of Herod 1758 (Tartar 1743).
Partisan, b. h., 1811.	Walton 1799 (Sir Peter T. 1784)—Parasol 1800 by Pot-8-o-'s 1773 (Eclipse 1764)—Prunella 1788 by Highflyer 1774 (Herod 1758)—Promise 1768 by Snap 1750 (Snip 1736)—Julia 1756 by Blank 1740 (God. Ar. 1724).
Pauline, b. m., 1826.	Moses 1819 ([Whalebone or] Seymour 1807)—Quadrille 1815 by Selim 1802 (Buzzard* 1787)—Canary Bird 1806 by [Whiskey or] Sorcerer 1796 (Trumpator 1782)—Canary 1797 by Coriander 1786 (Pot-8-o-'s 1773)—Miss Green 1787 by Highflyer 1774 (Herod 1758).
[Starch or] Voltaire, br. h., 1826.	Blacklock 1814 (Whitelock 1803)—Dau. 1816 of Phantom 1808 (Walton 1799)—Dau. 1802 of Overton 1788 (King Fergus 1775)—Dau. 1793 of Walnut 1786 (Highflyer 1774)—Dau. 1786 of Ruler 1777 (Y. Marske 1771).
Belinda, br. m., 1825.	Blacklock 1814 (Whitelock 1803)—Wagtail 1818 by Prime Minister 1810 (Sancho 1801)—Tranby*'s dam 1812 by Orville 1799 (Beningbrough 1791)—Miss Grimstone 1796 by Weazle 1776 (Herod 1758)—Dau. of Ancaster 1768 (Blank 1740).
Castrel, ch. h., 1801.	Buzzard* 1787 (Woodpecker 1773)—Dau. of Alexander 1782 (Eclipse 1764)—Dau. of Highflyer 1774 (Herod 1758)—Dau. of Alfred 1770 (Matchem 1748)—Dau. 1770 of Engineer 1756 (Sampson 1745).
Idalia, ch. m., 1815.	Peruvian 1806 (Sir Peter T. 1784)—Musidora 1804 by Meteor 1783 (Eclipse 1764)—Maid-of-all-Work 1786 by Highflyer 1774 (Herod 1758)—Sist. to Tandem 1771 by Syphon 1750 (Squirt 1732)—Dau. of Regulus 1739 (God. Ar. 1724).
Master Henry, b. h., 1815.	Orville 1799 (Beningbrough 1791)—Miss Sophia 1805 by Stamford 1794 (Sir Peter T. 1784)—Sophia 1798 by Buzzard* 1787 (Woodpecker 1773)—Huncamunca 1787 by Highflyer 1774 (Herod 1758)—Cypher 1772 by Squirrel 1754 (Traveller 1735).
Boadicea, b. m., 1807.	Alexander 1782 (Eclipse 1764)—Brunette by Amaranthus 1766 (Old England 1741)—Mayfly 1771 by Matchem 1748 (Cade 1734)—Dau. 1755 of Anc. Starling 1738 (Starling 1727)—Look-at-me-Lads 1731 by Brist. Grasshopper (Byerly Turk).
Brutandorf, b. h., 1821.	Blacklock 1814 (Whitelock 1803)—Mandane 1800 by Pot-8-o's 1773 (Eclipse 1764)—Y. Camilla 1787 by Woodpecker 1773 (Herod 1758)—Camilla 1778 by Trentham 1766 (T. Sweepstakes 1743)—Coquette 1765 by Compton Barb.
Primette, b. m., 1830.	Prime Minister 1810 (Sancho 1801)—Miss Paul 1811 by Sir Paul 1802 (Sir Peter T. 1784)—Miss Dunnington 1807 by Shuttle 1793 (Y. Marske 1771)—Miss Grimstone 1796 by Weazle 1776 (Herod 1756)—Dau. of Ancaster 1768 (Blank 1740).
Mulatto, b. h., 1823.	Catton 1809 (Golumpus 1802)—Desdemona 1811 by Orville 1799 (Beningbrough 1791)—Fanny 1796 by Sir Peter T. 1784 (Highflyer 1774)—Dau. 1788 of Diomed* 1777 (Florizel 1768)—Desdemona 1770 by Marske 1750 (Squirt 1732).
Linda, b. m., 1825.	Waterloo 1811 (Walton 1799)—Cressida 1807 by Whiskey 1789 (Saltram* 1780)—Y. Giantess 1790 by Diomed* 1777 (Florizel 1768—Giantess 1769 by Matchem 1748 (Cade 1734)—Molly Long Legs 1753 by Babraham 1740 (God. Ar. 1724).
Sir Hercules, blk. h., 1826.	Whalebone 1807 (Waxy 1790)—Peri 1822 by Wanderer 1811 (Gohanna 1790)—Thalestris 1809 by Alexander 1782 (Eclipse 1764)—Rival 1800 by Sir Peter T. 1784 (Highflyer 1774)—Hornet 1790 by Drone 1777 (Herod 1758).
Guiccioli, ch. m., 1823.	Bob Booty 1804 (Chanticleer 1787)—Flight 1809 by Ir. Escape 1802 (Commodore 1793)—Y. Heroine by Bagot 1780 (Herod 1758)—Heroine 1775 by Hero 1753 (Cade 1734)—Dau. 1766 of Snap 1750 (Snip 1736).
Clarion, b. h., 1836.	Sultan 1816 (Selim 1802)—Clara 1826 by Filho-da-Puta 1812 (Haphazard 1797)—Clari 1824 by Smolensko 1810 (Sorcerer 1796)—Dau. 1804 of Precipitate* 1787 (Mercury 1778)—Y. Juno 1788 by Highflyer 1774 (Herod 1758).
Anette, b. m., 1835.	Priam* 1827 (Emilius 1820)—The Potentate's dam b. m. 1823 by Don Juan 1814 (Orville 1799)—Moll-in-the-Wad br. m. 1810 by Hambletonian 1792 (King Fergus 1775)—Spitfire b. m. 1799 by Pipator 1786 (Imperator 1776)—Farewell b. m. 1792 by Slope 1782 (Highflyer 1774)—Dau. of Y. Marske 1771 (Marske 1750)—Dau. of Broth. to Silvio (Cade 1734)—Sist. to Stripling by Hutton's Spot 1728 (Hartley's Blind Horse).

IMPORTED ST. BLAISE.

WINNER OF THE TROY STAKES, 1882, AND THE DERBY, 1883.

The property of Messrs. Chas. Reed & Sons, Fairview Stud, Gallatin, Tenn. Yearling sales annually.

St. Blaise, chestnut horse, bred by Lord Alington, foaled 1880, imported by the late Hon. A. Belmont, by Hermit, son of Newminster by Touchstone, dam Fusee by Marsyas, son of Orlando by Touchstone, grandam Vesuvienne by Gladiator, son of Partisan and sire of Sweetmeat, out of Venus by Sir Hercules, son of Whalebone. A glance at his tabulated pedigree will show that St. Blaise is a highly and fashionably bred horse. Hermit, his sire, won the Derby and was the most popular stallion in England. Amongst the many race-horses he sired may be mentioned Tristan, Shotover (winner of the 2,000 Guineas and the Derby), Peter, Thebais (winner of the 1,000 Guineas and Oaks), St. Marguerite, St. Louis, St. Medard, Lonely (winner of the Oaks), etc. Marsyas sired George Frederick, the Derby winner of 1874, who was a success at the stud, among his get being Frontin (winner of the Grand Prize of Paris), Beau Brummel, etc. St. Blaise is inbred to Touchstone, and in his pedigree will be found the blood of Blacklock, Sultan, Gladiator and Sir Hercules.

St. Blaise was himself a good race-horse. As a two-year old started six times, won four races, including a walk-over and a dead heat. He made his *debut* in and won the Stockbridge Biennial, 6 furlongs, at the Bibury Club meeting, beating four others. In the Troy Stakes he had a walk-over; won the Hurstbourne Stakes, about 5 furlongs, both these events being at Stockbridge; he was defeated by Macheath, but was second, beating Mr. Houldsworth's Adriana, conceding her weight. In the Molecomb Stakes, at Goodwood, 6 furlongs, ran a dead heat with Elzevir; and in the Dewhurst Plate at Newmarket, 7 furlongs, he was not placed. Won the Newmarket Troy Stakes, 5 furlongs 140 yards, giving weight to everything that started. At three years old, in the 2,000 Guineas, won by Galliard, he ran unplaced. He turned the tables by winning the Derby from Highland Chief, second, Galliard third. St. Blaise's performance was very meritorious showing the greatest pluck and stamina. In the Grand Prize of Paris he vindicated his victory by running second to Frontin, who was thought the best three-year old of that year, only beaten a head; the good colt Farfadet was a poor third, three lengths away. In the Ascot Derby he conceded 10 lbs. to the winner, Ladislas, and 17 lbs. to Ossian, who ran second and afterward won the St. Leger. He closed his turf and three-year old career by walking over for the Stockbridge Biennial for three-year olds and the Winchester Biennial Stakes.

St. Blaise was a good class race-horse and has shown himself a phenomenally successful stallion. His sire, Hermit (winner of the Derby, etc.), was one of the best stallions England has ever known. Fusee (St. Blaise's dam) won as a two-year old the Chelmsford Nursery Handicap, beating a field of eight, and the Stratton Nursery Handicap, carrying top weight, defeating seven others. She was second in the Nursery Plate at Croydon and third in the Ramsay Abbey Nursery Handicap. At three years old she started eleven times and won nine races, including the Nursery Cup, 1 mile, at Chester; the Little John Stakes, 1 mile, at Nottingham; a Welter Cup Handicap, 1 mile, at Kingsbury, carrying top weight; the Kingsbury Handicap, 1¼ miles, again carrying top weight; the Queen's Plate, 2 miles, at Nottingham; the Queen's Plate, upwards of 3½ miles, at Newmarket; the Queen's Plate, 2 miles, at Winchester; the Queen's Plate, 3 miles, at Salisbury, and the Queen's Plate, 2½ miles, at Chelmsford. Thus proving a speedy and game animal and possessed of the highest stamina. In the stud Fusee has proved successful. She produced, to the cover of Blair Athol, Go Bang (winner of the Winchester Biennial Stakes at two years old), Friar Rush (a full brother to St. Blaise), a good and frequent winner. Next came St. Blaise, and, after missing one year, she produced Match Girl by Plebeian, who was a stake winner at two years old. Fusee produced another full brother to St. Blaise in Candlemas, a good performer, winner of the Epsom Grand Prize, the Zetland Biennial and the Chesterfield Cup. Fusee also produced Fuse (by Bend'or), that won the Kempton Park Two-Year Old Stakes.

St. Blaise is the sire of the winners Lady Agnes, Mamie B., Mr. Pelham, Padishah, St. Carlo, St. James, Adair, Admiral, Beauty, Belisarius, Chatham, Chesapeake, Clarendon, Coriolanus, Emily Carter, Faustina, Fearless, Flavia, Fritz, Her Highness, La Tosca (one of the best fillies ever foaled), Lizzie, My Queen, Polydora, Potomac (winner of the Futurity and Realization Stakes), St. Charles, St. Omer, St. Patrick, St. Denis, St. Mark, St. Carolus, St. Florian (a grand one), Schuylkill, Flattery, Chatham and others.

He is descended in direct female line from the Belgrade Turk mare, to which traces Taurus, Gladiator, Sting, Belladrum, New Holland, Boulevard, Lelaps, Wilful, etc.

Pot-8-o-'s 1773– Eclipse 1764– Marske 1750 Squirt 1732– Bartlett's Childers– Darley Arabian 1699.

ST. BLAISE.

CHESTNUT HORSE, FOALED 1880; BRED BY LORD ALINGTON, ENGLAND.

- HERMIT, ch. h., 1864.
  - Newminster, b. h., 1848.
    - Touchstone, br. h., 1831. (Camel, br. h., 1822. — Banter, br. m., 1826.)
    - Bee's-wing, b. m., 1833. (Dr. Syntax, br. h., 1811. — Tomboy's dam, ch. m., 1817.)
  - Seclusion, b. m., 1857.
    - Tadmor, br. h., 1846. (Ion, br. h., 1835. — Palmyra, br. m., 1838.)
    - Miss Sellon, b. m., 1851. (Cowl, b. h., 1842. — Belle Dame, ch. m., 1839.)
- FUSEE, b. m., 1867.
  - Marsyas, ch. h., 1851.
    - Orlando, b. h., 1841. (Touchstone, br. h., 1831. — Vulture, ch. m., 1833.)
    - Malibran, b. m., 1830. (Whisker, b. h., 1812. — Garcia, br. m., 1823.)
  - Vesuvienne, b. m., 1847.
    - Gladiator, ch. h., 1833. (Partisan, b. h., 1811. — Pauline, b. m., 1826.)
    - Venus, b. m., 1840. (Sir Hercules, bl. h., 1826. — Echo, b. m., 1828.)

Whalebone, br. h., 1807.	Waxy 1790 (Pot-8-o-'s 1773)—Penelope 1798 by Trumpator 1782 (Conductor 1767)—Prunella 1788 by Highflyer 1774 (Herod 1758)—Promise 1768 by Snap 1750 (Snip 1736)—Julia 1756 by Blank 1740 (God. Ar. 1724).
Daughter of, b. m., 1812.	Selim 1802 (Buzzard* 1787)—Maiden 1801 by Sir Peter T. 1784 (Highflyer 1774)—Dau. 1788 of Phenomenon 1780 (Herod 1758)—Matron 1782 by Florizel 1768 (Herod 1758)—Maiden 1770 by Matchem 1748 (Cade 1734).
Master Henry, b. h., 1815.	Orville 1799 (Beningbrough 1791)—Miss Sophia 1805 by Stamford 1794 (Sir Peter T. 1784)—Sophia 1798 by Buzzard* 1787 (Woodpecker 1773)—Huncamunca 1787 by Highflyer 1774 (Herod 1758)—Cypher 1772 by Squirrel 1754 (Traveller 1735).
Boadicea, b. m., 1807.	Alexander 1782 (Eclipse 1764)—Brunette by Amaranthus 1766 (Old England 1741)—Mayfly 1771 by Matchem 1748 (Cade 1734)—Dau. 1775 of Anc. Starling 1738 (Starling 1727)—Look-at-me-Lads 1731 by Brist. Grasshopper (Byerly T.).
Paynator, br. h., 1791.	Trumpator 1782 (Conductor 1767)—Dau. of Mark Anthony 1767 (Spectator 1749)—Signora 1767 by Snap 1750 (Snip 1736)—Miss Windsor 1754 by God. Ar. 1724—Sist. to Volunteer by Y. Belgrade (Belgrade Turk).
Daughter of, —, —.	Beningbrough 1791 (King Fergus 1775)—Jenny Mole 1787 by Carbuncle 1772 (Bab. Blank 1758)—Dau. 1773 of Prince T'Quassaw 1751 (Snap 1750)—Sultana 1759 by Regulus 1739 (God. Ar. 1724)—Dau. of Partner 1718 (Jigg).
Ardrossan, b. h., 1809.	John Bull 1789 (Fortitude 1777)—Miss Whip 1793 by Volunteer 1780 (Eclipse 1764)—Wimbleton 1785 by Evergreen 1769 (Herod 1758)—Sist. 1774 to Calash by Herod 1758 (Tartar 1743)—Teresa 1767 by Matchem 1748 (Cade 1734).
Lady Eliza, b. m., 1813.	Whitworth 1805 (Agonistes 1797)—X. Y. Z.'s dam 1798 by Spadille 1784 (Highflyer 1774)—Sylvia 1783 by Y. Marske 1771 (Marske 1750)—Ferret 1765 by Broth. to Silvio (Cade 1734)—Dau. of Regulus 1739 (God. Ar. 1724).
Cain, b. h., 1822.	Paulowitz 1813 (Sir Paul 1802)—Dau. 1800 of Paynator 1791 (Trumpator 1782)—Dau. 1785 of Delpini 1781 (Highflyer 1774)—Dau. 1785 of Y. Marske 1771 (Marske 1750)—Gentle Kitty 1774 by Silvio 1754 (Cade 1734).
Margaret, br. m., 1831.	Edmund 1824 (Orville 1799)—Medora 1811 by Selim 1802 (Buzzard* 1787)—Dau. 1802 of Sir Harry* 1795 (Sir Peter T. 1784)—Dau. 1798 of Volunteer 1780 (Eclipse 1764)—Dau. 1779 of Herod 1758 (Tartar 1743).
Sultan, b. h., 1816.	Selim 1802 (Buzzard* 1787)—Bacchante 1809 by W's Ditto 1800 (Sir Peter T. 1784)—Sist. 1791 to Calomel by Mercury 1778 (Eclipse 1764)—Dau. 1776 of Herod 1758 (Tartar 1743)—Folly 1771 by Marske 1750 (Squirt 1732).
Hester, br. m., 1832.	Camel 1822 (Whalebone 1807)—Monimia 1821 by Muley 1810 (Orville 1799)—Sist. 1816 to Petworth by Precipitate* 1787 (Mercury 1778)—Dau. 1787 of Woodpecker 1773 (Herod 1758)—Sist. 1773 to Juniper by Snap 1750 (Snip 1736).
Bay Middleton, b. h., 1833.	Sultan 1816 (Selim 1802)—Cobweb 1821 by Phantom 1808 (Walton 1799)—Filagree 1815 by Soothsayer 1808 (Sorcerer 1796)—Web 1808 by Waxy 1790 (Pot-8-o-'s 1773)—Penelope 1798 by Trumpator 1782 (Conductor 1767).
Crucifix, b. m., 1837.	Priam* 1827 (Emilius 1820)—Octaviana 1815 by Octavian 1807 (Stripling 1795)—Dau. 1807 of Shuttle 1793 (Y. Marske 1771)—Zara 1801 by Delpini 1781 (Highflyer 1774)—Flora 1789 by King Fergus 1775 (Eclipse 1764).
Belshazzar,* ch. h., 1830.	Blacklock 1814 (Whitelock 1803)—Mandane 1800 by Pot-8-o-'s 1773 (Eclipse 1764)—Y. Camilla 1787 by Woodpecker 1773 (Herod 1758)—Camilla 1778 by Trentham 1766 (T. Sweepstakes 1743).
Ellen, b. m., 1831.	Starch 1819 (Waxy Pope 1806)—Cuirass 1823 by Oiseau 1809 (Camillus 1803)—Castanea 1806 by Gohanna 1790 (Mercury 1778)—Grey Skim 1793 by Woodpecker 1773 (Herod 1758)—Dau. 1780 of Herod 1758 (Tartar 1743).
Camel, br. h., 1822.	Whalebone 1807 (Waxy 1790)—Dau. 1812 of Selim 1802 (Buzzard* 1787)—Maiden 1801 by Sir Peter T. 1784 (Highflyer 1774)—Dau. 1788 of Phenomenon 1780 (Herod 1758)—Matron 1782 by Florizel 1768 (Herod 1758).
Banter, br. m., 1826.	Master Henry 1815 (Orville 1799)—Boadicea 1807 by Alexander 1782 (Eclipse 1764)—Brunette by Amaranthus 1766 (Old England 1741)—Mayfly 1771 by Matchem 1748 (Cade 1734)—Dau. 1755 of Anc. Starling 1738 (Starling 1727).
Langar, ch. h., 1817.	Selim 1802 (Buzzard* 1787)—Dau. 1808 of Walton 1799 (Sir Peter T. 1784)—Y. Giantess 1790 by Diomed* 1777 (Florizel 1768)—Giantess 1769 by Matchem 1748 (Cade 1734)—Molly Long Legs 1753 by Babraham 1740 (God. Ar. 1724).
Kite, b. m., 1821.	Bustard 1813 (Castrel 1801)—Olympia 1815 by Sir Oliver 1800 (Sir Peter T. 1784)—Scotilla 1795 by Anvil 1777 (Herod 1758)—Scota 1783 by Eclipse 1764 (Marske 1750)—Harmony 1775 by Herod 1758 (Tartar 1743).
Waxy, b. h., 1790.	Pot-8-o-'s 1773 (Eclipse 1764)—Maria 1777 by Herod 1758 (Tartar 1743)—Lisette 1772 by Snap 1750 (Snip 1736)—Miss Windsor 1754 by God. Ar. 1724—Sist. to Volunteer by Y. Belgrade (Belgrade T.).
Penelope, b. m., 1798.	Trumpator 1782 (Conductor 1767)—Prunella 1788 by Highflyer 1774 (Herod 1758)—Promise 1768 by Snap 1750 (Snip 1736)—Julia 1756 by Blank 1740 (God. Ar. 1724)—Spectator's dam 1735 by Partner 1718 (Jigg).
Octavian, ch. h., 1807.	Stripling 1795 (Phenomenon 1780)—Dau. 1797 of Oberon 1790 (Highflyer 1774)—Sist. 1779 to Sharper by Ranthos 1763 (Matchem 1748)—Dau. of T. Sweepstakes 1743 (Bolt. Sweepstakes 1722)—Sist. 1748 to Careless by Spot 1728 (Hartley's Blind Horse).
Daughter of, b. m., 1806.	Shuttle 1793 (Y. Marske 1771)—Katharine 1798 by Delpini 1781 (Highflyer 1774)—Dau. 1786 of Paymaster 1766 (Blank 1740)—Dau. 1774 of Le Sang 1759 (Changeling 1747)—Dau. 1751 of Rib 1733 (Crab 1722).
Walton, b. h., 1799.	Sir Peter T. 1784 (Highflyer 1774)—Arethusa 1792 by Dungannon 1780 (Eclipse 1764)—Dau. 1777 of Prophet 1760 (Regulus 1739)—Virago 1764 by Snap 1750 (Snip 1736)—Dau. of Regulus 1739 (God. Ar. 1724).
Parasol, b. m., 1800.	Pot-8-o-'s 1773 (Eclipse 1764)—Prunella 1788 by Highflyer 1774 (Herod 1758)—Promise 1768 by Snap 1750 (Snip 1736)—Julia 1756 by Blank 1740 (God. Ar. 1724)—Spectator's dam 1735 by Partner 1718 (Jigg).
Moses, b. h., 1819.	[Whalebone or] Seymour 1807 (Delpini 1781)—Dau. 1807 of Gohanna 1790 (Mercury 1778)—Grey Skim 1793 by Woodpecker 1773 (Herod 1758)—Silver's dam 1780 by Herod 1758 (Tartar 1743)—Y. Hag 1761 by Skim 1746 (Starling 1728).
Quadrille, b. m., 1815.	Selim 1802 (Buzzard* 1787)—Canary Bird 1806 by [Whiskey or] Sorcerer 1796 (Trumpator 1782)—Canary 1797 by Coriander 1786 (Pot-8-o-'s 1773)—Miss Green 1787 by Highflyer 1774 (Herod 1758)—Harriet 1769 by Matchem 1748 (Cade 1734).
Whalebone, br. h., 1807.	Waxy 1790 (Pot-8-o-'s 1773)—Penelope 1798 by Trumpator 1782 (Conductor 1767)—Prunella 1788 by Highflyer 1774 (Herod 1758)—Promise 1768 by Snap 1750 (Snip 1736)—Julia 1756 by Blank 1740 (God. Ar. 1724).
Peri, b. m., 1822.	Wanderer 1811 (Gohanna 1790)—Thalestris 1809 by Alexander 1782 (Eclipse 1764)—Rival 1800 by Sir Peter T. 1784 (Highflyer 1774)—Hornet 1790 by Drone 1777 (Herod 1758)—Manilla 1777 by Goldfinder 1764 (Snap 1750).
Emilius, b. h., 1820.	Orville 1799 (Beningbrough 1791)—Emily 1810 by Stamford 1794 (Sir Peter T. 1784)—Dau. 1799 of Whiskey 1789 (Saltram* 1780)—Grey Dorimant 1781 by Dorimant 1772 (Otho 1760)—Dizzy 1757 by Blank 1740 (God. Ar. 1724).
Daughter of, b. m., 1820.	[Scud or] Pioneer 1804 (Whiskey 1789)—Canary Bird br. m. 1806 by [Whiskey or] Sorcerer 1796—Canary b. m. 1797 by Coriander 1786 (Pot-8-o-'s 1773)—Miss Green b. m. 1787 by Highflyer 1774 (Herod 1758)—Harriet b. m. 1769 by Matchem 1748 (Cade 1734)—Flora b. m. 1740 by Regulus 1739 (God. Ar. 1724)—Dau. of Bartlett's Childers (Darley Ar.

1699)—Dau. of Bay Bolton 1705 (Grey Hautboy)—Dau. of Belgrade Turk.

SALVATOR.

WINNER OF THE TITAN, MAPLE, REALIZATION, LORILLARD, CHAMPION AND TIDAL STAKES, THE SUBURBAN HANDICAP, AND OTHER PROMINENT EVENTS.

Property of J. B. Haggin, Rancho-del-Paso Stud, Sacramento, Cal. Yearlings sold annually.

SALVATOR, chestnut horse, foaled 1886, bred in the Elmendorf Stud, Ky., by imp. Prince Charlie, son of Blair Athol by Stockwell, dam Salina, dam of Danicheff, Duchess, etc., by Lexington, out of Lightsome, dam of Nevada (Luke Blackburn's dam), Crucifix (Semper Felix's dam), Sprightly (Volturno's dam), etc., by imp. Glencoe, tracing through Levity by imp. Trustee to Vandal's dam by imp. Tranby. Salvator is descended from one of the grandest racing and best producing families in the world. From it have come the grand race-horses and stallions Lexington (four miles in 7.19¾ and 7.23¾), Vandal, the best son of imp. Glencoe, Alaric, Rurie, Lever, Monitor, Volturno, Longstreet, The Bard, Monarchist, Aretino, Luke Blackburn, Tom Martin, Strathmore, Duke of Montrose, etc.

Salvator was a first-class race-horse. At two years old he won four races out of six in which he started, ran second to Proctor Knott in the Futurity Stakes, three-quarters of a mile, in 1.15½; won Flatbush Stakes, 7 furlongs, 100 lbs., in 1.30⅗; Maple Stakes, three-quarters of a mile, 120 lbs., in 1.17¾; Tuckahoe Stakes, Jerome Park, 1,400 yards, 120 lbs., in 1.24; Titan Stakes, 1,400 yards, 117 lbs., in 1.22¾. At three years old won seven out of eight races in which he started. Won the Tidal Stakes, 1 mile, 121 lbs., in 1.44⅗; won Realization Stakes, 1⅝ miles 122 lbs., in 2.51, Tenny, Long Dance, Longstreet and five others behind him; won Lorillard Stakes, 1½ miles, 118 lbs., in 2.37½, Longstreet, Centaur (Faverdale colt) and five others behind him; was third to Longstreet 118 lbs., and Proctor Knott, 120 lbs., in the Omnibus Stakes, 1½ miles, in 2.36¾, Salvator, 125 lbs., and four others behind him; won the Jersey Handicap, 1½ miles, 124 lbs., in 2.10¾, nine others behind him; won September Stakes, 1¾ miles, 129 lbs, in 3.05⅗, beating J. A. B. and Sorrento: won Sweepstakes, 1 1/16 miles, 127 lbs., in 1.50½; won purse, 1 1/16 miles, 113 lbs., in 2.01. At four years old won all his races. Won the Suburban Handicap, 1¼ miles, 127 lbs., in 2.06⅘, Cassius, 107 lbs., second, Tenny, 126 lbs., third, Reclare, Prince Royal, Longstreet, Firenzi, Raceland and others behind him: won the match against Tenny, each carrying 122 lbs., 1¼ miles, in 2.05; walked over for Monmouth Cup, 1¾ miles: won the Champion Stakes, 1½ miles, beating Tenny, each 122 lbs., in 2.35½; won the purse of $2,500 to beat the best time on record, 1.39¼, carrying 110 lbs.; he ran the distance in 1.35½, the fastest and best mile ever run in the world. He retired from the turf sound.

Salina, the dam of Salvator, was a good race mare. Won the Monmouth Oaks, Continental Hotel Stakes, and the Robins Stakes, 2 mile heats, Bugaman Stakes, and other good races. Prince Charlie his sire, was the Prince of the two-year old course, and combines in his pedigree all the most desirable strains of blood, Stockwell, Touchstone, Melbourne, and Priam through Crucifix. He must be a success in the stud.

He is decended in direct female line through old Montagu mare from the Royal mare, to which traces Conductor, O'Kelly's Eclipse, Filho-da-Puta, Voltaire, Sterling, Oxford, Springfield, Marsyas, Adventurer, Prince Charlie, Lexington, Vandal, Monarchist, Volturno, Luke Blackburn, Salvator, The Bard, Strathmore, Tom Martin, Grinstead, Duke of Montrose, Potomac, Tenny, and other noted horses from the Duchess family.

SALVATOR.

CHESTNUT HORSE, FOALED 1886; BRED BY D. SWIGERT, ELMENDORF STUD, KENTUCKY.

*PRINCE CHARLIE, ch. h., 1869.	Blair Athol, ch. h., 1861.	Stockwell, ch. h., 1849.	The Baron, ch. h., 1842.	Ir. Birdcatcher ch. h., 1833.	Sir Hercules 1826 (Whalebone 1807)—Guiccioli 1823 by Bob Booty 1804 (Chanticleer 1787)—Flight 1809 by Ir. Escape 1802 (Commodore 1793) Y. Heroine by Bagot 1780 (Herod 1758) Heroine 1775 by Hero 1753 (Cade 1734).
				Echidna, b. m., 1838.	Economist 1825 (Whisker 1812) Miss Pratt 1825 by Blacklock 1814 (Whitelock 1803) Gadabout 1812 by Orville 1799 (Beningbrough 1791) Minstrel 1803 by Sir Peter T. 1784 (Highflyer 1774)—Matron 1782 by Florizel 1768 (Herod 1758).
			Pocahontas, b. m., 1837.	Glencoe,* ch. h., 1831.	Sultan 1816 (Selim 1802) Trampoline 1825 by Tramp 1810 (Dick Andrews 1797) Web 1808 by Waxy 1790 (Pot-8-o-'s 1773) Penelope 1798 by Trumpator 1782 (Conductor 1767) Prunella 1788 by Highflyer 1774 (Herod 1758).
				Marpessa, b. m., 1830.	Muley 1810 (Orville 1799)—Clare 1824 by Marmion 1806 (Whiskey 1789)—Harpalice 1814 by Gohanna 1790 (Mercury 1778)—Amazon 1799 by Driver 1783 (Trentham 1766) Fractious 1792 by Mercury 1778 (Eclipse 1764).
		Blink Bonny, b. m., 1854.	Melbourne, br. h., 1834.	Humphrey Clinker, b. h., 1822.	Comus 1809 (Sorcerer 1796)—Clinkerina 1812 by Clinker 1805 (Sir Peter T. 1784) Pewet 1786 by Tandem 1773 (Syphon 1750) Termagant by Tantrum 1760 (Cripple 1750)—Cantatrice by Sampson 1745 (Blaze 1733).
				Daughter of, b. m., 1825.	Cervantes 1806 (Don Quixote 1784) Dau. 1818 of Golumpus 1802 (Gohanna 1790) Dau. 1810 of Paynator 1791 (Trumpator 1782) Circie 1802 by St. George 1789 (Highflyer 1774) Abigail 1788 by Woodpecker 1773 (Herod 1758).
			Queen Mary, b. m., 1843.	Gladiator, ch. h., 1833.	Partisan 1811 (Walton 1799)—Pauline 1826 by Moses 1819 ([Whalebone or] Seymour 1807) Quadrille 1815 by Selim 1802 (Buzzard* 1787)—Canary Bird 1806 by [Whiskey or] Sorcerer 1796 (Trumpator 1782)—Canary 1797 by Coriander 1786 (Pot-8-o-'s 1773).
				Daughter of, b. m., 1840.	Plenipotentiary 1831 (Emilius 1820) Myrrha 1830 by Whalebone 1807 (Waxy 1790) Gift 1818 by Y. Gohanna 1810 (Gohanna 1790) Sist. 1802 to Grazier by Sir Peter T. 1784 (Highflyer 1774)—Dau. 1788 of Trumpator 1782 (Conductor 1767).
	Eastern Princess, ch. m., 1858.	Surplice, b. h., 1845.	Touchstone, br. h., 1831.	Camel, br. h., 1822.	Whalebone 1807 (Waxy 1790)—Dau. 1812 of Selim 1802 (Buzzard* 1787) Maiden 1801 by Sir Peter T. 1784 (Highflyer 1774)—Dau. 1788 of Phenomenon 1780 (Herod 1758)—Matron 1782 by Florizel 1768 (Herod 1758).
				Banter, br. m., 1826.	Master Henry 1815 (Orville 1799)—Boadicea 1807 by Alexander 1782 (Eclipse 1764)—Brunette by Amaranthus 1766 (Old England 1741)—Mayfly 1771 by Matchem 1748 (Cade 1734)—Dau. 1755 of Anc. Starling 1738 (Starling 1727).
			Crucifix, b. m., 1837.	Priam,* b. h., 1827.	Emilius 1820 (Orville 1799)—Cressida 1807 by Whiskey 1789 (Saltram* 1780) Y. Giantess 1790 by Diomed* 1777 (Florizel 1768)—Giantess 1769 by Matchem 1748 (Cade 1734)—Molly Long Legs 1753 by Babraham 1740 (God. Ar. 1724).
				Octaviana, b. m., 1815.	Octavian 1807 (Stripling 1795) Dau. 1807 of Shuttle 1793 (Y. Marske 1771)—Zara 1801 by Delpini 1781 (Highflyer 1774)—Flora 1789 by King Fergus 1775 (Eclipse 1764) Atalanta 1769 by Matchem 1748 (Cade 1734).
		Tomyris, b. m., 1851.	Sesostris, b. h., 1845.	Slane, b. h., 1833.	Royal Oak 1823 (Catton 1809) Dau. 1819 of Orville 1799 (Beningbrough 1791)—Epsom Lass 1803 by Sir Peter T. 1784 (Highflyer 1774)—Alexina 1788 by King Fergus 1775 (Eclipse 1764)—Lardella 1780 by Y. Marske 1771 (Marske 1750).
				Palmyra, br. m., 1818.	Sultan 1816 (Selim 1802)—Hester 1832 by Camel 1822 (Whalebone 1807) Monimia 1821 by Muley 1810 (Orville 1799) Sist. 1796 to Petworth by Precipitate* 1787 (Mercury 1778)—Dau. 1787 of Woodpecker 1773 (Herod 1758).
			Daughter of, ch. m., 1842.	Glaucus, b. h., 1830.	Partisan 1811 (Walton 1799)—Nanine 1823 by Selim 1802 (Buzzard* 1787)—Bizarre 1811 by Peruvian 1806 (Sir Peter T. 1784)—Violante 1802 by John Bull 1789 (Fortitude 1777)—Dau. 1788 of Highflyer 1774 (Herod 1758).
				Io, ch. m., 1836.	Taurus 1826 ([Phantom or] Morisco 1819)—Arethissa 1824 by Quiz 1798 (Buzzard* 1787)—Persepolis 1803 by Alexander 1782 (Eclipse 1764)—Sist. 1784 to *Tickle Toby by Alfred 1770 (Matchem 1748) Cælia 1775 by Herod 1758 (Tartar 1743).
SALINA, b. m., 1868.	Lexington, b. h., 1850.	Boston, ch. h., 1833.	Timoleon, ch. h., 1813.	Sir Archy, b. h., 1805.	*Diomed 1777 (Florizel 1768)—*Castianira 1796 by Rockingham 1781 (Highflyer 1774)—Tabitha 1782 by Trentham 1766 (T. Sweepstakes 1743)—Dau. of Bosphorus 1754 (Babraham 1740) Dau. of W's Forester 1750 (Forester 1736).
				Daughter of, ch. m., —.	*Saltram 1780 (Eclipse 1764)—Dau. of Symmes' Wildair 1767 (*Fearnought 1755)—Dau. of Tyler's Driver 1760 (*Othello)—Dau. of *Fallower 1761 (Blank 1740)—Dau. of *Vampire 1757 (Regulus 1739).
			Daughter of, ch. m., 1814.	Ball's Florizel, ch. h., 1801.	*Diomed 1777 (Florizel 1768)—Dau. of *Shark 1771 (Marske 1750)—Dau. of Harris' Eclipse 1770 (*Fearnought 1755) Dau. of *Fearnought 1755 (Regulus 1739)—Dau. of *Jolly Roger 1741 (Roundhead 1733).
				Daughter of, b. m., 1799.	*Alderman 1787 (Pot-8-o-'s 1773)—Dau. of *Clockfast 1774 (Gimcrack 1760)—Dau. of Symmes' Wildair 1767 (*Fearnought 1755)—Y. Kitty Fisher 1767 by *Fearnought 1755 (Regulus 1739)—*Kitty Fisher 1755 by Cade 1734 (God. Ar. 1724).
		Alice Carneal, b. m., 1836.	*Sarpedon, br. h., 1828.	Emilius, b. h., 1820.	Orville 1799 (Beningbrough 1791)—Emily 1810 by Stamford 1794 (Sir Peter T. 1784) Dau. 1799 of Whiskey 1789 (Saltram* 1780)—Grey Dorimant 1781 by Dorimant 1772 (Otho 1760)—Dizzy 1757 by Blank 1740 (God. Ar. 1724).
				Icaria, b. m., 1824.	The Flyer 1814 (Vandyke Jr. 1808)—Parma 1818 by Dick Andrews 1797 (Joe Andrews 1778)—May 1804 by Beningbrough 1791 (King Fergus 1775) Primrose 1787 by Mambrino 1768 (Engineer 1756)—Cricket 1776 by Herod 1758 (Tartar 1743).
			Rowena, ch. m., 1826.	Sumpter, ch. h., 1818.	Sir Archy 1805 (*Diomed 1777)—Dau. of *Robin Redbreast 1796 (Sir Peter T. 1784)—Dau. 1788 of *Obscurity 1777 (Eclipse 1764)—Slamerkin 1768 by *Wildair 1753 (Cade 1734)—*Cub mare 1762 by Cub 1739 (Fox 1714).
				Lady Grey, b. m., 1817.	Robin Grey 1805 (*Royalist 1790)—Maria 1802 by Melzar 1791 (*Medley 1776) Dau. of *Highflyer 1784 (Highflyer 1774)—Dau. of *Fearnought 1755 (Regulus 1739) Dau. of Ariel 1756 (*Traveller 1747).
	Lightsome, ch. m., 1854.	*Glencoe, ch. h., 1831.	Sultan, b. h., 1816.	Selim, ch. h., 1802.	Buzzard* 1787 (Woodpecker 1773)—Dau. of Alexander 1782 (Eclipse 1764) Dau. of Highflyer 1774 (Herod 1758)—Dau. of Alfred 1770 (Matchem 1748)—Dau. 1770 of Engineer 1756 (Sampson 1745).
				Bacchante, br. m., 1809.	W's Ditto 1800 (Sir Peter T. 1784)—Dau. 1791 of Mercury 1778 (Eclipse 1764)—Dau. 1776 of Herod 1758 (Tartar 1743)—Folly 1771 by Marske 1750 (Squirt 1732)—Vixen 1773 by Regulus 1739 (God. Ar. 1724).
			Trampoline, ch. m., 1825.	Tramp, b. h., 1810.	Dick Andrews 1797 (Joe Andrews 1778)—Dau. 1803 of Gohanna 1790 (Mercury 1778)—Fraxinella 1793 by Trentham 1766 (T. Sweepstakes 1743)—Dau. 1785 of Woodpecker 1773 (Herod 1758)—Everlasting 1775 by Eclipse 1764 (Marske 1750).
				Web, b. m., 1808.	Waxy 1790 (Pot-8-o-'s 1773)—Penelope 1798 by Trumpator 1782 (Conductor 1767) Prunella 1788 by Highflyer 1774 (Herod 1758)—Promise 1768 by Snap 1750 (Snip 1736)—Julia 1756 by Blank 1740 (God Ar. 1724).
		Levity, b. m., 1845.	*Trustee, ch. h., 1829.	Catton, b. h., 1809.	Golumpus 1802 (Gohanna 1790)—Lucy Gray 1804 by Timothy 1794 (Delpini 1781)—Lucy 1789 by Florizel 1768 (Herod 1758)—Frenzy 1774 by Eclipse 1764 (Marske 1750)—Dau. of Engineer 1756 (Sampson 1745).
				Emma, ch. m., 1824.	Whisker 1812 (Waxy 1790)—Gibside Fairy 1811 by Hermes 1790 (Mercury 1778)—Vicissitude 1800 by Pipator 1786 (Imperator 1776)—Beatrice 1791 by Sir Peter T. 1784 (Highflyer 1774)—Pyrrha 1771 by Matchem 1748 (Cade 1734).
			Vandal's dam, b. m., 1837.	*Tranby, br. h., 1826.	Blacklock 1814 (Whitelock 1803) Dau. 1812 of Orville 1799 (Beningbrough 1791) Miss Grimstone 1796 by Weazle 1776 (Herod 1758)—Dau. of Ancaster 1768 (Blank 1740)—Dau. of Damascus Ar. 1754.
				Lucilla, b. m., —.	Trumpator 1823 (Sir Solomon 1805)—Lucy b. m. 1820 by Orphan 1810 (Ball's Florizel 1801)—Lady Grey b. m. 1817 by Robin Grey 1805 (*Royalist 1790)—Maria 1802 by Melzar 1791 (*Medley 1776)—Dau. of *Highflyer 1781 (Highflyer 1774)—Dau. of *Fearnought 1755 (Regulus 1739)—Dau. of Ariel 1756 (*Traveller 1747)—Dau. of *Jack of Diamonds (Cullen Ar. 1745)—*Duchess by Cullen Ar. 1745—Lady Thigh ch. m. 1781 by Partner 1718 (Jigg)—Dau. ch. m. 1723 of Greyhound (Chillaby)—Sophonisba's dam by Curwen Bay Barb—Dau. of D'Arcy Chestnut Ar.—Dau. of Whiteshirt—Lord Darcy's Montagu mare by Old Montagu—Dau. of Hautboy (D'Arcy White Turk)—Dau. of Brimmer (D'Arcy Yellow Turk)—Royal mare.

SENSATION.

WINNER OF THE JUVENILE AND NURSERY STAKES AT JEROME PARK, JULY, CRITERION AND AUGUST STAKES AT MONMOUTH PARK, SARATOGA AND FLASH STAKES AT SARATOGA, AND CENTRAL STAKES AT BALTIMORE, 1879.

Sensation is private stallion in the Rancocas Stud, Mr. P. Lorillard, proprietor, Jobstown, N. J.

SENSATION by imp. Leamington, son of Faugh-a-Ballagh, bred by Mr. A. Welch, Erdenheim Stud, Pa., foaled 1877, dam Susan Beane, dam of Stratford, Susquehanna, Onondaga, etc., by Lexington, son of Boston, out of Sally Lewis, dam of John Morgan, Hunter's Lexington, Acrobat, etc., by imp. Glencoe. Sensation was truly the sensation of his two-year old year. Started eight times and won all of his races. Jerome Park, May 31, won the Juvenile Stakes, half-mile, defeating Ethel and nine others, time, 0.50. Monmouth Park, July 8, won the July Stakes, 5 furlongs, defeating his stable companions, Grenada and Rosalie, who finished respectively second and third, and ten others, time, 1.07. Saratoga, July 22, won the Flash Stakes, half-mile, Grenada and Rosalie again finishing second and third, five others behind them, time, 0.49¾; same meeting, July 29, won the Saratoga Stakes, three-quarters of a mile, defeating Grenada, who finished second, Beata, Glidelia, Giroflé and By-and-By, time, 1.18. Monmouth Park, Aug. 23, won the August Stakes, three-quarters of a mile, defeating Grenada, who ran second, and seven others, time, 1.18¼; same meeting, Aug. 26, won the Criterion Stakes, three-quarters of a mile, carrying 5 lbs. penalty, and defeated Grenada, second, and five others, time, 1.22, track slow and heavy. Jerome Park, Oct. 2, won the Nursery Stakes, three-quarters of a mile, defeating Ferncliff, Grenada and four others, time, 1.18. Baltimore, Oct. 23, won the Central Stakes, 1 mile, defeating Grenada and Oden, time, 1.50½. Total winnings for the year, $20,250. He met and defeated all the best two-year olds of his year, winning from a half to one mile. He bursted his foot in the spring of 1880, which compelled his retirement. He is from one of the great racing families of the country. Acrobat and Hunter's Lexington, full brothers to his dam, were fine race-horses and could stay a distance. John Morgan, a half brother, was also a superior race-horse, and, for his chances, a very successful stallion. Motto, Nannie Lewis, Aldebaran were all good ones. Glenmore is also from the same family, and his dam traces back to the dam of the great race-horse Medoc, who was the most successful sire in any day on all sorts of mares. Sensation is a brown, full 16 hands high, with a white stripe down his face and a little white on the pastern of his left fore leg. He has a fine, well-placed shoulder and is symmetrically formed all over, with faultless action and good temper. Being a son of Leamington and tracing to a racing family, and an uncommon fine race-horse, he is a success in the stud. He has the Pantaloon and Whalebone crosses on the sire's side, and the Glencoe on the dam's, with the blood of Sir Archy and Diomed through its best sources.

Sensation has been used very little in the stud, but is the sire of the following winners: Banker, Carrie C., Lord Fauntleroy, Loantaka (winner of the Suburban Handicap), Centaur (Faverdale colt), My Own, Guarantee, Queen Elizabeth, Specialty, Raymond, Salvini, Chemise, Flitaway, Satisfaction, Prince George, Saunterer, Count, Sentiment, Vendetta and others.

Sensation traces to the same Royal mare through old Montagu mare to which trace in direct female line Conductor, Eclipse, Sterling, Weatherbit, Oxford, Kingston, Scottish Chief, Adventurer, Lexington, Vandal, Leonatus, Salvator, Onondaga, Stratford and all the Duchess family.

SENSATION.

BROWN HORSE, FOALED 1877; BRED BY MR. A. WELCH, CHESTNUT HILL STUD, PENNSYLVANIA.

Sire: *LEAMINGTON, br. h., 1853.
- Faugh-a-Ballagh, br. h., 1841.
  - Sir Hercules, blk. h., 1826.
    - Whalebone, br. h., 1807.
    - Peri, b. m., 1822.
  - Guiccioli, ch. m., 1823.
    - Bob Booty, ch. h., 1804.
    - Flight, ch. m., 1809.
- Daughter of, b. m., 1841.
  - Pantaloon, ch. h., 1824.
    - Castrel, ch. h., 1801.
    - Idalia, ch. m., 1815.
  - Daphne, br. m., 1837.
    - Laurel, br. h., 1824.
    - Maid of Honor, br. m., 1829.

Dam: SUSAN BEANE, ch. m., 1868.
- Lexington, b. h., 1850.
  - Boston, ch. h., 1833.
    - Timoleon, ch. h., 1813.
    - Sist. to Tuckahoe, ch. m., 1814.
  - Alice Carneal, b. m., 1836.
    - *Sarpedon, br. h., 1828.
    - Rowena, ch. m., 1826.
- Sally Lewis, ch. m., 1852.
  - *Glencoe, ch. h., 1831.
    - Sultan, b. h., 1816.
    - Trampoline, ch. m., 1825.
  - Motto, ch. m., 1839.
    - *Barefoot, ch. h., 1820.
    - Lady Thompkins, ch. m., 1831.

Waxy, b. h., 1790.	Pot-8-o-'s 1773 (Eclipse 1764)—Maria 1777 by Herod 1758 (Tartar 1743)—Lisette 1772 by Snap 1750 (Snip 1736)—Miss Windsor 1754 by God. Ar. 1724—Sist. to Volunteer by Y. Belgrade (Belgrade Turk).
Penelope, b. m., 1798.	Trumpator 1782 (Conductor 1767)—Prunella 1788 by Highflyer 1774 (Herod 1758)—Promise 1768 by Snap 1750 (Snip 1736)—Julia 1756 by Blank 1740 (God. Ar. 1724)—Spectator's dam 1735 by Partner 1718 (Jigg).
Wanderer, b. h., 1811.	Gohanna 1790 (Mercury 1778)—Catherine 1795 by Woodpecker 1773 (Herod 1758)—Camilla 1778 by Trentham 1766 (T. Sweepstakes 1743)—Coquette 1764 by Compton Barb—Sist. 1743 to Regulus by God. Ar. 1724.
Thalestris, blk. m., 1809.	Alexander 1782 (Eclipse 1764)—Rival 1800 by Sir Peter T. 1784 (Highflyer 1774)—Hornet 1790 by Drone 1777 (Herod 1758)—Manilla 1777 by Goldfinder 1764 (Snap 1750)—Dau. 1766 of Old England 1741 (God. Ar. 1724).
Chanticleer, ch. h., 1787.	Woodpecker 1773 (Herod 1758)—Dau. 1778 of Eclipse 1764 (Marske 1750)—Rosebud 1765 by Snap 1750 (Snip 1736)—Miss Belsea 1753 by Regulus 1739 (God. Ar. 1724)—Dau. of Bartlett's Childers (Darley Ar. 1699).
Ierne, ch. m., 1790.	Bagot 1780 (Herod 1758)—Dau. of Gamahoe (Bustard 1741)—Patty by Tim 1752 (Squirt 1732)—Miss Patch 1737 by Justice 1725 (H. C. Litton Ar.)—Ringtail Galloway by Curwen Bay Barb.
Ir. Escape, ch. h., 1802.	Commodore 1793 (Tom Tug 1777)—Moll-in-the-Wad 1791 by Highflyer 1774 (Herod 1758)—Shift 1779 by Sweetbriar 1769 (Syphon 1750)—Black Susan 1762 by Snap 1750 (Snip 1736)—Dau. of Cade 1734 (God. Ar. 1724).
Y. Heroine, —	Bagot 1780 (Herod 1758)—Heroine 1775 by Hero 1753 (Cade 1734)—Dau. 1766 of Snap 1750 (Snip 1736)—Sist. 1743 to Regulus by God. Ar. 1724—Grey Robinson 1723 by Bald Galloway (St. Victor Barb).
Buzzard,* ch. h., 1787.	Woodpecker 1773 (Herod 1758)—Misfortune 1775 by Dux 1761 (Matchem 1748)—Curiosity 1760 by Snap 1750 (Snip 1736)—Dau. 1749 of Regulus 1739 (God. Ar. 1724)—Dau. of Bartlett's Childers (Darley Ar. 1699).
Daughter of, b. m., —.	Alexander 1782 (Eclipse 1764)—Dau. of Highflyer 1774 (Herod 1758)—Dau. of Alfred 1770 (Matchem 1748)—Dau. 1770 of Engineer 1756 (Sampson 1745)—Bay Malton's dam 1756 by Cade 1734 (God. Ar. 1724).
Peruvian, b. h., 1806.	Sir Peter T. 1784 (Highflyer 1774)—Dau. 1788 of Boudrow 1777 (Eclipse 1764)—Dau. 1764 of Squirrel 1754 (Traveller 1735)—Sist. to Lowther's Babraham by Babraham 1740 (God. Ar. 1724)—Dau. 1750 of Golden Ball 1735 (Partner 1718).
Musidora, b. m., 1804.	Meteor 1783 (Eclipse 1764)—Maid-of-all-Work 1786 by Highflyer 1774 (Herod 1758)—Dau. 1771 of Syphon 1750 (Squirt 1732)—Dau. of Regulus 1739 (God. Ar. 1724)—Sist. 1758 to Judgement by Snip 1736 (Flying Childers 1715).
Blacklock, b. h., 1814.	Whitelock 1803 (Hambletonian 1792)—Dau. 1799 of Coriander 1786 (Pot-8-o-'s 1773)—Wildgoose 1792 by Highflyer 1774 (Herod 1758)—Coheiress 1786 by Pot-8-o-'s 1773 (Eclipse 1764)—Manilla 1777 by Goldfinder 1764 (Snap 1750).
Wagtail, b. m., 1818.	Prime Minister 1810 (Sancho 1801)—Trauby*'s dam 1812 by Orville 1799 (Beningbrough 1791)—Miss Grimstone 1796 by Weazle 1776 (Herod 1758)—Dau. of Ancaster 1768 (Blank 1740)—Dau. of Damascus Ar. 1754.
Champion, ch. h., 1812.	Selim 1802 (Buzzard* 1787)—Podagra 1805 by Gouty 1796 (Sir Peter T. 1784)—Jet 1785 by Magnet 1770 (Herod 1758)—Jewel 1775 by Squirrel 1754 (Traveller 1735)—Sophia 1764 by Blank 1740 (God. Ar. 1724).
Etiquette, br. m., 1820.	Orville 1799 (Beningbrough 1791)—Boadicea 1807 by Alexander 1782 (Eclipse 1764)—Brunette by Amaranthus 1766 (Old England 1741)—Mayfly 1771 by Matchem 1748 (Cade 1734)—Dau. 1755 of Starling 1738 (Starling 1727).
Sir Archy, b. h., 1805.	*Diomed 1777 (Florizel 1768)—*Castianira 1796 by Rockingham 1781 (Highflyer 1774)—Tabitha 1784 by Trentham 1766 (T. Sweepstakes 1743)—Dau. of Bosphorus 1754 (Babraham 1740)—Dau. of Forester 1750 (Forester 1736).
Daughter of, ch. m., —.	*Saltram 1780 (Eclipse 1764)—Dau. of Symmes' Wildair 1767 (*Fearnought 1755)—Dau. of Tyler's Driver 1760 (*Othello)—Dau. of *Fallower 1761 (Blank 1740)—Dau. of *Vampire 1757 (Regulus 1739).
Ball's Florizel, ch. h., 1801.	*Diomed 1777 (Florizel 1768)—Dau. of *Shark 1771 (Marske 1750)—Dau. of Harris' Eclipse 1770 (*Fearnought 1755)—Dau. of *Fearnought 1755 (Regulus 1739)—Dau. of *Jolly Roger 1741 (Roundhead 1733).
Daughter of, b. m., 1799.	*Alderman 1787 (Pot-8-o-'s 1773)—Dau. of *Clockfast 1774 (Gimcrack 1760)—Dau. of Symmes' Wildair 1767 (*Fearnought 1755)—Y. Kitty Fisher 1767 by *Fearnought 1755 (Regulus 1739)—*Kitty Fisher 1755 by Cade 1734 (God. Ar. 1724).
Emilius, b. h., 1820.	Orville 1799 (Beningbrough 1791)—Emily 1810 by Stamford 1794 (Sir Peter T. 1784)—Dau. 1799 of Whiskey 1789 (Saltram* 1780)—Grey Dorimant 1781 by Dorimant 1772 (Otho 1760)—Dizzy 1757 by Blank 1740 (God. Ar. 1724).
Icaria, b. m., 1824.	The Flyer 1814 (Vandyke Jr. 1808)—Parma 1818 by Dick Andrews 1797 (Joe Andrews 1778)—May 1804 by Beningbrough 1791 (King Fergus 1775)—Primrose 1787 by Mambrino 1768 (Engineer 1756)—Cricket 1776 by Herod 1758 (Tartar 1743).
Sumpter, ch. h., 1818.	Sir Archy 1805 (*Diomed 1777)—Dau. of *Robin Redbreast 1796 (Sir Peter T. 1784)—Dau. 1788 of *Obscurity 1777 (Eclipse 1764)—Slamerkin 1768 by *Wildair 1753 (Cade 1734)—*Cub mare 1762 by Cub 1739 (Fox 1714).
Lady Grey, b. m., 1817.	Robin Grey 1805 (*Royalist 1790)—Maria 1802 by Melzar 1791 (*Medley 1776)—Dau. of *Highflyer 1784 (Highflyer 1774)—Dau. of *Fearnought 1755 (Regulus 1739)—Dau. of Ariel 1756 (*Traveller 1747).
Selim, ch. h., 1802.	Buzzard* 1787 (Woodpecker 1773)—Dau. of Alexander 1782 (Eclipse 1764)—Dau. of Highflyer 1774 (Herod 1758)—Dau. of Alfred 1770 (Matchem 1748)—Dau. 1770 of Engineer 1756 (Sampson 1745).
Bacchante, br. m., 1800.	W's Ditto 1800 (Sir Peter T. 1784)—Dau. 1791 of Mercury 1778 (Eclipse 1764)—Dau. 1776 of Herod 1758 (Tartar 1743)—Folly 1771 by Marske 1750 (Squirt 1732)—Vixen 1753 by Regulus 1739 (God. Ar. 1724).
Tramp, b. h., 1810.	Dick Andrews 1797 (Joe Andrews 1778)—Dau. 1803 of Gohanna 1790 (Mercury 1778)—Fraxinella 1798 by Trentham 1766 (T. Sweepstakes 1743)—Dau. 1785 of Woodpecker 1773 (Herod 1758)—Everlasting 1775 by Eclipse 1764 (Marske 1750).
Web, b. m., 1808.	Waxy 1790 (Pot-8-o-'s 1773—Penelope 1798 by Trumpator 1782 (Conductor 1767)—Prunella 1788 by Highflyer 1774 (Herod 1758)—Promise 1768 by Snap 1750 (Snip 1736)—Julia 1756 by Blank 1740 (God Ar. 1724).
Tramp, b. h., 1810.	Dick Andrews 1797 (Joe Andrews 1778)—Dau. 1803 of Gohanna 1790 (Mercury 1778)—Fraxinella 1798 by Trentham 1766 (T. Sweepstakes 1743)—Dau. 1785 of Woodpecker 1773 (Herod 1758)—Everlasting 1775 by Eclipse 1764 (Marske 1750).
Rosamond, ch. m., 1798.	Buzzard* 1787 (Woodpecker 1773)—Roseberry 1792 by Phenomenon 1780 (Herod 1758)—Miss West 1777 by Matchem 1748 (Cade 1734)—Sist. to Favourite by Regulus 1739 (God. Ar. 1724—Y. Ebony 1742 by Crab 1722 (Alcock Ar.).
Am. Eclipse, ch. h., 1814.	Duroc 1806 (*Diomed 1777)—Miller's Damsel 1802 by *Messenger 1780 (Mambrino 1768)—*Mare 1792 by Pot-8-o-'s 1773 (Eclipse 1764)—Dau. 1778 of Gimcrack 1760 (Cripple 1750)—Snapdragon 1759 by Snap 1750 (Snip 1736).
Katy Ann, ch. m., 1825.	Ogle's Oscar 1800 (*Gabriel 1790)—Y. Maid-of-the-Oaks ch. m. 1817 by *Expedition 1795 (Pegasus 1784)—Maid-of-the-Oaks ch. m. 1801 by *Spread Eagle 1792 (Volunteer 1780)—Annette by *Shark 1771 (Marske 1750)—Dau. of Rockingham (Partner 1755)—Dau. of True Whig (Fitzhugh's Regulus 1765)—Dau. of Gallant 1770 (*Fearnought 1755)—Dau. of *Regulus 1747 (Regulus 1739)—*Duchess by Cullen Ar.—Lady Thigh ch. m. 1731 by Partner 1718 (Jigg)—Dau. ch. m. 1723 by Greyhound (Chillaby)—Sophonisba's dam by Curwen Bay Barb—Dau. of D'Arcy Chestnut Ar.—Dau. of Whiteshirt—Montagu mare by D'Arcy's old Montagu—Dau. of Hautboy (D'Arcy White Turk)—Dau. of Brimmer (D'Arcy Yellow Turk)—Royal mare.

IMPORTED SIDDARTHA.

The property of Dr. A. W. McAlester, Columbia, Mo. Terms annually. Yearlings sold annually.

SIDDARTHA, bay horse, foaled 1880, bred by Mr. W. I'Anson, by Pero Gomez, son of Beadsman, sire of Blue Gown and other good ones. Pero Gomez was beaten a head in the Derby by Pretender, and beat him in the St. Leger, which he won. Pero Gomez is the sire of Pontiac, Merrythought, Gil Blas, Harbinger Hidalgo, Espada, Titania, etc. Siddartha is out of The Pearl, the dam of Selby, winner of the Great Yorkshire Handicap, 1886, by Newminster, sire of the noted Hermit. He is from the grandest racing family in England. Caller Ou, his grandam, won 31 Queen's Plates, the Doncaster St. Leger and the Brighton and York Cups. Blink Bonny, out of Queen Mary, won the Derby and Oaks and was the dam of Borealis, Blair Athol (winner of the Derby and St. Leger), and Breadalbane (winner of the Prince of Wales and Gratwicke Stakes). Other famous winners in the family are Sir Bevys (winner of the Derby), Hampton (winner of the Epsom Gold Cup, Doncaster and Goodwood Cups and Goodwood Stakes), Broomielaw (winner of the Dee Stakes and Chesterfield Stakes), Blinkhoolie (winner of the Ascot Gold Vase and Alexandra Plate), Freeman (winner of the Great Northern Handicap twice, Goodwood Stakes, Chesterfield Cup and Alexandra Plate), Good Hope (winner of the Vienna Derby), Beauclerc (winner of the Middle Park Plate). Thrift, the dam of Tristan, for which $50,000 was refused, is from the same family. The family is potent in great race-horses, prolific brood mares and successful sires. The colts by Siddartha are said to be the best their dams ever produced. Siddartha never raced, but Mr. Jacob Pincus, who trained him, says he was a very speedy, good race-horse.

Siddartha is the sire of the winners Crispin, Joe Walton, Sid, Gracie, Ed. Dougherty, Chauntress, Millie R., etc. He has had little or no chance in the stud, but is a symmetrically formed horse and descended from one of the grandest and best racing and producing families in the world, and must succeed in the stud.

He traces in direct female line to the Childers mare, to which traces Postmaster (by Herod), Bustard (by Buzzard), Aimator, imp. Balrownie, imp. Bonnie Scotland, Hampton (by Lord Clifden), Blair Athol, Breadalbane, Pretender, Sir Bevys, Tristan, Carlton, Crafton, imp. Bernice and Belladonna, and all the Queen Mary family in England; Aurora Raby, Bill Bruce, Atilla, and all the imported Peggy family in America.

Dick Andrews 1797—Joe Andrews 1778—Eclipse 1764—Marske 1750—Squirt 1732—Bartlett's Childers—Darley Arabian 1699.

SIDDARTHA.

BAY HORSE, FOALED 1880; BRED BY MR. W. I'ANSON, ENGLAND.

PERO GOMEZ, br. h., 1866.	Beadsman, br. h., 1855.	Weatherbit, br. h., 1842.	Sheet Anchor, br. h., 1832.	Lottery, br. h., 1820.	Tramp 1810 (Dick Andrews 1797)—Mandane 1800 by Pot-8-o's 1773 (Eclipse 1764)—Y. Camilla 1787 by Woodpecker 1773 (Herod 1758)—Camilla 1778 by Trentham 1766 (T. Sweepstakes 1743)—Coquette 1765 by Compton Barb.
				Morgiana, blk. m., 1820.	Muley 1810 (Orville 1799)—Miss Stephenson 1814 by [Scud or] Sorcerer 1796 (Trumpator 1782)—Dau. 1795 of Precipitate* 1787 (Mercury 1778)—Dau. 1787 of Woodpecker 1773 (Herod 1758)—Sist. 1773 to Juniper by Snap 1750 (Snip 1736).
			Miss Letty, b. m., 1834.	Priam, b. h., 1827.	Emilius 1820 (Orville 1799)—Cressida 1807 by Whiskey 1789 (Saltram 1780)—Y. Giantess 1790 by Diomed 1777 (Florizel 1768)—Giantess 1769 by Matchem 1748 (Cade 1734)—Molly Long Legs 1753 by Babraham 1740 (God. Ar. 1724).
				Daughter of, br. m., 1815.	Orville 1799 (Beningbrough 1791)—Dau. 1800 of Buzzard* 1787 (Woodpecker 1773)—Hornpipe 1793 by Trumpator 1782 (Conductor 1767)—Luna 1779 by Herod 1758 (Tartar 1743)—Proserpine 1763 by Marske 1750 (Squirt 1732).
		Mendicant, br. m., 1843.	Touchstone, br. h., 1831.	Camel, br. h., 1822.	Whalebone 1807 (Waxy 1790)—Dau. 1812 of Selim 1802 (Buzzard* 1787)—Maiden 1801 by Sir Peter T. 1784 (Highflyer 1774)—Dau. 1788 of Phenomenon 1780 (Herod 1758)—Matron 1782 by Florizel 1768 (Herod 1758).
				Banter, br. m., 1826.	Master Henry 1815 (Orville 1799)—Boadicea 1807 by Alexander 1782 (Eclipse 1764)—Brunette by Amaranthus 1766 (Old England 1741)—Mayfly 1771 by Matchem 1748 (Cade 1734)—Dau. 1755 of Anc. Starling 1738 (Starling 1727).
			Lady Moore Carew, b. m., 1830.	Tramp, b. h., 1810.	Dick Andrews 1797 (Joe Andrews 1778)—Dau. 1803 of Gohanna 1790 (Mercury 1778)—Fraxinella 1793 by Trentham 1766 (T. Sweepstakes 1743)—Dau. 1785 of Woodpecker 1773 (Herod 1758)—Everlasting 1775 by Eclipse 1764 (Marske 1750).
				Kite, b. m., 1821.	Bustard 1813 (Castrel 1801)—Olympia 1815 by Sir Oliver 1800 (Sir Peter T. 1784)—Scotilla 1795 by Anvil 1777 (Herod 1758)—Scota 1783 by Eclipse 1764 (Marske 1750)—Harmony 1775 by Herod 1778 (Tartar 1743).
	Salamanca, b. m., 1859.	Student, ch. h., 1851.	Chatham, ch. h., 1839.	The Colonel, ch. h., 1825.	Whisker 1812 (Waxy 1790)—Dau. 1802 of Delpini 1781 (Highflyer 1774)—Tipple Cyder 1788 by King Fergus 1775 (Eclipse 1764)—Sylvia 1783 by Y. Marske 1771 (Marske 1750)—Ferret 1765 by Broth. to Silvio (Cade 1734).
				Hester, br. m., 1832.	Camel 1822 (Whalebone 1807)—Moulmia 1821 by Muley 1810 (Orville 1799)—Dau. 1796 of Precipitate* 1787 (Mercury 1778)—Dau. 1787 of Woodpecker 1773 (Herod 1758)—Sist. 1773 to Juniper by Snap 1750 (Snip 1736).
			Daughter of, b. m., 1840.	Laurel, br. h., 1824.	Blacklock 1814 (Whitelock 1803)—Wagtail 1818 by Prime Minister 1810 (Sancho 1801)—Tranby*'s dam 1812 by Orville 1799 (Beningbrough 1791)—Miss Grimstone 1796 by Weazle 1776 (Herod 1758)—Dau. of Ancaster 1768 (Blank 1740).
				Flight, br. m., 1831.	Velocipede 1825 (Blacklock 1814)—Miss Wilkes 1818 by Octavian 1807 (Stripling 1795)—Dau. 1807 of Remembrancer 1800 (Sir Peter T. 1784)—Mary 1797 by Y. Marske 1771 (Marske 1750)—Gentle Kitty 1774 by Silvio 1754 (Cade 1734).
		Bravery, br. m., 1852.	Gameboy, blk. h., 1842.	Tomboy, b. h., 1829.	Jerry 1821 (Smolensko 1810)—Dau. 1817 of Ardrossan 1809 (John Bull 1789)—Lady Eliza 1813 by Whitworth 1805 (Agonistes 1797)—Dau. 1793 of Spadille 1784 (Highflyer 1774)—Sylvia 1783 by Y. Marske 1771 (Marske 1750).
				Lady Moore Carew, b. m., 1830.	Tramp 1810 (Dick Andrews 1797)—Kite 1821 by Bustard 1813 (Castrel 1801)—Olympia 1815 by Sir Oliver 1800 (Sir Peter T. 1784)—Scotilla 1795 by Anvil 1777 (Herod 1758)—Scota 1783 by Eclipse 1764 (Marske 1750).
			Ennui, br. m., 1843.	Bay Middleton, b. h., 1833.	Sultan 1816 (Selim 1802)—Cobweb 1821 by Phantom 1808 (Walton 1799)—Filagree 1815 by Soothsayer 1808 (Sorcerer 1796)—Web 1808 by Waxy 1790 (Pot-8-o's 1773)—Penelope 1798 by Trumpator 1782 (Conductor 1767).
				Blue Devils, ch. m., 1837.	Velocipede 1825 (Blacklock 1814)—Care 1825 by Woful 1809 (Waxy 1790)—Dau. 1819 of Rubens 1805 (Buzzard* 1787)—Tippitywitchet 1808 by Waxy 1790 (Pot-8-o's 1773)—Hare 1794 by Sweetbriar 1769 (Syphon 1756).
THE PEARL, b. m., 1863.	Newminster, b. h., 1848.	Touchstone, br. h., 1831.	Camel, br. h., 1822.	Whalebone, br. h., 1807.	Waxy 1790 (Pot-8-o's 1773)—Penelope 1798 by Trumpator 1782 (Conductor 1767)—Prunella 1788 by Highflyer 1774 (Herod 1758)—Promise 1768 by Snap 1750 (Snip 1736)—Julia 1756 by Blank 1740 (God. Ar. 1724).
				Daughter of, b. m., 1812.	Selim 1802 (Buzzard* 1787)—Maiden 1801 by Sir Peter T. 1784 (Highflyer 1774)—Dau. 1788 of Phenomenon 1780 (Herod 1758)—Matron 1782 by Florizel 1768 (Herod 1758)—Maiden 1770 by Matchem 1748 (Cade 1734).
			Banter, br. m., 1826.	Master Henry, b. h., 1815.	Orville 1799 (Beningbrough 1791)—Miss Sophia 1805 by Stamford 1794 (Sir Peter T. 1784)—Sophia 1798 by Buzzard* 1787 (Woodpecker 1773)—Huncamunca 1787 by Highflyer 1774 (Herod 1758)—Cypher 1772 by Squirrel 1754 (Traveller 1735).
				Boadicea, b. m., 1807.	Alexander 1782 (Eclipse 1764)—Brunette by Amaranthus 1766 (Old England 1741)—Mayfly 1771 by Matchem 1748 (Cade 1734)—Dau. 1755 of Starling 1738 (Starling 1727)—Look-at-me-Lads 1731 by Grasshopper (Byerly Turk).
		Bee's-wing, b. m., 1833.	Dr. Syntax, br. h., 1811.	Paynator, br. h., 1791.	Trumpator 1782 (Conductor 1767)—Dau. of Mark Anthony 1767 (Spectator 1749)—Signora 1767 by Snap 1750 (Snip 1736)—Miss Windsor 1754 by God. Ar. 1724—Sist. to Volunteer by Y. Belgrade (Belgrade Turk).
				Daughter of, —,,—.	Beningbrough 1791 (King Fergus 1775)—Jenny Mole 1787 by Carbuncle 1772 (Babr. Blank 1758)—Dau. 1773 of Prince T'Quassaw 1751 (Snip 1736)—Sultana 1759 by Regulus 1739 (God. Ar. 1724)—Dau. of Partner 1718 (Jigg).
			Tomboy's dam, ch. m., 1817.	Ardrossan, b. h., 1809.	John Bull 1789 (Fortitude 1777)—Miss Whip 1793 by Volunteer 1780 (Eclipse 1764)—Wimbleton 1785 by Evergreen 1769 (Herod 1758)—Dau. 1774 of Herod 1758 (Tartar 1743)—Teresa 1767 by Matchem 1748 (Cade 1734).
				Lady Eliza, b. m., 1813.	Whitworth 1805 (Agonistes 1797)—Dau. 1793 of Spadille 1784 (Highflyer 1774)—Sylvia 1783 by Y. Marske 1771 (Marske 1750)—Ferret 1765 by Broth. to Silvio (Cade 1734)—Dau. of Regulus 1739 (God. Ar. 1724).
	Caller Ou, br. m., 1858.	Stockwell, ch. h., 1849.	The Baron, ch. h., 1842.	Ir. Birdcatcher, ch. h., 1833.	Sir Hercules 1826 (Whalebone 1807)—Guiccioli 1823 by Bob Booty 1804 (Chanticleer 1787)—Flight 1809 by Ir. Escape 1802 (Commodore 1793)—Y. Heroine by Bagot 1780 (Herod 1758)—Heroine 1775 by Hero 1753 (Cade 1734).
				Echidna, b. m., 1838.	Economist 1825 (Whisker 1812)—Miss Pratt 1825 by Blacklock 1814 (Whitelock 1803)—Gadabout 1812 by Orville 1799 (Beningbrough 1791)—Minstrel 1803 by Sir Peter T. 1784 (Highflyer 1774)—Matron 1782 by Florizel 1768 (Herod 1758).
			Pocahontas, b. m., 1837.	Glencoe,* ch. h., 1831.	Sultan 1816 (Selim 1802)—Trampoline 1825 by Tramp 1810 (Dick Andrews 1797)—Web 1808 by Waxy 1790 (Pot-8-o's 1773)—Penelope 1798 by Trumpator 1782 (Conductor 1767)—Prunella 1788 by Highflyer 1774 (Herod 1758).
				Marpessa, b. m., 1830.	Muley 1810 (Orville 1799)—Clare 1824 by Marmion 1806 (Whiskey 1789)—Harpalice 1814 by Gohanna 1790 (Mercury 1778)—Amazon 1799 by Driver 1783 (Trentham 1766)—Fractions 1792 by Mercury 1778 (Eclipse 1764).
		Haricot, br. m., 1847.	Lanercost, br. h., 1835.	Liverpool, b. h., 1828.	Tramp 1810 (Dick Andrews 1797)—Dau. 1822 of Whisker 1812 (Waxy 1790)—Mandane 1800 by Pot-8-o's 1773 (Eclipse 1764)—Y. Camilla 1787 by Woodpecker 1773 (Herod 1758)—Camilla 1778 by Trentham 1766 (T. Sweepstakes 1743).
				Otis, b. m., 1820.	Bustard 1801 (Buzzard* 1787)—Dau. 1813 of Election 1804 (Gohanna 1790)—Dau. 1791 of Highflyer 1774 (Herod 1758)—Dau. 1778 of Eclipse 1764 (Marske 1750)—Rosebud 1765 by Snap 1750 (Snip 1736).
			Queen Mary, [Mango or] b. m., 1843.	Gladiator, ch. h., 1833.	Partisan 1811 (Walton 1799)—Pauline 1826 by Moses 1819 ([Whalebone or] Seymour 1807)—Quadrille 1815 by Selim 1802 (Buzzard* 1787)—Canary Bird 1806 by [Whiskey or] Sorcerer 1796 (Trumpator 1782)—Canary 1797 by Coriander 1786 (Pot-8-o's 1773).
				Daughter of, b. m., 1840.	Plenipotentiary 1831 (Emilius 1820)—Myrrha b. m. 1830 by Whalebone 1807 (Waxy 1790)—Gift gr. m. 1818 by Y. Gohanna 1810 (Gohanna 1790)—Sist. to Grazier b. m. 1802 by Sir Peter T. 1784 (Highflyer 1774)—Sist. to Aimator b. m. 1788 by Trumpator 1782 (Conductor 1767)—Sist. b. m. 1774 to Postmaster by Herod 1758 (Tartar 1743)—Dau. of Snap 1750 (Snip 1736)—Dau. of Gower Stallion 1740 (God. Ar. 1724)—Dau. of Childers.

IMPORTED SILVERMINE.

The property of R. E. Roberts, Georgetown, Ky.

SILVERMINE, brown horse, foaled 1881, bred by Mr. W. Smith, by Silvio (winner of the Derby and St. Leger, Ascot Derby and other good races), son of Blair Athol by Stockwell. Blair Athol also won the Derby and St. Leger, and was the sire of imp. Prince Charlie (winner of the 2,000 Guineas) and other good horses. Silvermine's dam, Nuneaton by Orlando (winner of the Derby), son of Touchstone, and the sire of the dams of Camballo (winner of the 2,000 Guineas), Petrarch (winner of the 2,000 Guineas and St. Leger), Gang Forward (winner of the 2,000 Guineas), Gen. Peel (winner of the 2,000 Guineas), etc., grandam, Nun Appleton by Bay Middleton (winner of the 2,000 Guineas and Derby), son of Sultan, sire of imp. Glencoe. Bay Middleton is the sire of The Flying Dutchman (winner of both the Derby and St. Leger in 1849), sire also of the dam of Wild Dayrell (winner of the Derby) and of the dam of Sunbeam (winner of the St. Leger). His dam, Nuneaton, is own sister to Julia, dam of Julius by St. Albans, son of Stockwell. Silvermine is almost brother in blood to Julius, who won the Cesarewitch Handicap, Newmarket Biennial Stakes, Beaufort Cup, and defeated Hermit in a match. Julius was an excellent stallion, sire of many good winners, and full brother to Julius Cæsar (winner of the Westminster Stakes, second to Petrarch in the 2,000 Guineas, third to Kisber in the Derby, and third to Petrarch in the St. Leger, beating Kisber, won the City and Suburban Handicap and Royal Hunt Cup, and was a successful stallion). Silvermine, therefore, is not only a well-bred horse, but is from a racing and producing family of great merit.

Silvermine was regarded as a good race-horse. Started in three races at three years old, but won none. With a very limited chance in the stud, he is the sire of the winners Silver Charm, Silver Lake, Sunshine Whiskey, Mine Heart, Silvermint, etc.

He traces in direct female line to Daffodil, to which traces Vauxhall Snap, Mambrino (sire of imp. Messenger). Otterington, Grog, Lambinos, Ansel, Virgil and all the descendants of imp. Trumpetta in America.

Whalebone 1807 Waxy 1790—Pot-8-o-'s 1773 Eclipse 1764—Marske 1750 Squirt 1732—Bartlett's Childers Darley Ar. 1699.

SILVERMINE.

BROWN HORSE, FOALED 1881; BRED BY MR. W. SMITH, ENGLAND.

- SILVIO, b. h., 1874.
  - Blair Athol, ch. h., 1861.
    - Stockwell, ch. h., 1849.
      - The Baron, ch. h., 1842.
      - Pocahontas, b. m., 1837.
    - Blink Bonny, b. m., 1854.
      - Melbourne, br. h., 1834.
      - Queen Mary, b. m., 1843.
  - Silverhair, b. m., 1858.
    - Kingston, b. h., 1849.
      - Venison, br. h., 1833.
      - Queen Anne, b. m., 1833.
    - England's Beauty, ch. m., 1840.
      - Ir. Birdcatcher, ch. h., 1833.
      - The Prairie Bird, b. m., 1844.
- NUNEATON, b. m., 1857.
  - Orlando, b. h., 1841.
    - Touchstone, br. h., 1831.
      - Camel, br. h., 1822.
      - Banter, br. m., 1826.
    - Vulture, ch. m., 1833.
      - Langar, ch. h., 1817.
      - Kite, b. m., 1821.
  - Nun Appleton, b. m., 1845.
    - Bay Middleton, b. h., 1833.
      - Sultan, b. h., 1816.
      - Cobweb, b. m., 1821.
    - Miss Milner, br. m., 1832.
      - Malek, ch. h., 1824.
      - Daughter of, b. m., —.

Ir Birdcatcher ch. h., 1833.	Sir Hercules 1826 (Whalebone 1807) Guiccioli 1823 by Bob Booty 1804 (Chanticleer 1787)—Flight 1809 by Ir. Escape 1802 (Commodore 1793) Y. Heroine by Bagot 1780 (Herod 1758)—Heroine 1775 by Hero 1753 (Cade 1734).
Echidna, b. m., 1838.	Economist 1825 (Whisker 1812) Miss Pratt 1825 by Blacklock 1814 (Whitelock 1803) Gadabout 1812 by Orville 1799 (Beningbrough 1791) Minstrel 1803 by Sir Peter T. 1784 (Highflyer 1774) Matron 1782 by Florizel 1768 (Herod 1758).
Glencoe,* ch. h., 1831.	Sultan 1816 (Selim 1802) Trampoline 1825 by Tramp 1810 (Dick Andrews 1797)—Web 1808 by Waxy 1790 (Pot-8-o-'s 1773) Penelope 1798 by Trumpator 1782 Conductor 1767) Prunella 1788 by Highflyer 1774 (Herod 1758).
Marpessa, b. m., 1830.	Muley 1810 (Orville 1799)—Clare 1824 by Marmion 1806 (Whiskey 1789) Harpalice 1814 by Gohanna 1790 (Mercury 1778) Amazon 1799 by Driver 1783 (Trentham 1766) Fractious 1792 by Mercury 1778 (Eclipse 1764).
Humphrey Clinker, b. h., 1822.	Comus 1809 (Sorcerer 1796) Clinkerina 1812 by Clinker 1805 (Sir Peter T. 1784) Pewet 1786 by Tandem 1773 (Syphon 1750) Termagant by Tantrum 1760 (Cripple 1750)—Cantatrice by Sampson 1745 (Blaze 1733).
Daughter of, b. m., 1825.	Cervantes 1806 (Don Quixote 1784) Dau. 1818 of Golumpus 1802 (Gohanna 1790) Dau. 1810 of Paynator 1791 (Trumpator 1782)—Circle 1802 by St. George 1789 (Highflyer 1774) Abigail 1788 by Woodpecker 1773 (Herod 1758).
Gladiator, ch. h., 1833.	Partisan 1811 (Walton 1799) Pauline 1826 by Moses 1819 ([Whalebone or] Seymour 1807) Quadrille 1815 by Selim 1802 (Buzzard* 1787)—Canary Bird 1806 by [Whiskey or] Sorcerer 1796 (Trumpator 1782)—Canary 1797 by Coriander 1786 (Pot-8-o's 1773).
Daughter of, b. m., 1840.	Plenipotentiary 1831 (Emilius 1820) Myrrha 1830 by Whalebone 1807 (Waxy 1790) Gift 1818 by Y. Gohanna 1810 (Gohanna 1790)—Dau. 1802 of Sir Peter T. 1784 (Highflyer 1774)—Dau. 1788 of Trumpator 1782 (Conductor 1767).
Partisan, b. h., 1811.	Walton 1799 (Sir Peter T. 1784) Parasol 1800 by Pot-8-o-'s 1773 (Eclipse 1764) Prunella 1788 by Highflyer 1774 (Herod 1758) Promise 1768 by Snap 1750 (Snip 1736) Julia 1756 by Blank 1740 (God. Ar. 1724).
Fawn, br. m., 1823.	Smolensko 1810 (Sorcerer 1796) Jerboa 1803 by Gohanna 1790 (Mercury 1778) Camilla 1778 by Trentham 1766 (T. Sweepstakes 1743)—Coquette 1764 by Compton Barb—Sist. 1743 to Regulus by God. Ar. 1724.
Slane, b. h., 1833.	Royal Oak 1823 (Catton 1809) Dau. 1819 of Orville 1799 (Beningbrough 1791) Epsom Lass 1803 by Sir Peter T. 1784 (Highflyer 1774) Alexina 1788 by King Fergus 1775 (Eclipse 1764) Lardella 1780 by Y. Marske 1771 (Marske 1750).
Garcia, br. m., 1823.	Octavian 1807 (Stripling 1795) Dau. 1806 of Shuttle 1793 (Y. Marske 1771) Katharine 1798 by Delpini 1781 (Highflyer 1774) Dau. 1786 of Paymaster 1766 (Blank 1740)—Dau. 1774 of Le Sang 1759 (Changeling 1747).
Sir Hercules, blk. h., 1826.	Whalebone 1807 (Waxy 1790) Peri 1822 by Wanderer 1811 (Gohanna 1790) Thalestris 1809 by Alexander 1782 (Eclipse 1764)—Rival 1800 by Sir Peter T. 1784 (Highflyer 1774)—Hornet 1790 by Drone 1777 (Herod 1758).
Guiccioli, ch. m., 1823.	Bob Booty 1804 (Chanticleer 1787)—Flight 1809 by Ir. Escape 1802 (Commodore 1793)—Y. Heroine by Bagot 1780 (Herod 1758) Heroine 1775 by Hero 1753 (Cade 1734)—Dau. 1766 of Snap 1750 (Snip 1736).
Touchstone, br. h., 1831.	Camel 1822 (Whalebone 1807) Banter 1826 by Master Henry 1815 (Orville 1799) Boadicea 1807 by Alexander 1782 (Eclipse 1764)—Brunette by Amaranthus 1766 (Old England 1741)—Mayfly 1771 by Matchem 1748 (Cade 1734).
Zillah, ch. m., 1835.	Reveller 1815 (Comus 1809) Morisca 1826 by Morisco 1819 (Muley 1810)—Waltz 1822 by Election 1804 (Gohanna 1790) Penelope 1798 by Trumpator 1782 (Conductor 1767)—Prunella 1788 by Highflyer 1774 (Herod 1758).
Whalebone, br. h., 1807.	Waxy 1790 (Pot-8-o-'s 1773)—Penelope 1798 by Trumpator 1782 (Conductor 1767)—Prunella 1788 by Highflyer 1774 (Herod 1758)—Promise 1768 by Snap 1750 (Snip 1736)—Julia 1756 by Blank 1740 (God. Ar. 1724).
Daughter of, b. m., 1812.	Selim 1802 (Buzzard* 1787)—Maiden 1801 by Sir Peter T. 1784 (Highflyer 1774)—Dau. 1788 of Phenomenon 1780 (Herod 1758)—Matron 1782 by Florizel 1768 (Herod 1758)—Maiden 1770 by Matchem 1748 (Cade 1734).
Master Henry, b. h., 1815.	Orville 1799 (Beningbrough 1791)—Miss Sophia 1805 by Stamford 1794 (Sir Peter T. 1784)—Sophia 1798 by Buzzard* 1787 (Woodpecker 1773)—Huncamunca 1787 by Highflyer 1774 (Herod 1758) Cypher 1772 by Squirrel 1754 (Traveller 1735).
Boadicea, b. m., 1807.	Alexander 1782 (Eclipse 1764) Brunette by Amaranthus 1766 (Old England 1741) Mayfly 1771 by Matchem 1748 (Cade 1734) Dau. 1755 of Anc. Starling 1738 (Starling 1727)—Look-at-me-Lads 1731 by Bristol Grasshopper (Byerly Turk).
Selim, ch. h., 1802.	Buzzard* 1787 (Woodpecker 1773) Dau. of Alexander 1782 (Eclipse 1764) Dau. of Highflyer 1774 (Herod 1758)—Dau. of Alfred 1770 (Matchem 1748)—Dau. 1770 of Engineer 1756 (Sampson 1745).
Daughter of, ch. m., 1808.	Walton 1799 (Sir Peter T. 1784) Y. Giantess 1790 by Diomed* 1777 (Florizel 1768)—Giantess 1769 by Matchem 1748 (Cade 1734)—Molly Long Legs 1753 by Babraham 1740 (God. Ar. 1724) Dau. of Cole's Foxhunter 1727 (Brisk 1711).
Bustard, b. h., 1813.	Castrel 1801 (Buzzard* 1787) Miss Hap 1806 by Shuttle 1793 (Y. Marske 1771)—Dau. 1798 of Sir Peter T. 1784 (Highflyer 1774)—Miss Hervey 1775 by Eclipse 1764 (Marske 1750)—Clio 1769 by Y. Cade 1747 (Cade 1734).
Olympia, b. m., 1815.	Sir Oliver 1800 (Sir Peter T. 1784)—Scotilla 1795 by Anvil 1777 (Herod 1758)—Scota 1783 by Eclipse 1764 (Marske 1750)—Harmony 1775 by Herod 1758 (Tartar 1743)—Rutilia 1769 by Blank 1740 (God. Ar. 1724).
Selim, ch. h., 1802.	Buzzard* 1787 (Woodpecker 1773) Dau. of Alexander 1782 (Eclipse 1764) Dau. of Highflyer 1774 (Herod 1758) Dau. of Alfred 1770 (Matchem 1748)—Dau. 1770 of Engineer 1756 (Sampson 1745).
Bacchante, br. m., 1809.	W's Ditto 1800 (Sir Peter T. 1784) Dau. 1791 of Mercury 1778 (Eclipse 1764)—Dau. 1776 of Herod 1758 (Tartar 1743)—Folly 1771 by Marske 1750 (Squirt 1732)—Vixen 1753 by Regulus 1739 (God. Ar. 1724).
Phantom, b. h., 1808.	Walton 1799 (Sir Peter T. 1784)—Julia 1799 by Whiskey 1789 (Saltram* 1780)—Y. Giantess 1790 by Diomed* 1777 (Florizel 1768)—Giantess 1769 by Matchem 1748 (Cade 1734)—Molly Long Legs 1753 by Babraham 1740 (God. Ar. 1724).
Filagree, ch. m., 1815.	Soothsayer 1808 (Sorcerer 1796)—Web 1808 by Waxy 1790 (Pot-8-o-'s 1773) Penelope 1798 by Trumpator 1782 (Conductor 1767)—Prunella 1788 by Highflyer 1774 (Herod 1758)—Promise 1768 by Snap 1750 (Snip 1736).
Blacklock, b. h., 1814.	Whitelock 1803 (Hambletonian 1792)—Dau. 1799 of Coriander 1786 (Pot-8-o-'s 1773)—Wildgoose 1792 by Highflyer 1774 (Herod 1758)—Coheiress 1786 by Pot-8-o-'s 1773 (Eclipse 1764)—Manilla 1777 by Goldfinder 1764 (Snap 1750).
Velocipede's dam, —, 1817.	Juniper 1805 (Whiskey 1789)—Dau. 1810 of Sorcerer 1796 (Trumpator 1782)—Virgin 1801 by Sir Peter T. 1784 (Highflyer 1774)—Dau. 1791 of Pot-8-o-'s 1773 (Eclipse 1764)—Editha 1781 by Herod 1758 (Tartar 1743).
Whisker, b. h., 1812.	Waxy 1790 (Pot-8-o-'s 1773)—Penelope 1798 by Trumpator 1782 (Conductor 1767)—Prunella 1788 by Highflyer 1774 (Herod 1758)—Promise 1768 by Snap 1750 (Snip 1736)—Julia 1756 by Blank 1740 (God. Ar. 1724).
Daughter of, b. m., 1812.	Orville 1799 (Beningbrough 1791)—Otterington's dam b. m. 1805 by Expectation 1796 (Sir Peter T. 1784)—Calabria b. m. 1795 by Spadille 1784 (Highflyer 1774)—Grog's dam 1781 by Alfred 1770 (Matchem 1748)—Dau. of Locust 1744 (Crab 1722) Dau. b. m. 1758 of Changeling 1747 (Cade 1734)—Mambrino's dam 1751 by Cade 1734 (God. Ar. 1724)—Dau. of Bolton Little John 1731 (Partner 1718)—Durham's Favourite gr. m. 1728 by Son of Bald Galloway—Daffodil's dam by Sir T. Gascoigne's Foreign Horse.

IMPORTED SIMON MAGUS.

Is private stallion in the Rancocas Stud, property of Mr. P. Lorillard, Jobstown, N. J.

SIMON MAGUS, brown horse, foaled 1888, bred by the Duke of Portland, by St. Simon, son of Galopin by Vedette, dam Wheel of Fortune, dam of Oberon, Dodona, etc., by Adventurer, son of Newminster by Touchstone, grandam Queen Bertha, dam of Queen's Messenger, Spinaway, etc., by Kingston, son of Venison, out of Flax by Surplice, son of Touchstone, etc. St. Simon did not start in the Derby and St. Leger, in consequence of the death of his breeder and nominator, Prince Batthyany, but was never beaten during his turf career. Won the Epsom Gold Cup, Ascot Cup, Gold Cup at Newcastle, Goodwood Cup, and other good races. St. Simon is the sire of Semolina (winner of the 1,000 Guineas), Memoir (winner of the Oaks, St. Leger and Newmarket Stakes), Signorina (winner of the Select Stakes), St. Serf (winner of the Epsom Grand Prize, Rous Memorial Stakes, Sussex Stakes and other good races), La Flèche (winner of the 1,000 Guineas, Oaks, St. Leger and second in the Derby in 1892).

Simon Magus is brother to Oberon (winner of the Lincoln Handicap, etc.), and half brother to Dodona, a stake winner. These were the only living produce of Wheel of Fortune up to 1889. Wheel of Fortune was a good mare. At two years old won the Richmond Stakes at Goodwood, the Prince of Wales Stakes at York, walked over for the Wentworth Stakes at Doncaster, won the Buckenham Stakes at Newmarket, won the Triennal Produce Stakes, Newmarket, and the Middle Park Plate. At three years old Wheel of Fortune won the 1,000 Guineas, the Oaks, the Prince of Wales Stakes at Ascot and the Yorkshire Oaks. Queen Bertha, Simon Magus's grandam, was a first-class race-mare. At two years old she beat Blue Mantle and Falcon and was never unplaced. At three years old won the Oaks, ran second in the Ascot Derby Stakes, won the Ascot Triennial Stakes, ran second to Lord Clifden in the St. Leger, and second to Macaroni for the Doncaster Cup. In the stud Queen Bertha did well. She produced Gertrude, a good performer and winner of races, including the Yorkshire Oaks. She produced Charibert (winner of the 2,000 Guineas), Childeric (a good race-horse, second in the St. Leger, etc.), and the winner King Clovis. Her full sister, Blanchfleur, produced the two good winners Syringe and Choufleur. Spinaway, daughter of Queen Bertha, won the 1,000 Guineas, Oaks, etc., produced Busybody (winner of the 1,000 Guineas, the Oaks, etc.), Spinning Jenny (dam of Spinaway Top), and the winners Merry-go-round, Darnaway and Arcadian. Queen Bertha produced the race-horse Queen's Messenger and the winners Paladin and Great Carle. Flax, the great grandam (by Surplice, winner of the Derby, St. Leger, etc.), produced the winners Reginella, Hollyfox, Honiton (a good sire), Courtmantle, Linsey Woolsey, Traveller's Joy and Folengo. Of these Reginella produced Guy Dayrell and Cobham, and the winner Queen Gladys. Her daughter Zingarella produced the winners Simon and Donald Caird.

Simon Magus ran in three races and paid forfeit in a match to Imp. Masetto. He won no race.

He is descended in direct female line from the Tregonwell Natural Barb mare, to which traces Woodpecker, Whalebone, Whisker, Partisan, Rockingham, imp. Glencoe, Bay Middleton, Trumpeter, Melbourne, Bran, Bend'or, Robert the Devil, Speculum, Charibert, Silvio, imp. Mortemer, The Lambkin, Paradox, The Bard (by Petrarch), imp. Eclipse, in England: Uncas, Wanderer, and all the descendants of old Prunella.

Whitelock 1803-Hambletonian 1792-King Fergus 1775-Eclipse 1764-Marske 1750-Squirt 1732-Bart. Childers-Darley Ar. 1699.

SIMON MAGUS.

BROWN HORSE, FOALED 1888; BRED BY LORD FALMOUTH, ENGLAND.

ST. SIMON, b. h., 1881.	Galopin, b. h., 1872.	Vedette, br. h., 1854.	Voltigeur, br. h., 1847.	Voltaire, br. h., 1826.	Blacklock 1814 (Whitelock 1803)—Dau. 1816 of Phantom 1808 (Walton 1799)—Dau. 1802 of Overton 1788 (King Fergus 1775)—Dau. 1796 of Walnut 1786 (Highflyer 1774)—Dau. 1786 of Ruler 1777 (Y. Marske 1771).
				Martha Lynn, br. m., 1837.	Mulatto 1824 (Catton 1809)—Leda 1824 by Filho-da-Puta 1812 (Haphazard 1797)—Treasure 1809 by Camillus 1803 (Hambletonian 1792)—Dau. 1801 of Hyacinthus 1797 (Coriander 1786)—Flora 1789 by King Fergus 1775 (Eclipse 1764).
			Mrs. Ridgway, br. m., 1849.	Ir. Birdcatcher ch. h., 1833.	Sir Hercules 1826 (Whalebone 1807)—Guiccioli 1823 by Bob Booty 1804 (Chanticleer 1787)—Flight 1809 by Ir. Escape 1802 (Commodore 1793)—Y. Heroine by Bagot 1780 (Herod 1758)—Heroine 1775 by Hero 1757 (Cade 1734).
				Nan Darell, gr. m., 1844.	Inheritor 1831 (Lottery 1820)—Nell 1831 by Blacklock 1814 (Whitelock 1803)—Madame Vestris 1818 by Comus 1809 (Sorcerer 1796)—Lisette 1803 by Hambletonian 1792 (King Fergus 1775)—Constantia 1796 by Walnut 1786 (Highflyer 1774).
		Flying Duchess, b. m., 1853.	The Flying Dutchman, br. h., 1846.	Bay Middleton, b. h., 1833.	Sultan 1816 (Selim 1802)—Cobweb 1821 by Phantom 1808 (Walton 1799)—Filagree 1815 by Soothsayer 1808 (Sorcerer 1796)—Web 1808 by Waxy 1790 (Pot-8-o's 1773)—Penelope 1798 by Trumpator 1782 (Conductor 1767).
				Barbelle, b. m., 1836.	Sandbeck 1818 (Catton 1809)—Darioletta 1822 by Amadis 1807 (Don Quixote 1784)—Selima 1810 by Selim 1802 (Buzzard* 1787)—Dau. 1791 of Pot-8-o's 1773 (Eclipse 1764)—Editha 1781 by Herod 1758 (Tartar 1743).
			Merope, b. m., 1841.	Voltaire, br. h., 1826.	Blacklock 1814 (Whitelock 1803)—Dau. 1816 of Phantom 1808 (Walton 1799)—Dau. 1802 of Overton 1788 (King Fergus 1775)—Dau. 1796 of Walnut 1786 (Highflyer 1774)—Dau. 1786 of Ruler 1777 (Y. Marske 1771).
				Velocipede's dam, .. 1817.	Juniper 1805 (Whiskey 1789)—Dau. 1810 of Sorcerer 1796 (Trumpator 1782)—Virgin 1801 by Sir Peter T. 1784 (Highflyer 1774)—Dau. 1791 of Pot-8-o's 1773 (Eclipse 1764)—Editha 1781 by Herod 1758 (Tartar 1743).
	St. Angela, b. m., 1865.	King Tom, b. h., 1851.	Harkaway, ch. h., 1834.	Economist, b. h., 1825.	Whisker 1812 (Waxy 1790)—Floranthe 1818 by Octavian 1807 (Stripling 1795)—Caprice 1797 by Anvil 1777 (Herod 1758)—Madcap 1774 by Eclipse 1764 (Marske 1750)—Sappho 1763 by Blank 1740 (God. Ar. 1724).
				Fanny Dawson, ch. m., 1823.	Nabocklish 1811 (Rugantino 1803)—Miss Tooley 1808 by Teddy-the-Grinder 1798 (Asparagus 1787)—Lady Jane 1796 by Highflyer 1774 (Herod 1758)—Paulina 1778 by Florizel 1768 (Herod 1758)—Captive 1771 by Matchem 1748 (Cade 1734).
			Pocahontas, b. m., 1837.	Glencoe,* ch. h., 1831.	Sultan 1816 (Selim 1802)—Trampoline 1825 by Tramp 1810 (Dick Andrews 1797)—Web 1808 by Waxy 1790 (Pot-8-o's 1773)—Penelope 1798 by Trumpator 1782 (Conductor 1767)—Prunella 1788 by Highflyer 1774 (Herod 1758).
				Marpessa, b. m., 1830.	Muley 1810 (Orville 1799)—Clare 1824 by Marmion 1806 (Whiskey 1789)—Harpalice 1814 by Gohanna 1790 (Mercury 1778)—Amazon 1799 by Driver 1783 (Trentham 1766)—Fractious 1792 by Mercury 1778 (Eclipse 1764).
		Adeline, b. m., 1851.	Ion, br. h., 1835.	Cain, b. h., 1822.	Paulowitz 1813 (Sir Paul 1802)—Dau. 1810 of Paynator 1791 (Trumpator 1782)—Dau. 1805 of Delpini 1781 (Highflyer 1774)—Dau. 1795 of Y. Marske 1771 (Marske 1750)—Gentle Kitty 1774 by Silvio 1754 (Cade 1734).
				Margaret, br. m., 1831.	Edmund 1824 (Orville 1799)—Medora 1811 by Selim 1802 (Buzzard* 1787)—Dau. 1803 of Sir Harry* 1795 (Sir Peter T. 1784)—Dau. 1793 of Volunteer 1780 (Eclipse 1764)—Dau. 1779 of Herod 1758 (Tartar 1743).
			Little Fairy, ch. m., 1841.	Hornsea, ch. h., 1832.	Velocipede 1825 (Blacklock 1814)—Dau. 1820 of Cerberus 1802 (Gohanna 1790)—Miss Cranfield 1803 by Sir Peter T. 1784 (Highflyer 1774)—Dau. 1796 of Pegasus 1784 (Eclipse 1764)—Dau. 1783 of Paymaster 1766 (Blank 1740).
				Lacerta, b. m., 1816.	Zodiac 1801 (St. George 1789)—Jerboa 1803 by Gohanna 1790 (Mercury 1778)—Camilla 1778 by Trentham 1766 (T. Sweepstakes 1743)—Coquette 1764 by the Compton Barb—Sist. 1743 to Regulus by God. Ar. 1724.
WHEEL OF FORTUNE, b. m., 1876.	Adventurer, b. h., 1859.	Newminster, b. h., 1848.	Touchstone, br. h., 1831.	Camel, br. h., 1822.	Whalebone 1807 (Waxy 1790)—Dau. 1812 of Selim 1802 (Buzzard* 1787)—Maiden 1801 by Sir Peter T. 1784 (Highflyer 1774)—Dau. 1788 of Phenomenon 1780 (Herod 1758)—Matron 1782 by Florizel 1768 (Herod 1758).
				Banter, br. m., 1826.	Master Henry 1815 (Orville 1799)—Boadicea 1807 by Alexander 1782 (Eclipse 1764)—Brunette by Amaranthus 1766 (Old England 1741)—Mayfly 1771 by Matchem 1748 (Cade 1734)—Dau. 1755 of Anc. Starling 1738 (Starling 1727).
			Bee's-wing, b. m., 1833.	Dr. Syntax, br. h., 1811.	Paynator 1791 (Trumpator 1782)—Dau. of Beningbrough 1791 (King Fergus 1775)—Jenny Mole 1787 by Carbuncle 1772 (Babr. Blank 1758)—Dau. 1773 of Prince T'Quassaw 1751 (Snip 1736)—Sultana 1759 by Regulus 1739 (God. Ar. 1724).
				Daughter of, ch. m., 1817.	Ardrossan 1809 (John Bull 1789)—Lady Eliza 1813 by Whitworth 1805 (Agonistes 1797)—X. Y. Z.'s dam 1798 by Spadille 1784 (Highflyer 1774)—Sylvia 1783 by Y. Marske 1771 (Marske 1750)—Ferret 1765 by Broth. to Silvio (Cade 1734).
		Palma, br. m., 1840.	Emilius, b. h., 1820.	Orville, b. h., 1799.	Beningbrough 1791 (King Fergus 1775)—Evelina 1791 by Highflyer 1774 (Herod 1758)—Termagant by Tantrum 1760 (Cripple 1750)—Cantatrice by Sampson 1745 (Blaze 1733)—Dau. 1757 of Regulus 1739 (God. Ar. 1724).
				Emily, ch. m., 1810.	Stamford 1794 (Sir Peter T. 1784)—Dau. 1799 of Whiskey 1789 (Saltram* 1780)—Grey Dorimant 1781 by Dorimant 1772 (Otho 1760)—Dizzy 1757 by Blank 1740 (God. Ar. 1724)—Anc. Dizzy 1741 by Driver 1727 (Wynn Ar.).
			Francesca, br. m., 1829.	Partisan, b. h., 1811.	Walton 1799 (Sir Peter T. 1784)—Parasol 1800 by Pot-8-o's 1773 (Eclipse 1764)—Prunella 1788 by Highflyer 1774 (Herod 1758)—Promise 1768 by Snap 1750 (Snip 1736)—Julia 1756 by Blank 1740 (God. Ar. 1724).
				Daughter of, br. m., 1815.	Orville 1799 (Beningbrough 1791)—Golden Leg's dam 1800 by Buzzard* 1787 (Woodpecker 1773)—Hornpipe 1798 by Trumpator 1782 (Conductor 1767)—Luna 1779 by Herod 1758 (Tartar 1743)—Proserpine 1766 (Sist. to Eclipse) by Marske 1750 (Squirt 1732).
	Queen Bertha, b. m., 1860.	Kingston, b. h., 1849.	Venison, br. h., 1833.	Partisan, b. h., 1811.	Walton 1799 (Sir Peter T. 1784)—Parasol 1800 by Pot-8-o's 1773 (Eclipse 1764)—Prunella 1788 by Highflyer 1774 (Herod 1758)—Promise 1768 by Snap 1750 (Snip 1736)—Julia 1756 by Blank 1740 (God. Ar. 1724).
				Fawn, br. m., 1823.	Smolensko 1810 (Sorcerer 1796)—Jerboa 1803 by Gohanna 1790 (Mercury 1778)—Camilla 1778 by Trentham 1766 (T. Sweepstakes 1743)—Coquette 1764 by Compton Barb—Sist. 1743 to Regulus by God. Ar. 1724.
			Queen Anne, b. m., 1843.	Slane, b. h., 1833.	Royal Oak 1823 (Catton 1809)—Dau. 1819 of Orville 1799 (Beningbrough 1791)—Epsom Lass 1803 by Sir Peter T. 1784 (Highflyer 1774)—Alexina 1788 by King Fergus 1775 (Eclipse 1764)—Lardella 1780 by Y. Marske 1771 (Marske 1750).
				Garcia, br. m., 1823.	Octavian 1807 (Stripling 1795)—Dau. 1806 of Shuttle 1793 (Y. Marske 1771)—Katharine 1798 by Delpini 1781 (Highflyer 1774)—Dau. 1786 of Paymaster 1766 (Blank 1740)—Dau. 1774 of Le Sang 1759 (Changeling 1747).
		Flax, b. m., 1855.	Surplice, b. h., 1845.	Touchstone, br. h., 1831.	Camel 1822 (Whalebone 1807)—Banter 1826 by Master Henry 1815 (Orville 1799)—Boadicea 1807 by Alexander 1782 (Eclipse 1764)—Brunette by Amaranthus 1766 (Old England 1741)—Mayfly 1771 by Matchem 1748 (Cade 1734).
				Crucifix, b. m., 1837.	Priam* 1827 (Emilius 1820)—Octaviana 1815 by Octavian 1807 (Stripling 1795)—Dau. 1807 of Shuttle 1793 (Y. Marske 1771)—Zara 1801 by Delpini 1781 (Highflyer 1774)—Flora 1789 by King Fergus 1775 (Eclipse 1764).
			Odessa, ch. m., 1831.	Sultan, b. h., 1816.	Selim 1802 (Buzzard* 1787)—Bacchante 1809 by W's Ditto 1800 (Sir Peter T. 1784)—Sist. 1791 to Calomel by Mercury 1778 (Eclipse 1764)—Dau. 1776 of Herod 1758 (Tartar 1743)—Folly 1771 by Marske 1750 (Squirt 1732).
				Sister to Cobweb, ch. m., 1820.	Phantom 1808 (Walton 1799)—Filagree ch. m. 1815 by Soothsayer 1808 (Sorcerer 1796)—Web b. m. 1808 by Waxy 1790 (Pot-8-o's 1773)—Penelope b. m. 1798 by Trumpator 1782 (Conductor 1767)—Prunella b. m. 1788 by Highflyer 1774 (Herod 1758)—Promise br. m. 1768 by Snap 1750 (Snip 1736)—Julia b. m. 1756 by Blank 1740 (God. Ar. 1724)—

Spectator's dam b. m. 1735 by Partner 1718 (Jigg)—Bonny Lass b. m. 1723 by Bay Bolton 1705 (Grey Hautboy)—Dau. of Darley Ar. 1699—Dau. of Byerly Turk—Dau. of Taffolet Barb—Dau. of Place's White Turk—Natural Barb mare of Mr. Tregonwell's.

SIR DIXON.

WINNER OF CAMDEN, SELECT AND FLATBUSH STAKES AT TWO YEARS OLD, THE WITHERS, BELMONT, LORILLARD, TRAVERS, OMNIBUS AND OTHER STAKES AT THREE YEARS OLD.

In the Runnymede Stud, property of Messrs. Clay & Woodford, Paris, Ky. Yearling sales annually.

SIR DIXON, brother to Belvidere, bay horse, foaled 1885, bred in the Runnymede Stud, by imp. Billet, son of Voltigeur by Voltaire, dam Jaconet, own sister to Iroquois, by imp. Leamington, grandam Maggie B. B. by imp. Australian, great grandam Madeline by old Boston, out of Magnolia, dam of Kentucky, Daniel Boone, Gilroy, etc., by imp. Glencoe.

Sir Dixon is by imp. Billet, a good race-horse, winner of the Zetland Stakes at York, Engham Stakes at Engham, and other races. Billet's sire, Voltigeur, won both the Derby and St. Leger, and was by Voltaire, son of Blacklock. Voltigeur was also the sire of Vedette, sire of the noted Galopin, sire of the unbeaten St. Simon. Sir Dixon's dam, Jaconet, is full sister to Iroquois, one of the best race-horses, judged by his performances, ever bred in America, and the only horse, except Donovan, that ever won the Derby, Prince of Wales Stakes and the Doncaster St. Leger. Iroquois was second to Peregrine in the 2,000 Guineas and beat him in the Derby. Iroquois also won the Chesterfield Stakes at Newmarket, Levant Stakes at Goodwood, the Burwell Stakes, Newmarket Derby and other important events. In these races he beat such good ones as Peregrine, Ishmael, Scobel, Geologist, Tristan, Camilliard, Bal Gal, Lucy Glitters and others. Sir Dixon, therefore, is descended from one of the best bred and most potent racing families in America. His grandam, Maggie B. B., was the dam of Pera, Harold, Panique, Red and Blue (Sally McClelland's dam), etc. Harold had only one colt on the turf, Harefoot, an excellent race-horse, and Harold was own brother to Iroquois, a good index to the potency of the blood. Madeline, his great grandam, was by the great Boston, sire of Lexington, Lecompte, Bostona, Nina and a host of good ones. Magnolia was the dam of Kentucky, the best son of Lexington, Daniel Boone, Magic, Charley Ball, Madonna, Skedaddle, Gilroy, Victory, etc. Skedaddle is the dam of Saucebox, Scramble, Florence B., Sly Boots, Joe S., The Slashes and Squeez'em (the dam of Day Star, winner of the Kentucky Derby, Blue Ribbon Stakes and other fast races) and Sylph (the dam of La Sylphide, 1¼ miles, in 2.07½, and dam of Semper Fidelo). Kaloolah, Kincsem, etc. Sly Boots is the dam of Leveller, Sly Dance, Sachem and Savanac. Myrtle was by Mameluke (winner of the Derby). Bobadilla won the Ascot Gold Cup and Drawing Room Stakes.

Sir Dixon was a superior race-horse. At two years old won the Camden Stakes, three-quarters of a mile, 109 lbs., in 1.16, defeating a field of nine; won the Select Stakes, three-quarters of a mile, 111 lbs., in 1.17, and the Flatbush Stakes, 7 furlongs, 115 lbs., in 1.29, beating such good ones as Tea Tray, King Fish, Reclare, etc.; was placed third in the Junior Champion Stakes. At three years old won the Analostan Stakes, 1 mile, 118 lbs., in 1.46; the Carlton Stakes, 1 mile, 119 lbs., beating Raceland, in 1.56¼; ran second to Emperor of Norfolk in Brooklyn Derby, 1½ miles, in 2.08¾, beating Prince Royal and Raceland; ran second to The Bard in St. James Hotel Stakes, 1¼ miles, in 2.08½; won the Withers Stakes, 1 mile, 118 lbs., in 1.47¾, track muddy, Prince Royal second, Tea Tray third; won Belmont Stakes, 1½ miles, 118 lbs., in 2.40¼, beating Prince Royal; won the Lorillard Stakes, 1½ miles, 118 lbs., in 2.37½, beating Los Angeles, Prince Royal, etc.; won the Travers Stakes, 1¾ miles, 118 lbs., in 3.07¾, beating Los Angeles and two others; ran second to Taragon in the Omnibus Stakes, 1½ miles, in 2.41, giving him 11 lbs.; ran second to Geo. Oyster in Choice Stakes, 1½ miles, in 2.42, giving him 10 lbs., Taragon and two others behind him. Sir Dixon did not race at four years old. At five years old won the St. James Hotel Stakes, 1¼ miles, 126 lbs., defeating Taragon, etc. He was placed second twice and third five times. Sir Dixon was a good race-horse and is from a grand racing and producing family.

He is descended in direct female line from the Layton Barb mare, to which trace Young Greyhound, Cartouch, Matchem, Young Cade, Drone by Herod, Audubon, Phiammon, St. Honorat, Kentucky, Daniel Boone, Gilroy, Iroquois and all the Magnolia family, Powhattan, Wild Irishman, Frankfort, Prince George, Virgilian, Ferida and all the imp. Cub mare's descendents.

King Fergus 1775—Eclipse 1764—Marske 1750 Squirt 1732 Bartlett's Childers—Darley Arabian 1699.

SIR DIXON.

BAY HORSE, FOALED 1885; BRED IN THE RUNNYMEDE STUD.

*BILLET, b. h., 1865.	Voltigeur, br. h., 1847.	Voltaire, br. h., 1826.	Blacklock, b. h., 1814.	Whitelock, b. h., 1803.	Hambletonian 1792 (King Fergus 1775)—Rosalind 1788 by Phenomenon 1780 (Herod 1758)—Atalanta 1769 by Matchem 1748 (Cade 1734)—Y. Lass of the Mill 1753 by Oroonoko 1745 (Crab 1722)—Dau. 1744 of Traveler 1735 (Partner 1718).
				Daughter of, —, 1799.	Coriander 1786 (Pot-8-o-'s 1773)—Wildgoose 1792 by Highflyer 1774 (Herod 1758)—Coheiress 1786 by Pot-8-o-'s 1773 (Eclipse 1764)—Manilla 1777 by Goldfinder 1764 (Snap 1750)—Dau. 1764 of Old England 1741 (God. Ar. 1724).
			Daughter of, b. m., 1816.	Phantom, b. h., 1808.	Walton 1799 (Sir Peter T. 1784)—Julia 1790 by Whiskey 1789 (Saltram* 1780)—Y. Giantess 1790 by Diomed* 1777 (Florizel 1768)—Giantess 1769 by Matchem 1748 (Cade 1734)—Molly Long Legs 1753 by Babraham 1740 (God. Ar. 1724).
				Daughter of, b. m., 1802.	Overton 1788 (King Fergus 1775)—Dau. 1793 of Walnut 1786 (Highflyer 1774)—Dau. 1786 of Ruler 1777 (Y. Marske 1771)—Piracantha 1772 by Matchem 1748 (Cade 1734)—Prophetess 1758 by Regulus 1739 (God. Ar. 1724).
		Martha Lynn, br. m., 1837.	Mulatto, b. h., 1823.	Catton, b. h., 1809.	Golumpus 1802 (Gohanna 1790)—Lucy Gray 1804 by Timothy 1794 (Delpini 1781)—Lucy 1789 by Florizel 1768 (Herod 1758)—Frenzy 1774 by Eclipse 1764 (Marske 1750)—Dau. of Engineer 1756 (Sampson 1745).
				Desdemona, br. m., 1811.	Orville 1799 (Beningbrough 1791)—Fanny 1793 by Sir Peter T. 1784 (Highflyer 1774)—Dau. 1788 of Diomed* 1777 (Florizel 1768)—Desdemona 1777 by Marske 1750 (Squirt 1732)—Y. Hag 1761 by Skim 1746 (Starling 1727).
			Leda, b. m., 1824.	Filho-da-Puta, br. h., 1812.	Haphazard 1797 (Sir Peter T. 1784)—Mrs. Barnet 1801 by Waxy 1790 (Pot-8-o-'s 1773)—Dau. 1788 of Woodpecker 1773 (Herod 1758)—Heinel 1771 by Squirrel 1754 (Traveler 1735)—Principessa 1762 by Blank 1740 (God. Ar. 1724).
				Treasure, ch. m., 1809.	Camillus 1803 (Hambletonian 1792)—Dau. 1804 of Hyacinthus 1797 (Coriander 1786)—Flora 1789 by King Fergus 1775 (Eclipse 1764)—Atalanta 1769 by Matchem 1748 (Cade 1734)—Y. Lass of the Mill 1753 by Oroonoko 1745 (Crab 1722).
	Calcutta, b. m., 1858.	Flatcatcher, br. h., 1845.	Touchstone, br. h., 1831.	Camel, br. h., 1822.	Whalebone 1807 (Waxy 1790)—Dau. 1812 of Selim 1802 (Buzzard* 1787)—Maiden 1801 by Sir Peter T. 1784 (Highflyer 1774)—Dau. 1788 of Phenomenon 1780 (Herod 1758)—Matron 1782 by Florizel 1768 (Herod 1758).
				Banter, br. m., 1826.	Master Henry 1815 (Orville 1799)—Boadicea 1807 by Alexander 1782 (Eclipse 1764)—Brunette by Amaranthus 1766 (Old England 1741)—Mayfly 1771 by Matchem 1748 (Cade 1734)—Dau. 1755 of Anc. Starling 1738 (Starling 1727).
			Decoy, b. m., 1830.	Filho-da-Puta, br. h., 1812.	Haphazard 1797 (Sir Peter T. 1784)—Mrs. Barnet 1803 by Waxy 1790 (Pot-8-o-'s 1773)—Dau. 1788 of Woodpecker 1773 (Herod 1758)—Heinel 1771 by Squirrel 1754 (Traveler 1735)—Principessa 1762 by Blank 1740 (God. Ar. 1724).
				Finesse, b. m., 1815.	Peruvian 1806 (Sir Peter T. 1784)—Violante 1802 by John Bull 1789 (Fortitude 1777)—Dau. 1788 of Highflyer 1774 (Herod 1758)—Everlasting 1775 by Eclipse 1764 (Marske 1750)—Hyæna 1762 by Snap 1750 (Snip 1736).
		Miss Martin, br. m., 1844.	St. Martin, br. h., 1835.	Actæon, ch. h., 1822.	Scud 1804 (Beningbrough 1791)—Diana 1808 by Stamford 1794 (Sir Peter T. 1784)—Dau. 1799 of Whiskey 1789 (Saltram* 1780)—Grey Dorimant 1781 by Dorimant 1772 (Otho 1760)—Dizzy 1757 by Blank 1740 (God. Ar. 1724).
				Galena, br. m., 1823.	Walton 1799 (Sir Peter T. 1784)—Comedy 1817 by Comus 1809 (Sorcerer 1796)—Dau. 1803 of Star* 1785 (Highflyer 1774)—Dau. 1794 of Y. Marske 1771 (Marske 1750)—Emma 1781 by Telemachus 1770 (Herod 1758).
			Wagtail, br. m., 1829.	Whisker, b. h., 1812.	Waxy 1790 (Pot-8-o-'s 1773)—Penelope 1798 by Trumpator 1782 (Conductor 1767)—Prunella 1788 by Highflyer 1774 (Herod 1758)—Promise 1768 by Snap 1750 (Snip 1736)—Julia 1756 by Blank 1740 (God. Ar. 1724).
				Daughter of, br. m., 1814.	Sorcerer 1796 (Trumpator 1782)—Dau. 1808 of Sir Solomon 1793 (Sir Peter T. 1784)—Dau. 1794 of Y. Marske 1771 (Marske 1750)—Dau. 1788 of Phenomenon 1780 (Herod 1758)—Calliope 1763 by Slouch 1747 (Cade 1734).
JACONET (full sist. to Iroquois), ch. m., 1875.	*Leamington, br. h., 1853.	Faugh-a-Ballagh, br. h., 1841.	Sir Hercules, blk. h., 1826.	Whalebone, br. h., 1807.	Waxy 1790 (Pot-8-o-'s 1773)—Penelope 1798 by Trumpator 1782 (Conductor 1767)—Prunella 1788 by Highflyer 1774 (Herod 1758)—Promise 1768 by Snap 1750 (Snip 1736)—Julia 1756 by Blank 1740 (God. Ar. 1724).
				Peri, b. m., 1822.	Wanderer 1811 (Gohanna 1790)—Thalestris 1809 by Alexander 1782 (Eclipse 1764)—Rival 1800 by Sir Peter T. 1784 (Highflyer 1774)—Hornet 1790 by Drone 1777 (Herod 1758)—Manilla 1777 by Goldfinder 1764 (Snap 1750).
			Guiccioli, ch. m., 1823.	Bob Booty, ch. h., 1804.	Chanticleer 1787 (Woodpecker 1773)—Ierne 1790 by Bagot 1780 (Herod 1758)—Dau. of Gamahoe (Bustard 1741)—Patty by Tim 1752 (Squirt 1732)—Miss Patch 1737 by Justice 1725 (H. C. Litton Ar.).
				Flight, ch. m., 1809.	Ir. Escape 1802 (Commodore 1793)—Y. Heroine by Bagot 1780 (Herod 1758)—Heroine 1775 by Hero 1758 (Cade 1734)—Dau. 1766 of Snap 1750 (Snip 1736)—Sist. 1743 to Regulus by God. Ar. 1724.
		Daughter of, b. m., 1841.	Pantaloon, ch. h., 1824.	Castrel, ch. h., 1801.	Buzzard* 1787 (Woodpecker 1773)—Dau. of Alexander 1782 (Eclipse 1764)—Dau. of Highflyer 1774 (Herod 1758)—Dau. of Alfred 1770 (Matchem 1748)—Dau. 1770 of Engineer 1756 (Sampson 1745).
				Idalia, ch. m., 1815.	Peruvian 1806 (Sir Peter T. 1784)—Musidora 1804 by Meteor 1783 (Eclipse 1764)—Maid of all Work 1786 by Highflyer 1774 (Herod 1758)—Sist. 1771 to Tandem by Syphon 1750 (Squirt 1732)—Dau. of Regulus 1739 (God. Ar. 1724).
			Daphne, br. m., 1837.	Laurel, br. h., 1824.	Blacklock 1814 (Whitelock 1803)—Wagtail 1818 by Prime Minister 1810 (Sancho 1801)—Tranby*'s Dam 1812 by Orville 1799 (Beningbrough 1791)—Miss Grimstone 1796 by Weazle 1776 (Herod 1758)—Dau. of Ancaster 1768 (Blank 1740).
				Maid of Honor, br. m., 1829.	Champion 1812 (Selim 1802)—Etiquette 1820 by Orville 1799 (Beningbrough 1791)—Boadicea 1807 by Alexander 1782 (Eclipse 1764)—Brunette by Amaranthus 1766 (Old England 1741)—Mayfly 1771 by Matchem 1748 (Cade 1734).
	Maggie B. B., ch. m., 1867.	*Australian, ch. h., 1858.	West Australian, b. h., 1850.	Melbourne, br. h., 1834.	Humprey Clinker 1822 (Comus 1809)—Dau. 1825 of Cervantes 1806 (Don Quixote 1781)—Dau. 1818 of Golumpus 1802 (Gohanna 1790)—Dau. 1810 of Paynator 1791 (Trumpator 1782)—Circle 1802 by St. George 1789 (Highflyer 1774).
				Mowerina, b. m., 1843.	Touchstone 1831 (Camel 1822)—Emma 1824 by Whisker 1812 (Waxy 1790)—Gibside Fairy 1811 by Hermes 1790 (Mercury 1778)—Vicissitude 1800 by Pipator 1786 (Imperator 1776)—Beatrice 1791 by Sir Peter T. 1784 (Highflyer 1774).
			Emilia,* b. m., 1840.	Y. Emilius, b. h., 1831.	Emilius 1820 (Orville 1799)—Shoveler 1816 by Scud 1804 (Beningbrough 1791)—Goosander 1805 by Hambletonian 1792 (King Fergus 1775)—Rally 1790 by Trumpator 1782 (Conductor 1767)—Fancy 1780 (Sist. to *Diomed) by Florizel 1768 (Herod 1758).
				Persian, b. m., 1829.	Whisker 1812 (Waxy 1790)—Variety 1816 by [Selim or] Soothsayer 1808 (Sorcerer 1796)—Sprite 1807 by Bobtail 1795 (Precipitate* 1787)—Catherine 1795 by Woodpecker 1773 (Herod 1758)—Camilla 1778 by Trentham 1766 (T. Sweepstakes 1743).
		Madeline, ch. m., 1849.	Boston, ch. h., 1833.	Timoleon, ch. h., 1813.	Sir Archy 1805 (*Diomed 1777)—Dau. of *Saltram 1780 (Eclipse 1764)—Dau. of Symme's Wildair 1767 (*Fearnought 1755)—Dau. of Tyler's Driver 1760 (*Othello)—Dau. of *Fallower 1761 (Blank 1740).
				Sister to Tuckahoe, ch. m., 1814.	Ball's Florizel 1801 (*Diomed 1777)—Dau. 1799 of *Alderman 1787 (Pot-8-o-'s 1773)—Dau. of *Clockfast 1771 (Gimcrack 1760)—Dau. of Symme's Wildair 1767 (*Fearnought 1755)—Y. Kitty Fisher 1767 by *Fearnought 1755 (Regulus 1739).
			Magnolia, ch. m., 1841.	*Glencoe, ch. h., 1831.	Sultan 1816 (Selim 1802)—Trampoline 1825 by Tramp 1810 (Dick Andrews 1797)—Web 1808 by Waxy 1790 (Pot-8-o-'s 1773)—Penelope 1798 by Trumpator 1782 (Conductor 1767)—Prunella 1788 by Highflyer 1774 (Herod 1758).
				*Myrtle, ch. m., 1834.	Mameluke 1824 (Partisan 1811)—Bobadilla br. m. 1825 by Bobadil 1813 (Rubens 1805)—Pythoness blk. m. 1813 by Sorcerer 1796 (Trumpator 1782)—Princess blk. m. 1796 by Sir Peter T. 1784 (Highflyer 1774)—Dau. br. m. 1789 of Dungannon 1780 (Eclipse 1764)—Dau. ch. m. 1773 of Turf 1760 (Matchem 1748)—Dau. b. m. 1779 of Herod 1758 (Tartar 1743)—Golden Grove ch. m. 1760 by Blank 1740 (God. Ar. 1724)—Spinster (the Widdrington mare) ch. m. 1735 by Partner 1718 (Jigg)—Bay Bloody Buttocks b. m. 1729 by Bloody Buttocks—Sist. 1722 to Guy by Greyhound (Chillaby)—Brown Farewell br. m. 1710 by Makeless (Oglethorpe Ar.)—Dau. of Brimmer (D'Arcy Yellow Turk)—Dau. of Place's White Turk—Dau. of Dodsworth—Layton Barb mare.

IMPORTED SIR MODRED.

WINNER OF CANTERBURY DERBY, DUNEDIN AND CANTERBURY CUP. NEW ZEALAND.

Is private stallion in Rancho-del-Paso Stud, J. B. Haggin, proprietor, Sacramento, Cal. Yearlings sold annually.

SIR MODRED, bay horse, foaled 1877, bred by the Middle Park Stud Co., New Zealand, by Traducer, son of The Libel by Pantaloon, dam Idalia by Cambuscan, son of Newminster by Touchstone, grandam Dulcibella by Voltigeur, son of Voltaire. His sire, Traducer was bred in England, by The Libel (son of Pantaloon) and Arethusa. Traducer was the most popular and successful stallion that has ever stood in New Zealand. From 1867 to 1881 he got nine winners of the Canterbury Derby, the great event in that country, viz: 1867, Scandal; 1870, Envy; 1871, Defamation; 1872, Calumny; 1877, Trump Card; 1878, Natator; 1880, Sir Modred; 1881, The Dauphin; 1882, Cheviot. In three of these years his get ran first, second and third, and in two others first and second. He got Vanguard (winner of the New Zealand Cup), Welcome Jack (winner of the Canterbury Cup), and Lurline, a great mare, out of imp. Mermaid. Lurline is the dam of imp. Darebin, winner of the Victoria Derby of 1881, and the best horse of his day in Australia. Idalia, his dam, is full sister to Onslow (winner of the Prince of Wales Stakes at York, defeating Cremorne). She is half sister to Mabelle, dam of Beau Brummell (winner of the Woodcote and Hopeful Stakes), and The Child of the Mist (Mr. M. Daly's), winner of the Chesterfield and Ascot Biennial Stakes and the Ascot Derby. Cambuscan, her sire, is the sire of Camballo, winner of the 2,000 Guineas. Dulcibella, his grandam, won the Cesarewitch Handicap, beating Optimist, in 1860, and Brighton Cup. Her sire, Voltigeur, won both the Derby and St. Leger and was the sire of Vedette, sire of the noted Galopin. Priestess, his great grandam, was a superior race-mare, winner of the Northhampshire Cup, Caledonia St. Leger, Caledonia Cup, Gold Cup at Warwick, and several Queen's Plates, and by The Doctor, one of the gamest horses of his day. Priestess is half sister to The Gem, dam of Regalia (winner of the Oaks), and dam of Verneuil (winner of the Ascot Gold Cup) and Zut (winner of the French Derby, etc.). Idalia was a grand brood mare. She produced, in 1876, Betrayer; in 1877, Sir Modred; 1878, Idalium; 1879, Cheviot; 1880, July, all by Traducer; in 1881, Liverpool by King of Clubs; 1883, chestnut filly by Apremont, son of our imported Mortemer; 1884, a bay filly by Apremont. Betrayer won the Canterbury Champagne Stakes, Canterbury Cup, Wanganal Cup and Timaru Cup. Sir Modred won the Dunedin Champagne Stakes, Canterbury Champagne Stakes, Dunedin Cup, Canterbury Derby, Timaru Cup, Dunedin Birthday Handicap, Canterbury Cup, Christchurch Plate, and the Sydney Great Metropolitan. July ran second for the Dunedin Cup. Liverpool won the Welcome Stakes. Cheviot won the Canterbury Derby, Midsummer Handicap, and Craven Stakes, and only Sir Modred, his brother, could beat him for the Canterbury Cup.

Sir Modred is a success in the stud, sire of the winners Ballarat, Sir John, Marie Lovell, Fannie J., Tournament (winner of the Realization Stakes), Dr. Hasbrouck, Dr. Helmuth, Masterlode, Mary Stone, Gloaming Beverwyck, Discount, Enid, Sykeston, Gold Dollar, Grand Prix, Sir Matthew, Lizetta, Motto, Dr. Wilcox, Prince Charming, Sir Launcelot, Comanche, Sir Roy, Wheeler, Uncle Jess, Fairplay, Courtship, Gladiator, Shellbark, Reginald, Integrity and many others.

He is descended in direct female line through the Warlock Galloway from the Byerly Turk mare, to which traces Cub, Sancho, Young Blacklock, Pantaloon, Prime Warden, Yattendon (Australian), Cymbal, Beau Brummell, Child of the Mist, imp. Sovereign, imp. Citizen and other noted horses.

SIR MODRED.

BAY HORSE, FOALED 1877; BRED BY THE MIDDLE PARK STUD COMPANY, NEW ZEALAND.

†TRADUCER, br. h., 1857.	The Libel, br. h., 1842.	Pantaloon, ch. h., 1824.	Castrel, ch. h., 1801.	Buzzard,* ch. h., 1787.	Woodpecker 1773 (Herod 1758) Misfortune 1775 by Dux 1761 (Matchem 1748) Curiosity 1760 by Snap 1750 (Snip 1736) Dau. 1741 of Regulus 1739 (God. Ar. 1724) Dau. of Bartlett's Childers (Darley Ar. 1699).
				Daughter of, b. m., .	Alexander 1782 (Eclipse 1764)—Dau. of Highflyer 1774 (Herod 1758) Dau. of Alfred 1770 (Matchem 1748) Dau. 1770 of Engineer 1756 (Sampson 1745) Bay Malton's dam 1756 by Cade 1734 (God. Ar. 1724).
			Idalia, ch. m., 1815.	Peruvian, b. h., 1806.	Sir Peter T. 1784 (Highflyer 1774)—Dau. 1788 of Boudrow 1777 (Eclipse 1764) Escape's dam 1768 by Squirrel 1754 (Traveller 1735) Sist. to Lowther's Babraham by Babraham 1740 (God. Ar. 1724) Dau. 1750 of Golden Ball 1735 (Partner 1718).
				Musidora, b. m., 1804.	Meteor 1783 (Eclipse 1764) Maid-of-all-Work 1786 by Highflyer 1774 (Herod 1758) Sist. 1771 to Tandem by Syphon 1750 (Squirt 1732) Dau. of Regulus 1739 (God. Ar. 1724) Sist. 1758 to Judgement by Snip 1736 (Flying Childers 1715).
		Pasquinade, br. m., 1830.	Camel, br. h., 1822.	Whalebone, br. h., 1807.	Waxy 1790 (Pot-8-o-'s 1773)—Penelope 1798 by Trumpator 1782 (Conductor 1767) Prunella 1788 by Highflyer 1774 (Herod 1758) Promise 1768 by Snap 1750 (Snip 1736) Julia 1756 by Blank 1740 (God. Ar. 1724).
				Daughter of, b. m., 1812.	Selim 1802 (Buzzard* 1787)—Maiden 1801 by Sir Peter T. 1784 (Highflyer 1774) Dau. 1788 of Phenomenon 1780 (Herod 1758) Matron 1782 by Florizel 1768 (Herod 1758) Maiden 1770 by Matchem 1748 (Cade 1734).
			Banter, br. m., 1826.	Master Henry, b. h., 1815.	Orville 1799 (Beningbrough 1791)—Miss Sophia 1805 by Stamford 1794 (Sir Peter T. 1784) Sophia 1798 by Buzzard* 1787 (Woodpecker 1773) Huncamunca 1787 by Highflyer 1774 (Herod 1758) Cypher 1772 by Squirrel 1754 (Traveller 1735).
				Boadicea, b. m., 1807.	Alexander 1782 (Eclipse 1764) Brunette by Amaranthus 1766 (Old England 1741) Mayfly 1771 by Matchem 1748 (Cade 1734)—Dau. 1755 of Anc. Starling 1738 (Starling 1727) Look-at-me-Lads 1731 by Grasshopper (Byerly Turk).
	Arethusa, b. m., 1839.	Elis, ch. h., 1833.	Langar, ch. h., 1817.	Selim, ch. h., 1802.	Buzzard* 1787 (Woodpecker 1773)—Dau. of Alexander 1782 (Eclipse 1764) Dau. of Highflyer 1774 (Herod 1758)—Dau. of Alfred 1770 (Matchem 1748) Dau. 1770 of Engineer 1756 (Sampson 1745).
				Daughter of ch. m., 1808.	Walton 1799 (Sir Peter T. 1784)—Y. Giantess 1790 by Diomed* 1777 (Florizel 1768)—Giantess 1769 by Matchem 1748 (Cade 1734) Molly Long Legs 1753 by Babraham 1740 (God. Ar. 1724)—Dau. of Cole's Foxhunter 1727 (Brisk 1711).
			Olympia, b. m., 1815.	Sir Oliver, b. h., 1800.	Sir Peter T. 1784 (Highflyer 1774)—Fanny 1790 by Diomed* 1777 (Florizel 1768)—Ambrosia 1783 by Woodpecker 1773 (Herod 1758) Ruth 1761 by Blank 1740 (God. Ar. 1724)—Sist. 1751 to South by Regulus 1739 (God. Ar. 1724).
				Scotilla, b. m., 1795.	Anvil 1777 (Herod 1758) Scota 1783 by Eclipse 1764 (Marske 1750) Harmony 1775 by Herod 1758 (Tartar 1743) Rutilia 1769 by Blank 1740 (God. Ar. 1724) Sist. 1751 to South by Regulus 1739 (God. Ar. 1724).
		Languid, b. m., 1831.	Cain, b. h., 1822.	Paulowitz, br. h., 1813.	Sir Paul 1802 (Sir Peter T. 1784)—Evelina 1791 by Highflyer 1774 (Herod 1758)—Termagant by Tantrum 1760 (Cripple 1750)—Cantatrice by Sampson 1745 (Blaze 1733)—Dau. 1757 of Regulus 1739 (God. Ar. 1724).
				Daughter of, —, —, 1810.	Paynator 1791 (Trumpator 1782)—Dau. 1805 of Delpini 1781 (Highflyer 1774)—Dau. 1795 of Y. Marske 1771 (Marske 1750)—Dau. of Dorimond 1758 (Dormouse 1738) - Portia 1758 by Regulus 1739 (God. Ar. 1724).
			Lydia, b. m., 1822.	Poulton, b. h., 1805.	Sir Peter T. 1784 (Highflyer 1774)—Fanny 1790 by Diomed* 1777 (Florizel 1768)—Ambrosia 1783 by Woodpecker 1773 (Herod 1758)—Ruth 1761 by Blank 1740 (God. Ar. 1724)—Sist. 1751 to South by Regulus 1739 (God. Ar. 1724).
				Variety, b. m., 1808.	Hyacinthus 1797 (Coriander 1786)—Sist. 1790 to Swordsman by Weazle 1776 (Herod 1758)—Dau. of Turk 1763 (Regulus 1739)—Dau. of Locust 1744 (Crab 1722)—Dau. of 1758 Changeling 1747 (Cade 1734).
†IDALIA, b. m., 1870.	Cambuscan, ch. h., 1861.	Newminster, b. h., 1848.	Touchstone, br. h., 1831.	Camel, br. h., 1822.	Whalebone 1807 (Waxy 1790)—Dau. 1812 of Selim 1802 (Buzzard* 1787)—Maiden 1801 by Sir Peter T. 1784 (Highflyer 1774)—Dau. 1788 of Phenomenon 1780 (Herod 1758)—Matron 1782 by Florizel 1768 (Herod 1758).
				Banter, br. m., 1826.	Master Henry 1815 (Orville 1799)—Boadicea 1807 by Alexander 1782 (Eclipse 1764)—Brunette by Amaranthus 1766 (Old England 1741)—Mayfly 1771 by Matchem 1748 (Cade 1734)—Dau. 1755 of Anc. Starling 1738 (Old Starling 1727).
			Bee's-wing, b. m., 1833.	Dr. Syntax, br. h., 1811.	Paynator 1791 (Trumpator 1782)—Dau. of Beningbrough 1791 (King Fergus 1775) Jenny Mole 1787 by Carbuncle 1772 (Babr. Blank 1758)—Dau. 1773 of Prince T'Quassaw 1751 (Snip 1736)—Sultana 1759 by Regulus 1739 (God. Ar. 1724).
				Daughter of, ch. m., 1817.	Ardrossan 1809 (John Bull 1789)—Lady Eliza 1813 by Whitworth 1805 (Agonistes 1797)—X. Y. Z.'s dam 1794 by Spadille 1784 (Highflyer 1774)—Sylvia 1783 by Y. Marske 1771 (Marske 1750)—Ferret 1765 by Broth. to Silvio (Cade 1734).
		The Arrow, b. m., 1850.	Slane, b. h., 1833.	Royal Oak, br. h., 1823.	Catton 1809 (Golumpus 1802)—Dau. 1818 of Smolensko 1810 (Sorcerer 1796) Lady Mary 1810 by Beningbrough 1791 (King Fergus 1775)—Dau. of Highflyer 1774 (Herod 1758)—Sist. to Miss Bell by Marske 1750 (Squirt 1732).
				Daughter of, b. m., 1819.	Orville 1799 (Beningbrough 1791)—Epsom Lass 1803 by Sir Peter T. 1784 (Highflyer 1774)—Alexina 1788 by King Fergus 1775 (Eclipse 1764)—Lardella 1780 by Y. Marske 1771 (Marske 1750)—Dau. of Y. Cade 1747 (Cade 1734).
			Southdown, b. m., 1835.	Defence, b. h., 1824.	Whalebone 1807 (Waxy 1790)—Defiance 1816 by Rubens 1805 (Buzzard* 1787)—Little Folly 1806 by Highland Fling 1798 (Spadille 1784)—Harriet 1799 by Volunteer 1780 (Eclipse 1764) Dau. 1785 of Alfred 1770 (Matchem 1748).
				Feltona, b. m., 1819.	X. Y. Z. 1808 (Haphazard 1797)—Janetta 1803 by Beningbrough 1791 (King Fergus 1775)—Dau. 1797 of Drone 1777 (Herod 1758)—Contessina 1787 by Y. Marske 1771 (Marske 1750) Tuberose 1772 by Herod 1758 (Tartar 1743).
	Dulcibella, br. m., 1857.	Voltigeur, br. h., 1847.	Voltaire, br. h., 1826.	Blacklock, b. h., 1814.	Whitelock 1803 (Hambletonian 1792)—Dau. 1799 of Coriander 1786 (Pot-8-o-'s 1773)—Wildgoose 1792 by Highflyer 1774 (Herod 1758)—Coheiress 1786 by Pot-8-o-'s 1773 (Eclipse 1764)—Manilla 1777 by Goldfinder 1764 (Snap 1750).
				Daughter of, b. m., 1816.	Phantom 1808 (Walton 1799)—Dau. 1802 of Overton 1788 (King Fergus 1775)—Gratitude's dam 1793 by Walnut 1786 (Highflyer 1774)—Dau. 1786 of Ruler 1777 (Y. Marske 1771)—Piracantha 1772 by Matchem 1748 (Cade 1734).
			Martha Lynn, br. m., 1837.	Mulatto, b. h., 1823.	Catton 1809 (Golumpus 1802)—Desdemona 1811 by Orville 1799 (Beningbrough 1791)—Fanny 1796 by Sir Peter T. 1784 (Highflyer 1774)—Dau. 1788 of Diomed* 1777 (Florizel 1768)—Desdemona 1770 by Marske 1750 (Squirt 1732).
				Leda, b. m., 1824.	Filho-da-Puta 1812 (Haphazard 1797)—Treasure 1809 by Camillus 1803 (Hambletonian 1792)—Dau. 1804 of Hyacinthus 1797 (Coriander 1786)—Flora 1789 by King Fergus 1775 (Eclipse 1764) Atalanta 1769 by Matchem 1748 (Cade 1734).
		Priestess, bl. m., 1846.	The Doctor, bl. h., 1834.	Dr. Syntax, br. h., 1811.	Paynator 1791 (Trumpator 1782)—Dau. of Beningbrough 1791 (King Fergus 1775)—Jenny Mole 1787 by Carbuncle 1772 (Babr. Blank 1758)—Dau. 1773 of Prince T'Quassaw 1751 (Snip 1736)—Sultana 1759 by Regulus 1739 (God. Ar. 1724).
				Daughter of, br. m., 1828.	Lottery 1820 (Tramp 1810)—Elizabeth 1821 by Walton 1799 (Sir Peter T. 1784)—Trulla 1815 by Sorcerer 1796 (Trumpator 1782)—Dau. 1793 of Weathercock 1785 (Highflyer 1774)—Cora 1777 by Matchem 1748 (Cade 1734).
			The Biddy, ch. m., 1833.	Bran, ch. h., 1831.	Humprey Clinker 1822 (Comus 1809)—Velvet 1823 by Oiseau 1809 (Camillus 1803)—Wire 1811 by Waxy 1790 (Pot-8-o-'s 1773)—Penelope 1798 by Trumpator 1782 (Conductor 1767)—Prunella 1788 by Highflyer 1774 (Herod 1758).
				Idalia, ch. m., 1815.	Peruvian 1806 (Sir Peter T. 1784)—Musidora b. m. 1804 by Meteor 1783 (Eclipse 1764)—Maid-of-all-Work b. m. 1786 by Highflyer 1774 (Herod 1758)—Sist. b. m. 1771 to Tandem by Syphon 1750 (Squirt 1732)—Dau. b. m. of Regulus 1739 (God. Ar. 1724) Sist. 1758 to Judgement by Snip 1736 (Flying Childers 1715)—Dau. b. m. 1745 of Cottingham 1735 (Hartley's Blind Horse)—The Warlock Galloway ch. m. 1728 by Snake (Lister Turk)—Old Lady ch. m. by the Bald Galloway (St. Victor Barb)—The Wharton mare by the Carlisle Turk—Dau. of Byerly Turk.

SPENDTHRIFT.

WINNER OF THE SANFORD STAKES FOR TWO-YEAR OLDS AT LOUISVILLE, YOUNG AMERICA STAKES No. 1 AT NASHVILLE, BELMONT AND LORILLARD STAKES AT JEROME PARK, CHAMPION STAKES AND JERSEY DERBY AT MONMOUTH PARK.

Located in Hartland Stud, Versailles, Ky. Yearlings sold annually.

SPENDTHRIFT by imp. Australian, son of West Australian, bred in the Woodburn Stud, Ky., foaled 1876, dam Aerolite, dam of Mozart, Rutherford, Fellowcraft, Miser, etc., by Lexington, out of Florine, dam of Idlewild by imp. Glencoe. Aerolite, Spendthrift's dam, was the great brood mare of the Woodburn Stud. Her full sister, Idlewild, was the best mare of her day, ran four miles when five years old, 117 lbs., in 7.26¼, and is the dam of Wildidle, that ran four miles in 7.25½. Spendthrift was the phenomenal two-year old of 1878; won all his races, five in number. At Lexington, Ky., won Sweepstakes for two-year olds, 1 mile, in 1.58½, beating Montreal, Scully and three others. At Louisville, Ky., won the Sanford Stakes for two-year olds, 1 mile, in 1.46¼, beating Montreal, Trinidad and five others, among which was Goodnight and Strathmore; same place, won a Sweepstakes for two-year olds, 1 mile, in 1.45. At Nashville, Tennessee, won the Young America Stakes No. 1 for two-year olds, three-quarters of a mile, in 1.16¼, beating Lord Murphy, Charlemagne and Cal. Morgan; at same place, won a Sweepstakes for two-year olds, 1 mile, in 1.44½, beating Lord Murphy, the only other starter. Season of 1879 he started eight times, won five races, and was second in three. Spendthrift made his first appearance at Jerome Park, May 31. He ran second in the Withers Stakes, 1 mile, won by his stable companion, Dan Sparling, by sufferance, Report was third, Harold, Plevna and Mulrooney unplaced, time, 1.48; June 5, won the Belmont Stakes, 1½ miles, defeating Monitor, Jericho, Pawnee, Rochester and Harold, time, 2.42¾; June 10, won the rich Lorillard Stakes, 1½ miles, defeating Harold, Magnetism, Report, Dan Sparling, Monitor, Rochester, Jericho, Boardman, Plevna and Eunice. This race was the grandest achievement ever accomplished by a three-year old in America. Spendthrift carried 5 lbs. penalty for winning the Belmont Stakes, making his weight 123 lbs., and was almost left at the post. Harold delayed the start by his ugly temper, and gave Spendthrift a vicious kick just as the signal for the start was given. Harold bolted off in the lead, and Spendthrift was the last to leave the starting point, fully fifty yards in his rear. Harold turned into the home-stretch with fully thirty yards lead on Spendthrift. Spendthrift gradually gained on Harold, and the game son of Australian came on and won easily by two lengths, Harold second, Magnetism third, time, 2.25¾. Monmouth Park, July 8, won the Jersey Derby, 1½ miles, defeating Wilful with ease, time, 2.53. Saratoga, N. Y., July 9, Spendthrift met the son of Enquirer, Falsetto, for the first time. The race was for the Travers Stakes, 1¾ miles. Falsetto won, Spendthrift was a good second, Harold third, Jericho and Dan Sparling unplaced, time, 3.09¼; Aug. 12, Falsetto again defeated Spendthrift, who finished second, in the Kenner Stakes, 2 miles, Jericho was third, Monitor and Harold unplaced, time, 3.35¼. Monmouth Park, Aug. 23, Spendthrift won the Champion Stakes for all ages, 1½ miles, defeating Report, Bramble, Volturno, Harold and Bonnie Oaks, time, 2.41¼, the track was heavy and slow. Jerome Park, N. S., Oct. 2, Spendthrift, unfit to run, started in the Jerome Stakes, 1¾ miles, Monitor won, Spendthrift second, Report third, Jericho fourth, time, 3.12. Spendthrift, at four years old, was sent to England, but when amiss ran unplaced for the Cambridgeshire. Ran two races at Jerome Park when unfit at five years old: was second to Barrett, 1 mile, in 1.44½, and was unplaced in the other race.

Spendthrift is the sire of a number of good winners, including the famous sprinter Kingston, Bankrupt, Carrie G., Stockton, Thriftless, Touch Pas, Shasta, Golden Reel, Defaulter, Alarm Bell, America, Extravagance, Speedwell, Victor Sass, Prodigal, Speedwest, Bill Barnes, Kingsbridge, Kingstock, Marty B., Pall Mall, Pickpocket, Queenston, Harlem, Lamplighter (1½ miles in 2.32¾, best on record), Maria Stoops, Stowaway and many others.

Spendthrift has the Touchstone, Whisker and Whalebone blood on the sire's side, the Glencoe on the dam's side, and is much inbred to Sir Archy and Diomed, with the Gimcrack blood through imp. Medley.

Trumpator 1782 Conductor 1767 Matchem 1748 Cade 1734—Godolphin Arabian 1724.

SPENDTHRIFT.

CHESTNUT HORSE, FOALED 1876; BRED IN THE WOODBURN STUD, SPRING STATION, KY.

AEROLITE (Sist. to Idlewild), ch. m., 1862. — Florine, ch. m., 1851. (Melody, ch. m., 1837: Rodolph's dam, br. m., 1824; Medoc, ch. h., 1829. — *Glencoe, ch. h., 1831: Trampoline, ch. m., 1825; Sultan, b. h., 1816.) — Lexington, b. h., 1850. (Alice Carneal, b. m., 1836: Rowena, ch. m., 1826; *Sarpedon, br. h., 1828. — Boston, ch. h., 1833: Sist. to Tuckahoe, ch. m., 1814; Timoleon, ch. h., 1813.)

AUSTRALIAN, ch. h., 1858. — Emilia, b. m., 1840. (Persian, b. m., 1829: Variety, ch. m., 1816; Whisker, b. h., 1812. — Y. Emilius, b. h., 1833: Shoveler, b. m., 1816; Emilius, b. h., 1820.) — West Australian, b. h., 1850. (Mowerina, b. m., 1843: Emma, ch. m., 1824; Touchstone, br. h., 1831. — Melbourne, br. h., 1834: Daughter of, b. m., 1825; Humphrey Clinker, b. h., 1822.)

Comus, ch. h., 1809.	Sorcerer 1796 (Trumpator 1782) Epsom Lass 1803 by Sir Peter T. 1784 (Highflyer 1774) Alexina 1788 by King Fergus 1775 (Eclipse 1764) Lardella 1780 by Y. Marske 1771 (Marske 1750) Dau. of Y. Cade 1747 (Cade 1734).
Clinkerina, br. m., 1812.	Clinker 1805 (Sir Peter T. 1784) Pewet 1786 by Tandem 1773 (Syphon 1750) Termagant by Tantrum 1760 (Cripple 1750) Cantatrice by Sampson 1745 (Blaze 1733) Dau. 1757 of Regulus 1739 (God. Ar. 1724).
Cervantes, b. h., 1806.	Don Quixote 1784 (Eclipse 1764) Evelina 1791 by Highflyer 1774 (Herod 1758) Termagant by Tantrum 1760 (Cripple 1750) Cantatrice by Sampson 1745 (Blaze 1733) Dau. 1757 of Regulus 1739 (God. Ar. 1724).
Daughter of, br. m., 1818.	Golumpus 1802 (Gohanna 1790) Dau. 1810 of Paymator 1791 (Trumpator 1782) Circle 1802 by St. George 1789 (Highflyer 1774) Abigail 1788 by Woodpecker 1773 (Herod 1758) Firetail 1772 by Eclipse 1764 (Marske 1750).
Camel, br. h., 1822.	Whalebone 1807 (Waxy 1790) Dau. 1812 of Selim 1802 (Buzzard* 1787) Maiden 1801 by Sir Peter T. 1784 (Highflyer 1774) Dau. 1788 of Phenomenon 1780 (Herod 1758)—Matron 1782 by Florizel 1768 (Herod 1758).
Banter, br. m., 1826.	Master Henry 1815 (Orville 1799) Boadicea 1807 by Alexander 1782 (Eclipse 1764)—Brunette by Amaranthus 1766 (Old England 1741) Mayfly 1771 by Matchem 1748 (Cade 1734)—Dau. 1755 of Anc. Starling 1738 (Starling 1727).
Whisker, b. h., 1812.	Waxy 1790 (Pot-8-o-'s 1773) Penelope 1798 by Trumpator 1782 (Conductor 1767) Prunella 1788 by Highflyer 1774 (Herod 1758) Promise 1768 by Snap 1750 (Snip 1736) Julia 1756 by Blank 1740 (God. Ar. 1724).
Gibside Fairy, b. m., 1811.	Hermes 1790 (Mercury 1778) Vicissitude 1800 by Pipator 1786 (Imperator 1776)—Beatrice 1791 by Sir Peter T. 1784 (Highflyer 1774)—Pyrrha 1771 by Matchem 1748 (Cade 1734)—Duchess 1748 by Whitenose 1742 (God. Ar. 1724).
Orville, b. h., 1799.	Beningbrough 1791 (King Fergus 1775) Evelina 1791 by Highflyer 1774 (Herod 1758)—Termagant by Tantrum 1760 (Cripple 1750)—Cantatrice by Sampson 1745 (Blaze 1733)—Dau. 1757 of Regulus 1739 (God. Ar. 1724).
Emily, ch. m., 1810.	Stamford 1794 (Sir Peter T. 1784) Dau. 1799 of Whiskey 1789 (Saltram* 1780) Grey Dorimant 1781 by Dorimant 1772 (Otho 1760) Dizzy 1757 by Blank 1740 (God Ar. 1724)—Dizzy 1741 by Driver 1727 (Wynn Ar.).
Scud, b. h., 1804.	Beningbrough 1791 (King Fergus 1775) Eliza 1791 by Highflyer 1774 (Herod 1758)—Augusta 1781 by Eclipse 1764 (Marske 1750)—Hardwicke's dam by Herod 1758 (Tartar 1748)—Nettle 1763 by Bajazet 1740 (God. Ar. 1724).
Goosander, b. m., 1805.	Hambletonian 1792 (King Fergus 1775)—Rally 1790 by Trumpator 1782 (Conductor 1767) Fancy 1780 by Florizel 1768 (Herod 1758)—Dau. 1763 of Spectator 1749 (Crab 1722)—Horatia 1758 by Blank 1740 (God. Ar. 1724).
Waxy, b. h., 1790.	Pot-8-o-'s 1773 (Eclipse 1764) Maria 1777 by Herod 1758 (Tartar 1748)—Lisette 1772 by Snap 1750 (Snip 1736)—Miss Windsor 1754 by God. Ar. 1724—Sist. to Volunteer by Y. Belgrade (Belgrade Turk).
Penelope, b. m., 1798.	Trumpator 1782 (Conductor 1767)—Prunella 1788 by Highflyer 1774 (Herod 1758)—Promise 1768 by Snap 1750 (Snip 1736)—Julia 1756 by Blank 1740 (God. Ar. 1724)—Spectator's dam 1735 by Partner 1718 (Jigg).
[Selim or] Soothsayer, ch. h., 1808.	Sorcerer 1796 (Trumpator 1782)—Golden Locks 1793 by Delpini 1781 (Highflyer 1774)—Violet 1787 by Shark* 1771 (Marske 1750)—Dau. 1772 of Syphon 1750 (Squirt 1732)—Charlotte 1756 by Blank 1740 (God. Ar. 1724).
Sprite, ch. m., 1807.	Bobtail 1795 (Precipitate* 1787)—Catherine 1795 by Woodpecker 1773 (Herod 1758)—Camilla 1778 by Trentham 1766 (T. Sweepstakes 1743)—Coquette 1764 by Compton Barb—Sist. 1743 to Regulus by God. Ar. 1724.
Sir Archy, b. h., 1805.	*Diomed 1777 (Florizel 1768)—*Castianira 1796 by Rockingham 1781 (Highflyer 1774)—Tabitha 1782 by Trentham 1766 (T. Sweepstakes 1743) Dau. of Bosphorus 1754 (Babraham 1740)—Dau. of W's Forester 1750 (Forester 1736).
Daughter of, ch. m., —.	*Saltram 1780 (Eclipse 1764)—Dau. of Symmes' Wildair 1767 (*Fearnought 1755)—Dau. of Tyler's Driver 1760 (*Othello)—Dau. of *Fallower 1761 (Blank 1740)—Dau. of *Vampire 1757 (Regulus 1739).
Ball's Florizel, ch. h., 1801.	*Diomed 1777 (Florizel 1768)—Dau. of *Shark 1771 (Marske 1750)—Dau. of Harris's Eclipse 1770 (*Fearnought 1755) Dau. of *Fearnought 1755 (Regulus 1739)—Dan. of *Jolly Roger 1741 (Roundhead 1733).
Daughter of, b. m, 1791.	*Alderman 1787 (Pot-8-o-'s 1773)—Dau. of *Clockfast 1771 (Gimcrack 1760)—Dau. of Symmes' Wildair 1767 (*Fearnought 1755)—Y. Kitty Fisher by *Fearnought 1755 (Regulus 1739) *Kitty Fisher 1755 by Cade 1734 (God. Ar. 1724).
Emilius, b. h., 1820.	Orville 1799 (Beningbrough 1791)—Emily 1810 by Stamford 1794 (Sir Peter T. 1784)—Dau. 1799 of Whiskey 1789 (Saltram* 1780)—Grey Dorimant 1781 by Dorimant 1772 (Otho 1760) Dizzy 1757 by Blank 1740 (God. Ar. 1724).
Icaria, b. m., 1824.	The Flyer 1814 (Vandyke Jr. 1808)—Parma 1813 by Dick Andrews 1797 (Joe Andrews 1778)—May 1804 by Beningbrough 1791 (King Fergus 1775) Primrose 1787 by Mambrino 1768 (Engineer 1756) Cricket 1776 by Herod 1758 (Tartar 1743).
Sumpter, ch. h., 1818.	Sir Archy 1805 (*Diomed 1777)—Dau. of *Robin Redbreast 1796 (Sir Peter T. 1784)—Dau. 1788 of *Obscurity 1777 (Eclipse 1764)—Slamerkin 1768 by *Wildair 1753 (Cade 1734)—*Cub mare 1762 by Cub 1739 (Fox 1714).
Lady Grey, b. m., 1817.	Robin Grey 1805 (*Royalist 1790)—Maria 1802 by Melzar 1791 (*Medley 1773)—Dau. of *Highflyer 1784 (Highflyer 1774)—Dau. of *Fearnought 1755 (Regulus 1739)—Dau. of Ariel 1753 (*Traveller 1747).
Selim, ch. h., 1802.	Buzzard* 1787 (Woodpecker 1773)—Dau. of Alexander 1782 (Eclipse 1764)—Dan. of Highflyer 1774 (Herod 1758)—Dau. of Alfred 1770 (Matchem 1748)—Dau. 1770 of Engineer 1756 (Sampson 1745).
Bacchante, br. m., 1809.	W's Ditto 1800 (Sir Peter T. 1784)—Dau. 1791 of Mercury 1778 (Eclipse 1764)—Dau. 1776 of Herod 1758 (Tartar 1743)—Folly 1771 by Marske 1750 (Squirt 1732)—Vixen 1753 by Regulus 1739 (God. Ar. 1724).
Tramp, b. h., 1810.	Dick Andrews 1797 (Joe Andrews 1778)—Dau. 1803 of Gohanna 1790 (Mercury 1778)—Fraxinella 1793 by Trentham 1766 (T. Sweepstakes 1743)—Dau. 1785 of Woodpecker 1773 (Herod 1758) Everlasting 1775 by Eclipse 1764 (Marske 1750).
Web, b. m., 1808.	Waxy 1790 (Pot-8-o-'s 1773)—Penelope 1798 by Trumpator 1782 (Conductor 1767) Prunella 1788 by Highflyer 1774 (Herod 1758)—Promise 1768 by Snap 1750 (Snip 1736)—Julia 1756 by Blank 1740 (God Ar. 1724).
Am. Eclipse, ch. h., 1814.	Duroc 1806 (*Diomed 1777)—Miller's Damsel 1802 by *Messenger 1780 (Mambrino 1768)—*Dau. 1792 of Pot-8-o-'s 1773 (Eclipse 1764)—Dau. 1778 of Gimcrack 1760 (Cripple 1750) Snapdragon 1759 by Snap 1750 (Snip 1736).
Y. Maid of the Oaks, ch. m., 1817.	*Expedition 1795 (Pegasus 1784)—Maid of the Oaks 1801 by *Spread Eagle 1792 (Volunteer 1780)—Annette by *Shark 1771 (Marske 1750)—Dau. of Rockingham (Partner 1755) Dau. of True Whig (Fitzhugh's Regulus 1765).
Haxall's Moses, b. h., 1816.	*Sir Harry 1795 (Sir Peter T. 1784)—*Mermaid by Waxy 1790 (Pot-8-o-'s 1773) *Promise 1803 by Buzzard* 1787 (Woodpecker 1773)—Dau. 1795 of Precipitate* 1787 (Mercury 1778)—Lady Harriet 1773 by Mark Anthony 1767 (Spectator 1749).
Daughter of, —, —.	Cook's Whip 1804 (*Whip 1794) Hart's Maria by Craig's Alfred (*Medley 1776)—The Pryor mare [a good race-mare] by Tayloe's Bellair 1786 (*Medley 1776)—Thoroughbred mare of Mr. T. D. Owings.

SPRINGBOK.

WINNER OF THE BELMONT STAKES AT JEROME PARK, SARATOGA CUP AT SARATOGA, AND MANY OTHER RACES.

In the Megibben-Edgewater Stud. Application to F. S. Ashbrook, secretary, Cynthiana, Ky.

SPRINGBOK by imp. Australian, son of West Australian, bred in the Woodburn Stud, Ky., foaled 1870, dam Hester by Lexington, second dam Heads I Say, dam of Maggie Hunter, Hazard, etc., by imp. Glencoe, out of imp. Heads or Tails by Lottery. Springbok came out as a two-year old; started in two races. Jerome Park, won a dash of 5 furlongs for two-year olds in 1.05½, beating Minnie Mac, Katy Pease and others, and was unplaced in the Saratoga Stakes, at Saratoga, three-quarters of a mile, won by Catesby in 1.17¼. As a three-year old started in seven races, of which he won three. At Jerome Park, won the Belmont Stakes for three-year olds, 1 mile 5 furlongs, beating Count D'Orsay, imp. Strachino, Waverly and six others, among which were Fellowcraft and imp. Ill-Used; same place, won a race of mile heats in 1.45½, 1.44½, beating imp. Buckden and two others; won mile heats in 1.46½, 1.48, 1.48¼. Minnie Mac won the first heat. At Monmouth Park, was second in the Jersey Derby for three-year olds, 1½ miles, won by Tom Bowling in 2.45¾. At Jerome Park, was second in the Jerome Stakes for three-year olds, 2 miles, won by Tom Bowling, no official time given. At Saratoga, was unplaced in the Travers Stakes for three-year olds, 1¾ miles, won by Tom Bowling in 3.09¾, and was third in the Kenner Stakes for three-year olds, 2 miles, won by imp. Ill-Used in 3.39. As a four-year old started in nine races, of which he won eight. Jerome Park, won a selling race, 1¾ miles, in 3.14¼, carrying 115 lbs., and beating Mate, Fellowcraft and others; won a Handicap Sweepstakes, 1 mile and 5 furlongs, in 2.53. At Utica, N. Y., won the Citizens' Handicap Stakes for all ages, 2 miles, in 3.36¾, beating Artist and Katy Pease; same place, won the Hotel Purse, mile heats, in 1.45, 1.42¾. At Saratoga, won a dash of 1¼ miles, in 2.09¾, beating Mate, Survivor and Dublin; won the Saratoga Cup, 2¼ miles, in 4.11¾, beating Preakness, Katy Pease and Wanderer; won a dash of 2 miles and a furlong, in 3.56; won a dash of 3 miles, in 5.42¼, beating Preakness and Fellowcraft. Jerome Park, was second in the Jockey Club Handicap, 2 miles, won by Preakness in 3.42. As a five year old he started in six races, winning four and running dead heat in one. At Jerome Park, won a dash of 1½ miles, in 2.43½, beating Mate, the only other starter; won a race of mile heats in 1.44¼, 1.46¼, beating Spindrift, Big Fellow and Bill Bruce; won a dash of 1 mile and a furlong in 1.58¼, beating Bob Woolley, Mate and four others. At Saratoga, ran dead heat with Preakness in the Saratoga Cup, 2¼ miles, in 3.56¼, the fastest race ever ran at the distance. At Jerome Park, won a dash of 1¾ miles, in 3.12, beating Milner, Big Fellow and two others. At Saratoga, was second in a Sweepstakes, 1¼ miles, won by Grinstead in 2.08¾. As a six-year old started once in a race of mile heats and was distanced in the first heat. In this race Freebooter was the winner, with 101 lbs., while Springbok carried 124 lbs. Springbok was a first-class race-horse, and is descended from a racing family on both sides. He is much inbred to Waxy through Whisker, Whalebone and Web, and traces on his dam's side to Eleanor by Whiskey, the first mare to win both the Derby and Oaks. In addition to his Diomed blood through Lexington, his dam traces to Diomed's best daughter in England, Young Giantess, the dam of Sorcerer, and a sister to Roxana, who first brought the great Godolphin Arabian into notice to a daughter of Spanker.

Springbok is the sire of the winners Almeda, Huntress, Laura Garrison, Winona, Lila May, Lute Arnold, Elgin, Markland, Aurelian, Ascoli, Kasson, Miss Charmer, Loveland, Minola, Easterbok, Eberlee, The Elk, W. G. Morris, Vallera, Harry Kuhl, Blue Jeans, Birdella M., Rook Laidly, Ben March, Spectator, Flower Dellis, Maj. Thornton, Neva C., Julia May, W. L. Munson, Springaway and many others.

Springbok descends in direct female line through Old Bald Peg from the Barb mare, to which traces Lath, Cade, Flying Childers, Diomed, Sorcerer, Priam, Phantom, Langar, Muley, Plenipotentiary, Turnus, See Saw, Zinganee, Troubadour, Harry of the West and all the imp. Sylva family.

Springbok is a bright chesnut, 16 hands high, with star and snip over his left nostril, and white spot between the nostrils, and black spot over his right eye, left fore leg white over the ankle, both hind feet white nearly to the hocks. He is a horse of great substance and power, highly finished all over, without a particle of lumber, and has made his mark in the stud.

SPRINGBOK.

CHESTNUT HORSE, FOALED 1870; BRED BY A. J. ALEXANDER, WOODBURN STUD, KENTUCKY.

Trumpator 1782—Conductor 1767—Matchem 1748—Cade 1734—Godolphin Arabian 1724.

*AUSTRALIAN, ch. h., 1858.	West Australian, b. h., 1850.	Melbourne, br. h., 1834.	Humphrey Clinker, b. h., 1822.	**Comus, ch. h., 1809.** Sorcerer 1796 (Trumpator 1782)—Epsom Lass 1801 by Sir Peter T. 1784 (Highflyer 1774)—Alexina 1788 by King Fergus 1775 (Eclipse 1764)—Lardella 1780 by Y. Marske 1771 (Marske 1750)—Dau. of Y. Cade 1747 (Cade 1734).
				Clinkerina, br. m., 1812. Clinker 1805 (Sir Peter T. 1784)—Pewet 1786 by Tandem 1773 (Syphon 1750)—Termagant by Tantrum 1760 (Cripple 1750)—Cantatrice by Sampson 1745 (Blaze 1733)—Dau. 1757 of Regulus 1739 (God. Ar. 1724).
			Daughter of, b. m., 1825.	**Cervantes, b. h., 1806.** Don Quixote 1784 (Eclipse 1764)—Evelina 1791 by Highflyer 1774 (Herod 1758)—Termagant by Tantrum 1760 (Cripple 1750)—Cantatrice by Sampson 1745 (Blaze 1733)—Dau. 1757 of Regulus 1739 (God. Ar. 1724).
				Daughter of, br. m., 1818. Golumpus 1802 (Gohanna 1790)—Dau. 1810 of Paynator 1791 (Trumpator 1782)—Circle 1802 by St. George 1789 (Highflyer 1774)—Abigail 1788 by Woodpecker 1773 (Herod 1758)—Firetail 1772 by Eclipse 1764 (Marske 1750).
		Mowerina, b. m., 1843.	Touchstone, br. h., 1831.	**Camel, br. h., 1822.** Whalebone 1807 (Waxy 1790)—Dau. 1812 of Selim 1802 (Buzzard* 1787)—Maiden 1801 by Sir Peter T. 1784 (Highflyer 1774)—Dau. 1788 of Phenomenon 1780 (Herod 1758)—Matron 1782 by Florizel 1768 (Herod 1758).
				Banter, br. m., 1826. Master Henry 1815 (Orville 1799)—Boadicea 1807 by Alexander 1782 (Eclipse 1764)—Brunette by Amaranthus 1766 (Old England 1741)—Mayfly 1771 by Matchem 1748 (Cade 1734)—Dau. 1755 of Anc. Starling 1738 (Starling 1727).
			Emma, ch. m., 1824.	**Whisker, b. h., 1812.** Waxy 1790 (Pot-8-o's 1773)—Penelope 1798 by Trumpator 1782 (Conductor 1767)—Prunella 1788 by Highflyer 1774 (Herod 1758)—Promise 1768 by Snap 1750 (Snip 1736)—Julia 1756 by Blank 1740 (God. Ar. 1724).
				Gibside Fairy, b. m., 1811. Hermes 1790 (Mercury 1778)—Vicissitude 1800 by Pipator 1786 (Imperator 1776)—Beatrice 1791 by Sir Peter T. 1784 (Highflyer 1774)—Pyrrha 1771 by Matchem 1748 (Cade 1734)—Duchess 1748 by Whitenose 1742 (God. Ar. 1724).
	Emilia,* b. m., 1840.	Y. Emilius, b. h., 1833.	Emilius, b. h., 1820.	**Orville, b. h., 1799.** Beningbrough 1791 (King Fergus 1775)—Evelina 1791 by Highflyer 1774 (Herod 1758)—Termagant by Tantrum 1760 (Cripple 1750)—Cantatrice by Sampson 1745 (Blaze 1733)—Dau. 1757 of Regulus 1739 (God. Ar. 1724).
				Emily, ch. m., 1810. Stamford 1794 (Sir Peter T. 1784)—Dau. 1799 of Whiskey 1789 (Saltram* 1780)—Grey Dorimant 1781 by Dorimant 1772 (Otho 1760)—Dizzy 1757 by Blank 1740 (God. Ar. 1724)—Dizzy 1741 by Driver 1727 (Wynn Ar.).
			Shoveler, b. m., 1816.	**Scud, b. h., 1804.** Beningbrough 1791 (King Fergus 1775)—Eliza 1791 by Highflyer 1774 (Herod 1758)—Augusta 1781 by Eclipse 1764 (Marske 1750)—Hardwicke's dam by Herod 1758 (Tartar 1743)—Nettle 1763 by Bajazet 1740 (God. Ar. 1724).
				Goosander, b. m., 1805. Hambletonian 1792 (King Fergus 1775)—Rally 1790 by Trumpator 1782 (Conductor 1767)—Fancy 1780 by Florizel 1768 (Herod 1758)—Dau. 1768 of Spectator 1749 (Crab 1722)—Horatia 1758 by Blank 1740 (God. Ar. 1724).
		Persian, b. m., 1829.	Whisker, b. h., 1812.	**Waxy, b. h., 1790.** Pot-8-o's 1773 (Eclipse 1764)—Maria 1777 by Herod 1758 (Tartar 1743)—Lisette 1772 by Snap 1750 (Snip 1736)—Miss Windsor 1754 by God. Ar. 1724—Sist. to Volunteer by Y. Belgrade (Belgrade Turk).
				Penelope, b. m., 1798. Trumpator 1782 (Conductor 1767)—Prunella 1788 by Highflyer 1774 (Herod 1758)—Promise 1768 by Snap 1750 (Snip 1736)—Julia 1756 by Blank 1740 (God. Ar. 1724)—Spectator's dam 1735 by Partner 1718 (Jigg).
			Variety, ch. m., 1816.	**[Selim or] Soothsayer, ch. h., 1808.** Sorcerer 1796 (Trumpator 1782)—Golden Locks 1793 by Delpini 1781 (Highflyer 1774)—Violet 1787 by Shark* 1771 (Marske 1750)—Dau. 1772 of Syphon 1750 (Squirt 1732)—Charlotte 1756 by Blank 1740 (God. Ar. 1724).
				Sprite, ch. m., 1807. Bobtail 1795 (Precipitate* 1787)—Catherine 1795 by Woodpecker 1773 (Herod 1758)—Camilla 1778 by Trentham 1766 (T. Sweepstakes 1743)—Coquette 1764 by Compton Barb—Sist. 1743 to Regulus by God. Ar. 1724.
HESTER, b. m., 1866.	Lexington, b. h., 1850.	Boston, ch. h., 1833.	Timoleon, ch. h., 1813.	**Sir Archy, b. h., 1805.** *Diomed 1777 (Florizel 1768)—*Castianira 1796 by Rockingham 1781 (Highflyer 1774)—Tabitha 1782 by Trentham 1766 (T. Sweepstakes 1743)—Dau. of Bosphorus 1754 (Babraham 1740)—Dau. of W's Forester 1750 (Forester 1736).
				Daughter of, ch. m., —. *Saltram 1780 (Eclipse 1764)—Dau. of Symmes' Wildair 1767 (*Fearnought 1755)—Dau. of Tyler's Driver 1760 (*Othello)—Dau. of *Fallower 1761 (Blank 1740)—Dau. of *Vampire 1757 (Regulus 1739).
			Sist. to Tuckahoe, ch. m., 1814.	**Ball's Florizel, ch. h., 1801.** *Diomed 1777 (Florizel 1768)—Dau. of *Shark 1771 (Marske 1750)—Dau. of Harris's Eclipse 1770 (*Fearnought 1755)—Dau. of *Fearnought 1755 (Regulus 1739)—Dau. of *Jolly Roger 1741 (Roundhead 1733).
				Daughter of, b. m., 179[illegible]. *Alderman 1787 (Pot-8-o's 1773)—Dau. of *Clockfast 1774 (Gimcrack 1760)—Dau. of Symmes' Wildair 1767 (*Fearnought 1755)—Y. Kitty Fisher by *Fearnought 1755 (Regulus 1739)—*Kitty Fisher 1755 by Cade 1734 (God. Ar. 1724).
		Alice Carneal, b. m., 1836.	*Sarpedon, br. h., 1828.	**Emilius, b. h., 1820.** Orville 1799 (Beningbrough 1791)—Emily 1810 by Stamford 1794 (Sir Peter T. 1784)—Dau. 1799 of Whiskey 1789 (Saltram* 1780)—Grey Dorimant 1781 by Dorimant 1772 (Otho 1760)—Dizzy 1757 by Blank 1740 (God. Ar. 1724).
				Icaria, b. m., 1824. The Flyer 1814 (Vandyke Jr. 1808)—Parma 1813 by Dick Andrews 1797 (Joe Andrews 1778)—May 1804 by Beningbrough 1791 (King Fergus 1775)—Primrose 1787 by Mambrino 1768 (Engineer 1756)—Cricket 1776 by Herod 1758 (Tartar 1743).
			Rowena, ch. m., 1826.	**Sumpter, ch. h., 1818.** Sir Archy 1805 (*Diomed 1777)—Dau. of *Robin Redbreast 1796 (Sir Peter T. 1784)—Dau. 1788 of *Obscurity 1777 (Eclipse 1764)—Slamerkin 1768 by *Wildair 1753 (Cade 1734)—*Cub mare 1762 by Cub 1739 (Fox 1714).
				Lady Grey, b. m., 1817. Robin Grey 1805 (*Royalist 1790)—Maria 1802 by Melzar 1791 (*Medley 1776)—Dau. of *Highflyer 1781 (Highflyer 1774)—Dau. of *Fearnought 1755 (Regulus 1739)—Dau. of Ariel 1753 (*Traveller 1747).
	Heads-I-Say, b. m., 1851.	*Glencoe, ch. h., 1831.	Sultan, b. h., 1816.	**Selim, ch. h., 1802.** Buzzard* 1787 (Woodpecker 1773)—Dau. of Alexander 1782 (Eclipse 1764)—Dau. of Highflyer 1774 (Herod 1758)—Dau. of Alfred 1770 (Matchem 1748)—Dau. 1770 of Engineer 1756 (Sampson 1745).
				Bacchante, br. m., 1809. W's Ditto 1800 (Sir Peter T. 1784)—Dau. 1791 of Mercury 1778 (Eclipse 1764)—Dau. 1776 of Herod 1758 (Tartar 1743)—Folly 1771 by Marske 1750 (Squirt 1732)—Vixen 1753 by Regulus 1739 (God. Ar. 1724).
			Trampoline, ch. m., 1825.	**Tramp, b. h., 1810.** Dick Andrews 1797 (Joe Andrews 1778)—Dau. 1803 of Gohanna 1790 (Mercury 1778)—Fraxinella 1793 by Trentham 1766 (T. Sweepstakes 1743)—Dau. 1785 of Woodpecker 1773 (Herod 1758)—Everlasting 1775 by Eclipse 1764 (Marske 1750).
				Web, b. m., 1808. Waxy 1790 (Pot-8-o's 1773)—Penelope 1798 by Trumpator 1782 (Conductor 1767)—Prunella 1788 by Highflyer 1774 (Herod 1758)—Promise 1768 by Snap 1750 (Snip 1736)—Julia 1756 by Blank 1740 (God Ar. 1724).
		*Heads-or-Tails, b. m., 1831.	Lottery, br. h., 1820.	**Tramp, b. h., 1810.** Dick Andrews 1797 (Joe Andrews 1778)—Dau. 1803 of Gohanna 1790 (Mercury 1778)—Fraxinella 1793 by Trentham 1766 (T. Sweepstakes 1743)—Dau. 1785 of Woodpecker 1773 (Herod 1758)—Everlasting 1775 by Eclipse 1764 (Marske 1750).
				Mandane, ch. m., 1800. Pot-8-o's 1773 (Eclipse 1764)—Y. Camilla 1787 by Woodpecker 1773 (Herod 1758)—Camilla 1778 by Trentham 1766 (T. Sweepstakes 1743)—Coquette 1765 by Compton Barb—Sist. 1743 to Regulus by God. Ar. 1724.
			Active, b. m., 1820.	**Partisan, b. h., 1811.** Walton 1799 (Sir Peter T. 1784)—Parasol 1800 by Pot-8-o's 1773 (Eclipse 1764)—Prunella 1788 by Highflyer 1774 (Herod 1758)—Promise 1768 by Snap 1750 (Snip 1736)—Julia 1756 by Blank 1740 (God. Ar. 1724).
				Eleanor, b. m., 1798. Whiskey 1789 (Saltram* 1780)—Y. Giantess b. m. 1790 by Diomed* 1777 (Florizel 1768)—Giantess b. m. 1769 by Matchem 1748 (Cade 1734)—Molly Long Legs b. m. 1753 by Babraham 1740 (God. Ar. 1724)—Dau. of Cole's Foxhunter 1727 (Brisk 1711)—Sist. to Cato by Partner 1718 (Jigg)—Sist. ch. m. 1719 to Roxana by Bald Galloway (St. Victor Barb)—Sist. to Chaunter by Akaster Turk—Cream Cheeks by Leedes Ar.—Dau. 1680 by Spanker (D'Arcy Yellow Turk)—Old Morocco mare by Fairfax's Morocco Barb—Old Bald Peg by an Arabian—A Barb mare.

IMPORTED STALWART.

Imported and owned by the Estate of the late D. D. Withers, Holmdel, N. J. Sales annually.

STALWART, bay horse, foaled 1883, bred by the Yardley Stud, by Sterling, son of Oxford by Birdcatcher, dam Sea Mark by Adventurer, son of Newminster, out of Sea Gull by Lifeboat, son of Sir Hercules. Stalwart received an injury crossing the ocean and was never trained.

Sterling, the sire of Stalwart, one of the most popular stallions in England, started only a few times. He won the Liverpool Autumn Cup, beating Louise Victoria and twelve others, and was third to Montargis and Walnut in the Cambridgeshire, with 133 lbs. He won the Ascot Biennial Stakes, the Summer Stakes at Newmarket, the Derby Free Handicap and Craven Stakes. He was second to Boswell in the 2,000 Guineas, a defeat which was one of the most glaring flukes in the annals of the turf, for it was admitted by everyone that he could concede quite 42 lbs. to the winner. He was not in the Derby. His value is best shown in the stud, having sired such good ones as Isonomy, who himself is the sire of Riviera, one of the favorites for her year's Derby, Formidable, another good two-year old, Antibes. Common, winner of 2,000 Guineas, Derby and St. Leger, etc., etc., and one of the best race-horses of his time. His get this year won over $100,000, and he was standing at $1,000, with his season full. Harvester, who ran a dead heat with St. Gatien, and divided the stakes for the Derby, and was third in the 2,000 Guineas. Paradox, Enterprise and Enthusiast, all three winners of the 2,000 Guineas, the former only beaten a head in the Derby and the latter defeating the famous Donovan. Energy, winner of sixteen races out of thirty seven and one of the speediest horses of his day. Beaudesert, the winner of the Middle Park Plate and sire of many winners last year. Caxtonian, Discount, Post Orbit, Geologist, El Dorado (another conqueror of Donovan) and Cherry all come from Sterling's loins. Stalwart's dam, Sea Mark, is the dam of Astronomy, Beacon and the fine young stallion Top Gallant. His grandam, Sea Gull, was the dam of Beaudesert, winner of the Middle Park Plate, Queenstown, The Admiral, Discordant, etc. Wild Cherry was the dam of Grace Darling, Cherry Ripe, Sugar Plum, Heliotrope, Reciprocity, etc. Bridget was the dam of The Baronet, Wild Strawberry, Pomona, Miss Fanny, etc. Stalwart traces through Ringlet, a daughter of Whisker, Clinkerina, the dam of Humphrey Clinker, sire of the noted Melbourne by Clinker, Pewet, a winner of the St. Leger, by tandem, to the noted Blacklegs mare, dam of Marske, the sire of Eclipse. He combines in his pedigree a double cross of Sir Hercules, upon a large infusion of Touchstone blood, through such excellent sources as Adventurer, Flatcatcher and Surplice, and traces on his sire's and dam's side to the noted daughters of Priam. Crucifix and Miss Letty. Many of the best race-horses and stallions in England are descended from the family. He must be a valuable acquisition to the blood stock of the country. He is descended in direct female line from the Bustler mare, to which traces Merlin (by Castrel), Marske, sire of Eclipse, Orville, Cervantes. Cain. Buzzard (by Blacklock), Humphrey Clinker, Newminster, Strafford, Mariner, Melton, etc.

Pot-8-o's 1773 Eclipse 1764—Marske 1750 Squirt 1732 Bartlett's Childers Darley Arabian 1699.

STALWART.

BAY HORSE, FOALED 1883; BRED AT THE YARDLEY STUD, ENGLAND.

STERLING, b. h., 1868.	Oxford, ch. h., 1857.	Ir. Birdcatcher, ch. h., 1833.	Sir Hercules, blk. h., 1826.	Whalebone, br. h., 1807.	Waxy 1790 (Pot-8-o's 1773)—Penelope 1798 by Trumpator 1782 (Conductor 1767)—Prunella 1788 by Highflyer 1774 (Herod 1758)—Promise 1768 by Snap 1750 (Snip 1736)—Julia 1756 by Blank 1740 (God. Ar. 1724).
				Peri, b. m., 1822.	Wanderer 1811 (Gohanna 1790)—Thalestris 1809 by Alexander 1782 (Eclipse 1764)—Rival 1800 by Sir Peter T. 1784 (Highflyer 1774)—Hornet 1790 by Drone 1777 (Herod 1758)—Manilla 1777 by Goldfinder 1764 (Snap 1750).
			Guiccioli, ch. m., 1823.	Bob Booty, ch. h., 1804.	Chanticleer 1787 (Woodpecker 1773)—Ierne 1790 by Bagot 1780 (Herod 1758)—Dau. of Gamahoe (Bustard 1741)—Patty by Tim 1752 (Squirt 1732)—Miss Patch 1737 by Justice 1725 (H. C. Litton Ar.).
				Flight, ch. m., 1809.	Ir. Escape 1802 (Commodore 1793)—Y. Heroine by Bagot 1780 (Herod 1758)—Heroine 1775 by Hero 1758 (Cade 1734)—Dau. 1766 of Snap 1750 (Snip 1736)—Sist. 1743 to Regulus by God. Ar. 1724.
		Honey Dear, b. m., 1844.	Plenipotentiary, ch. h., 1831.	Emilius, b. h., 1820.	Orville 1799 (Beningbrough 1791)—Emily 1810 by Stamford 1794 (Sir Peter T. 1784)—Dau. 1799 of Whiskey 1789 (Saltram* 1780)—Grey Dorimant 1781 by Dorimant 1772 (Otho 1760)—Dizzy 1757 by Blank 1740 (God. Ar. 1724).
				Harriet, blk. m., 1819.	Pericles 1809 (Evander 1801)—Dau. 1812 of Selim 1802 (Buzzard* 1787)—Pipylina 1803 by Sir Peter T. 1784 (Highflyer 1774)—Rally 1790 by Trumpator 1782 (Conductor 1767)—Fancy 1780 by Florizel 1768 (Herod 1758).
			My Dear, b. m., 1841.	Bay Middleton, b. h., 1833.	Sultan 1816 (Selim 1802)—Cobweb 1821 by Phantom 1808 (Walton 1799)—Filagree 1815 by Soothsayer 1808 (Sorcerer 1796)—Web 1808 by Waxy 1790 (Pot-8-o's 1773)—Penelope 1798 by Trumpator 1782 (Conductor 1767).
				Miss Letty, b. m., 1834.	Priam* 1827 (Emilius 1820)—Dau. 1815 of Orville 1799 (Beningbrough 1791)—Dau. 1800 of Buzzard* 1787 (Woodpecker 1773)—Hornpipe 1793 by Trumpator 1782 (Conductor 1767)—Luna 1779 by Herod 1758 (Tartar 1743).
	Whisper, b. m., 1857.	Flatcatcher, br. h., 1845.	Touchstone, br. h., 1831.	Camel, br. h., 1822.	Whalebone 1807 (Waxy 1790)—Dau. 1812 of Selim 1802 (Buzzard* 1787)—Maiden 1801 by Sir Peter T. 1784 (Highflyer 1774)—Dau. 1788 of Phenomenon 1780 (Herod 1758)—Matron 1782 by Florizel 1768 (Herod 1758).
				Banter, br. m., 1826.	Master Henry 1815 (Orville 1799)—Boadicea 1807 by Alexander 1782 (Eclipse 1764)—Brunette by Amaranthus 1766 (Old England 1741)—Mayfly 1771 by Matchem 1748 (Cade 1734)—Dau. 1755 of Anc. Starling 1738 (Starling 1727).
			Decoy, b. m., 1830.	Filho-da-Puta, br. h., 1812.	Haphazard 1797 (Sir Peter T. 1784)—Mrs. Barnet 1806 by Waxy 1790 (Pot-8-o's 1773)—Dau. 1788 of Woodpecker 1773 (Herod 1758)—Heinel 1771 by Squirrel 1754 (Traveller 1735)—Principessa 1762 by Blank 1740 (God. Ar. 1724).
				Finesse, b. m., 1815.	Peruvian 1806 (Sir Peter T. 1784)—Violante 1802 by John Bull 1789 (Fortitude 1777)—Dau. 1788 of Highflyer 1774 (Herod 1758)—Everlasting 1775 by Eclipse 1764 (Marske 1750)—Hyæna 1762 by Snap 1750 (Snip 1736).
		Silence, b. m., 1848.	Melbourne, br. h., 1834.	Humphrey Clinker, b. h., 1822.	Comus 1809 (Sorcerer 1796)—Clinkerina 1812 by Clinker 1805 (Sir Peter T. 1784)—Pewet 1786 by Tandem 1773 (Syphon 1750)—Termagant by Tantrum 1760 (Cripple 1750)—Cantatrice by Sampson 1745 (Blaze 1733).
				Daughter of, b. m., 1825.	Cervantes 1806 (Don Quixote 1784)—Dau. 1818 of Golumpus 1802 (Gohanna 1790)—Dau. 1810 of Paynator 1791 (Trumpator 1782)—Circle 1802 by St. George 1789 (Highflyer 1774)—Abigail 1788 by Woodpecker 1773 (Herod 1758).
			Secret, ch. m., 1841.	Hornsea, ch. h., 1832.	Velocipede 1825 (Blacklock 1814)—Dau. 1820 of Cerberus 1802 (Gohanna 1790)—Miss Cranfield 1803 by Sir Peter T. 1784 (Highflyer 1774)—Dau. 1796 of Pegasus 1784 (Eclipse 1764)—Dau. 1783 of Paymaster 1766 (Blank 1740).
				Solace, b. m., 1830.	Longwaist 1821 (Whalebone 1807)—Dulcamara 1818 by Waxy 1790 (Pot-8-o's 1773)—Witchery 1814 by Sorcerer 1796 (Trumpator 1782)—Cobbêa 1802 by Skyscraper 1786 (Highflyer 1774)—Dau. 1788 of Woodpecker 1773 (Herod 1758).
SEA MARK, ch. m., 1873.	Adventurer, b. h., 1859.	Newminster, b. h., 1848.	Touchstone, br. h., 1831.	Camel, br. h., 1822.	Whalebone 1807 (Waxy 1790)—Dau. 1812 of Selim 1802 (Buzzard* 1787)—Maiden 1801 by Sir Peter T. 1784 (Highflyer 1774)—Dau. 1788 of Phenomenon 1780 (Herod 1758)—Matron 1782 by Florizel 1768 (Herod 1758).
				Banter, br. m., 1826.	Master Henry 1815 (Orville 1799)—Boadicea 1807 by Alexander 1782 (Eclipse 1764)—Brunette by Amaranthus 1766 (Old England 1741)—Mayfly 1771 by Matchem 1748 (Cade 1734)—Dau. 1735 of Anc. Starling 1738 (Old Starling 1727).
			Bee's-wing, b. m., 1833.	Dr. Syntax, br. h., 1811.	Paynator 1791 (Trumpator 1782)—Dau. of Beningbrough 1791 (King Fergus 1775)—Jenny Mole 1787 by Carbuncle 1772 (Babr. Blank 1758)—Dau. 1778 of Prince T'Quassaw 1751 (Snip 1736)—Sultana 1759 by Regulus 1739 (God. Ar. 1724).
				Daughter of, ch. m., 1817.	Ardrossan 1809 (John Bull 1789)—Lady Eliza 1813 by Whitworth 1805 (Agonistes 1797)—N. Y. Z.'s dam 1793 by Spadille 1784 (Highflyer 1774)—Sylvia 1783 by Y. Marske 1771 (Marske 1750)—Ferret 1765 by Broth. to Silvio (Cade 1734).
		Palma, br. m., 1840.	Emilius, b. h., 1820.	Orville, b. h., 1799.	Beningbrough 1791 (King Fergus 1775)—Evelina 1791 by Highflyer 1774 (Herod 1758)—Termagant by Tantrum 1760 (Cripple 1750)—Cantatrice by Sampson 1745 (Blaze 1733)—Dau. 1757 of Regulus 1739 (God. Ar. 1724).
				Emily, ch. m., 1810.	Stamford 1794 (Sir Peter T. 1784)—Dau. 1799 of Whiskey 1789 (Saltram* 1780)—Grey Dorimant 1781 by Dorimant 1772 (Otho 1760)—Dizzy 1757 by Blank 1740 (God. Ar. 1724)—Anc. Dizzy 1741 by Driver 1727 (Wynn Ar.)
			Francesca, br. m., 1821.	Partisan, b. h., 1811.	Walton 1799 (Sir Peter T. 1784)—Parasol 1800 by Pot-8-o's 1773 (Eclipse 1764)—Prunella 1788 by Highflyer 1774 (Herod 1758)—Promise 1768 by Snap 1750 (Snip 1736)—Julia 1756 by Blank 1740 (God. Ar. 1724).
				Miss Fanny's dam, br. m., 1815.	Orville 1799 (Beningbrough 1791)—Dau. 1800 of Buzzard* 1787 (Woodpecker 1773)—Hornpipe 1793 by Trumpator 1782 (Conductor 1767)—Luna 1779 by Herod 1758 (Tartar 1743)—Proserpine 1766 (Eclipse's sist.) by Marske 1750 (Squirt 1732).
	Sea Gull, blk. m., 1866.	Lifeboat, br. h., 1855.	Sir Hercules, blk. h., 1826.	Whalebone, br. h., 1807.	Waxy 1790 (Pot-8-o's 1773)—Penelope 1798 by Trumpator 1782 (Conductor 1767)—Prunella 1788 by Highflyer 1774 (Herod 1758)—Promise 1768 by Snap 1750 (Snip 1736)—Julia 1756 by Blank 1740 (God. Ar. 1724).
				Peri, b. m., 1822.	Wanderer 1811 (Gohanna 1790)—Thalestris 1809 by Alexander 1782 (Eclipse 1764)—Rival 1800 by Sir Peter T. 1784 (Highflyer 1774)—Hornet 1790 by Drone 1777 (Herod 1758)—Manilla 1777 by Goldfinder 1764 (Snap 1750).
			Yard-Arm, b. m., 1843.	Sheet Anchor, br. h., 1832.	Lottery 1820 (Tramp 1810)—Morgiana 1820 by Muley 1810 (Orville 1799)—Miss Stephenson 1814 by [Scud or] Sorcerer 1796 (Trumpator 1782)—Sist. 1793 to Petworth by Precipitate* 1787 (Mercury 1778)—Dau. 1787 of Woodpecker 1773 (Herod 1758).
				Fanny Kemble, b. m., 1827.	Paulowitz 1813 (Sir Paul 1802)—Loyalty 1817 by Rubens 1805 (Buzzard* 1787)—Pennyroyal 1805 by Coriander 1786 (Pot-8-o's 1773)—Peppermint 1787 by Highflyer 1774 (Herod 1758)—Promise 1768 by Snap 1750 (Snip 1736).
		Wild Cherry, blk. m., 1854.	Surplice, b. h., 1845.	Touchstone, br. h., 1831.	Camel 1822 (Whalebone 1807)—Banter 1826 by Master Henry 1815 (Orville 1799)—Boadicea 1807 by Alexander 1782 (Eclipse 1764)—Brunette by Amaranthus 1766 (Old England 1741)—Mayfly 1771 by Matchem 1748 (Cade 1734).
				Crucifix, b. m., 1837.	Priam* 1827 (Emilius 1820)—Octaviana 1815 by Octavian 1807 (Stripling 1795)—Dau. 1807 of Shuttle 1793 (Y. Marske 1771)—Zara 1801 by Delpini 1781 (Highflyer 1774)—Flora 1789 by King Fergus 1775 (Eclipse 1764).
			Bridget, br. m., 1846.	[Rocket or] Cardinal Puff, br. h., 1834.	Pantaloon 1824 (Castrel 1801)—Puff 1823 by Waterloo 1811 (Walton 1799)—Blowing 1808 by Buzzard* 1787 (Woodpecker 1773)—Dau. 1796 of Pot-8-o's 1773 (Eclipse 1764)—Maid-of-all-Work 1786 by Highflyer 1774 (Herod 1758).
				Ringlet, br. m., 1826.	Whisker 1812 (Waxy 1790)—Clinkerina br. m. 1812 by Clinker 1805 (Sir Peter T. 1784)—Pewet b. m. 1786 by Tandem 1773 (Syphon 1750)—Termagant by Tantrum 1760 (Cripple 1750)—Cantatrice by Sampson 1745 (Blaze 1733)—Dau. br. m. 1757 by Regulus 1739 (God. Ar. 1724)—The Raby mare (Marske's dam) by Hutton's

Blacklegs 1723 (Mulso Turk)—Dau. of Bay Bolton 1705 (Grey Hautboy)—Dau. of Fox Cub 1714 (Clumsy)—Dau. of Coneyskins 1702 (Lister Turk)—Dau. of Hutton's Grey Barb 1700—Dau. of Hutton's Royal Colt 1700 (Helmsley Turk)—Dau. of Byerly Turk—Dau. of Bustler (Helmsley Turk).

IMPORTED STONEHENGE.

WINNER OF THE COTTAGERS' CUP AT LONG BRANCH, 1873.

In the Brookdale Stud, Estate of D. D. Withers, proprietor, near Holmdel, N. J. Private stallion. Sales annually.

STONEHENGE by Blair Athol, best son of Stockwell, bred by Mr. W. Blenkiron, foaled 1870, dam Coimbra, dam Light of Wine, etc., by Kingston, son of Venison, out of Calcavella, dam of Queen Bee, etc., by Irish Birdcatcher. His sire, Blair Athol, won the double events, Derby and St. Leger, and his sire, Stockwell, was the greatest and grandest of all the English stallions. Kingston, the sire of his dam, traveled all over England, and was winner in the best of company. Irish Birdcatcher by Sir Hercules, son of Whalebone, was not only a good race-horse but he got The Baron, winner of the St. Leger, and he got Stockwell and Rataplan. Birdcatcher was also sire of Daniel O'Rourke, a Derby winner, Songstress, winner of the Oaks, Knight of St. George and Warlock, winners of the St. Leger. Faugh-a-Ballagh, the sire of Leamington, was own brother to Birdcatcher. Stonehenge made his debut as a two-year old in the Nursery Stakes at Jerome Park, but was unplaced. This was his only start in his two-year old form. When three years old he started six times, won one race, the Cottagers' Cup, at Long Branch, 1½ miles, in 2.44½; he was third to Tom Bowling in the Robins Stake, with Lizzie Lucas and The Minstrel behind him; was third to Survivor and Fellowcraft in a race of 1¼ miles, with four others behind. He was unplaced in his other races. His son, Marsh Redon, showed himself a good two-year old. He won the Surf Stakes at Sheepshead Bay, 5 furlongs, in 1.03¼, beating Onondaga and Memento; won August Stakes at Long Branch, three-quarters of a mile, in 1.16, with 115 lbs., beating Wyoming, Onondaga and Memento; was second in 1882 to Forester, in the Withers Stakes; second to Runnymede, in the Tidal Stakes at Sheepshead Bay, 1 mile, in 1.43¾; won Club Purse, 1 mile, in 1.43¼, beating Barrett and Fellowplay. Buckstone (2) won the Moët and Chandon Stakes, three-quarters of a mile, in 1.21, track heavy, defeating a field of five; won the Homebred Produce Stakes, three-quarters of a mile, in 1.18½, both at Long Branch. He is a blood bay, with blaze face and snip, near hind leg roan and white to the hock, is 15 hands 2½ inches high, and weighs 1,140 lbs. With a Glencoe cross, through Stockwell, and a double cross of Whalebone, through Irish Birdcatcher, fortified by the blood of Eclipse and Herod, he only needed a chance to distinguish himself as a sire.

Stonehenge is the sire of the winners Hopeless, Judge Griffith, Frolic, Jack, Salisbury, Gold Star, Alice, Druidess, Elmstone, Richard K. Fox. Overstone, Adamant, Rosa, Rose, Stonemason, Genevieve, Soho, Necktie, Stonenell, etc.

Stonehenge descends in direct female line from the Sister to Stripling, to which traces Burgundy, Ambrose, Dundee, Brown Bread, Van Amburgh, Tibthorpe, St. Gatien, Ormonde, Kendal, Muncaster, imp. Stylites, imp Rossington, and all the Agnes family in England.

STONEHENGE.

BAY HORSE, FOALED 1870; BRED BY MR. BLENKIRON, ENGLAND.

Waxy 1790 Pot-8-o's 1773 Eclipse 1764 Marske 1750 Squirt 1732 Bartlett's Childers Darley Arabian 1699.

BLAIR ATHOL, ch. h., 1861.	Stockwell, ch. h., 1849.	The Baron, ch. h., 1842.	Ir. Birdcatcher, ch. h., 1833.	Sir Hercules, blk. h., 1826.	Whalebone 1807 (Waxy 1790) Peri 1822 by Wanderer 1811 (Gohanna 1790)—Thalestris 1809 by Alexander 1782 (Eclipse 1764) Rival 1800 by Sir Peter T. 1784 (Highflyer 1774) Hornet 1790 by Drone 1777 (Herod 1758).
				Guiccioli, ch. m., 1823.	Bob Booty 1804 (Chanticleer 1787)—Flight 1809 by Ir. Escape 1802 (Commodore 1793) Y. Heroine by Bagot 1780 (Herod 1758) Heroine 1775 by Hero 1753 (Cade 1734) Dau. 1766 of Snap 1750 (Snip 1736).
			Echidna, b. m., 1838.	Economist, b. h., 1825.	Whisker 1812 (Waxy 1790)—Floranthe 1818 by Octavian 1807 (Stripling 1795) Caprice 1797 by Anvil 1777 (Herod 1758)—Madcap 1774 by Eclipse 1764 (Marske 1750) Sappho 1763 by Blank 1740 (God. Ar. 1724).
				Miss Pratt, b. m., 1825.	Blacklock 1814 (Whitelock 1803) Gadabout 1812 by Orville 1799 (Beningbrough 1791) Minstrel 1803 by Sir Peter T. 1784 (Highflyer 1774)—Matron 1782 by Florizel 1768 (Herod 1758)—Maiden 1770 by Matchem 1748 (Cade 1734).
		Pocahontas, b. m., 1837.	Glencoe*, ch. h., 1831.	Sultan, b. h., 1816.	Selim 1802 (Buzzard 1787) Bacchante 1809 by W's Ditto 1800 (Sir Peter T. 1784) Dau. 1791 of Mercury 1778 (Eclipse 1764)—Dau. 1776 of Herod 1758 (Tartar 1743) Folly 1771 by Marske 1750 (Squirt 1732).
				Trampoline, ch. m., 1825.	Tramp 1810 (Dick Andrews 1797) Web 1808 by Waxy 1790 (Pot-8-o's 1773) Penelope 1798 by Trumpator 1782 (Conductor 1767) Prunella 1788 by Highflyer 1774 (Herod 1758) Promise 1768 by Snap 1750 (Snip 1736).
			Marpessa, b. m., 1830.	Muley, b. h., 1810.	Orville 1799 (Beningbrough 1791) Eleanor 1798 by Whiskey 1789 (Saltram 1780) Y. Giantess 1790 by Diomed* 1777 (Florizel 1768) Giantess 1769 by Matchem 1748 (Cade 1734) Molly Long Legs 1753 by Babraham 1740 (God. Ar. 1724).
				Clare, b. m., 1824.	Marmion 1806 (Whiskey 1789) Harpalice 1814 by Gohanna 1790 (Mercury 1778) Amazon 1799 by Driver 1783 (Trentham 1766)—Fractious 1792 by Mercury 1778 (Eclipse 1764) Dau. 1785 of Woodpecker 1773 (Herod 1758).
	Blink Bonny, b. m., 1854.	Melbourne, br. h., 1834.	Humphrey Clinker, b. h., 1822.	Comus, ch. h., 1809.	Sorcerer 1796 (Trumpator 1782) Houghton Lass 1801 by Sir Peter T. 1784 (Highflyer 1774) Alexina 1788 by King Fergus 1775 (Eclipse 1764) Lardella 1780 by Y. Marske 1771 (Marske 1750) Dau. of Y. Cade 1747 (Cade 1734).
				Clinkerina, br. m., 1812.	Clinker 1805 (Sir Peter T. 1784)—Pewet 1786 by Tandem 1773 (Syphon 1750) Termagant by Tantrum 1760 (Cripple 1750)—Cantatrice by Sampson 1745 (Blaze 1733) Dau. 1757 of Regulus 1739 (God. Ar. 1724).
			Daughter of, b. m., 1825.	Cervantes, b. h., 1806.	Don Quixote 1784 (Eclipse 1764) Evelina 1791 by Highflyer 1774 (Herod 1758)—Termagant by Tantrum 1760 (Cripple 1750) Cantatrice by Sampson 1745 (Blaze 1733)—Dau. 1757 of Regulus 1739 (God. Ar. 1724).
				Daughter of, br. m., 1818.	Golumpus 1802 (Gohanna 1790)—Dau. 1810 of Paynator 1791 (Trumpator 1782)—Circle 1802 by St. George 1789 (Highflyer 1774)—Abigail 1788 by Woodpecker 1773 (Herod 1758)—Firetail 1772 by Eclipse 1764 (Marske 1750).
		Queen Mary, b. m., 1843.	Gladiator, ch. h. 1833.	Partisan, b. h., 1811.	Walton 1799 (Sir Peter 1784) Parasol 1800 by Pot-8-o's 1773 (Eclipse 1764) Prunella 1788 by Highflyer 1774 (Herod 1758)—Promise 1768 by Snap 1750 (Snip 1736) Julia 1756 by Blank 1740 (God. Ar. 1724).
				Pauline, b. m., 1826.	Moses 1819 ([Whalebone or] Seymour 1807)—Quadrille 1815 by Selim 1802 (Buzzard* 1787)—Canary Bird 1806 by [Whiskey or] Sorcerer 1796 (Trumpator 1782)—Canary 1797 by Coriander 1786 (Pot-8-o's 1773)—Miss Green 1787 by Highflyer 1774 (Herod 1758).
			Daughter of, b. m., 1840.	Plenipotentiary, ch. h., 1831.	Emilius 1820 (Orville 1799)—Harriet 1819 by Pericles 1809 (Evander 1801)—Dau. 1812 of Selim 1802 (Buzzard* 1787) Pipylina 1803 by Sir Peter T. 1784 (Highflyer 1774) Rally 1790* by Trumpator 1782 (Conductor 1767).
				Myrrha, b. m., 1830.	Whalebone 1807 (Waxy 1790)—Gift 1818 by Y. Gohanna 1810 (Gohanna 1790)—Dau. 1802 of Sir Peter T. 1784 (Highflyer 1774)—Dau. 1788 of Trumpator 1782 (Conductor 1767)—Sist. 1774 to Postmaster by Herod 1758 (Tartar 1743).
COIMBRA, b. m., 1861.	Kingston, b. h., 1849.	Venison, br. h., 1833.	Partisan, b. h., 1811.	Walton, b. h., 1799.	Sir Peter T. 1784 (Highflyer 1774)—Arethusa 1792 by Dungannon 1780 (Eclipse 1764)—Dau. 1777 of Prophet 1760 (Regulus 1739)—Virago 1764 by Snap 1750 (Snip 1736)—Dau. of Regulus 1739 (God. Ar. 1724).
				Parasol, b. m., 1800.	Pot-8-o's 1773 (Eclipse 1764)—Prunella 1788 by Highflyer 1774 (Herod 1758)—Promise 1768 by Snap 1750 (Snip 1736)—Julia 1756 by Blank 1740 (God. Ar. 1724)—Spectator's dam 1735 by Partner 1718 (Jigg).
			Fawn, br. m., 1823.	Smolensko, blk. h., 1810.	Sorcerer 1796 (Trumpator 1782) Wowski 1797 by Mentor 1784 (Justice 1774) Maria 1777 by Herod 1758 (Tartar 1743)—Lisette 1772 by Snap 1750 (Snip 1736)—Miss Windsor 1754 by God. Ar. 1724.
				Jerboa, b. m., 1803.	Gohanna 1790 (Mercury 1778)—Camilla 1778 by Trentham 1766 (T. Sweepstakes 1743)—Coquette 1764 by Compton Barb Sist. 1743 to Regulus by God. Ar. 1724—Grey Robinson 1723 by Bald Galloway (St. Victor Barb).
		Queen Anne, b. m., 1843.	Slane, b. h., 1833.	Royal Oak, br. h., 1823.	Catton 1809 (Golumpus 1802)—Dau. 1818 of Smolensko 1810 (Sorcerer 1796)—Lady Mary 1800 by Beningbrough 1791 (King Fergus 1775)—Dau. of Highflyer 1774 (Herod 1758)—Sist. to Miss Bell by Marske 1750 (Squirt 1732).
				Daughter of, b. m., 1819.	Orville 1799 (Beningbrough 1791)—Epsom Lass 1803 by Sir Peter T. 1784 (Highflyer 1774)—Alexina 1788 by King Fergus 1775 (Eclipse 1764)—Lardella 1780 by Y. Marske 1771 (Marske 1750)—Dau. of Y. Cade 1747 (Cade 1734).
			Garcia, br. m., 1822.	Octavian, ch. h., 1807.	Stripling 1795 (Phenomenon 1780)—Dau. 1797 of Oberon 1790 (Highflyer 1774)—Sist. 1779 to Sharper by Ranthos 1763 (Matchem 1748)—Dau. of T. Sweepstakes 1743 (Bolt. Sweepstakes 1722)—Dau. 1748 of Hutton's Spot 1728 (Hartley's Blind Horse).
				Daughter of, b. m., 1806.	Shuttle 1793 (Y. Marske 1771)—Katharine 1798 by Delpini 1781 (Highflyer 1774)—Dau. 1786 of Paymaster 1766 (Blank 1740)—Dau. 1774 of Le Sang 1759 (Changeling 1747)—Dau. 1751 of Rib 1736 (Crab 1722)
	Calcavella, b. m., 1844.	Ir. Birdcatcher, ch. h., 1833.	Sir Hercules, blk. h., 1826.	Whalebone, br. h., 1807.	Waxy 1790 (Pot-8-o's 1773)—Penelope 1798 by Trumpator 1782 (Conductor 1767)—Prunella 1788 by Highflyer 1774 (Herod 1758) Promise 1768 by Snap 1750 (Snip 1736)—Julia 1756 by Blank 1740 (God. Ar. 1724).
				Peri, b. m., 1822.	Wanderer 1811 (Gohanna 1790)—Thalestris 1809 by Alexander 1782 (Eclipse 1764)—Rival 1800 by Sir Peter T. 1784 (Highflyer 1774)—Hornet 1790 by Drone 1777 (Herod 1758)—Manilla 1777 by Goldfinder 1764 (Snap 1750).
			Guiccioli, ch. m., 1823.	Bob Booty, ch. h., 1804.	Chanticleer 1787 (Woodpecker 1773)—Ierne 1790 by Bagot 1780 (Herod 1758) Dau. of Gamahoe (Bustard 1741)—Patty by Tim 1752 (Squirt 1732)—Miss Patch 1737 by Justice 1725 (H. C. Litton Ar.).
				Flight, ch. m., 1809.	Ir. Escape 1802 (Commodore 1793)—Y. Heroine by Bagot 1780 (Herod 1758)—Heroine 1775 by Hero 1753 (Cade 1734)—Dau. 1766 of Snap 1750 (Snip 1736) Sist. 1743 to Regulus by God. Ar. 1724.
		Caroline, b. m., 1836.	Ir. Drone, gr. h., 1823.	Master Robert, gr. h., 1811.	Buffer 1798 (Prizefighter 1781) Spinster 1805 by Shuttle 1793 (Y. Marske 1771) Dau. 1797 of Sir Peter T. 1784 (Higflyer 1774) Bab 1787 by Bourdeaux 1774 (Herod 1758)—Speranza 1778 by Eclipse 1764 (Marske 1750).
				Daughter of, br. m., 1817.	Sir Walter Raleigh 1801 (Waxy 1790)—Miss Tooley 1808 by Teddy the Grinder 1798 (Asparagus 1787)—Lady Jane 1796 by Sir Peter T. 1784 (Highflyer 1774)—Paulina 1778 by Florizel 1768 (Herod 1758)—Captive 1771 by Matchem 1748 (Cade 1734).
			The Potentate's dam, b. m., 1823.	Don Juan, b. h., 1814.	Orville 1799 (Beningbrough 1791)—Peterea 1804 by Sir Peter T. 1784 (Highflyer 1774)—Mary Grey 1781 by Friar 1768 (Hero 1753)—Timante by Tim 1752 (Squirt 1732)—Noblessa 1769 by Gamahoe (Bustard 1741).
				Moll-in-the-Wad, br. m., 1810.	Hambletonian 1792 (King Fergus 1775) Spitfire b. m. 1799 by Pipator 1786 (Imperator 1768) Farewell b. m. 1792 by Slope 1782 (Highflyer 1774)—Dau. of Y. Marske 1771 (Marske 1750)—Dau. of Broth. to Silvio (Cade 1734)—Sist. to Stripling by Hutton's Spot 1728 (Hartley's Blind Horse).

STRATFORD.

Will be used as private stallion in the Chesterbrook Stud, Berwyn, Pa., the property of A. J. Cassatt, Esq.

STRATFORD (Leamington 2d) by imp. Leamington, son of Faugh-a-Ballagh, bred by A. Welch, Erdenheim Stud, Pa., foaled 1873, dam Susan Beane, dam of Sensation by Lexington, son of Boston, out of Sally Lewis, dam of Hunter's Lexington, John Morgan, Acrobat, etc., by imp. Glencoe. Stratford was an exceedingly promising horse as a two-year old, but, having been taken with a severe attack of distemper, was totally unable to start, and the disease left him in such a poor plight, having injured his lungs, so that whilst he had speed he was unable to stay a distance. At three years old he started in fifteen races, winning two, running second in five, third in one, and unplaced in the others. Point Breeze Park, Pa., won purse, three-quarters of a mile, in 1.17½, beating Durango, Paladin, and four others; same place, won purse, heats of half a mile, in 0.49¾, 0.49½, 0.49¼, beating First Chance, who won first heat, Hobkirk and Grey Lag, the fastest three heats ever run to that date. Jerome Park, ran second to Ambush, 1 mile, in 1.48, beating Woodlands. Long Branch, ran second to Romney, three-quarters of a mile, in 1.18, Faithless, Hattie F. and three others behind him. Point Breeze Park, ran second to First Chance, three-quarters of a mile, in 1.15, beating Culpepper and two others; ran second to Waco, 1 mile, in 1.44, beating Hobkirk, Emma G. and another. Washington, D. C., ran second to Burgoo, 1¼ miles, in 2.14, beating First Chance, Fadladeen and one other. As a four-year, ran unplaced in three races. Stratford is a highly bred horse, own brother to the great Sensation and Onondaga, is bred from one of the great racing families of America, a family which were almost uniformly large and profitable winners on the turf and very successful as stallions and brood mares, the family which produced Sensation, John Morgan, Acrobat, Hunter's Lexington, Motto, Nannie Lewis, Glenmore and a host of others, possessing speed, conformation, racing lineage and good blood.

Stratford is a brown, 15¾ hands, with a Leamington blaze in his face and black legs; he has magnificent shoulders, good middle-piece, good back, hip and loin, and excellent legs and feet. He much resembles his sire in general conformation and appearance.

Stratford is the sire of the winners Columbine, Connemara, Lansdown, Now or Never, Tarragon, Bessie, Lady Archer, Leopold, Felicia II., Shamrock, Seymour, Emma J., Foxford, Francis S., Greylock, Josie W., Judge Mitchell, Nellie Bly, Stratagem, Octagon, Tartar, Donohue, Baltimore II., Sir Richard, Sweet Alice, etc. A success, considering that he is a private stallion.

Stratford is descended in direct female line from the old Montagu mare and a Royal mare, to which trace Young Marske, Conductor, O'Kelly's Eclipse, Voltaire, Sterling, Oxford, Kingston, Marsyas, Scottish Chief, The Duke, Adventurer, Prince Charlie, Lexington, Vandal, Luke Blackburn, Leonatus, Salvator, Sensation, The Bard, Chesapeake, Potomac, Longstreet, John Morgan, Hunter's Lexington, and all the Levity and Lady Thompkins families.

STRATFORD.

BROWN HORSE, FOALED 1873; BRED BY MR. A. WELCH, CHESTNUT HILL STUD, PENNSYLVANIA.

*LEAMINGTON, br. h., 1853.
- Faugh-a-Ballagh, br. h., 1841.
  - Sir Hercules, blk. h., 1826.
    - Whalebone, br. h., 1807.
    - Peri, b. m., 1822.
  - Guiccioli, ch. m., 1823.
    - Bob Booty, ch. h., 1804.
    - Flight, ch. m., 1809.
- Daughter of, b. m., 1841.
  - Pantaloon, ch. h., 1824.
    - Castrel, ch. h., 1801.
    - Idalia, ch. m., 1815.
  - Daphne, br. m., 1837.
    - Laurel, br. h., 1824.
    - Maid of Honor, br. m., 1829.

SUSAN BEANE, ch. m., 1868.
- Lexington, b. h., 1850.
  - Boston, ch. h., 1833.
    - Timoleon, ch. h., 1813.
    - Sist. to Tuckahoe, ch. m., 1814.
  - Alice Carneal, b. m., 1836.
    - *Sarpedon, br. h., 1828.
    - Rowena, ch. m., 1826.
- Sally Lewis, ch. m., 1852.
  - *Glencoe, ch. h., 1831.
    - Sultan, b. h., 1816.
    - Trampoline, ch. m., 1825.
  - Motto, ch. m., 1839.
    - *Barefoot, ch. h., 1820.
    - Lady Thompkins, ch. m., 1831.

Waxy, b. h., 1790.	Pot-8-o-'s 1773 (Eclipse 1764)—Maria 1777 by Herod 1758 (Tartar 1743)—Lisette 1772 by Snap 1750 (Snip 1736)—Miss Windsor 1754 by God. Ar. 1724—Sist. to Volunteer by Y. Belgrade (Belgrade Turk).
Penelope, b. m., 1798.	Trumpator 1782 (Conductor 1767)—Prunella 1788 by Highflyer 1774 (Herod 1758)—Promise 1768 by Snap 1750 (Snip 1736)—Julia 1756 by Blank 1740 (God. Ar. 1724)—Spectator's dam 1735 by Partner 1718 (Jigg).
Wanderer, b. h., 1811.	Gohanna 1790 (Mercury 1778)—Catherine 1795 by Woodpecker 1773 (Herod 1758)—Camilla 1778 by Trentham 1766 (T. Sweepstakes 1743)—Coquette 1764 by Compton Barb—Sist. 1743 to Regulus by God. Ar. 1724.
Thalestris, blk. m., 1809.	Alexander 1782 (Eclipse 1764)—Rival 1800 by Sir Peter T. 1784 (Highflyer 1774)—Hornet 1790 by Drone 1777 (Herod 1758)—Manilla 1777 by Goldfinder 1764 (Snap 1750)—Dau. 1766 of Old England 1741 (God. Ar. 1724).
Chanticleer, ch. h., 1787.	Woodpecker 1773 (Herod 1758)—Dau. 1778 of Eclipse 1764 (Marske 1750)—Rosebud 1765 by Snap 1750 (Snip 1736)—Miss Belsea 1753 by Regulus 1739 (God. Ar. 1724)—Dau. of Bartlett's Childers (Darley Ar. 1699).
Ierne, ch. m., 1790.	Bagot 1780 (Herod 1758)—Dau. of Gamahoe (Bustard 1741)—Patty by Tim 1752 (Squirt 1732)—Miss Patch 1737 by Justice 1725 (H. C. Litton Ar.)—Ringtail Galloway by Curwen Bay Barb.
Ir. Escape, ch. h., 1802.	Commodore 1793 (Tom Tug 1777)—Moll-in-the-Wad 1791 by Highflyer 1774 (Herod 1758)—Shift 1779 by Sweetbriar 1769 (Syphon 1750)—Black Susan 1762 by Snap 1750 (Snip 1736)—Dau. of Cade 1734 (God. Ar. 1724).
Y. Heroine, . .	Bagot 1780 (Herod 1758)—Heroine 1775 by Hero 1753 (Cade 1734)—Dau. 1766 of Snap 1750 (Snip 1736)—Sist. 1743 to Regulus by God. Ar. 1724—Grey Robinson 1723 by Bald Galloway (St. Victor Barb).
Buzzard,* ch. h., 1787.	Woodpecker 1773 (Herod 1758)—Misfortune 1775 by Dux 1761 (Matchem 1748)—Curiosity 1760 by Snap 1750 (Snip 1736)—Dau. 1749 of Regulus 1739 (God. Ar. 1724)—Dau. of Bartlett's Childers (Darley Ar. 1699).
Daughter of, b. m., .	Alexander 1782 (Eclipse 1764)—Dau. of Highflyer 1774 (Herod 1758)—Dau. of Alfred 1770 (Matchem 1748)—Dau. 1770 of Engineer 1756 (Sampson 1745)—Bay Malton's dam 1756 by Cade 1734 (God. Ar. 1724).
Peruvian, b. h., 1806.	Sir Peter T. 1784 (Highflyer 1774)—Dau. 1788 of Boudrow 1777 (Eclipse 1764)—Dau. 1764 of Squirrel 1754 (Traveller 1735)—Sist. to Lowther's Babraham by Babraham 1740 (God. Ar. 1724)—Dau. 1750 of Golden Ball 1735 (Partner 1718).
Musidora, b. m., 1804.	Meteor 1783 (Eclipse 1764)—Maid-of-all-Work 1786 by Highflyer 1774 (Herod 1758)—Dau. 1771 of Syphon 1750 (Squirt 1732)—Dau. of Regulus 1739 (God. Ar. 1724)—Sist. 1758 to Judgement by Snip 1736 (Flying Childers 1715).
Blacklock, b. h., 1814.	Whitelock 1803 (Hambletonian 1792)—Dau. 1799 of Coriander 1786 (Pot-8-o-'s 1773)—Wildgoose 1792 by Highflyer 1774 (Herod 1758)—Coheiress 1786 by Pot-8-o-'s 1773 (Eclipse 1764)—Manilla 1777 by Goldfinder 1764 (Snap 1750).
Wagtail, b. m., 1818.	Prime Minister 1810 (Sancho 1801)—Tranby*'s dam 1812 by Orville 1799 (Beningbrough 1791)—Miss Grimstone 1796 by Weazle 1776 (Herod 1758)—Dau. of Ancaster 1768 (Blank 1740)—Dau. of Damascus Ar. 1754.
Champion, ch. h., 1812.	Selim 1802 (Buzzard* 1787)—Podagra 1805 by Gouty 1793 (Sir Peter T. 1784)—Jet 1785 by Magnet 1770 (Herod 1758)—Jewel 1775 by Squirrel 1754 (Traveller 1735)—Sophia 1754 by Blank 1740 (God. Ar. 1724).
Etiquette, br. m., 1820.	Orville 1799 (Beningbrough 1791)—Boadicea 1807 by Alexander 1782 (Eclipse 1764)—Brunette by Amaranthus 1766 (Old England 1741)—Mayfly 1771 by Matchem 1748 (Cade 1734)—Dau. 1755 of Starling 1738 (Starling 1727).
Sir Archy, b. h., 1805.	*Diomed 1777 (Florizel 1768)—*Castianira 1796 by Rockingham 1781 (Highflyer 1774)—Tabitha 1782 by Trentham 1766 (T. Sweepstakes 1743)—Dau. of Bosphorus 1754 (Babraham 1740)—Dau. of Forester 1750 (Forester 1736).
Daughter of, ch. m., —.	*Saltram 1780 (Eclipse 1764)—Dau. of Symmes' Wildair 1767 (*Fearnought 1755)—Dau. of Tyler's Driver 1760 (*Othello)—Dau. of *Fallower 1761 (Blank 1740)—Dau. of *Vampire 1757 (Regulus 1739).
Ball's Florizel, ch. h., 1801.	*Diomed 1777 (Florizel 1768)—Dau. of *Shark 1771 (Marske 1750)—Dau. of Harris' Eclipse 1770 (*Fearnought 1755)—Dau. of *Fearnought 1755 (Regulus 1739)—Dau. of *Jolly Roger 1741 (Roundhead 1733).
Daughter of, b. m., 1799.	*Alderman 1787 (Pot-8-o-'s 1773)—Dau. of *Clockfast 1774 (Gimcrack 1760)—Dau. of Symmes' Wildair 1767 (*Fearnought 1755)—Y. Kitty Fisher 1767 by *Fearnought 1755 (Regulus 1739)—*Kitty Fisher 1755 by Cade 1734 (God. Ar. 1724).
Emilius, b. h., 1820.	Orville 1799 (Beningbrough 1791)—Emily 1810 by Stamford 1794 (Sir Peter T. 1784)—Dau. 1799 of Whiskey 1789 (Saltram* 1780)—Grey Dorimant 1781 by Dorimant 1772 (Otho 1763)—Dizzy 1757 by Blank 1740 (God. Ar. 1724).
Icaria, b. m., 1824.	The Flyer 1814 (Vandyke Jr. 1808)—Parma 1813 by Dick Andrews 1797 (Joe Andrews 1778)—May 1804 by Beningbrough 1791 (King Fergus 1775)—Primrose 1787 by Mambrino 1768 (Engineer 1756)—Cricket 1776 by Herod 1758 (Tartar 1743).
Sumpter, ch. h., 1818.	Sir Archy 1805 (*Diomed 1777)—Dau. of *Robin Redbreast 1796 (Sir Peter T. 1784)—Dau. 1788 of *Obscurity 1777 (Eclipse 1764)—Slamerkin 1768 by *Wildair 1753 (Cade 1734)—*Cub mare 1762 by Cub 1739 (Fox 1714).
Lady Grey, b. m., 1817.	Robin Grey 1805 (*Royalist 1790)—Maria 1802 by Melzar 1791 (*Medley 1776)—Dau. of *Highflyer 1784 (Highflyer 1771)—Dau. of *Fearnought 1755 (Regulus 1739)—Dau. of Ariel 1756 (*Traveller 1747).
Selim, ch. h., 1802.	Buzzard* 1787 (Woodpecker 1773)—Dau. of Alexander 1782 (Eclipse 1764)—Dau. of Highflyer 1774 (Herod 1758)—Dau. of Alfred 1770 (Matchem 1748)—Dau. 1770 of Engineer 1756 (Sampson 1745).
Bacchante, br. m., 1809.	W's Ditto 1800 (Sir Peter T. 1784)—Dau. 1791 of Mercury 1778 (Eclipse 1764)—Dau. 1776 of Herod 1758 (Tartar 1743)—Folly 1771 by Marske 1750 (Squirt 1732)—Vixen 1753 by Regulus 1739 (God. Ar. 1724).
Tramp, b. h., 1810.	Dick Andrews 1797 (Joe Andrews 1778)—Dau. 1803 of Gohanna 1790 (Mercury 1778)—Fraxinella 1793 by Trentham 1766 (T. Sweepstakes 1743)—Dau. 1785 of Woodpecker 1773 (Herod 1758)—Everlasting 1775 by Eclipse 1764 (Marske 1750).
Web, b. m., 1808.	Waxy 1790 (Pot-8-o-'s 1773)—Penelope 1798 by Trumpator 1782 (Conductor 1767)—Prunella 1788 by Highflyer 1774 (Herod 1758)—Promise 1768 by Snap 1750 (Snip 1736)—Julia 1756 by Blank 1740 (God Ar. 1724).
Tramp, b. h., 1810.	Dick Andrews 1797 (Joe Andrews 1778)—Dau. 1803 of Gohanna 1790 (Mercury 1778)—Fraxinella 1793 by Trentham 1766 (T. Sweepstakes 1743)—Dau. 1785 of Woodpecker 1773 (Herod 1758)—Everlasting 1775 by Eclipse 1764 (Marske 1750).
Rosamond, ch. m., 1798.	Buzzard* 1787 (Woodpecker 1773)—Roseberry 1792 by Phenomenon 1780 (Herod 1758)—Miss West 1777 by Matchem 1748 (Cade 1734)—Sist. to Favourite by Regulus 1739 (God. Ar. 1724)—Y. Ebony 1742 by Crab 1722 (Alcock Ar.).
Am. Eclipse, ch. h., 1814.	Duroc 1806 (*Diomed 1777)—Miller's Damsel 1802 by *Messenger 1780 (Mambrino 1768)—*Mare 1792 by Pot-8-o-'s 1773 (Eclipse 1764)—Dau. 1778 of Gimcrack 1760 (Cripple 1750)—Snapdragon 1759 by Snap 1750 (Snip 1736).
Katy Ann, ch. m., 1825.	Ogle's Oscar 1800 (*Gabriel 1790)—Y. Maid-of-the-Oaks ch. m. 1817 by *Expedition 1795 (Pegasus 1784)—Maid-of-the-Oaks ch. m. 1801 by *Spread Eagle 1792 (Volunteer 1780)—Annette by *Shark 1771 (Marske 1750)—Dau. of Rockingham (Partner 1755)—Dau. of True Whig (Fitzhugh's Regulus 1765)—Dau. of Gallant 1770 (*Fearnought 1755)—Dau. of *Regulus 1747 (Regulus 1739)—*Duchess by Cullen Ar.—Lady Thigh ch. m. 1731 by Partner 1718 (Jigg)—Dau. ch. m. 1723 by Greyhound (Chillaby)—Sophonisba's dam by Curwen Bay Barb—Dau. of D'Arcy Chestnut Ar.—Dau. of Whiteshirt—Montagu mare by D'Arcy's old Montagu—Dau. of Hautboy (D'Arcy White Turk)—Dau. of Brimmer (D'Arcy Yellow Turk)—Royal mare.

THE BARD.

WINNER OF THE RED BANK, BOUQUET AND CAPITOL STAKES AT TWO YEARS OLD, THE PREAKNESS, SPINDRIFT, BARNEGAT, FREEHOLD, OMNIBUS, CHOICE, SEPTEMBER, JEROME, DIXIE, BRECKENRIDGE AND POTOMAC STAKES AT THREE YEARS OLD, SPECIAL CONEY ISLAND CUP TWICE, MONMOUTH CUP, BROOKLYN CUP AND OTHER GOOD RACES.

Is private stallion in the Chesterbrook Stud, Mr. A. J. Cassatt, proprietor, Berwyn, Chester Co., Pa. Yearlings sold annually.

THE BARD, bay horse, bred by Messrs. Chas. Reed & Sons, Fairview Stud, Tenn., by Longfellow, son of imp. Leamington, dam Brademante by War Dance, son of Lexington, grandam Brenna by imp. Knight of St. George, son of Birdcatcher, out of Levity by imp. Trustee. The Bard was unquestionably one of the grandest race-horses ever bred in America, and the best son of his great sire Longfellow. Brademante, his dam, was a superior race-mare, winner of two-mile heats in 3.32¼, 3.29, dam also of Vivacity (dam of Tenacity) and the winners Father John and Joseph. From the grandam, Brenna, came Athalaric (dam of the good horse Tom Martin), Strathmore (a successful young sire), and the winners Kinkead, Sir Walter, Long Stop and Violante, the last named of which won five-eighths of a mile at Westebester, in 0.59½. Great grandam Levity, whose descendants exceed in brilliancy those of any other mare in the world. Amongst the prominent ones may be named, Rurie, Lightsome (dam of Salina, Sprightly, Crucifix, Nevada and Gulnare), Mildred (dam of Ringlet, Nora Worth, Minx, Monarchist, Stamps, Sultana and Frederick the Great), Lever (sire of Leveller, Mahlstick, Apollo, Lou Lanier, etc.), Legatee (a good race-horse) and Sister to Rurie (dam of Grinstead, Viceroy, Kelpie, Ermengarde and Gilly Flower). From the grandaughters of Levity have come Salvator, Danicheff, Volturno, Elias Lawrence, Avalon, Aretino, Fairplay, Semper Felix (dam of Leonatus, Warder, Leopold, Felicia and Semper Vive, dam of Fordham, Folsom, and grandam of Longstreet) Blanch J., Cripple, St. Augustine, Luke Blackburn, Greenland, Ecuador, Bradford, Bravo, Sierra Nevada (dam of Mountain Range, Monte Rosa's dam), Gulnare (dam of Costello and Satisfaction), Restless, Wandering, Force, Startle, Tramp, Finality, Monitor, Mirth, Minnock, Louisette, Preciosa, Bertha, Start, Loch Tanna, Katie Pearce (dam of Lizzie S., Ballard, Redstone and Katrina), Janet (dam of Sir Peter, Golden Reel, Sinaloa, Volunteer's dam, and Spaldie, Lizzie D.'s dam), Miss Austine (dam of Tenton), Duke of Montrose, Annette, Kelp (dam of Onaway), Ovation, Overture, Baltic, Brocade (dam of Burch, Jennie B. and Bertha B., the dam of Lizzie Baker and Judge Murray), She, Padisha, Louisette (dam of Prince George), and any number of others.

As a two-year old started fourteen times, won three races, was second once and third in four races. Ran second to Portland in the Great Post Stakes, three-quarters of a mile, 115 lbs, in 1.17¾; ran third in the Seabright Stakes, three-quarters of a mile, Electric first, Buffalo second; won Red Bank Stakes, three-quarters of a mile, 108 lbs., in 1.16¼ Buffalo second, Charity third; ran third in Flatbush Stakes, 120 lbs., seven-eighths of a mile, in 1.31¼ Charity, 97 lbs., won, Dew Drop, 107 lbs., second; won Bouquet Stakes, seven eighths of a mile, 108 lbs., in 1.31¾, Inspector B. second, Florence Fonso third; ran third in the Eelectic Stakes, 1 mile, 110 lbs., 1.44¼, Dew Drop, 107 lbs., won, Winfred, 100 lbs., second; ran third in the Arlington Stakes, three-quarters of a mile, 118 lbs., 1.17, Bess, 108 lbs., won, Elkwood, 105 lbs., second. He wound up his two-year old career by winning the Capitol Stakes, 1 mile 115 lbs., in 1.48, beating Bess, Biggonet, etc. At three years old he started seventeen times, won eleven races, was five times second and once unplaced. He won Preakness Stakes, 1½ miles, 118 lbs., 2.45, Eurus second, Elkwood third; ran second in the Belmont Stakes, 1½ miles, 118 lbs., 2.41, Inspector B. won, Linden third; unplaced in the Green Grass Stakes, 1⅛ miles, won by Dry Monopole; ran second to Inspector B. in the Stud Stakes, 1⅝ miles, 118 lbs., in 2.23¾, beating Ban Fox and Buffalo. Ran second by a short head to Winfred 108½ lbs., in the Emporium, 1½ miles, 118 lbs., 2.43; ran dead heat with Dew Drop, 115 lbs., in the Spendthrift Stakes, 1¼ miles, 114 lbs., in 2.09½, the Bard walked over for the Stakes; won Barnegat Stakes, 1½ miles, 113 lbs., in 2.45½, in a canter from Quito, Winfred, etc.; ran second to Dew Drop, 118 lbs., in the Stevens Stakes, 1⅜ miles, 118 lbs., in 2.34½, beating Quito, Linden and Inspector B.; ran second to Charity, 106 lbs., in the Raritan Stakes, 1¼ miles, 118 lbs., 2.14, beating Linden, Pontico and Winfred. After this race he scored eight consecutive winnings, in the Freehold Stakes, 1½ miles, 100 lbs., 2.45½; Omnibus Stakes, 1½ miles, 118 lbs., in 2.39, beating by four lengths Dew Drop, Ben Ali, Charity, etc.; Choice Stakes, 1½

(Continued on page 268.)

THE BARD.

BAY HORSE, FOALED 1883; BRED BY CHAS. REED, & SONS, FAIRVIEW STUD, TENNESSEE.

LONGFELLOW, br. h., 1867.
- *Leamington, br. h., 1853.
  - Faugh-a-Ballagh, br. h., 1841.
    - Sir Hercules, blk. h., 1823.
    - Guiccioli, ch. m., 1823.
  - Daughter of, b. m., 1841.
    - Pantaloon, ch. h., 1824.
    - Daphne, br. m., 1837.
- Nantura, b. m., 1855.
  - Brawner's Eclipse, ch. h., 1833.
    - Am Eclipse, ch. h., 1814.
    - Daughter of, ch. m., 1834.
  - Quiz, b. m., 1836.
    - Bertrand, b. h., 1821.
    - Lady Fortune, ch. m., —.

BRADEMANTE, ch. m., 1874.
- War Dance, ch. h., 1859.
  - Lexington, b. h., 1850.
    - Boston, ch. h., 1833.
    - Alice Carneal, b. m., 1836.
  - Reel, gr. m., 1838.
    - *Glencoe, ch. h., 1831.
    - *Galopade, gr. m., 1828.
- Brenna, b. m., 1861.
  - *Knight of St. George, b. h., 1851.
    - Ir. Birdcatcher, ch. h., 1833.
    - Maltese, b. m., 1845.
  - Levity, b. m., 1845.
    - *Trustee, ch. h., 1829.
    - Vandal's dam, b. m., 1837.

Whalebone, br. h., 1807.	Waxy 1790 (Pot-8-o-'s 1773)—Penelope 1798 by Trumpator 1782 (Conductor 1767)—Prunella 1788 by Highflyer 1774 (Herod 1758)—Promise 1768 by Snap 1750 (Snip 1736)—Julia 1756 by Blank 1740 (God. Ar. 1724).
Peri, b. m., 1822.	Wanderer 1811 (Gohanna 1790)—Thalestris 1809 by Alexander 1782 (Eclipse 1764)—Rival 1800 by Sir Peter T. 1784 (Highflyer 1774)—Hornet 1790 by Drone 1777 (Herod 1758)—Manilla 1777 by Goldfinder 1764 (Snap 1750).
Bob Booty, ch. h., 1804.	Chanticleer 1787 (Woodpecker 1773)—Ierne 1790 by Bagot 1780 (Herod 1758)—Dau. of Gamahoe (Bustard 1741)—Patty by Tim 1752 (Squirt 1732)—Miss Patch 1752 by Justice 1745 (H. C. Litton Ar.).
Flight, ch. m., 1809.	Ir. Escape 1802 (Commodore 1793)—Y. Heroine by Bagot 1780 (Herod 1758)—Heroine 1775 by Hero 1753 (Cade 1734)—Dau. 1766 of Snap 1750 (Snip 1736)—Sist. 1743 to Regulus 1739 (God. Ar. 1724).
Castrel, ch. h., 1801.	Buzzard* 1787 (Woodpecker 1773)—Dau. of Alexander 1782 (Eclipse 1764)—Dau. of Highflyer 1774 (Herod 1758)—Dau. of Alfred 1770 (Matchem 1748)—Dau. 1770 of Engineer 1756 (Sampson 1745).
Idalia, ch. m., 1815.	Peruvian 1806 (Sir Peter T. 1784)—Musidora 1804 by Meteor 1783 (Eclipse 1764)—Maid-of-all-Work 1786 by Highflyer 1774 (Herod 1758)—Dau. 1771 of Syphon 1750 (Squirt 1732)—Dau. of Regulus 1739 (God. Ar. 1724).
Laurel, br. h., 1824.	Blacklock 1814 (Whitelock 1803)—Wagtail 1818 by Prime Minister 1810 (Sancho 1801)—Tranby*'s dam 1812 by Orville 1799 (Beningbrough 1791)—Miss Grimstone 1796 by Weazle 1776 (Herod 1758)—Dau. of Ancaster 1768 (Blank 1740).
Maid of Honor, br. m., 1829.	Champion 1812 (Selim 1802)—Etiquette 1820 by Orville 1799 (Beningbrough 1791)—Boadicea 1807 by Alexander 1782 (Eclipse 1764)—Brunette by Amaranthus 1766 (Old England 1741)—Mayfly 1771 by Matchem 1748 (Cade 1734).
Duroc, ch. h., 1806.	*Diomed 1777 (Florizel 1768)—Amanda 1800 by Grey Diomed 1786 (*Medley 1776)—Dau. of Va. Cade (Partner 1755)—Dau. of Hickman's Independence 1776 (*Fearnought 1755)—Dau. of *Badger (Bosphorus 1754).
Miller's Damsel, gr. m., 1802.	*Messenger 1780 (Mambrino 1768)—*Dau. 1792 of Pot-8-o-'s 1773 (Eclipse 1764)—Dau. 1778 of Gimcrack 1760 (Cripple 1750)—Snapdragon 1759 by Snap 1750 (Snip 1736)—Dau. 1749 of Regulus 1739 (God. Ar. 1724).
Henry, ch. h., 1819.	Sir Archy 1805 (*Diomed 1777)—Dau. of *Diomed 1777 (Florizel 1768)—Bellona by Bellair 1786 (*Medley 1776)—Indian Queen by Pilgrim 1771 (*Fearnought 1755)—Dau. of *Valiant 1756 (Dormouse 1738).
Y. Romp, ch. m., 1815.	Duroc 1806 (*Diomed 1777)—Old Romp 1805 by *Messenger 1780 (Mambrino 1768)—*Dau. 1792 of Pot-8-o-'s 1773 (Eclipse 1764)—Dau. 1778 of Gimcrack 1760 (Cripple 1750)—Snapdragon 1759 by Snap 1750 (Snip 1736).
Sir Archy, b. h., 1805.	*Diomed 1777 (Florizel 1768)—*Castianira 1796 by Rockingham 1781 (Highflyer 1774)—Tabitha 1782 by Trentham 1766 (T. Sweepstakes 1743)—Dau. of Bosphorus 1754 (Babraham 1740)—Sist. to Grecian Princess by Forester 1750 (Forester 1736).
Eliza, ch. m., 1804.	*Bedford 1792 (Dungannon 1780)—*Mambrina 1785 by Mambrino 1768 (Engineer 1756)—Sist. to Sally by Blank 1740 (God. Ar. 1724)—Dau. of Ward 1731 (Crab 1722)—Dau. of Merlin (Bustler).
[Brimmer or] Blue Beard, ch. h., 1802.	*Stirling 1791 (Volunteer 1780)—Brilliant 1779 by *Matchem (Bosphorus 1754)—*Dau. of Brilliant 1750 (Crab 1722)—Dau. of Tartar 1743 (Partner 1718)—Dau. of Son of Flying Childers 1715.
Woodpecker's dam, ch. m., 1809.	*Buzzard 1787 (Woodpecker 1773)—The Faun by Craig's Alfred (*Medley 1776)—Dau. of *Traveller 1747 (Partner 1718)—Dau. of *Whittington (Cullen Ar.)—Dau. of *Childers (Blaze 1733).
Timoleon, ch. h., 1813.	Sir Archy 1805 (*Diomed 1777)—Dau. of *Saltram 1780 (Eclipse 1764)—Dau. of Symmes' Wildair 1767 (*Fearnought 1755)—Dau. of Tyler's Driver 1760 (*Othello)—Dau. of *Fallower 1761 (Blank 1740).
Sister to Tuckahoe, ch. m., 1814.	Ball's Florizel 1801 (*Diomed 1777)—Dau. 1799 of *Alderman 1787 (Pot-8-o-'s 1773)—Dau. of *Clockfast 1774 (Gimcrack 1760)—Dau. of Symmes' Wildair 1767 (*Fearnought 1755)—Y. Kitty Fisher 1767 by *Fearnought 1755 (Regulus 1739).
*Sarpedon, br. h., 1828.	Emilius 1820 (Orville 1799)—Icaria 1824 by The Flyer 1814 (Vandyke Jr. 1808)—Parma 1813 by Dick Andrews 1797 (Joe Andrews 1778)—May 1804 by Beningbrough 1791 (King Fergus 1775)—Primrose 1787 by Mambrino 1768 (Engineer 1756).
Rowena, ch. m., 1826.	Sumpter 1818 (Sir Archy 1805)—Lady Grey 1817 by Robin Grey 1805 (*Royalist 1790)—Maria 1802 by Melzar 1791 (*Medley 1776)—Dau. of *Highflyer 1784 (Highflyer 1774)—Dau. of *Fearnought 1755 (Regulus 1739).
Sultan, b. h., 1816.	Selim 1802 (Buzzard* 1787)—Bacchante 1809 by W's Ditto 1800 (Sir Peter T. 1784)—Dau. 1791 of Mercury 1778 (Eclipse 1764)—Dau. 1776 of Herod 1758 (Tartar 1743)—Folly 1771 by Marske 1750 (Squirt 1732).
Trampoline, ch. m., 1825.	Tramp 1810 (Dick Andrews 1797)—Web 1808 by Waxy 1790 (Pot-8-o-'s 1773)—Penelope 1798 by Trumpator 1782 (Conductor 1767)—Prunella 1788 by Highflyer 1774 (Herod 1758)—Promise 1768 by Snap 1750 (Snip 1736).
Catton, b. h., 1809.	Golumpus 1802 (Gohanna 1790)—Lucy Grey 1804 by Timothy 1794 (Delpini 1781)—Lucy 1789 by Florizel 1768 (Herod 1758)—Frenzy 1774 by Eclipse 1764 (Marske 1750)—Dau. of Engineer 1756 (Sampson 1745).
Camillina, gr. m., 1822.	Camillus 1803 (Hambletonian 1792)—Dau. 1818 of Smolensko 1810 (Sorcerer 1796)—Miss Cannon 1811 by Orville 1799 (Beningbrough 1791)—Dau. 1796 of Weathercock 1785 (Highflyer 1774)—Cora 1777 by Matchem 1748 (Cade 1734).
Sir Hercules, blk. h., 1826.	Whalebone 1807 (Waxy 1790)—Peri 1822 by Wanderer 1811 (Gohanna 1790)—Thalestris 1809 by Alexander 1782 (Eclipse 1764)—Rival 1800 by Sir Peter T. 1784 (Highflyer 1774)—Hornet 1790 by Drone 1777 (Herod 1758).
Guiccioli, ch. m., 1823.	Bob Booty 1804 (Chanticleer 1787)—Flight 1809 by Ir. Escape 1802 (Commodore 1793)—Y. Heroine by Bagot 1780 (Herod 1758)—Heroine 1775 by Hero 1753 (Cade 1734)—Dau. 1766 of Snap 1750 (Snip 1736).
Hetman Platoff, b. h., 1836.	Brutandorf 1821 (Blacklock 1814)—Dau. 1820 of Comus 1809 (Sorcerer 1796)—Marciana 1809 by Stamford 1794 (Sir Peter T. 1784)—Marcia 1797 by Coriander 1786 (Pot-8-o-'s 1773)—Faith 1779 by Pacolet 1763 (Blank 1740).
Waterwitch, blk. m., 1833.	Sir Hercules 1826 (Whalebone 1807)—Mary Anne 1820 by Waxy Pope 1806 (Waxy 1790)—Witch 1811 by Sorcerer 1796 (Trumpator 1782)—Miss Buckle 1801 by Precipitate* 1787 (Mercury 1778)—Diddler's dam 1788 by Highflyer 1774 (Herod 1758).
Catton, b. h., 1809.	Golumpus 1802 (Gohanna 1790)—Lucy Gray 1804 by Timothy 1794 (Delpini 1781)—Lucy 1789 by Florizel 1768 (Herod 1758)—Frenzy 1774 by Eclipse 1764 (Marske 1750)—Dau. of Engineer 1756 (Sampson 1745).
Emma, ch. m., 1824.	Whisker 1812 (Waxy 1790)—Gibside Fairy 1811 by Hermes 1790 (Mercury 1778)—Vicissitude 1800 by Pipator 1786 (Imperator 1776)—Beatrice 1791 by Sir Peter T. 1784 (Highflyer 1774)—Pyrrha 1771 by Matchem 1748 (Cade 1734).
*Tranby, br. h., 1826.	Blacklock 1814 (Whitelock 1803)—Dau. 1812 by Orville 1799 (Beningbrough 1791)—Miss Grimstone 1796 by Weazle 1776 (Herod 1758)—Dau. of Ancaster 1768 (Blank 1740)—Dau. of Damascus Ar. 1751.
Lucilla, b. m., —.	Trumpator 1823 (Sir Solomon 1805)—Lucy b. m. 1821 by Orphan 1810 (Ball's Florizel 1801)—Lady Grey b. m. 1817 by Robin Grey 1805 (*Royalist 1790)—Maria 1802 by Melzar 1791 (*Medley 1776)—Dau. of *Highflyer 1784 (Highflyer 1774)—Dau. of *Fearnought 1755 (Regulus 1739)—Dau. of Ariel 1756 (*Traveller 1747)—Dau. of *Jack of Diamonds (Cullen Ar. 1745)—*Duchess by Cullen Ar. 1745—Lady Thig ch. m. 1731 by Partner 1718 (Jigg)—Dau. ch. m. 1723 of Greyhound (Chillaby)—Sophonisba's dam by Curwen Bay Barb—Dau. of D'Arcy Chestnut Ar.—Dau. of Whiteshirt—Lord D'Arcy's Montagu mare by Montagu—Dau. of Hautboy (D'Arcy White Turk)—Dau. of Brimmer (D'Arcy Yellow Turk)—Royal mare.

THE BARD—Continued.

miles, 118 lbs., in 2.39¼, Dew Drop, 113 lbs., second, Ben Ali and Blue Wing beaten off; September Stakes, 1¾ miles, 125 lbs., in 3.05, beating Ferona, 107 lbs., Linden, 110 lbs., etc; Jerome Stakes, 1¾ miles, 125 lbs., no time taken, beating Elkwood, Mollie McCarthy's Last, etc.: Dixie Stakes, 2 miles, 118 lbs., in 3.33; Breckenridge Stakes, 2 miles, 123 lbs, in 3.35, and Potomac Stakes, 1½ miles, 118 lbs., 2:27, winning in a canter by eight lengths, Elkwood second. At four years old started eight times, won six races, was second twice. Won June Special, 1½ miles, 113 lbs., in 1.55; St. James Hotel Stakes, 1¼ miles, 118 lbs., 2.08¾, beating Hidalgo, Barnum, Elkwood; Coney Island Stakes, 1¼ miles, 118 lbs., in 1.55, by three lengths from Troubadour; Coney Island Cup, 1¾ miles, 118 lbs., in 3.03, winning in a canter from Barnum, Elkwood, Rataplan; ran second to Troubadour in the Ocean Stakes, 1½ miles, 118 lbs., in 1.56¼, and in the Monmouth Cup, 1¾ miles, 123 lbs., in 3.04. In the Freehold Stakes, 1½ miles, 118 lbs., in 2.39¾, he beat Troubadour and Barnum, and won the Eatontown Stakes, 1 mile, 118 lbs., in 1.43¾. At five years old he won seven races out of eight starts, only beaten in his last race by Firenzi, when she ran in record time, 1½ miles in 2:31. He won the Brooklyn Handicap, 1¼ miles, 125 lbs., in 2.13, Hanover, 125 lbs., second, Exile, 114 lbs., third, eight others beaten off; St. James Hotel Stakes, 1¼ miles 122 lbs., in 2.08½, beating Sir Dixon, 104 lbs; Brooklyn Cup, 1½ miles, 123 lbs., in 2.42½, by ten lengths from Volante, Hanover, Fenelon. He had only to walk over for the Second Special, 1⅜ miles, and for the Coney Island Stakes, 1½ miles. Won the Coney Island Cup, 1¾ miles, 122 lbs., beating Hidalgo and Elkwood; the Ocean Stakes, 1⅛ miles, 120 lbs., in 1.55, Kingston, 118 lbs., second, Firenzi, 113 lbs., third. In the Freehold Stakes, 1½ miles, with 120 lbs., he ran second to Firenzi, 113 lbs., in 2.31 (record time).

The Bard is untried in the stud, yearlings being the oldest of his get, and they are grand ones. He is an extraordinary fine horse, possessing speed, capacity to carry weight, and could stay over a distance of ground.

He is descended in female line from the same Royal mare to which traces Eclipse, Voltaire, Sterling, Oxford, Kingston, Scottish Chief, Lexington, Vandal, Leonatus, Sensation, Salvator, Strathmore, Monarchist, Grinstead, Longstreet, Potomac and all the Duchess family.

IMPORTED THE JACOBITE.

The property of J. H. McGavock, Max Meadows, Va. Terms of service advertised annually.

THE JACOBITE, chestnut horse, foaled 1876, bred by Her Majesty, by Prince Charlie, son of Blair Athol, dam sister to Little Lady by Orlando (winner of the Derby), son of Touchstone, grandam sister to Voltigeur and Barnton by Voltaire, son of Blacklock, out of Martha Lynn by Mulatto, etc. Prince Charlie was the winner of the 2,000 Guineas and was regarded as one of the speediest horses of his day, sire of Princess Catherine, the dam of Clover and Clamart, both winners of the French Derby, Prestonpans (winner of the Liverpool Autumn Cup). In this country, Salvator, winner of the fastest Suburban ever run, in 2.06⅘, and the fastest mile ever run in the world, 1.35½, Senorita (winner of the Monmouth Oaks), etc. The Jacobite is full brother to Lady Charlie (winner of the Croydon Spring Stakes and Juvenile Plate at Hampton, and dam of Madame d'Albany, winner of the Liverpool Autumn Cup, 1891) and Lady Jacobite (winner of the British Dominion Two-year Old Stakes. The Jacobite's dam is full sister to Little Lady, who won as a yearling Anglesey Stakes, one-quarter of a mile, and won afterwards a good number of stakes and plates, and was dam of Camballo (winner of the 2,000 Guineas) by Orlando, winner of the Derby, sire of Fitz Roland, Fazzoletto and Diophantus (winners of the 2,000 Guineas), Teddington (winner of the Derby), and Imperieuse (winner of the 1,000 Guineas and St. Leger), she being almost full sister in blood to Little Lady, being out of Eulogy, half sister to Volley, which produced, to the cover of Orlando, Eurydice, winner of the Cambridgeshire Handicap. Volley, The Jacobite's grandam, was full sister to Barnton and the great Voltigeur (winner of the Derby, St. Leger and Doncaster Cup), and is the grandam of the good sire Lord Clifden, winner of the St. Leger.

The Jacobite is a highly bred horse and from a grand racing family. He has had but little chance in the stud, but is the sire of the winners; Ella, Chiswell, etc.

The Jacobite descends in direct female line from the Burton Barb mare, to which traces Whiskey, Selim, Castrel, Rubens, Catton, Blacklock, Sir Hercules, Harkaway, Cremorne, Voltigeur, Surplice, Lord Clifden, Pell Mell, Brag, Sheen, Gold, imp. Ainderby, imp. Billet, imp. Hurrah, imp. Margrave, imp. Yorkshire, and all the descendants of imp. Mary Grey.

THE JACOBITE.

CHESTNUT HORSE, FOALED 1876; BRED BY HER MAJESTY, ENGLAND.

PRINCE CHARLIE,* ch. h., 1869.	Blair Athol, ch. h., 1861.	Stockwell, ch. h., 1849.	The Baron, ch. h., 1842.	Ir. Birdcatcher ch. h., 1833.	Sir Hercules 1826 (Whalebone 1807)—Guiccioli 1823 by Bob Booty 1804 (Chanticleer 1787)—Flight 1809 by Ir. Escape 1802 (Commodore 1793)—Y. Heroine by Bagot 1780 (Herod 1758)—Heroine 1775 by Hero 1753 (Cade 1734).
				Echidna, b. m., 1838.	Economist 1825 (Whisker 1812)—Miss Pratt 1825 by Blacklock 1814 (Whitelock 1803)—Gadabout 1812 by Orville 1799 (Beningbrough 1791)—Minstrel 1803 by Sir Peter T. 1784 (Highflyer 1774)—Matron 1782 by Florizel 1768 (Herod 1758).
			Pocahontas, b. m., 1837.	Glencoe,* ch. h., 1831.	Sultan 1816 (Selim 1802)—Trampoline 1825 by Tramp 1810 (Dick Andrews 1797)—Web 1808 by Waxy 1790 (Pot-8-o's 1773)—Penelope 1798 by Trumpator 1782 (Conductor 1767)—Prunella 1788 by Highflyer 1774 (Herod 1758).
				Marpessa, b. m., 1830.	Muley 1810 (Orville 1799)—Clare 1824 by Marmion 1806 (Whiskey 1789)—Harpalice 1814 by Gohanna 1790 (Mercury 1778)—Amazon 1799 by Driver 1783 (Trentham 1766)—Fractious 1792 by Mercury 1778 (Eclipse 1764).
		Blink Bonny, b. m., 1854.	Melbourne, br. h., 1834.	Humphrey Clinker, b. h., 1822.	Comus 1809 (Sorcerer 1796)—Clinkerina 1812 by Clinker 1805 (Sir Peter T. 1784)—Pewet 1786 by Tandem 1773 (Syphon 1750)—Termagant by Tantrum 1760 (Cripple 1750)—Cantatrice by Sampson 1745 (Blaze 1733).
				Daughter of, b. m., 1825.	Cervantes 1806 (Don Quixote 1784)—Dau. 1818 of Golumpus 1802 (Gohanna 1790)—Dau. 1810 of Paynator 1791 (Trumpator 1782)—Circle 1802 by St. George 1789 (Highflyer 1774)—Abigail 1788 by Woodpecker 1773 (Herod 1758).
			Queen Mary, b. m., 1843.	Gladiator, ch. h., 1833.	Partisan 1811 (Walton 1799)—Pauline 1826 by Moses 1819 ([Whalebone or] Seymour 1807)—Quadrille 1815 by Selim 1802 (Buzzard* 1787)—Canary Bird 1806 by [Whiskey or] Sorcerer 1796 (Trumpator 1782)—Canary 1797 by Coriander 1786 (Pot-8-o's 1773).
				Daughter of, b. m., 1840.	Plenipotentiary 1831 (Emilius 1820)—Myrrha 1830 by Whalebone 1807 (Waxy 1790)—Gift 1818 by Y. Gohanna 1810 (Gohanna 1790)—Sist. 1802 to Grazier by Sir Peter T. 1784 (Highflyer 1774)—Dau. 1788 of Trumpator 1782 (Conductor 1767).
	Eastern Princess, ch. m., 1858.	Surplice, b. h., 1845.	Touchstone, br. h., 1831.	Camel, br. h., 1822.	Whalebone 1807 (Waxy 1790)—Dau. 1812 of Selim 1802 (Buzzard* 1787)—Maiden 1801 by Sir Peter T. 1784 (Highflyer 1774)—Dau. 1788 of Phenomenon 1780 (Herod 1758)—Matron 1782 by Florizel 1768 (Herod 1758).
				Banter, br. m., 1823.	Master Henry 1815 (Orville 1799)—Boadicea 1807 by Alexander 1782 (Eclipse 1764)—Brunette by Amaranthus 1766 (Old England 1741)—Mayfly 1771 by Matchem 1748 (Cade 1734)—Dau. 1755 of Anc. Starling 1738 (Starling 1727).
			Crucifix, b. m., 1837.	Priam,* b. h., 1827.	Emilius 1820 (Orville 1799)—Cressida 1807 by Whiskey 1789 (Saltram* 1780)—Y. Giantess 1790 by Diomed* 1777 (Florizel 1768)—Giantess 1769 by Matchem 1748 (Cade 1734)—Molly Long Legs 1753 by Babraham 1740 (God. Ar. 1724).
				Octaviana, b. m., 1815.	Octavian 1807 (Stripling 1795)—Dau. 1807 of Shuttle 1793 (Y. Marske 1771)—Zara 1801 by Delpini 1781 (Highflyer 1774)—Flora 1789 by King Fergus 1775 (Eclipse 1764)—Atalanta 1769 by Matchem 1748 (Cade 1734).
		Tomyris, b. m., 1851.	Sesostris, b. h., 1845.	Slane, b. h., 1833.	Royal Oak 1823 (Catton 1809)—Dau. 1819 of Orville 1799 (Beningbrough 1791)—Epsom Lass 1803 by Sir Peter T. 1784 (Highflyer 1774)—Alexina 1788 by King Fergus 1775 (Eclipse 1764)—Lardella 1780 by Y. Marske 1771 (Marske 1750).
				Palmyra, br. m., 1838.	Sultan 1816 (Selim 1802)—Hester 1832 by Camel 1822 (Whalebone 1807)—Monimia 1821 by Muley 1810 (Orville 1799)—Sist. 1796 to Petworth by Precipitate* 1787 (Mercury 1778)—Dau. 1787 of Woodpecker 1773 (Herod 1758).
			Daughter of, ch. m., 1842.	Glaucus, b. h., 1830.	Partisan 1811 (Walton 1799)—Nanine 1823 by Selim 1802 (Buzzard* 1787)—Bizarre 1811 by Peruvian 1806 (Sir Peter T. 1784)—Violante 1802 by John Bull 1789 (Fortitude 1777)—Dau. 1788 of Highflyer 1774 (Herod 1758).
				Io, ch. m., 1836.	Taurus 1824 ([Phantom or] Morisco 1819)—Arethusa 1824 by Quiz 1798 (Buzzard* 1787)—Persepolis 1808 by Alexander 1782 (Eclipse 1764)—Sist. 1784 to *Tickle Toby by Alfred 1770 (Matchem 1748)—Celia 1775 by Herod 1758 (Tartar 1743).
SISTER TO LITTLE LADY, b. m., 1857.	Orlando, b. h., 1841.	Touchstone, br. h., 1831.	Camel, br. h., 1822.	Whalebone, br. h., 1807.	Waxy 1790 (Pot-8-o's 1773)—Penelope 1798 by Trumpator 1782 (Conductor 1767)—Prunella 1788 by Highflyer 1774 (Herod 1758)—Promise 1768 by Snap 1750 (Snip 1736)—Julia 1756 by Blank 1740 (God. Ar. 1724).
				Daughter of, b. m., 1812.	Selim 1802 (Buzzard* 1787)—Maiden 1801 by Sir Peter T. 1784 (Highflyer 1774)—Dau. 1788 of Phenomenon 1780 (Herod 1758)—Matron 1782 by Florizel 1768 (Herod 1758)—Maiden 1770 by Matchem 1748 (Cade 1734).
			Banter, br. m., 1826.	Master Henry, b. h., 1815.	Orville 1799 (Beningbrough 1791)—Miss Sophia 1805 by Stamford 1794 (Sir Peter T. 1784)—Sophia 1798 by Buzzard* 1787 (Woodpecker 1773)—Huncamunca 1787 by Highflyer 1774 (Herod 1758)—Cypher 1772 by Squirrel 1754 (Traveller 1735).
				Boadicea, b. m., 1807.	Alexander 1782 (Eclipse 1764)—Brunette by Amaranthus 1766 (Old England 1741)—Mayfly 1771 by Matchem 1748 (Cade 1734)—Dau. 1755 of Anc. Starling 1738 (Starling 1727)—Look-at-me-Lads 1731 by Bristol Grasshopper (Byerly Turk).
		Vulture, ch. m., 1833.	Langar, ch. h., 1817.	Selim, ch. h., 1802.	Buzzard* 1787 (Woodpecker 1773)—Dau. of Alexander 1782 (Eclipse 1764)—Dau. of Highflyer 1774 (Herod 1758)—Dau. of Alfred 1770 (Matchem 1748)—Dau. 1770 of Engineer 1756 (Sampson 1745).
				Daughter of, ch. m., 1808.	Walton 1799 (Sir Peter T. 1784)—Y. Giantess 1790 by Diomed* 1777 (Florizel 1768)—Giantess 1769 by Matchem 1748 (Cade 1734)—Molly Long Legs 1753 by Babraham 1740 (God. Ar. 1724)—Dau. of Cole's Foxhunter 1727 (Brisk 1711).
			Kite, b. m., 1821.	Bustard, b. h., 1813.	Castrel 1801 (Buzzard* 1787)—Miss Hap 1806 by Shuttle 1793 (Y. Marske 1771)—Dau. 1798 of Sir Peter T. 1784 (Highflyer 1774)—Miss Hervey 1775 by Eclipse 1764 (Marske 1750)—Clio 1760 by Y. Cade 1747 (Cade 1734).
				Olympia, b. m., 1815.	Sir Oliver 1800 (Sir Peter T. 1784)—Scotilla 1795 by Anvil 1777 (Herod 1758)—Scota 1783 by Eclipse 1764 (Marske 1750)—Harmony 1775 by Herod 1758 (Tartar 1743)—Rutilia 1769 by Blank 1740 (God. Ar. 1724).
	Volley, b. m., 1845.	Voltaire, br. h., 1826.	Blacklock, b. h., 1814.	Whitelock, b. h., 1803.	Hambletonian 1792 (King Fergus 1775)—Rosalind 1788 by Phenomenon 1780 (Herod 1758)—Atalanta 1769 by Matchem 1748 (Cade 1734)—Y. Lass-of-the-Mill 1756 by Oroonoko 1745 (Crab 1722)—Dau. 1744 of Traveller 1735 (Partner 1718).
				Daughter of, —, 1799.	Coriander 1786 (Pot-8-o's 1773)—Wildgoose 1792 by Highflyer 1774 (Herod 1758)—Coheiress 1786 by Pot-8-o's 1773 (Eclipse 1764)—Manilla 1777 by Goldfinder 1764 (Snap 1750)—Dau. 1766 of Old England 1741 (God. Ar. 1724).
			Daughter of, b. m., 1816.	Phantom, b. h., 1808.	Walton 1799 (Sir Peter T. 1784)—Julia 1799 by Whiskey 1789 (Saltram* 1780)—Y. Giantess 1790 by Diomed* 1777 (Florizel 1768)—Giantess 1769 by Matchem 1748 (Cade 1734)—Molly Long Legs 1753 by Babraham 1740 (God. Ar. 1724).
				Daughter of, b. m., 1802.	Overton 1788 (King Fergus 1775)—Dau. 1796 of Walnut 1786 (Highflyer 1774)—Dau. 1786 of Ruler 1777 (Y. Marske 1771)—Piracantha 1772 by Matchem 1748 (Cade 1734)—Prophetess 1758 by Regulus 1739 (God. Ar. 1724).
		Martha Lynn, br. m., 1837.	Mulatto, b. h., 1823.	Catton, b. h., 1809.	Golumpus 1802 (Gohanna 1790)—Lucy Grey 1804 by Timothy 1794 (Delpini 1781)—Lucy 1789 by Florizel 1768 (Herod 1758)—Frenzy 1774 by Eclipse 1764 (Marske 1750)—Dau. of Engineer 1756 (Sampson 1745).
				Desdemona, br. m., 1811.	Orville 1799 (Beningbrough 1791)—Fanny 1796 by Sir Peter T. 1784 (Highflyer 1774)—Dau. 1788 of Diomed* 1777 (Florizel 1768)—Desdemona 1770 by Marske 1750 (Squirt 1732)—Y. Hag 1761 by Skim 1746 (Starling 1727).
			Leda, b. m., 1824.	Filho-da-Puta, br. h., 1812.	Haphazard 1797 (Sir Peter T. 1784)—Mrs. Barnet 1806 by Waxy 1790 (Pot-8-o's 1773)—Dau. 1788 of Woodpecker 1773 (Herod 1758)—Heinel 1771 by Squirrel 1754 (Traveller 1735)—Principessa 1762 by Blank 1740 (God. Ar. 1724).
				Treasure, ch. m., 1809.	Camillus 1803 (Hambletonian 1792)—Dau. ch. m. 1804 of Hyacinthus 1797 (Coriander 1786)—Flora ch. m. 1789 by King Fergus 1775 (Eclipse 1764)—Atalanta ch. m. 1769 by Matchem 1749 (Cade 1734)—Y. Lass-of-the-Mill b. m. 1756 by Oroonoko 1745 (Crab 1722)—Dau. 1744 of Traveller 1735 (Partner 1718)—Miss Makeless b. m. 1737 by Y. Greyhound 1723 (Greyhound)—Dau. 1731 of Partner 1718 (Jigg)—Brown Woodcock br. m. 1717 by D'Arcy Woodcock (Bustler)—Lusty Thornton ch. m. by Croft's Bay Barb (Chillaby)—Chestnut Thornton ch. m. by Makeless (Oglethorpe Ar.)—Old Thornton by Brimmer (D'Arcy Yellow Turk)—Dau. of Dickey Pierson (Dodsworth)—Burton Natural Barb mare.

IMPORTED THE SAILOR PRINCE.

WINNER OF THE CAMBRIDGESHIRE AND LEWES SPRING HANDICAPS AND PORTLAND STAKES.

Is private stallion in Rancocas Stud, Mr. P. Lorillard, proprietor, Jobstown, N. J.

THE SAILOR PRINCE, bay horse, foaled 1880, bred by Mr. B. Norrington, by Albert Victor, son of Marsyas by Orlando, dam Hermita by Hermit, son of Newminster, grandam Affection by Lifeboat, son of Sir Hercules, out of Rose d'Amour by Wild Dayrell, etc. The Sailor Prince is own brother to Trixie, winner of the Alexandra Plate, out of Hermita (winner of the Grand Stand, Kentish and Camden Park Handicaps and other races) by the great Hermit (winner of the Derby), sire of imp. St. Blaise (winner of the Derby), Shotover (winner of the 2,000 Guineas and Derby), Thebais (winner of the 1,000 Guineas and Oaks), St. Marguerite (winner of the 2,000 Guineas, dam of Seabreeze, winner of the Oaks and St. Leger), Tristan and imp. Mr. Pickwick. His grandam, Affection, is by Lifeboat, son of Sir Hercules and Yard Arm by Sheet Anchor (winner of the Great Metropolitan Stakes, Chester Cup and many other races). His great grandam, Rose d'Amour by Wild Dayrell (winner of the Derby) and sire of Hurricane (winner of the 1,000 Guineas) and the dam of Atlantic (winner of the 2,000 guineas) and his full sister, Atlantis, dam of imp. King Ban, by Thormanby, and sire of Buccaneer, the sire of Kisber, winner of the Derby and Grand Prize of Paris.

The Sailor Prince was a good race-horse at two years old. He ran third to Rookery and Sigmophone in the Windsor Castle Stakes at Ascot; second to Britomarte in the Champagne Stakes at Bibury, six others behind him; won the Four Oaks Two-year Old Plate, with eight others behind him; was beaten a neck by Sarah Bernhardt in the Great Witley Stakes at Worcester, giving her 10 lbs.; won the Royal Two-year Old Plate at Kempton, with ten others behind him, and was second to Hornpipe in the Thames Plate. At three years old ran three times. Was third in Prince of Wales Handicap. At four years old won a Handicap Sweepstakes, 5 furlongs, at Bibury. At five years old won Royal Plate Handicap, 1 mile, at Sandown; ran third in the Goodwood Corinthian Plate, 1 mile, won by Reine Blanche, Wild Thyme second; ran second in the Daviss Park Free Welter Handicap, 138 lbs., Cintra, 108 lbs., won; ran second to Castor, 110 lbs., in the Autumn Plate at Manchester, he carrying 146 lbs.; ran second to Cohort in the Great Shropshire Handicap at Shrewsbury; ran a dead heat and divided the stakes in the Great Lancashire Handicap with Kimbolton at Liverpool; ran second to Alb, 102 lbs., in the Holmby Handicap Plate at Northampton, carrying 131 lbs.; ran second to Verano, 140 lbs., he carrying 166 lbs., in the Coventry Welter Plate at Warwick. At six years old, won the Lewes Spring Handicap, 1 mile, 121 lbs., Mohawk second, Cavalier third; was third to Middlethorpe and Perdita II. in the Liverpool Cup, with eleven others behind him; won the Cambridgeshire Handicap, 1 mile 240 yards, 105 lbs., with St. Mirin, Carlton, Melton and twelve others behind him. At seven years old, won the Portland Stakes, 1 mile, with 131 lbs., with three behind him. He retired sound.

The Sailor Prince is a handsomely bred horse, and from a racing family that not only races but produces race-horses.

He is descended in direct female line from the old Woodcock mare, to which trace Skim, Reveller (by Comus) Heron, Alarm, Vindex, Monarque, Vedette (sire of Galopin), Cambuscan, Vespasian, Isonomy, Fernandez, King Lud, in England: Lisbon, imp. Clockfast, Exile (by Mortemer), etc., etc.

Waxy 1790 Pot-8-o's 1773—Eclipse 1764—Marske 1750—Squirt 1732—Bartlett's Childers—Darley Arabian 1690.

THE SAILOR PRINCE.

BAY HORSE, FOALED 1880; BRED BY MR. R. NORRINTON, ENGLAND.

ALBERT VICTOR, ch. h., 1868.	Marsyas, ch. h., 1851.	Orlando, b. h., 1841.	Touch-stone, br. h., 1831.	Camel, br. h., 1822.	Whalebone 1807 (Waxy 1790)—Dau. 1812 of Selim 1802 (Buzzard* 1787)—Maiden 1801 by Sir Peter T. 1784 (Highflyer 1774)—Dau. 1788 of Phenomenon 1780 (Herod 1758)—Matron 1782 by Florizel 1768 (Herod 1758).
				Banter, br. m., 1826.	Master Henry 1815 (Orville 1799)—Boadicea 1807 by Alexander 1782 (Eclipse 1764)—Brunette by Amaranthus 1766 (Old England 1741)—Mayfly 1771 by Matchem 1748 (Cade 1734)—Dau. 1755 of Starling 1738 (Starling 1727).
			Vulture, ch. m., 1833.	Langar, ch. h., 1817.	Selim 1802 (Buzzard 1787)—Dau. 1808 of Walton 1799 (Sir Peter T. 1784)—Y. Giantess 1790 by Diomed* 1777 (Florizel 1768)—Giantess 1769 by Matchem 1748 (Cade 1734)—Molly Long Legs 1753 by Babraham 1740 (God. Ar. 1724).
				Kite, b. m., 1821.	Bustard 1813 (Castrel 1801)—Olympia 1815 by Sir Oliver 1800 (Sir Peter T. 1784)—Scotilla 1795 by Anvil 1777 (Herod 1758)—Scota 1783 by Eclipse 1764 (Marske 1750)—Harmony 1775 by Herod 1758 (Tartar 1743).
		Malibran, b. m., 1840.	Whisker, b. h., 1812.	Waxy, b. h., 1790.	Pot-8-o's 1773 (Eclipse 1764)—Maria 1777 by Herod 1758 (Tartar 1743)—Lisette 1772 by Snap 1750 (Snip 1736)—Miss Windsor 1751 by God. Ar. 1724—Dau. of Y. Belgrade (Belgrade Turk).
				Penelope, b. m., 1798.	Trumpator 1782 (Conductor 1767)—Prunella 1788 by Highflyer 1774 (Herod 1758)—Promise 1768 by Snap 1750 (Snip 1736)—Julia 1756 by Blank 1740 (God. Ar. 1724)—Spectator's Dam 1735 by Partner 1718 (Jigg).
			Garcia, br. m., 1823.	Octavian, ch. h., 1807.	Stripling 1795 (Phenomenon 1780)—Dau. 1797 of Oberon 1790 (Highflyer 1774)—Sist. 1779 to Sharper by Ranthos 1763 (Matchem 1748)—Dau. of T. Sweepstakes 1743 (B. Sweepstakes 1722)—Dau. 1748 of Hutton's Spot 1728 (Bartley's Blind Horse).
				Daughter of, b. m., 1806.	Shuttle 1793 (Y. Marske 1771)—Katharine 1798 by Delpini 1781 (Highflyer 1774)—Dau. 1788 of Paymaster 1766 (Blank 1740)—Dau. 1774 of Le Sang 1759 (Changeling 1747)—Dau. 1751 of Rib 1736 (Crab 1722).
	Princess of Wales, ch. m., 1862.	Stockwell, ch. h., 1849.	The Baron, ch. h., 1842.	Ir. Birdcatcher, ch. h., 1833.	Sir Hercules 1826 (Whalebone 1807)—Guiccioli 1823 by Bob Booty 1804 (Chanticleer 1787)—Flight 1809 by Ir. Escape 1802 (Commodore 1793)—Y. Heroine by Bagot 1780 (Herod 1758)—Heroine 1775 by Hero 1753 (Cade 1734).
				Echidna, b. m., 1838.	Economist 1825 (Whisker 1812)—Miss Pratt 1825 by Blacklock 1814 (Whitelock 1803)—Gadabout 1812 by Orville 1799 (Beningbrough 1791)—Minstrel 1803 by Sir Peter T. 1784 (Highflyer 1774)—Matron 1782 by Florizel 1768 (Herod 1758).
			Pocahontas, b. m., 1837.	Glencoe,* ch. h., 1831.	Sultan 1816 (Selim 1802)—Trampoline 1825 by Tramp 1810 (Dick Andrews 1797)—Web 1808 by Waxy 1790 (Pot-8-o's 1773)—Penelope 1798 by Trumpator 1782 (Conductor 1767)—Prunella 1788 by Highflyer 1774 (Herod 1758).
				Marpessa, b. m., 1830.	Muley 1810 (Orville 1799)—Clare 1824 by Marmion 1806 (Whiskey 1789)—Harpalice 1814 by Gohanna 1790 (Mercury 1778)—Amazon 1799 by Driver 1783 Trentham 1766)—Fractions 1792 by Mercury 1778 (Eclipse 1764).
		The Bloomer, b. m., 1850.	Melbourne, br. h., 1834.	Humphrey Clinker, b. h., 1822.	Comus 1809 (Sorcerer 1796)—Clinkerina 1812 by Clinker 1805 (Sir Peter T. 1784)—Pewet 1786 by Tandem 1773 (Syphon 1750)—Termagant by Tantrum 1760 (Cripple 1750)—Cantatrice by Sampson 1745 (Blaze 1733).
				Daughter of, b. m., 1825.	Cervantes 1806 (Don Quixote 1784)—Dau. 1818 of Golumpus 1802 (Gohanna 1790)—Dau. 1810 of Paynator 1791 (Trumpator 1782)—Circle 1802 by St. George 1789 (Highflyer 1774)—Abigail 1788 by Woodpecker 1773 (Herod 1758).
			Lady Sarah, ch. m., 1841.	Velocipede, ch. h., 1825.	Blacklock 1814 (Whitelock 1803)—Dau. 1817 of Juniper 1805 (Whiskey 1789)—Dau. 1810 of Sorcerer 1796 (Trumpator 1782)—Virgin 1801 by Sir Peter T. 1784 (Highflyer 1774)—Dau. 1791 of Pot-8-o's 1773 (Eclipse 1764).
				Lady Moore Carew, b. m., 1830.	Tramp 1810 (Dick Andrews 1797)—Kite 1821 by Bustard 1813 (Castrel 1801)—Olympia 1815 by Sir Oliver 1800 (Sir Peter T. 1784)—Scotilla 1795 by Anvil 1777 (Herod 1758)—Scota 1783 by Eclipse 1764 (Marske 1750).
HERMITA, b. m., 1871.	Hermit, ch. h., 1864.	Newminster, b. h., 1848.	Touchstone, br. h., 1831.	Camel, br. h., 1822.	Whalebone 1807 (Waxy 1790)—Dau. 1812 of Selim 1802 (Buzzard* 1787)—Maiden 1801 by Sir Peter T. 1784 (Highflyer 1774)—Dau. 1788 of Phenomenon 1780 (Herod 1758)—Matron 1782 by Florizel 1768 (Herod 1758).
				Banter, br. m., 1826.	Master Henry 1815 (Orville 1799)—Boadicea 1807 by Alexander 1782 (Eclipse 1764)—Brunette by Amaranthus 1766 (Old England 1741)—Mayfly 1771 by Matchem 1748 (Cade 1734)—Dau. 1755 of Anc. Starling 1738 (Old Starling 1727).
			Bee's-wing, b. m., 1833.	Dr. Syntax, br. h., 1811	Paynator 1791 (Trumpator 1782)—Dau. of Beningbrough 1791 (King Fergus 1775)—Jenny Mole 1787 by Carbuncle 1772 (Babr. Blank 1758)—Dau. 1773 by Prince T'Quassaw 1751 (Snip 1736)—Sultana 1759 by Regulus 1739 (God. Ar. 1724).
				Daughter of, ch. m., 1817.	Ardrossan 1809 (John Bull 1789)—Lady Eliza 1813 by Whitworth 1805 (Agonistes 1797)—X. Y. Z.'s dam 1793 by Spadille 1784 (Highflyer 1774)—Sylvia 1783 by Y. Marske 1771 (Marske 1750)—Ferret 1765 by Bro. to Silvio (Cade 1734).
		Seclusion, b. m., 1857.	Tadmor, br. h., 1846.	Ion, br. h., 1835.	Cain 1822 (Paulowitz 1813)—Margaret 1831 by Edmund 1824 (Orville 1799)—Medora 1811 by Selim 1802 (Buzzard* 1787)—Dau. 1803 of Sir Harry* 1795 (Sir Peter T. 1784)—Dau. 1790 of Volunteer 1780 (Eclipse 1764).
				Palmyra, br. m., 1838.	Sultan 1816 (Selim 1802)—Hester 1832 by Camel 1822 (Whalebone 1807)—Monimia 1821 by Muley 1810 (Orville 1799)—Sist. 1796 to Petworth by Precipitate 1787 (Mercury 1778)—Dau. 1787 of Woodpecker 1773 (Herod 1758).
			Miss Sellon, b. m., 1851.	Cowl, b. h., 1842.	Bay Middleton 1833 (Sultan 1816)—Crucifix 1837 by Priam* 1827 (Emilius 1820)—Octaviana 1815 by Octavian 1807 (Stripling 1795)—Dau. 1807 of Shuttle 1793 (Y. Marske 1771)—Zara 1801 by Delpini 1781 (Highflyer 1774).
				Belle Dame, ch. m., 1839.	Belshazzar* 1830 (Blacklock 1814)—Ellen 1831 by Starch 1819 (Waxy Pope 1806)—Cuirass 1823 by Oiseau 1809 (Camillus 1803)—Castanea 1806 by Gohanna 1790 (Mercury 1778)—Grey Skim 1793 by Woodpecker 1773 (Herod 1758).
	Affection, br. m., 1865.	Lifeboat, br. h., 1855.	Sir Hercules, blk. h., 1826.	Whalebone, br. h., 1807.	Waxy 1790 (Pot-8-o's 1773)—Penelope 1798 by Trumpator 1782 (Conductor 1767)—Prunella 1788 by Highflyer 1774 (Herod 1758)—Promise 1768 by Snap 1750 (Snip 1736)—Julia 1756 by Blank 1740 (God. Ar. 1724).
				Peri, b. m., 1822.	Wanderer 1811 (Gohanna 1790)—Thalestris 1809 by Alexander 1782 (Eclipse 1764)—Rival 1800 by Sir Peter T. 1784 (Highflyer 1774)—Hornet 1790 by Drone 1777 (Herod 1758)—Manilla 1777 by Goldfinder 1764 (Snap 1750).
			Yard Arm, b. m., 1843.	Sheet Anchor, br. h., 1832.	Lottery 1820 (Tramp 1810)—Morgiana 1820 by Muley 1810 (Orville 1799)—Miss Stephenson 1814 by [Scud or] Sorcerer 1796 (Trumpator 1782)—Sist. 1795 to Petworth by Precipitate* 1787 (Mercury 1778)—Dau. 1787 of Woodpecker 1773 (Herod 1758).
				Fanny Kemble, b. m., 1827.	Paulowitz 1813 (Sir Paul 1802)—Loyalty 1817 by Rubens 1805 (Buzzard* 1787)—Pennyroyal 1805 by Coriander 1786 (Pot-8-o's 1773)—Peppermint 1787 by Highflyer 1774 (Herod 1758)—Promise 1768 by Snap 1750 (Snip 1736).
		Rose d'Amour, br. m., 1850.	Wild Dayrell, br. h., 1852.	Ion, br. h., 1835.	Cain 1822 (Paulowitz 1813)—Margaret 1831 by Edmund 1824 (Orville 1799)—Medora 1811 by Selim 1802 (Buzzard* 1787)—Dau. 1803 of Sir Harry* 1795 (Sir Peter T. 1784)—Dau. 1790 of Volunteer 1780 (Eclipse 1764).
				Ellen Middleton, br. m., 1846.	Bay Middleton 1833 (Sultan 1816)—Myrrha 1831 by Malek 1824 (Blacklock 1814)—Bessy 1815 by Y. Gouty 1805 (Gouty 1793)—Grandiflora 1810 by Sir Harry Dimsdale 1809 (Sir Peter T. 1784)—Dau. 1799 of Pipator 1786 (Imperator 1776).
			Rosa, ch. m., 1842.	Cain, b. h., 1822.	Paulowitz 1813 (Sir Paul 1802)—Dau. 1819 of Paynator 1791 (Trumpator 1782)—Dau. 1805 of Delpini 1781 (Highflyer 1774)—Sist. 1795 to Mary by Y. Marske 1771 (Marske 1750)—Gentle Kitty 1774 by Silvio 1754 (Cade 1734).
				Daughter of, ch. m., 1833.	Muley 1810 (Orville 1799)—Rosalia br. m. 1825 by Walton 1799 (Sir Peter T. 1784)—Rosanne b. m. 1811 by Dick Andrews 1797 (Joe Andrews 1778)—Rosette br. m. 1803 by Beningbrough 1791 (King Fergus 1775)—Rosamond b. m. 1788 by Tandem 1773 (Syphon 1750)—Tuberose gr. m. 1772 by Herod 1758 (Tartar 1743)—Grey Starling gr. m. 1745 by Bolton Starling 1727 (Bay Bolton 1705)—Coughing Polly b. m. 1736 by Bartlett's Childers (Darley Ar.)—Counsellor M. by Wood's Counsellor 1694 (D'Arcy Counsellor)—Dau. of Snake (Lister Turk)—Dau. of Luggs (D'Arcy White Turk)—Dau. of Davill's Old Woodcock.

TOM OCHILTREE.

In the Bowlingbrook Stud, Md. Application to R. W. Walden, Middleburg, Md.

TOM OCHILTREE by Lexington, son of Boston, bred by A. J. Alexander, Woodburn Stud, near Spring Station, Ky., foaled 1872, dam Katona, the dam of Metarie and Sandford, by Voucher, son of Wagner, out of Countess by imp. Margrave, winner of the St. Leger in 1832. Wagner, the sire of Voucher, he the sire of Katona, was closely inbred to Sir Archy. He defeated the great Grey Eagle in the two memorable contests, four-mile heats, at Louisville, Kentucky, in 1839. Voucher was a fine race-horse and was held in high esteem in his day. Margrave, the sire of his grandam, was a large, fine horse and a very successful sire; sired Brown Dick, whose three miles in 5.30, 5.28¼, has not often been excelled; he was also sire of Blue Dick, a famous horse, and of Mollie Jackson, whose famous third heat of three miles in 5.28¾ has never been equalled. Tom Ochiltree did not run in his two-year old form, but scored his first victory at Baltimore, and won four out of the nine races in which he started at three years old. Baltimore, won dash of three-quarters of a mile in 1.24¾, beating Audubon and seven others; won the Preakness Stakes, a Sweepstakes for three-year olds, 1½ miles, in 2.43½, beating Viator and seven others. Jerome Park, won the Annual Sweepstakes for three-year olds, 2 miles and a furlong, in 4.09½, beating Chesapeake and Ascension. Baltimore, Md., won the Dixie Stakes for three-year olds, 2 miles, in 3.42½, beating Viator, Aristides, Rhadamanthus and other good horses; was third in the Breckenridge Stakes for three-year olds, 2 miles, won by Aristides, Viator second, in 3.36½, he carrying a penalty of five pounds, and was unplaced in his other races. Four years old, started in ten races and won eight. Baltimore, Md., won the Baltimore Cup, 2¼ miles, in 4.09, beating Stampede, Viator and four others. Jerome Park, won the Jockey Club Handicap, 2 miles, in 3.41¾, beating Kildare, Chesapeake and five others; won the Centennial Stakes for all ages, 2¾ miles, in 5.09¼, beating Acrobat and Olitipa. Monmouth Park, won the Monmouth Cup, 2½ miles, in 4.48½, Stampede the only other starter; won the Capital Stakes for four-year olds, 3 miles, in 5.35¾, beating Chesapeake and Ascension. At Saratoga, won the Saratoga Cup, 2¼ miles, in 4.06½, beating Parole, Big Sandy and Madge. Jerome Park, won the Maturity Stakes for four-year olds, 3 miles, in 5.43½, beating Chesapeake, Matty A. and Grey Nun; won the Centennial Cup, 4 miles, in 7.36, beating Acrobat, Big Sandy and D'Artagnan. At Saratoga, was second in the All-aged Sweepstakes, 1¼ miles, won by Parole in 2.12¾; was unplaced in race of two-mile heats, won by Add in 3.47¼, 3.48¾, Ochiltree won first heat, but was distanced in the second. At five years old, started fourteen times and was first in nine races. Baltimore, Md., won the Baltimore Cup, 2¼ miles, in 4.14; won race of two-mile heats in 3.43, distanced the field first heat. Jerome Park, won the Westchester Cup, 2¼ miles, in 4.09¾, beating Athlene and Fellowcraft; won a dash of 2½ miles, in 4.36½, Athlene the only other starter. Saratoga, won a dash of 2 miles, in 3.39½, beating Athlene and Aunt Betsy; won a Handicap dash of 2 miles, in 3.42¼, beating Cloverbrook, Oriole and two others, Ochiltree carried 120 lbs. Jerome Park, won the Grand National Handicap, 2¼ miles, in 4.18½, carrying 124 lbs., beating Parole, Inspiration and Galway; won dash of 1¾ miles, in 3.14, beating Virginius and Warlock, track heavy. Saratoga, was second in the All-aged Stakes, 1¼ miles, won by Vera Cruz (3), Ochiltree carried 124 lbs.; was second in the Saratoga Cup, 2¼ miles, won by Parole in 4.04¼; was second in a dash of 2¼ miles, won by Whisper in 4.02; was second in a dash of 1¾ miles, won by Vera Cruz in 3.17. At Baltimore, was third in a dash of 2¼ miles, won by Parole in 4.37¾, track very heavy. Tom Ochiltree is one of Lexington's best sons; he could run fast and stay under heavy weights. His great size, coupled with symmetry of form, should commend him to breeders. He is deeply inbred to Diomed through his many Archy crosses, and the Eclipse cross is an addition through a staying family, fortified with a triple cross of Orville through his sons, Muley and Emilius. Ochiltree is a bay, 16 hands 2½ inches high, no white except a small star. He is one of the truest and best shaped big horses in the world, all bone and muscle, without a particle of lumber.

Tom Ochiltree is the sire of the winners Major Domo (winner of the Brookdale Handicap, 1¼ miles, in 1.54; Parkway Handicap, 1⅛ miles, in 1.47, 1 mile, in 1.39¾ and 1.41½), King Tom, Tattler, Omaha, Maggie, Mitchell, Letretia, Little Barefoot, Constellation, Cynosure, Orator, Sluggard, Tom Boy, Waterloo, Crutches, J. O. C., Sandstone, Morrisania, Jester, Disappointment, Exotic, Citizen, etc. He has covered only a few good mares.

TOM OCHILTREE.

BAY HORSE, FOALED 1873; BRED BY MR. A. J. ALEXANDER, WOODBURY STUD, KENTUCKY.

- LEXINGTON, b. h., 1850.
  - Boston, ch. h., 1833.
    - Timoleon, ch. h., 1813.
      - Sir Archy, b. h., 1805.
      - Daughter of, ch. m., —.
    - Sist. to Tuckahoe, ch. m., 1814.
      - Ball's Florizel, ch. h., 1801.
      - Daughter of, b. m., 1799.
  - Alice Carneal, b. m., 1836.
    - *Sarpedon, br. h., 1828.
      - Emilius, b. h., 1820.
      - Icaria, b. m., 1824.
    - Rowena, ch. m., 1826.
      - Sumpter, ch. h., 1818.
      - Lady Grey, b. m., 1817.
- KATONA, br. m., 1857.
  - Voucher, b. h., 1845.
    - Wagner, ch. h., 1834.
      - Sir Charles, ch. h., 1816.
      - Maria West, b. m., 1827.
    - *Britannia, br. m., 1834.
      - Muley, b. h., 1810.
      - Nancy, b. m., 1813.
  - Countess, ch. m., 1847.
    - *Margrave, ch. h., 1829.
      - Muley, b. h., 1810.
      - Daughter of, ch. m., 1815.
    - Daughter of, —.
      - Am. Eclipse, ch. h., 1814.
      - Daughter of, —.

*Diomed, ch. h., 1777.	Florizel 1768 (Herod 1758) Sist. 1763 to Juno by Spectator 1749 (Crab 1722)—Horatia 1758 by Blank 1740 (God. Ar. 1724)—Dau. of Flying Childers 1715 (Darley Ar. 1699) Miss Belvoir by Grey Grantham (Brownlow Turk).
*Castianira, br. m., 1796.	Rockingham 1781 (Highflyer 1774) Tabitha 1782 by Trentham 1766 (T. Sweepstakes 1743) Dau. of Bosphorus 1754 (Babraham 1740) Sist. to Grecian Princess by W's Forester 1750 (Forester 1736) Dau. of Coalition Colt 1730 (God. Ar. 1724).
Saltram, br. h., 1780.	Eclipse 1764 (Marske 1750) Virago 1764 by Snap 1750 (Snip 1736) Dau. of Regulus 1739 (God. Ar. 1724) Sist. 1750 to Othello by Crab 1722 (Alcock Ar.) Miss Slamerkin 1729 by Y. True Blue 1718 (Honeywood's Ar.).
Daughter of, —.	Symmes' Wildair 1767 (*Fearnought 1755) Dau. of Tyler's Driver 1760 (*Othello) Dau. of *Fallower 1761 (Blank 1740)—Dau. of *Vampire 1757 (Regulus 1739). [The Saltram mare was dam of Jenny Cockracy.]
*Diomed, ch. h., 1777.	Florizel 1768 (Herod 1758) Sist. 1763 to Juno by Spectator 1749 (Crab 1722)—Horatia 1758 by Blank 1740 (God. Ar. 1724)—Dau. of Flying Childers 1715 (Darley Ar. 1699) Miss Belvoir by Grey Grantham (Brownlow Turk).
Daughter of, —.	*Shark 1771 (Marske 1750)—Dau. of Harris' Eclipse 1770 (*Fearnought 1755) Dau. of *Fearnought 1755 (Regulus 1739) Dau. of *Jolly Roger 1741 (Roundhead 1733) Dau. of *Sober John 1748 (Rib 1736).
*Alderman, b. h., 1787.	Pot-8-o-'s 1773 (Eclipse 1764) Lady Bolingbroke 1766 by Squirrel 1754 (Traveller 1735) Cypron 1750 by Blaze 1733 (Flying Childers 1715) Salome 1733 by Bethell's Ar.—Dau. of Graham's Champion 1707 (Harpur's Ar.).
Daughter of, ab. 1791.	*Clockfast 1774 (Gimcrack 1760) Dau. of Symmes' Wildair 1767 (*Fearnought 1755) Y. Kitty Fisher 1767 by *Fearnought 1755 (Regulus 1739) *Kitty Fisher 1755 by Cade 1734 (God. Ar. 1724)—Dau. of Somerset Ar.
Orville, b. h., 1799.	Beningbrough 1791 (King Fergus 1775) Evelina 1791 by Highflyer 1774 (Herod 1758)—Termagant by Tantrum 1760 (Cripple 1750)—Cantatrice by Sampson 1745 (Blaze 1733)—Dau. 1757 of Regulus 1739 (God. Ar. 1724).
Emily, ch. m., 1810.	Stamford 1794 (Sir Peter T. 1784)—Dau. 1799 of Whiskey 1789 (Saltram* 1780) Grey Dorimant 1781 by Dorimant 1772 (Otho 1760)—Dizzy 1757 by Blank 1740 (God. Ar. 1724)—Dizzy 1741 by Driver 1727 (Wynn Ar.).
The Flyer, br. h., 1814.	Vandyke Jr. 1808 (Walton 1799)—Azalia 1804 by Beningbrough 1791 (King Fergus 1775)—Gilliflower 1787 by Highflyer 1774 (Herod 1758) Preference 1778 by Goldfinder 1764 (Snap 1750)—Dau. 1771 of Marske 1750 (Squirt 1732).
Parma, br. m., 1813.	Dick Andrews 1797 (Joe Andrews 1778) May 1804 by Beningbrough 1791 (King Fergus 1775)—Primrose 1787 by Mambrino 1768 (Engineer 1756) Cricket 1776 by Herod 1758 (Tartar 1743)—Sophia 1767 by Blank 1740 (God. Ar. 1724).
Sir Archy, b. h., 1805.	*Diomed 1777 (Florizel 1768)—*Castianira 1796 by Rockingham 1781 (Highflyer 1774)—Tabitha 1782 by Trentham 1766 (T. Sweepstakes 1743)—Dau. of Bosphorus 1754 (Babraham 1740)—Dau. of Forester 1750 (Forester 1736).
Robin mare, ch. m., —.	*Robin Redbreast 1793 (Sir Peter T. 1784)—Dau. 1788 of *Obscurity 1777 (Eclipse 1764)—Slamerkin 1768 by *Wildair 1753 (Cade 1734)—*Cub mare 1762 by Cub 1739 (Fox 1714)—Dau. 1755 of Second 1732 (Flying Childers 1715).
Robin Grey, gr. h., 1805.	*Royalist 1790 (Saltram* 1780) Belle Maria by Grey Diomed 1786 (*Medley 1776)—Queen by *St. George 1771 (Dragon 1757)—Dau. of Cassius 1778 (Black-and all-Black 1773)—Primrose by *Dove 1756 (Y. Cade 1747).
Maria, —. m., 1802.	Melzar 1791 (*Medley 1776)—Dau. of *Highflyer 1784 (Highflyer 1774)—Dau. of *Fearnought 1755 (Regulus 1739)—Dau. of Ariel 1756 (*Traveller 1747)—Dau. of *Jack of Diamonds (Cullen Ar.).
Sir Archy, b. h., 1805.	*Diomed 1777 (Florizel 1768)—*Castianira 1796 by Rockingham 1781 (Highflyer 1774)—Tabitha 1782 by Trentham 1766 (T. Sweepstakes 1743)—Dau. of Bosphorus 1754 (Babraham 1740)—Dau. of Forester 1750 (Forester 1736).
Daughter of, ch. m., —.	*Citizen 1785 (Pacolet 1763)—Dau. of Commutation (Symmes' Wildair 1767)—Dau. of *Dare Devil 1787 (Magnet 1770)—Sally Shark by *Shark 1771 (Marske 1750) Betsey Pringle by *Fearnought 1755 (Regulus 1739).
Marion, b. h., 1820.	Sir Archy 1805 (*Diomed 1777)—Dau. of *Citizen 1785 (Pacolet 1763)—Dau. of *Alderman 1787 (Pot-8-o-'s 1773)—Dau. of Ashe's Roebuck (*Sweeper) Dau. of Haine's King Herod 1770 (*Fearnought 1755).
Ella Crump, br. m., —.	*Citizen 1785 (Pacolet 1763)—Dau. 1795 of Huntsman 1780 (*Monsetrap 1771)—Dau. of Symmes' Wildair 1767 (*Fearnought 1755)—Dau. of *Fearnought 1755 (Regulus 1739)—Dau. of *Janus 1746 (Janus 1739).
Orville, b. h., 1799.	Beningbrough 1791 (King Fergus 1775)—Evelina 1791 by Highflyer 1774 (Herod 1758)—Termagant by Tantrum 1760 (Cripple 1750)—Cantatrice by Sampson 1745 (Blaze 1733)—Dau. 1757 of Regulus 1739 (God. Ar. 1724).
Eleanor, b. m., 1798.	Whiskey 1789 (Saltram* 1780)—Y. Giantess 1790 by Diomed* 1777 (Florizel 1768)—Giantess 1769 by Matchem 1748 (Cade 1734)—Molly Long Legs 1753 by Babraham 1740 (God. Ar. 1724)—Dau. of Cole's Foxhunter 1727 (Brisk 1711).
Dick Andrews, b. h., 1797.	Joe Andrews 1778 (Eclipse 1764) Dau. 1790 of Highflyer 1774 (Herod 1758)—Dau. 1777 of Cardinal Puff 1760 (Babraham 1740)—Dau. 1768 of Tatler 1754 (Blank 1740)—Bay Snip 1746 by Snip 1736 (Flying Childers 1715).
Spitfire, b. m., 1800.	Beningbrough 1791 (King Fergus 1775)—Dau. of Y. Sir Peter 1775 (Doge 1762)—Dau. of Engineer 1756 (Sampson 1745)—Dau. of Wilson Ar.—Dau. of Hutton's Spot 1728 (Hartley's Blind Horse).
Orville, b. h., 1799.	Beningbrough 1791 (King Fergus 1775)—Evelina 1791 by Highflyer 1774 (Herod 1758)—Termagant by Tantrum 1760 (Cripple 1750)—Cantatrice by Sampson 1745 (Blaze 1733)—Dau. 1757 of Regulus 1739 (God. Ar. 1724).
Eleanor, b. m., 1798	Whiskey 1789 (Saltram* 1780)—Y. Giantess 1790 by Diomed* 1777 (Florizel 1768)—Giantess 1769 by Matchem 1748 (Cade 1734)—Molly Long Legs 1753 by Babraham 1740 (God. Ar. 1724)—Dau. of Foxhunter 1727 (Brisk 1711).
Election, ch. h., 1804.	Gohanna 1790 (Mercury 1778—Chestnut Skim 1794 by Woodpecker 1773 (Herod 1758)—Silver*'s dam 1780 by Herod 1758 (Tartar 1743)—Y. Hag 1761 by Skim 1746 (Starling 1727)—Hag 1744 by Crab 1722 (Alcock Ar.).
Fair Helen, gr. m., 1808.	Hambletonian 1792 (King Fergus 1775)—Helen 1801 by Delpini 1781 (Highflyer 1774)—Rosalind 1788 by Phenomenon 1780 (Herod 1758) Atalanta 1769 by Matchem 1748 (Cade 1734)—Y. Lass-of-the-Mill 1756 by Oroonoko 1745 (Crab 1722).
Duroc, ch. h., 1806.	*Diomed 1777 (Florizel 1768) Amanda 1800 by Grey Diomed 1786 (*Medley 1776) Dau. of Va. Cade (Partner 1755)—Dau. of Hickman's Independence 1776 (*Fearnought 1755)—Dau. of *Badger (Bosphorus 1754).
Miller's Damsel, gr. m., 1802.	*Messenger 1780 (Mambrino 1768)—*Mare 1792 by Pot-8-o-'s 1773 (Eclipse 1764) Dau. 1778 of Gimcrack 1760 (Cripple 1750)—Snapdragon 1759 by Snap 1750 (Snip 1736)—Dau. 1749 of Regulus 1739 (God. Ar. 1724).
Rattler, ch. h., 1816.	Sir Archy 1805 (*Diomed 1777) Robin mare by *Robin Redbreast 1796 (Sir Peter T. 1784)—Dau. 1788 of *Obscurity 1777 (Eclipse 1764)—Slamerkin 1768 by *Wildair 1753 (Cade 1734)—*Cub mare 1762 by Cub 1739 (Fox 1714).
Daughter of, —, —.	Thornton's Diomed (Ball's Florizel 1801)—Dau. of Tiger 1812 (Cook's Whip 1804)—Dau. of *Shark 1771 (Marske 1750)—Dau. of *Fearnought 1755 (Regulus 1739).

IMPORTED TOP GALLANT.

Property of J. B. Ewing, Willamette Stud, Nashville, Tenn. Terms and yearling sales annually.

STERLING, the sire of Top Gallant, one of the most popular stallions in England, started only a few times. He won the Liverpool Autumn Cup, beating Louise Victoria and twelve others, and was third to Montargis and Walnut in the Cambridgeshire, with 133 lbs. He won the Ascot Biennial Stakes, the Summer Stakes at Newmarket, the Derby Free Handicap and Craven Stakes. He was second to Boswell in the 2,000 Guineas, a defeat which was one of the most glaring flukes in the annals of the turf, for it was admitted by every one that he could concede quite 42 lbs. to the winner; he was not in the Derby. His value is best shown in the stud, having sired such good ones as Isonomy, who himself is the sire of Riviera, one of the favorites for the Derby in 1890, Formidable, another good two-year old in 1890, Antibes, Common, winner of the 2,000 Guineas, the Derby and St. Leger in 1891, etc., and one of the best race-horses of his time. His get last year won over $100,000, and he stood at $1,000, with his season full in advance. Harvester, who ran a dead heat with St. Gatien and divided the stakes for the Derby, and was third in the 2,000 Guineas. Paradox, Enterprise and Euthusiast, all three winners of the 2,000 Guineas, the former only beaten a head in the Derby, and the latter defeating the famous Donovan. Energy, winner of sixteen races out of thirty-seven, and one of the speediest horses of his day. Beaudesert, the winner of the Middle Park Plate and sire of many winners. Caxtonian, Discount, Post Orbit, Geologist, El Dorado (another conquerer of Donovan) and Cherry all come from Sterling's loins. Top Gallant's dam, Sea Mark, is the dam of Astronomy, Beacon, and Mr. D. D. Withers' fine young stallion Stalwart, who, from an injury received in crossing the ocean, was never trained. His grandam, Sea Gull, was the dam of Beaudesert, winner of the Middle Park Plate, Queenstown, The Admiral, Discordant, etc. Wild Cherry was the dam of Grace Darling, Cherry Ripe, Sugar Plum, Heliotrope, Reciprocity, etc. Bridget was the dam of The Baronet, Wild Strawberry, Pomona, Miss Fanny, etc. Top Gallant traces through Ringlet, a daughter of Whisker, Clinkerina, the dam of Humphrey Clinker, sire of the noted Melbourne by Clinker, Pewet, a winner of the St. Leger, by Tandem, to the noted Blacklegs mare, dam of Marske, the sire of Eclipse. He combines in his pedigree a double cross of Sir Hercules upon a large infusion of Touchstone blood through such excellent sources as Adventurer, Flatcatcher and Surplice, and traces on his sire's side and dam's side to the noted daughters of Priam, Crucifix and Miss Letty. From the Coneyskin's mare, the fourteenth dam of Top Gallant, descend in direct female line the following noted race-horses and successful stallions: Marske, sire of Eclipse, Orville, St. Leger winner, Cervantes, Cain, Humphrey Clinker, Tomboy, Octavian, St. Leger winner, The Colonel, St. Leger winner, Sultan, sire of Glencoe and Bay Middleton, Sir Tatton Sykes, St. Leger winner, Newminster, St. Leger winner, and sire of the famous Hermit, Andover, Derby winner, Rifleman, Strafford, Cathedral, Mariner, Melton, winner of both Derby and St. Leger, and many other of the best race-horses and stallions in England. He is one of the few sons of the great Sterling in America accessible to breeders, and he must be a valuable acquisition to the blood stock of the country.

Top Gallant is a horse of rare individual merit. He carried off the first prize at the National Horse Show, in 1890, over a choice lot of stallions.

Pot-8-o's 1773 Eclipse 1764—Marske 1750 Squirt 1732 Bartlett's Childers · Darley Arabian 1699.

TOP GALLANT.

BAY HORSE, FOALED 1884; BRED AT THE YARDLEY STUD, ENGLAND.

STERLING, b. h., 1868.

- Oxford, ch. h., 1857.
  - Ir. Birdcatcher, ch. h., 1833.
    - Sir Hercules, blk. h., 1826.
    - Guiccioli, ch. m., 1823.
  - Honey Dear, b. m., 1844.
    - Plenipotentiary, ch. h., 1831.
    - My Dear, b. m., 1841.
- Whisper, b. m., 1857.
  - Flatcatcher, br. h., 1845.
    - Touchstone, br. h., 1831.
    - Decoy, b. m., 1830.
  - Silence, b. m., 1848.
    - Melbourne, br. h., 1834.
    - Secret, ch. m., 1841.

SEA MARK, ch. m., 1873.

- Adventurer, b. h., 1859.
  - Newminster, b. h., 1848.
    - Touchstone, br. h., 1831.
    - Bee's-wing, b. m., 1833.
  - Palma, br. m., 1840.
    - Emilius, b. h., 1820.
    - Francesca, br. m., 1823.
- Sea Gull, blk. m., 1866.
  - Lifeboat, br. h., 1855.
    - Sir Hercules, blk. h., 1826.
    - Yard-Arm, b. m., 1843.
  - Wild Cherry, blk. m., 1853.
    - Surplice, b. h., 1845.
    - Bridget, br. m., 1846.

Whalebone, br. h., 1807.	Waxy 1790 (Pot-8-o's 1773) Penelope 1798 by Trumpator 1782 (Conductor 1767) Prunella 1788 by Highflyer 1774 (Herod 1758) Promise 1768 by Snap 1750 (Snip 1736) Julia 1756 by Blank 1740 (God. Ar. 1724).
Peri, b. m., 1822.	Wanderer 1811 (Gohanna 1790)—Thalestris 1809 by Alexander 1782 (Eclipse 1764) Rival 1800 by Sir Peter T. 1784 (Highflyer 1774) Hornet 1790 by Drone 1777 (Herod 1758) Manilla 1777 by Goldfinder 1764 (Snap 1750).
Bob Booty, ch. h., 1804.	Chanticleer 1787 (Woodpecker 1773)—Ierne 1790 by Bagot 1780 (Herod 1758)—Dau. of Gamahoe (Bustard 1741) Patty by Tim 1752 (Squirt 1732)—Miss Patch 1737 by Justice 1725 (H. C. Litton Ar.).
Flight, ch. m., 1809.	Ir. Escape 1802 (Commodore 1793) Y. Heroine by Bagot 1780 (Herod 1758)—Heroine 1775 by Hero 1753 (Cade 1734)—Dau. 1756 of Snap 1750 (Snip 1736) Sist. 1743 to Regulus by God. Ar. 1724.
Emilius, b. h., 1820.	Orville 1799 (Beningbrough 1791)—Emily 1810 by Stamford 1794 (Sir Peter T. 1784) Dau. 1799 of Whiskey 1789 (Saltram* 1780)—Grey Dorimant 1781 by Dorimant 1772 (Otho 1760)—Dizzy 1757 by Blank 1740 (God. Ar. 1724).
Harriet, blk. m., 1819.	Pericles 1809 (Evander 1801)—Dau. 1812 of Selim 1802 (Buzzard* 1787)—Pipylina 1803 by Sir Peter T. 1784 (Highflyer 1774)—Rally 1790 by Trumpator 1782 (Conductor 1767)—Fancy 1780 by Florizel 1768 (Herod 1758).
Bay Middleton, b. h., 1833.	Sultan 1816 (Selim 1802)—Cobweb 1821 by Phantom 1808 (Walton 1799) Filagree 1815 by Soothsayer 1808 (Sorcerer 1796)—Web 1808 by Waxy 1790 (Pot-8-o's 1773)—Penelope 1798 by Trumpator 1782 (Conductor 1767).
Miss Letty, b. m., 1834.	Priam* 1827 (Emilius 1820)—Dau. 1815 of Orville 1799 (Beningbrough 1791)—Dau. 1800 of Buzzard* 1787 (Woodpecker 1773) Hornpipe 1793 by Trumpator 1782 (Conductor 1767)—Luna 1779 by Herod 1758 (Tartar 1743).
Camel, br. h., 1822.	Whalebone 1807 (Waxy 1790) Dau. 1812 of Selim 1802 (Buzzard* 1787) Maiden 1801 by Sir Peter T. 1784 (Highflyer 1774)—Dau. 1788 of Phenomenon 1780 (Herod 1758)—Matron 1782 by Florizel 1768 (Herod 1758).
Banter, br. m., 1826.	Master Henry 1815 (Orville 1799)—Boadicea 1807 by Alexander 1782 (Eclipse 1764)—Brunette by Amaranthus 1766 (Old England 1741) Mayfly 1771 by Matchem 1748 (Cade 1734)—Dau. 1755 of Anc. Starling 1738 (Starling 1727).
Filho-da-Puta br. h., 1812.	Haphazard 1797 (Sir Peter T. 1784)—Mrs. Barnet 1806 by Waxy 1790 (Pot-8-o's 1773)—Dau. 1788 of Woodpecker 1773 (Herod 1758)—Heinel 1771 by Squirrel 1754 (Traveller 1735) Principessa 1762 by Blank 1740 (God. Ar. 1724).
Finesse, b. m., 1815.	Peruvian 1806 (Sir Peter T. 1784)—Violante 1802 by John Bull 1789 (Fortitude 1777)—Dau. 1788 of Highflyer 1774 (Herod 1758) Everlasting 1775 by Eclipse 1764 (Marske 1750)—Hyæna 1762 by Snap 1750 (Snip 1736).
Humprey Clinker, b. h., 1824.	Comus 1809 (Sorcerer 1796)—Clinkerina 1812 by Clinker 1805 (Sir Peter T. 1784)—Pewet 1786 by Tandem 1773 (Syphon 1750)—Termagant by Tantrum 1760 (Cripple 1750)—Cantatrice by Sampson 1745 (Blaze 1733).
Daughter of, b. m., 1825.	Cervantes 1806 (Don Quixote 1784) Dau. 1818 of Golumpus 1802 (Gohanna 1790) Dau. 1810 of Paynator 1791 (Trumpator 1782) Circle 1802 by St. George 1789 (Highflyer 1774)—Abigail 1788 by Woodpecker 1773 (Herod 1758).
Hornsea, ch. h., 1832.	Velocipede 1825 (Blacklock 1814) Dau. 1820 of Cerberus 1802 (Gohanna 1790) Miss Cranfield 1803 by Sir Peter T. 1784 (Highflyer 1774)—Dau. 1796 of Pegasus 1784 (Eclipse 1764)—Dau. 1783 of Paymaster 1766 (Blank 1740).
Solace, b. m., 1830.	Longwaist 1821 (Whalebone 1807) Dulcamara 1818 by Waxy 1790 (Pot-8-o's 1773)—Witchery 1814 by Sorcerer 1796 (Trumpator 1782)—Cobbĕa 1802 by Skyscraper 1786 (Highflyer 1774)—Dau. 1788 of Woodpecker 1773 (Herod 1758).
Camel, br. h., 1822.	Whalebone 1807 (Waxy 1790)—Dau. 1812 of Selim 1802 (Buzzard* 1787)—Maiden 1801 by Sir Peter T. 1784 (Highflyer 1774)—Dau. 1788 of Phenomenon 1780 (Herod 1758)—Matron 1782 by Florizel 1768 (Herod 1758).
Banter, br. m., 1826.	Master Henry 1815 (Orville 1799)—Boadicea 1807 by Alexander 1782 (Eclipse 1764)—Brunette by Amaranthus 1766 (Old England 1741)—Mayfly 1771 by Matchem 1748 (Cade 1734)—Dau. 1755 of Anc. Starling 1738 (Old Starling 1727).
Dr. Syntax, br. h., 1811.	Paynator 1791 (Trumpator 1782)—Dau. of Beningbrough 1791 (King Fergus 1775)—Jenny Mole 1787 by Carbuncle 1772 (Babr. Blank 1758) Dau. 1773 of Prince T'Quassaw 1751 (Snip 1736)—Sultana 1759 by Regulus 1739 (God. Ar. 1724).
Daughter of, ch. m., 1817.	Ardrossan 1809 (John Bull 1789)—Lady Eliza 1813 by Whitworth 1805 (Agonistes 1797)—X. Y. Z.'s dam 1798 by Spadille 1784 (Highflyer 1774)—Sylvia 1783 by Y. Marske 1771 (Marske 1750)—Ferret 1765 by Broth. to Silvio (Cade 1734).
Orville, b. h., 1799.	Beningbrough 1791 (King Fergus 1775)—Evelina 1791 by Highflyer 1774 (Herod 1758)—Termagant by Tantrum 1760 (Cripple 1750)—Cantatrice by Sampson 1745 (Blaze 1733)—Dau. 1757 of Regulus 1739 (God. Ar. 1724).
Emily, ch. m., 1810.	Stamford 1794 (Sir Peter T. 1784)—Dau. 1799 of Whiskey 1789 (Saltram* 1780)—Grey Dorimant 1781 by Dorimant 1772 (Otho 1760) Dizzy 1757 by Blank 1740 (God. Ar. 1724)—Anc. Dizzy 1741 by Driver 1727 (Wynn Ar.)
Partisan, b. h., 1811.	Walton 1799 (Sir Peter T. 1784)—Parasol 1800 by Pot-8-o's 1773 (Eclipse 1764)—Prunella 1788 by Highflyer 1774 (Herod 1758)—Promise 1768 by Snap 1750 (Snip 1736)—Julia 1756 by Blank 1740 (God. Ar. 1724).
Miss Fanny's dam, br. m., 1815.	Orville 1799 (Beningbrough 1791)—Dau. 1800 of Buzzard* 1787 (Woodpecker 1773)—Hornpipe 1793 by Trumpator 1782 (Conductor 1767)—Luna 1779 by Herod 1758 (Tartar 1743)—Proserpine 1756 (Eclipse's sist.) by Marske 1750 (Squirt 1732).
Whalebone, br. h., 1807.	Waxy 1790 (Pot-8-o's 1773)—Penelope 1798 by Trumpator 1782 (Conductor 1767)—Prunella 1788 by Highflyer 1774 (Herod 1758)—Promise 1768 by Snap 1750 (Snip 1736)—Julia 1756 by Blank 1740 (God. Ar. 1724).
Peri, b. m., 1822.	Wanderer 1811 (Gohanna 1790)—Thalestris 1809 by Alexander 1782 (Eclipse 1764)—Rival 1800 by Sir Peter T. 1784 (Highflyer 1774) Hornet 1790 by Drone 1777 (Herod 1758)—Manilla 1777 by Goldfinder 1764 (Snap 1750).
Sheet Anchor, br. h., 1832.	Lottery 1820 (Tramp 1810)—Morgiana 1820 by Muley 1810 (Orville 1799)—Miss Stephenson 1814 by [Scud or] Sorcerer 1796 (Trumpator 1782) Sist. 1795 to Petworth by Precipitate* 1787 (Mercury 1778)—Dau. 1787 of Woodpecker 1773 (Herod 1758).
FannyKemble b. m., 1827.	Paulowitz 1813 (Sir Paul 1802) Loyalty 1817 by Rubens 1805 (Buzzard* 1787) Pennyroyal 1805 by Coriander 1786 (Pot-8-o's 1773)—Peppermint 1787 by Highflyer 1774 (Herod 1758)—Promise 1768 by Snap 1750 (Snip 1736).
Touchstone, br. h., 1831.	Camel 1822 (Whalebone 1807)—Banter 1826 by Master Henry 1815 (Orville 1799) Boadicea 1807 by Alexander 1782 (Eclipse 1764)—Brunette by Amaranthus 1766 (Old England 1741)—Mayfly 1771 by Matchem 1748 (Cade 1734).
Crucifix, b. m., 1837.	Priam* 1827 (Emilius 1820)—Octaviana 1815 by Octavian 1807 (Stripling 1795)—Dau. 1807 of Shuttle 1793 (Y. Marske 1771)—Zara 1801 by Delpini 1781 (Highflyer 1774)—Flora 1789 by King Fergus 1775 (Eclipse 1764).
[Rocket or] Cardinal Puff. br. h., 1834.	Pantaloon 1824 (Castrel 1801)—Puff 1823 by Waterloo 1814 (Walton 1799) Blowing 1803 by Buzzard* 1787 (Woodpecker 1773)—Dau. 1796 of Pot-8-o's 1773 (Eclipse 1764)—Maid-of-all-Work 1786 by Highflyer 1774 (Herod 1758).
Ringlet, br. m., 1826.	Whisker 1812 (Waxy 1790)—Clinkerina br. m. 1812 by Clinker 1805 (Sir Peter T. 1784)—Pewet b. m. 1786 by Tandem 1773 (Syphon 1750)—Termagant by Tantrum 1760 (Cripple 1750)—Cantatrice by Sampson 1745 (Blaze 1733) Dau. br. m. 1757 by Regulus 1739 (God. Ar. 1724)—The Ruby mare (Marske's dam) by Hutton's Blacklegs 1725 (Mulso Turk)—Dau. of Bay Bolton 1705 (Grey Hautboy)—Dau. of Fox Cub 1714 (Clumsy)—Dau. of Coneyskins 1712 (Lister Turk)—Dau. of Hutton's Grey Barb 1700—Dau. of Hutton's Royal Colt 1700 (Helmsley Turk)—Dau. of Byerly Turk—Dau. of Bustler (Helmsley Turk).

TREMONT.

WINNER OF THE JUVENILE, FOAM, SURF, JUNE, GREAT POST, ATLANTIC, TYRO, JUNIOR CHAMPION AND OTHER STAKES AT TWO YEARS OLD.

The property of Gen. W. H. Jackson, Belle Meade Stud, Nashville, Tenn. Terms annually. Yearling sales in June each year.

TREMONT, black horse, foaled 1884, bred in the Elmendorf Stud, Ky., by Virgil, son of Vandal by imp. Glencoe, dam Ann Fief by Alarm, son of imp. Eclipse by Orlando, grandam Kate Walker by Embry's Lexington, son of Lexington and Bellamira by imp. Monarch, son of imported Priam, etc. Tremont was a phenomenally fast horse and never met defeat during his racing career, which ended at two years old. Virgil, sire of Tremont, was by Vandal, the best son of imp. Glencoe, whose dam was the Tranby mare, that also produced Levity, Volga and Vistula. Virgil's dam was Hymenia by imp. Yorkshire. She produced Daisy by Cracker, Alaric, Ansel, Lady Petry and Mollie Wood. Virgil was the sire of not only the unbeaten Tremont, but of the great Hindoo (sire of Hanover), Portland, Vigil, Vagrant, Vera Cruz, Virginius, Vanguard, Virgilian, Fairplay, Katrina, Memento, Isaac Murphey, Carley B., Ben Ali, Richmond, Kingfish, Himalaya, etc., etc. Tremont inherited much of his speed from Ann Fief, the daughter of Alarm by imp. Eclipse. Ann Fief was out of Kate Walker by Embry's Lexington, son of the great Lexington and grandson of the stout Boston. Kate Walker, the dam of Ann Fief, was the dam of Bend'or, one of the best race-horses this country has ever produced for speed, stamina and weight-carrying capacity. As a four-year old Bend'or started in twelve races, of which he won nine. He won at heats of a mile and a sixteenth, capturing the second and third heats; won at 1½ miles, over a heavy track, at Chicago, carrying a 10 lb. penalty; he won the Board of Trade Handicap at the same distance; with 149 lbs., he ran a mile at Saratoga in 1.45; won at 1¼ miles, in 2.35½, defeating such a grand mare as Thora, and again in 2.36¾, with 123 lbs., conceding General Monroe 22 lbs.; while in the Grand Prize at Saratoga, 1¾ miles, he carried 122 lbs. and ran second to Glidelia, 116 lbs., in 3.01, the fastest and best race at the distance up to that date. Besides Bend'or, Kate Walker produced the winners Maggie H. and Hattie D. H. Ann Fief is the dam of Spanish King, Aurania (winner of the New and Spring Stakes as a two-year old and other races), Anna B. (winner of the Debutante Stakes) and Westchester (winner of the Seabright Stakes). Tremont won the Juvenile Stakes, Jerome Park, half a mile, 115 lbs., in 0.48¾, fourteen starters; won the Sequence Stakes, Jerome Park, 5 furlongs, 115 lbs., in 1.02½, seven starters; won the Foam Stakes, Sheepshead Bay, 5 furlongs, 115 lbs., in 1.04½, ten starters; won the Surf Stakes, Sheepshead Bay, 5 furlongs, 122 lbs., in 1.02 nine starters; won the Zephyr Stakes, Sheepshead Bay, three-quarters of a mile, 112 lbs., in 1.17½, four starters; won the Paddock Stakes, Sheepshead Bay, three-quarters of a mile, 112 lbs., in 1.17½, two starters; won the Spring Stakes, Sheepshead Bay, three-quarters of a mile, 114 lbs., in 1.16½; won the June Stakes, Sheepshead Bay, three-quarters of a mile, 117 lbs., in 1.18½, four starters; won the Great Post Stakes, Sheepshead Bay, three-quarters of a mile, 115 lbs., in 1.19¾, four starters; won the Good-Bye Stakes, Sheepshead Bay, three-quarters of a mile, 115 lbs., in 1.15, six starters; won the Atlantic Stakes, Monmouth Park, three-quarters of a mile, 115 lbs., in 1.16¾, six starters; won the Tyro Stakes, Monmouth Park, three-quarters of a mile, 115 lbs., in 1.19½, five starters; won the Junior Champion Stakes, Monmouth Park, three-quarters of a mile, 115 lbs., in 1.17½. In this last race he defeated Kingston and Fitzroy. He met and defeated all the best two-year olds of his year.

Tremont must be regarded as a success in the stud. The following winners are most all the get of his first season in the stud: Dagonet, Fremont, Fillide, Fleurette, Jessica, Lavish, Temple, Tringle, Marmont, Chicago, Seabright, Lovelace, etc., creditable performers.

TREMONT.

BLACK HORSE, FOALED 1884; BRED BY D. SWIGERT, ELMENDORF STUD, LEXINGTON, KY.

VIRGIL, br. h., 1864.
- Hymenia, b. m., 1851.
  - Little Peggy, ch. m., 1837.
    - Peggy Stewart, b. m., 1821.
    - Cripple, b. h., 1826.
  - *Yorkshire, b. h., 1834.
    - Miss Rose, b. m., 1826.
    - St. Nicholas, br. h., 1827.
- Vandal, b. h., 1850.
  - Alaric's dam, b. m., 1837.
    - Lucilla, b. m., —.
    - *Tranby, br. h., 1826.
  - *Glencoe, ch. h., 1831.
    - Trampoline, ch. m., 1825.
    - Sultan, b. h., 1816.

ANN FIEF, br. m., 1876.
- Kate Walker, br. m., 1868.
  - Carrie D., b. m., 1861.
    - Romance, b. m., 1831.
    - Don Juan, blk. h., 1857.
  - Embry's Lexington, b. h., 1858.
    - Bellamira, b. m., 1842.
    - Lexington, b. h., 1850.
- Alarm, b. h., 1869.
  - *Maud, b. m., 1859.
    - Countess of Albemarle, b. m., 1847.
    - Stockwell, ch. h., 1849.
  - *Eclipse, b. h., 1855.
    - Gaze, b. m., 1842.
    - Orlando, b. h., 1841.

Sultan	Selim, ch. h., 1802.	Buzzard* 1787 (Woodpecker 1773) Dau. of Alexander 1782 (Eclipse 1764) Dau. of Highflyer 1774 (Herod 1758) Dau. of Alfred 1770 (Matchem 1748) Dau. 1770 of Engineer 1756 (Sampson 1745).
	Bacchante, br. m., 1809.	W's Ditto 1800 (Sir Peter T. 1784) Dau. 1791 of Mercury 1778 (Eclipse 1764) Dau. 1776 of Herod 1758 (Tartar 1743) Folly 1771 by Marske 1750 (Squirt 1732) Vixen 1753 by Regulus 1739 (God. Ar. 1724).
Trampoline	Tramp, b. h., 1810.	Dick Andrews 1797 (Joe Andrews 1778) Dau. 1803 of Gohanna 1790 (Mercury 1778) Fraxinella 1793 by Trentham 1766 (T. Sweepstakes 1743) Dau. 1785 of Woodpecker 1773 (Herod 1758) Everlasting 1775 by Eclipse 1764 (Marske 1750).
	Web, b. m., 1808.	Waxy 1790 (Pot-8-o-'s 1773) Penelope 1798 by Trumpator 1782 (Conductor 1767) Prunella 1788 by Highflyer 1774 (Herod 1758)—Promise 1768 by Snap 1750 (Snip 1736)—Julia 1756 by Blank 1740 (God. Ar. 1724).
*Tranby	Blacklock, b. h., 1814.	Whitelock 1803 (Hambletonian 1792) Dau. 1799 of Coriander 1786 (Pot-8-o-'s 1773) Wildgoose 1792 by Highflyer 1774 (Herod 1758) Coheiress 1786 by Pot-8-o-'s 1773 (Eclipse 1764) Manilla 1777 by Goldfinder 1764 (Snap 1750).
	Miss Bowe's dam, br. m., 1812.	Orville 1799 (Beningbrough 1791) Miss Grimstone 1796 by Weazle 1776 (Herod 1758) Dau. of Ancaster 1768 (Blank 1740) Dau. of Damascus Ar. 1754 Dau. of Sampson 1745 (Blaze 1733).
Lucilla	Davenport's Trumpator, br. h., 1823.	Sir Solomon 1805 (*Tickle Toby 1786) Dau. of Hickory 1801 (Whip 1794)—Trumpetta 1797 by Trumpator 1782 (Conductor 1767) Sist. 1786 to Lambinos by Highflyer 1774 (Herod 1758) Dau. 1775 of Eclipse 1764 (Marske 1750).
	Lucy, b. m., 1821.	Orphan 1810 (Ball's Florizel 1801) Lady Grey 1817 by Robin Grey 1805 (*Royalist 1790) Maria 1802 by Melzar 1791 (*Medley 1776) Dau. of *Highflyer 1784 (Highflyer 1774) Dau. of *Fearnought 1755 (Regulus 1739).
St. Nicholas	Emilius, b. h., 1820.	Orville 1799 (Beningbrough 1791) Emily 1810 by Stamford 1794 (Sir Peter T. 1784)—Dau. 1799 of Whiskey 1789 (*Saltram 1780) Grey Dorimant 1781 by Dorimant 1772 (Otho 1760) Dizzy 1757 by Blank 1740 (God. Ar. 1724).
	Sea Mew, b. m., 1815.	Scud 1804 (Beningbrough 1791)—Goosander 1805 by Hambletonian 1792 (King Fergus 1775)—Rally 1790 by Trumpator 1782 (Conductor 1767)—Fancy 1780 by Florizel 1768 (Herod 1758) Dau. 1768 of Spectator 1749 (Crab 1722).
Miss Rose	Tramp, b. h., 1810.	Dick Andrews 1797 (Joe Andrews 1778)—Dau. 1803 of Gohanna 1790 (Mercury 1778)—Fraxinella 1793 by Trentham 1766 (T. Sweepstakes 1743) Dau. 1785 of Woodpecker 1773 (Herod 1758) Everlasting 1775 by Eclipse 1764 (Marske 1750).
	Daughter of, ch. m., 1810.	Sancho 1801 (Don Quixote 1784)—Blacklock's dam 1799 by Coriander 1786 (Pot-8-o-'s 1773) Wildgoose 1792 by Highflyer 1774 (Herod 1758)—Coheiress 1786 by Pot-8-o-'s 1773 (Eclipse 1764)—Manilla 1777 by Goldfinder 1764 (Snap 1750).
Cripple	Medoc, ch. h., 1829.	Am. Eclipse 1814 (Duroc 1806) Y. Maid-of-the-Oaks 1817 by *Expedition 1795 (Pegasus 1784) Maid-of-the-Oaks 1801 by *Spread Eagle 1792 (Volunteer 1780) Annette by *Shark 1771 (Marske 1750)—Dau. of Rockingham (Partner 1755).
	Grecian Princess, b. m., 1816.	Cook's Whip 1804 (*Whip 1794)—Jane Hunt 1796 by Hampton's Paragon 1788 (*Flimnap 1765) Moll by *Figure 1757 (Figure 1747)—Slamerkin 1768 by *Wildair 1772 (Cade 1734) *Cub mare 1762 by Cub 1739 (Fox 1714).
Peggy Stewart	Cook's Whip, b. h., 1804.	*Whip 1794 (Saltram* 1780) Speckleback 1790 by Randolph's Celer (Meade's Celer 1774) Old Speckleback by Meade's Celer 1774 (*Janus 1746) Dau. of *Sober John 1748 (Rib 1736) Thoroughbred mare of Mr. Wade.
	Mary Bedford, b. m., 1816.	Duke of Bedford (*Bedford 1792) Dau. 1813 of *Speculator 1795 (Dragon* 1787)—Dau. 1806 of *Dare Devil 1787 (Magnet 1770) *Trumpetta 1797 by Trumpator 1782 (Conductor 1767) Sist. 1786 to Lambinos by Highflyer 1774 (Herod 1758).
Orlando	Touchstone, br. h., 1831.	Camel 1822 (Whalebone 1807)—Banter 1826 by Master Henry 1815 (Orville 1799)—Boadicea 1807 by Alexander 1782 (Eclipse 1764)—Brunette by Amaranthus 1766 (Old England 1741)—Mayfly 1771 by Matchem 1748 (Cade 1734).
	Vulture, ch. m., 1833.	Langar 1817 (Selim 1802)—Kite 1821 by Bustard 1818 (Castrel 1801)—Olympia 1815 by Sir Oliver 1800 (Sir Peter T. 1784)—Scotilla 1795 by Anvil 1777 (Herod 1758)—Scota 1783 by Eclipse 1764 (Marske 1750).
Gaze	Bay Middleton, b. h., 1833.	Sultan 1816 (Selim 1802)—Cobweb 1821 by Phantom 1808 (Walton 1799) Filagree 1815 by Soothsayer 1808 (Sorcerer 1796)—Web 1808 by Waxy 1790 (Pot-8-o-'s 1773)—Penelope 1798 by Trumpator 1782 (Conductor 1767).
	Flycatcher, ch. m., 1831.	Godolphin 1818 (Partisan 1811)—Sist. 1820 to Cobweb by Phantom 1808 (Walton 1799) Filagree 1815 by Soothsayer 1808 (Sorcerer 1796) Web 1808 by Waxy 1790 (Pot-8-o-'s 1773)—Penelope 1798 by Trumpator 1782 (Conductor 1767).
Stockwell	The Baron, ch. h., 1842.	Ir. Birdcatcher 1833 (Sir Hercules 1826)—Echidna 1838 by Economist 1825 (Whisker 1812)—Miss Pratt 1825 by Blacklock 1814 (Whitelock 1803) Gadabout 1812 by Orville 1799 (Beningbrough 1791)—Minstrel 1803 by Sir Peter T. 1784 (Highflyer 1774).
	Pocahontas, b. m., 1837.	Glencoe* 1831 (Sultan 1816)—Marpessa 1830 by Muley 1810 (Orville 1799) Clare 1824 by Marmion 1806 (Whiskey 1789) Harpalice 1814 by Gohanna 1790 (Mercury 1778)—Amazon 1799 by Driver 1783 (Trentham 1766).
Countess of Albemarle	Lanercost, br. h., 1835.	Liverpool 1828 (Tramp 1810)—Otis 1820 by Bustard 1801 (Buzzard* 1787)—Dau. 1813 of Election 1804 (Gohanna 1790)—Sist. 1791 to Skysweeper by Highflyer 1774 (Herod 1758)—Chanticleer's dam 1778 by Eclipse 1764 (Marske 1750).
	Sister to Hornsea, b. m., 1837.	Velocipede 1825 (Blacklock 1814)—Dau. 1820 of Cerberus 1802 (Gohanna 1790)—Miss Cranfield 1803 by Sir Peter T. 1784 (Highflyer 1774)—Dau. 1796 of Pegasus 1784 (Eclipse 1764)—Dau. 1783 of Paymaster 1766 (Blank 1740).
Lexington	Boston, ch. h., 1833.	Timoleon 1813 (Sir Archy 1805)—Dau. 1814 of Ball's Florizel 1801 (*Diomed 1777)—Dau. 1799 of *Alderman 1787 (Pot-8-o-'s 1773)—Dau. of *Clockfast 1774 (Gimcrack 1760)—Dau. of Symmes' Wildair 1767 (*Fearnought 1755).
	Alice Carneal, b. m., 1836.	*Sarpedon 1828 (Emilius 1820)—Rowena 1826 by Sumpter 1818 (Sir Archy 1805)—Lady Grey 1817 by Robin Grey 1805 (*Royalist 1790)—Maria 1802 by Melzar 1791 (*Medley 1776)—Dau. of *Highflyer 1784 (Highflyer 1774).
Bellamira	*Monarch, b. h., 1834.	*Priam 1827 (Emilius 1820)—*Delphine 1825 by Whisker 1812 (Waxy 1790) My Lady* 1818 by Comus 1809 (Sorcerer 1796)—The Colonel's dam 1802 by Delpini 1781 (Highflyer 1774)—Tipple Cyder 1788 by King Fergus 1775 (Eclipse 1764).
	Kitty Heath, ch. m., 1833.	Am. Eclipse 1814 (Duroc 1806)—Am. Pomona 1820 by Sir Alfred 1803 (*Sir Harry 1795)—Dau. of *Sir Harry 1795 (Sir Peter T. 1784)—*Pomona 1802 by Worthy 1795 (Pot-8-o-'s 1773)—Comedy 1799 by Buzzard* 1787 (Woodpecker 1773).
Don Juan	Berthune, b. h., 1837.	Sidi Hamet 1825 (Virginian 1815)—Susette 1828 by Aratus 1820 (Director 1811) Jenny Cockracy 1814 by Potomac 1805 (*Diomed 1777)—Timoleon's dam by *Saltram 1780 (Eclipse 1764)—Dau. of Symmes' Wildair 1767 (*Fearnought 1755).
	Ariel, br. m., 1844.	*Sarpedon 1828 (Emilius 1820)—Lancess 1835 by Lance 1821 (Am. Eclipse 1814)—Aurora 1827 by Aratus 1820 (Director 1811) Paragon 1807 by *Buzzard 1787 (Woodpecker 1773) Indiana 1802 by Columbus (*Pantaloon 1778).
Romance	Sir Leslie, b. h., 1828.	Sir William of Transport 1820 (Sir Archy 1805) Miss Dowden 1809 by *Buzzard 1787 (Woodpecker 1773)—Dau. of *Speculator 1795 (Dragon* 1787)—Dau. of Damon (*Fearnought 1755)—Dau. of Fearnought.
	Arian, b. m., 1834.	Davenport's Trumpator 1823 (Sir Solomon 1805) Mary Jones b. m. 1827 by Kosciusko 1815 (Sir Archy 1805)—Multiflora's grandam by Y. Bedford (*Bedford 1792)—Dau. of Arion (Polyphemus) Dan. of Brilliant (Burwell's Traveller, son of *Traveller 1747).

TROUBADOUR.

WINNER OF THE SENSATION, CRITERION, KIMBALL AND OTHER STAKES AT TWO YEARS OLD, THE CHARLES GREEN, PALMER HOUSE AND ST. LEGER STAKES AT THREE YEARS OLD, SUBURBAN HANDICAP, OCEAN STAKES AND MONMOUTH CUP, ETC.

The property of Capt. S. S. Brown, Wildwood Stud, Pa.

TROUBADOUR, bay horse, foaled 1882 bred in the Woodburn Stud, Ky., by Lisbon son of imp. Phaeton by King Tom, dam Glenluine, dam of Windrush by imp. Glenelg, son of Caterer by Stockwell, grandam Lute, dam of Virginius by Lexington, out of Lulu Horton by imp. Albion. Troubadour was a grand race-horse and is from a great racing and producing family. Glenluine was a grand producer. She is the dam of School Girl and her sister, Patrimony, the dam of Lisimony (a first-class race-horse), and Patron (winner of the Great American Stakes), Windrush (winner of the Illinois Derby, etc.), and Lizzie Krepps, a good two-year old winner. Lute, the grandam, produced that good horse Virginius, and Lute Edgar (the dam of Shiloh), Torrent (a two-year old winner), and Tenerine (dam of Leo and Louise, a two-year winner in 1891). The great grandam, Lulu Horton, produced Kildare and Amy Farley (dam of Mattie Amelia, dam of Hemlock, Borealis and Sallie Taylor, Terese. Fonsie's dam, Pathfinder and Blue Grass). Blue Grass won in England the Burwell Stakes, Newmarket, the Ascot Biennial Stakes, and the Racing Stakes, Goodwood, beating Ossian, winner of the St. Leger, etc. Other good ones from Lulu Horton were Spartan, Pequot and Patula, dam of Ceawood (winner of the Clay Stakes, etc.). The Stockholder mare, the fifth dam, produced Compromise, Invincible, Laura (dam of Lily Ward, Harry of the West, Fannie Cheatham, Judge Durrell and Lerna) and Iodine. From these came the good ones Laura Faris, Ascension, Buckette, Hattie Hugar, Fanny Brown, Ada Hayes and Memento. Troubadour at two years old ran fourteen times, won seven races, was second in three and third in one. Ran second to Favor in McGrath Stakes five-eighths of a mile, 110 lbs., in 1.03¼; ran second to O.B. in Alexander Stakes, five-eighths of a mile, 110 lbs., in 1.02½; won purse, five-eighths of a mile, 105 lbs., in 1.02¾; won Sensation Stakes, three-quarters of a mile, 110 lbs., in 1.22¼, beating Ten Stone, Freeman, etc.; won Criterion Stakes, three-quarters of a mile, 110 lbs., in 1.21¾; ran third in Kenwood Stakes, five-eighths of a mile, 112 lbs., in 1.01½; ran second to Ten Stone for Purse, five-eighths of a mile, 110 lbs., in 1.03½; won purse, three-quarters of a mile, 110 lbs., in 1.18¼; won purse, 1 mile, 109 lbs., in 1.50¾; won Kimball Stakes, three-quarters of a mile, 115 lbs., in 1.17; won Barrett Stakes, 1 mile, 118 lbs., in 1.45. At three years old ran twenty-two races, won five, was second in three and third in nine. Ran third in Distillers' Stakes, 1½ miles, 102 lbs., in 2.08¾, won by Modesty; ran third in Phoenix Hotel Stakes, 1¼ miles, 110 lbs., in 2.03¾; won purse, 1¼ miles, 102 lbs., in 2.08¾; ran second to Bersan in Clark Stakes, 1½ miles, 110 lbs., in 2.09½; ran second to Aretino for purse, 1⅜ miles, 105 lbs., in 2.54½; won Charles Green Stakes, 1¼ miles, 118 lbs., in 2.13¼, beating Thistle, Favor, etc.; ran third in American Derby 1½ miles, 123 lbs., in 2.49½, Volante won, Favor second; ran third in Boulevard Stakes 1¼ miles, 112 lbs., in 2.09; ran third in Sheridan Stakes, 1¼ miles, 123 lbs., in 2.09¼; ran third in Drexel Stakes, 1 mile, 123 lbs., in 1.50; won Palmer House Stakes, 1¼ miles, 106 lbs., in 2.10¾; won St. Louis St. Leger Stakes, 2 miles, 123 lbs, in 3.48; ran third in Kaskaskia Stakes, 1½ miles, 114 lbs., in 2.44; ran third in Hotel Stakes, 1½ miles, 125 lbs., in 2.45½; ran third in Brewers' Stakes, 1¾ miles, 113 lbs., in 3.24; ran second to Editor in Falsetto Stakes, 2 miles, 125 lbs., in 3.43¾, beating Volo, Ten Booker, etc. At four years old ran three times and won all his races. Won Handicap purse, 1½ miles, 110 lbs., in 1.57; won Suburban, 1¼ miles, 115 lbs., in 2.12½, beating Richmond, Barnum, Lizzie Dwyer, Ban Fox, Charity, Favor, etc.; won special race against Miss Woodford, 1¼ miles, 118 lbs., in 2.08¾. At five years old ran six races, won four, was second in one and third in the other. Won purse, 1 mile, 122 lbs., in 1.42½; ran second to The Bard in Coney Island Stakes, 1½ miles, 122 lbs, in 1.55; won Sweepstakes, 1 mile, 115 lbs., in 1.41; won Ocean Stakes, 1½ miles, 120 lbs., 1.56½, beating The Bard; won Monmouth Cup, 1¾ miles, 121 lbs., in 3.04, beating The Bard; ran third in Freehold Stakes, 1½ miles, 118 lbs., in 2.39¾, The Bard won, Barnum was second and Troubadour broke down.

Troubadour is young in the stud. Is sire of the winners Arnold, Lookout, Too Quick, La Cigale, about all that have been trained.

Troubadour is descended in direct female line through imp. Silver (Sylva) from the Barb mare, to which trace Lath, Cade, Flying Childers, imp. Diomed, Sorcerer, imp. Priam, Phantom, Langar, Muley, The Cure, Plenipotentiary, Turnus, See-Saw, imp. Zinganee, in England; Windrush, Tyrant, Virginius, Blue Grass, Epsilon, Hidalgo, Voltigeur, etc., etc., in America.

Waxy 1790 Pot-8-o-'s 1773—Eclipse 1764 Marske 1750 Squirt 1732 Bartlett's Childers Darley Arabian 1699.

TROUBADOUR.

BAY HORSE, FOALED 1882; BRED BY A. J. ALEXANDER, WOODBURN STUD, KENTUCKY.

- **LISBON, b. h., 1874.**
  - *Phaeton, b. h., 1865.
    - King Tom, b. h., 1851.
      - Harkaway, ch. h., 1834.
        - Economist, b. h., 1825. — Whisker 1812 (Waxy 1790) Floranthe 1818 by Octavian 1807 (Stripling 1795)—Caprice 1797 by Anvil 1777 (Herod 1758) Madcap 1774 by Eclipse 1764 (Marske 1750) Sappho 1763 by Blank 1740 (God. Ar. 1724).
        - Fanny Dawson, ch. m., 1823. — Nabocklish 1811 (Rugantino 1803) Miss Tooly 1808 by Teddy-the Grinder 1798 (Asparagus 1787) Lady Jane 1796 by Sir Peter T. 1784 (Highflyer 1774)—Paulina 1778 by Florizel 1768 (Herod 1758)—Captive 1774 by Matchem 1748 (Cade 1734).
      - Pocahontas, b. m., 1837.
        - Glencoe*, ch. h., 1831. — Sultan 1816 (Selim 1802) Trampoline 1825 by Tramp 1810 (Dick Andrews 1797) Web 1808 by Waxy 1790 (Pot-8-o-'s 1773)—Penelope 1798 by Trumpator 1782 (Conductor 1767)—Prunella 1788 by Highflyer 1774 (Herod 1758).
        - Marpessa, b. m., 1830. — Muley 1810 (Orville 1799) Clare 1824 by Marmion 1806 (Whiskey 1789)—Harpalice 1814 by Gohanna 1790 (Mercury 1778) Amazon 1799 by Driver 1783 (Trentham 1766)—Fractions 1792 by Mercury 1778 (Eclipse 1764).
    - Merry Sunshine, ch. m., 1855.
      - Storm, b. h., 1848.
        - Touchstone, br. h., 1831. — Camel 1822 (Whalebone 1807) Banter 1826 by Master Henry 1815 (Orville 1799)—Boadicea 1807 by Alexander 1782 (Eclipse 1764) Brunette by Amaranthus 1766 (Old England 1741) Mayfly 1771 by Matchem 1748 (Cade 1734).
        - Ghuznee, b. m., 1838. — Pantaloon 1824 (Castrel 1801) Languish 1830 by Cain 1822 (Paulowitz 1813) Lydia 1822 by Poulton 1805 (Sir Peter T. 1784) Variety 1808 by Hyacinthus 1797 (Coriander 1786) Dau. 1790 of Weazle 1776 (Herod 1758).
      - Daughter of, b. m., 1849.
        - Falstaff, br. h., 1842. — Touchstone 1831 (Camel 1822) Decoy 1830 by Filho-da-Puta 1812 (Haphazard 1797)—Finesse 1815 by Peruvian 1806 (Sir Peter T. 1784) Violante 1802 by John Bull 1789 (Fortitude 1777) Sist. 1788 to Skyscraper by Highflyer 1774 (Herod 1758).
        - Sister to Pompey, b. m., 1843. — Emilius 1820 (Orville 1799) Variation 1827 by Bustard 1813 (Castrel 1801)—Johanna Southcote 1811 by Beningbrough 1791 (King Fergus 1775)—Lavinia 1802 by Pipator 1786 (Imperator 1776) Dick Andrew's dam 1790 by Highflyer 1774 (Herod 1758).
  - *Lady Love, br. m., 1865.
    - [Stockwell or] Caterer, b. h., 1859.
      - Stockwell, ch. h., 1849.
        - The Baron, ch. h., 1842. — Ir. Birdcatcher 1833 (Sir Hercules 1826)—Echidna 1838 by Economist 1825 (Whisker 1812) Miss Pratt 1825 by Blacklock 1814 (Whitelock 1803) Gadabout 1812 by Orville 1799 (Beningbrough 1791) Minstrel 1803 by Sir Peter T. 1784 (Highflyer 1774).
        - Pocahontas, b. m., 1837. — Glencoe* 1831 (Sultan 1816)—Marpessa 1830 by Muley 1810 (Orville 1799) Clare 1824 by Marmion 1806 (Whiskey 1789) Harpalice 1814 by Gohanna 1790 (Mercury 1778) Amazon 1799 by Driver 1783 (Trentham 1766).
      - Selina, b. m., 1851.
        - Orlando, b. h., 1841. — Touchstone 1831 (Camel 1822)—Vulture 1833 by Langar 1817 (Selim 1802) Kite 1821 by Bustard 1813 (Castrel 1801)—Olympia 1815 by Sir Oliver 1800 (Sir Peter T. 1784)—Scotilla 1795 by Anvil 1777 (Herod 1758).
        - The Ladye of the Silverkeld Well, ch. m., 1839. — Velocipede 1825 (Blacklock 1814)—Emma 1824 by Whisker 1812 (Waxy 1790)—Gibside Fairy 1811 by Hermes 1790 (Mercury 1778)—Vicissitude 1800 by Pipator 1786 (Imperator 1776)—Beatrice 1791 by Sir Peter T. 1784 (Highflyer 1774).
    - Chère Amie, br. m., 1857.
      - Sweetmeat, br. h., 1842.
        - Gladiator, ch. h., 1833. — Partisan 1811 (Walton 1799)—Pauline 1826 by Moses 1819 ([Whalebone or] Seymour 1807)—Quadrille 1815 by Selim 1802 (Buzzard* 1787)—Canary Bird 1806 by [Whiskey or] Sorcerer 1796 (Trumpator 1782)—Canary 1797 by Coriander 1786 (Pot-8-o-'s 1773).
        - Lollypop, blk. m., 1836. — [Starch or] Voltaire 1826 (Blacklock 1814)—Belinda 1825 by Blacklock 1814 (Whitelock 1803)—Wagtail 1818 by Prime Minister 1810 (Sancho 1801)—Tranby*'s dam 1812 by Orville 1799 (Beningbrough 1791)—Miss Grimstone 1796 by Weazle 1776 (Herod 1758).
      - Blanne, b. m., 1851.
        - Touchstone, br. h., 1831. — Camel 1822 (Whalebone 1807)—Banter 1826 by Master Henry 1815 (Orville 1799)—Boadicea 1807 by Alexander 1782 (Eclipse 1764)—Brunette by Amaranthus 1766 (Old England 1741) Mayfly 1771 by Matchem 1748 (Cade 1734).
        - Maid of Lyme, br. m., 1843. — Tomboy 1829 (Jerry 1821)—Lapwing 1837 by Bustard 1813 (Castrel 1801)—Dau. 1830 of Muley 1810 (Orville 1799)—Rosanne 1811 by Dick Andrews 1797 (Joe Andrews 1778)—Rosette 1803 by Beningbrough 1791 (King Fergus 1775).
- **GLENLUINE, b. m., 1874.**
  - *Glenelg, b. h., 1866.
    - Citadel, ch. h., 1859.
      - Stockwell, ch. h., 1849.
        - The Baron, ch. h., 1842. — Ir. Birdcatcher 1833 (Sir Hercules 1826)—Echidna 1838 by Economist 1825 (Whisker 1812)—Miss Pratt 1825 by Blacklock 1814 (Whitelock 1803)—Gadabout 1812 by Orville 1799 (Beningbrough 1791)—Minstrel 1803 by Sir Peter T. 1784 (Highflyer 1774).
        - Pocahontas, b. m., 1837. — Glencoe* 1831 (Sultan 1816)—Marpessa 1830 by Muley 1810 (Orville 1799)—Clare 1824 by Marmion 1806 (Whiskey 1789)—Harpalice 1814 by Gohanna 1790 (Mercury 1778)—Amazon 1799 by Driver 1783 (Trentham 1766).
      - Sortie, b. m., 1851.
        - Melbourne, br. h., 1834. — Humprey Clinker 1822 (Comus 1809)—Dau. 1825 of Cervantes 1806 (Don Quixote 1784)—Dau. 1818 of Golumpus 1802 (Gohanna 1790)—Dau. 1810 of Paynator 1791 (Trumpator 1782)—Circle 1802 by St. George 1789 (Highflyer 1774).
        - Escalade, b. m., 1846. — Touchstone 1831 (Camel 1822)—Ghuznee 1838 by Pantaloon 1824 (Castrel 1801)—Languish 1830 by Cain 1822 (Paulowitz 1813)—Lydia 1822 by Poulton 1805 (Sir Peter T. 1784)—Variety 1803 by Hyacinthus 1797 (Coriander 1786).
    - *Babta, b. m., 1858.
      - Kingston, b. h., 1849.
        - Venison, br. h., 1833. — Partisan 1811 (Walton 1799)—Fawn 1823 by Smolensko 1810 (Sorcerer 1796)—Jerboa 1803 by Gohanna 1790 (Mercury 1778)—Camilla 1778 by Trentham 1766 (T. Sweepstakes 1743)—Coquette 1764 by Compton Barb.
        - Queen Anne, b. m., 1843. — Slane 1833 (Royal Oak 1823)—Garcia 1823 by Octavian 1807 (Stripling 1795)—Dau. 1806 of Shuttle 1793 (Y. Marske 1771)—Katharine 1798 by Delpini 1781 (Highflyer 1774)—Dau. 1786 of Paymaster 1766 (Blank 1740).
      - Alice Lowe, ch. m., 1841.
        - Defence, b. h., 1824. — Whalebone 1807 (Waxy 1790)—Defiance 1816 by Rubens 1805 (Buzzard* 1787)—Little Folly 1806 by Highland Fling 1798 (Spadille 1784)—Harriet 1790 by Volunteer 1780 (Eclipse 1764)—Dau. 1785 of Alfred 1770 (Matchem 1748).
        - Pet, b. m., 1826. — Gainsborough 1813 (Rubens 1805)—Dau. 1818 of Topsy Turvey 1805 (St. George 1789)—Agnes 1805 by Shuttle 1793 (Y. Marske 1771)—Dau. 1788 of Highflyer 1774 (Herod 1758)—Dau. 1779 of Goldfinder 1764 (Snap 1750).
  - Lute, blk. m., 1865.
    - Lexington, b. h., 1850.
      - Boston, ch. h., 1833.
        - Timoleon, ch. h., 1813. — Sir Archy 1805 (*Diomed 1777)—Dau. of *Saltram 1780 (Eclipse 1764)—Dau. of Symmes' Wildair 1767 (*Fearnought 1755)—Dau. of Tyler's Driver 1760 (*Othello)—Dau. of Fallower 1761 (Blank 1740).
        - Sister to Tuckahoe, ch. m., 1814. — Ball's Florizel 1801 (*Diomed 1777)—Dau. of *Clockfast 1774 (Gimcrack 1760)—Dau. of Symmes' Wildair 1767 (*Fearnought 1755)—Y. Kitty Fisher 1767 by *Fearnought 1755 (Regulus 1739)—*Kitty Fisher 1755 by Cade 1734 (God. Ar. 1724).
      - Alice Carneal, b. m., 1836.
        - *Sarpedon, br. h., 1828. — Emilius 1820 (Orville 1799)—Icaria 1824 by The Flyer 1814 (Vandyke Jr. 1808) Parma 1813 by Dick Andrews 1797 (Joe Andrews 1778)—May 1804 by Beningbrough 1791 (King Fergus 1775)—Primrose 1787 by Mambrino 1768 (Engineer 1756).
        - Rowena, ch. m., 1826. — Sumpter 1818 (Sir Archy 1805)—Lady Grey 1817 by Robin Grey 1805 (*Royalist 1790)—Maria 1802 by Melzar 1791 (*Medley 1771)—Dau. of *Highflyer 1784 (Highflyer 1774)—Dau. of *Fearnought 1755 (Regulus 1739).
    - Lulu Horton, br. m., 1857.
      - *Albion, blk. h., 1837.
        - [Cain or] Actæon, ch. h., 1822. — Scud 1804 (Beningbrough 1791)—Diana 1808 by Stamford 1794 (Sir Peter T. 1784)—Dau. 1799 of Whiskey 1789 (Saltram* 1780)—Grey Dorimant 1781 by Dorimant 1772 (Otho 1760)—Dizzy 1757 by Blank 1740 (God. Ar. 1724).
        - Panthea, br. m., 1821. — [Comus or] Blacklock 1814 (Whitelock 1803)—Manuella 1809 by Dick Andrews 1797 (Joe Andrews 1778)—Mandane 1800 by Pot-8-o-'s 1773 (Eclipse 1764) Y. Camilla 1787 by Woodpecker 1773 (Herod 1758)—Camilla 1778 by Trentham 1766 (T. Sweepstakes 1743).
      - Martha Dunn, b. m., 1849.
        - *Sovereign, b. h., 1836. — Emilius 1820 (Orville 1799)—Fleur de Lis 1822 by Bourbon 1811 (Sorcerer 1796)—Lady Rachel 1803 by Stamford 1794 (Sir Peter T. 1784)—Y. Rachel 1799 by Volunteer 1780 (Eclipse 1764)—Rachel 1790 by Highflyer 1774 (Herod 1758).
        - Daughter of, b. m., 1833. — Stockholder (Sir Archy 1805)—Dau. ch. m. 1822 of Pacolet 1806 (*Citizen 1785)—Nell Saunders ch. m. 1815 by Wilkes' Wonder 1800 (*Diomed 1777)—Julietta by *Dare Devil 1787 (Magnet 1770)—Rosetta ch. m. 1785 by *Centinel 1758 (Blank 1740)—Diana ch. m. by Clodius 1778 (*Janus 1746)—Sally Painter by *Sterling 1762 (Bellsize Ar.* 1756)—*Silver 1762 by Bellsize Ar.* 1756—Sist. to Cato by Partner 1718 (Jigg)—Sist. ch. m. 1719 to Roxana by Bald Galloway (St. Victor Barb)—Sist. to Chanuter by Akaster Turk—Cream Cheeks by Leedes Ar.—Dau. 1690 of Spanker (D'Arcy Yellow Turk)—Old Morocco mare by Lord Fairfax's Morocco mare—Old Bald Peg by an Arabian—A Barb mare.

TURCO.

WINNER OF THE RARITAN STAKES AND OTHER RACES.

The property of Mr. L. O. Appleby, near Eatontown, N. J. Yearlings sold annually.

TURCO, chestnut horse, foaled 1879, bred in the Nursery Stud, by Kingfisher, son of Lexington by Boston, dam imp. Touceques, dam of La Tosca by Monarque, son of The Emperor by Defence, grandam La Touceques by The Baron, sire of Stockwell and Rataplan, great grandam Tapestry by Melbourne, son of Humphrey Clinker by Comus, out of Stitch by Hornsea, son of Velocipede by Blacklock, etc. Turco was a good but unfortunate race-horse. At two years old ran four races. Won purse, three-quarters of a mile, 100 lbs., in 1.15¾, beating Vampire, Bayadere, etc.; ran second to Hiawasse, three-quarters of a mile, in 1.16, with six others behind him; ran third to Wyoming and Olivia, 5 furlongs, in 1.03¼, with five others behind him. At three years old, won two races out of three starts; 1 mile, 118 lbs., in 1.46¾, beating Free Gold, Hilarity and three others; won Raritan Stakes, 1⅜ miles, 118 lbs., in 2.29, beating Macbeth, Duplex and three others. His sire, Kingfisher, was a first-class race-horse, winner of the Belmont, Champion, Jerome and Travers Stakes. Imp. Eltham Lass was half sister to Viridis, dam of Springfield (winner of the Stockbridge Cup) by Kingston, son of Venison and winner of the Goodwood Cup. Kingfisher is the sire of Lady Rosebery (winner of the Champagne Stakes), Rica, Prince Royal (winner of eighteen races, including the Midsummer Handicap, 1 mile, at Monmouth, in which he conceded Eurus and Badge 9 lbs. each and won in 1.40: Rancho-del-Paso Handicap, 1$\frac{1}{16}$ miles, 115 lbs., in 1.46½, which was the best on record to that date; the Harvest Handicap, 1¼ miles, 125 lbs., in 2.07, an excellent race). Imp. Touceques, his dam, won the Filly Sapling Stakes, and was by Monarque (winner of the French Derby and sire of Gladiateur, winner of the 2,000 Guineas, Derby, St. Leger, Grand Prize of Paris and Ascot Gold Cup, the only horse that ever won all these events). Touceques is the dam also of the Tarbouche, Turk and the great filly La Tosca. His grandam, La Touceques (winner of the French Oaks, French Derby and Prix Royal Oak, 1 mile 7 furlongs) by The Baron, son of Birdcatcher and sire of the great Stockwell. Great grandam, Tapestry, own sister to Volante, dam of the good horse Thunder (winner of the Epsom Cup, Craven Stakes, Goodwood and Alexandra Plate) by Melbourne, sire of West Australian (winner of the 2,000 Guineas, Derby and St. Leger), Blink Bonny (winner of the Oaks and Derby) and dam of Blair Athol (winner of the Derby and St. Leger), out of Stitch, half sister to Lady Evelyn (winner of the Oaks) by Hornsea, she out of Industry (winner of the Oaks) by imp. Priam. Turco is handsomely and fashionably bred from a grand racing and producing family on both paternal and maternal sides.

Turco traces through the best racing blood to the Burton Barb mare, from which descend in direct female line Whiskey, Selim, Rubens, Castrel, Catton, Whitelock, Blacklock, Sir Hercules, imp. Margrave, Harkaway, Voltigeur, Surplice, St. Albans, Lord Clifden, Mask, Pell Mell, Brag, Sheen, Gold, imp. Hurrah, imp. Ainderby, imp. Billet, imp. Eagle and Spread Eagle, Waverly (by imp. Australian), etc.

Florizel 1768—Herod 1758—Tartar 1743 Partner 1718 Jigg Byerly Turk, about 1685.

TURCO.

CHESTNUT HORSE, FOALED 1879; BRED BY MR. A. BELMONT, NURSERY STUD, LONG ISLAND.

KINGFISHER, b. h., 1867.	Lexington, b. h., 1850.	Boston, ch. h., 1833.	Timoleon, ch. h., 1813.	Sir Archy, b. h., 1805.	*Diomed 1777 (Florizel 1768)—*Castianira 1796 by Rockingham 1781 (Highflyer 1774)—Tabitha 1782 by Trentham 1766 (T. Sweepstakes 1743)—Dau. of Bosphorus 1751 (Babraham 1740)—Dau. of Forester 1770 (Forester 1736).
				Daughter of, ch. m., —.	*Saltram 1780 (Eclipse 1764)—Dau. of Symmes' Wildair 1767 (*Fearnought 1755)—Dau. of Tyler's Driver 1760 (Othello)—Dau. of *Fallower 1761 (Blank 1740)—Dau. of *Vampire 1757 (Regulus 1739).
			Sist. to Tuckahoe, ch. m., 1814.	Ball's Florizel, ch. h., 1801.	*Diomed 1777 (Florizel 1768)—Dau. of *Shark 1771 (Marske 1750)—Dau. of Harris' Eclipse 1770 (*Fearnought 1755)—Dau. of *Fearnought 1755 (Regulus 1739)—Dau. of *Jolly Roger 1741 (Roundhead 1733).
				Daughter of, b. m., 1799.	*Alderman 1787 (Pot-8-o-'s 1773)—Dau. 1791 of *Clockfast 1774 (Gimcrack 1760)—Dau. of Symmes' Wildair 1767 (*Fearnought 1755)—Y. Kitty Fisher 1767 by *Fearnought 1755 (Regulus 1739)—*Kitty Fisher 1755 by Cade 1734 (God. Ar. 1724).
		Alice Carneal, b. m., 1836.	*Sarpedon, br. h., 1828.	Emilius, b. h., 1820.	Orville 1799 (Beningbrough 1791)—Emily 1810 by Stamford 1794 (Sir Peter T. 1784)—Dau. 1799 of Whiskey 1789 (Saltram* 1780)—Grey Dorimant 1781 by Dorimant 1772 (Otho 1760)—Dizzy 1757 by Blank 1740 (God. Ar. 1724).
				Icaria, b. m., 1824.	The Flyer 1814 (Vandyke Jr. 1808)—Parma 1813 by Dick Andrews 1797 (Joe Andrews 1778)—May 1804 by Beningbrough 1791 (King Fergus 1775)—Primrose 1787 by Mambrino 1768 (Engineer 1756)—Cricket 1776 by Herod 1758 (Tartar 1743).
			Rowena, ch. m., 1826.	Sumpter, ch. h., 1818.	Sir Archy 1805 (*Diomed 1777)—Dau. of *Robin Redbreast 1796 (Sir Peter T. 1784)—Dau. 1788 of *Obscurity 1777 (Eclipse 1764)—Slamerkin 1768 by *Wildair 1753 (Cade 1734)—*Cub mare 1762 by Cub 1739 (Fox 1714).
				Lady Grey, b. m., 1817.	Robin Grey 1805 (*Royalist 1790)—Maria 1802 by Melzar 1791 (*Medley 1776)—Dau. of *Highflyer 1784 (Highflyer 1774)—Dau. of *Fearnought 1755 (Regulus 1739)—Dau. of Ariel 1756 (*Traveller 1747).
	*Eltham Lass, b. m., 1859.	Kingston, b. h., 1849.	Venison, br. h., 1833.	Partisan, b. h., 1811.	Walton 1799 (Sir Peter T. 1784)—Parasol 1800 by Pot-8-o-'s 1773 (Eclipse 1764)—Prunella 1788 by Highflyer 1774 (Herod 1758)—Promise 1768 by Snap 1750 (Snip 1736)—Julia 1756 by Blank 1740 (God. Ar. 1724).
				Fawn, br. m., 1823.	Smolensko 1810 (Sorcerer 1796)—Jerboa 1803 by Gohanna 1790 (Mercury 1778)—Camilla 1778 by Trentham 1766 (T. Sweepstakes 1743)—Coquette 1764 by Compton Barb—Sist. 1743 to Regulus by God. Ar. 1724.
			Queen Anne, b. m., 1841.	Slane, b. h., 1833.	Royal Oak 1823 (Catton 1809)—Dau. 1819 of Orville 1799 (Beningbrough 1791)—Epsom Lass 1803 by Sir Peter T. 1784 (Highflyer 1774)—Alexina 1788 by King Fergus 1775 (Eclipse 1764)—Lardella 1780 by Y. Marske 1771 (Marske 1750).
				Garcia, br. m., 1823.	Octavian 1807 (Stripling 1795)—Dau. 1806 of Shuttle 1793 (Y. Marske 1771)—Katharine 1798 by Delpini 1781 (Highflyer 1774)—Dau. 1786 of Paymaster 1766 (Blank 1740)—Dau. 1774 of Le Sang 1759 (Changeling 1747).
		Maid of Palmyra, b. m., 1855.	Pyrrhus I., ch. h., 1842.	Epirus, ch. h., 1834.	Langar 1817 (Selim 1802)—Olympia 1815 by Sir Oliver 1800 (Sir Peter T. 1784)—Scotilla 1795 by Anvil 1777 (Herod 1758)—Scota 1783 by Eclipse 1764 (Marske 1750)—Harmony 1775 by Herod 1758 (Tartar 1743).
				Fortress, b. m., 1833.	Defence 1824 (Whalebone 1807)—Jewess 1827 by Moses 1819 ([Whalebone or] Seymour 1807)—Calendulæ 1815 by Camerton 1808 (Hambletonian 1792)—Snowdrop 1806 by Highland Fling 1798 (Spadille 1784)—Daisy 1793 by Buzzard* 1787 (Woodpecker 1773).
			Palmyra, br. m., 1838.	Sultan, b. h., 1816.	Selim 1802 (Buzzard* 1787)—Bacchante 1809 by W's Ditto 1800 (Sir Peter T. 1784)—Dau. 1791 of Mercury 1778 (Eclipse 1764)—Dau. 1776 of Herod 1758 (Tartar 1743)—Folly 1771 by Marske 1750 (Squirt 1732).
				Hester, br. m., 1832.	Camel 1822 (Whalebone 1807)—Monimia 1821 by Muley 1810 (Orville 1799)—Sist. 1796 to Petworth by Precipitate* 1787 (Mercury 1778)—Dau. 1787 of Woodpecker 1773 (Herod 1758)—Dau. 1773 of Snap 1750 (Snip 1736).
*TOUCQUES, ch. m., 1868.	Monarque, b. h., 1852.	[The Baron, Sting or] The Emperor, ch. h., 1841.	Defence, b. h., 1824.	Whalebone, br. h., 1807.	Waxy 1790 (Pot-8-o-'s 1773)—Penelope 1798 by Trumpator 1782 (Conductor 1767)—Prunella 1788 by Highflyer 1774 (Herod 1758)—Promise 1768 by Snap 1750 (Snip 1736)—Julia 1756 by Blank 1740 (God. Ar. 1724).
				Defiance, ch. m., 1816.	Rubens 1805 (Buzzard* 1787)—Little Folly 1806 by Highland Fling 1798 (Spadille 1784)—Harriet 1799 by Volunteer 1780 (Eclipse 1764)—Dau. 1785 of Alfred 1770 (Matchem 1748)—Magnolia 1771 by Marske 1750 (Squirt 1732).
			Daughter of, ch. m., 1831.	Reveller, b. h., 1815.	Comus 1809 (Sorcerer 1796)—Rosette 1803 by Beningbrough 1791 (King Fergus 1775)—Rosamond 1788 by Tandem 1773 (Syphon 1750)—Tuberose 1772 by Herod 1758 (Tartar 1743)—Grey Starling 1745 by Starling 1727 (Bay Bolton 1705).
				Design,* ch. m., 1827.	Tramp 1810 (Dick Andrews 1797)—Defiance 1816 by Rubens 1805 (Buzzard* 1787)—Little Folly 1806 by Highland Fling 1798 (Spadille 1784)—Harriet 1799 by Volunteer 1780 (Eclipse 1764)—Dau. 1785 of Alfred 1770 (Matchem 1748).
		Poetess, b. m., 1838.	Royal Oak, br. h., 1823.	Catton, b. h., 1809.	Golumpus 1802 (Gohanna 1790)—Lucy Gray 1804 by Timothy 1794 (Delpini 1781)—Lucy 1789 by Florizel 1768 (Herod 1758)—Frenzy 1774 by Eclipse 1764 (Marske 1750)—Dau. of Engineer 1756 (Sampson 1745).
				Daughter of, b. m., 1818.	Smolensko 1810 (Sorcerer 1796)—Lady Mary 1800 by Beningbrough 1791 (King Fergus 1775)—Dau. of Highflyer 1774 (Herod 1758)—Sist. to Miss Bell by Marske 1750 (Squirt 1732)—Dau. of Bajazet 1740 (God. Ar. 1724).
			Ada, b. m., 1834.	Whisker, b. h., 1812.	Waxy 1790 (Pot-8-o-'s 1773)—Penelope 1798 by Trumpator 1782 (Conductor 1767)—Prunella 1788 by Highflyer 1774 (Herod 1758)—Promise 1768 by Snap 1750 (Snip 1736)—Julia 1756 by Blank 1740 (God. Ar. 1724).
				Anna Bella, b. m., 1807.	Shuttle 1793 (Y. Marske 1771)—Dau. 1797 of Drone 1777 (Herod 1758)—Contessina 1787 by Y. Marske 1771 (Marske 1750)—Tuberose 1772 by Herod 1758 (Tartar 1743)—Grey Starling 1745 by Starling 1727 (Bay Bolton 1705).
	La Toucques, ch. m., 1860.	The Baron, ch. h., 1842.	Ir. Birdcatcher, ch. h., 1833.	Sir Hercules, blk. h., 1826.	Whalebone 1807 (Waxy 1790)—Peri 1822 by Wanderer 1811 (Gohanna 1790)—Thalestris 1809 by Alexander 1782 (Eclipse 1764)—Rival 1800 by Sir Peter T. 1784 (Highflyer 1774)—Hornet 1790 by Drone 1777 (Herod 1758).
				Guiccioli, ch. m., 1823.	Bob Booty 1804 (Chanticleer 1787)—Flight 1809 by Ir. Escape 1802 (Commodore 1793)—Y. Heroine by Bagot 1780 (Herod 1758)—Heroine 1775 by Hero 1753 (Cade 1734)—Dau. 1766 of Snap 1750 (Snip 1736).
			Echidna, b. m., 1838.	Economist, b. h., 1825.	Whisker 1812 (Waxy 1790)—Floranthe 1818 by Octavian 1807 (Stripling 1795)—Caprice 1797 by Anvil 1777 (Herod 1758)—Madcap 1774 by Eclipse 1764 (Marske 1750)—Sappho 1768 by Blank 1740 (God. Ar. 1724).
				Miss Pratt, b. m., 1825.	Blacklock 1814 (Whitelock 1803)—Gadabout 1812 by Orville 1799 (Beningbrough 1791)—Minstrel 1803 by Sir Peter T. 1784 (Highflyer 1774)—Matron 1782 by Florizel 1768 (Herod 1758)—Maiden 1770 by Matchem 1748 (Cade 1724).
		Tapestry, b. m., 1853.	Melbourne, br. h., 1834.	Humphrey Clinker, b. h., 1822.	Comus 1809 (Sorcerer 1796)—Clinkerina 1812 by Clinker 1805 (Sir Peter T. 1784)—Pewet 1786 by Tandem 1773 (Syphon 1750)—Termagant by Tantrum 1760 (Cripple 1750)—Cantatrice by Sampson 1745 (Blaze 1733).
				Daughter of, b. m., 1825.	Cervantes 1806 (Don Quixote 1784)—Dau. 1818 of Golumpus 1802 (Gohanna 1790)—Dau. 1810 of Paynator 1791 (Trumpator 1782)—Circle 1802 by St. George 1789 (Highflyer 1774)—Abigail 1788 by Highflyer 1774 (Herod 1758).
			Stitch, ch. m., 1842.	Hornsea, ch. h., 1832.	Velocipede 1825 (Blacklock 1814)—Dau. 1820 of Cerberus 1802 (Gohanna 1790)—Miss Cranfield 1803 by Sir Peter T. 1784 (Highflyer 1774)—Dau. 1796 of Pegasus 1784 (Eclipse 1764)—Dau. 1783 of Paymaster 1766 (Blank 1740).
				Industry, br. m., 1835.	Priam* 1827 (Emilius 1820)—Arachne br. m. 1822 by Filho-da-Puta 1812 (Haphazard 1797)—Treasure ch. m. 1809 by Camillus 1803 (Hambletonian 1792)—Dau. ch. m. 1804 of Hyacinthus 1797 (Coriander 1786)—Flora ch. m. 1789 by King Fergus 1775 (Eclipse 1764)—Atalanta ch. m. 1769 by Matchem 1748 (Cade 1734)—Y. Lass-of-the-Mill b. m. 1756 by Oroonoko 1745 (Crab 1722)—Dau. 1744 of Traveller 1735 (Partner 1718)—Miss Makeless b. m. 1737 by Y. Greyhound 1723 (Greyhound)—Dau 1731 of Partner 1718 (Jigg)—Brown Woodcock br. m. 1717 by D'Arcy Woodcock (Bustler)—Lusty Thornton ch. m. by Croft's Bay Barb (Chillaby)—Chestnut Thornton ch. m. by Makeless (Oglethorpe Ar.)—Old Thornton by Brimmer (D'Arcy Yellow Turk)—Dau. of Dickey Pierson (Dodsworth)—Burton Natural Barb mare.

IMPORTED UHLAN.

WINNER OF THE ASCOT AND GOODWOOD STAKES, BRIGHTON AND DONCASTER CUPS AND FOUR QUEEN'S PLATES.

Is private stallion in Chesterbrook Stud, Mr. A. J. Cassatt, proprietor, Berwyn, Chester Co., Pa. Yearlings sold annually.

UHLAN, brown horse, foaled 1869, bred by Mr. Savile, by The Ranger, son of Voltigeur, by Voltaire, dam La Méchante by Turnus, son of Taurus by Phantom, grandam Traviata by The Libel, son of Pantaloon, out of Fernande by Slane, son of Royal Oak. Uhlan's sire, The Ranger, won the Grand Prize of Paris in 1863. He was a full brother to Skirmisher (winner of the Ascot Gold Cup) and by Voltigeur (winner of the Derby and St. Leger), the sire of imp. Billet and grandsire of Galopin. Hence the male line is the most popular one at present, both here and in England. The Ranger's dam was the grandam of Cremorne (winner of the Derby, Grand Prize of Paris and Ascot Gold Cup). La Méchante, his dam, won the Grand Stand Plate at Nottingham and Handicap Plate, beating Dalesman, Blue Mantle, etc., and Town Plate at Lewes, and was by Turnus, sire of the Oaks winner Butterfly; his grandam Traviata by The Libel, the sire of Traducer, who got the popular sire and race-horse, imp. Sir Modred, and his brother, imp. Cheviot. Zinc, his fifth dam, won the 1,000 Guineas and Oaks.

Uhlan was a great race-horse. He won the Ascot Stakes, the Goodwood Stakes, Brighton and Doncaster Cups, four Queen's Plates and many other races. Uhlan stood sixth on the list of winning sires in France for the year 1885, with a credit of sixty-four races won, and $46,654.90, a remarkable showing, considering he had not been in that country since 1881, and consequently the youngest of his get in France were that season three-year olds. Amongst his principal winners in France that season were Blond II., a three-year old filly, who won the Prix de Lilas, 1¼ miles, at Paris, also the Prix de la Societe d'Encouragement (handicap), 1⅞ miles, when she carried 133 lbs., beating Coulon, three years, 121 lbs., Bourges, three years old, 121 lbs., and three others. She also won the Prix National, 2½ miles, with 122 lbs. La Nonaille by Uhlan won the Prix de Petit Cercle, 2¼ miles. Sceptique won the Prix de la Ferme, at Montauban, also the Prix de Biarritz, 1⅞ miles, with 120 lbs., conceding 11 lbs. to Bernay, the second horse, and 19 lbs. to third. Sceptique also won the Prix de la Ville, at Bordeaux. Sourire by Uhlan won the Prix de Compiegne, at Paris, and won the Grand Prix de la Ville, 1⅝ miles, at Lyons, beating the famous Relusiant, who had just won the French Derby and the Prix de Chemin de Fer at Deauville. Anatole by Uhlan won the Prix de Parc, at Vincennes, Prix de Garches, at Paris, and Prix de Bievre, at Vincennes. Don Gigandas by Uhlan won the Prix de Fervacques, 1⅞ miles, at Deauville.

Uhlan has had little chance in America, but is the sire of Grateful, Sir Abner, Mount Vernon, Rita, Expense and Tragedienne filly, etc.

He is descended in direct female line from the old Scarboro' mare, to which trace Comus, Slane, Young Melbourne, North Lincoln, Sefton, in England; Planet, Exchequer, and all the Nina family.

King Fergus 1775—Eclipse 1764—Marske 1750 Squirt 1732 Bartlett's Childers Darley Arabian 1689.

UHLAN.

BROWN HORSE, FOALED 1869; BRED BY MR. SAVILLE, ENGLAND.

Parent	Grandparent	Great-grandparent	G.-g.-grandparent	G.-g.-g.-grandparent	Pedigree
THE RANGER, br. h., 1860.	Voltigeur, br. h., 1847.	Voltaire, br. h., 1826.	Blacklock, b. h., 1814.	Whitelock, b. h., 1803.	Hambletonian 1792 (King Fergus 1775)—Rosalind 1788 by Phenomenon 1780 (Herod 1758)—Atalanta 1769 by Matchem 1748 (Cade 1734)—Y. Lass-of-the-Mill 1756 by Oroonoko 1745 (Crab 1722)—Dau. 1741 of Traveller 1735—Partner 1718).
				Daughter of, [illegible], 1799.	Coriander 1786 (Pot-8-o-'s 1773)—Wildgoose 1792 by Highflyer 1774 (Herod 1758)—Coheiress 1786 by Pot-8-o-'s 1773 (Eclipse 1764)—Manilla 1777 by Goldfinder 1764 (Snap 1750)—Dau. 1766 of Old England 1741 (God. Ar. 1724).
			Daughter of, b. m., 1816.	Phantom, b. h., 1808.	Walton 1799 (Sir Peter T. 1784)—Julia 1799 by Whiskey 1789 (Saltram* 1780)—Y. Giantess 1790 by Diomed* 1777 (Florizel 1768)—Giantess 1769 by Matchem 1748 (Cade 1734)—Molly Long Legs 1753 by Babraham 1740 (God. Ar. 1724).
				Daughter of, b. m., 1802.	Overton 1788 (King Fergus 1775)—Dau. 1796 of Walnut 1786 (Highflyer 1774)—Dau. 1786 of Ruler 1777 (Y. Marske 1771)—Piracantha 1772 by Matchem 1748 (Cade 1734)—Prophetess 1758 by Regulus 1739 (God. Ar. 1724).
		Martha Lynn, br. m., 1837.	Mulatto, b. h., 1823.	Catton, b. h., 1809.	Golumpus 1802 (Gohanna 1790)—Lucy Gray 1804 by Timothy 1794 (Delpini 1781)—Lucy 1789 by Florizel 1768 (Herod 1758)—Frenzy 1774 by Eclipse 1764 (Marske 1750)—Dau. of Engineer 1756 (Sampson 1745).
				Desdemona, br. m., 1811.	Orville 1799 (Beningbrough 1791)—Fanny 1796 by Sir Peter T. 1784 (Highflyer 1774)—Dau. 1788 of Diomed* 1777 (Florizel 1768)—Desdemona 1770 by Marske 1750 (Squirt 1732)—Y. Hag 1761 by Skim 1746 (Starling 1727).
			Leda, b. m., 1824.	Filho-da-Puta, br. h., 1812.	Haphazard 1797 (Sir Peter T. 1784)—Mrs. Barnet 1806 by Waxy 1790 (Pot-8-o-'s 1773)—Dau. 1788 of Woodpecker 1773 (Herod 1758)—Heinel 1771 by Squirrel 1754 (Traveller 1735)—Principessa 1762 by Blank 1740 (God. Ar. 1724).
				Treasure, ch. m., 1809.	Camillus 1803 (Hambletonian 1792)—Dau. 1804 of Hyacinthus 1797 (Coriander 1786)—Flora 1789 by King Fergus 1775 (Eclipse 1764)—Atalanta 1769 by Matchem 1748 (Cade 1734)—Y. Lass-of-the-Mill 1756 by Oroonoko 1745 (Crab 1722).
	Skirmisher's dam, ., 1845.	Gardham, b. h., 1831.	Falcon, gr. h., 1822.	Interpreter, b. h., 1815.	Soothsayer 1808 (Sorcerer 1796)—Blowing 1803 by Buzzard* 1787 (Woodpecker 1773)—Dau. 1796 of Pot-8-o-'s 1773 (Eclipse 1764)—Maid-of-all-Work 1786 by Highflyer 1774 (Herod 1758)—Dau. 1771 of Syphon 1750 (Squirt 1732).
				Miss Newton, gr. m., 1804.	Delpini 1781 (Highflyer 1774)—Tipple Cyder 1788 by King Fergus 1775 (Eclipse 1764)—Sylvia 1783 by Y. Marske 1771 (Marske 1750)—Ferret 1765 by Broth. to Silvio (Cade 1734)—Dau. of Regulus 1739 (God. Ar. 1724).
			Muta, b. m., 1819.	Tramp, b. h., 1810.	Dick Andrews 1797 (Joe Andrews 1778)—Dau. 1803 of Gohanna 1790 (Mercury 1778)—Fraxinella 1793 by Trentham 1766 (T. Sweepstakes 1743)—Dau. 1785 of Woodpecker 1773 (Herod 1758)—Everlasting 1775 by Eclipse 1764 (Marske 1750).
				Mandane, ch. m., 1800.	Pot-8-o-'s 1773 (Eclipse 1764)—Y. Camilla 1787 by Woodpecker 1773 (Herod 1758)—Camilla 1778 by Trentham 1766 (T. Sweepstakes 1743)—Coquette 1765 by Compton Barb—Sist. 1743 to Regulus by God. Ar. 1724.
		Daughter of, br. m., 1837.	Langar, ch. h., 1817.	Selim, ch. h., 1802.	Buzzard* 1787 (Woodpecker 1773)—Dau. of Alexander 1782 (Eclipse 1764)—Dau. of Highflyer 1774 (Herod 1758)—Dau. of Alfred 1770 (Matchem 1748)—Dau. 1770 of Engineer 1756 (Sampson 1745).
				Daughter of, ch. m., 1803.	Walton 1799 (Sir Peter T. 1784)—Y. Giantess 1790 by Diomed* 1777 (Florizel 1767)—Giantess 1769 by Matchem 1748 (Cade 1734)—Molly Long Legs 1753 by Babraham 1740 (God. Ar. 1724)—Dau. of Foxhunter 1727 (Brisk 1711).
			Sist. to Busto, br. m., 1816.	Clinker, br. h., 1805.	Sir Peter T. 1784 (Highflyer 1774)—Hyale 1797 by Phenomenon 1780 (Herod 1758)—Rally 1790 by Trumpator 1782 (Conductor 1767)—Fancy 1780 (Sist. to Diomed*) by Florizel 1768 (Herod 1758)—Dau. 1763 of Spectator 1749 (Crab 1722).
				Bronze, br. m., 1803.	Buzzard* 1787 (Woodpecker 1773)—Selim's dam by Alexander 1782 (Eclipse 1764)—Dau. of Highflyer 1774 (Herod 1758)—Dau. of Alfred 1770 (Matchem 1748)—Dau. 1770 of Engineer 1756 (Sampson 1745).
LA MÉCHANTE, blk. m., 1862.	Turnus, b. h., 1846.	Taurus,‡ ch. h., 1826.	Morisco, b. h., 1819.	Muley, b. h., 1810.	Orville 1799 (Beningbrough 1791)—Eleanor 1798 by Whiskey 1789 (Saltram* 1780)—Y. Giantess 1790 by Diomed* 1777 (Florizel 1768)—Giantess 1769 by Matchem 1748 (Cade 1734)—Molly Long Legs 1753 by Babraham 1740 (God. Ar. 1724).
				Aquilina, b. m., 1807.	Eagle* 1796 (Volunteer 1780)—Dau. 1796 of Precipitate* 1787 (Mercury 1778)—Dau. 1787 of Woodpecker 1773 (Herod 1758)—Sist. 1773 to Juniper by Snap 1750 (Snip 1736)—Y. Marske's dam 1759 by Blank 1740 (God. Ar. 1724).
			Katherine, ch. m., 1821.	Soothsayer, ch. h., 1808.	Sorcerer 1796 (Trumpator 1782)—Golden Locks 1798 by Delpini 1781 (Highflyer 1774)—Violet 1787 by Shark* 1771 (Marske 1750)—Dau. 1772 of Syphon 1750 (Squirt 1732)—Charlotte 1754 by Blank 1740 (God. Ar. 1724).
				Quadrille, b. m., 1815.	Selim 1802 (Buzzard* 1787)—Canary Bird 1806 by [Whiskey or] Sorcerer 1796 (Trumpator 1782)—Canary 1797 by Coriander 1786 (Pot-8-o-'s 1773)—Miss Green 1787 by Highflyer 1774 (Herod 1758)—Harriet 1769 by Matchem 1748 (Cade 1734).
		Clarissa, br. m., 1835.	Defence, b. h., 1824.	Whalebone, br. h., 1807.	Waxy 1790 (Pot-8-o-'s 1773)—Penelope 1798 by Trumpator 1782 (Conductor 1767)—Prunella 1788 by Highflyer 1774 (Herod 1758)—Promise 1768 by Snap 1750 (Snip 1736)—Julia 1756 by Blank 1740 (God. Ar. 1724).
				Defiance, ch. m., 1816.	Rubens 1805 (Buzzard* 1787)—Little Folly 1806 by Highland Fling 1798 (Spadille 1784)—Harriet 1799 by Volunteer 1780 (Eclipse 1764)—Dau. 1785 of Alfred 1770 (Matchem 1748)—Magnolia 1771 by Marske 1750 (Squirt 1732).
			Clara, b. m., 1829.	Filho-da-Puta, br. h., 1812.	Haphazard 1797 (Sir Peter T. 1784)—Mrs. Barnet 1806 by Waxy 1790 (Pot-8-o-'s 1773)—Dau. 1788 of Woodpecker 1773 (Herod 1758)—Heinel 1771 by Squirrel 1754 (Traveller 1735)—Principessa 1762 by Blank 1740 (God. Ar. 1724).
				Clari, br. m., 1824.	Smolensko 1810 (Sorcerer 1796)—Dau. 1804 of Precipitate* 1787 (Mercury 1778)—Y. Juno 1788 by Highflyer 1774 (Herod 1758)—Juno 1764 by Spectator 1749 (Crab 1722)—Horatia 1758 by Blank 1740 (God. Ar. 1724).
	Traviata, b. m., 1856.	The Libel, br. h., 1842.	Pantaloon, ch. h., 1821.	Castrel, ch. h., 1801.	Buzzard* 1787 (Woodpecker 1773)—Dau. of Alexander 1782 (Eclipse 1764)—Dau. of Highflyer 1774 (Herod 1758)—Dau. of Alfred 1770 (Matchem 1748)—Dau. 1770 of Engineer 1756 (Sampson 1745).
				Idalia, ch. m., 1815.	Peruvian 1806 (Sir Peter T. 1784)—Musidora 1804 by Meteor 1783 (Eclipse 1764)—Maid-of-all-Work 1786 by Highflyer 1774 (Herod 1758)—Dau. 1771 of Syphon 1750 (Squirt 1732)—Dau. of Regulus 1739 (God. Ar. 1724).
			Pasquinade, br. m., 1829.	Camel, br. h. 1822.	Whalebone 1807 (Waxy 1790)—Dau. 1812 of Selim 1802 (Buzzard* 1787)—Maiden 1801 by Sir Peter T. 1784 (Highflyer 1774)—Dau. 1788 of Phenomenon 1780 (Herod 1758)—Matron 1782 by Florizel 1768 (Herod 1758).
				Banter, br. m., 1826.	Master Henry 1815 (Orville 1799)—Boadicea 1807 by Alexander 1782 (Eclipse 1764)—Brunette by Amaranthus 1766 (Old England 1741)—Mayfly 1771 by Matchem 1748 (Cade 1734)—Dau. 1755 of Starling 1738 (Starling 1727).
		Fernande, br. m., 1847.	Slane, b. h., 1833.	Royal Oak, br. h., 1823.	Catton 1809 (Golumpus 1802)—Dau. 1818 of Smolensko 1810 (Sorcerer 1796)—Lady Mary 1800 by Beningbrough 1791 (King Fergus 1775)—Dau. of Highflyer 1774 (Herod 1758)—Sist. to Miss Bell by Marske 1750 (Squirt 1732).
				Daughter of, b. m., 1819.	Orville 1799 (Beningbrough 1791)—Epsom Lass 1803 by Sir Peter T. 1784 (Highflyer 1774)—Alexina 1788 by King Fergus 1775 (Eclipse 1764)—Lardella 1780 by Y. Marske 1771 (Marske 1750)—Dau. of Y. Cade 1747 (Cade 1734).
			Elf, b. m., 1835.	Shakespeare, br. h., 1823.	Smolensko 1810 (Sorcerer 1796)—Charming Molly 1813 by Rubens 1805 (Buzzard* 1787)—Comedy 1804 by Beningbrough 1791 (King Fergus 1775)—Mrs. Jordan 1784 by Highflyer 1774 (Herod 1758)—Harriet 1769 by Matchem 1748 (Cade 1734).
				Zink, br. m., 1820.	Woful 1809 (Waxy 1790)—Zaida blk. m. 1806 by Sir Peter T. 1784 (Highflyer 1774)—Alexina ch. m. 1788 by King Fergus 1775 (Eclipse 1764)—Lardella b. m. 1780 by Y. Marske 1771 (Marske 1750)—Dau. of Y. Cade 1747 (Cade 1734)—Beaufremont's dam gr. m. by Broth. to Fearnought (Bay Bolton 1705)—Miss Wyndham by Wyndham 1719 (Hautboy)—Dau. of Belgrade Turk—Old Scarboro' mare 1715 by Makeless (Oglethorpe Ar.)—Dau. of Brimmer (D'Arcy Yellow Turk).

‡ Taurus is by Phantom or Morisco. Morisco's pedigree is accepted.

UNCAS.

WINNER OF THE KENTUCKY STAKES AT SARATOGA AND WESTCHESTER CUP AT JEROME PARK, N. Y.

A private stallion in the Brookdale Stud, Monmouth County, near Holmdel, N. J. Estate of D. D. Withers, proprietors.

UNCAS by Lexington, son of Boston, bred in Woodburn Stud, Kentucky, foaled 1876, dam Coral, dam of Wanderer by Vandal, son of Glencoe, out of Cairn Gorme by Cotherstone, winner of the 2,000 Guineas and Derby, son of Touchstone, winner of the St. Leger, Doncaster Cups in 1835 and 1836, and Ascot Gold Cups in 1836 and 1837. Uncas only started three times at two years old; won the Kentucky Stakes at Saratoga, three-quarters of a mile, in 1.20½, beating Harold, Idler, Dan Sparling, etc.; he ran second for the Saratoga Stakes to Harold, and second in the Nursery at Jerome Park to the King Ernest gelding. Uncas was then shipped to England, and in 1879 ran unplaced in the 2,000 Guineas, won by Charibert, he behaved very badly at the post and delayed the start. He was returned to America, and at four years old started fifteen times, won five races, was second in six, third in three and unplaced in one. Jerome Park, June 5, won the Westchester Cup, 2¼ miles, in 4.09½, beating Ferida, Monitor and Franklin, a capital race. Gravesend, June 23, ran second to Monitor in Coney Island Cup, 2¼ miles, 4.36½, beating Ferida and Danicheff; June 26, ran second to Glenmore in Stirrup Cup, 2 miles and a furlong, in 3.48½, beating Jim Beck, an excellent race. Long Branch, July 5, ran second to Report (4), 108 lbs., Uncas (4), 114 lbs., in Shrewsbury Handicap, 1¾ miles, in 3 12½, beating Ferida (4), 105 lbs., and Danicheff (5), 100 lbs.; was third to Report, 115 lbs., and Monitor, 120 lbs., Uncas, 121 lbs., in Monmouth Cup, 2¼ miles, in 2.34, the fastest race ever run at the distance to that date; August 21, won Handicap Sweepstake, 1¼ miles, in 2.16½, Uncas 112, lbs., beating One Dime (4), 115 lbs. Gravesend, September 4, ran third to Luke Blackburn and Monitor in Great Challenge Stakes, 1½ miles, in 2.38, beating One Dime; September 7, ran second to Glenmore (5), 107 lbs., Uncas, 107 lbs., 1¾ miles, in 3.07, beating Monitor, 109 lbs., etc.; September 14, was beaten by Luke Blackburn, a match, each carrying 108 lbs., 1½ miles, in 2.42½; September 23, won Handicap Sweepstakes, 1 mile and 3 furlongs, 107 lbs., in 2.21¾, beating Mary Anderson (3), 78 lbs., and Checkmate (5), 114 lbs. This is the fastest and best race ever run at the distance to that date. Jerome Park, October 7, won Grand National Handicap, 2¼ miles, carrying 113 lbs., in 4.05½, beating Monitor, 116 lbs., Ferida, 114 lbs., and Franklin (5), 102 lbs. Baltimore, October 20, was beaten by Monitor in Pimlico Stakes, 2 miles and a furlong, in 3.44½, the fastest race ever run at the distance. Jerome Park, November 2, won Handicap Sweepstakes, 1¾ miles, in 3.16½, 119 lbs., beating Harlequin (4), 89 lbs., and Krupp Gun (5), 100 lbs. At five years old, started five times, won no race. Uncas, for his chances, is a success in the stud, is the sire of the winners Dunboyne, Laggard (a first-class race-horse), Nat Goodwin, Cascade, Matawan, Chieftain, Sorcerer, Surplus, Bel Demonio, McCarthy, St. Michael, Copyright, Pactolus and others. Uncas is a blood bay, stands 15½ hands high, with his right fore-foot white, and a small star in his forehead. He is quite a compact, well and evenly-formed horse all over, having a neat head and neck, with good shoulders and depth of girth, with extra good body, hips, stifles, and excellent legs and feet. The pedigree is one of the best in the Stud Book, and from a distinguished racing family. His dam is by Glencoe's best son, and he traces back through an own sister to Glencoe to a natural Barb mare. If the blood of Lexington and Glencoe is to be preserved to the country, it is through just such pure channels as Uncas presents.

Uncas traces in direct female line to the natural Barb mare of Mr. Tregonwell, to which traces Woodpecker (by Herod), Waxy Pope, Whalebone, Whisker, Partisan, imp. Glencoe, Bay Middleton, Melbourne, Bend'or, Robert the Devil, imp. Mortemer, Nutbourne, Speculum and all the Prunella family.

Herod 1758—Tartar 1743—Partner 1718—Jigg—Byerly Turk, about 1685.

UNCAS.

BAY HORSE, FOALED 1876; BRED BY A. J. ALEXANDER, WOODBURN STUD, KENTUCKY.

Sire: LEXINGTON, b. h., 1850.
- Boston, ch. h., 1833.
  - Timoleon, ch. h., 1813.
    - Sir Archy, b. h., 1805.
    - Daughter of, ch. m., —.
  - Sist. to Tuckahoe, ch. m., 1814.
    - Ball's Florizel, ch. h., 1801.
    - Daughter of, b. m., 1799.
- Alice Carneal, b. m., 1836.
  - *Sarpedon, br. h., 1828.
    - Emilius, b. h., 1820.
    - Icaria, b. m., 1824.
  - Rowena, ch. m., 1826.
    - Sumpter, ch. h., 1818.
    - Lady Grey, b. m., 1817.

Dam: CORAL, br. m., 1864.
- Vandal, b. h., 1850.
  - *Glencoe, ch. h., 1831.
    - Sultan, b. h., 1816.
    - Trampoline, ch. m., 1825.
  - Alaric's dam, b. m., 1837.
    - *Tranby, br. h., 1826.
    - Lucilla, b. m., .
- *Cairngorm, b. m., 1852.
  - Cotherstone, b. h., 1840.
    - Touchstone, br. h., 1831.
    - Emma, ch. m., 1824.
  - Glenluce, b. m., 1843.
    - Slane, b. h., 1833.
    - Glencairne, b. m., 1828.

*Diomed, ch. h., 1777.	Florizel 1768 (Herod 1758) Sist. 1763 to Juno by Spectator 1749 (Crab 1722) Horatia 1758 by Blank 1740 (God. Ar. 1724) Dau. of Flying Childers 1715 (Darley Ar. 1699) Miss Belvoir by Grey Grantham (Brownlow Turk).
*Castianira, br. m., 1796.	Rockingham 1781 (Highflyer 1774) Tabitha 1782 by Trentham 1766 (T. Sweepstakes 1743) Dau. of Bosphorus 1754 (Babraham 1740) Sist. to Grecian Princess by W's Forester 1750 (Forester 1736) Dau. of Coalition Colt 1729 (God. Ar. 1724).
*Saltram, br. h., 1780.	Eclipse 1764 (Marske 1750) Virago 1764 by Snap 1750 (Snip 1736) Dau. of Regulus 1739 (God. Ar. 1724)—Sist. 1759 to Othello by Crab 1722 (Alcock Ar.)—Miss Slamerkin 1729 by Y. True Blue 1718 (Honeywood's Ar.).
Daughter of, —.	Symmes' Wildair 1767 (*Fearnought 1755) Dau. of Tyler's Driver 1760 (Othello) Dau. of *Fallower 1761 (Blank 1740) Dau. of Vampire 1757 (Regulus 1739). [The Saltram mare was dam of Jenny Cockracy].
*Diomed, ch. h., 1777.	Florizel 1768 (Herod 1758)—Sist. 1763 to Juno by Spectator 1749 (Crab 1722) Horatia 1758 by Blank 1740 (God. Ar. 1724) Dau. of Flying Childers 1715 (Darley Ar. 1699) Miss Belvoir by Grey Grantham (Brownlow Turk).
Daughter of, —.	*Shark 1771 (Marske 1750) Dau. of Harris's Eclipse 1770 (Fearnought 1755) Dau. of *Fearnought 1755 (Regulus 1739) Dau. of *Jolly Roger 1741 (Roundhead 1733) Dau. of *Sober John 1748 (Rib 1736).
*Alderman, b. h., 1787.	Pot-8-o's 1773 (Eclipse 1764) Lady Bolingbroke 1766 by Squirrel 1754 (Traveller 1735) Cypron 1750 by Blaze 1733 (Flying Childers 1715)—Salome 1733 by Bethell's Ar. Dau. of Graham's Champion 1707 (Harpur's Ar.).
Daughter of, ab. 1791.	*Clockfast 1774 (Gimcrack 1760) Dau. of Symmes' Wildair 1767 (*Fearnought 1755) Y. Kitty Fisher 1767 by *Fearnought 1755 (Regulus 1739)—*Kitty Fisher 1755 by Cade 1734 (God. Ar. 1724) Dau. of Somerset Ar.
Orville, b. h., 1799.	Beningbrough 1791 (King Fergus 1775) Evelina 1791 by Highflyer 1774 (Herod 1758) Termagant by Tantrum 1760 (Cripple 1750)—Cantatrice by Sampson 1745 (Blaze 1733) Dau. 1757 of Regulus 1739 (God. Ar. 1724).
Emily, ch. m., 1810.	Stamford 1794 (Sir Peter T. 1784)—Dau. 1799 of Whiskey 1789 (Saltram* 1780)—Grey Dorimant 1781 by Dorimant 1772 (Otho 1760) Dizzy 1757 by Blank 1740 (God. Ar. 1724) Anc. Dizzy 1741 by Driver 1727 (Wynn Ar.).
The Flyer, br. h., 1814.	Vandyke Jr. 1808 (Walton 1799) Azalia 1804 by Beningbrough 1791 (King Fergus 1775) Gilliflower 1787 by Highflyer 1774 (Herod 1758)—Preference 1778 by Goldfinder 1764 (Snap 1750)—Dau. 1774 of Marske 1750 (Squirt 1732).
Parma, br. m., 1813.	Dick Andrews 1797 (Joe Andrews 1778) May 1801 by Beningbrough 1791 (King Fergus 1775)—Primrose 1787 by Mambrino 1768 (Engineer 1756) Cricket 1776 by Herod 1758 (Tartar 1743) Sophia 1764 by Blank 1740 (God. Ar. 1724).
Sir Archy, b. h., 1805.	*Diomed 1777 (Florizel 1768)—Castianira 1796 by Rockingham 1781 (Highflyer 1774)—Tabitha 1782 by Trentham 1766 (T. Sweepstakes 1743) Dau. of Bosphorus 1754 (Babraham 1740)—Dau. of Forester 1750 (Forester 1736).
Robin mare, ch. m., —.	*Robin Redbreast 1796 (Sir Peter T. 1784) Dau. 1788 of *Obscurity 1777 (Eclipse 1764) Slamerkin 1768 by *Wildair 1753 (Cade 1734)—*Cub mare 1762 by Cub 1739 (Fox 1714)—Dau. 1755 of Second 1732 (Flying Childers 1715).
Robin Grey, gr. h., 1805.	*Royalist 1790 (Saltram* 1780) Belle Maria by Grey Diomed 1786 (*Medley 1776) Queen by *St. George 1771 (Dragon 1757) Dau. of Cassius 1778 (Black-and-all-Black 1773)—Primrose by *Dove (Y. Cade 1747).
Maria, —. m., 1802.	Melzar 1791 (*Medley 1776)—Dau. of *Highflyer 1784 (Highflyer 1774)—Dau. of *Fearnought 1755 (Regulus 1739) Dau. of Ariel 1753 (*Traveller 1747)—Dau. of *Jack of Diamonds (Cullen Ar.).
Selim, ch. h., 1802.	Buzzard* 1787 (Woodpecker 1773)—Dau. of Alexander 1782 (Eclipse 1764) Dau. of Highflyer 1774 (Herod 1758)—Dau. of Alfred 1770 (Matchem 1748) Dau. 1770 of Engineer 1756 (Sampson 1745).
Bacchante, br. b., 1809.	W's Ditto 1800 (Sir Peter T. 1784)—Dau. 1791 of Mercury 1778 (Eclipse 1764)—Dau. 1776 of Herod 1758 (Tartar 1743)—Folly 1771 by Marske 1750 (Squirt 1732)—Vixen 1753 by Regulus 1739 (God. Ar. 1724).
Tramp, b. h., 1810.	Dick Andrews 1797 (Joe Andrews 1778) Dau. 1803 of Gohanna 1790 (Mercury 1778)—Fraxinella 1793 by Trentham 1766 (T. Sweepstakes 1743)—Dau. 1785 of Woodpecker 1773 (Herod 1758)—Everlasting 1775 by Eclipse 1764 (Marske 1750).
Web, b. m., 1808.	Waxy 1790 (Pot-8-o's 1773)—Penelope 1798 by Trumpator 1782 (Conductor 1767) Prunella 1788 by Highflyer 1774 (Herod 1758)—Promise 1768 by Snap 1750 (Snip 1736)—Julia 1756 by Blank 1740 (God. Ar. 1724).
Blacklock, b. h., 1814.	Whitelock 1803 (Hambletonian 1792)—Dau. 1799 of Coriander 1786 (Pot-8-o's 1773)—Wildgoose 1792 by Highflyer 1774 (Herod 1758)—Coheiress 1786 by Pot-8-o's 1773 (Eclipse 1764)—Manilla 1777 by Goldfinder 1764 (Snap 1750).
Miss Bowe's dam, br. m., 1812.	Orville 1799 (Beningbrough 1791)—Miss Grimstone 1793 by Weazle 1776 (Herod 1758)—Dau. of Ancaster 1768 (Blank 1740)—Dau. of Damascus Ar. 1751 Dau. of Sampson 1745 (Blaze 1733).
Trumpator, br. h., 1823.	Sir Solomon 1805 (*Tickle Toby 1786)—Dau. of Hickory 1804 (*Whip 1794)—*Trumpetta 1797 by Trumpator 1782 (Conductor 1767) Sist. to Lambinos 1786 by Highflyer 1774 (Herod 1758)—Dau. 1775 of Eclipse 1764 (Marske 1750).
Lucy, b. m., 1821.	Orphan 1810 (Ball's Florizel 1801)—Lady Grey 1817 by Robin Grey 1805 (*Royalist 1790)—Maria 1802 by Melzar 1791 (*Medley 1776) Dau. of *Highflyer 1784 (Highflyer 1774)—Dau. of *Fearnought 1755 (Regulus 1739).
Camel, br. h., 1822.	Whalebone 1807 (Waxy 1790)—Dau. 1812 of Selim 1802 (Buzzard* 1787) Maiden 1801 by Sir Peter T. 1784 (Highflyer 1774)—Dau. 1788 of Phenomenon 1780 (Herod 1758)—Matron 1782 by Florizel 1768 (Herod 1758).
Banter, br. m., 1826.	Master Henry 1815 (Orville 1799)—Boadicea 1807 by Alexander 1782 (Eclipse 1764) Brunette by Amaranthus 1766 (Old England 1741)—Mayfly 1771 by Matchem 1748 (Cade 1734)—Dau. 1755 of Anc. Starling 1738 (Starling 1727).
Whisker, b. h., 1812.	Waxy 1790 (Pot-8-o's 1773)—Penelope 1798 by Trumpator 1782 (Conductor 1767) Prunella 1788 by Highflyer 1774 (Herod 1758)—Promise 1768 by Snap 1750 (Snip 1736)—Julia 1756 by Blank 1740 (God. Ar. 1724).
Gibside Fairy, b. m., 1811.	Hermes 1790 (Mercury 1778)—Vicissitude 1800 by Pipator 1786 (Imperator 1776) Beatrice 1791 by Sir Peter T. 1784 (Highflyer 1774)—Pyrrha 1771 by Matchem 1748 (Cade 1734) Duchess 1748 by Whitenose 1742 (God. Ar. 1724).
Royal Oak, br. h., 1823.	Catton 1809 (Golumpus 1802)—Dau. 1818 of Smolensko 1810 (Sorcerer 1796)—Lady Mary 1800 by Beningbrough 1791 (King Fergus 1775)—Dau. of Highflyer 1774 (Herod 1758)—Sist. to Miss Bell by Marske 1750 (Squirt 1732).
Daughter of, b. m., 1819.	Orville 1799 (Beningbrough 1791)—Epsom Lass 1803 by Sir Peter T. 1784 (Highflyer 1774)—Alexina 1788 by King Fergus 1775 (Eclipse 1764)—Lardella 1780 by Y. Marske 1771 (Marske 1750)—Dau. of Y. Cade 1747 (Cade 1734).
Sultan, b. h., 1816.	Selim 1802 (Buzzard* 1787) Bacchante 1809 by W's Ditto 1800 (Sir Peter T. 1784)—Dau. 1791 of Mercury 1778 (Eclipse 1764)—Dau. 1776 of Herod 1758 (Tartar 1743) Folly 1771 by Marske 1750 (Squirt 1732).
Trampoline, ch. m., 1825.	Tramp 1810 (Dick Andrews 1797)—Web b. m. 1808 by Waxy 1790 (Pot-8-o's 1773)—Penelope b. m. 1798 by Trumpator 1782 (Conductor 1767)—Prunella b. m. 1788 by Highflyer 1774 (Herod 1758)—Promise br. m. 1768 by Snap 1750 (Snip 1736) Julia b. m. 1756 by Blank 1740 (God. Ar. 1724)—Spectator's dam. b. m. 1735 by Partner 1718

(Jigg)—Bonny Lass b. m. 1723 by Bay Bolton 1705 (Grey Hautboy)—Dau. of Darley Ar. 1699—Dau. of Byerly Turk—Dau. of Taffolet Barb—Dau. of Place's White Turk—Natural Barb mare of Mr. Tregonwell's.

VENTILATOR.

Is a private stallion in the Brookdale Stud, near Holmdel, N. J., Estate of D. D. Withers, proprietors. Yearling sales annually.

VENTILATOR by Vandal, best son of Imp. Glencoe. bred by Gen. W. G. Harding. Belle Meade Stud, near Nashville, Tenn., foaled 1872, dam Carolin, the dam of Bounce, By-the-Way, Ella, etc., by imp. Scythian, son of Orlando by Touchstone. grandam Rosette, dam of Quartermaster by imp. Yorkshire, out of Picayune, the dam of Doubloon, Florin, Louis d'Or, etc., by Medoc, son of American Eclipse. Ventilator made his first appearance as a two-year old in the Young America Stakes No. 1, 1 mile, at Nashville, Tenn., running second to Katie Pearce in 1.45½, with Elemi Leona, Bob Woolley and six others behind him. Ran dead heat with Elemi for the Young America Stakes No. 2, 1 mile, in 1.46¾; Elemi beat him the deciding heat in 1.49¾; behind him in this were Katie Pearce, Bob Woolley and five others. At three years old was unplaced in the Clark Stakes at Louisville, won by his stable companion, Voltigeur. In 1876 and 1877 he was in retirement. In 1878 he came out in poor hands, ran second in one race and unplaced in seven. In 1879 started in twenty-one races. all over timber, won four, was second in three, third in five and unplaced in the others. He ran 2¼ miles over the Steeplechase Course, at Sheepshead Bay, in 4.27½, and won the second heat in a hurdle race of heats, 1¼ miles, in 2.23. In 1880 started in fourteen races, won seven, was second in five, and unplaced in two. His races were all over timber, and some of them fast ones, beating such horses as Judith, Bertha, Pomeroy, Chimneysweep, etc. During his retirement he covered three mares, and got Mary Anderson, Bagdad and Ventriloquist, all of them winners in good company. The family is a famous racing one on both sides. On the sire's side belong Lexington, Vandal, Monarchist, Voltnruo, Ruric, Alaric, Luke Blackburn, Hindoo. Monitor, The Bard and others of the best horses of the American turf. On the side of his dam the family is perhaps equally famous, such horses as Doubloon, Florin. Louis d'Or, the best horses of their day, as also Quartermaster, Vagrant, Bounce, Raceland, Madge, Brown Duke, Red Dick, Bay Dick, Plenipo, Sherrod, etc.

Ventilator, in addition to those mentioned above, is the sire of the winners Airshaft, Airtight, Airplant, Punka, Bass Viol, Mirth, Wizard, Wind Sail, John Finn, Futurity, Drizzle, Single Stone, Oragense, etc., etc.

Ventilator is a dark chestnut, with a small leaf-shaped star in the forehead, 15¼ hands high, but very stout, well shaped and muscular. He weighs 1,055 lbs. He has gotten and will get race-horses with a fair chance.

VENTILATOR.

CHESTNUT HORSE, FOALED 1872; BRED BY GEN. W. G. HARDING, BELLE MEADE STUD, NASHVILLE, TENN.

VANDAL, b. h., 1850.	*Glencoe, ch. h., 1831.	Sultan, b. h., 1816.	Selim, ch. h., 1802. Buzzard*, ch. h., 1787.	Woodpecker 1773 (Herod 1758) Misfortune 1775 by Dux 1761 (Matchem 1748) Curiosity 1760 by Snap 1750 (Snip 1736) Dau. 1749 of Regulus 1739 (God. Ar. 1724) Dau. of Bartlett's Childers (Darley Ar. 1689).
			Daughter of, b. m., — .	Alexander 1782 (Eclipse 1764) Dau. of Highflyer 1774 (Herod 1758) Dau. of Alfred 1770 (Matchem 1748)—Dau. 1770 of Engineer 1756 (Sampson 1745) Bay Malton's dam 1756 by Cade 1734 (God. Ar. 1724).
			Bacchante, br. m., 1809. Williamson's Ditto, b. h., 1800.	Sir Peter T. 1784 (Highflyer 1774) Arethusa 1792 by Dungannon 1780 (Eclipse 1764)—Dau. 1777 of Prophet 1760 (Regulus 1739)—Virago 1764 by Snap 1750 (Snip 1736)—Dau. of Regulus 1739 (God. Ar. 1724).
			Sister to Calomel, b. m., 1791.	Mercury 1778 (Eclipse 1764)—Dau. 1776 of Herod 1758 (Tartar 1743)—Folly 1771 by Marske 1750 (Squirt 1732)—Vixen 1753 by Regulus 1739 (God. Ar. 1724) Dau. of Hutton's Spot 1738 (Hartley's Blind Horse).
		Trampoline, ch. m., 1825.	Tramp, b. h., 1810. Dick Andrews, b. h., 1797.	Joe Andrews 1778 (Eclipse 1764)—Dau. 1790 of Highflyer 1774 (Herod 1758) Dau. 1777 of Cardinal Puff 1760 (Babraham 1740) Dau. 1768 of Tatler 1751 (Blank 1740) Bay Snip 1746 by Snip 1736 (Fl. Childers 1715).
			Daughter of, b. m., 1803.	Gohanna 1790 (Mercury 1778) Fraxinella 1793 by Trentham 1766 (T. Sweepstakes 1743) Dau. 1785 of Woodpecker 1773 (Herod 1758) Everlasting 1775 by Eclipse 1764 (Marske 1750)—Hyæna 1762 by Snap 1750 (Snip 1736).
			Web, b. m., 1808. Waxy, b. h., 1790.	Pot-8-o-'s 1773 (Eclipse 1764) Maria 1777 by Herod 1758 (Tartar 1743) Lisette 1772 by Snap 1750 (Snip 1736) Miss Windsor 1754 by God. Ar. 1724 Sist. to Volunteer by Y. Belgrade (Belgrade Turk).
			Penelope, b. m., 1798.	Trumpator 1782 (Conductor 1767) Prunella 1788 by Highflyer 1774 (Herod 1758) Promise 1768 by Snap 1750 (Snip 1736)—Julia 1756 by Blank 1740 (God. Ar. 1724)—Spectator's dam 1735 by Partner 1718 (Jigg).
	Alaric's dam, b. m., 1837.	*Tranby, br. h., 1826.	Blacklock, b. h., 1814. Whitelock, b. h., 1803.	Hambletonian 1792 (King Fergus 1775)—Rosalind 1788 by Phenomenon 1780 (Herod 1758)—Atalanta 1769 by Matchem 1748 (Cade 1734) Y. Lass-of-the-Mill 1756 by Oroonoko 1745 (Crab 1722)—Dau. 1744 of Traveller 1735 (Partner 1718).
			Daughter of, —, —, 1799.	Coriander 1786 (Pot-8-o-'s 1773)—Wildgoose 1792 by Highflyer 1774 (Herod 1758)—Coheiress 1786 by Pot-8-o-'s 1773 (Eclipse 1764)—Manilla 1777 by Goldfinder 1764 (Snap 1750)—Dau. 1766 of Old England 1741 (God. Ar. 1724).
			Miss Bowe's dam, br. m., 1812. Orville, b. h., 1799.	Beningbrough 1791 (King Fergus 1775)—Evelina 1791 by Highflyer 1774 (Herod 1758) Termagant by Tantrum 1760 (Cripple 1750)—Cantatrice by Sampson 1745 (Blaze 1733)—Dau. 1757 of Regulus 1739 (God. Ar. 1724).
			Miss Grimstone, b. m., 1796.	Weazle 1776 (Herod 1758)—Dau. of Ancaster 1768 (Blank 1740)—Dau. of Damascus Ar. 1754—Dau. of Sampson 1745 (Blaze 1733)—Stolk's dam 1757 by Oroonoko 1745 (Crab 1722).
		Lucilla, b. m., — .	Trumpator, br. h., 1823. Sir Solomon, br. h., 1805.	*Tickle Toby 1786 (Alfred 1770)—Vesta by Dreadnought (Celer 1774)—Bandy by *Clockfast 1774 (Gimcrack 1760)—Dau. of Americus (*Fearnought 1755) Dau. of *Fearnought 1755 (Regulus 1739).
			Daughter of, —, —.	Hickory 1804 (*Whip 1794)—*Trumpetta 1797 by Trumpator 1782 (Conductor 1767)—Dau. 1786 of Highflyer 1774 (Herod 1758) Dau. 1775 of Eclipse 1764 (Marske 1750)—Vauxhall's dam by Y. Cade 1747 (Cade 1734).
			Lucy, b. m., 1821. Orphan, ch. h., 1810.	Ball's Florizel 1801 (*Diomed 1777) Fair Rachel by *Diomed 1777 (Florizel 1768)—Susan Jones by *Shark 1771 (Marske 1750)—Dau. of Thornton's Wildair (Symmes' Wildair 1767).
			Lady Grey, b. m., 1817.	Robin Grey 1805 (*Royalist 1790)—Maria 1802 by Melzar 1791 (*Medley 1776) Dau. of *Highflyer 1784 (Highflyer 1774)—Dau of *Fearnought 1755 (Regulus 1739)—Dau. of Ariel 1756 (*Traveller 1747)
CAROLIN, b. m., 1858.	*Scythian, b. h., 1851.	Orlando, b. h., 1841.	Touchstone, br. h., 1831. Camel, br. h., 1822.	Whalebone 1807 (Waxy 1790)—Dau. 1812 of Selim 1802 (Buzzard* 1787)—Maiden 1801 by Sir Peter T. 1784 (Highflyer 1774)—Dau. 1788 of Phenomenon 1780 (Herod 1758)—Matron 1782 by Florizel 1768 (Herod 1758).
			Banter, br. m., 1826.	Master Henry 1815 (Orville 1799)—Boadicea 1807 by Alexander 1782 (Eclipse 1764)—Brunette by Amaranthus 1766 (Old England 1741) Mayfly 1771 by Matchem 1748 (Cade 1734)—Dau. 1755 of Anc. Starling 1738 (Starling 1727).
			Vulture, ch. m., 1833. Langar, ch. h., 1817.	Selim 1802 (Buzzard* 1787)—Dau. 1808 of Walton 1799 (Sir Peter T. 1784)—Y. Giantess 1790 by Diomed* 1777 (Florizel 1768)—Giantess 1769 by Matchem 1748 (Cade 1734)—Molly Long Legs 1753 by Babraham 1740 (God. Ar. 1724).
			Kite, b. m., 1821.	Bustard 1813 (Castrel 1801)—Olympia 1815 by Sir Oliver 1800 (Sir Peter T. 1784) Scotilla 1795 by Anvil 1777 (Herod 1758)—Scota 1783 by Eclipse 1764 (Marske 1750)—Harmony 1775 by Herod 1758 (Tartar 1743).
		Scythia, b. m., 1846.	Hetman Platoff, b. h., 1836. Brutandorf, b. h., 1821.	Blacklock 1814 (Whitelock 1803)—Mandane 1800 by Pot-8-o's 1773 (Eclipse 1764)—Y. Camilla 1787 by Woodpecker 1773 (Herod 1758)—Camilla 1778 by Trentham 1766 (T. Sweepstakes 1743) Coquette 1765 by Compton Barb.
			Daughter of, gr. m., 1820.	Comus 1809 (Sorcerer 1796)—Marciana 1809 by Stamford 1794 (Sir Peter T. 1784)—Marcia 1797 by Coriander 1786 (Pot-8-o-'s 1773) Faith 1779 by Pacolet 1763 (Blank 1740)—Atalanta 1769 by Matchem 1748 (Cade 1734).
			The Princess, ch. m., 1841. Slane, b. h., 1833.	Royal Oak 1823 (Catton 1809)—Dau. 1819 of Orville 1799 (Beningbrough 1791)—Epsom Lass 1803 by Sir Peter T. 1784 (Highflyer 1774)—Alexina 1788 by King Fergus 1775 (Eclipse 1764)—Lardella 1780 by Y. Marske 1771 (Marske 1750).
			Sister to Cobweb, ch. m., 1823.	Phantom 1808 (Walton 1799)—Filagree 1815 by Soothsayer 1808 (Sorcerer 1796)—Web 1808 by Waxy 1790 (Pot-8-o-'s 1773)—Penelope 1798 by Trumpator 1782 (Conductor 1767)—Prunella 1788 by Highflyer 1774 (Herod 1758).
	Rosette, ch. m., 1851.	*Yorkshire, b. h., 1834.	St. Nicholas, br. h., 1827. Emilius, b. h., 1820.	Orville 1799 (Beningbrough 1791)—Emily 1810 by Stamford 1794 (Sir Peter T. 1784)—Dau. 1799 of Whiskey 1789 (Saltram* 1780) Grey Dorimant 1781 by Dorimant 1772 (Otho 1760)—Dizzy 1757 by Blank 1740 (God. Ar. 1724).
			Sea Mew, b. m., 1815.	Scud 1804 (Beningbrough 1791)—Goosander 1805 by Hambletonian 1792 (King Fergus 1775)—Rally 1790 by Trumpator 1782 (Conductor 1767)—Fancy 1780 by Florizel 1768 (Herod 1758) Dau. 1763 by Spectator 1735 (Crab 1722).
			Miss Rose, b. m., 1825. Tramp, b. h., 1810.	Dick Andrews 1797 (Joe Andrews 1778)—Dau. 1803 of Gohanna 1790 (Mercury 1778)—Fraxinella 1793 by Trentham 1766 (T. Sweepstakes 1743)—Dau. 1785 of Woodpecker 1773 (Herod 1758)—Everlasting 1775 by Eclipse 1764 (Marske 1750).
			Daughter of, ch. m., 1810.	Sancho 1801 (Don Quixote 1784) Blacklock's dam 1799 by Coriander 1786 (Pot-8-o-'s 1773)—Wildgoose 1792 by Highflyer 1774 (Herod 1758)—Coheiress 1786 by Pot-8-o-'s 1773 (Eclipse 1764)—Manilla 1777 by Goldfinder 1764 (Snap 1750).
		Picayune, ch. m., 1833.	Medoc, ch. h., 1829. American Eclipse, ch. h., 1814.	Duroc 1806 (*Diomed 1777)—Miller's Damsel 1802 by *Messenger 1780 (Mambrino 1768)—*Pot-8-o-'s mare 1792 by Pot-8-o-'s 1773 (Eclipse 1764) Dau. 1778 of Gimcrack 1760 (Cripple 1750)—Snapdragon 1759 by Snap 1750 (Snip 1736).
			Y. Maid of the Oaks, ch. m., 1817.	*Expedition 1795 (Pegasus 1784)—Maid of the Oaks 1801 by *Spread Eagle 1792 (Volunteer 1780)—Anette by *Shark 1771 (Marske 1750)—Dau. of Rockingham (Partner 1755)—Dau. of True Whig (*Fitzhugh's Regulus 1765).
			Sally Howe, ch. m., —. Sir William, b. h., 1820.	Sir Archy 1805 (*Diomed 1777) Transport 1812 by Virginius 1805 (*Diomed 1777)—Nancy Air 1799 by *Bedford 1792 (Dungannon 1780)—Annette by *Shark 1771 (Marske 1750)—Dau. of Rockingham (Partner 1755).
			Lady Robin, gr. m., —.	Robin Grey 1805 (*Royalist 1790)—Dau. of Quicksilver 1789 (*Medley 1776)—Dau. of Meade's Celer 1774 (*Janus 1746).

VOCALIC.

In the Iroquois Stud, James B. Clay, proprietor, Lexington, Ky. Terms advertised and yearling sales annually.

VOCALIC, black horse, foaled 1881, bred in the Elmendorf Stud, by Virgil, son of Vandal by imp. Glencoe, dam Acoustic by imp. Australian, son of West Australian by Melbourne, grandam Nemesis, dam of Rhadamanthus by imp. Eclipse, out of Echo by Lexington, etc. Virgil, sire of Vocalic, was a good race-horse and winner of many races. He was by the great race-horse Vandal, the best son of imported Glencoe, and whose dam was the Trauby mare, that also produced Levity, Volga and Vistula. Virgil's dam was Hymenia by imp. Yorkshire, and she also produced Daisy by Cracker, Alaric, Ansel, Lady Petry and Mollie Wood. Virgil is the sire of the unbeaten Tremont, the great Hindoo (sire of Hanover), Portland, Vigil, Vagrant, Vera Cruz, Virginius, Vanguard, Virgilian, Fairplay, Katrina, Memento, Carley B., Ben Ali, Richmond, Kingfish, Himalaya, etc., etc. His get were potent factors on the turf for many years, and nothing commends Vocalic more highly than that he is a son of Virgil. His dam, Acoustic, is by imp. Australian, son of West Australian and Emilia by Young Emilius, sire of Waverly, Spendthrift, Fellowcraft, Maggie B. B., the dam of Iroquois, Farfalletta, dam of Falsetto, Ivy Leaf, Bramble's dam, Madge, Lizzie Lucas, dam of Cerise, Morello's dam, Perhaps, dam of Chance and Reckon, etc. His grandam, Nemesis, is the dam of Rhadamanthus, the horse that laid the foundation of the turf fame and fortune of the Dwyer Brothers, by imp. Eclipse, son of Orlando by Touchstone. Naphtha, the full sister to Nemesis, produced Golden Gate (dam of Oro and Laura Winston) and Explosion, the dam of Dew Drop, the best filly of her day, that sold at auction for $29,000 and won upwards of $40,000 before her untimely death at three years of age. Maria Innis was the dam of Bulletin, Meteor and Kathleen, the dam of the great racer George Kinney, Kennesaw, John Happy, Kern and other winners. Ann Innis was the dam of Hebron, Jim Barton and La Grande Duchesse, dam of Aniella, Metells, Talisman, Uberto and Christine, dam of Bonnie Australian, Montana Regent, Julia L. and High Tariff. Vocalic won at Sheepshead Bay, 1 mile, in 1.43¼, at two years old, defeating Dutch Roller, Blast, Rataplan, Economy and other good ones, and a Handicap, 1⅛ miles, at Jerome Park, in 2.17; he was placed second and third in several good races.

Although young in the stud, nearly everything by him thus far trained has proved a winner, including Bellevue, Jay Q. L., the Alice Ward colt, Vortex, Vocalite, Vocaletta, Blanche's Last, Vocal, etc.

He comes from the same family as Maiden, the best mare of her day, winner of the Travers and other great stakes, dam of James A., Parole, Paw Paw, Pappoose, Perfection, Pawnee and the promising sire Powhattan, Kitty Clark, the dam of Charlotte Buford (dam of Florence I., the dam of Powhattan III., Florence, Fonso and Florio), Kate Clark, Rosa Clark and La Henderson (dam of Ferida, Aella and Ferona). Miss Obstinate was the dam of Mary Morris and Ann Innis, from whom Falsetto, Lizzie Lucas, Geo. Kinney, Rhadamanthus, Eachus, Dew Drop, Morello and many great ones have descended. The family is one of the grandest racing and producing ones in the world.

Woodpecker 1773 Herod 1758 Tartar 1743—Partner 1718—Jigg Byerly Turk, about 1685.

VOCALIC.

BLACK HORSE, FOALED 1881; BRED BY MR. D. SWIGERT, ELMENDORF STUD, KENTUCKY.

- VIRGIL, br. h., 1864.
  - Vandal, b. h., 1850.
    - *Glencoe, ch. h., 1831.
      - Sultan, b. h., 1816.
      - Trampoline, ch. m., 1825.
    - Alaric's dam, b. m., 1837.
      - *Tranby, br. h., 1826.
      - Lucilla, b. m., [illegible]
  - Hymenia, b. m., 1851.
    - *Yorkshire, b. h., 1834.
      - St. Nicholas, br. h., 1827.
      - Miss Rose*, b. m., 1826.
    - Little Peggy, ch. m., 1847.
      - Cripple, b. h., 1836.
      - Peggy Stewart, b. m., 1821.
- ACOUSTIC, br. m., 1875.
  - *Australian, ch. h., 1858.
    - West Australian, b. h., 1850.
      - Melbourne, br. h., 1834.
      - Mowerina, b. m., 1843.
    - Emilia,* b. m., 1840.
      - Y. Emilius, b. h., 1833.
      - Persian, b. m., 1829.
  - Nemesis, blk. m., 1865.
    - *Eclipse, b. h., 1855.
      - Orlando, b. h., 1841.
      - Gaze, b. m., 1842.
    - Echo, b. m., 1859.
      - Lexington, b. h., 1850.
      - Maria Innis, blk. m., 1853.

Selim, ch. h., 1802. — Buzzard 1787 (Woodpecker 1773) Dau. of Alexander 1782 (Eclipse 1764) Dau. of Highflyer 1774 (Herod 1758) Dau. of Alfred 1770 (Matchem 1748) Dau. 1770 of Engineer 1756 (Sampson 1745).

Bacchante, br. m., 1809. — W's Ditto 1800 (Sir Peter T. 1784) Dau. 1791 of Mercury 1778 (Eclipse 1764) Dau. 1776 of Herod 1758 (Tartar 1743) Folly 1771 by Marske 1750 (Squirt 1732) Vixen 1753 by Regulus 1739 (God. Ar. 1724).

Tramp, b. h., 1810. — Dick Andrews 1797 (Joe Andrews 1778) Dau. 1803 of Gohanna 1790 (Mercury 1778) Fraxinella 1793 by Trentham 1766 (T. Sweepstakes 1743) Dau. 1785 of Woodpecker 1773 (Herod 1758) Everlasting 1775 by Eclipse 1764 (Marske 1750).

Web, b. m., 1808. — Waxy 1790 (Pot-8-o-'s 1773) Penelope 1798 by Trumpator 1782 (Conductor 1767) Prunella 1788 by Highflyer 1774 (Herod 1758)—Promise 1768 by Snap 1750 (Snip 1736)—Julia 1756 by Blank 1740 (God. Ar. 1724).

Blacklock, b. h., 1814. — Whitelock 1803 (Hambletonian 1792) Dau. 1799 of Coriander 1786 (Pot-8-o-'s 1773) Wildgoose 1792 by Highflyer 1774 (Herod 1758) Coheiress 1786 by Pot-8-o-'s 1773 (Eclipse 1764) Manilla 1777 by Goldfinder 1764 (Snap 1750).

Miss Bowe's dam, br. m., 1812. — Orville 1799 (Beningbrough 1791) Miss Grimstone 1796 by Weazle 1776 (Herod 1758) Dau. of Ancaster 1768 (Blank 1740) Dau. of Damascus Ar. 1754 Dau. of Sampson 1745 (Blaze 1733).

Davenport's Trumpator, br. h., 1824. — Sir Solomon 1805 (*Tickle Toby 1786) Dau. of Hickory 1801 (*Whip 1794)—*Trumpetta 1797 by Trumpator 1782 (Conductor 1767) Sist. 1786 to Lambinos by Highflyer 1774 (Herod 1758) Dau. 1775 of Eclipse 1764 (Marske 1750).

Lucy, b. m., 1821. — Orphan 1810 (Ball's Florizel 1801)—Lady Grey 1817 by Robin Grey 1805 (*Royalist 1790) Maria 1802 by Melzar 1791 (*Medley 1776) Dau. of *Highflyer 1784 (Highflyer 1774) Dau. of *Fearnought 1755 (Regulus 1739).

Emilius, b. h., 1820. — Orville 1799 (Beningbrough 1791)—Emily 1810 by Stamford 1794 (Sir Peter T. 1784)—Dau. 1799 of Whiskey 1789 (*Saltram 1780) Grey Dorimant 1781 by Dorimant 1772 (Otho 1760) - Dizzy 1757 by Blank 1740 (God. Ar. 1724).

Sea Mew, b. m., 1815. — Send 1804 (Beningbrough 1791)—Goosander 1805 by Hambletonian 1792 (King Fergus 1775)—Rally 1790 by Trumpator 1782 (Conductor 1767)—Fancy 1780 by Florizel 1768 (Herod 1758) Dau. 1763 of Spectator 1749 (Crab 1722).

Tramp, b. h., 1810. — Dick Andrews 1797 (Joe Andrews 1778) - Dau. 1803 of Gohanna 1790 (Mercury 1778)—Fraxinella 1793 by Trentham 1766 (T. Sweepstakes 1743) Dau. 1785 of Woodpecker 1773 (Herod 1758)- Everlasting 1775 by Eclipse 1764 (Marske 1750).

Daughter of, ch. m., 1810. — Sancho 1801 (Don Quixote 1784)- Blacklock's dam 1799 by Coriander 1786 (Pot-8-o-'s 1773) Wildgoose 1792 by Highflyer 1774 (Herod 1758)—Coheiress 1786 by Pot-8-o-'s 1773 (Eclipse 1764) - Manilla 1777 by Goldfinder 1764 (Snap 1750).

Medoc, ch. h., 1829. — Am. Eclipse 1814 (Duroc 1806)—Y. Maid-of-the-Oaks 1817 by *Expedition 1795 (Pegasus 1784) Maid-of-the-Oaks 1801 by *Spread Eagle 1792 (Volunteer 1780) Annette by *Shark 1771 (Marske 1750) - Dau. of Rockingham (Partner 1755).

Grecian Princess, b. m., 1816. — Cook's Whip 1804 (*Whip 1794)—Jane Hunt 1796 by Hampton's Paragon 1788 (*Flimnap 1765)—Moll by *Figure 1757 (Figure 1747)—Slamerkin 1768 by *Wildair 1753 (Cade 1734) - *Cub mare 1762 by Cub 1739 (Fox 1714).

Cook's Whip, b. h., 1804. — *Whip 1794 (Saltram* 1780)- Speckleback 1790 by Randolph's Celer (Meade's Celer 1774)—Old Speckleback by Meade's Celer 1774 (*Janus 1746)—Dau. of *Sober John 1748 (Rib 1736) Thoroughbred mare of Mr. Wade.

MaryBedford, b. m., 1816. — Duke of Bedford (*Bedford 1792) Dau. 1813 of *Speculator 1795 (Dragon* 1787) Dau. 1806 of *Dare Devil 1787 (Magnet 1770)—*Trumpetta 1797 by Trumpator 1782 (Conductor 1767) Sist. 1786 to Lambinos by Highflyer 1774 (Herod 1758).

Humprey Clinker, b. h., 1822. — Comus 1809 (Sorcerer 1796)—Clinkerina 1812 by Clinker 1805 (Sir Peter T. 1784)—Pewet 1786 by Tandem 1773 (Syphon 1750) Termagant by Tantrum 1760 (Cripple 1750)—Cantatrice by Sampson 1745 (Blaze 1733).

Daughter of, b. m., 1825. — Cervantes 1806 (Don Quixote 1784)—Dau. 1818 of Golumpus 1802 (Gohanna 1790) Dau. 1810 of Paynator 1791 (Trumpator 1782)—Circle 1802 by St. George 1789 (Highflyer 1774) - Abigail 1788 by Woodpecker 1773 (Herod 1758).

Touchstone, br. h., 1831. — Camel 1822 (Whalebone 1807) - Banter 1826 by Master Henry 1815 (Orville 1799) Boadicea 1807 by Alexander 1782 (Eclipse 1764)- Brunette by Amaranthus 1766 (Old England 1741)—Mayfly 1771 by Matchem 1748 (Cade 1734).

Emma, ch. m., 1824. — Whisker 1812 (Waxy 1790) - Gibside Fairy 1811 by Hermes 1790 (Mercury 1778)—Vicissitude 1800 by Pipator 1786 (Imperator 1776) - Beatrice 1791 by Sir Peter T. 1784 (Highflyer 1774)- Pyrrha 1771 by Matchem 1748 (Cade 1734).

Emilius, b. h., 1820. — Orville 1799 (Beningbrough 1791)—Emily 1810 by Stamford 1794 (Sir Peter T. 1784)—Dau. 1799 of Whiskey 1789 (Saltram* 1780)—Grey Dorimant 1781 by Dorimant 1772 (Otho 1760) - Dizzy 1757 by Blank 1740 (God. Ar. 1724).

Shoveler, b. m., 1816. — Send 1804 (Beningbrough 1791)—Goosander 1805 by Hambletonian 1792 (King Fergus 1775) Rally 1790 by Trumpator 1782 (Conductor 1767) Fancy 1780 by Florizel 1768 (Herod 1758) - Sist. 1763 to Juno by Spectator 1749 (Crab 1722).

Whisker, b. h., 1812. — Waxy 1790 (Pot-8-o-'s 1773) Penelope 1798 by Trumpator 1782 (Conductor 1767) - Prunella 1788 by Highflyer 1774 (Herod 1758)—Promise 1768 by Snap 1750 (Snip 1736)—Julia 1756 by Blank 1740 (God. Ar. 1724).

Variety, ch. m., 1816. — [Selim or] Soothsayer 1808 (Sorcerer 1796)—Sprite 1807 by Bobtail 1795 (Precipitate* 1787) - Catherine 1795 by Woodpecker 1773 (Herod 1758) - Camilla 1778 by Trentham 1766 (T. Sweepstakes 1743)- Coquette 1765 by Compton Barb.

Touchstone, br. h., 1831. — Camel 1822 (Whalebone 1807) Banter 1826 by Master Henry 1815 (Orville 1799) Boadicea 1807 by Alexander 1782 (Eclipse 1764) Brunette by Amaranthus 1766 (Old England 1741)—Mayfly 1771 by Matchem 1748 (Cade 1734).

Vulture, ch. m., 1833. — Langar 1817 (Selim 1802) - Kite 1821 by Bustard 1813 (Castrel 1801)—Olympia 1815 by Sir Oliver 1800 (Sir Peter T. 1784) - Scotilla 1795 by Anvil 1777 (Herod 1758)—Scota 1783 by Eclipse 1764 (Marske 1750).

Bay Middleton, b. h., 1833. — Sultan 1816 (Selim 1802) Cobweb 1821 by Phantom 1808 (Walton 1799) Filagree 1815 by Soothsayer 1808 (Sorcerer 1796)—Web 1808 by Waxy 1790 (Pot-8-o-'s 1773) Penelope 1798 by Trumpator 1782 (Conductor 1767).

Flycatcher, ch. m., 1831. — Godolphin 1818 (Partisan 1811)—Sist. 1820 to Cobweb by Phantom 1808 (Walton 1799)—Filagree 1815 by Soothsayer 1808 (Sorcerer 1796) - Web 1808 by Waxy 1790 (Pot-8-o-'s 1773) Penelope 1798 by Trumpator 1782 (Conductor 1767).

Boston, ch. h., 1833. — Timoleon 1813 (Sir Archy 1805)—Sist. 1814 to Tuckahoe by Ball's Florizel 1801 (*Diomed 1777)—Dau. 1799 of *Alderman 1787 (Pot-8-o-'s 1773)—Dau. of *Clockfast 1771 (Gimcrack 1760) Dau. of Symmes' Wildair 1767 (*Fearnought 1775)

Alice Carneal, b. m., 1836. — *Sarpedon 1828 (Emilius 1820)—Rowena 1826 by Sumpter 1818 (Sir Archy 1805)- Lady Grey 1817 by Robin Grey 1805 (*Royalist 1790)—Maria 1802 by Melzar 1791 (*Medley 1776) - Dau. of *Highflyer 1784 (Highflyer 1774).

Yorkshire, b. h., 1834. — St. Nicholas 1827 (Emilius 1820)—Miss Rose 1826 by Tramp 1810 (Dick Andrews 1797)—Dan. 1810 of Sancho 1801 (Don Quixote 1784)—Blacklock's dam 1799 by Coriander 1786 (Pot-8-o-'s 1773)—Wildgoose 1792 by Highflyer 1774 (Herod 1758).

Ann Innis, blk. m., 1838. — Am. Eclipse 1814 (Duroc 1806)—Miss Obstinate b. m. 1829 by Sumpter 1818 (Sir Archy 1805)—Jenny Slamerkin b. m. 1823 by Tiger 1812 (Cook's Whip 1804)—Paragon b. m. 1807 by Buzzard 1787 (Woodpecker 1773)—Indiana b. m. 1802 by Butler's Columbus (*Pantaloon 1778)—Jane Hunt b. m. 1796 by Hampton's Paragon 1788 (*Flimnap 1765)—Moll b. m. by *Figure 1757 (Figure 1747)—Slamerkin b. m. 1768 by *Wildair 1753 (Cade 1734)—*Cub mare b. m. 1762 by Cub 1739 (Fox 1714)—Dau. of b. m. 1755 by Second 1732 (Flying Childers 1715)—Sist. 1746 to Leedes by Starling 1727 (Bay Bolton 1705)—Sist. b. m. 1740 to Vane's Little Partner by Partner 1718 (Jigg)—Sist. 1722 to Guy by Greyhound (Chillaby)—Brown Farewell br. m. 1710 by Makeless (Oglethorpe Ar.)—Dau. of Brimmer (D'Arcy Yellow Turk)—Dau. of Place's White Turk—Dau. of Dodsworth—Layton Barb mare.

VOLANTE.

WINNER OF THE FLASH, AMERICAN DERBY, SEQUEL, WOODLAWN, COLUMBIA, EXCELSIOR AND CHAMPION STAKES, AND BROOKLYN CUP, ETC.

The property of Nicholas Finzer, Louisville, Ky. Services annually.

VOLANTE, bay horse, foaled 1882, bred by E. J. Baldwin, by Grinstead, son of Gilroy by Lexington, dam Sister Anne by imp. Glenelg, son of Citadel by Stockwell, out of The Nun, sister to Norfolk by Lexington, son of Boston. Grinstead, the sire of Volante, is from the grand racing family known as the Levity family, from which have come so many famous race-horses, potent brood mares and successful stallions. Grinstead was a good race-horse; won the Mansion House All-aged Sweepstakes and the Summer Handicap at Saratoga. Volante was a superior race-horse; won two out of four races at two years old, including Flash Stakes at Saratoga, half a mile, in 0.49½, with 110 lbs., was second in one and third in one. At three years old started eighteen times, won six races, was second in eight and third in three. Won the Street Railway Stakes, 1¾ miles, 118 lbs., in 2.25; won Bankers and Brokers Stakes, 1¼ miles, 132 lbs., in 2.12; won the American Derby, 1½ miles, 128 lbs., in 2.49½, track heavy; won First Sweepstakes, 1 mile, 105 lbs., in 1.43¼; won Sequel Stakes, 1⅝ miles, 123 lbs., in 3.00; won Woodlawn Stakes, 1½ miles, 124 lbs., in 2.41. At four years old started twenty times, won ten races, was second in six and third in one. Won Cuyama Stakes, 1¼ miles, 118 lbs., in 2.17; won Handicap, $1\frac{1}{16}$ miles, 118 lbs., in 1.50; received forfeit in match, 1½ miles, from Tyrant; won purse, 1½ miles, 118 lbs., in 2.36½; won Columbia Stakes, 1¾ miles, 118 lbs., in 3.04; won Excelsior Stakes, 1¼ miles, 118 lbs., in 2.13½; won Saratoga Cup, 2¼ miles, 118 lbs., in 4.25; won Champion Stakes, 1½ miles, 118 lbs., in 2.45; won Free Handicap, 1¼ miles, 119 lbs., in 2.12½; won the Autumnal Handicap, 1½ miles, 124 lbs., in 2.36½. At five years old started twenty-eight times, won thirteen races, was second in ten and third in four. Won purse, 1⅛ miles, 114 lbs., in 1.56½; won Merchants' Stakes at Latonia, 1⅛ miles, 122 lbs., in 1.58½; won Merchants' Stakes at Louisville, Ky., 1⅛ miles, 127 lbs., in 1.57½; won Ellis Wainwright Stakes, 1¼ miles, 122 lbs., in 2.18; won Columbia Stakes, 1¾ miles, 116 lbs., in 3.06½; won purse, $1\frac{1}{16}$ miles, 116 lbs., in 1.51½; won Beverwyck Stakes, 1 mile 500 yards, 120 lbs., in 2.15½; won Handicap, 1⅜ miles, 120 lbs., in 2.30; won First Special Stakes, 1¼ miles, 118 lbs., in 2.12; won Handicap, 1 mile, 121 lbs., in 1.47½; won Autumnal Handicap, 1¼ miles, 122 lbs., in 2.09½; won Brokers' Handicap, 1¼ miles, 123 lbs., in 2.11; won Brewers' Cup, St. Louis, 1⅝ miles, 111 lbs., in 3.07½. At six years old started eleven times, won two races, was second in four and third in two. Won Handicap, 1 mile, 119 lbs., in 1.41½; won Boulevard Stakes, 1¼ miles, 120 lbs., in 2.08½; his last season. Volante won with high weights, met the best horses and defeated the best of them. Volante is from a grand racing and producing family. His dam, Sister Anne, is full sister to Clara D. (the dam of Winona, Santa Anna, Santiago) by the great Glenelg. His grandam, The Nun, was own sister to the unbeaten Norfolk, winner of 3-mile heats in 5.27½, 5.29½, the best ever run, and he one of the most successful stallions. Chloe, Anderson, the third dam, was grand in the stud. She produced Nora (the dam of Bombay, Persia, Sovereign Pat and Grimaldi) and Kate Anderson that gave us Adelgiza (dam of Sarah B.), the grand mare Novice (dam of Norfolk, sire of El Rio Rey, Emperor of Norfolk, The Czar, etc., and one of Lexington's best sons, The Nun, Notice and Notre Dame) and Sally Anderson (dam of Amanda, Buford, Carrie Anderson and Florence Anderson). From the above-mentioned grandaughters of Chloe Anderson came the following winners and producers, viz.: Avondale (dam of Ben Edwards, A. O. H. and Go Lucky), Faugh-a-Ballagh, Sister Anne (dam of Estrella, another good one, and Marie Stuart, Caliente and San Pueblo's dam), Nota Bene (dam of Vindex, Geo. C. Burnett and Take Notice), Eugenia Pickens (dam of Major), Bahama, Gambetta, Susie Linwood (dam of Volta, Red Light's dam), Sangaree (dam of Value), Roderick Random, Full Sail, etc. Belle Anderson, the fourth dam, produced the famous Zenith, the best racer of his day. Volante, on his grand qualities as a race-horse, should get race-horses.

VOLANTE.

BAY HORSE, FOALED 1882; BRED BY E. J. BALDWIN, SANTA ANITA STUD, CALIFORNIA.

Diomed* 1777 Florizel 1768—Herod 1758—Tartar 1743 Partner 1718—Jigg—Byerly Turk, about 1685.

GRINSTEAD, b. h., 1871.	Gilroy, b. h., 1862.	Lexington, b. h., 1850.	Boston, ch. h., 1833.	Timoleon, ch. h., 1813.	Sir Archy 1805 (*Diomed 1777) Dau. of *Saltram 1780 (Eclipse 1764) Dau. of Symmes' Wildair 1767 (*Fearnought 1755) Dau. of Tyler's Driver 1760 (*Othello) Dau. of *Fallower 1761 (Blank 1740).
				Daughter of, ch. m., 1814.	Ball's Florizel 1801 (*Diomed 1777) Dau. 1790 of *Alderman 1787 (Pot-8-o's 1773) Dau. of *Clockfast 1774 (Gimcrack 1760)—Dau. of Symmes' Wildair 1767 (*Fearnought 1755) Y. Kitty Fisher 1767 by *Fearnought 1755 (Regulus 1739).
			AliceCarneal, b. m., 1836.	*Sarpedon, br. h., 1828.	Emilius 1820 (Orville 1799) Icaria 1824 by The Flyer 1814 (Vandyke Jr. 1808) Parma 1813 by Dick Andrews 1797 (Joe Andrews 1778) May 1801 by Beningbrough 1791 (King Fergus 1775) Primrose 1787 by Mambrino 1768 (Engineer 1756).
				Rowena, ch. m., 1826.	Sumpter 1818 (Sir Archy 1805) Lady Grey 1817 by Robin Grey 1805 (*Royalist 1790)—Maria 1802 by Melzar 1791 (*Medley 1776)—Dau. of *Highflyer 1784 (Highflyer 1774)—Dau. of *Fearnought 1755 (Regulus 1739).
		Magnolia, ch. m., 1841.	*Glencoe, ch. h., 1831.	Sultan, b. h., 1816.	Selim 1802 (Buzzard* 1787)—Bacchante 1809 by W's Ditto 1800 (Sir Peter T. 1784)—Dau. 1791 of Mercury 1778 (Eclipse 1764) Dau. 1776 of Herod 1758 (Tartar 1743) Folly 1771 by Marske 1750 (Squirt 1732)—Vixen 1753 by Regulus 1739 (God. Ar. 1724).
				Trampoline, ch. m., 1825.	Tramp 1810 (Dick Andrews 1797) Web 1808 by Waxy 1790 (Pot-8-o's 1773) Penelope 1798 by Trumpator 1782 (Conductor 1767)—Prunella 1788 by Highflyer 1774 (Herod 1758) Promise 1768 by Snap 1750 (Snip 1736).
			*Myrtle, ch. m., 1831.	Mameluke, b. h., 1824.	Partisan 1811 (Walton 1799) Miss Sophia 1805 by Stamford 1794 (Sir Peter T. 1784)—Sophia 1798 by Buzzard* 1787 (Woodpecker 1773)—Huncamunca 1787 by Highflyer 1774 (Herod 1758) Cypher 1772 by Squirrel 1754 (Traveller 1735)
				Bobadilla, br. m., 1825.	Bobadil 1813 (Rubens 1805)—Pythoness 1813 by Sorcerer 1796 (Trumpator 1782) Princess 1795 by Sir Peter T. 1784 (Highflyer 1774) Dau. 1789 by Dungannon 1780 (Eclipse 1764) Dau. 1783 of Turf 1760 (Matchem 1748).
	Sister to Ruric, b. m., 1862.	*Sovereign, b. h., 1836.	Emilius, b. h., 1820.	Orville, b. h., 1799.	Beningbrough 1791 (King Fergus 1775) Evelina 1791 by Highflyer 1774 (Herod 1758)—Termagant by Tantrum 1760 (Cripple 1750) Cantatrice by Sampson 1745 (Blaze 1733)—Dau. 1757 of Regulus 1739 (God. Ar. 1724).
				Emily, ch. m., 1810.	Stamford 1794 (Sir Peter T. 1784) Dau. 1799 of Whiskey 1789 (Saltram* 1780)—Grey Dorimant 1781 by Dorimant 1772 (Otho 1760) Dizzy 1757 by Blank 1740 (God. Ar. 1724)—Dizzy 1741 by Driver 1727 (Wynn Ar.)
			Fleur-de-Lis, b. m., 1822.	Bourbon, b. h., 1811.	Sorcerer 1796 (Trumpator 1782) Dau. 1797 of Precipitate 1787 (Mercury 1778)—Y. Tiffany 1787 by Highflyer 1774 (Herod 1758) Tiffany 1775 by Eclipse 1764 (Marske 1750) Y. Hag 1761 by Skim 1746 (Starling 1727).
				Lady Rachel, b. m., 1805.	Stamford 1794 (Sir Peter T. 1784) Y. Rachel 1790 by Volunteer 1780 (Eclipse 1764)—Rachel 1790 by Highflyer 1774 (Herod 1758)—Dau. 1771 of Syphon 1750 (Squirt 1732)—Dau. of Regulus 1739 (God. Ar. 1724).
		Levity, b. m., 1845.	*Trustee, ch. h., 1829.	Catton, b. h., 1809.	Golumpus 1802 (Gohanna 1790) Lucy Gray 1804 by Timothy 1794 (Delpini 1781) Lucy 1789 by Florizel 1768 (Herod 1758)—Frenzy 1774 by Eclipse 1764 (Marske 1750)—Dau. of Engineer 1756 (Sampson 1745).
				Emma, ch., m., 1824.	Whisker 1812 (Waxy 1790) Gibside Fairy 1811 by Hermes 1790 (Mercury 1778) Vicissitude 1800 by Pipator 1786 (Imperator 1776)—Beatrice 1791 by Sir Peter T. 1784 (Highflyer 1774)—Pyrrha 1771 by Matchem 1748 (Cade 1734).
			Vandal's dam, b. m., 1837.	*Trauby, br. h., 1826.	Blacklock 1814 (Whitelock 1803)—Dau. 1812 by Orville 1799 (Beningbrough 1791) Miss Grimstone 1796 by Weazle 1776 (Herod 1758)—Dau. of Ancaster 1768 (Blank 1740)—Dau. of Damascus Ar. 1754.
				Lucilla, b. m., —.	Trumpator 1823 (Sir Solomon 1805)—Lucy 1821 by Orphan 1810 (Ball's Florizel 1801)—Lady Grey 1817 by Robin Grey 1805 (*Royalist 1790) Maria 1802 by Melzar 1791 (*Medley 1776)—Dau. of *Highflyer 1784 (Highflyer 1774).
SISTER ANNE, b. m., 1876.	*Glenelg, b. h., 1866.	Citadel, ch. h., 1859.	Stockwell, ch. h., 1849.	The Baron, ch. h., 1842.	Ir. Birdcatcher 1833 (Sir Hercules 1826)—Echidna 1838 by Economist 1825 (Whisker 1812)—Miss Pratt 1825 by Blacklock 1814 (Whitelock 1803) Gadabout 1812 by Orville 1799 (Beningbrough 1791)—Minstrel 1803 by Sir Peter T. 1784 (Highflyer 1774).
				Pocahontas, b. m., 1837.	Glencoe* 1831 (Sultan 1816)—Marpessa 1830 by Muley 1810 (Orville 1799)—Clare 1824 by Marmion 1806 (Whiskey 1789) Harpalice 1814 by Gohanna 1790 (Mercury 1778)—Amazon 1799 by Driver 1783 (Trentham 1766).
			Sortie, b. m., 1851.	Melbourne, br. h., 1834.	Humprey Clinker 1822 (Comus 1809)—Dau. 1825 of Cervantes 1806 (Don Quixote 1784)—Dau. 1818 of Golumpus 1802 (Gohanna 1790)—Dau. 1810 of Paynator 1791 (Trumpator 1782)—Circle 1802 by St. George 1789 (Highflyer 1774).
				Escalade, b. m., 1846.	Touchstone 1831 (Camel 1822)—Ghuznee 1838 by Pantaloon 1824 (Castrel 1801)—Languish 1830 by Cain 1822 (Paulowitz 1813)—Lydia 1822 by Poulton 1805 (Sir Peter T. 1784)—Variety 1808 by Hyacinthus 1797 (Coriander 1786).
		*Babta, b. m., 1858.	Kingston, b. h., 1849.	Venison, br. h., 1833.	Partisan 1811 (Walton 1799)—Fawn 1823 by Smolensko 1810 (Sorcerer 1796)—Jerboa 1803 by Gohanna 1790 (Mercury 1778) Camilla 1778 by Trentham 1766 (T. Sweepstakes 1743)—Coquette 1765 by Compton Barb.
				Queen Anne, b. m., 1843.	Slane 1833 (Royal Oak 1823)—Garcia 1823 by Octavian 1807 (Stripling 1795)—Dau. 1806 of Shuttle 1793 (Y. Marske 1771)—Katharine 1798 by Delpini 1781 (Highflyer 1774)—Dau. 1785 of Paymaster 1766 (Blank 1740).
			Alice Lowe, ch. m., 1841.	Defence, b. h., 1824.	Whalebone 1807 (Waxy 1790)—Defiance 1816 by Rubens 1805 (Buzzard* 1787)—Little Folly 1806 by Highland Fling 1798 (Spadille 1784)—Harriet 1799 by Volunteer 1780 (Eclipse 1764)—Dau. 1785 of Alfred 1770 (Matchem 1748).
				Pet, b. m., 1826.	Gainsborough 1813 (Rubens 1805) Dau. 1818 of Topsy Turvey 1805 (St. George 1789)—Agnes 1805 by Shuttle 1793 (Y. Marske 1771)—Dau. 1788 of Highflyer 1774 (Herod 1758)—Dau. 1779 of Goldfinder 1764 (Snap 1750).
	The Nun, ch. m., 1866.	Lexington, b. h., 1850.	Boston, ch. h., 1833.	Timoleon, ch. h., 1813.	Sir Archy 1805 (*Diomed 1777) Dau. of *Saltram 1780 (Eclipse 1764) Dau. of Symmes' Wildair 1767 (*Fearnought 1755)—Dau. of Tyler's Driver 1760 (*Othello) Dau. of *Fallower 1761 (Blank 1740).
				Sister to Tuckahoe, ch. m., 1814.	Ball's Florizel 1801 (*Diomed 1777)—Dau. 1790 of *Alderman 1787 (Pot-8-o's 1773) Dau. of *Clockfast 1774 (Gimcrack 1760)—Dau. of Symmes' Wildair 1767 (*Fearnought 1755) Y. Kitty Fisher 1767 by *Fearnought 1755 (Regulus 1739).
			AliceCarneal, b. m., 1836.	*Sarpedon, br. h., 1828.	Emilius 1820 (Orville 1799)—Icaria 1824 by The Flyer 1814 (Vandyke Jr. 1808) Parma 1813 by Dick Andrews 1797 (Joe Andrews 1778) May 1804 by Beningbrough 1791 (King Fergus 1775)—Primrose 1787 by Mambrino 1768 (Engineer 1756).
				Rowena, ch. m., 1826.	Sumpter 1818 (Sir Archy 1805)—Lady Grey 1817 by Robin Grey 1805 (*Royalist 1790)—Maria 1802 by Melzar 1791 (*Medley 1776)—Dau. of *Highflyer 1784 (Highflyer 1774)—Dau. of *Fearnought 1755 (Regulus 1739).
		Novice, b. m., 1853.	*Glencoe, ch. h., 1831.	Sultan, b. h., 1816.	Selim 1802 (Buzzard* 1787)—Bacchante 1809 by W's Ditto 1800 (Sir Peter T. 1784) Dau. 1791 of Mercury 1778 (Eclipse 1764)—Dau. 1776 of Herod 1758 (Tartar 1743)—Folly 1771 by Marske 1750 (Squirt 1732).
				Trampoline, ch. m., 1825.	Tramp 1810 (Dick Andrews 1797)—Web 1808 by Waxy 1790 (Pot-8-o's 1773)—Penelope 1798 by Trumpator 1782 (Conductor 1767)—Prunella 1788 by Highflyer 1774 (Herod 1758)—Promise 1768 by Snap 1750 (Snip 1736).
			Chloe Anderson, b. m., 1839.	Rodolph, b. h., 1831.	Sir Archy Montorio 1823 (Sir Archy 1805)—Melody's dam 1824 by Haxall's Moses 1816 (*Sir Harry 1795)—Dau. of Cook's Whip 1804 (*Whip 1794)—Hart's Maria by Craig's Alfred (*Medley 1776)—Pryor mare by Tayloe's Bellair 1786 (*Medley 1776).
				Belle Anderson, b. m., 1830.	Sir William of Transport 1820 (Sir Archy 1805)—Butterfly ch. m. 1825 by Sumpter 1818 (Sir Archy 1805) Dau. of *Buzzard 1787 (Woodpecker 1773)—Dau. of Dandridge's Fearnought (*Fearnought 1755)—Dau. of *Janus 1746 (Janus 1789). [The Dandridge's Fearnought mare was a good race-mare in Kentucky about 1790.]

IMPORTED WAGNER.

The property of Messrs. L. B. Field & Bro., Lexington, Ky. Terms advertised annually.

WAGNER, bay horse, foaled 1882, bred by Mr. T. Stevens, by Prince Charlie, son of Blair Athol by Stockwell, dam Duchess of Malfi by Elland, son of Rataplan by The Baron, grandam The Duchess by St. Albans, son of Stockwell by The Baron, out of Bay Celia by Orlando, son of Touchstone. Prince Charlie, the sire of Wagner, was one of the speediest horses ever in England, and was called the "Prince of the Two-year old Course." He won the 2,000 Guineas, Middle Park Plate, Criterion, Fernhill, Drawing Room, Don, Cheveley and other good stakes; sire of Prestonpans (winner of the Liverpool Autumn Cup), Eos, Pride of the Highlands, Reconciliation, etc., in England; Salvator (winner of the fastest Suburban ever run, 2.06⅘, beat Tenny in the match, 1¼ miles, in 2.05, and ran the fastest mile ever run in the world, 1.35½), Senorita (winner of the Monmouth Oaks) and other good ones in America. Wagner's dam, Duchess of Malfi, a winner, by Elland (winner of the Beaufort and Liverpool Autumn Cups, Gold Vase at Ascot, Claret Stakes and Queen's Plate). He beat such good ones as Breadalbane, The Duke, John Davis, Broomielaw, etc. Elland was almost full brother to imp. Ill Used, sire of His Highness. Ill Used being by Breadalbane, son of Stockwell, and Elland by Rataplan, the own brother to Stockwell, both out of Ellermire by Chanticleer, son of Birdcatcher, Ellermire being half sister to the Derby winner Ellington. Grandam The Duchess by St. Albans, winner of the St. Leger and sire of Springfield (winner of the July Cup and Champion Stakes), out of Bay Celia, dam of the Duke, winner of the Goodwood Cup and grandsire of Robert the Devil (winner of the St. Leger) and of The Earl (winner of the Grand Prize of Paris), she by Orlando (winner of the Derby), whose daughters are noted in the stud as the dams of Gang Forward, Camballo and General Peel, winners of the 2,000 Guineas, and of Petrarch (winner of the 2,000 Guineas and St. Leger). Wagner is thus by a very excellent race-horse and potent sire, and from a racing and producing family on the side of his dam. Wagner ran only once in England, won the Wilton Park Stakes, 5 furlongs, 121 lbs., in a field of four others. He is young in the stud but is the sire of the speedy Minnehaha, May T., etc.

Wagner is descended in direct female line through the old Montagu mare, from the Royal mare, to which traces Conductor, Eclipse (O'Kelly's), Voltaire, Sterling, Weatherbit, Springfield, Scottish Chief, Adventurer, Prince Charlie, in England; Lexington, Vandal, The Bard, Leonatus, Salvator, Longstreet, Sensation, Onondaga, Stratford, Tenny, Luke Blackburn, and all the Duchess family in America.

Whalebone 1807—Waxy 1790 Pot-8-o's 1773 Eclipse 1764—Marske 1750 Squirt 1732 Bartlett's Childers Darley Ar. 1699.

WAGNER.

BAY HORSE, FOALED 1882; BRED BY MR. T. STEVENS, ENGLAND.

PRINCE CHARLIE,* ch. h., 1869.	Blair Athol, ch. h., 1861.	Stockwell, ch. h., 1849.	The Baron, ch. h., 1842.	Ir. Birdcatcher, ch. h., 1833.	Sir Hercules 1826 (Whalebone 1807)—Guiccioli 1823 by Bob Booty 1804 (Chanticleer 1787)—Flight 1809 by Ir. Escape 1802 (Commodore 1793)—Y. Heroine by Bagot 1780 (Herod 1758)—Heroine 1775 by Hero 1753 (Cade 1734).
				Echidna, b. m., 1838.	Economist 1825 (Whisker 1812)—Miss Pratt 1825 by Blacklock 1814 (Whitelock 1803)—Gadabout 1812 by Orville 1799 (Beningbrough 1791)—Minstrel 1803 by Sir Peter T. 1784 (Highflyer 1774)—Matron 1782 by Florizel 1768 (Herod 1758).
			Pocahontas, b. m., 1837.	Glencoe,* ch. h., 1831.	Sultan 1816 (Selim 1802)—Trampoline 1825 by Tramp 1810 (Dick Andrews 1797)—Web 1808 by Waxy 1790 (Pot-8-o's 1773)—Penelope 1798 by Trumpator 1782 (Conductor 1767)—Prunella 1788 by Highflyer 1774 (Herod 1758).
				Marpessa, b. m., 1830.	Muley 1810 (Orville 1799)—Clare 1824 by Marmion 1806 (Whiskey 1789)—Harpalice 1814 by Gohanna 1790 (Mercury 1778)—Amazon 1799 by Driver 1783 (Trentham 1766)—Fractious 1792 by Mercury 1778 (Eclipse 1764).
		Blink Bonny, b. m., 1854.	Melbourne, br. h., 1834.	Humprey Clinker, b. h., 1822.	Comus 1809 (Sorcerer 1796)—Clinkerina 1812 by Clinker 1805 (Sir Peter T. 1784)—Pewet 1786 by Tandem 1773 (Syphon 1750)—Termagant by Tantrum 1760 (Cripple 1750)—Cantatrice by Sampson 1745 (Blaze 1733).
				Daughter of, b. m., 1825.	Cervantes 1806 (Don Quixote 1784)—Dau. 1818 of Golumpus 1802 (Gohanna 1790)—Dau. 1810 of Paynator 1791 (Trumpator 1782)—Circle 1802 by St. George 1789 (Highflyer 1774)—Abigail 1788 by Woodpecker 1773 (Herod 1758).
			Queen Mary, b. m., 1843.	Gladiator, ch. h., 1833.	Partisan 1811 (Walton 1799)—Pauline 1826 by Moses 1819 ([Whalebone or] Seymour 1807)—Quadrille 1815 by Selim 1802 (Buzzard* 1787)—Canary Bird 1806 by [Whiskey or] Sorcerer 1796 (Trumpator 1782)—Canary 1797 by Coriander 1786 (Pot-8-o's 1773).
				Daughter of, b. m., 1840.	Plenipotentiary 1831 (Emilius 1820)—Myrrha 1830 by Whalebone 1807 (Waxy 1790)—Gift 1818 by Y. Gohanna 1810 (Gohanna 1790)—Sist. 1802 to Grazier by Sir Peter T. 1784 (Highflyer 1774)—Dau. 1788 of Trumpator 1782 (Conductor 1767).
	Eastern Princess, ch. m., 1858.	Surplice, b. h., 1845.	Touchstone, br. h., 1831.	Camel, br. h., 1822.	Whalebone 1807 (Waxy 1790)—Dau. 1812 of Selim 1802 (Buzzard* 1787)—Maiden 1801 by Sir Peter T. 1784 (Highflyer 1774)—Dau. 1788 of Phenomenon 1780 (Herod 1758)—Matron 1782 by Florizel 1768 (Herod 1758).
				Banter, br. m., 1826.	Master Henry 1815 (Orville 1799)—Boadicea 1807 by Alexander 1782 (Eclipse 1764)—Brunette by Amaranthus 1766 (Old England 1741)—Mayfly 1771 by Matchem 1748 (Cade 1734)—Dau. 1755 of Anc. Starling 1738 (Starling 1727).
			Crucifix, b. m., 1837.	Priam,* b. h., 1827.	Emilius 1820 (Orville 1799)—Cressida 1807 by Whiskey 1789 (Saltram* 1780)—Y. Giantess 1790 by Diomed* 1777 (Florizel 1768)—Giantess 1769 by Matchem 1748 (Cade 1734)—Molly Long Legs 1753 by Babraham 1740 (God. Ar. 1724).
				Octaviana, b. m., 1815.	Octavian 1807 (Stripling 1795)—Dau. 1807 of Shuttle 1793 (Y. Marske 1771)—Zara 1801 by Delpini 1781 (Highflyer 1774)—Flora 1789 by King Fergus 1775 (Eclipse 1764)—Atalanta 1769 by Matchem 1748 (Cade 1734).
		Tomyris, b. m., 1851.	Sesostris, b. h., 1845.	Slane, b. h., 1833.	Royal Oak 1823 (Catton 1809)—Dau. 1819 of Orville 1799 (Beningbrough 1791)—Epsom Lass 1803 by Sir Peter T. 1784 (Highflyer 1774)—Alexina 1788 by King Fergus 1775 (Eclipse 1764)—Lardella 1780 by Y. Marske 1771 (Marske 1750).
				Palmyra, br. m., 1838.	Sultan 1816 (Selim 1802)—Hester 1832 by Camel 1822 (Whalebone 1807)—Monimia 1821 by Muley 1810 (Orville 1799)—Sist. 1796 to Petworth by Precipitate* 1787 (Mercury 1778)—Dau. 1787 of Woodpecker 1773 (Herod 1758).
			Daughter of, ch. m., 1842.	Glaucus, b. h., 1830.	Partisan 1811 (Walton 1799)—Nanine 1823 by Selim 1802 (Buzzard* 1787)—Bizarre 1811 by Peruvian 1806 (Sir Peter T. 1784)—Violante 1802 by John Bull 1789 (Fortitude 1777)—Dau. 1783 of Highflyer 1774 (Herod 1758).
				Io, ch. m., 1836.	Taurus 1826 ([Phantom or] Morisco 1819)—Arethissa 1824 by Quiz 1798 (Buzzard* 1787)—Persepolis 1808 by Alexander 1782 (Eclipse 1764)—Sist. 1784 to *Tickle Toby by Alfred 1770 (Matchem 1748)—Cœlia 1775 by Herod 1758 (Tartar 1743).
DUCHESS OF MALFI, b. m., 1873.	Elland, b. h., 1862.	Rataplan, ch. h., 1850.	The Baron, ch. h., 1842.	Ir Birdcatcher, ch. h., 1833.	Sir Hercules 1826 (Whalebone 1807)—Guiccioli 1823 by Bob Booty 1804 (Chanticleer 1787)—Flight 1809 by Ir. Escape 1802 (Commodore 1793)—Y. Heroine by Bagot 1780 (Herod 1758)—Heroine 1775 by Hero 1753 (Cade 1734).
				Echidna, b. m., 1838.	Economist 1825 (Whisker 1812)—Miss Pratt 1825 by Blacklock 1814 (Whitelock 1803)—Gadabout 1812 by Orville 1799 (Beningbrough 1791)—Minstrel 1803 by Sir Peter T. 1784 (Highflyer 1774)—Matron 1782 by Florizel 1768 (Herod 1758).
			Pocahontas, b. m., 1837.	Glencoe,* ch. h., 1831.	Sultan 1816 (Selim 1802)—Trampoline 1825 by Tramp 1810 (Dick Andrews 1797)—Web 1808 by Waxy 1790 (Pot-8-o's 1773)—Penelope 1798 by Trumpator 1782 (Conductor 1767)—Prunella 1788 by Highflyer 1774 (Herod 1758).
				Marpessa, b. m., 1830.	Muley 1810 (Orville 1799)—Clare 1824 by Marmion 1806 (Whiskey 1789)—Harpalice 1814 by Gohanna 1790 (Mercury 1778)—Amazon 1799 by Driver 1783 (Trentham 1766)—Fractious 1792 by Mercury 1778 (Eclipse 1764).
		Ellermire, b. m., 1852.	Chanticleer, gr. h., 1843.	Ir Birdcatcher, ch. h., 1833.	Sir Hercules 1826 (Whalebone 1807)—Guiccioli 1823 by Bob Booty 1804 (Chanticleer 1787)—Flight 1809 by Ir. Escape 1802 (Commodore 1793)—Y. Heroine by Bagot 1780 (Herod 1758)—Heroine 1775 by Hero 1753 (Cade 1734).
				Whim, gr. m., 1832.	Ir. Drone 1823 (Master Robert 1811)—Kiss 1827 by Waxy Pope 1806 (Waxy 1790)—Dau. 1810 of Champion 1797 (Pot-8-o's 1773)—Brown Fanny 1799 by Maximin 1785 (Evergreen 1769)—Dau. 1782 of Highflyer 1774 (Herod 1758).
			Ellerdale, br. m., 1841.	Lanercost, br. h., 1835.	Liverpool 1828 (Tramp 1810)—Otis by 1820 Bustard 1801 (Buzzard* 1787)—Gayhurst's dam 1813 by Election 1804 (Gohanna 1790)—Sist. 1791 to Skysweeper by Highflyer 1774 (Herod 1758)—Dau. 1778 of Eclipse 1764 (Marske 1750).
				Daughter of, br. m., 1838.	Tomboy 1829 (Jerry 1821)—Tesane 1830 by Whisker 1812 (Waxy 1790)—Lady of the Tees 1822 by Octavian 1807 (Stripling 1795)—Dau. 1810 of Sancho 1801 (Don Quixote 1784)—Miss Furey 1798 by Trumpator 1782 (Conductor 1767).
	The Duchess, br. m., 1864.	St. Albans, ch. h., 1857.	Stockwell, ch. h., 1849.	The Baron, ch. h., 1842.	Ir. Birdcatcher 1833 (Sir Hercules 1826)—Echidna 1838 by Economist 1825 (Whisker 1812)—Miss Pratt 1825 by Blacklock 1814 (Whitelock 1803)—Gadabout 1812 by Orville 1799 (Beningbrough 1791)—Minstrel 1803 by Sir Peter T. 1784 (Highflyer 1774).
				Pocahontas, b. m., 1837.	Glencoe* 1831 (Sultan 1816)—Marpessa 1830 by Muley 1810 (Orville 1799)—Clare 1824 by Marmion 1806 (Whiskey 1789)—Harpalice 1814 by Gohanna 1790 (Mercury 1778)—Amazon 1799 by Driver 1783 (Trentham 1766).
			Bribery, ch. m., 1851.	The Libel, br. h., 1842.	Pantaloon 1824 (Castrel 1801)—Pasquinade 1829 by Camel 1822 (Whalebone 1807)—Banter 1826 by Master Henry 1815 (Orville 1799)—Boadicea 1807 by Alexander 1782 (Eclipse 1764)—Brunette by Amaranthus 1766 (Old England 1741).
				Splitvote, ch. m., 1841.	St. Luke 1833 (Bedlamite 1823)—Electress 1819 by Election 1804 (Gohanna 1790)—Dau. 1805 of Stamford 1794 (Sir Peter T. 1784)—Miss Judy 1784 by Alfred 1770 (Matchem 1748)—Manilla 1777 by Goldfinder 1764 (Snap 1750).
		Bay Celia, b. m., 1851.	Orlando, b. h., 1841.	Touchstone, br. h., 1831.	Camel 1822 (Whalebone 1807)—Banter 1826 by Master Henry 1815 (Orville 1799)—Boadicea 1807 by Alexander 1782 (Eclipse 1764)—Brunette by Amaranthus 1766 (Old England 1741)—Mayfly 1771 by Matchem 1748 (Cade 1734).
				Vulture, ch. m., 1833.	Langar 1817 (Selim 1802)—Kite 1821 by Bustard 1813 (Castrel 1801)—Olympia 1815 by Sir Oliver 1800 (Sir Peter T. 1784)—Scotilla 1795 by Anvil 1777 (Herod 1758)—Scota 1783 by Eclipse 1764 (Marske 1750).
			Hersey, b. m., 1842.	Glaucus, b. h., 1830.	Partisan 1811 (Walton 1799)—Nanine 1823 by Selim 1802 (Buzzard* 1787)—Bizarre 1811 by Peruvian 1806 (Sir Peter T. 1784)—Violante 1802 by John Bull 1789 (Fortitude 1777)—Sist. 1788 to Skyscraper by Highflyer 1774 (Herod 1758).
				Hester, br. m., 1832.	Camel 1822 (Whalebone 1807)—Monimia b. m. 1821 by Muley 1810 (Orville 1799)—Sist. ch. m. 1796 to Petworth by Precipitate* 1787 (Mercury 1778)—Dau. b. m. 1787 of Woodpecker 1773 (Herod 1758)—Sist. b. m. 1773 to Juniper by Snap 1750 (Snip 1736)—Y. Marske's dam. ch. m. 1759 by Blank 1740 (God. Ar. 1724)—Bay Starling b. m. ab. 1749 by Bolton Starling 1727 (Bay Bolton 1705)—Meynell ch. m. 1736 by Partner 1718 (Jigg)—Sist. ch. m. 1723 to Sampson by Greyhound (Chillaby)—Sophonisba's dam by Curwen Bay Barb—Dau. of D'Arcy Chestnut Arab—Dau. of Whiteshirt—Lord D'Arcy's Montagu mare by his Old Montagu—Dau. of Hautboy (D'Arcy White Turk)—Dau. of Brimmer (D'Arcy Yellow Turk)—A Royal mare.

WANDERER.

WINNER OF THE RAILROAD STAKES AT NASHVILLE, 1872, THE MONMOUTH CUP AT LONG BRANCH, AND WESTCHESTER CUP AT JEROME PARK, 1873.

The property of Dr. J. R. McKee. Application to Dr. J. R. McKee, Spring Station, Ky.

WANDERER by Lexington, son of Boston, bred in the Woodburn Stud, Ky., foaled 1868, dam Coral, dam of Uncas and Rambler by Vandal, son of Glencoe, out of imp. Cairngorme by Cotherstone, winner of the 2,000 Guineas and Derby, son of Touchstone, winner of the St. Leger.

Wanderer is descended from one of the most noted racing families of England, tracing back through an own sister to imp. Glencoe to the famous Web by Waxy. The family has furnished some of the most noted race-horses and successful sires of the English turf. From it came Whalebone, winner of the Derby in 1810, Whisker, winner of the same event in 1815, Cobweb, winner of the Oaks and 1,000 Guineas in 1824, Riddlesworth, winner of the 2,000 Guineas and Derby in 1836, Glencoe, winner of the 2,000 Guineas in 1831, Bay Middleton, winner of the 2,000 Guineas and Goodwood Cup in 1834, Blue Gown, winner of the Derby in 1868, Silvio, winner of both Derby and St. Leger in 1877, and a host of others.

Wanderer made his first appearance as a three-year old, and was unplaced in the Belmont Stakes at Jerome Park and the Jersey Derby at Long Branch, both events having been won by his half brother, Harry Bassett. At Long Branch he won a Club Purse, mile heats, in 1.48¾, 1.48½, defeating a field of seven. In his four-year old form, at New Orleans, he won a race of mile heats in 1.51, 1.47½, 1.47½, beating Frank Ross, winner of the first heat, Niagara and Glenrose. Won the Railroad Stakes at Nashville, Tenn., 2 mile heats, in 3.41½, 3.38½, beating Hollywood, Frogtown, etc., and walked over for Club Purse, 2 mile heats. As a five-year old he won the Monmouth Cup at Long Branch, 2¼ miles, in 4.34¼, beating Preakness, Hubbard and others; won the Westchester Cup at Jerome Park, 2¼ miles, in 4.04, beating True Blue, Eolus and others. At Saratoga ran second to Hubbard, 3 miles, in 5.34, beating Harry Bassett and King Henry; same meeting, ran second to Arizona, 1¼ miles, in 2.38, beating Boss Tweed and Eolus. As a six-year old won Club Purse at Savannah, Ga., 2 miles, in 3.43½, beating Granger and four others; and also, same meeting, a Club Purse, 1½ miles, in 2.18¾, beating Ortolan and Tabitha. Nashville, Tenn., won the Johnson Stakes, 2¼ miles, in 4.06½, beating Planchette, St. George and two others. At Jerome Park ran second to Shylock, son of Lexington, in the Westchester Cup, 2¼ miles, in 4.13, beating Lizzie Lucas, Abd-el-Koree and two others, track heavy. Saratoga, won Club Purse, 2¼ miles, in 4.00½, beating Fellowcraft, Jack Frost, Galway and Katie Pease; same place, ran second to Fellowcraft, 4 miles, in 7.19½, the fastest race at the distance ever run in a contested race. In this race Wanderer beat Katie Pease and was timed the distance in 7.20.

Wanderer has been a success in the stud. He sired One Dime, a winner, 1½ miles in 1.55½, 1¼ miles in 2.09¼, mile heats in 1.44½, 1.44½; Elkhorn Stakes, 1¾ miles, in 3.05½. Minnie C. (Mrs. Chubbs), winner of the Filly Stakes at Lexington. Juanita, 1¼ miles in 2.10. Lizzy S., Filly Stakes at Lexington, half-mile, in 0.49; Coquette Stakes at St. Louis, three-quarters of a mile, in 1.18½; 5 furlongs in 1.02; three-quarters of a mile in 1.16½; 1 mile in 1.43. Mary Lamphier, Farragut, Nomad, Wakefield, Cash Clay, Waterford, Prophet, Rambler, Wandering, winner 1 mile in 1.45¼, and Capital Stakes, 1 mile, in 1.44¾, Cruiser, J. J. Healey, May Queen, Danville, Lottie Wall, Pearl L., Red Stone, Wanderment, Wanderoo, Warder, Loup, Tourist, Annie Queen and others, all winners. His colts are the best the dams have produced. He is a rich golden chestnut, with the marks of his sire, is handsomely shaped and of very blood-like appearance, muscular and highly finished. His sire was the best race-horse and stallion this country has produced; his dam is by the best son of imp. Glencoe, and traces through an own sister to Glencoe and Web by Waxy to a natural Barb mare. This is one of the pure sources through which the blood of Lexington and Glencoe should be preserved. (See Uncas, his brother.)

WANDERER.

CHESTNUT HORSE, FOALED 1868; BRED BY A. J. ALEXANDER, WOODBURN STUD, KENTUCKY.

- LEXINGTON, b. h., 1850.
  - Boston, ch. h., 1833.
    - Timoleon, ch. h., 1813.
      - Sir Archy, b. h., 1805.
      - Daughter of, ch. m., —
    - Sist. to Tuckahoe, ch. m., 1814.
      - Ball's Florizel, ch. h., 1801.
      - Daughter of, b. m., 1799.
  - Alice Carneal, b. m., 1836.
    - *Sarpedon, br. h., 1828.
      - Emilius, b. h., 1820.
      - Icaria, b. m., 1824.
    - Rowena, ch. m., 1826.
      - Sumpter, ch. h., 1818.
      - Lady Grey, b. m., 1817.
- CORAL, br. m., 1864.
  - Vandal, b. h., 1850.
    - *Glencoe, ch. h., 1831.
      - Sultan, b. h., 1816.
      - Trampoline, ch. m., 1825.
    - Alaric's dam, b. m., 1837.
      - *Tranby, br. h., 1826.
      - Lucilla, b. m., —
  - *Cairngorm, b. m., 1859.
    - Cotherstone, b. h., 1840.
      - Touchstone, br. h., 1831.
      - Emma, ch. m., 1824.
    - Glenluce, b. m., 1849.
      - Slane, b. h., 1833.
      - Glencairne, b. m., 1838.

*Diomed, ch. h., 1777.	Florizel 1768 (Herod 1758) Sist. 1763 to Juno by Spectator 1749 (Crab 1722) Horatia 1758 by Blank 1740 (God. Ar. 1724) Dau. of Flying Childers 1715 (Darley Ar. 1689) Miss Belvoir by Grey Grantham (Brownlow Turk).
*Castianira, br. m., 1796.	Rockingham 1781 (Highflyer 1774) Tabitha 1782 by Trentham 1766 (T. Sweepstakes 1743) Dau. of Bosphorus 1754 (Babraham 1740) Sist. to Grecian Princess by W's Forester 1750 (Forester 1736) Dau. of Coalition Colt 1739 (God. Ar. 1724).
*Saltram, br. h., 1780.	Eclipse 1764 (Marske 1750) Virago 1764 by Snap 1750 (Snip 1736) Dau. of Regulus 1739 (God. Ar. 1724)—Sist. 1750 to Othello by Crab 1722 (Alcock Ar.)—Miss Slamerkin 1729 by Y. True Blue 1718 (Honeywood's Ar.).
Daughter of, —.	Symmes' Wildair 1767 (*Fearnought 1755)—Dau. of Tyler's Driver 1760 (Othello) Dau. of *Fallower 1761 (Blank 1740) Dau. of Vampire 1757 (Regulus 1739). [The Saltram mare was dam of Jenny Cockracy].
*Diomed, ch. h., 1777.	Florizel 1768 (Herod 1758)—Sist. 1763 to Juno by Spectator 1749 (Crab 1722) Horatia 1758 by Blank 1740 (God. Ar. 1724) Dau. of Flying Childers 1715 (Darley Ar. 1689)—Miss Belvoir by Grey Grantham (Brownlow Turk).
Daughter of, —.	*Shark 1771 (Marske 1750) Dau. of Harris's Eclipse 1770 (*Fearnought 1755) Dau. of *Fearnought 1755 (Regulus 1739)—Dau. of *Jolly Roger 1741 (Roundhead 1733) Dau. of *Sober John 1748 (Rib 1736).
*Alderman, b. h., 1787.	Pot-8-o-'s 1773 (Eclipse 1764) Lady Bolingbroke 1766 by Squirrel 1754 (Traveller 1735)—Cypron 1750 by Blaze 1733 (Flying Childers 1715)—Salome 1733 by Bethell's Ar.—Dau. of Graham's Champion 1707 (Harpur's Ar.).
Daughter of, ab. 1791.	*Clockfast 1774 (Gimcrack 1760) Dau. of Symmes' Wildair 1767 (*Fearnought 1755)—Y. Kitty Fisher 1767 by *Fearnought 1755 (Regulus 1739)—*Kitty Fisher 1755 by Cade 1734 (God. Ar. 1724) Dau. of Somerset Ar.
Orville, b. h., 1799.	Beningbrough 1791 (King Fergus 1775) Evelina 1791 by Highflyer 1774 (Herod 1758)—Termagant by Tantrum 1760 (Cripple 1750)—Cantatrice by Sampson 1745 (Blaze 1733)—Dau. 1757 of Regulus 1739 (God. Ar. 1724).
Emily, ch. m., 1810.	Stamford 1794 (Sir Peter T. 1784)—Dau. 1799 of Whiskey 1789 (Saltram* 1780) Grey Dorimant 1781 by Dorimant 1772 (Otho 1760)—Dizzy 1757 by Blank 1740 (God. Ar. 1724)—Anc. Dizzy 1741 by Driver 1727 (Wynn Ar.).
The Flyer, br. h., 1814.	Vandyke Jr. 1808 (Walton 1799)—Azalia 1804 by Beningbrough 1791 (King Fergus 1775) Gilliflower 1787 by Highflyer 1774 (Herod 1758)—Preference 1778 by Goldfinder 1764 (Snap 1750)—Dau. 1774 of Marske 1770 (Squirt 1732).
Parma, br. m., 1813.	Dick Andrews 1797 (Joe Andrews 1778) May 1804 by Beningbrough 1791 (King Fergus 1775) Primrose 1787 by Mambrino 1768 (Engineer 1756) Cricket 1776 by Herod 1758 (Tartar 1743)—Sophia 1764 by Blank 1740 (God. Ar. 1724).
Sir Archy, b. h., 1805.	*Diomed 1777 (Florizel 1768)—*Castianira 1796 by Rockingham 1781 (Highflyer 1774)—Tabitha 1782 by Trentham 1766 (T. Sweepstakes 1743)—Dau. of Bosphorus 1754 (Babraham 1740)—Dau. of Forester 1750 (Forester 1736).
Robin mare, ch. m., .	*Robin Redbreast 1796 (Sir Peter T. 1784) Dau. 1788 of *Obscurity 1777 (Eclipse 1764)—Slamerkin 1768 by *Wildair 1753 (Cade 1734)—*Cub mare 1762 by Cub 1739 (Fox 1714)—Dau. 1755 of Second 1732 (Flying Childers 1715).
Robin Grey, gr. h., 1805.	*Royalist 1790 (Saltram* 1780)—Belle Maria by Grey Diomed 1786 (*Medley 1776)—Queen by *St. George 1771 (Dragon 1757) Dau. of Cassius 1778 (Black-and-all-Black 1773)—Primrose by *Dove (Y. Cade 1747).
Maria, —. m., 1802.	Melzar 1791 (*Medley 1776)—Dau. of *Highflyer 1784 (Highflyer 1774)—Dau. of *Fearnought 1755 (Regulus 1739)—Dau. of Ariel 1756 (*Traveller 1747)—Dau. of *Jack of Diamonds (Cullen Ar.).
Selim, ch. h., 1802.	Buzzard* 1787 (Woodpecker 1773)—Dau. of Alexander 1782 (Eclipse 1764)—Dau. of Highflyer 1774 (Herod 1758)—Dau. of Alfred 1770 (Matchem 1748) Dau. 1770 of Engineer 1756 (Sampson 1745).
Bacchante, br. h., 1801.	W's Ditto 1800 (Sir Peter T. 1784) Dau. 1791 of Mercury 1778 (Eclipse 1764)—Dau. 1776 of Herod 1758 (Tartar 1743)—Folly 1771 by Marske 1750 (Squirt 1732)—Vixen 1753 by Regulus 1739 (God. Ar. 1724).
Tramp, b. h., 1810.	Dick Andrews 1797 (Joe Andrews 1778) Dau. 1803 of Gohanna 1790 (Mercury 1778)—Fraxinella 1793 by Trentham 1766 (T. Sweepstakes 1743)—Dau. 1785 of Woodpecker 1773 (Herod 1758)—Everlasting 1775 by Eclipse 1764 (Marske 1750).
Web, b. m., 1808.	Waxy 1790 (Pot-8-o-'s 1773)—Penelope 1798 by Trumpator 1782 (Conductor 1767)—Prunella 1788 by Highflyer 1774 (Herod 1758)—Promise 1768 by Snap 1750 (Snip 1736)—Julia 1756 by Blank 1740 (God. Ar. 1724).
Blacklock, b. h., 1814.	Whitelock 1803 (Hambletonian 1792)—Dau. 1799 of Coriander 1786 (Pot-8-o-'s 1773)—Wildgoose 1792 by Highflyer 1774 (Herod 1758)—Coheiress 1786 by Pot-8-o-'s 1773 (Eclipse 1764)—Manilla 1777 by Goldfinder 1764 (Snap 1750).
Miss Bowe's dam, br. m., 1812.	Orville 1799 (Beningbrough 1791)—Miss Grimstone 1796 by Weazle 1776 (Herod 1758)—Dau. of Ancaster 1768 (Blank 1740)—Dau. of Damascus Ar. 1754—Dau. of Sampson 1745 (Blaze 1733).
Trumpator, br. h., 1823.	Sir Solomon 1805 (*Tickle Toby 1786)—Dau. of Hickory 1804 (*Whip 1794)—*Trumpetta 1797 by Trumpator 1782 (Conductor 1767) Sist. to Lambinos 1786 by Highflyer 1774 (Herod 1758)—Dau. 1775 of Eclipse 1764 (Marske 1750).
Lucy, b. m., 1821.	Orphan 1810 (Ball's Florizel 1801) Lady Grey 1817 by Robin Grey 1805 (*Royalist 1790)—Maria 1802 by Melzar 1791 (*Medley 1776)—Dau. of *Highflyer 1784 (Highflyer 1774)—Dau. of *Fearnought 1755 (Regulus 1739).
Camel, br. h., 1822.	Whalebone 1807 (Waxy 1790)—Dau. 1812 of Selim 1802 (Buzzard* 1787) Maiden 1801 by Sir Peter T. 1784 (Highflyer 1774)—Dau. 1788 of Phenomenon 1780 (Herod 1758)—Matron 1782 by Florizel 1768 (Herod 1758).
Banter, br. m., 1821.	Master Henry 1815 (Orville 1799)—Boadicea 1807 by Alexander 1782 (Eclipse 1764)—Brunette by Amaranthus 1766 (Old England 1741)—Mayfly 1771 by Matchem 1748 (Cade 1734)—Dau. 1755 of Anc. Starling 1738 (Starling 1727).
Whisker, b. h., 1812.	Waxy 1790 (Pot-8-o-'s 1773)—Penelope 1798 by Trumpator 1782 (Conductor 1767)—Prunella 1788 by Highflyer 1774 (Herod 1758)—Promise 1768 by Snap 1750 (Snip 1736)—Julia 1756 by Blank 1740 (God. Ar. 1724).
Gibside Fairy, b. m., 1811.	Hermes 1790 (Mercury 1778) Vicissitude 1800 by Pipator 1786 (Imperator 1776)—Beatrice 1791 by Sir Peter T. 1784 (Highflyer 1774)—Pyrrha 1771 by Matchem 1748 (Cade 1734) Duchess 1748 by Whitenose 1742 (God. Ar. 1724).
Royal Oak, br. h., 1823.	Catton 1809 (Golumpus 1802)—Dau. 1818 of Smolensko 1810 (Sorcerer 1796)—Lady Mary 1800 by Beningbrough 1791 (King Fergus 1775)—Dau. of Highflyer 1774 (Herod 1758)—Sist. to Miss Bell by Marske 1750 (Squirt 1732).
Daughter of, b. m., 1819.	Orville 1799 (Beningbrough 1791) Epsom Lass 1803 by Sir Peter T. 1784 (Highflyer 1774)—Alexina 1788 by King Fergus 1775 (Eclipse 1764)—Lardella 1780 by Y. Marske 1771 (Marske 1750) Dau. of Y. Cade 1747 (Cade 1734).
Sultan, b. h., 1816.	Selim 1802 (Buzzard* 1787) Bacchante 1809 by W's Ditto 1800 (Sir Peter T. 1784)—Dau. 1791 of Mercury 1778 (Eclipse 1764)—Dau. 1776 of Herod 1758 (Tartar 1743) Folly 1771 by Marske 1750 (Squirt 1732).
Trampoline, ch. m., 1825.	Tramp 1810 (Dick Andrews 1797)—Web b. m. 1808 by Waxy 1790 (Pot-8-o-'s 1773)—Penelope b. m. 1798 by Trumpator 1782 (Conductor 1767)—Prunella b. m. 1788 by Highflyer 1774 (Herod 1758)—Promise br. m. 1768 by Snap 1750 (Snip 1736)—Julia b. m. 1756 by Blank 1740 (God. Ar. 1724)—Spectator's dam, b. m. 1735 by Partner 1718

(Jigg)—Bonny Lass b. m. 1723 by Bay Bolton 1705 (Grey Hautboy)—Dau. of Darley Ar. 1699—Dau. of Byerly Turk—Dau. of Taffolet Barb—Dau. of Place's White Turk—Natural Barb mare of Mr. Tregonwell's.

WHISPER.

WINNER OF NASHVILLE CUP, LOUISVILLE CUP AT LOUISVILLE, KY., AND OTHER GOOD RACES.

In the Fleetwood Stud. Application to L. P. Tarlton, proprietor, Fleetwood Stud, Frankfort, Ky.

WHISPER by Planet, son of Revenue, bred in the Woodburn Stud, Ky., foaled 1870, dam Mattie Gross, dam of Mate, Grenada, etc., by Lexington, out of Dick Doty's dam by American Eclipse. Whisper did not run as a two-year old. At three years old was unplaced in the Phœnix Hotel Stakes, 1¼ miles, won by Artist in 2.12½. Monmouth Park, unplaced in the Jersey Derby, 1½ miles, won by Tom Bowling in 2.45¾; was second in the Jersey Jockey Club Purse, mile heats, best three in five, won by Arizona in 1.46, 1.48¾, 1.47¼, 1.53½, 1.55, Whisper won the first two heats; was third in the Cottagers' Cup, 1½ miles, won by imp. Stonehenge in 2.44½; unplaced in a Consolation Purse, 1½ miles, won by Shylock in 2.44½. Saratoga, unplaced in the Kenner Stakes, 2 miles, won by imp. Ill Used in 2.39. At (4) Jerome Park, walked over for purse of $700, 1¾ miles. Monmouth Park, was second in the Monmouth Cup, 2½ miles, won by Tom Bowling in 4.42½; was second in a dash of four miles, won by Fellowcraft in 7.43. Jerome Park, was unplaced in a dash of 1½ miles, won by Grinstead in 2.40¾; was third in a dash of 2½ miles, won by Acrobat in 4.33¾. Baltimore, distanced first heat of the Bowie Stakes, 4 mile heats, won by Jack Frost in 7.33, 7.41, 8.11, Bessie Lee won the first heat. At (5) Louisville Ky., unplaced in a dash of 1 mile and a furlong, won by Calvin in 1.56½; unplaced in Louisville Club Stakes, 2¼ miles, won by Ballinkeel, in 4.01½; won a Handicap mile heats, in 1.52¾, 1.53¾. At (6) Louisville, Ky., was third in a dash of 1¼ miles, won by Brakesman in 2.11; was third in a dash of 2 miles and a furlong, won by Elemi in 3.49½; was second in a race of mile heats, won by Kilburn in 1.45½, 1.44¼, 1.48¼, Whisper won the first heat. Lexington, Ky., Fall Meeting, was third in a dash of 1½ miles, won by Phillis in 2.37¼; won a Selling Race, 1 mile and a furlong, in 1.58. Louisville, Ky., won a Handicap Sweepstakes, 1¾ miles, in 3.09. Nashville, Tennessee, won dash of 1½ miles for all ages, in 2.44½. At (7) Nashville, Tenn., won the Nashville Cup, 2¼ miles, in 4.08; was second in a dash of 1½ miles, won by McWhirter in 2.38½. Louisville, Ky., won a dash of 2 miles, in 3.36¾; won the Louisville Cup, 2¼ miles, in 3.59¼. Saratoga, won Handicap for all ages, 1¾ miles, in 3.12¾; was third in a dash of 1 mile and a furlong, won by Vera Cruz, in 1.57½; won a dash of 2¼ miles in 4.02. Louisville Fall Meeting, unplaced in a Handicap, 1¾ miles, won by Longbow in 3.07¾; Whisper was third in a dash of 2 miles, won by Ten Broeck in 3.36. At (8) Louisville, Ky., was second in the Louisville Cup, 2¼ miles, won by Mahlstick in 4.07½; won a Handicap dash of 4 miles in 7.36. St. Louis, Mo., won the Blow Stakes, 3 mile heats, in 5.39, 5.35½. Whisper is by the best son of Revenue, Planet, a winner at all distances over the best horses of his day. Whisper's dam, Mattie Gross, is the dam of Mate and Grenada, and his grandam was the dam of Dick Doty, the famous four miler, whilst his great grandam was the dam of Eliza Jenkins and the Big Archy mare, which produced the great race-horse Creath, who beat Alice Carneal and won a second heat of 3 miles in 5.43, in 1841. Whisper is a bay, full 16 hands high, with a star and right hind leg white above the hoof. He is one of the finest and best shaped horses in America, and could not only run fast but stay a distance. In addition to being a grandson of imp. Trustee, he is much inbred to Sir Archy and Diomed, having ten crosses of that most desirable blood. He has also a double cross of Whisker.

Whisper is the sire of the winners Clarion, Canto, Clamor and others.

WHISPER.

BAY HORSE, FOALED 1870; BRED BY A. J. ALEXANDER, WOODBURN STUD, KENTUCKY.

PLANET, ch. h., 1855.

Revenue, b. h., 1843.

*Trustee, ch. h., 1829.

Catton, b. h., 1809.

Golumpus, b. h., 1802. — Gohanna 1790 (Mercury 1778)—Catherine 1795 by Woodpecker 1773 (Herod 1758)—Camilla 1778 by Trentham 1766 (T. Sweepstakes 1743)—Coquette 1765 by Compton Barb—Sist. 1743 to Regulus by God. Ar. 1724.

Lucy Gray, ch. m., 1804. — Timothy 1794 (Delpini 1781)—Lucy 1789 by Florizel 1768 (Herod 1758)—Frenzy 1774 by Eclipse 1764 (Marske 1750)—Dau. of Engineer 1756 (Sampson 1745)—Dau. 1763 of Blank 1740 (God. Ar. 1724).

Emma, ch. m., 1824.

Whisker, b. h., 1812. — Waxy 1790 (Pot-8-o's 1773)—Penelope 1798 by Trumpator 1782 (Conductor 1767)—Prunella 1788 by Highflyer 1774 (Herod 1758)—Promise 1768 by Snap 1750 (Snip 1736)—Julia 1756 by Blank 1740 (God. Ar. 1724).

Gibside Fairy, b. m., 1811. — Hermes 1790 (Mercury 1778)—Vicissitude 1800 by Pipator 1786 (Imperator 1776)—Beatrice 1791 by Sir Peter T. 1784 (Highflyer 1774)—Pyrrha 1771 by Matchem 1748 (Cade 1734)—Duchess 1748 by Whitenose 1742 (God. Ar. 1724).

Rosalie Somers, b. m., 1831.

Sir Charles, ch. h., 1816.

Sir Archy, b. h., 1805. — *Diomed 1777 (Florizel 1768)—*Castianira 1796 by Rockingham 1781 (Highflyer 1774)—Tabitha 1782 by Trentham 1766 (T. Sweepstakes 1743)—Dau. of Bosphorus 1754 (Babraham 1740)—Dau. of Forester 1750 (Forester 1736).

Daughter of, ch. m., —. — *Citizen 1785 (Pacolet 1763)—Dau. of Commutation (Symmes' Wildair 1767)—Dau. of *Dare Devil 1787 (Magnet 1770)—Sally Shark by *Shark 1771 (Marske 1750)—Betsey Pringle by *Fearnought 1755 (Regulus 1739).

Mischief, b. m., 1826.

Virginian, b. h., 1815. — Sir Archy 1805 (*Diomed 1777)—Meretrix by Magog 1799 (Chanticleer)—Narcissa 1796 by *Shark 1771 (Marske 1750)—Rosetta 1785 by *Centinel 1758 (Blank 1740)—Diana by Clodius 1778 (*Janus 1746).

Trifle's grandam, gr. m., —. — *Bedford 1792 (Dungannon 1780)—Dau. of Bellair 1786 (*Medley 1776)—Dau. of *Shark 1771 (Marske 1750)—Dau. of Symmes' Wildair 1767 (*Fearnought 1755)—Dau. of Lexington 1774 (Washington).

Nina, b. m., 1848.

Boston, ch. h., 1833.

Timoleon, ch. h., 1813.

Sir Archy, b. h., 1805. — *Diomed 1777 (Florizel 1768)—*Castianira 1796 by Rockingham 1781 (Highflyer 1774)—Tabitha 1782 by Trentham 1766 (T. Sweepstakes 1743)—Dau. of Bosphorus 1754 (Babraham 1740)—Dau. of Forester 1750 (Forester 1736).

Daughter of, ch. m., —. — *Saltram 1780 (Eclipse 1764)—Dau. of Symmes' Wildair 1767 (*Fearnought 1755)—Dau. of Tyler's Driver 1760 (*Othello)—Dau. of *Fallower 1761 (Blank 1740)—Dau. of *Vampire 1757 (Regulus 1739).

Sist. to Tuckahoe, ch. m., 1814.

Ball's Florizel, ch. h., 1801. — *Diomed 1777 (Florizel 1768)—Dau. of *Shark 1771 (Marske 1750)—Dau. of Harris' Eclipse 1770 (*Fearnought 1755)—Dau. of *Fearnought 1755 (Regulus 1739)—Dau. of *Jolly Roger 1741 (Roundhead 1733).

Daughter of, b. m., 1799. — *Alderman 1787 (Pot-8-o's 1773)—Dau. of *Clockfast 1774 (Gimcrack 1760)—Dau. of Symmes' Wildair 1767 (*Fearnought 1755)—Y. Kitty Fisher 1767 by *Fearnought 1755 (Regulus 1739)—*Kitty Fisher 1755 by Cade 1734 (God. Ar. 1724).

*Frolicksome Fanny, b. m., 1832.

Lottery, br. h., 1820.

Tramp, b. h., 1810. — Dick Andrews 1797 (Joe Andrews 1778)—Dau. 1803 of Gohanna 1790 (Mercury 1778)—Fraxinella 1793 by Trentham 1766 (T. Sweepstakes 1743)—Dau. 1785 of Woodpecker 1773 (Herod 1758)—Everlasting 1775 by Eclipse 1764 (Marske 1750).

Mandane, ch. m., 1800. — Pot-8-o's 1773 (Eclipse 1764)—Y. Camilla 1787 by Woodpecker 1773 (Herod 1758)—Camilla 1778 by Trentham 1766 (T. Sweepstakes 1743)—Coquette 1765 by Compton Barb—Sist. 1743 to Regulus by God. Ar. 1724.

Sist. to Catterick, b. m., 1821.

Whisker, b. h., 1812. — Waxy 1790 (Pot-8-o's 1773)—Penelope 1798 by Trumpator 1782 (Conductor 1767)—Prunella 1788 by Highflyer 1774 (Herod 1758)—Promise 1768 by Snap 1750 (Snip 1736)—Julia 1756 by Blank 1740 (God. Ar. 1724).

Daughter of, —, —. — Bay Trophonius 1807 (Beningbrough 1791)—Dau. 1791 of Slope 1782 (Highflyer 1774)—Lardella 1780 by Y. Marske 1771 (Marske 1750)—Dau. of Y. Cade 1747 (Cade 1734)—Beaufremont's dam by Broth. to Fearnought (Bay Bolton 1705).

MATTIE GROSS, b. m., 1860.

Lexington, b. h., 1850.

Boston, ch. h., 1833.

Timoleon, ch. h., 1813.

Sir Archy, b. h., 1805. — *Diomed 1777 (Florizel 1768)—*Castianira 1796 by Rockingham 1781 (Highflyer 1774)—Tabitha 1782 by Trentham 1766 (T. Sweepstakes 1743)—Dau. of Bosphorus 1754 (Babraham 1740)—Dau. of Forester 1750 (Forester 1736).

Daughter of, ch. m., —. — *Saltram 1780 (Eclipse 1764)—Dau. of Symmes' Wildair 1767 (*Fearnought 1755)—Dau. of Tyler's Driver 1760 (*Othello)—Dau. of *Fallower 1761 (Blank 1740)—Dau. of *Vampire 1757 (Regulus 1739).

Sist. to Tuckahoe, ch. m., 1814.

Ball's Florizel, ch. h., 1801. — *Diomed 1777 (Florizel 1768)—Dau. of *Shark 1771 (Marske 1750)—Dau. of Harris' Eclipse 1770 (*Fearnought 1755)—Dau. of *Fearnought 1755 (Regulus 1739)—Dau. of *Jolly Roger 1741 (Roundhead 1733).

Daughter of, b. m., 1799. — *Alderman 1787 (Pot-8-o's 1773)—Dau. of *Clockfast 1774 (Gimcrack 1760)—Dau. of Symmes' Wildair 1767 (*Fearnought 1755)—Y. Kitty Fisher 1767 by *Fearnought 1755 (Regulus 1739)—*Kitty Fisher 1755 by Cade 1734 (God. Ar. 1724).

Alice Carneal, b. m., 1836.

*Sarpedon, br. h., 1828.

Emilius, b. h., 1820. — Orville 1799 (Beningbrough 1791)—Emily 1810 by Stamford 1794 (Sir Peter T. 1784)—Dau. 1799 of Whiskey 1789 (Saltram* 1780)—Grey Dorimant 1781 by Dorimant 1772 (Otho 1760)—Dizzy 1757 by Blank 1740 (God. Ar. 1724).

Icaria, b. m., 1824. — The Flyer 1814 (Vandyke Jr. 1808)—Parma 1813 by Dick Andrews 1797 (Joe Andrews 1778)—May 1804 by Beningbrough 1791 (King Fergus 1775)—Primrose 1787 by Mambrino 1768 (Engineer 1756)—Cricket 1776 by Herod 1758 (Tartar 1743).

Rowena, ch. m., 1826.

Sumpter, ch. h., 1818. — Sir Archy 1805 (*Diomed 1777)—Dau. of *Robin Redbreast 1796 (Sir Peter T. 1784)—Dau. 1788 of *Obscurity 1777 (Eclipse 1764)—Slamerkin 1768 by *Wildair 1753 (Cade 1734)—*Cub mare 1762 by Cub 1739 (Fox 1714).

Lady Grey, b. m., 1817. — Robin Grey 1805 (*Royalist 1790)—Maria 1802 by Melzar 1791 (*Medley 1776)—Dau. of *Highflyer 1784 (Highflyer 1774)—Dau. of *Fearnought 1755 (Regulus 1739)—Dau. of Ariel 1756 (*Traveller 1747).

Dick Doty's dam, ch. m., 1844.

American Eclipse, ch. h., 1814.

Duroc, ch. h., 1806.

*Diomed, ch. h., 1777. — Florizel 1768 (Herod 1758)—Dau. 1763 of Spectator 1749 (Crab 1722)—Horatia 1758 by Blank 1740 (God. Ar. 1724)—Dau. of Flying Childers 1715 (Darley Ar. 1699)—Miss Belvoir by Grey Grantham (Brownlow Turk).

Amanda, ch. m., 1800. — Grey Diomed 1786 (*Medley 1776)—Dau. of Va. Cade (Partner 1755)—Dau. of Hickman's Independence 1776 (*Fearnought 1755)—Dau. of *Badger (Bosphorus 1754).

Miller's Damsel, gr. m., 1802.

*Messenger, gr. h., 1780. — Mambrino 1768 (Engineer 1756)—Dau. 1774 of Turf 1760 (Matchem 1748)—Dau. 1761 of Regulus 1739 (God. Ar. 1724)—Dau. 1753 of Starling 1727 (Bay Bolton 1705)—Snap's dam 1739 by Fox 1714 (Clumsey).

*Pot-8-o's mare, ch. m., 1792. — Pot-8-o's 1773 (Eclipse 1764)—Dau. 1778 of Gimcrack 1760 (Cripple 1750)—Snapdragon 1759 by Snap 1750 (Snip 1736)—Dau. 1749 of Regulus 1739 (God. Ar. 1724)—Dau. of Bartlett's Childers (Darley Ar. 1699).

Nell, ch. m., —.

Orphan, ch. h., 1810.

Ball's Florizel, ch. h., 1801. — *Diomed 1777 (Florizel 1768)—Dau. of *Shark 1771 (Marske 1750)—Dau. of Harris' Eclipse 1770 (*Fearnought 1755)—Dau. of *Fearnought 1755 (Regulus 1739)—Dau. of *Jolly Roger 1741 (Roundhead 1733).

Fair Rachel, —, —. — *Diomed 1777 (Florizel 1768)—Susan Jones by *Shark 1771 (Marske 1750)—Dau. of Thornton Wildair (Symmes' Wildair 1767).

Daughter of, —, —.

*Buzzard, ch. h., 1787. — Woodpecker 1773 (Herod 1758)—Misfortune 1775 by Dux 1761 (Matchem 1748)—Curiosity 1760 by Snap 1750 (Snip 1736)—Dau. 1740 of Regulus 1739 (God. Ar. 1724)—Dau. of Bartlett's Childers (Darley Ar. 1699).

Daughter of, —, —. — Silvertail 1773 (*Fearnought 1755)—Dau. of *Dove (Y. Cade 1747).

WILDIDLE.

WINNER OF THE NURSERY STAKES, FORDHAM AND JOCKEY CLUB HANDICAPS AT JEROME PARK.

In the Wildidle Stud, J. C. Judson, proprietor, Santa Clara, Cal. Terms annually.

WILDIDLE by imp. Australian, son of West Australian, bred in the Woodburn Stud, Ky., foaled 1870, dam Idlewild by Lexington, out of Florine by imp. Glencoe. Wildidle made his bow to the public by winning the Nursery Stakes. 1 mile, at Jerome Park, in 1.48¾, defeating Long Branch, Catesby and eleven others. He was out of condition and ran unplaced in one race as a three-year old. At four years old was second to Catesby in Maturity Stakes, 3 miles, in 5.36, beating Carver and Ransom; was unplaced in purse, 1¾ miles, won by Grinstead in 3.10; was second to Acrobat in purse, 2¼ miles, in 4.33¾, beating Whisper, all run at Jerome Park. At five years old, 104 lbs., Jerome Park, won the Fordham Handicap, 1¼ miles, in 2.12, beating Spindrift, aged, 118 lbs., Preakness, aged, 131 lbs., and nine others; won the Jockey Club Handicap, 111 lbs., 2 miles, in 3.38½, beating Preakness, 130 lbs., Grinstead (4), 110 lbs., and Tubman (aged). 115 lbs. Monmouth Park, was beaten by Aaron Pennington and Ballenkeel in Monmouth Cup, 2¼ miles, in 4.34; was second to Rutherford in 4 mile dash, run in 7.34¾, Big Fellow and Bessie Lee behind him; won Club Purse, 2 mile heats, 3.40¾. 3.47¼, beating Dublin. Saratoga, was unplaced in Saratoga Cup, 2¼ miles, in 3.56¼, dead heat, and stakes divided between Preakness and Springbok: this is the fastest race ever run at the distance; was second to Rutherford in Club Purse, 3 miles, in 5.38, beating Madge; was second to Grinstead (4), 110 lbs., in the Summer Handicap, 2 miles, in 3.37½, Wildidle (5), 114 lbs., beating Mattie W. (3), 93 lbs.; won the Club Purse, 4 miles, in 7.39, beating Rutherford. San Francisco, Cal., won the Wise Plate, 4 miles, in 7.25½, beating Grinstead, Sherman and Revenue, Jr. Wildidle comes honestly by his racing qualities. His dam was second to no mare ever raised in America. She ran four miles over the Centreville Course, L. I., 1863, when five years old, with 117 lbs., in 7.26¼, and this was the best time ever made by a mare until beaten by Ferida, in 1880, when she ran a first heat in 7.23¼.

Wildidle, for his chances, has been a very successful sire. The best of his get are Tillie C., May D., Record, Jack Douglass, Ellen Douglass, Belshaw, Jim Douglas (winner of ten out of fourteen races in 1881), Wildidler, Lottie J., Jack Brady, Carmen, Not Idle, Patti, Lady R., Neilson, Brown Idle, King Idle, Alfaretta, Ella Hill, Fanny S., Flambeau, Gambo, Muta, Philander, Raindrop, Rosemead, Telephone, Wild Oats, Garcia, Rosebud, Rover, Stufax, Tearless, Folly, Gorget, Nomad, Wild Rose, Edith, Lord of the Harem, Jessie Carr, Nero, Fanny F., Oscar, etc., etc.

Wildidle is inbred to Waxy on the sire's side, and has much of the blood of Sir Archy and Diomed, in addition to a cross of Eclipse, through his famous son Medoc, and three crosses of imp. Medley, son of Gimcrack by Cripple, son of the Godolphin Arabian. His family has always been a grand racing and producing one. He is brother in blood to Spendthrift, sire of Kingston, Lamplighter, etc., and to Miser, sire of Yorkville Belle, and to Fellowcraft (4 miles in 7.19½).

WILDIDLE.

BAY HORSE, FOALED 1870; BRED IN THE WOODBURN STUD, SPRING STATION, KY.

*AUSTRALIAN, ch. h., 1858.	West Australian, b. h., 1850.	Melbourne, br. h., 1834.	Humphrey Clinker, b. h., 1822.	Comus, ch. h., 1809.	Sorcerer 1796 (Trumpator 1782)—Epsom Lass 1803 by Sir Peter T. 1784 (Highflyer 1774)—Alexina 1788 by King Fergus 1775 (Eclipse 1764)—Lardella 1780 by Y. Marske 1771 (Marske 1750)—Dau. of Y. Cade 1747 (Cade 1734).
				Clinkerina, br. m., 1812.	Clinker 1805 (Sir Peter T. 1784)—Pewet 1786 by Tandem 1773 (Syphon 1750)—Termagant by Tantrum 1760 (Cripple 1750)—Cantatrice by Sampson 1745 (Blaze 1733)—Dau. 1757 of Regulus 1739 (God. Ar. 1724).
			Daughter of, b. m., 1825.	Cervantes, b. h., 1806.	Don Quixote 1784 (Eclipse 1764)—Evelina 1791 by Highflyer 1774 (Herod 1758)—Termagant by Tantrum 1760 (Cripple 1750)—Cantatrice by Sampson 1745 (Blaze 1733)—Dau. 1757 of Regulus 1739 (God. Ar. 1724).
				Daughter of, br. m., 1818.	Golumpus 1802 (Gohanna 1790)—Dau. 1810 of Paynator 1791 (Trumpator 1782)—Circle 1802 by St. George 1789 (Highflyer 1774)—Abigail 1788 by Woodpecker 1773 (Herod 1758)—Firetail 1772 by Eclipse 1764 (Marske 1750).
		Mowerina, b. m., 1843.	Touch-stone, br. h., 1831.	Camel, br. h., 1822.	Whalebone 1807 (Waxy 1790)—Dau. 1812 of Selim 1802 (Buzzard 1787)—Maiden 1801 by Sir Peter T. 1784 (Highflyer 1774)—Dau. 1788 of Phenomenon 1780 (Herod 1758)—Matron 1782 by Florizel 1768 (Herod 1758).
				Banter, br. m., 1826.	Master Henry 1815 (Orville 1799)—Boadicea 1807 by Alexander 1782 (Eclipse 1764)—Brunette by Amaranthus 1766 (Old England 1741)—Mayfly 1771 by Matchem 1748 (Cade 1734)—Dau. 1755 of Anc. Starling 1738 (Starling 1727).
			Emma, ch. m., 1824.	Whisker, b. h., 1812.	Waxy 1790 (Pot-8-o-'s 1773)—Penelope 1798 by Trumpator 1782 (Conductor 1767)—Prunella 1788 by Highflyer 1774 (Herod 1758)—Promise 1768 by Snap 1750 (Snip 1736)—Julia 1756 by Blank 1740 (God. Ar. 1724).
				Gibside Fairy, b. m., 1811.	Hermes 1790 (Mercury 1778)—Vicissitude 1800 by Pipator 1786 (Imperator 1776)—Beatrice 1791 by Sir Peter T. 1784 (Highflyer 1774)—Pyrrha 1771 by Matchem 1748 (Cade 1734)—Duchess 1748 by Whitenose 1742 (God. Ar. 1724).
	Emilia,* b. m., 1840.	Y. Emilius, b. h., 1833.	Emilius, b. h., 1820.	Orville, b. h., 1799.	Beningbrough 1791 (King Fergus 1775)—Evelina 1791 by Highflyer 1774 (Herod 1758)—Termagant by Tantrum 1760 (Cripple 1750)—Cantatrice by Sampson 1745 (Blaze 1733)—Dau. 1757 of Regulus 1739 (God. Ar. 1724).
				Emily, ch. m., 1810.	Stamford 1794 (Sir Peter T. 1784)—Dau. 1799 of Whiskey 1789 (Saltram* 1780)—Grey Dorimant 1781 by Dorimant 1772 (Otho 1760)—Dizzy 1757 by Blank 1740 (God Ar. 1724)—Dizzy 1741 by Driver 1727 (Wynn Ar.).
			Shoveler, b. m., 1816.	Scud, b. h., 1804.	Beningbrough 1791 (King Fergus 1775)—Eliza 1791 by Highflyer 1774 (Herod 1758)—Augusta 1781 by Eclipse 1764 (Marske 1750)—Hardwicke's dam by Herod 1758 (Tartar 1743)—Nettle 1763 by Bajazet 1740 (God. Ar. 1724).
				Goosander, b. m., 1805.	Hambletonian 1792 (King Fergus 1775)—Rally 1790 by Trumpator 1782 (Conductor 1767)—Fancy 1780 by Florizel 1768 (Herod 1758)—Dau. 1763 of Spectator 1749 (Crab 1722)—Horatia 1758 by Blank 1740 (God. Ar. 1724).
		Persian, b. m., 1829.	Whisker, b. h., 1812.	Waxy, b. h., 1790.	Pot-8-o-'s 1773 (Eclipse 1764)—Maria 1777 by Herod 1758 (Tartar 1743)—Lisette 1772 by Snap 1750 (Snip 1736)—Miss Windsor 1754 by God. Ar. 1724—Sist. to Volunteer by Y. Belgrade (Belgrade Turk).
				Penelope, b. m., 1798.	Trumpator 1782 (Conductor 1767)—Prunella 1788 by Highflyer 1774 (Herod 1758)—Promise 1768 by Snap 1750 (Snip 1736)—Julia 1756 by Blank 1740 (God. Ar. 1724)—Spectator's dam 1735 by Partner 1718 (Jigg).
			Variety, ch. m., 1816.	[Selim or] Soothsayer, ch. h., 1808.	Sorcerer 1796 (Trumpator 1782)—Golden Locks 1793 by Delpini 1781 (Highflyer 1774)—Violet 1787 by Shark* 1771 (Marske 1750)—Dau. 1772 of Syphon 1750 (Squirt 1732)—Charlotte 1756 by Blank 1740 (God. Ar. 1724).
				Sprite, ch. m., 1807.	Bobtail 1795 (Precipitate* 1787)—Catherine 1795 by Woodpecker 1773 (Herod 1758)—Camilla 1778 by Trentham 1766 (T. Sweepstakes 1743)—Coquette 1764 by Compton Barb—Sist. 1743 to Regulus by God. Ar. 1724.
IDLEWILD (Sist. to Aerolite), b. m., 1859.	Lexington, b. h., 1850.	Boston, ch. h., 1833.	Timoleon, ch. h., 1813.	Sir Archy, b. h., 1805.	*Diomed 1777 (Florizel 1768)—*Castianira 1796 by Rockingham 1781 (Highflyer 1774)—Tabitha 1782 by Trentham 1766 (T. Sweepstakes 1743)—Dau. of Bosphorus 1754 (Babraham 1740)—Dau. of W's Forester 1750 (Forester 1736).
				Daughter of, ch. m., —.	*Saltram 1780 (Eclipse 1764)—Dau. of Symmes' Wildair 1767 (*Fearnought 1755)—Dau. of Tyler's Driver 1760 (*Othello)—Dau. of *Fallower 1761 (Blank 1740)—Dau. of *Vampire 1757 (Regulus 1739).
			Sist. to Tuckahoe, ch. m., 1814.	Ball's Florizel, ch. h., 1801.	*Diomed 1777 (Florizel 1768)—Dau. of *Shark 1771 (Marske 1750)—Dau. of Harris's Eclipse 1770 (*Fearnought 1755)—Dau. of *Fearnought 1755 (Regulus 1739)—Dau. of *Jolly Roger 1741 (Roundhead 1733).
				Daughter of, b. m., 1799.	*Alderman 1787 (Pot-8-o-'s 1773)—Dau. of *Clockfast 1774 (Gimcrack 1760)—Dau. of Symmes' Wildair 1767 (*Fearnought 1755)—Y. Kitty Fisher by *Fearnought 1755 (Regulus 1739)—*Kitty Fisher 1755 by Cade 1734 (God. Ar. 1724).
		Alice Carneal, b. m., 1836.	*Sarpedon, br. h., 1828.	Emilius, b. h., 1820.	Orville 1799 (Beningbrough 1791)—Emily 1810 by Stamford 1794 (Sir Peter T. 1784)—Dau. 1799 of Whiskey 1789 (Saltram* 1780)—Grey Dorimant 1781 by Dorimant 1772 (Otho 1760)—Dizzy 1757 by Blank 1740 (God. Ar. 1724).
				Icaria, b. m., 1824.	The Flyer 1814 (Vandyke Jr. 1808)—Parma 1813 by Dick Andrews 1797 (Joe Andrews 1778)—May 1804 by Beningbrough 1791 (King Fergus 1775)—Primrose 1787 by Mambrino 1768 (Engineer 1756)—Cricket 1776 by Herod 1758 (Tartar 1743).
			Rowena, ch. m., 1826.	Sumpter, ch. h., 1818.	Sir Archy 1805 (*Diomed 1777)—Dau. of *Robin Redbreast 1796 (Sir Peter T. 1784)—Dau. 1788 of *Obscurity 1777 (Eclipse 1764)—Slamerkin 1768 by *Wildair 1753 (Cade 1734)—*Cub mare 1762 by Cub 1739 (Fox 1714).
				Lady Grey, b. m., 1817.	Robin Grey 1805 (*Royalist 1790)—Maria 1802 by Melzar 1791 (*Medley 1773)—Dau. of *Highflyer 1784 (Highflyer 1774)—Dau. of *Fearnought 1755 (Regulus 1739)—Dau. of Ariel 1754 (*Traveller 1747).
	Florine, ch. m., 1851.	*Glencoe, ch. h., 1831.	Sultan, b. h., 1816.	Selim, ch. h., 1802.	Buzzard* 1787 (Woodpecker 1773)—Dau. of Alexander 1782 (Eclipse 1764)—Dau. of Highflyer 1774 (Herod 1758)—Dau. of Alfred 1770 (Matchem 1748)—Dau. 1770 of Engineer 1756 (Sampson 1745).
				Bacchante, br. m., 1809.	W's Ditto 1800 (Sir Peter T. 1784)—Dau. 1791 of Mercury 1778 (Eclipse 1764)—Dau. 1776 of Herod 1758 (Tartar 1743)—Folly 1771 by Marske 1750 (Squirt 1732)—Vixen 1753 by Regulus 1739 (God. Ar. 1724).
			Trampoline, ch. m., 1825.	Tramp, b. h., 1810.	Dick Andrews 1797 (Joe Andrews 1778)—Dau. 1803 of Gohanna 1790 (Mercury 1778)—Fraxinella 1793 by Trentham 1766 (T. Sweepstakes 1743)—Dau. 1785 of Woodpecker 1773 (Herod 1758)—Everlasting 1775 by Eclipse 1764 (Marske 1750).
				Web, b. m., 1808.	Waxy 1790 (Pot-8-o-'s 1773)—Penelope 1798 by Trumpator 1782 (Conductor 1767)—Prunella 1788 by Highflyer 1774 (Herod 1758)—Promise 1768 by Snap 1750 (Snip 1736)—Julia 1756 by Blank 1740 (God Ar. 1724).
		Melody, ch. m., 1837.	Medoc, ch. h., 1829.	Am. Eclipse, ch. h., 1814.	Duroc 1806 (*Diomed 1777)—Miller's Damsel 1802 by *Messenger 1780 (Mambrino 1768)—*Dau. 1792 of Pot-8-o-'s 1773 (Eclipse 1764)—Dau. 1778 of Gimcrack 1760 (Cripple 1750)—Snapdragon 1759 by Snap 1750 (Snip 1736).
				Y. Maid of the Oaks, ch. m., 1817.	*Expedition 1795 (Pegasus 1784)—Maid of the Oaks 1801 by *Spread Eagle 1792 (Volunteer 1780)—Annette by *Shark 1771 (Marske 1750)—Dau. of Rockingham (Partner 1755)—Dau. of True Whig (Fitzhugh's Regulus 1765).
			Rodolph's dam, br. m., 1824.	Haxall's Moses, b. h., 1816.	*Sir Harry 1795 (Sir Peter T. 1784)—*Mermaid by Waxy 1790 (Pot-8-o-'s 1773)—*Promise 1803 by Buzzard* 1787 (Woodpecker 1773)—Dau. 1795 of Precipate* 1787 (Mercury 1778)—Lady Harriet 1773 by Mark Anthony 1767 (Spectator 1749).
				Daughter of, —, —.	Cook's Whip 1814 (*Whip 1794)—Hart's Maria by Craig's Alfred (*Medley 1776)—The Pryor mare [a good race-mare] by Tayloe's Bellair 1786 (*Medley 1776)—Thoroughbred mare of Mr. T. D. Owings.

WILFUL.

WINNER OF THE JOCKEY CLUB HANDICAP AT JEROME PARK, SEASIDE HANDICAP AT PROSPECT PARK, AND HANDICAP SWEEPSTAKES AT MONMOUTH PARK, IN 1879.

Is private stallion in Foxhill Stud, Mr. R. Ellis, Philadelphia, Pa., proprietor. Yearlings sold annually.

WILFUL, foaled 1876, bred by John Hunter, N. Y., by imp. Australian, son of West Australian, dam imp. Pussy, dam of Bangweola and Lelaps by Diophantus, out of Agapemone by Bay Middleton, son of Sultan. Wilful ran unplaced in Champagne Stakes at Jerome Park when two years old. As a three-year old started eleven times, won three races, was second in four, third in one and unplaced in three. Jerome Park, June 7, made his maiden effort in the Jockey Club Handicap, 2 miles, defeating his stable companion, Lou Lanier, who finished second, Gen Philips, Bonnie Wood, Warfield and Bertha, time 3.42¾. Prospect Park, L. I., Inaugnration Meeting, Coney Island Jockey Club, June 24, won the Seaside Handicap, 2 miles, defeating Gen. Philips, Gov. Hampton and Bramble, time 3.31¾, this was one of the best contested races of the year; same meeting, June 26, ran second to Kenney in the Brighton Hotel Purse, 1¾ miles, time 3.07½, Gov. Hampton and Gen. Philips also ran. Monmouth Park, July 5, won Handicap Sweepstakes, 1½ miles, defeating Kenney, Pilot, Lou Lanier and Belle, time 2.41; same place, July 8, ran second to Spendthrift in the Jersey Derby, 1½ miles, time 2.53; same place, July 10, finished second in the Shrewsbury Handicap, 1¾ miles, won by his stable companion, Lou Lanier, time 3.11¾, Wakefield, Terror and Zoo-Zoo also ran. Saratoga, July 24, ran second to Bramble in the Saratoga Cup, 2¼ miles, Lou Lanier third, Belle and Danicheff unplaced, time 4.11¾. This closed his racing career. Imp. Australian, his sire, is the sire of Spendthrift (sire of Kingston, Lamplighter, etc.), Fellowcraft, the first horse to beat Lexington's four-mile time (7.19¾), Fellowcraft running in 7.19½, Miser (sire of Yorkville Belle), Rutherford, Wildidle (4 miles in 7.25½) and of Ivy Leaf, dam of Bramble, Lizzie Lucas, dam of Cerise (Morrello's dam), Maggie B. B., dam of the grand racehorse and successful stallion Iroquois, Farfaletta, dam of Falsetto and others.

Wilful is a very finely bred horse, having a double cross of Touchstone, and is much inbred to Waxy through his best sons Whisker and Whalebone, and to Godolphin Arabian through Regulus and Matchem. Lelaps, his half brother, has done well in the stud, and Wilful is sire of the winners Wilfred, Dick Turpin, Meade, Parthian, Wild Cherry, Wayward, Westmoreland, Wickham, Maiden Hair, etc. He is a dark chestnut, without white, except a small star in his forehead, is 16 hands high, and highly finished all over. His colts, from what we hear, are very promising and fine.

He is descended in direct female line through Flora by Regulus from the Belgrade Turk mare, to which trace Taurus, Gladiator, Sting, Belladrum, imp. St. Blaise, Mainstay, New Holland, Boulevard, in England; and the good colt Victory, etc.

WILFUL.

CHESTNUT HORSE, FOALED 1876; BRED BY JOHN HUNTER, NEW YORK.

Parent	Grandparent	Great-grandparent	Great-great-grandparent	Ancestor	Pedigree
*AUSTRALIAN, ch. h., 1858.	West Australian, b. h., 1850.	Melbourne, br. h., 1834.	Humphrey Clinker, b. h., 1822.	Comus, ch. h., 1809.	Sorcerer 1796 (Trumpator 1782) Epsom Lass 1803 by Sir Peter T. 1784 (Highflyer 1774) Alexina 1788 by King Fergus 1775 (Eclipse 1764) Lardella 1780 by Y. Marske 1771 (Marske 1750) Dau. of Y. Cade 1747 (Cade 1734).
				Clinkerina, br. m., 1812.	Clinker 1805 (Sir Peter T. 1784) Pewet 1786 by Tandem 1773 (Syphon 1750) Termagant by Tantrum 1760 (Cripple 1750) Cantatrice by Sampson 1745 (Blaze 1733) Dau. 1757 of Regulus 1739 (God. Ar. 1724).
			Daughter of, b. m., 1825.	Cervantes, b. h., 1806.	Don Quixote 1784 (Eclipse 1764) Evelina 1791 by Highflyer 1774 (Herod 1758) Termagant by Tantrum 1760 (Cripple 1750) Cantatrice by Sampson 1745 (Blaze 1733) Dau. 1757 of Regulus 1739 (God. Ar. 1724).
				Daughter of, br. m., 1818.	Golumpus 1802 (Gohanna 1790) Dau. 1810 of Paynator 1791 (Trumpator 1782) Circle 1802 by St. George 1789 (Highflyer 1774) Abigail 1788 by Woodpecker 1773 (Herod 1758) Firetail 1772 by Eclipse 1764 (Marske 1750).
		Mowerina, b. m., 1843.	Touchstone, br. h., 1831.	Camel, br. h., 1822.	Whalebone 1807 (Waxy 1790) Dau. 1812 of Selim 1802 (Buzzard* 1787) Maiden 1801 by Sir Peter T. 1784 (Highflyer 1774) Dau. 1788 of Phenomenon 1780 (Herod 1758)—Matron 1782 by Florizel 1768 (Herod 1758).
				Banter, br. m., 1826.	Master Henry 1815 (Orville 1799) Boadicea 1807 by Alexander 1782 (Eclipse 1764) Brunette by Amaranthus 1766 (Old England 1741) Mayfly 1771 by Matchem 1748 (Cade 1734)—Dau. 1755 of Anc. Starling 1738 (Starling 1727).
			Emma, ch. m., 1824.	Whisker, b. h., 1812.	Waxy 1790 (Pot-8-o's 1773) Penelope 1798 by Trumpator 1782 (Conductor 1767) Prunella 1788 by Highflyer 1774 (Herod 1758) Promise 1768 by Snap 1750 (Snip 1736) Julia 1756 by Blank 1740 (God. Ar. 1724).
				Gibside Fairy, b. m., 1811.	Hermes 1790 (Mercury 1778) Vicissitude 1800 by Pipator 1786 (Imperator 1776)—Beatrice 1791 by Sir Peter T. 1784 (Highflyer 1774)—Pyrrha 1771 by Matchem 1748 (Cade 1734) Duchess 1748 by Whitenose 1742 (God. Ar. 1724).
	Emilia,* b. m., 1840.	Y. Emilius, b. h., 1833.	Emilius, b. h., 1820.	Orville, b. h., 1799.	Beningbrough 1791 (King Fergus 1775) Evelina 1791 by Highflyer 1774 (Herod 1758)—Termagant by Tantrum 1760 (Cripple 1750)—Cantatrice by Sampson 1745 (Blaze 1733)—Dau. 1757 of Regulus 1739 (God. Ar. 1724).
				Emily, ch. m., 1810.	Stamford 1794 (Sir Peter T. 1784) Dau. 1790 of Whiskey 1789 (Saltram* 1780)—Grey Dorimant 1781 by Dorimant 1772 (Otho 1760) Dizzy 1757 by Blank 1740 (God. Ar. 1724)—Dizzy 1741 by Driver 1727 (Wynn Ar.).
			Shoveler, b. m., 1816.	Scud, b. h., 1804.	Beningbrough 1791 (King Fergus 1775) Eliza 1791 by Highflyer 1774 (Herod 1758)—Augusta 1781 by Eclipse 1764 (Marske 1750)—Hardwicke's dam by Herod 1758 (Tartar 1743) Nettle 1763 by Bajazet 1740 (God. Ar. 1724).
				Goosander, b. m., 1805.	Hambletonian 1792 (King Fergus 1775)—Rally 1790 by Trumpator 1782 (Conductor 1767)—Fancy 1780 by Florizel 1768 (Herod 1758)—Dau. 1763 of Spectator 1749 (Crab 1722)—Horatia 1758 by Blank 1740 (God. Ar. 1724).
		Persian, b. m., 1829.	Whisker, b. h., 1812.	Waxy, b. h., 1790.	Pot-8-o's 1773 (Eclipse 1764)—Maria 1777 by Herod 1758 (Tartar 1743)—Lisette 1772 by Snap 1750 (Snip 1736)—Miss Windsor 1754 by God. Ar. 1724—Sist. to Volunteer by Y. Belgrade (Belgrade Turk).
				Penelope, b. m., 1798.	Trumpator 1782 (Conductor 1767)—Prunella 1788 by Highflyer 1774 (Herod 1758)—Promise 1768 by Snap 1750 (Snip 1736) Julia 1756 by Blank 1740 (God. Ar. 1724)—Spectator's dam 1735 by Partner 1718 (Jigg).
			Variety, ch. m., 1814.	[Selim or] Soothsayer, ch. h., 1808.	Sorcerer 1796 (Trumpator 1782)—Golden Locks 1793 by Delpini 1781 (Highflyer 1774)—Violet 1787 by Shark* 1771 (Marske 1750)—Dau. 1772 of Syphon 1750 (Squirt 1732)—Charlotte 1756 by Blank 1740 (God. Ar. 1724).
				Sprite, ch. m., 1807.	Bobtail 1795 (Precipitate* 1787)—Catherine 1795 by Woodpecker 1773 (Herod 1758)—Camilla 1778 by Trentham 1766 (T. Sweepstakes 1743)—Coquette 1765 by Compton Barb—Sist. 1748 to Regulus by God. Ar. 1724.
*PUSSY, b. m., 1864.	Diophantus, ch. h., 1858.	Orlando, b. h., 1841.	Touchstone, br. h., 1831.	Camel, br. h., 1822.	Whalebone 1807 (Waxy 1790)—Dau. 1812 of Selim 1802 (Buzzard* 1787)—Maiden 1801 by Sir Peter T. 1784 (Highflyer 1774)—Dau. 1788 of Phenomenon 1780 (Herod 1758)—Matron 1782 by Florizel 1768 (Herod 1758).
				Banter, br. m., 1826.	Master Henry 1815 (Orville 1799)—Boadicea 1807 by Alexander 1782 (Eclipse 1764)—Brunette by Amaranthus 1766 (Old England 1741) Mayfly 1771 by Matchem 1748 (Cade 1734)—Dau. 1755 of Anc. Starling 1738 (Starling 1727).
			Vulture, ch. m., 1833.	Langar, ch. h., 1817.	Selim 1802 (Buzzard* 1787)—Dau. 1808 of Walton 1799 (Sir Peter T. 1784)—Y. Giantess 1790 by Diomed* 1777 (Florizel 1768)—Giantess 1769 by Matchem 1748 (Cade 1734)—Molly Long Legs 1753 by Babraham 1740 (God. Ar. 1724).
				Kite, b. m., 1821.	Bustard 1813 (Castrel 1801)—Olympia 1815 by Sir Oliver 1800 (Sir Peter T. 1784)—Scotilla 1795 by Anvil 1777 (Herod 1758)—Scota 1783 by Eclipse 1764 (Marske 1750)—Harmony 1775 by Herod 1758 (Tartar 1743).
		Equation, b. m., 1839.	Emilius, b. h., 1820.	Orville, b. h., 1799.	Beningbrough 1791 (King Fergus 1775)—Evelina 1791 by Highflyer 1774 (Herod 1758)—Termagant by Tantrum 1760 (Cripple 1750)—Cantatrice by Sampson 1745 (Blaze 1733)—Dau. 1757 of Regulus 1739 (God. Ar. 1724).
				Emily, ch. m., 1810.	Stamford 1794 (Sir Peter T. 1784)—Dau. 1790 of Whiskey 1789 (Saltram* 1780)—Grey Dorimant 1781 by Dorimant 1772 (Otho 1760)—Dizzy 1757 by Blank 1740 (God. Ar. 1724)—Anc. Dizzy 1741 by Driver 1727 (Wynn Ar.).
			Maria, b. m., 1827.	Whisker, b. h., 1812.	Waxy 1790 (Pot-8-o's 1773)—Penelope 1798 by Trumpator 1782 (Conductor 1767)—Prunella 1788 by Highflyer 1774 (Herod 1758)—Promise 1768 by Snap 1750 (Snip 1736)—Julia 1756 by Blank 1740 (God. Ar. 1724).
				Gibside Fairy, b. m., 1811.	Hermes 1790 (Mercury 1778) Vicissitude 1800 by Pipator 1786 (Imperator 1776)—Beatrice 1791 by Sir Peter T. 1784 (Highflyer 1774)—Pyrrha 1771 by Matchem 1748 (Cade 1734)—Duchess 1748 by Whitenose 1742 (God. Ar. 1724).
	*Agapemone, b. m., 1851.	Bay Middleton, b. h., 1833.	Sultan, b. h., 1816.	Selim, ch. h., 1802.	Buzzard* 1787 (Woodpecker 1773)—Dau. of Alexander 1782 (Eclipse 1764)—Dau. of Highflyer 1774 (Herod 1758)—Dau. of Alfred 1770 (Matchem 1748)—Dau. 1770 of Engineer 1756 (Sampson 1745).
				Bacchante, br. m., 1809.	W's Ditto 1800 (Sir Peter T. 1784)—Dau. 1791 of Mercury 1778 (Eclipse 1764) Dau. 1776 of Herod 1758 (Tartar 1743)—Folly 1771 by Marske 1750 (Squirt 1732)—Vixen 1753 by Regulus 1739 (God. Ar. 1724).
			Cobweb, b. m., 1821.	Phantom, b. h., 1808.	Walton 1799 (Sir Peter T. 1784)—Julia 1790 by Whiskey 1789 (Saltram* 1780)—Y. Giantess 1790 by Diomed* 1777 (Florizel 1768)—Giantess 1769 by Matchem 1748 (Cade 1734)—Molly Long Legs 1753 by Babraham 1740 (God. Ar. 1724).
				Filagree, ch. m., 1815.	Soothsayer 1808 (Sorcerer 1796)—Web 1808 by Waxy 1790 (Pot-8-o's 1773) Penelope 1798 by Trumpator 1782 (Conductor 1767) Prunella 1788 by Highflyer 1774 (Herod 1758)—Promise 1768 by Snap 1750 (Snip 1736).
		Venus, b. m., 1840.	Sir Hercules, bl. h., 1826.	Whalebone, br. h., 1807.	Waxy 1790 (Pot-8-o's 1773)—Penelope 1798 by Trumpator 1782 (Conductor 1767)—Prunella 1788 by Highflyer 1774 (Herod 1758)—Promise 1768 by Snap 1750 (Snip 1736)—Julia 1756 by Blank 1740 (God. Ar. 1724).
				Peri, b. m., 1822.	Wanderer 1811 (Gohanna 1790)—Thalestris 1809 by Alexander 1782 (Eclipse 1764)—Rival 1800 by Sir Peter T. 1784 (Highflyer 1774)—Hornet 1790 by Drone 1777 (Herod 1758)—Manilla 1777 by Goldfinder 1764 (Snap 1750).
			Echo, b. m., 1828.	Emilius, b. h., 1820.	Orville 1799 (Beningbrough 1791)—Emily 1810 by Stamford 1794 (Sir Peter T. 1784)—Dau. 1790 of Whiskey 1789 (Saltram* 1780)—Grey Dorimant 1781 by Dorimant 1772 (Otho 1760)—Dizzy 1757 by Blank 1740 (God. Ar. 1724).
				Daughter of, b. m., 1820.	[Scud or] Pioneer 1804 (Whiskey 1789)—Canary Bird br. m. 1806 by [Whiskey or] Sorcerer 1796 (Trumpator 1782)—Canary b. m. 1797 by Coriander 1786 (Pot-8-o's 1773)—Miss Green b. m. 1787 by Highflyer 1774 (Herod 1758)—Harriet b. m. 1769 by Matchem 1748 (Cade 1734)—Flora b. m. 1749 by Regulus 1739 (God. Ar. 1724)—Dau. of Bartlett's Childers (Darley Ar. 1699)—Dau. of Bay Bolton 1705 (Grey Hautboy) Dau. of Belgrade Turk.

IMPORTED WOOD MOSS.

Bred and owned by Dr. A. W. McAlester, Columbia, Mo. Terms of Service and sales annually.

WOOD MOSS bay horse, foaled 1887, by Creancier, son of Le Sarrazin by Monarque, dam imp. Lady Nar by Macaroni, grandam Queen of Scots by Blair Athol, out of East Sheen by Kingston, son of Venison by Partisan. Creancier won the Trial Stakes at Epsom and other races. Le Sarrazin, the sire of Creancier, was full brother to La Favorite, the dam of Flageolet (winner of the Goodwood Cup and sire of imp. Rayon-d'Or) by Monarque, the sire of Gladiateur (winner of the 2.000 Guineas, Derby, St. Leger, Grand Prize of Paris and Prix de Royal Oak), Creancier's dam by Vermouth (winner of the Grand Prize of Paris and a successful stallion). Lady Nar, Wood Moss's dam, was a good race-mare, won the Winchester Biennial Stakes for three-year olds, 1 mile; was second in Welter Handicap, 1½ miles, beaten by a head. As an individual she is unsurpassed. She is by the great horse Macaroni (winner of the Derby and the 2,000 Guineas). The Macaroni mares are more sought after in England than any others. The dams of the great horses Ormonde, Galliard, Bonavista, Busy Body, etc., are Macaroni mares. Macaroni was by Sweetmeat, a grand race-horse, started twenty-four times, won twenty-two and a half, among them the Queen's Vase at Ascot. Queen o' Scots, dam of Lady Nar, was by Blair Athol (Derby and St. Leger winner), he the sire of Prince Charlie (winner of the 2,000 Guineas), of Silvio (winner of both Derby and St. Leger), and Craig Millar (winner of the St. Leger). East Sheen, third dam of Wood Moss, was by Kingston, a winner at all distances, from two to five years old, carrying the heaviest weights, against the best horses of his day, from one to four miles, and the sire of Caractacus (winner of the Derby); Queen Bertha (winner of the Oaks and dam of Spinaway and Wheel of Fortune, winners of the 1.000 Guineas and Oaks), imp. Eltham Lass, dam of Kingfisher, and Majestic, dam of the good race-horse Eurus.

Wood Moss's breeding is gilt-edged. The most successful cross of modern times is the double cross of Whalebone, which is secured by intermixing the descendants of Whalebone through Sir Hercules, Birdcatcher, The Baron, Stockwell, Rataplan, Camel, Touchstone, Newminster and Adventurer. Isonomy by Sterling, son of Oxford by Birdcatcher, one of the best race and best bred horses of the century, had no less than four strains of Whalebone and seven of Waxy in his pedigree, besides one of Shuttle, the rarest of all. Measured by this standard, Wood Moss should make a successful stock getter. He has nine strains of Whalebone, two of Whisker, full brother to Whalebone, thirteen of Waxy and two of Shuttle.

He is a rich, dark bay with black points, no marks, 16 hands high, highly finished and symmetrical. He is of the Leamington type, with more bone than the Leamingtons usually show.

Wood Moss descends in direct line from the same Royal mare to which traces Woodpecker, Waxy Pope, Whalebone, Whisker, Partisan, Bay Middleton, imp. Glencoe, Melbourne, Robert the Devil, Bend'or, imp. Eclipse (Morris), Uncas, Wanderer and all the Prunella family.

Waxy 1790—Pot-8-o's 1773 Eclipse 1764—Marske 1750 Squirt 1732 Bartlett's Childers—Darley Arabian 1695.

WOOD-MOSS.

BAY HORSE, FOALED 1887 (IMPORTED IN UTERO); BRED BY DR. McALESTER, COLUMBIA, MO.

CREANCIER, br. h., 1878.

Cravache, br. m., 1869. — Le Sarrazin, b. h., 1865.

Mademoiselle Cravachon, b. m., 1856. — Vermouth, b. h., 1861. — Constance, ch. m., 1848. — Monarque,‡ b. h., 1852.

LolaMontez, b. m., 1845. — Loadstone, br. h., 1845. — Vermeille, ch. m., 1843. — The Nabob, blk. h., 1849. — Lanterne, b. m., 1841. — Gladiator, ch. h., 1833. — Poetess, b. m., 1838. — TheEmperor, ch. h., 1841.

Defence, b. h., 1824. Whalebone 1807 (Waxy 1790)—Defiance 1816 by Rubens 1805 (Buzzard* 1787)—Little Folly 1806 by Highland Fling 1798 (Spadille 1784)—Harriet 1793 by Volunteer 1780 (Eclipse 1764)—Dau. 1785 of Alfred 1770 (Matchem 1748).

Daughter of, ch. m., 1831. Reveller 1815 (Comus 1809)—Design 1827 by Tramp 1810 (Dick Andrews 1797)—Defiance 1816 by Rubens 1805 (Buzzard* 1787)—Little Folly 1806 by Highland Fling 1798 (Spadille 1784)—Harriet 1793 by Volunteer 1780 (Eclipse 1764).

Royal Oak, br. h., 1823. Catton 1809 (Golumpus 1802)—Dau. 1818 of Smolensko 1810 (Sorcerer 1796)—Lady Mary 1800 by Beningbrough 1791 (King Fergus 1775)—Dau. of Highflyer 1774 (Herod 1758)—Sist. to Miss Bell by Marske 1750 (Squirt 1732).

Ada, b. m., 1824. Whisker 1812 (Waxy 1790)—Anna Bella 1807 by Shuttle 1793 (Y. Marske 1771)—Dau. 1797 of Drone 1777 (Herod 1758)—Contessina 1787 by Y. Marske 1771 (Marske 1750)—Tuberose 1772 by Herod 1758 (Tartar 1743).

Partisan, b. h., 1811. Walton 1799 (Sir Peter T. 1784)—Parasol 1800 by Pot-8-o's 1773 (Eclipse 1764)—Prunella 1788 by Highflyer 1774 (Herod 1758)—Promise 1768 by Snap 1750 (Snip 1736)—Julia 1756 by Blank 1740 (God. Ar. 1724).

Pauline, b. m., 1826. Moses 1819 ([Whalebone or] Seymour 1807)—Quadrille 1815 by Selim 1802 (Buzzard* 1787)—Canary Bird 1806 by [Whiskey or] Sorcerer 1796 (Trumpator 1782)—Canary 1797 by Coriander 1786 (Pot-8-o's 1773)—Miss Green 1787 by Highflyer 1774 (Herod 1758).

Hercule, ch. h., 1839. Rainbow 1808 (Walton 1799)—Aimable 1822 by Election 1804 (Gohanna 1790)—Dau. 1808 of Y. Whiskey 1801 (Whiskey 1789)—Dau. 1800 of Walnut 1786 (Highflyer 1774)—Bay Javelin 1793 by Javelin 1772 (Eclipse 1764).

Elvira, b. m., 1829. Eryx 1816 (Milo 1802)—Coral 1816 by Orville 1799 (Beningbrough 1791)—Fairing 1807 by Waxy 1790 (Pot-8-o's 1773)—Rattle 1793 by Trumpator 1782 (Conductor 1767)—Fancy 1780 (Sist. to *Diomed) by Florizel 1768 (Herod 1758).

The Nob, b. h., 1838. Glaucus 1830 (Partisan 1811)—Octave 1830 by Emilius 1820 (Orville 1799)—Whizgig 1819 by Rubens 1805 (Buzzard* 1787)—Penelope 1798 by Trumpator 1782 (Conductor 1767)—Prunella 1788 by Highflyer 1774 (Herod 1758).

Hester, br. m., 1832. Camel 1822 (Whalebone 1807)—Monimia 1821 by Muley 1810 (Orville 1799)—Sist. 1796 to Petworth by Precipitate* 1787 (Mercury 1778)—Dau. 1787 of Woodpecker 1773 (Herod 1758)—Sist. 1773 to Juniper by Snap 1750 (Snip 1736).

The Baron, ch. h., 1842. Ir. Birdcatcher 1833 (Sir Hercules 1826)—Echidna 1838 by Economist 1825 (Whisker 1812)—Miss Pratt 1825 by Blacklock 1814 (Whitelock 1803)—Gadabout 1812 by Orville 1799 (Beningbrough 1791)—Minstrel 1803 by Sir Peter T. 1784 (Highflyer 1774).

Fair Helen, b. m., 1837. Priam* 1827 (Emilius 1820)—Dirce 1830 by Partisan 1811 (Walton 1799)—Antiope 1817 by Whalebone 1807 (Waxy 1790)—Amazon 1799 by Driver 1783 (Trentham 1766)—Fractious 1792 by Mercury 1778 (Eclipse 1764).

Touchstone, br. h., 1831. Camel 1822 (Whalebone 1807)—Banter 1826 by Master Henry 1815 (Orville 1799)—Boadicea 1807 by Alexander 1782 (Eclipse 1764)—Brunette by Amaranthus 1766 (Old England 1741)—Mayfly 1771 by Matchem 1748 (Cade 1734).

Latitude, b. m., 1835. Langar 1817 (Selim 1802)—Olympia 1815 by Sir Oliver 1800 (Sir Peter T. 1784)—Scotilla 1795 by Anvil 1777 (Herod 1758)—Scota 1783 by Eclipse 1764 (Marske 1750)—Harmony 1775 by Herod 1758 (Tartar 1743).

Slane, b. h., 1833. Royal Oak 1823 (Catton 1809)—Dau. 1819 of Orville 1799 (Beningbrough 1791)—Epsom Lass 1803 by Sir Peter T. 1784 (Highflyer 1774)—Alexina 1788 by King Fergus 1775 (Eclipse 1764)—Lardella 1780 by Y. Marske 1771 (Marske 1750).

Hester, br. m., 1832. Camel 1822 (Whalebone 1807)—Monimia 1821 by Muley 1810 (Orville 1799)—Sist. 1796 to Petworth by Precipitate* 1787 (Mercury 1778)—Dau. 1787 of Woodpecker 1773 (Herod 1758)—Sist. 1773 to Juniper by Snap 1750 (Snip 1736).

*LADY NAR, ch. m., 1877.

Queen o' Scots, b. m., 1869. — Macaroni, b. h., 1860.

East Sheen, br. m., 1858. — Blair Athol, ch. h., 1861. — Jocose, b. m., 1843. — Sweetmeat, br. h., 1842.

Miss Slick, br. m., 1843. — Kingston, b. h., 1849. — BlinkBonny, b. m., 1851. — Stockwell, ch. h., 1849. — Banter, br. m., 1826. — Pantaloon, ch. h., 1824. — Lollypop, blk. m., 1836. — Gladiator, ch. h., 1833.

Partisan, b. h., 1811. Walton 1799 (Sir Peter T. 1784)—Parasol 1800 by Pot-8-o's 1773 (Eclipse 1764)—Prunella 1788 by Highflyer 1774 (Herod 1758)—Promise 1768 by Snap 1750 (Snip 1736)—Julia 1756 by Blank 1740 (God. Ar. 1724).

Pauline, b. m., 1826. Moses 1819 ([Whalebone or] Seymour 1807)—Quadrille 1815 by Selim 1802 (Buzzard* 1787)—Canary Bird 1806 by [Whiskey or] Sorcerer 1796 (Trumpator 1782)—Canary 1797 by Coriander 1786 (Pot-8-o's 1773)—Miss Green 1787 by Highflyer 1774 (Herod 1758).

[Starch or] Voltaire, br. h., 1826. Blacklock 1814 (Whitelock 1803)—Dau. 1816 of Phantom 1808 (Walton 1799)—Dau. 1802 of Overton 1788 (King Fergus 1775)—Dau. 1796 of Walnut 1786 (Highflyer 1774)—Dau. 1783 of Ruler 1771 (Y. Marske 1771).

Belinda, br. m., 1825. Blacklock 1814 (Whitelock 1803)—Wagtail 1818 by Prime Minister 1810 (Sancho 1801)—Tranby*'s dam 1812 by Orville 1799 (Beningbrough 1791)—Miss Grimstone 1796 by Weazle 1773 (Herod 1758)—Dau. of Ancaster 1768 (Blank 1740).

Castrel, ch. h., 1801. Buzzard* 1787 (Woodpecker 1773)—Dau. of Alexander 1782 (Eclipse 1764)—Dau. of Highflyer 1774 (Herod 1758)—Dau. of Alfred 1770 (Matchem 1748)—Dau. 1770 of Engineer 1756 (Sampson 1745).

Idalia, ch. m., 1815. Peruvian 1806 (Sir Peter T. 1784)—Musidora 1801 by Meteor 1783 (Eclipse 1764)—Maid-of-all-Work 1786 by Highflyer 1774 (Herod 1758)—Dau. 1771 of Syphon 1750 (Squirt 1732)—Dau. of Regulus 1739 (God. Ar. 1724).

Master Henry, b. h., 1815. Orville 1799 (Beningbrough 1791)—Miss Sophia 1805 by Stamford 1794 (Sir Peter T. 1784)—Sophia 1798 by Buzzard* 1787 (Woodpecker 1773)—Huncamunca 1787 by Highflyer 1774 (Herod 1758)—Cypher 1772 by Squirrel 1754 (Traveller 1735).

Boadicea, b. m., 1807. Alexander 1782 (Eclipse 1764)—Brunette by Amaranthus 1766 (Old England 1741)—Mayfly 1771 by Matchem 1748 (Cade 1734)—Dau. 1755 of Anc. Starling 1738 (Starling 1727)—Look-at-me-Lads 1731 by Grasshopper (Byerly Turk).

The Baron, ch. h., 1842. Ir. Birdcatcher 1833 (Sir Hercules 1826)—Echidna 1838 by Economist 1825 (Whisker 1812)—Miss Pratt 1825 by Blacklock 1814 (Whitelock 1803)—Gadabout 1812 by Orville 1799 (Beningbrough 1791)—Minstrel 1803 by Sir Peter T. 1784 (Highflyer 1774).

Pocahontas, b. m., 1837. Glencoe* 1831 (Sultan 1816)—Marpessa 1830 by Muley 1810 (Orville 1799)—Clare 1824 by Marmion 1806 (Whiskey 1789)—Harpalice 1814 by Gohanna 1790 (Mercury 1778)—Amazon 1799 by Driver 1783 (Trentham 1766).

Melbourne, br. h., 1834. Humphrey Clinker 1822 (Comus 1809)—Dau. 1825 of Cervantes 1806 (Don Quixote 1784)—Dau. 1818 of Golumpus 1802 (Gohanna 1790)—Dau. 1810 of Paynator 1791 (Trumpator 1782)—Circle 1802 by St. George 1789 (Highflyer 1774).

Queen Mary, b. m., 1843. Gladiator 1833 (Partisan 1811)—Dau. 1840 of Plenipotentiary 1831 (Emilius 1820)—Myrrha 1830 by Whalebone 1807 (Waxy 1790)—Gift 1818 by Y. Gohanna 1810 (Gohanna 1790)—Dau. 1802 of Sir Peter T. 1784 (Highflyer 1774).

Venison, br. h., 1833. Partisan 1811 (Walton 1799)—Fawn 1823 by Smolensko 1810 (Sorcerer 1796)—Jerboa 1803 by Gohanna 1790 (Mercury 1778)—Camilla 1778 by Trentham 1766 (T. Sweepstakes 1743)—Coquette 1764 by Compton Barb.

Queen Anne, b. m., 1843. Slane 1833 (Royal Oak 1823)—Garcia 1823 by Octavian 1807 (Stripling 1795)—Dau. 1806 of Shuttle 1793 (Y. Marske 1771)—Katharine 1798 by Delpini 1781 (Highflyer 1774)—Dau. 1786 of Paymaster 1766 (Blank 1740).

Muley Moloch, br. h., 1830. Muley 1810 (Orville 1799)—Nancy 1813 by Dick Andrews 1797 (Joe Andrews 1778)—Spitfire 1800 by Beningbrough 1791 (King Fergus 1775)—Dau. of Y. Sir Peter 1775 (Doge 1762)—Dau. of Engineer 1756 (Sampson 1745).

Daughter of, br. m., 1832. Whisker 1812 (Waxy 1790)—Dau. ch. m. 1826 of Sam 1815 (Scud 1804)—Morel ch. m. 1805 by Sorcerer 1796 (Trumpator 1782)—Hornby Lass b. m. 1795 by Buzzard* 1787 (Woodpecker 1773)—Puzzle b. m. 1778 by Matchem 1748 (Cade 1734)—Princess b. m. 1769 by Herod 1758 (Tartar 1743)—Julia b. m. 1756 by Blank 1740 (God. Ar. 1724)—Spectator's dam b. m. 1735 by Partner 1718 (Jigg)—Bonny Lass b. m. 1723 by Bay Bolton 1705 (Grey Hautboy)—Dau. of Darley Ar. 1699—Dau. of Byerly Turk—Dau. of Taffolet Barb—Dau. of Place's White Turk—Natural Barb mare of Mr. Tregonwell's.

‡ Monarque is by The Baron, Sting or The Emperor. Emperor's pedigree is accepted.

MATCHEM,

HEROD,

ECLIPSE,

THE FOUNDERS OF THE THREE GREAT LINES OF BLOOD.

MATCHEM.

WINNER OF THE JOCKEY CLUB PLATE AT NEWMARKET, 1756, ETC.

MATCHEM, a bay horse, by Cade, was foaled in 1748, bred by Mr. John Holme, of Carlisle, and was the third produce of Sister to Miss Partner (likewise dam of Miss Roundhead by Roundhead, Changeling by Cade, etc.) by Crofts' Partner, who was bred in 1718 by Mr. Crofts, of Barforth, Yorkshire. Matchem, who was purchased by Mr. William Fenwick, of Bywell, Northumberland, ran his first race at York in 1753, for the great Subscription Purse of 160 guineas, for five-year olds, which he won, beating Mr. Shafto's Barforth Billy by Forester and Mr. Watson's Bold by Cade, and the same year he won a Plate of £50 at Morpeth, beating Mr. Shafto's Blameless. In 1754 Matchem won the Ladies' Plate of 126 guineas, four mile heats, at York, beating Sedbury in two smartly contested heats, and he also won the Ladies' Plate of 80 guineas, at Lincoln, in two heats, from the Duke of Ancaster's Martin and Mr. Smith's Skim, while at Morpeth he walked over for a £50 Plate. In 1755 Matchem won the £50 Plate for six-year olds, 8 st. 7 lbs., aged, 9 stones, at Newmarket, over the Beacon Course, beating Mr. Bowle's Trajan by Regulus. The race was run in 7.20, carrying 8 st. 7 lbs. each, and two others were distanced. Trajan showed the better speed but was unable to maintain it, owing, it was thought, to his indifferent condition; consequently, on Mr. Fenwick offering to run Matchem for the Whip in the following April against any horse in the world, Mr. Bowles accepted the challenge with Trajan. In August Matchem walked over at York for £20, no horse entering. The pair met accordingly on April 11th, 1756, for a Sweepstakes of 200 guineas each and the Whip, 10 stone each, over the Beacon Course. Matchem, who was ridden by John Singleton, permitted Trajan, who was a fretful, pulling horse, to make the running, and he held such a long lead over the Flat that 5 to 1 was betted on him; but at the Turn of the Lands 100 to 1 was betted on Matchem, who there had managed to close with his antagonist, and finally won easily. But, notwithstanding his defeat, Sir Richard Grosvenor gave 250 guineas for Trajan immediately after the race. Matchem was subsequently beaten that year by Spectator for the Jockey Club Plate, but won the Plate of 60 guineas at Newcastle, beating Draweansir and Full Moon. In 1758 Matchem again ran for the Jockey Club Plate, which was decided in one heat, the Stewards having so ordered it, owing to the difficulty of placing the horses in the previous year, an arrangement, however, that failed to prevent the usual heavy wagering on their placing. The race was won with great difficulty by Mirza by the Godolphin Arabian, Matchem second, Jason third, Feather fourth and Forester last. Feather was the favorite at evens, 6 to 4 against Jason, 6 to 1 against Mirza, and 10 to 1 against Matchem. Immense sums were won on Mirza, and also on Matchem being placed second, respecting which it was observed "that the friends of Matchem not only *combed the Golden Fleece,* but *dressed the Feathers very handsomely.*" Matchem only started once subsequently, viz., at Scarborough, where he won the £50 Plate, 9 stone each, beating Foxhunter and Sweetlips. In 1763 Matchem was put to the stud and became the leading sire in the North of England. He at first covered at the low fee of 5 guineas, which was raised in 1765 to 10 guineas, and in 1770 to 20 guineas. In 1775 his fee was further increased to 50 guineas for twenty-five mares, including those belonging to his owner, Mr. Fenwick, who is believed to have realized £17,000 by Matchem's services as a sire. The progeny of Matchem won on the turf, during the twenty-three years they were running, £151,097.

Matchem got Conductor (the sire of Imperator and Trumpator). Imperator got Pipator and Remembrancer, and Trumpator got Sorcerer. Sorcerer got Comus, he Humphrey Clinker, he Melbourne, the founder of the stout Melbourne line, which is so well represented in this country through Melbourne's grandson, imp. Australian, one of the best and luckiest importations ever made. Matchem got Alfred, who got imp. Tickle Toby, who got Sir Solomon, sire of Trumpator (Davenport's).

Matchem got Teetotum (winner of the Oaks in 1780), Hollandaise (winner of the St. Leger in 1778), the dam of Eager (winner of the Derby in 1791), Rarity (dam of Maid of the Oaks, winner of the Oaks in 1783), Miss West (dam of Quiz, winner of the St. Leger in 1801). He got also Giantess, that produced a mare by Highflyer (the dam of Cælia, winner of the Oaks, and the Fidget colt, winner of the Derby); and Young Giantess by Diomed, that produced Sorcerer; Eleanor, the first mare to win the Derby and the Oaks, and dam of Muley; Julia, dam of the Derby winner, Phantom; Cressida, dam of imp. Priam, winner of Derby, 2,000 Guineas and Goodwood Cup, and Antar, winner of the 2,000 Guineas; a Walton mare, dam of Nicolo, winner of the 2,000 Guineas, and Langar.—Atalanta, that produced Faith, the dam of Camillus, and Rosalind, the dam of Whitelock and Hyacinthus.—Captive, that produced Paulina, dam of the Oaks winner, Hermione, which was in turn dam of the Oaks winner, Brisels, and she produced Corinne, winner of the 1,000 Guineas and Oaks.—Teresa, that produced Calash, the dam of Whiskey and the St. Leger winner, Paragon.—Monimia, that produced a Highflyer mare, the dam of Hambletonian, winner of the St. Leger.—Puzzle, that produced Woodbine, the dam of the two Oaks winners, Music and Minuet, and Hornby Lass, the dam of the Oaks winner, Morel, which was the founder of a great family.—Maiden, that produced a mare by Herod, the dam of Gohanna and Precipitate; Walnut and Matron, the dam of Sir Solomon.—Pyrrha, that produced a mare by Herod, dam of the St. Leger winner, Beningbrough, and Beatrice, the dam of Vicissitude.—Mayfly, that produced Brunette by Amaranthus, the grandam of Banter, Touchstone's dam.—Marotte, the dam of Bagot.—Purity, the dam of Rockingham, the sire of imp. Castianira.

MATCHEM (MATCH'EM).

BAY HORSE, FOALED 1748; BRED BY MR. JOHN HOLME AND SOLD TO WILLIAM FENWICK. DIED FEB. 21, 1781, AGED 33 YEARS.

‡CADE, b. h., 1734.	Godolphin Arabian, br. h., 1724.				The most noted of all the Eastern sires, about whom opinions are divided as to whether he was an Arab or Barb, the best judges inclining to the opinion that he was a Barb. He was purchased in Paris by Mr. Coke in 1728, who gave him to Mr. R. Williams, by whom he was presented to Earl Godolphin. He was a brown bay, 15 hands high, with some white on the off hind-heel. Died 1753.
	Roxana, ch. m., 1718.	Bald Galloway, —, —.	St. Victor Barb, —, —.		Brought from France by Capt. Rider, of Whittlebury.
			Grey Whynot, —, —.	Old Whynot, —, —.	Fenwick Barb—A Royal mare.
				A Royal Mare, —, —.	
		Sister to Chaunter, —, —.	Akaster Turk, —, —.		Sire of Chaunter, Terror, Thwait's dun mare, dam of Driver etc.
			Cream Cheeks, —, —.	Leedes Arabian, —, —.	Sire of Leedes, Bay Peg, Dyer's Dimple, etc.
				Daughter of, —, 1680.	Spanker (D'Arcy Yellow Turk—Old Morocco mare)—Old Morocco mare by Morocco Barb—Old Bald Peg by an Arabian—A Barb mare.
CHANGELING'S DAM (Sister to Miss Partner), —, 1735.	(Crofts') Partner, ch. h., 1718.	Sir Mostyn's Jigg, —, —.	Byerly Turk, —, —.		Byerly Turk was Capt. Byerly's charger in Ireland in all King William's wars, 1689. Sire of Basto, Jigg, etc.
			Daughter of, —, —.	Spanker, —, —.	D'Arcy Yellow Turk—Old Morocco mare by Morocco Barb—Old Bald Peg by an Arabian—Barb mare.
		Sist. to Mixbury, —, —.	CurwenBayBarb, —, —.		Was presented to Louis XIV., King of France (1638-1715), by Muley Ishmael, King of Morocco, and brought to England by Mr. Curwen. Sire of Mixbury, Tantivy, etc.
			Daughter of, —, —.	The Curwen Old Spot, —, —.	Selaby Turk, property of Mr. Marshall. Sire of the dams of Wyndham, Spot, White-neck, etc.
				Daughter of, —, —.	White-legged Lowther Barb—Vintner mare.
	Brown Farewell, br. m., 1710.	Makeless, —, —.	OglethorpeArabian, —, —.		This may be the horse described as Mr. Oglethorpe's son of D'Arcy Yellow Turk. Sire of Bald Frampton, etc.
			Daughter of, —, —.		
		§Daughter of, —, —.	Brimmer, —, —.	D'Arcy Yellow Turk, —, —.	Brimmer was bred in Yorkshire by the D'Arcy family. His name is in all the best pedigrees.
				A Royal Mare, —, —.	
			Trumpet's dam, —, —.	Place's White Turk, —, —.	Was the property of Mr. Place, studmaster to Oliver Cromwell (died 1658), when Protector.
				Daughter of, —, —.	Dodsworth (though foaled in England, was a Natural Barb, his cover was in Tangiers, out of a Royal mare [a Natural Barb, the dam of Vixen])—Layton (Violet) Barb mare.

‡ Got his name while Roxana died, leaving him about 10 days old, and he was fed with cow's milk.

§ Mare given to Mr. Crofts by Queen Anne. Trumpet's dam was called the D'Arcy Yellow mare.

HEROD (KING HEROD).

HEROD, a bay horse, by Tartar, was foaled in 1758, bred by his Royal Highness the Duke of Cumberland, and was the fourth produce of Cypron (likewise dam of Dapper by Y. Cade, Hollyhock by Y. Cade, Protector by Matchem, etc.), who was bred in 1750 by Sir Wm. St. Quintin by Blaze, son of Flying Childers, out of Salome (Selima) by Bethell's Arabian. Herod (first called King Herod) ran his first race in October, 1763, when five years old, at Newmarket, where he beat the Duke of Ancaster's Roman by Blank, five years, 8 st. 7 lbs. each, over the Beacon Course, for a match of 500 guineas a side; and in the following April, 1764, he won a Sweepstakes of 300 sovereigns each (10 stone, nine subscriptions), Beacon Course, beating Sir John Moore's (Mr. Wildman's) Tartar by Tartar, out of Miss Meredith. At Ascot he won his match for 1,000 guineas a side, giving 6 lbs. to Lord Rockingham's Tom Tinker, four miles; and at Newmarket, in the October Meeting, he next gave 3 lbs. to the Duke of Grafton's Antinous, 8 st. 8 lbs. over the Beacon Course, for 500 guineas a side, 6 to 4 being betted on the loser. Not content with this defeat of Antinous, the Duke of Grafton matched him to run in the following May, 1765, for 1,000 guineas a side, Herod to give him 9 lbs., but the result was the same, although Antinous was the favourite at 7 to 5 on. In October, 1765, Herod, carrying 9 stone, suffered his first defeat, for a match of 1,000 guineas a side, against Sir Jas. Lowther's Ascham, six years, 8 stone, over the Beacon Course. As much as 3 to 1 was betted on Herod, who was beaten easily. On the death of His Royal Highness the Duke of Cumberland, which happened that year, Herod was purchased by Sir John Moore, who matched him at the following Newmarket April Meeting, 1766, to give Lord Bolingbroke's Turf (five years) 6 lbs., over Beacon Course, for 1,000 guineas a side. The betting was 7 to 4 on Herod, but again he suffered an easy defeat. At York he started for the Great Subscription Purse against Bay Malton, Beaufremont and several others, but, a blood vessel breaking in his head, he was beaten off. In 1767 King Herod ran second to Bay Malton for a Sweepstakes of 500 guineas a side, 8 st. 7 lbs. each, over the Beacon Course, in which Lord Bolingbroke's Turf and Mr. Shafto's Ascham were likewise defeated. This race, says a chronicler of the time, brought together a greater number of noblemen, gentlemen, sportsmen and people of all ranks, from all parts of the kingdom, than were ever seen before at Newmarket, and those from Yorkshire backed Bay Malton freely and won thousands. The betting was 6 to 4 against Turf, 7 to 4 against Bay Malton, 4 to 1 against Ascham, and 5 to 1 against Herod, and immense sums changed hands on the result. Among other bets, Lord Rockingham took 500 to 20 that he placed the whole party and won. In the following May, 1767, Herod received 7 lbs. from Ascham, and beat him over the Beacon Course for 1,000 guineas a side, which was his last race. Herod was put to the stud in 1770, and during the nineteen years his stock were running they won £201,505, besides several cups and 44 hogsheads of claret. The most distinguished of Herod's progeny on the turf and at the stud include the three Oaks winners, Bridget, Faith and Maid-of-the-Oaks, and the St. Leger winner Phenomenon (which died after landing in America in 1803), Anvil, Bagot, Bourdeaux, Drone, Evergreen, Florizel (sire of imp. Diomed), Fortitude (sire of John Bull), Highflyer (sire of Delpini, Rockingham, Sir Peter, Walnut, etc.), Justice, Rosalba, Rover (afterwards Tug in Ireland), Telemachus, Tuberose, Weazle, Woodpecker (sire of imp. Buzzard, the sire of Selim, Castrel, Rubens, etc.). Herod got also the unnamed daughter, dam of the Derby winner Aimwell—another, the dam of the St. Leger winner Beningbrough (he the sire of Orville)—another, the dam of the Oaks winner Platina—another, the dam of the Oaks winner Portia—Hip, the dam of the Oaks winner Hippolyta—Cytheréa, the dam of Oaks winner Nightshade—Maria, the dam of the Derby winner Waxy (one of the grandest sires) and Wowski, the dam of Smolensko (winner of the 2,000 Guineas and Derby), and Thunderbolt, and Jemima, the grandam of imp. Miss Shipton—Calash, the dam of the St. Leger winner Paragon, of Whiskey and Lady Charlotte, the dam of the 2,000 Guineas winner Cwrw—Aspasia, dam of the Derby winner Sergeant—Perdita, dam of the Oaks winner, the Yellow mare, and her sister, Deceit (the dam of the Oaks winner Parisot)—Expectation, the dam of Anticipation, the founder of the great Hawthorne and Manganese families—Princess, the dam of Puzzle—Tuberose, the dam of Contessina and Rosamond—Cælia, the dam of imp. Tickle Toby, Highland Fling and a mare by King Fergus (the dam of the Oaks winner Scotia)—Harmony, the dam of Scota—Peggy (sister to Postmaster), the dam of imp. Peggy.

HEROD (Called also KING HEROD).

BAY HORSE, FOALED 1758; BRED BY H. R. H. WILLIAM, DUKE OF CUMBERLAND. DIED, MAY 12, 1780, AGED 22 YEARS.

TARTAR (Partner), ch. h., 1743.

(Croft's) Partner, ch. h., 1718. | Meliora, ch. m., 1729.

Sir R. Mostyn's Jigg, —, . | Sist. to Mixbury, —, —. | Fox, b. h., 1714. | Witty's Milkmaid, b. m., 1720.

Daughter of, Byerly Turk, —, —. | Daughter of, Curwen Bay Barb, ab. 1703. —, —. | Clumsey, —, —. | Bay Peg, —, . | Daughter of, OldSnail(Snales), ab. 1712. —, —.

	Byerly Turk was Capt. Byerly's charger in Ireland in King William's wars, 1689.
Spanker, —, .	D'Arcy Yellow Turk—Old Morocco mare by Morocco Barb Old Bald Peg by an Arabian Barb mare.
	Was presented to Louis XIV., King of France (1638-1715), by Muly Ishmael, King of Morocco, and brought to England by Mr. Curwen.
The Curwen Old Spot, —, —.	Selaby Turk, owned by Mr. Marshall, sire of the dam of Wyndham.
Daughter of, —, .	White-legged Lowther Barb—Vintner mare.
Hautboy, —, —.	D'Arcy's White Turk—A Royal mare.
Miss D'Arcy's Pet Mare, —, —.	No sire given—A Sedbury Royal mare.
Leedes Arabian, —, .	Was a famous Eastern Sire, imported by Mr. Leedes, of North Milford, a large breeder.
Young Bald Peg, —, .	Leedes Arabian—Old Morocco mare by Lord Fairfax's Morocco Barb—Old Bald Peg by an Arabian—Barb mare.
Whynot, —, —.	The Fenwick Barb—A Royal mare.
Wilkinson's Mare, Daughter of, —, —.	Wilkinson's Bay Arabian—Natural Barb mare.
Mr. Curwen's Shields Galloway, —, .	Bred by Mr. Curwen, of Workington. Pedigree unknown.

CYPRON, b. m., 1750.

Blaze, b. h., 1733. | Salome (Selima), blk. m., 1738.

Flying Childers, ch. h., 1715. | Confederate Filly, —, 1719. | Bethell's Arabian, —, . | Daughter of, —, .

Darley Arabian, —, 1699. | Betty Leedes, —, . | Daughter of, Grey Grantham, —, —. | Daughter of, Graham's Champion, b. h., 1707. | —, —.

	Probably a Turk or Syrian horse; was brought over from Smyrna by a brother of Mr. Darley, hence his name.
Old Careless, ab. 1696.	Spanker (D'Arcy Yellow Turk—Old Morocco mare)—Barb mare.
(Sist. to Leedes) Cream Cheeks, —, —.	Leedes Arabian—Daughter 1690 of Spanker (Darcy Yellow Turk—Old Morocco mare)—Old Morocco mare by Morocco Barb Old Bald Peg by an Arabian—A Barb mare.
The Brownlow Turk, —, —.	Grey Grantham, sired Miss Belvoir, Shadow, the Grantham Filly, Green's Galloway, etc.
Unknown.	
Duke of Rutland's Black Barb, —, —.	The Rutland Black Barb was imported by the Duke of Rutland and used as private stallion.
Bright's Roan, —, —.	Bred by Mr. Leedes, and said to be by the Arabian sire of Leedes.
	This Arabian was brought to England by Mr. Bethell, of Rise, in Holderness, a district in the East Riding of York, famous for its breed of horses. Mr. Bethell bred some horses of note, Ruffler, Castaway, Woodcock, etc.
Harpur's Arabian, —, —.	This Arabian was brought to England by Sir J. Harpur, of Yorkshire, hence his name.
Almanzor's dam, Daughter of —, .	Old Hautboy (D'Arcy's White Turk—Royal mare).
Darley Arabian, —, 1690.	Was purchased by Mr. Darley, a merchant, who presented him to his brother, J. B. Darley, of Buttercrambe (now Aldby Park), about 1702.
Daughter of, —, —.	Old Merlin (Bustler, son of the Helmsley Turk). Remainder of this pedigree unknown.

ECLIPSE (O'KELLY'S).

WINNER OF ELEVEN KING'S PLATES AND NEVER BEATEN.

Eclipse, chestnut horse, foaled during the great eclipse in 1764, hence his name, bred by H. R. H. William, Duke of Cumberland, and purchased after His Highness' death by Mr. Wildman, who sold him to Captain Dennis O'Kelly, by Marske, son of Squirt by Bartlett's Childers; dam Spiletta, the dam of Proserpine, Garrick and Briseïs by Regulus, son of the Godolphin Arabian; grandam Mother Western by Smith's Son of Snake, son of Lister's Snake, etc. Eclipse made his debut at Epsom, May 3, 1769. Won £50 for horses that never won £30, five-year olds 8 stone, six-year olds 9 st. 3 lbs., four mile heats, beating Gower, Chance, Trial and Plume. The horses in this race were all bunched at the end of three miles, when Eclipse drew away and distanced them all under the hardest pull. Mr. O'Kelly had laid large bets that he placed the horses. When called upon to do so, he replied, "Eclipse first, the others no place." Won £50 at Ascot Heath for five-year olds, 9 st. 3 lbs., two mile heats, beating Cream-de-Barbade. Won King's Plate for six-year olds (he only five years old), 12 stone, beating Slouch (winner of the King's Plate at Guilford), Chigger, etc.; walked over for £50; walked over for the King's Plate at Salisbury; won City Silver Bowl, with £30 added, 10 stone, beating Sulphur and distancing Forester; walked over for the King's Plate at Canterbury for six-year olds, 12 stone; won the King's Plate for six-year olds, 12 stone, at Lewes, beating Kingston; won King's Plate at Lichfield, 8 st. 7 lbs., three mile heats; at Newmarket, 1770, beat Bucephalus, same age, 8 st. 7 lbs. each, Beacon Course, purchased after this race by Mr. O'Kelly: won King's Plate, 12 stone, Round Course, four mile heats, beating Pensioner, Diana and Chigger, latter drawn after first heat, balance distanced; walked over for the King's Plates at Guilford, Nottingham and York, 12 stone; won £319:10s., 8 st. 7 lbs., beating Tortoise, aged, 9 stone, and Bellario, aged, four miles; walked over for the King's Plate at Lincoln; at Newmarket won 150 guineas and upward (paying 100 guineas entrance) for six-year olds, 8 st. 7 lbs., Beacon Course, four miles, beating Corsican, 70 to 1 on Eclipse; next day walked over for the King's Plate, Round Course, 12 stone. This ended his turf career. Won eleven King's Plates, in ten of which the weight was 12 stone (168 lbs). Previous to running for the King's Plate at Winchester in 1769, Mr. O'Kelly paid Mr. Wildman 650 guineas for half of the horse, and in April, 1770, paid 1,100 guineas for the other half.

Eclipse covered, at Clay Hill, near Epsom, Surrey, in 1771 at 50 guineas, in 1772, 1773 and 1774 at 25 guineas; afterwards, by subscription, 40 mares at 30 guineas, in 1779 at 30 guineas, in 1780 at 20 guineas. In 1781, 1782, 1783, 1784 he was stinted to 50 mares at 20 guineas; in 1785 and 1786 to 40 mares at 30 guineas; in 1787 and 1788 to 25 mares at 30 guineas, after which Eclipse was advertised to cover at the same price at Cannons, Middlesex, to which he was removed from Epsom in his carriage, drawn by two horses, with his groom as inside passenger. When, like other travelers, he chose to take a glass of gin or anniseed for himself, he was ordered to furnish Eclipse with a lock of hay and a drop out of the pail.

His progeny continued to distinguish themselves on the turf for twenty-three years, and, besides various Cups, won £158,047. Eclipse's most distinguished progeny on the race-course and at the stud include three Derby winners, Y. Eclipse, Saltram (imp.) and Sergeant, and Annette, winner of the Oaks, also Alexander, Boudrow, Don Quixote, Dungannon, Harmonia, Hermes, Horatia, Javelin, Joe Andrews (sire of Jack Andrews), Dick Andrews, Jupiter, King Fergus (sire of Beningbrough, Overton, Hambletonian, etc.), Laura, Luna, Madcap, Maria, Mercury (sire of Calomel, Precipitate, Gohanna, etc.), Meteor, Miss Hervey, imp. Obscurity, Pegasus, Satellite, Pot-8-o-'s (sire of Waxy and Coriander), Soldier, Sister to Soldier, Spitfire, Ticklepitcher, Volunteer (sire of imp. Eagle and Spread Eagle), Zara and Zodiac, etc.

Eclipse got also Augusta, the dam of imp. Anvelina and Eliza, the dam of Seud—Frenzy, the dam of the St. Leger winner, Phenomenon, and Lucy, the grandam of Catton—Everlasting, the dam of the Derby winner, Skyscraper, and his full sister, the dam of Violante, also a mare by Woodpecker, which was the dam of Fractions and Fraxinella—an unnamed daughter, that produced Chanticleer, the sire of Bob Booty—Speranza, the grandam of imp. Psyche—Scota, the dam of Scotilla, that produced Olive, winner of the 2,000 Guineas, and the great Olympia, the dam of the St. Leger winner, Elis, Epirus and Kite—Queen Mab (sister to Mercury and Volunteer), the dam of the St. Leger winner, Remembrancer—Venus (full sister to Queen Mab), dam of the Oaks winner, Tag—Xantippe (full sister to Alexander), the dam of the Derby winner, John Bull—Miss Hervey, the dam of Haphazard—Madcap, the dam of Caprice, grandam of Economist—Tiffany, the dam of Y. Tiffany, the grandam of Bourbon and imp. Popinjay mare—Horatia, the dam of the Derby winners, imp. Archduke and Paris—Anna, the dam of the Oaks winner, Bellina—Bobtail, the dam of the Oaks winner, Ephemera—Fanny, the dam of the Oaks winner, Volante—Ruth, the dam of the St. Leger winner Tartar.

Eclipse was high in his hips and short in his forehand. His heart weighed, when taken out, 13 lbs. To such an extraordinary weight he was supposed to owe his natural courage. At his interment ale and cake were given, as at that of the Godolphin Arabian.

EPITAPH ON ECLIPSE.

Praise to departed worth! Illustrious steed!
    Not the fam'd Phernicus of Pindar's ode
O'er thee, ECLIPSE, possess'd transcendent speed,
    When by a keen Newmarket jockey rode.

Tho' from the hoof of Pegasus arose,
    Inspiring Hippocrene, a fount divine,
A richer stream superior merit shows:
    Thy matchless foot produc'd O'Kelly wine.

True, o'er the tomb in which this fav'rite lies,
    No vaunting boast appears of lineage good;
Yet the Turf Register's bright page defies
    The race of Herod to show better blood.

ECLIPSE.

FOALED DURING THE GREAT ECLIPSE, 1764. DIED FEB. 27, 1789, AGED 25 YEARS.

Chestnut horse, foaled 1764; bred by H. R. H. the Duke of Cumberland, and purchased after the Duke's death by Mr. W. Wildman, who sold him to Dennis O'Kelly.

MARSKE, br. h., 1750.
Squirt, ch. h., 1732. — Bartlett's Childers, — Darley Arabian, 1702. — Betty Leedes, Clumsey's Sister, — Sist. to Old Country Wench, — Lister's Snake, — Grey Wilkes.
The Ruby Mare, — Hutton's Blacklegs, 1725. — [Hutton's or] Mulso Bay Turk. — Daughter of, — Daughter of, — Bay Bolton, br. h., 1705. — Daughter of, — Daughter of,

	Probably a Turk or Syrian horse. Was brought over from Smyrna by a brother of Mr. Darley, hence his name.
Old Careless, ab. 1680.	Spanker (D'Arcy Yellow Turk—Old Morocco mare)—A Barb mare.
(Sist. to Leedes) Cream Cheeks, —, —.	Leedes Arabian—Dau. 1680 of Spanker (D'Arcy Yellow Turk—Old Morocco mare)—Old Morocco mare by Morocco Barb—Old Bald Peg by an Arabian—A Barb mare.
Lister Turk,	Was brought into England by the Duke of Berwick from the siege of Buda, in the reign of James II. (1633-1701).
Daughter of,	Hautboy (D'Arcy White Turk—A Royal mare).
Hautboy,	D'Arcy White Turk—A Royal mare.
Miss B. D'Arcy's Pet Mare,	Daughter of a Sedbury Royal mare.
	An Eastern horse, imported about 1720 by Mr. Hutton, of Marské, York. Sire also of Black Chance, the best horse of his day with high weights.
Coneyskins, gr. h., 1712.	Lister Turk—Daughter of Jigg (Byerly Turk—Daughter of Spanker).
Old Club Foot,	Hautboy (D'Arcy White Turk—A Royal mare).
Grey Hautboy, —, —.	Old Hautboy (D'Arcy White Turk—A Royal mare).
Mr. Pierson's Daughter of, blk. m., —.	Makeless (Oglethorpe Arabian)—Daughter of Brimmer (D'Arcy Yellow Turk—A Royal mare)—Daughter of D'Arcy's Diamond (The Helmsley Turk)—A full sister to Old Merlin by Bustler.
Fox Cub, b. h., 1714.	Clumsey (Hautboy—D'Arcy's Pet mare)—Charming Jenny by Leedes Arabian—Dau. 1690 of Spanker (D'Arcy Yellow Turk—Old Morocco mare)—Old Morocco mare by Morocco Barb—Old Bald Peg by an Arabian—A Barb mare.
Daughter of, —, —.	Coneyskins gr. h. 1712 (Lister Turk—Dau. of Jigg)—Dau. of Hutton's Grey Barb—Dau. of Hutton's Royal Colt ab. 1700 (The Helmsley Turk—Sedbury Royal mare)—Dau. of Byerly Turk—Dau. of Bustler (The Helmsley Turk).

SPILETTA, b. m., 1749.
Regulus, b. h., 1739. — Godolphin Arabian, br. h., 1724. — Grey Robinson, gr. m., 1723. — Bald Galloway, — Sist. to Old Country Wench.
Mother Western, —. — Smith's son of Snake, b. h., 1721. — Lister's Snake, — Daughter of, — Daughter of, — Lord D'Arcy's Old Montagu, — Daughter of,

	The most noted of all the Eastern sires, about whom opinions are divided whether he was an Arab or Barb. The best judges inclined to the opinion that he was a Barb. He was purchased in Paris by Mr. Coke in 1728, who gave him to Mr. R. Williams, by whom he was presented to Earl Godolphin. He was a brown bay, 15 hands high, with some white on off hind-heel. He died in 1753.
St. Victor Barb, —, —.	Brought from France by Capt. Rider, of Whittlebury.
Grey Whynot, gr. m., —.	Old Whynot (Fenwick Barb—A Royal mare)—A Royal mare.
Lister's Snake, —, —.	Lister Turk—Dau. of Hautboy (D'Arcy White Turk—Royal mare).
Grey Wilkes, —, —.	Hautboy (D'Arcy White Turk—A Royal mare)—Miss B. D'Arcy's Pet mare—A daughter of a Sedbury Royal mare.
The Lister Turk, —, —.	Was brought into England by the Duke of Berwick from the siege of Buda, in the reign of James II. (1633-1701).
Daughter of, —, —.	Hautboy (D'Arcy White Turk—A Royal mare).
Akaster Turk, —, —.	Sire of Chaunter, Thwaite's Dun mare and the dam of Roxana, the dam of Lath and Cade.
Daughter of, —, —.	Son of Pulleine's Chestnut Arabian—Dau. of Brimmer (D'Arcy Yellow Turk—A Royal mare).
	Was in Lord D'Arcy's stud. Said to have been bred by Lord Montagu, a nobleman in Charles II.'s reign noted for his fine breed of horses.
Hautboy, —, —.	D'Arcy White Turk—A Royal mare.
Daughter of, —, —.	Brimmer (D'Arcy Yellow Turk—A Royal mare)—A Royal mare.

TABLE SHOWING THE FEMALE DESCENT FROM AN ORIGINAL MARE (FOUNDER OF A FAMILY).

IMP. DUCHESS BY CULLEN ARABIAN.

- Mare by imp. Regulus.
  - Mare by Gallant.
  - Mare by True Whig.
  - Mare by Rockingham.
  - Annette by imp. Shark.
  - Maid of the Oaks by imp. Spread Eagle.
  - Y. Maid of the Oaks by imp. Expedition (MEDOC'S dam).
  - Katy Ann by Ogle's Oscar.
  - Lady Thompkins by Am. Eclipse.
    - Blue Filly by imp. Hedgford.
      - Blue Bell by Chorister.
      - Julia Mattingly by John Morgan.
      - Belle of Maywood by Hunter's Lexington (TENNY'S dam).
    - Motto by imp. Barefoot.
      - Sally Lewis by imp. Glencoe (JOHN MORGAN'S dam).
      - Susan Beane by Lexington (SENSATION'S dam).
      - Susquehanna by imp. Leamington (POTOMAC'S dam).
- Mare by imp. Jack of Diamonds.
  - Mare by Ariel.
  - Mare by imp. Fearnought.
  - Mare by imp. Highflyer.
  - Maria by Melzar.
  - Lady Grey by Robin Grey.
    - Rowena by Sumpter.
      - Alice Carneal by imp. Sarpedon (LEXINGTON'S dam).
    - Lucy by Orphan.
      - Lucilla by Trumpator.
      - Mare by imp. Tranby (VANDAL'S dam).
        - Volga by imp. Glencoe.
          - Evadne by Lexington.
          - Blink Bonnie by imp. Bonnie Scotland.
          - Tullahoma by imp. Great Tom (TAMMANY'S dam).
        - Levity by imp. Trustee.
          - Brenna by imp. Knight of St. George.
            - Brademante by War Dance (THE BARD'S dam).
          - Lightsome by imp. Glencoe
            - Salina by Lexington (SALVATOR'S dam).
            - Nevada (LUKE BLACKBURN'S dam).

www.ingramcontent.com/pod-product-compliance
Lightning Source LLC
Chambersburg PA
CBHW030728230426
43667CB00007B/635

* 9 7 8 3 3 3 7 1 4 4 9 1 3 *